THE VASCULAR FLORA OF OHIO

Volume Two

THE

Dicotyledoneae of Ohio

PART 2. LINACEAE THROUGH CAMPANULACEAE

*This work is a project of the Ohio Flora Committee
of The Ohio Academy of Science, sponsored by the Academy
and supported by the Research Council of Kent State University
and by the Ohio Department of Natural Resources,
with publication costs partially underwritten
by the ODNR's Division of Natural Areas and Preserves.*

TOM S. COOPERRIDER

THE *Dicotyledoneae* of Ohio

PART 2. LINACEAE THROUGH CAMPANULACEAE

Original drawings by Michael Allen Lewis, Jennifer Dye, Paul Sahre, and Ronald C. Perruchon

OHIO STATE UNIVERSITY PRESS: Columbus

Endpaper source: Miami University, Oxford, Ohio.

Copyright © 1995 by the Ohio State University Press.
All rights reserved.

Library of Congress Cataloging-in-Publication Data
(Revised for vol. 2)

The Dicotyledoneae of Ohio.

(The Vascular flora of Ohio ; v. 2)
"A project of the Ohio Flora Committee of the Ohio Academy of Science, sponsored by the Academy."
Includes bibliographical references and indexes.
Contents: — pt. 2. Linaceae through Campanulaceae / Tom S. Cooperrider ; original drawings by Michael Allen Lewis . . . [et al.] — pt. 3. Asteraceae / T. Richard Fisher ; original drawings by Sharon Ames Glett.
1. Dicotyledons—Ohio—Identification. 2. Dicotyledons—Ohio—Geographical distribution —Maps. 3. Botany—Ohio. 4. Phytogeography—Ohio —Maps. I. Fisher, T. Richard, 1921– . II. Cooperrider, Tom S. III. Ohio Academy of Science. Ohio Flora Committee.
QK180.V35 vol. 2 581.9771 87-31257
583'.09771

ISBN 0-8142-0446-5 (pt. 3)
ISBN 0-8142-0628-X (pt. 2 : alk. paper)

Type set in Goudy Old Style by The Composing Room of Michigan, Inc., Grand Rapids, MI.
Printed by Braun-Brumfield, Ann Arbor, MI.

The paper in this book meets the guidelines for permanence and durability of the Committee on Production Guidelines for Book Longevity of the Council on Library Resources. ∞

9 8 7 6 5 4 3 2

This book is dedicated to

Frederick H. Deering
Member, Ohio House of Representatives, 1973–1992

In appreciation of his support
of the Ohio Flora Project

FOREWORD

The families Linaceae through Campanulaceae, covered in this book, include some of the best-known plants in Ohio. The systematic text treats the native members of these families, the alien plants of these families that have become established or that appear occasionally in the state's flora, and to a lesser extent the families' major cultivated plants.

Wildflowers compose a prominent part of this plant group's native herbaceous species. Harbinger of spring, blooming in February, leads off Ohio's annual wildflower show. The display expands in March to include other members of the group such as trailing arbutus and blue violets, and blows wide in April and May with the addition of yellow violets, white violets, dwarf ginseng, lousewort, bluebells, blue-eyed Mary, shooting star or pride of Ohio, Jacob's ladder, bluets or Quaker ladies, wild geranium, hoary puccoon, and Indian paintbrush. Some of summer's showiest wildflowers bloom next: foxglove beard-tongue, sundrops, wild bergamot, butterfly-weed, milkweed, bee-balm, spotted touch-me-not or jewel-weed, rose-pink, and swamp candles; then those of late summer and autumn: rose-mallow, evening-primrose, cardinal flower, great blue lobelia, turtlehead, and finally bottle gentian.

The wildflowers are paralleled in Ohio gardens by cultivated relatives such as creeping phlox, garden pansy, geranium, petunia, snapdragon, foxglove, salvia, obedience, garden phlox, scarlet sage, purple loosestrife, and hollyhock.

This group of families includes a number of favorite trees and ornamental shrubs commonly cultivated around Ohio homes. Among the trees are the dogwoods, maples, ashes, buckeyes, and horse-chestnuts; among the shrubs, forsythia, lilac, azalea, rhododendron, mountain laurel, privet, shrub-honeysuckle, rose of Sharon, burning-bush or winged wahoo, red-osier dogwood, and snowberry. The families include also the ornamental flowering vines, Japanese honeysuckle, trumpet-vine, and morning glory; the wall climbers, English ivy and Boston ivy; and two common ground covers, pachysandra and periwinkle or myrtle.

Grapes, blueberries, celery, parsnips, carrots, potatoes, sweet potatoes, tomatoes, cucumbers, squash, muskmelon, and watermelon are produced by species of this group, as are seasonings such as dill, parsley, sage, rosemary, thyme, oregano, sweet basil, peppermint, and spearmint.

However, not all members of these families are attractive or benign. Poison ivy and poison sumac are harmful to humans; poison-hemlock, water-hemlock, castor bean, and several species of the nightshades are poisonous. Bindweed is sometimes a troublesome weed in cultivated fields, and Jimson-weed is an ill-smelling and obnoxious plant of barnyards and feedlots.

Other noteworthy representatives of the group are native plants of thicket, woodland, and forest habitats. The white Indian pipe, the brownish beechdrops and squawroot, and the orange-stemmed dodders are interesting parasitic plants. Partridge-berry and wintergreen or teaberry are small evergreens of the forest floor. Bittersweet, Virginia creeper, and several species of wild grapes are climbing vines. Shrubs of these habitats include smooth sumac, staghorn sumac, elderberry, and several species each of dogwoods, viburnum or arrow-wood, and blueberries. Well known small trees within the group are American holly, bladdernut, flowering dogwood, sourwood, fringe-tree, and silver bell. The families include also a few of Ohio's large native trees, namely sour-gum or tupelo and the basswoods, ashes, maples, and buckeyes.

The text provides descriptions, illustrations, and county dot-distribution maps for some 700 species. Of these about 75 percent are regarded as native to Ohio; the remaining 25 percent are considered to be aliens that have come from other areas, chiefly Eurasia.

The Ohio maps show varied and interesting distribution patterns. Some native species are limited to a particular part of the state. *Galium palustre*, northern marsh bedstraw (p. 524), is known only from Ashtabula County in the northeasternmost part of Ohio, and *Euphorbia polygonifolia*, seaside spurge (p. 52), only from the northernmost counties on the beaches bordering Lake Erie. *Bignonia capreolata*, cross-vine (p. 495), and *Spermacoce glabra*, smooth button-weed (p. 529), grow only in the southernmost counties, respectively on the wooded slopes and the open shores along the Ohio River. *Paxistima canbyi*, cliff-green (p. 71), has been found at only two sites in the state, both "in the drainage basin of the preglacial Teays River whose old valleys are prominent features of southern Ohio" (Braun, 1961). *Cornus stolonifera*, red-osier dogwood (p. 242), grows only in the northern half of Ohio, *Passiflora lutea*, yellow passion-flower (p. 162), only in the southern half.

Other native species delineate particular phytogeographic regions. *Oxydendrum arboreum*, sourwood (p. 259), and *Aureolaria laevigata*, entire-leaved false foxglove (p. 488), occur exclusively in, but throughout, Ohio's southeastern Allegheny Plateau region (Silberhorn, 1970). The ranges of *Kalmia latifolia*, mountain laurel (p. 258), and *Cunila origanoides*, dittany (p. 415), mark precisely the unglaciated portion of Ohio, which occupies the east-central and southeastern parts of the state (Cusick and

Silberhorn, 1977). The distributions of *Shepherdia canadensis*, buffalo-berry (p. 166), *Gentiana clausa*, blind gentian (p. 300), and *Viburnum alnifolium*, hobblebush (p. 555), provide examples of native species limited to the glaciated Allegheny Plateau region of northeastern Ohio (Andreas, 1989). The maps of *Hibiscus laevis*, halberd-leaved rose-mallow (p. 112), *Fraxinus quadrangulata*, blue ash (p. 290), and *Veronica catenata*, water speedwell (p. 477), show species limited to the calcareous soils and waters of the western half of Ohio, while that of *Epigaea repens*, trailing arbutus (p. 260), shows a species limited to the acidic soils of eastern Ohio.

Two species provide interesting examples of north/south climatic discontinuities occurring within the boundaries of the state. The first instance involves a northern species with a disjunct population in southern Ohio, the second a southern species with disjunct populations in northern Ohio. *Viola rotundifolia*, round-leaved violet (p. 151), is a boreal plant ranging southward into counties of northeastern Ohio and even farther south at higher altitudes in the Appalachian Mountains. In Ohio, it is one of a group of northern species (Cooperrider, 1982b) with outlying southern populations in the cool microclimates of the hills of Hocking County. Conversely, *Linum striatum*, ridged yellow flax (p. 8), is a southern species inhabiting the "Zone 6"[1] area of southern Ohio and the narrow Zone 6 band in northern counties bordering and in winter warmed by Lake Erie, but absent from the large intervening colder section of northern Ohio classed as Zone 5.

Some native species grow throughout Ohio. This is especially true of plants such as *Rhus glabra*, smooth sumac (p. 60), and *Toxicodendron radicans*, poison ivy (p. 62), whose weedy tendencies enable them to compete successfully in disturbed habitats. But a statewide distribution is also true of some nonweedy survivors of Ohio's primeval forests such as *Acer saccharum*, sugar maple (p. 77), and *Fraxinus americana*, white ash (p. 286).

The distribution patterns of alien species established in the flora range from that of *Vinca major*, greater periwinkle (p. 306), known from a single small colony near the Ohio River in southwestern Ohio, to those of *Ailanthus altissima*, tree-of-heaven (p. 25), and *Verbascum thapsus*, common mullein (p. 446), both naturalized in all of Ohio's eighty-eight counties. Other alien species have ranges intermediate between those extremes. *Rhamnus frangula*, glossy buckthorn (p. 89), and *Epilobium hir-*

1. For a map of these hardiness zones, see the inside covers of *Hortus Third* (Bailey Hortorium Staff, 1976).

sutum, hairy willow-herb (p. 180), are naturalized only in northern, especially northeastern, Ohio; *Croton monanthogynus*, prairie-tea (p. 35), and *Galium pedemontanum*, Piedmont bedstraw (p. 528), only in southern Ohio. *Ludwigia decurrens*, erect primrose-willow (p. 176), and *Ludwigia leptocarpa*, hairy primrose-willow (p. 178), are more restricted invaders, growing only along the shores of the Ohio River on the state's southern boundary.

This book is one of a series of three covering the dicotyledons of Ohio; all are a part of the Ohio Flora Project. The Project originated in 1950, under the leadership of the late E. Lucy Braun and the ægis of The Ohio Academy of Science (Cooperrider, 1984a). It has served as a stimulus for an extensive amount of field work in the state, producing thousands of new herbarium specimens and distribution records (Cooperrider, 1961, 1992). The result of this enterprise is a detailed knowledge of plant occurrence that places Ohio's flora among the best known of any of the fifty states.

Dr. Braun prepared two volumes on the flora, *The Woody Plants of Ohio* (1961; reprint 1989) and *The Monocotyledoneae* (1967). In addition to Braun's books, more than fifty journal articles and theses on various aspects of the flora have resulted from the research projects of other botanists. Many of these are cited in the review paper, "Ohio's Herbaria and the Ohio Flora Project" (Cooperrider, 1984a).

Two major publications on the flora have appeared since Braun's 1967 volume. Both are Ohio Biological Survey Bulletins: *The Vascular Plants of Unglaciated Ohio*, by Allison W. Cusick and Gene M. Silberhorn (1977), and *The Flora of the Glaciated Allegheny Plateau Region of Ohio*, by Barbara K. Andreas (1989).

Other recent research has focused on Ohio's endangered plant species. The many new herbarium specimens collected since 1950 for the Ohio Flora Project and since 1980 for the Ohio Natural Heritage Program of the Ohio Department of Natural Resources (ODNR) have provided not only basic data on the flora itself, but have also formed a sound database for the determinations of which plant species are in danger of becoming extirpated from the state. *Endangered and Threatened Plants of Ohio* (Cooperrider, 1982a), another publication of the Ohio Biological Survey, included individual sections written by William Adams, Marvin L. Roberts, Ray E. Showman, Jerry A. Snider, Ronald L. Stuckey, and myself. Together, the sections treated all of the flora, both vascular and nonvascular plants, except for the algae and the fungi. In addition to identifying those species in jeopardy, the authors presented valuable lists of excluded species, mostly ones that had been incorrectly attributed to the Ohio flora by past workers. In 1984, *Ohio Endangered and Threatened Vascular Plants*, a collection of comprehensive abstracts for 357 species, edited by Robert M. McCance, Jr., and James F. Burns, was published by the ODNR's Division of Natural Areas and Preserves. That agency has also in recent alternate years (1980, 1982, 1984, etc.) issued precise status lists of extirpated, endangered, threatened, and potentially threatened species, each revised list incorporating new data resulting from the field work of the previous biennium.

A long-term goal of the Ohio Flora Project has been the preparation of three books on Ohio's dicotyledons (dicots) to complement Braun's volume on the monocotyledons (monocots), thereby completing coverage of all the flowering plants. Publication of the dicot series began with *The Dicotyledoneae of Ohio: Part 3. Asteraceae* by T. Richard Fisher (1988). It continues with the present book, *Part 2. Linaceae through Campanulaceae*. The final book, *Part 1. Saururaceae through Fabaceae (Leguminosae)*, is under preparation by John J. Furlow. Plans are for Dr. Furlow's book to present an introduction to all the dicotyledons, including a discussion of their position in the flora and their overall relationship to Ohio's physiography, along with a key to the families, similar to Braun's (1961, 1967) introductions to the woody plants and the monocotyledons. All books follow the family sequence of Fernald (1950).

I thank Lynn Edward Elfner, of The Ohio Academy of Science, for the support of the Academy, and the members of the Academy's Ohio Flora Committee: Barbara K. Andreas, Will H. Blackwell, Philip D. Cantino, T. Richard Fisher, John J. Furlow, J. Arthur Herrick, Ronald L. Stuckey, Tod F. Stuessy, and Michael A. Vincent, for their support and service. I thank also David B. MacLean, John H. Olive, David E. Todt, Ralph W. Wisniewski, and the late Carson Stauffer for their timely help.

I thank especially Charles C. King, of the Ohio Biological Survey, and Guy L. Denny, Richard E. Moseley, Jr., and Ralph E. Ramey, of the Ohio Department of Natural Resources, for valuable assistance each gave the Ohio Flora Project in recent years. I appreciate the cooperation of the members of the 113th (1979–80), 115th (1983–84), 116th (1985–86), and 119th (1991–92) Ohio General Assemblies in providing financial support for the Project, even under tight budget conditions.

The support of all these individuals made this work on Ohio's dicotyledons possible.

Tom S. Cooperrider
Chair, Ohio Flora Committee
The Ohio Academy of Science

PREFACE

The text, illustrations, and maps for this book follow the plan and style established by E. Lucy Braun (1967) for *The Vascular Flora of Ohio, Volume 1, The Monocotyledoneae*, and continued by T. Richard Fisher (1988) in *Volume 2, The Dicotyledoneae of Ohio. Part 3. Asteraceae*. As was the case with both those authors, my goal has been to achieve completeness and accuracy, combined in a work that can be used as a source of information about a particular group of Ohio plants and as a means of identifying a specimen whose identity is unknown.

For support that provided the time required to prepare this book, I express my gratitude to the Division of Natural Areas and Preserves of the Ohio Department of Natural Resources and to the Kent State University Research Council.

Many individuals have assisted me along the way, and it is a pleasure to have this opportunity to acknowledge their contribution.

I thank my wife and fellow botanist, Miwako K. Cooperrider, for her help in countless matters, not the least of which has been the day-to-day supervision of the Kent State University Herbarium (Cooperrider and Cooperrider, 1994).

I am grateful also to the botanists who reviewed the manuscript for this book: Carol C. Baskin, University of Kentucky; Jerry M. Baskin, University of Kentucky; Craig C. Freeman, Kansas Biological Survey; James S. Pringle, Royal Botanical Gardens, Hamilton, Ontario; and Edward G. Voss, University of Michigan. Their careful criticisms and thoughtful suggestions improved the book in many ways. I thank Charles C. King, Ohio Biological Survey, and Ronald L. Stuckey, Ohio State University, both of whom made valuable suggestions regarding the introductory sections. I thank Dr. King also for his editorial services during the manuscript review.

I express my thanks to Peter John Givler, director of the Ohio State University Press, and to the talented Press staff. Lynne M. Bonenberger, Marla Bucy, Peg Carr, John Delaine, Charlotte Dihoff, Alex Holzman, Ruth Melville, and Ellen Satrom did much to make this book a reality.

I am deeply indebted to Linda L. Matz, of Kent State University, for excellent word processing, skillfully converting reams of rough or revised copy into a finished manuscript. I appreciate Julie A. Cooperrider's help in the proofreading of the final copy.

I thank Richard E. Moseley, Jr., and Ralph E. Ramey, of the Ohio Department of Natural Resources, and Carol M. Toncar, of Kent State University, for their support and help throughout the past several years. I thank William R. Carr and James S. McCormac, and especially Allison

W. Cusick, of the Ohio Department of Natural Resources, for a steady supply of specimens that provided many valuable new Ohio distribution records. I appreciate greatly the efforts of Barbara K. Andreas, of Cuyahoga Community College, and Jeffrey D. Knoop, of The Nature Conservancy, in procuring helpful financial support.

The Kent State University Audio-Visual Services staff has provided quick and always competent help; I thank Cathy E. Brown, Lawrence M. Rubens, and Paul Sisamis for their graphic and photographic work. I thank Scott D. Robinson and Tana Smith for preparing preliminary distribution maps, and John A. Cooperrider, Paul Sahre, and Wendy Shindler for help with other graphic work.

Jennifer Dye drew the illustrations for the families Rubiaceae and Caprifoliaceae. Paul Sahre prepared those for the Valerianaceae, Dipsacaceae, Cucurbitaceae, and Campanulaceae, as well as the drawings of the fruits of *Linum*, the habit sketch of *Oxalis stricta*, and the drawings of *Acer negundo*. The habit sketches of *Androsace occidentalis* and *Moneses uniflora* and the drawings of the fruits of *Diospyros virginiana* and *Malva neglecta* were made by Ronald C. Perruchon. The illustration of *Aralia spinosa* by Elizabeth Dalvé is reprinted by permission from Braun (1961). All other drawings were done by the principal illustrator for the book, Michael Allen Lewis. My thanks go to the artists both for their talent and for their patience in our efforts to make the illustrations accurate.

The project has been aided by fellow herbarium curators, who assisted in my study of their specimens, and by many other fellow biologists, who collected special specimens to fill distribution gaps or provided high quality recent material, answered questions about plants of which they had special knowledge, and helped in other important ways. For assistance in such matters, I thank W. Preston Adams, Dennis M. Anderson, Barbara K. Andreas, John R. Baird, Harvey E. Ballard, Jr., Marjorie Becus, Virginia L. Bielert, James K. Bissell, Will H. Blackwell, Orland J. Blanchard, Jr., David E. Boufford, Richard K. Brummitt, James F. Burns, Philip D. Cantino, William R. Carr, Carl F. Chuey, Barbara H. Costelloe, Richard W. Couch, Allison W. Cusick, Jane M. Decker, Guy L. Denny, Judith Doty Dumke, N. William Easterly, David P. Emmitt, Dan K. Evans, T. Richard Fisher, John J. Furlow, Alwyn H. Gentry, Charles A. Hammer, E. Dennis Hardin, Raymond M. Harley, Peter C. Hoch, Barbara Hosta, David M. Johnson, George T. Jones, Susan R. Kephart, Jeffrey D. Knoop, Thomas G. Lammers, Donald H. Les, Jimmy R. Massey, James S. McCormac, Steven M. McKee, Landon E. McKinney, Susan J. Meades, Robert H. Mohlenbrock, C. Susan Munch, Lytton J. Musselman, James E. Nellessen, Marilyn W. Ortt, Wilson T. Rankin, Peter H. Raven, Robert

H. Read, Marvin L. Roberts, Clark G. Schaack, Gregory J. Schneider, Jerry A. Snider, Victor G. Soukup, Alvin E. Staffan, Stanley J. Stine, Jr., Warren P. Stoutamire, Ronald L. Stuckey, John W. Thieret, John D. Usis, Michael A. Vincent, Edward G. Voss, Warren L. Wagner, David A. Webb, David H. Webb, Clara G. Weishaupt, Jean D. Wistendahl, and Warren A. Wistendahl.

I am grateful to the editors of *Castanea*, Journal of the Southern Appalachian Botanical Society, for publishing preliminary studies of several individual families and genera. The stimulus and professional sustenance provided by these publications helped the work go forward. In this regard, I am indebted especially to Earl L. Core, Jesse F. Clovis, and Audrey Mellichamp.

Several Kent State students have done special studies that provided specimens or other valuable aid; they are Judith Doty Dumke, Donald W. Dunkle, Richard J. Goldsworth, Sally A. Hamlin, William J. Humphrey, Frederick J. Peabody, John M. Tomci, Karl R. Valley, Thomas J. Zavortink, and Thomas S. Zuppke. Students in my plant taxonomy course have added many important specimens from their class projects.

I thank especially the students who have completed graduate degrees under my direction, each of whom has contributed to knowledge of the Ohio flora: Joyce E. Amann, Sandra M. Anderson, Barbara K. Andreas, Gregory D. Bentz, Judy S. Bradt-Barnhart, Bruce L. Brockett, James F. Burns, Robert J. Cline, Tammy E. Cook, Allison W. Cusick, David P. Emmitt, Edward J. P. Hauser, William D. Hawver, George A. McCready, Lynne D. Miller, Paul L. "Larry" Pusey, Robert F. Sabo, Gene M. Silberhorn, Lewis W. Tandy, Paula L. Van Natta, and Hugh D. Wilson. The botanical adventures we shared have been a highlight of my professional career.

Department of Biological Sciences, Kent State University

CONTENTS

INTRODUCTION

This book covers the families Linaceae through Campanulaceae. It treats the taxa native to Ohio, the alien taxa naturalized in the Ohio flora, and to a lesser extent those aliens appearing from time to time as adventives. The text takes note also of the families' members that are cultivated in Ohio. Original illustrations are provided for nearly all the native and naturalized species, and for each there is also a county dot-distribution map.

In the course of this work, I examined all Ohio holdings of the following herbaria:

The University of Akron,
Bowling Green State University,
University of Cincinnati,
Kent State University,
Miami University,
Oberlin College,
The Ohio State University,
Ohio University, and
Youngstown State University.

I also examined some but not all Ohio specimens from the herbaria of Antioch College, Carnegie-Mellon University, the Cleveland Museum of Natural History, the Dayton Museum of Natural History, Denison University, the University of Michigan, Missouri Botanical Garden, Muskingum College, Ohio Wesleyan University, and the Smithsonian Institution.

In constructing the maps, I used only records from herbarium specimens (some 80,000 in number) that I actually examined. In previous publications of mine and my students, we have on occasion accepted records from scientific literature that we regarded as reliable. Because no literature records were used in the preparation of the present maps, some county records in those earlier publications do not appear here. However, these additional records from the literature are cited in the text. The records used for constructing the maps were assembled on 5-inch × 7-inch index cards, which indicate the herbaria housing the county record specimens. After publication of this book, the cards will be deposited in the herbarium of The Ohio State University, Columbus.

SYSTEMATIC TEXT

In the text below, the names of species and interspecific hybrids that are native in all or part of their Ohio range are printed in **boldface**. Names of naturalized alien species and names of frequently occurring adventive alien species are printed in CAPITALS AND SMALL CAPITALS. Rare adventives and marginally established aliens are noted in the introductory statement of their family or genus and printed in *italics*. Common or vernacular names are printed in CAPITALS AND SMALL CAPITALS.

In the taxonomic and nomenclatural synonymy, printed in *italics*, I have attempted to account for all other names applied to these plants in any of the following major publications:

1. *Gray's Manual of Botany*, eighth edition (Fernald, 1950);
2. *The New Britton and Brown Illustrated Flora of the Northeastern United States and Adjacent Canada* (Gleason, 1952a, 1952b);
3. *The Woody Plants of Ohio* (Braun, 1961);
4. *Manual of Vascular Plants of Northeastern United States and Adjacent Canada* (Gleason and Cronquist, 1963, and second edition, 1991);
5. *Flora Europaea* (Tutin et al., 1964–1980);
6. *Vascular Plants of Ohio*, third edition (Weishaupt, 1971);
7. *Hortus Third* (Bailey Hortorium Staff, 1976);
8. *The Vascular Plants of Unglaciated Ohio* (Cusick and Silberhorn, 1977);
9. *Endangered and Threatened Plants of Ohio* (Cooperrider, 1982a);
10. *Ohio Endangered and Threatened Vascular Plants* (McCance and Burns, 1984); and
11. *The Flora of the Glaciated Allegheny Plateau Region of Ohio* (Andreas, 1989).

Other works that provided significant taxonomic or descriptive data are those of Schaffner (1932), Deam (1940), Lawrence (1951), G. N. Jones and G. D. Fuller (1955), Steyermark (1963), Rickett (1966, 1967), Radford, Ahles, and Bell (1968), Polunin (1969), Petrides (1972), Strausbaugh and Core (1978), Kartesz and Kartesz (1980), Cronquist (1981), Soper and Heimburger (1982), Voss (1985b), Beal and Thieret (1986), and Mohlenbrock (1986).

The treatments of all woody taxa are based at least in part on Braun (1961). Data on champion Ohio trees, assembled by the Ohio Forestry Association (Anonymous, 1989), provided valuable information on maximum sizes attained by individual species of trees in Ohio and were also interesting reading.

The nomenclature and circumscription of families follow Cronquist (1981). The taxonomy of genera and taxa of lower rank generally follows one or more of the works listed above. In other cases it is based on the published research cited in the discussion of the taxon in question. Most abbreviations of authors' surnames are those used by Fernald (1950, pp. liii–lviii); names of other authors are presented in the citation style of Meikle (1980).

The only major infraspecific rank used is that of variety; equivalent names at the rank of subspecies are given in synonymy only in cases where the subspecific epithet differs from the varietal epithet. In most instances in which the only variety found in Ohio is the type variety, no varietal name is listed, e.g., *Linum sulcatum* instead of *Linum sulcatum* var. *sulcatum* —because of the sometimes uncertain validity of the other varieties. When two or more varieties occur within the state, they are mapped separately except in those species in which for one reason or another a significant number of specimens are unassignable; in such cases all infraspecific elements are combined on a single species map.

At the rank of form, I have included those named elements that are distinctive and clearly differentiated from the typical plants of the species in corolla color, corolla shape, or leaf shape. I have not included forms in instances where there is a great amount of intergradation, e.g., the varying shades of corolla color in *Impatiens*, *Phlox*, *Polygala*, and *Prunella*, or those forms based on varying kinds and degrees of vestiture.

Major, and some minor, interspecific hybrids are designated by name, e.g., *Lysimachia* × *producta*, as well as by formula, e.g., *Lysimachia quadrifolia* × *L. terrestris*. Other minor hybrids are designated by formula only.

Distribution of the distinctive forms, of the minor interspecific hybrids, and of the infrequent adventive and rarely established alien species is given by listing the counties from which they have been collected, rather than by separate maps.

Descriptions of taxa are abbreviated, with emphasis on those characters that give an idea of the overall appearance of the plants and the characters that distinguish members of the taxon from other Ohio plants. The data in the descriptions and keys apply to Ohio plants; they may or may not be accurate and complete for these same taxa in other parts of their range. The illustrations are likewise designed to show diagnostic characters and the general aspect of the plants.

Ohio flowering dates, obtained from herbarium specimens and field observation, are given at the conclusion of each species description.

Significant publications on the taxon under discussion are cited in the text; these are generally limited to works published since 1950. The cited literature includes publications through 1991. It also includes a few post-1991 papers that I read in manuscript form while preparation of this text was in progress.

Linaceae. FLAX FAMILY

A small cosmopolitan family whose principal genus, *Linum*, is represented in the Ohio flora by several species. Robertson (1971) studied the family in southeastern United States.

1. Linum L. FLAX

Annual or perennial herbs; leaves simple, alternate or opposite; flowers perfect, in terminal, much-branched paniculate or corymbose cymes; calyx of 5 separate or nearly separate sepals imbricate in two close series, with 2 or 3 inner sepals and the remainder outer; corolla actinomorphic, of 5 separate petals; stamens 5, united below; carpels 5, united to form a compound pistil with 5 separate styles (styles united basally in *L. sulcatum*); ovary superior; fruit a septicidal capsule. A genus of perhaps 200 species, widespread in distribution, but centered in the temperate and subtropical regions of the Northern Hemisphere.

Three members of the genus are occasionally cultivated in Ohio, and adventives of each have been collected at various times. *Linum grandiflorum* Desf., FLOWERING FLAX, with a large, showy, red corolla to 4 cm across, has been collected from Auglaize and Defiance counties, both in northwestern Ohio. Collections of the blue-flowered *Linum perenne* L. (incl. *L. lewisii* Pursh), PERENNIAL FLAX, have been made from Ashtabula, Erie, Highland, and Lake counties. COMMON FLAX, the more frequent blue-flowered adventive, is discussed below.

The treatment here is based in large part on studies by C. M. Rogers (1957, 1963, 1984).

a. Petals blue or red, 10–25 mm long; fruiting pedicels 10–75 mm long.
  b. Corolla red; sepals exceeding capsules; see above. . . . . *Linum grandiflorum*
  bb. Corolla blue; sepals shorter than to nearly as long as the capsules.
    c. Leaves 3-nerved, the 2 lateral nerves sometimes faint; inner sepals ciliate-serrulate . . . . 1. *Linum usitatissimum*
    cc. Leaves essentially 1-nerved, sometimes faintly 3-nerved at base; inner sepals with entire margins; see above . . . . *Linum perenne*
aa. Petals yellow, 3–10 mm long; fruiting pedicels mostly 0.5–6 mm long, rarely to 11 mm long.
  b. Stems with two dark stipular glands at leaf bases, these glands remaining on stem after the leaves fall; outer sepals mostly more than 4 mm long, margins of both outer and inner sepals markedly glandular-serrate; petals 5–10 mm long; styles united at base for 1/4 their length or more; plants annual . . . . 2. *Linum sulcatum*

bb. Stems lacking stipular glands; outer sepals less than 4 mm long, margins of outer sepals entire, those of the inner sepals glandular-serrate to merely glandular or entire; petals 2.5–8 mm long; styles separate; plants perennial.

c. Margins of inner sepals, especially near the sepal apex, glandular-serrate or with stalked glands; inflorescence more or less corymbose in aspect, but with stiffly ascending branches . . . 3. *Linum medium*

cc. Margins of inner sepals with sessile glands or entire; inflorescence either corymbose in aspect and with spreading branches or paniculate in aspect and with ascending branches.

d. Inflorescence corymbose with spreading branches, nearly or quite as wide as long, the lowest branches of the inflorescence much exceeding subtending leaves; outer surface of mature carpels flat or slightly concave; upper internodes low-striate angled or terete . . . . 4. *Linum virginianum*

dd. Inflorescence paniculate with ascending branches, longer than wide, the lowest branches of the inflorescence not or only slightly exceeding leaves; outer surface of mature carpels slightly convex; upper internodes markedly striate-angled . . . . . 5. *Linum striatum*

1. Linum usitatissimum L. Common Flax

Erect annual, to 10 dm tall; leaves alternate, linear-lanceolate, the major leaves 3-nerved, the 2 lateral nerves sometimes faint; the usually solitary stem branched apically to produce an open, few- to several-flowered panicle, but with the flowers borne in axils of reduced leaves and appearing nearly solitary; pedicels to 20 mm long; sepals shorter than to nearly as long as the capsules, the outer sepals entire, the inner ciliate-serrulate; petals light blue, rarely white, corolla about 15 mm across. A once infrequent, but now rare, adventive, probably of Asian origin; disturbed habitats, especially along roads and railroads. June–Sept.

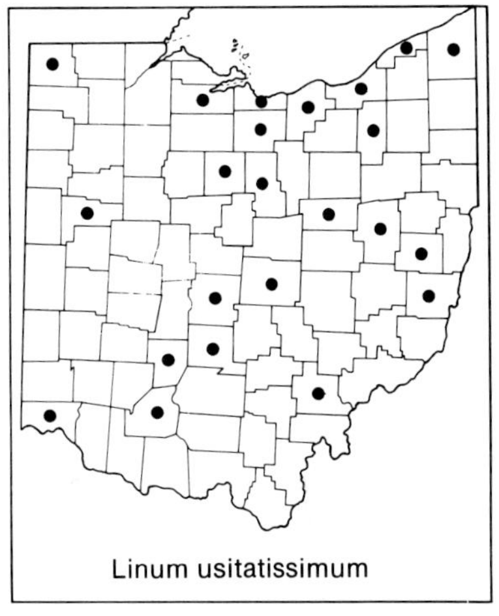
Linum usitatissimum

Linum usitatissimum is cultivated both for its long stem fibers, from which linen is produced, and for its seeds, which yield the linseed oil used in paints and other sealants and in the manufacture of linoleum. Linseed oil meal was used in the past as a dietary protein supplement for cattle. Flax was once a minor farm crop in Ohio, but now is seen mainly in demonstration plots of historical plants. Most Ohio herbarium specimens were collected from 1877 to 1910. Few agricultural sights can equal the beauty of a field of flax in bloom, the sea of thousands of blue flowers presenting an unforgettable image.

2. **Linum sulcatum** Riddell Grooved Yellow Flax

Rather stiffly erect annual, to ca. 7 dm tall; leaves alternate throughout or sometimes opposite below, linear-lanceolate, each leaf bearing two dark

Linum sulcatum

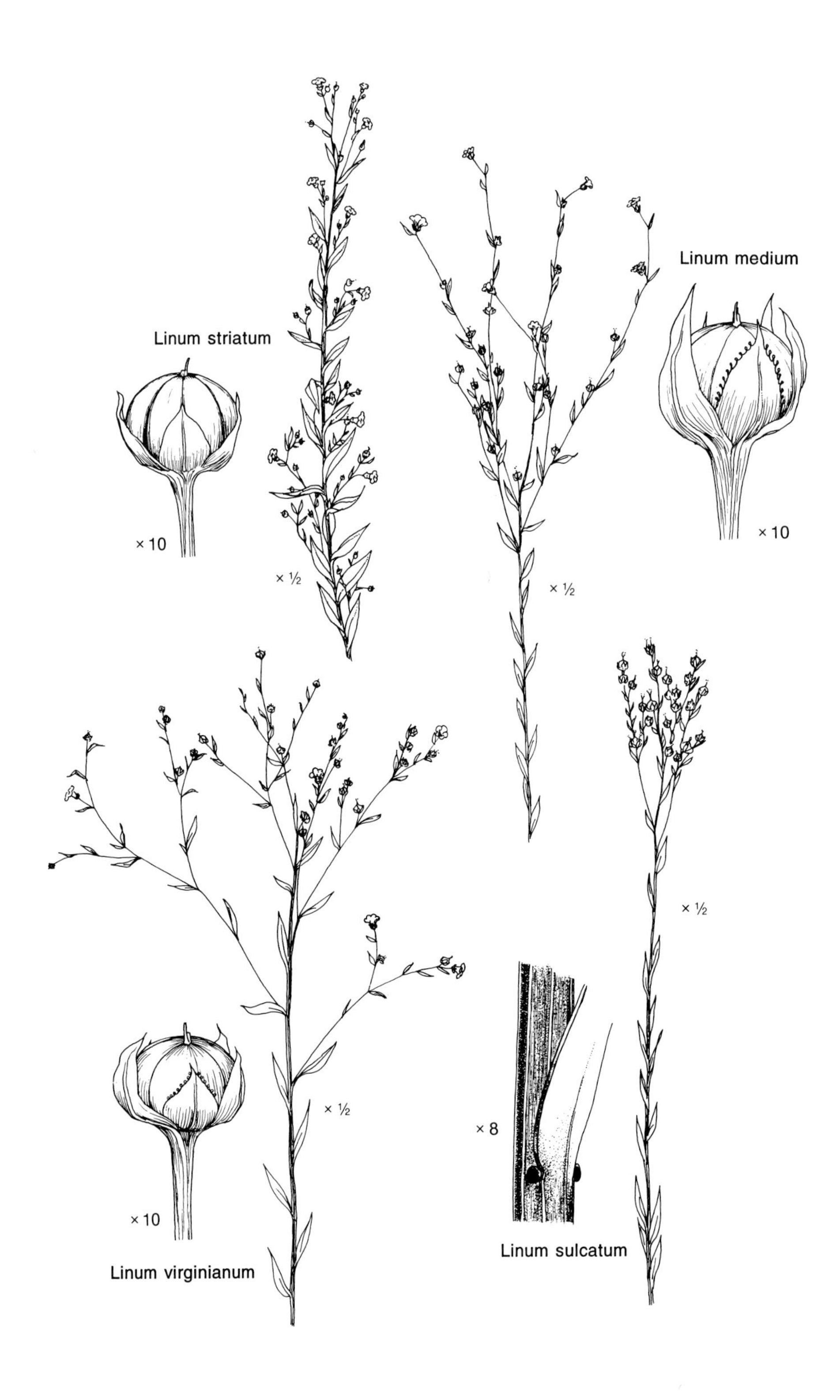
Linum striatum
× 10
× ½
Linum medium
× 10
× ½
× ½
× ½
× 8
× 10
Linum virginianum
Linum sulcatum

stipular glands at its base, these glands persisting on the stem after the leaves fall; the stem branched above to form a many-flowered paniculate inflorescence; pedicels to 4 mm long; outer sepals mostly 4–6 mm long, margins of both inner and outer sepals markedly glandular-serrate; petals yellow, 5–10 mm long; styles united at base for 1/4 their length or more. In Ohio, known from a few sites mostly in western counties; dry, sandy or gravelly calcareous soils. Late June–August.

C. M. Rogers (1984) notes that the type specimen (in the U.S. National Herbarium at the Smithsonian Institution) was collected by Riddell near Dover, Ohio, on 3 August 1835. I have not seen nor mapped this specimen, presumably from Tuscarawas County, where Dover is located.

3. **Linum medium** (Planch.) Britt. STIFF YELLOW FLAX

Erect perennial, to 6 or 7 dm tall; leaves mostly alternate, the lowest opposite, narrowly lanceolate to narrowly oblanceolate; stem branched above to form a many-flowered, more or less corymbose inflorescence with stiffly ascending branches; pedicels to 5 mm long, but often shorter; outer sepals entire, to 3.5 mm long, inner sepals, especially near the apex, glandular-serrate or with stalked glands; petals yellow, to 8 mm long. In northern and in southern Ohio; moist roadsides, open fields, woodland openings. June–early August.

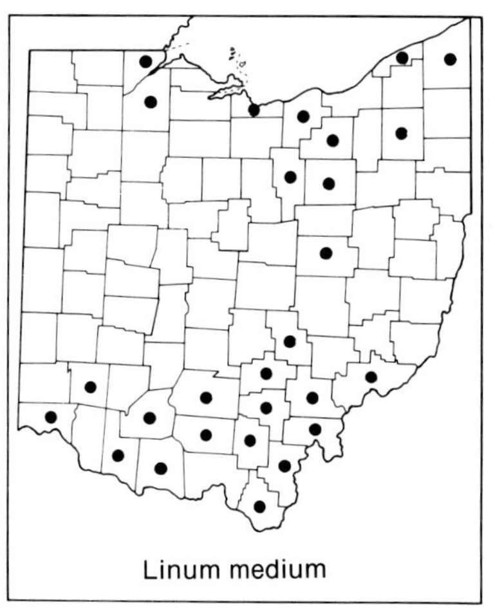

Linum medium

Ohio plants belong to the diploid var. *texanum* (Planch.) Fern. The tetraploid var. *medium*, more northern in distribution, with the inner sepals either entire or less glandular-serrate, was attributed to northern Ohio by Fernald (1950). C. M. Rogers' (1963) map shows a locality for var. *medium* in northwestern Pennsylvania, very near Ashtabula County; but in a later study, Harris (1968) found the tetraploids to be limited to Ontario. Variety *medium* is therefore excluded from the Ohio flora.

It is of interest that this species is sometimes merged with the next, *Linum virginianum*, in which case var. *medium* becomes *L. virginianum* var. *medium* Planch., and the Ohio variety becomes *L. virginianum* var. *texanum* Planch. In his treatment of *Linum* for North and South Carolina, Ahles (1968) went one step farther, reducing var. *texanum* to synonymy under *L. virginianum* var. *medium*.

4. **Linum virginianum** L. SLENDER YELLOW FLAX

Erect perennial, to 6 or 7 dm tall; leaves oblanceolate or lanceolate, opposite on the lower part of the stem, alternate on the upper; the stem branched above to form a wide corymbose inflorescence with spreading branches, the lowest branches of the inflorescence much exceeding their subtending leaves; pedicels to 9 (–11) mm long; outer sepals entire, to 4 mm long, inner sepals usually with sessile glands, especially near the sepal apex; petals yellow, to 5 mm long; the outer surface of mature carpels flat

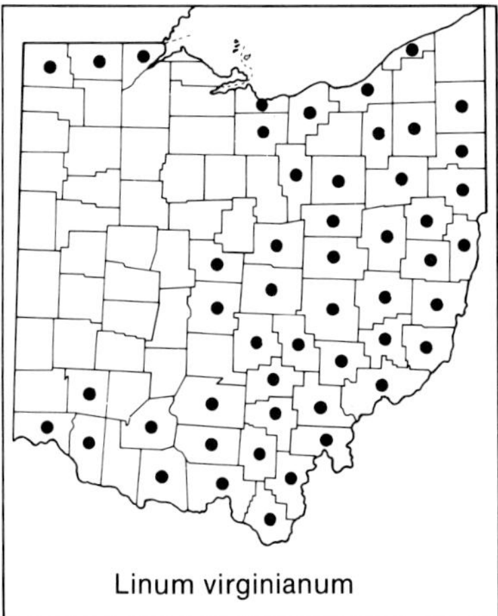

Linum virginianum

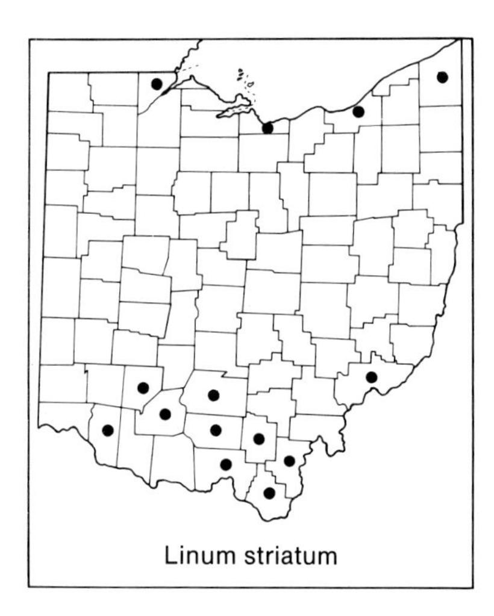
Linum striatum

or slightly concave. Frequent in eastern counties, scarce or absent westward; dry woodland openings and dry fields. Late June–early August.

See comments above under *Linum medium* on the relationship between these two species.

5. **Linum striatum** Walt. Ridged Yellow Flax
Erect perennial, to 10 dm tall; leaves narrowly elliptic, opposite on the lower part of the stem and alternate on the upper; the upper internodes markedly striate-angled; the stem branched above to form a long, leafy, rather narrow, paniculate inflorescence with ascending branches, the lowest branches of the inflorescence not or only slightly exceeding their subtending leaves; pedicels to 4 mm long; outer sepals entire, to 3 mm long, the inner also entire or nearly so; petals yellow, to 4.5 mm long; the outer surface of the mature carpels slightly convex. In southern counties and a few northern counties bordering Lake Erie; moist ditches, roadsides, fields. Late June–August.

Oxalidaceae. WOOD-SORREL FAMILY

A chiefly tropical family of seven genera and about 1,000 species; most of the species are members of *Oxalis*, the only genus occurring in Ohio.

1. Oxalis L. WOOD-SORREL

Perennial, or rarely annual, herbs; leaves trifoliolate, alternate and cauline, or basal, the leaflets obcordate; flowers perfect, in few- to several-flowered cymes or umbellate cymes, or rarely the flowers solitary; calyx of 5 separate sepals; corolla actinomorphic, of 5 separate petals; stamens 10, of two lengths: 5 longer and 5 shorter, all basally fused; carpels 5, fused to form a compound pistil with 5 styles, the styles of the same length within a flower, but varying between flowers, the individual flowers with long, short, or midlength styles; ovary superior; fruit a capsule. A genus of perhaps 800 species, widely distributed but centered in South Africa and Latin America. The leaves are noted for their "sleep movements," the leaflets drooping at night (and on hot sunny afternoons).

Wingo (1962) made a study of the Oxalidaceae of Ohio. Other valuable works are those of Robertson (1975) on the Oxalidaceae of southeastern United States, and Young (1958) on *Oxalis* in the British Isles.

This small group has had more than its share of classification and nomenclatural problems, several of which remain unsettled and will likely continue to cause confusion for some time. The taxonomy of our yellow-flowered species, all assigned to *Oxalis* section *Corniculatae* DC., was studied by Eiten (1963) and later by Lourteig (1979). Eiten's taxonomy is followed here; it was also adopted by Gleason and Cronquist (1963, 1991) and by Lovett Doust, Lovett Doust, and Cavers (1981). Eiten's application of the name *Oxalis stricta*, a key element in which his taxonomy differs from that of Lourteig, was recently supported by Watson (1989).

a. Plants acaulescent, lacking aerial stems, the leaves and peduncles arising from a rhizome or a bulbous stem either underground or at soil surface; corolla white, pink, violet, or purple.

b. Leaves and peduncles arising from slender scaly rhizomes; leaves ciliate, the blades shallowly lobed at apex; peduncles with a single flower; petals white to pale pink or pinkish-purple, often with evident darker veins, slightly notched at apex; sepals green or purplish at apex . . . . 1. *Oxalis acetosella*

bb. Leaves and peduncles arising from a thick-scaled bulbous stem; leaf-margins glabrous, the blades subtruncate to broadly emarginate at apex; peduncles with 2–several flowers; petals usually violet to purple, with cream-colored base, rarely white throughout, rounded at apex; sepals with bright orange callus at apex. . . . 2. *Oxalis violacea*

aa. Plants caulescent with leaves and peduncles arising from aerial stems; corolla yellow.

b. Petals 12–16 mm long; leaflets of the principal leaves (20–) 25–45 mm wide, usually with narrow purplish border; inflorescence cymose; fruiting pedicels erect or ascending . . . . 6. *Oxalis grandis*

bb. Petals 4–11 mm long; leaflets 4–20 (–25) mm wide, usually without purplish border; inflorescence cymose to umbellate; fruiting pedicels erect to reflexed.

c. Leaf bases with stipules as wide at apex as at base or widest at apex and tapering to base; main stem prostrate and creeping, rooting at the nodes; fruiting pedicels deflexed . . . . 3. *Oxalis corniculata*

cc. Leaf bases with stipules narrow at apex and tapering to wider base; main stem erect or ascending, not rooting at the nodes; fruiting pedicels deflexed, ascending, or erect.

d. Stems, petioles, and peduncles usually pubescent, the hairs mostly appressed, not septate; fruiting pedicels usually deflexed, sometimes merely divergent; inflorescence umbellate . . . . 4. *Oxalis dillenii*

dd. Stems, petioles, and peduncles occasionally glabrous, more often with spreading hairs, these usually with septae or constrictions that become tan or brown upon drying; fruiting pedicels erect or ascending; inflorescence cymose . . . . 5. *Oxalis stricta*

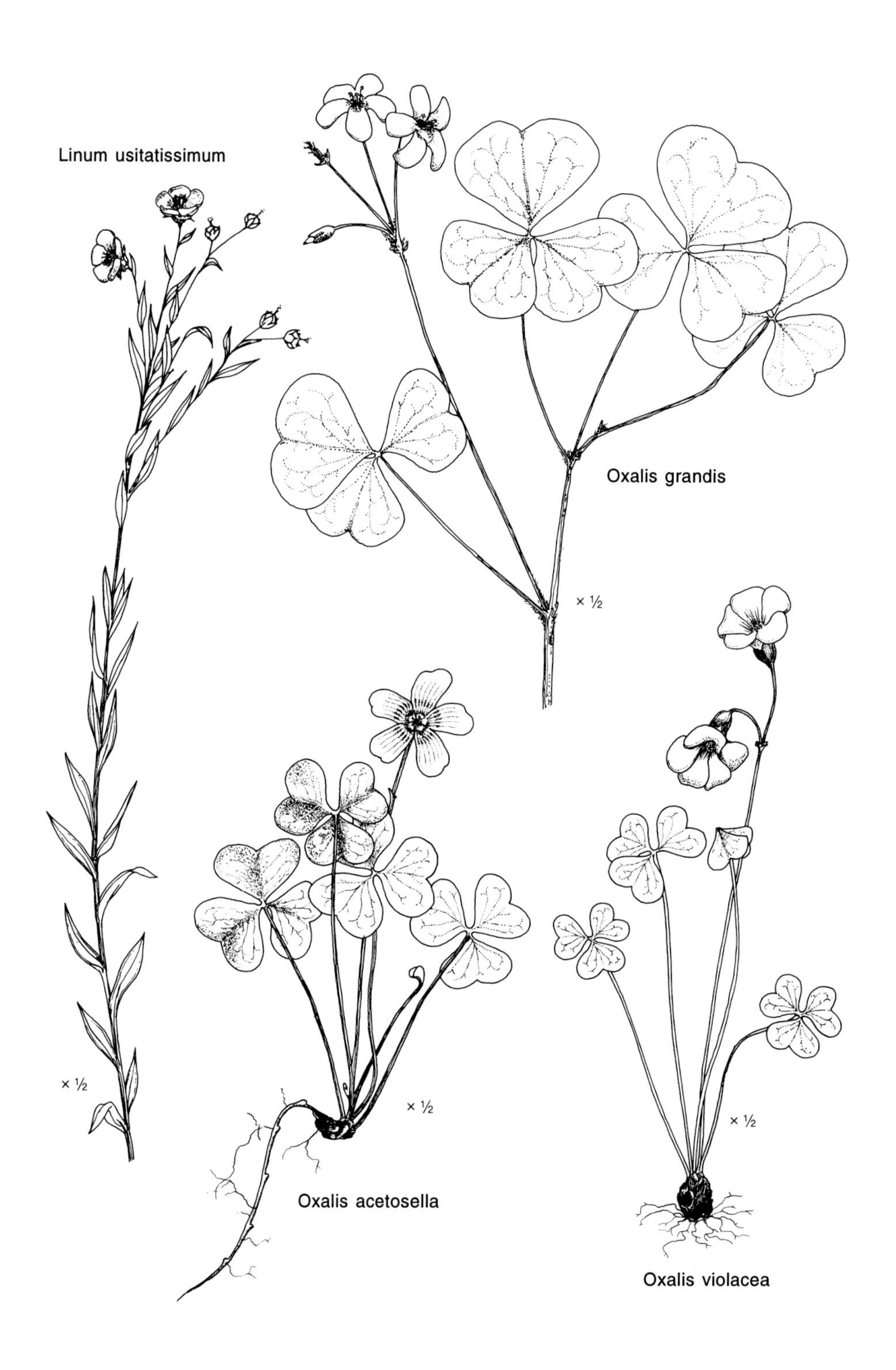
Linum usitatissimum
Oxalis grandis
× ½
× ½
× ½
Oxalis acetosella
× ½
Oxalis violacea

1. **Oxalis acetosella** L. White Wood-sorrel
incl. *O. montana* Raf.

Acaulescent perennial with slender scaly rhizome; leaves basal, petioles to 10 cm long, petioles and leaflets sparsely pubescent and the leaflets short-ciliate; flowers solitary on peduncles to 10 cm or more in length; petals white with pink veins or pinkish-purple with darker veins, 8–15 mm long, shallowly and asymmetrically notched at the apex. The plants produce cleistogamous flowers on slender peduncles late in the season; the varying ratios of cleistogamous and chasmogamous flowers are reported in a recent study by Jasieniuk and Lechowicz (1987). Known from two eastern Ohio counties; rich moist woodlands. Late May–July.

As defined here in the broad sense (following Knuth, 1930), the species is circumboreal. The plants of eastern North America may be distinguished as *Oxalis montana* Raf. or as *O. acetosella* subsp. *montana* (Raf.) D. Löve.

Oxalis acetosella

2. **Oxalis violacea** L. Violet Wood-sorrel

Acaulescent perennial with bulbous, scaly, subterranean base; leaves basal, the petioles to 14 cm long, delicate at base and easily broken from their underground attachment, the petioles and leaves usually glabrous; flowers in an umbellate cluster of 2–several flowers, the peduncles to 25 cm long, and also delicately attached at base; sepals with a distinctive orange callus at tip; petals usually violet to purple and with cream-colored base, rarely white throughout, about 15 mm long, rounded at apex. Throughout Ohio, except northwestern and west-central counties; open woods, grassy woodland borders. April–June.

Denton (1973) recognizes no varieties within this species and places var. *trichocarpa* Fassett in synonymy. A white-flowered specimen, the rare forma *albida* Fassett, was collected from Washington County, in 1976.

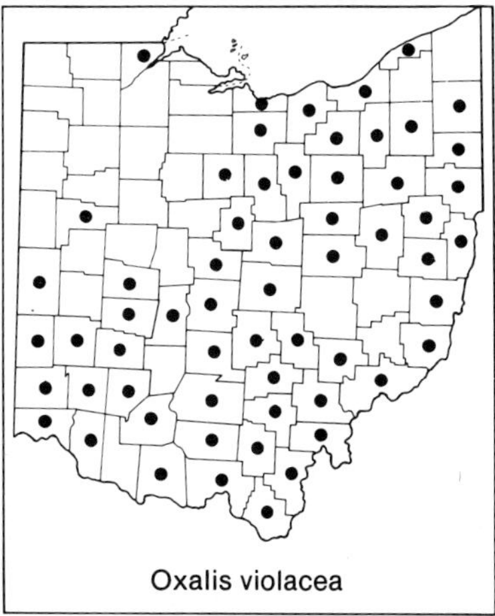
Oxalis violacea

3. Oxalis corniculata L. Creeping Wood-sorrel. Creeping Lady's-sorrel
O. repens Thunb.

Much-branched perennial with the main stem prostrate and creeping, and rooting at the nodes; leaves caulescent, green or sometimes purplish, the leaflets varying in size but generally smaller than in other species of the genus in Ohio, the smallest only 4 mm wide, the leaf bases with stipules either as wide at the apex as at the base or widest at the apex and tapering to the base; flowers 1–few in umbellate cymes; corolla yellow, 4–8 mm long; fruiting pedicels deflexed. A weed of European origin, naturalized at scattered sites in Ohio, probably more frequent than collections indicate; lawns and gardens, roadsides, also a greenhouse weed. May–August.

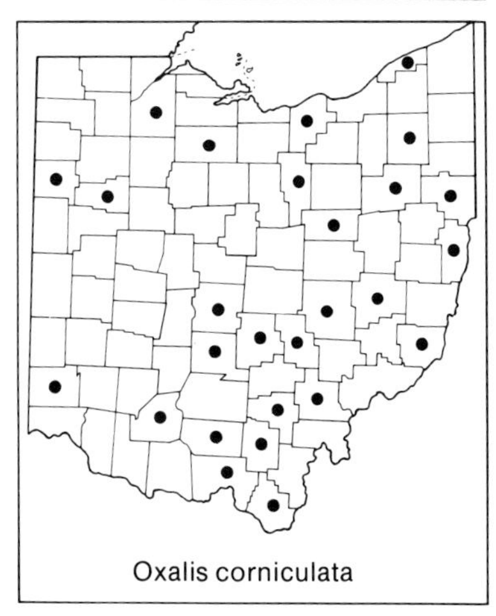
Oxalis corniculata

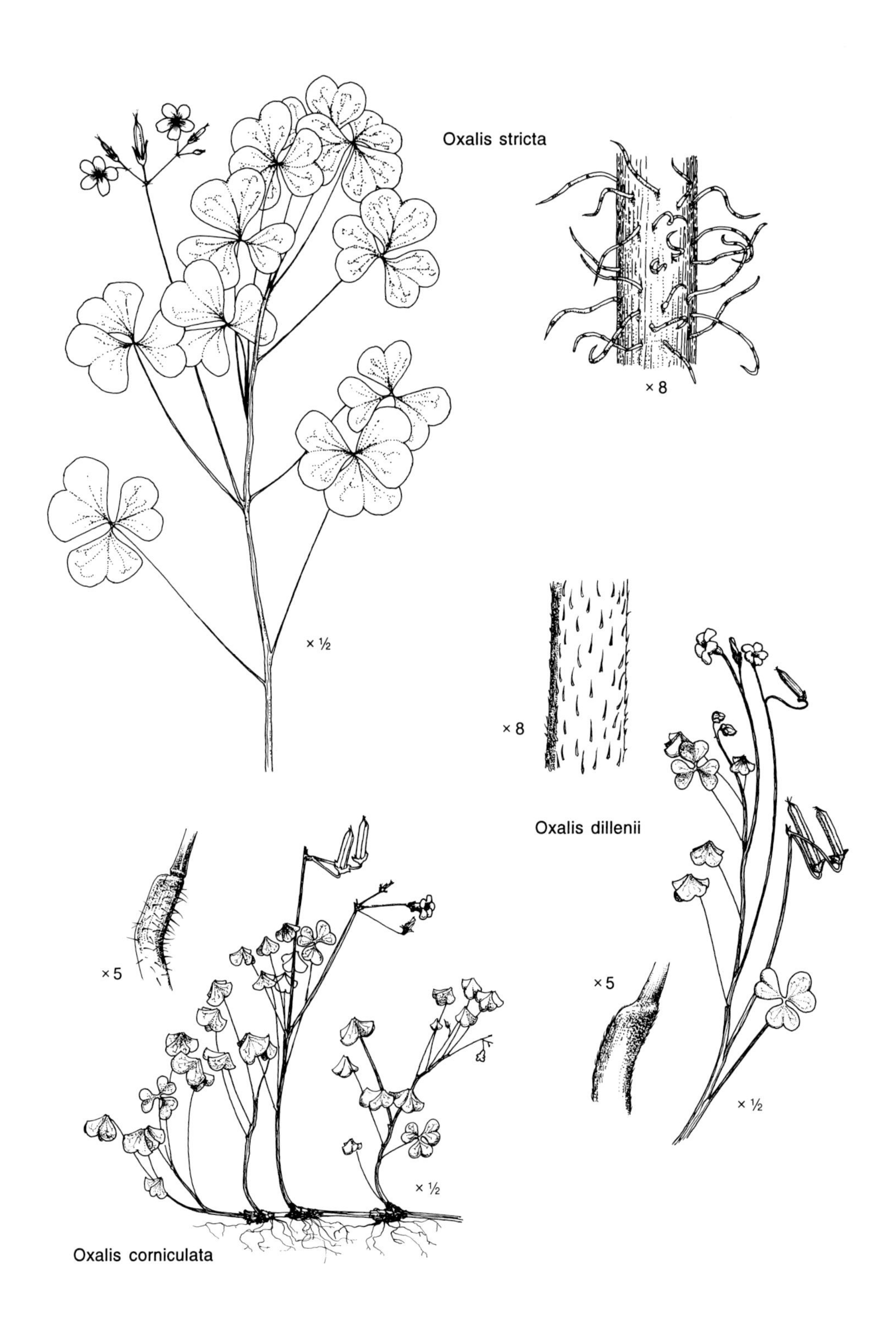
Oxalis stricta
× 8
× ½
× 8
Oxalis dillenii
× 5
× 5
× ½
× ½
Oxalis corniculata

Lourteig (1979) recognizes two varieties from plants of northeastern North America: *O. corniculata* var. *corniculata* and var. *purpurea* Planch., the latter variety with purplish leaves. Eiten (1963) places this variety in synonymy under the species, a decision that seems sensible to me.

The fertility relationships among this species and the next two, *O. dillenii* and *O. stricta*, were the subject of experimental work by Lovett Doust, Lovett Doust, and Cavers (1981).

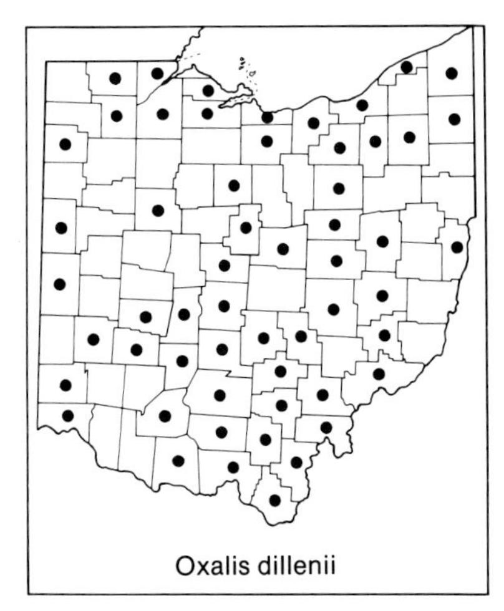

Oxalis dillenii

4. **Oxalis dillenii** Jacq. SOUTHERN YELLOW WOOD-SORREL
   incl. *O. florida* Salisb.
   O. stricta of authors, incl. Fernald (1950), Gleason (1952a), and Lourteig (1979), not L.

Erect or later decumbent perennial, stems not rooting at nodes; leaves caulescent, the leaflets variable in size, but most from 8–15 mm wide, the leaf base with stipules narrow at the apex and tapering to a wider base; stems, petioles, and peduncles usually pubescent with upwardly appressed, nonseptate hairs; flowers 1–few in umbellate cymes; corolla yellow, 7–11 mm long; fruiting pedicels usually deflexed, sometimes merely divergent. A weedy species occurring throughout Ohio, but less common than the next species, *O. stricta*; roadsides, along railroads, in parking lots, and other usually dry disturbed habitats. May–July (–September).

Lourteig (1979) assigns the name *Oxalis stricta* L. to this species, as did Fernald (1950) and Gleason (1952a), placing *O. dillenii* Jacq. in synonymy. The treatment here follows Eiten's (1955) conclusions.

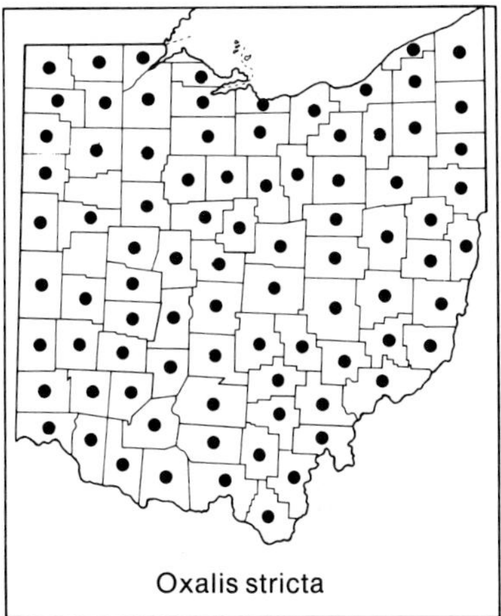

Oxalis stricta

5. **Oxalis stricta** L. COMMON YELLOW WOOD-SORREL
   O. fontana Bunge; *O. europaea* Jord.

Erect to later decumbent annual or perennial; leaves caulescent, the leaflets variable in size, but most 8–25 mm wide, the leaf bases with stipules narrow at the apex and tapering to a wider base; stems, petioles, and peduncles occasionally glabrous, but usually with spreading hairs, the hairs usually with septae or constrictions that become tan or brown upon drying; flowers few to several in cymes; corolla yellow, 7–11 mm long; fruiting pedicels erect or ascending. Throughout Ohio, in all kinds of disturbed habitats, but more often in moist or shaded ones than is *O. dillenii*. June–August (–October).

Applying the name *O. stricta* to the previous species, Lourteig (1979) calls this species *O. fontana* Bunge, a name that antedates *O. europaea* Jord. used by Fernald (1950) and Gleason (1952a). Application of the name *O. stricta* to this species follows the conclusions of Eiten (1955) and Watson (1989).

Fernald (1950) recognized two varieties and several forms based on amount and kind of pubescence. These seem of little merit to me.

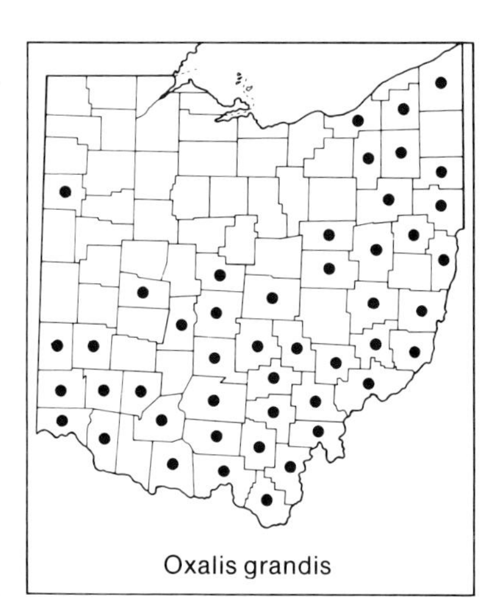
Oxalis grandis

6. **Oxalis grandis** Small GREAT YELLOW WOOD-SORREL
Erect perennial, 2.5–6 (–10) dm tall, the tallest of the Ohio members of this genus; stems and petioles with soft spreading pubescence; leaves cauline, the leaflets variable in size, but most from 25–45 mm wide, and usually with a narrow purplish border; flowers 1–few in cymes; corolla yellow, 12–16 mm long; fruiting pedicels erect or spreading. Chiefly Appalachian in its distribution; eastern and southern Ohio; open woods and woodland borders. May–early July.

Geraniaceae. GERANIUM FAMILY

Annual, biennial, or perennial herbs; leaves opposite, compound or deeply dissected; flowers perfect, in corymbose or umbellate inflorescences; calyx of 5 separate sepals, persistent; corolla actinomorphic or nearly so, of 5 separate petals; stamens 5 or 10; carpels 5, fused to form a compound pistil with 5 styles, the styles separate below but fused apically to the top of a long central column; fruit a schizocarp separating into 5 mericarps, the separation beginning in the ovary and progressing toward the apex, where the 5 mericarps ultimately remain attached to the top of the column, the styles (called "awns" or "beaks") twisting and curling as the upward separation progresses. A family of 11 genera and 700 or more species, nearly half of which are in the genus *Geranium*.

Cultivars of the South African genus, *Pelargonium*, "GERANIUM," are among the most familiar of all Ohio house plants, popular because they often bloom indoors in the winter months, producing large umbellate clusters of bright red, pink, or white flowers. During summer months they are widely cultivated outdoors in flower beds and boxes.

The treatment here is based in large part on the recent study of Ohio Geraniaceae by McCready and Cooperrider (1984). Hanks and Small (1907) published a comprehensive study of the family in North America; Robertson (1972) studied the family in southeastern United States.

a. Leaves palmately lobed or palmately compound; flowers with 10 fertile stamens (except G. *pusillum* with only 5 stamens, some of which are occasionally sterile). . . . . 1. *Geranium*
aa. Leaves pinnately compound, the leaflets pinnately cleft; flowers with 5 fertile and 5 sterile stamens. . . . . 2. *Erodium*

1. Geranium L. CRANE'S-BILL

Perennial, biennial, annual, or winter annual herbs; leaves palmately veined and either palmately lobed or palmately compound. All species except *Geranium pusillum* have 10 fertile stamens. Accurate identification is in some cases dependent on seeing the type of, or lack of, pubescence on the ovary, but the ovary is often obscured by the persistent sepals, which must be bent back or removed. A genus of nearly 300 species, widespread in north temperate regions.

Geranium dissectum L., CUT-LEAVED CRANE'S-BILL, a European species with tiny purplish-pink flowers, occurring rarely in Ohio as an adventive weed, has been collected from Brown, Fairfield, Hamilton, and Lake counties.

a. Plants perennial, with thick horizontal rhizome; petals (10–) 14–25 mm long.

b. Aerial stems erect, usually unbranched below inflorescence, most with only two cauline leaves (in addition to basal leaves originating from rhizome); the cauline leaves 5–15 cm wide . . . . 1. *Geranium maculatum*

bb. Aerial stems erect or decumbent and spreading, branched, with several to many cauline leaves; the cauline leaves 3–5 cm wide . . . . 2. *Geranium sanguineum*

aa. Plants annual or biennial, with slender roots, rhizome lacking; petals 2–10 (–12) mm long.

b. Leaves palmately compound with 3–5 leaflets, the leaflets deeply pinnately lobed, the terminal leaflet stalked. . . . . 8. *Geranium robertianum*

bb. Leaves moderately to deeply palmately lobed, but not divided into leaflets, the lobes in turn variously lobed or cleft.

c. Sepals acuminate with awned tips, the awns 1–3 mm long; outer sepals 6–12 mm long.

d. Some or all pedicels (2–) 3 or more times length of calyx.

e. Pedicels densely glandular-villous; ovaries and mericarps with spreading glandless hairs . . . . 4. *Geranium bicknellii*

ee. Pedicels with appressed, retrorse, glandless hairs; ovaries and mericarps glabrous or with few minute glandular hairs. . . . . 3. *Geranium columbinum*

dd. Pedicels 1–2 times length of calyx, or less.

e. Ovaries and mericarps with spreading to antrorse, glandless hairs 1–1.5 mm long, hairs of style sometimes glandular. . . . . 5. *Geranium carolinianum*

ee. Ovaries, mericarps, and styles with spreading glandular hairs 0.5 mm long or less; see above. . . . . *Geranium dissectum*

cc. Sepals acute to obtuse, without awns but some terminating in a tiny callous tip less than 1 mm long; outer sepals 2–4 mm long.

d. Ovaries and mericarps pubescent; stamens 5 .... 6. *Geranium pusillum*

dd. Ovaries and mericarps glabrous; stamens 10 .... 7. *Geranium molle*

1. **Geranium maculatum** L. WILD GERANIUM. WILD CRANE'S-BILL

Perennial with a thick rhizome producing a few long-petiolate basal leaves, the petioles to 20 cm long, and with an aerial stem bearing two or more leaves; leaves 5–15 (–20) cm wide, 5- (or 7-) lobed; peduncles 1–several, each bearing 1–several pairs of flowers, in all forming an umbellate inflorescence; petals rose-purple, 15–25 mm long, the flowers 3.2–3.8 cm across. One of Ohio's most prized spring wildflowers; throughout the state; moist woods, especially wooded floodplains, less often in drier and more open sites in fields and thickets. Agren and Willson (1991) studied seed production in this species. Late April–June (–August).

The rare white-flowered forma *albiflorum* (Raf.) House was collected from Pickaway County in 1931.

2. GERANIUM SANGUINEUM L. BLOOD-RED CRANE'S-BILL

Branched perennial with a thick rhizome; aerial stems erect or more often decumbent and spreading, with several to many cauline leaves 3–5 cm wide and deeply 5–7-lobed; flowers axillary, 3.3–3.5 cm across, petals crimson-purple, the petals (10–) 15–20 mm long. An introduced Eurasian species; frequently cultivated in Ohio, occasionally escaping and becoming naturalized mostly in southern counties; disturbed open sites. May–August.

3. GERANIUM COLUMBINUM L. LONG-STALKED CRANE'S-BILL

Annual with erect or ascending branches; leaves 2–5 cm wide, deeply 5- or 7-lobed; peduncles axillary, each bearing usually a pair of pedicellate

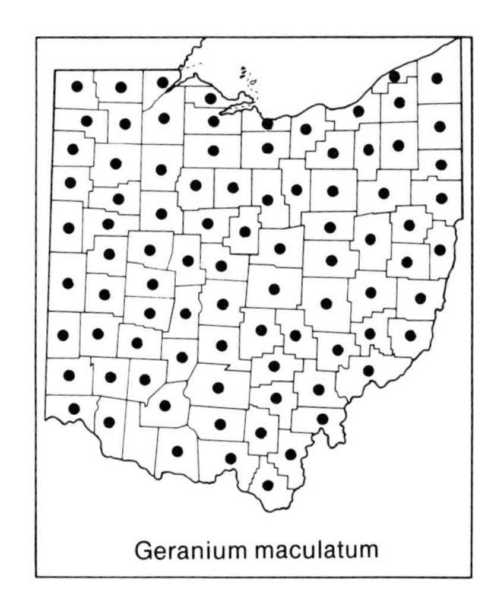

Geranium maculatum

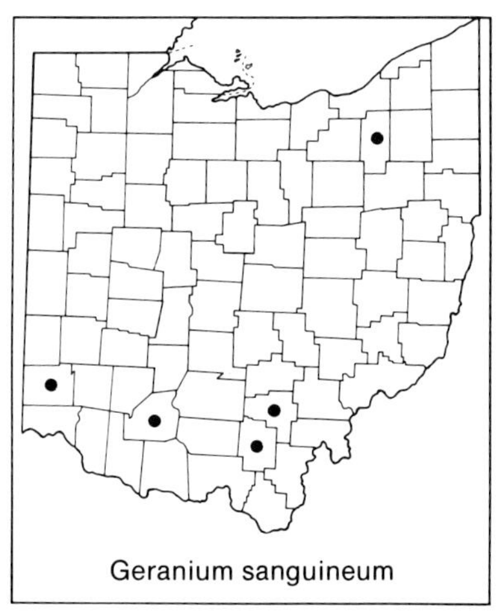

Geranium sanguineum

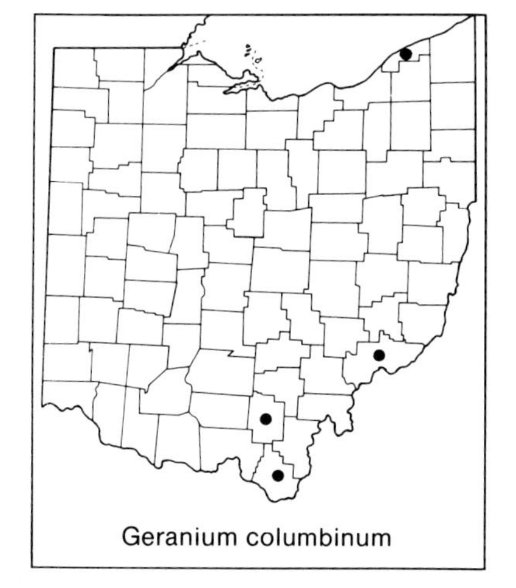

Geranium columbinum

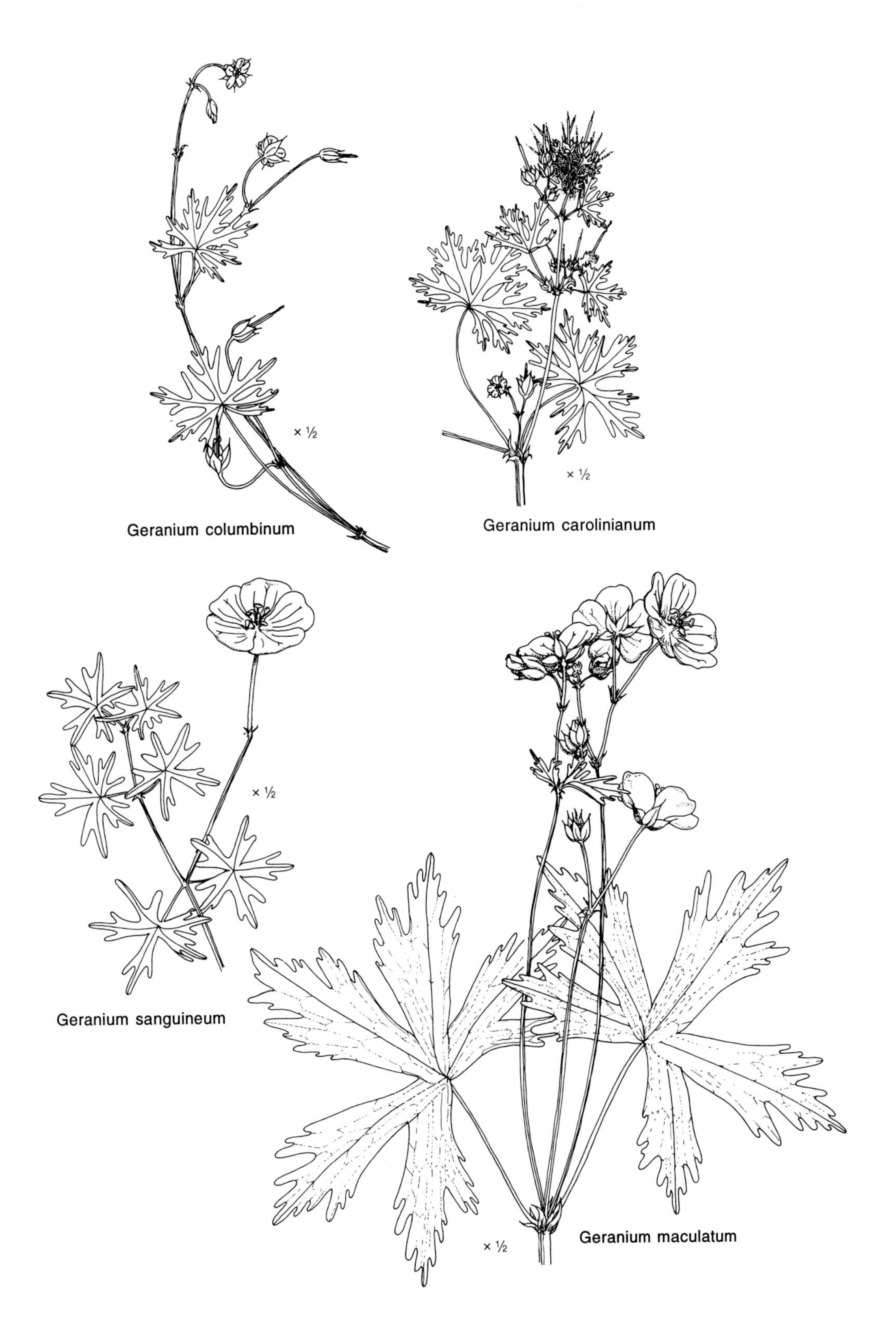
× ½
Geranium columbinum
× ½
Geranium carolinianum
× ½
Geranium sanguineum
× ½
Geranium maculatum

flowers; some or all of the pedicels 3 times the length of the calyx or more and with appressed, retrorse, glandless hairs; flowers 1.6–2.0 cm wide; sepals acuminate with awned tips, the outer sepals 10–12 mm long; petals purplish-pink, 7–9 mm long; ovaries and mericarps glabrous or with few minute glandular hairs. A European species that is naturalized, or perhaps merely adventive, at a few sites, mostly in southern counties; open disturbed ground. June–September.

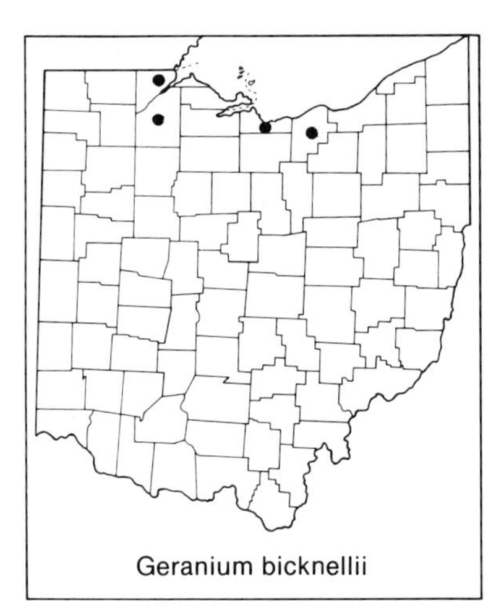

Geranium bicknellii

4. **Geranium bicknellii** Britt. Bicknell's Crane's-bill

Annual or biennial, erect with spreading branches; leaves 3–7 cm wide, deeply 5-lobed; peduncles axillary, each usually bearing a pair of pedicellate flowers; pedicels densely glandular-villous, some or all 3 times the length of the calyx or more; flowers about 1.2 cm across; sepals acuminate with awned tips, the outer sepals 7.5–8.5 mm long; petals pinkish-purple, about 8 mm long; ovaries and mericarps with spreading glandless hairs. Known from a few counties in northern Ohio; open, disturbed, sandy or gravelly sites; thought to have been extirpated from Ohio (Cooperrider, 1982b), but recently discovered in Lucas County (E. D. Hardin, 1985). July–September.

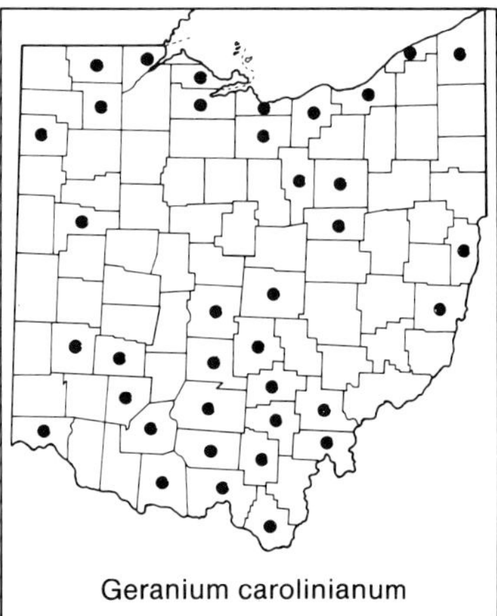

Geranium carolinianum

5. **Geranium carolinianum** L. Carolina Crane's-bill

incl. var. *carolinianum* and var. *confertiflorum* Fern.

Annual with ascending branches; leaves 3–6 (–7) cm wide, 5–7-lobed; flowers few to many in a terminal corymb; pedicels at most twice as long as calyx; flowers 6–8.5 mm across; sepals acuminate with awned tips, the outer sepals 6–10 mm long; petals pink or pale pink, about 6 mm long; ovaries and mericarps with spreading to antrorse glandless hairs, the hairs 1.0–1.5 mm long, hairs of the style sometimes glandular. Widespread in Ohio, but chiefly in south-central counties and northern ones bordering Lake Erie; in a variety of habitats, mostly dry, especially at roadsides, along railroads, and in fields and openings. Late April–June.

Fernald recognized the two varieties listed in synonymy above, but after a detailed study, McCready and Cooperrider (1984) concluded that, in "the absence of any useful discontinuities in morphology or geographic range, it seems inappropriate to recognize varieties among Ohio members of the species, a course of action also adopted by Radford, Ahles, and Bell (1968) for the Carolinas."

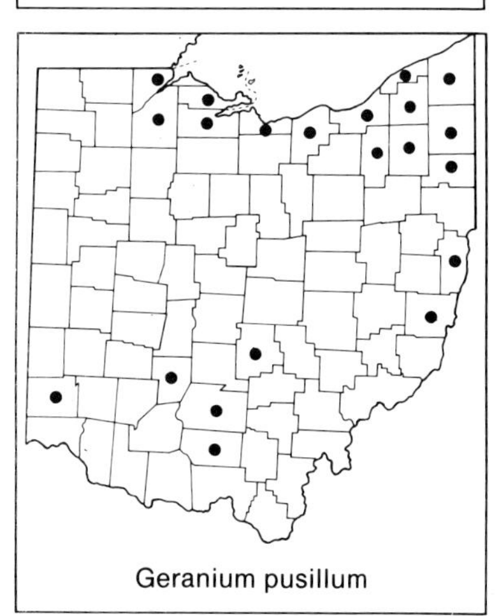

Geranium pusillum

6. Geranium pusillum L. Small-flowered Crane's-bill

Annual with decumbent or prostrate branches; leaves 1–4 (–6) cm wide, 3–7-lobed; peduncles axillary, each usually bearing a pair of pedicellate flowers; flowers 3–5 mm across; sepals acute to obtuse, but without awns, the outer sepals 2.5–4 mm long; petals pale violet, 2–4 mm long; stamens 5; ovaries and mericarps pubescent. A European species, infrequently nat-

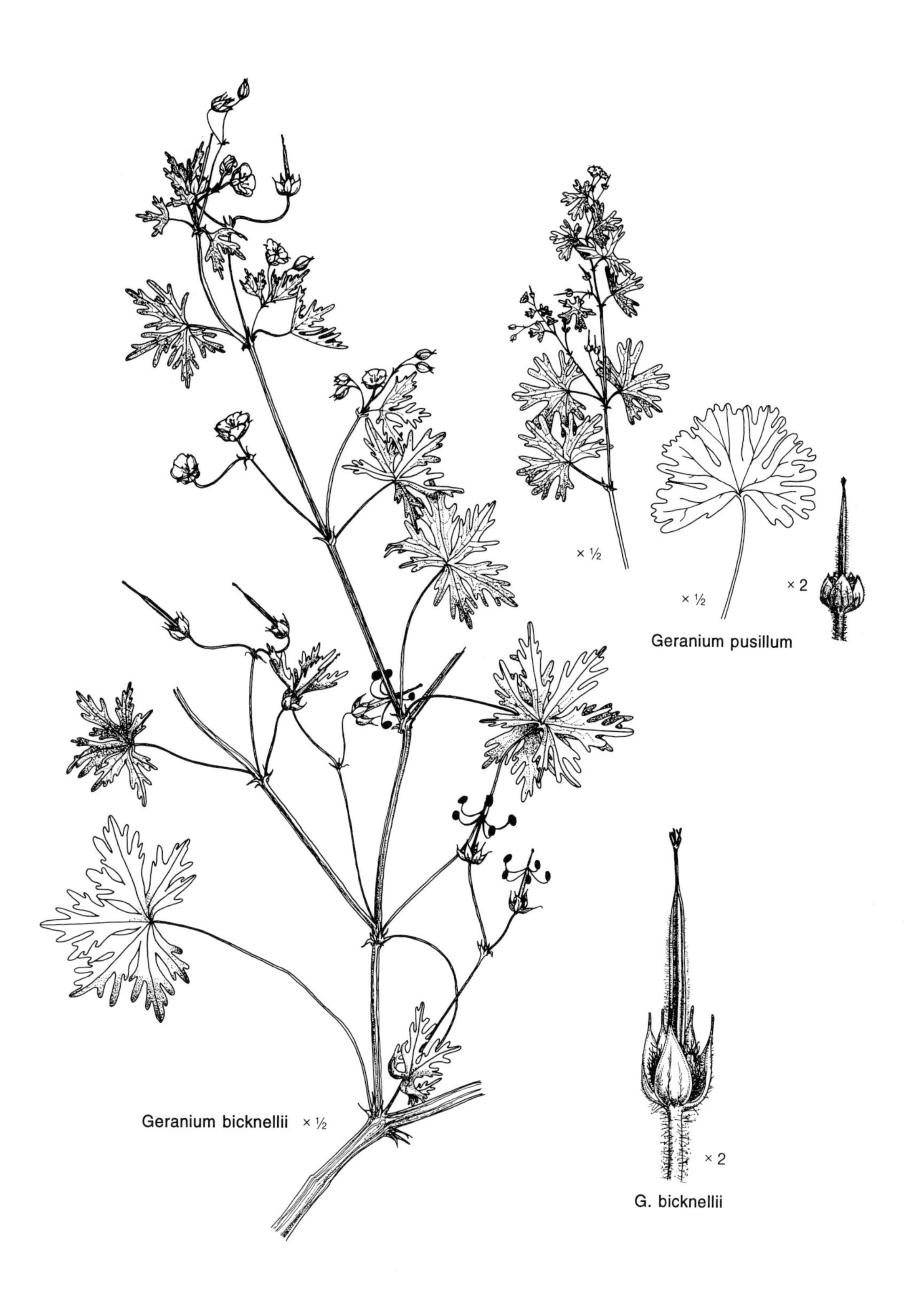
× ½
× ½
× 2
Geranium pusillum
Geranium bicknellii × ½
× 2
G. bicknellii

uralized, mostly in northeastern and north-central counties; lawns and other open, disturbed, weedy places. May–September.

7. Geranium molle L. Dove's-foot Crane's-bill

Annual with spreading or decumbent branches; leaves 2–4 (–6) cm wide, 5–7-lobed; peduncles borne in axils of full-sized or reduced leaves, each usually bearing a pair of pedicellate flowers; flowers 5–7 mm across; sepals acute to obtuse, but without awns, the outer sepals 3–4 mm long; petals pinkish-purple, 3–7 mm long; ovaries and mericarps glabrous. A European species infrequently naturalized, mostly in northern counties; lawns and other open disturbed sites. May–August.

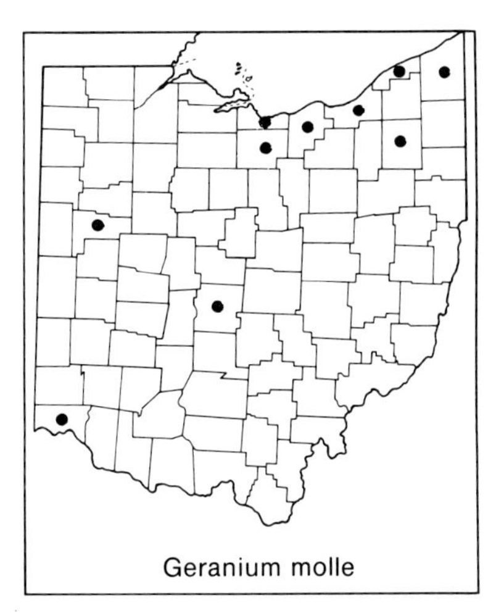

Geranium molle

8. **Geranium robertianum** L. Herb Robert

Branched, strong-scented annual with weak, pubescent, erect to procumbent stems; leaves 3–8 cm wide, palmately compound with 3–5 leaflets, the leaflets deeply pinnately lobed, the terminal leaflet stalked; peduncles axillary, each usually bearing two pedicellate flowers; flowers 11–15 mm across; sepals 6–8 mm long, with filiform tips; petals bright pink to purplish, 9–12 mm long, clawed. Mostly in northern counties; moist woods and borders, and on moist, shaded, usually calcareous, rock exposures. May–September.

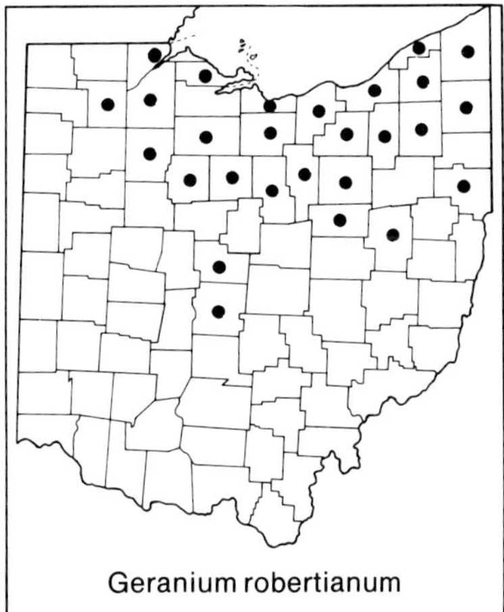

Geranium robertianum

2. Erodium Ait. STORK'S-BILL

A genus of the Mediterranean region with some 60 or more species, one of which has become naturalized in Ohio.

1. Erodium cicutarium (L.) L'Hér. Alfilaria. Red-stemmed Filaree

Low, somewhat sprawling, winter annual or biennial herb; leaves pinnately compound, the leaflets further pinnately dissected; flowers purplish-pink, 6–9 mm across; with 5 fertile and 5 sterile stamens. Naturalized from Europe at scattered sites; open disturbed ground, especially lawns, cultivated ground, and roadsides. The species is apparently increasing in frequency in Ohio. March–October.

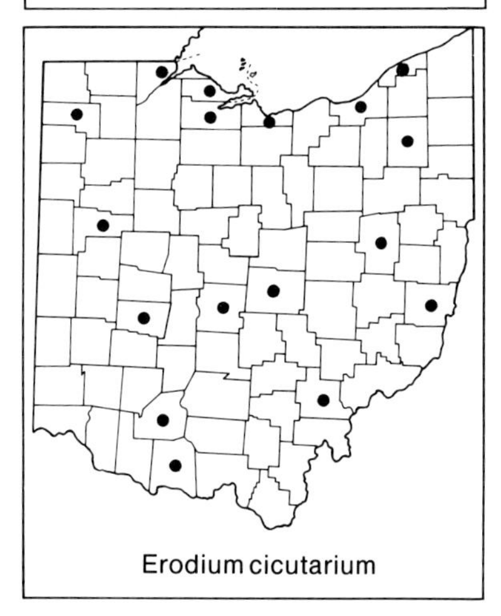

Erodium cicutarium

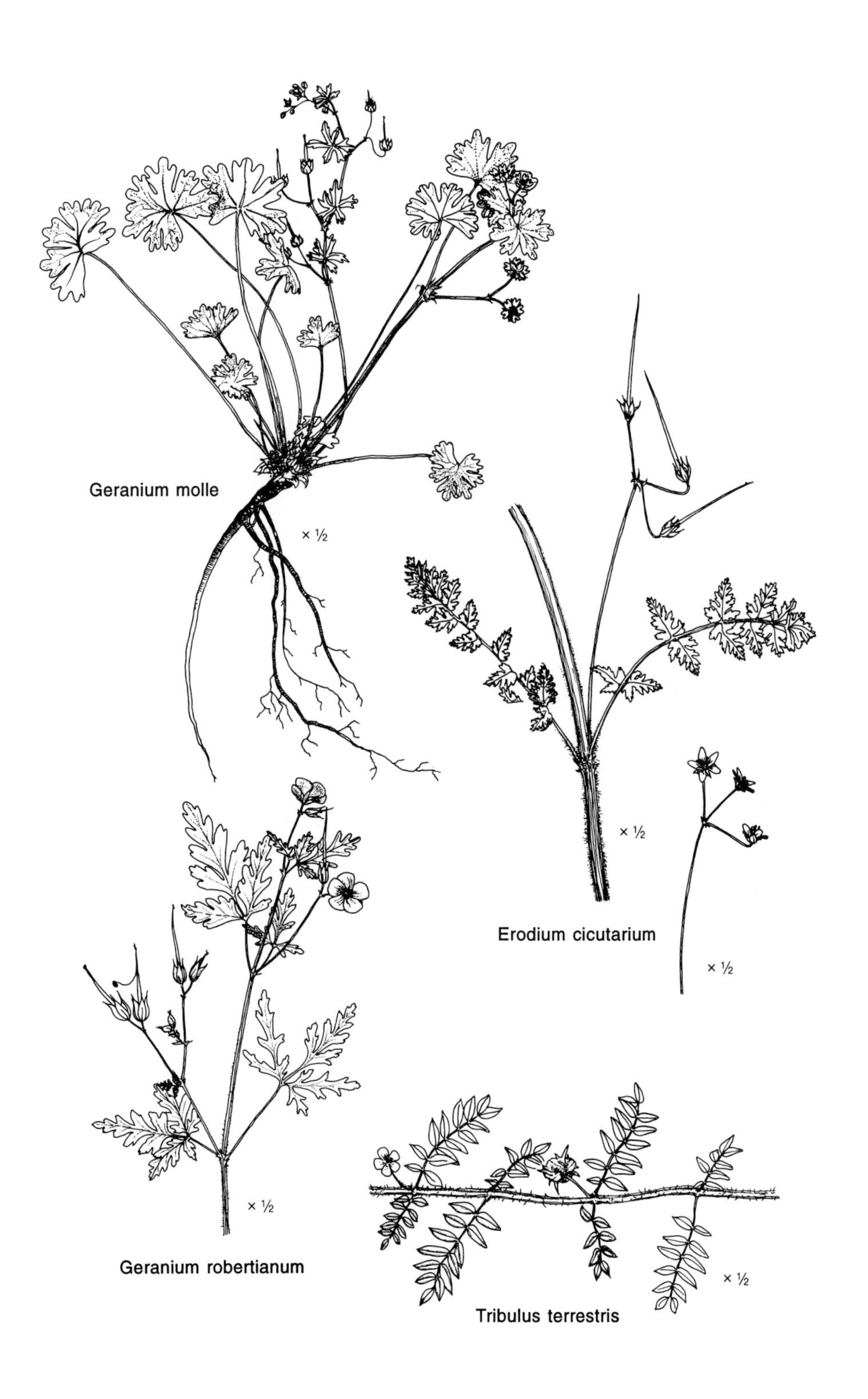
Geranium molle
× ½
× ½
Erodium cicutarium
× ½
× ½
Geranium robertianum
× ½
Tribulus terrestris

Zygophyllaceae. CALTROP FAMILY. CREOSOTE-BUSH FAMILY

A small, chiefly tropical family with one genus represented in the Ohio flora. *Larrea tridentata* (DC.) Cov., also a member of the family, is the CREOSOTE-BUSH of southwestern United States. Porter (1972) studied the family in southeastern United States.

1. Tribulus L. CALTROP

A small genus of 12 or more species, mostly in tropical or subtropical regions, with a single species in Ohio.

1. TRIBULUS TERRESTRIS L. PUNCTURE-VINE. PUNCTURE-WEED

Small, much-branched, prostrate annual with hirsute stems; leaves opposite, pinnately compound, the two leaves at each node unequal in length; flowers perfect, solitary in leaf axils, on short peduncles; calyx of 5 separate, early deciduous sepals; corolla actinomorphic, 1 cm or less across, of 5 separate, pale yellow petals; stamens 10, in two series of 5, slightly fused at base; carpels 5, fused to form a compound pistil with a single style; ovary superior; fruit a spiny schizocarp, ca. 1 cm in diameter, separating at maturity into 5 hard mericarps, each bearing two stout divergent spines and a central ridge with smaller spines. Native of the Mediterranean region; naturalized, or perhaps merely adventive, at a few scattered sites in Ohio; disturbed ground, especially along railroad tracks. June–September.

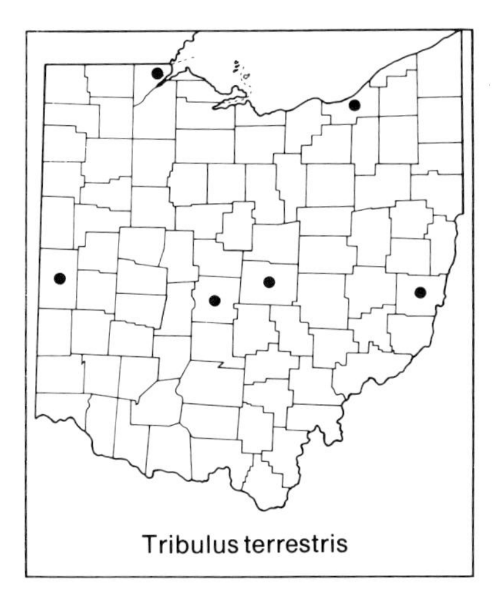
Tribulus terrestris

Rutaceae. RUE FAMILY

Deciduous shrubs and small trees; leaves alternate, compound; flowers small, actinomorphic or nearly so, perfect or imperfect, greenish to whitish, in axillary or terminal cymes; fruit a samara or subfleshy follicle. A family of about 150 genera and 1,500 species, widely distributed around the world, with distribution centers in South Africa and Australia. *Citrus*, the source of citrus fruits, is the best-known genus. Trees of the genera *Evodia* and *Phellodendron* are occasionally seen in cultivation in Ohio. Brizicky (1962a) treats the family as it is represented in the flora of southeastern United States.

a. Leaves pinnately compound with 5–11 leaflets; stems with prickles . . . . 1. *Zanthoxylum*
aa. Leaves with 3 leaflets; stems without prickles . . . . 2. *Ptelea*

1. Zanthoxylum L. PRICKLY-ASH

A chiefly tropical genus of about 200 species, one of which is native to Ohio. Porter (1976) revised the genus in North America, north of Mexico. The generic name is sometimes spelled *Xanthoxylum*.

1. **Zanthoxylum americanum** Mill. PRICKLY-ASH

Dioecious shrubs, or very rarely small trees; stems to 5.5 m (18 feet) tall, but often much shorter, with stout paired prickles at the nodes (very stout in young growth) and red-rusty pubescent winter buds; leaves alternate, odd-pinnately compound, the leaflets mostly 7 or 9, ovate-elliptic to narrowly elliptic, very shallowly toothed to subentire, the petiole and rachis sometimes with a few prickles; fruit a tiny, dull, subfleshy follicle; the fruits and foliage pleasantly aromatic when crushed. Throughout Ohio, but more frequent westward; often forming clonal thickets (Reinartz and Popp, 1987); open moist woodlands and on wooded floodplains, terraces, and steep banks. April–May.

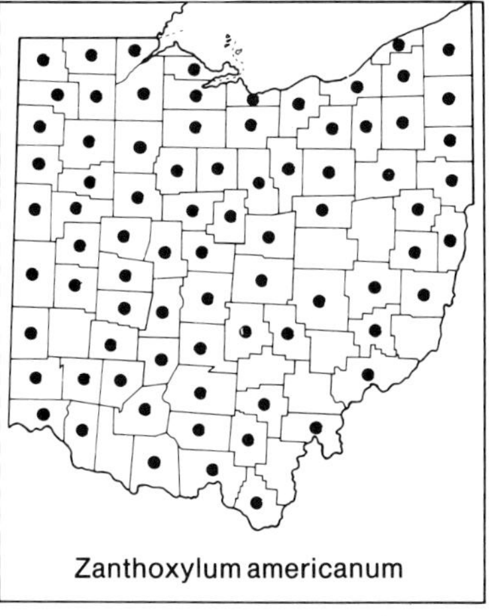

Zanthoxylum americanum

2. Ptelea L. HOP-TREE

A North American genus of three or more species, one of which is native to Ohio. Bailey (1962) prepared the most recent revision of the genus.

1. **Ptelea trifoliata** L. HOP-TREE. WAFER-ASH

Erect shrubs or small trees, to 8 m (26 feet) tall; leaves alternate, trifoliolate, the terminal leaflet very short-stalked, the lateral two sessile and more or less oblique at base, the leaflets entire or subentire, and ovate to elliptic or rhombic in shape, the foliage unpleasantly aromatic when crushed; fruit a distinctive, nearly circular, samara. Throughout Ohio, but

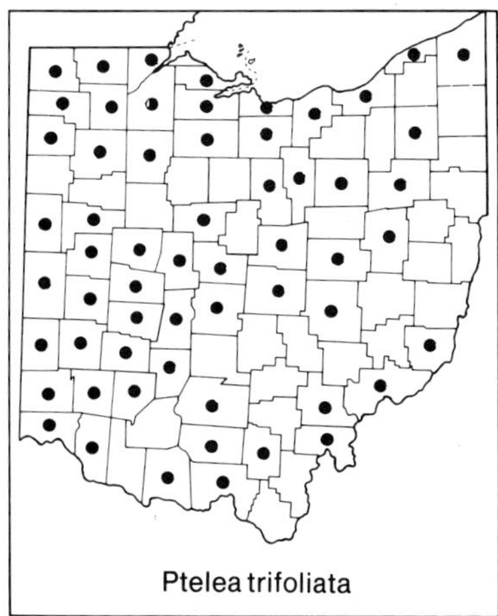

Ptelea trifoliata

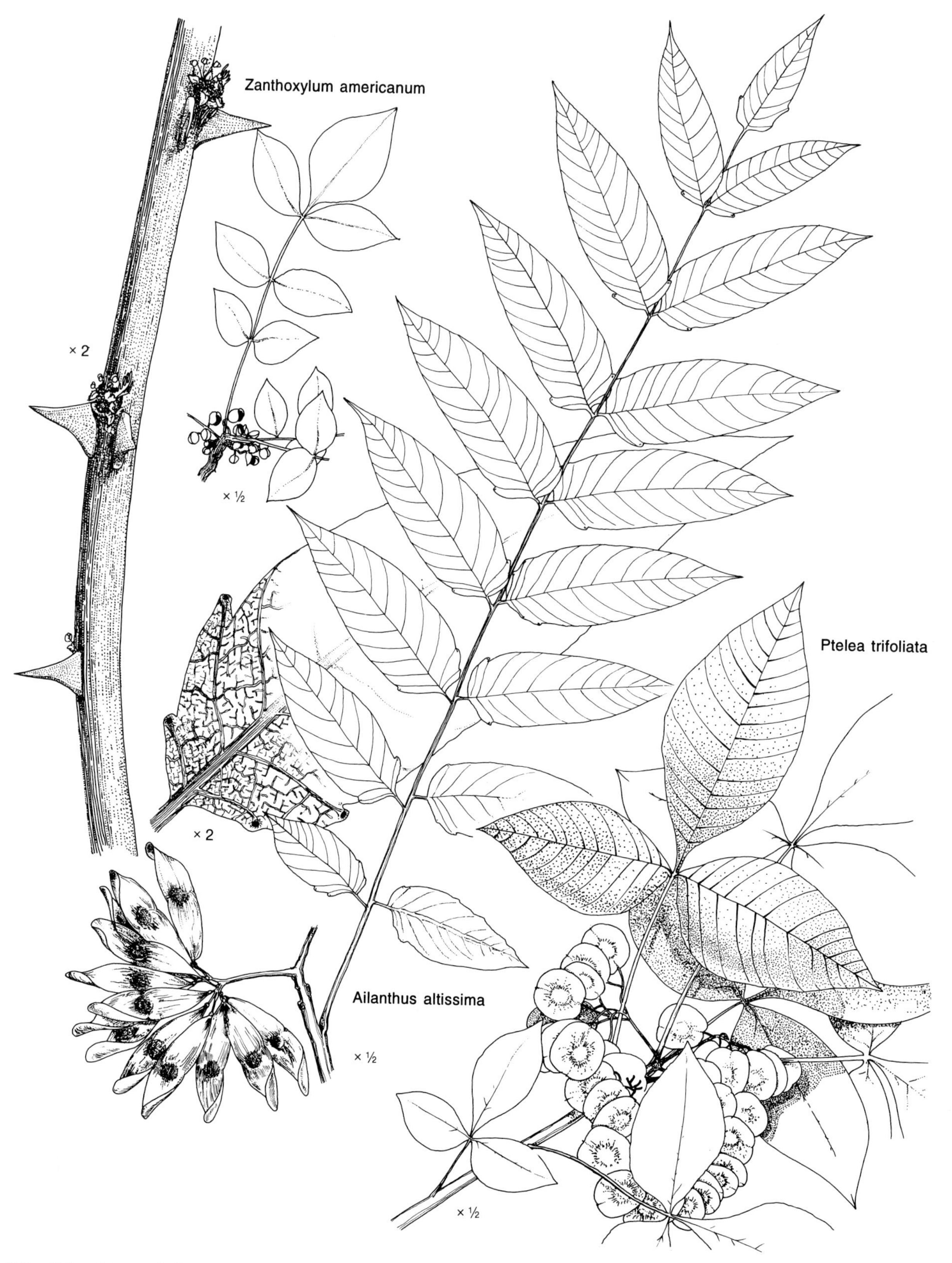
Zanthoxylum americanum
× 2
× ½
× 2
Ptelea trifoliata
Ailanthus altissima
× ½
× ½

more frequent in western counties and in northern ones bordering Lake Erie; moist woods and thickets. Bailey, Herlin, and Bailey (1970) examined variation among populations of this species throughout eastern United States. Late May–June.

Although Ohio plants are mostly glabrous, a form persistently pilose on the undersurface of leaflets, which occurs in some western counties (Braun, 1961), has been recognized as forma *pubescens* (Pursh) Fern. Bailey (1962) does not recognize this form. She assigns all Ohio plants to *Ptelea trifoliata* subsp. *trifoliata* var. *trifoliata*.

Simaroubaceae. QUASSIA FAMILY

A small, chiefly tropical family with a single genus in the Ohio flora. Brizicky (1962b) treats the family in southeastern United States.

1. Ailanthus Desf.

A genus of eastern and southeastern Asia with 15 species, one of which is established in Ohio.

1. Ailanthus altissima (Mill.) Swingle Tree-of-heaven

Tree, to 22.5 m (74 feet) tall, usually much shorter; stems emitting a foul odor when broken; leaves alternate, odd-pinnately compound, with 11–25 or more leaflets, each leaflet with 1–4 low callous-tipped teeth near the base of the otherwise entire margin, the individual leaves 2–6 dm or more in length; flowers small, yellowish to greenish, actinomorphic, in large, terminal, many-flowered panicles; flowers mostly imperfect, the species functionally dioecious, the flowers of staminate trees emitting a foul odor; fruit a slightly twisted samara, produced in abundance by the pistillate trees. A Chinese species, introduced into North America in 1784 (Wyman, 1965), this tree is now naturalized in disturbed habitats throughout Ohio. It is abundant in southern counties. June.

A rapidly growing plant, its fragile wood fractures easily. The frequently seen stout sucker shoots grow as much as 3 m in a single season, and display very large shield-shaped leaf scars in winter.

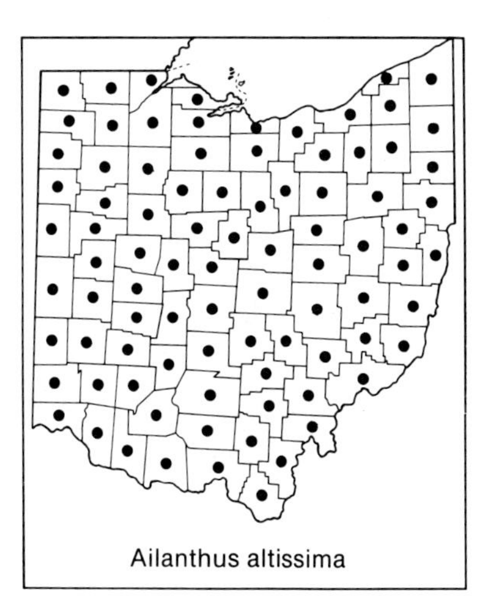
Ailanthus altissima

Polygalaceae. MILKWORT FAMILY

A cosmopolitan family of about 10 genera and 1,000 species, represented in the Ohio flora by a single genus. N. G. Miller (1971) prepared a study of the family in southeastern United States.

1. Polygala L. MILKWORT

Annual or perennial herbs; leaves simple; flowers perfect, in many-flowered racemes, or in spike-like or head-like racemes, or infrequently the flowers few or solitary; calyx zygomorphic, of 5 sepals: 3 outer ones small and green, and 2 inner ones (called "wings") larger and petal-like in color; corolla zygomorphic, of 3 petals: 2 upper ones alike, and the lower one keeled, the 3 variously fused to each other or to the staminal sheath; stamens 8, less often 6 or 7, basally fused to form a sheath split along the upper side; carpels 2, fused to form a compound pistil with a single, but apically 2-lobed, style; ovary superior; fruit a loculicidal capsule. A widely distributed genus of more than 500 species.

The treatment here is based in part on the recent study of Ohio species of *Polygala* by Burns (1986). Saulmon (1971) treated the genus as it is represented in the flora of southeastern United States. J. M. Gillett (1968) covered the genus in Canada and has a set of fine illustrations for each species, including all the Ohio species except *P. cruciata* and *P. curtissii.*

a. Flowers large, 12–20 mm long, purplish-rose or rarely white, solitary or in clusters of 2, 3, or 4; the few to several upper leaves clustered below flowers, the lower leaves abruptly much reduced and scale-like; plants perennial . . . . 1. *Polygala paucifolia*

aa. Flowers small, 1.5–10 mm long, purple, rose, white, or greenish, in many-flowered racemes, the racemes sometimes spike-like or head-like; leaves more or less evenly spaced, the lower not, or only gradually, reduced in size toward base of stem; plants annual or perennial.

b. Lower and medial leaves whorled or sometimes only opposite; plants annual.

c. Racemes 10–15 mm wide, cylindric, abruptly apiculate at apex; the inflorescence immediately subtended by full-sized leaves; the two inner sepals ("wings") deltoid to triangular, tapering to an acuminate apex. . . . . 7. *Polygala cruciata*

cc. Racemes 2–6 mm wide, slenderly cylindric or slenderly conic, gradually tapering toward apex; the peduncle naked or with small bracts below inflorescence; the two inner sepals ("wings") oblong-elliptic, mostly rounded at apex. . . . . 8. *Polygala verticillata* bb. Leaves alternate; plants annual or perennial.

c. Plants perennial, with (1–) 2–several stems arising from fibrous or thickened roots; inflorescence loose or dense, elongate, mostly 3–8 times as long as wide.

d. The plants with thick, often knotted, roots; inflorescence either dense throughout or dense above and loose below; pedicels 0.2–0.6 mm long; flowers white or greenish . . . . . . 3. *Polygala senega*

dd. The plants with fibrous roots; inflorescence loose and open; pedicels 1–4 mm long; flowers pink or purplish-pink, rarely white . . . . 2. *Polygala polygama*

cc. Plants annual, with a single, simple or branched stem and a slender taproot; inflorescence dense or very dense, mostly 1–4 times as long as wide.

d. Leaves narrowly linear, 0.3–0.6 mm wide, early deciduous, with many fallen by anthesis; inflorescence spicate, the pedicels less than 1 mm long . . . . 4. *Polygala incarnata*

dd. Leaves broadly linear to oblong-linear or very narrowly oblanceolate, 0.8–4.5 mm wide, persistent; inflorescence racemose-capitate, the pedicels 1–2 mm long.

e. Inflorescence very dense, obtuse at apex, purple to rose or greenish or whitish; largest inner sepals ("wings") deltoid to broadly rhombic or orbicular-elliptic, (2.5-) 3.0–3.5 mm wide, truncate to obtuse at base . . . . 5. *Polygala sanguinea*

ee. Inflorescence dense but somewhat loose below, acute at apex, purplish-pink; largest inner sepals ("wings") oblong to oblanceolate, 1.5–2.5 (–3.0) mm wide, tapering and acute at base . . . . 6. *Polygala curtissii*

1. **Polygala paucifolia** Willd. FRINGED MILKWORT. GAY-WINGS
Perennial with underground rhizomes originating from small tubers; aerial stems erect, often bearing minute cleistogamous flowers ca. 2 mm long on short branches near the stem base; stems with 3–7 full-sized leaves clustered near stem apex, the lower leaves abruptly much-reduced and scale-like; leaves alternate, elliptic to narrowly ovate; flowers rose-purple to white, solitary or in clusters of 2–4 at stem tip, 12–20 mm long (much larger than flowers of the other Ohio species of *Polygala*); stamens 6. A rare species known from a few sites in northern Ohio; open sandy soils, woodland openings. LaFrankie (1983) describes the growth habit of this species. May–June.

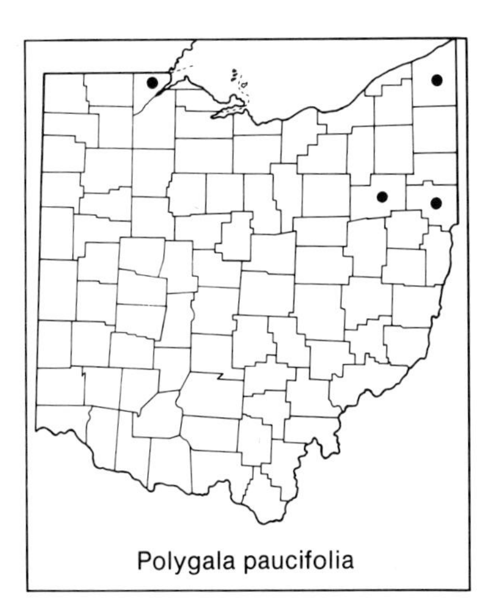
Polygala paucifolia

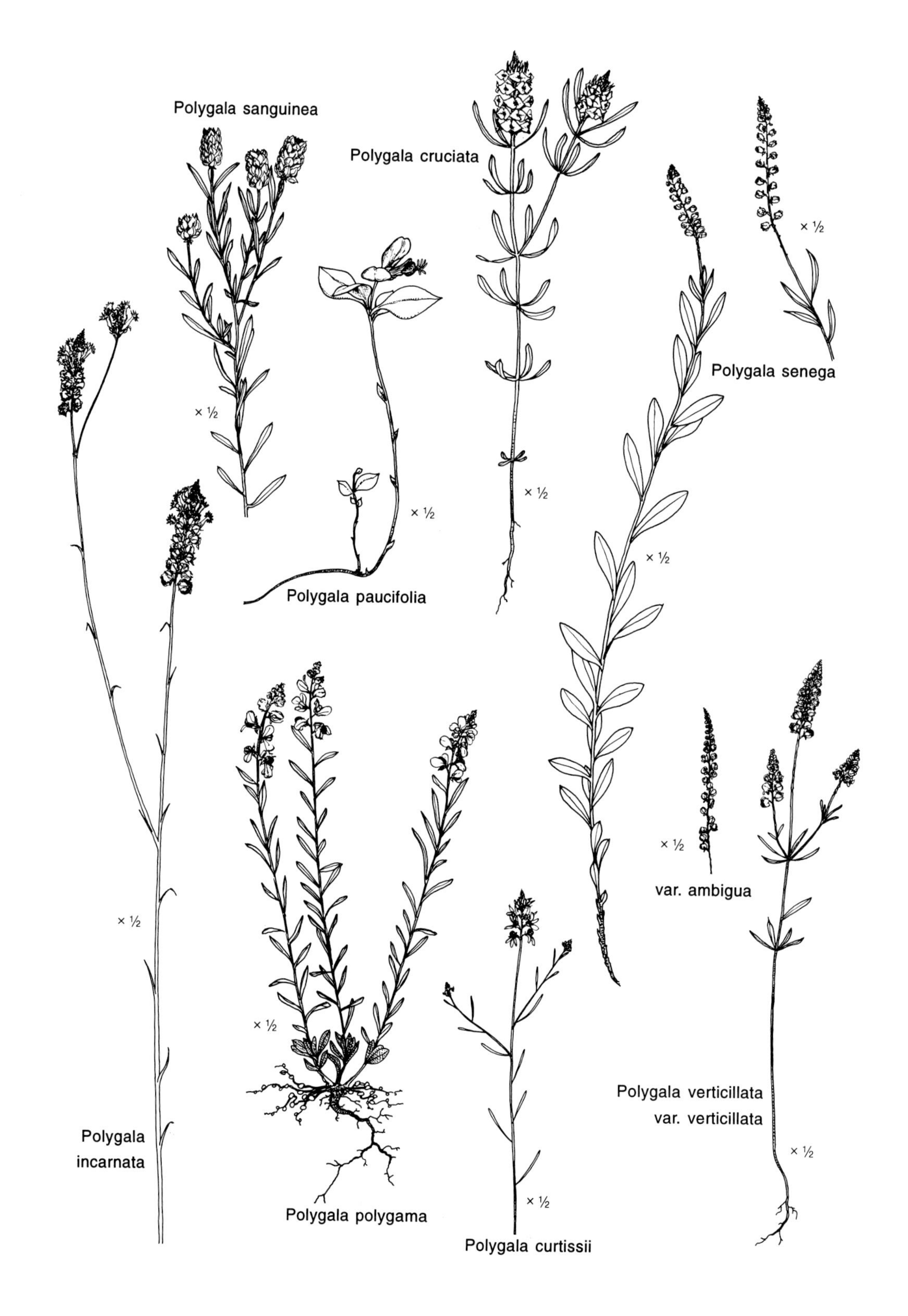
Polygala sanguinea
Polygala cruciata
× ½
Polygala senega
× ½
× ½
× ½
Polygala paucifolia
× ½
× ½
× ½
var. ambigua
× ½
× ½
Polygala verticillata
var. verticillata
× ½
Polygala
incarnata
× ½
Polygala polygama
Polygala curtissii

2. **Polygala polygama** Walt. RACEMED MILKWORT

incl. var. *polygama* and var. *obtusata* Chodat

Perennial; few to several erect stems originating from a crown, the crown also producing short underground racemes with minute cleistogamous flowers; leaves numerous, alternate, linear to narrowly oblanceolate or very narrowly oblong; flowers pink or purplish-pink, rarely white, 4–6 mm long, borne on short pedicels 1–4 mm long in loose open racemes; stamens 7. A few northern Ohio counties; open, usually sandy, sites. June–early August.

In Fernald's (1950) treatment, all or most Ohio plants would be assigned to var. *obtusata.* James (1957), however, found no merit in the recognition of varieties within this species and placed var. *obtusata* and var. *ramulosa* Farwell in synonymy under the species. Voss (1985a) reduced the latter variety to the rank of *forma.*

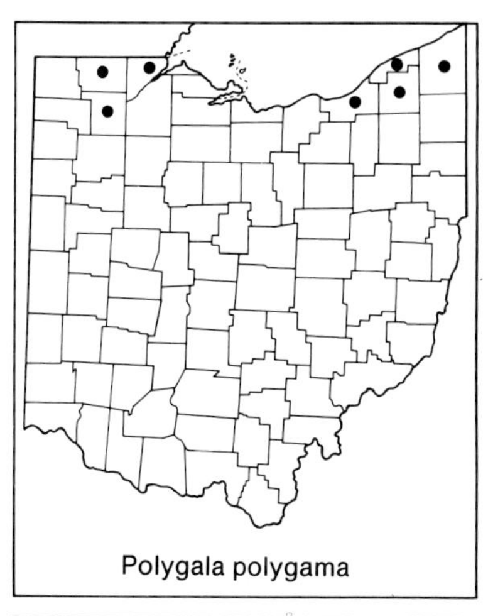
Polygala polygama

3. **Polygala senega** L. SENECA SNAKEROOT

incl. var. *senega* and var. *latifolia* T. & G.

Perennial with thick, often knotted, roots; few to several erect, puberulent stems originating from a crown; leaves numerous, alternate, the lowest reduced and scale-like, the others highly variable, ranging from elliptic-linear to lanceolate to elliptic, rhombic, or ovate; flowers white or greenish-white, 3–3.5 mm long, subsessile on very short pedicels, in a dense to somewhat open raceme. Known from scattered localities in the western half of Ohio and some northeastern counties, but absent from most parts of the unglaciated Allegheny Plateau; open woodlands, prairies, and other openings. May–early July.

Ohio specimens show great variation in leaf width, from 3 to 20 mm, with a complete series of transitional forms from narrow to broad leaves. The broadest are separated as var. *latifolia* by Fernald (1950), but in view of the complete series of intergrades, such a separation seems of little value. Burns (1986) reports a specimen from Adams County of a single plant on which one stem has broad leaves and another stem narrow leaves.

The dried roots have been used medicinally as an emetic and as a stimulant. The common name derives from its use by Native Americans in the treatment of snakebite.

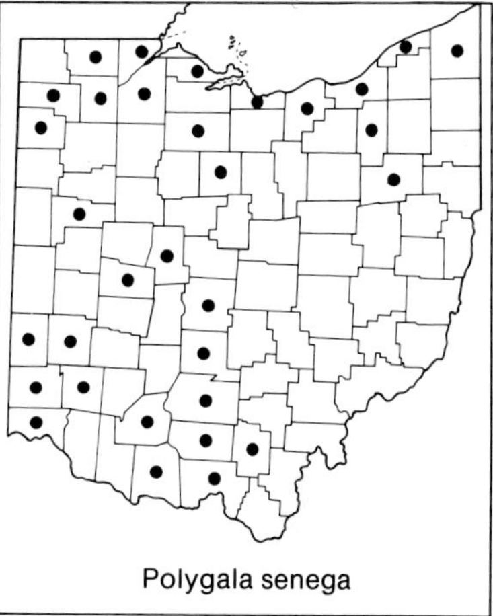
Polygala senega

4. **Polygala incarnata** L. PINK MILKWORT

Annual with slender glabrous stems, to ca. 4 dm tall; leaves alternate, few, linear, and to only ca. 1 cm long, early deciduous or becoming tan and scale-like later in the season, the stems thus appearing nearly leafless; flowers pale pinkish-purple, 7–10 mm long, subsessile in dense spicate racemes terminating the stems. Known only from a few south-central counties; open, often disturbed, habitats, such as fields, roadsides, and prairies. June–September (–October).

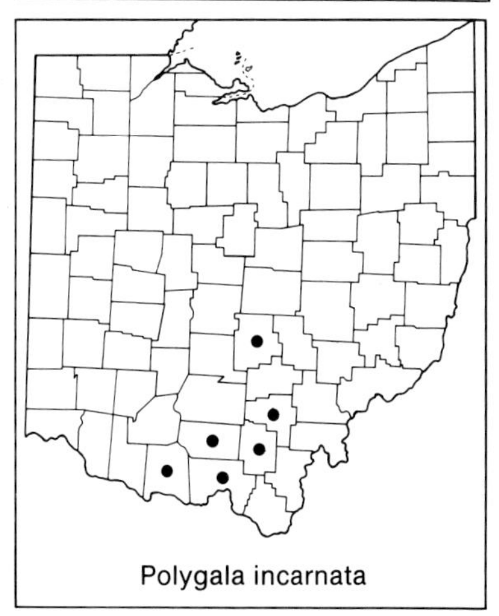
Polygala incarnata

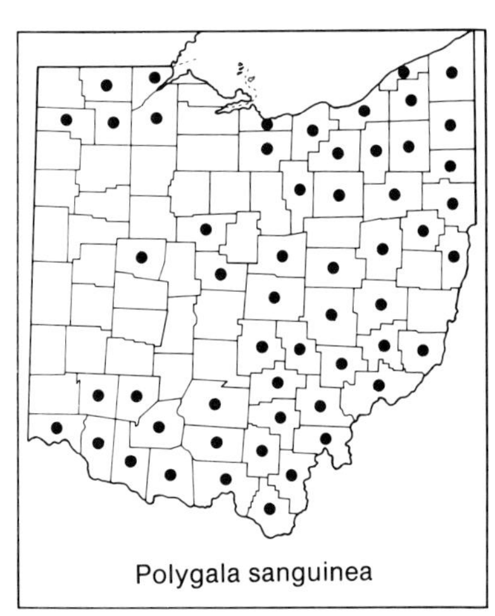
Polygala sanguinea

Polygala curtissii

Polygala cruciata

5. **Polygala sanguinea** L. PURPLE MILKWORT. FIELD MILKWORT
Annual with slender glabrous stems, to ca. 3 dm tall, some flowering stems less than 1 dm; leaves alternate, numerous, broadly linear to oblong-linear; flowers variable in color, ranging from rose-purple to greenish or nearly white, to 4 mm long, subsessile in densely compact head-like racemes; largest inner sepals ("wings") deltoid to broadly rhombic or orbicular-elliptic, (2.5–) 3.0–3.5 mm wide, truncate to obtuse at base. Throughout most of Ohio, except in central and west-central counties, where it is scarce or absent; dry to moist fields, roadsides, openings. Late June–early October.

Voss (1985a) reiterates that *P. sanguinea*, rather than *P. viridescens* L., is the correct name for the species.

6. **Polygala curtissii** A. Gray CURTISS' MILKWORT
Delicate annual with slender glabrous stems, to 3 dm tall; leaves alternate, numerous, broadly linear to very narrowly oblanceolate; flowers rose-purple, to 4 mm long, subsessile in densely compact, head-like racemes; largest inner sepals ("wings") oblong to oblanceolate, 1.5–2.5 (–3.0) mm wide, tapering and acute at base. A species of the southeastern United States, reaching a northern distributional limit in Jackson County, in south-central Ohio; dry, open, sometimes disturbed habitats. August–September.

This species is very difficult to separate from the much more common *P. sanguinea*. In addition to the key characters used above, Burns (1986) reports that the aril/seed length ratio is reliable: the aril of *P. curtissii* is about 1/4 the length of the seed, that of *P. sanguinea* about 3/4 the length of the seed. Burns also notes that earlier published reports on the occurrence of *P. curtissii* in Adams and Perry counties proved to be based on misidentified specimens of *P. sanguinea*, and that the known Ohio distribution of *P. curtissii* is limited to Jackson County.

7. **Polygala cruciata** L. CROSS-LEAVED MILKWORT
Annual with slender, glabrous, usually branched stems, to ca. 2 dm tall; leaves all or mostly in whorls of 3 or 4, narrowly oblanceolate; flowers pale rose-purple to greenish-white, to 5 mm long, subsessile in densely compact, head-like racemes; racemes 10–15 mm wide, cylindric, abruptly apiculate at apex, and immediately subtended by full-sized leaves; the two inner sepals ("wings") deltoid to triangular, tapering to an acuminate apex. Known from Gallia County in extreme southern Ohio and from Lucas and Wood counties in northernmost Ohio; in sand or peat in moist, open or partially shaded sites. July–September.

In Fernald's (1950) treatment, Ohio plants are assigned to var. *aquilonia* Fern. & Schub., with var. *cruciata* limited to the southeastern states.

Gleason and Cronquist (1963, 1991) recognize no varieties within the species, and the value of recognizing var. *aquilonia* is debatable.

8. **Polygala verticillata** L. WHORLED MILKWORT

Delicate annual with slender, glabrous, usually branched stems, to ca. 2 dm tall, often much shorter; lower leaves whorled or opposite, the upper subopposite to alternate; leaves linear to elliptic-linear; flowers white to greenish-white, infrequently pink-tinged, to 3.5 mm long, subsessile in dense to somewhat open racemes; racemes 2–6 mm wide, slenderly cylindric to slenderly conic, gradually tapering toward apex, the racemes on a naked or slightly bracteate peduncle; the two inner sepals ("wings") oblong-elliptic, mostly rounded at apex. Throughout Ohio, except in some counties of the western half of the state; openings in woodlands and thickets, fields, roadsides, and other disturbed open places. Late June–early October.

Burns (1986) presents a thorough summary of the numerous infraspecific taxa that have been recognized in this species. His conclusion that only two varieties are meaningful among Ohio plants is adopted in the key to varieties and their synonymy given below. *Polygala verticillata* var. *verticillata* occurs throughout the Ohio range of the species; var. *ambigua* is limited to south-central and a few southeastern counties.

a. Inflorescences congested throughout, with tightly packed flowers. . . . . var. *verticillata*
incl. var. *isocycla* Fern. and var. *sphenostachya* Pennell

aa. Inflorescences congested above but open at base, with flowers in lower portion loosely disposed. . . . . var. *ambigua* (Nutt.) Alph. Wood
P. ambigua Nutt.

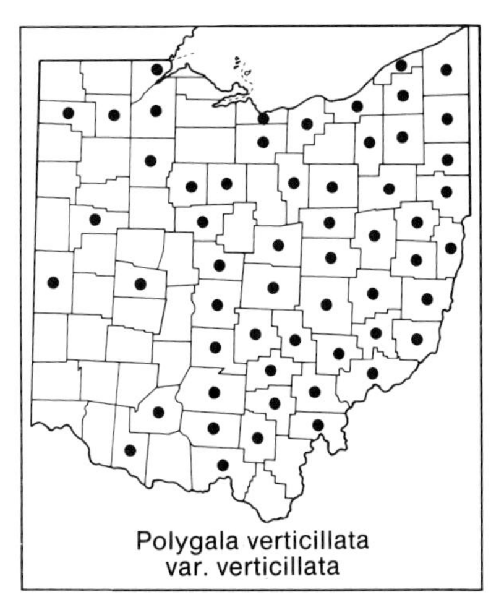
Polygala verticillata var. verticillata

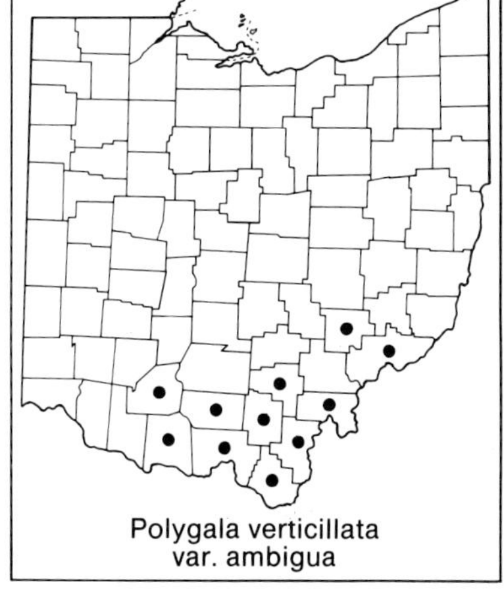
Polygala verticillata var. ambigua

Euphorbiaceae. SPURGE FAMILY

Annual or perennial herbs, often with milky sap; leaves simple; inflorescences various; flowers imperfect, the species either monoecious or dioecious; calyx and corolla minute or absent; stamens 1 to several; carpels mostly 3, fused to form a compound pistil with 3 separate or partially fused styles; ovary superior; fruit a distinctive, usually 3-lobed capsule. A major plant family, with 300 genera and 7,500 species, widely distributed around the earth. Several members of the family are poisonous.

Mercurialis annua L., HERB MERCURY or BOYS-AND-GIRLS, is a small dioecious European annual, rarely adventive in Ohio. There are collections from Franklin and Lake counties, both made in the early part of the present century. CASTOR BEAN, a more frequent adventive, is discussed below.

Two tropical species are noteworthy plants of commerce. The sap of *Hevea brasiliensis* (Willd.) Muell. Arg. is the main natural source of rubber. Tapioca is derived from the roots of *Manihot esculenta* Crantz.

Wisniewski (1967) prepared an earlier study of the Euphorbiaceae of Ohio. Other significant references are the studies of the family in West Virginia by Corbett (1973), and in southeastern United States by Webster (1967).

a. Plants with milky sap, monoecious; flowers lacking both calyx and corolla; staminate flowers consisting of a single stamen, pistillate of a single pistil; inflorescence (the cyathium) consisting of few to several staminate flowers surrounding a single pistillate flower and borne within a green cup-shaped involucre, in all resembling a single flower, the involucre sometimes with perianth-like appendages .... 5. *Euphorbia*

aa. Plants with watery or rarely colored but nonmilky sap, monoecious or dioecious; flowers with calyx, with or without corolla; staminate flowers with 3–20 or more stamens, pistillate flowers with a single pistil; flowers in axillary pairs or in axillary or terminal spikes, racemes, or panicles, not borne within a cup-shaped involucre and not in an inflorescence resembling a single flower.

b. Leaves palmately lobed, alternate .... 3. *Ricinus*

bb. Leaves entire or dissected but not palmately lobed, alternate or opposite.

c. Stems and leaves with stellate pubescence; leaves alternate or subopposite .... 1. *Croton*

cc. Stems and leaves glabrous or with simple hairs; leaves alternate or opposite.

d. Leaves opposite; stems glabrous; plants dioecious; styles 2, capsules 2–loculate; see above .... *Mercurialis* dd. Leaves alternate or sometimes subopposite; stems glabrous or pubescent; plants monoecious; styles 3, capsules 3–loculate.

e. Each flower or cluster of flowers subtended by a wide, conspicuous, lobed or cleft bract; stems usually pubescent, rarely glabrous; leaves mostly crenate, serrate, or dentate, less often entire . . . . 2. *Acalypha*

ee. Flowers without bracts; stems glabrous; leaves entire . . . . 4. *Phyllanthus*

1. Croton L. CROTON

Monoecious annuals with nonmilky sap; leaves alternate to subopposite; stems and leaves stellate-pubescent. A chiefly tropical genus of some 700 or more species.

a. Leaf margin serrate; leaves with 1 or 2 yellowish, cup-shaped glands at junction of petiole and blade . . . . 1. *Croton glandulosus*

aa. Leaf margin entire or shallowly crenate; leaves without glands at junction of petiole and blade.

b. Leaves mostly lanceolate; principal leaves with blades 4–6 cm long, and apex acuminate and scarcely apiculate; styles 3, each 2- or 3-branched; capsule usually with 3 seeds . . . . . . 2. *Croton capitatus*

bb. Leaves mostly broadly oblong to oblong or narrowly ovate; principal leaves with blades 1.5–3 (–4) cm long, and apex obtuse to rounded and short-apiculate; styles 2, each 2-branched; capsule usually with 1 seed. . . . . 3. *Croton monanthogynus*

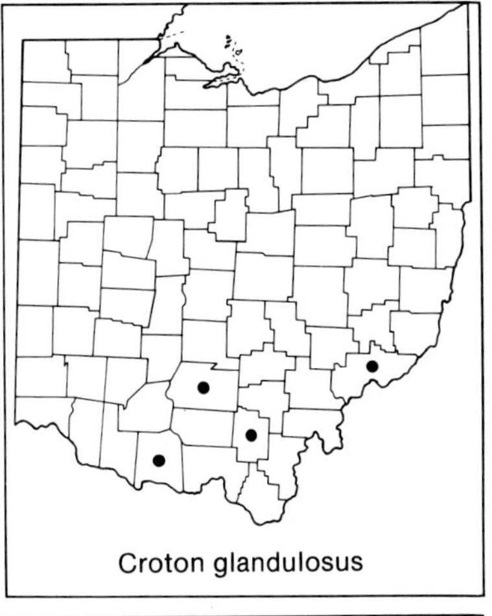

Croton glandulosus

Croton capitatus

1. **Croton glandulosus** L. GLANDULAR CROTON. SAND CROTON

Pubescent several-branched annual, to 6 dm tall; leaves alternate to more often subopposite, more or less oblong, coarsely serrate and with 1 or 2 yellowish cup-shaped glands at junction of petiole and blade; styles 3, each deeply 2-lobed. A New World species ranging from subtropical parts of South America northward to east-central United States, and adventive at sites farther north. The few stations in southern Ohio are at the northern boundary of the species' native range; the Bowling Green State University Herbarium includes a specimen of an adventive collected from Lucas County in northwestern Ohio (not mapped). Open, sandy or dry disturbed ground. July–October.

Ohio plants are assigned to var. *septentrionalis* Muell. Arg.; var. *glandulosus* grows farther south in subtropical and tropical regions.

2. CROTON CAPITATUS Michx. WOOLLY CROTON. HOGWORT

Woolly, 1–few-branched annual, to ca. 1 m tall; leaves alternate, entire or subentire, mostly lanceolate, blades 4–6 cm long, apex acuminate and scarcely apiculate; styles 3, each 2- or 3-branched; capsule usually with 3

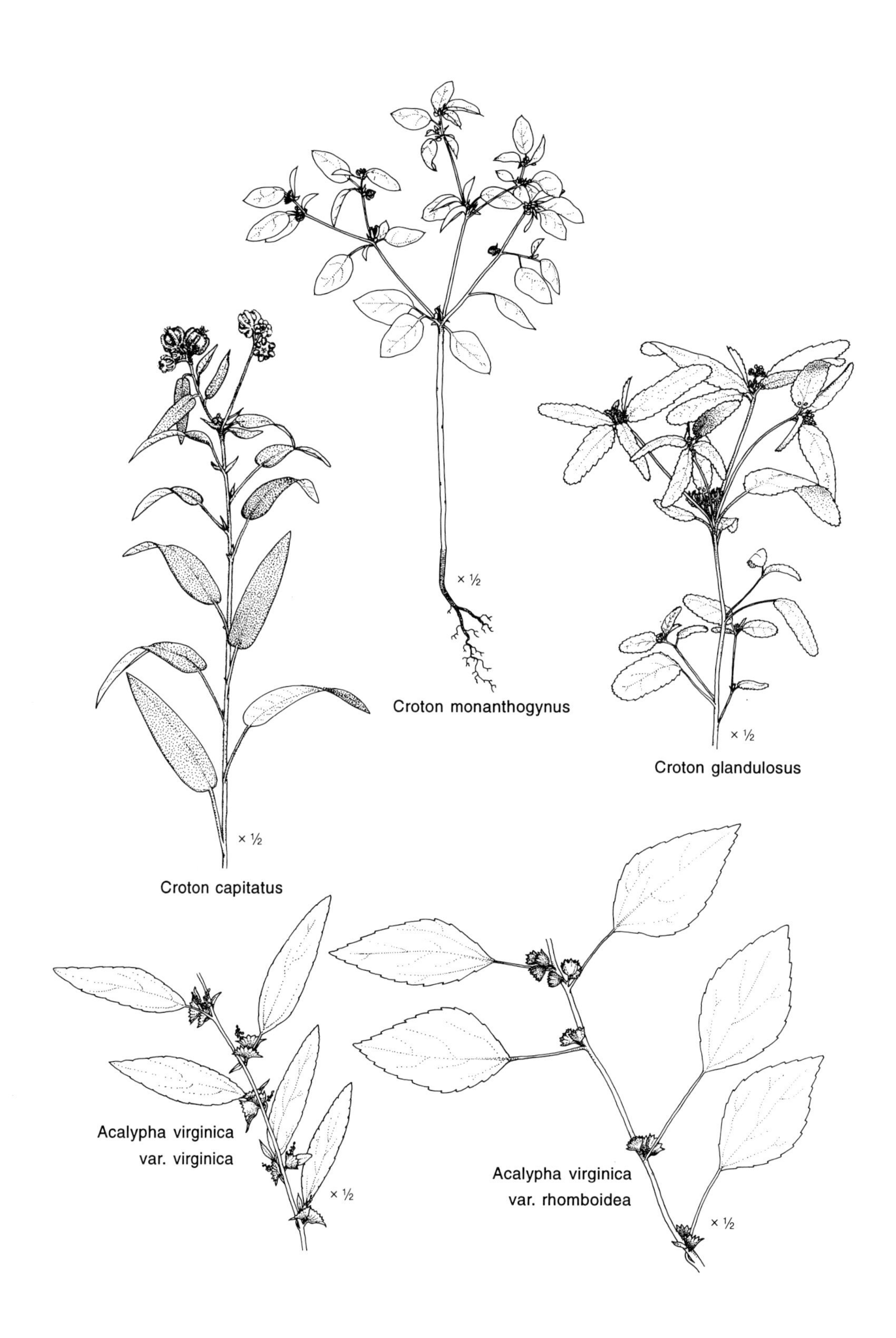
× ½
Croton monanthogynus
× ½
Croton glandulosus
× ½
Croton capitatus
Acalypha virginica
var. virginica
× ½
Acalypha virginica
var. rhomboidea
× ½

seeds. A species of southeastern United States and the southern Great Plains states, adventive or perhaps naturalized at a few sites in southern Ohio counties; dry disturbed ground. July–October.

Croton capitatus was treated as an Ohio native by Fernald (1950) and Gleason (1952a). Roberts and Cooperrider (1982), following the lead of Deam (1940), who regarded the Indiana records as "adventive from the south," concluded that this species was not a native member of the Ohio flora. Gleason noted some adventive occurrences of the species in eastern states.

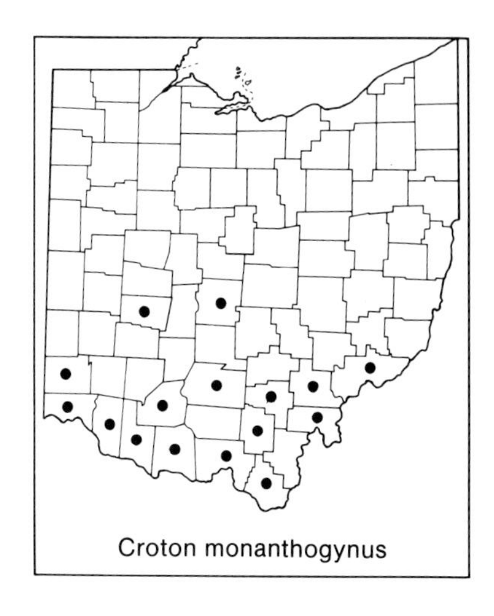
Croton monanthogynus

3. Croton monanthogynus Michx. Prairie-tea

Somewhat rough, glandular-pubescent, widely branched annual, to ca. 4 dm tall; leaves alternate to subopposite, entire, mostly broadly oblong to oblong or narrowly ovate, principal leaves with blades 1.5–3 (–4) cm long, and apex obtuse to rounded and short-apiculate; styles 2, each 2-branched; capsule usually with 1 seed. A species of Mexico and south-central and southeastern United States, naturalized in southern Ohio; dry disturbed habitats along roads, railroads, and the like. July–October.

Croton monanthogynus was thought by Fernald (1950) and Gleason (1952a) to be an Ohio native. As in the case of the preceding species, Roberts and Cooperrider (1982) followed the conclusion of Deam (1940), who regarded plants of *C. monanthogynus* as non-native to the Indiana flora.

2. Acalypha L. THREE-SEEDED MERCURY

Monoecious annuals with nonmilky sap; stems usually pubescent to rarely glabrous; leaves mostly alternate, dentate or serrate to crenate or rarely entire; flowers or clusters of flowers subtended by a wide, conspicuous, lobed or cleft bract. A chiefly tropical genus of more than 400 species. L. A. W. Miller (1964) prepared a study of *Acalypha* in the United States.

Acalypha gracilens A. Gray, Slender Three-seeded Mercury, adventive from the southeastern United States and the Atlantic coastal region, has been collected at disturbed sites in Hamilton, Hocking, and Lake counties. Some of the few Ohio specimens known approach *A. virginica*, suggesting that *A. gracilens* might best be treated as a variety of *A. virginica*. Gleason (1952a) observed that "well grown plants of [*A. gracilens*] are often difficult to distinguish from depauperate plants of *A. virginica*."

a. Leaves shallowly cordate at base; capsules bearing numerous, short, soft prickles. . . . . 2. *Acalypha ostryifolia*

aa. Leaves tapering to broad or narrow base, not cordate; capsules glabrous or pubescent, but without prickles.

b. Bracts subtending pistillate flowers with 9–13 very shallow, broadly deltoid lobes, the sinuses extending only about 1/5 the distance to base of bract; petioles mostly about 1 cm long, less than 1/5 the length of the blade; hairs on stem all short and incurved; see above . . . . *Acalypha gracilens*

bb. Bracts subtending pistillate flowers with 5–15 linear to lanceolate, oblong-lanceolate, or narrowly triangular lobes, the sinuses mostly extending more than 1/3 the distance to base of bract; petioles mostly more than 1 cm long, 1/5 as long as to nearly equalling length of the blade; hairs on the stem either short and incurved or longer and spreading. . . . . 1. *Acalypha virginica*

1. **Acalypha virginica** L.

Much-branched, more or less pubescent annual, to ca. 5 dm tall; leaves tapering to a broad or narrow base, not cordate; capsules glabrous or pubescent, but without prickles. Throughout Ohio; usually open, dry or moist, mostly disturbed habitats. July–October.

As treated here, three varieties occur among the Ohio plants of this species; each of the varieties has been treated at the rank of species (Mohlenbrock, 1982a; Webster, 1967). Plants assignable to or approaching var. *virginica* are more common in southern counties but are also known from northern ones, while plants assignable to or approaching var. *rhomboidea* occur throughout Ohio; intergrades between these two are numerous. In a recent discussion (Cooperrider, 1984b), I noted that in addition to the problems posed by intergradation, "no single reliable diagnostic character or combination of characters" separates the two taxa. The rarest of the

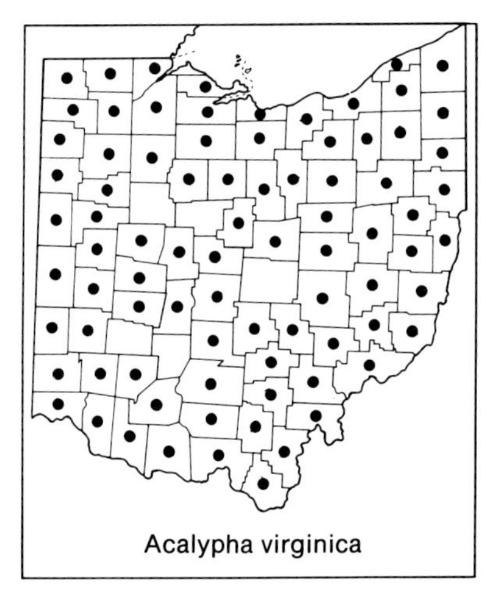
Acalypha virginica

three varieties, var. *deamii*, is known from a few sites in southern counties along the Ohio River.

a. Seeds mostly 2.5 mm or longer; principal leaves 7–11 cm long . . . . var. *deamii* Weath.
   A. deamii (Weath.) Ahles
   DEAM'S THREE-SEEDED MERCURY

aa. Seeds mostly 1.5–1.8 mm long; principal leaves 2–9 cm long.

b. Leaves mostly ovate to broadly rhombic, the petioles 1/2 to nearly as long as the blades; bracts subtending pistillate flowers mostly with 5–9 lobes, these often bearing stalked glands and also rarely straight divergent hairs; stems usually with all hairs short and incurved. . . . . var. *rhomboidea* (Raf.) Cooperr.
   A. rhomboidea Raf.
   RHOMBIC THREE-SEEDED MERCURY

bb. Leaves mostly broadly lanceolate to narrowly rhombic, the petioles ca. 1/5–1/2 the length of the blade; bracts subtending pistillate flowers mostly with 9–15 lobes, these usually with many straight divergent hairs and rarely also with stalked glands; stems, especially above, often with few to many straight divergent hairs as well as short incurved ones . . . . var. *virginica*
   VIRGINIA THREE-SEEDED MERCURY

2. ACALYPHA OSTRYIFOLIA Riddell   HORNBEAM THREE-SEEDED MERCURY

Branched, very short-puberulent annual, to ca. 6 dm tall; leaves shallowly cordate at base; capsules bearing numerous, short, soft prickles. A species ranging from Mexico across south-central and southeastern United States; naturalized or adventive in southwestern and south-central counties of Ohio; river banks, in cultivated fields, and other disturbed sites. August–October.

Deam (1940) concluded that the species was not indigenous to southern Indiana. On the basis of Deam's treatment and the fact that Ohio representatives of the species are limited to disturbed habitats, Roberts and Cooperrider (1982) concluded that it was also not native to Ohio. Similarly, Radford, Ahles, and Bell (1968) regarded it as "doubtfully native" to North and South Carolina.

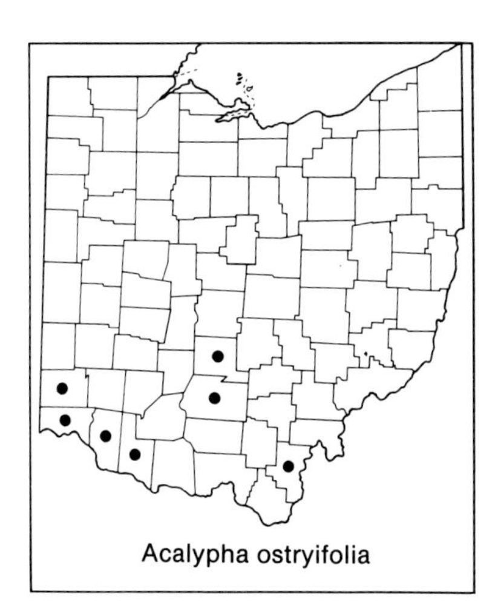
Acalypha ostryifolia

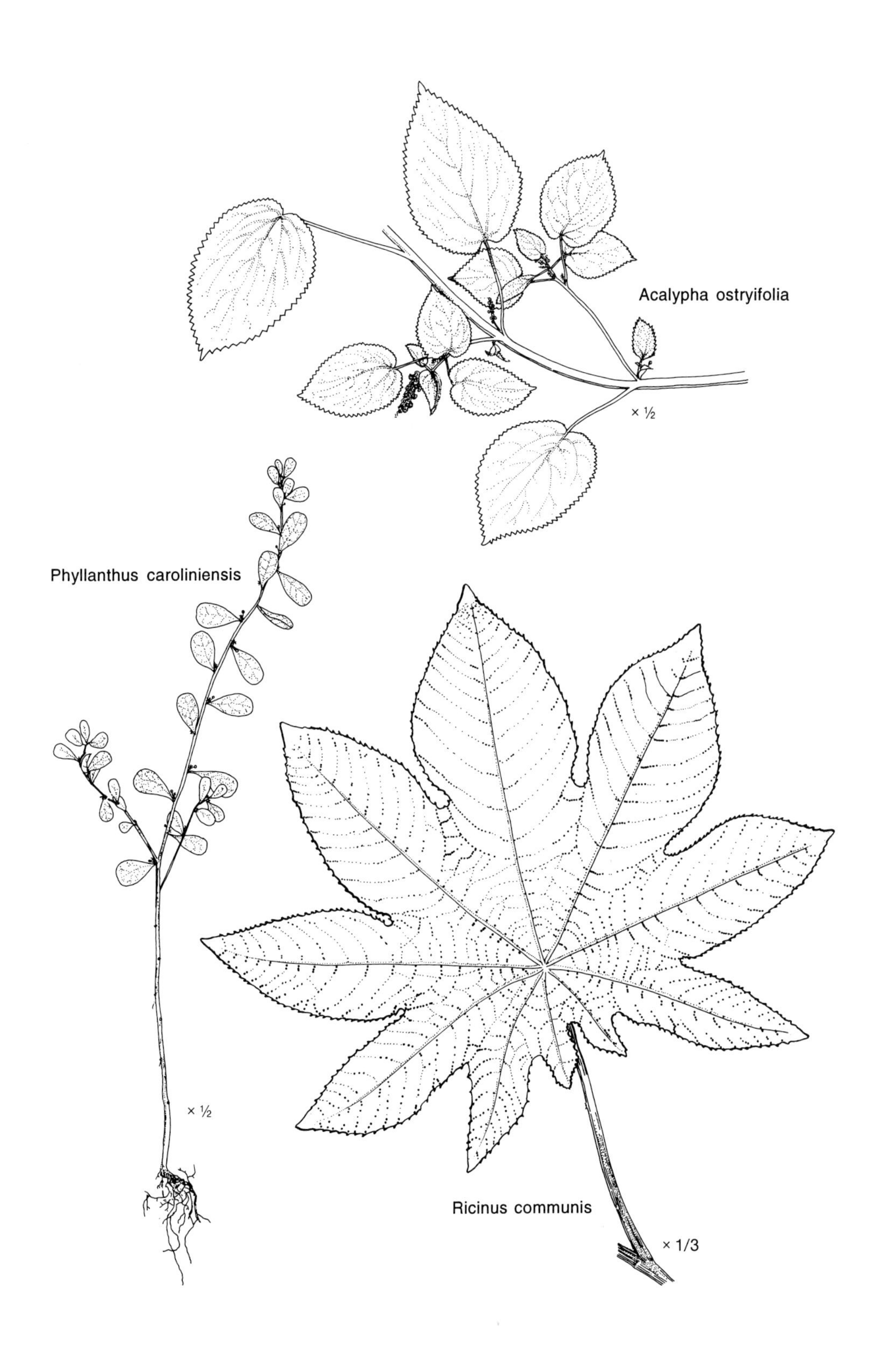
Acalypha ostryifolia
× ½
Phyllanthus caroliniensis
× ½
Ricinus communis
× 1/3

3. Ricinus L.

A genus of a single species.

1. Ricinus communis L. Castor Bean

Tall, coarse, monoecious herb, reaching heights of ca. 4 m (13 feet) during its single season; leaves alternate, palmately veined and palmately lobed, sometimes nearly 1 m wide; capsules spiny, borne in panicles. A distinctive but variable species from tropical Africa, becoming a shrub or small tree in the tropics. In temperate regions it is grown as an ornamental annual, prized for its height and huge leaves. Adventive in disturbed areas, because of its size probably undercollected and more frequent than collections indicate. July–October.

The seeds, although deadly poisonous, yield the extract "castor oil," used medicinally. Once prescribed by parents as a general cure-all for internal ailments of the young, castor oil was in the not-too-distant past a vile part of the lives of many Ohio children, including myself.

Ricinus communis

4. Phyllanthus L.

A genus of some 750 species, chiefly of the Old World tropics, but also with numerous species in the American tropics; a single species occurs in the Ohio flora. Webster (1970) revised the genus as it is represented in the flora of continental United States.

1. **Phyllanthus caroliniensis** Walt. Carolina Leaf-flower

Small, glabrous, monoecious annual, about 1–3 dm tall; leaves alternate, entire, more or less oblanceolate. South-central and southwestern Ohio counties; moist, more or less open, often sandy habitats. July–October.

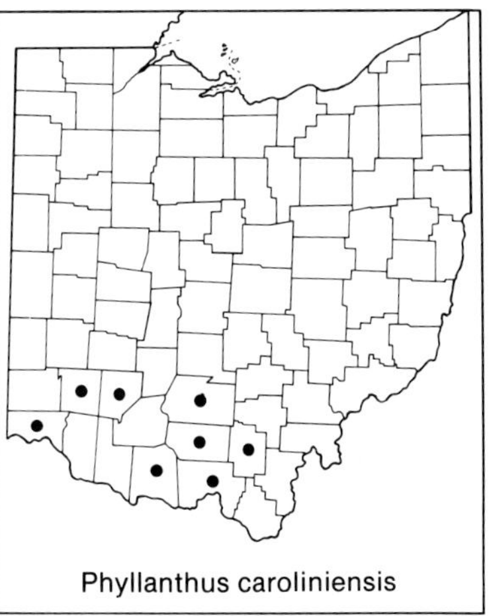
Phyllanthus caroliniensis

5. Euphorbia L. spurge

Monoecious annual, biennial, or perennial herbs with milky sap; leaves opposite or alternate; the flowers borne in a small unique inflorescence (called the "cyathium") that resembles a single flower and includes 1–several staminate flowers (each consisting of a single stamen) and one pistillate flower (consisting only of a single pistil), all borne in a green cup-shaped involucre which appears somewhat like a calyx-tube, the involucre sometimes bearing petaloid glands at its apex. Widely distributed and with more than 1,500 species, *Euphorbia* is one of the largest genera of flowering plants in the world.

Euphorbia helioscopia L., Sun Spurge, a European species, frequent in southern Ontario, was collected twice as an adventive weed from Lake County, in northern Ohio, once in 1892 and again in 1902. *Euphorbia exigua* L., Dwarf Spurge, another European adventive, was attributed to

Ohio by Gleason (1952a), but I have seen no specimens other than one from Lake County marked "cultivated." *Euphorbia glyptosperma* Engelm., a transcontinental species ranging across northern North America from New Brunswick and Maine to British Columbia and in the west southward to Mexico, was attributed to Ohio by Fernald (1950), but I have seen no specimens. Ohio lies outside the range of this species as presented by Wheeler (1941). *Euphorbia pulcherrima* Willd. is the familiar POINSETTIA of the winter holiday season.

Richardson (1968) treats the species of *Euphorbia* occurring in several north-central states.

Wheeler (1941) and Burch (1966) have published, respectively, a revision and commentary upon *Euphorbia* subgenus *Chamaesyce* Raf., which includes species nos. 14–20, below. This group of species is sometimes placed in the segregate genus *Chamaesyce* S. F. Gray (e.g., by Mohlenbrock, 1982a; and by Webster, 1967, who notes, however, that *Chamaesyce* is "accepted diffidently as a distinct genus"). Mohlenbrock and others also give generic rank to *Poinsettia* Graham; see notes under *E. dentata* and *E. cyathophora*, species nos. 1 and 2 below.

a. Erect plants with their cyathia (cup-like involucres) bearing conspicuous, white, petal-like appendages, the appendages 2 mm or more in width.

b. Plants perennial; leaves green throughout, the leaf apex obtuse to rounded or retuse; stems mostly glabrous . . . 13. *Euphorbia corollata*

bb. Plants annual; upper or uppermost leaves with narrow to very broad white border, the leaf apex usually apiculate or acuminate; stems more or less pubescent, at least above. . . . . 12. *Euphorbia marginata*

aa. Erect, ascending, or prostrate plants with their cyathia (cup-like involucres) either without petal-like appendages or with inconspicuous, white to reddish appendages, the appendages 0.5 mm or less in width.

b. Principal leaves with petioles 6–20 (–25) mm long; plants erect, often branching below the middle.

c. The principal leaves opposite or subopposite, more or less regularly serrate to round-dentate, slightly scabrous-pubescent above; stems scabrous or otherwise pubescent; bracts immediately beneath inflorescence usually whitish-green at base . . . . . . 1. *Euphorbia dentata*

cc. The principal leaves alternate, subentire or with few irregularly spaced teeth or shallow lobes, glabrous above; stems glabrous; leaves immediately beneath inflorescence often reddish at base . . . . 2. *Euphorbia cyathophora* bb. Principal leaves sessile or with petioles 1–4 (–5) mm long; plants mostly either erect and branching (if at all) above the middle, or prostrate to ascending and branching near the base.

c. Stem leaves narrowly linear to linear-oblong or linear-lanceolate, 1–15 cm long, 1–15 mm wide, (7–) 10 or more times as long as wide.

d. Leaves of main stem mostly alternate-scattered, linear, 1–2 (–3) mm wide, 1–3 (–3.5) cm long; stems often branched near base.

e. Plants perennial; flowering stems terminated by a multi-rayed umbel, the stems infrequently also with a few axillary umbels immediately below main umbel; cyathia ca. 3 mm long; seeds not tuberculate . . . . 6. *Euphorbia cyparissias*

ee. Plants annual; flowering stems with axillary and terminal umbels, most with 3–5 rays; cyathia ca. 1 mm long; seeds tuberculate; see above . . . . *Euphorbia exigua*

dd. Leaves of main stem alternate or opposite, linear or linear-oblong to narrowly lanceolate or narrowly oblanceolate, (2–) 3–10 mm wide, 3–25 cm long; stems unbranched or branched above the middle.

e. Principal cauline leaves alternate; floral leaves ca. 1 cm long; cyathia 2.5–3 mm long. 7. *Euphorbia esula*

ee. Principal cauline leaves opposite; floral leaves 2–6 cm long; cyathia ca. 4 mm long . . . . 11. *Euphorbia lathyris*

cc. Stem leaves variously shaped, but not linear to linear-oblong or linear-lanceolate, only 1.5–6 times as long as wide.

d. Stems erect, either unbranched or branching above—or mostly above—the middle; lower leaves often alternate, the upper opposite or alternate; leaves mostly equilateral at base.

e. Principal stem leaves 5–15 cm long or longer; cyathia ca. 4 mm long; capsules 6–9 mm long.

f. Cauline leaves obtuse to rounded, and minutely apiculate, at apex; floral leaves varying from nearly as wide as long to wider than long, rounded or retuse-rounded at apex . . . . 3. *Euphorbia purpurea*

ff. Cauline leaves acute to acuminate at apex; floral leaves longer than wide, acute to acuminate at apex. . . 11. *Euphorbia lathyris*

ee. Principal stem leaves 1–5 cm long; cyathia 1–2 mm long; capsules 2–3 mm long.

f. Leaves serrulate or finely serrulate.

g. The leaves inequilateral at base, one or more on each plant usually marked with red; cyathia with minute, white, petal-like appendages . . . . 19. *Euphorbia nutans*

gg. The leaves equilateral at base, green throughout; cyathia lacking white petal-like appendages.

h. Capsules smooth or nearly so, without wart-like protuberances; both stem and floral leaves narrowed at base; see above . . . . *Euphorbia helioscopia*

hh. Capsules beset with wart-like protuberances; stem leaves narrowed at base, floral leaves with broad, truncate to subcordate base.

i. Undersurface of leaves pubescent; leaf apices acute . . . . 4. *Euphorbia platyphylla*

ii. Undersurface of leaves glabrous; leaf apices obtuse to rounded . . . . 5. *Euphorbia obtusata*

ff. Leaves entire.

g. Stem leaves sessile or subsessile, acute at apex; floral leaves longer than to slightly longer than wide, with subulate tip 0.5–1 mm in length. . . . . 10. *Euphorbia falcata*

gg. Stem leaves sessile or petiolate, obtuse or retuse at apex; floral leaves either longer than wide or wider than long, but either without subulate tip or (rarely) with a tip less than 0.5 mm in length.

h. Floral leaves wider than long; medial stem leaves sessile, the lowermost sometimes petiolate . . . . 8. *Euphorbia commutata*

hh. Floral leaves longer than wide; medial and lowermost stem leaves petiolate . . . . 9. *Euphorbia peplus*

dd. Stems prostrate to ascending, branching near the base; all leaves opposite; leaves often inequilateral at base.

e. Both the upper internodes and the capsules glabrous; plants mostly prostrate.

f. Leaves entire.

g. The leaves linear-oblong to narrowly oblong-lanceolate, mostly 8–15 mm long; seeds 2–2.5 mm long . . . . 17. *Euphorbia polygonifolia*

gg. The leaves ovate-orbicular to broadly oblong, mostly 2–7 mm long; seeds ca. 1 mm long. . . . . 18. *Euphorbia serpens*

ff. Leaves serrulate across the rounded summit and usually for a short distance down sides of blade; the leaves narrowly oblong to linear-oblong, mostly 4–8 mm long; see above. . . . . *Euphorbia glyptosperma* ee. Upper internodes more or less pubescent, the capsules either glabrous or pubescent; plants prostrate to ascending.

f. Capsules pubescent; stems all or mostly prostrate.

g. Pubescence of capsules mostly appressed; principal leaves 8–15 (–17) mm long.

h. Principal leaves linear to oblong, 1–5 (–6) mm wide, most 2.5–3 times as long as wide; seeds with a few low transverse ridges . . . . 14. *Euphorbia maculata*

hh. Principal leaves elliptic to oblong-ovate or ovate, 6–10 mm wide, most 1.5–2 times as long as wide; seeds smooth or slightly roughened. . . . . 15. *Euphorbia humistrata*

gg. Pubescence of capsules divergent from capsule wall; principal leaves mostly 5–7 mm long . . . . 16. *Euphorbia prostrata*

ff. Capsules glabrous; stems prostrate to ascending.

g. Stems ascending or nearly erect; stem pubescence sparse (sometimes only a single line of hairs) and appressed; principal leaves 1.5–3.5 cm long . . . . 19. *Euphorbia nutans*

gg. Stems prostrate or slightly ascending; stem pubescence moderate to heavy, spreading; principal leaves 0.5–1.8 cm long. . . . . 20. *Euphorbia vermiculata*

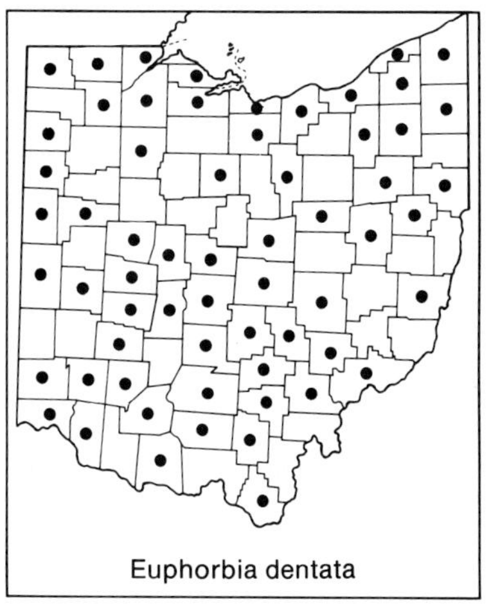
Euphorbia dentata

1. Euphorbia dentata Michx. Toothed Spurge. Wild Poinsettia

Erect annual, to ca. 6 dm tall, often branching below the middle; stems scabrous or otherwise pubescent; leaves opposite or subopposite; petioles hispid-pubescent, to 20 (–25) mm long; blades slightly scabrous-pubescent above, highly variable in shape (two extremes are illustrated), ranging from very narrowly long-lanceolate to ovate or broadly ovate, the margins ranging from more or less regularly serrate to round-dentate; bracts immediately beneath the inflorescence usually whitish-green at base. A species of Central America, Mexico, and southwestern to eastern United States, naturalized throughout Ohio; mostly along railroads, also at roadsides and in other disturbed habitats. Late June–early October.

In his synopsis, Dressler (1961) assigns this species to the segregate genus *Poinsettia* Graham.

Euphorbia cyathophora

2. Euphorbia cyathophora Murr. Fire-on-the-mountain. Painted-leaf

E. heterophylla of authors, incl. Fernald (1950) and Gleason (1952a), not L.

Erect annual, to ca. 6 dm tall, often branching below the middle; stems glabrous or essentially so; principal leaves alternate; petioles glabrous, to 15 mm long; blades glabrous or nearly so, varying from ovate to fiddle-shaped to sublinear, with few irregularly spaced teeth or shallow lobes to

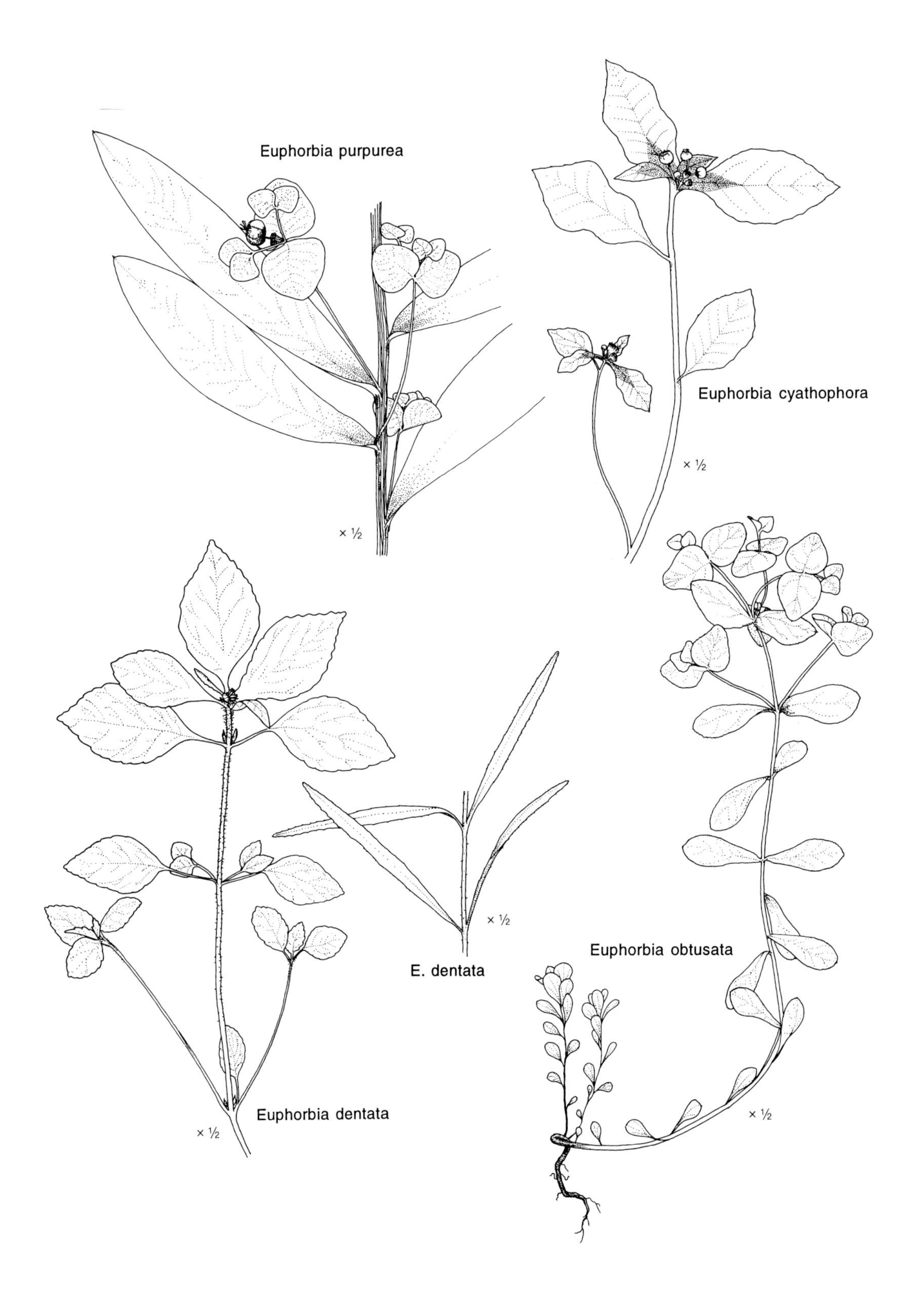
Euphorbia purpurea
× ½
Euphorbia cyathophora
× ½
× ½
E. dentata
Euphorbia obtusata
Euphorbia dentata
× ½
× ½

nearly entire; bracts immediately beneath inflorescence often reddish at base. A species ranging from northern Mexico to central and southeastern United States; adventive at a few disturbed sites in southwestern Ohio. Summer.

Dressler (1961) separates this species from *Euphorbia heterophylla* L., native to tropical and subtropical America, assigning both to the segregate genus *Poinsettia*.

3. **Euphorbia purpurea** (Raf.) Fern. GLADE SPURGE
Erect, glabrous, somewhat fleshy perennial, to ca. 1 m tall, branched only in the inflorescence; stem leaves alternate, sessile or subsessile, entire, oblanceolate to elliptic-lanceolate, the upper ovate; cauline leaves to 10 cm or longer, obtuse to rounded at apex, and minutely apiculate; the floral leaves varying from nearly as wide as long to wider than long, rounded or retuse-rounded at apex; capsules 6–9 mm long. A species of east-central United States, known in Ohio from a few stations in southern counties, where it is evidently at the western limit of its range; dry to moist wooded slopes. June–August.

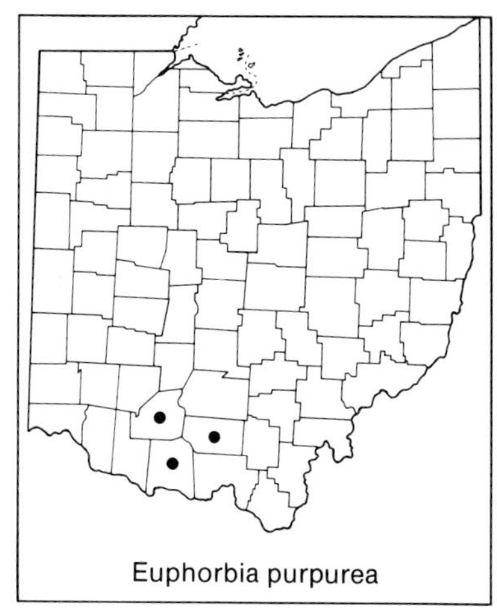
Euphorbia purpurea

Thought to have been extirpated from Ohio, the species was recently discovered in Pike County (Knoop, 1990). In addition to that record and valid earlier records from Adams and Highland counties, a specimen at The University of Cincinnati Herbarium is marked "Columbus, O.," but has no other label data. While the species might once have grown in Franklin County, the record seems too doubtful to warrant a dot on the distribution map.

4. EUPHORBIA PLATYPHYLLA L. BROAD-LEAVED SPURGE
Erect branched annual, to ca. 5 dm tall; stems glabrous or nearly so; stem leaves sessile or subsessile, pubescent beneath, to 4 or 5 cm long, oblanceolate, acute at apex, the upper half of the leaf finely serrate, tapering below to a narrowed, entire, cordate base; floral leaves with broad, truncate to subcordate base; capsules 2.5–3 mm long, beset with wart-like tubercles. A European species naturalized in northern Ohio counties; disturbed habitats. June–August.

Euphorbia platyphylla

5. **Euphorbia obtusata** Pursh BLUNT-LEAVED SPURGE
Erect, glabrous, few- to much-branched annual, to ca. 6 dm tall; stem leaves sessile, glabrous beneath, to 4 cm long, oblanceolate to oblong-spatulate, obtuse to rounded at apex, the upper 1/2–2/3 of the leaf finely serrate, tapering to a narrowed, entire, cordate, slightly clasping base; floral leaves with broad truncate to subcordate base; capsules 3 mm long, beset with wart-like tubercles. At scattered Ohio sites, mostly in south-central counties; moist openings. May–June.

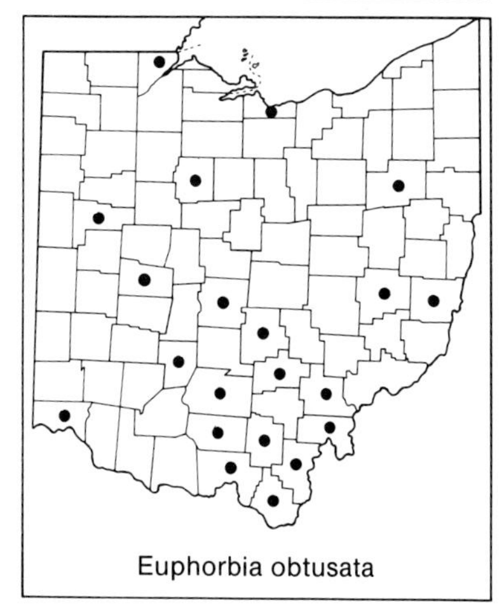
Euphorbia obtusata

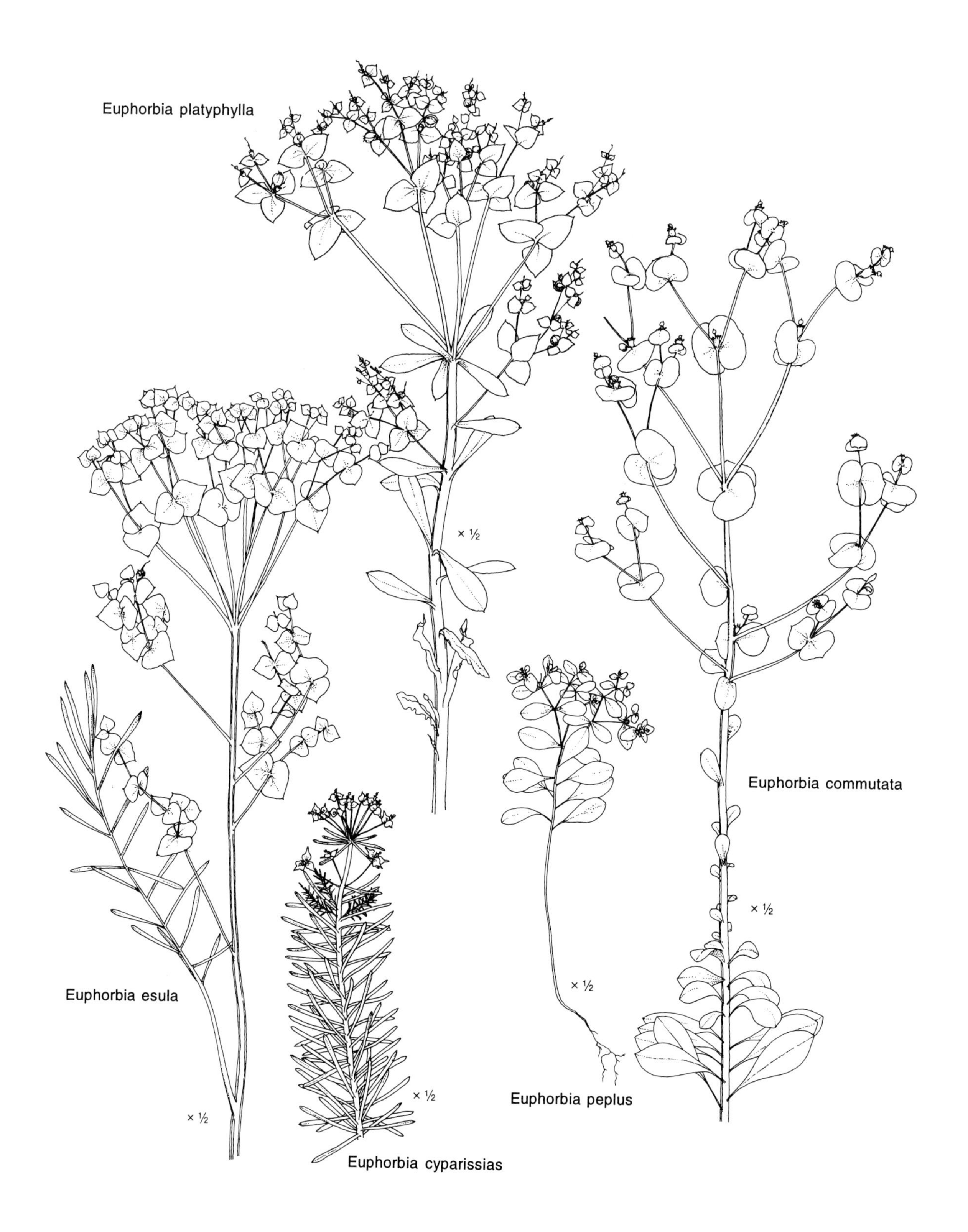
Euphorbia platyphylla
× ½
Euphorbia commutata
× ½
Euphorbia esula
× ½
× ½
Euphorbia cyparissias
× ½
Euphorbia peplus

6. Euphorbia cyparissias L. Cypress Spurge
Erect, glabrous, few-branched, leafy perennial, mostly 2–3 dm tall; stem leaves numerous, crowded-alternate, sessile, linear, to 3 cm or slightly longer; flowering stems terminated by a multirayed umbel, the stems infrequently also with a few axillary umbels; the small floral leaves subtending the cyathia broadly triangular-ovate, yellowish when young; capsules 3 mm long. A Eurasian species cultivated, more in the past than at present, in rock gardens and at flower garden borders; escaped and naturalized throughout Ohio, disturbed areas, especially along roadsides and near graveyards. April–July.

R. J. Moore and Frankton (1969) report naturally occurring hybrids between this species and the next, *E. esula*, growing in Ontario.

7. Euphorbia esula L. Leafy Spurge
Erect, glabrous, branched perennial, to 6 (–10) dm tall; stem leaves sessile or essentially so, mostly alternate, broadly linear to narrowly lanceolate or narrowly oblanceolate, to 7 (–8) cm long; flowering stems terminating in a rather open umbel, the stem also with several, lower, axillary flowering branches; the floral leaves broadly triangular-ovate to reniform, ca. 1 cm long; cyathia 2.5–3 mm long; capsules warty, 2.5–3 mm long. A weed of Eurasian origin established across northern United States and southern Canada. It has become in some areas, especially in the north-central states, a troublesome weed in pastures and cultivated fields. Best et al. (1980) describe the species, its history in North America, and control measures. Known from four Ohio collections, the first made in 1968; roadsides and other disturbed ground. June–July.

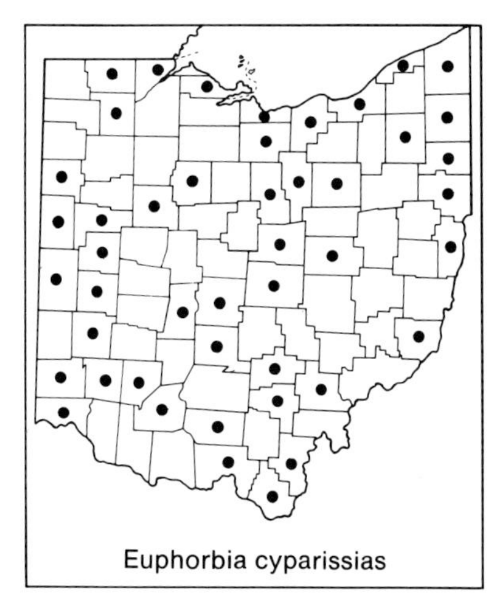

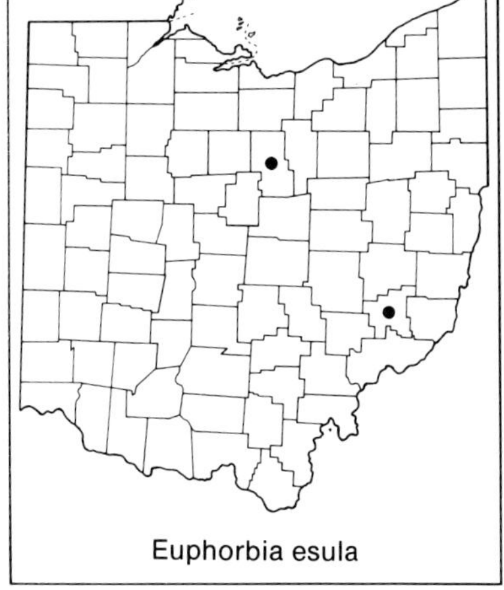

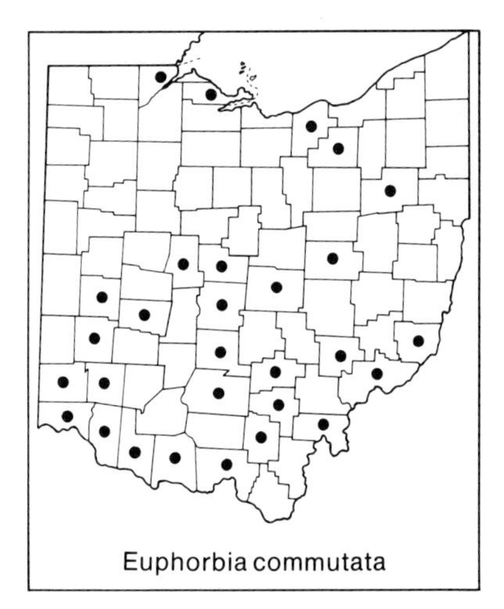
Euphorbia commutata

8. **Euphorbia commutata** Engelm. Wood Spurge
Erect, glabrous, branched, winter annual (or perhaps perennial), to ca. 3 dm tall; stem leaves alternate, entire, mostly obovate, to 15 mm long, the lower petiolate, the medial and upper sessile; the floral leaves hemispheric-reniform, sometimes connate and forming a circular blade; capsules 3 mm long. Mostly in the southern half of Ohio, rare or absent in most northern counties; mesic to moist woods, especially on wooded slopes. April–July.

9. Euphorbia peplus L. Petty Spurge
Erect, glabrous, branched annual, to 3 dm tall; stem leaves alternate, entire, obovate to elliptic or roundish, to 2 cm long, all but the upper petiolate; the floral leaves ovate to elliptic, longer than wide; capsules 2 mm long. A Eurasian species, naturalized mostly in northeastern counties; shaded, moist, disturbed ground. June–November.

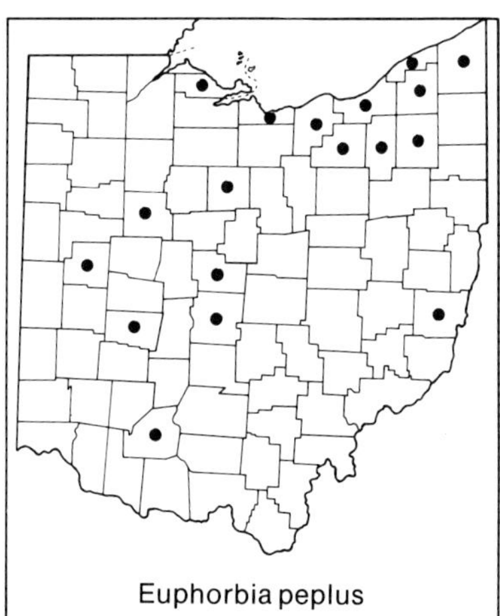
Euphorbia peplus

10. Euphorbia falcata L. Falcate Spurge
Erect, glabrous, branched annual, to ca. 3 dm tall; stem leaves alternate, entire, oblanceolate but acute at apex, to 3 cm long, sessile or subsessile; the floral leaves ovate, longer than wide—sometimes only slightly so, with a subulate tip 0.5–1 mm long; capsules 2 mm long. A Eurasian species naturalized, or perhaps merely adventive, in a few southwestern counties; disturbed sites. July–September.

11. Euphorbia lathyris L. Caper Spurge
Erect glabrous annual or biennial, to ca. 1 m tall; stem leaves mostly opposite, entire, linear-oblong to narrowly lanceolate, to 15 cm or more in length, sessile with slightly clasping base; the floral leaves ovate-lanceolate to lanceolate, 2–6 cm long; capsules to 13 mm long. A Eurasian species

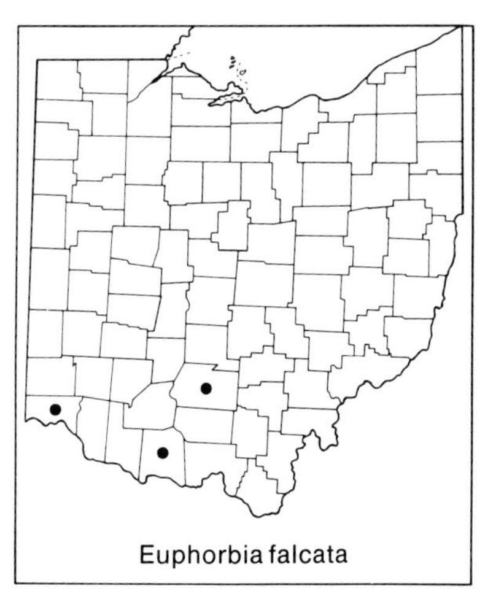
Euphorbia falcata

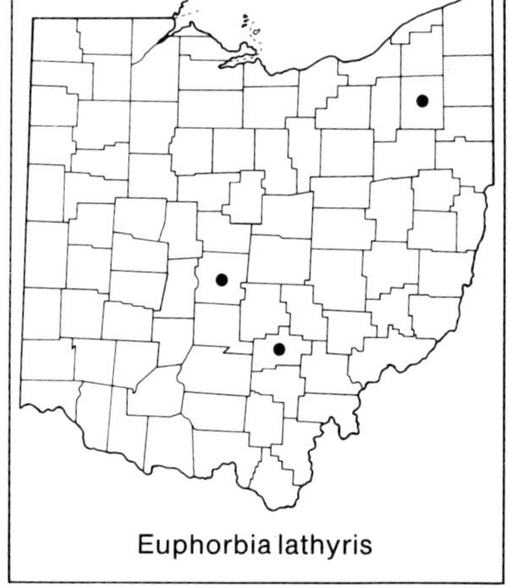
Euphorbia lathyris

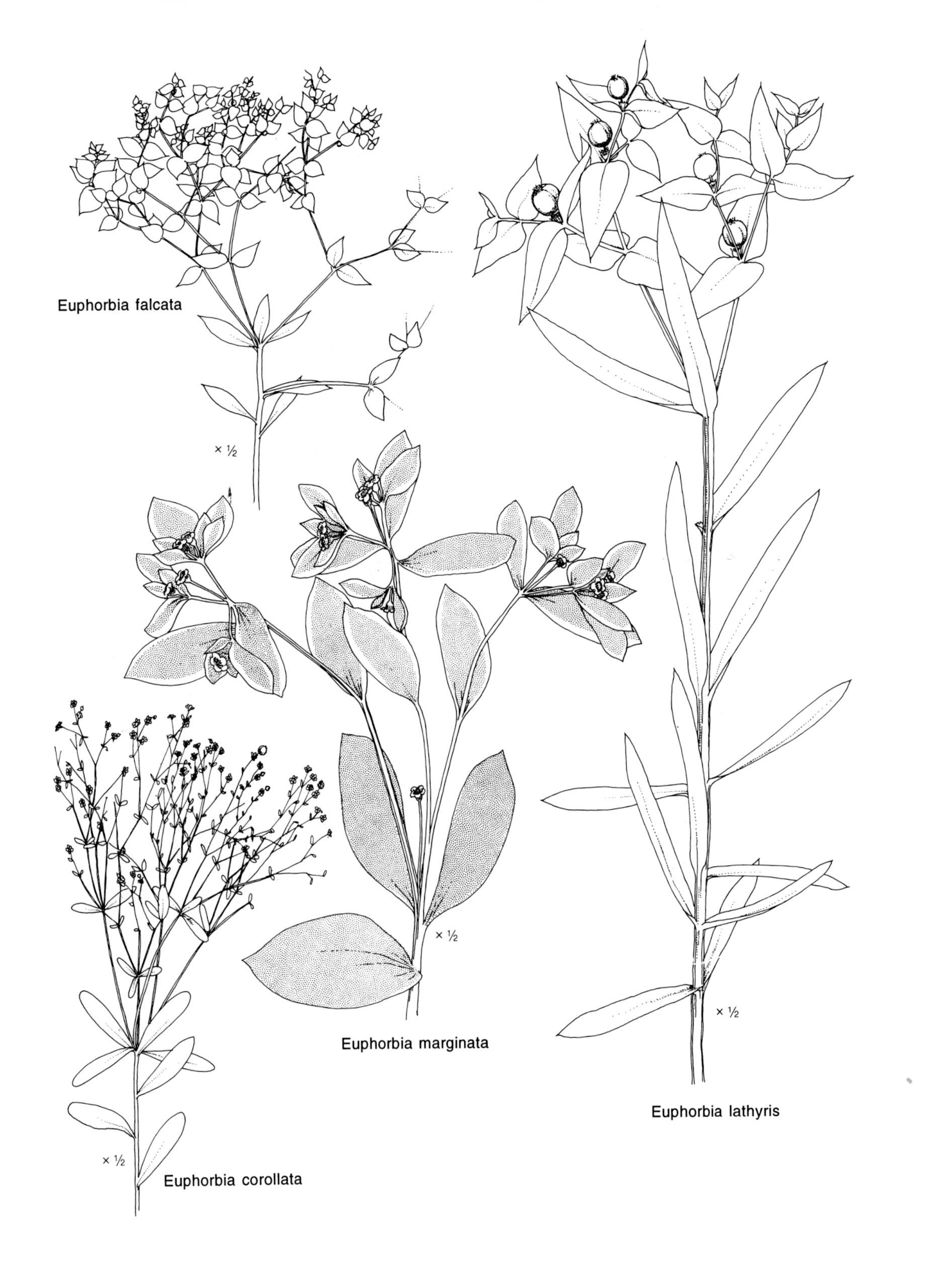
Euphorbia falcata
× ½
× ½
Euphorbia marginata
× ½
Euphorbia lathyris
× ½
Euphorbia corollata

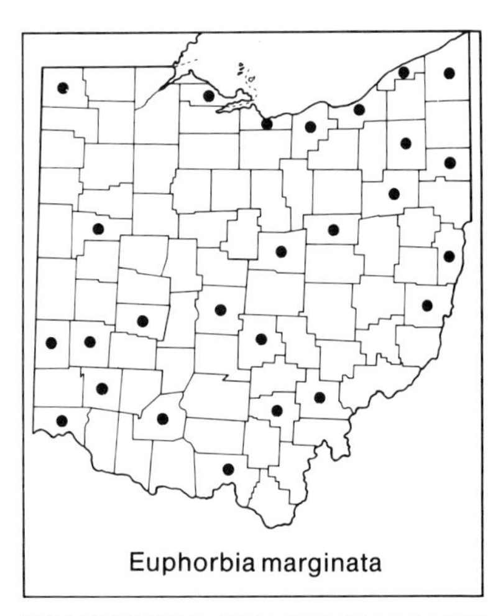

Euphorbia marginata

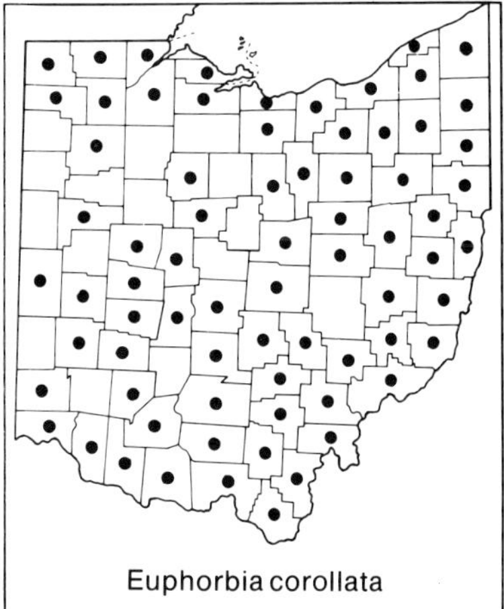

Euphorbia corollata

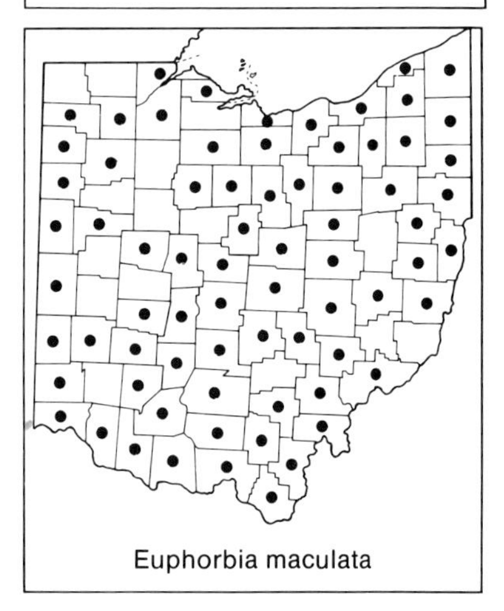

Euphorbia maculata

known from three scattered Ohio counties; adventive or perhaps locally naturalized in disturbed sites. May–September.

The epithet is sometimes incorrectly spelled *lathyrus*.

12. EUPHORBIA MARGINATA Pursh SNOW-ON-THE-MOUNTAIN

Erect few-branched annual, to ca. 9 dm tall; stem pubescent above and in the inflorescence; stem leaves alternate, entire, sessile, elliptic to oblong, to 8 cm long, the apex usually apiculate or acuminate, the upper leaves or at least the uppermost with a narrow to very broad white border; cyathia with conspicuous, white, petaloid appendages; capsules hirsute, to 6 mm long. Native to central United States, often cultivated and escaped eastward; at scattered sites more or less throughout Ohio; adventive or perhaps locally naturalized in disturbed habitats. July–October.

The milky sap of this plant is extremely caustic.

13. **Euphorbia corollata** L. FLOWERING SPURGE

Erect perennial, to 1 m tall, branched above in the inflorescence; stems mostly glabrous, sometimes with a few scattered hairs above; stem leaves alternate, entire, sessile or very short-petiolate, broadly linear to elliptic-oblong or elliptic-oblanceolate, to 6 cm long, the apex obtuse to rounded or retuse; cyathia numerous and flower-like, with conspicuous white petaloid appendages, making the plant somewhat showy when in bloom; capsules 3–4 mm long. Throughout Ohio but infrequent in west-central counties; dry openings, prairies, and a variety of weedy habitats, such as disturbed grassy fields, roadsides, railroad banks, and the like; sometimes cultivated. Late June–early September.

The pubescent southern phase, called var. *mollis* Millsp. by Fernald (1950), evidently does not reach as far north as Ohio.

14. **Euphorbia maculata** L. PROSTRATE SPURGE

E. supina Raf., the name used by Fernald (1950).

Prostrate and sprawling to ascending annual, much-branched from near the base; stems villous, at least above; leaves opposite, minutely serrulate to subentire, typically with a red blotch on upper surface, the principal leaves linear to oblong, 8–15 (–17) mm long, 1–5 (–6) mm wide, most 2.5–3 times as long as wide; capsules pubescent, the hairs mostly appressed; seeds with a few low transverse ridges. Throughout Ohio but less frequent westward; open sandy sites, and at roadsides, along railroads, in gardens, lawns, sidewalks, parking lots, and similar disturbed habitats. June–early October.

I follow Burch (1966) in using the name *Euphorbia maculata* for this species, which some authors call *E. supina*.

See notes under the next species, *E. humistrata*, on the difficulty of separating that species from this.

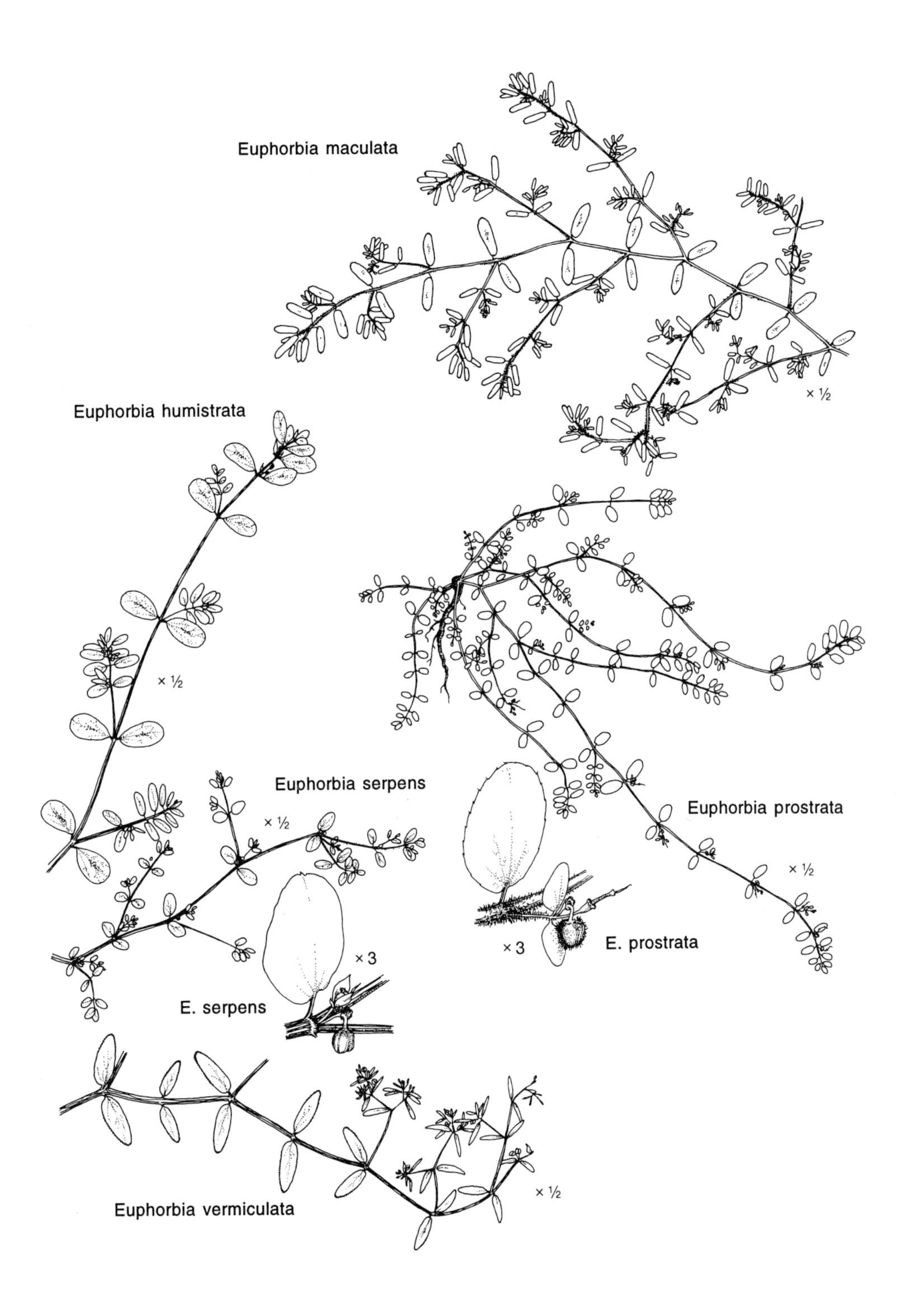
Euphorbia maculata
× ½
Euphorbia humistrata
× ½
Euphorbia serpens
× ½
× 3
E. serpens
× 3
E. prostrata
Euphorbia prostrata
× ½
Euphorbia vermiculata
× ½

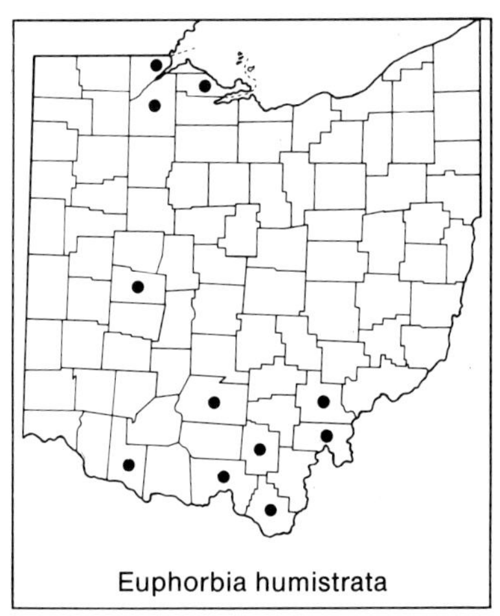
Euphorbia humistrata

15. **Euphorbia humistrata** Engelm. Spreading Spurge

Prostrate and sprawling to ascending annual, much-branched from near base; stems more or less pubescent; leaves opposite, the principal leaves elliptic to oblong-ovate or ovate, minutely serrulate to subentire, 6–10 mm wide and 8–14 mm long, most 1.5–2 times as long as wide; capsules pubescent, the hairs mostly appressed; seeds smooth or slightly wrinkled. A species of the middle and lower Mississippi Valley, evidently native to Ohio, where it is known chiefly from southern counties, but adventive in eastern United States, and possibly adventive at some Ohio sites; river flats and disturbed areas. August–October.

Wheeler (1941) notes the difficulty of separating some specimens from the previous species, *E. maculata*. Richardson (1968) has a comprehensive chart listing the contrasting characters he found most useful in separating the two species. In Ohio, *E. maculata* is much more common than *E. humistrata*.

Euphorbia prostrata

16. Euphorbia prostrata Ait. Ground-fig Spurge

E. chamaesyce of authors, incl. Fernald (1950) and Gleason (1952a), not L.

Prostrate and sprawling to ascending annual, much-branched from near base; stems crisp-pubescent, at least above; leaves opposite, the principal leaves oblong to elliptic or ovate, serrulate, 5–7 mm long; capsules pubescent, the hairs divergent from capsule wall. Native of tropical America; specimens have been collected from three sites in western Ohio, two along railroad tracks and the other in a lawn, all found since 1977—indicating that the species is perhaps becoming established in the state. July–October.

I have adopted Burch's (1966) conclusion in applying the name *E. prostrata* to this species.

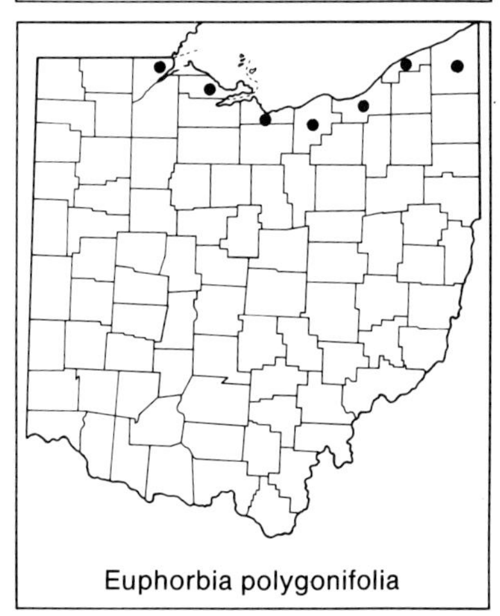
Euphorbia polygonifolia

17. **Euphorbia polygonifolia** L. Seaside Spurge

Prostrate and sprawling to ascending, glabrous annual, much-branched from near base; leaves opposite, entire, short-petiolate, linear-oblong to narrowly oblong-lanceolate, mostly 8–15 mm long; capsules glabrous, 3–3.5 mm long; seeds 2–2.5 mm long. A North American species restricted to dunes and beaches along the Atlantic Ocean and those bordering the Great Lakes (except Lake Superior). Guire and Voss (1963) have an interesting map of its Great Lakes distribution, including that in Ohio, where it is limited to northern counties bordering Lake Erie. July–October.

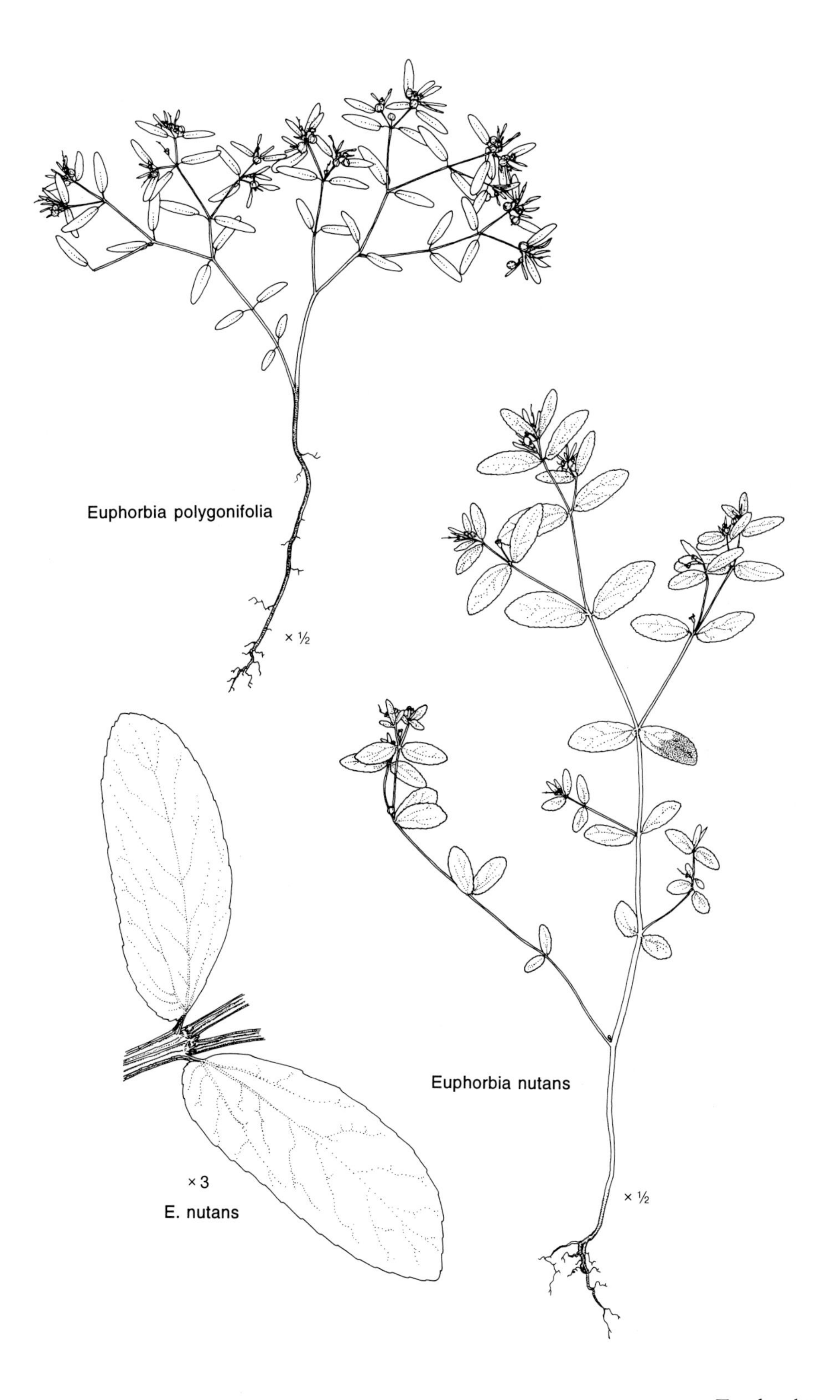
Euphorbia polygonifolia
× ½
Euphorbia nutans
× ½
× 3
E. nutans

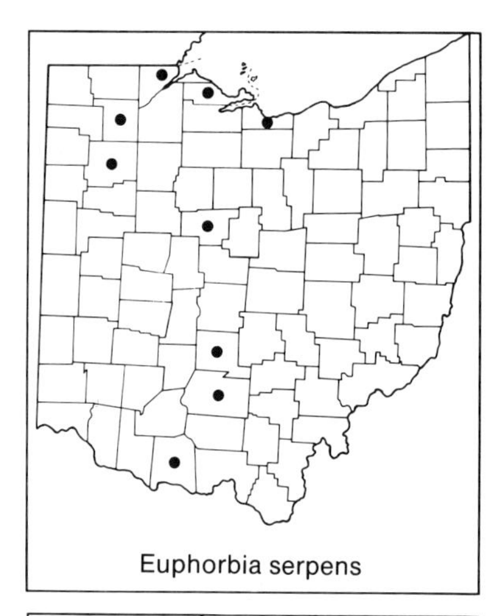
Euphorbia serpens

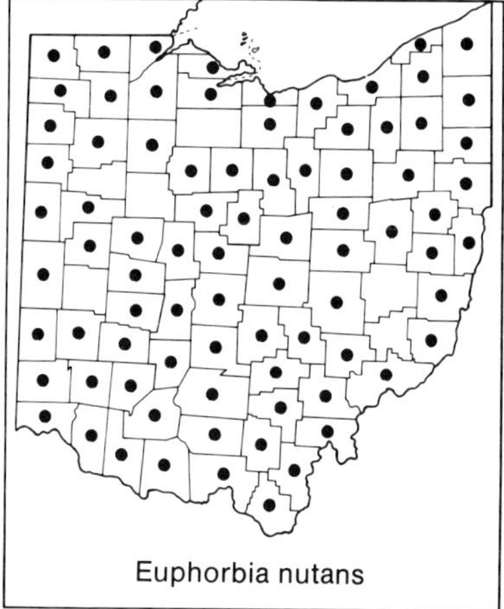
Euphorbia nutans

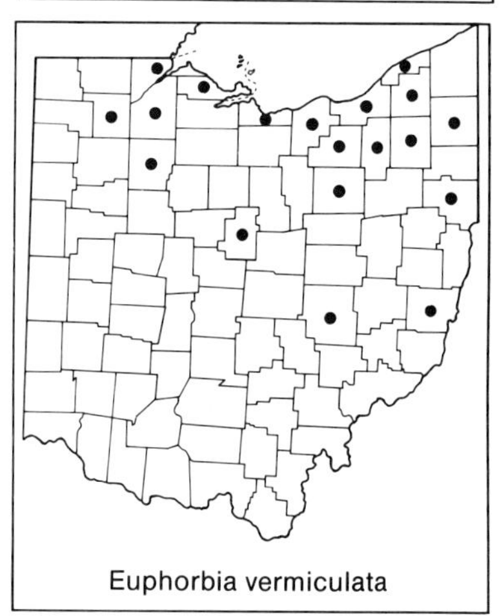
Euphorbia vermiculata

18. **Euphorbia serpens** HBK. ROUND-LEAVED SPURGE
Prostrate glabrous annual, much-branched from near base; leaves opposite, entire, ovate-orbicular to broadly oblong, mostly 2–7 mm long; capsules glabrous, ca. 1.2 mm long; seeds ca. 1 mm long. Known from a few sites in the western half of Ohio; moist alluvial or sandy soil, sometimes in disturbed habitats. July–October.

19. **Euphorbia nutans** Lag. SPOTTED SPURGE. NODDING SPURGE
E. preslii Guss.; *E. maculata* of authors, incl. Fernald (1950), not L. Erect or ascending, several-branched annual; stems sparsely appressed-pubescent at least above; leaves opposite, often slightly curved, serrulate, usually one or more on each plant marked with red, more or less oblong, markedly inequilateral at base; the principal leaves 1.5–3.5 cm long; cyathia with minute, white petaloid appendages; capsules glabrous, ca. 2 mm long. Throughout Ohio; openings, woodland borders, and especially along roads and railroads, and in similar disturbed habitats. June–early October.

20. **Euphorbia vermiculata** Raf. HAIRY SPURGE
Prostrate to slightly ascending annual, much-branched from near base; stems with spreading pubescence; leaves opposite, serrulate, ovate to oblong, the principal leaves 0.5–1.8 cm long; capsules glabrous, 1.5–2 mm long. Mostly in northern, especially northeastern counties; edges of fields, roadsides, along railroads. August–October.

Callitrichaceae. WATER-STARWORT FAMILY

A family with only one genus.

1. Callitriche L. WATER-STARWORT. WATER-CHICKWEED

Small, delicate, monoecious, annual herbs of aquatic or wet terrestrial habitats; leaves opposite, simple; flowers imperfect, usually solitary in leaf axils; calyx and corolla absent, but the flowers usually subtended by a pair of tiny horn-shaped bracts; staminate flowers with a single stamen; pistillate flowers with 2 carpels fused to form a compound pistil with 2 separate styles; fruit a tiny schizocarp separating into 2 or 4 mericarps. A cosmopolitan genus of 35–40 species. Fassett (1951) prepared a revision of the New World taxa.

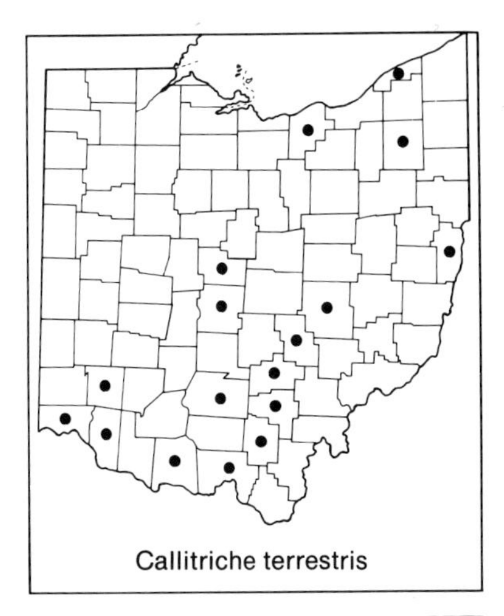
Callitriche terrestris

a. Fruits short-stalked, wider than long, deeply notched at apex; length of styles less than length of fruit; plants terrestrial on moist sites, with spatulate to oblanceolate leaves more or less uniform in size throughout plant . . . . 1. *Callitriche terrestris*

aa. Fruits sessile or subsessile, about equally as wide as long or longer than wide, shallowly or scarcely notched at apex; length of styles less than to greater than length of fruit; plants usually aquatic, sometimes on wet terrestrial sites, with either linear submersed leaves, or broadly spatulate to obovate or oblanceolate floating leaves, or both.

b. Fruits subellipsoid, longer than wide, and with marginal wing or keel spreading across upper margins and top of fruit; floating leaves thin and translucent, oblanceolate with a rounded apex; length of styles about equalling or less than length of fruit . . . 2. *Callitriche palustris*

bb. Fruits suborbicular, about as wide as long, and at maturity with rounded, nonwinged and nonkeeled margins; floating leaves thick and opaque, obovate to broadly spatulate, rounded to subtruncate at apex; length of styles greater than length of fruit. . . . . 3. *Callitriche heterophylla*

1. **Callitriche terrestris** Raf. TERRESTRIAL WATER-STARWORT
C. *deflexa* A. Br. var. *austinii* (Engelm.) Hegelm.

Annual; leaves oblanceolate to spatulate, more or less uniform in shape and size throughout plant; fruits short-stalked, wider than long, deeply notched at apex; length of styles less than length of fruit. At scattered sites in Ohio, but absent from northwestern and west-central counties; mud flats, moist ditches, and similar open moist places. April–July.

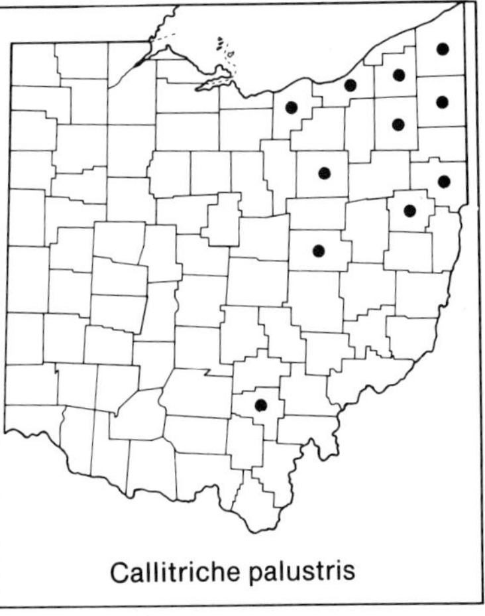
Callitriche palustris

2. **Callitriche palustris** L. VERNAL WATER-STARWORT
incl. C. *verna* L.

Annual; submersed leaves linear, floating leaves thin and translucent, oblanceolate, rounded at apex; fruits subellipsoid, longer than wide, shallowly or scarcely notched at apex, a wing or keel running along the upper margin and top of the fruit; length of styles about equalling or less than length of fruit. Widespread in cooler parts of the Northern Hemisphere; known from several Ohio counties, most in the northeastern quarter of the state; streams, ponds, temporary pools, and rarely in very moist terrestrial places. April–September.

There is disagreement as to which of the two Linnaean names listed above is correct for this species. My use of C. *palustris* rather than C. *verna* follows *Flora Europaea* (Tutin et al., 1964–1980).

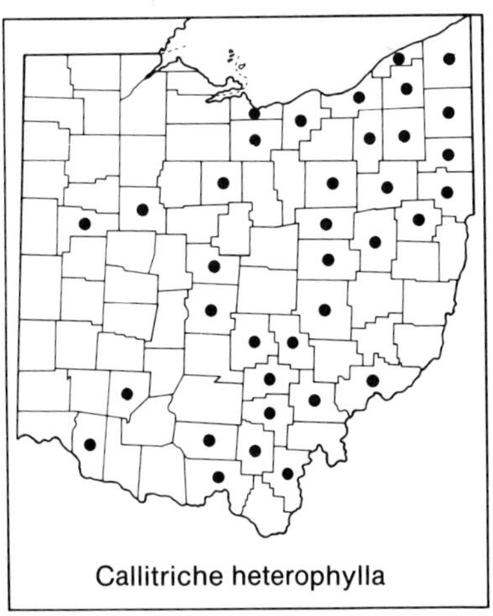
Callitriche heterophylla

3. **Callitriche heterophylla** Pursh LARGER WATER-STARWORT

Annual; submersed leaves linear, floating leaves thick and opaque, obovate to broadly spatulate, rounded to subtruncate at apex; fruits suborbicu-

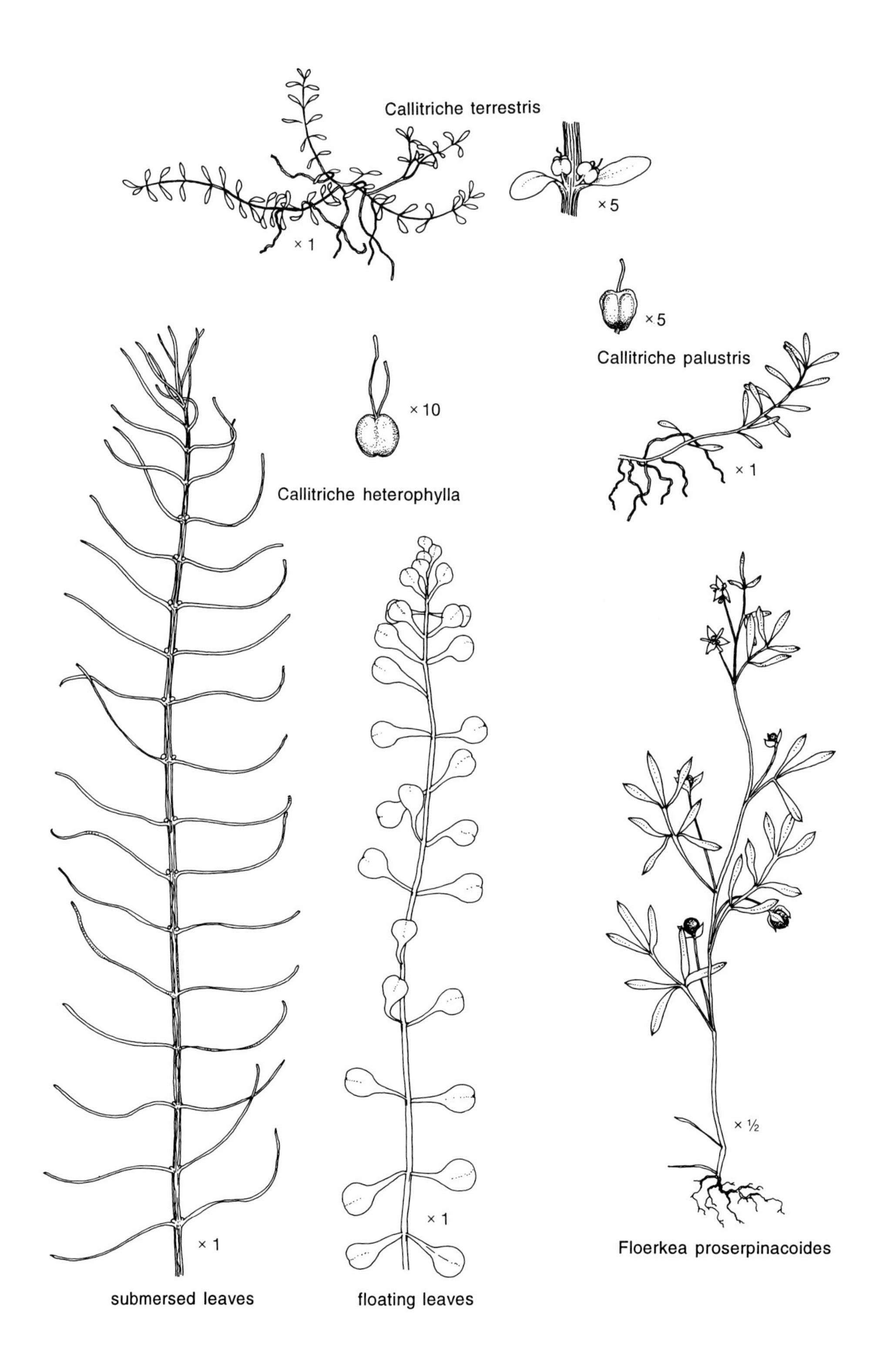
Callitriche terrestris
× 1
× 5
× 5
Callitriche palustris
× 10
× 1
Callitriche heterophylla
× ½
× 1
× 1
Floerkea proserpinacoides
submersed leaves
floating leaves

lar, about as wide as long, at maturity with rounded, nonwinged and nonkeeled, margins, shallowly or scarcely at all notched at apex; length of styles greater than length of fruits. Chiefly in the eastern half of Ohio; in water of ditches, canals, ponds, and streams, or rarely on moist ground in marshes or floodplains. Deschamp and Cooke (1985, and earlier papers cited therein) have studied the causal mechanisms controlling leaf dimorphism in this species. April–September.

Buxaceae. BOXWOOD FAMILY

A small chiefly subtropical family, three species of which are cultivated in Ohio. *Buxus sempervirens* L., COMMON BOXWOOD, of Old World origin, is a tall shrub with puberulent branches and opposite, entire, evergreen, aromatic leaves that are widest below the middle. Along with the southern magnolia, it is a legendary ornamental of old homes and plantations in southeastern United States. Only a few cultivars of this species are hardy in Ohio, where it is seen most often in southern counties, and these do not attain the heights and magnificence of plants in the South. There is a single collection of this species from Middle Bass Island, in Ottawa County, that may be from an adventive plant, or perhaps from one merely persisting from cultivation. *Buxus microphylla* Sieb. & Zucc., ORIENTAL BOXWOOD, a similar but shorter shrub with glabrous branches and leaves that are widest above the middle is also cultivated in Ohio.

Pachysandra terminalis Sieb. & Zucc., JAPANESE PACHYSANDRA, is a procumbent, stoloniferous, herbaceous-appearing plant with simple, alternate, dentate, evergreen leaves, usually crowded at the ends of low erect branches. Hardy in all parts of Ohio, it forms dense stands and is a popular and widely planted evergreen ground cover. Collections of adventive plants or of ones that have spread from cultivation have been made in Cuyahoga and Hamilton counties, and doubtless could be made in many other counties.

Limnanthaceae. FALSE MERMAID FAMILY. MEADOW-FOAM FAMILY

A small North American family with two genera, one of which is native to Ohio.

1. Floerkea Willd. FALSE MERMAID

A genus with a single species.

1. **Floerkea proserpinacoides** Willd. FALSE MERMAID

Small, weak-stemmed, ephemeral annual, 0.6–2 (–3) dm tall; leaves alternate, pinnately compound or deeply pinnately lobed, the leaflets or

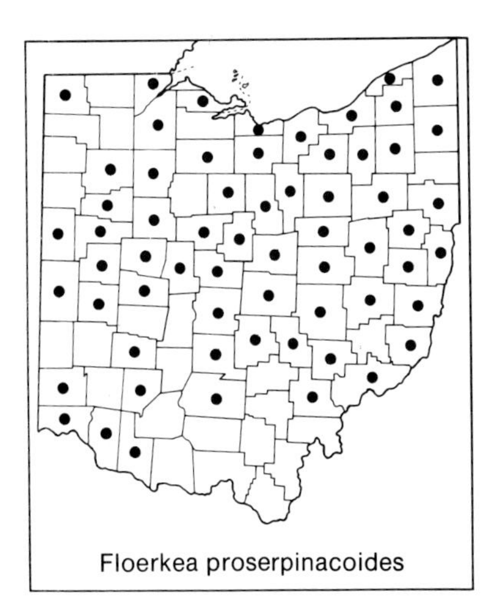
Floerkea proserpinacoides

lobes 3–7, entire or sometimes again lobed; flowers perfect, solitary in leaf axils; calyx of 3 separate sepals; corolla actinomorphic, of 3 separate, white petals, shorter than the sepals; stamens 6; carpels 2–3, only partially fused at base, but with a single style; ovary superior; fruit a small dark, subglobose, subfleshy achene, tuberculate near the apex. Throughout most of Ohio, but absent from south-central and some western counties; rich moist woodlands. Baskin, Baskin, and McCann (1988) studied the germination ecology. April–May.

Anacardiaceae. SUMAC FAMILY. CASHEW FAMILY

A small family of woody plants, chiefly tropical but extending into the temperate regions, with two weakly separated genera native to Ohio. *Cotinus coggygria* Scop., SMOKE-TREE, is frequently cultivated in Ohio; herbarium specimens have been collected from adventive plants in Ashtabula, Highland, Jefferson, and Trumbull counties. The cashew nut comes from the Brazilian tree, *Anacardium occidentale* L. The tropical mango is the fruit of *Mangifera indica* L.; although delicious, it causes dermatitis in some individuals. Barkley (1937) revised *Rhus* (incl. *Toxicodendron*) in North America. Brizicky (1962c) covers the genera in southeastern United States.

a. Leaves simple; fruits asymmetrical, with lateral styles; see above . . . . *Cotinus*
aa. Leaves trifoliolate or pinnately compound; fruits symmetrical, with terminal styles.
b. Inflorescences dense; fruits red and pubescent . . . . 1. *Rhus*
bb. Inflorescences open; fruits whitish and glabrous or nearly so. . . . . 2. *Toxicodendon*

1. Rhus L. SUMAC

Shrubs and small trees with milky sap; leaves alternate, compound; flowers small, white, greenish, or yellowish, actinomorphic, in dense inflorescences; flowers mostly imperfect, the species mostly functionally dioecious; fruit a red, pubescent, and dryish drupe. A widely distributed genus of about 100 species.

a. Leaves pinnately compound, leaflets 5–31, rarely a leaf or two with only 3 leaflets.
b. Rachis winged between all or most leaflets . . . . . . 3. *Rhus copallina*
bb. Rachis not winged.
c. Young branches densely soft-pubescent . . . . 1. *Rhus typhina*
cc. Young branches glabrous . . . . 2. *Rhus glabra*
aa. Leaves trifoliolate, all with 3 leaflets. . . . . 4. *Rhus aromatica*

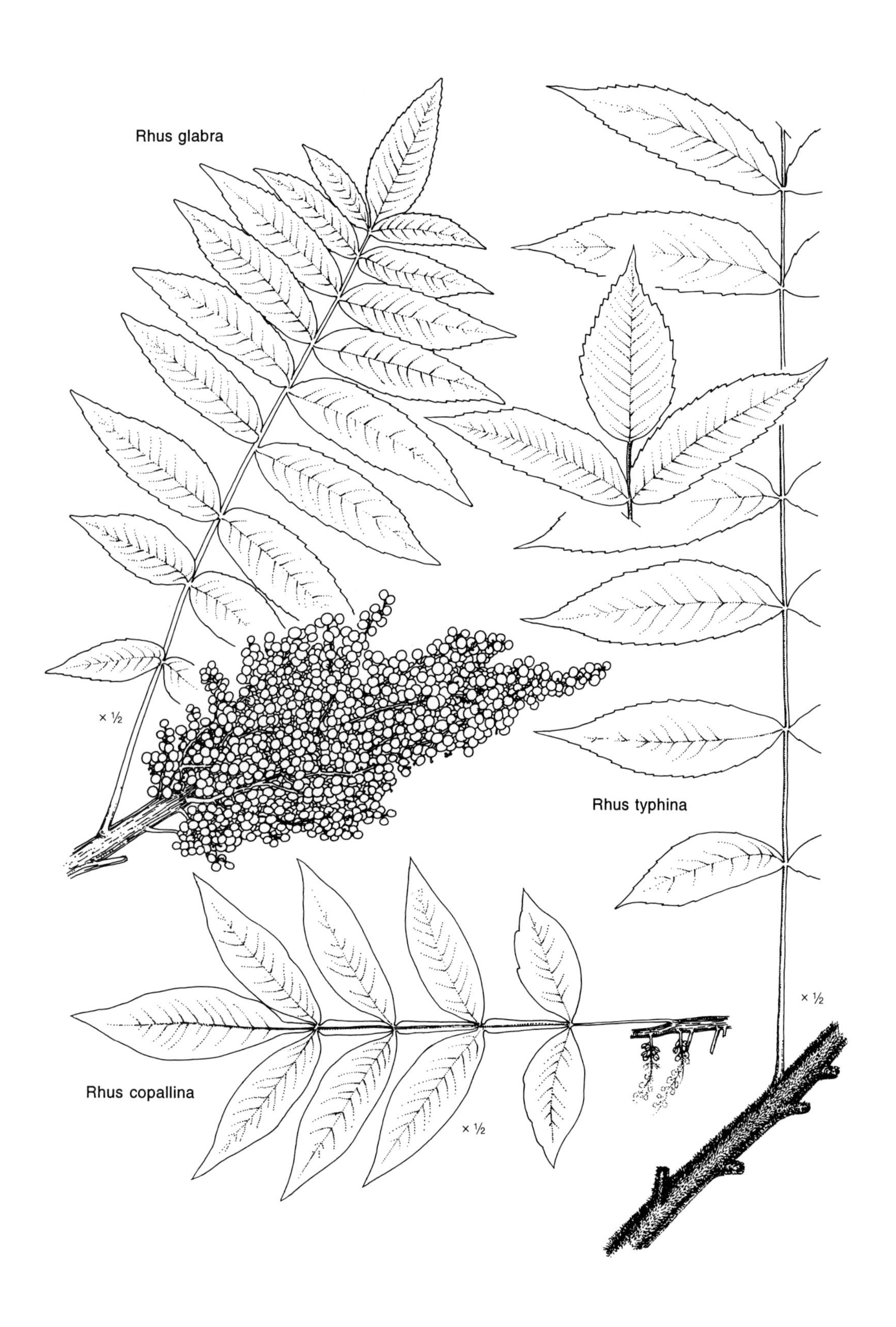
Rhus glabra
× ½
Rhus typhina
× ½
Rhus copallina
× ½

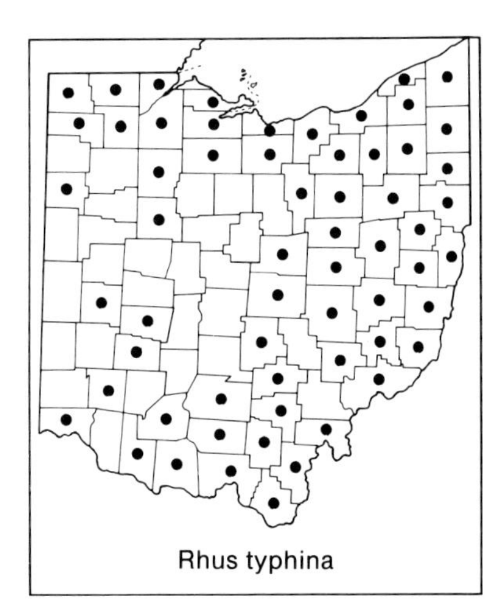
Rhus typhina

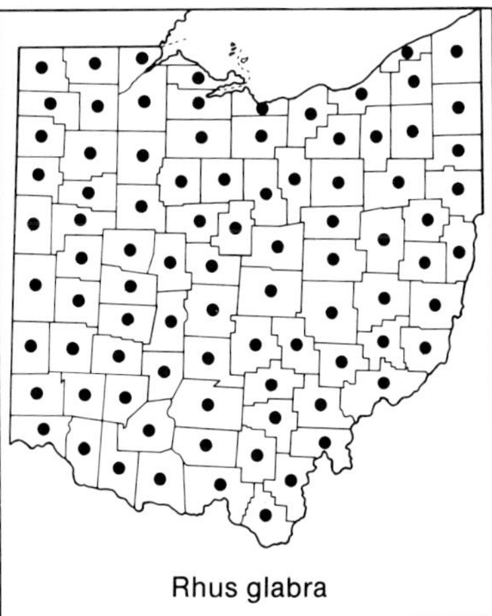
Rhus glabra

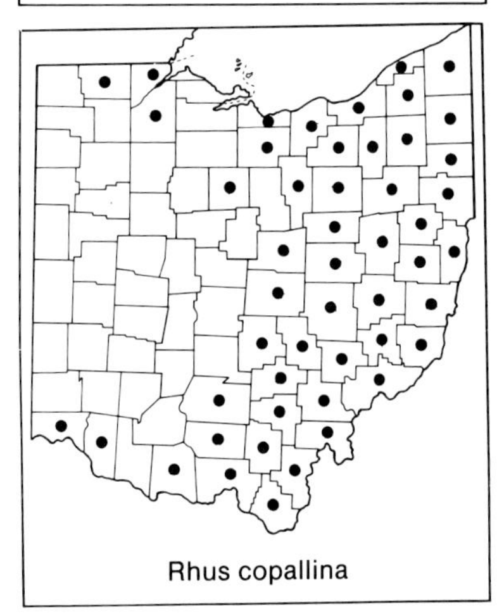
Rhus copallina

1. **Rhus typhina** L. Staghorn Sumac
Shrub or small tree, to 9 m (30 feet) tall, but usually about half that height; young branches brown and densely soft-pubescent (when leafless, in winter, reminiscent of a stag's horns); leaves odd-pinnately compound with 9–31 leaflets, the leaflets more or less oblong but acuminate at apex, pale green beneath, the margins variably shallowly to deeply serrate; fruits red and pubescent. Throughout Ohio except in some central and some west-central counties; open woods, thickets, roadsides, disturbed habitats. June–early July.

Reveal (1991) recently concluded that the correct name for this species is *R. hirta* (L.) Sudworth.

A specimen of *R. typhina* forma *dissecta* Rehd., with leaflets irregularly pinnately lobed or divided, was collected from Stark County in 1960.

Rhus × pulvinata E. Greene, the hybrid of *Rhus glabra* × *R. typhina*, has been collected from Ashtabula and Erie counties, both in northern Ohio. J. W. Hardin and L. L. Phillips (1985), who made successful artificial crosses between the two species, called the hybrid *R.* × *borealis* (Britt.) E. Greene, but Voss (1985b) believes the correct name to be *R.* × *pulvinata*. Hardin and Phillips also report naturally occurring hybrid swarms and introgressant populations.

2. **Rhus glabra** L. Smooth Sumac
Shrub or rarely a small tree, to 7.5 m (25 feet) tall, but usually about half that height; young branches tan to medium brown, glabrous and often squarish; leaves odd-pinnately compound with 9–31 leaflets, the leaflets more or less oblong but acuminate at apex, pale green beneath, the margins serrate; fruits red and minutely pubescent. Throughout Ohio; woodland borders, openings, thickets, roadsides, along railroads, and in similar mostly dry open habitats. Late June–July.

See note above under *Rhus typhina* on the naturally occurring hybrid between these two species.

3. **Rhus copallina** L. Shining Sumac. Winged Sumac
Shrub or very rarely a small tree, to 7.5 m (25 feet) tall, but usually half that height or less; young branches often squarish and with short crisp pubescence; leaves odd-pinnately compound with (3–) 5–13 leaflets, the leaflets varying from lanceolate to elliptic to ovate, and from acute to subobtuse at apex, the margins usually entire but sometimes with few to several teeth near the leaflet apex, the rachis with a conspicuous wing connecting the upper leaflets, the wing narrowing below; fruits red and minutely pubescent. Chiefly in the eastern half of Ohio; dry openings, thickets, and fields. July–August.

In Fernald's (1950) infraspecific classification, which seems to have merit, Ohio plants are assigned to the western var. *latifolia* Engler. The more eastern var. *copallina* does not occur in Ohio.

4. **Rhus aromatica** Ait. FRAGRANT SUMAC

Shrub, 0.3–1.5 m (1–5 feet) tall, young branches minutely puberulent to glabrous, pungently aromatic when cut (another infrequently encountered common name, POLECAT BUSH, suggesting that some find the "fragrance" less than pleasant); leaves trifoliolate, the terminal leaflet tapering to a sessile base or to a stalk no more than 4 mm long; fruits red and conspicuously pubescent. April–May.

In the taxonomy adopted here, which is the same as that of Fernald (1950) and Braun (1961), Ohio representatives of this species may be assigned to the two varieties keyed below.

a. Terminal leaflet ovate to rhombic-elliptic, widest at about the middle, most 4 cm or longer; apex of most leaflets acute, a few approaching obtuse. . . . . var. *aromatica*

aa. Terminal leaflet broadly obovate, widest above middle, less than 4 cm long; apex of most leaflets obtuse to rounded . . . . var. *arenaria* (E. Greene) Fern.

R. arenaria (E. Greene) G. Jones; *R. trilobata* Nutt. var. *arenaria* (E. Greene) F. Barkley

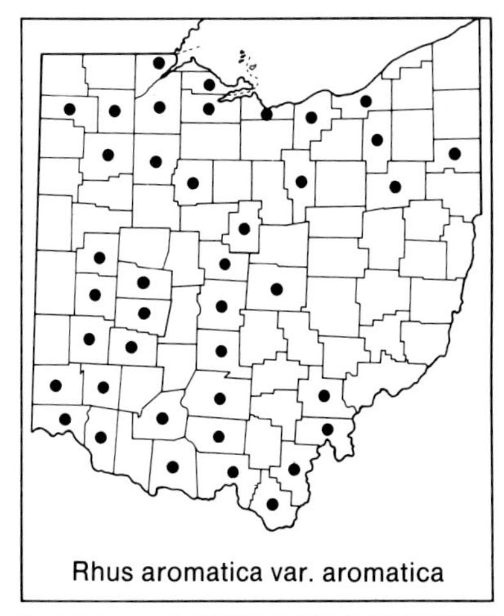
Rhus aromatica var. aromatica

Rhus aromatica var. arenaria

Variety *aromatica* ranges diagonally from New England and southeastern Canada to eastern Texas; in Ohio it grows chiefly in counties of the western half of the state, in prairies, fields, and other openings, usually on calcareous soils. Variety *arenaria* is limited to the dunes along the Great Lakes in Ohio, Indiana, and Illinois. The only bona fide Ohio collection of this species was made from a small colony found in 1931 near the Lake Erie shore in Ashtabula County. A few specimens assigned to var. *aromatica* show intergradation toward var. *arenaria*. As noted in the synonymy above, var. *arenaria* has sometimes been classified as a part of *Rhus trilobata*; but I believe the instances of intergradation between the two Ohio taxa support their being placed within the same species.

2. Toxicodendron Mill.

Woody vines, shrubs, and small trees with milky sap; leaves alternate, compound; flowers small and greenish, in open axillary inflorescences; flowers imperfect, the species dioecious; fruit a dry whitish drupe, glabrous or nearly so. A genus of ca. 10 species of the Western Hemisphere and eastern Asia, often and perhaps better treated as a subgenus of *Rhus* (Brizicky, 1963a). The true POISON-OAK of southeastern United States, *Toxicodendron pubescens* Mill. (*Rhus toxicodendron* L.), does not reach Ohio; Ohio plants called "poison-oak" are forms of *T. radicans*.

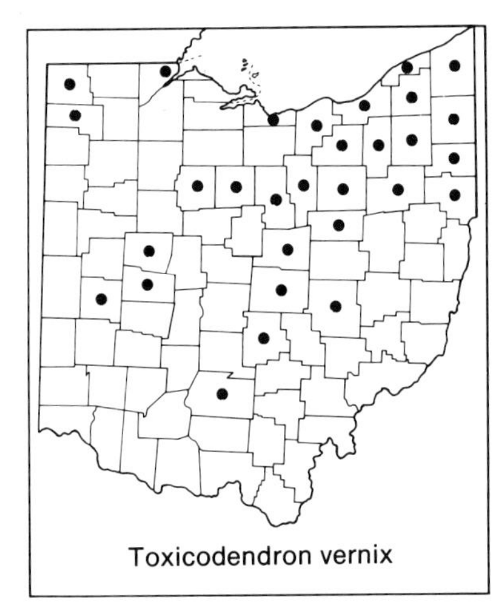

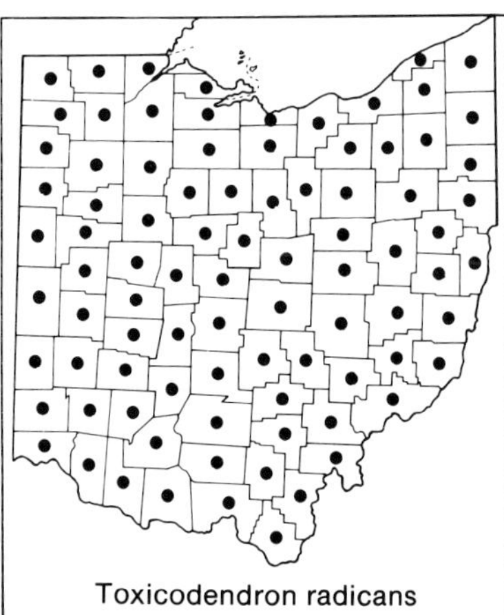

a. Leaves pinnately compound, with 7–13 leaflets .... 1. *Toxicodendron vernix*
aa. Leaves trifoliolate..................... 2. *Toxicodendron radicans*

1. **Toxicodendron vernix** (L.) Ktze. Poison Sumac
Rhus vernix L.

Shrub with few to several main stems, to 7.5 m (25 feet) tall, but usually about half that height; young branches glabrous; leaves odd-pinnately compound with 7–13 leaflets, the leaflets more or less oblong to elliptic or obovate, acutish at apex, the margins entire; fruits whitish. Chiefly in counties of the northeastern quarter of the state, at scattered sites elsewhere; sphagnum bogs, fens, swamps. June–early July.

This species is like poison ivy in causing a skin reaction in persons sensitive to its toxic oils. Cases of dermatitis resulting from contact with this species are fewer (although usually more severe) than those acquired from poison ivy because of the more restricted and remote habitat of poison sumac. Although the species is easy to find in northeastern Ohio, probably only a small fraction of the people living there have ever seen the plant, and even fewer have brushed against it.

2. **Toxicodendron radicans** (L.) Ktze. Poison Ivy
incl. var. *radicans* (*Rhus radicans* L.) and var. *rydbergii* (Rydb.) Erskine (*T. rydbergii* (Rydb.) Greene)

Trailing to climbing or scrambling vine, when climbing, often with long ascending to diverging side branches projecting outward from the supporting pole or tree or wall, or infrequently a thin-stemmed, erect or semierect, usually unbranched shrub; leaves trifoliolate, the terminal leaflet long-stalked, the stalk 1–4 cm long; the leaflets highly variable, most ranging in shape from narrowly ovate to ovate or rhombic, the two lateral leaflets often markedly asymmetric, the margins ranging from irregularly serrate or dentate to entire or rarely the lateral leaflets shallowly lobed; fruits whitish. Widely distributed in temperate North America, this species is found throughout Ohio, sometimes abundantly so, in a variety of open or shaded habitats. Late May–June.

Of the Ohio plants harmful to humans, this is by far the most familiar in terms of name recognition because of the many people who have had the dermatitis caused by contact with the plant or who know of others who have. Unfortunately, despite its widespread occurrence in Ohio, the species is recognized at sight by only a very small percentage of the population. This is due in part to its many "look-alikes," e.g., Boston ivy (p. 93), box-elder (p. 79), fragrant sumac (p. 61), and Virginia creeper (p. 93), causing some persons to mistrust them all, including many harmless plants. It is in part due also to the unusual amount of variation within

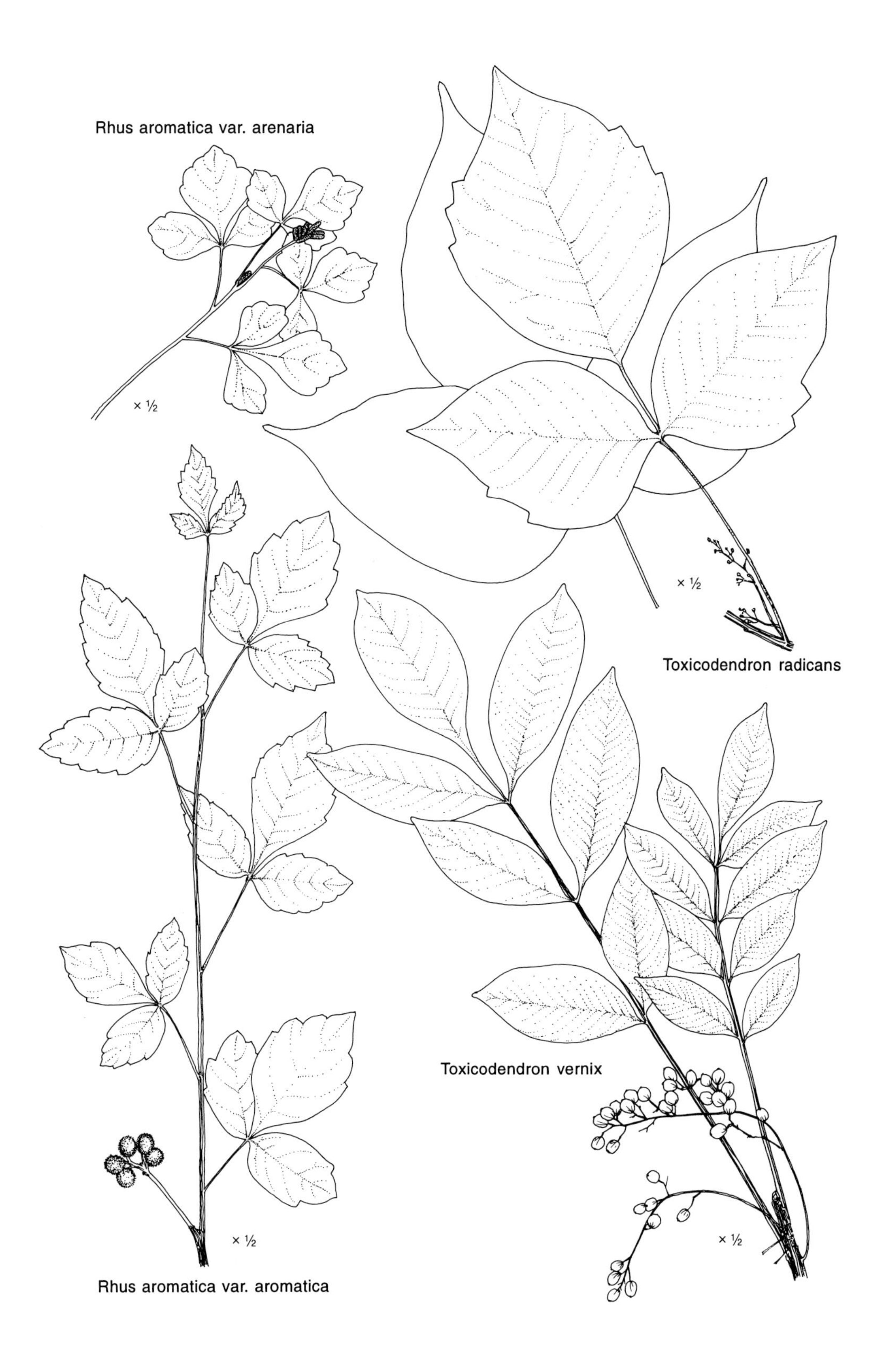
Rhus aromatica var. arenaria
× ½
× ½
Toxicodendron radicans
Toxicodendron vernix
× ½
× ½
Rhus aromatica var. aromatica

the species, giving it many guises one must learn if it is to be successfully avoided.

The great amount of variation in habit and leaf form and size, noted above, has given rise to a complicated taxonomy. The erect or suberect forms are often segregated as var. *rydbergii* or as the species *T. rydbergii*. Agreeing with Voss' (1985b) treatment of the species in Michigan, I believe Ohio plants are best treated as belonging to a polymorphic species, without recognized infraspecific categories. For a full account of the various infraspecific names published, see the revision by Gillis (1971).

Aquifoliaceae. HOLLY FAMILY

Shrubs or small trees; leaves alternate, simple; flowers small, greenish or whitish, actinomorphic, in axillary cymose clusters; flowers mostly imperfect, the species dioecious, or at least functionally so; fruits in Ohio species bright red, or rarely yellow, drupes (holly "berries"). A rather small family of woody plants with two genera in the Ohio flora. Brizicky (1964a) studied the family in southeastern United States.

a. Leaves either conspicuously serrate or spine-toothed; petals fused basally . . . . 1. *Ilex*

aa. Leaves entire or with few low inconspicuous serrations; petals separate . . . . 2. *Nemopanthus*

1. Ilex L. HOLLY

Shrubs or small trees; leaves serrate or spine-toothed; petals fused basally; stamens epipetalous. A genus of more than 300 species with scattered centers of distribution in temperate and tropical regions of Asia and the Americas.

a. Leaves evergreen, with spine-toothed margins; small tree . . . . 1. *Ilex opaca*

aa. Leaves deciduous, with serrate but not spine-toothed margins; shrub . . . . 2. *Ilex verticillata*

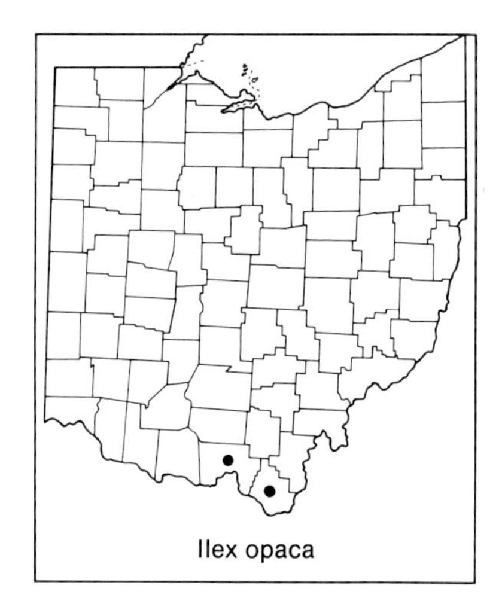

1. **Ilex opaca** Ait. AMERICAN HOLLY

Tree, to 14 m (45 feet) tall, but usually much shorter; leaves evergreen, elliptic to elliptic-oblong or narrowly obovate, margins spine-toothed. A species ranging across southeastern United States and northward along the Atlantic coast to Massachusetts and inland to southern Ohio and southern Illinois. South-central counties; moist woods, disturbed woods and thickets. May–June.

The two trees from which the southernmost Ohio collections were made, one each in Lawrence and Scioto counties, are thought to have

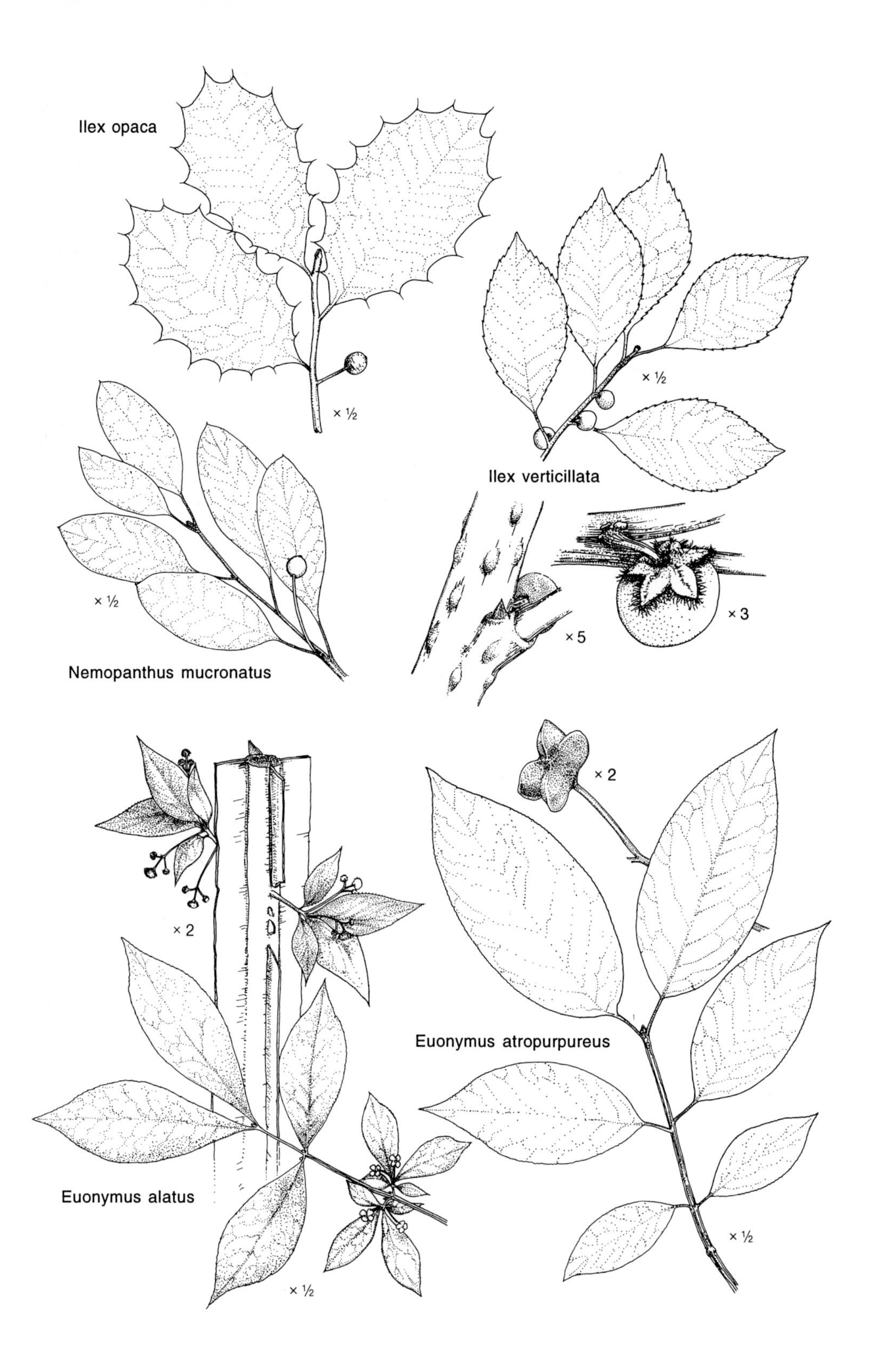
Ilex opaca
× ½
× ½
Ilex verticillata
× ½
Nemopanthus mucronatus
× 5
× 3
× 2
× 2
Euonymus atropurpureus
Euonymus alatus
× ½
× ½

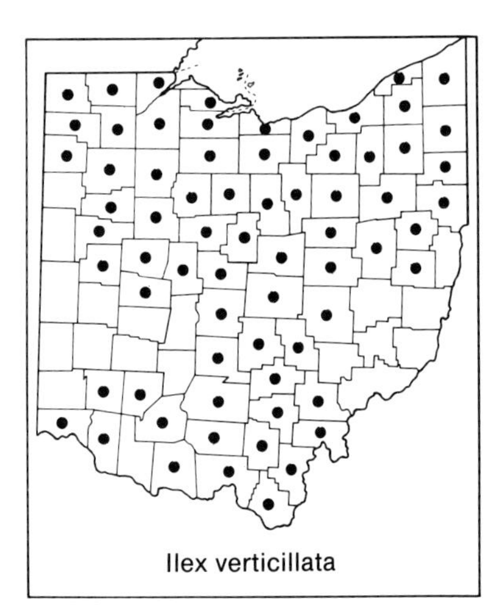
Ilex verticillata

been natives, slightly disjunct from the northern boundary of the more continuous range of the species that reaches as far north as Lewis County, Kentucky, directly across the Ohio River from Scioto County, Ohio (Braun, 1961). Specimens of adventive escapes from cultivation have been collected from Fairfield, Jackson, Licking, Meigs, and Tuscarawas counties.

Ilex opaca and hybrids between it and related species have given rise to a large number of cultivars, including one named 'Beautiful Ohio'. Many of the cultivars are on display at the Secrest Arboretum of the Ohio Agricultural Research and Development Center at Wooster. In general, these hollies are taller and more vigorous in southern counties and northern counties bordering Lake Erie than elsewhere in Ohio. This is the holly most often used for Christmas decorations although cut branches from other cultivars, especially those of ENGLISH HOLLY, *Ilex aquifolium* L., are often shipped to Ohio from warmer climates during the winter holiday season. *Ilex opaca* is noted also for its heartwood of unusual whiteness, used for inlays and for carvings.

2. **Ilex verticillata** (L.) A. Gray WINTERBERRY
incl. var. *verticillata*, var. *padifolia* (Willd.) T. & G., and var. *tenuifolia* (Torr.) S. Wats.

Shrub, to 4.5 m (15 feet) tall; leaves deciduous, varying in shape from oblanceolate to obovate to elliptic, rhombic-elliptic, or rarely subcircular, margins serrate to sharply serrate, sometimes slightly double-serrate. Throughout much of Ohio; marshes, open swamps, lake borders, and similar moist habitats; sometimes cultivated. June–July.

Several varieties, those listed in synonymy above and others, have been recognized on the basis of variation in leaf shape and amounts of pubescence. Ohio plants vary continuously in these features and designation of varieties among them seems of little value, a conclusion also reached by Voss (1985b) in studying Michigan plants.

Herbarium specimens of this species are frequently confused with those of *Rhamnus alnifolia* (p. 87) and *R. lanceolata* (p. 87). They may be distinguished by the presence in *I. verticillata* of small, dark, pointed, persistent stipules and of ciliate calyx lobes, persisting beneath the fruit. In *Rhamnus* there are no small, dark, persistent stipules, and the fruits are subtended by a glabrous, saucer-shaped hypanthium, sometimes once-cleft, but otherwise unlobed.

2. Nemopanthus Raf. MOUNTAIN-HOLLY

A genus of a single species. *Nemopanthus* is merged under *Ilex* by Ahles (1968) but accepted as a distinct genus by most American floristic workers.

Baas (1984) found characters supporting each point of view, but believing further study was needed did not make a judgment in the case.

1. **Nemopanthus mucronatus** (L.) Loesener MOUNTAIN-HOLLY. CATBERRY

Shrub, to 3 m (10 feet) tall; leaves deciduous, oblong to elliptic, entire or with a few, low, usually inconspicuous serrations; petals separate; stamens free. Mostly in northeastern counties; sphagnum bogs and sometimes in wet woods. A specimen in The Ohio State University Herbarium, collected by Wm. C. Werner in 1890 from "North Columbus, Franklin County," and originally identified as *Aronia melanocarpa*, is not mapped because of the possibility of error in the stated location. May.

Voss (1985a) discusses Loesener's authorship of this name.

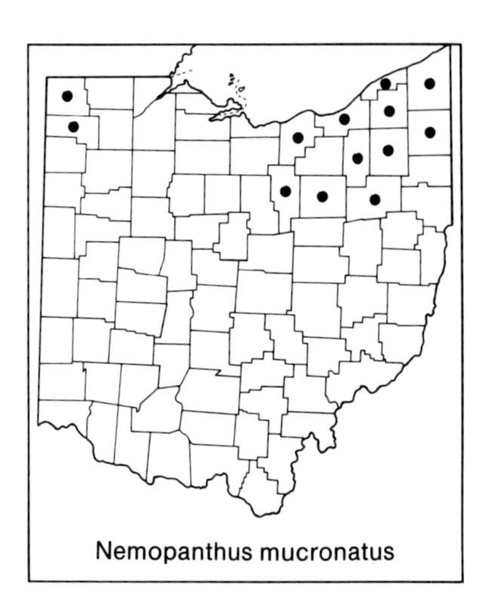
Nemopanthus mucronatus

Celastraceae. BITTERSWEET FAMILY. STAFF-TREE FAMILY

Shrubs or woody vines; leaves simple, opposite or alternate; flowers small, greenish or dark red to purplish, actinomorphic, perfect or imperfect, solitary or in axillary cymes or terminal racemes; fruit a capsule, often colored, which upon opening reveals seeds encased within a fleshy, and in most instances bright-colored, aril. A widely distributed family of more than 50 genera but only about 800 species. Brizicky (1964a) studied the family in southeastern United States.

a. Leaves opposite, deciduous or evergreen; stems erect, trailing or climbing.
  b. Erect shrubs with deciduous leaves or trailing or climbing shrubs with evergreen or deciduous leaves; leaves mostly 3–12 cm long; fruit 3–5-valved, seeds with red or orange aril . . . . 1. *Euonymus*
  bb. Low, erect to decumbent shrubs to 0.3 m tall, with evergreen leaves; leaves mostly 1–2 cm long; fruit 2-valved, seeds with white aril . . . . 2. *Paxistima*
aa. Leaves alternate, deciduous; stems twining and climbing . . . . 3. *Celastrus*

1. Euonymus L. SPINDLE-TREE

Erect to trailing, or climbing and vine-like, shrubs or small trees; leaves opposite. A genus of more than 150 species, most native to Asia.

a. Branches with conspicuous corky wings, the wings 2–5 mm wide; sepals and petals 4 . . . . 4. *Euonymus alatus*
aa. Branches sometimes ridged, or rarely with low wings less than 1 mm wide, or without wings or ridges; sepals and petals 4 or 5.

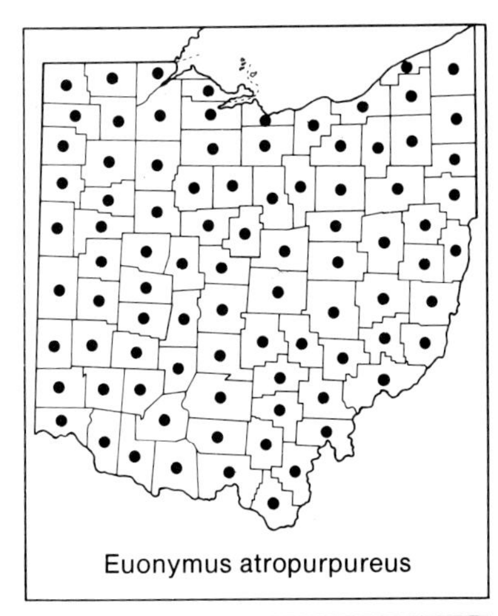
Euonymus atropurpureus

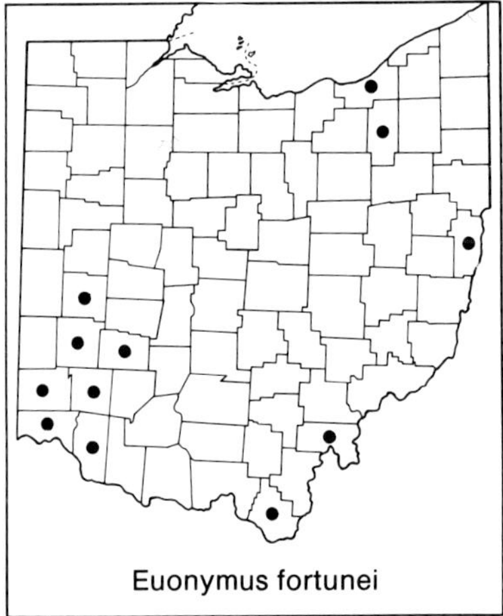
Euonymus fortunei

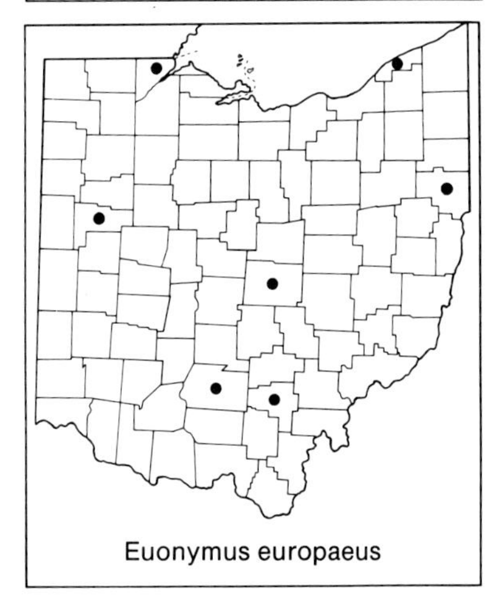
Euonymus europaeus

b. Trailing shrub or climbing (by rootlets) vines with evergreen leaves . . . . 2. *Euonymus fortunei*

bb. Erect or trailing shrubs with deciduous leaves.

c. Principal leaves with petioles usually (8–) 10–15 mm long; petals and sepals 4; capsules not tuberculate.

d. Undersurface of leaves densely short-pubescent; flowers purplish . . . . 1. *Euonymus atropurpureus*

dd. Undersurface of leaves glabrous; flowers greenish-yellow to white . . . . 3. *Euonymus europaeus*

cc. Principal leaves with petioles 1–4 (–6) mm long; petals and sepals 5; capsules tuberculate.

d. Erect to straggly shrubs; leaves elliptic-lanceolate to ovate-lanceolate, gradually acuminate at apex, subsessile on petioles 1–3 mm long. . . . . 5. *Euonymus americanus*

dd. Trailing to decumbent shrubs, stems often rooting; leaves obovate, obtuse to rounded at apex or abruptly short-acute, on petioles 2–6 mm long . . . . 6. *Euonymus obovatus*

1. **Euonymus atropurpureus** Jacq. Wahoo. Burning-bush

Erect shrub or small tree, to ca. 4 m (13 feet) tall; leaves deciduous, long-elliptic to ovate-elliptic, acuminate at apex, densely short-pubescent beneath, petioles 8–15 (–20) mm long; sepals 4; petals 4, purplish; capsules not tuberculate. Collected from every Ohio county, common but not abundant; woods, borders, thickets, floodplains. Late May–early July.

2. Euonymus fortunei (Turcz.) Hand.-Mazz. Chinese Wintercreeper

Trailing shrub or woody vine climbing by aerial rootlets; leaves evergreen, the midrib and principal lateral veins usually whitish, elliptic to narrowly ovate or elliptic-lanceolate, varying greatly in size from plant to plant. An Asian species from which a host of cultivars have been derived, several escaped from cultivation and established locally, mostly in southwestern counties; disturbed woodlands. Not illustrated. Rarely flowering.

3. Euonymus europaeus L. European Spindle-tree

Shrub or small tree, to 5 m (16 feet) tall; leaves deciduous, varying greatly in shape from plant to plant: broadly ovate or broadly elliptic to elliptic or rather narrowly oblong. A Eurasian species from which several cultivars have been derived, infrequently escaped and established at widely scattered Ohio sites; open disturbed areas. Not illustrated. May–June.

4. Euonymus alatus (Thunb.) Sieb. Winged Wahoo. Winged Spindle-tree

Erect shrub, to 2.5 m (8 feet) tall; branches conspicuously and widely to narrowly corky-winged; leaves deciduous, oblanceolate to obovate or rhombic-elliptic, petioles 1–3 mm long; sepals and petals 4; capsules not tuberculate. A species of eastern Asia cultivated as an ornamental, the foliage a bright rosy-red in autumn, the distinctive 4-winged branches attractive in winter. Escaped at scattered sites around Ohio, and perhaps becoming naturalized at some; most collections are recent. Ebinger (1983) reports the species to be naturalized with increasing frequency in Illinois. May.

5. **Euonymus americanus** L. American Strawberry-bush. Hearts-bursting-open-with-love

Low, somewhat straggling to erect shrub, to 2 m (6 feet) tall; leaves deciduous, elliptic-lanceolate to ovate-lanceolate, gradually acuminate at apex, subsessile on petioles 1–3 mm long; petals and sepals 5; capsules tuberculate. A species of eastern United States ranging from Florida to Texas and north to New York and Missouri, the Ohio distribution limited to a few south-central counties where the species is at a northern limit of its range; moist woods, hemlock slopes. May–June.

6. **Euonymus obovatus** Nutt. Running Strawberry-bush

Trailing to decumbent shrub, the stems often rooting; leaves deciduous, obovate, obtuse to rounded at apex or abruptly short-acute, petioles 2–6 mm long; petals and sepals 5; capsules tuberculate. Throughout Ohio, except in some of the southernmost counties; moist to mesic woods, wooded floodplains, terraces, and slopes. Late April–early June.

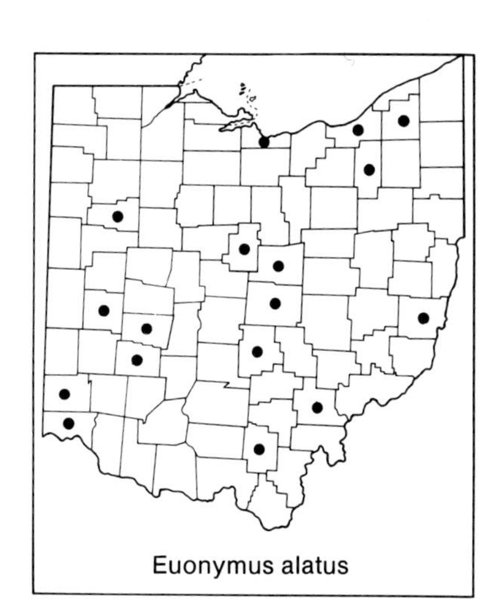
Euonymus alatus

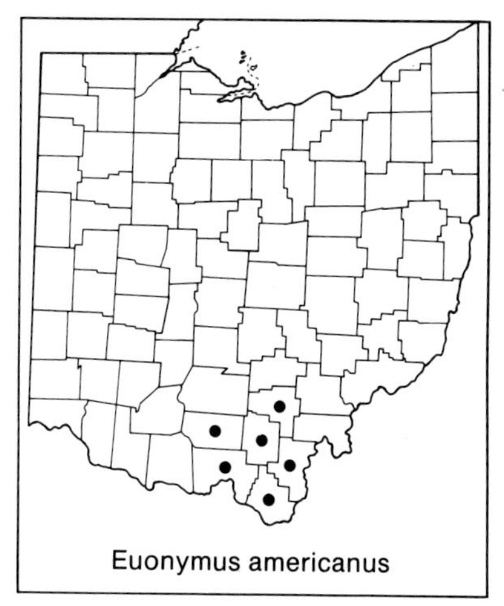
Euonymus americanus

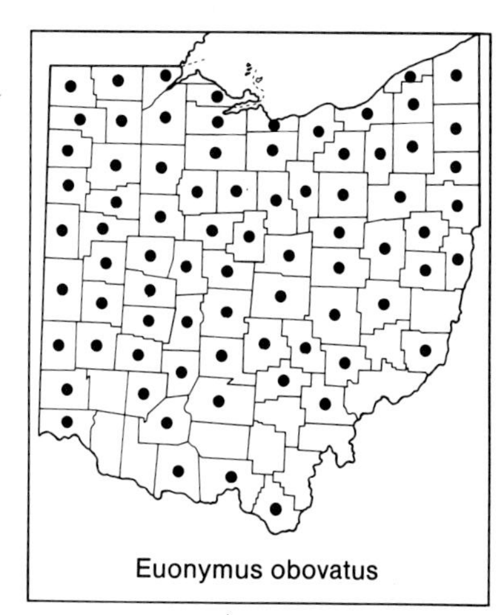
Euonymus obovatus

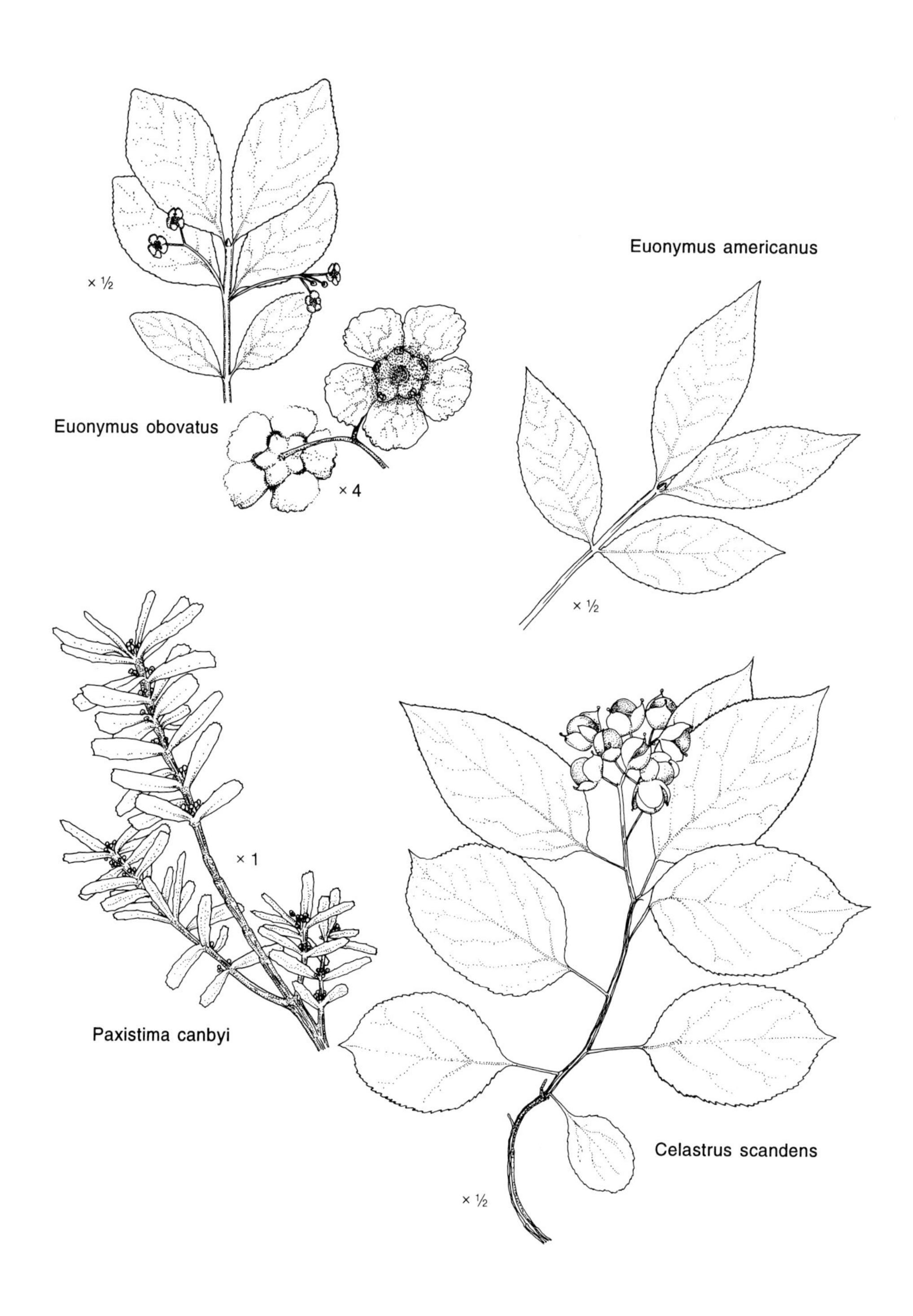
× ½
Euonymus americanus
Euonymus obovatus
× 4
× ½
× 1
Paxistima canbyi
Celastrus scandens
× ½

2. Paxistima Raf.

A genus of two species, the other native to western North America. The generic name is sometimes spelled *Pachistima* or *Pachystima*; Wheeler (1943) discussed the orthography. The genus was recently revised by Navaro and Blackwell (1990).

1. **Paxistima canbyi** A. Gray CLIFF-GREEN. CANBY'S MOUNTAIN-LOVER

Low, erect to decumbent shrub, to 0.3 m (1 foot) tall; leaves leathery and evergreen, opposite, mostly narrowly oblong, obtuse to truncate at apex, 1–2 cm long; sepals and petals 4; capsules 2-valved, arils white. Endemic to a limited area in Ohio, Kentucky, West Virginia, Virginia, Tennessee, and Pennsylvania; in Ohio known from only two southern sites, both "on rocky (dolomite) knolls near Ohio Brush Creek" (Braun, 1961), located in Adams and Highland counties. Braun (1961) and McCance and Burns (1984) give detailed information on this species and its occurrence in Ohio. Late March–early April.

Paxistima canbyi

3. Celastrus L. BITTERSWEET

Twining woody vines; leaves alternate, deciduous; sepals and petals 5; capsules dull orange, arils scarlet. A widely scattered genus of some 30 or more species, revised by Hou (1955). One native and one alien species occur in Ohio. Both are sometimes cultivated, and fruits of each are often seen in decorative arrangements of dried plant materials.

a. Flowers and fruits usually 6 or more in terminal panicles; leaves variable but mostly oblong-elliptic to ovate . . . . 1. *Celastrus scandens*

aa. Flowers and fruits 2 or 3 in axillary cymes, the cymes few to numerous; leaves orbicular to suborbicular or broadly obovate . . . . 2. *Celastrus orbiculatus*

1. **Celastrus scandens** L. BITTERSWEET. WAXWORK

Leaves variable in shape, mostly oblong-elliptic to ovate, but some varying to lanceolate and others to broadly elliptic or obovate; flowers and fruits usually 6 or more in terminal panicles. Throughout Ohio; woodland borders and thickets, sometimes climbing to considerable heights on supporting fence-posts, poles, and trees. Late May–June.

Celastrus scandens

2. CELASTRUS ORBICULATUS Thunb. ORIENTAL BITTERSWEET

Leaves orbicular to suborbicular or broadly obovate; flowers and fruits 2 or 3 in axillary cymes. An Asian species, escaped from cultivation and naturalized at scattered sites in southern counties; open disturbed areas, thickets; most collections are recent. Not illustrated. May–June.

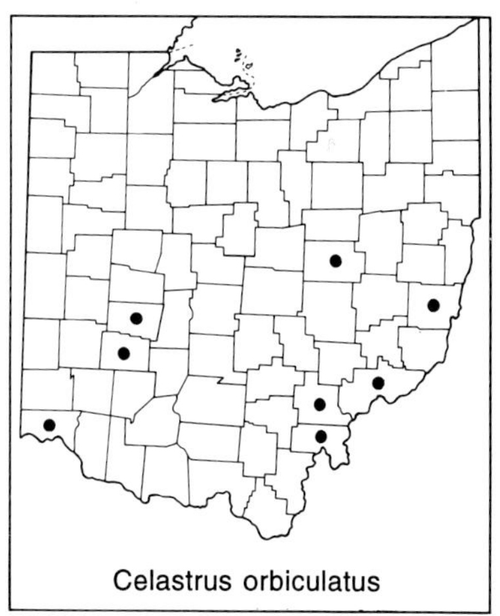
Celastrus orbiculatus

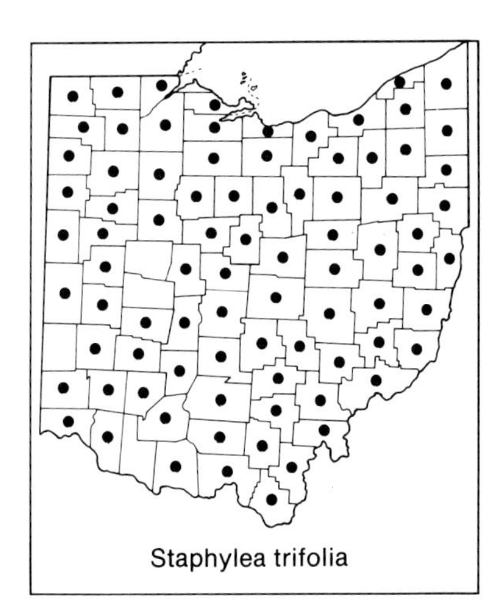
Staphylea trifolia

Staphyleaceae. BLADDERNUT FAMILY

A small family of woody plants, chiefly of the Northern Hemisphere, with a single genus in Ohio. Spongberg (1971) studied the family in southeastern United States.

1. Staphylea L. BLADDERNUT

A genus of temperate regions of Eurasia and North America with 10–15 species, one of which is native to Ohio.

1. **Staphylea trifolia** L. BLADDERNUT

Shrub or small tree, to 8 m (26 feet) tall; the trunk and older branches with whitish stripes; leaves opposite, trifoliolate, the 3 leaflets ovate, acuminate at apex, and finely serrate; flowers perfect, actinomorphic, white or whitish, in drooping racemose panicles; fruit a large, inflated, apically 3-lobed capsule, 3–6 cm long. Throughout Ohio, but not abundant; moist to mesic woodlands; wooded floodplains, terraces, and slopes. Late April–June.

Aceraceae. MAPLE FAMILY

A small family consisting chiefly of the genus *Acer*; a second genus, *Dipteronia*, has but two species, both native to China. Brizicky (1963b) studied the family in southeastern United States.

1. Acer L. MAPLE

Large to small trees (infrequently tall shrubs, in *A. pensylvanicum* and *A. spicatum*); leaves opposite, simple or compound; flowers actinomorphic, greenish to yellow or red, mostly imperfect, borne in various types of multiflowered inflorescences; fruit a distinctive schizocarp, splitting at maturity into two winged mericarps (called "keys" or "samaras"). A genus of about 150 species of woody plants, most native to the North Temperate Zone.

In addition to the species keyed and described below, many of which are taken into cultivation, the following species also are frequently seen as ornamentals in Ohio and may occasionally escape near plantings: two species from Europe and western Asia, *Acer campestre* L., HEDGE MAPLE, and *A. tataricum* L., TATARIAN MAPLE; and three species from eastern Asia, *Acer ginnala* Maxim., AMUR MAPLE, *A. griseum* (Franch.) Pax, PAPERBARK MAPLE, and *A. palmatum* Thunb., JAPANESE MAPLE. An escape of *A. palmatum* has recently been collected from Geauga County, and one of *A. tataricum* from Hamilton County.

a. Leaves trifoliolate or pinnately compound, with 3–7 (–9) leaflets; plants dioecious; staminate flowers on drooping slender pedicels in umbellate fascicles; pistillate flowers in racemes; petals absent .... 7. *Acer negundo*

aa. Leaves simple, variously palmately lobed; plants monoecious or with perfect flowers; inflorescences various; petals present or absent.

b. Petioles with milky sap; leaves with 5 (–7) lobes, each lobe with few to several spinulose-acuminate teeth, the leaf color varying from dark green to bronze-purple; flowers with conspicuous petals and arranged in erect corymbs; wings of schizocarps widely divergent, forming nearly a straight line................ 3. *Acer platanoides*

bb. Petioles with watery sap; leaves with 3–5 (–7) lobes, teeth (if present) acute or acuminate but not spinulose-acuminate, the leaf color mostly green; flowers with or without petals and in various types of inflorescenses; wings of schizocarps less widely divergent, forming a narrow or broad V-shape.

c. The three major lobes of leaf either entire-margined, or undulate, or with 1–3 broad teeth along sides of lobe; flowers on drooping pedicels in umbellate clusters; petals lacking... 4. *Acer saccharum*

cc. The three major lobes of leaf with several to many serrations or dentations along margins; flowers either in erect to divergent clusters (not drooping) or in narrow panicles or narrow racemes; petals present or not.

d. Small trees or shrubs; leaves with 3 (–5) lobes, the leaf margins more or less uniformly serrate or dentate; flowers in slender elongated panicles or racemes, the flowers with petals.

e. Margin of leaves coarsely serrate or dentate; flowers in narrow, slightly drooping to ascending or erect panicles .... 1. *Acer spicatum*

ee. Margins of leaves finely serrate; flowers in pendulous racemes .... 2. *Acer pensylvanicum*

dd. Large trees; leaves with (3–) 5 lobes, the leaf margins irregularly dentate and/or serrate; flowers in short rounded clusters, the flowers with or without petals.

e. Sinuses on either side of terminal lobe rarely or not extending to middle of blade; the terminal lobe widest at base; flowers with petals; both members of each pair of fruits usually fully developed .... 5. *Acer rubrum*

ee. Sinuses on either side of terminal lobe extending below middle of blade; the terminal lobe widest at about the middle and tapering at its base; flowers without petals; one member of each pair of fruits often small and poorly developed.... 6. *Acer saccharinum*

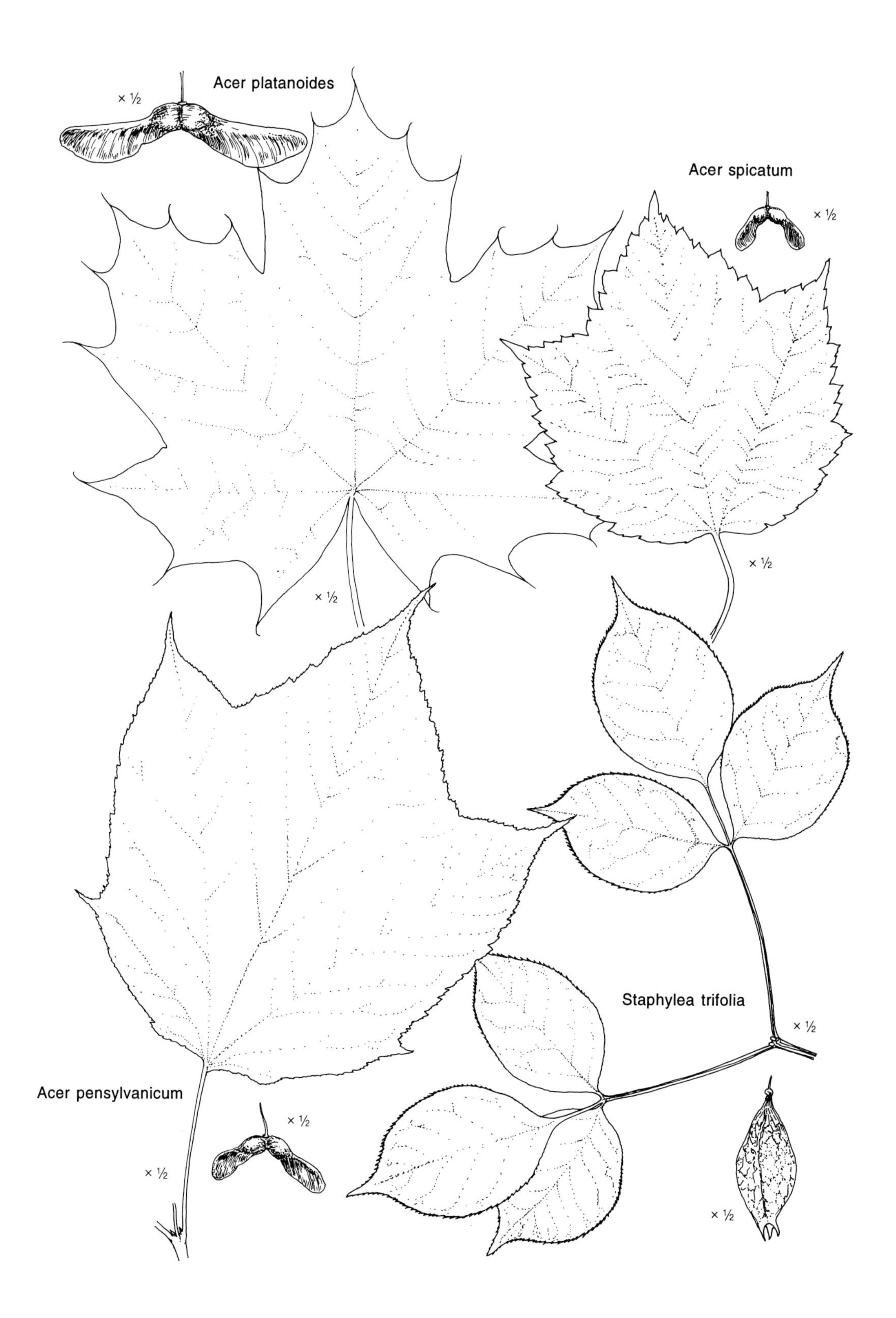
Acer platanoides
× ½
Acer spicatum
× ½
× ½
× ½
× ½
Staphylea trifolia
× ½
Acer pensylvanicum
× ½
× ½
× ½

1. **Acer spicatum** Lam. MOUNTAIN MAPLE
Small tree or large shrub, to 3.5 m (11 feet) tall; leaves simple, 3- (5-) lobed, the margins coarsely serrate or dentate; flowers borne in narrow, slightly drooping to ascending or erect panicles. Mostly in northeastern counties of Ohio; moist wooded gorges and north-facing slopes. May–June.

2. **Acer pensylvanicum** L. STRIPED MAPLE. MOOSEWOOD
Small tree or large shrub, to 5 m (16 feet) tall; bark of mature branches dark green with vertical whitish stripes; leaves simple, 3- (5-) lobed, the margins finely serrate; flowers borne in pendulous racemes. A species of southeastern Canada and northeastern United States, extending south in the Appalachians; in Ohio found only in the northeasternmost section of the state, in Ashtabula County; rich moist woodlands. May–June.

3. ACER PLATANOIDES L. NORWAY MAPLE
Tree, to 24 m (79 feet) tall; leaves simple, dark green to bronze-purple, 5- (7-) lobed, each lobe with few to several spinulose-acuminate teeth; flowers greenish-yellow, larger than those of native maples, borne in erect, moderately showy corymbs; wings of schizocarp widely divergent, forming nearly a straight line. An ornamental shade tree of Eurasian origin from which a host of cultivars, diagnosed mostly on variation in leaf color, have been derived; escaped and locally established at a few disturbed sites in northern, mostly northeastern, counties. April–May.

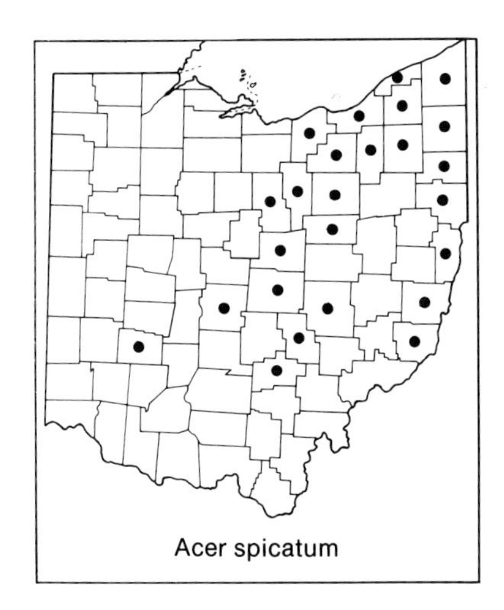
Acer spicatum

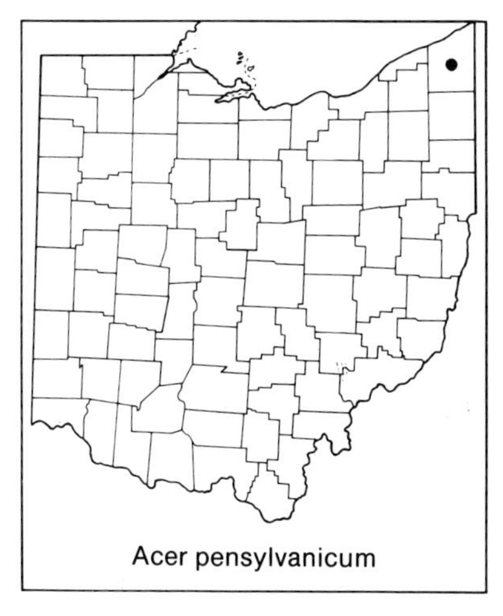
Acer pensylvanicum

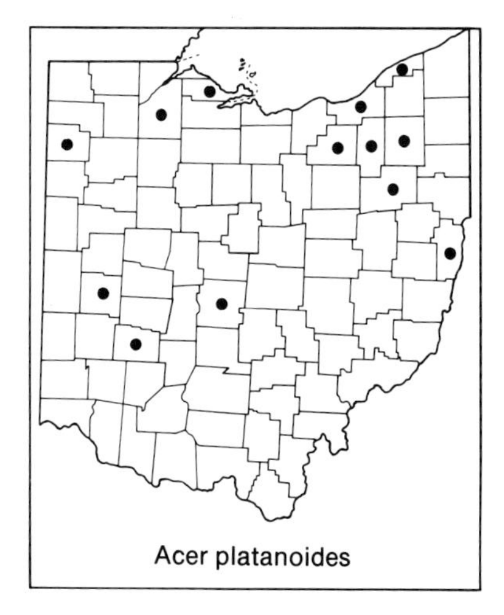
Acer platanoides

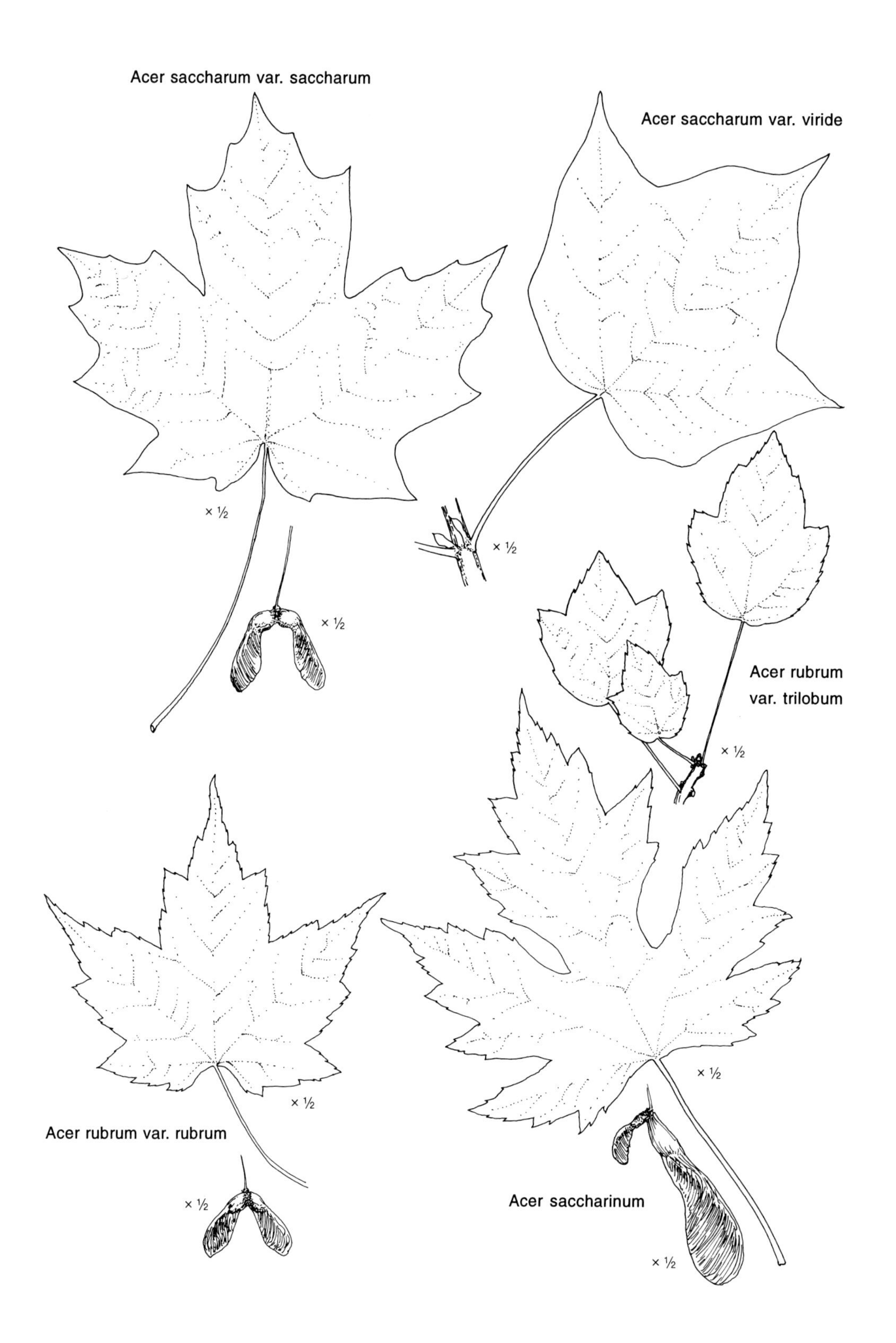
Acer saccharum var. saccharum
Acer saccharum var. viride
× ½
× ½
× ½
Acer rubrum
var. trilobum
× ½
× ½
× ½
Acer rubrum var. rubrum
× ½
Acer saccharinum
× ½

4. **Acer saccharum** Marsh.

Tree, to 30 m (100 feet) tall; leaves simple, 3–5 (–7)-lobed, the lobes with 0-few teeth, the teeth, when present, mostly acute, not spinulose-acuminate; flowers without petals, borne in umbellate clusters on drooping pedicels. Throughout Ohio; rich woods. Late April–May.

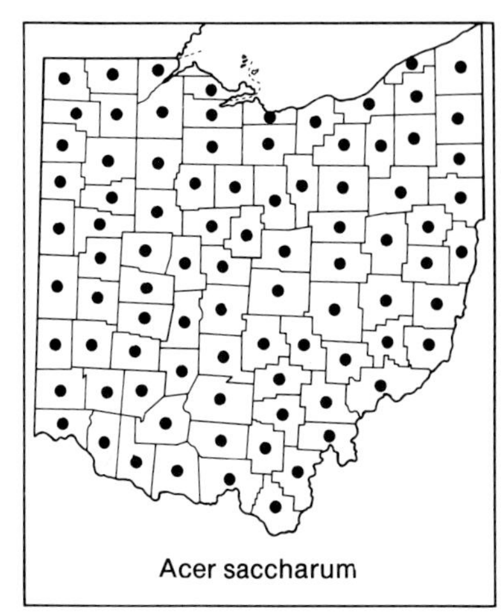
Acer saccharum

The two varieties keyed below have often been treated as separate species, e.g., by Fernald (1950), Gleason (1952a), and Braun (1961). In an extensive study, including the collection of 22,718 leaves from as many trees, Desmarais (1952) concluded that both varieties, and a few other taxa not found in Ohio, are best treated as part of a single variable species. Both var. *saccharum* and var. *viride*, and intermediates between them, occur throughout Ohio. Both are valued for their sap, which is made into maple syrup and maple sugar. Both are prized also for their heartwood, some decorative forms of which are called "Bird's-eye Maple" and "Curly Maple." They are also popular shade trees.

Within *Acer saccharum* sensu stricto, Fernald (1950) recognizes the trees with 3-lobed leaves as forma *rugelii* (Pax) Palmer & Steyerm., which Braun (1961) reports from a few sites in southwestern Ohio. Given the great variation found among the leaves of *A. saccharum* sensu lato, I see little merit in recognition of this form.

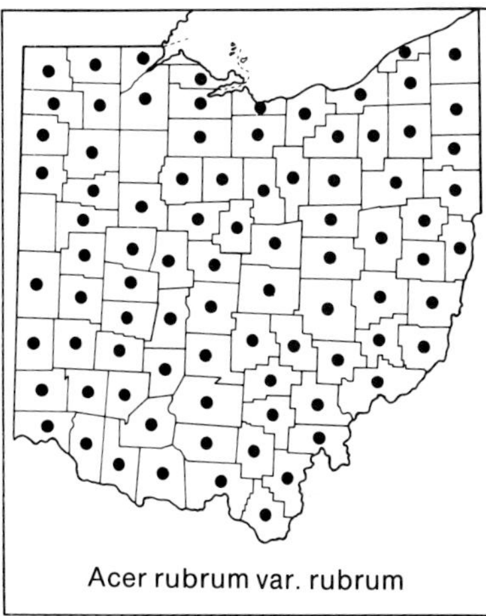
Acer rubrum var. rubrum

a. Leaf blades flat, undersurface pale green, basal sinuses evident; the leaves lacking stipules . . . . var. *saccharum*
Sugar Maple

aa. Leaf blades curved downwards and drooping at margins, undersurface yellowish-green, basal sinuses lacking or nearly so; the leaves often with stipules . . . . var. *viride* (Schmidt) E. Murray
A. nigrum Michx. f. (*A. saccharum* var. *nigrum* (Michx. f.) Britt.)
Black Maple

5. **Acer rubrum** L. Red Maple

Tree, to 25 m (83 feet) tall; bark on younger branches and young trunks smooth and light gray, that of older trunks dark and shallowly furrowed; leaves simple, 3–5-lobed, the terminal lobe widest at its base, the sinuses on either side of the terminal lobe only moderately deep, rarely extending to middle of blade, the margins irregularly dentate and/or serrate; flowers bright red, borne in short rounded clusters. Throughout Ohio; mesic woodlands, swamps, floodplain woods, and (stunted trees) in bogs. The species is cultivated and prized for its brilliant red foliage in autumn, and some Ohio specimens may represent escapes from cultivation. March–April.

Acer rubrum var. trilobum

The two varieties keyed below are usually rather distinctive in Ohio and easily separated. Variety *rubrum* occurs throughout Ohio; variety *trilobum*,

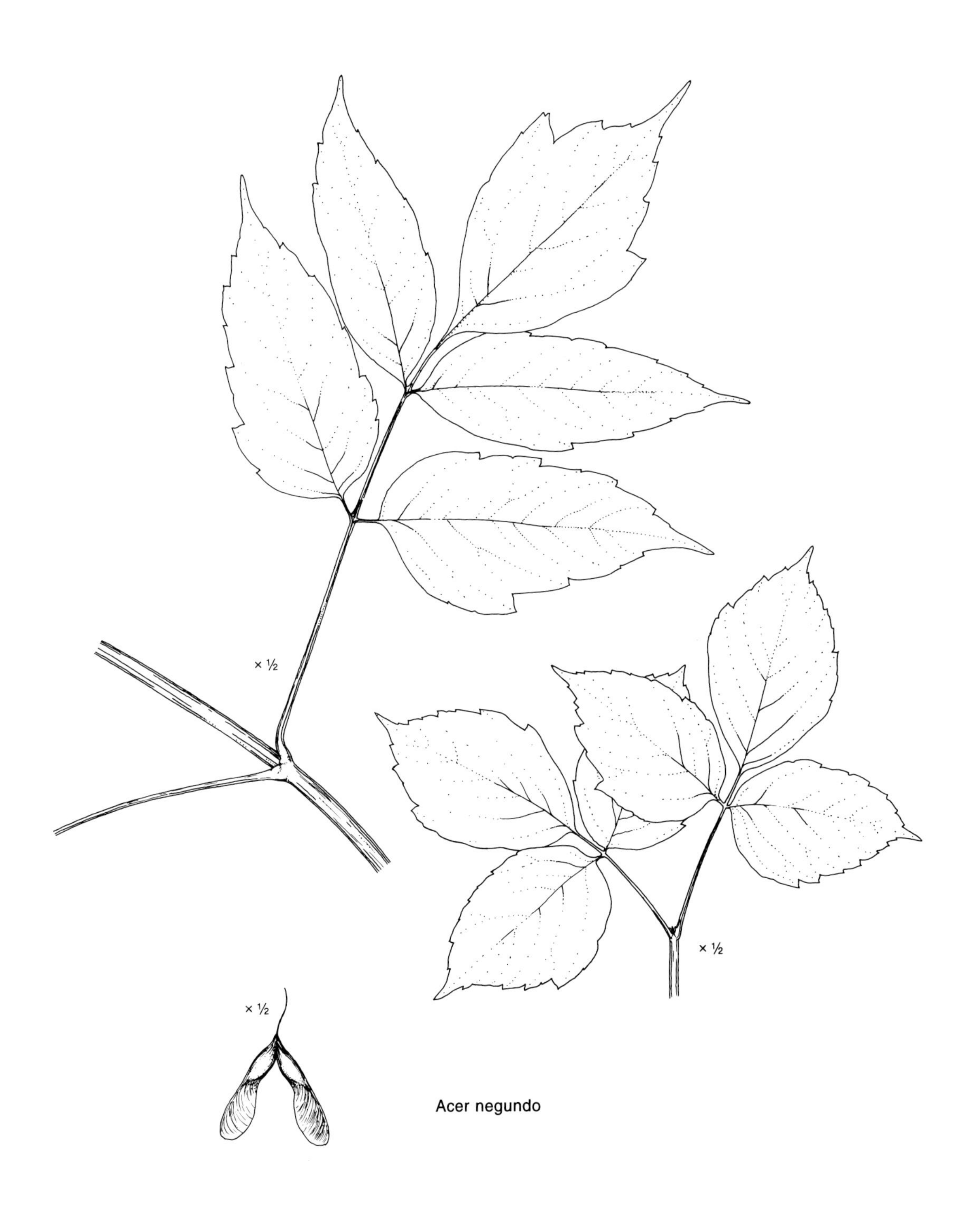

Acer negundo

chiefly of southeastern United States, is known in Ohio from only a few sites in south-central counties. In addition, Braun (1961) noted specimens from Clermont County "approaching var. *drummondii* (H. & A.) Sarg.," but that variety does not occur in Ohio.

a. Leaves 5-lobed, with 3 principal lobes near apex and two smaller ones near base of blade . . . . var. *rubrum*

aa. Leaves 3-lobed, the 3 lobes all on upper part of blade, the base unlobed . . . . var. *trilobum* K. Koch

var. *tridens* Alph. Wood

Additional record: Jackson County (Ellis, 1963).

In a revision of the red maple complex, Ellis (1963) noted specimens from Hancock and Licking counties that he determined as putative hybrids, *A. rubrum* × *A. saccharinum*. Occasional specimens are difficult to assign with certainty to one species or the other.

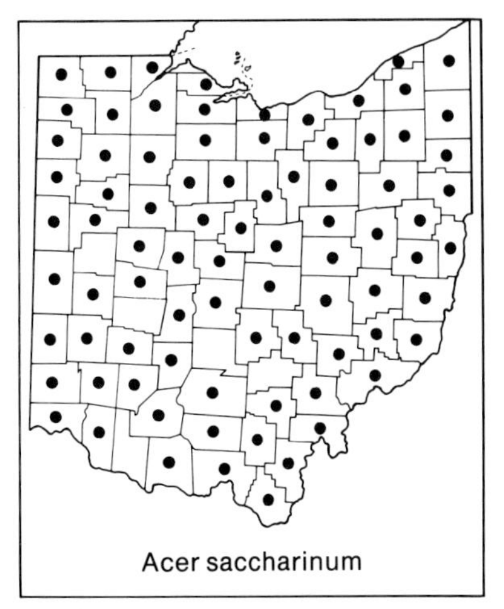
Acer saccharinum

Acer negundo

6. **Acer saccharinum** L. Silver Maple

Tree, to 33.5 m (110 feet) tall; leaves simple, 5-lobed, the terminal lobe widest at about the middle and tapering to a narrower base, the sinuses on either side of the terminal lobe deep, extending below the middle of the blade, the margins irregularly dentate and/or serrate; flowers greenish-yellow to reddish, borne in short rounded clusters; one mericarp of the schizocarp-pair usually small and poorly developed (see illustration). Throughout Ohio; floodplains and stream banks. Late February–April.

Often cultivated, especially in western Ohio, the tree is fast-growing but somewhat fragile. See comments under the previous species, *Acer rubrum*, regarding putative hybrids between the two. Unlike *Acer rubrum*, the leaves of *A. saccharinum* are yellow in autumn.

7. **Acer negundo** L. Box-elder. Ash-leaved Maple

Tree, to 19 m (62 feet) tall; young branches purple to green, and glaucous; leaves compound, either pinnately compound with 5–7 (–9) leaflets or trifoliolate; leaflets ovate to lanceolate, irregularly toothed; dioecious, the staminate flowers borne in umbellate fascicles on drooping slender pedicels, the pistillate flowers in racemes. Throughout Ohio; river banks, floodplains, and in moist to mesic disturbed weedy habitats; often cultivated, either directly or indirectly by allowing volunteers to persist. Wagner (1975) has notes on the floral biology of this species. April–early May.

Ohio plants belong to var. *negundo* (incl. var. *violaceum* (Kirch.) Jaeg.). *Acer negundo* var. *texanum* Pax was attributed to southern Ohio by Fernald (1950), but no specimens have been found in this study.

Forms with three leaflets, especially those produced on short sprouts in spring, often resemble poison ivy (p. 62), giving rise to the interesting but infrequently heard common name, Poison-ivy-tree. The two species can be readily distinguished by leaf arrangement; the trifoliolate leaves are alternate in poison ivy, opposite in box-elder.

Hippocastanaceae. HORSE-CHESTNUT FAMILY

A small family of two genera, one occurring in Ohio. J. W. Hardin (1957a) studied the American members of the family, and Brizicky (1963b) studied the family in southeastern United States.

1. Aesculus L. HORSE-CHESTNUT. BUCKEYE

Trees; leaves opposite, palmately compound; flowers zygomorphic, imperfect or perfect, borne in large, showy, terminal panicles or thyrses; fruit a fleshy to subleathery capsule with 1–3 very large shiny brown seeds. A genus of about 13 species native to the North Temperate Zone; Beatley (1979) examined the distribution of buckeyes in Ohio.

Aesculus hippocastanum L., HORSE-CHESTNUT, native to Europe and frequently cultivated in Ohio, is noted for its large "buckeye" seeds produced in a bristly capsule. Collections of adventives of *A. hippocastanum* have been made from Fairfield, Medina, Ottawa, Portage, and Ross counties. J. W. Hardin (1960) studied this and other Old World species.

a. Leaves mostly with 7 leaflets, rarely fewer; the leaflets markedly obovate, widest well above the middle, and abruptly narrowed to a short-acuminate apex; terminal bud dark reddish-brown, gummy and sticky; capsules prickly; see above . . . . *Aesculus hippocastanum*

aa. Leaves mostly with 5 leaflets, rarely more; the leaflets elliptic to slightly obovate and gradually narrowed to a long-acuminate apex; terminal bud tan to medium-brown, not gummy and sticky; capsules prickly or smooth.

b. Capsules prickly, especially when young; pedicels pubescent and with no or few short-stalked glands; petals yellowish-green, approximately equal in size; stamens conspicuously exserted; leaflets widest at about the middle, their undersurface often with tufts of pale hair in vein axils . . . . 1. *Aesculus glabra*

bb. Capsules without prickles; pedicels pubescent and with numerous long-stalked glands; petals yellow to red or purplish, unequal in size; stamens included or only slightly exserted; leaflets widest above the middle, at about 2/3 the distance from base to apex, their undersurface often with tufts of reddish to dark brown hair in vein axils . . . . 2. *Aesculus flava*

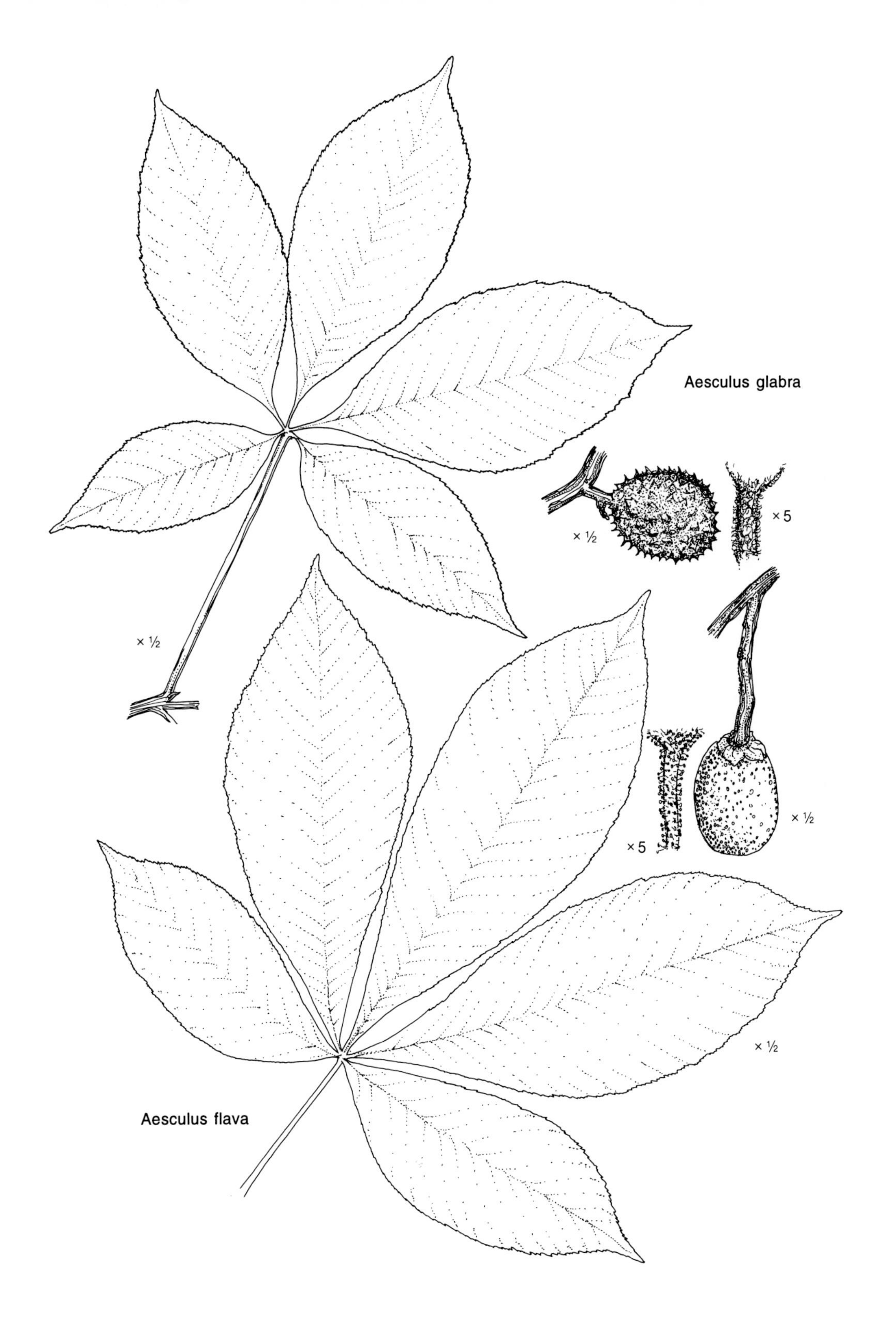
Aesculus glabra
× ½
× 5
× ½
× ½
× 5
× ½
Aesculus flava

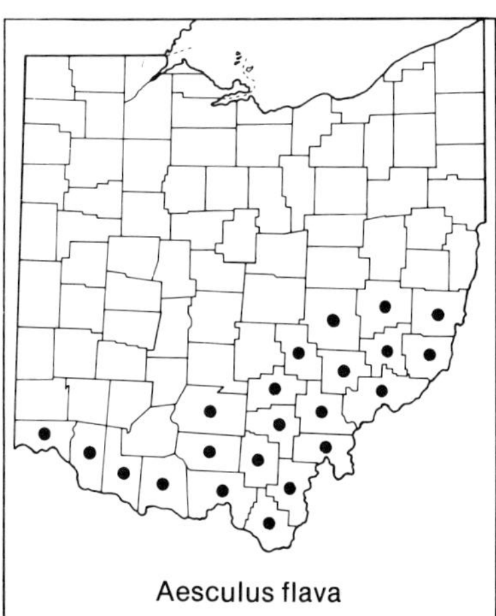

1. **Aesculus glabra** Willd. OHIO BUCKEYE
incl. var. *sargentii* Rehd.

Tree, to 35 m (116 feet) tall, but usually much shorter; leaflets widest at about the middle, their undersurface often with tufts of pale hair in vein axils; pedicels pubescent but with no or only few short-stalked glands; petals yellowish-green, approximately equal in size; stamens 7, conspicuously exserted; capsules prickly, especially so when young. Common in the calcareous soils of western Ohio, but less common in eastern counties (see discussion below); river banks, bottomlands, slopes, and thickets. April–May.

J. W. Hardin (1957b) reports several Ohio specimens of naturally occurring hybrids of *Aesculus glabra* × *A. octandra* [=*A. flava*]. He writes, "A few F_1 hybrids have been found; most of the intermediates, however, represent backcrosses or various recombinants. Since intermediates are detected by floral characteristics only, some sterile specimens annotated and cited as *A. glabra* may actually be" hybrids between the two species. Similarly, Beatley (1979) notes the presence of "hybridization and introgression between the species" and the necessity of having flowering or fruiting material for accurate identification.

A related problem concerns the Ohio distribution of *A. glabra.* Beatley (1979) concluded that its natural range in Ohio is limited primarily to the glaciated parts of the state and that elsewhere the species is "restricted to [calcareous] outwash deposits and limestone bedrock areas beyond the limits of glaciation." Even within the glaciated region, it is scarce or absent in parts of the glaciated Allegheny Plateau of northeastern Ohio. She believed misidentified sterile specimens and collections from cultivated plants (not so indicated on the specimen labels) to be responsible for past reports of the species being common in the unglaciated portion of the state, the east-central and southeastern counties. The species is often cultivated in Ohio, and it is also possible that some specimens from eastern counties were collected from adventive individuals. The reader should view the map presented here with these caveats in mind, and may wish to consult Beatley's interesting generalized distribution map and discussion.

2. **Aesculus flava** Soland. SWEET BUCKEYE. YELLOW BUCKEYE
A. octandra Marsh.

Tree, to 30 m (98 feet) tall; leaflets widest above the middle, at about 2/3 the distance from base to apex, the undersurface often with tufts of reddish to dark brown hair in vein axils; pedicels pubescent and with numerous long-stalked glands; petals yellow to red or purplish, unequal in size; stamens 7 or 8, included or only slightly exserted; capsules without prickles. An American species, chiefly of the Appalachians; in southeastern and south-central Ohio counties and extending westward across southernmost Ohio "in the deeply dissected terrain bordering the Ohio River" (Beatley, 1979); rich wooded slopes. May–June.

Meyer and Hardin (1987) recently determined that *Aesculus flava* is the correct name of this species, long known as *A. octandra.* See notes above under *A. glabra* concerning intermediates between that species and this.

Sapindaceae. SOAPBERRY FAMILY

A chiefly tropical family, two species of which are often cultivated as ornamentals in Ohio. The herbaceous *Cardiospermum halicacabum* L., BALLOON-VINE, the common name coming from its inflated fruit, has been collected as an adventive escape from Hamilton and Highland counties, both in southwestern Ohio. *Koelreuteria paniculata* Laxm., GOLDEN-RAINTREE, a tree with large showy sprays of yellow flowers, is widely cultivated. It has been collected as an escape in Portage County, in northeastern Ohio. Brizicky (1963b) studied the family in southeastern United States.

Balsaminaceae. BALSAM FAMILY. TOUCH-ME-NOT FAMILY

A family of two genera, all species but one belonging to the genus *Impatiens.* Wood (1975) studied the family in southeastern United States.

1. Impatiens L. BALSAM. JEWEL-WEED. TOUCH-ME-NOT

Erect, tender, annual herbs with succulent, easily damaged, watery stems; leaves alternate, simple; flowers perfect, axillary, solitary or in few-flowered clusters; calyx zygomorphic, of 3 separate sepals, the 2 upper small and green, the lower one large, petal-like in color, saccate, and spurred; corolla zygomorphic, of 5 petals, those of the two lateral pairs fused to the other member of the pair, the uppermost, fifth petal shorter than the laterals, wider than long, curved and somewhat keeled; stamens 5, partially fused; carpels 5, fused to form a compound pistil with 1 style and 5 stigmas; ovary superior; fruit a fleshy green capsule, at maturity exploding when touched and propelling the seeds for a short distance from the plant. Plants also bear cleistogamous flowers that do not open, but that do produce seeds. A genus of about 450 species, chiefly of the tropics and subtropics, with distribution centered in Asia and Africa. Rust (1977) did research on pollination in the two species native to Ohio.

The Asian species *Impatiens balsamina* L., GARDEN BALSAM, with white or yellow to red or purple flowers, is frequently seen in cultivation, usually

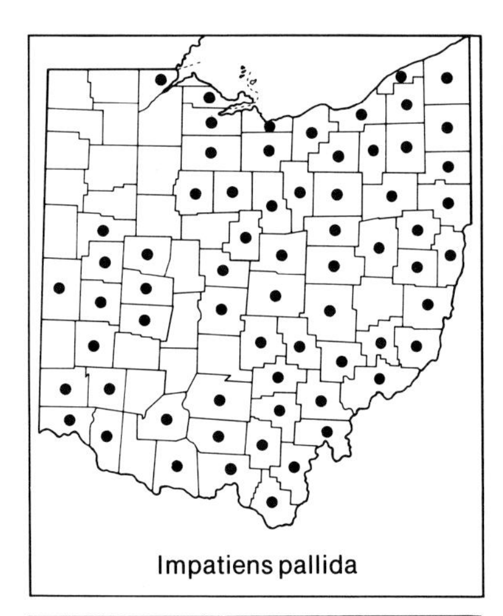

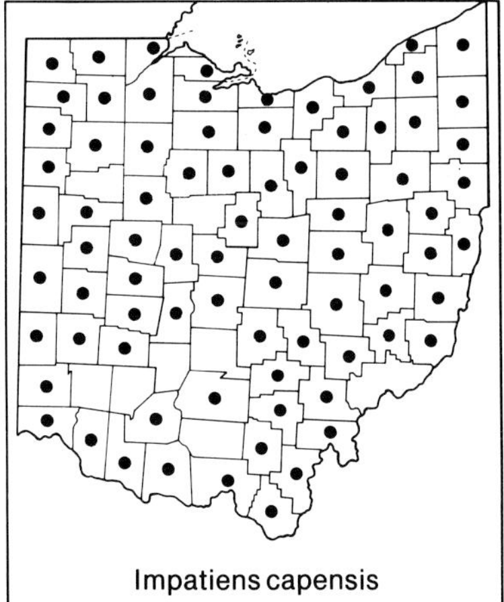

in its double-flowered forms. Adventive plants of this species have been collected from Butler, Highland, Holmes, Lawrence, and Washington counties, mostly in southern Ohio. Also cultivated is the African species, *Impatiens walleriana* Hook. f., ZANZIBAR BALSAM or SULTANA, a favorite for shaded sites, where it prospers. Each plant produces scores of white, pink, red, or purple flowers throughout the summer and fall, but dies at the first touch of frost.

a. Leaf margins crenate or crenate-serrate, base acute to broadly acute above a long petiole; flowers orange, yellow, or rarely white, in few-flowered axillary inflorescences.

b. Flowers usually yellow, rarely white; calyx spur 4–5 mm long, mostly diverging from sepal sac at 45°–90° angle . . . . . . 1. *Impatiens pallida*

bb. Flowers usually orange or yellow-orange, very rarely yellow or white; calyx spur 7–9 mm long, mostly curved sharply forward at an angle less than 45° and lying alongside the sepal sac . . . . 2. *Impatiens capensis*

aa. Leaf margins sharply serrate, either with base long-tapering to a short petiole or the leaf subsessile; flowers white, pink, red, or purple, solitary in leaf axils; see above. . . . . *Impatiens balsamina*

1. **Impatiens pallida** Nutt. PALE TOUCH-ME-NOT. JEWEL-WEED

Erect branched annual, to 2 m tall; flowers yellow or rarely white; calyx spur 4–5 mm long, mostly diverging from sepal sac at 45°–90° angle. Frequent in the eastern half of Ohio, less frequent or absent westward; moist shaded ditches and woodland borders. Late June–September.

2. **Impatiens capensis** Meerb. SPOTTED TOUCH-ME-NOT. JEWEL-WEED
I. biflora Walt.

Erect branched annual, to 1.5 m tall; flowers usually orange or yellow-orange (rarely yellow or white); calyx spur 7–9 mm long, mostly curved sharply forward at an angle less than 45° and lying alongside the sepal sac. Throughout Ohio; in a wide variety of moist habitats ranging from swamps to wet ditches, wet woodland borders, and moist thickets and roadsides; more weedy and more frequent than *I. pallida*, but patches of the two species are occasionally seen growing side by side. Late June–September.

Rhamnaceae. BUCKTHORN FAMILY

Shrubs or small trees; leaves simple, usually alternate; flowers small, white to green, actinomorphic, solitary or clustered; flowers perfect or imperfect; fruit a drupe or capsule-like drupe. A cosmopolitan family of 55 genera and 900 species. Brizicky (1964c) studied the family in southeastern United States.

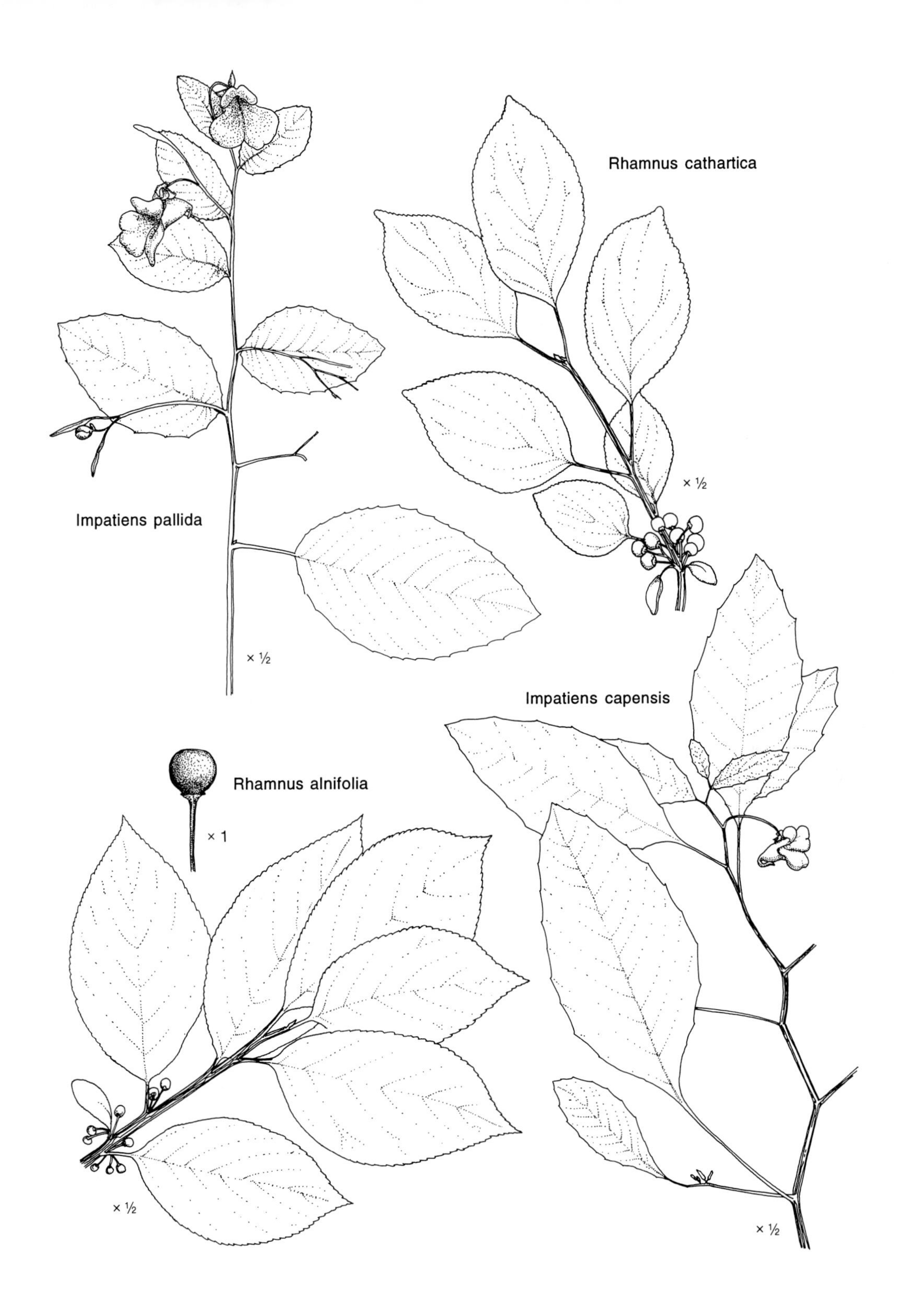
Rhamnus cathartica
× ½
Impatiens pallida
× ½
Impatiens capensis
Rhamnus alnifolia
× 1
× ½
× ½

Berchemia scandens (Hill) K. Koch, Supple-jack, was attributed to Ohio by C. H. Jones (1945), but no specimens have been found in this study. The attribution was probably based on misidentified specimens of *Rhamnus frangula*.

a. Leaf blades with one main vein originating from apex of petiole; flowers green or greenish-white, solitary or in few-flowered axillary clusters; fruit a drupe . . . . 1. *Rhamnus*

aa. Leaf blades with 3 main veins originating from apex of petiole; flowers white, numerous in terminal and axillary umbellate clusters; fruit dry and capsule-like . . . . 2. *Ceanothus*

1. Rhamnus L. BUCKTHORN

Shrubs or small trees; leaves pinnately veined; flowers green to greenish-white, solitary or in few-flowered clusters, axillary; fruit a small, blackish drupe. A genus of more than 100 species, chiefly in the North Temperate Zone but also with some tropical members.

a. Leaves opposite or subopposite, blades mostly with 2–3 prominent lateral veins on either side of midrib; sepals 4; petals 4; drupes with (3 or) 4 pits . . . . 2. *Rhamnus cathartica*

aa. Leaves alternate, blades with 4–9 prominent lateral veins on either side of midrib; sepals 4 or 5; petals 0, 4, or 5; drupes with 2 or 3 pits.

b. Leaf margins distinctly serrate or crenate-serrate; flowers either with sepals and petals 4, or with sepals 5 and petals 0.

c. Shrubs, at most 1 m tall; leaves mostly with 5 (sometimes 4 or 6) prominent lateral veins on either side of midrib; sepals 5; petals 0; drupes with 3 pits . . . . 1. *Rhamnus alnifolia*

cc. Shrubs, usually 1–2 m tall; leaves mostly with 7 (rarely with 6) prominent lateral veins on either side of midrib; sepals 4; petals 4; drupes with 2 pits . . . . 3. *Rhamnus lanceolata*

bb. Leaf margins entire, subentire, or with very low teeth on upper half of blade; flowers with sepals 5 and petals 5.

c. Leaf margins with numerous, extremely small, rounded serrations; leaves elliptic to elliptic-oblanceolate, principal blades mostly 8–12 cm long, leaf apex gradually to subabruptly acute; clusters of flowers and fruits on short peduncles 2–5 mm long, both peduncle and pedicels roughly short-pubescent . . . . 4. *Rhamnus caroliniana*

cc. Leaf margins either entire with shallow undulations, or with very few rounded teeth; leaves elliptic to elliptic-obovate, principal blades mostly 2–6 (–9) cm long, leaf apex either rounded or rounded and abruptly short-acute to short-acuminate; clusters of flowers and fruits pedicellate but without peduncles, pedicels glabrous or becoming so . . . . 5. *Rhamnus frangula*

1. **Rhamnus alnifolia** L'Hér. ALDER-LEAVED BUCKTHORN

Low shrub, to 0.8 (–1.0) m (3 feet) tall; leaves alternate, mostly with 5, sometimes 4 or 6, prominent lateral veins on either side of midrib, margins distinctly serrate to crenate-serrate; sepals 5; petals 0; drupes with 3 pits. A northern species ranging across Canada and southward into parts of the United States; the Ohio stations, mostly in northeastern counties, are at a southern boundary of the species' range; fens, wet sedge meadows, swampy places. May–June.

See note under *Ilex verticillata* (p. 66) on separating that species from this one.

2. RHAMNUS CATHARTICA L. EUROPEAN BUCKTHORN

Shrub or small tree, to 7 m (23 feet) tall; branches sometimes spine-tipped; leaves opposite or subopposite, mostly with 2–3 prominent lateral veins on either side of midrib; margins finely crenate-serrate; sepals 4; petals 4; drupes with (3 or) 4 pits. A European species naturalized at several Ohio sites; disturbed habitats along railroads, in thickets, and at woodland borders. May–June.

3. **Rhamnus lanceolata** Pursh LANCE-LEAVED BUCKTHORN

incl. var. *lanceolata* and var. *glabrata* Gleason

Shrub, to 2 m (7 feet) tall; leaves alternate, mostly with 7 (rarely 6) prominent lateral veins on either side of the midrib, margins distinctly but finely serrate to crenate-serrate; sepals 4; petals 4; drupes with 2 pits. Chiefly in calcareous soils of western Ohio; open or shaded banks, ravines, and woodland borders. May.

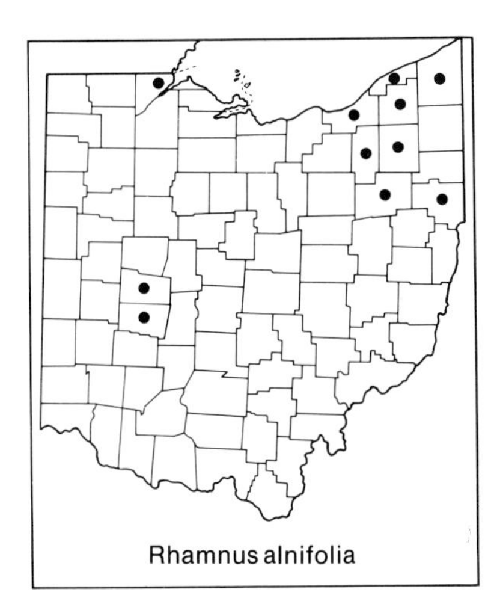
Rhamnus alnifolia

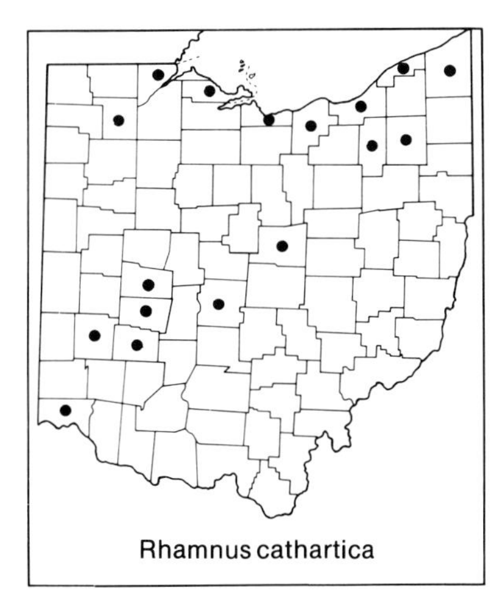
Rhamnus cathartica

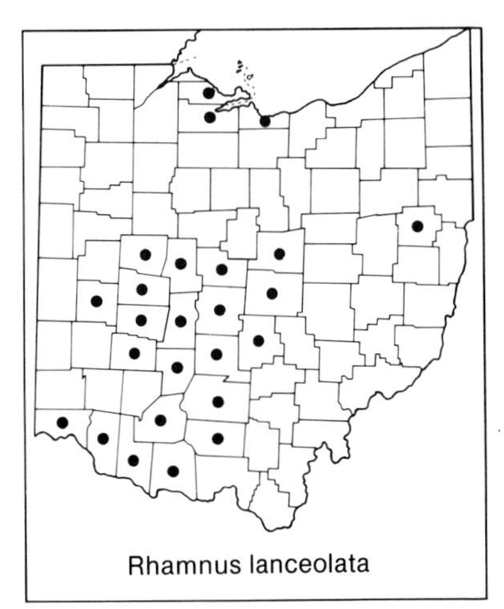
Rhamnus lanceolata

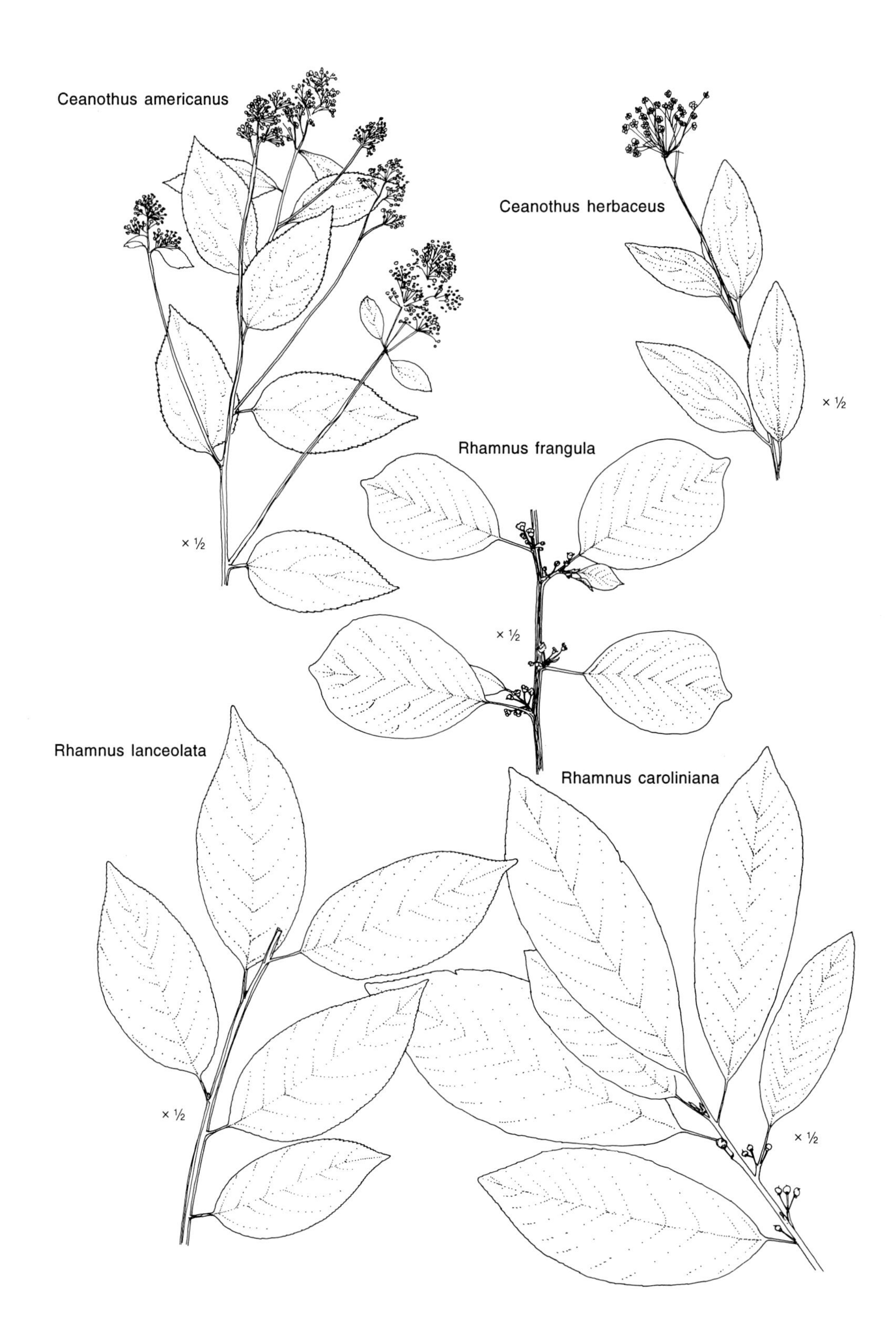
Ceanothus americanus
Ceanothus herbaceus
× ½
Rhamnus frangula
× ½
× ½
Rhamnus lanceolata
Rhamnus caroliniana
× ½
× ½

Most Ohio plants have some pubescence on the under-leaf surface and could, therefore, be assigned to var. *lanceolata*; a few without the pubescence could be assigned to var. *glabrata*; others are intermediate in this character. I concur with L. A. Johnston (1975), who regarded these as trivial pubescence forms not meriting varietal rank.

See note under *Ilex verticillata* (p. 66) on separating that species from this one.

Rhamnus caroliniana

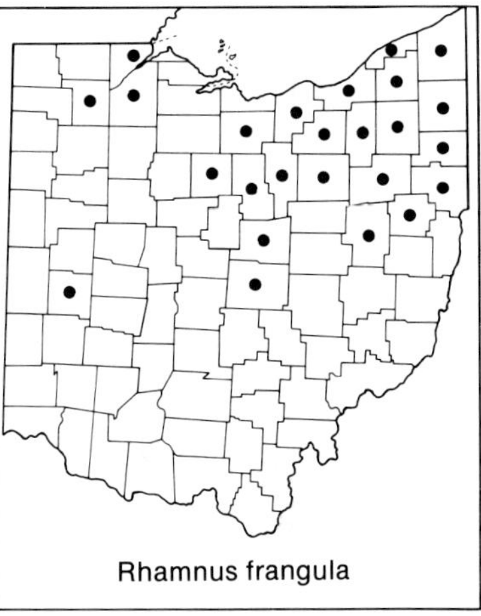

Rhamnus frangula

4. **Rhamnus caroliniana** Walt. CAROLINA BUCKTHORN. INDIAN CHERRY

incl. var. *caroliniana* and var. *mollis* Fern.

Shrub or small tree, to 5 m (16 feet) tall; leaves alternate, elliptic to elliptic-oblanceolate, blades of principal leaves 8–12 cm long, apex gradually to subabruptly acute, the margins with numerous, extremely fine, rounded serrations; clusters of flowers and fruits on short peduncle 2–5 mm long, both peduncle and pedicels roughly short-pubescent; sepals 5; petals 5. A southern species reaching a northern boundary of its range in southernmost Ohio; Braun (1961) reports it from "dolomite ledges, dry rocky south slopes, and borders of prairie patches." May–June.

In describing the plants of southeastern United States, Brizicky (1964c) writes, "Two variants which occur throughout this range (although one may be prevalent locally) have sometimes been recognized as var. *caroliniana*, with the leaves glabrous or glabrescent beneath, and var. *mollis* Fern., with the leaves permanently pubescent ('velvety') on the lower surface." As Braun (1961) noted, both elements and intermediates between them occur in Ohio. There seems little merit in recognition of these forms at the rank of variety.

5. RHAMNUS FRANGULA L. GLOSSY BUCKTHORN. ALDER BUCKTHORN

Frangula alnus Mill.

Shrub or small tree, to 6 m (20 feet) tall; branches ill-smelling when cut or broken; leaves alternate, elliptic to elliptic-obovate, blades of principal leaves mostly 2–6 (–9) cm long, apex either rounded or rounded and abruptly short-acute to short-acuminate, the margins entire and with shallow undulations or with very few rounded teeth; flowers and fruits clustered but without peduncles, pedicels glabrous or becoming so; sepals 5; petals 5. Naturalized from Eurasia in disturbed moist to mesic habitats in northern, especially northeastern, counties. Howell and Blackwell (1977) write that, "In all probability, *R. frangula* entered the Ohio flora in the early 1920's with the westward spread of the European starling . . ." May–September.

Although with few obvious attractive features, the species is sometimes cultivated in Ohio. Escapes of the strikingly different forma *asplenifolia*

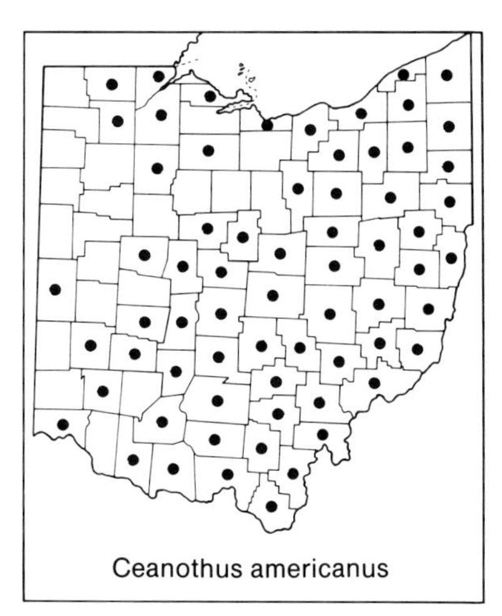
Ceanothus americanus

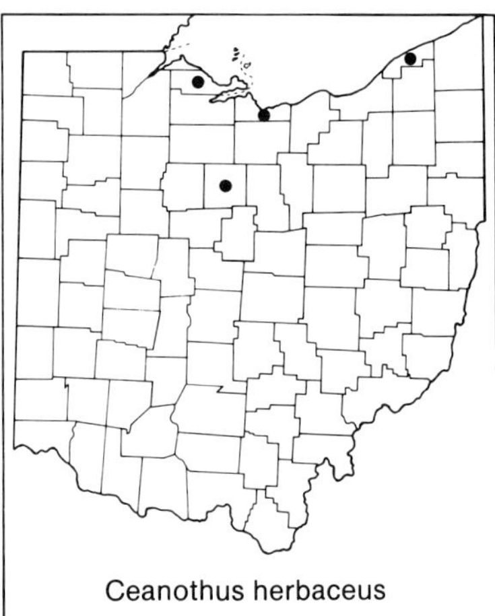
Ceanothus herbaceus

(Dippel) Beissner, with narrow, almost linear leaves, 10 or more times longer than wide, and irregularly crenate margins, were collected from Summit County in 1979, and from Cuyahoga County in 1986.

2. Ceanothus L.

Shrubs; leaves with three main veins at the base of the blade; flowers white and numerous in terminal and axillary umbellate clusters; fruit dry and capsule-like. A North American genus of about 50 species.

a. Leaves elliptic-ovate to ovate, leaf apex acuminate or infrequently merely acute; inflorescences on axillary peduncles, the peduncles progressively shorter toward shoot apex . . . . 1. *Ceanothus americanus*

aa. Leaves lanceolate to elliptic-lanceolate, leaf apex obtuse or infrequently acutish; inflorescences terminal on leafy shoots, all on short peduncles . . . . 2. *Ceanothus herbaceus*

1. **Ceanothus americanus** L. New Jersey-tea

Shrub, to 1 m (3 feet) tall; leaves elliptic-ovate to ovate, apex acuminate or infrequently merely acute; the moderately showy inflorescences borne on axillary peduncles, the lower peduncles long, the others progressively shorter toward shoot apex. Frequent in eastern and southern counties, infrequent or absent elsewhere in Ohio; dry roadside banks, woodland borders and openings, dry disturbed areas. June–July.

Fernald (1950), Gleason (1952a), and Gleason and Cronquist (1963, 1991) recognize three varieties within this species: var. *americanus*, var. *pitcheri* T. & G., and var. *intermedius* (Pursh) T. & G., the last southern and probably not reaching Ohio according to Braun (1961). Braun notes that Ohio lies in a region where there is extensive intergradation between the more eastern var. *americanus* with the under-leaf surface glabrous or nearly so, and the more western var. *pitcheri* with the under-leaf surface "velvety pubescent." Judging from Ohio specimens, I believe an attempted separation on the basis of this character state is meaningless for our plants. All have at least some hairs along the midrib; some have hairs along secondary veins as well; others have hairs also along minor veins; and still others have additional hairs between the veins. Accordingly, Ohio plants are assigned to var. *americanus* (incl. var. *pitcheri* T. & G.)

2. **Ceanothus herbaceus** Raf. Narrow-leaved New Jersey-tea. Eastern Redroot

C. *ovatus* of authors, not Desf.; see Brizicky (1964d).

Much-branched shrub, to 1 m (3 feet) tall, but usually shorter; leaves lanceolate to elliptic-lanceolate, apex obtuse or infrequently acutish; inflorescences all with short peduncles, terminal on leafy shoots. Known from only a few sites in northern Ohio; open, dry, sandy or gravelly habitats. Moseley (1931) believed that Native Americans may have introduced this species into northern Ohio. June–July.

Vitaceae. GRAPE FAMILY. VINE FAMILY

Tendril-bearing, climbing, woody vines; leaves alternate, simple or compound; flowers perfect or imperfect, very small, more or less greenish, actinomorphic, in cymose or paniculate inflorescences; fruit a dark-colored berry. A chiefly tropical family of 12 genera and 700 species. Brizicky (1965a) studied the family in southeastern United States. Some sections of the keys below are adapted from Braun (1961).

a. Leaves palmately compound with 5 leaflets, or rarely with 3, 4, 6, or 7 leaflets, the leaflets toothed but not lobed . . . . 2. *Parthenocissus*

aa. Leaves simple, sometimes very deeply 3–5-lobed, when deeply 5-lobed, the lobes often again lobed.

b. Inflorescences and fruit-clusters wider than long; corolla opening normally, the petals not fused at apex, falling separately. . . . . 1. *Ampelopsis*

bb. Inflorescences and fruit-clusters longer than wide; corolla not opening, the petals fused at apex and falling as a unit . . . . 3. *Vitis*

1. Ampelopsis Michx.

Leaves unlobed to shallowly or deeply lobed; flowers and fruits in broad, cymose clusters. A genus of 20–25 species of Asia and North America.

Ampelopsis arborea (L.) Koehne, Pepper Vine, with twice-pinnately compound leaves, native to southeastern United States, is cultivated in southern Ohio.

a. Leaves unlobed or with two very shallow, shoulder-like lobes . . . . 1. *Ampelopsis cordata*

aa. Leaves moderately to deeply 3–5-lobed. . . . . 2. *Ampelopsis brevipedunculata*

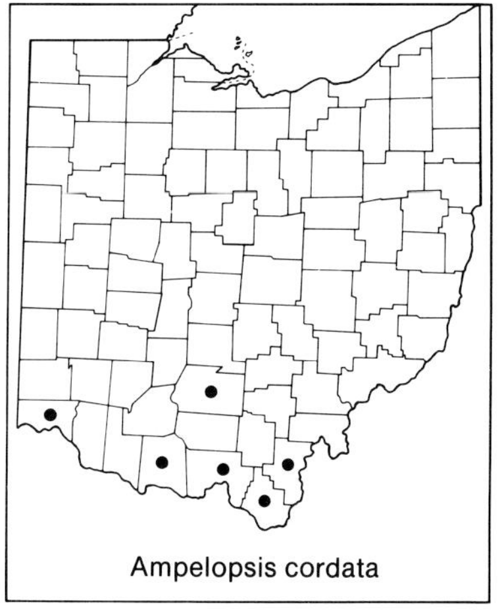
Ampelopsis cordata

Ampelopsis brevipedunculata

1. **Ampelopsis cordata** Michx. Raccoon Grape

High-climbing woody vine; leaves unlobed or with two very shallow, shoulder-like lobes; clusters of flowers and fruits wider than long. Sterile specimens are easily confused with some forms of *Vitis vulpina* (p. 97). A generally southern species at a northern boundary of its range in southern Ohio; alluvial woods and low thickets. May–June.

2. Ampelopsis brevipedunculata (Maxim.) Trautv.

Climbing woody vine; leaves moderately to deeply lobed, the lobes coarsely toothed or sometimes again lobed. An ornamental of Asian origin, cul-

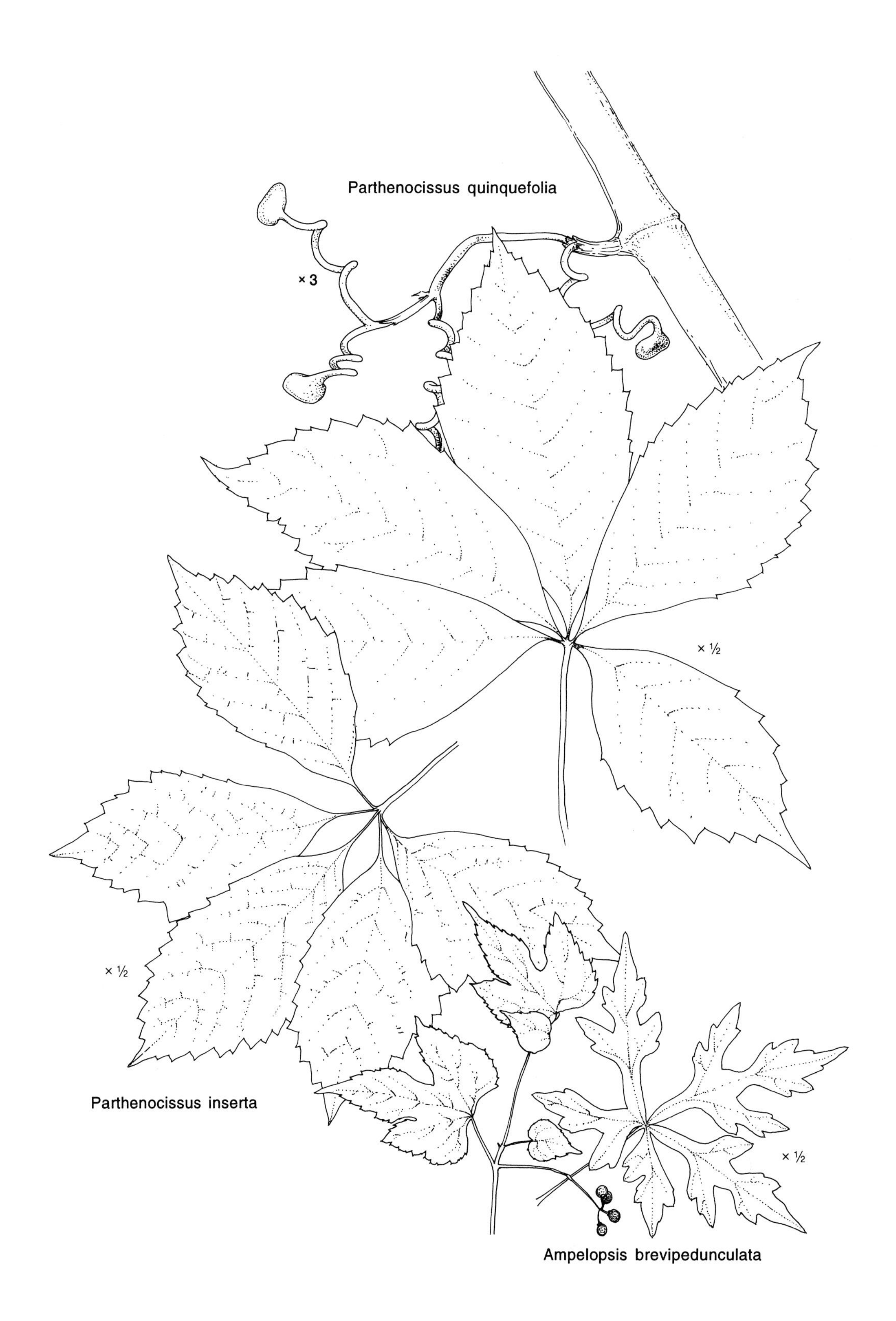
Parthenocissus quinquefolia
× 3
× ½
× ½
Parthenocissus inserta
× ½
Ampelopsis brevipedunculata

tivated for its interesting and variable leaf shapes. Adventive or perhaps locally naturalized at a few sites in Ohio; thickets along roads and railroads; fencerows. July.

2. Parthenocissus Planch. VIRGINIA CREEPER. WOODBINE

Leaves palmately compound with usually 5 (rarely 3, 4, 6, or 7) leaflets. A genus of 10–15 species of Asia and North America. D. A. Webb (1967) discusses the nomenclatural history of Ohio's two native species.

On most college campuses in Ohio, the ivy on "ivy-covered walls" is Boston Ivy, *Parthenocissus tricuspidata* (Siebold & Zucc.) Planch. Of Asian origin despite its common name, Boston ivy climbs by means of tendrils with adhesive discs. Most leaves are simple and shallowly 3-lobed, but branches near the base of the building next to the ground often have, in addition, some leaves with both sinuses cut all the way to the petiole (making the leaf trifoliolate) and others with only one of the sinuses cut to the petiole (making the leaf asymmetrically bifoliolate). Aside from occasional stands of poison ivy (p. 62), the other ivy seen on college buildings is English ivy (p. 204), more frequent in southern than in northern Ohio. In contrast to Boston ivy, both English ivy and poison ivy climb by means of aerial rootlets.

a. Blades of leaflets moderately tapering to long-tapering at base, the leaflets nearly sessile or with petiolules 1–10 mm long; tips of tendrils terminating in adhesive discs; inflorescences more or less paniculate, with an elongate central axis and smaller side branches . . . . 1. *Parthenocissus quinquefolia*

aa. Blade of leaflets short-tapering at base, the leaflets with petiolules 5–20 (–30) mm long; tips of tendrils sometimes enlarged but without adhesive discs; inflorescences usually forked near base and usually with two main branches approximately equal in length and diameter . . . . 2. *Parthenocissus inserta*

1. **Parthenocissus quinquefolia** (L.) Planch. Virginia Creeper

Woody vine, climbing by means of adhesive discs on the tips of tendril branches; leaflets subsessile or with petiolules 1–10 mm long, blades of leaflets moderately tapering to long-tapering at base. Throughout Ohio in woods, thickets, and fencerows, and on walls; sometimes casually cultivated. June–August.

Some herbarium specimens are difficult to separate from those of the next species, *P. inserta*. The most reliable character is the presence in this species of adhesive discs at the tendril tips, but as noted by D. A. Webb (1967), the discs of *P. quinquefolia* "are formed only as a response to con-

Parthenocissus quinquefolia

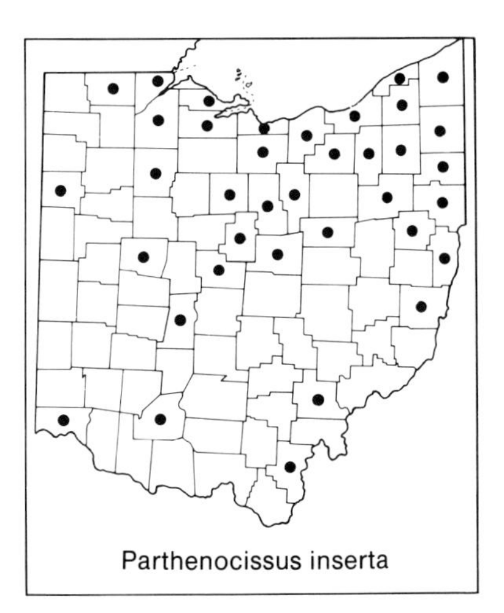
Parthenocissus inserta

tact, and this means that although they are very easily seen on a wall, they are often poorly represented in herbarium specimens, which will tend to show only young, unattached tendrils. On the other hand, it must be realized that the tendril branches of [*P. inserta*] can swell up inside a crevice to give a swollen tip which, if not carefully examined, might be mistaken for a disc."

2. **Parthenocissus inserta** (A. Kerner) Fritsch   THICKET CREEPER
   P. vitacea (Knerr) Hitchc.

Woody vine, sometimes climbing, but more often trailing or scrambling than *P. quinquefolia*; tips of tendril branches sometimes enlarged, but not ending in adhesive discs; leaflets with petiolules 5–20 (–30) mm long, blades of leaflets short-tapering at base. Chiefly in northeastern and north-central counties; fields, fencerows, thickets, railroad embankments, walls, and the like. June–August.

See notes above, under *P. quinquefolia*, on the difficulty of separating that species from this.

3. Vitis L. GRAPE

Leaves simple, usually 3-lobed but sometimes unlobed; flowers and fruits in elongate paniculate clusters. A genus of 50–60 species with distribution centered in temperate regions of Asia and North America. Cultivars of the European species, *Vitis vinifera* L., WINE GRAPE, and cultivars derived from hybrids between that species and native American grapes, principally *V. labrusca* and *V. riparia*, are raised in the vineyards supplying Ohio's wineries.

The treatment here follows that of M. O. Moore (1991). As noted in commentary below, intergrades between species are frequent in Ohio.

a. Leaves at maturity densely pubescent on undersurface with rust-colored to grayish hairs, the pubescence forming a felt-like covering that completely obscures the leaf surface; leaves mostly shallowly 3-lobed, some unlobed and others more deeply 3-lobed; stems with either tendrils or inflorescences at three or more consecutive nodes; fruits large, 1.3 cm or more in diameter . . . . 1. *Vitis labrusca*

aa. Leaves at maturity glabrous or with cobwebby pubescence, or the pubescence in scattered patches, but not obscuring the leaf surface (except on juvenile leaves); leaves unlobed or lobed; stems with tendrils and inflorescences at not more than two consecutive nodes; fruits 0.3–1.2 cm in diameter.

b. Young stems, petioles, and veins on leaf undersurface all densely beset with short, stiffish, diverging hairs ca. 0.2 mm long; leaves unlobed or with two very short lateral lobes or shoulders . . . . 3. *Vitis cinerea*

bb. Young stems glabrous or pubescent, but lacking dense, short stiffish hairs (except sometimes near the nodes); petioles and veins on under-leaf surface sometimes with dense, short, stiffish hairs, but more often either glabrous or with longer lax or cobwebby hairs; leaves unlobed or lobed.

c. Undersurface of leaves either greenish with cobwebby, and usually rust-colored, pubescence or strongly glaucous. . . . 2. *Vitis aestivalis*

cc. Undersurface of leaves greenish and either glabrous or with short noncobwebby hairs.

d. Leaves with two lateral lobes, the lobes acute or acuminate at the apex, pointing more or less apically, their sinuses forming acute angles; teeth of leaf margins sharply acute, sides of teeth usually straight or slightly concave . . . . 4. *Vitis riparia*

dd. Leaves unlobed or with two short, shoulder-like lateral lobes, the lobes short-acute to obtusish, pointing outward, their sinuses forming obtuse angles; teeth of leaf margins broadly acute or obtuse, sides of teeth usually convex. . . . . 5. *Vitis vulpina*

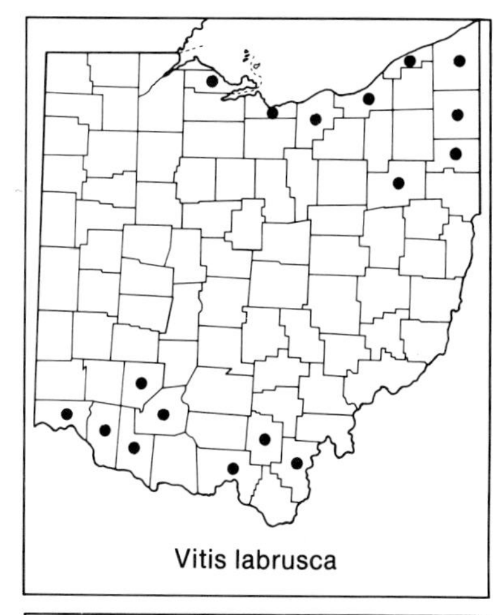
Vitis labrusca

1. **Vitis labrusca** L. Fox Grape
incl. *V. labruscana* L. Bailey

Woody vine; stems with either tendrils or inflorescences at three or more consecutive nodes; leaves mostly shallowly 3-lobed, sometimes unlobed or more deeply 3-lobed; leaves at maturity densely pubescent beneath, with rust-colored to grayish hairs forming a thick felt-like covering that completely obscures the leaf surface; fruits 1.3 cm or more in diameter, the largest fruits of any Ohio grape species. Native in northeastern and extreme north-central counties and native, or perhaps naturalized, in south-central and southwestern counties; thickets and woodland borders. June.

Several cultivars have been derived from this species; the most common in Ohio is the familiar 'Concord' grape.

Some intergrades occur between this species and the next, *V. aestivalis*.

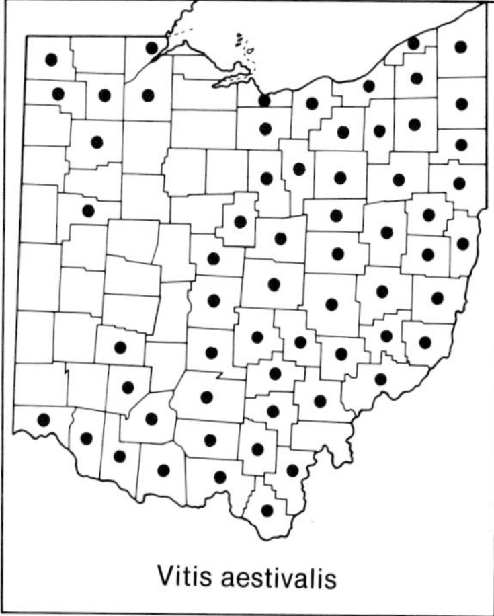
Vitis aestivalis

2. **Vitis aestivalis** Michx. Summer Grape

Woody vine; stems with tendrils and inflorescences at not more than any two consecutive nodes; leaves shallowly to deeply 3- (rarely 5-) lobed. Common in the eastern half of Ohio, but infrequent or absent westward; woods, woodland borders, thickets, roadsides. June–July.

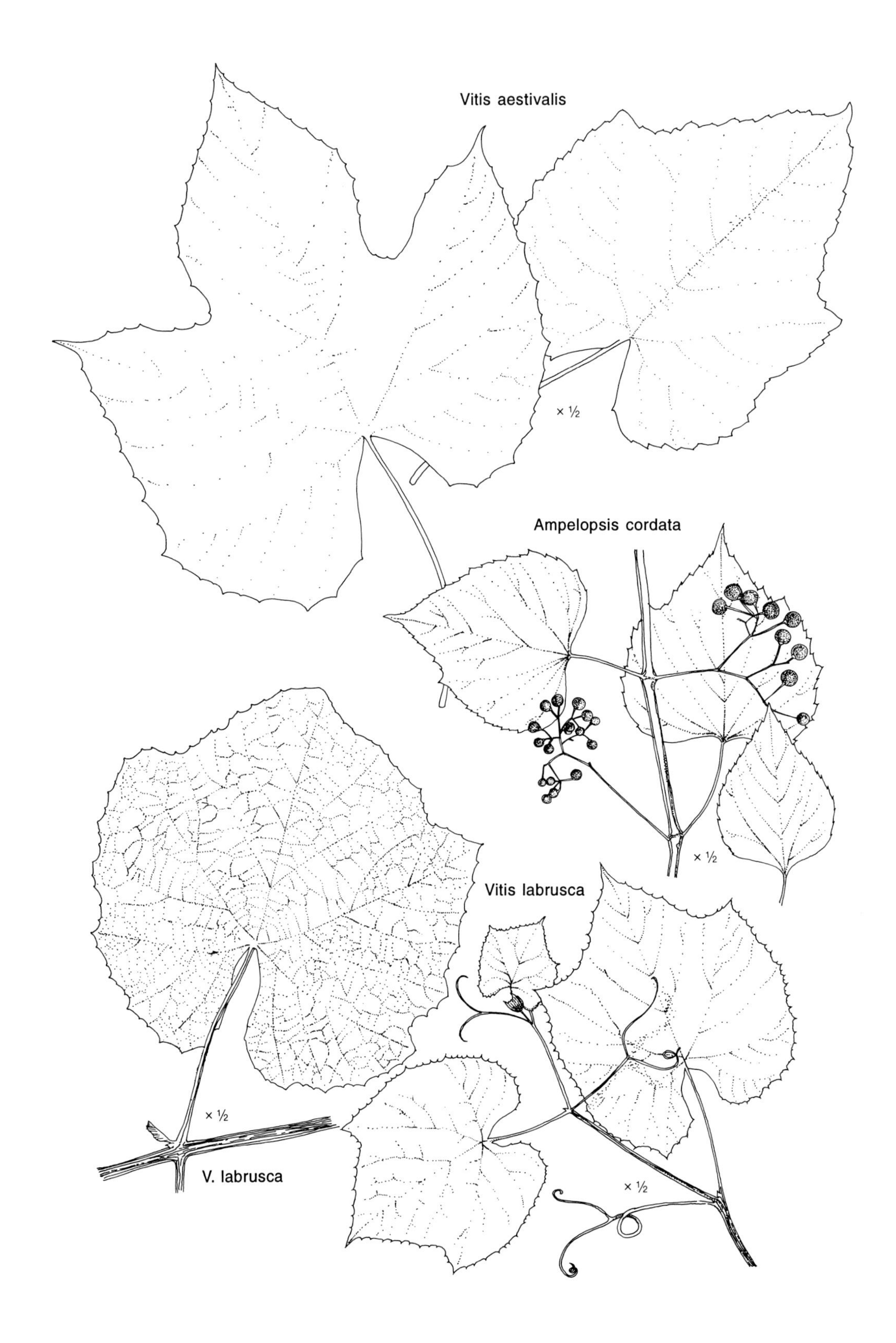
Vitis aestivalis
× ½
Ampelopsis cordata
× ½
Vitis labrusca
× ½
V. labrusca
× ½

Two varieties have been recognized among Ohio members of this species.

a. Lower leaf surface with cobwebby, rust-colored pubescence, not or only slightly glaucous . . . . var. *aestivalis*

aa. Lower leaf surface glabrous or somewhat pubescent, strongly glaucous . . . . var. *bicolor* Deam

var. *argentifolia* (Munson) Fern.

The two varieties intergrade freely in Ohio, and both have approximately the same range in the state, although var. *bicolor* is more common in northern counties.

Some Ohio specimens appear to be intergrades between *Vitis aestivalis* and *V. cinerea*, and others intergrades between *V. aestivalis* and *V. vulpina*.

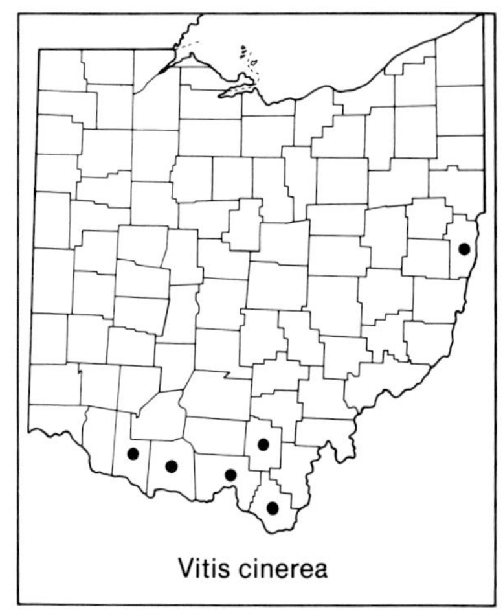
Vitis cinerea

3. **Vitis cinerea** Millard PIGEON GRAPE

Woody vine; stems with tendrils and inflorescences at not more than any two consecutive nodes; leaves unlobed or with two very short lateral lobes or mere shoulders; young stems, petioles, and veins on leaf undersurface all densely beset with short, stiffish, diverging hairs ca. 0.2 mm long. Southern, mostly extreme south-central, counties; bottomland thickets, stream banks. June.

Some Ohio specimens are intermediates between *V. cinerea* and *V. aestivalis*. Most can be assigned to one species or the other, but a few cannot be satisfactorily placed in either.

In Moore's (1991) treatment, adopted here, Ohio plants are assigned to var. *baileyana* (Munson) Comeaux (*V. baileyana* Munson); var. *cinerea* occurs to the southwest of Ohio.

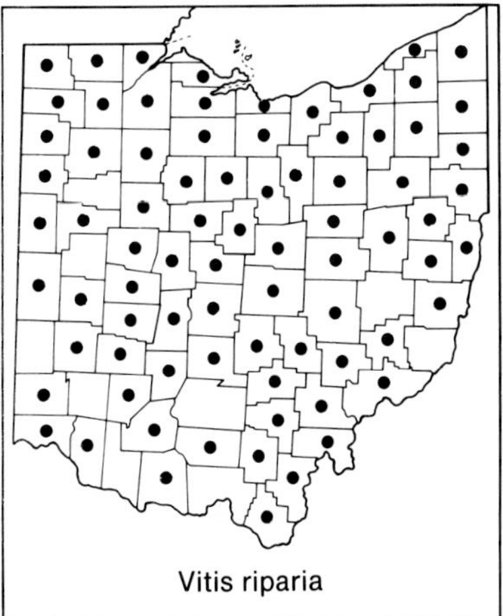
Vitis riparia

4. **Vitis riparia** Michx. RIVERBANK GRAPE

incl. var. *syrticola* (Fern. & Wieg.) Fern.

Woody vine; stems with tendrils and inflorescences at not more than any two consecutive nodes; leaves green beneath, the undersurface glabrous or with short hairs, the leaves with two usually shallow, lateral lobes, each lobe acute or acuminate at the apex, the lobes pointing more or less apically, each of the two sinuses forming an acute angle; teeth of leaf margins sharply acute (more so than in any other Ohio grape), sides of the teeth usually straight or slightly concave. Ohio's most common grape species, growing throughout the state; river banks and moist thickets. May–June.

Occasional specimens are difficult to separate from *V. vulpina*.

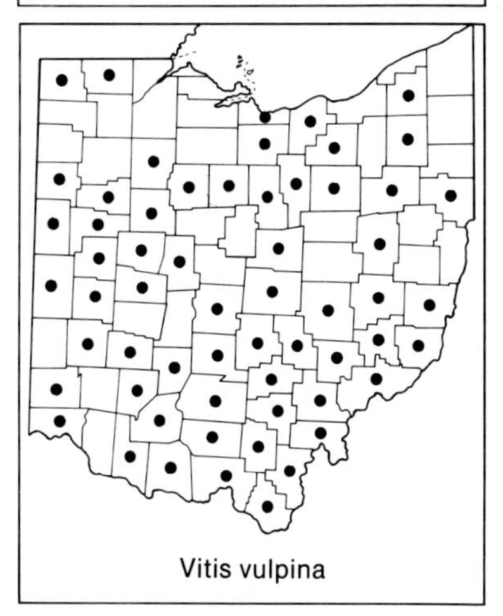
Vitis vulpina

5. **Vitis vulpina** L. FROST GRAPE

Woody vine; stems with tendrils at not more than any two consecutive nodes; leaves green beneath, the undersurface glabrous or with short hairs, the leaves unlobed or with two short, shoulder-like, lateral lobes, the lobes

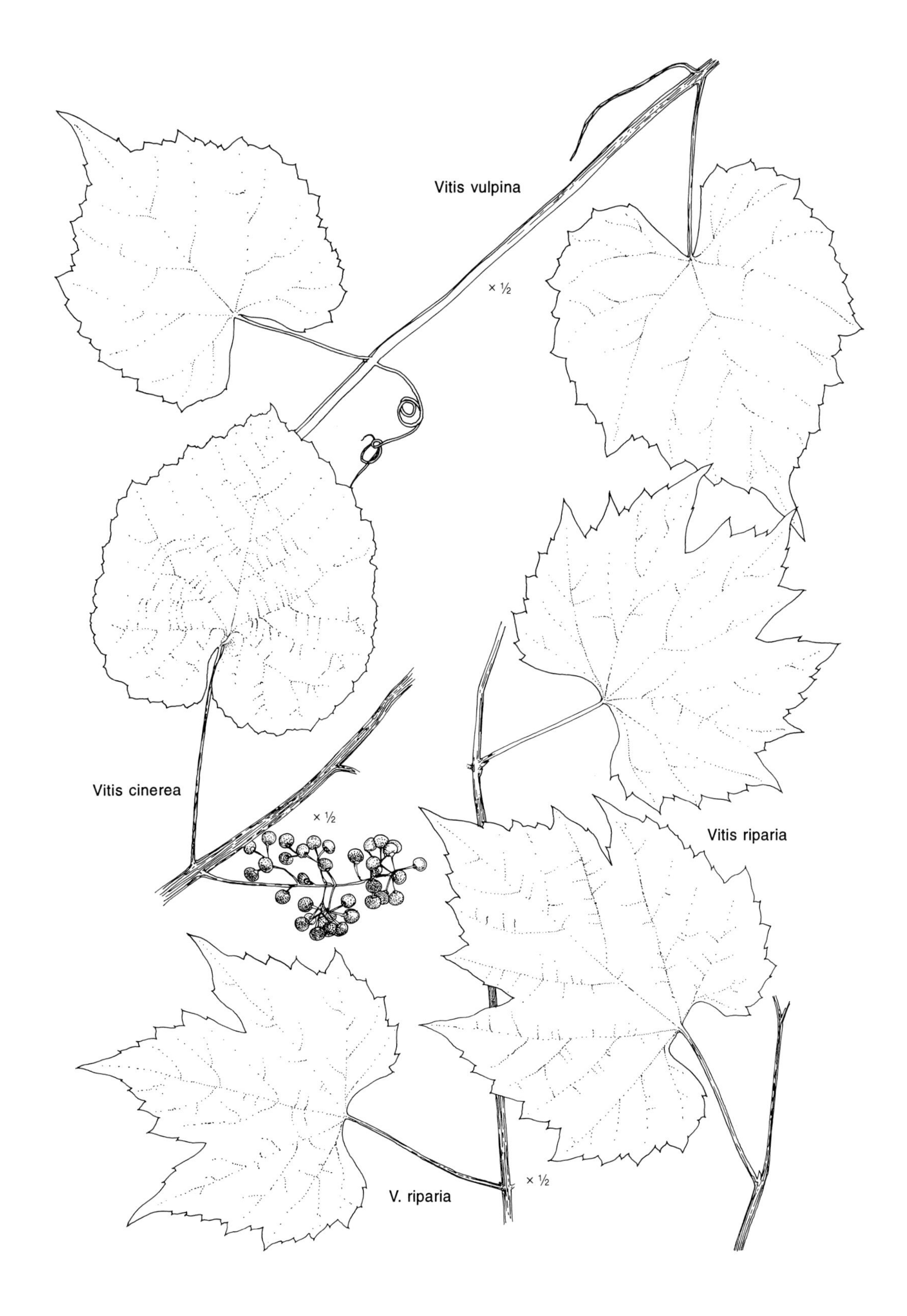

Vitis vulpina
× ½
Vitis cinerea
× ½
Vitis riparia
V. riparia
× ½

pointing outward, each of the two sinuses forming an obtuse angle; teeth of leaf margins broadly acute or obtuse, sides of the teeth usually convex. Widespread in Ohio but chiefly in the southern half of the state; not as common as *V. riparia*; moist woods and thickets, shores, river banks. May–June.

In Ohio, this species intergrades with both *V. aestivalis* and *V. riparia*.

Tiliaceae. LINDEN FAMILY

A chiefly tropical and subtropical family with some 50 genera, one of which is represented in the Ohio flora. Brizicky (1965b) studied the family in southeastern United States.

1. Tilia L. BASSWOOD. LINDEN

Large trees with simple alternate leaves; flowers perfect, sweetly fragrant, yellowish to off-white, actinomorphic, in pedunculate, corymbose to cymose inflorescences, the lower half of the peduncle fused to a unique, narrowly oblong, foliaceous bract, the upper half diverging from near the center of the bract; fruit small, globose, nutlike. A genus of some 30 species of the Northern Hemisphere, chiefly in temperate regions. G. N. Jones (1968) prepared the most recent revision of the American species of *Tilia*. His taxonomy, adopted here, places the Ohio trees in the two species discussed below. However, several other workers, including Ashby (1964), Hickok and Anway (1972), and J. W. Hardin (1990), have concluded that *Tilia heterophylla* should be merged under *T. americana* sensu lato.

Tilia floridana (V. Engler) Small was attributed to Ohio by Braun (1960, 1961), but G. N. Jones (1968) determined that Ohio specimens so identified were in fact misidentified forms of *T. americana*.

In addition to our two native species, *Tilia cordata* Mill., SMALL-LEAVED EUROPEAN LINDEN, is frequently cultivated in Ohio.

a. Undersurface of leaves green, glabrous or with small amounts of pubescence, young leaves sometimes with stellate pubescence, but this pubescence soon falling .... 1. *Tilia americana*

aa. Undersurface of leaves densely and permanently covered with fine, white to brownish, stellate pubescence .... 2. *Tilia heterophylla*

1. **Tilia americana** L. BASSWOOD

incl. *T. neglecta* Spach

Tree, to 36 m (122 feet) tall, but usually much shorter; base of leaves cordate to markedly oblique, the leaf undersurface green, glabrous or with small amounts of pubescence, young leaves sometimes with stellate pubes-

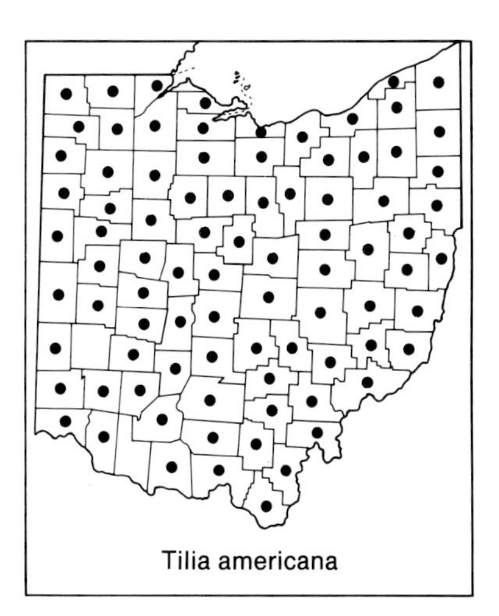

Tilia americana

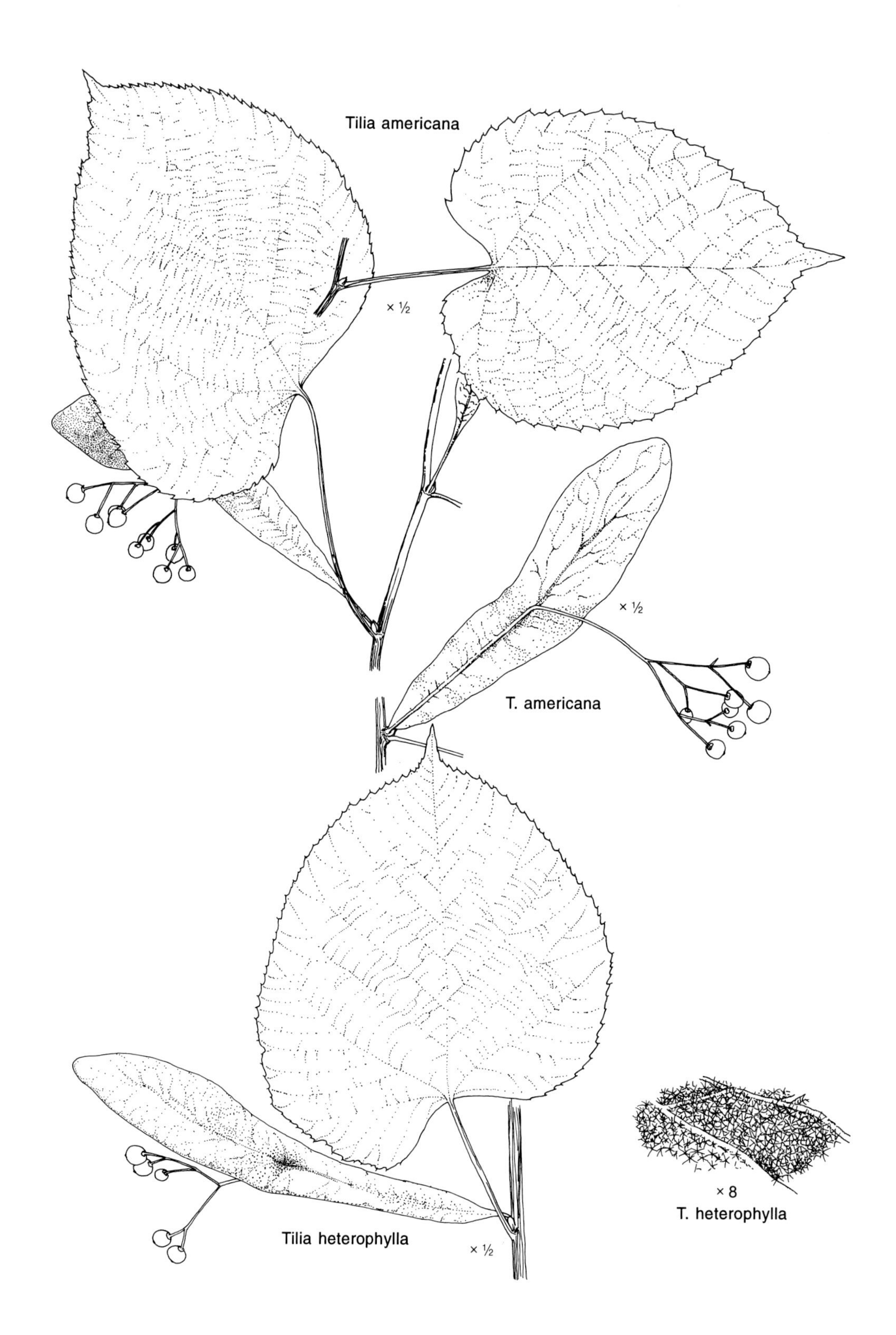
Tilia americana
× ½
× ½
T. americana
× 8
T. heterophylla
Tilia heterophylla
× ½

cence, but this soon shed. Throughout Ohio; rich moist woods, especially wooded bottomlands, terraces, and stream banks. June–July.

G. N. Jones (1968) concluded that specimens identified as *T. neglecta* were in nearly all cases ones taken from sucker shoots of *T. americana*; J. W. Hardin (1990) suggests that they may be introgressants between this taxon and the next.

2. **Tilia heterophylla** Vent. WHITE BASSWOOD

T. americana L. var. *heterophylla* (Vent.) Loud.

Tree, to 23 m (75 feet) tall; base of leaves cordate to markedly oblique, the leaf undersurface densely and permanently covered with fine, white to brownish, stellate pubescence. Southern and east-central counties; rich moist woodlands. June–July.

J. W. Hardin (1990) treats this species as a variety of *T. americana*; some Ohio specimens are intermediate between the two species.

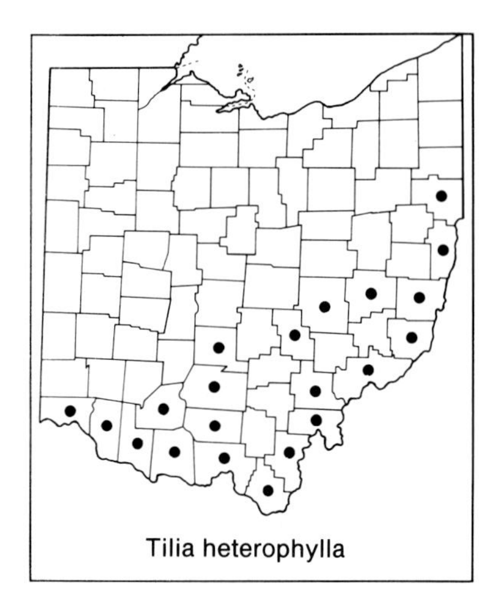
Tilia heterophylla

Malvaceae. MALLOW FAMILY

Herbs (or rarely shrubs in *Hibiscus*); leaves alternate, simple; flowers perfect, mostly solitary in leaf axils, less often arranged in panicles or corymbs; calyx of 5 sepals, separate or slightly fused at the base; corolla of 5 separate petals; stamens numerous and fused by their filaments (monadelphous) to form a unique, conspicuous, staminal tube; carpels 5–many, and fused to form a compound pistil with one long style that passes through, but is not fused to, the staminal tube, the stigmas separate and often conspicuous; ovary superior; fruit a capsule. An important and widespread family of about 1,500 species. Kearney (1951) presents a synopsis of the genera of the Malvaceae in the New World.

Cotton is obtained from species of the genus *Gossypium* L. Okra or gumbo is the fruit of *Abelmoschus esculentus* (L.) Moench.

Callirhoe involucrata (T. & G.) A. Gray, POPPY-MALLOW, a low red-flowered perennial, was reported by Schaffner (1932) and Fernald (1950) to be adventive in Ohio, but I have seen no specimens other than ones probably collected from cultivated plants.

a. Flowers with whorls of sepal-like bracts immediately subtending calyx.
  b. Bracts 3.
    c. Calyx 10–20 mm long, the lobes ovate; petals shallowly notched at apex . . . . 1. *Malva*
    cc. Calyx 3–10 mm long, the lobes triangular-lanceolate; petals truncate and denticulate at apex; see above . . . . *Callirhoe*
  bb. Bracts 5–14.

c. Styles 5; carpels united at maturity to form a capsule. . . . . 8. *Hibiscus*

cc. Styles usually 15 or more; carpels separating at maturity and falling individually.

d. Petals 1.5–2.0 cm long; bracts subtending calyx narrowly lanceolate . . . . 2. *Althaea*

dd. Petals 3–6 cm long; bracts subtending calyx triangular to broadly triangular . . . . 3. *Alcea*

aa. Flowers without bracts subtending calyx.

b. Leaves deeply 5–9 (–11)-palmately lobed, the lobes again shallowly and irregularly lobed or coarsely toothed; the leaves mostly as wide as to wider than long; flowers white, imperfect, the plants dioecious . . . . 4. *Napaea*

bb. Leaves either unlobed or 3–5 (–7)-palmately lobed, if lobed the lobe margins dentate or serrate but not again lobed; the leaves mostly as long as to longer than wide; flowers white or other colors, perfect.

c. Flowers solitary, borne on long pedicels equal, or approximately equal, to the length of subtending petiole; petals pale blue to lavender. . . . . 5. *Anoda*

cc. Flowers either solitary or in clusters, if solitary the length of the pedicel much less than length of subtending petiole; petals white, yellow, or orange-yellow.

d. Leaves deeply 3–5 (–7)-lobed; corolla white. . . . . 6. *Sida*

dd. Leaves unlobed; corolla yellow to orange-yellow.

e. The leaves oblong-lanceolate to lanceolate, rounded at base and acutish at apex; carpels with a single seed . . . . 6. *Sida*

ee. The leaves broadly ovate to suborbicular, cordate at base and acuminate at apex; carpels with (2–) 3–6 (–9) seeds . . . . 7. *Abutilon*

1. Malva L. MALLOW

Annual, biennial, or perennial herbs; flowers sometimes showy, subtended by 3 bracts. A genus of about 30 species, native to Eurasia and northern Africa.

Malva alcea L., VERVAIN MALLOW, an adventive escape from cultivation, has been collected from Ashland, Lake, and Lorain counties, all in northeastern Ohio. *Malva verticillata* L., WHORLED MALLOW, also an adventive escape, has been collected from Delaware and Lorain counties.

a. Leaves on upper half of stem unlobed or with shallow rounded lobes, the sinuses not extending beyond the middle of the blade; flowers clustered in leaf axils and usually distributed along entire length of stem.

b. Bracts subtending calyx oblong-lanceolate or oblong-elliptic to narrowly oblong-ovate; corolla reddish-purple; largest leaves to 15 cm wide. . . . . 1. *Malva sylvestris*

bb. Bracts subtending calyx linear to linear-lanceolate; corolla white, pale lilac, or pale blue; largest leaves to 7 cm wide.

c. Petals 5–12 mm long, about twice the length of the calyx; carpels 11–15, their outer faces obscurely or not reticulate, the margins with no or with a very low hyaline ridge; stems prostrate, ascending, or erect.

d. Stems prostrate or ascending at tip; flowers and fruits on pedicels 3–20 (–30) mm long . . . . 2. *Malva neglecta*

dd. Stems erect; flowers subsessile, fruiting pedicels less than 10 mm long; see above . . . . *Malva verticillata*

cc. Petals 3–4.5 mm long, shorter than to slightly exceeding the calyx; carpels 8–12, their outer faces conspicuously rugose-reticulate, the margins with a distinct hyaline ridge; stems ascending or erect . . . . 3. *Malva rotundifolia*

aa. Leaves on upper half of stem deeply palmately lobed, the sinuses extending nearly to base of the blade, the lobes often again lobed; flowers borne near apex of stem, some forming a terminal inflorescence, the others solitary in axils of upper leaves.

b. Upper stem pubescent with simple divergent hairs; bracts subtending calyx linear to narrowly lanceolate or narrowly elliptic . . . . 4. *Malva moschata*

bb. Upper stem stellate-pubescent; bracts subtending calyx ovate to ovate-oblong; see above . . . . *Malva alcea*

1. Malva sylvestris L. High Mallow

Biennial, to 0.8 m tall; largest leaves to 15 cm wide, upper leaves with shallow rounded lobes; flowers in axillary clusters; bracts subtending calyx 3, oblong-lanceolate or oblong-elliptic to narrowly oblong-ovate; petals 12–30 mm long, reddish-purple. A European species, sometimes cultivated in Ohio and infrequently escaped; adventive or perhaps locally naturalized at scattered disturbed sites. July–August (–October).

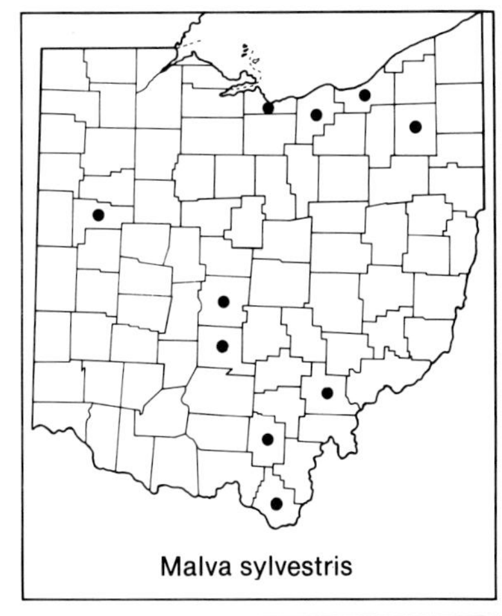
Malva sylvestris

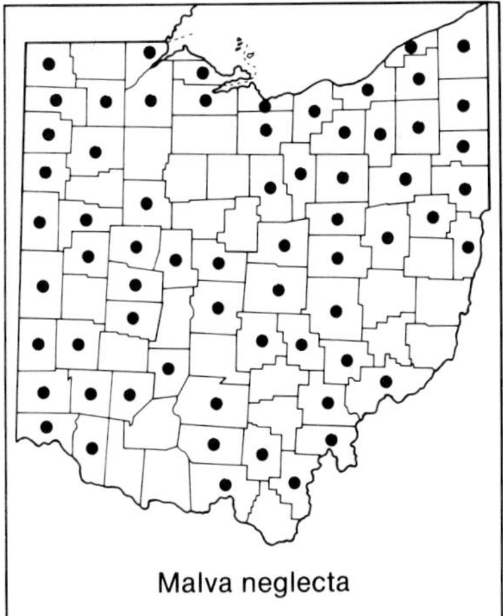
Malva neglecta

2. Malva neglecta Wallr. Common Mallow. Cheeses

Annual; stems prostrate or ascending at tips; leaves with very shallow, rounded lobes, the largest to 7 cm wide; flowers in axillary clusters; bracts

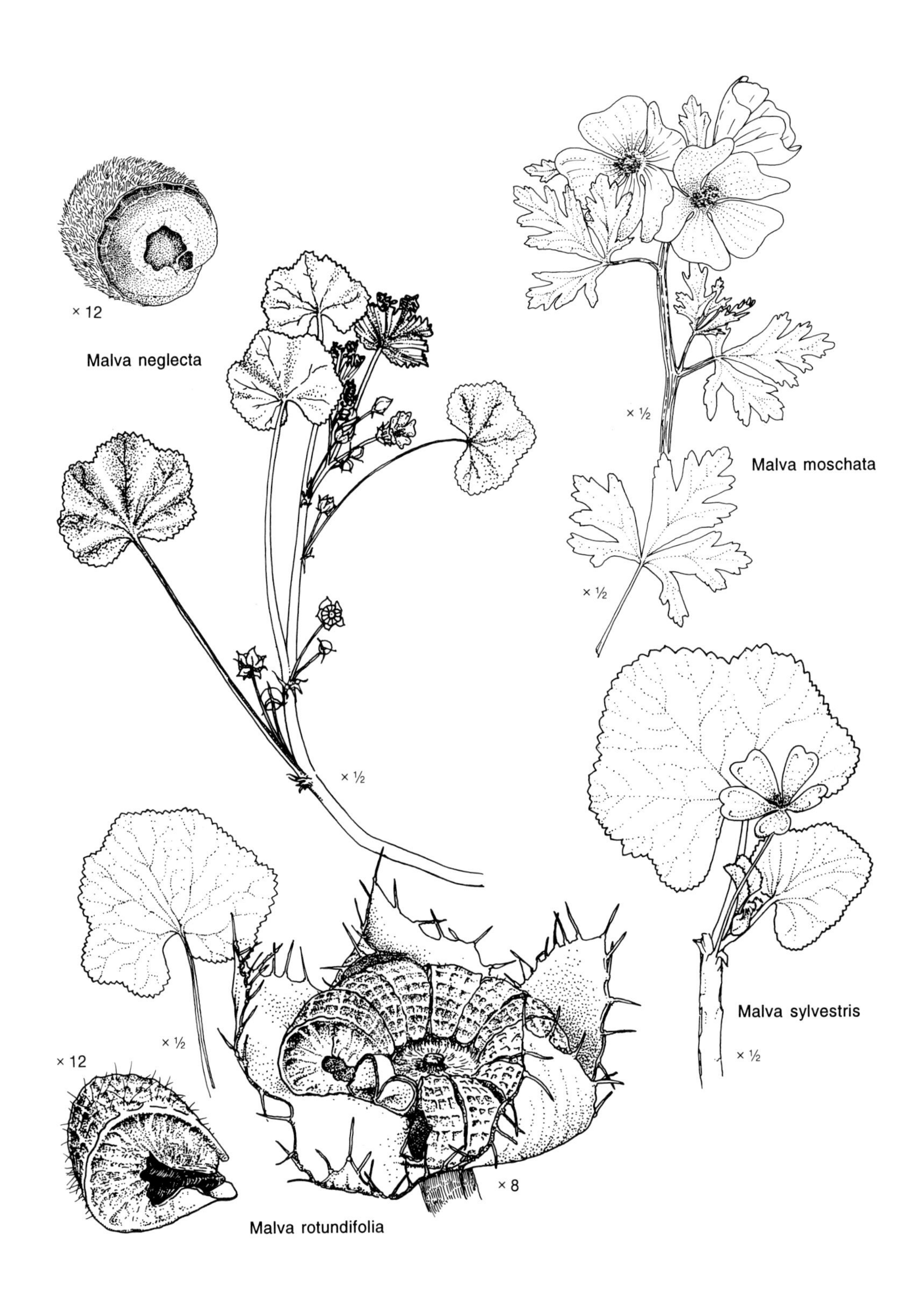
× 12
Malva neglecta
× ½
Malva moschata
× ½
× ½
× ½
× 12
Malva sylvestris
× ½
× 8
Malva rotundifolia

subtending calyx 3, broadly linear; petals pale lilac to whitish, 9–12 mm long, about twice as long as the calyx; carpels 11–15, their outer faces obscurely or not reticulate. A Eurasian species, naturalized throughout Ohio; in waste places near homes and buildings, in barnyards, at roadsides, and in other disturbed sites. Often confused with the next species, M. *rotundifolia*. May–October.

3. MALVA ROTUNDIFOLIA L. ROUND-LEAVED MALLOW. DWARF MALLOW

M. *pusilla* Sm.

Annual; stems ascending or erect; leaves with very shallow, rounded lobes, the largest to 7 cm wide; flowers in axillary clusters; bracts subtending calyx 3, broadly linear; petals pale lilac to pale blue, 3.0–4.5 mm long, shorter than to slightly exceeding calyx; carpels 8–12, their outer faces conspicuously rugose-reticulate, and the margins with a distinct hyaline ridge. A European species found in disturbed habitats at a few scattered sites in Ohio, often confused with M. *neglecta*, but much less frequent in Ohio than that species; naturalized or perhaps merely adventive. Makoswki and Morrison (1989) discuss the biology of this species. July–September.

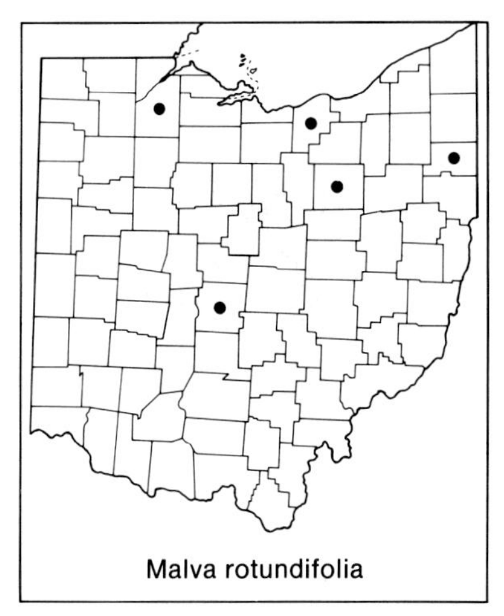
Malva rotundifolia

4. MALVA MOSCHATA L. MUSK MALLOW

Perennial, to 1 m tall; stems pubescent, erect; leaves on upper stem deeply palmately lobed, the sinuses extending nearly to base of blade, and the lobes often again lobed; flowers borne near stem apex, some in a terminal inflorescence, others solitary in axils of upper leaves; bracts subtending calyx 3, narrowly elliptic; corolla showy, the petals white, pink, rose, or pale purple, to 3 cm long. A species of Europe and northern Africa, cultivated in Ohio; escaped, especially in northeastern counties, and locally naturalized at some sites, adventive at others; fields and roadsides. June–September.

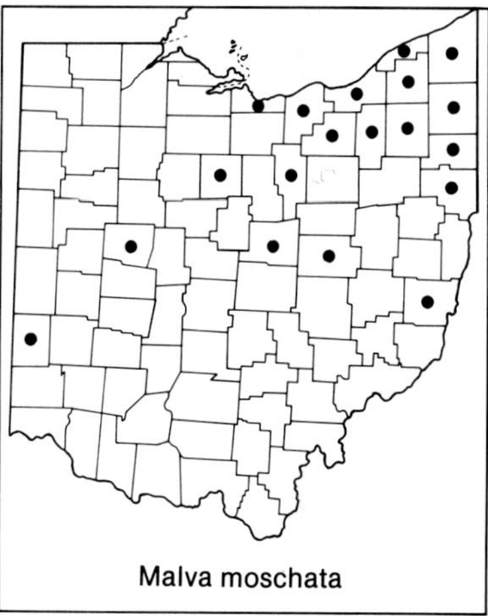
Malva moschata

2. Althaea L.

A Eurasian genus of 12 species, with a single species in Ohio.

1. ALTHAEA OFFICINALIS L. MARSH-MALLOW

Velvety pubescent perennial, to ca. 1 m tall; leaves not or only shallowly lobed; inflorescences of few to several flowers, axillary; bracts 6–9, narrowly lanceolate; corolla moderately showy, the petals 1.5–2 cm long, pale pink or pale rose. A European species, infrequently cultivated as an ornamental in Ohio; escaped into disturbed habitats and locally naturalized at a few scattered sites, probably only adventive at others. June–September.

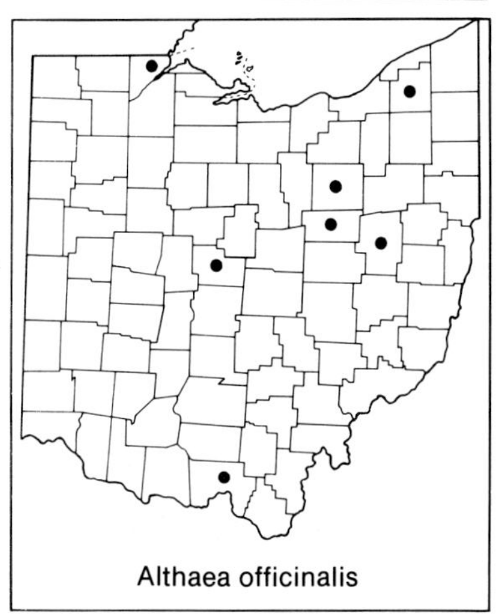
Althaea officinalis

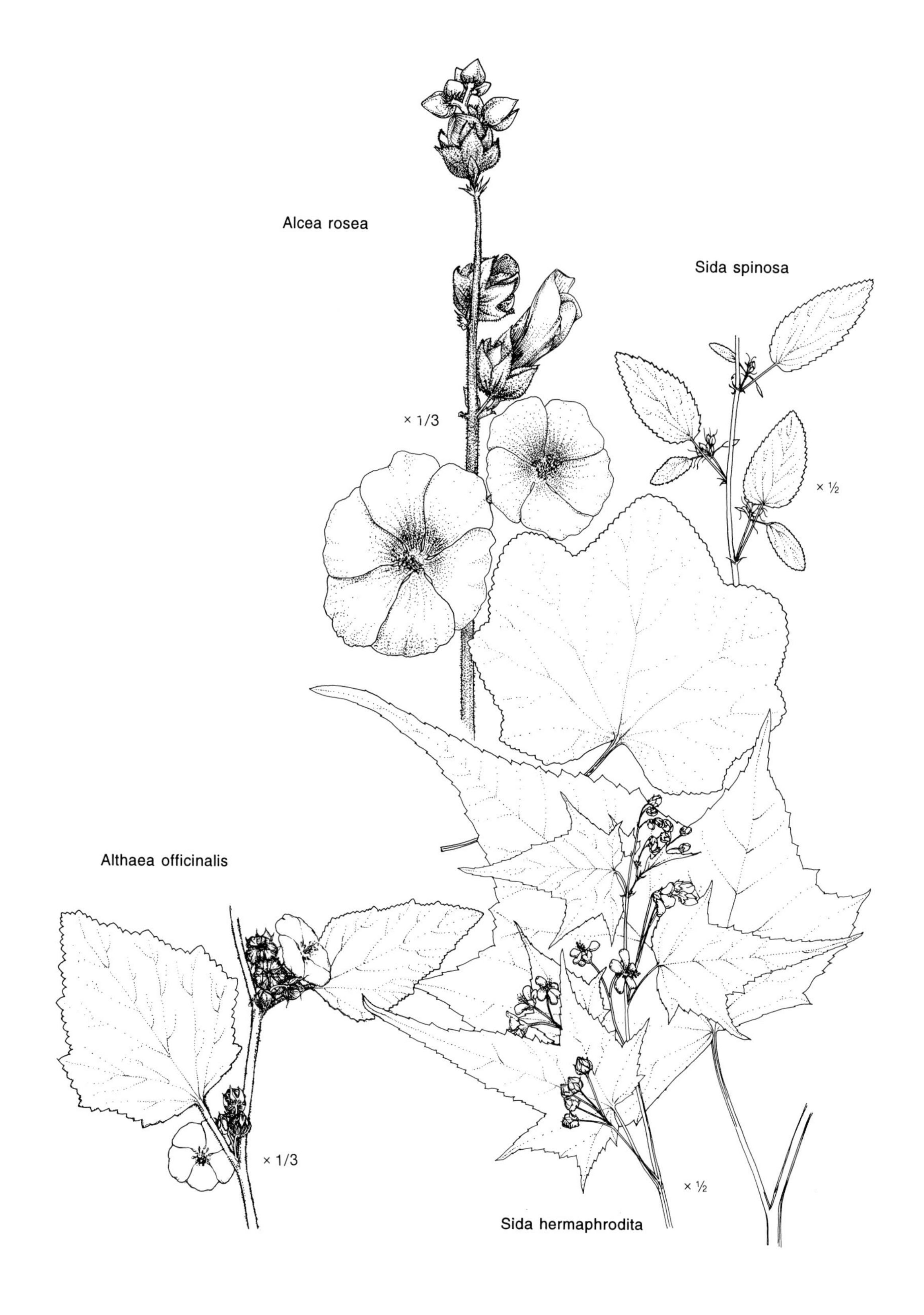
Alcea rosea
Sida spinosa
× 1/3
× ½
Althaea officinalis
× 1/3
× ½
Sida hermaphrodita

3. Alcea L. HOLLYHOCK

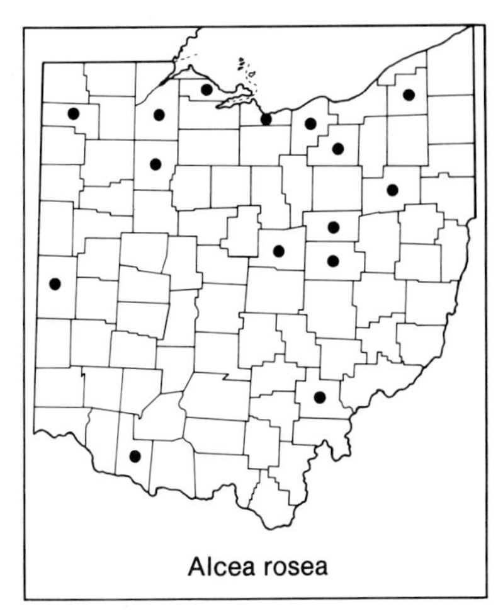
Alcea rosea

A Eurasian genus of 60 species, one of which is cultivated in Ohio.

1. ALCEA ROSEA L. HOLLYHOCK
   Althaea rosea (L.) Cav.

Pubescent biennial, to 2 (or 3) m tall; leaves shallowly lobed; flowers solitary in leaf axils and in terminal racemes; bracts usually 5 or 6, triangular to broadly triangular; corolla showy, the petals 5 or 6, 3–6 cm long, in a variety of colors, mostly white to pink, red, or purple. A Eurasian species, perhaps of hybrid origin, in the past a very popular ornamental and still today widely, if sometimes only casually, cultivated; at scattered Ohio sites; adventive on roadsides, along railroads, and in other disturbed sites. Some cultivars have "double" flowers. June–October.

4. Napaea L. GLADE-MALLOW

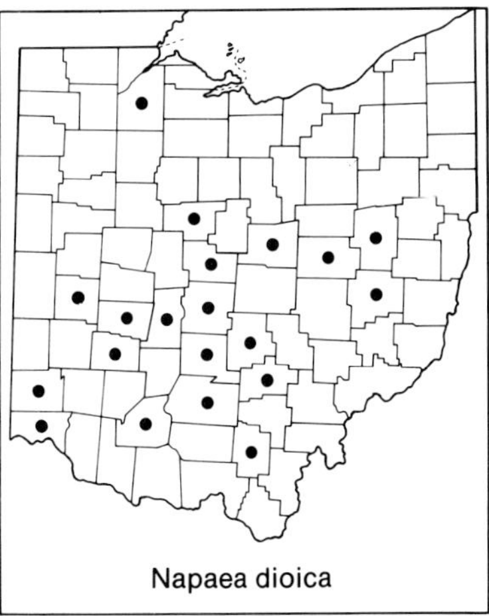
Napaea dioica

A genus with a single species.

1. **Napaea dioica** L. GLADE-MALLOW

Dioecious perennial, to 2.5 m tall; leaves large, to 3.5 dm wide, mostly as wide as to wider than long, deeply 5–9 (–11)-palmately lobed, the lobes again shallowly and irregularly lobed or coarsely toothed; flowers unisexual, numerous in large terminal or axillary panicles; calyx without subtending bracts; corolla white, 3–9 mm long. Native to north-central United States, reaching in Ohio an eastward boundary of the species' range; mostly in the southern two-thirds of Ohio; wet prairies, moist borders, and moist openings, terraces, and floodplains. As noted by Iltis (1963), plants of the western half of Ohio are mostly natives, while those of the eastern half are adventives growing in disturbed habitats beyond the boundary of the species' natural range. Additional record: Fayette County (Iltis, 1963). June–August.

5. Anoda Cav.

Anoda cristata

A New World genus of some 15 species, one of which grows in Ohio.

1. ANODA CRISTATA (L.) Schlecht. SPURRED ANODA

Villous-hirsute annual; stems more or less erect, to 1 m long; leaves coarsely toothed and sometimes shallowly lobed; flowers long-penduncculate, solitary from leaf axils; calyx without subtending bracts; corolla pale blue to lavender, to 2 cm long. A locally frequent adventive from southwestern United States and Mexico; in a few southern counties; waste places. August–September.

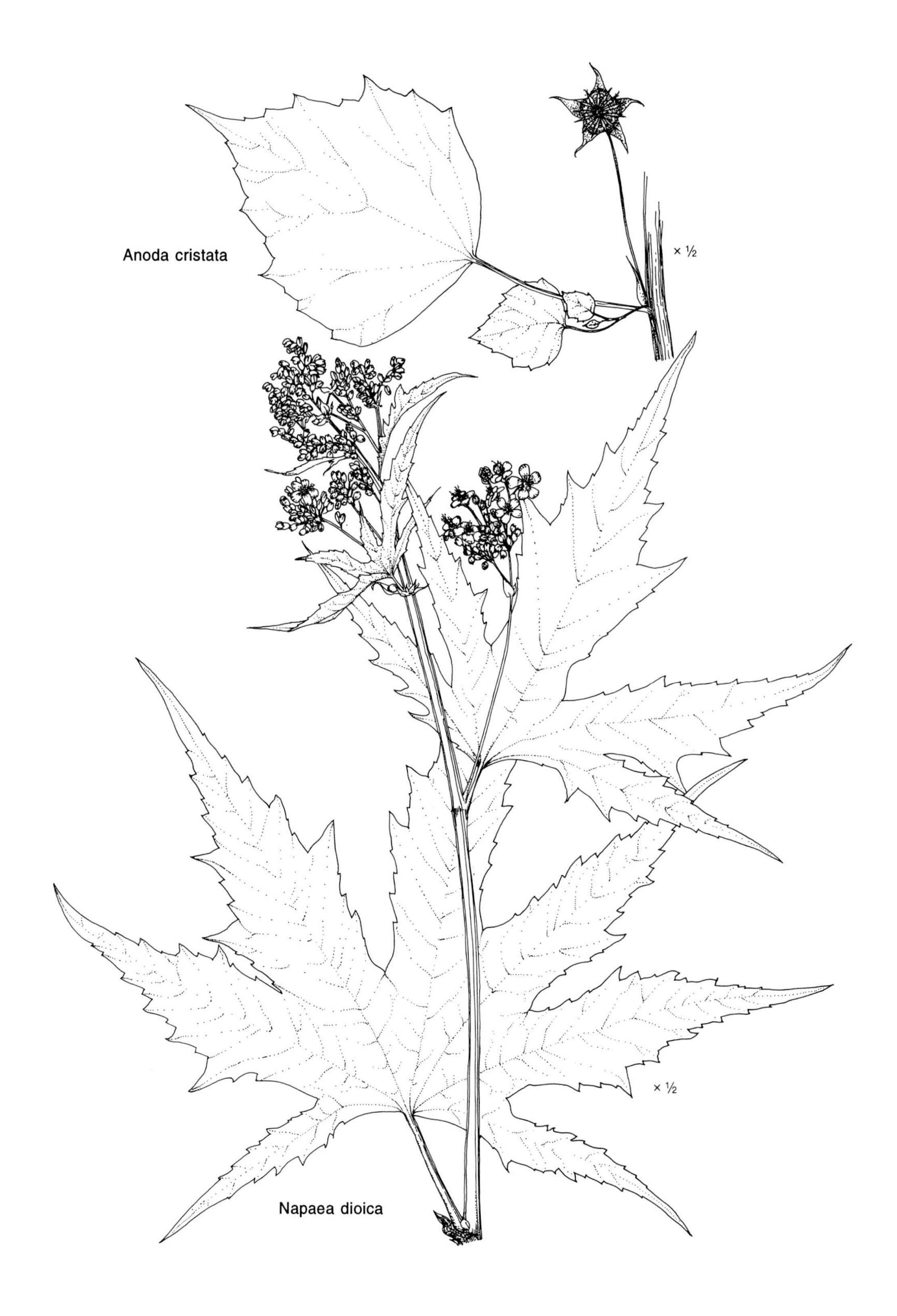
Anoda cristata
× ½
× ½
Napaea dioica

6. Sida L.

A widely distributed genus of some 150 species. The two species in Ohio, both bearing bractless flowers, are otherwise markedly dissimilar in appearance. Fryxell (1985) made a study of the species of North and Central America.

a. Leaves deeply 3–5 (–7)-palmately lobed; corolla white . . . . 1. *Sida hermaphrodita*
aa. Leaves unlobed; corolla yellow . . . . 2. *Sida spinosa*

1. **Sida hermaphrodita** (L.) Rusby TALL SIDA
Perennial, to 3 m tall; leaves mostly 3–5-lobed, the terminal lobe much the longest; flowers few to several in axillary clusters; calyx without subtending bracts; petals white, 10–15 mm long. Moist, open to partially open sites along streams and rivers, and at moist to mesic roadsides; in several southern counties bordering the Ohio River, and in two northwestern counties. Spooner et al. (1985) speculate that the species is native in both parts of Ohio; some collections in each area may be from adventive or naturalized stands. Late June–October.

Clement (1957) made a taxonomic study of this and other species of *Sida*.

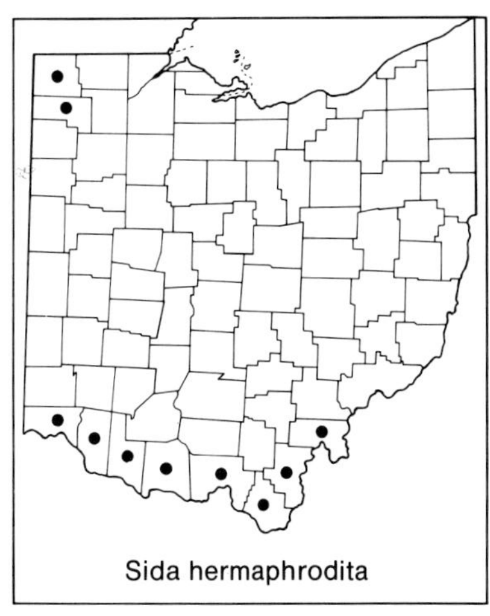
Sida hermaphrodita

2. SIDA SPINOSA L. PRICKLY SIDA
Annual, to ca. 6 dm tall, usually much branched from near base of main stem; leaves crenate to serrate, but unlobed, each usually with a tiny spine-like process (scarcely, if at all, prickly) at base of petiole; flowers in few-flowered axillary clusters; calyx without subtending bracts; corolla pale yellow, the petals 5–7 mm long. A native of more tropical regions, naturalized mostly in the southern half of Ohio; along roadsides and in other weedy places. Baskin and Baskin (1984e) studied seed germination in this species. July–September.

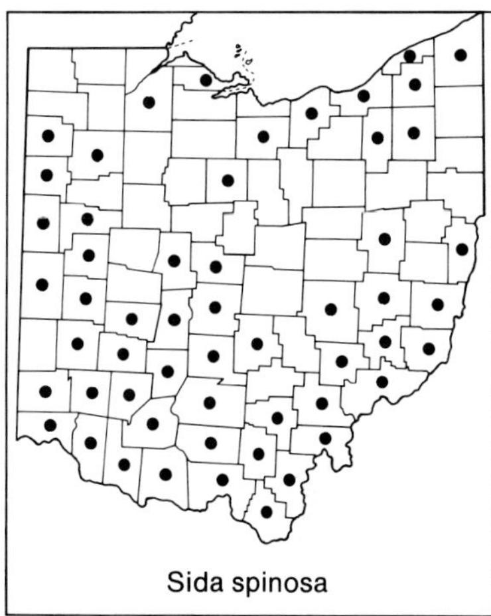
Sida spinosa

7. Abutilon Mill.

A widely distributed genus of about 150 species, one of which is a common Ohio weed.

1. ABUTILON THEOPHRASTI Medic. VELVET-LEAF
Pubescent annual, to 1.5 m tall; leaves unlobed, margins low crenate-serrate, cordate at base and usually long-acuminate at apex; flowers solitary or in few-flowered clusters in leaf axils; calyx without subtending bracts; corolla orange-yellow, the petals 10–15 mm long. A species of Asian origin, "introduced into the New World before 1750 as a potential fiber crop for the American colonies" (Spencer, 1984); naturalized in disturbed sites throughout Ohio, and a sometimes troublesome weed of cultivated fields and gardens. Late June–October.

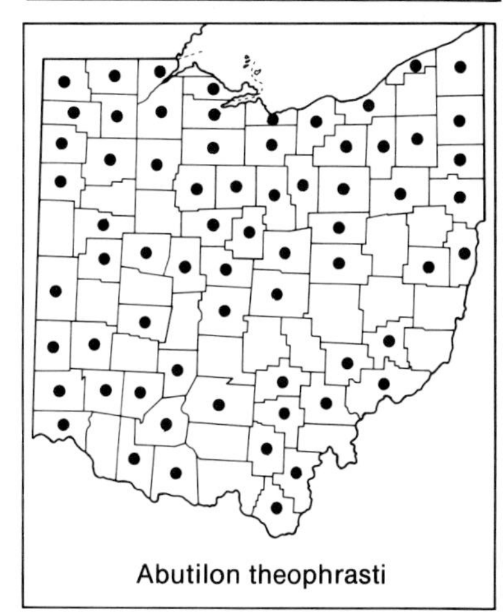
Abutilon theophrasti

8. Hibiscus L. ROSE-MALLOW

A genus of about 250 species, chiefly of warm and tropical climates. The Ohio species bear showy, and in some cases very large, flowers subtended by numerous narrow bracts.

Hibiscus syriacus L., the cultivated Rose of Sharon, has been collected as an adventive escape from Franklin, Highland, and Lawrence counties, in the southern half of the state. *Hibiscus rosa-sinensis* L., Chinese Hibiscus or Hawaiian Hibiscus, grown indoors in Ohio, produces the large, red, yellow, orange, and white hibiscus flowers of the tropics, noted for their bright-colored petals and the long-exserted staminal column. Bates' (1965) taxonomic study of the species of *Hibiscus* cultivated in the United States includes the two species named above as well as the three below.

a. Perennial herbs or shrubs; leaves unlobed or when lobed with sinuses at most 2/3 of way to base of blade; calyx not inflated; petals of various colors, 4–10 cm long.

b. Shrubs; leaves mostly 3-lobed, the lateral lobes obtuse to rounded; petals mostly 4–6 cm long; see above .... *Hibiscus syriacus*

bb. Tall, coarse perennial herbs; leaves either unlobed or if lobed the lateral lobes acuminate; petals 6–12 cm long.

c. Upper internodes of stem and under-leaf surfaces pubescent; leaves unlobed or very shallowly lobed.......... 1. *Hibiscus moscheutos*

cc. Upper internodes of stem and under-leaf surfaces glabrous; most leaves markedly 3-lobed, more or less hastate, the lateral lobes widely divergent.......................... 2. *Hibiscus laevis*

aa. Annual herbs; leaves deeply 3–5-lobed, the sinuses extending nearly to base of blade, the lobes again lobed; calyx inflated in fruit; petals yellow with dark purple base, 2–4 cm long ..... 3. *Hibiscus trionum*

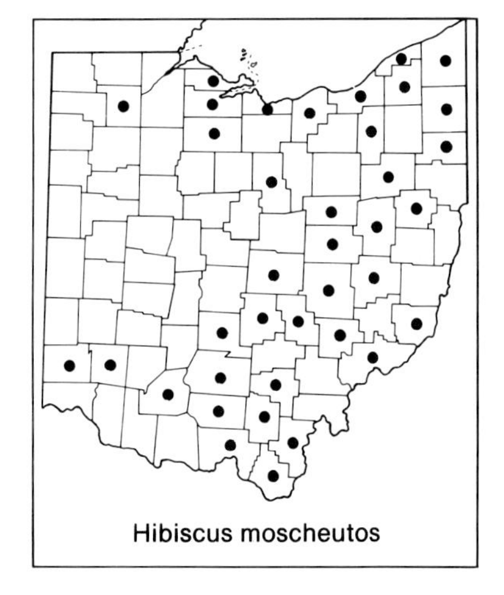
Hibiscus moscheutos

1. **Hibiscus moscheutos** L. Swamp Rose-mallow. Common Rose-mallow

incl. *H. palustris* L.

Coarse, perennial, shrub-like herb, to 2.5 m tall; upper internodes pubescent; leaves with dentate-serrate margins, unlobed or very shallowly lobed, pubescent beneath; flowers solitary in upper axils; bracts 10–14, narrowly linear-lanceolate; corolla white to pink or red, sometimes darker at the center, and large—to 16 cm across, the individual petals 8–12 cm long. Mostly in the eastern half of Ohio; native at some sites, naturalized at others; lake shores, marshy areas, wet ditches, and similar wet open habitats; often cultivated. Additional record: Lucas County (Blanchard, 1976). Late July–September.

When *Hibiscus palustris* L. (*H. moscheutos* subsp. *palustris* (L.) R. T. Clausen) is treated as a distinct taxon, its range is generally said to include

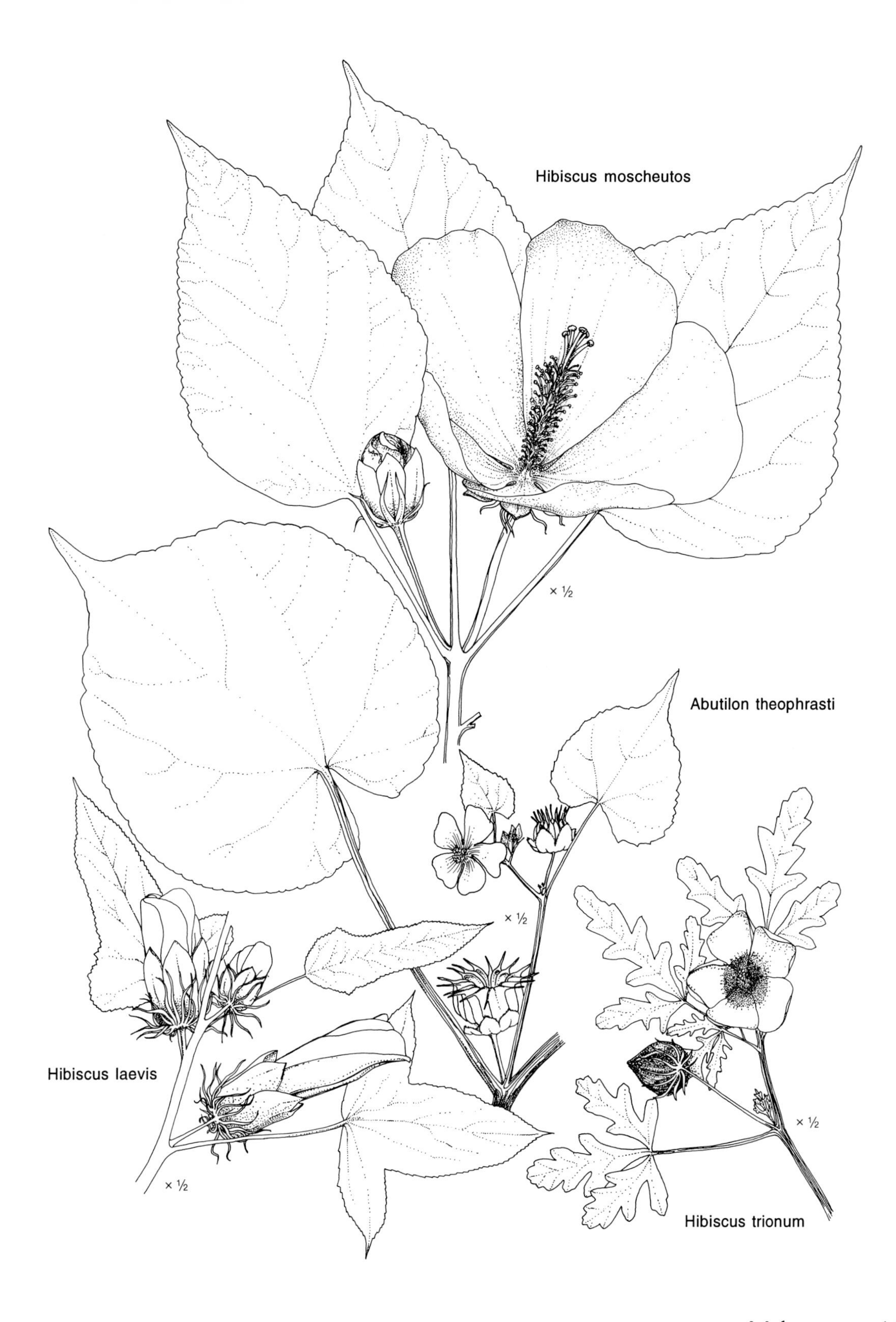
Hibiscus moscheutos
× ½
Abutilon theophrasti
× ½
Hibiscus laevis
× ½
× ½
Hibiscus trionum

northern Ohio, while that of *H. moscheutos* sensu stricto is said to include southern Ohio (Fernald, 1950; Gleason, 1952a). I have found no consistent way to separate them and, following Blanchard (1976) and Gleason and Cronquist (1991), place *H. palustris* in synonymy without rank.

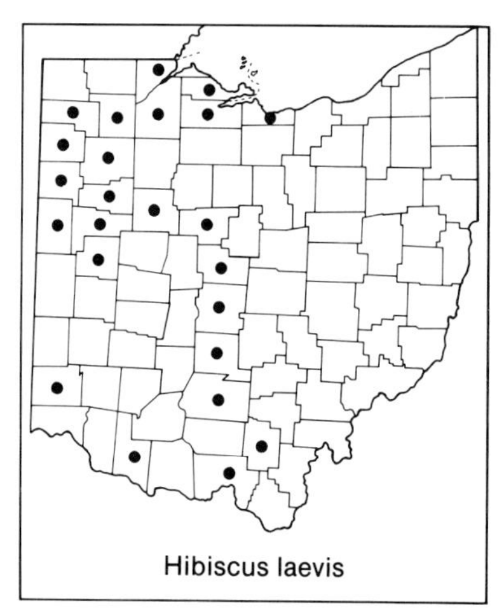

Hibiscus laevis

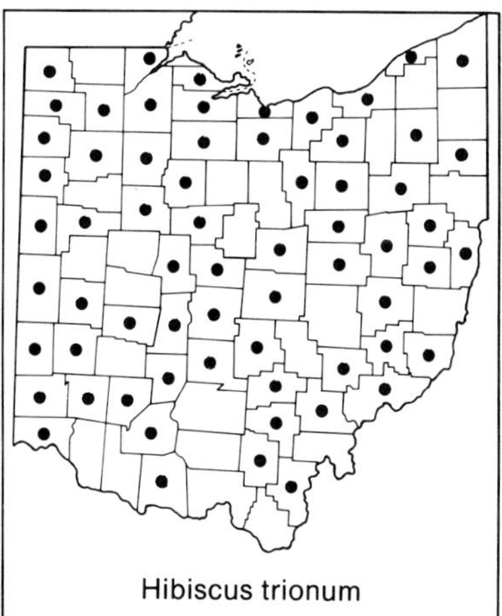

Hibiscus trionum

2. **Hibiscus laevis** All. HALBERD-LEAVED ROSE-MALLOW. SOLDIER ROSE-MALLOW
   H. militaris Cav.

Coarse, perennial, stiffly erect herb, to 2 m tall; upper internodes glabrous; leaves mostly markedly 3-lobed, more or less hastate, the terminal lobe much the longest, the lateral lobes divergent; flowers solitary in upper axils; bracts 10–14, linear to narrowly linear-lanceolate; corolla pink to off-white, sometimes darker at the center, the petals 6–8 cm long. Rather sharply limited to the western half of Ohio; wet woods and marshes. Additional record: Gallia County (Blanchard, 1976). July–September.

In his revision, Blanchard (1976) adopted the name *H. laevis* for this species, which had long been known as *H. militaris*.

3. HIBISCUS TRIONUM L. FLOWER-OF-AN-HOUR

Low, pubescent, more or less erect annual, to 5 dm tall, often branched from near base of main stem; leaves deeply 3–5-lobed, the sinuses extending nearly to center of blade, and the lobes again lobed; flowers short-lived, solitary in leaf axils; bracts 7–12, linear; calyx inflated in fruit; corolla yellow to pale yellow, purple in the center, the petals 2–4 cm long. A Eurasian species, naturalized throughout Ohio; cultivated fields and gardens, along railroads, and in similar disturbed areas; attractive when blooming but in some places an unwelcome weed. June–September.

Clusiaceae or Guttiferae. MANGOSTEEN FAMILY. ST. JOHN'S-WORT FAMILY

Perennial or annual herbs or, less often, shrubs; leaves opposite, simple, punctate; flowers perfect, mostly in cymose inflorescences, sometimes solitary; calyx of 4 or 5 separate, persistent, subequal sepals; corolla actinomorphic, of 4 or 5 separate petals, yellow to pinkish, often showy; stamens 5–many, often slightly fused at base and sometimes organized into 3–5 loose clusters (fascicles); carpels 2–5, fused to form a compound pistil with as many separate or basally fused styles; ovary superior; fruit a capsule.

Ohio's genera, *Hypericum* and *Triadenum*, along with a few additional genera, have sometimes been assigned to a segregate family, the Hypericaceae (Gleason, 1952a; Gleason and Cronquist, 1963). I follow Cronquist (1968, 1981) and Thorne (1968, 1976, 1983) in retaining these genera within the Clusiaceae or Guttiferae sensu lato. Treated thus, it is a family of 50 genera and more than 1,000 species. The family has two scientific names; the use of either is equally correct. W. P. Adams (1973) and Wood and Adams (1976) studied the family in southeastern United States; Gillett and Robson (1981) prepared a treatment of the family as it is represented in the flora of Canada. The text below is based on my recent study (Cooperrider, 1989) of the family in Ohio.

a. Corolla yellow or orangish-yellow; stamens 5–100 or more, separate or slightly connate at base, sometimes arranged in 3–5 clusters; flowers without glands; principal leaves 1 mm–10 cm long, green (rarely reddish), rounded to acuminate at apex . . . . 1. *Hypericum*

aa. Corolla purplish-pink or yellowish-pink; stamens 9, in 3 fascicles of 3 basally connate stamens each; flowers with 3 large orange glands alternating with fascicles of stamens; principal leaves 3–15 cm long, often purplish, mostly obtuse, rounded, or retuse at apex . . . 2. *Triadenum*

1. Hypericum L. ST. JOHN'S-WORT

Perennial or annual herbs or, less often, shrubs; leaves green or rarely reddish, varying from rounded to acuminate at apex, varying from 1 mm to 10 cm in length; corolla yellow or orangish-yellow; stamens 5–100 or more, separate or in 3–5 clusters. A genus of about 300 species, widely distributed around the world. W. P. Adams' (1962) taxonomic study of *Hypericum* section *Myriandra* includes species nos. 4–8 below. Rodríguez Jiménez' (1973) study of *Hypericum* section *Brathys* subsection *Spachium* includes species nos. 9–16 below; D. H. Webb (1980) did biosystematic research on the same set of species in eastern North America. Webb reports several instances of naturally occurring interspecific hybrids among species of this group; as noted below, only one has been found in Ohio: *H. canadense* × *H. mutilum*.

Adopting the conclusions of W. P. Adams and N. B. K. Robson (1961), *Hypericum* is defined here to include the segregate genus *Ascyrum* L.

Hypericum calycinum L., CREEPING ST. JOHN'S-WORT or GOLDFLOWER, an attractive low (to 4 dm tall) shrub from Asia Minor with showy yellow flowers ca. 5 cm across, is sometimes cultivated in Ohio as a ground cover or as a border plant.

a. Sepals 4, the outer pair broad, the inner pair narrow; petals 4; low, scarcely woody shrubs, to 3 dm tall. . . . . . . 8. *Hypericum hypericoides*

aa. Sepals 5, approximately equal in size; petals 5; upright shrubs or perennial or annual herbs.

b. Shrubs, 3–20 dm tall.

c. Styles and locules in ovary 5; principal leaves 1.5–4 (–5) cm long; shrubs 3–6 (–10) dm tall; young branches 4-angled, the angles prominent and all with low wings . . . . . . 4. *Hypericum kalmianum*

cc. Styles and locules in ovary 3 or rarely 4; principal leaves 3–8 cm long; shrubs (3–) 6–20 dm tall; young branches either 2-angled and with low wings, or 4-angled with 2 of the angles prominent and with low wings and 2 angles obscure and lacking wings. . . . . 5. *Hypericum prolificum*

bb. Herbs, 1–10 (–15) dm tall.

c. Flowers ca. 6 cm across; styles 5, locules of ovary 5; capsules 15–30 mm long; plants (7–) 8–15 dm tall; leaves 4–10 cm long . . . . 1. *Hypericum pyramidatum*

cc. Flowers to 2.5 cm across; styles 3, locules of ovary 1–3; capsules 3–12 mm long; plants 1–8 (-10) dm tall; leaves 0.1–4.0 (–7.0) cm long.

d. Leaves linear to narrowly scale-like, 0.3–1.5 mm wide, mostly ascending or subappressed, some spreading.

e. The leaves evident, 5–25 mm long; mature capsules nearly equalling to exceeding sepals.

f. Apex of leaves subacute and bluntish; the leaves 10–30 mm long, some usually spreading, others ascending; mature capsules exceeding sepals. . . . . 14. *Hypericum canadense*

ff. Apex of leaves acuminate and usually with a tiny callous tip; the leaves 5–10 (–15) mm long, mostly ascending; mature capsules nearly equalling or slightly exceeding sepals . . . . 15. *Hypericum drummondii*

ee. The leaves scarcely visible, narrowly scale-like, 1–2 mm long; mature capsules much exceeding sepals. . . . . 16. *Hypericum gentianoides*

dd. Leaves broader, 2–15 (–20) mm wide, spreading.

e. Plants with black dots or lines (as well as translucent dots) on one or more of these parts: upper leaves, upper internodes, sepals, and/or petals.

f. Principal leaves 1–2 (–2.5) cm long, 3–7 (–10) mm wide; upper internodes sharply ridged; styles 4–6 mm long . . . . 2. *Hypericum perforatum*

ff. Principal leaves (2–) 3–7 cm long, (8–) 10–25 mm wide; upper internodes without conspicuous ridges, rounded or nearly so; styles ca. 2 mm long. . . . . 3. *Hypericum punctatum*

ee. Plants with translucent dots only, lacking black dots and black lines.

f. Petals (4–) 5–10 mm long, flowers 8–15 mm across; stamens numerous, 20 or more.

g. Principal leaves 3–7 cm long, linear-lanceolate to elliptic-oblong, and acute to less often obtuse at apex . . . . 7. *Hypericum sphaerocarpum*

gg. Principal leaves 1–3 (–4) cm long, narrowly oblong to elliptic, and acutish, obtuse, or rounded at apex.

h. Leaves mostly obtuse to rounded at apex; styles united at base, the united portion persisting as a beak on capsule; capsules 5–6 mm long . . . . 6. *Hypericum ellipticum*

hh. Leaves acutish at apex; styles separate to base, not persisting as a beak on capsule; capsules 3–5 mm long . . . . 9. *Hypericum denticulatum*

ff. Petals 1–4 (–5) mm long, flowers mostly 3–7 mm across; stamens 5–15 (–20 in *H. canadense* and *H. majus*).

g. Leaves linear to linear-oblanceolate, mostly 1–2 mm wide . . . . 14. *Hypericum canadense*

gg. Leaves lanceolate to elliptic, oblong, ovate, or deltoid-ovate, 3–10 mm wide.

h. The leaves deltoid-ovate, 1.5–2 times as long as wide . . . . 12. *Hypericum gymnanthum*

hh. The leaves elliptic to oblong, ovate, or lanceolate, 2–5 times as long as wide.

i. Leaves mostly lanceolate, 2.0–4.5 cm long, apex of all or at least the upper leaves acute . . . . 13. *Hypericum majus*

ii. Leaves mostly elliptic, oblong, or narrowly ovate, 1–3 (–4) cm long, apex broadly acute to obtuse or rounded.

j. Bracts subtending individual flowers (or ultimate clusters of 3 flowers) leaf-like, 3–5 mm long, 1–2 mm wide, and obtuse or rounded at apex . . . . 10. *Hypericum boreale*

jj. Bracts subtending individual flowers (or ultimate clusters of 3 flowers) subulate, ca. 1 mm long, 0.1–0.2 mm wide, and long acuminate at apex . . . . 11. *Hypericum mutilum*

1. **Hypericum pyramidatum** Ait. Great St. John's-wort

Large perennial, to 1.5 m tall; principal leaves 4–10 cm long; flowers showy, ca. 6 cm across; sepals and petals 5; stamens more than 100, united at base into 5 groups; styles 5; capsules 1.5–3 cm long. A species of northeastern North America, reaching a southern boundary of its range in Ohio; chiefly in northeastern counties; moist thickets and fields, and other open moist habitats. July–August.

Hypericum pyramidatum

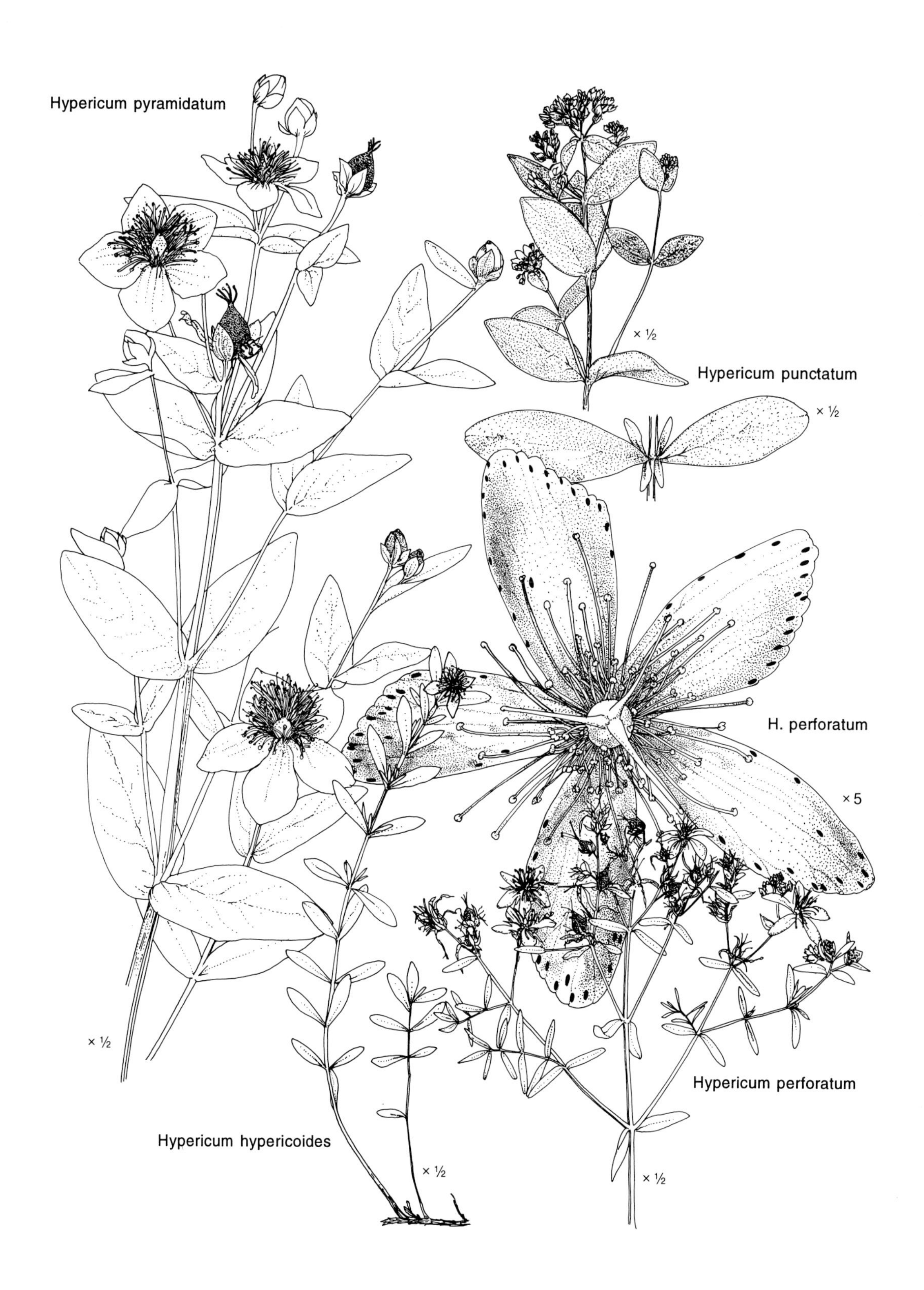
Hypericum pyramidatum
× ½
Hypericum punctatum
× ½
× ½
H. perforatum
× 5
Hypericum perforatum
× ½
Hypericum hypericoides
× ½

This species is merged under *H. ascyron* L., a taxon otherwise native only to eastern Asia, by the Bailey Hortorium Staff (1976), Gillett and Robson (1981), and others. Pending a detailed study of the complex, I continue to treat the American plants as a separate species.

2. Hypericum perforatum L. Common St. John's-wort

Tough perennial, to 8 dm tall; plants with black dots on one or more of these parts: upper internodes, upper leaves, sepals, and/or petals; principal leaves 1–2 (–2.5) cm long; flowers 2–2.5 cm across, in several- to many-flowered, somewhat showy cymes; sepals and petals 5; stamens numerous; styles 3, 4–6 mm long. A Eurasian species, naturalized throughout most of Ohio in a variety of disturbed, usually dry, habitats, especially at roadsides, along railroads, and in parking lots. June–August (–October).

3. **Hypericum punctatum** Lam. Spotted St. John's-wort

Perennial, to 1 m tall; plants with black dots on upper internodes and leaves, the leaves sometimes reddish, more or less long-elliptic to oblanceolate, the principal leaves (2–) 3–7 cm long, showing considerable variation in leaf size between plants; flowers ca. 1 cm across, in dense, several-flowered cymes; sepals and petals 5, streaked with black lines; stamens numerous; styles 3, ca. 2 mm long. Throughout most of Ohio, but scarce or absent in west-central counties; open woods and woodland margins, and in disturbed areas, especially along roadsides. July–August.

4. **Hypericum kalmianum** L. Kalm's St. John's-wort

Small, branched shrub, to 6 (–10) dm (3 feet) tall; young branches prominently 4-angled, the angles with low wings; principal leaves 1.5–4 (–5) cm long; flowers rather showy, 1.5–2.5 cm across, in few-flowered cymes; se-

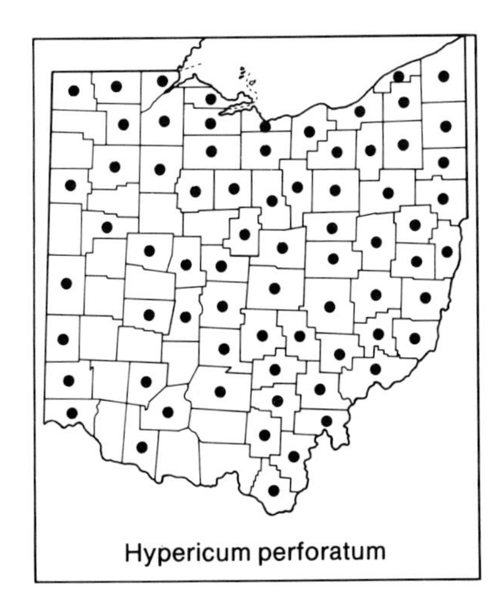
Hypericum perforatum

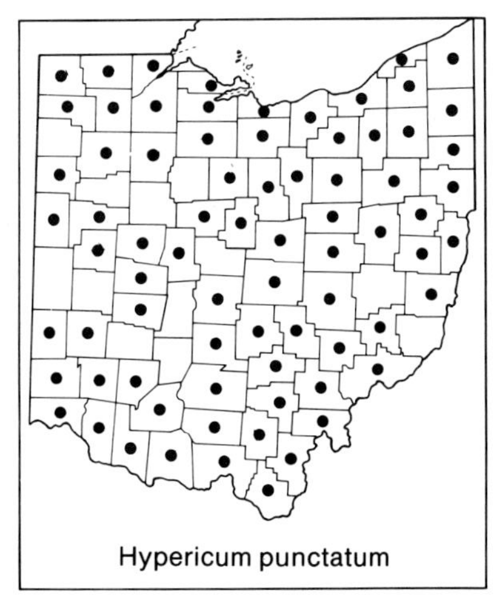
Hypericum punctatum

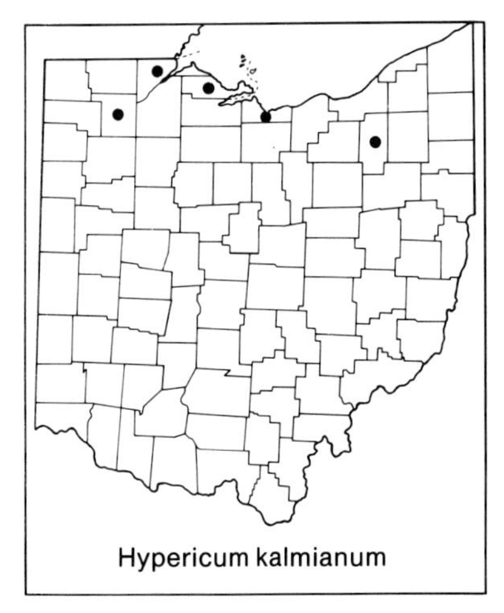
Hypericum kalmianum

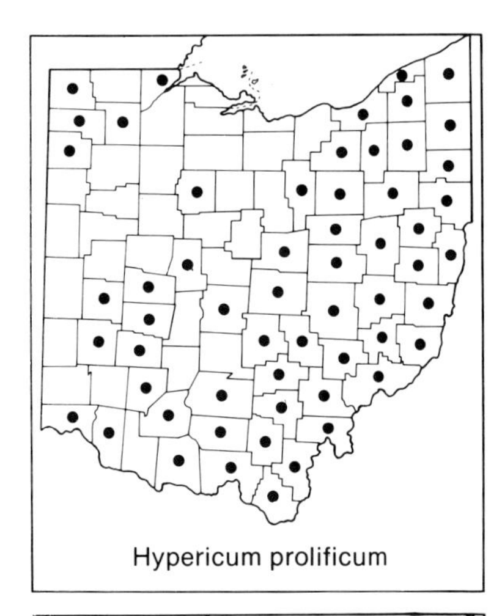
Hypericum prolificum

Hypericum ellipticum

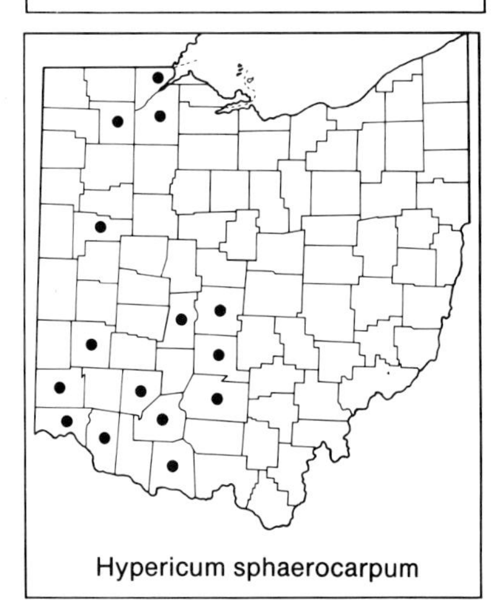
Hypericum sphaerocarpum

pals and petals 5; stamens numerous; styles 5, separate in fruit, but usually not separated in flower. A species of the Great Lakes region (for a distribution map, see Guire and Voss, 1963), at a southern limit of its range in northern Ohio counties; open sandy or gravelly sites chiefly near Lake Erie, and sometimes in moist calcareous inland sites. July–September.

5. **Hypericum prolificum** L. SHRUBBY ST. JOHN'S-WORT
H. spathulatum (Spach) Steud.

Bushy shrub, to 2 m (6.5 feet) tall; young branches mostly 2-angled, each angle with low wings, or if 4-angled then two angles prominent and low-winged and two obscure and lacking wings; principal leaves 3–8 cm long; flowers showy, 1.5–2.5 cm across, in few-flowered cymes; sepals and petals 5; stamens numerous; styles 3, separate in fruit, but usually not separated in flower. Frequent in eastern and southernmost counties, scarce or absent in the till plains of western Ohio; woodland borders and openings, fields, thickets. Late June–early September.

W. P. Adams (1959) established that *H. prolificum* is the correct name for this species.

6. **Hypericum ellipticum** Hook. ELLIPTIC-LEAVED ST. JOHN'S-WORT

Perennial, to 5 dm tall; leaves elliptic to oblong-elliptic, mostly obtuse to rounded at apex, the principal leaves 1.0–3.5 cm long; flowers ca. 1.2 cm across, in few-flowered cymes; sepals and petals 5; stamens numerous; styles 3; capsules 5–6 mm long, beaked. A species of northeastern North America, extending south into the Appalachians and westward into the northeasternmost counties of Ohio; damp shores and fields. June–July.

7. **Hypericum sphaerocarpum** Michx. ROUND-FRUITED ST. JOHN'S-WORT

Perennial, to 6 dm tall; leaves linear-lanceolate to elliptic-oblong, acute to less often obtuse at apex, the principal leaves 3–7 cm long; flowers ca. 8.0 mm across, in many-flowered cymes; sepals and petals 5; stamens numerous; styles 3. A species of the interior United States, at a northeastern boundary of its range in western Ohio; calcareous openings, banks, and prairies. July–September.

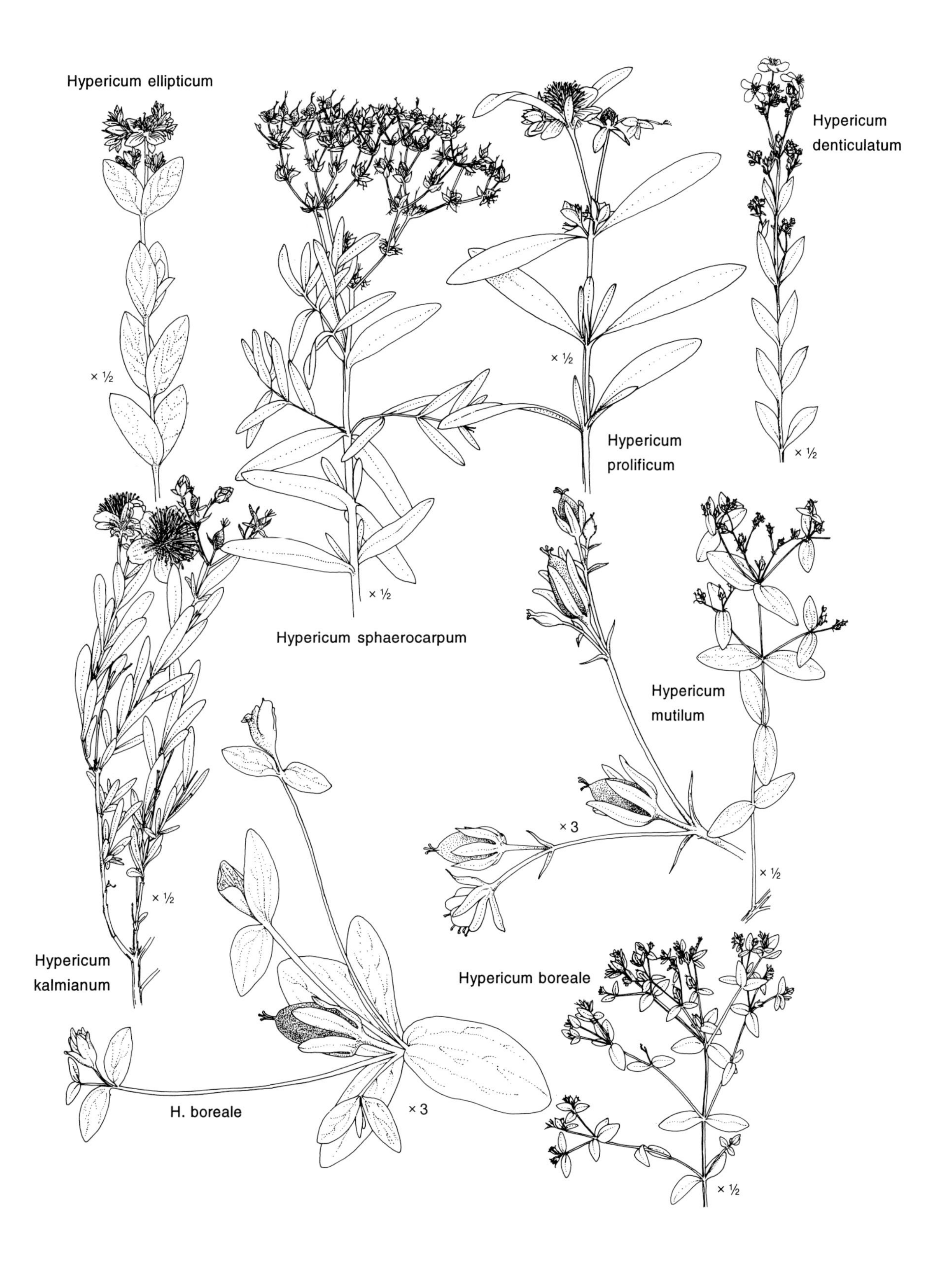
Hypericum ellipticum
× ½
Hypericum denticulatum
× ½
× ½
Hypericum prolificum
× ½
Hypericum sphaerocarpum
Hypericum mutilum
× 3
× ½
× ½
Hypericum kalmianum
Hypericum boreale
H. boreale
× 3
× ½

8. **Hypericum hypericoides** (L.) Crantz St. Andrew's-cross
Low, much-branched, scarcely woody shrub, somewhat sprawling or erect, to ca. 3 dm tall; principal leaves 1.0–2.5 cm long, both surfaces with numerous tiny translucent dots; flowers ca. 1.5 cm across; sepals 4, the outer pair broad, the inner pair narrow; petals 4; stamens numerous; styles 2. A plant of southeastern United States, reaching a northern limit of its range in southern Ohio; dry open woodlands, thickets, and woodland borders. July–August.

Cooperrider (1989) gives details of recent changes in taxonomy centering about this taxon. Ohio plants are assigned to var. *multicaule* (Michx.) Fosberg (*Ascyrum hypericoides* L. var. *multicaule* (Michx.) Fern.; *H. stragulum* W. Adams and N. Robson).

9. **Hypericum denticulatum** Walt. Coppery St. John's-wort
Perennial, to 6 dm tall; leaves oblanceolate to oblong-elliptic, acutish at apex, principal leaves 1–3 (–4) cm long; flowers ca. 1.2 cm across; sepals and petals 5; stamens numerous; styles 3. Variety *denticulatum*, which does not occur in Ohio, is a plant chiefly of the mid-Atlantic seaboard. Variety *acutifolium* (Ell.) S. F. Blake (incl. var. *recognitum* Fern. and Schub.), from more inland regions of southeastern United States, reaches a northern limit of its range in southern Ohio; sandy soils in shaded to open fields. July–August.

The infraspecific classification used here is that of D. H. Webb (1980).

10. **Hypericum boreale** (Britt.) Bickn. Northern St. John's-wort
Much-branched perennial, to 4 dm tall; leaves mostly 1–2 cm long, narrowly elliptic to oblong, rounded to obtuse or broadly acute at apex, bracts

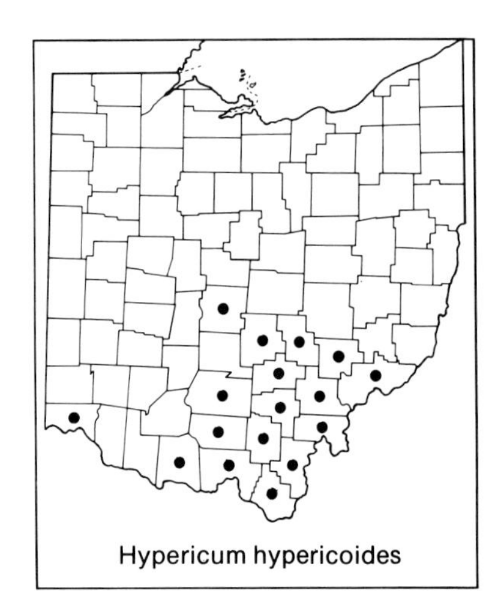

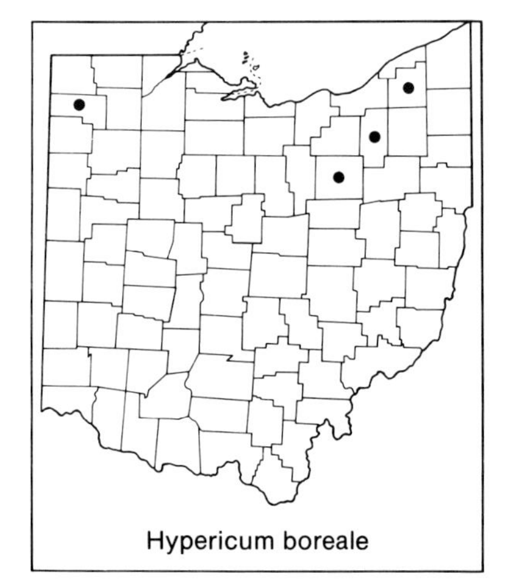

subtending individual flowers (or ultimate clusters of 3 flowers) leaf-like, 3–5 mm long, 1–2 mm wide, and obtuse or rounded at apex; flowers numerous, tiny, 4–5 mm across; sepals and petals 5; stamens 8–15; styles 3. A species chiefly of more northern regions of eastern North America, reaching a southern boundary of its range in northern Ohio; wet shores, bogs, and other open moist habitats. July–August.

J. M. Gillett (1979) merged this species under the next, calling it *H. mutilum* subsp. *boreale* (Britt.) J. M. Gillett. Following D. H. Webb (1980), as well as my own judgment, I retain it at the rank of species.

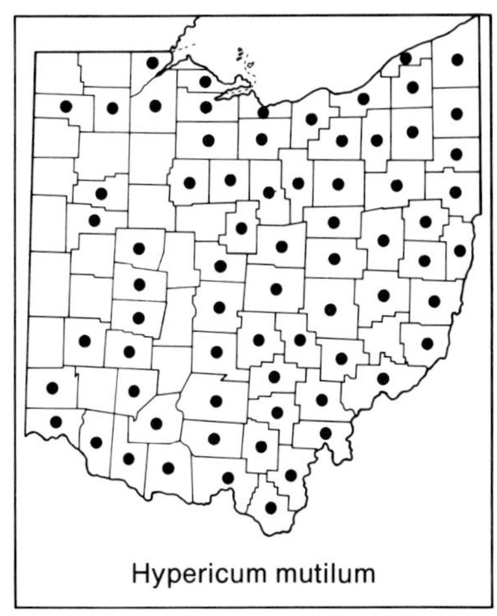
Hypericum mutilum

11. **Hypericum mutilum** L. SMALL-FLOWERED ST. JOHN'S-WORT. DWARF ST. JOHN'S-WORT

incl. var. *mutilum* and var. *parviflorum* (Willd.) Fern.

Perennial or annual, to 6 dm tall; leaves 1–3 (–4) cm long, elliptic to lanceolate-elliptic or narrowly ovate, apex rounded to obtuse or broadly acute; bracts subtending individual flowers (or ultimate clusters of 3 flowers) subulate, ca. 1 mm long, 0.1–0.2 mm wide, and long-acuminate at apex; flowers numerous, tiny, 3–4 mm across; sepals and petals 5; stamens 5–15; styles 3. Throughout much of Ohio, but scarce or absent in westernmost counties; wet banks, shores, ditches, and fields; also in moist shaded gardens and similar moist weedy habitats. July–September.

When the two varieties listed in synonymy above are recognized, most Ohio plants either are or approach var. *parviflorum*, but following D. H. Webb (1980) no infraspecific taxa are recognized here. A few specimens were difficult to separate from *H. majus*.

I have determined one specimen from Erie County and two from Lucas County as *H. canadense* × *H. mutilum*; both counties border Lake Erie. Although D. H. Webb (1980) does not report this hybrid from Ohio, he does report it from seven other eastern states.

Hypericum gymnanthum

12. **Hypericum gymnanthum** Engelm. & A. Gray CLASPING-LEAVED ST. JOHN'S-WORT

Annual, to 7 dm tall; leaves 1.0–2.5 cm long, deltoid-ovate, clasping at base, acute to subacute at apex; flowers ca. 5 mm across; sepals and petals 5; stamens 10–15; styles 3. A species chiefly of the mid-Atlantic states with scattered inland stations, reaching a northern boundary of its range in Ohio; open wet sites. Late June–September.

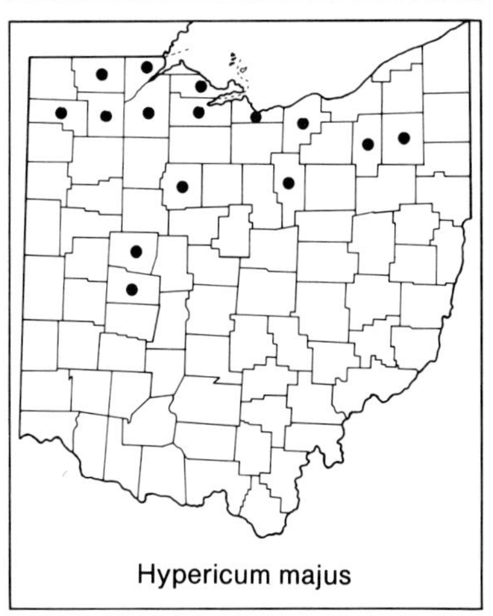
Hypericum majus

13. **Hypericum majus** (A. Gray) Britt. TALL ST. JOHN'S-WORT

Perennial, to 6 dm tall; leaves 2–4.5 cm long, lanceolate to narrowly elliptic, apex of all leaves, or at least the upper, acute; flowers ca. 7 mm across; sepals and petals 5; stamens 15–20; styles 3. A species of northeastern and north-central United States and adjacent Canada, at a southern boundary of its range in the northern half of Ohio; wet ditches, pond margins, dis-

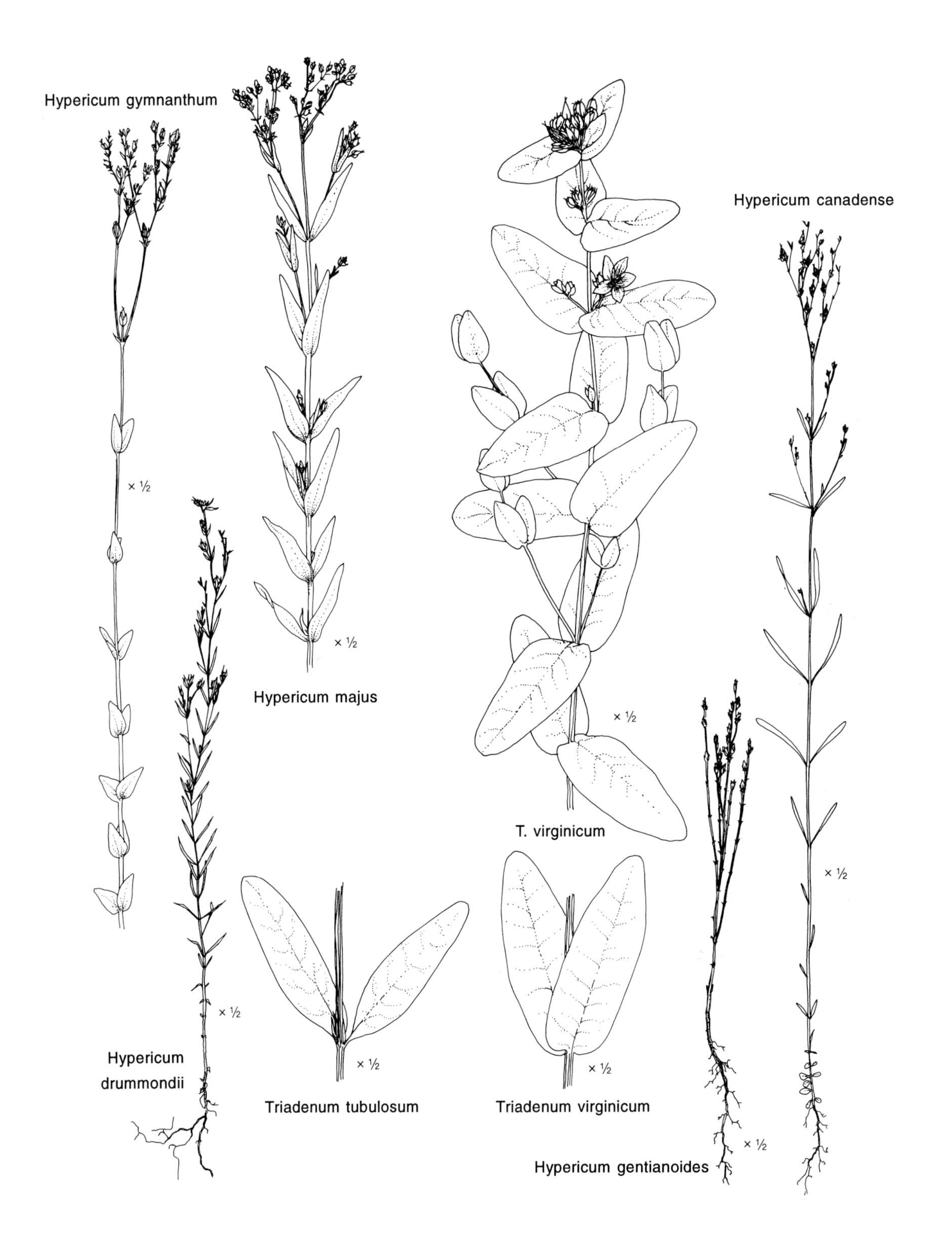
Hypericum gymnanthum
Hypericum canadense
× ½
× ½
Hypericum majus
× ½
T. virginicum
× ½
× ½
Hypericum
drummondii
× ½
Triadenum tubulosum
× ½
Triadenum virginicum
× ½
Hypericum gentianoides
× ½

turbed bogs. Additional record: Montgomery County (D. H. Webb, 1980). July–September.

Some specimens are difficult to separate from *H. mutilum*.

14. **Hypericum canadense** L. CANADIAN ST. JOHN'S-WORT

Annual or perennial, to 4 dm tall; leaves 1–3 cm long, linear to linear-oblanceolate, but subacute and blunt at apex, some leaves usually ascending, others spreading; flowers ca. 3–4 mm across; sepals and petals 5; stamens ca. 15–20; styles 3; capsules exceeding the calyx. At a few scattered sites in Ohio; open moist fields, marsh and lake margins. July–September.

See note under *H. mutilum*, above, regarding hybrids between that species and this.

Hypericum canadense

15. **Hypericum drummondii** (Grev. & Hook.) T. & G. NITS-AND-LICE

Annual, few- to much-branched, to 5 dm tall; leaves 5–10 (–15) mm long, linear, acuminate and usually with a tiny callous tip, mostly ascending; flowers ca. 8 mm across; sepals and petals 5; stamens 10–20; styles 3; capsules nearly equalling or only slightly exceeding the calyx. A species of southeastern and south-central United States, reaching a northern boundary of its range in southern and eastern counties of Ohio; dry open ground, fallow fields, abandoned roads. Additional record: Warren County (D. H. Webb, 1980). July–September.

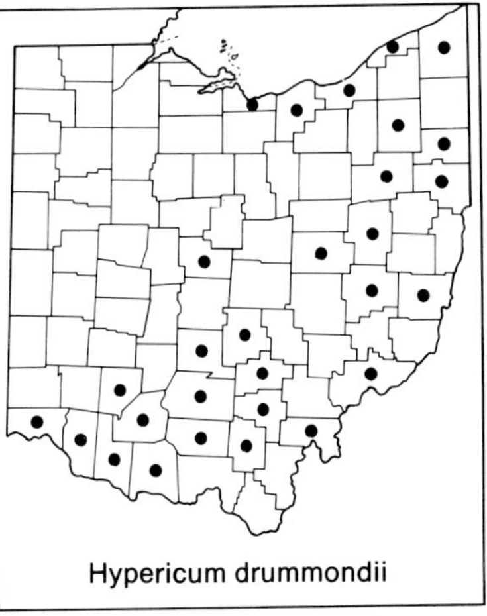
Hypericum drummondii

16. **Hypericum gentianoides** (L.) BSP. ORANGE-GRASS. PINEWEED

Annual, few- to much-branched, to 3 dm tall; leaves minute, only 1–2 mm long, narrow and scale-like and often scarcely visible; flowers ca. 3–4 mm across; sepals and petals 5; stamens 5–10; styles 3; mature capsules much exceeding calyx. South-central counties and at scattered sites northward, where the species is at a boundary of its range; open sandy ground, dry fields, and other dry disturbed sites. July–September.

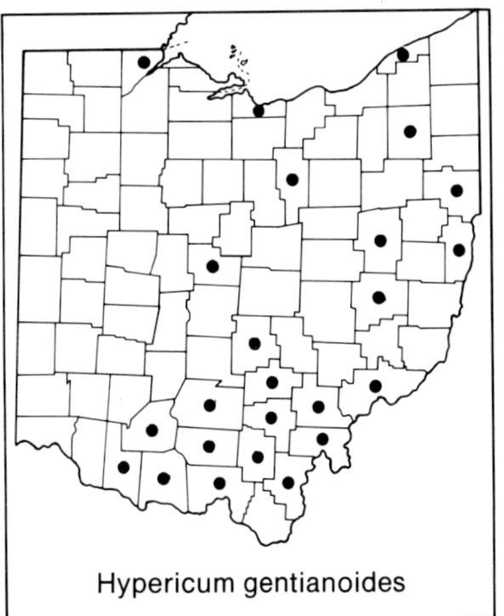
Hypericum gentianoides

2. Triadenum Raf. MARSH ST. JOHN'S-WORT

Perennial herbs; leaves green or often purplish-tinged, mostly obtuse, rounded, or retuse at apex, the principal leaves 3–15 cm long; sepals and petals 5; corolla purplish-pink to yellowish-pink; stamens 9, in 3 fascicles of 3 stamens each, the fascicles alternating with 3 large, orange glands; carpels and styles 3. A small North American genus consisting of only the four taxa below. As indicated by her annotations of specimens at The Ohio State University Herbarium, Susan J. Meades found all four to be a part of the Ohio flora.

Gleason (1947) presents reasons for recognizing this as a genus distinct from *Hypericum*.

a. Leaves sessile, cordate, and more or less clasping at base . . . . 1. *Triadenum virginicum*

aa. Leaves, at least the lower, narrowed to base, either sessile or petiolate . . . . 2. *Triadenum tubulosum*

1. **Triadenum virginicum** (L.) Raf. MARSH ST. JOHN'S-WORT

   Hypericum virginicum L.

Perennial, to 6 dm tall; leaves sessile, cordate and more or less clasping. In northeastern counties and scattered elsewhere; bogs, marshes, wet open woodlands, and similar moist habitats. July–early September.

Ohio plants may be separated into the following two varieties, which are often treated at the rank of species.

a. Sepals mostly 5–9 mm long, acute to acuminate at apex; styles 2–3 mm long . . . . var. *virginicum*

   Hypericum virginicum var. *virginicum*

aa. Sepals mostly 3–5 mm long, obtuse to rounded at apex; styles 0.5–1.5 mm long . . . . var. *fraseri* (Spach) Cooperr.

   Triadenum fraseri (Spach) Gleason; *Hypericum virginicum* var. *fraseri* (Spach) Fern.

2. **Triadenum tubulosum** (Walt.) Gleason LARGE MARSH ST. JOHN'S-WORT

   Hypericum tubulosum Walt.

Perennial, to 6 dm tall; leaves, at least the lower, narrowed at base, either sessile or petiolate. Scattered sites in eastern, especially southeastern, counties; marshes, swamp borders, and other open wet places. July–September.

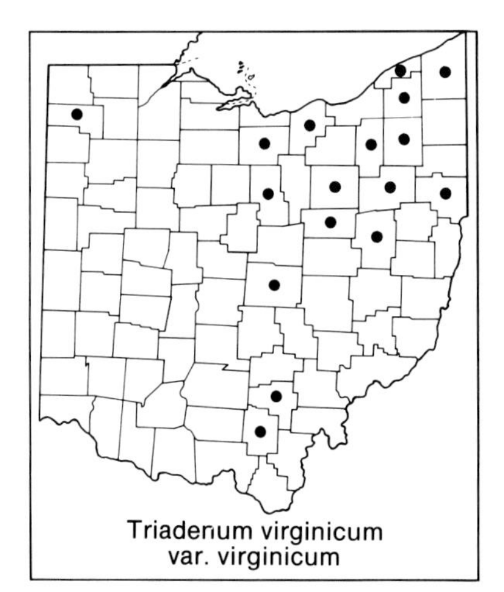

Triadenum virginicum
var. virginicum

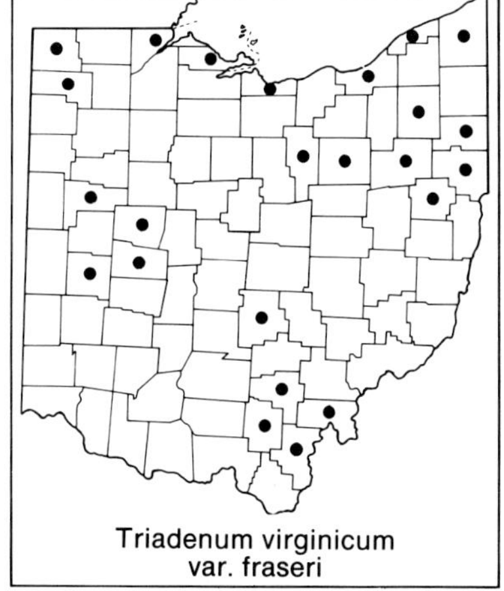

Triadenum virginicum
var. fraseri

Ohio plants may be separated into the following two varieties, which are sometimes treated at the rank of species.

a. Lower leaves sessile, with few dots on undersurface; sepals acute at apex. . . . . var. *tubulosum*
   Hypericum tubulosum var. *tubulosum*

aa. Lower leaves petiolate, with numerous dots on undersurface; sepals obtuse at apex . . . . var. *walteri* (Gmel.) Cooperr.
   Triadenum walteri (Gmel.) Gleason; *Hypericum tubulosum* var. *walteri* (Gmel.) Lott

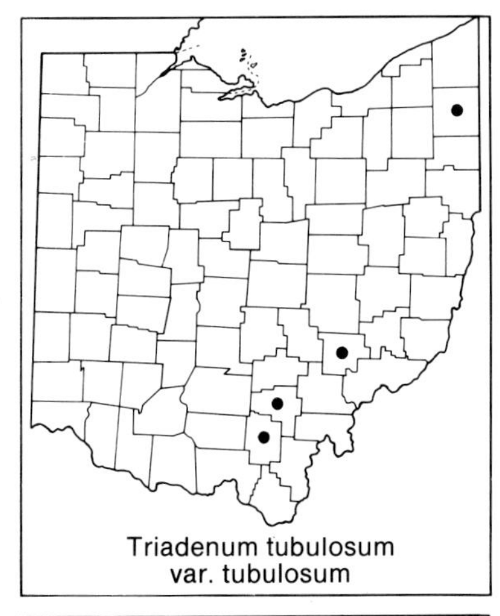

Triadenum tubulosum var. tubulosum

Triadenum tubulosum var. walteri

Elatine triandra

Elatinaceae. WATERWORT FAMILY

A very small family of two genera. G. C. Tucker (1986) made a study of the family in southeastern United States.

1. Elatine L. WATERWORT

A genus of about 20 species, widely distributed in aquatic or wet terrestrial habitats, chiefly in the Northern Hemisphere. A single species has been collected in Ohio. W. H. Duncan (1964) discusses the genus as it is represented in the flora of southeastern United States.

1. **Elatine triandra** Schkuhr WATERWORT

Small, delicate, creeping, branched annual; leaves opposite, simple; flowers quite inconspicuous, solitary or in pairs in leaf axils; flowers perfect; calyx of 3 separate sepals; corolla actinomorphic, of 3 separate pinkish petals; stamens 3; carpels 3, fused to form a compound pistil with 3 separate short styles; ovary superior; fruit a capsule. July–September.

The only known Ohio collections of this species were made by Floyd Bartley and Leslie Pontius, in November 1935, from the margin of a small pond in Pickaway Township, Pickaway County. The identity of one of their specimens was confirmed by Fassett (1939) and that of another by Robert H. Read, a current student of the genus. Although Fernald (1950) accepted the Ohio record as a native occurrence, Stuckey and Roberts (1977) treated the species as an alien member of the Ohio flora. Awaiting a modern distributional study, I tentatively follow Fernald's conclusion and treat it as a rare native species, perhaps often overlooked because of its tiny size and inconspicuous appearance.

Ohio plants belong to var. *brachysperma* (A. Gray) Fassett (*E. brachysperma* A. Gray).

Tamaricaceae. TAMARISK FAMILY

A few collections of *Tamarix chinensis* Lour., CHINESE TAMARISK, presumably from adventive individuals, have been made at sites near Lake Erie, in Ottawa County. The plants are shrubs or small trees with tiny scale-like leaves on pendulous branches. Baum (1967) discussed the introduced and naturalized tamarisks of the United States and Canada.

Cistaceae. ROCK-ROSE FAMILY

Herbs or shrubs; leaves small, simple, opposite or alternate; plants sometimes producing cleistogamous as well as chasmogamous flowers, the chasmogamous flowers small, perfect, solitary or in cymose or paniculate inflorescences; calyx of 3–5 separate sepals; corolla actinomorphic, of 3–5 separate petals; stamens 3–many; carpels 3, fused to form a compound pistil with a single style; ovary superior; fruit a small capsule, either dull and leathery or shining and bony. A small but widespread family, chiefly of the Northern Hemisphere. Brizicky (1964b) studied the family in southeastern United States. The text below is based in part on O'Connor and Blackwell's (1974) treatment of the family in Ohio.

a. Plants low, spreading, bushy shrubs; leaves scale-like, broadest at base, 1–4 mm long, 1 mm wide or less, appressed, densely gray-pubescent . . . . 2. *Hudsonia*

aa. Plants erect perennials; leaves linear to oblanceolate or elliptic, mostly tapering to a narrowed base, (5–) 10–30 mm long, 1–7 mm wide, spreading, pubescent to glabrous.

b. Petals either 5, yellow, and falling early, or petals none in late flowers; style short, stigma capitate; plants without overwintering basal shoots . . . . 1. *Helianthemum*

bb. Petals 3, reddish, persisting after withering; style none, stigma 3-branched; plants often with overwintering basal shoots . . . . 3. *Lechea*

1. Helianthemum Mill. FROSTWEED. ROCK-ROSE

Perennial herbs; stems and leaves with all or some of the pubescence stellate; leaves alternate, simple, more or less elliptic; chasmogamous flowers with 5 sepals, the 2 outer narrow and the 3 inner broad; petals 5, or none in late flowers. A widespread genus of more than 100 species. The most recent revision of the North American taxa is that of Daoud and Wilbur (1965).

a. Upper surface of leaves with numerous short stellate hairs and several long simple hairs; chasmogamous flowers usually solitary or sometimes in clusters of 2 or 3, their peduncles with both stellate pubescence and conspicuous, simple, diverging hairs 0.5–1 mm long . . . . 1. *Helianthemum canadense*

aa. Upper surface of leaves with numerous short stellate hairs only; chasmogamous flowers in clusters of (2–) 6–10, their peduncles with short stellate pubescence only, the hairs less than 0.5 mm long . . . . 2. *Helianthemum bicknellii*

1. **Helianthemum canadense** (L.) Michx. CANADA FROSTWEED

Perennial, to 5 dm tall; upper leaf surface with numerous short stellate hairs and also several long simple hairs; chasmogamous flowers usually solitary or sometimes in clusters of 2 or 3, their peduncles with both stellate pubescence and conspicuous, simple diverging hairs 0.5–1.0 mm long. In Ohio, known mostly from northern counties; open sand and other open dry habitats. May–June.

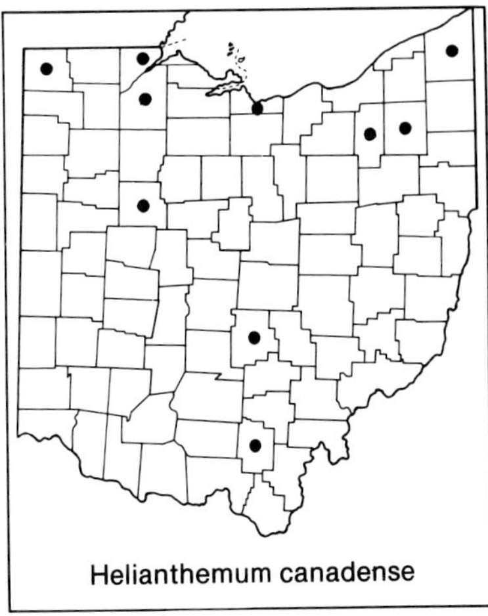
Helianthemum canadense

2. **Helianthemum bicknellii** Fern. PLAINS FROSTWEED

Perennial, to 5 dm tall; upper leaf surface with numerous short stellate hairs only; chasmogamous flowers in clusters of (2–) 6–10, their peduncles with short stellate pubescence only, the hairs less than 0.5 mm long. At scattered Ohio sites, mostly in northern counties; open sandy soil and other open dry habitats. June–July.

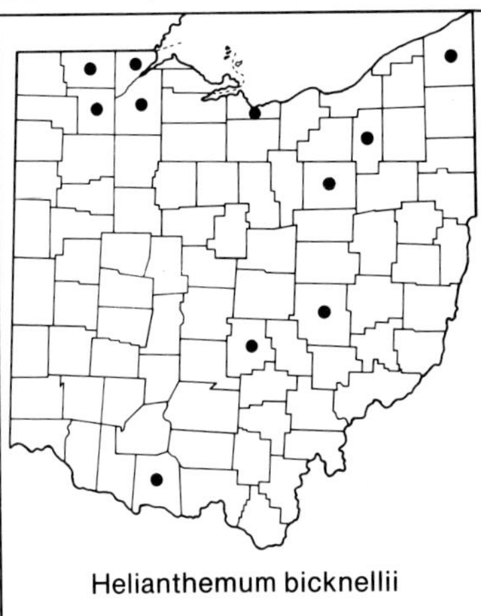
Helianthemum bicknellii

2. Hudsonia L. BEACH-HEATHER. FALSE HEATHER

A small genus of eastern North America. Fernald (1950), Gleason (1952a), and other authors recognize three species within the genus, only one of which, *H. tomentosa*, is found in Ohio. Skog and Nickerson (1972) treat all plants of the genus as members of a single species, *Hudsonia ericoides* L. sensu lato, composed of four subspecies.

1. **Hudsonia tomentosa** Nutt. BEACH-HEATHER. POVERTY-GRASS

H. ericoides subsp. *tomentosa* (Nutt.) Nickerson & J. Skog

Low, spreading, bushy shrub, to ca. 2 dm tall; young stems and leaves gray-pubescent; leaves minute, scale-like and appressed, broadest at base, 1–4 mm long, 1 mm wide or less; sepals 3; petals 5. Dry, open, sandy sites; known from two northwestern counties. May–July.

Hudsonia tomentosa

3. Lechea L. PINWEED

Perennial herbs with all pubescence of simple hairs; leaves simple, the cauline leaves mostly alternate, small, and narrowly elliptic; chasmogamous flowers with 5 sepals, the 2 outer narrow and sometimes inconspicuous, the 3 inner broad; petals 3. A North American genus of 17 species. The most recent comprehensive revision is that of Hodgdon (1938). Wilbur and Daoud (1961) studied the taxa of this genus in southeastern United States.

Lechea stricta Britt., Prairie Pinweed, attributed to Ohio on the basis of misidentified specimens of *L. racemulosa*, was deleted from the flora by Roberts and Cooperrider (1982).

a. Stems with most pubescence divergent or spreading; inner and outer sepals approximately equal in length, the inner conspicuously keeled; capsules globose or subglobose . . . . 1. *Lechea villosa*

aa. Stems with pubescence appressed or closely ascending; inner and outer sepals generally of different lengths, neither keeled; capsules globose or otherwise.

b. The narrow outer sepals usually longer than the broad inner sepals, rarely a few flowers with the outer sepals merely equalling the inner.

c. Principal leaves of flowering stems narrowly elliptic to lanceolate, 3–9 mm long, 1–3 mm wide, most about 3 times as long as wide; capsule and inner sepals (considered together) ellipsoid to obovate, longer than wide . . . . 2. *Lechea minor*

cc. Principal leaves of flowering stems linear, 8–20 mm long, 1 (–1.5) mm wide, most about 10 times as long as wide; capsule and inner sepals (considered together) subglobose, about as long as wide . . . . 3. *Lechea tenuifolia*

bb. The narrow outer sepals mostly half to three-fourths as long as the broad inner sepals, occasionally a few flowers with the outer sepals nearly or approximately equalling the inner.

c. Capsule and inner sepals (considered together) narrowly ellipsoid to obpyriform, about twice as long as wide; mature fruiting calyx brown above, tan at base, the outer sepals about 3/4 as long as to nearly equalling the inner; leaves of basal shoots (when present) about 2–3 times as long as wide . . . . 4. *Lechea racemulosa*

cc. Capsule and inner sepals (considered together) subglobose or obovoid, less than twice as long as wide, sometimes wider than long; mature fruiting calyx medium to light brown, the color uniform throughout, the outer sepals at most half as long as the inner; leaves of basal shoots (when present) about 4–6 times as long as wide.

d. Capsules obovoid, tapering to a narrowed base; seeds mostly 2–3 per capsule. . . . . 5. *Lechea leggettii*

dd. Capsules more or less globose, rounded at base; seeds 4–6 per capsule . . . . 6. *Lechea intermedia*

Helianthemum canadense
Helianthemum bicknellii
× ½
× ½
Lechea villosa
× 10
× ½
× 5
Hudsonia tomentosa
× 1
Elatine triandra

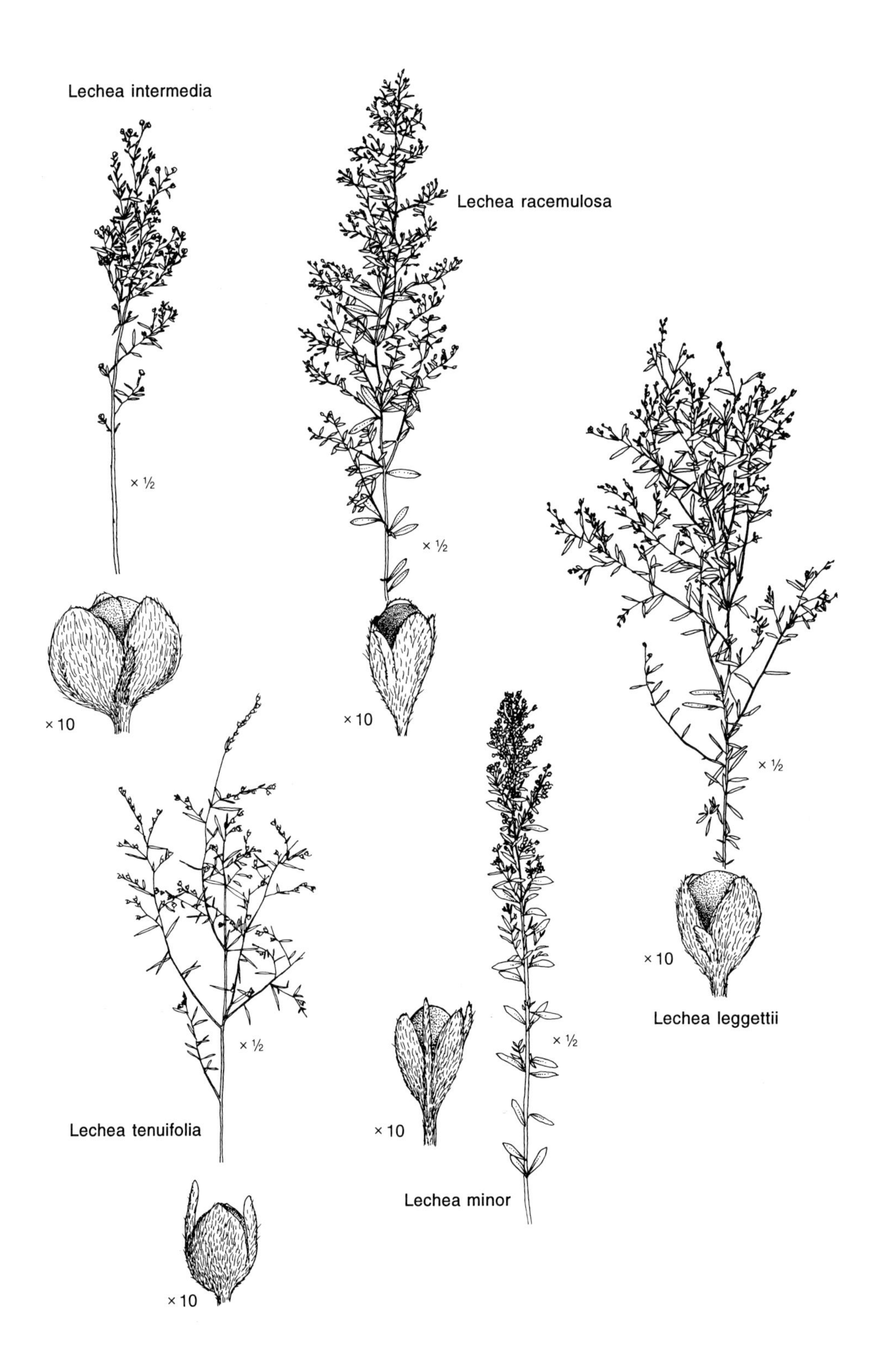
Lechea intermedia
Lechea racemulosa
× ½
× ½
× 10
× 10
× ½
× 10
Lechea leggettii
× ½
× ½
Lechea tenuifolia
× 10
Lechea minor
× 10

1. **Lechea villosa** Ell. HAIRY PINWEED
(?*L. mucronata* Raf.)
Perennial, to 8 dm tall; stems pubescent with divergent or spreading hairs; the narrow outer sepals and broader inner sepals approximately equal in length, the inner conspicuously keeled; capsules globose or subglobose. Scattered sites, mostly in northern counties; dry woodland openings and borders, sandy fields. July–August.

Lechea mucronata Raf. antedates *L. villosa* Ell. and would therefore be the correct name if, as Wilbur (1966) claims, the name *L. mucronata* could be clearly established as applying to this species. Hodgdon (1938), studying the same situation, found Rafinesque's description too vague to be applied with certainty and rejected *L. mucronata* as a name for this species.

2. **Lechea minor** L. THYME-LEAVED PINWEED
Perennial, to 5 dm tall; stems pubescent with subappressed to slightly divergent hairs; principal leaves of flowering stems narrowly elliptic to lanceolate, 3–9 mm long, 1–3 mm wide, most about 3 times as long as wide; the narrow outer sepals equalling to more often longer than the broader inner sepals; capsule and inner sepals (considered together) ellipsoid to obovate, longer than wide. Known from a few scattered localities, mostly in northwestern counties; open, usually dry, sandy habitats. July–August.

Reveal (1986) concluded that the correct name for this species is *L. thymifolia* Michx. Pending further study of the situation, I continue to call it *L. minor*, as do Gleason and Cronquist (1991).

3. **Lechea tenuifolia** Michx. NARROW-LEAVED PINWEED
Perennial, to 4 dm tall; stems pubescent with appressed hairs; principal leaves of flowering stems linear, 8–20 mm long, 1 (–1.5) mm wide, most

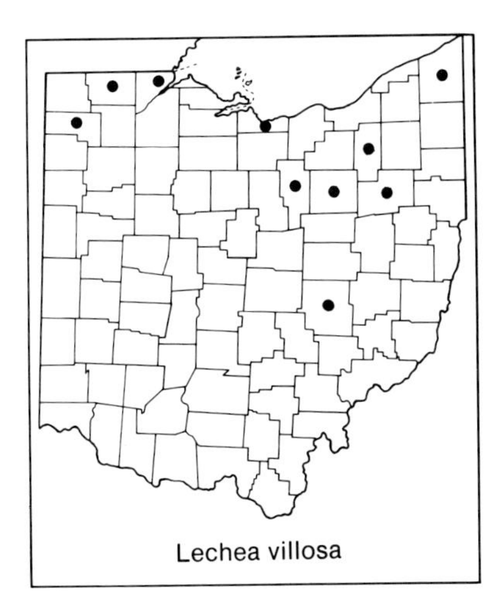
Lechea villosa

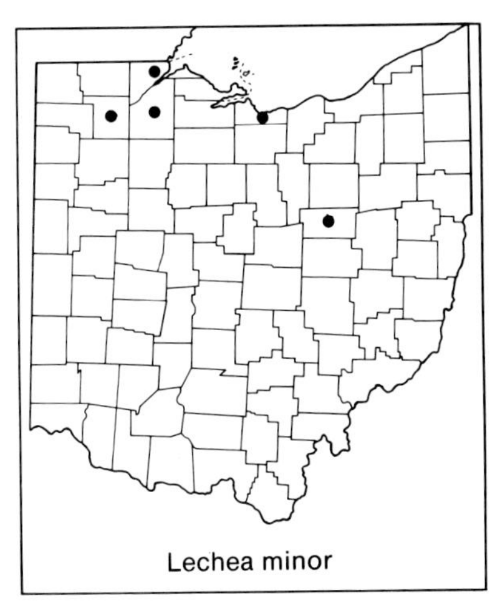
Lechea minor

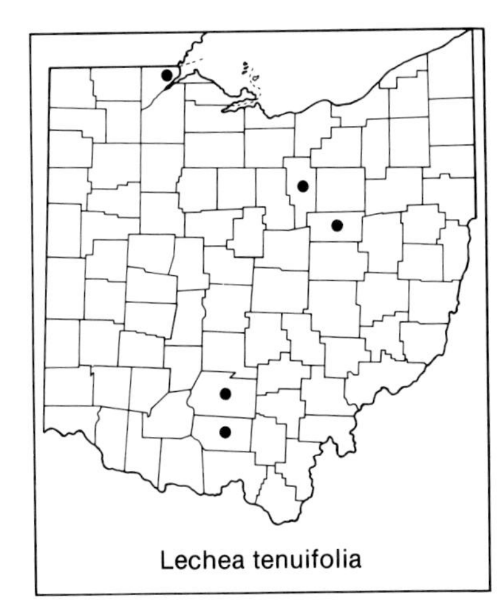
Lechea tenuifolia

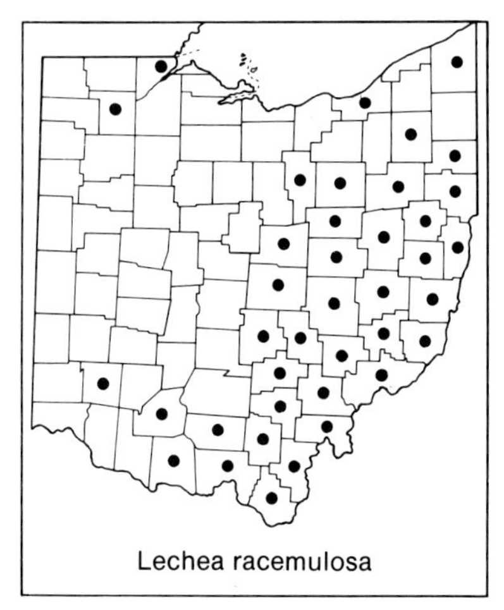
Lechea racemulosa

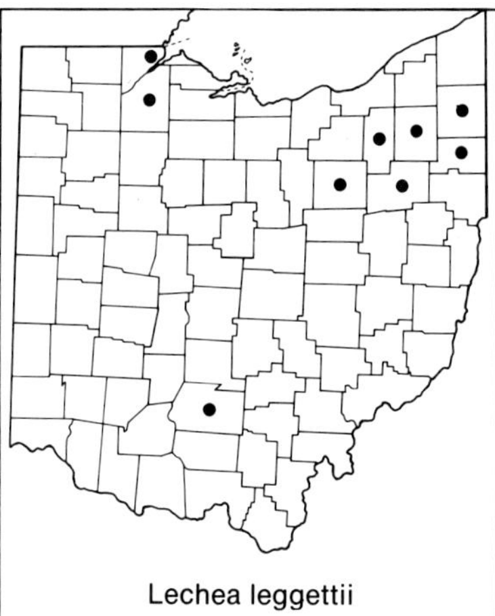
Lechea leggettii

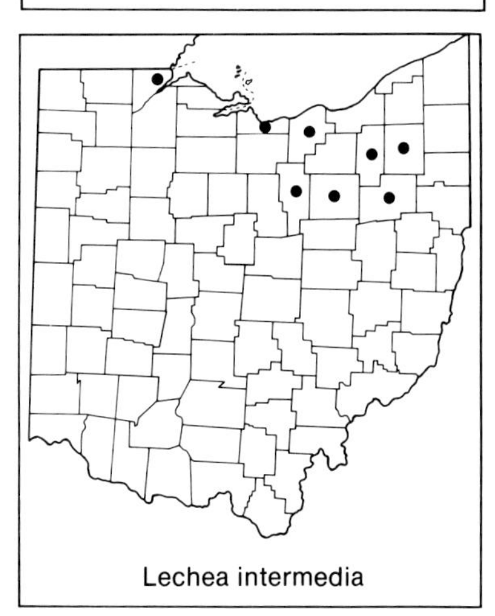
Lechea intermedia

about 10 times as long as wide; the narrow outer sepals equalling to more often longer than the broader inner sepals; capsule and inner sepals (considered together) subglobose, about as long as wide. Known in Ohio from a few scattered sites; dry open woods and fields. July–August.

4. **Lechea racemulosa** Michx. OBLONG-FRUITED PINWEED
Perennial, to 4 dm tall; stems pubescent with subappressed hairs; the narrow outer sepals about 3/4 as long as to nearly equalling length of broader inner sepals; capsule and inner sepals (considered together) narrowly ellipsoid to obpyriform, about twice as long as wide; mature fruiting calyx brown above, tan at base. Throughout the unglaciated counties of east-central and southeastern Ohio and at scattered sites elsewhere in the state; dry open woodlands, dry slopes, and dry roadside banks. July–August.

5. **Lechea leggettii** Britt. & Hollick LEGGETT'S PINWEED
incl. var. *moniliformis* (Bickn.) Hodgdon (?*L. pulchella* Raf.)
Perennial, to 5 dm tall; stems pubescent with appressed hairs; the narrow outer sepals about half (or less than half) as long as the broader inner sepals; capsule obovoid, tapering to a narrowed base, but less than twice as long as wide; fruiting calyx medium to dark brown, uniform in color throughout; seeds mostly 2–3 per capsule. Known from a few scattered sites in Ohio, mostly in northern counties; dry to less often moist open habitats, sandy woodland borders and openings, stream and pond margins. July–August.

In Hodgdon's (1938) treatment of this species, two varieties occur in Ohio, var. *leggettii* and var. *moniliformis*. Gleason (1952a) notes that the two intergrade and have the same range. I have found them scarcely if at all separable in Ohio, and treat all our plants as var. *leggettii* sensu lato.

Wilbur (1966) calls this species *L. pulchella* Raf., a name which antedates *L. leggettii* Britt. & Hollick, but the case he makes for Rafinesque's name applying to this species seems to me weak, and I continue to use *L. leggettii* (cf. discussion under *L. villosa* above).

6. **Lechea intermedia** Britt. ROUND-FRUITED PINWEED
Perennial, to 6 dm tall; stems pubescent with appressed hairs; the narrow outer sepals about half (or less than half) as long as the broader inner sepals; capsule more or less globose, rounded at base; fruiting calyx medium to dark brown, uniform in color throughout; seeds 4–6 per capsule. Northern counties; dry fields and slopes, sandy fields. July–August.

Violaceae. VIOLET FAMILY

Herbs; leaves simple, sometimes deeply lobed (rarely trifoliolate), alternate, often basal; flowers perfect, solitary or in small clusters; calyx of 5 separate persistent sepals; corolla often showy, slightly to markedly zygomorphic, with 5 separate petals, the lowest usually spurred, sometimes conspicuously so; stamens 5, with short flat filaments; carpels 3, fused to form a compound pistil with a single style; ovary superior; fruit a capsule. A cosmopolitan family of about 800 species. Important past studies of the family include those for southeastern United States (Brizicky, 1961b) and western Pennsylvania (Henry, 1953). The treatment here is based in large part on the study of the family in Ohio, prepared by L. D. Miller (1976).

a. Perennials with erect, leafy stems; corolla greenish-white, very small, 3–4 mm wide, slightly zygomorphic, the lowest petal slightly enlarged and saccate at base; sepals without auricles at base .... 1. *Hybanthus*

aa. Perennials or annuals with or without erect leafy stems; corolla variously blue, purple, white, and/or yellow, but not greenish-white, larger, 5–70 mm wide, markedly zygomorphic, the lowest petal distinctly enlarged and spurred; sepals with auricles at base, but these sometimes minute.. 2. *Viola*

1. Hybanthus Jacq. GREEN VIOLET

A chiefly tropical genus of about 80 species. The one species occurring in Ohio is segregated into the monotypic genus *Cubelium* Raf. by Gleason (1952a) and Gleason and Cronquist (1963).

1. **Hybanthus concolor** (T. F. Forst.) Spreng. GREEN VIOLET
   Cubelium concolor (T. F. Forst.) Raf.

Perennial, to ca. 1 m tall; leaves elliptic to obovate, but long-tapering at both apex and base, usually entire, occasionally with a few irregular teeth; flowers solitary or few in axillary clusters, very small, 3–4 mm wide; sepals without auricles; corolla greenish-white, slightly zygomorphic, the lowest petal slightly enlarged and saccate at base; capsules large, 1.5–2.0 cm long. Frequent in southern counties, scattered elsewhere in Ohio; rich woodlands. Late April–May.

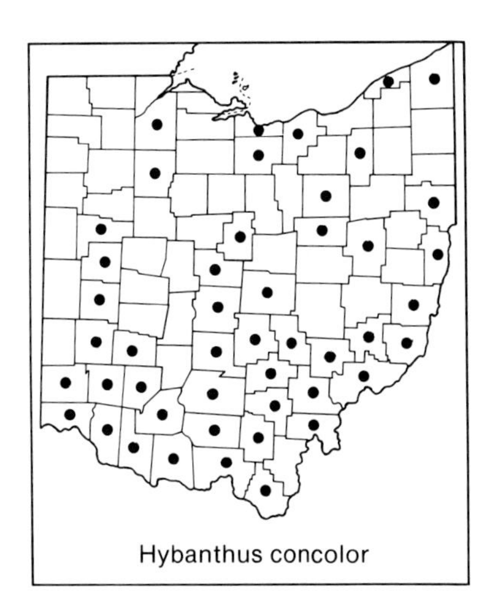
Hybanthus concolor

2. Viola L. VIOLET

Perennial or annual herbs; leaves either basal and originating from a rhizome (the plants called "acaulescent"), or cauline and originating from an aerial stem (the plants called "caulescent"); chasmogamous flowers usually small, 5–25 mm across, and somewhat showy; sepals with auricles, but these sometimes small or minute; corolla variously blue or purple to white and/or yellow, zygomorphic, with 2 upper petals, 2 lateral petals, and a lowest petal that is more or less conspicuously spurred, many species also producing rather inconspicuous, apetalous, cleistogamous flowers late in the season; capsules 4–15 mm long. A widely distributed genus of about 500 species. V. B. Baird (1942) and Klaber (1976) have published illustrated accounts of American violets. Grover (1939) has notes on the violets of Ohio. Russell's (1965) major study of the violets of central and eastern United States, as well as the work by L. D. Miller (1976), noted in the family commentary above, were used in preparing the text for this genus. Landon E. McKinney kindly provided me with an early draft of his manuscript on aculescent blue-flowered violets. Harvey E. Ballard, Jr., checked the identity of several violet hybrids and provided me with a draft of his key to the violets of eastern North America.

Many of Ohio's wild violets, especially *Viola sororia*, are often casually cultivated. *Viola* × *wittrockiana* Gams, Garden Pansy, with large flowers in a variety of colors and color combinations, is more intentionally cultivated; adventive escapes of this hybrid have been collected from Ashtabula, Highland, and Lucas counties. Also cultivated and also with a variety of bright-colored forms is the bedding pansy *Viola cornuta* L., Horned Violet or Viola. *Viola villosa* Walt., Southern Woolly Violet, was attributed to Ohio by Russell (1965), but excluded by L. D. Miller (1976), who determined the specimens in question to be young or depauperate plants of *V. fimbriatula* and *V. affinis*.

a. Plants acaulescent, the stalks of leaves and flowers arising from underground rhizomes, a few species also stoloniferous, the stolons sometimes with 1–few leaves.
 b. Corolla yellow or white.
  c. Corolla yellow; leaves at first inrolled or folded, later expanding and becoming ovate to orbicular, to 13 cm long and to 10 cm wide . . . . 17. *Viola rotundifolia*
  cc. Corolla white; leaves with various shapes.
   d. Most leaves 3–5-lobed. . . . . 10. *Viola triloba*
   dd. All leaves unlobed.
    e. Blades glabrous or rarely with 1–few scattered hairs, often however with pubescent petioles.
     f. Leaves narrowly to broadly lanceolate or rarely oblanceolate, their blades mostly 5 or more times as long as wide . . . . 15. *Viola lanceolata*
     ff. Leaves ovate, orbicular, reniform, or subdeltoid, their blades 1–2 times as long as wide or slightly wider than long.

g. Leaves ovate to broadly oblong or obovate, lower part of blade obtuse, truncate, or rounded and then tapering to base, not cordate .... 16. *Viola primulifolia*

gg. Leaves orbicular, ovate, reniform, or subdeltoid, base of blade cordate.

h. Corollas 0.7–1.5 cm across; plants sometimes with slender stolons; leaf blades ovate to orbicular, 1–8 cm long.

i. Largest leaves mostly 1–3 cm long .... 13. *Viola macloskeyi*

ii. Largest leaves mostly 4–8 cm long .... 14. *Viola blanda*

hh. Corollas 1.5–2.5 (–3.0) cm across; plants nonstoloniferous; leaf blades ovate, reniform, or subdeltoid, 2–12 cm long .... 6. *Viola sororia*

ee. Blades pubescent, at least on base of upper surface, the petioles pubescent or glabrous.

f. Leaves ovate to broadly oblong or obovate, lower part of blade obtuse, truncate, or rounded, and then tapering to base, not cordate........................... 16. *Viola primulifolia*

ff. Leaves ovate to orbicular or reniform, base cordate.

g. Style hooked at tip; petioles and peduncles pubescent .... 18. *Viola odorata*

gg. Style more or less straight, not hooked at tip; petioles and peduncles glabrous or nearly so .... 14. *Viola blanda*

bb. Corolla blue, purple, or violet.

c. Some or all leaves lobed or divided.

d. Blades 1.5–3 times as long as wide.

e. Blades glabrous to pubescent on both surfaces; early blades shallowly toothed near base and crenate above, later blades deeply toothed near base—or sometimes hastately lobed—and shallowly toothed along middle and upper margins; later blades shallowly cordate to sagittate at base; petioles glabrous or slightly pubescent, those of later leaves as long as to twice as long as the blades................................ 9. *Viola sagittata*

ee. Blades pubescent on both surfaces, usually villous; early blades mostly crenate throughout, later blades crenate throughout or the lower part of blade sometimes toothed or rarely shallowly hastately lobed; later blades truncate to shallowly cordate at base; petioles pubescent, those of later leaves shorter than to scarcely exceeding length of blades .... 8. *Viola fimbriatula*

dd. Blades mostly about as wide as long or wider than long, rarely to 1.5 times as long as wide.

e. Leaves all deeply lobed, the primary lobes further deeply to shallowly lobed one or more times, the lobes and ultimate segments uniform in appearance, more or less linear, mostly entire-margined, or with few apical teeth; base of blade truncate to cuneate; corolla flattened or not; lateral petals bearded or not.

f. Corolla 2–3.5 cm wide, concolorous lilac-purple or bicolorous with the two upper petals dark purple and the lower three pinkish-purple; petals beardless; sepals 9–12 mm long .... 1. *Viola pedata*

ff. Corolla 1.5–2.5 cm wide, concolorous violet-colored; lateral and spurred petals bearded; sepals 5–8 mm long............. .... 12. *Viola pedatifida*

ee. Leaves shallowly to deeply lobed, earliest leaves sometimes unlobed, lobes of various shapes and sizes, rarely if ever uniform in appearance, rarely linear, their margins entire or more commonly crenate or toothed; base of blade mostly cordate, sometimes subtruncate; corolla not flattened, petals projecting forward; lateral petals bearded.

f. Earliest leaves unlobed, later leaves with 3–5 main lobes, sinuses very shallow to deep, the lobes sometimes deeply toothed or rarely again shallowly lobed .... 10. *Viola triloba*

ff. All leaves shallowly to deeply 5–11-lobed, the lobes entire, crenate or toothed......................... 11. *Viola palmata*

cc. Leaves unlobed.

d. Blades elongate, 1.5–3 times as long as wide.

e. Blades glabrous to pubescent on both surfaces; early blades shallowly toothed near base and crenate above, later blades deeply toothed near base—or sometimes hastately lobed—and shallowly toothed along mid and upper margins; later blades shallowly cordate to sagittate at base; petioles glabrous or slightly pubescent, those of later leaves as long as to twice as long as the blades................................ 9. *Viola sagittata*

ee. Blades pubescent on both surfaces, usually villous; early blades mostly crenate throughout, later blades crenate throughout or the lower part of blade sometimes toothed or rarely shallowly hastately lobed; later blades truncate to shallowly cordate at base; petioles pubescent, those of later leaves shorter than to scarcely exceeding length of blades .... 8. *Viola fimbriatula*

dd. Blades mostly about as long as wide, or wider than long, rarely to 1.5 times as long as wide.

e. Leaf blades glabrous or at most slightly pubescent on upper surface; petioles glabrous or slightly pubescent.

f. Spurred petal bearded.

g. Blades more or less triangular, lower half wide, upper half somewhat abruptly narrow-triangular, the upper half usually with only 1–3 teeth on either margin, apex acute; blades glabrous . . . . 3. *Viola missouriensis*

gg. Blades narrowly ovate to reniform, upper half not abruptly narrow-triangular, apex acuminate, acute, or obtuse; blades either with few scattered short stiff hairs on upper surface or glabrous.

h. Leaves ovate to broadly reniform, apex acute to obtuse; leaf texture firm to subleathery . . . . 4. *Viola nephrophylla*

hh. Leaves narrowly ovate to ovate or less often broadly ovate, apex mostly acuminate to acute, some leaves subreniform and obtuse at apex; leaf texture thin . . . . 5. *Viola affinis*

ff. Spurred petal beardless or nearly so.

g. Leaves more or less triangular, apex acute; upper half of blade somewhat abruptly narrow-triangular . . . . 3. *Viola missouriensis*

gg. Leaves ovate to reniform or deltoid, apex acute to obtuse or rounded; upper half of blade not somewhat abruptly narrow-triangular.

h. Lateral petals bearded with short stout hairs, the hairs 0.1–0.5 (–0.7) mm long; corolla pale to medium blue or violet; peduncles usually much elongated, 0.5–3 (–4.5) dm long, the flowers held above or well above the leaves; sepal auricles obvious, 1.5–3 (–6) mm long . . . . 2. *Viola cucullata*

hh. Lateral petals bearded with long slender hairs, some or most of them 1.0–1.5 (–2.2) mm long; corolla purple, purplish-blue, or mixed purple and white; peduncles shorter, 0.4–1.2 (–3) dm long, the flowers held below, even with, or slightly above the leaves; sepal auricles small, 1 mm or less in length . . . . 6. *Viola sororia*

ee. Leaf blades slightly to conspicuously pubescent; petioles pubescent or glabrous.

f. Style hooked at tip; petioles and peduncles pubescent; plants stoloniferous, forming extensive mats . . . . 18. *Viola odorata*

ff. Style more or less straight, not hooked at tip; petioles and peduncles pubescent or glabrous; plants not stoloniferous.

g. Upper surface of blade with moderate to dense silvery pubescence, undersurface glabrous or with few hairs, the petioles glabrous; upper surface of blade mottled green in color and sometimes with numerous dark dots, undersurface often purplish and with many dark dots; leaf veins and petioles often purplish; spurred petal glabrous or slightly bearded . . . . 7. *Viola hirsutula*

gg. Upper surface of blade more or less pubescent, undersurface and petioles pubescent or glabrous; upper and undersurfaces green, not mottled and without dark dots; leaf veins and petioles green; spurred petal beardless.

h. Lateral petals bearded with short stout hairs 0.1–0.5 (–0.7) mm long; corolla pale to medium blue or violet; peduncles usually much elongated, 0.5–4 (–4.5) dm long, the flowers held above or well above the leaves; sepal auricles obvious, 1.5–3 (–6) mm long. . . . . 2. *Viola cucullata*

hh. Lateral petals bearded with long slender hairs, some or most of them 1.0–1.5 (–2.2) mm long; corolla purple, purplish-blue, or mixed purple and white; peduncles shorter, 0.4–1.2 (–3) dm long, the flowers held below, even with, or slightly above the leaves; sepal auricles small, 1 mm or less in length . . . . 6. *Viola sororia*

aa. Plants caulescent, the leaves and flowers arising from erect to prostrate stems, or sometimes from stolons.

b. Corolla large, (2–) 3–7 cm wide, in a variety of colors and variegated color combinations; rare escape from cultivation; see above . . . . *Viola* × *wittrockiana*

bb. Corolla smaller, 0.5–2.5 cm wide.

c. Corolla violet, purple, or blue, the colors sometimes pale.

d. Plants annual; early leaves orbicular, later leaves linear, elliptic, or oblanceolate; stipules deeply lobed, the lobes mostly elliptic; corolla flattened; spur 0.5–3 mm long. . . . . . 29. *Viola rafinesquii*

dd. Plants perennial; all leaves ovate, orbicular, or reniform; stipules deeply or sharply toothed, but not lobed; corolla not flattened; spur 3–16 mm long.

e. Spur prominent, 7–16 mm long; lateral petals beardless . . . . 26. *Viola rostrata*

ee. Spur less prominent, 3–6 mm long; lateral petals bearded.

f. Stems at first short and erect, then becoming prostrate; plants stoloniferous and forming extensive mats; stems and leaves puberulent . . . . 25. *Viola walteri* ff. Stems erect or ascending; plants nonstoloniferous; stems and leaves glabrous, or sometimes upper leaf surface slightly pubescent . . . . 24. *Viola conspersa*

cc. Corolla white to yellow, or variegated with combinations of violet and white and/or yellow.

d. Petals white or cream-colored, sometimes tinged with blue.

e. Plants annual; earliest leaves orbicular to ovate, later leaves linear to elliptic, oblanceolate, or ovate; stipules deeply lobed; corolla flattened.

f. Terminal lobe of stipule much larger than adjacent lobes, cuneate at base, often long-stalked, its blade elliptic to ovate and resembling a leaf blade; sepals as long as or longer than the petals; early leaves orbicular to ovate, later leaves ovate to broadly or narrowly elliptic; petals yellow to cream-colored, sometimes purple-tinged or purple-tipped . . . . 28. *Viola arvensis*

ff. Terminal lobe of stipules more or less elliptic, not stalked, not or only slightly resembling leaf blade, usually not more than twice the length of the adjacent lobes, the two uppermost lobes often—in combination—appearing mitten-shaped; sepals much shorter than to nearly equalling the petals; early leaves orbicular, later leaves linear, elliptic, or oblanceolate; petals white or bluish, often turning somewhat yellow soon after drying . . . . 29. *Viola rafinesquii*

ee. Plants perennial; earliest leaves suborbicular or subreniform, later leaves ovate to broadly ovate or suborbicular; stipules entire or toothed, not lobed; corolla not flattened.

f. Stipules entire, lanceolate, scarious throughout, 0.7–1.7 cm long; corolla white with a yellow center, petals tinged with purple on back, especially in age . . . . 22. *Viola canadensis*

ff. Stipules deeply toothed, ovate to long-lanceolate, green, 1–3 cm long; corolla white or cream-colored, without a yellow center, petals not tinged with purple . . . . 23. *Viola striata*

dd. Petals all bright yellow, or variegated with combinations of violet, white, and yellow.

e. Petals yellow, sometimes tinged with blue, purple, or red-purple, or the spur sometimes blue; corolla flattened or not.

f. Stipules entire or shallowly toothed, not lobed; cauline leaves clustered near summit of stem; corolla with petals projecting forward, not flattened.

g. All or most leaves either deeply parted into 3 elongate lobes or trifoliolate . . . . 19. *Viola tripartita* gg. Leaves unlobed, simple.

h. Leaf blades broadly ovate or broadly triangular to nearly reniform, mostly as wide as long or wider than long, rarely to 1.5 times as long as wide; one to several basal leaves often present; corolla 1–2 cm wide; petals sometimes tinged with purple on back; sepals 6–11 mm long . . . 21. *Viola pubescens*

hh. Leaf blades ovate, triangular, long-triangular, or subrhomboid, mostly 1.5–2 times as long as wide; basal leaves none; corolla 0.5–1.5 cm wide; upper or all petals often tinged with purple or red-purple on back; sepals 4–8 mm long.

i. Stems 0.4–1 (–1.5) dm tall at anthesis, green or purplish; leaves usually variegated with silvery-green zones on upper surface, especially in age; veins and undersurface of leaves purplish; stipules small, 1–4 mm long; upper petals beardless, lateral petals slightly bearded . . . . 20. *Viola hastata*

ii. Stems 1–3 dm tall at anthesis, green; leaves uniformly green; veins uniformly green; stipules larger, 4–12 mm long; upper petals slightly bearded, lateral petals bearded . . . . 19. *Viola tripartita*

ff. Stipules deeply lobed; cauline leaves more or less evenly distributed on stem, not clustered near summit; corolla flattened.

g. Terminal lobe of stipule much larger than adjacent lobes, cuneate at base, often long-stalked, its blade elliptic to ovate and resembling a leaf blade; sepals as long as or longer than the petals; early leaves orbicular to ovate, later leaves ovate to broadly or narrowly elliptic; petals yellow to cream-colored, sometimes purple-tinged or purple-tipped . . . . 28. *Viola arvensis*

gg. Terminal lobe of stipules more or less elliptic, not stalked, not or only slightly resembling leaf blade, usually not more than twice the length of adjacent lobes, the two uppermost lobes often—in combination—appearing mitten-shaped; sepals much shorter than to nearly equalling the petals; early leaves orbicular, later leaves linear, elliptic, or oblanceolate; petals white or bluish, often turning somewhat yellow soon after drying . . . . 29. *Viola rafinesquii*

ee. Petals variegated; corolla flattened.

f. Sepals shorter than the petals; corolla usually with the two upper petals purple, the two lateral petals white or pale yellow, and the spurred petal bright yellow, the lateral and spurred petals sometimes lavender or purple-tinged . . . . 27. *Viola tricolor*

ff. Sepals equalling or longer than the petals; corolla with all petals usually pale yellow to cream-colored, sometimes purple-tinged or purple-tipped, occasionally the two upper petals purple throughout, the spur sometimes bluish . . . . . 28. *Viola arvensis*

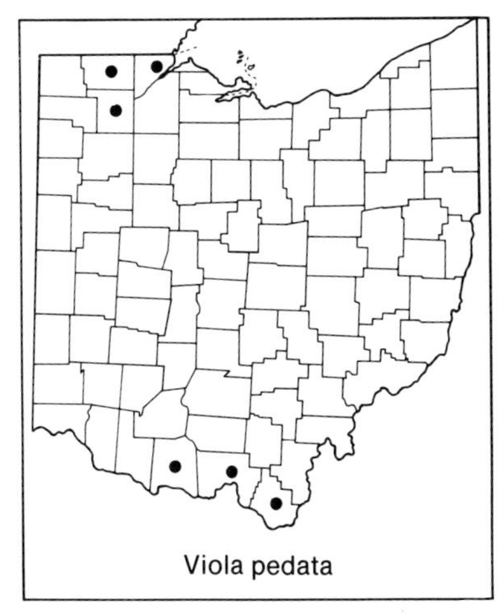

Viola pedata

1. **Viola pedata** L. Birdfoot Violet

incl. var. *pedata* and var. *lineariloba* DC.

Acaulescent perennial, nonstoloniferous; leaf blades lobed, reniform to obovate in outline, ciliate but otherwise glabrous or with short stiff hairs along the veins; leaves with deep sinuses separating 3 main lobes, the central lobe usually linear, sometimes again 3-lobed, entire except for a few teeth near the apex, the two lateral lobes cleft into 2–5 linear segments, these segments entire or with a few teeth near the apex, base of blade mostly broadly to narrowly cuneate; sepals 9–12 mm long; corolla 2–3.5 cm across, either concolorous lilac-purple or bicolorous with the upper two petals dark purple and the lower three pinkish-purple; all petals beardless. At a few Ohio sites, mostly in northwestern and south-central counties; open sand and open dry rocky habitats, roadsides, fields. A specimen in the Youngstown State University Herbarium collected from Carroll County, in east-central Ohio, is presumably an adventive (not mapped). April–May (August).

Russell (1965) believes that the two varieties listed above in synonymy may be mere environmental forms; I follow his treatment in recognizing no varieties within the species.

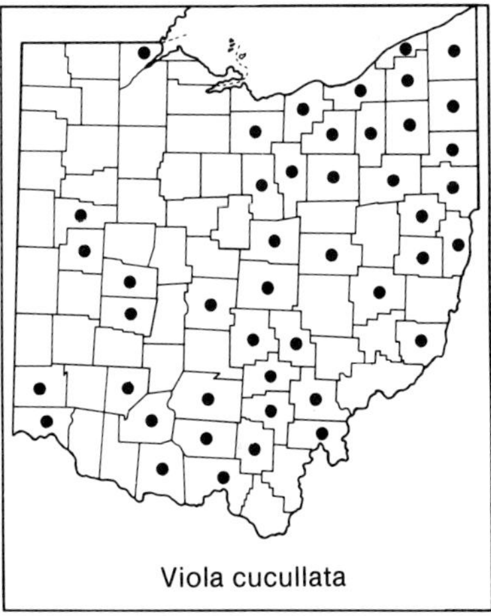

Viola cucullata

2. **Viola cucullata** Ait. Marsh Blue Violet

(?*V. obliqua* Hill)

Acaulescent perennial, nonstoloniferous; leaf blades ovate to deltoid or subreniform, about as wide as long, glabrous or with a few short stiff hairs on upper surface—mostly at base of blade, apex acute to obtuse, margins crenate-serrate, base cordate; peduncles usually much elongated, 0.5–3.0 (–4.5) dm long, the flowers held above or well above the leaves; sepal auricles obvious, 1.5–3.0 (–6.0) mm long; corolla pale to medium blue or violet, 1.0–2.5 cm across; spurred petal beardless, lateral petals bearded with short stout hairs 0.1–0.5 (–0.7) mm long. Frequent in eastern Ohio counties, but scarce or absent westward; swamps, bogs, wet alluvial woods, wet thickets, shaded marshes, wet fields. April–early June (October).

In *Flora Europaea*, Tutin et al. (1964–80) call this species *V. obliqua* Hill, a name that has been applied by various authors to six different spe-

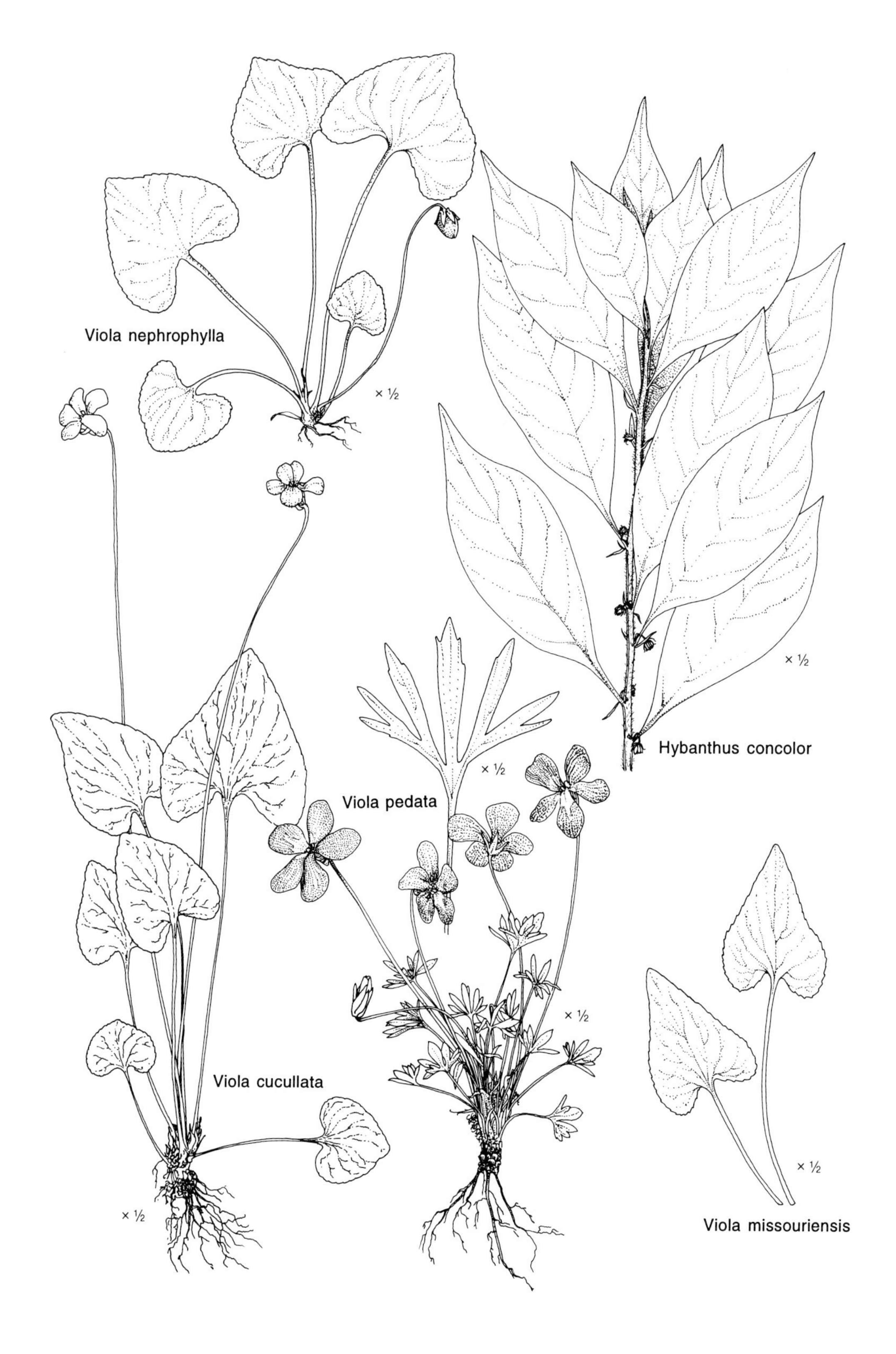
Viola nephrophylla
× ½
Hybanthus concolor
× ½
× ½
Viola pedata
× ½
Viola cucullata
× ½
× ½
Viola missouriensis

cies of American violets. It is uncertain whether or not that name, which antedates *V. cucullata* Ait., applies to this species. Commentary on the problem can be found in Brainerd (1907), House (1924, p. 503), and Voss (1985b, p. 601).

3. **Viola missouriensis** E. Greene MISSOURI VIOLET

Acaulescent perennial, nonstoloniferous; leaf blades more or less triangular in outline, lower half of blade wide, the upper half somewhat abruptly narrowly triangular, glabrous, apex acute, margins low crenate-serrate, usually only 1–3 teeth on each side of narrowed upper half of blade, base cordate to subtruncate; corolla pale violet, 1.0–1.5 cm across, spurred petal glabrous or slightly bearded, lateral petals bearded. A species (sensu Russell, 1965) of central United States, in Ohio known only from Brown County, in southwestern Ohio, where it is at the easternmost point of its range; dry woodland borders. April–May.

Viola missouriensis

Gleason and Cronquist (1991) merge this species without rank under *V. sororia*. McKinney (1992) treats it as a variety of that species, calling it *V. sororia* var. *missouriensis* (E. Greene) McKinney. He reports it from Licking and Scioto counties in Ohio, and from many locations in other states eastward to the Atlantic Coast.

4. **Viola nephrophylla** E. Greene NORTHERN BOG VIOLET

Acaulescent perennial, nonstoloniferous; leaves firm to subleathery, leaf blades ovate to broadly reniform, glabrous or rarely with few scattered hairs on upper surface, apex acute to obtuse, margin crenate-serrate, base cordate to broadly cordate or broadly subsagittate; corolla medium violet with a white center, 1.5–2.5 cm across, spurred petal slightly bearded, lateral petals heavily bearded. Known from a few north-central counties; mostly on wet calcareous rocks and cliffs along or near Lake Erie. May–early June.

Viola nephrophylla

McKinney (1992) merges this species under *Viola sororia* var. *sororia*, assigning it no rank.

5. **Viola affinis** LeConte THIN-LEAVED VIOLET. LECONTE'S VIOLET

Acaulescent perennial, nonstoloniferous; leaf blades thin in texture, mostly narrowly ovate to ovate, less often broadly ovate to subreniform, glabrous except for a few scattered short stiff hairs on upper surface, apex mostly acuminate to acute, obtuse in subreniform leaves, margin crenate-dentate, base cordate; corolla violet, 1.5–2.5 cm across; spurred petal slightly bearded, lateral petals heavily bearded. Throughout Ohio; moist woodlands, woodland openings and borders. March–May.

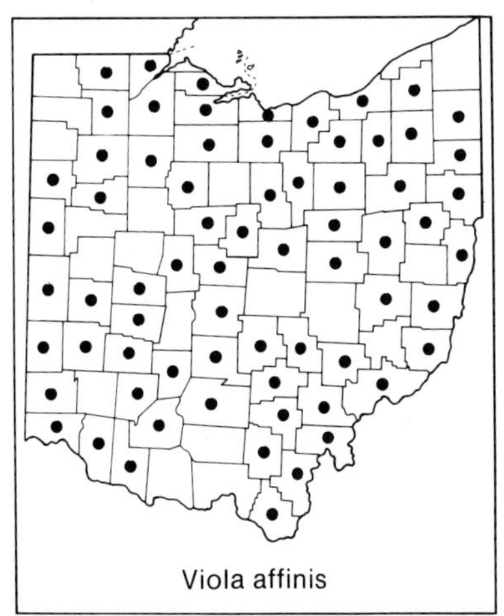

Viola affinis

Gleason and Cronquist (1991) merge this species without rank under *V. sororia*. McKinney (1992) treats it as *V. sororia* var. *affinis* (LeConte) McKinney.

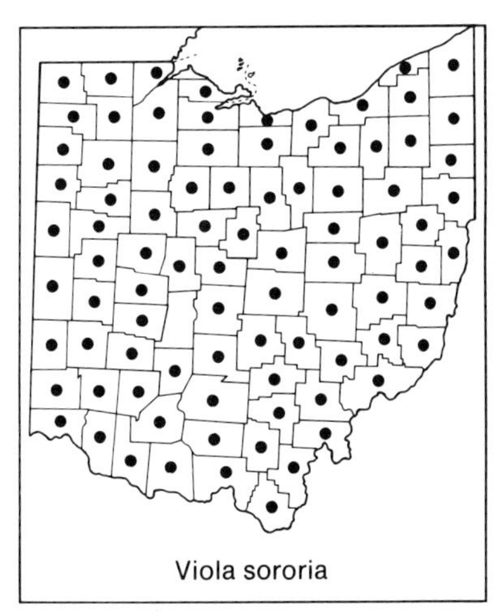
Viola sororia

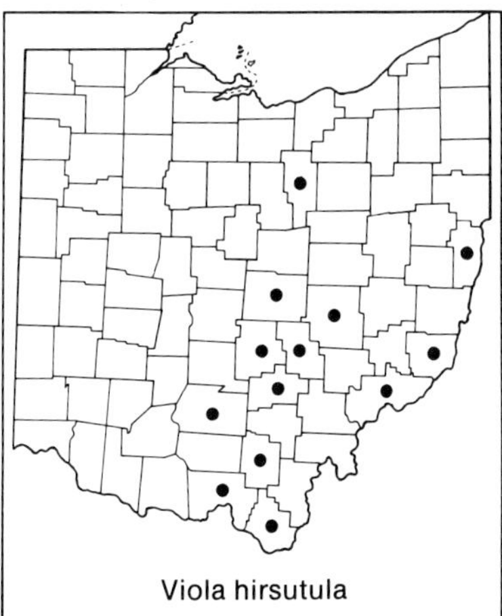
Viola hirsutula

6. **Viola sororia** Willd. Common Blue Violet
incl. *V. papilionacea* Pursh

Acaulescent perennial, nonstoloniferous; leaf blades ovate or subdeltoid to reniform, variable in pubescence, ranging from both surfaces villous to both surfaces glabrous, apex acute to less often obtuse or rounded, margin crenate, base cordate; flowers held below, even with, or slightly above the leaves; sepal auricles small, 0.5–1.0 mm long; corolla purple to purplish-blue (or other colors in the forms keyed below), 1.5–2.5 (–3.0) cm across; spurred petal beardless or nearly so, lateral petals bearded with long, slender, slightly knobbed hairs. Common and locally abundant throughout Ohio; wooded floodplains and slopes, moist open woods and fields, lawns and other disturbed grassy sites. This often weedy species is sometimes a volunteer in flower gardens and sometimes purposely brought into cultivation. Late March–early June (September–October).

Following Russell's (1965) classification, plants usually identified as *Viola papilionacea* are treated as glabrous or glabrate forms of *V. sororia*, and assigned no taxonomic rank. Many intermediates exist between these glabrous plants and the villous forms constituting *V. sororia* sensu stricto. This species is merged under *V. palmata*, as *V. palmata* var. *sororia* (Willd.) Pollard, by Ahles (1968).

Three significant color forms are found in Ohio.

a. Corolla purplish-blue or purple throughout . . . . forma *sororia*

aa. Corolla partly or wholly white or gray.

b. Corolla white throughout, or the lowest petal with a few purple lines . . . . forma *beckwithiae* House
incl. *V. papilionacea* forma *albiflora* Grover
Erie and Tuscarawas counties.

bb. Corolla gray, with purple veins near center . . . . forma *priceana* (Pollard) Cooperr.
V. papilionacea var. *priceana* (Pollard) E. J. Alex.
Confederate Violet
Franklin, Fulton, Lawrence, and Portage counties; for commentary, see Cooperrider (1984b).

7. **Viola hirsutula** Brainerd Southern Wood Violet

Acaulescent perennial, nonstoloniferous; petioles and leaf veins often purplish; leaf blades ovate to reniform or nearly orbicular, the upper surface mottled green and with moderate to dense silvery pubescence, the undersurface frequently purplish and glabrous or with few scattered hairs, the undersurface and sometimes the upper with many dark dots, base cordate; corolla violet or pinkish-violet, 1–2 cm across; spurred petal glabrous or slightly bearded, lateral petals bearded. East-central and southeastern

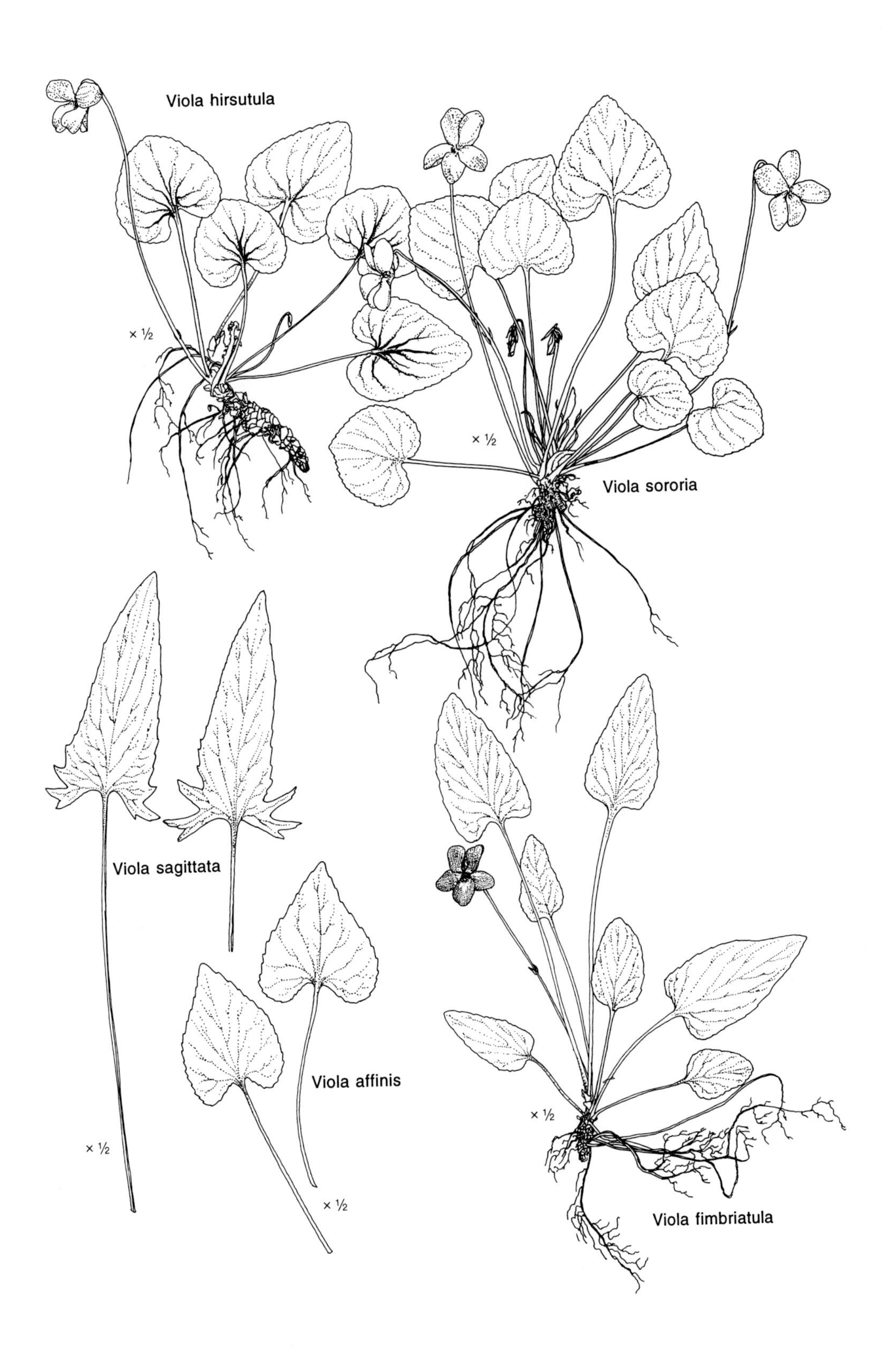
Viola hirsutula
× ½
× ½
Viola sororia
Viola sagittata
Viola affinis
× ½
× ½
× ½
Viola fimbriatula

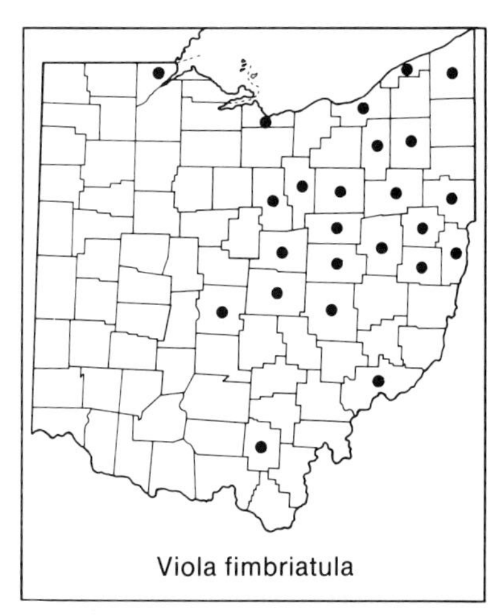

Viola fimbriatula

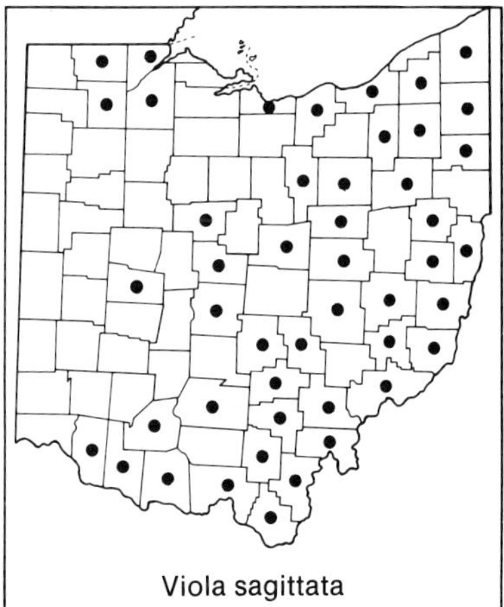

Viola sagittata

counties, sharply confined to the unglaciated Allegheny Plateau (a record from Geauga County in northeastern Ohio reported by Russell (1965) was based on a misidentified specimen of *V. rotundifolia*); dry woodlands, open wooded slopes and bottomlands. Ohio stations are at a northwestern boundary of the species' range. Late April–May.

Gleason and Cronquist (1991) treat this species as a part of *V. villosa* Walt.

8. **Viola fimbriatula** Sm. OVATE-LEAVED VIOLET. NORTHERN DOWNY VIOLET

Acaulescent perennial, nonstoloniferous; petioles pubescent, those of early leaves often longer than blades, those of later leaves shorter than to scarcely exceeding length of blades; leaf blades ovate-orbicular to ovate or ovate-oblong in early leaves, and oblong-ovate to oblong-lanceolate in later leaves, both surfaces pubescent, usually villous; margins shallowly crenate in early leaves, and crenate and sometimes toothed or rarely shallowly lobed on lower part of blade in later leaves; base of early and late blades truncate to shallowly cordate; flowers on pubescent peduncles; corolla violet with a white center, 1–2 cm across; spurred petal slightly bearded, lateral petals heavily bearded. Mostly in the eastern half of Ohio, especially in east-central counties; dry, often sandy, soils, woodland borders and openings, thickets, open fields, roadside banks. Late April–May.

Viola fimbriatula is generally thought to be a close relative of the next species, *V. sagittata*, and is merged under *V. sagittata* by Voss (1985b) and Gleason and Cronquist (1991). McKinney (1992) treats it as *V. sagittata* var. *ovata* (Nutt.) T. & G.

9. **Viola sagittata** Ait. ARROW-LEAVED VIOLET

Acaulescent perennial, nonstoloniferous; petioles glabrous or slightly pubescent, those of early leaves as long as, and those of later leaves to twice as long as, the blades; leaf blades lanceolate, oblong-lanceolate or long-triangular, glabrous to pubescent on both surfaces, sometimes becoming reddish in autumn; margins mostly shallowly toothed near base and crenate above in early leaves, and deeply toothed or sometimes hastately lobed near base and shallowly toothed along mid and upper margins in later leaves; base truncate to shallowly cordate in early leaves, and shallowly cordate to sagittate in later leaves; flowers on glabrous peduncles; corolla purple-violet with a white center, 1–2 cm across; spurred and lateral petals bearded. Chiefly, but not exclusively, in the eastern half of Ohio; dry open ground of fields, slopes, and roadside banks; less often in moist or wooded places. April–early June.

10. **Viola triloba** Schwein. THREE-LOBED VIOLET

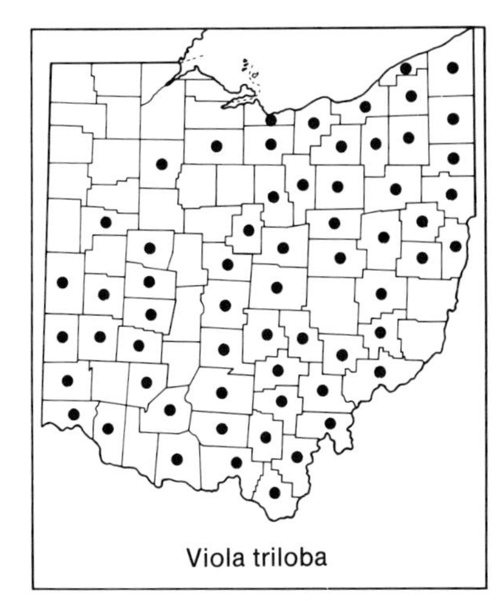
Viola triloba

Acaulescent perennial, nonstoloniferous; blades of earliest leaves more or less reniform, unlobed, and sometimes purplish beneath; blades of later leaves triangular to broadly ovate or subreniform in outline, the upper surface glabrous to moderately pubescent, the undersurface glabrous to villous, margins crenate or variously toothed, rarely subentire, the leaves with 3–5 main lobes, the sinuses very shallow to very deep, the central lobe often the widest and the basal lobes often lunate, the lobes sometimes deeply toothed or rarely again lobed; base of leaves cordate to subtruncate; corolla violet, 1.5–2.5 cm across; spurred petal glabrous or slightly bearded, lateral petals bearded. Throughout most of eastern and southern Ohio, scarce or absent in northwestern counties; dry upland woodlands, woodland openings, grassy woodland borders. April–June.

All Ohio specimens I have seen are assignable to var. *triloba*; those I have seen identified as var. *dilatata* (Ell.) Brainerd proved to be specimens of *V. palmata*.

This species is merged under *V. palmata* as *V. palmata* var. *triloba* (Schwein.) DC. by Ahles (1968), and some Ohio specimens are difficult to separate from *V. palmata*. Gleason and Cronquist (1991) and McKinney (1992) merge it without rank under *V. palmata*.

11. **Viola palmata** L. EARLY BLUE VIOLET. PALMATE-LEAVED VIOLET

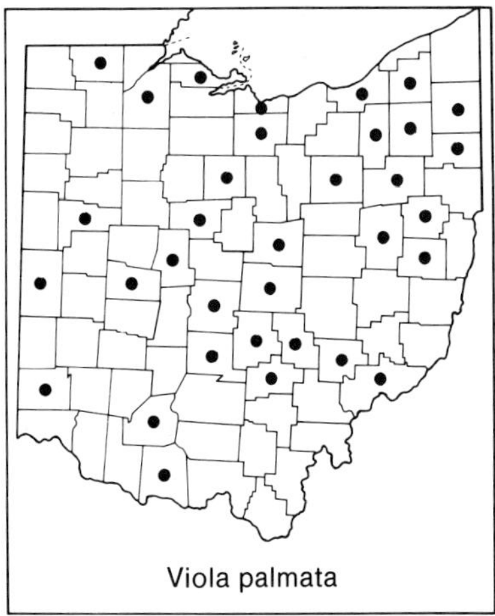
Viola palmata

Acaulescent perennial, nonstoloniferous; leaf blades reniform, orbicular, or ovate to narrowly triangular in outline, upper surface moderately pubescent to glabrous, undersurface villous to glabrous, all leaves shallowly to deeply 5–11-lobed, the margins of the lobes entire, crenate, or toothed; base of blade cordate to cordate-hastate; corolla violet, 1.5–2.5 cm across; spurred petal glabrous or slightly bearded, lateral petals bearded. Scattered localities, mostly in eastern and central counties; dry to mesic open woodlands, rich woodlands. Late April–May.

12. **Viola pedatifida** G. Don PRAIRIE VIOLET. LARKSPUR VIOLET

V. palmata L. var. *pedatifida* (G. Don) Cronq.

Viola pedatifida

Acaulescent perennial, nonstoloniferous; leaf blades broadly ovate to reniform in outline, upper surface glabrous or becoming so, undersurface pubescent with short stiff hairs, especially near margins and along veins; leaves deeply 3-parted, nearly or quite trifoliolate, each division again 3-parted, often deeply so, and each of these parts further cut into 2–4 elongate segments, the ultimate lobes or segments linear with mostly entire margins; base of blade subtruncate to cuneate; sepals 5–8 mm long; corolla violet, 1.5–2.5 cm across; spurred and lateral petals bearded. A

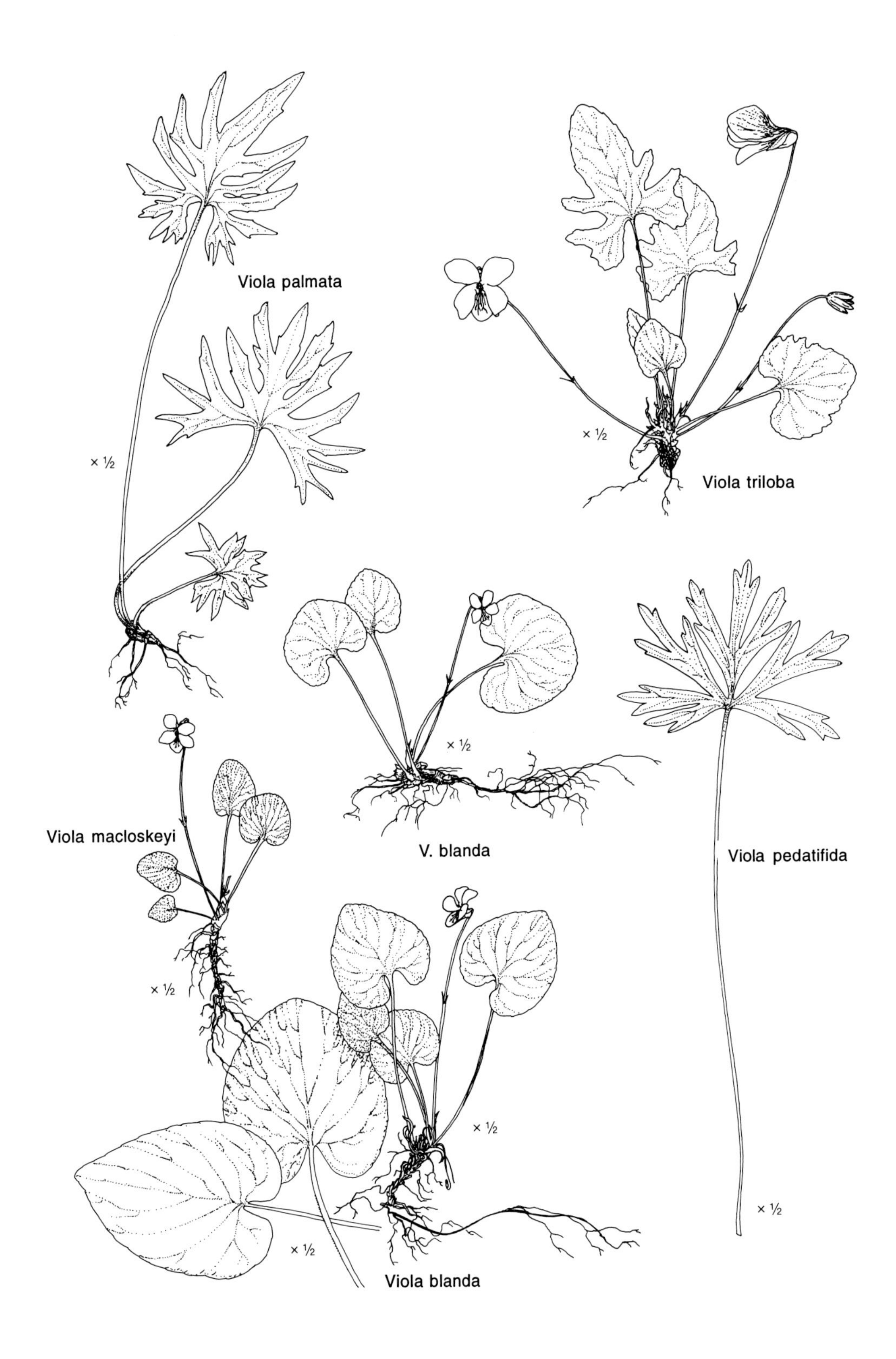
Viola palmata
× ½
Viola triloba
× ½
× ½
V. blanda
Viola macloskeyi
× ½
Viola pedatifida
× ½
× ½
× ½
Viola blanda

prairie species of the northern plains reaching an eastern boundary of its range in a few counties of western Ohio; dry prairie remnants, dry open fields. May.

This species has been attributed to Adams and Highland counties, but all so-identified specimens from those counties that I have examined proved to be forms of *V. palmata* with deeply lobed leaves (Cooperrider, 1982a). In this regard, however, it is noteworthy that Gleason and Cronquist (1991) treat *V. pedatifida* as a variety of *V. palmata*. McKinney (1992) maintains it at the rank of species.

13. **Viola macloskeyi** F. Lloyd NORTHERN WHITE VIOLET
incl. *V. pallens* (DC.) Brainerd

Acaulescent perennial, sometimes producing slender stolons in summer; leaf blades ovate to orbicular, 1–3 (–4) cm long, glabrous or rarely with 1–few scattered hairs on upper surface, base cordate; corolla white, 0.7–1.5 cm across, the lateral petals and especially the spurred petal with dark purple veins; spurred petal beardless, lateral petals slightly bearded. Northeastern counties and at scattered sites elsewhere, mostly in northern Ohio; sphagnum bogs, wet meadows, low wet woodlands, sometimes on moss-covered logs. Late April–May.

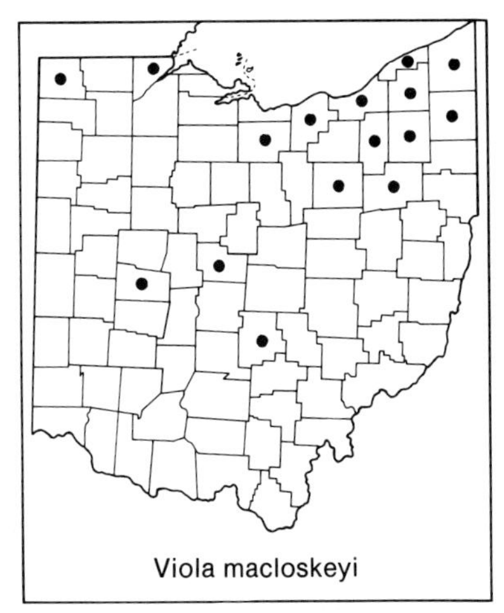
Viola macloskeyi

Ohio plants belong to var. *pallens* (DC.) C. L. Hitchc. M. S. Baker (1953a, 1953b) discusses the merger of *Viola pallens* under *V. macloskeyi*.

14. **Viola blanda** Willd. SWEET WHITE VIOLET
incl. *V. incognita* Brainerd

Acaulescent perennial, with stolons sometimes to 30 cm long; leaf blades narrowly to broadly ovate or orbicular, or the earliest subreniform, 1–5 (–8.5) cm long, upper surface with few to many short stiff hairs either over the entire surface or more often only at base of blade, undersurface glabrous or with few scattered hairs, base deeply cordate; corolla white, 0.7–1.5 cm across, the lateral petals and especially the spurred petal with reddish-purple to purple veins; spurred petal beardless, lateral petals beardless or slightly bearded. Chiefly in the eastern half of Ohio; cool moist hemlock or deciduous forests, shaded rocky ravines and slopes, among mosses and liverworts on rock ledges or crevices, stream banks, swamps. April–early June.

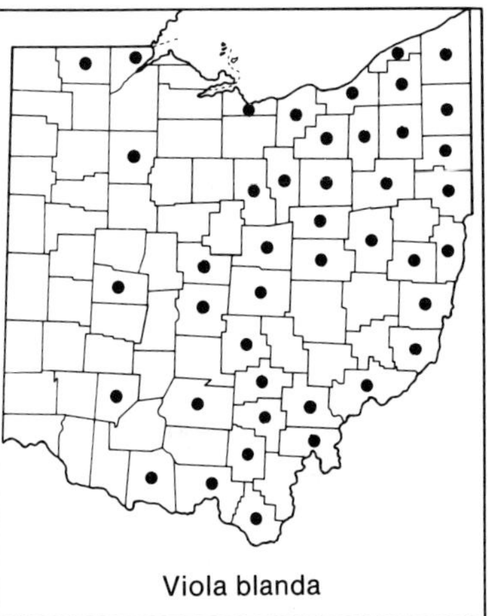
Viola blanda

Viola incognita is merged without rank under *V. blanda* following the treatment of Gleason and Cronquist (1991).

15. **Viola lanceolata** L. LANCE-LEAVED VIOLET

Acaulescent perennial, stoloniferous in summer, forming small mats; leaf blades narrowly to broadly lanceolate or rarely oblanceolate, to 14 cm long, glabrous, base long-tapering to petiole; corolla white with a greenish-yellow center, 0.7–1.3 cm across, the lateral petals and especially

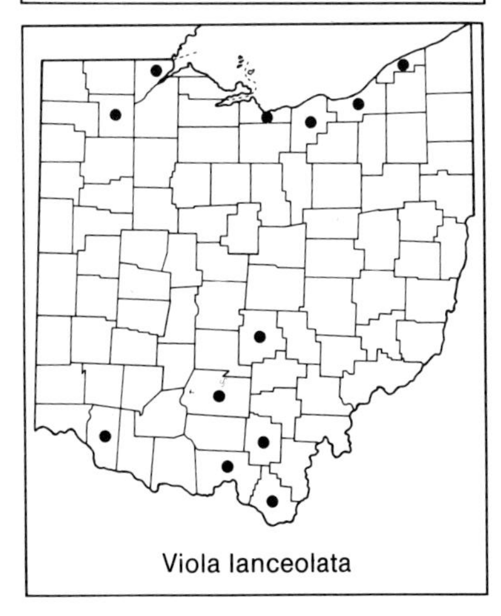
Viola lanceolata

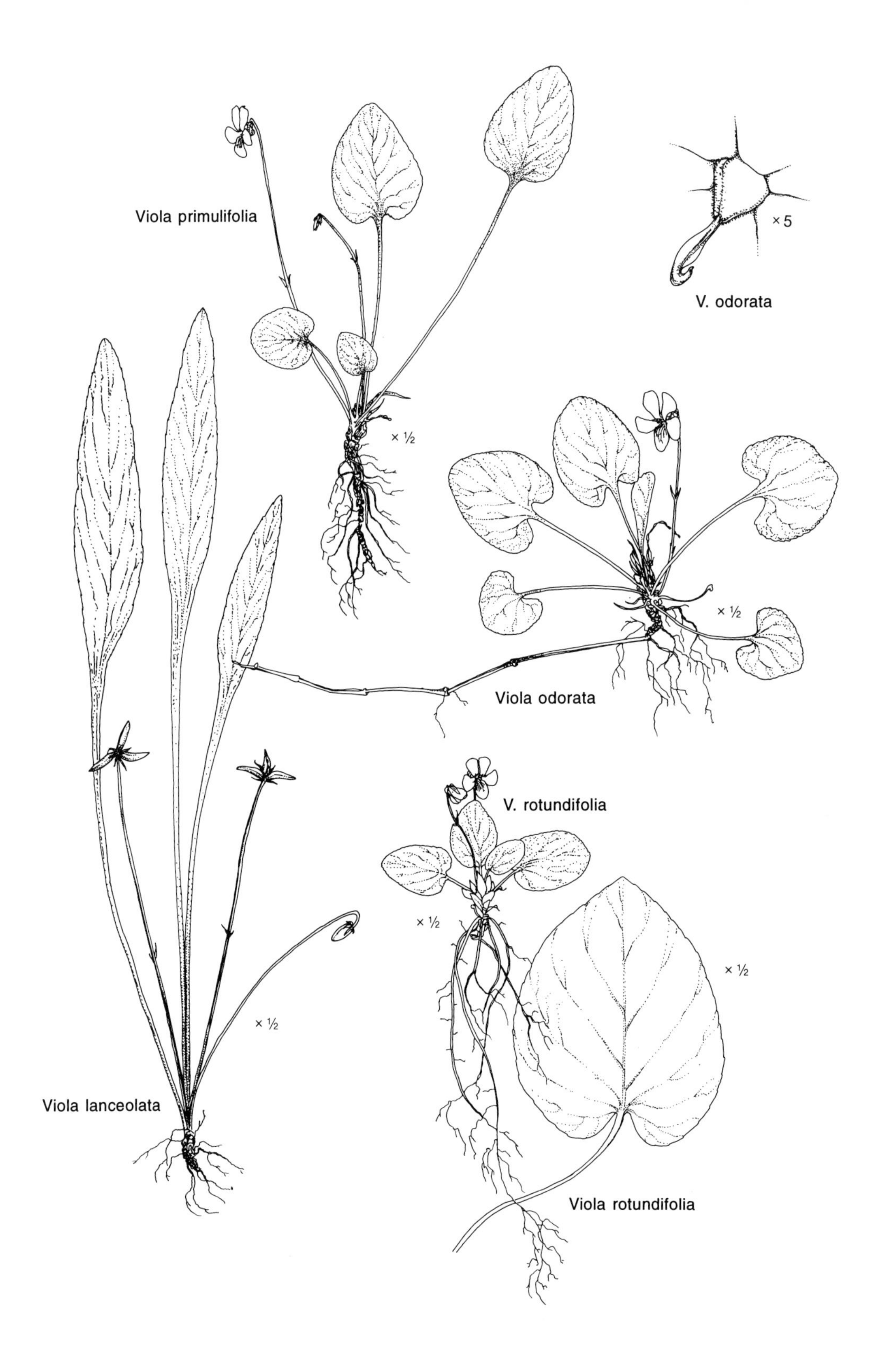
Viola primulifolia
V. odorata
×5
× ½
× ½
Viola odorata
V. rotundifolia
× ½
× ½
× ½
Viola lanceolata
Viola rotundifolia

the spurred petal with dark purple veins; all petals beardless. At scattered sites in northern and south-central counties; moist open, usually sandy, habitats, often along ponds and streams. May–early June.

16. **Viola primulifolia** L. PRIMROSE-LEAVED VIOLET
incl. var. *primulifolia* and var. *acuta* (Bigel.) T. & G.

Acaulescent perennial, producing stolons in summer; leaf blades ovate to ovate-oblong, to 8 cm long, upper surface glabrous, undersurface pubescent or sometimes glabrous, base obtuse, truncate, or rounded, and then abruptly tapering to petiole; corolla white, 0.7–1.4 cm across, the lateral and spurred petals with purple veins, spurred petal beardless, lateral petals glabrous or slightly bearded. Known from one northeastern and two south-central counties; moist open sites along ponds and streams, bogs and marshes. May–early June.

Viola primulifolia

Voss (1985b) treats this taxon as an interspecific hybrid, *Viola* × *primulifolia*, regarding it as a cross of *V. lanceolata* × either *V. blanda* (incl. *V. incognita*) or *V. macloskeyi*. If it is of hybrid origin, the latter would seem the more likely parent because *V. lanceolata* and *V. macloskeyi* are reported to have the same chromosome number, $2n=24$, while that reported for *V. blanda* is $2n=44$. Russell (1955), however, treats *V. primulifolia* as a species and recognizes as well naturally occurring interspecific hybrids between *V. lanceolata* and *V. macloskeyi*, and between *V. lanceolata* and *V. primulifolia*.

Following Russell (1965), no infraspecific taxa are recognized in this species.

17. **Viola rotundifolia** Michx. ROUND-LEAVED VIOLET. EARLY YELLOW VIOLET

Acaulescent perennial, nonstoloniferous; leaf blades ovate to orbicular, at anthesis small and at first inrolled or folded, at maturity to 13 cm long and to 10 cm wide, upper surface with scattered pubescence, undersurface glabrous or slightly pubescent, base cordate; corolla bright yellow, 0.5–1.7 cm across, the lateral and spurred petals with reddish-brown veins; spurred petal glabrous, lateral petals bearded. Limited to northeastern counties except for one disjunct station in Hocking County in southeastern Ohio; steep, rocky, wooded slopes, often associated with hemlock. Bouchard and Maycock (1970) present a detailed study of the species' range and habitat. April–early May.

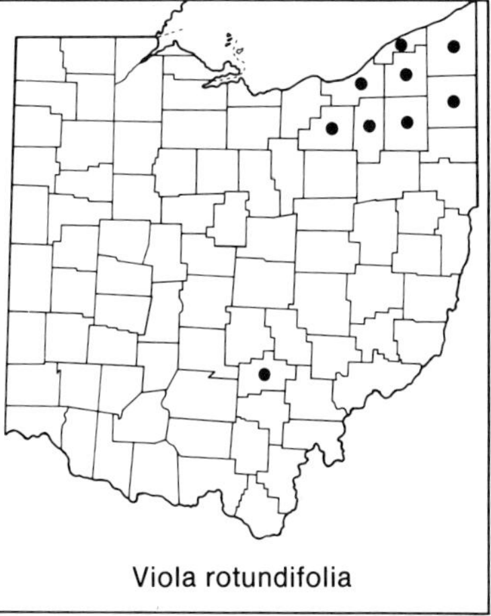
Viola rotundifolia

18. VIOLA ODORATA L. SWEET VIOLET

Acaulescent perennial, with many stolons forming extensive mats; leaf blades ovate to orbicular or subreniform, base cordate; flowers fragrant; corolla either violet or white with a blue spur, 1.0–1.5 cm across; spurred petal beardless, lateral petals usually bearded; the style distinctly hook-shaped at apex. A Eurasian species cultivated as an ornamental and occasionally es-

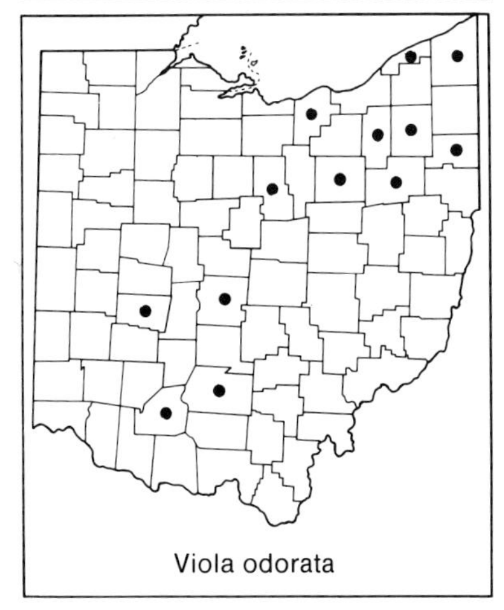
Viola odorata

caping, naturalized in northeastern counties and at a few scattered sites elsewhere in Ohio; lawns, roadsides, fields, and sometimes in open woodlands. March–April.

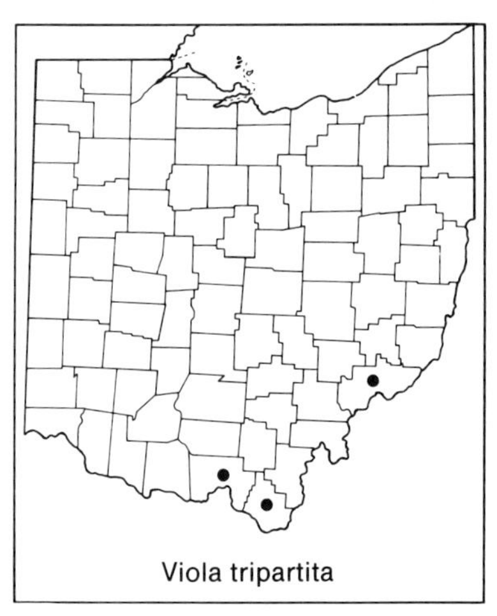
Viola tripartita

19. **Viola tripartita** Ell.

Caulescent perennial, 1–3 (–4.5) dm tall, nonstoloniferous; stems green; leaves either unlobed or deeply 3-lobed to trifoliolate, glabrous to moderately pubescent on both surfaces; base in unlobed leaves broadly cuneate to truncate; stipules 4–12 mm long; sepals 4–7 mm long; corolla yellow, 0.5–1.5 cm across, the lateral and spurred petals with dark reddish-brown veins, the two upper petals often tinged on back with red to purple, upper petals slightly bearded, lateral petals bearded. A species of southeastern United States, reaching a northern boundary of its range in southern Ohio; dry woodlands. Late April–May.

The two taxa keyed below are often treated at the rank of variety; but following Fernald (1950) and Russell (1965) I treat them here at the rank of *forma*. Forma *tripartita* has been found only in Lawrence County, forma *glaberrima* only in Scioto and Washington counties.

a. Leaves deeply 3-lobed or trifoliolate . . . . forma *tripartita*
   V. tripartita var. *tripartita*
   Three-parted Violet

aa. Leaves unlobed, simple . . . . forma *glaberrima* (DC.) Fern.
   V. tripartita var. *glaberrima* (DC.) Harper
   Southern Appalachian Yellow Violet

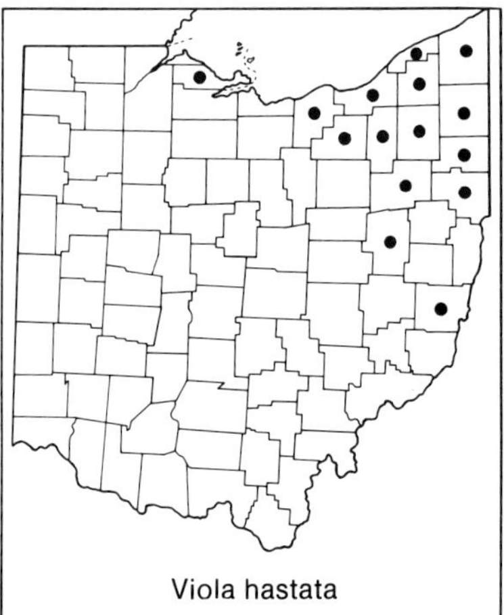
Viola hastata

20. **Viola hastata** Michx. Halberd-leaved Violet

Caulescent perennial, 0.4–2 dm tall, nonstoloniferous, one or two stems rising from a deeply buried white rhizome; stems green or sometimes purplish above; leaf blades ovate to triangular or long-triangular, glabrous or slightly pubescent, usually variegated with green and silvery-green zones on upper surface—especially in age, veins and undersurface purplish, base shallowly cordate to truncate; stipules small, 1–4 mm long; sepals 5–8 mm long; corolla pale yellow, 0.5–1.5 cm across, the lateral and spurred petals with dark purple-brown veins, all petals tinged with purple on back, upper petals beardless, lateral petals bearded. An Appalachian species reaching a northwestern boundary of its range in Ohio; mostly in northeastern counties; rich mesic woodlands, often in beech-maple forests. April–May.

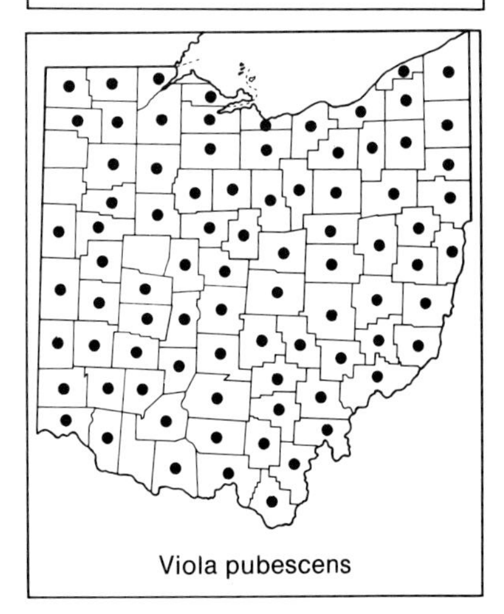
Viola pubescens

21. **Viola pubescens** Ait. Common Yellow Violet

Caulescent perennial, 0.5–3.0 (–4.0) dm tall, nonstoloniferous; stems green; leaf blades broadly ovate or broadly triangular to occasionally subreniform, pubescent to nearly glabrous on both surfaces, base cordate or truncate or sometimes broadly cuneate; stipules 5–16 mm long; sepals 6–11 mm long; corolla yellow, 1–2 cm across, sometimes purplish tinged on

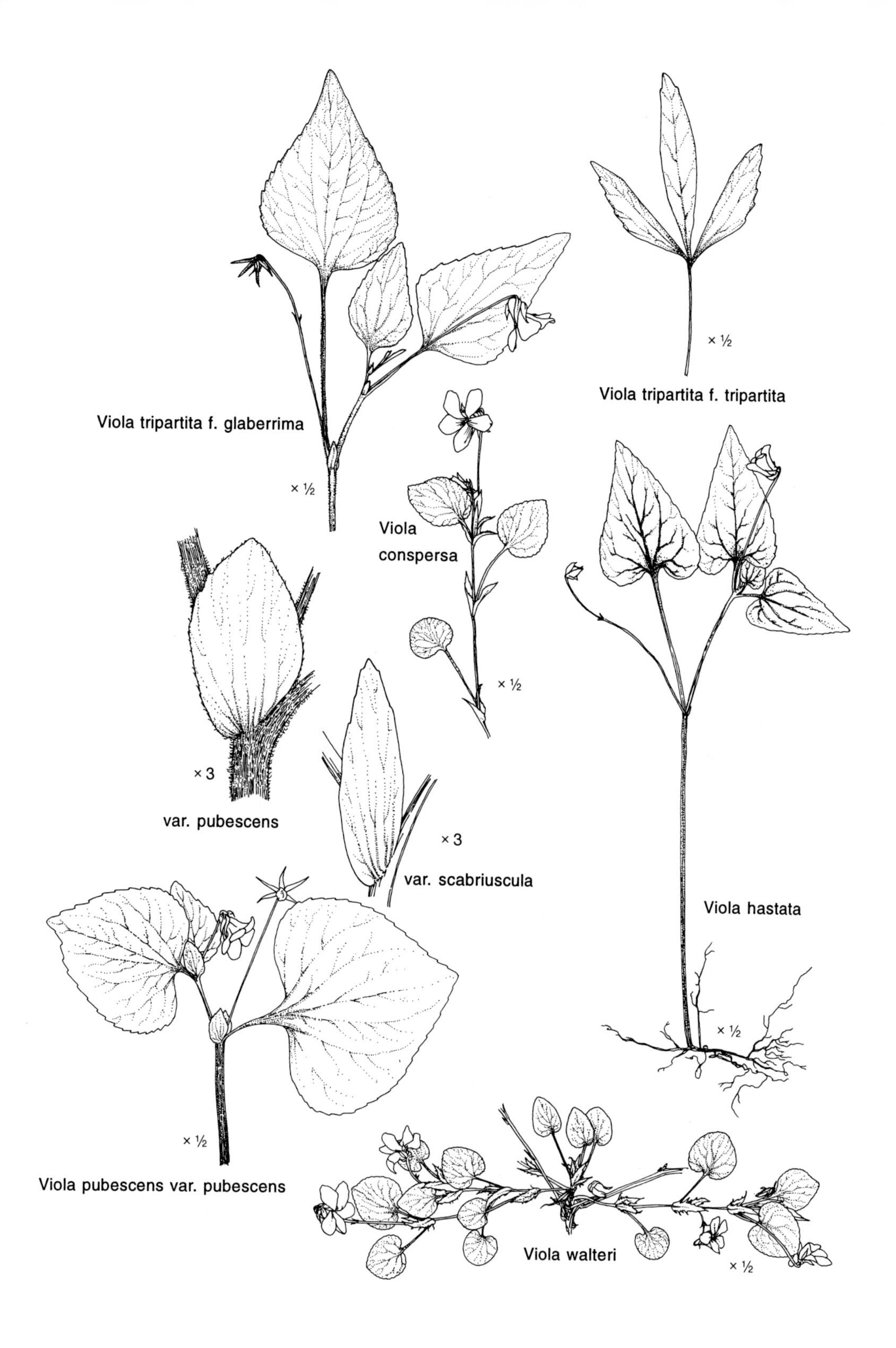
Viola tripartita f. glaberrima
× ½
Viola tripartita f. tripartita
× ½
Viola conspersa
× ½
× 3
var. pubescens
× 3
var. scabriuscula
Viola hastata
× ½
× ½
Viola pubescens var. pubescens
Viola walteri
× ½

back, the lateral and spurred petals with purple-brown veins, lateral petals bearded. Throughout Ohio; moist to dry woodlands. April–May.

Ohio plants may be separated into two intergrading varieties, var. *scabriuscula*, the more common of the two, which grows throughout the state, and var. *pubescens*, found chiefly in northeastern counties. Cain (1967) studied hybrid swarms between the two taxa. In their taxonomic study of stemmed yellow violets, Lévesque and Dansereau (1966) treat these varieties at the rank of species: *Viola pubescens* sensu stricto and *V. eriocarpa* Schwein.; I follow Russell (1965) in treating them as varieties of a single species, *V. pubescens* sensu lato. Gleason and Cronquist (1991) merge the two, recognizing no varieties within *V. pubescens*.

a. Flowering stem usually solitary, villous; basal leaves 0, 1, or 2; most cauline leaves with approximately 30–45 teeth; stipules broadly ovate to ovate-oblong, 5–13 mm long, 3–7 mm wide . . . . . . var. *pubescens*
   incl. *V. pubescens* var. *eriocarpa* Nuttall (1818, p. 150); *V. pensylvanica* Michx. (see G. N. Jones, 1959).
   Downy Common Yellow Violet

aa. Flowering stems one to several, finely pubescent to glabrous; basal leaves several; cauline leaves with approximately 15–30 teeth; stipules usually lanceolate, 9–16 mm long, 3–6 mm wide . . . . var. *scabriuscula* Torr. & A. Gray
   V. pubescens var. *eriocarpa* (Schwein.) N. Russell (1965; see Waterfall, 1971); *V. eriocarpa* Schwein.; *V. pensylvanica* of authors, incl. Fernald (1950), not Michx.
   Smooth Common Yellow Violet

Viola canadensis

22. **Viola canadensis** L. Canada Violet

Caulescent perennial, (1–) 2–4 dm tall, nonstoloniferous; leaf blades ovate to broadly ovate or suborbicular, basal leaves sometimes subreniform, upper surface slightly pubescent, undersurface glabrous, base cordate or on the uppermost leaves sometimes truncate; stipules entire, lanceolate, scarious throughout, 0.7–1.7 cm long; flowers slightly fragrant; corolla white with a yellow center, 1.5–2.5 cm across, the lateral and spurred petals with dark brown-purple veins, all petals purple-tinged on back and becoming darker in age, lateral petals bearded. Mostly in eastern counties; rich moist woodlands. April–early June (and July–September).

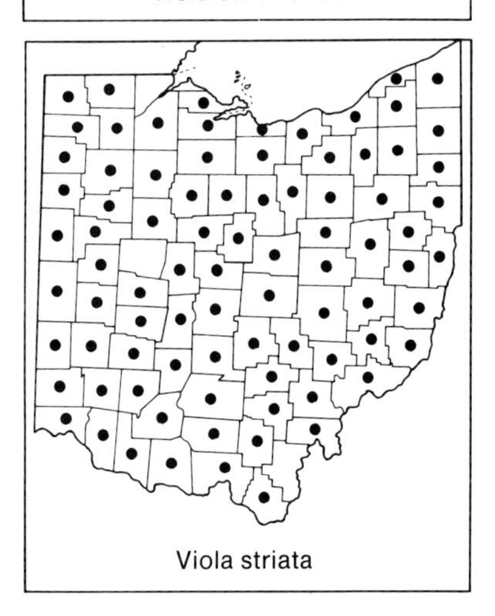
Viola striata

23. **Viola striata** Ait. Common White Violet. Striped Violet

Caulescent perennial, 1–4 (–7) dm tall, nonstoloniferous; blades of earliest leaves suborbicular to subreniform, those of later leaves ovate, both surfaces glabrous or with a few, scattered, short, stiff hairs, base cordate; stipules deeply toothed, long-lanceolate to ovate; corolla white or cream-colored, 1.0–2.3 cm across, the lateral and spurred petals striped with black-purple veins, the upper two petals sometimes slightly bearded, lateral petals heavily bearded. Throughout Ohio; low moist woodlands, stream banks, wet fields and ditches, roadside banks. April–June (and August–October).

24. **Viola conspersa** Reichenb. AMERICAN DOG VIOLET

Caulescent perennial, nonstoloniferous; stems ascending, 0.5–1.5 dm long; leaf blades reniform to orbicular or broadly ovate, upper surface glabrous or slightly pubescent, undersurface glabrous, base cordate to subtruncate; stipules lanceolate, usually sharply toothed, often with large teeth on the outer margin and shallower teeth on the inner margin, and the upper portion nearly entire; corolla lilac-violet to bluish-violet, 0.5–1.5 cm across, spur 3–6 mm long, lateral petals bearded. Chiefly in northern, especially northeastern, counties; wet or moist woodlands, floodplains. Late April–May.

25. **Viola walteri** House WALTER'S VIOLET

Low caulescent perennial, stoloniferous and forming extensive mats; stems 0.1–1.0 (–1.5) dm long, erect at first, later becoming prostrate; leaf blades broadly ovate to orbicular, usually puberulent on both surfaces, somewhat variegated with two shades of green, the veins and undersurface sometimes purplish, base cordate; stipules oblong-lanceolate, deeply toothed; corolla pinkish-violet, 1.0–1.5 cm across, spur 3–5 mm long, lateral petals heavily bearded. Known from only a few sites in south-central counties, where the species is at the northern limit of its range; calcareous rocky ledges in dry woodland openings. April–May.

Viola conspersa

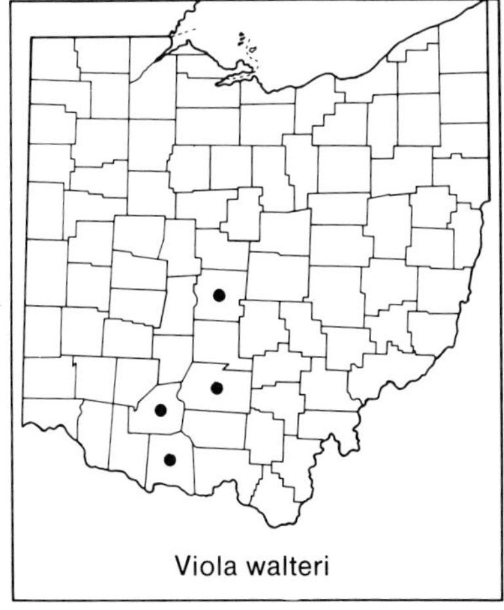
Viola walteri

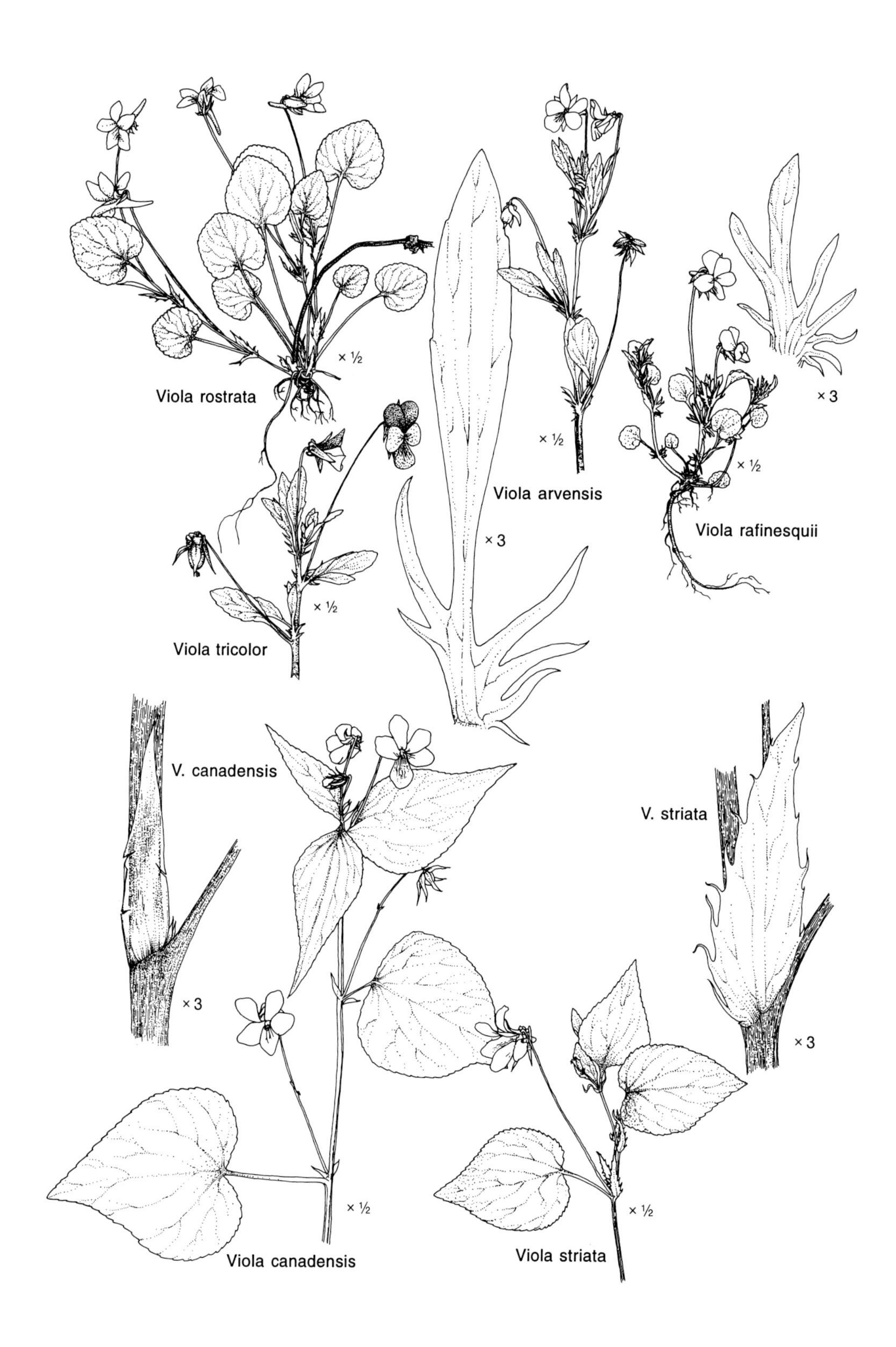
× ½
Viola rostrata
× 3
× ½
Viola arvensis
× 3
× ½
Viola rafinesquii
× ½
Viola tricolor
V. canadensis
× 3
V. striata
× 3
× ½
Viola canadensis
× ½
Viola striata

26. **Viola rostrata** Pursh LONG-SPURRED VIOLET

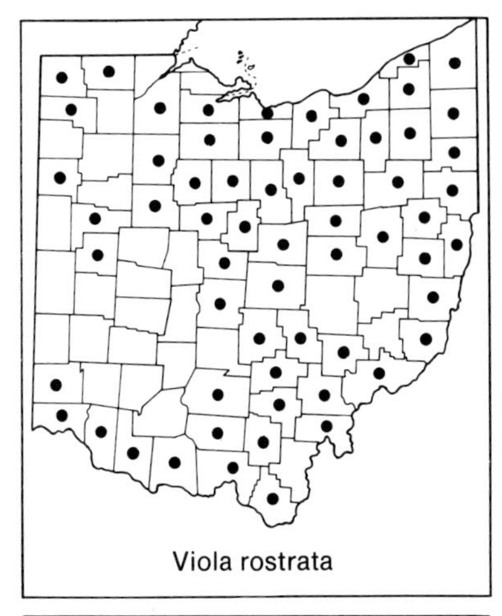
Viola rostrata

Caulescent perennial, nonstoloniferous; stems 0.2–4.0 dm long; blades of earliest leaves orbicular to reniform, those of later leaves ovate, upper surface glabrous or with few short stiff hairs, undersurface glabrous, base cordate; stipules lanceolate, sharply toothed; corolla pale lilac to darker lilac near center, 1–2 cm across, the two upper petals frequently reflexed and lying against the spur, the spur unusually long, 7–16 mm, straight or curving upward (the lateral petals also spurred in a rare form); petals beardless. Throughout eastern Ohio, but scarce westward and unknown from some western counties; rich moist woods, floodplains. April–early June.

a. Corolla with a single spurred petal . . . . forma *rostrata*

aa. Corolla with three spurred petals . . . . forma *trirostrata* Grover

The only Ohio record is from Lorain County (Grover, 1939).

27. VIOLA TRICOLOR L. JOHNNY-JUMP-UP

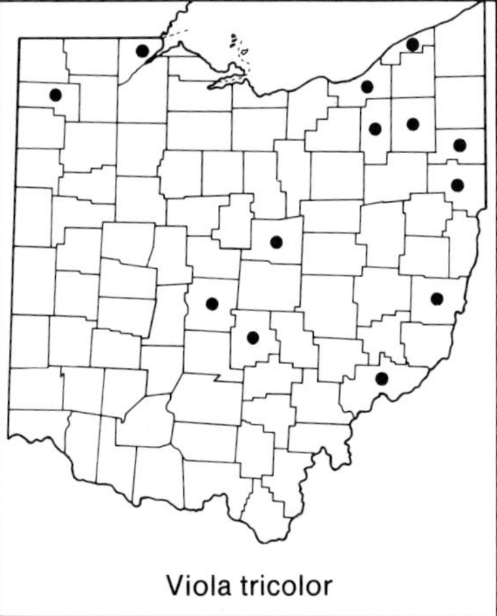
Viola tricolor

Caulescent annual, 1–4 dm tall, nonstoloniferous; blades of early leaves ovate to orbicular, later leaves oblong, elliptic, or linear, glabrous or puberulent on both surfaces, base cuneate to attenuate; stipules large, 1–4 cm long, deeply pinnately lobed, the terminal lobe oblanceolate and crenate, often resembling a smaller version of the leaf blade, not or scarcely stalked, much larger than adjacent lobes; sepals 5–14 mm long, equalling or longer than petals; corolla 1.0–2.5 cm across, nearly flat, variegated, usually with upper petals purple, lateral petals white or pale yellow, and spurred petal bright yellow, the lateral and spurred petals with wide, dark, brown-purple veins, occasional forms with both lateral and spurred petals purple-tinged; spur 2–4 mm long; lateral and spurred petals bearded. A Eurasian species sporadically escaped from cultivation, mostly in northeastern counties; adventive or perhaps locally naturalized in lawns, gardens, and other cultivated ground. April–September.

28. VIOLA ARVENSIS Murr. EUROPEAN FIELD PANSY. EUROPEAN WILD PANSY

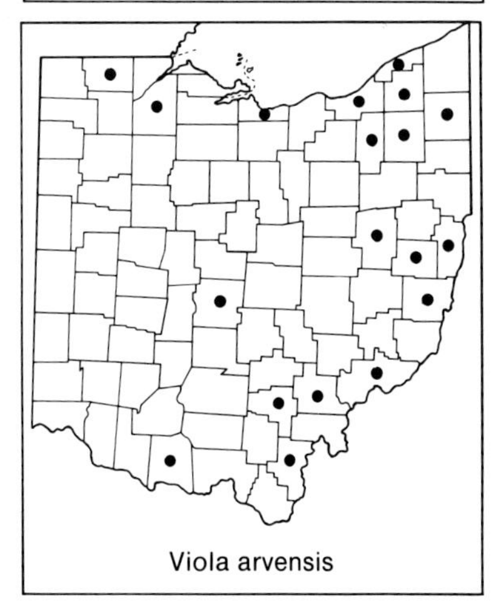
Viola arvensis

Caulescent annual, 0.7–4 dm tall, nonstoloniferous; blades of early basal leaves ovate to orbicular, blades of later leaves ovate to broadly or narrowly elliptic, more or less puberulent on both surfaces, base cuneate to truncate; stipules large, 1–5 cm long, deeply pinnatifid, the terminal lobe often long-stalked, its blade elliptic to ovate and resembling a leaf blade, much larger than other lobes; sepals 7–15 mm long, as long as or longer than the petals; corolla 0.7–2 cm across, flattened, pale yellow to cream-colored, the spurred petal bright yellow within at base, petals sometimes purple-tipped or purple-tinged or occasionally the two upper petals purple throughout, lateral and spurred petals sometimes with purple veins, spur 1–3 mm long, sometimes bluish; lateral petals bearded. A Eurasian spe-

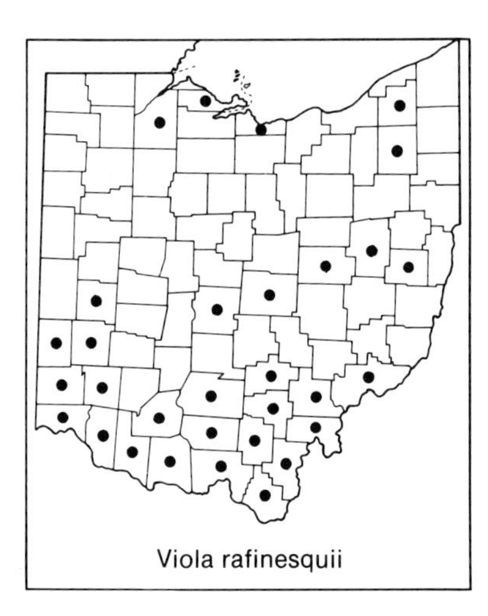
Viola rafinesquii

cies, sometimes casually cultivated; naturalized, or adventive at some sites, chiefly in eastern counties; cultivated fields, grassy fields, cemeteries, roadsides, and similar disturbed habitats. May–September.

29. **Viola rafinesquii** E. Greene WILD PANSY. FIELD PANSY

V. kitaibeliana R. & S. var. *rafinesquii* (E. Greene) Fern.

Caulescent annual, 0.3–5.0 dm tall, nonstoloniferous; blades of early leaves more or less orbicular, later leaves linear to elliptic or oblanceolate, upper surface glabrous, undersurface glabrous or puberulent, base cuneate to attenuate; stipules large, 0.5–1.5 (–3.5) cm long, deeply lobed, the terminal lobe stalkless, more or less elliptic, somewhat larger than other lobes, it and the lobe nearest it—in combination—appearing mitten-shaped; sepals 4–11 mm long, much shorter than to equalling petals; corolla 0.5–1.5 cm across, flattened, white or infrequently bluish (sometimes yellowish after drying), lateral and spurred petals often with blue-purple veins; spurred petal slightly bifid at apex, spur 0.5–2.0 mm long, sometimes blue; lateral petals bearded. Fernald (1950) regarded this taxon as naturalized from Eurasia, but Clausen, Channell, and Nur (1964) and Russell (1965) concluded that it is a native of eastern North America. Southwestern and south-central Ohio counties and scattered elsewhere; pastures, dry fields, cemeteries, roadsides. Baskin and Baskin (1972) studied the seed germination requirements of this species. March–July.

Kartesz and Gandhi (1990) conclude that the correct name for this species is *V. bicolor* Pursh.

Putative Hybrid Violets Native to Ohio

As is the case with Ohio's oaks (Braun, 1961, pp. 131–132) and sunflowers (Fisher, 1988, pp. 154–156), interspecific hybridization is a common occurrence among Ohio's native violets [see Brainerd (1924) for an account of violet hybrids in North America]. In describing the causes of this phenomenon, Russell (1954) notes that many violet species were originally separated by ecological barriers and that where habitats are unaltered by human or natural disturbance, closely related violet species, even though sympatric, do not interbreed. He writes:

> Man has, however, very greatly altered the natural habitats of plants over the greater part of the eastern United States, and, in consequence, many species, formerly separated by considerable distances, now occur side by side.
>
> Plowed and now fallow fields, lumbering roads, drained swamps, etc., have brought into contiguity edaphic situations that were formerly separated by large stretches of intermediate habitats. In addition, entirely new habitats have been created. Perhaps most important, competition pressure has been released in habitats formerly closed, allowing species hybrids to become established where before this was rarely possible.

The following naturally occurring, putative hybrids have been found in Ohio. Most of the county distribution records listed are based on identifications made by Russell in the preparation of his 1965 publication and by L. D. Miller for her 1976 study.

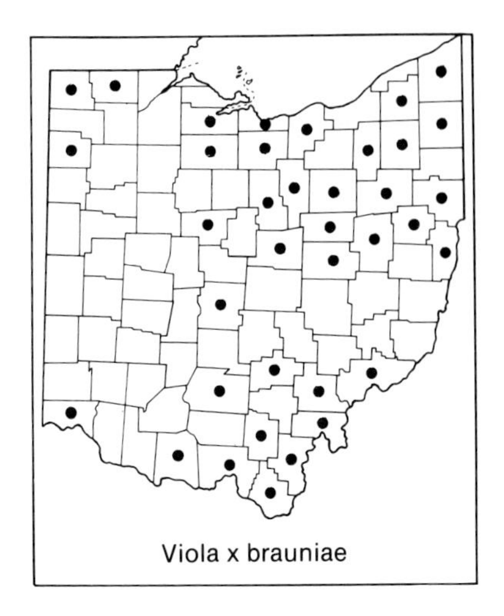
Viola x brauniae

1. **Viola × aberrans** E. Greene
   incl. *V.* × *fernaldii* House
   Viola fimbriatula Sm. × *V. sororia* Willd.
   Lake County.
2. **Viola × bissellii** House
   incl. *V.* × *conturbata* House
   Viola cucullata Ait. × *V. sororia* Willd.
   Columbiana, Erie, Fayette, Gallia, Geauga, Guernsey, Holmes, Stark, Tuscarawas, and Washington counties.
3. **Viola × brauniae** Cooperr. BRAUN'S HYBRID VIOLET
   Viola rostrata Pursh × *V. striata* Ait.
   See separate map for distribution of this, Ohio's most common hybrid violet, and Cooperrider (1986b) for details of its history and nomenclature.
4. **Viola × cooperrideri** H. Ballard COOPERRIDER'S HYBRID VIOLET
   Viola striata Ait. × *V. walteri* House
   Reported by Ballard (1993) from Adams and Highland counties.
5. **Viola × dissita** House
   Viola hirsutula Brainerd × *V. triloba* Schwein.
   Lawrence County.
6. **Viola × emarginata** (Nutt.) LeConte TRIANGLE-LEAVED VIOLET
   Viola affinis LeConte × *V. sagittata* Ait.
   Brown, Erie, Greene, and Henry counties.
   This hybrid was studied by Russell and Risser (1960).
7. **Viola × festata** House
   Viola cucullata Ait. × *V. sagittata* Ait.
   Wayne County.
8. **Viola × filicetorum** E. Greene
   incl. *V.* × *consona* House
   Viola affinis LeConte × *V. sororia* Willd.
   Athens, Columbiana, Crawford, Erie, Lorain, and Scioto counties.
9. **Viola × malteana** House
   Viola conspersa Reichenb. × *V. rostrata* Pursh
   Reported by Ballard (1990) from Scioto County.
10. **Viola × milleri** Moldenke
    Viola affinis LeConte × *V. triloba* Schwein.
    Huron, Jackson, Knox, Lorain, and Wyandot counties.

11. **Viola × populifolia** E. Greene
    Viola sororia Willd. × *V. triloba* Schwein.
    Clinton, Holmes, Lorain, Perry, Ross, and Scioto counties.
12. **Viola × porteriana** Pollard
    Viola cucullata Ait. × *V. fimbriatula* Sm.
    Fairfield County.

In addition, *Viola* × *eclipes* H. Ballard, reported from Michigan (Ballard, 1989), is to be expected in northern Ohio, where its parents, *V. conspersa* and *V. striata*, are sympatric.

Passifloraceae. PASSION-FLOWER FAMILY

A chiefly tropical family of about 600 species; most are members of *Passiflora*, the only genus represented in the Ohio flora. Brizicky (1961a) studied the family in southeastern United States.

1. Passiflora L. PASSION-FLOWER

Herbaceous vines climbing by means of tendrils; leaves alternate, simple, 3-lobed; flowers solitary or few in leaf axils, perfect and with a short tube formed from the fused bases of the calyx, corolla, and corona; calyx of 5 separate sepals; corolla actinomorphic, of 5 separate petals; the corona (an additional unique and ornate floral structure located just inward from the corolla) composed of numerous, long, colored filamentous processes; stamens 5, fused basally to form a staminal tube, that tube in turn fused to a stalk (the androgynophore) bearing the ovary, the 5 filaments diverging from the staminal tube just beneath the ovary; carpels 3, fused to form a compound pistil with 3 separate styles; ovary superior, elevated above the corona and perianth by the androgynophore; fruit a berry. A tropical, chiefly American, genus of about 400 species.

The berry of the tropical plant *Passiflora edulis* Sims provides the passion-fruit juice of tropical beverages.

a. Flowers 1.5–2.5 cm wide; leaves shallowly 3-lobed, the sinuses extending halfway or less than halfway to base of blade, the lobes obtuse to rounded at apex, margins of lobes entire . . . . 1. *Passiflora lutea*

aa. Flowers 4–8 cm wide; leaves deeply 3- (4- or 5-) lobed, the sinuses between the principal lobes extending more than halfway to base of blade, the lobes acuminate at apex, margins of lobes serrate . . . . 2. *Passiflora incarnata*

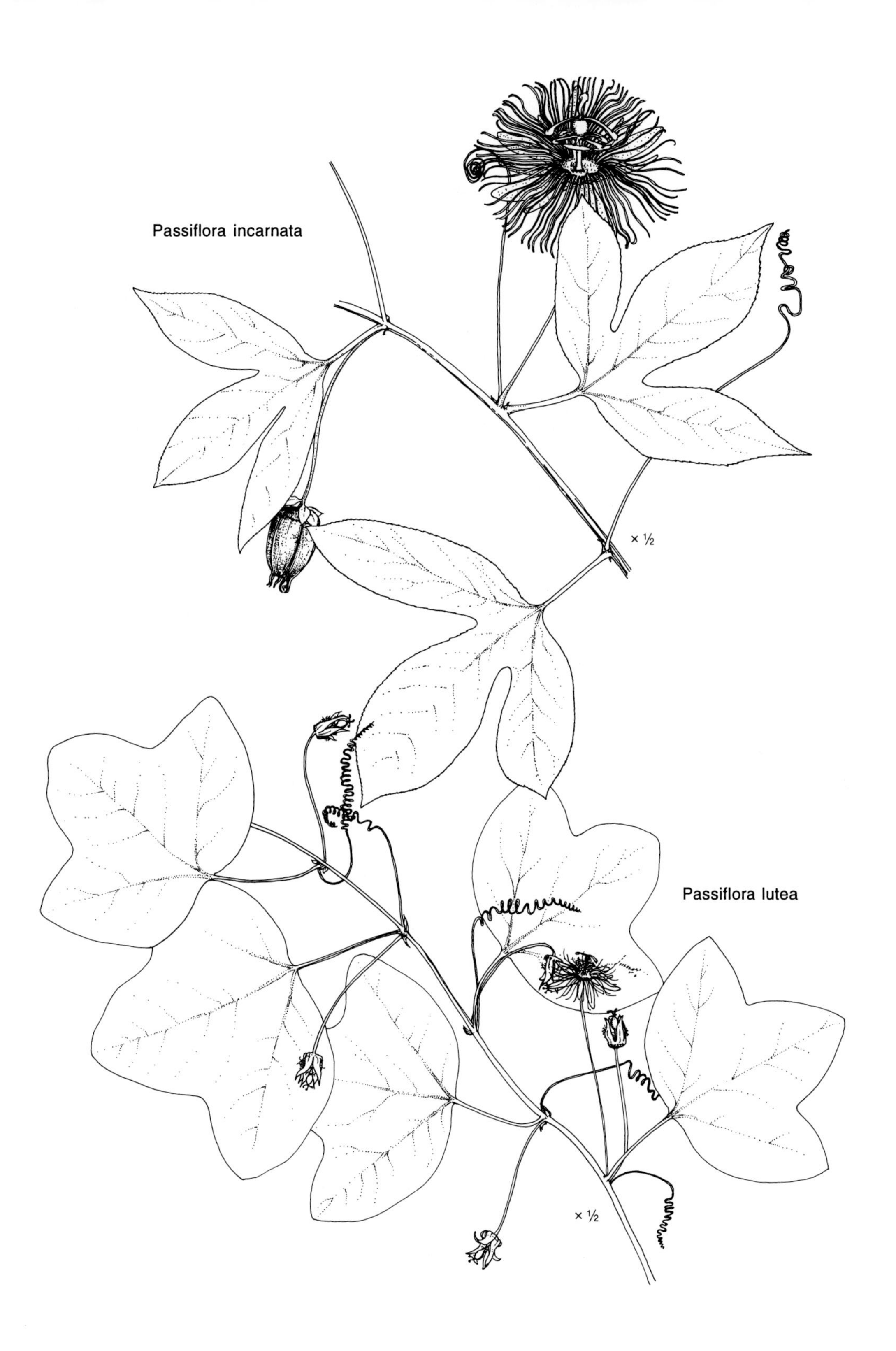
Passiflora incarnata
× 1/2
Passiflora lutea
× 1/2

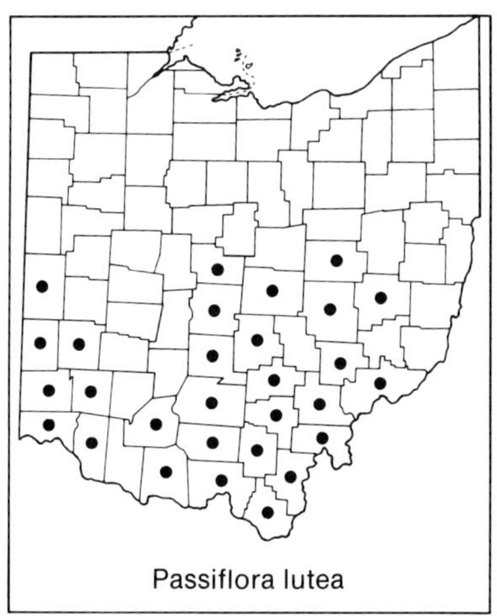
Passiflora lutea

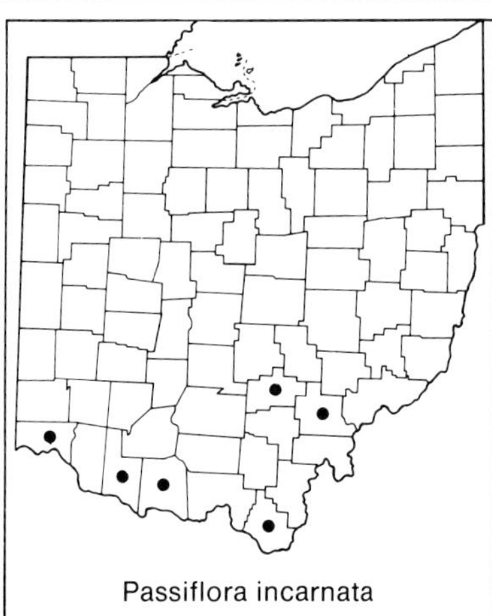
Passiflora incarnata

1. **Passiflora lutea** L. Yellow Passion-flower
incl. var. *lutea* and var. *glabriflora* Fern.

Perennial, herbaceous vine; leaves shallowly 3-lobed, the sinuses extending halfway or less than halfway to base of blade, the lobes obtuse to rounded at apex, margins of lobes entire; flowers small, 1.5–2.5 cm across, greenish-yellow, outer fringe of corona yellowish; fruit purple to blackish, more or less globose, ca. 1.2 cm in diameter. A species of southeastern and south-central United States, reaching a northern boundary of its range in the southern counties of Ohio; woodland borders and openings, moist thickets, roadsides. July–September.

Fernald (1950), Gleason (1952a), and Gleason and Cronquist (1991) recognize the two varieties listed in synonymy above, the pilose var. *lutea*, and the glabrous var. *glabriflora*. In Ohio, the two intergrade to such an extent that their recognition seems pointless.

2. **Passiflora incarnata** L. Maypop

Perennial, herbaceous vine; leaves deeply 3- (or 4- or 5-) lobed, the sinuses between the principal lobes extending more than halfway to base of blade, the lobes acuminate at apex, margins of the lobes serrate; flowers large, 4–8 cm across, greenish-white; fruit yellow, ovoid-ellipsoid, ca. 5 cm long. A species of southeastern and south-central United States, reaching a northern boundary of its range in southern Ohio; open wooded areas, thickets. July–September.

Cactaceae. CACTUS FAMILY

A famous family of New World succulents with 1,200 or more species, two of which are members of the Ohio flora. Benson's (1982) major study covers the family in the United States and Canada. Noelle and Blackwell (1972) studied the Ohio taxa.

1. Opuntia Mill. PRICKLY-PEAR

Succulent, essentially leafless, spiny and/or bristly perennials with flattish segmented and jointed stems, their longer spines, when present, and short bristles (called "glochids") clustered in the axils (called "areoles") of much-reduced or missing leaves; flowers solitary and showy, perfect, actinomorphic, calyx and corolla of numerous separate sepals and petals and some intermediate transitional structures, their bases fused with those of the stamens to form a short hypanthium above the inferior ovary; stamens numerous, separate; carpels several, fused to form a compound pistil with a single style and several stigmas; ovary inferior and sunken into, surrounded by, and fused with the stem tip, the sides therefore bearing vestigial leaves and areoles with tiny bristles; fruit a berry. A genus of nearly 300 species, ranging throughout much of North and South America, but most abundant in arid regions. The nomenclature and synonymy of both species below follow Benson (1982).

a. Areoles producing no major spines, or one major spine, or rarely 2; roots fibrous but without tuberous thickenings . . 1. *Opuntia humifusa*

aa. Areoles producing 3–4 (–7) major spines; roots fibrous and with tuberous thickenings . . . . 2. *Opuntia macrorhiza*

1. **Opuntia humifusa** (Raf.) Raf. COMMON PRICKLY-PEAR

incl. O. *compressa* (Salisb.) Macbr.; O. *vulgaris* of authors, not Mill.

Low, succulent, mat-forming perennial; roots fibrous, without tuberous thickenings; stem with one to several broadly elliptic to subcircular segments; aeroles with bristles and 0–1 (–2) longer spines 1–3.5 cm long; flowers yellow, showy, 6–7 cm across; fruit dark red. In several southern and a few north-central counties; open, dry, usually sandy, sites. Moseley (1931) believed that this species may have been introduced into northern Ohio by Native Americans. June–July.

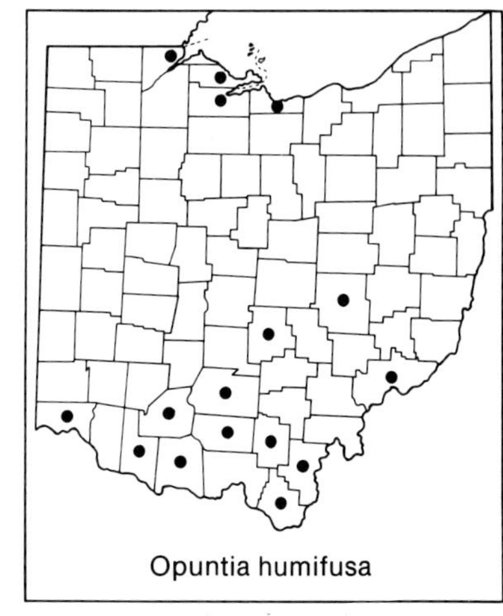
Opuntia humifusa

2. OPUNTIA MACRORHIZA Engelm.

incl. O. *tortispina* Engelm.

Low, succulent, clump-forming perennial; roots fibrous and with tuberous thickenings; stem of one to several broadly elliptic to subcircular segments; areoles producing 3–4 (–7) longer spines of varying lengths, the longest to 5 cm long; flowers yellow or reddish at base, 5–6 cm across; fruit dark red. A species of south-central and southwestern United States, naturalized in Hamilton County, in southwesternmost Ohio. Noelle and Blackwell (1972) report it to be locally established in dry disturbed sites, at "several closely spaced stations," and speculate that it may have been introduced into Ohio between 1916 and 1918. Not illustrated. June–July.

Opuntia macrorhiza

Thymelaeaceae. MEZEREUM FAMILY

A widely distributed family of about 500 species, represented in the Ohio flora by one native species and the following two alien adventives. The Asian species, *Daphne mezereum* L., MEZEREON, an ornamental shrub with fragrant, white to pale purple flowers, was collected from a swamp in Ashtabula County, the northeasternmost county of Ohio, in the 1930s. More recently, *Thymelaea passerina* (L.) Coss. & Germ., a small inconspicuous weed of European origin, was discovered on a disturbed gravelly slope in Montgomery County in southwestern Ohio. Vincent and Thieret (1987) report this discovery and provide an illustration of the species. For an additional illustration, see Mohlenbrock (1982a, p. 147). Nevling (1962) studied this family in southeastern United States.

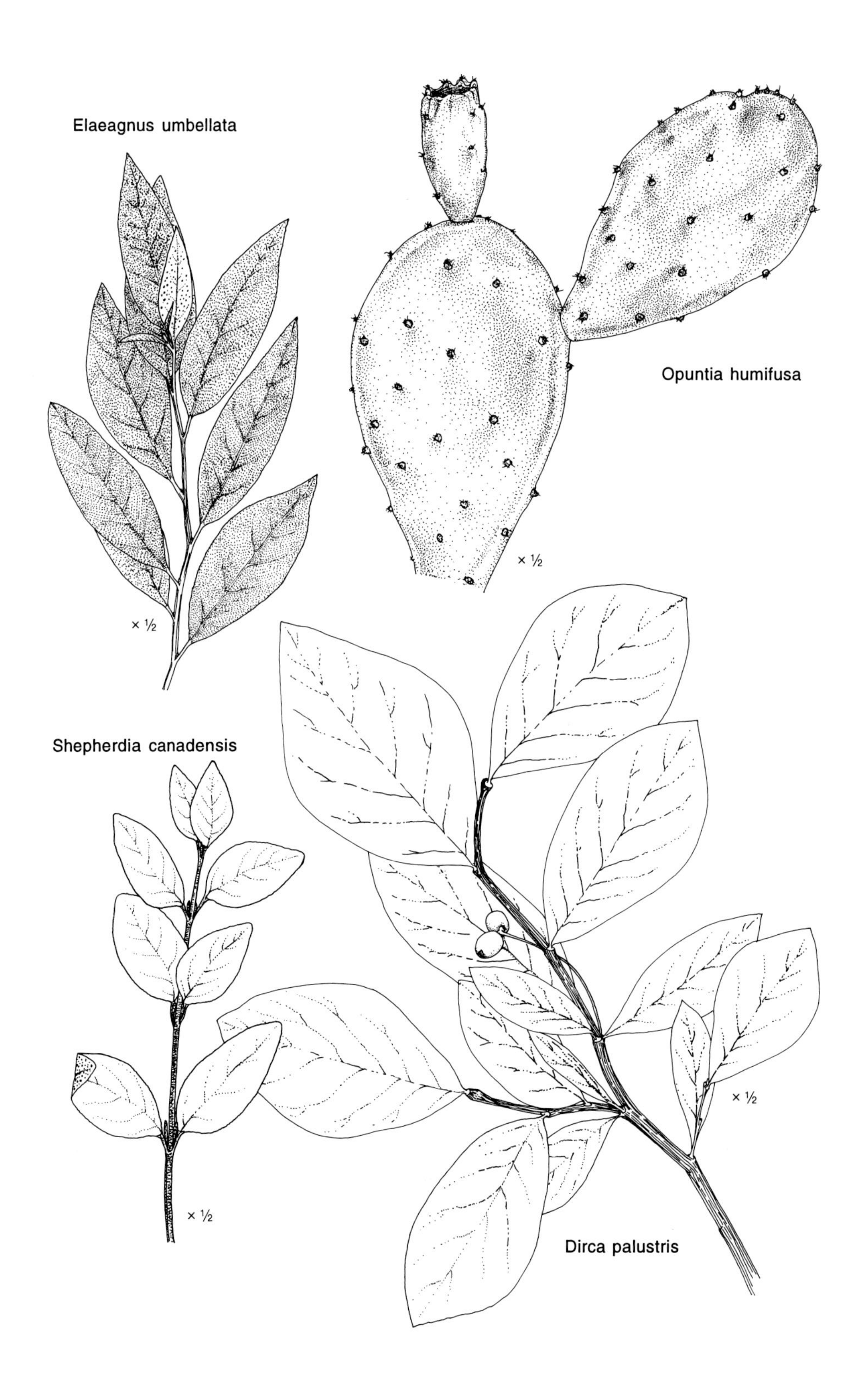
Elaeagnus umbellata
Opuntia humifusa
× ½
× ½
Shepherdia canadensis
× ½
× ½
Dirca palustris

a. Plants herbaceous annuals; leaves 1–2 mm wide; see above .... Thymelaea
aa. Plants woody perennial shrubs; leaves (8–) 10 mm or more wide.
b. Leaves broadly elliptic to broadly subrhombic or broadly obovate, principal leaves 3–6 cm wide, more than half as wide as long; sepals minute; stamens and style exserted .... 1. *Dirca*
bb. Leaves narrowly elliptic to oblanceolate, principal leaves 1–2 cm wide, less than half as wide as long; sepals evident, spreading and petal-like; stamens and style included; see above .... *Daphne*

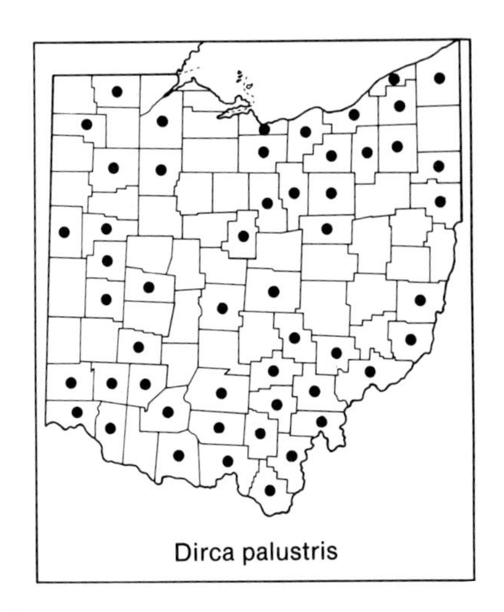
Dirca palustris

1. Dirca L. LEATHERWOOD

A North American genus of two species, the other, *D. occidentalis* A. Gray, is native to a very restricted area in California.

1. **Dirca palustris** L. LEATHERWOOD
Much-branched shrub, to 3 m (10 feet) tall; the young branches and their bark unusually supple and leathery; leaves alternate, simple, entire, elliptic-rhombic; flowers yellowish, appearing before the leaves, actinomorphic, perfect, in few-flowered clusters; fruit a greenish drupe. In much of northeastern and southern Ohio, and at scattered sites elsewhere, but not common; rich mesic to dryish woodlands. April.

Elaeagnaceae. OLEASTER FAMILY

Shrubs or small trees; leaves simple, alternate or opposite; branches and leaves with scattered to dense lepidote and/or stellate hairs; flowers small, yellow or yellowish, actinomorphic, solitary or in few-flowered racemose or umbellate inflorescences, calyx tubular and 4-lobed; corolla none; flowers perfect or imperfect, if the latter the species functionally dioecious; fruits drupelike. A small family of three genera and about 50 species, chiefly of Eurasia and North America. Graham (1964b) studied the family in southeastern United States.

a. Leaves alternate; stamens 4 .... 1. *Elaeagnus*
aa. Leaves opposite; stamens 8 .... 2. *Shepherdia*

1. Elaeagnus L. OLEASTER

Shrubs or small trees; leaves alternate. A genus of perhaps 40 species, chiefly of Eurasia. The Asian species *Elaeagnus multiflora* Thunb. was collected in 1938 from a disturbed woods in Jackson County, where it was adventive or perhaps locally naturalized.

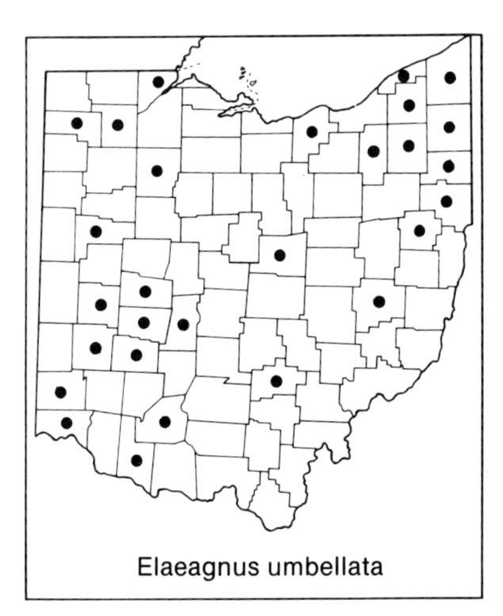
Elaeagnus umbellata

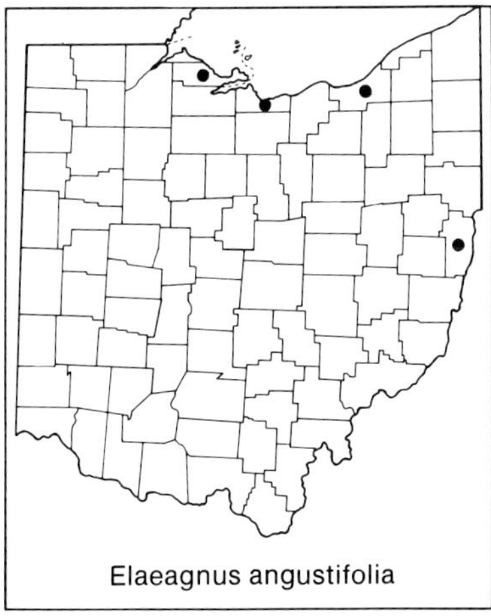
Elaeagnus angustifolia

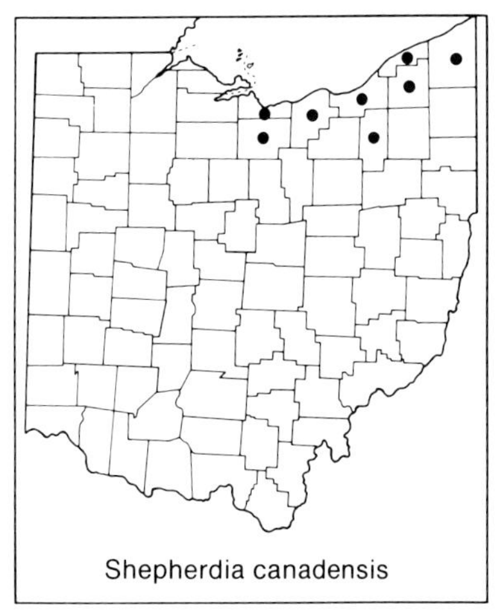
Shepherdia canadensis

a. Leaves lanceolate or narrowly lanceolate; leaves and young branches with silvery scales only; calyx tube ca. 1½ times as long as calyx lobes; fruits yellow, with silvery scales; stalk of fruit 2–4 mm long . . . . 2. *Elaeagnus angustifolia*

aa. Leaves elliptic to ovate-oblong or ovate; leaves and young branches with both silvery and brown scales or with brown scales only; calyx tube equalling to much longer than lobes; fruits red at maturity; stalk of fruit 3–25 mm long.

b. Stalk of fruit 3–5 (–8) mm long; calyx tube 2–3 times as long as the lobes. . . . . 1. *Elaeagnus umbellata*

bb. Stalk of fruit (5–) 9–25 mm long; calyx tube 1–2 times as long as lobes; see above . . . . *Elaeagnus multiflora*

1. Elaeagnus umbellata Thunb. Autumn-olive

Shrub, to 3 m (10 feet) tall; branches with both silvery and brown scales; leaves silvery on undersurface, green or soon becoming green on upper surface, elliptic to ovate, ovate-oblong or elliptic-lanceolate, to 7 cm long, petioles with silvery and brown scales; flowers not or only slightly fragrant; calyx tube twice (or more) as long as calyx lobes; fruit red at maturity, ellipsoid to ovoid-globose, ca. 8–9 mm long; stalk of fruit short, 3–5 (–8) mm long. An Asian species, escaping with increasing frequency from plantings and now becoming naturalized throughout much of Ohio; roadsides and other disturbed areas. Ebinger and Lehnen (1981) and Ebinger (1983) describe the spread of this species in Illinois; Voss (1985b) discusses its spread in Michigan. May–June.

2. Elaeagnus angustifolia L. Russian Olive. Oleaster

Shrub or small tree, to 6 m (20 feet) tall; branches with dense silvery scales; leaves silvery on both surfaces or greenish above, lanceolate to narrowly lanceolate, to 10 cm long, petioles silvery; flowers pungently fragrant; calyx tube ca. 1½ times as long as calyx lobes; fruits yellow with silvery scales, subglobose to ovoid, to 12 mm long; stalk of fruit short, 3–4 mm long. A Eurasian species escaped and either adventive or naturalized at a few Ohio sites, mostly in northern counties; moist disturbed habitats. Not illustrated. May–June.

2. Shepherdia Nutt. BUFFALO-BERRY

A North American genus of three species, one of which is a native member of the Ohio flora.

1. **Shepherdia canadensis** (L.) Nutt. Buffalo-berry

Shrub, to 2 m (7 feet) tall; branches with dense brown scales; leaves opposite, variable in shape, ranging from ovate, rhombic, or broadly elliptic to oblongish or narrowly elliptic, the undersurface silvery and with reddish-brown scales, upper surface green; flowers small and inconspicuous, opening in early spring before the leaves appear; fruits orangish. Transcontinental across northern North America, at a southern boundary of its range in northeastern and north-central counties of Ohio; woodland openings, crests of ridges and bluffs. April–May.

Lythraceae. LOOSESTRIFE FAMILY

Annual or perennial herbs or rarely shrubs; leaves simple; flowers perfect, solitary or in few-flowered clusters in leaf axils or in terminal spikes; bases of sepals, petals, and stamens fused to form an often conspicuously parallel-ribbed or strongly veined hypanthium; calyx and corolla of 4, 5, or 6 sepals and petals, separate above the hypanthium; stamens few to many; carpels 2–5, fused to form a compound pistil with a single style; ovary superior; fruit a capsule. A chiefly tropical family of about 500 species. Graham (1964a, 1975) studied the family in southeastern United States. The treatment here is based in large part on Blackwell's (1970) study of the family as it is represented in the Ohio flora.

Perhaps the best known member of the family is the cultivated ornamental shrub, *Lagerstroemia indica* L., CRAPE-MYRTLE. It survives Ohio winters only in the southernmost counties.

Didiplis diandra (DC.) Alph. Wood (*Peplis diandra* DC.), WATER-PURSLANE, was attributed to Ohio by Fernald (1950), Gleason (1952a), and others, but excluded by Roberts and Cooperrider (1982) for lack of substantiating specimens.

a. Hypanthium subglobose or globose, as wide or nearly as wide as long.
  b. Plants perennial; stems woody, at least at base, usually more than 1 m long; leaves mostly opposite or in whorls of 3 or 4, sometimes subopposite, or rarely alternate; flowers and fruits on pedicels (2–) 4–10 mm long; stamens 10 (5 long and 5 short); petals 5, typically 5–15 mm long; sepals 5–7. . . . . 3. *Decodon*
  bb. Plants annual; stems herbaceous, less than 1 m long; leaves opposite; flowers and fruits sessile or on pedicels to 1.5 mm long; stamens 1–8; petals 4, minute, 1–3 mm long and early deciduous, or petals lacking; sepals 4.
    c. Leaves strongly narrowed at base, more or less petiolate; flowers solitary in leaf axils . . . . 1. *Rotala*
    cc. Leaves slightly widened and somewhat clasping at base, sessile; flowers 1–7 in each axil. . . . . 2. *Ammannia*

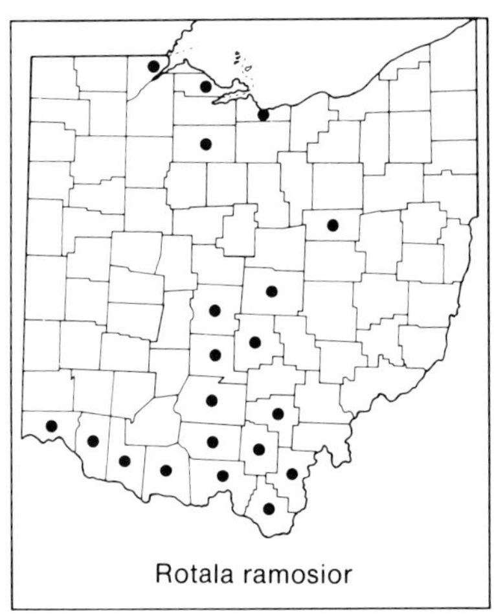
Rotala ramosior

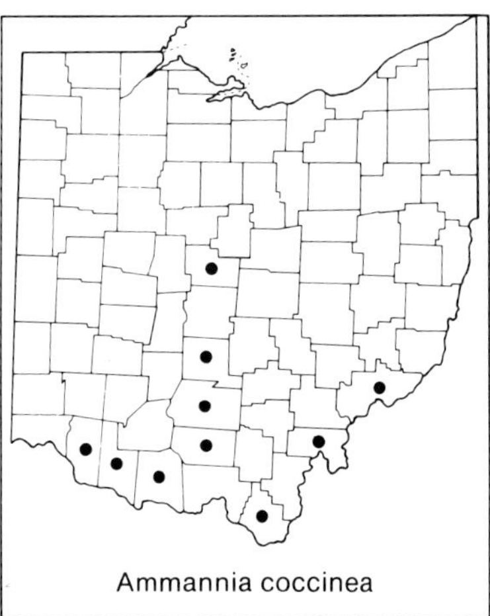
Ammannia coccinea

aa. Hypanthium cylindrical, urn-shaped, or vase-shaped, at least twice as long as wide.

b. Plants glabrous or if pubescent, not viscid; leaves sessile or nearly so; hypanthium symmetric; corolla actinomorphic, the petals equal in size . . . . 4. *Lythrum*

bb. Plants viscid-pubescent; principal leaves with petioles 5–12 mm long; hypanthium asymmetric, gibbous near base; petals markedly unequal in size, two larger and four smaller . . . . 5. *Cuphea*

1. Rotala L.

A genus of some 40 species, chiefly of tropical regions in Asia and Africa; one species is a member of the Ohio flora.

1. **Rotala ramosior** (L.) Koehne   TOOTH-CUP
   incl. var. *ramosior* and var. *interior* Fern. & Grisc.

Low annual, to 3.5 dm tall; leaves opposite, narrowed at base and more or less petiolate; flowers and fruits solitary in leaf axils, sessile, hypanthium globose to subglobose, as wide as or nearly as wide as long; sepals 4; petals 4; stamens 4. Southern and a few northern counties; muddy ground on banks of streams and ponds, wet ditches, and other damp sites. July–early October.

Fernald (1950) and Gleason (1952a) assign Ohio plants to var. *interior*; I follow Blackwell (1970) in recognizing no varieties within this species.

2. Ammannia L.

Annuals; leaves opposite; flowers actinomorphic, tetramerous, usually in axillary clusters, the hypanthium more or less globose. A widely distributed genus of about 25 species. Some data below are from Graham's (1985) revision of the genus in the Western Hemisphere.

a. Flowers and fruits on peduncles 1–4 mm long; capsules 3–4 mm wide; flowers 1–7 in each axil . . . . 1. *Ammannia coccinea*

aa. Flowers and fruits subsessile, on peduncles 0.3–0.8 mm long; capsules 4–6 mm wide; flowers 1–5 in each axil . . . . 2. *Ammannia robusta*

1. **Ammannia coccinea** Rottb.   LONG-LEAVED AMMANNIA

Annual, to 1 m tall; leaves broadly linear, clasping at base; flowers and fruits 1–7 in each axil, on peduncles 1–4 mm long; capsules more or less globose, 3–4 mm wide. A species ranging from South America northward throughout much of southeastern and south-central United States, reaching a northern boundary of its range in southern and central Ohio; mud flats, pond and river margins, and other wet habitats. July–October.

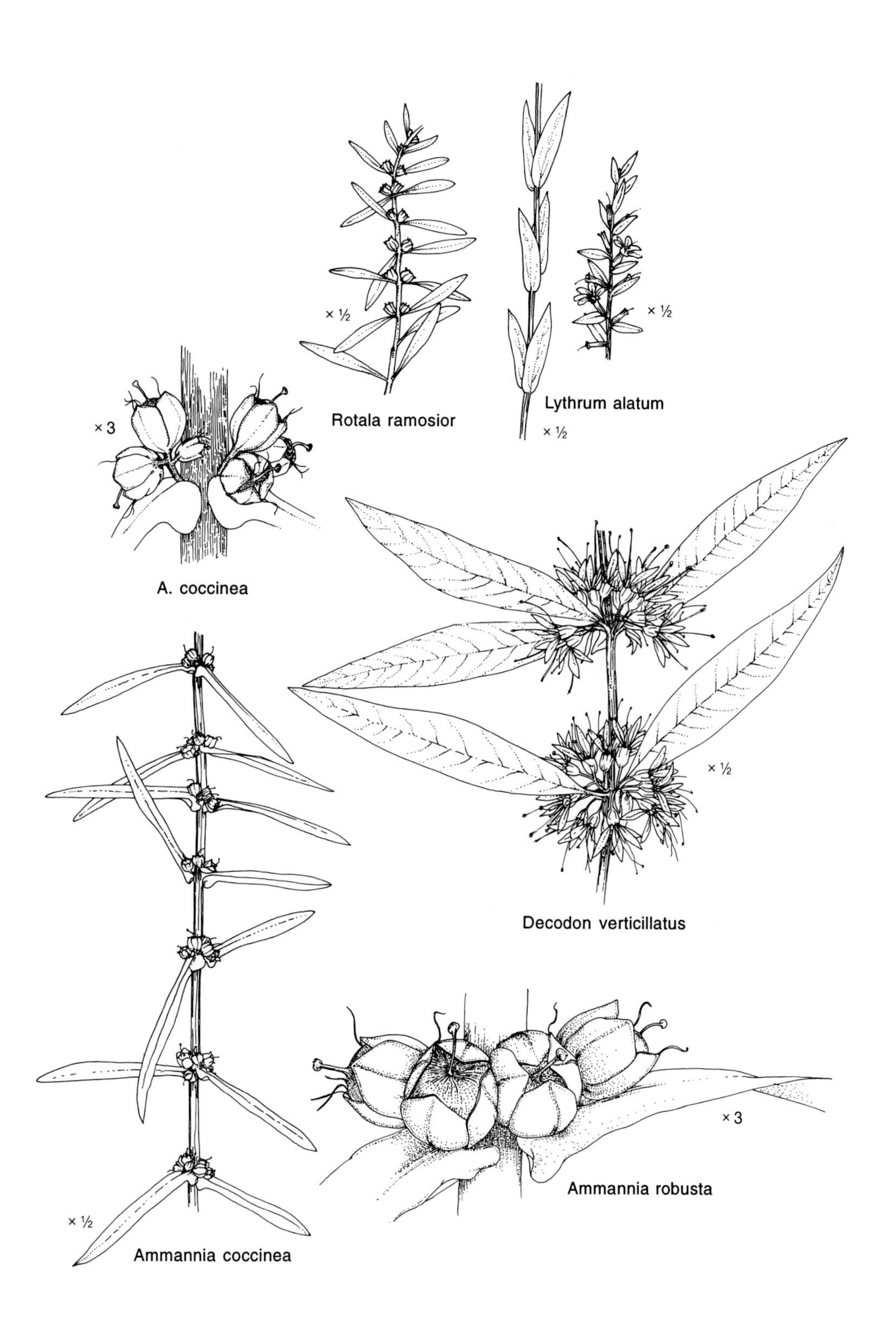
× ½
Rotala ramosior
Lythrum alatum
× ½
× ½
× 3
A. coccinea
× ½
Decodon verticillatus
× 3
Ammannia robusta
× ½
Ammannia coccinea

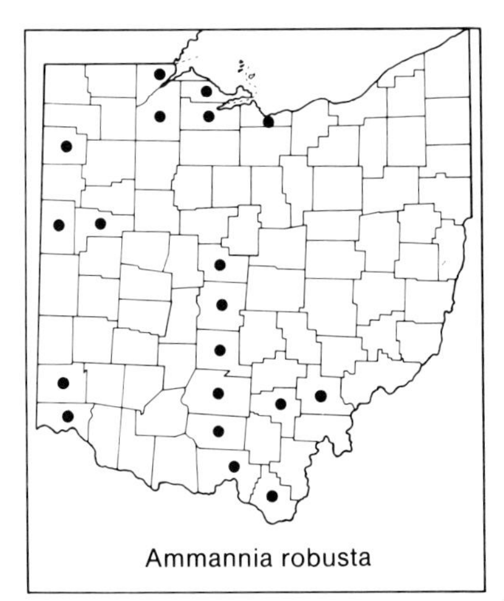
Ammannia robusta

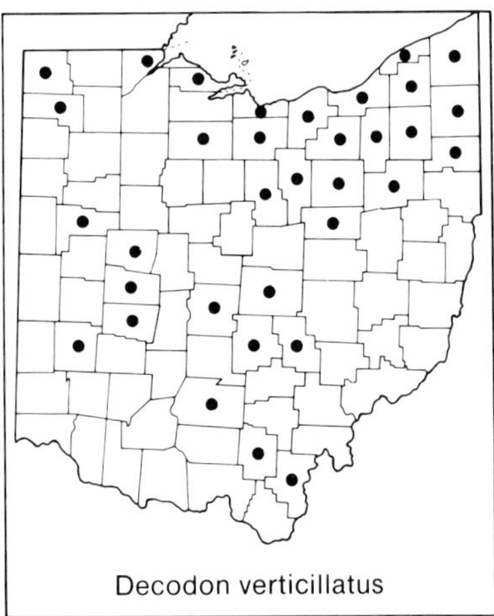
Decodon verticillatus

2. **Ammannia robusta** Heer & Regel

Annual, to 1 m tall; leaves broadly linear, clasping at base; flowers and fruits 1–5 in each axil, subsessile on peduncles 0.3–0.8 mm long; capsules more or less globose, 4–6 mm wide. South-central and northwestern counties, native or perhaps naturalized; mud flats, pond and river margins, and other wet habitats. July–October.

Graham (1979) demonstrated the validity of this species, previously hidden within the circumscription of A. *coccinea*.

3. Decodon J. F. Gmel. SWAMP LOOSESTRIFE

A genus with a single species.

1. **Decodon verticillatus** (L.) Ell. SWAMP LOOSESTRIFE. WATER-WILLOW

incl. var. *verticillatus* and var. *laevigatus* T. & G.

Much-branched, somewhat scrambling shrub, the branches 1–3 m long, arching, rooting at the tips, and forming dense tangled thickets; leaves to 15 cm long, opposite or in whorls of 3 or 4 per node, sometimes subopposite and rarely alternate, elliptic-oblong to lanceolate, tapering at apex and base, margin entire; flowers in showy axillary clusters on pedicels (2–) 4–10 mm long; hypanthium subglobose; sepals 5–7; petals 5, pinkish-purple, 5–15 mm long; stamens 10 (5 long and 5 short). Northeastern and north-central counties and at scattered sites elsewhere; in shallow water at margins of ponds, lakes, and rivers, and in marshes. Late July–September.

Fernald (1950) and Gleason (1952a) assign Ohio plants to var. *laevigatus*; however Graham (1975) favored recognition of no varieties within the species, the taxonomy adopted here.

4. Lythrum L. LOOSESTRIFE

Perennials or annuals; flowers actinomorphic; hypanthium cylindric; petals 6. A genus of about 30 species, chiefly of North America, but widely scattered elsewhere in the world. *Lythrum hyssopifolia* L., HYSSOP-LEAVED LOOSESTRIFE, an annual of the Atlantic coastal region, was collected several times from 1935 to 1955, in a wet field in Pickaway County; it seems best treated as an adventive in the Ohio flora. The key below is adapted from that of Blackwell (1970).

a. Stems pubescent, at least above; leaves of main stem opposite or sometimes in whorls of 3; flowers several in each of many axillary clusters, together forming a narrow, racemose, almost spike-like panicle; stamens 12 . . . . 2. *Lythrum salicaria*

aa. Stems glabrous; the upper leaves of main stem usually alternate, the lower usually opposite; flowers one or two in each axil; stamens 6.

b. Plants annual, simple or few-branched; principal leaves narrowed at the base; hypanthium at most only faintly nerved; stamens and style included; petals 2 to 3 mm long; see above .... *Lythrum hyssopifolia*

bb. Plants perennial, usually several-branched; principal leaves rounded, truncate, or slightly clasping at base, not narrowed; hypanthium prominently nerved; either stamens or style exserted (flowers dimorphic); petals 3–6 mm long .... 1. *Lythrum alatum*

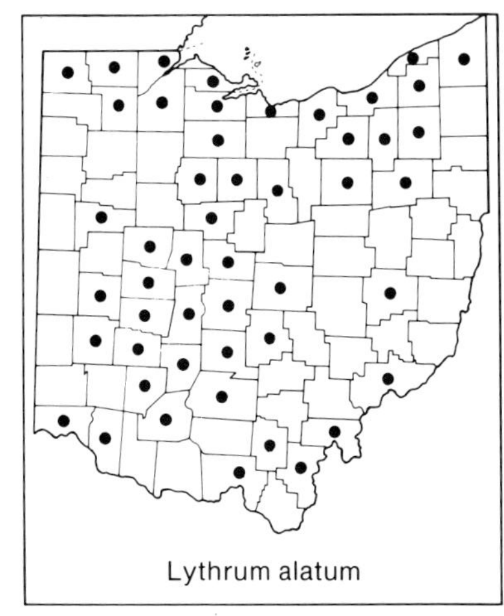
Lythrum alatum

1. **Lythrum alatum** Pursh   WING-ANGLED LOOSESTRIFE

Perennial, to 1 m tall, usually with several branches; stems glabrous; lower leaves opposite, the upper alternate, principal leaves rounded, truncate, or slightly clasping at base; flowers solitary in axils of reduced leaves; hypanthium cylindric, prominently nerved; petals 6, pinkish-purple, 3–6 mm long. Chiefly in northern and southwestern counties; moist open ground in fields and bogs, dry calcareous sites. Late June–September.

Following Graham (1975), Ohio plants are assigned to var. *alatum* (including *L. dacotanum* Nieuwl.).

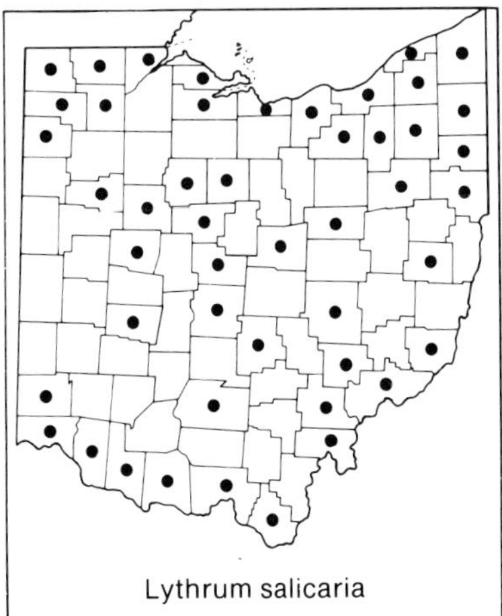
Lythrum salicaria

2. LYTHRUM SALICARIA L.   PURPLE LOOSESTRIFE. SPIKED LOOSESTRIFE

Erect branched perennial, to 1.5 m tall; stems pubescent throughout or at least above and in the inflorescence; leaves opposite or sometimes in whorls of 3, sessile and more or less clasping at base; flowers several in each of many axillary clusters, together forming a dense, showy, narrow, racemose, almost spike-like, multiflowered panicle; hypanthium cylindric, conspicuously nerved; petals 6, reddish-purple, 7–9 mm long. A frequently cultivated Eurasian species, escaped and naturalized throughout much of Ohio; wet ditches and shores, moist roadsides and other moist, open, disturbed sites. Unfortunately, this attractive species is an aggressive weed, now becoming abundant at some stations and crowding out native wetland plants; Stuckey (1980) details its spread across North America. Balogh and Bookhout (1989) studied its occurrence in Ohio's Lake Erie marshes. Thompson, Stuckey, and Thompson (1987) discuss the impact and control of this weed. June–September.

Following the treatment in *Flora Europaea* (Tutin et al., 1964–80), no varieties are recognized within this species.

5. Cuphea P. Br.

A chiefly tropical genus of the Western Hemisphere with perhaps 250 species. Graham (1988) revised *Cuphea* section *Heterodon*, which includes the one species below.

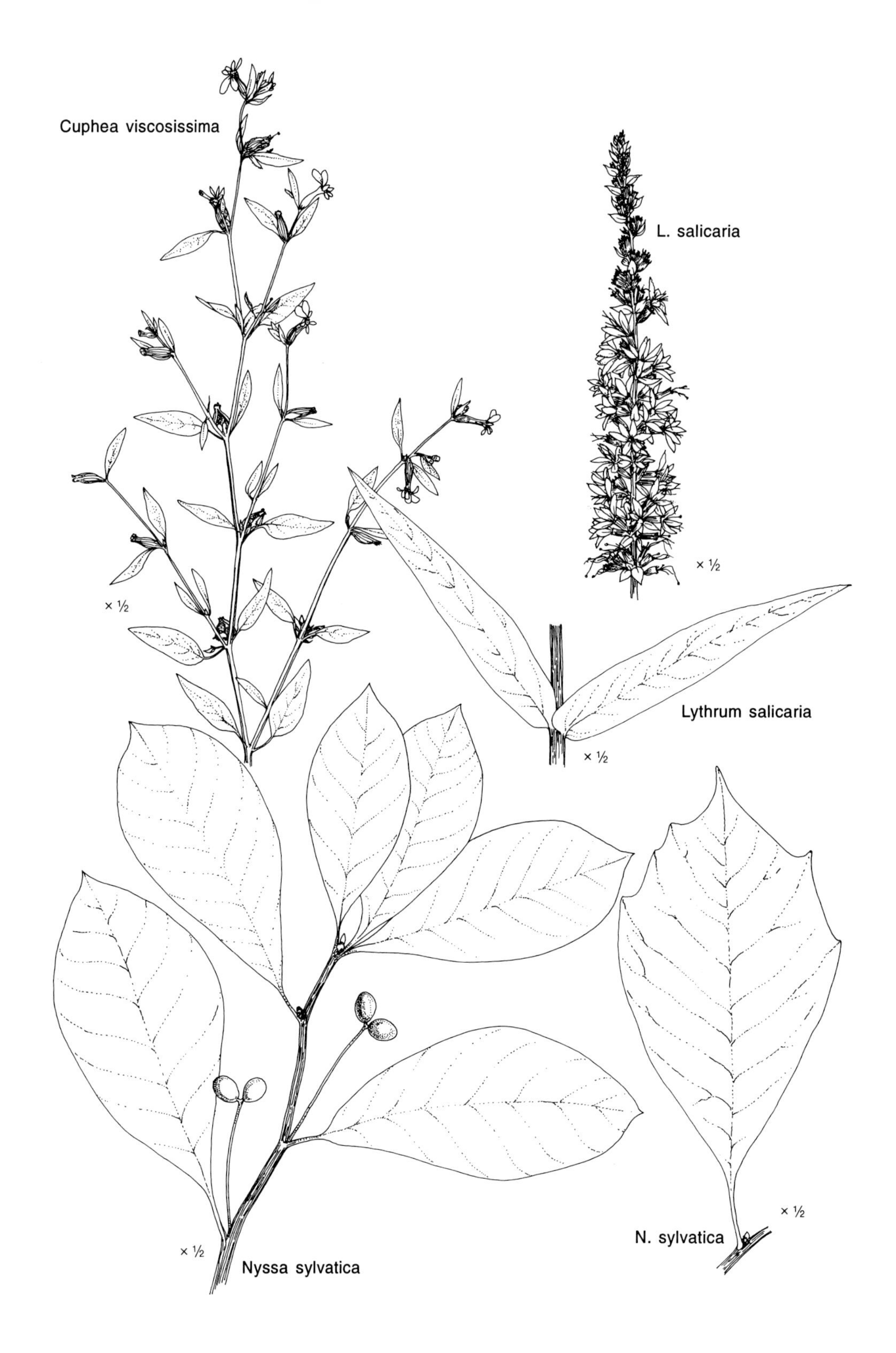
Cuphea viscosissima
L. salicaria
× ½
× ½
Lythrum salicaria
× ½
× ½
N. sylvatica
× ½
Nyssa sylvatica

1. **Cuphea viscosissima** Jacq. CLAMMY CUPHEA. BLUE WAXWEED
C. *petiolata* (L.) Koehne

Branched annual, to 6 dm tall; stems viscid-pubescent and sometimes reddish above; principal leaves with petioles 5–12 mm long and blades 1.5–6.0 cm long, the blades varying from narrowly to broadly lanceolate; flowers mostly solitary in leaf axils; hypanthium cylindric, gibbous near base; petals 6, purple, markedly unequal in size with two larger than the other four. Southern and eastern counties; dry fields and roadside banks, woodland openings and borders, sometimes in mesic to moist habitats. Late June–September.

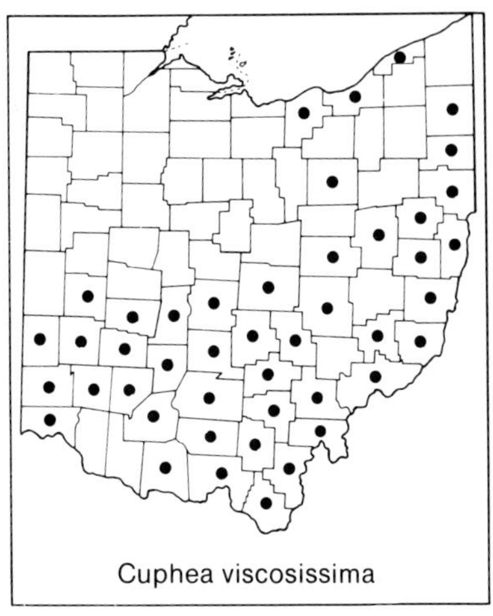

Cuphea viscosissima

Nyssaceae. SOUR-GUM FAMILY

A small family of Asia and North America with three genera and about 10 species, one of which is native to Ohio. Eyde (1966) studied the family in southeastern United States.

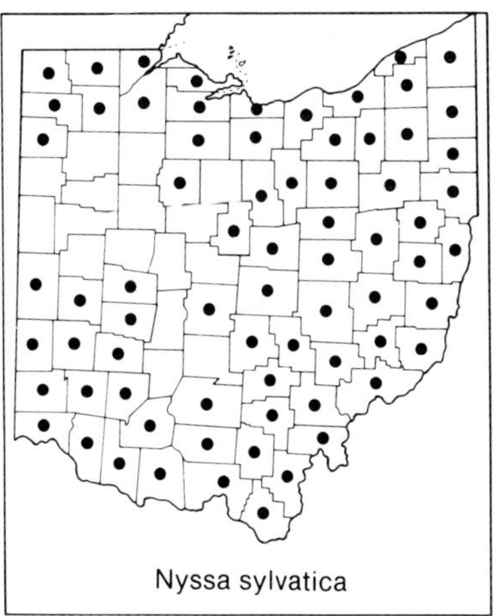

Nyssa sylvatica

1. Nyssa L. TUPELO

A genus of about seven species, native to North America and southeastern Asia. This genus is placed in the family Cornaceae by Gleason (1952a) and Gleason and Cronquist (1963, 1991), but maintained as the type genus of a separate family by Cronquist (1968, 1981).

1. **Nyssa sylvatica** Marsh. BLACK-GUM. SOUR-GUM. TUPELO
incl. var. *sylvatica* and var. *caroliniana* (Poir.) Fern.

Tree, mostly to ca. 18 m (60 feet) tall, but mature specimens sometimes attaining heights of ca. 30 m (100 feet); trunks of mature trees with thick, deeply checkered bark; leaves alternate, simple, mostly entire but occasionally with one or a few broad teeth near apex; flowers minute, greenish, in few-flowered clusters; flowers imperfect, the species dioecious; fruit a dark blue drupe, usually borne in pairs at the end of a peduncle 2–5 cm long. Throughout Ohio, except in some west-central counties; mesic to dry or moist woodlands. The shining scarlet leaves of this tree are the first harbinger of autumn, a few changing color in midsummer. May.

There is considerable variation in leaf shape, but I follow Braun (1961) and Eyde (1966) in recognizing no varieties within Ohio material.

Some herbarium specimens have been confused with *Diospyros virginiana* (p. 281). Most can be separated using one or more of the following characters.

Nyssa sylvatica: pith white and evenly diaphragmed; leaves usually widest above the middle, sometimes with a few coarse teeth, the ultimate veinlets on undersurface of leaves not conspicuously orange-brown.

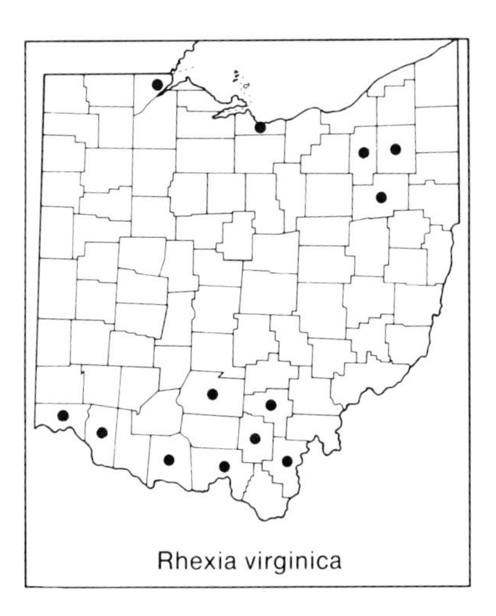
Rhexia virginica

Diospyros virginiana: pith often brownish and chambered with irregular and usually incomplete cross-partitions; leaves usually widest at or below the middle, always lacking teeth, the ultimate veinlets on undersurface of leaves conspicuously orange-brown.

Melastomataceae. MELASTOME FAMILY

One of the earth's major families of dicots, with about 150 genera and 4,000 species. Chiefly tropical, its distribution is centered in South America. One species is a part of the Ohio flora. Wurdack and Kral (1982) studied the family in southeastern United States.

1. Rhexia L. MEADOW-BEAUTY. DEERGRASS

A North American genus of 12–15 species, most of them in southeastern United States. James (1956) and Kral and Bostick (1969) have published revisions of the genus.

1. **Rhexia virginica** L. Virginia Meadow-beauty

Perennial, to 6 dm tall, with few or no branches; stems erect, quadrangular and with low wings; leaves simple, opposite, spinulose-serrate, with 3 main more or less parallel veins; flowers perfect, in few- to several-flowered cymes; bases of sepals, petals, and stamens fused to form a globose hypanthium, contracted above into a small neck; calyx of 4 sepals, separate above the hypanthium; corolla actinomorphic, of 4 showy rosy-purple petals, separate above the hypanthium; stamens 8, bright yellow, with long, curved, and more or less crescent-shaped anthers; carpels 4, fused to form a compound pistil with a single style and stigma; ovary partly inferior; fruit a capsule. Known from southern and a few northern counties; open fields, often in moist sandy soil. Late July–early September.

Onagraceae. EVENING-PRIMROSE FAMILY

Annual, biennial, or perennial herbs; leaves simple; flowers perfect, either solitary in leaf axils or in spikes, racemes, or panicles; calyx of 2–7 sepals, usually 4; corolla actinomorphic, of 2–7 petals, usually 4, or corolla absent; stamens 2–12, usually 8; in most species the bases of the sepals, petals, and stamens fused to form a tubular hypanthium topping the inferior ovary; carpels 2–several, usually 4, united to form a compound pistil with a single style; ovary inferior; fruit usually a capsule. A family of about 700 species, with distribution centered in temperate regions of the Western Hemisphere. Much of the text below is based on Munz' (1965) taxonomic treatment of the family in North America.

Clarkia pulchella Pursh, Farewell-to-spring, an ornamental annual with pink to purplish flowers, was collected once, in 1891, as an adventive in Franklin County.

The tropical Fuchsia, *Fuchsia magellanica* Lam., and cultivars derived from it and its hybrids with other species are often seen in greenhouses and in summertime hanging baskets. Forms with red sepals and purple petals and others with pink sepals and white petals are popular.

a. Petals lacking; sepals 4; stamens 4 . . . . 1. *Ludwigia*
aa. Petals present; sepals 2–7; stamens 2–12.
  b. Flowers with 2 sepals, 2 petals, and 2 stamens . . . . 5. *Circaea*
  bb. Flowers with 4 or more sepals, petals, and stamens.
    c. Hypanthium above ovary lacking or very short, 0–2 mm long.
      d. Petals deep yellow to whitish-yellow; sepals persistent on fruit apex; seeds lacking hairs . . . . 1. *Ludwigia*
      dd. Petals purple, rose, pink, or white; sepals deciduous; seeds comose, with cluster of hairs at one end . . . . 2. *Epilobium*
    cc. Hypanthium above ovary (3–) 4–100 (–150) mm long.
      d. Fruit indehiscent; petals white, pink, or red, and 3–7 mm long . . . . 4. *Gaura*
      dd. Fruit dehiscent; petals in most species yellow and 5–30 mm long, in the others white, pink, or lavender, and 10–40 mm long.
        e. Petals 3-lobed, pink to lavender; filaments attached near base of anther; see above . . . . *Clarkia*
        ee. Petals unlobed, yellow in most species, white to pink in others; filaments attached near middle of anther . . . . 3. *Oenothera*

1. Ludwigia L.

A genus of about 75 species, widely distributed in tropical and temperate regions. Following Munz (1965)—with some misgivings—*Ludwigia* is treated sensu lato, so as to include *Jussiaea* L. Peng (1989) made a systematic study of section *Microcarpium*, which includes species no. 5 below.

a. Capsules 12–50 mm long, 4 or more times as long as wide; sepals 4–7; petals 4–7; stamens 8–14.
  b. Sepals 4; capsules broadly cylindric to obpyramidal, 12–20 mm long . . . . 1. *Ludwigia decurrens*
  bb. Sepals 5 (–7); capsules cylindric or narrowly cylindric, 20–50 mm long.
    c. Peduncles 2–6 cm long; fully developed leaves obovate to oblanceolate, with 6–9 principal veins on each side of midrib, apex subacute to obtuse or rounded . . . . 2. *Ludwigia peploides*

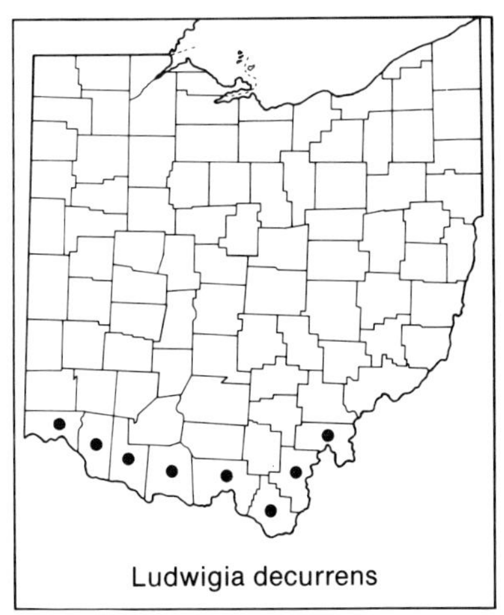

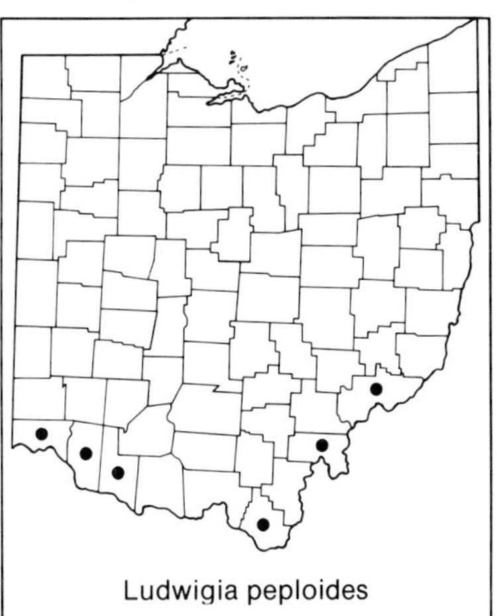

cc. Peduncles 0.5–1.5 cm long; fully developed leaves lanceolate to oblanceolate, with 11–16 principal veins on each side of midrib, apex acute to narrowly obtuse . . . . 3. *Ludwigia leptocarpa*

aa. Capsules 2–6 mm long, 1–2 times as long as wide; sepals 4; petals 4 or more; stamens 4.

b. Leaves opposite; stems either creeping and ascending or floating; flowers and fruits sessile or nearly so; petals none. . . . . 6. *Ludwigia palustris*

bb. Leaves alternate; stems erect or nearly so; flowers and fruits sessile or short-pedicellate; petals present or not.

c. Flowers and fruits on pedicels 2–7 mm long; petals yellow, 8–10 mm long; sepals 7–11 mm long; capsule 5–7 mm thick . . . . 4. *Ludwigia alternifolia*

cc. Flowers and fruits sessile; petals either green and minute (less than 1 mm long) or lacking; sepals 2.5–3.5 mm long; capsule 3–5 mm thick . . . . 5. *Ludwigia polycarpa*

1. Ludwigia decurrens Walt. Erect Primrose-willow
Jussiaea decurrens (Walt.) DC.

Erect branched annual, to ca. 1 m tall; leaves alternate, sessile, the principal leaves lanceolate, to 12 cm long; flowers solitary in axils of upper leaves; sepals 4; petals 4, yellow; stamens 8; capsules broadly cylindric to obpyramidal, conspicuously 4-angled, 12–20 mm long. Mud flats and wet ground along the Ohio River, in southwestern and south-central counties. Ranging from South America to North America, the species is perhaps native to Ohio, but is more likely a naturalized element in the flora (so regarded by Stuckey and Roberts, 1977). August–October.

2. Ludwigia peploides (HBK.) Raven Creeping Primrose-willow

Creeping, mat-forming, branched perennial; leaves alternate, petioles 2–4 cm long, blades obovate to oblanceolate, with 6–9 principal veins on each side of midrib, apex subacute to obtuse or rounded, to 10 cm long; flowers solitary in leaf axils; peduncles 2–6 cm long; sepals 5; petals 5, yellow; stamens 10; capsules long-cylindric, more or less 5-angled, 25–40 mm long. Mud flats along or near the Ohio River, in the southernmost tier of counties; adventive, or perhaps becoming naturalized, from southeastern United States. The earliest Ohio collection I have seen was made in 1981. Not illustrated. July–October.

Ohio plants belong to var. *glabrescens* (Ktze.) Shinners (*Jussiaea repens* L. var. *glabrescens* Ktze.). The species has also been recently added (as a presumed adventive) to the flora of West Virginia, from sites along or near the Ohio River (Brant, 1987).

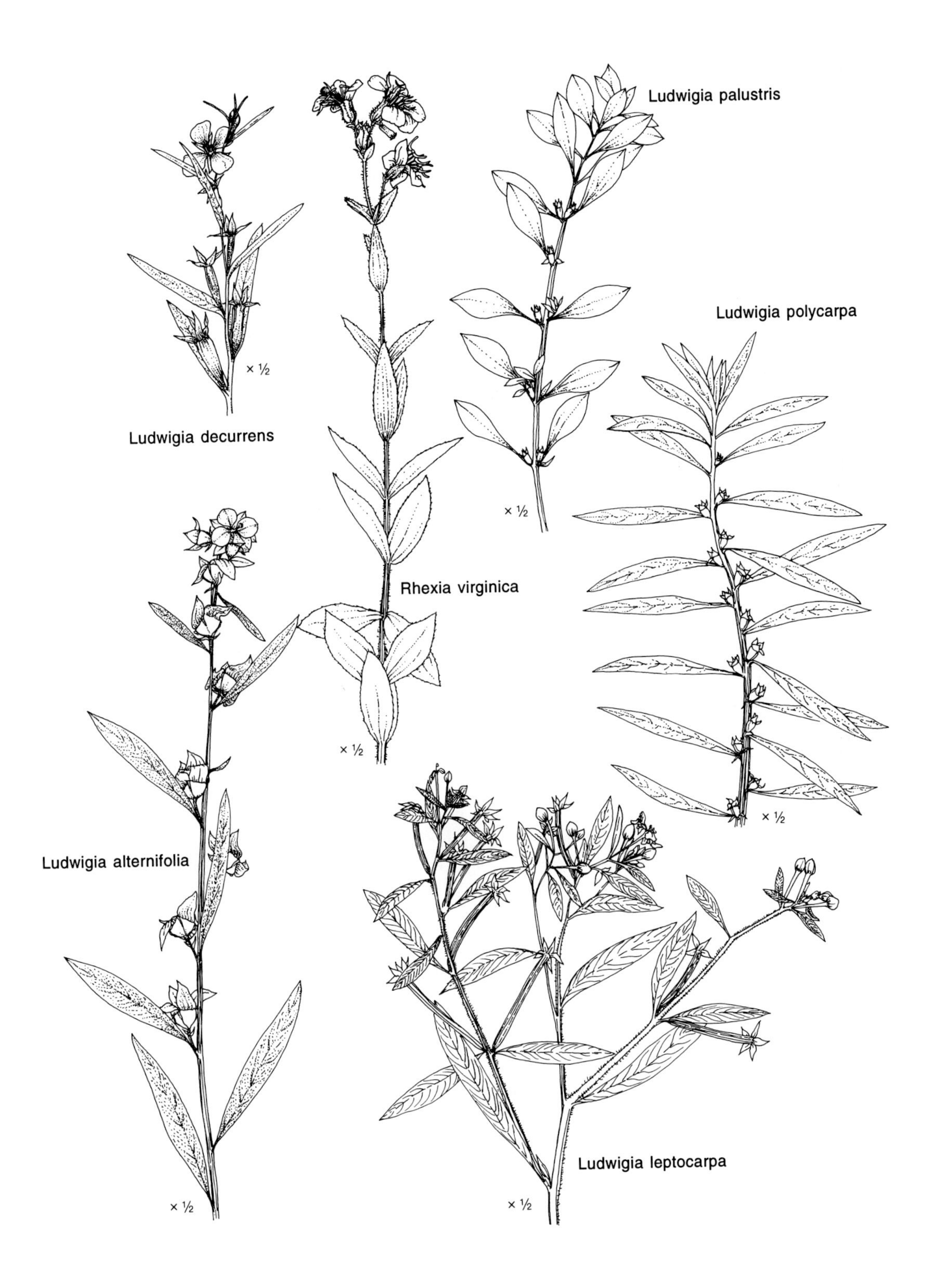

Ludwigia palustris
Ludwigia decurrens
× ½
Ludwigia polycarpa
× ½
Rhexia virginica
× ½
Ludwigia alternifolia
× ½
Ludwigia leptocarpa
× ½
× ½

3. LUDWIGIA LEPTOCARPA (Nutt.) Hara HAIRY PRIMROSE-WILLOW
Jussiaea leptocarpa Nutt.
Erect branched annual or perennial, to ca. 1 m tall; stems densely pubescent above; leaves alternate, lanceolate to oblanceolate, to 15 cm long, with 11–16 principal veins on each side of midrib, tapering to short petiole at base, apex acute to narrowly obtuse; flowers solitary in leaf axils; peduncles 0.5–1.5 cm long; sepals 5 or 6 (or 7); petals 5 or 6 (or 7), yellow; stamens 10 or 12 (sometimes 14); capsules long-cylindric, 10–14 nerved, 15–50 mm long. Mud flats and wet weedy shores of the Ohio River, in the southernmost tier of counties; naturalized from southern states. August–October.

4. **Ludwigia alternifolia** L. SEEDBOX
Erect branched perennial, to ca. 1 m tall; leaves alternate, lanceolate to linear-lanceolate, blades to 10 cm long, tapering at base to very short petiole; flowers solitary in leaf axils, flowers and fruits on short pedicels 2–7 mm long; sepals 4, 7–11 mm long; petals 4, yellow, 8–10 mm long; stamens 4; capsules squarish, 5–6 mm long, 5–7 mm thick. Northernmost, eastern, and southern counties, but absent from much of the western half of Ohio; moist open ground, ditches, shores, swamps. July–August.

5. **Ludwigia polycarpa** Short & R. Peter FALSE LOOSESTRIFE
Erect branched perennial, to 8 dm tall; leaves alternate, lance-linear, to 12 cm long, tapering at base to a very short petiole; flowers and fruits solitary in leaf axils, sessile; sepals 4, 2.5–3.5 mm long; petals none or minute and green; stamens 4; capsules very short-cylindric, 3–4 mm long, 3–5 mm thick. Chiefly in northwestern counties, and from a few scattered sites

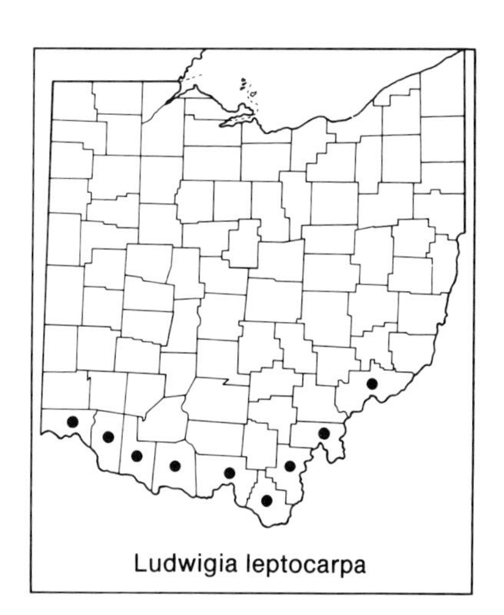
Ludwigia leptocarpa

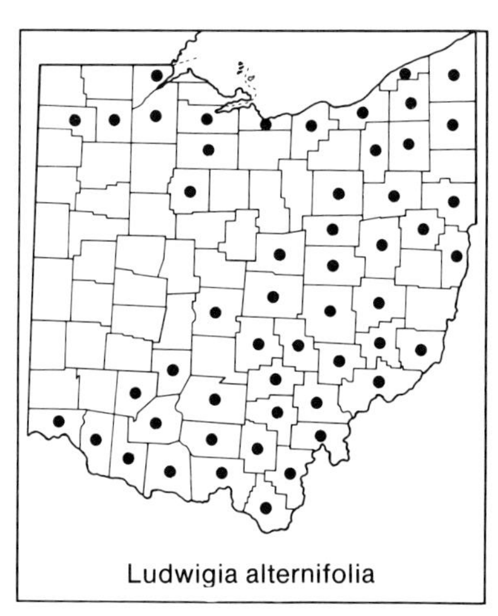
Ludwigia alternifolia

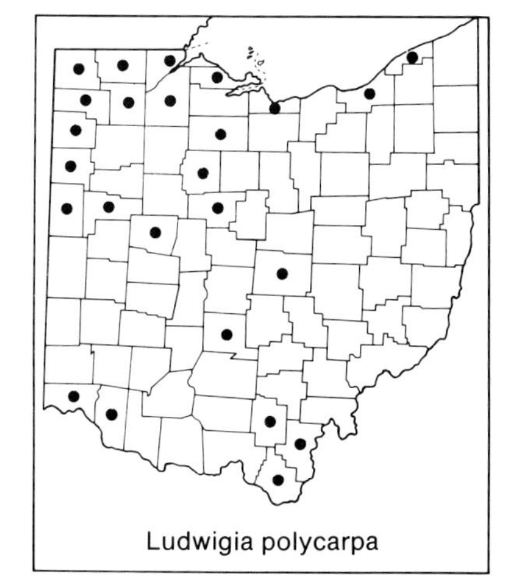
Ludwigia polycarpa

elsewhere in Ohio; wet ground, swamps, ditches, fields, shores. Additional record: Butler County (Peng, 1989). August–September.

6. **Ludwigia palustris** (L.) Ell. WATER-PURSLANE. MARSH-PURSLANE
Annual or perennial herbs; stems either creeping and ascending, or partially submersed; leaves opposite, petiolate, blades more or less elliptic, to 3 cm long; flowers and fruits solitary in leaf axils, sessile or nearly so; sepals 4, ca. 1 mm long; petals none; stamens 4; capsules short-cylindric, 2–4 mm long and about as thick. Throughout Ohio, except for a few west-central counties; wet ditches and shores, shallow water of ponds. July–September.

I follow Munz (1965) in treating Ohio plants as var. *palustris* (incl. var. *americana* (DC.) Fern. & Grisc.).

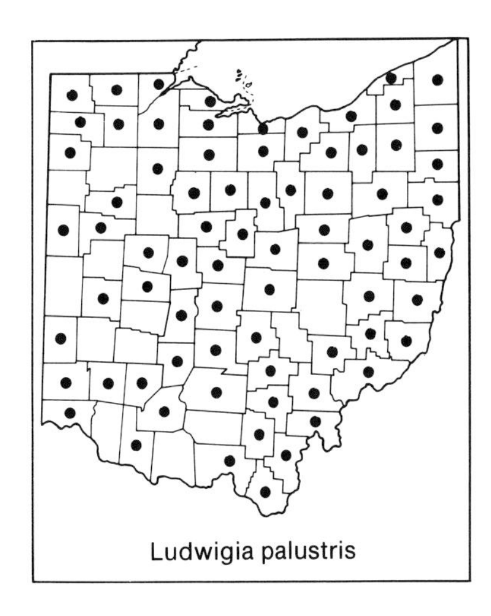
Ludwigia palustris

2. Epilobium L. WILLOW-HERB

Perennial herbs; sepals and petals 4; stamens 8; seeds comose, with a cluster of hairs at one end. A genus of more than 100 species, of temperate regions.

a. Petals 10–20 mm long; stigma 4-lobed.
 b. Stems glabrous below, puberulent above; principal leaves alternate; flowers in well-defined racemes, all but the lowest subtended by only minute bracts; petals medium-purple to pinkish, not notched at apex . . . . 1. *Epilobium angustifolium*
 bb. Stems pubescent to densely villous; principal leaves opposite; flowers solitary in axils of leaves or in weakly defined racemes, all subtended by slightly reduced leaves; petals purple or purplish, notched at apex . . . . 2. *Epilobium hirsutum*
aa. Petals 2–7 (–9) mm long; stigma either unlobed or 4-lobed (only in *E. parviflorum*).
 b. Stigma 4-lobed; petals (4–) 6–7 (–9) mm long; stems densely pubescent to villous with long, mostly spreading hairs . . . . 3. *Epilobium parviflorum*
 bb. Stigma unlobed; petals 2–6 mm long; stems variously pubescent, but not villous with long, mostly spreading hairs.
  c. Principal leaves linear to linear-lanceolate, 1–6 (–8) mm wide, their margins entire or very remotely and shallowly denticulate; stems terete, without decurrent ridges or lines of pubescence running downward from bases of the leaves.
   d. Stems with short, straight, spreading hairs; leaves mostly 3–6 (–8) mm wide, the lateral veins apparent . . . . 4. *Epilobium strictum*

dd. Stems with appressed or incurved hairs; leaves mostly 1–3 mm wide, the lateral veins not or scarcely visible . . . . 5. *Epilobium leptophyllum*

cc. Principal leaves lanceolate to long-lanceolate or narrowly ovate, 5–25 mm wide, their margins conspicuously serrulate; stems 4-angled, with either decurrent ridges or lines of pubescence running downward from bases of the leaves.

d. Principal leaves long-lanceolate, 5–15 cm long, with acuminate apices; some or all leaves with distinct petioles 3–6 mm long; some or all internodes with low pubescence running in lines down the stem, the uppermost internodes sometimes with scattered pubescence, the lower glabrous; sepal tips in bud forming a beak mostly 0.2–0.4 mm long, the individual sepal tips somewhat divergent and spreading from this beak; mature seeds with brown coma (pale when immature) . . . . 6. *Epilobium coloratum*

dd. Principal leaves lanceolate to narrowly ovate, 3–7 cm long, with acute apices; most leaves sessile or narrowed into a very short winged petiole less than 2 mm long; internodes with pubescence scattered more or less evenly across stem surface or some internodes glabrous; sepal tips in bud scarcely forming a beak 0.1 mm long or less, the individual sepal tips held together, not spreading; mature seeds with whitish coma . . . . 7. *Epilobium ciliatum*

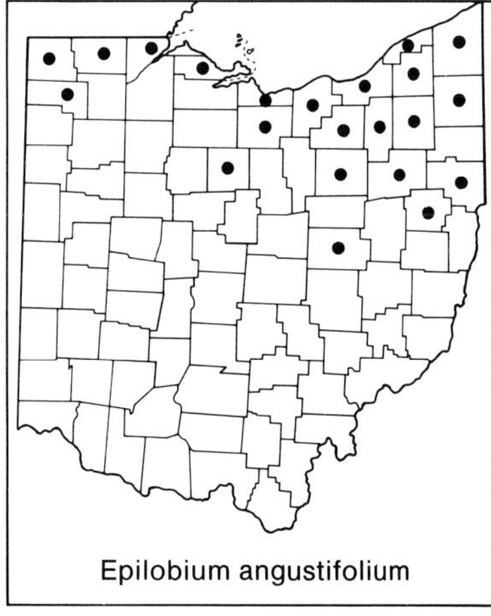
Epilobium angustifolium

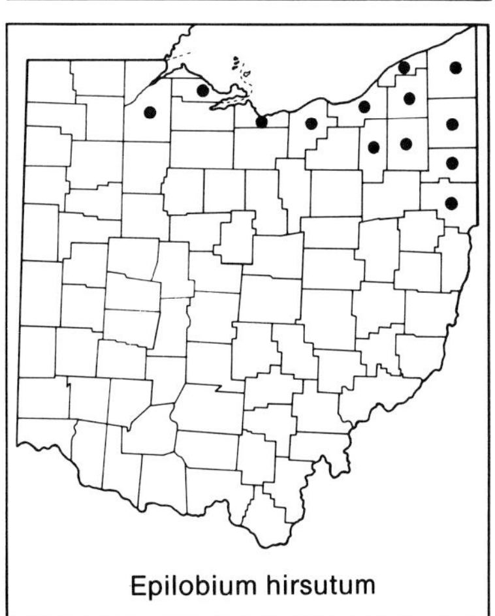
Epilobium hirsutum

1. **Epilobium angustifolium** L. FIREWEED

Erect, usually unbranched perennial, to 1.5 m tall; stems glabrous below, puberulent above; principal leaves alternate, lanceolate, to 17 cm long; flowers showy, in well-defined racemes, all but the lowest subtended by only minute bracts, petals medium-purple to pinkish, 10–20 mm long, not notched at apex; stigma 4-lobed. A circumboreal species at a southern boundary of its range in counties of northern Ohio; moist, open, weedy fields and clearings. Mosquin (1966) studied the taxonomy of this species. June–August.

Ohio plants belong to var. *platyphyllum* (Daniels) Fern. (*E. angustifolium* subsp. *circumvagum* Mosquin); var. *angustifolium* (incl. var. *canescens* (Wood) Britt.) occurs to the north of Ohio.

2. EPILOBIUM HIRSUTUM L. HAIRY WILLOW-HERB

Erect perennial, to 1.5 m tall; stems pubescent to densely villous above; principal leaves opposite, often softly pubescent, clasping, more or less oblong, to 10 cm long; flowers showy, either solitary in leaf axils or in weakly defined racemes, subtended by only slightly reduced leaves; petals purple or purplish, 12–20 mm long, notched at apex; stigma 4-lobed. A Eurasian species naturalized in northeastern and north-central counties; wet

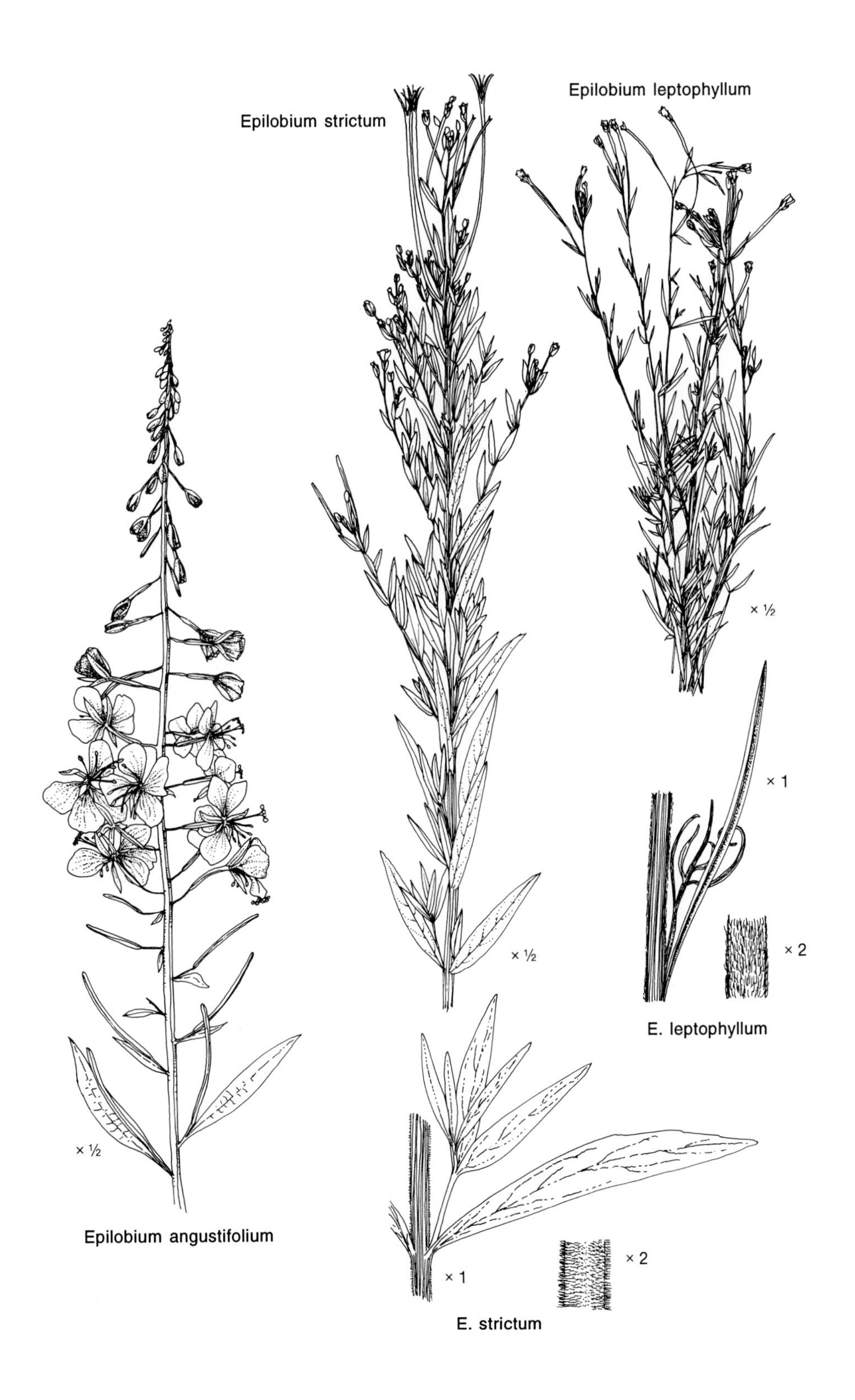
Epilobium strictum
Epilobium leptophyllum
× ½
× 1
× 2
E. leptophyllum
× ½
× ½
Epilobium angustifolium
× 1
× 2
E. strictum

Epilobium parviflorum

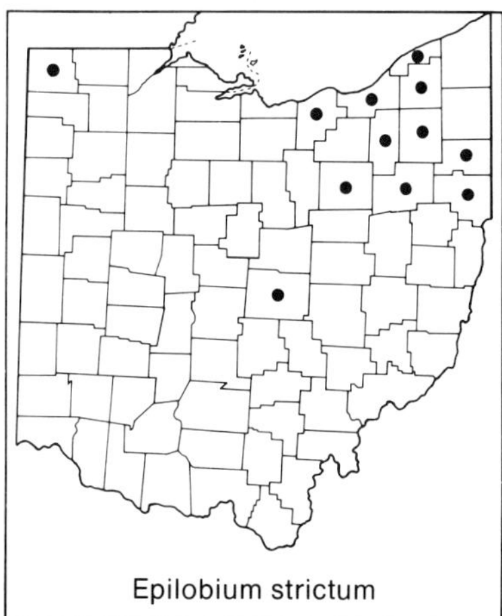
Epilobium strictum

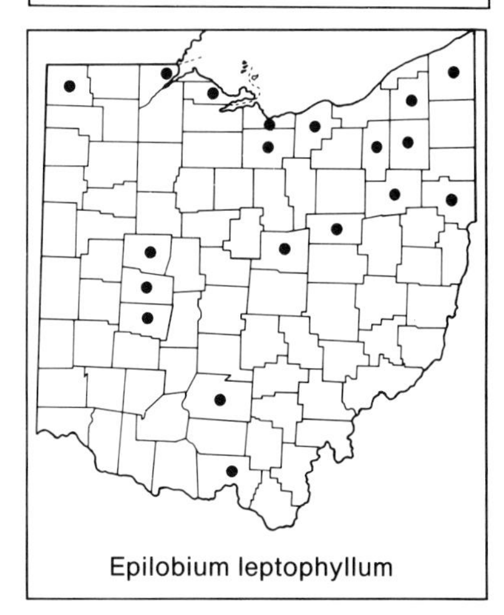
Epilobium leptophyllum

ditches, shores and gravel bars in streams, marshes. Stuckey (1970) describes the introduction of this species into North America and its spread across northeastern and midwestern United States. Late June–September.

3. Epilobium parviflorum Schreb. Small-flowered Hairy Willow-herb

Erect perennial, to 1 (–2) m tall; stems densely pubescent to villous, with long, mostly spreading hairs; leaves opposite throughout or alternate above, softly pubescent, sessile but not clasping, narrowly oblong to lanceolate, to 10 cm long; flowers in weakly defined racemes; petals pinkish-purple, (4–) 6–7 (–9) mm long, notched at apex; stigma 4-lobed. A Eurasian species currently spreading in northeastern North America, first collected in Michigan in 1966 (Voss, 1985b) and in Ontario in 1968 (Purcell, 1976). The first Ohio collections were made in 1979 (Cooperrider and Andreas, 1991); northeastern counties; wet banks and ditches, marshes and fens. Not illustrated. July–September.

4. **Epilobium strictum** Muhl. Downy Willow-herb. Simple Willow-herb

Erect perennial, to 1 m tall; stems terete, pubescent with short, straight, spreading hairs; principal leaves mostly opposite, narrowly lance-oblong to broadly linear, to 5 cm long, 3–5 (–8) mm wide, lateral veins apparent; flowers solitary in axils of upper leaves or, when numerous, appearing paniculate; petals pink, 4–6 mm long; stigma unlobed. A species of southeastern Canada and northeastern United States, at a southern boundary of its range in Ohio, mostly in northeastern counties; sedge meadows, marshes, bogs, swamps. Late July–early September.

5. **Epilobium leptophyllum** Raf. Linear-leaved Willow-herb

Erect perennial, to 1 m tall; stems terete, pubescent with appressed or incurved hairs; principal leaves alternate or opposite, linear to rarely broadly linear, to 5 cm long, mostly 1–3 mm wide, the lateral veins not or scarcely visible; flowers solitary in axils of upper leaves; petals pale pink, 4–6 mm long; stigma unlobed. Mostly in northern counties, absent from most of unglaciated Ohio; fens, swamps, marshes. Late July–early September.

This species was merged under *E. palustre* L. by Gleason and Cronquist (1963) but recognized as distinct by Gleason and Cronquist (1991).

6. **Epilobium coloratum** Biehler PURPLE-LEAVED WILLOW-HERB

Erect perennial, to 1 m tall; stems 4-angled, some or all internodes with low pubescence running in lines down the stem, the uppermost internodes sometimes also with scattered pubescence, the lower glabrous; some or all leaves with a distinct petiole, principal leaves long-lanceolate, 5–15 cm long, apex acuminate, older leaves often becoming irregularly blotched with red or purple; flowers in a bracteate panicle; sepal tips in bud forming a beak mostly 0.2–0.4 mm long, the individual sepal tips somewhat divergent and spreading from this beak; petals whitish to pale pink, 3–5 mm long; stigma unlobed; mature seeds with brown coma (coma pale when immature). Throughout Ohio; floodplains, shores and banks, marshes, wet ditches, swamps, wet fields. July–September.

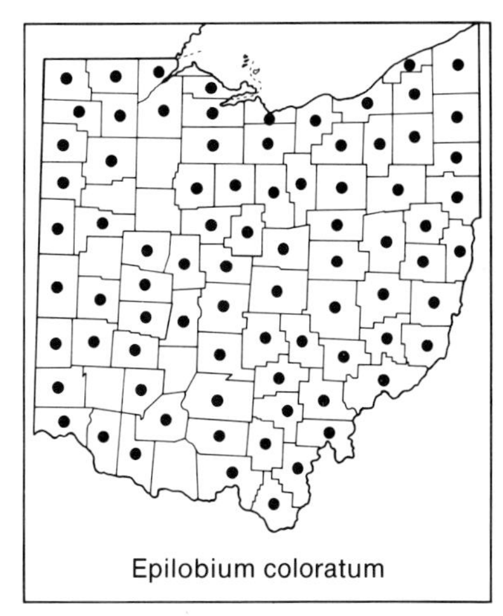

Epilobium coloratum

Epilobium × wisconsinense Ugent, the naturally occurring hybrid of *Epilobium ciliatum × E. coloratum*, has been collected from Geauga, Hamilton, Henry, and Lorain counties. Ugent (1963) has a discussion and an illustration of this hybrid.

7. **Epilobium ciliatum** Raf. NORTHERN WILLOW-HERB

E. adenocaulon Haussk.; *E. glandulosum* Lehm. var. *adenocaulon* (Haussk.) Fern.

Erect perennial, to 1 m tall; stems somewhat 4-angled with decurrent ridges running downward from leaf bases, internodes with pubescence scattered more or less evenly across stem surface or some internodes glabrous; most leaves either sessile or narrowed at base to a very short, winged petiole less than 2 mm long, principal leaves lanceolate to narrowly ovate, 3–7 cm long, apex acute; flowers in a bracteate panicle; sepal tips in bud scarcely forming a beak 0.1 mm long or less, the individual sepal tips held together, not spreading from the beak; petals white to pale pink, 2–6 mm long; stigma unlobed; mature seeds with whitish coma. Mostly in northern counties, but scattered elsewhere; wet, sometimes rocky, habitats. June–August.

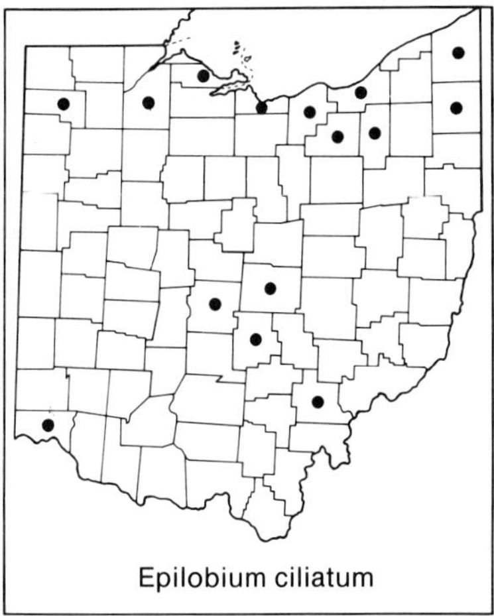

Epilobium ciliatum

The nomenclature and synonymy above follow Hoch (1978), who treats Ohio plants as *E. ciliatum* subsp. *ciliatum*. See discussion above for comments on the hybrid between this species and *E. coloratum*.

3. Oenothera L. EVENING-PRIMROSE

Perennial, or sometimes biennial or annual herbs; hypanthium 4 mm–10 cm long; sepals and petals 4; stamens 8. A variable and taxonomically difficult genus of about 150 species of the Western Hemisphere.

Dietrich and Wagner's (1988) study of *Oenothera* section *Oenothera* subsection *Raimannia* includes species no. 2 below. Straley (1977) has prepared a systematic study of *Oenothera* section *Kneiffia*, which includes species nos. 3–5 below. Warren L. Wagner and Peter C. Hoch kindly sent me Ohio records from their work and made helpful comments on the native or non-native status of several Ohio *Oenothera* taxa.

Several of the species described below, especially *Oenothera fruticosa* and *O. pilosella*, are cultivated in Ohio. Also cultivated is the striking species *O. missouriensis* Sims, OZARK SUNDROPS, a decumbent to slightly erect perennial with large, showy, yellow flowers ca. 8 cm across, and petals to 5 cm long. *Oenothera albicaulis* Pursh, PRAIRIE EVENING-PRIMROSE, was attributed to Ohio in error by C. H. Jones (1941), on the basis of a misidentified specimen of *O. speciosa*.

a. Petals white to pink, 25–40 mm long . . . . 7. *Oenothera speciosa*

aa. Petals yellow or orangish-yellow (drying red in *O. laciniata*), 5–30 mm long.

b. Plants without aerial stems, leaves oblanceolate, in a basal cluster; leaf margins coarsely and shallowly toothed to deeply lobed, nearly to the midrib . . . . 6. *Oenothera triloba*

bb. Plants with aerial leafy stems; leaf margins various, but not deeply lobed.

c. Capsules club-shaped, narrowed at base, the basal portion bearing no seeds and appearing like a stalk; ovary and fruit tetragonal and more or less winged, especially above.

d. Petals 5–10 mm long; stigma borne on same plane as anthers; inflorescence usually nodding prior to flowering . . . . 5. *Oenothera perennis*

dd. Petals (8–) 15–30 mm long; stigma borne well above the anthers; inflorescence erect.

e. Upper internodes with spreading hairs 1–2 mm long; free tips of sepals in bud, 1–4 mm long; principal cauline leaves 5–10 cm long . . . . 3. *Oenothera pilosella*

ee. Upper internodes either glabrous or with short ascending or diverging hairs about 0.5 mm long; free tips of sepals in bud to 1 mm long; principal cauline leaves 2–5 (–7) cm long . . . . 4. *Oenothera fruticosa*

cc. Capsules cylindric, not narrowed at base, sessile, the basal portion bearing seeds; ovary and fruit terete or nearly so, not winged.

d. Cauline leaves remotely dentate, denticulate, or subentire . . . . 1. *Oenothera biennis*

dd. Cauline leaves evenly and sharply dentate or sharply lobed . . . . 2. *Oenothera laciniata*

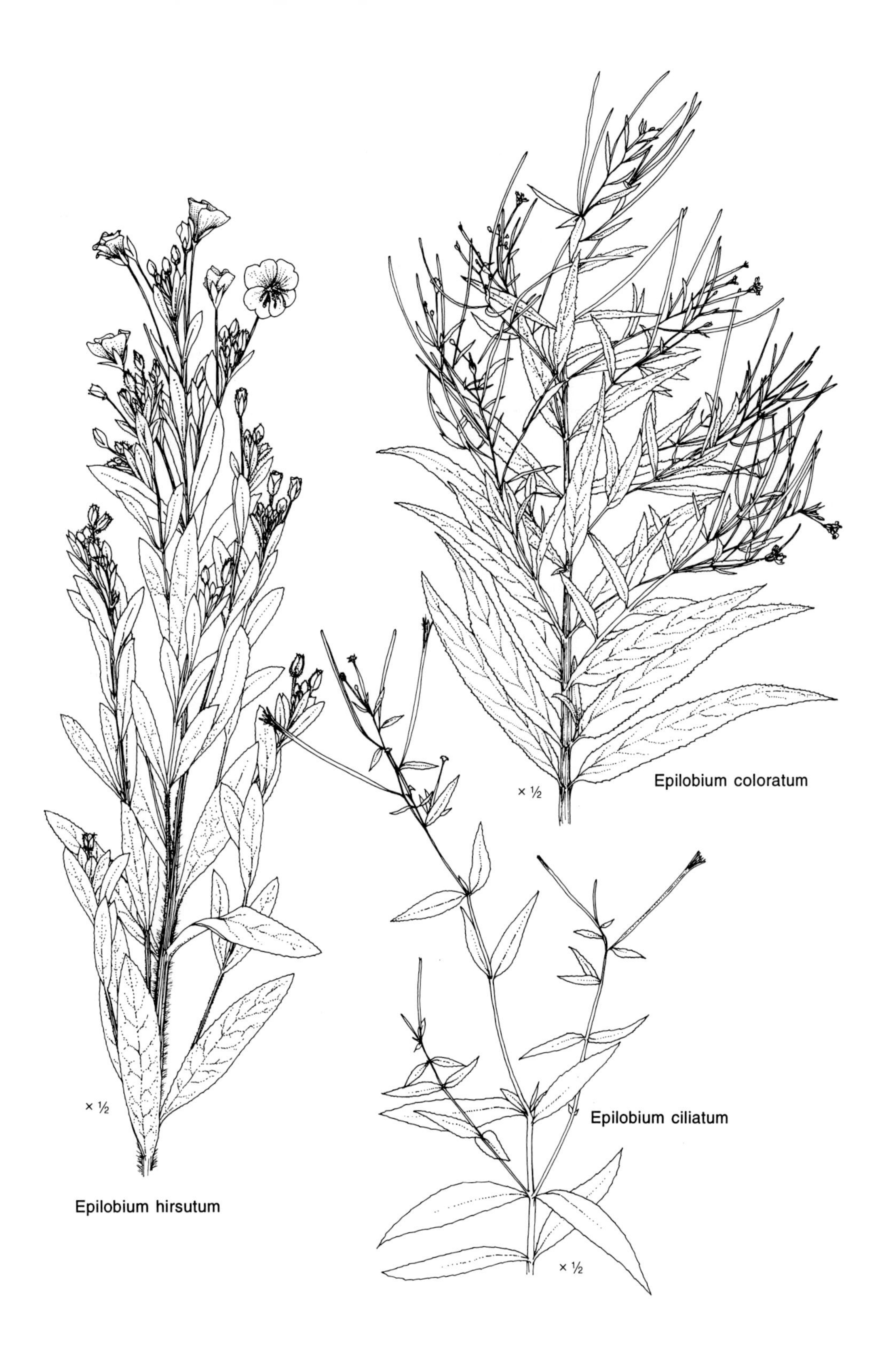
Epilobium coloratum
× ½
Epilobium ciliatum
× ½
× ½
Epilobium hirsutum

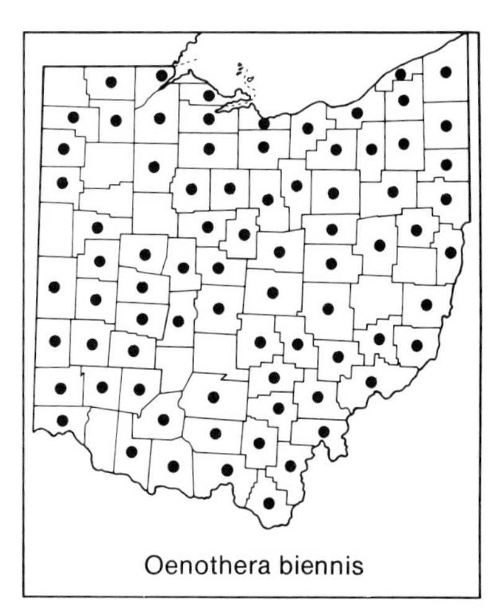
Oenothera biennis

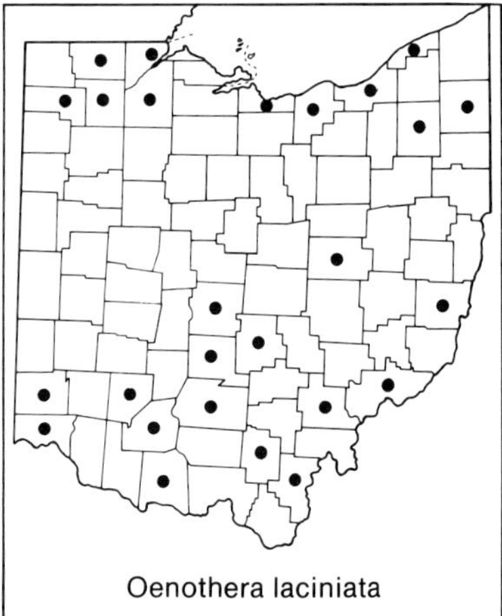
Oenothera laciniata

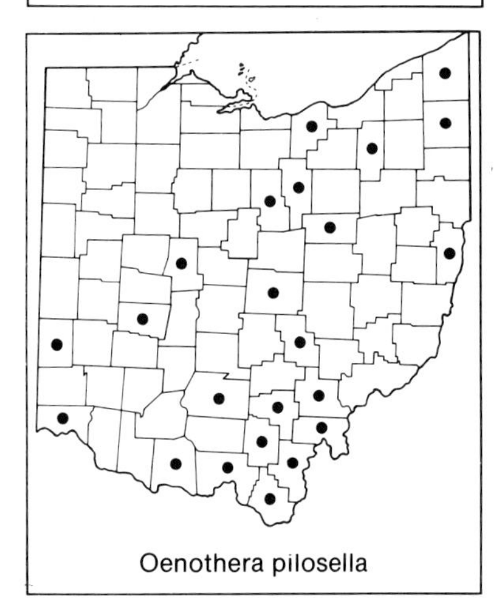
Oenothera pilosella

1. **Oenothera biennis** L. COMMON EVENING-PRIMROSE
Erect biennial, to 2 m tall; leaves lanceolate, margins subentire to denticulate or remotely dentate; flowers in showy terminal racemes; petals yellow, to 2.5 cm long; ovary terete; capsule cylindric, sessile, not narrowed at base, the basal portion bearing seeds. Throughout Ohio; fields, roadsides, open disturbed habitats. July–October.

Lacking a modern systematic treatment of *Oenothera* section *Oenothera* subsection *Oenothera*, of which the following taxa are a part, Ohio members of this complex are treated here as belonging to a polymorphic species with four varieties, often scarcely or not at all separable. Occasional individual specimens are assignable to (1) var. *canescens* T. & G. (*O. villosa* Thunb.), with dense, closely appressed pubescence above, widespread and probably originally native; (2) var. *nutans* (Atkinson & Bartlett) Wieg. (*O. nutans* Atkinson & Bartlett), with a more or less glabrous calyx, collected from a few scattered locations mostly in the northern half of Ohio, and perhaps naturalized from more eastern states; (3) var. *oakesiana* A. Gray (*O. oakesiana* (A. Gray) Wats. & Coult.; *O. parviflora* L. var. *oakesiana* (A. Gray) Fern.), a plant of the Atlantic coastal region ranging inland around the Great Lakes, collected in counties bordering Lake Erie, where it is probably native; or (4) var. *biennis* (*O. biennis* sensu stricto, incl. var. *pycnocarpa* (Atkinson & Bartlett) Wieg.), with varying amounts of pubescence, widespread and native. This treatment is based in part on data from W. L. Wagner and P. C. Hoch, who, however, treat each of the four taxa at the rank of species. Raven, Dietrich, and Stubbe (1979) and Voss (1985b) have additional commentary on this "*Oenothera biennis* complex." A population of the closely related **Oenothera parviflora** L. [var. *parviflora*], SMALL-FLOWERED EVENING-PRIMROSE, was recently reported from Erie County (Ramey, 1992).

2. **Oenothera laciniata** Hill CUT-LEAVED EVENING-PRIMROSE
Erect to suberect annual, to 4 dm tall; leaves oblanceolate to oblong-lanceolate, all—or at least the middle and upper—evenly and sharply dentate or sharply lobed; flowers solitary in leaf axils; petals pale yellow (becoming red or marked with red when dry), to 2 cm long; ovary terete; capsule cylindric, sessile, not narrowed at base, the basal portion bearing seeds. In southern and northern Ohio counties, but absent from a broad band of central counties; open dry habitats, usually in sand or gravel; probably adventive at some sites. May–September.

3. **Oenothera pilosella** Raf. MEADOW SUNDROPS
Erect perennial, to 8 dm tall; upper internodes with conspicuous, soft, spreading hairs 1–2 mm long; leaves lanceolate to narrowly ovate, margins

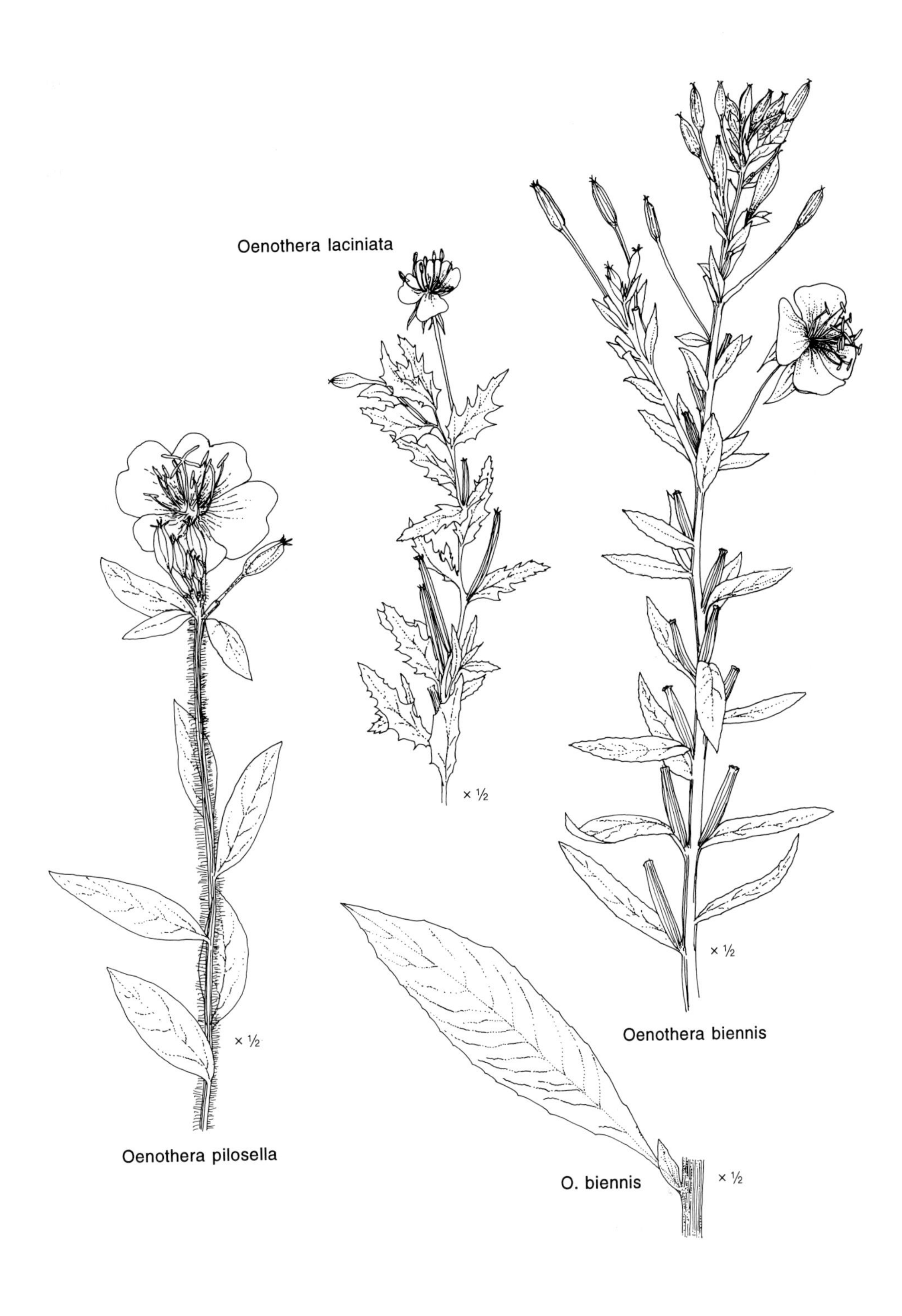
Oenothera laciniata
× ½
× ½
Oenothera biennis
× ½
Oenothera pilosella
O. biennis
× ½

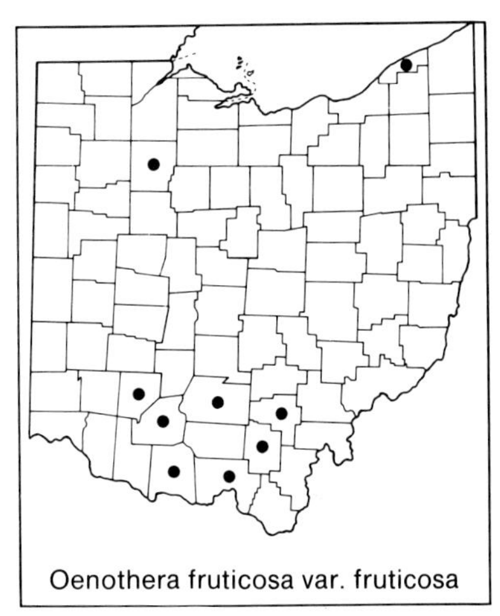
Oenothera fruticosa var. fruticosa

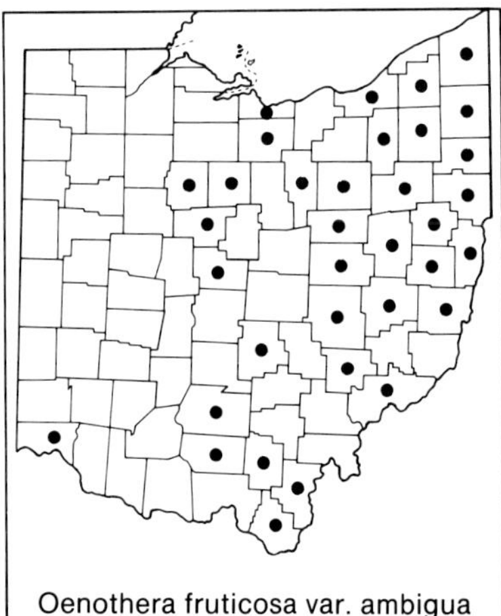
Oenothera fruticosa var. ambigua

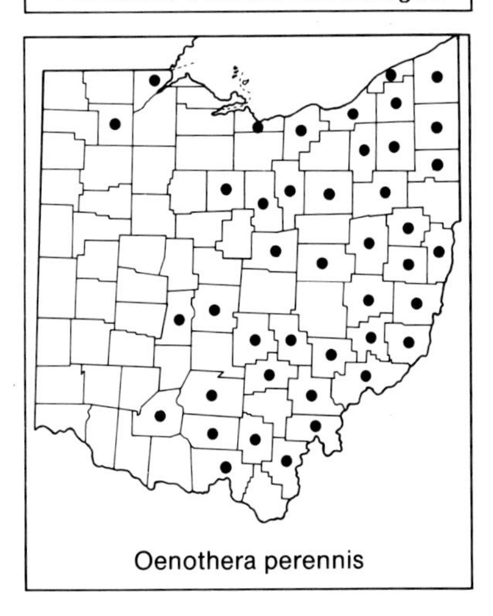
Oenothera perennis

subentire, principal cauline leaves 5–10 cm long; flowers several, yellow, large and showy, to 5 cm across, in a short erect raceme at stem tip; free tips of sepals in bud, 1–4 mm long; petals 15–30 mm long; stigma borne well above the anthers; capsule winged, club-shaped, narrowed and stalk-like at base, the basal portion bearing no seeds. South-central counties and widely scattered elsewhere, but absent from northwestern Ohio; roadside banks, moist fields, open woodlands. June–July.

4. **Oenothera fruticosa** L. Common Sundrops
incl. O. *tetragona* Roth

Erect perennial, to 8 dm tall; upper internodes either glabrous or with short ascending to diverging hairs about 0.5 mm long; leaves very narrowly elliptic and sublinear to lanceolate or oblanceolate, subentire to very low dentate, principal cauline leaves 2–5 (–7) cm long; flowers yellow, large and showy, to 4 cm across, in a short erect raceme at stem tip; free tips of sepals in bud only to 1 mm long; petals (8–) 15–25 mm long; stigma borne well above the anthers; capsule angled or winged, club-shaped, narrowed and stalk-like at base, the basal portion bearing no seeds. Mostly in eastern and southern counties; moist to dry fields, shores, open woodlands. June–July.

Following Straley (1977), Ohio plants are divided into two infraspecific elements, treated here at the rank of variety, but by Straley at the rank of subspecies. The key below is adapted from that of Straley.

a. Upper internodes with short ascending or divergent hairs; leaves usually pubescent, entire or nearly so; capsules widest above the middle; pubescence of ovaries and capsules mostly eglandular . . . . var. *fruticosa*
incl. var. *linearis* (Michx.) S. Wats.

aa. Upper internodes glabrous; leaves glabrous or sparsely pubescent, nearly entire to more or less dentate; capsules widest at about the middle; pubescence of ovaries and capsules mostly glandular. . . . . var. *ambigua* Nutt.
O. *fruticosa* subsp. *glauca* (Michx.) Straley; O. *tetragona* Roth

5. **Oenothera perennis** L. Small Sundrops

More or less erect perennial, to 4 dm tall; upper internodes with short, ascending to diverging hairs to 0.5 mm long; leaves mostly narrowly elliptic, sublinear in some cases, subentire to very low dentate, principal cauline leaves 3–7 cm long; flowers yellow, small, in a short raceme at stem tip, the inflorescence usually nodding prior to anthesis; petals 5–10 mm long; stigma borne on same plane as anthers; capsule winged, club-shaped, narrowed and stalk-like at base, the basal portion bearing no seeds.

Oenothera triloba
× ½
Gaura biennis
× ½
× ½
Oenothera perennis
× ½
Oenothera fruticosa

Oenothera triloba

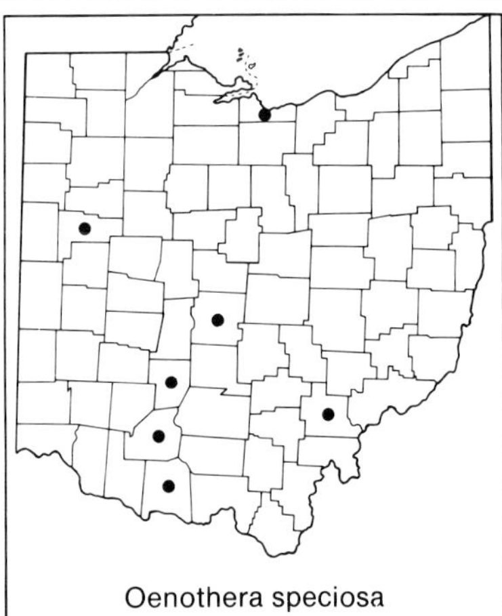
Oenothera speciosa

Throughout most of the eastern two-thirds of Ohio; moist to dry fields, pastures, and woodland borders. Late May–August.

6. Oenothera triloba Nutt. Sessile Evening-primrose

Nearly acaulescent biennial; leaves in a basal cluster, oblanceolate, margins coarsely and shallowly toothed to deeply lobed and dissected nearly to the midrib; flowers showy, yellow; petals to 2.5 cm long. A species ranging chiefly to the south and west of Ohio, with a single record from Montgomery County in southwestern Ohio; Deam (1940) regards the species as native to Indiana, but Warren L. Wagner (personal communication) believes that the species may have been naturalized or was perhaps merely adventive at the Montgomery County locale. May.

7. Oenothera speciosa Nutt. White Evening-primrose

More or less erect perennial, to 5 dm tall; leaves linear to oblong, shallowly dentate to (the lower) pinnately lobed; flowers showy; corolla white to pink, the petals 2.5–4.0 cm long. A species of the southern plains states, cultivated in Ohio, escaped, and perhaps locally established at a few sites, mostly in southern counties; open disturbed ground. Not illustrated. May–July.

4. Gaura L.

A North American genus of 21 species, one of which is native to Ohio. The most recent revision is that of Raven and Gregory (1972). *Gaura longiflora* Spach (*G. biennis* var. *pitcheri* T. & G.; *G. filiformis* Small), a species native from Illinois and Iowa south to Texas, was collected once, in 1960, as an adventive weed in Lorain County. Adventive specimens of *Gaura parviflora* Dougl., also native to the west of Ohio, have been collected from Erie, Jackson, and Stark counties, but mostly before 1910. *Gaura lindheimeri* Engelm. & A. Gray, White Gaura, an attractive species from south-central United States, is frequently cultivated. Carr et al. (1990) published a cladistic analysis of the genus.

a. Inflorescences usually unbranched spike-like racemes; sepals 2–3 mm long; petals 2–3 mm long; fruit 7–11 mm long; see above .... *Gaura parviflora*

aa. Inflorescences usually branched, usually paniculate; sepals 8–13 mm long; petals 3–15 mm long; fruit 5–9 mm long.

b. Stems with both incurved and spreading hairs or with spreading hairs only; petals 3–7 mm long .... 1. *Gaura biennis*

bb. Stems with all hairs appressed or incurved; petals 7–15 mm long; see above .... *Gaura longiflora*

1. **Gaura biennis** L. BIENNIAL GAURA

Winter annual or biennial, to 1.5 m tall; stems pubescent with both incurved and spreading hairs or with spreading hairs only; leaves narrowly long-elliptic, margins subentire to low dentate; flowers borne in a panicle or subpanicle or sometimes in a spike-like raceme (long-bracteate in a rare form from southern Ohio); hypanthium 6–12 mm long; sepals 4, 8–13 mm long; corolla somewhat zygomorphic, petals 4, white to pink or red, 3–7 mm long; stamens 8; fruit indehiscent, 5–9 mm long. Throughout Ohio; wet fields, banks and ditches along roads and railroads, and other disturbed usually moist habitats. Late July–October.

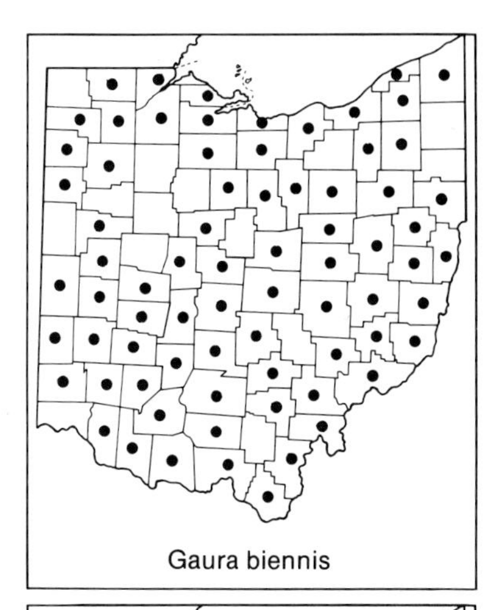
Gaura biennis

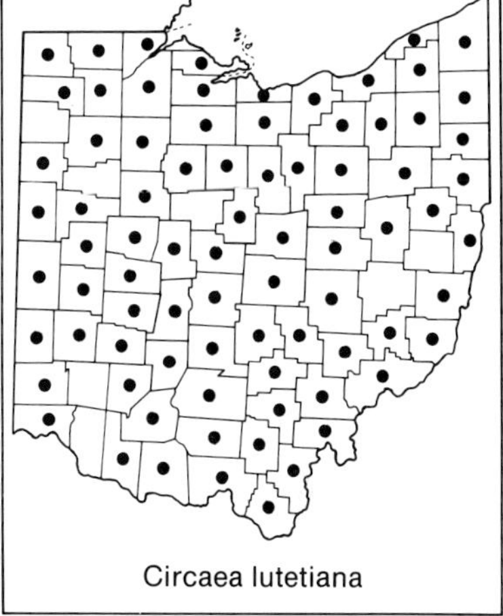
Circaea lutetiana

5. Circaea L. ENCHANTER'S-NIGHTSHADE

Perennial herbs; leaves opposite, mostly ovate, margins low-dentate; flowers in a raceme or panicle; sepals and petals 2; floral disk sometimes present; stamens 2; carpels 2; fruit indehiscent. A genus of seven species in temperate parts of the Northern Hemisphere. The classification and nomenclature used here are those adopted by Boufford (1982) in his comprehensive study of the systematics and evolution of *Circaea*. Boufford et al. (1990) made a cladistic analysis of the genus.

a. Stems firm, strictly erect, 2.5–10 dm tall; fruit bearing numerous stout hooked bristles; fruit width (including the bristles) 4–6 mm; anthers 1.0–1.2 mm long when fresh, 0.6–1.0 mm long when dry; floral disc 0.4–0.6 mm high when fresh. . . . . 1. *Circaea lutetiana*

aa. Stems weak, not strictly erect, 0.5–5 dm tall; fruit bearing numerous soft flexuous bristles; fruit width (including the bristles) 1–2 mm; anthers 0.3–0.8 mm long when fresh, 0.2–0.5 mm long when dry; floral disc lacking or to 0.2 mm high when fresh.

b. Plants not producing mature fruit, the ovaries 2–loculate, abscising soon after flowering; floral disc 0.1–0.2 mm high; anthers 0.5–0.8 mm long when fresh, 0.3–0.5 mm long when dry. . . . . 2. *Circaea* × *intermedia*

bb. Plants producing mature fruit, the ovaries 1–loculate, persisting long after flowering; floral disc lacking; anthers 0.3–0.5 mm long when fresh, 0.2–0.3 mm long when dry . . . . 3. *Circaea alpina*

1. **Circaea lutetiana** L. COMMON ENCHANTER'S-NIGHTSHADE

Erect perennial, to 1 m tall; stems firm; leaves mostly ovate, margins low-dentate; flowers in a raceme or panicle; floral disc 0.4–0.6 mm high; anthers 1.0–1.2 mm long when fresh, 0.6–1.0 mm long when dry; fruit bearing numerous stout hooked bristles, fruit (incl. bristles) 4–6 mm wide.

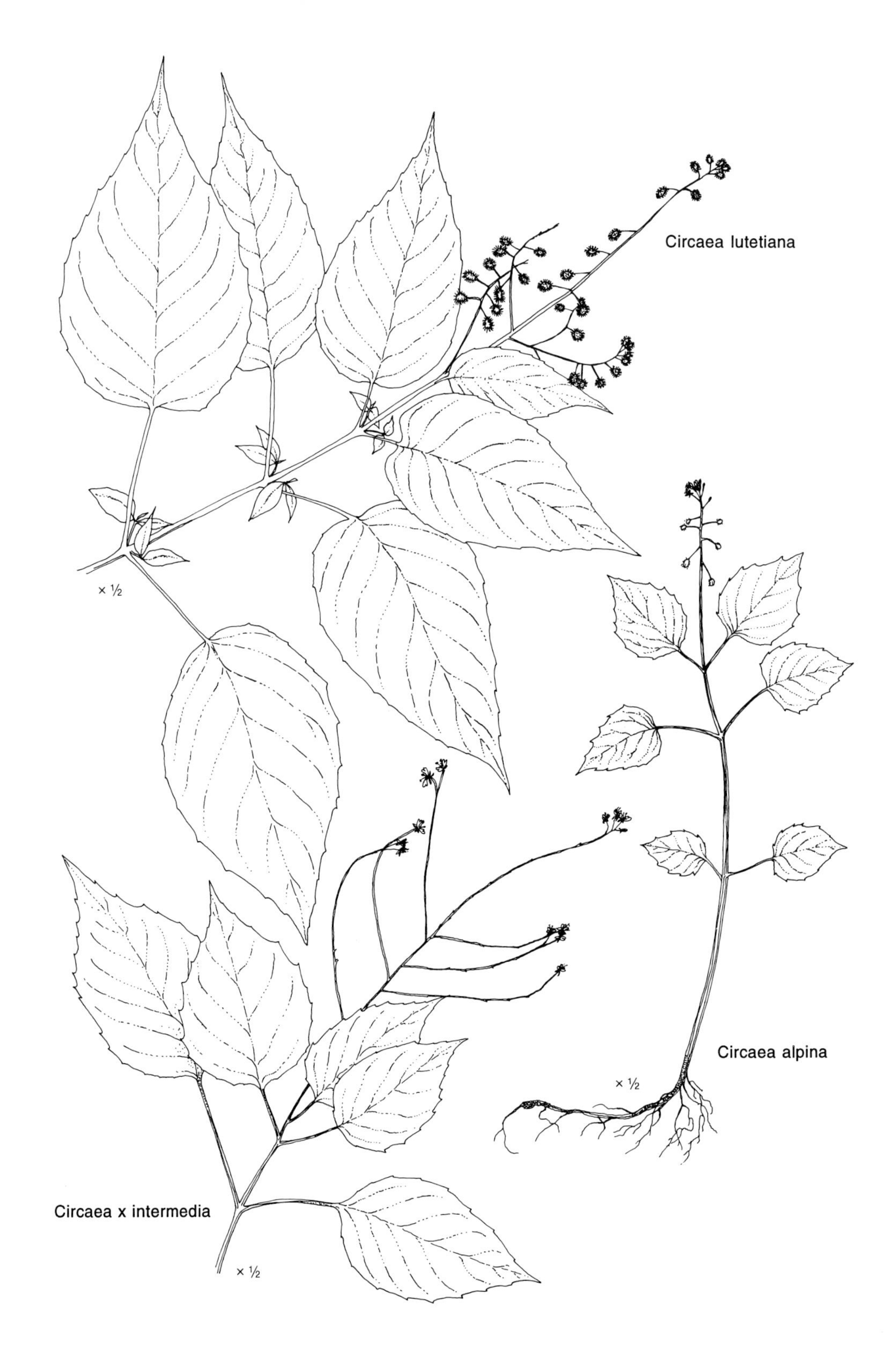
Circaea lutetiana
× ½
Circaea alpina
× ½
Circaea x intermedia
× ½

Throughout Ohio; moist to dry, disturbed, open or partially open woodlands. June–August.

Ohio plants belong to var. *canadensis* L. (*C. quadrisulcata* of authors, incl. Fernald (1950), Gleason (1952a), and Gleason and Cronquist (1963), not (Maxim.) Franch. & Sav.). Boufford et al. (1990) have recently proposed segregating the American members of this species under the name *C. canadensis* (L.) Hill subsp. *canadensis*.

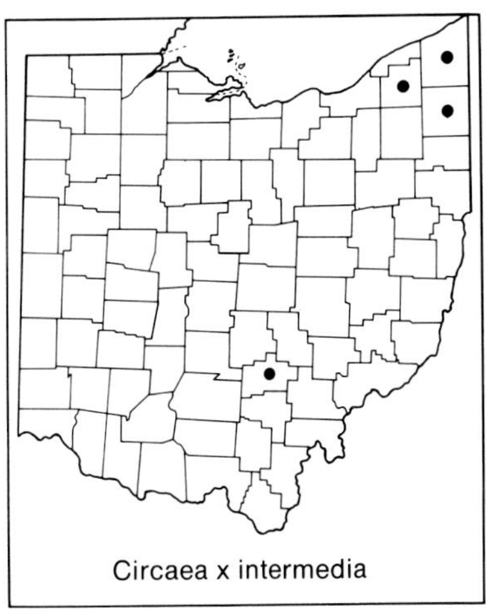
Circaea x intermedia

2. **Circaea × intermedia** Ehrh. INTERMEDIATE ENCHANTER'S-NIGHTSHADE

C. canadensis of authors, incl. Fernald (1950), Gleason (1952a), and Gleason and Cronquist (1963), not (L.) Hill.

The sterile interspecific hybrid of *Circaea alpina* × *C. lutetiana*, treated as a taxonomic species (Fernald, 1950; Gleason, 1952a; Gleason and Cronquist, 1963) before its hybrid status was determined (Cooperrider, 1962). In morphology, intermediate between the parental species; plants never producing mature fruit, the ovaries abscising soon after flowering, and the flowering continuing until frost. Known from a few Ohio sites, usually along streams, always in the vicinity of both parents. Additional record: Huron County (Boufford, 1982). Haber (1977) made an extensive study of this hybrid. June–October (–November).

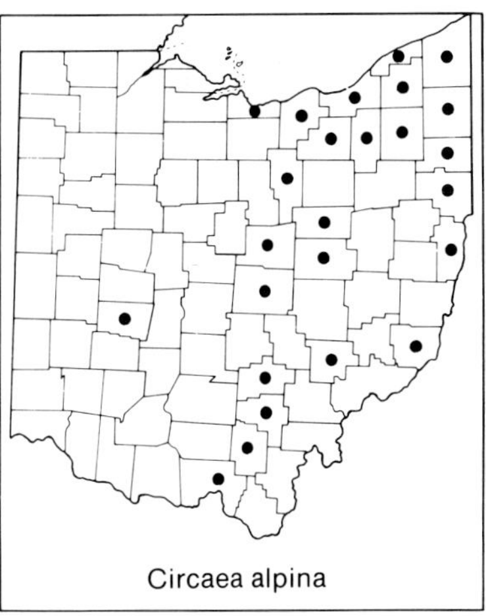
Circaea alpina

3. **Circaea alpina** L. SMALL ENCHANTER'S-NIGHTSHADE

Delicate perennial, to 2 dm tall; stems weak; floral disc lacking; anthers minute, 0.3–0.5 mm long when fresh, 0.2–0.3 mm long when dry; fruit bearing soft flexuous bristles, fruit (incl. bristles) 1–2 mm wide. Chiefly in counties of the eastern half of Ohio; usually on wet, dripping, mossy rocks and cliffs. Additional records: Franklin and Huron counties (Boufford, 1982). June–August.

Haloragaceae. WATER-MILFOIL FAMILY

Perennial herbs; leaves alternate, opposite, or whorled, often highly dissected; flowers either perfect or imperfect (the plants monoecious), small and inconspicuous, either solitary to few in leaf axils or numerous in terminal spikes; calyx of 3 or 4 sepals, sometimes minute; corolla actinomorphic, of 4 small petals, or corolla absent; stamens 3, 4, or 8; carpels 3 or 4, fused to form a compound pistil with 1 or 3 styles; ovary inferior; fruit small and nutlike. A widely distributed family of about 100 species of aquatic and wet terrestrial habitats. Crow and Hellquist (1983) prepared a treatment of the family as it is represented in the New England flora, and Meyer and Mohlenbrock (1966) a study of the Illinois species.

a. Carpels 4; stamens 4 or 8; plants either lacking emersed leaves or with emersed leaves usually less than 2 cm in length .... 1. *Myriophyllum*

aa. Carpels 3; stamens 3; plants usually with some or all leaves emersed, these narrowly lanceolate or oblanceolate, 2–6 (–8) cm long. .... 2. *Proserpinaca*

1. Myriophyllum L. WATER-MILFOIL

Perennials; submersed leaves finely dissected, emersed leaves either none or small and bract-like; some species producing turions (winter buds), specialized structures that break free from the parent plant and serve as a propagating unit capable of surviving over winter. A cosmopolitan genus of some 40 species. In addition to the species in the text below, three others have been collected as adventives in Ohio. *Myriophyllum aquaticum* (Vellozo) Verdcourt (*M. brasiliense* Camb.), Parrot's-feather, a South American species sometimes used in aquaria, was collected from two different ponds in Scioto County, first in 1949, and again in 1959. *Myriophyllum humile* (Raf.) Morong, Lowly Water-milfoil, a species of the Atlantic coastal region, was collected in 1976, from a lake in Butler County. A southern species, *Myriophyllum pinnatum* (Walt.) BSP., Pinnate Water-milfoil, was collected in 1937, in Lake County. The identifications of the specimens of *M. humile* and *M. pinnatum* were both confirmed by E. N. Nelson and R. W. Couch, current students of the genus. Couch (in personal communication) expressed the opinion that both were adventive in Ohio; he wrote, "I'm convinced that many aquatic plant species get transported by bird migration (and other means), [become] established for a few years, then die out for one reason or another." Aiken (1981) published a conspectus of *Myriophyllum* in North America.

a. Submersed leaves alternate to subopposite, with 3–8 pairs of segments; terminal flowers alternate.

b. Fruits tuberculate, borne on emersed spikes; see above .... *Myriophyllum pinnatum*

bb. Fruits smooth or nearly so, borne in axils of submersed leaves or on emersed spikes; see above.................. *Myriophyllum humile*

aa. Submersed leaves whorled, with 4–24 pairs of segments; terminal flowers whorled.

b. Flowers originating from axils of submersed leaves; submersed leaves with 10–20 pairs of filiform segments; emersed leaves mostly 2.5–3.5 cm long, with segments 4–8 mm long; see above .... *Myriophyllum aquaticum* bb. Flowers borne on emersed spikes; submersed leaves with 4–24 pairs of filiform segments; emersed leaves mostly 0.5–2.5 cm long, with segments 1–2 mm long or none.

c. Mid and upper bracts of inflorescence deeply pinnatifid; lower bracts mostly more than twice the length of pistillate flowers; stamens 8; submersed leaves with 8–13 pairs of segments .... 3. *Myriophyllum verticillatum*

cc. Mid and upper bracts of inflorescence with entire or serrate margins, not pinnatifid; lower bracts mostly twice the length of pistillate flowers or less; stamens 8 or 4; submersed leaves with 4–24 pairs of segments.

d. Mid and upper bracts of inflorescence with entire margins; lower bracts twice the length of the pistillate flowers or less; stamens 8.

e. Leaves with 4–13 pairs of segments; plants often producing turions in autumn.................... 1. *Myriophyllum sibiricum*

ee. Leaves with (5–) 14–24 pairs of segments; plants not producing turions......................... 2. *Myriophyllum spicatum*

dd. Mid and upper bracts of inflorescence with serrate margins; lower bracts mostly twice as long as pistillate flowers; stamens 4 .... 4. *Myriophyllum heterophyllum*

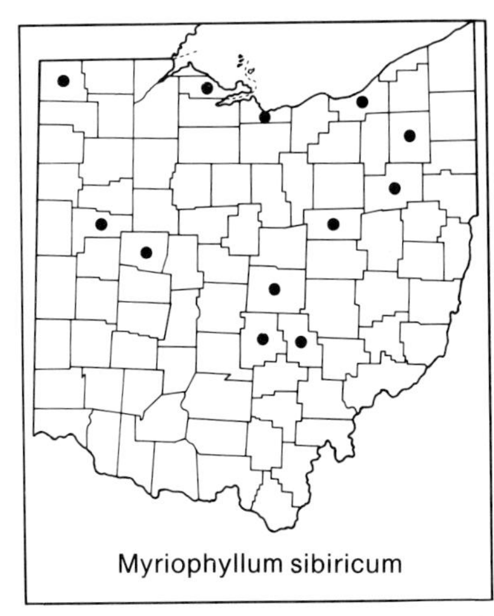

Myriophyllum sibiricum

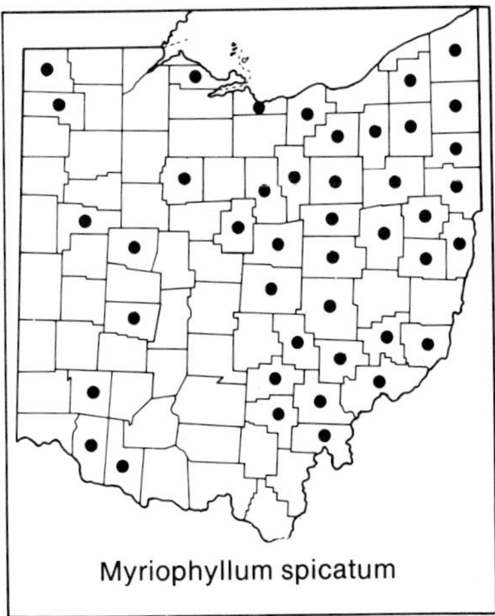

Myriophyllum spicatum

1. **Myriophyllum sibiricum** V. Komarov NORTHERN WATER-MILFOIL. AMERICAN WATER-MILFOIL

M. exalbescens Fern.; *M. spicatum* var. *exalbescens* (Fern.) Jepson; *M. spicatum* var. *squamosum* (Hartman f.) Hartman f.

Perennial, usually aquatic, often producing turions in autumn; submersed leaves in whorls of 3–4 (–5), 1.5–3.0 cm long, with 4–13 pairs of filiform segments, emersed leaves and bracts minute, 1–3 mm long, margins of the lowest deeply dentate to serrate, of the mid and upper leaves entire; stamens 8. At scattered Ohio sites, chiefly in northern counties, now scarce; lakes, ponds, quiet streams. Aiken and Walz (1979) and Weber (1972) describe turion formation. June–October.

Ceska and Ceska (1986) discuss the nomenclature of this species. Some Ohio specimens are difficult to separate from the next species, *M. spicatum*. A significant number of workers, among them: the Bailey Hortorium Staff (1976), Gleason and Cronquist (1963), Jones and Fuller (1955), Nichols (1975), Orchard (1981), Patten (1954), and Taylor and MacBryde (1977), have merged *M. sibiricum* under *M. spicatum* sensu lato.

2. MYRIOPHYLLUM SPICATUM L. EURASIAN WATER-MILFOIL. SPIKED WATER-MILFOIL

Perennial, usually aquatic; submersed leaves in whorls of 3–4 (–5), 1.5–3.0 cm long, with (5–) 14–24 pairs of filiform segments, emersed leaves

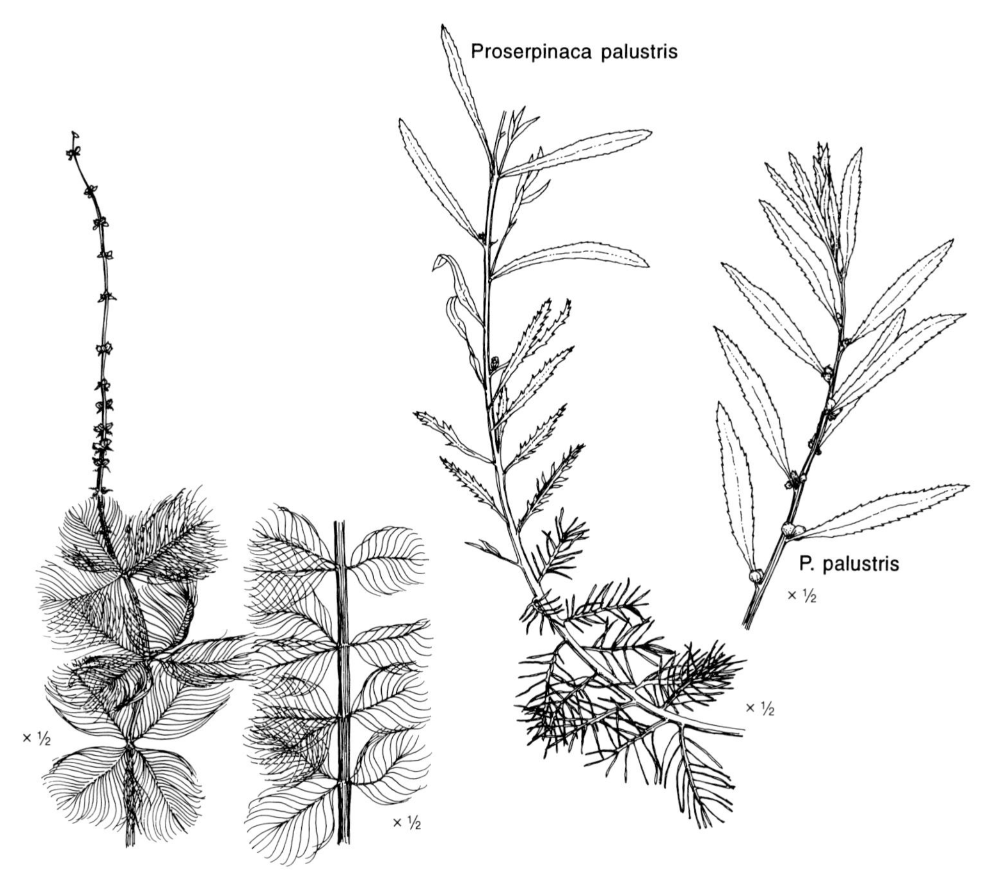
Proserpinaca palustris
P. palustris
× ½
× ½
× ½
× ½
Myriophyllum spicatum
Myriophyllum sibiricum

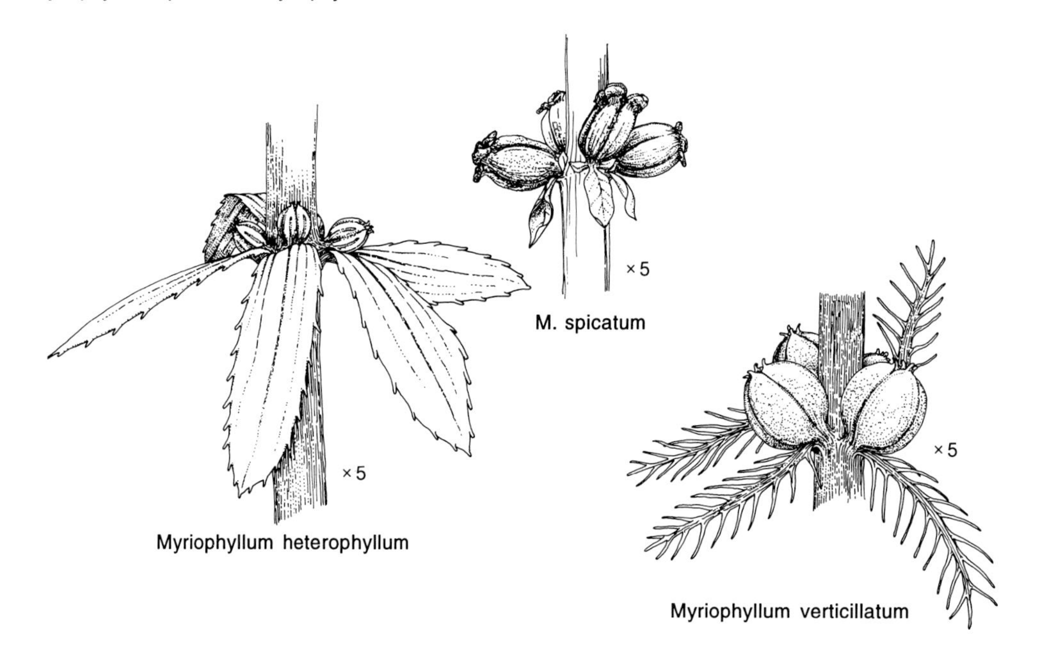
× 5
M. spicatum
× 5
Myriophyllum heterophyllum
× 5
Myriophyllum verticillatum

and bracts minute, 1–3 mm long, margins of the lowest leaves deeply dentate to serrate, of the middle and upper leaves entire; stamens 8. A Eurasian species, naturalized in eastern Ohio, and at scattered sites in the western half of the state; lakes, ponds, quiet streams. June–October.

These plants are aggressive aquatic weeds in the Midwest today, perhaps replacing the native M. *sibiricum* (B. T. Coffey and C. D. McNabb, 1974; Nichols, 1975; Voss, 1985b). All Ohio collections of this species that I have seen date since 1950. Aiken, Newroth, and Wile (1979) studied the species in Canada.

See the discussion under M. *sibiricum*, above, on the relationship between that species and this.

Myriophyllum verticillatum

3. **Myriophyllum verticillatum** L. WHORLED WATER-MILFOIL
   incl. var. *verticillatum* and var. *pectinatum* Wallr.

Perennial, usually aquatic; submersed leaves in whorls of 4–5, 1.5–4.0 cm long, with 8–13 pairs of filiform segments; emersed leaves and bracts to 6 mm or more in length, margins deeply pinnatifid; stamens 8. Known only from a few Ohio sites; lakes, swamps. Additional record: Summit County (Cusick, 1991). Weber and Noodén (1974) studied turion formation in these plants. June–September.

Following Aiken (1981), no varieties are recognized in this species.

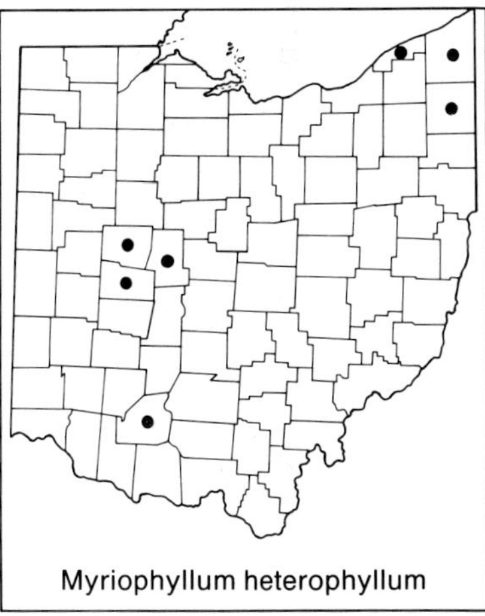
Myriophyllum heterophyllum

4. **Myriophyllum heterophyllum** Michx. BROAD-LEAVED WATER-MILFOIL

Perennial, usually aquatic; submersed leaves in whorls of 4–6, to 6 cm long, with 5–12 pairs of filiform segments; emersed leaves and bracts to 15 mm long, margins of the lower somewhat pinnatifid, of the mid and upper serrate; stamens 4. At a few sites in northeastern and southwestern counties; ponds, lakes, quiet streams. June–September.

2. Proserpinaca L. MERMAID-WEED

An American genus of three species, one of which occurs in Ohio.

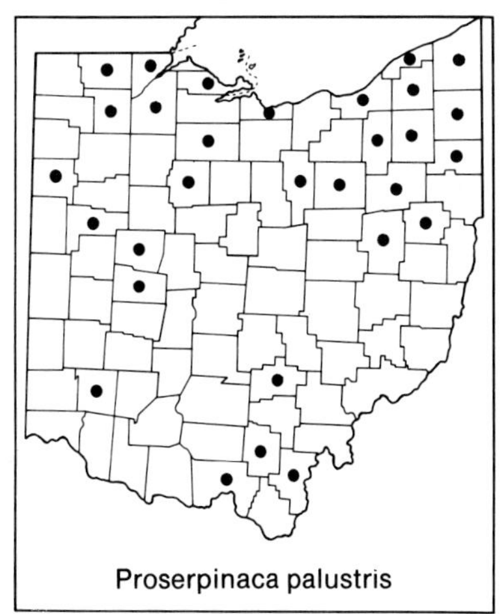
Proserpinaca palustris

1. **Proserpinaca palustris** L. MERMAID-WEED

Perennial, aquatic or terrestrial; leaves alternate, plants usually with some or all leaves emersed, these narrowly lanceolate or oblanceolate, 2–6 (–8) cm long; submersed leaves, when present, pinnately dissected with filiform segments, some plants with transitional forms between the two leaf types; carpels 3; stamens 3. Mostly in northern counties; ponds and muddy banks, marshes. Schmidt and Millington (1968) studied leaf variation and its controlling mechanisms in this species. June–October.

Ohio plants belong to var. *crebra* Fern. & Grisc.; var. *palustris* is mostly more southern in range.

Araliaceae. GINSENG FAMILY

Perennial herbs, shrubs, small trees, or woody vines; leaves alternate, compound or simple; flowers small, in umbellate inflorescences, perfect or less often imperfect; calyx of 5 sepals, often tiny or nearly absent; corolla more or less actinomorphic, of 5 separate petals; stamens 5; carpels 2–several, fused to form a compound pistil with as many styles; ovary inferior; fruit a berry. A chiefly tropical family of 700 or more species. Graham (1966) studied the family in southeastern United States, and Kerrigan and Blackwell (1973) the distribution of the Ohio species.

a. Climbing or creeping vines; leaves simple, either shallowly to moderately 3- (or 5-) lobed or unlobed . . . . 3. *Hedera*

aa. Erect herbs, shrubs, or small trees; leaves compound.

b. Leaves basal or cauline, ternately decompound or twice (or thrice) pinnately compound; plants bearing 2 or more umbels; carpels (and styles) usually 5; fruit black or nearly black. . . . . 1. *Aralia*

bb. Leaves cauline, in a single whorl, palmately compound; plants bearing a single umbel; carpels (and styles) 2 or 3; fruit yellow or red . . . . 2. *Panax*

1. Aralia L.

Herbs, shrubs, or small trees; leaves basal or cauline, ternately decompound or twice (or thrice) pinnately compound; flowers and fruits borne in 2–many umbels; carpels and styles usually 5; fruit black or nearly black. A genus of some 30 species of Asia and North America.

a. Shrubs or small trees, to 5 m tall; stems armed with stout coarse prickles; leaves twice-pinnately compound, most with 50 or more leaflets of approximately equal size.

b. Veins of leaflets curving apically and not reaching margin; major leaflets short-petiolulate . . . . 1. *Aralia spinosa*

bb. Veins of leaflets not or only slightly curving, running nearly to marginal teeth; major leaflets sessile or subsessile; see comments below under *A. spinosa* . . . . *Aralia elata*

aa. Herbaceous plants to 3 m tall; stems smooth throughout or prickly only at base; leaves variously compound, with fewer than 50 leaflets, the leaflets either equal or unequal in size.

b. Plants appearing acaulescent, the single compound leaf and the inflorescence arising basally from a long woody root . . . . 4. *Aralia nudicaulis*

bb. Plants with leaves and inflorescense borne on aerial stem.

c. Stems without prickles, smooth throughout; leaflets ovate to broadly ovate, most obliquely cordate or obliquely truncate at base, the largest 10–20 cm long, 5–15 cm wide; umbels numerous, more than 10 . . . . 2. *Aralia racemosa*

cc. Stems with prickles at base, otherwise smooth; leaflets narrowly ovate to lanceolate, tapering acutely at base, the largest 2.5–5 (–10) cm long, 1–3 (–5) cm wide; umbels 2–10 . . . . 3. *Aralia hispida*

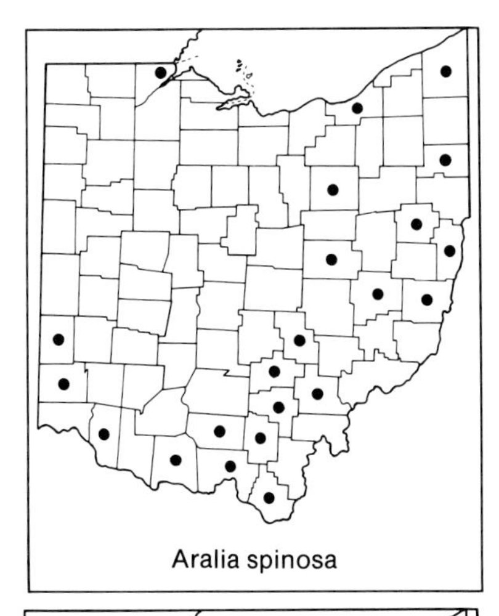
Aralia spinosa

1. **Aralia spinosa** L. Devil's Walkingstick. Hercules' Club

Shrub or small tree, to 5 m (15 feet) tall; stems armed with stout coarse prickles; leaves large, to 1 m or more long, twice-pinnately compound and often bearing prickles, most with 50 or more leaflets of approximately equal size; umbels in a long compound panicle. Native in southern Ohio, where it is at a northern boundary of the species' range; sometimes cultivated, escaped, and naturalized in northern counties; rich to moist woodlands. Smith (1982) describes leaf differences between juvenile and adult plants. July–September.

In addition to specimens from the counties mapped, there are others from Monroe, Portage, and Stark counties originally identified as *A. spinosa* that are probably specimens of the morphologically similar *A. elata* (Miq.) Seem., Japanese Angelica, escaped from cultivation.

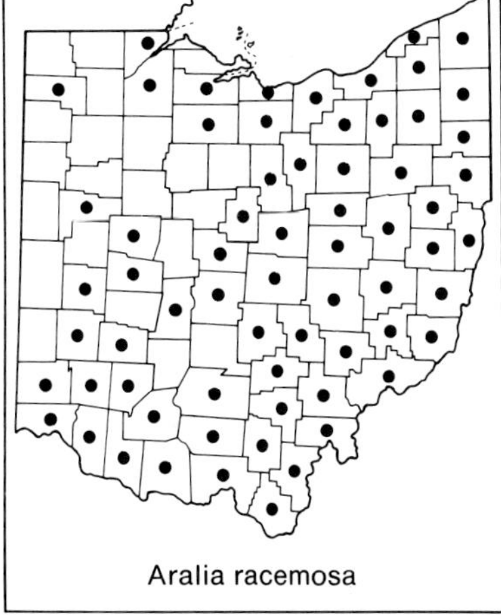
Aralia racemosa

2. **Aralia racemosa** L. Spikenard

Herbaceous perennial, sometimes shrub-like in appearance, to 2 m tall; stems without prickles; leaves ternately decompound, leaflets ovate to broadly ovate, mostly obliquely cordate or obliquely truncate at base, the largest 10–20 cm long, 5–15 cm wide; umbels numerous (more than 10) in a usually raceme-like panicle. Throughout eastern and southern Ohio, and at scattered northwestward localities; rocky wooded slopes, moist to mesic woodlands. Late June–August.

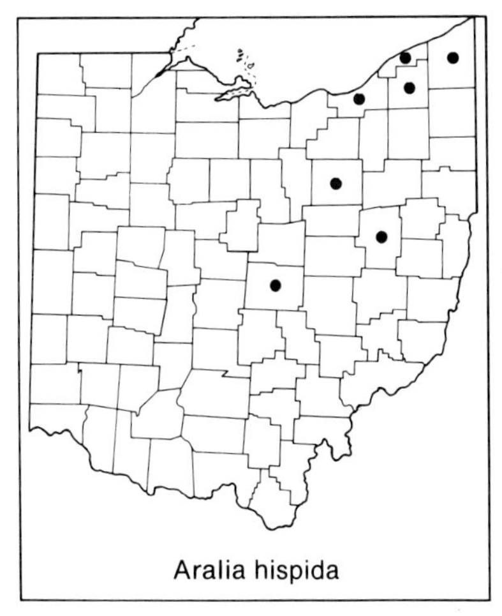
Aralia hispida

3. **Aralia hispida** Vent. Bristly Sarsaparilla

Herbaceous perennial, somewhat shrub-like in appearance, to 1 m tall; stems with prickles at base only, smooth above; leaves twice-pinnately compound, leaflets narrowly ovate to lanceolate, tapering acutely at base, the largest 2.5–5 (–10) cm long, 1–3 (–5) cm wide; umbels 2–10 in a paniculate cluster. Known from a few Ohio counties, mostly northeastern; dry, rocky, woodland openings and borders, clearings. June–August.

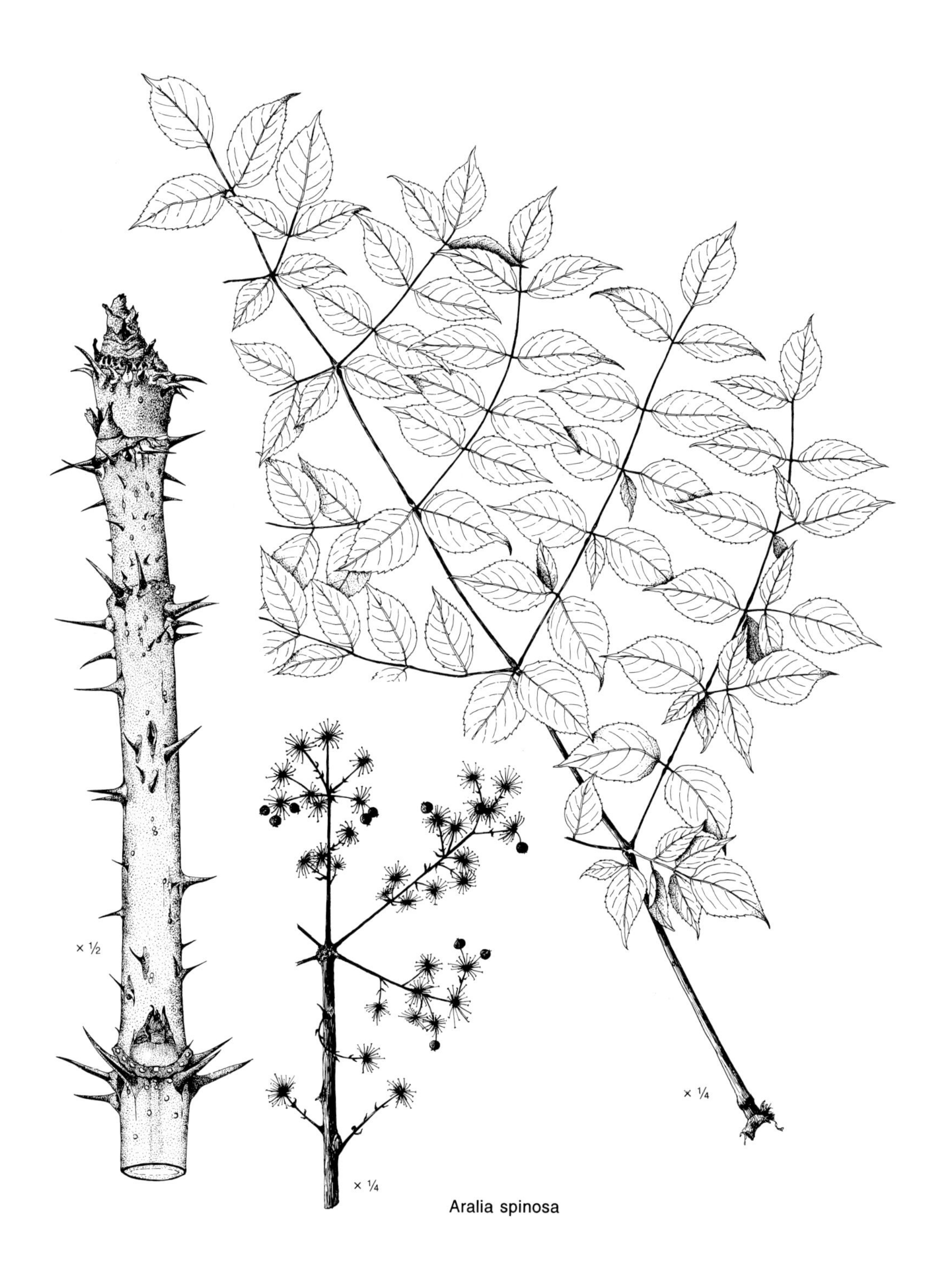

Aralia spinosa

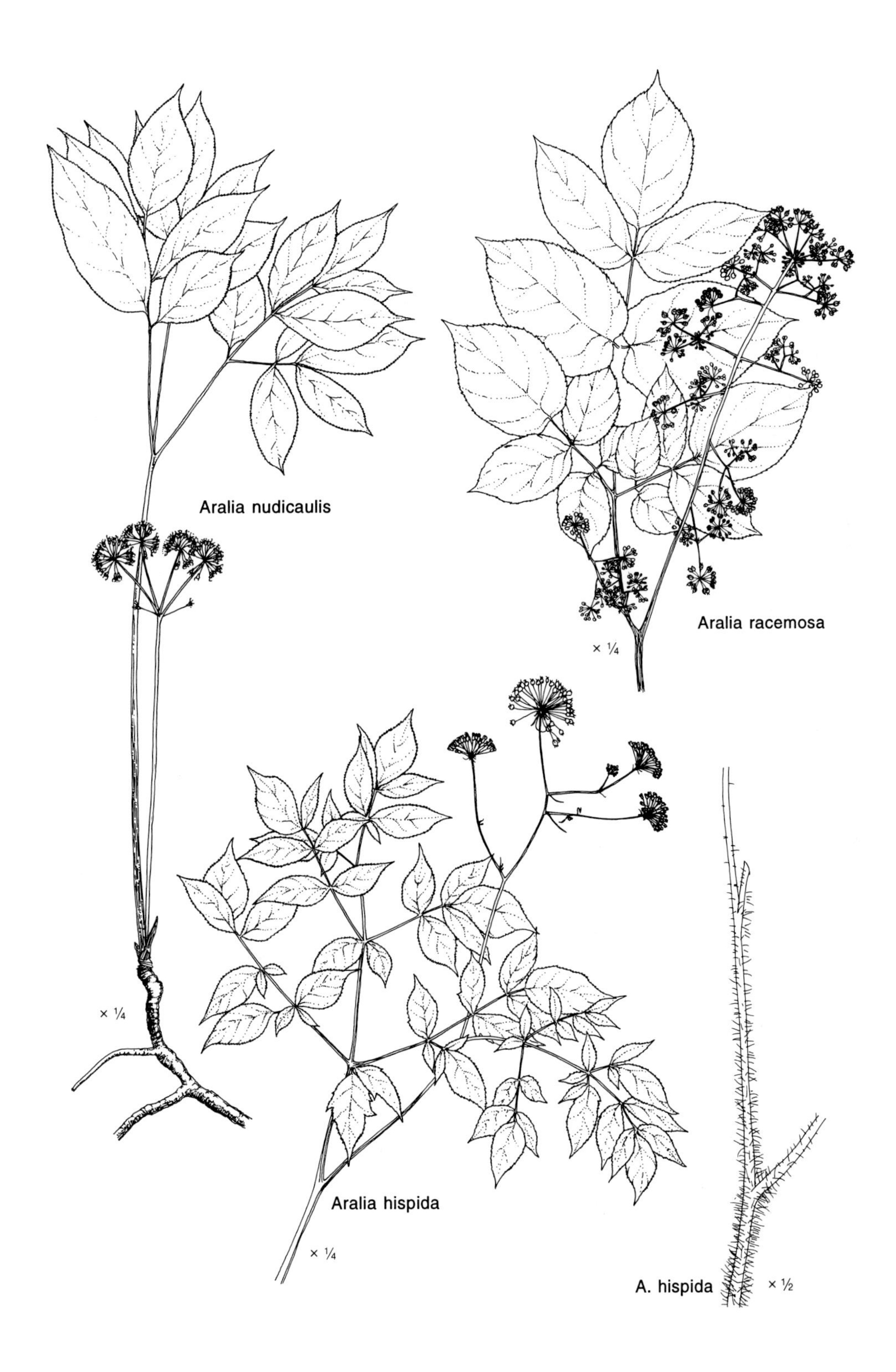
Aralia nudicaulis
Aralia racemosa
× ¼
× ¼
Aralia hispida
× ¼
A. hispida
× ½

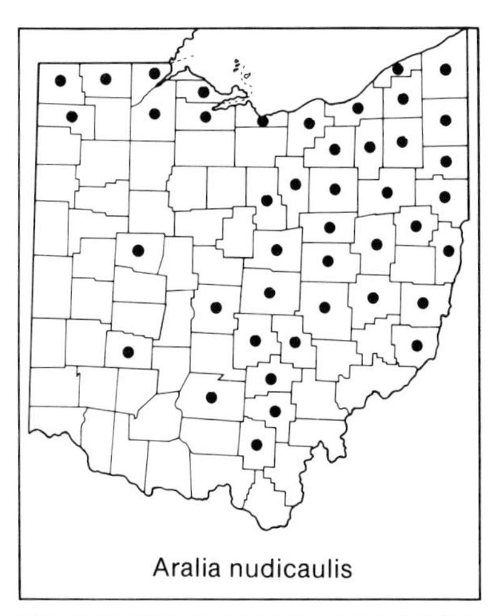
Aralia nudicaulis

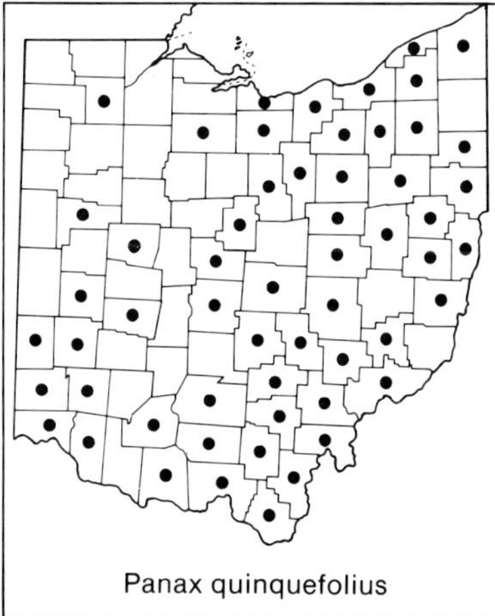
Panax quinquefolius

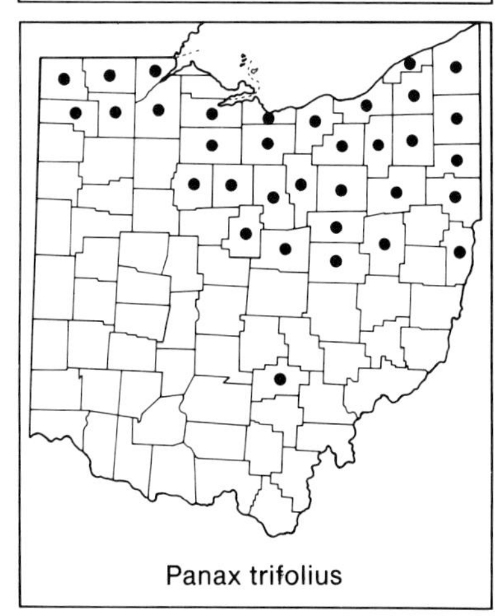
Panax trifolius

4. **Aralia nudicaulis** L. Wild Sarsaparilla

Perennial, without prickles, appearing acaulescent, the single leaf and inflorescence arising basally, the plants with a long woody root; leaf ternately compound, leaflets elliptic to ovate, often oblique at base, umbels 3–5 (–7), in an umbellate cluster on a long scape. Chiefly in northeastern counties, but also extending to northwesternmost Ohio and to some south-central counties; moist to mesic woodlands. Flanagan and Bain (1988) studied the biology of this species. Late April–June.

2. Panax L. GINSENG

Perennial herbs; leaves palmately compound; flowers and fruits in a single umbel; carpels and styles 2 or 3; fruit yellow or red. A genus of six or seven species, those below native to eastern North America, the others to eastern Asia. The best known of the Asian species is *Panax pseudoginseng* Wallich (*P. ginseng* C. A. Mey.), Ginseng, the source of the famed "ginseng root," a folk medicine prized in Asia as a general tonic for the mind and body, and as an aphrodisiac.

a. Leaflets 5, all or the upper 3 long-stalked; the leaflets elliptic-oblong to broadly obovate, the largest 7–12 cm long, 4–7 cm wide, acuminate at apex . . . . 1. *Panax quinquefolius*

aa. Leaflets 3–5, sessile; the leaflets lanceolate to narrowly elliptic or narrowly oblong, the largest 1.5–5 cm long, 0.5–1.5 cm wide, obtuse or acutish at apex . . . . 2. *Panax trifolius*

1. **Panax quinquefolius** L. American Ginseng

Perennial, to 6 dm tall, from a fusiform, often branched, root; stems erect; leaves (2) 3 or 4, in a single whorl, each leaf palmately compound with 5 leaflets, the 3 central leaflets long-stalked, the two outer short-stalked to occasionally subsessile; the leaflets elliptic-oblong to broadly obovate, acuminate at apex, the largest 7–12 cm long and 4–7 cm wide; fruits bright red, borne in a simple umbel at the tip of a long peduncle. Mostly in eastern and southern Ohio; woodlands. June–August.

This species is extensively harvested from the wild, mostly for shipment to Asia, where its dried root is used as a substitute for the Asian ginseng, noted above. The species has also been the subject of much research; for a study of floral biology and references to other recent publications, see Schlessman (1985).

2. **Panax trifolius** L. Dwarf Ginseng

Perennial, to 2 dm tall, but often shorter, from a globose bulb-shaped root; stems erect; leaves usually 3, occasionally 4, each leaf palmately compound with 3–5 sessile leaflets, the leaflets lanceolate to narrowly elliptic

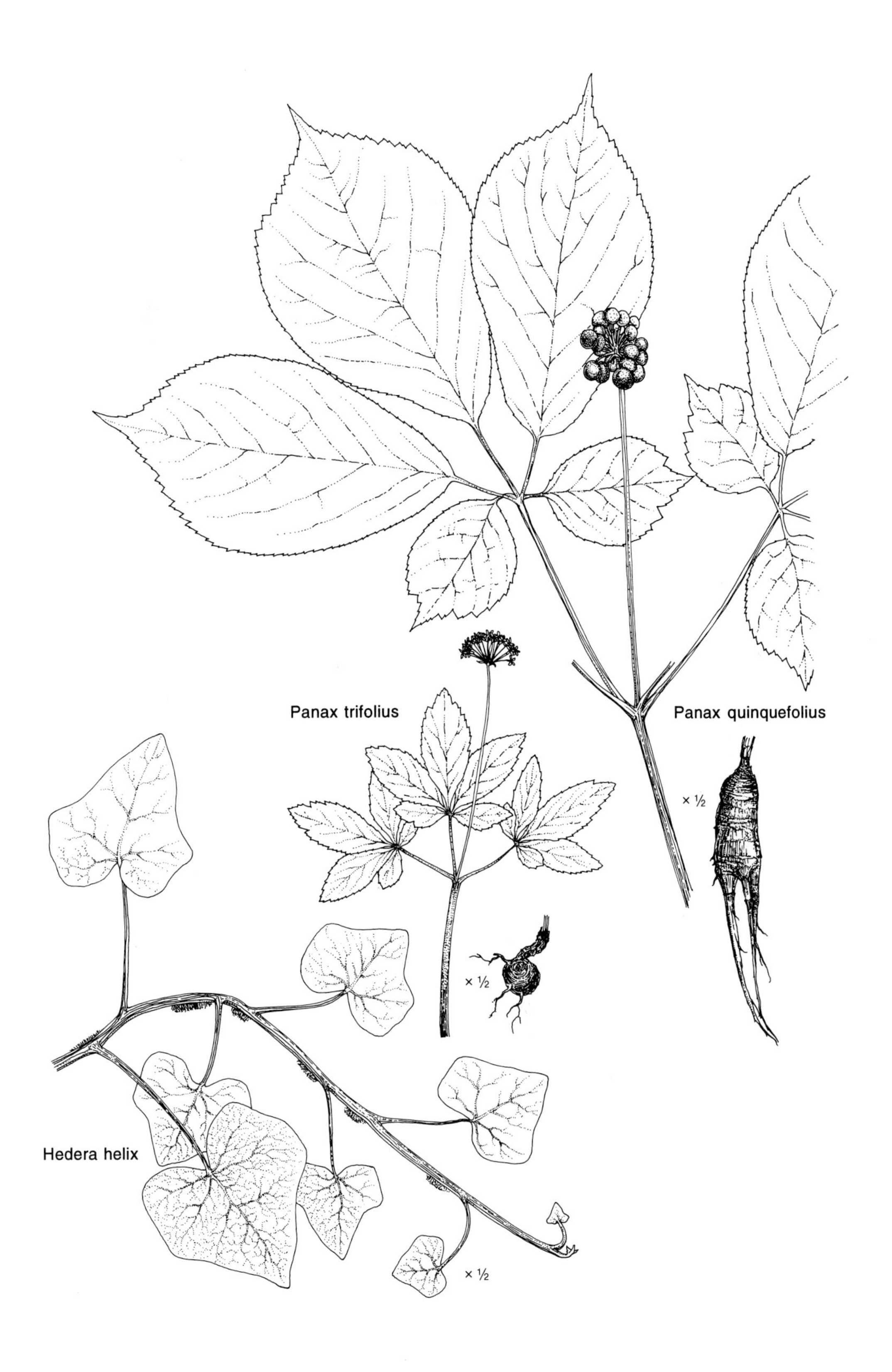
Panax trifolius
Panax quinquefolius
× ½
× ½
Hedera helix
× ½

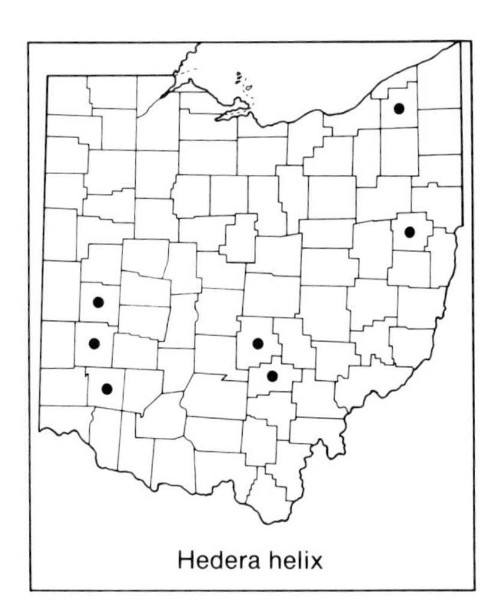
Hedera helix

or narrowly oblong, the largest 1.5–5 cm long, 0.5–1.5 cm wide, obtuse or acutish at apex; fruits dull yellow, borne in a simple umbel at the tip of a long peduncle. Chiefly in northeastern and north-central counties; moist to mesic woodlands. Philbrick (1983) studied the species' reproductive biology. April–May.

3. Hedera L. IVY

A genus of 1–few species native to the Old World.

1. HEDERA HELIX L. ENGLISH IVY

Woody vine, creeping, or climbing walls and trees by means of aerial rootlets; leaves evergreen, often variegated with white veins and/or margins, alternate, simple, shallowly to moderately 3- (or 5-) lobed or sometimes unlobed. Introduced from Eurasia and—with a host of cultivars—widely planted, usually as a ground cover; sometimes escaped and naturalized, probably more often than the few records indicate. Flowering only infrequently. August–October.

Apiaceae or Umbelliferae. PARSLEY FAMILY. CARROT FAMILY

Perennial, biennial, or annual herbs; leaves alternate, usually compound, the petioles widened and sheathing at the base; flowers small, perfect, usually in well-defined, simple or compound umbels; calyx of 5 minute sepals or none; corolla more or less actinomorphic, of 5 separate petals; stamens 5; carpels 2, fused to form a compound pistil with 2 styles, the styles enlarged at the base to form a unique structure, the stylopodium, which sits atop the inferior ovary; fruit a schizocarp splitting at maturity into two mericarps. A major and widely distributed family of about 3,000 species. The family has two scientific names; the use of either is equally correct. Much of the text below is based on Mathias and Constance' (1944–1945) study of the family in North America.

Many members of the family are cultivated as food seasonings. These and a few other weedy species occurring in Ohio as occasional adventives are represented by the following collections. *Aethusa cynapium* L., FOOL'S PARSLEY, a poisonous Eurasian plant, from Clark, Darke, Fayette, Lake, Miami, and Ottawa counties. *Anethum graveolens* L., DILL, often cultivated for use in making pickles, from Athens, Erie, Franklin, and Ottawa counties. *Apium graveolens* L., CELERY, from Cuyahoga and Highland counties. Two species of *Bupleurum*: *B. lancifolium* Hornem. (*B. subovatum* Link), a Mediterranean species distinguished from the next by its broader, more rounded, and abruptly apiculate bractlets and its wrinkled or strongly tuberculate fruit, from a flower bed in Franklin County, where it may have been introduced in bird seed; and *B. rotundifolium* L., THOROUGHWAX, a Eurasian weed with bractlets more ovate and acuminate, and fruit smooth, from Adams and Warren counties. *Carum carvi* L., CARAWAY, from a number of sites, mostly in northern Ohio, and most prior to 1950: Ashland, Auglaize, Columbiana, Cuyahoga, Fulton, Hamilton, Lake, Licking, Lorain, Ottawa, and Trumbull counties. *Coriandrum sativum* L., CORIANDER, from Allen County. *Foeniculum vulgare* Mill., FENNEL, from Erie County in northern Ohio, and from four southern counties: Highland, Hocking, Pike, and Scioto. *Levisticum officinale* Koch, EUROPEAN LOVAGE, a tall herb grown both for use as a seasoning and as an ornamental, from Cuyahoga and Highland counties. *Oenanthe aquatica* (L.) Poir., WATER-FENNEL, from a cemetery pond in Franklin County. *Petroselinum crispum* (Mill.) A. W. Hill, PARSLEY, from Highland and Madison counties. *Scandix pecten-veneris* L., VENUS' COMB or SHEPHERD'S NEEDLE, a Mediterranean weed, from Lake County.

Imperatoria ostruthium L. (*Peucedanum ostruthium* (L.) Koch), MASTERWORT, was attributed to Ohio by Cusick and Silberhorn (1977), but I have determined the specimens cited to be *Heracleum maximum*. *Pimpinella saxifraga* L., BURNET-SAXIFRAGE, was attributed to Ohio by Gleason (1952a); I have seen no specimens other than ones marked "cultivated."

a. Flowers in head-like clusters, the pedicels scarcely or not visible (except staminate flowers in *Sanicula* on pedicels to 5 mm long).
  b. Leaves palmately compound or deeply palmately lobed; fruits covered with numerous hooked bristles. . . . . 2. *Sanicula*
  bb. Leaves simple, unlobed; fruits without bristles.
    c. Leaves long-petiolate, the blades suborbicular to reniform . . . . 1. *Hydrocotyle*
    cc. Leaves without evident petioles, broadly linear to long-lanceolate . . . . 3. *Eryngium*
aa. Flowers in umbels or compound umbels and with evident pedicels.
  b. Cauline leaves manifestly simple.
    c. Leaf margins crenate to lobed; umbels simple . . . . . 1. *Hydrocotyle*
    cc. Leaf margins entire; umbels compound; see above. . . . . *Bupleurum*
  bb. Cauline leaves all or mostly compound or so deeply lobed or dissected as to appear compound (basal leaves sometimes simple).
    c. Ovary, and later the fruit, either pubescent or bearing spines or bristles.
      d. Leaves either once-pinnately or once-palmately compound.
        e. Petioles forming a broad sheath 1 cm or more in width; principal leaflets 11–25 cm or more in width. . . . . 23. *Heracleum*

ee. Petioles forming an inconspicuous sheath 0.5 cm or less in width; principal leaflets 1–8 cm wide.

f. Leaves palmately compound . . . . 2. *Sanicula*

ff. Leaves pinnately compound . . . . 7. *Torilis*

dd. Leaves twice or more compound.

e. Fruit short-pubescent, lacking spines or bristles.

f. Stems weak to somewhat firm, to 0.8 m tall; leaflets with narrow segments, the margins of the segments entire; fruit flattened laterally, ribbed but not winged . . . . 4. *Chaerophyllum*

ff. Stems stout, to 2 m tall; leaflets with low serrate-dentate margins, unlobed; fruit flattened dorsally, conspicuously winged . . . . 20. *Angelica*

ee. Fruit bearing spines or bristles.

f. Body of fruit (excluding beak) 5–25 mm long, with straight, either appressed or ascending bristles.

g. Fruit with short beak to 3.5 mm long. . . . . 6. *Osmorhiza*

gg. Fruit with conspicuous beak 20–70 mm long; see above . . . . *Scandix*

ff. Body of fruit (excluding beak) 2–4 mm long, with divergent, either hooked or curved bristles.

g. Inflorescence subtended by conspicuous, deeply lobed bracts . . . . 24. *Daucus*

gg. Inflorescence subtended by inconspicuous linear bracts or none.

h. Rays of umbels glabrous; sheaths of upper leaves ciliate . . . . 5. *Anthriscus*

hh. Rays of umbels hispid; sheaths of upper leaves lacking cilia . . . . 7. *Torilis*

cc. Ovary, and later the fruit, glabrous—although sometimes bearing small tubercles.

d. Principal leaves with most ultimate segments (5–) 10 mm or more in width.

e. Principal cauline leaves either trifoliolate or once-palmately compound or once-pinnately compound.

f. Petioles forming a broad sheath 1 cm or more in width; flowers white. . . . . 23. *Heracleum*

ff. Petioles forming a sheath less than 1 cm in width; flowers white, pink, purple, or yellow.

g. Leaves trifoliolate, the leaflets sometimes deeply lobed.

h. Most leaflets palmately veined; flowers white; see above . . . . *Apium* hh. Most leaflets more or less pinnately veined, with one main midrib; flowers yellow, purple, or white.

i. Flowers white; margins of most leaflets coarsely doubly serrate. . . . . 12. *Cryptotaenia*

ii. Flowers yellow or purple; margins of leaflets mostly simply serrate.

j. Fruit ribbed, wingless or with low wings; central flower or fruit in each umbellet sessile or subsessile on pedicel 0.1–0.5 mm long . . . . 10. *Zizia*

jj. Fruit with prominent wings; central flower or fruit in each umbellet on short pedicel 1.0 mm or longer. . . . . 18. *Thaspium*

gg. Leaves with five or more leaflets.

h. Leaflets lanceolate to linear, oblong, or elliptic; flowers white.

i. Margins of leaflets with numerous teeth . . . . 16. *Sium*

ii. Margins of leaflets entire or with 1–3 (–6) teeth along each side . . . . 21. *Oxypolis*

hh. All or some leaflets ovate to suborbicular in outline; flowers white or yellow.

i. Flowers yellow; fruit winged. . . . . 22. *Pastinaca*

ii. Flowers white; fruit without wings or with very narrow wings.

j. Leaf margin serrate with blunt teeth; see above . . . . *Apium*

jj. Leaf margin serrate with acute teeth; see above . . . . *Pimpinella*

ee. Principal cauline leaves twice or more compound.

f. Leaflets entire and unlobed; flowers yellow. . . . . . . 13. *Taenidia*

ff. Leaflets serrate or lobed; flowers yellow, purple, or white.

g. Flowers yellow or purple.

h. Inflorescences subtended by an involucre of numerous lanceolate bracts; see above . . . . *Levisticum*

hh. Inflorescences subtended by an involucre of few, inconspicuous, linear bracts or none.

i. Central flower or fruit in each umbellet sessile or subsessile on pedicel 0.1–0.5 mm long. . . . . 10. *Zizia*

ii. Central flower or fruit in each umbellet on short pedicel 1.0 mm or longer.

j. Fruit with low ribs but lacking wings; sepals none; see above . . . . *Petroselinum* jj. Fruit with conspicuous wings; sepals present, small .... 18. *Thaspium*

gg. Flowers white.

h. Petioles of upper leaves forming a broad sheath 1 cm or more in width when flattened.................. 20. *Angelica*

hh. Petioles of upper leaves forming a narrower sheath less than 1 cm in width when flattened.

i. Major branches from leaflet midveins leading nearly or quite to sinuses between marginal teeth, the vein then abruptly diverted and ending in an adjacent tooth ...... 11. *Cicuta*

ii. Major branches from leaflet midveins leading directly to marginal teeth.

j. Principal leaves twice- or thrice-pinnately compound, with numerous ultimate segments.............. 5. *Anthriscus*

jj. Principal leaves ternately compound, mostly with 9 leaflets.

k. Leaflets irregularly doubly serrate, with 20 or more teeth on each side; flowers lacking sepals .... 14. *Aegopodium*

kk. Leaflets singly serrate, usually with 10–17 teeth on each side; flowers with minute sepals .... 17. *Ligusticum*

dd. Principal leaves with ultimate segments narrowly elliptic to linear or filiform, 5 mm or less in width.

e. Ultimate segments of leaves filiform, to 0.5 mm wide.

f. Plants with some leaf segments filiform and others linear to narrowly lanceolate or ovate; petals white, pink, or lavender; fruit globose or subglobose.

g. Upper leaves with filiform segments, the lower with ovate but deeply dissected segments; plants often fruiting, never bearing bulblets; see above .... *Coriandrum*

gg. Leaves with an irregular pattern and a scattered combination of filiform or more often linear to lanceolate leaflets, the latter with dentate margins; plants rarely fruiting, often with bulblets in axils of upper leaves.................... 11. *Cicuta*

ff. Plants with all leaf segments filiform; petals yellow; fruit oblong to elliptic.

g. Fruit more or less terete, ribs low and scarcely evident, not winged; fresh foliage with anise scent when crushed; see above .... *Foeniculum*

gg. Fruit somewhat flattened, ribs evident, the lateral ribs winged; fresh foliage with dill scent when crushed; see above........ .... *Anethum* ee. Ultimate segments of leaves linear or broader, 1–5 mm wide.

f. Umbellets subtended by conspicuous bractlets, these broadly lanceolate to oblong or ovate, 1–2 mm wide.

g. Primary umbels with 1–5 rays.

h. Plants weak, lax to erect, 2–8 dm tall, annual with taproot; anthers pale; fruit longer than wide . . . . . 4. *Chaerophyllum*

hh. Plants delicate, erect, 0.5–1.5 dm tall, perennial from subglobose tuber; anthers dark; fruit wider than long . . . . 8. *Erigenia*

gg. Primary umbels with 6–15 or more rays.

h. Petiolar sheaths ciliate . . . . 5. *Anthriscus*

hh. Petiolar sheaths glabrous, without cilia . . . . 9. *Conium*

ff. Umbellets subtended by inconspicuous bractlets 0.5 mm or less in width, or bractlets none.

g. Flowers yellow.

h. Sepals none; fruit 2–3 mm long; see above . . . . *Petroselinum*

hh. Sepals present; fruit 3–4 mm long. . . . . 18. *Thaspium*

gg. Flowers white, pink, or lavender.

h. Sepals present, lanceolate to triangular.

i. Ultimate leaf segments usually irregularly dentate; plants rarely fruiting, often with bulblets in axils of upper leaves; fruits, when produced, 1.5–2 mm long . . . . 11. *Cicuta*

ii. Ultimate leaf segments entire; plants often fruiting, never bearing bulblets; fruits 3–4 mm long.

j. Ultimate segments 2–3 mm long, mucronate-tipped; see above. . . . . *Oenanthe*

jj. Ultimate segments 5–30 mm long, obtuse at tip . . . . 15. *Perideridia*

hh. Sepals none.

i. Principal leaves with blades 8–20 cm long; rays of inflorescence 15–30; fruit more or less oblong, 4–6 mm long . . . . 19. *Conioselinum*

ii. Principal leaves with blades 6–15 cm long; rays of inflorescence 7–20; fruit ellipsoid to ovoid or subglobose, 3–4 mm long.

j. Bractlets subtending umbellets minute or none; fruit more or less ellipsoid, slightly flattened, the ribs inconspicuous; pedicels varying markedly in length, the longest to 10–14 mm long; ultimate leaf segments narrowly linear; see above . . . . *Carum*

jj. Bractlets subtending umbellets 2–5, conspicuous, congregated on one side of umbellet; fruit broadly ovoid or subglobose, nearly terete, with prominent corky ribs; pedicels more or less uniform in length, the longest to 5 mm long; ultimate leaf segments ovate to narrowly ovate, many times cleft; see above . . . . *Aethusa*

1. Hydrocotyle L. WATER-PENNYWORT

Perennial herbs; leaves simple, long-petiolate, the blades more or less orbicular to reniform; flowers white or greenish, in head-like clusters or simple umbels; fruit glabrous. A cosmopolitan genus of some 60 or more species.

a. Leaves peltate . . . . 1. *Hydrocotyle umbellata*

aa. Leaves not peltate, the blades cleft to the point of petiole attachment, the petioles therefore attached at margin of blade.

b. Blades of principal leaves 5–12 mm wide; flowers and fruits individually sessile at the tip of a peduncle 5–20 mm long. . . . . 3. *Hydrocotyle sibthorpioides*

bb. Blades of principal leaves 15–60 mm wide; flowers and fruits individually short-pedicellate.

c. Leaves moderately lobed, the deeper sinuses extending 1/3–1/2 of way toward point of petiole attachment; pedicels of flowers and fruits borne at tip of a peduncle 4–60 mm long. . . . . 2. *Hydrocotyle ranunculoides*

cc. Leaves shallowly lobed, the sinuses extending at most 1/4 of way toward point of petiole attachment; pedicels of flowers and fruits either attached directly to stem, or attached to a petiole, or borne at tip of a short peduncle less than 1 mm long . . . . 4. *Hydrocotyle americana*

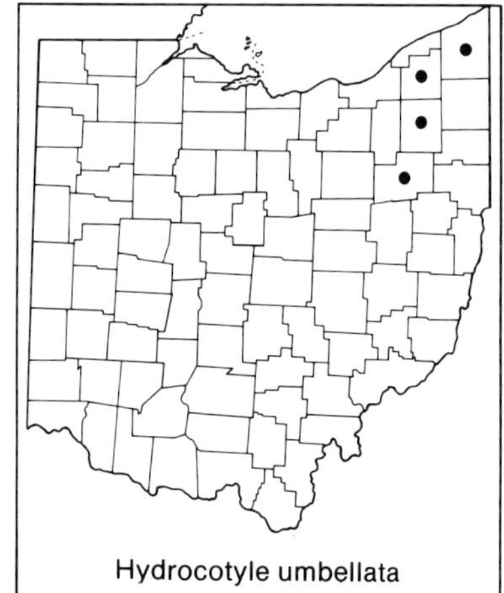
Hydrocotyle umbellata

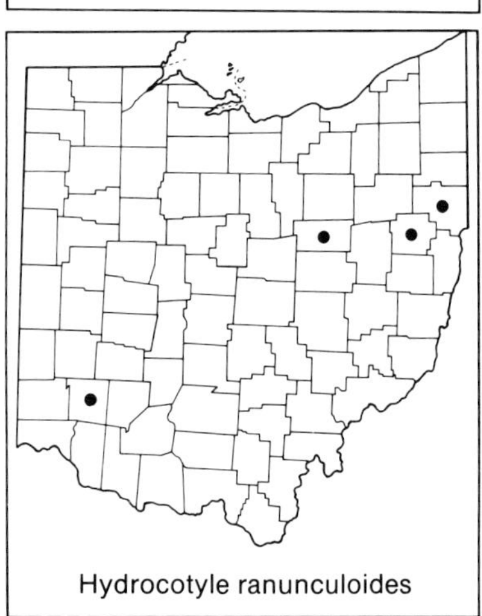
Hydrocotyle ranunculoides

1. **Hydrocotyle umbellata** L. UMBELLATE WATER-PENNYWORT
Creeping perennial; leaves peltate, orbicular, to 6 cm or more in diameter, margin crenate. A few localities in northeastern counties; shallow water of lakes and ponds, muddy shores. July–September.

2. HYDROCOTYLE RANUNCULOIDES L. f. FLOATING WATER-PENNYWORT
Creeping perennial; leaves orbicular-reniform, to 6 cm wide, moderately lobed, the deeper sinuses extending 1/3–1/2 of the way to point of petiole attachment; flowers and fruits individually short-pedicellate at tip of a peduncle 4–60 mm long. A few sites in scattered counties; probably naturalized from areas to the east and south of Ohio, but perhaps native; shallow water of ponds and marshes, muddy shores. June–August.

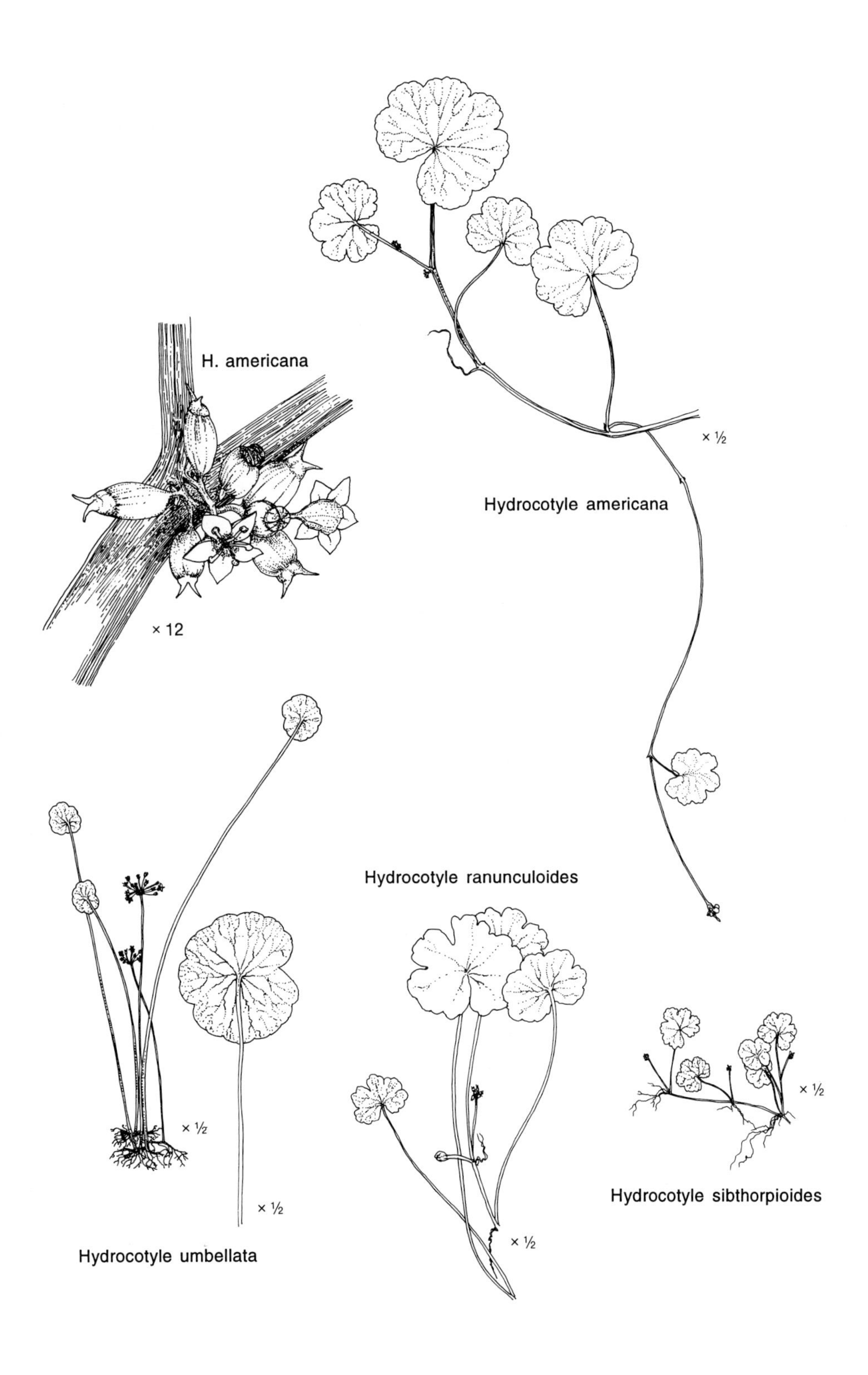
H. americana
× 12
× ½
Hydrocotyle americana
Hydrocotyle ranunculoides
× ½
× ½
× ½
× ½
Hydrocotyle sibthorpioides
Hydrocotyle umbellata

Hydrocotyle sibthorpioides

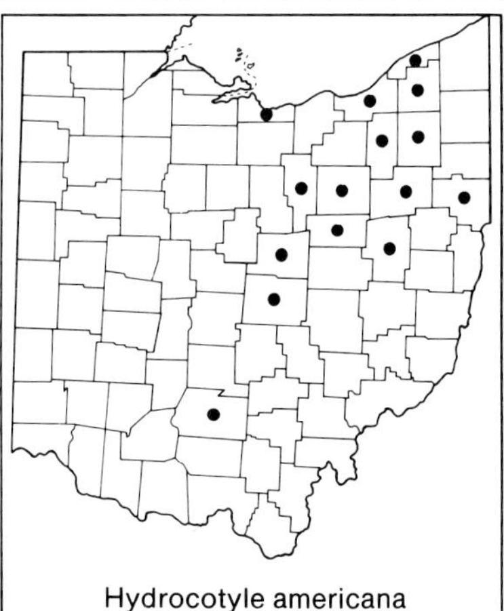
Hydrocotyle americana

3. Hydrocotyle sibthorpioides Lam. Lawn Water-pennywort
Creeping perennial; leaves nearly orbicular, to 12 mm in diameter, shallowly crenately lobed; flowers and fruits individually sessile at tip of a peduncle 5–20 mm long. An Asian species, naturalized in southernmost Ohio; lawns and moist, shaded, disturbed sites. June–August.

4. **Hydrocotyle americana** L. American Water-pennywort
Creeping perennial; leaves nearly orbicular, to 5 cm in diameter, shallowly lobed, the sinuses extending at most 1/4 of the way to point of petiole attachment; pedicels of flowers and fruits either attached directly to stem, or attached to a petiole, or attached to tip of a short peduncle less than 1 mm long. Mostly in the northeastern quarter of Ohio; seeping ledges, wet thickets, bogs, fens, marshes, swamps. June–September.

2. Sanicula L. SNAKEROOT. SANICLE

Perennial, or in some species perhaps biennial, herbs; leaves palmately compound or deeply palmately lobed; flowers white to greenish-white or greenish-yellow, in dense head-like clusters (simple umbels); inflorescences either with 3 perfect and 1–many staminate flowers or with staminate flowers only; fruits covered with numerous hooked bristles. A widely distributed genus of some 30 species, it was revised by Shan and Constance (1951). Phillippe's (1978) biosystematic study of *Sanicula* section *Sanicula* includes all the Ohio species. Pryer and Phillippe (1989) published a synopsis of the genus as it is represented in the flora of eastern Canada.

In addition to the four species below, Phillippe (1978) also lists *Sanicula smallii* Bickn. for Ohio, citing a specimen from Hocking County. This Ohio site is greatly disjunct from the species' range, in southeastern United States, shown by Shan and Constance (1951) and is considerably disjunct from the nearest sites, in southern West Virginia and east-central Kentucky, recorded by Phillippe. The specimen, which I have studied, was collected early in the season, and its fruits are less than mature. It may be merely an immature plant of *S. trifoliata.*

a. Inflorescences with 10–20 or more staminate flowers; persistent styles conspicuously exserted and recurved, longer than the bristles of the fruit.
  b. Sepals of staminate flowers narrowly lanceolate, 0.9–1.4 mm long, and with rigid subulate tip; fruit 4–6 mm long. . . . . 1. *Sanicula marilandica*
  bb. Sepals of staminate flowers triangular to narrowly triangular, 0.4–0.6 mm long, and with soft or only slightly hardened acute apex; fruit 3–5 mm long . . . . 2. *Sanicula gregaria*

aa. Inflorescences with 1–5 (–8) staminate flowers; persistent styles erect but shorter than to scarcely equalling length of the bristles of the fruit.

b. Pedicels of staminate flowers (2.5–) 3–5 mm long; fruit ovoid to elliptic-ovoid at maturity, 6–7 mm long (including bristles and terminal calyx), subsessile, borne on pedicels to 0.5 mm long; the fruit with a persistent calyx exceeding length of bristles and forming a conspicuous hard tuft at fruit apex . . . . 3. *Sanicula trifoliata*

bb. Pedicels of staminate flowers 1–2 mm long; fruit globose to slightly wider than long at maturity, 2–5 mm long (including bristles and terminal calyx), borne on pedicels 0.8–1.2 mm long; the fruit with a persistent calyx shorter than, and more or less hidden by, the bristles at fruit apex . . . . 4. *Sanicula canadensis*

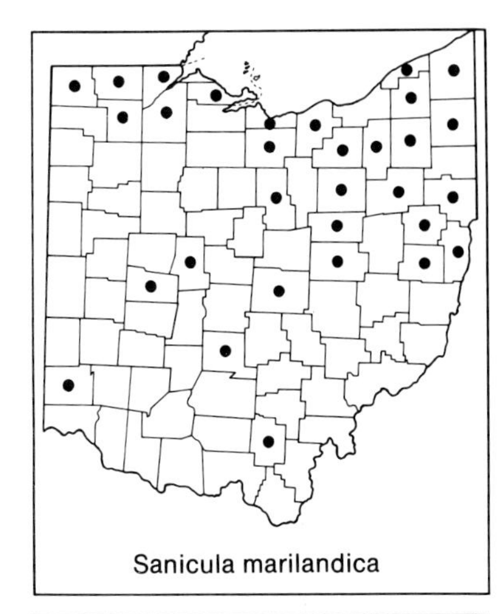
Sanicula marilandica

1. **Sanicula marilandica** L. Black Snakeroot

Perennial, to ca. 1 m tall; leaves 3–5-parted, cut nearly or quite to top of petiole, the two outer divisions sometimes again deeply lobed; inflorescences with 10–many staminate flowers, sepals of staminate flowers narrowly lanceolate, 0.9–1.4 mm long, and with a rigid subulate tip; fruit 4–6 mm long, the persistent styles evident, exserted and recurved, longer than the bristles of the fruit. Northern, especially northeastern counties, and scattered elsewhere; rich woods, wooded slopes, thickets. May–June.

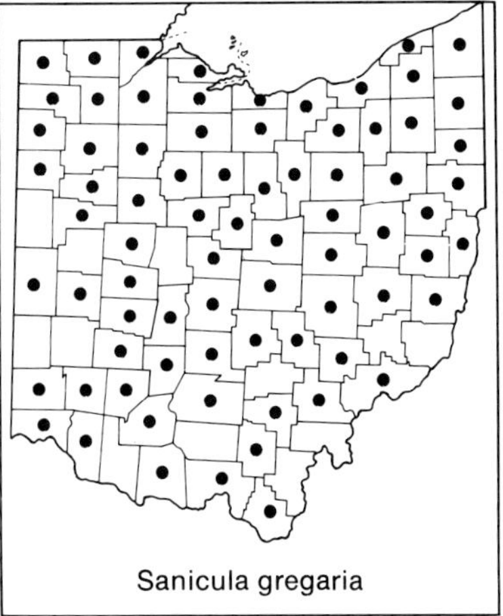
Sanicula gregaria

2. **Sanicula gregaria** Bickn. Clustered Snakeroot

Perennial, to ca. 7 dm tall; leaves 3–5-parted, cut nearly or quite to petiole; inflorescence with 12–25 or more staminate flowers, sepals of staminate flowers triangular to narrowly triangular, 0.4–0.6 mm long, and with a soft or only slightly hardened acute apex; fruit 3–5 mm long, the persistent styles exserted and recurved, longer than the bristles of the fruit. Throughout Ohio; moist to dry woodlands, thickets, roadsides. May–June.

Pryer and Phillippe (1989) have recently published the combination S. *odorata* (Raf.) Pryer & Phillippe, proposed as the correct name for this species.

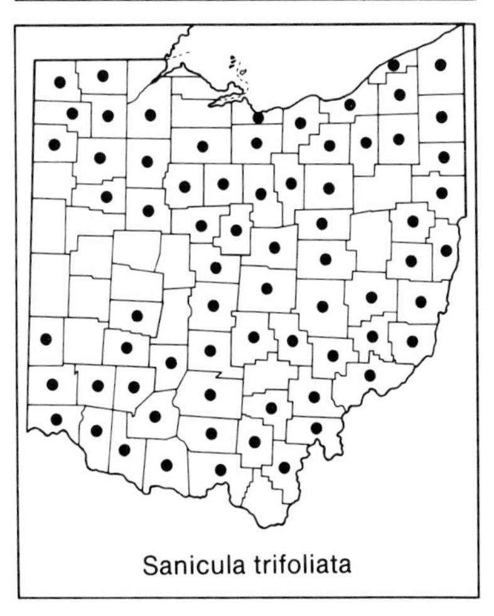
Sanicula trifoliata

3. **Sanicula trifoliata** Bickn. Large-fruited Snakeroot

Perennial or perhaps biennial, to ca. 7 dm tall; leaves 3-parted, cut to petiole, one or both lateral divisions sometimes lobed; inflorescences with (1–) 4–5 (–8) staminate flowers, pedicels of staminate flowers (2.5–) 3–5 mm long; fruit ovoid to elliptic-ovoid at maturity, 6–7 mm long (including bristles and terminal calyx), subsessile, borne on pedicels to 0.5 mm long, the fruit with a persistent calyx exceeding length of bristles and forming a conspicuous hard tuft at fruit apex, the persistent styles erect but shorter than to scarcely equalling length of bristles of the fruit. Throughout Ohio,

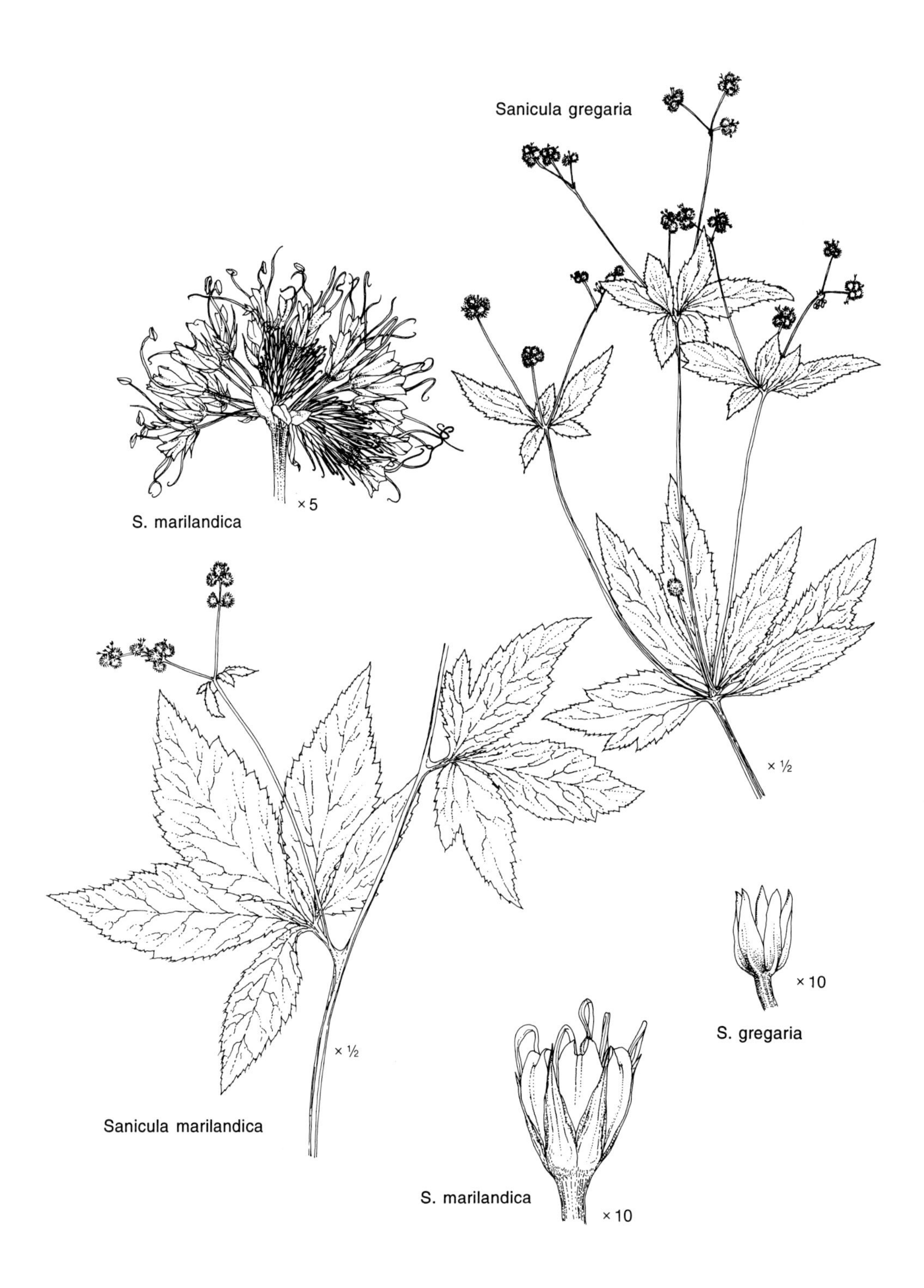
Sanicula gregaria
× 5
S. marilandica
× ½
× ½
× 10
S. gregaria
Sanicula marilandica
S. marilandica
× 10

except west-central counties; moist to mesic or dry woodlands. Additional record: Hocking County (Phillippe, 1978). June.

4. **Sanicula canadensis** L. SHORT-STYLED SNAKEROOT

incl. var. *canadensis* and var. *grandis* Fern.

Perennial or perhaps biennial, to ca. 1 m tall; leaves 3-parted, the two lateral divisions often lobed; inflorescences usually with 1–3 staminate flowers, pedicels of staminate flowers 1–2 mm long; fruit globose to slightly wider than long at maturity, 2–5 mm long (including bristles and terminal calyx), borne on pedicels 0.8–1.2 mm long; the fruit with a persistent calyx shorter than, and more or less hidden by, the bristles at fruit apex, the persistent styles erect but shorter than to scarcely equalling length of bristles of the fruit. Throughout Ohio; moist to dry woodlands, thickets. June.

Following Shan and Constance (1951), the species is defined here to include, without rank, the larger coarser plants recognized by Fernald (1950), Phillippe (1978), and Pryer and Phillippe (1989) as var. *grandis*.

3. Eryngium L. ERYNGO

A cosmopolitan genus of more than 200 species, one of which is a member of the Ohio flora. Bell (1963) studied the species of southeastern United States.

1. **Eryngium yuccifolium** Michx. RATTLESNAKE-MASTER

Tough perennial, to ca. 1 m tall; leaves coarse, simple, sessile, unlobed, parallel-veined, broadly linear to long-lanceolate, to ca. 5 dm long and to 3 cm wide, margins with weak spines; flowers white, in dense heads, the heads arranged in a cymose cluster; fruits with small flat scales on the angles. Chiefly in western counties; openings, prairies. July.

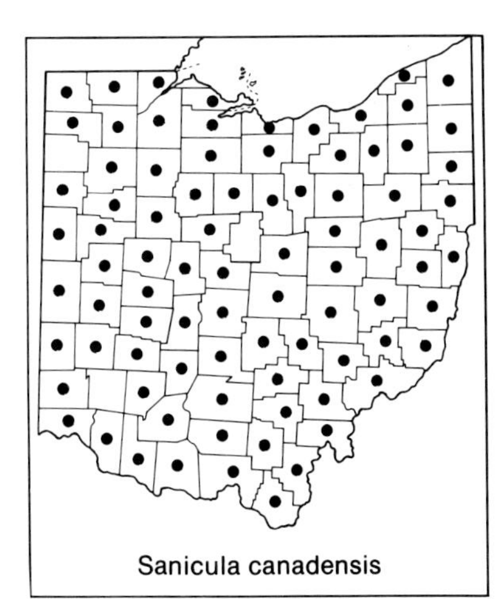

Sanicula canadensis

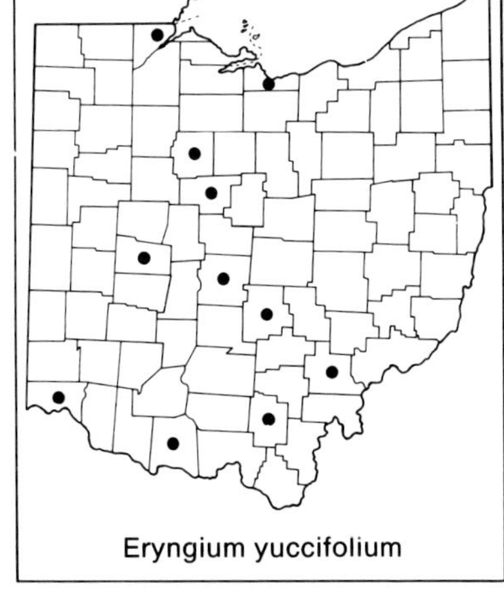

Eryngium yuccifolium

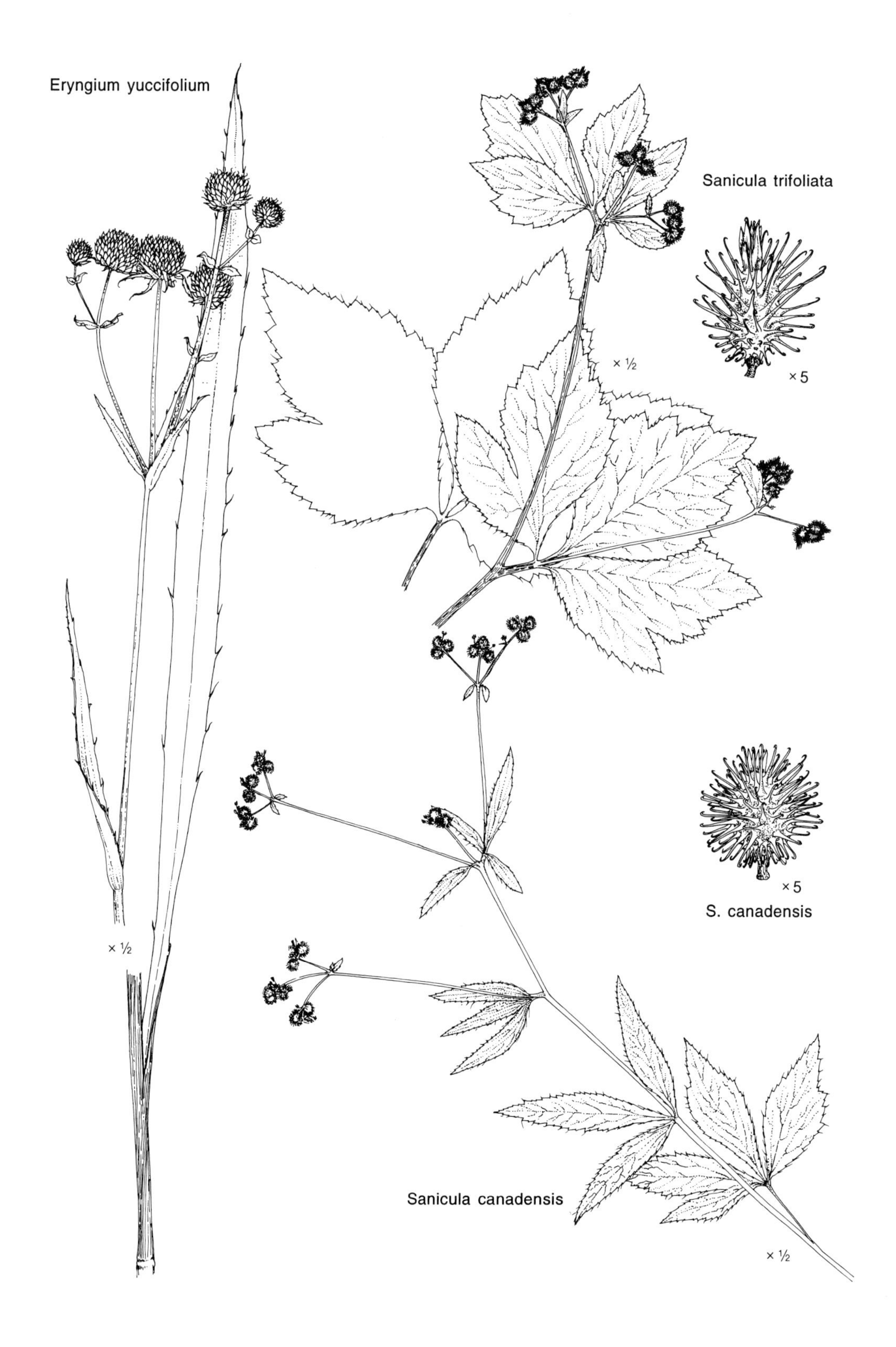
Eryngium yuccifolium
× ½
Sanicula trifoliata
× ½
× 5
× 5
S. canadensis
Sanicula canadensis
× ½

4. Chaerophyllum L.

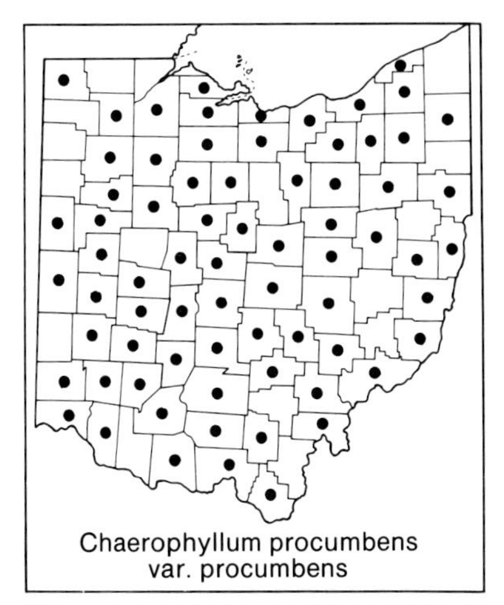
Chaerophyllum procumbens var. procumbens

A Northern Hemisphere genus of about 35 species, one of which is native to Ohio. *Chaerophyllum tainturieri* Hook. was attributed to Ohio by Fernald (1950) and others, but no specimens have been found in this study. All Ohio specimens encountered that were so-identified or so-annotated proved to be *C. procumbens*.

1. **Chaerophyllum procumbens** (L.) Crantz   Spreading Chervil

Weak, lax to semi-erect annual, to 6 (–8) dm tall, loosely branched at base; leaves more or less ternate, the divisions twice-pinnately compound, the ultimate segments narrow, entire-margined, and blunt at apex; flowers white, in simple or compound umbels; primary umbels with (1–) 2–3 rays, umbellets subtended by several conspicuous, but short, broadly lanceolate bractlets; anthers pale; fruit longer than wide, glabrous or pubescent. Throughout most of Ohio, scarce in northwestern counties; wooded flood-plains and terraces, mesic woodlands. April–May.

The two varieties keyed below are seemingly of minor significance, but are usually separable with little or no difficulty. Variety *procumbens* occurs throughout the species' Ohio range; var. *shortii* is found chiefly in south-central and southwestern counties.

a. Fruit glabrous . . . . var. *procumbens*
aa. Fruit pubescent . . . . var. *shortii* T. & G.
Short's Chervil

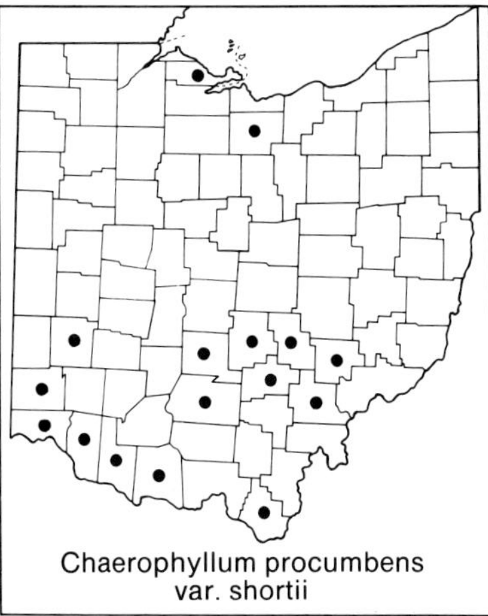
Chaerophyllum procumbens var. shortii

5. Anthriscus Persoon   CHERVIL

Tall weedy plants; leaves 2–3 times pinnately compound; flowers white, in compound umbels; fruits either hispid or glabrous. A Eurasian genus of about 12 species.

a. Fruit hispid, lanceolate-ovoid, 4 mm long . . . . . 1. *Anthriscus caucalis*
aa. Fruit smooth, lanceolate, 5–6 mm long . . . . . . 2. *Anthriscus sylvestris*

1. Anthriscus caucalis Bieb.   Bur Chervil
A. scandicina (Web.) Mansf.

Anthriscus caucalis

Annual, to 6 (–9) dm tall; rays of primary umbels 2–6; fruit lanceolate-ovoid, ca. 4 mm long, hispid with stiff bristles. Thieret (1976a), who reported the discovery of this Eurasian plant in Ohio, found it locally common "in a fallow field and in an adjacent thicket" on North Bass Island in Lake Erie, Ottawa County. Not illustrated. May–June.

Anthriscus sylvestris

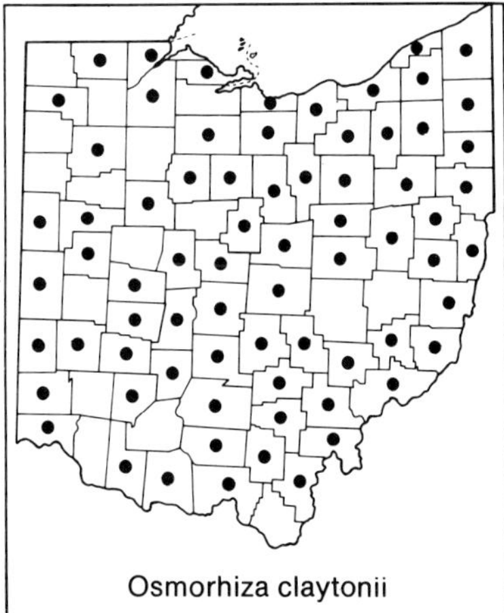
Osmorhiza claytonii

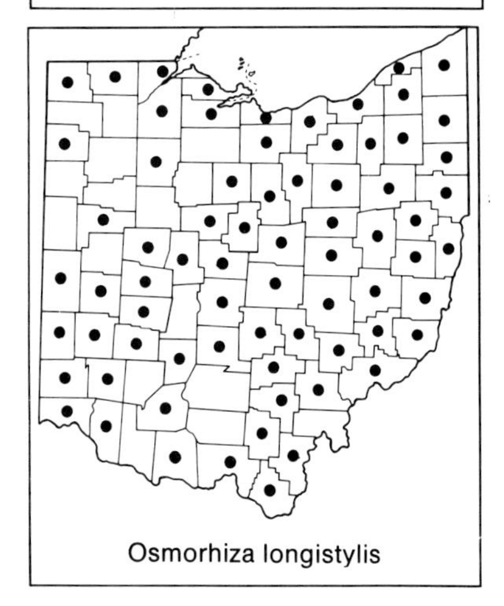
Osmorhiza longistylis

2. Anthriscus sylvestris (L.) Hoffm. Wild Chervil

Annual or perhaps biennial or perennial, to 1 m tall; rays of primary umbels 6–10; fruit lanceolate, 5–6 mm long, smooth. A Eurasian species, naturalized in a few Ohio localities; disturbed weedy habitats. May–June.

6. Osmorhiza Raf. SWEET CICELY

Perennial herbs; stems and roots aromatic when broken; leaves ternate, the divisions once or twice pinnately compound; flowers white, in compound umbels; fruits long and with appressed bristles. A genus of 10 species native to the Western Hemisphere and Asia, it was revised by Lowry and Jones (1984). Ostertag and Jensen (1980) studied variation in Ohio populations.

a. Combined length of stylopodium and style 0.7–1 mm in flower, 1–1.3 mm in fruit; each umbellet with 4–8 flowers; freshly cut stems with odor of carrot roots . . . . 1. *Osmorhiza claytonii*

aa. Combined length of stylopodium and style 2–4 mm in flower and in fruit; each umbellet with 7–16 flowers; freshly cut stems with odor of anise . . . . 2. *Osmorhiza longistylis*

1. **Osmorhiza claytonii** (Michx.) C. B. Clarke Woolly Sweet Cicely

Perennial, to 8 dm tall; freshly cut stems with odor of carrot roots; umbellets with 4–8 flowers; combined length of stylopodium and style 0.7–1.0 mm in flower, 1.0–1.3 mm in fruit. Throughout Ohio; woodlands. Baskin and Baskin (1991) studied seed dormancy in this species. May–June.

2. **Osmorhiza longistylis** (Torr.) DC. Smooth Sweet Cicely
incl. var. *longistylis*, var. *brachycoma* S. F. Blake, and var. *villicaulis* Fern.

Perennial, to 1 m tall; freshly cut stems with odor of anise; umbellets with 7–16 flowers; combined length of stylopodium and style 2–4 mm in flower and fruit. Throughout Ohio; woods and thickets. Baskin and Baskin (1984b) studied the species' germination ecophysiology. Late April–June.

Following Lowry and Jones (1984), no varieties are recognized.

7. Torilis Adans. HEDGE-PARSLEY

Weedy annuals; leaves 1–3 times pinnately compound; flowers white to pinkish, in compound umbels; fruits bristly. An Old World genus of some 20 species.

a. Major primary umbels subtended by 1–2 (–3) bracts or none; bristles on fruit divergent or slightly curved forward, gland-tipped when young; leaves terminating in a lanceolate or narrowly lanceolate segment . . . . 1. *Torilis arvensis*

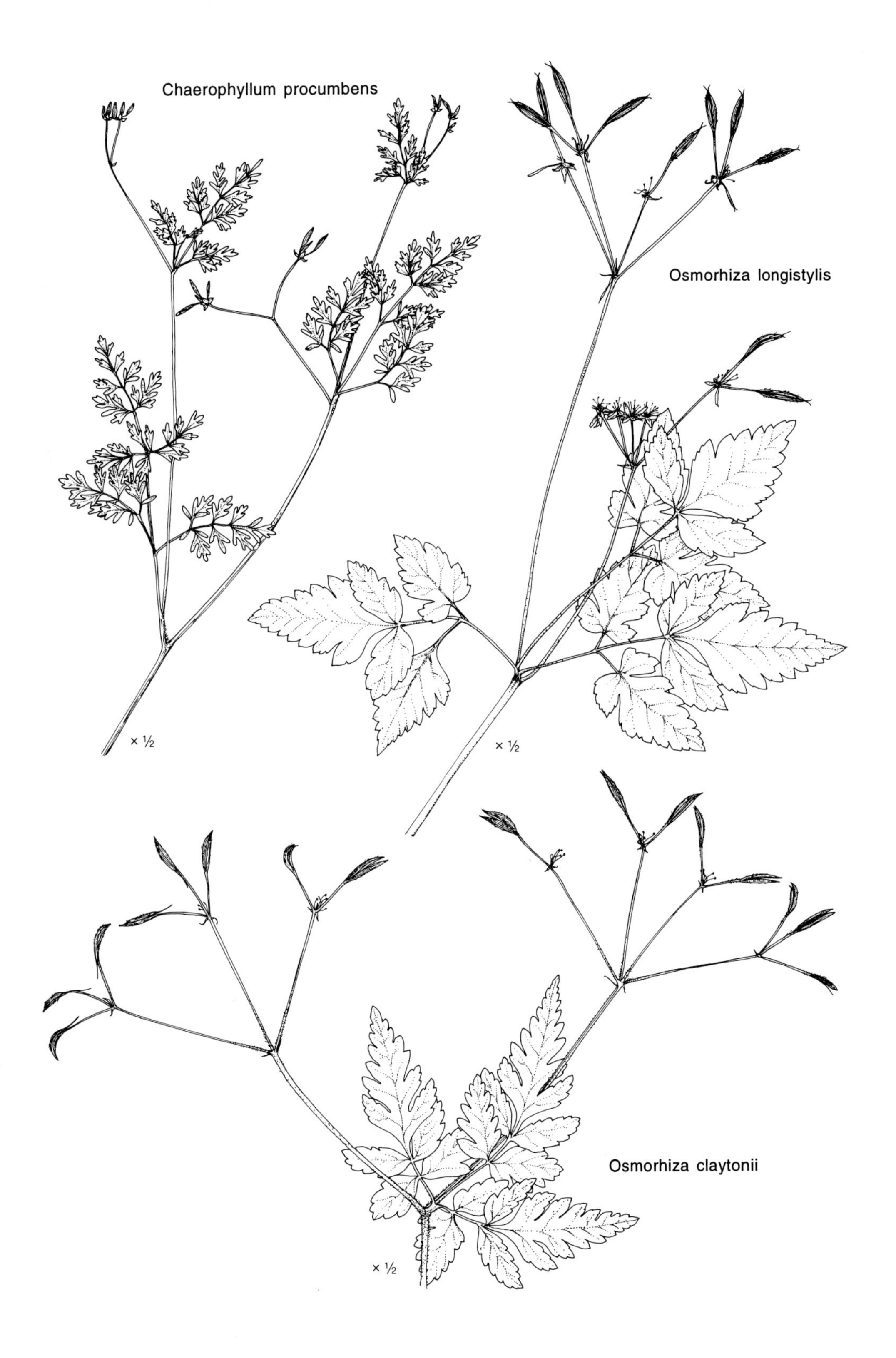
Chaerophyllum procumbens
Osmorhiza longistylis
× ½
× ½
Osmorhiza claytonii
× ½

aa. Major primary umbels subtended by 4–6 (–12) bracts; bristles on fruit hooked sharply forward at tip, not gland-tipped; leaves terminating in a narrowly ovate to ovate segment . . . . 2. *Torilis japonica*

1. TORILIS ARVENSIS (Huds.) Link FIELD HEDGE-PARSLEY

Annual, to 1 m tall; leaves 2–3 times pinnately compound, terminating in a lanceolate or narrowly lanceolate segment; major primary umbels subtended by 1–2 (–3) bracts or none; bristles on fruit divergent or curved slightly forward, gland-tipped when young. A Eurasian species, naturalized in the southwestern quarter of Ohio; along roads and railroads and in other disturbed weedy habitats. July–August.

2. TORILIS JAPONICA (Houtt.) DC. JAPANESE HEDGE-PARSLEY

Annual, to 8 dm tall; leaves 1–3 times pinnately compound, terminating in a narrowly ovate to ovate segment; major primary umbels subtended by 4–6 (–12) bracts; bristles on fruit hooked sharply forward at tip, not gland-tipped. Naturalized from Eurasia, mostly in northeastern counties but scattered elsewhere; woodland margins, along roads and railroads, and in other disturbed weedy habitats. July–August.

8. Erigenia Nutt.

A genus of a single species.

1. **Erigenia bulbosa** (Michx.) Nutt. HARBINGER OF SPRING. PEPPER-AND-SALT

Delicate perennial, to 1.5 (–2.0) dm tall; stems erect or nearly so, arising from a solid globose tuber; leaves ternate, the divisions 1–2 times pinnately compound, the ultimate segments narrow, entire-margined, and blunt at apex; flowers white, in compound umbels, primary umbels with

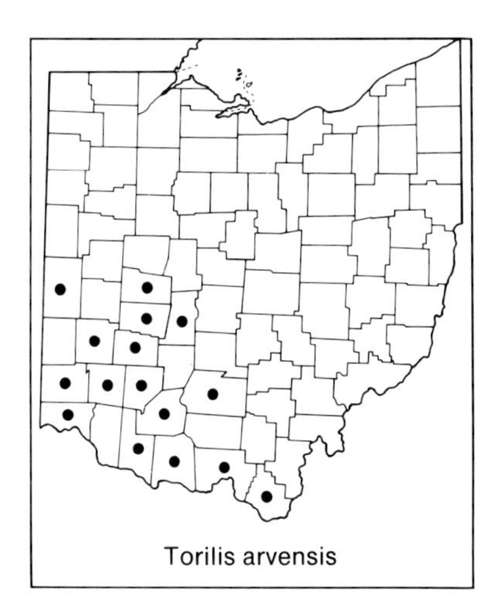
Torilis arvensis

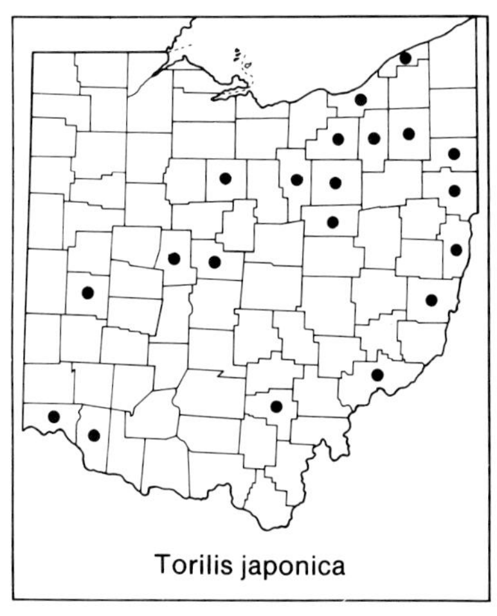
Torilis japonica

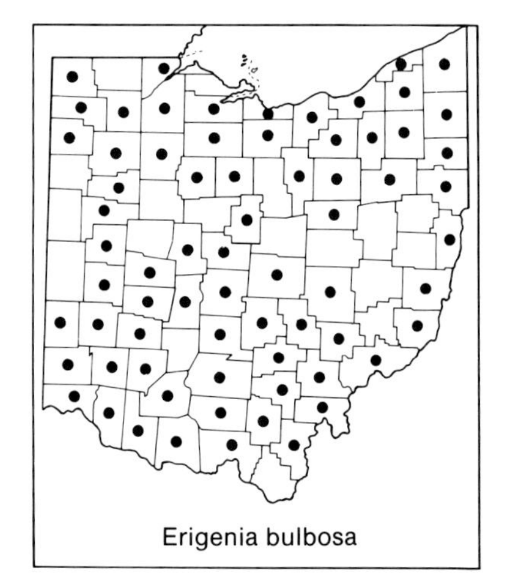
Erigenia bulbosa

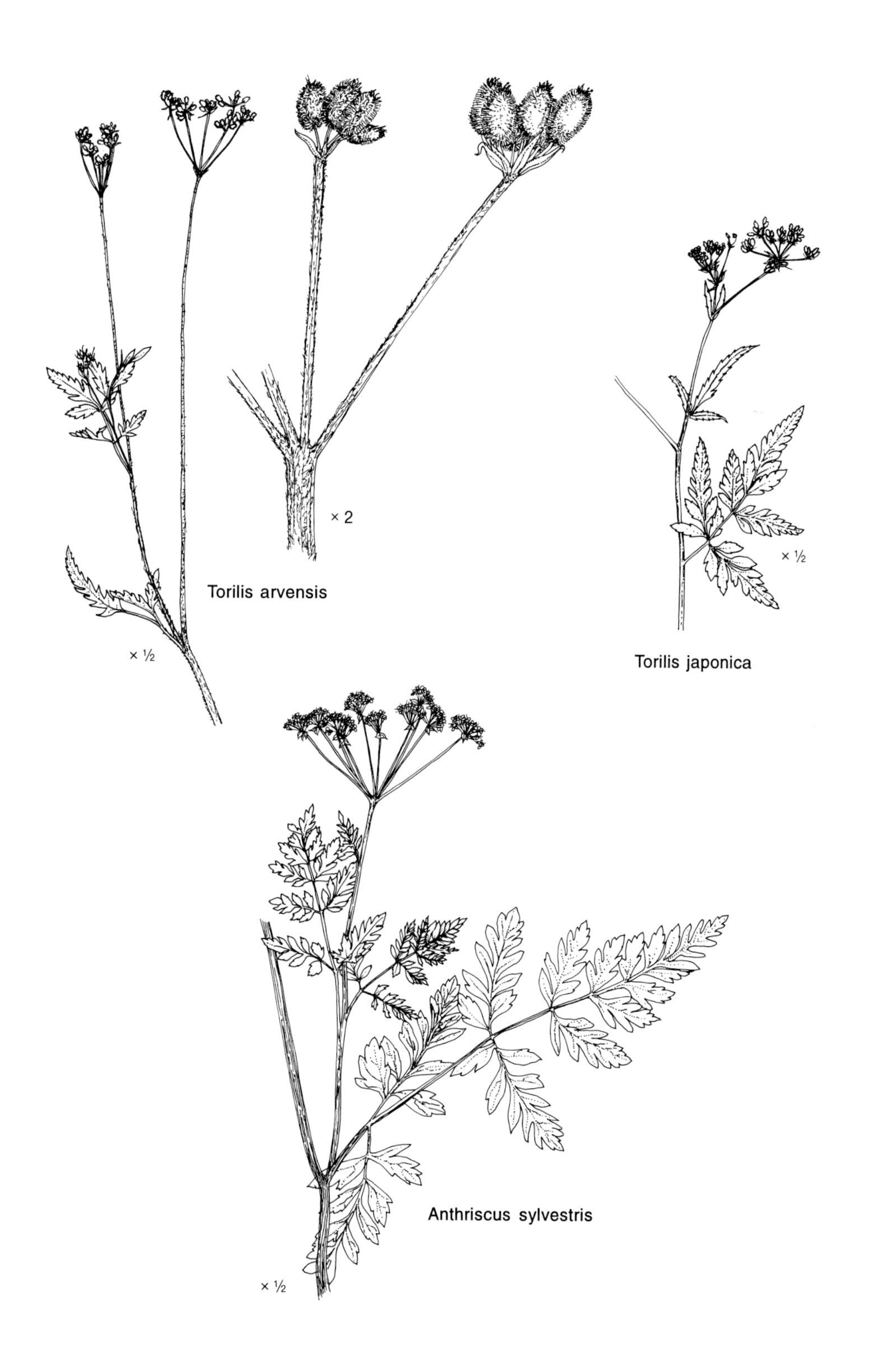
× 2
Torilis arvensis
× ½
× ½
Torilis japonica
Anthriscus sylvestris
× ½

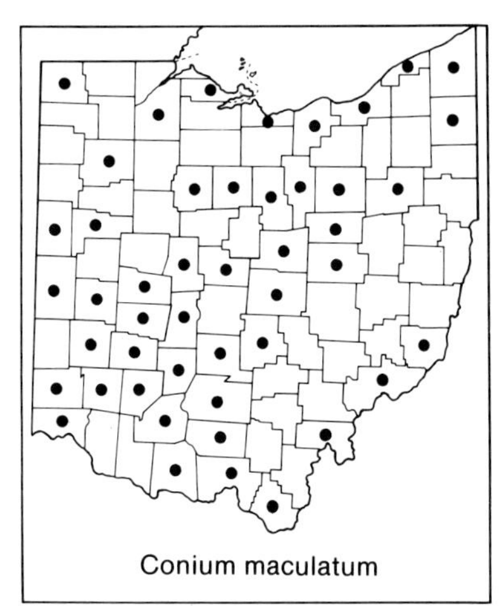

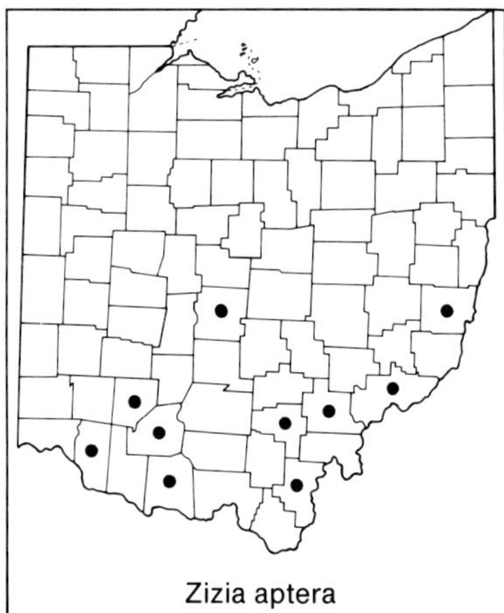

(1–) 3 (–5) rays; umbellets subtended by several oblong bractlets; anthers dark; fruit glabrous, wider than long. Throughout most of Ohio; wooded floodplains, terraces, and slopes; one of Ohio's earliest-blooming wildflowers. Late February–April.

9. Conium L.

A Eurasian genus of two species, both deadly poisonous.

1. Conium maculatum L. Poison-hemlock

Large, stout, bushy-topped biennial, to 2.5 m tall; leaves often large and elaborately dissected, to 4 dm long and 5 dm wide, 3–4 times pinnately compound, the ultimate segments oblong and incised; flowers white, in compound umbels, rays of primary umbel 10–15; fruit glabrous. Naturalized from Eurasia throughout much of Ohio, especially central, south-central, and southwestern counties; along roads and railroads, and in other disturbed weedy habitats. May–August.

Extremely poisonous, this species was used in classical Greece as a means of execution, the criminal being required to drink a cup of juice expressed from the stems and leaves. It was by this means that the philosopher Socrates was put to death.

10. Zizia Koch

Perennial herbs; flowers yellow, in compound umbels, central flower or fruit in each umbellet sessile or subsessile on pedicel 0.1–0.5 mm long; fruits glabrous, their ribs wingless or with low wings. A North American genus of four species, the two in Ohio often confused with species of *Thaspium* (p. 230). Lindsey (1984) and Lindsey and Bell (1985) did research on the reproductive biology of the two genera. The text below is based on my previous study (Cooperrider, 1985).

a. Basal leaves simple, rarely 3-lobed or trifoliolate, cauline leaves simple to mostly trifoliolate, margins of leaves or leaflets crenate-dentate . . . . 1. *Zizia aptera*

aa. Basal leaves mostly ternately decompound with 5–9 leaflets, rarely with only 3 leaflets, cauline leaves with 9 or more leaflets, margins of leaflets mostly sharply serrate . . . . 2. *Zizia aurea*

1. **Zizia aptera** (A. Gray) Fern. Heart-leaved Meadow-parsnip

Perennial, to 6 dm tall; basal leaves simple, rarely 3-lobed or trifoliolate, cordate at base; cauline leaves simple to mostly trifoliolate; margins of leaves or leaflets crenate-dentate. In several southern counties; prairie patches, woodland openings and borders. April–June.

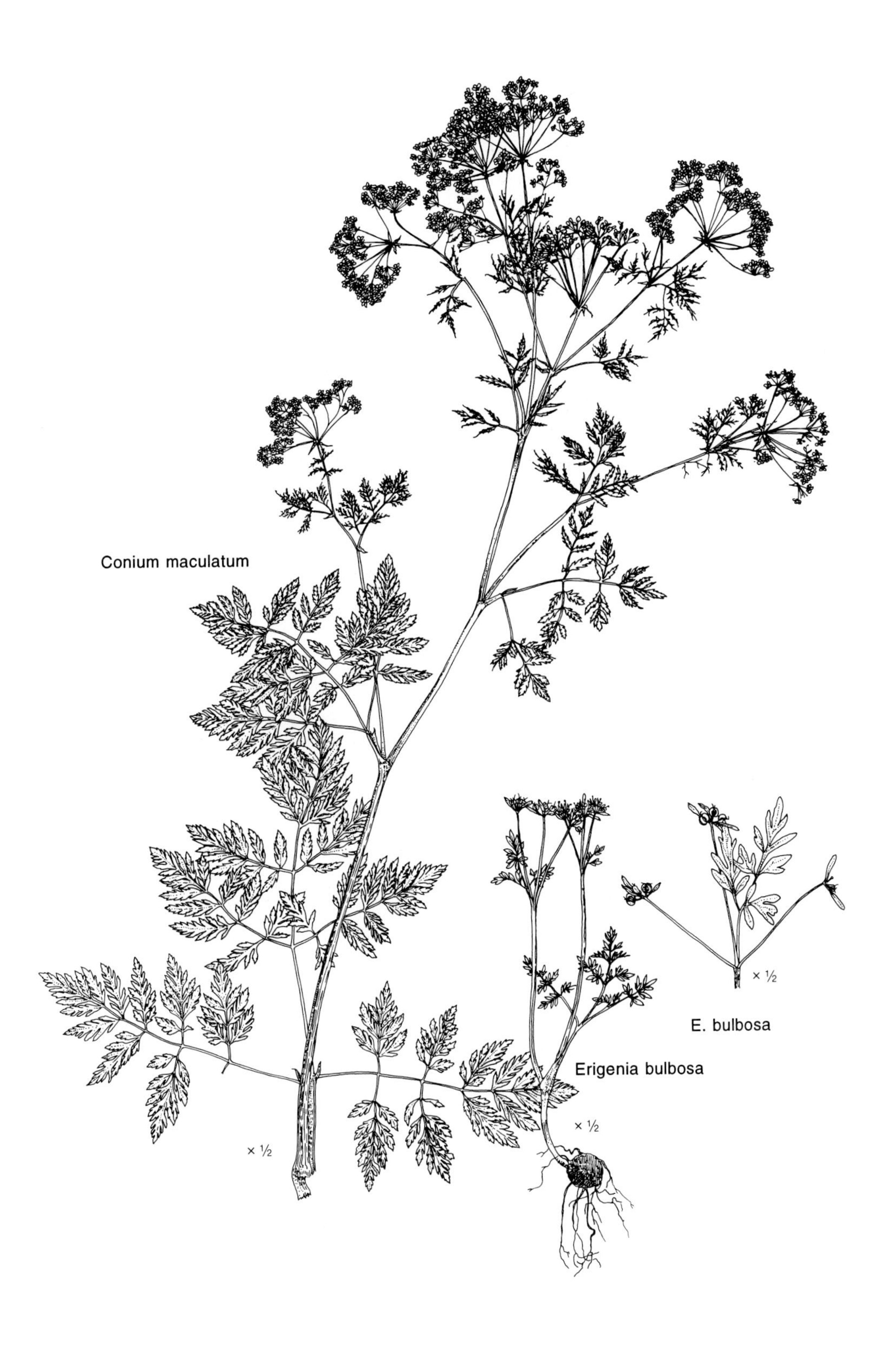
Conium maculatum
E. bulbosa
× 1/2
Erigenia bulbosa
× 1/2
× 1/2

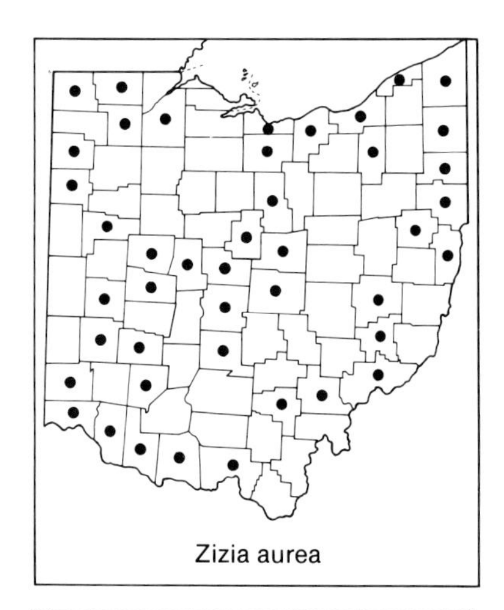
Zizia aurea

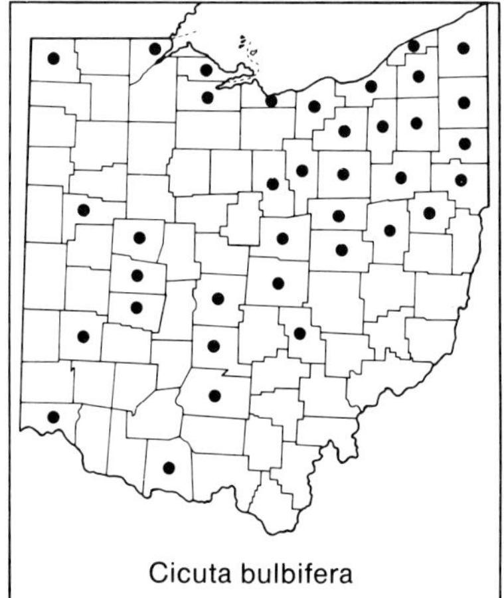
Cicuta bulbifera

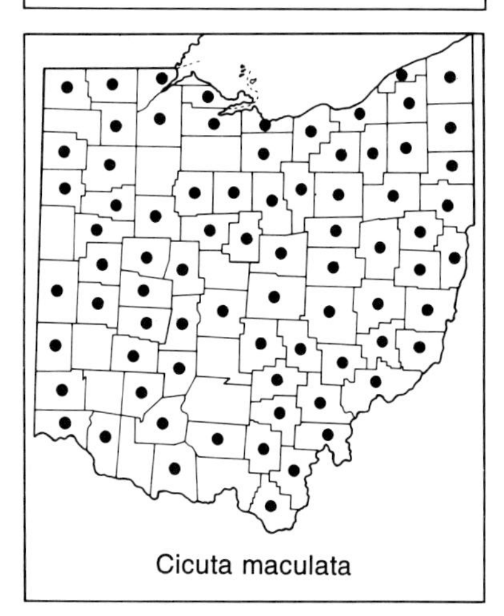
Cicuta maculata

Often confused with *Thaspium trifoliatum* (p. 230) and sometimes with the next species, *Zizia aurea*.

2. **Zizia aurea** (L.) Koch GOLDEN ALEXANDERS. EARLY MEADOW-PARSNIP

Perennial, to 8 dm tall; basal leaves mostly ternately decompound with 5–9 leaflets, rarely with only 3 leaflets; cauline leaves similar to the basal, with 9 or more leaflets; margins of leaflets mostly sharply serrate. Chiefly in western counties, scattered eastward; openings, fields, roadsides, floodplains. May–June.

Sometimes confused with *Thaspium barbinode* (p. 232).

11. Cicuta L.

Tall, perennial, deadly poisonous herbs; flowers white, in compound umbels; fruits glabrous, globose or ovoid. A genus of about 15 species, native to North America and Eurasia. Mulligan (1980) studied the North American species.

a. Leaflets linear, their margins irregularly toothed or incised or sometimes subentire; plants often with bulblets produced in axils of upper leaves . . . . 1. *Cicuta bulbifera*

aa. Leaflets lanceolate, their margins regularly and coarsely serrate or incised; plants not producing bulblets. . . . . 2. *Cicuta maculata*

1. **Cicuta bulbifera** L. BULBLET-BEARING WATER-HEMLOCK

Perennial, to 1 m tall, often producing bulblets in axils of upper leaves; leaves irregular in pattern, with a scattered combination of filiform segments and linear leaflets, their margins irregularly toothed or incised or sometimes subentire; plants sometimes flowering, but rarely producing fruit. Northeastern counties and scattered elsewhere in Ohio; margins of lakes and ponds, wet ditches, swamps, bogs. Late July–September.

2. **Cicuta maculata** L. WATER-HEMLOCK. SPOTTED WATER-HEMLOCK. SPOTTED COWBANE

Perennial, to 1.2 m tall, not producing bulblets; leaflets lanceolate, their margins regularly and coarsely serrate or incised, the major branches from leaflet midvein leading nearly or quite to a sinus, then abruptly diverging and terminating in an adjacent tooth; plants producing both flowers and fruits. Throughout Ohio; shores of lakes and ponds, marshes, bogs, wet ditches. June–July.

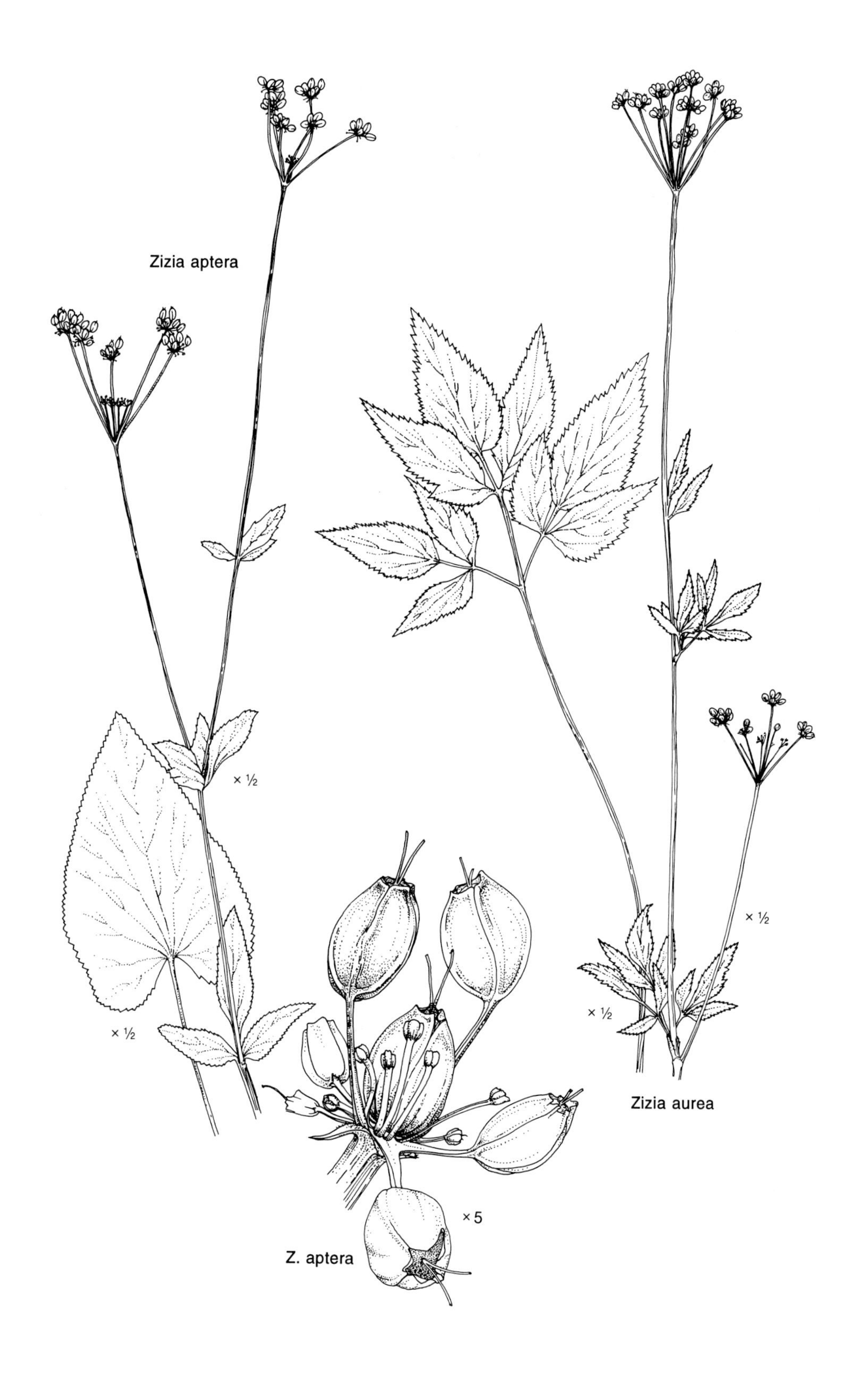
Zizia aptera
× ½
× ½
Zizia aurea
× ½
× ½
× 5
Z. aptera

C. bulbifera
× 1
× ½
Cicuta bulbifera
× ½
Cicuta maculata

12. Cryptotaenia DC. HONEWORT

A rather widespread genus of five species, one of which occurs in Ohio.

1. **Cryptotaenia canadensis** (L.) DC. Honewort

Perennial, to 9 dm tall; leaves trifoliolate, the leaflets sometimes deeply lobed, the margins of most leaflets coarsely doubly serrate; flowers white, in compound umbels; fruit glabrous, longer than wide, often curved. Throughout Ohio; wooded slopes, wooded ravines and floodplains, thickets. Baskin and Baskin (1988) studied the ecological life cycle. Late May–July (–August).

13. Taenidia (T. & G.) Drude

A genus of eastern North America with two species (as interpreted by Cronquist, 1982), one a member of the Ohio flora.

1. **Taenidia integerrima** (L.) Drude Yellow Pimpernel

Perennial, to 8 dm tall; leaves twice or more pinnately compound, the first division sometimes ternate, the leaflets elliptic and entire-margined; flowers yellow, in a compound umbel; fruit glabrous. Throughout Ohio; woodland openings and borders, roadside banks, prairies. May–June.

14. Aegopodium L.

A Eurasian genus of about five species, one of which occurs in Ohio.

1. Aegopodium podagraria L. Goutweed

Stout perennial, to 9 dm tall; leaves mostly ternate, each of the 3 divisions with 3 leaflets, the leaflets irregularly doubly serrate, with 20 or more teeth on each side of blade; flowers white, in compound umbels; sepals none;

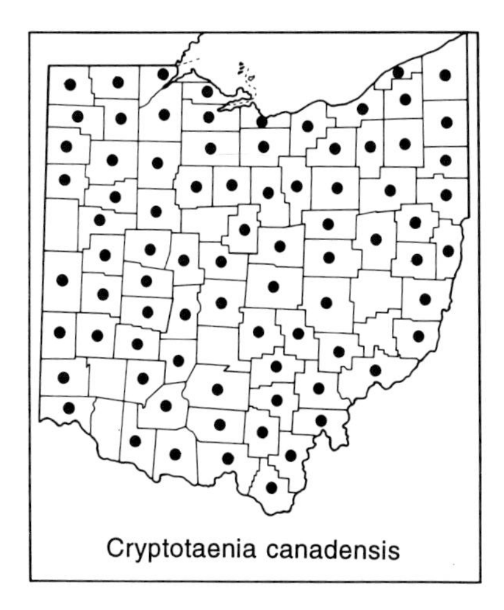
Cryptotaenia canadensis

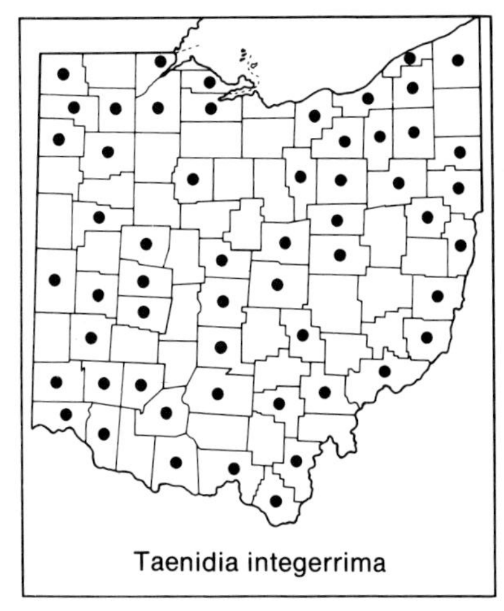
Taenidia integerrima

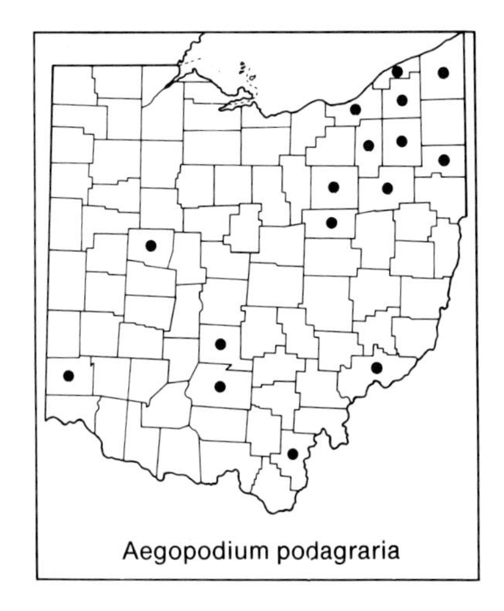
Aegopodium podagraria

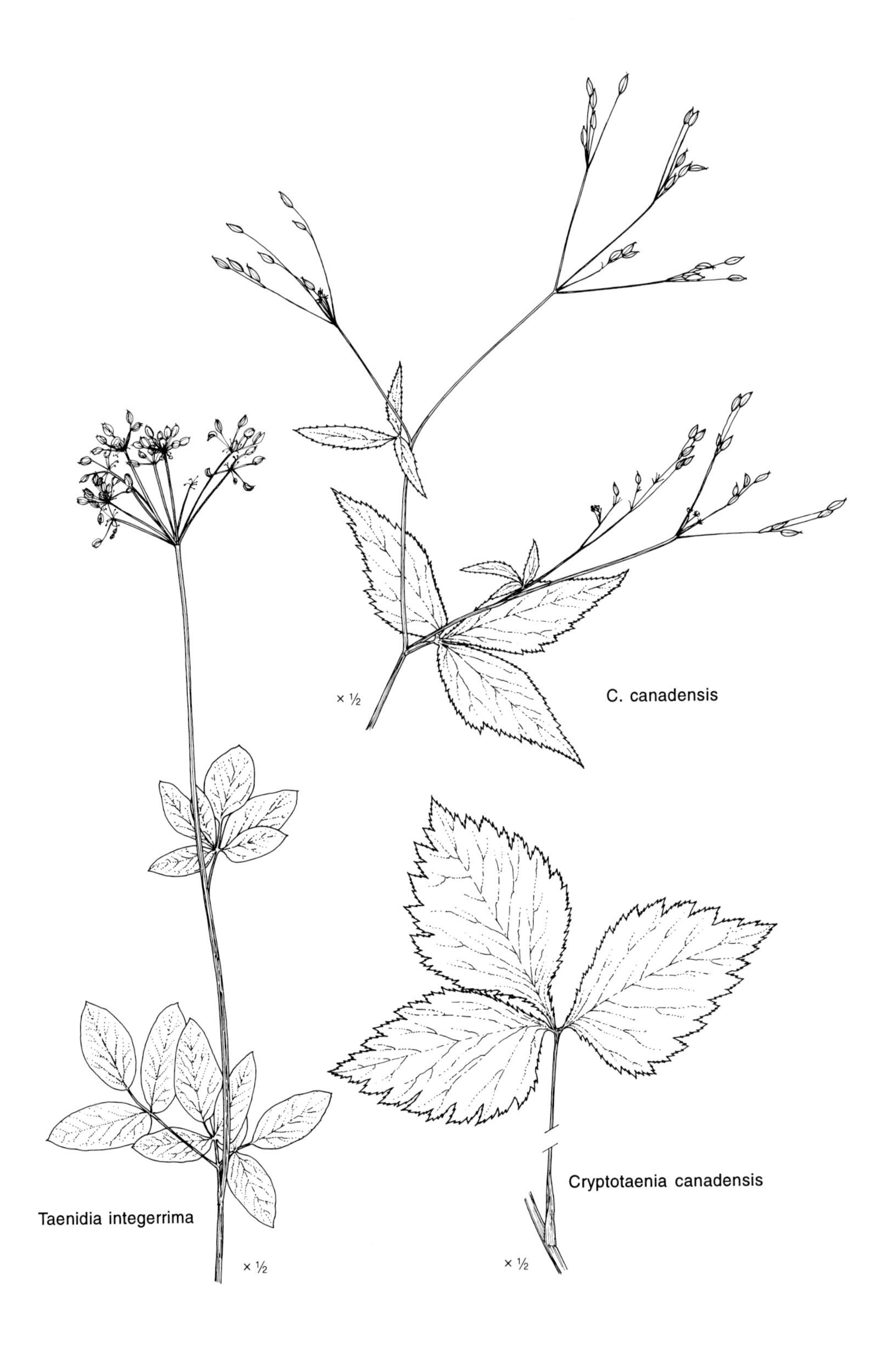
× ½
C. canadensis
Cryptotaenia canadensis
Taenidia integerrima
× ½
× ½

fruit glabrous. Naturalized from Eurasia, in northeastern counties and at scattered sites elsewhere in Ohio; disturbed moist woods, roadsides, and other disturbed habitats. June–July.

A cultivated form with white-margined leaves is seen frequently in Ohio, especially around older homes.

Perideridia americana

15. Perideridia Reichenb. YAMPAH

A genus with nine species of central and western North America, one extending eastward to Ohio. Chuang and Constance (1969) prepared a systematic study of this genus.

1. **Perideridia americana** (DC.) Reichenb. PERIDERIDIA

Perennial, to ca. 1 m tall; leaves mostly twice-pinnately compound, the leaflets once or twice deeply lobed into narrow oblong to linear, entire-margined segments; flowers white, in compound umbels; fruit glabrous. A species of the interior United States, ranging from Kansas to central Ohio, where, in Pickaway County, it reaches the eastern boundary of its range; Chuang and Constance (1969) report a record from Franklin County, also in central Ohio; low wet woodlands. June.

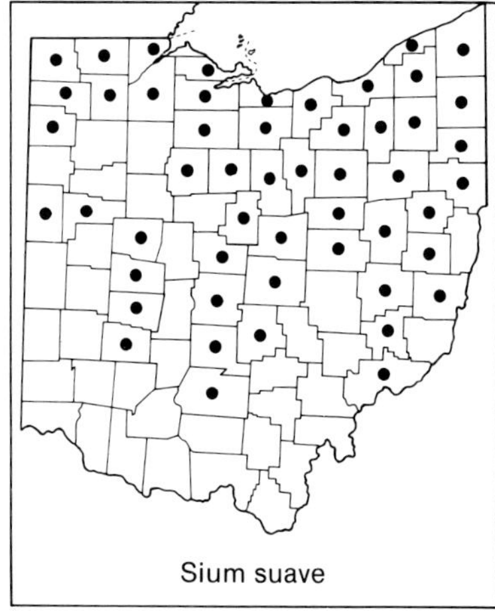
Sium suave

16. Sium L.

A rather widespread genus of ten species, one of them a member of the Ohio flora.

1. **Sium suave** Walt. WATER-PARSNIP

Perennial, to ca. 1 m tall; leaves once pinnately compound, margins of the leaflets sharply serrate with numerous teeth (aquatic phase leaves, found on plants rooted in water, highly dissected, 3–4 times pinnately compound with narrow ultimate segments); flowers white, in compound umbels; fruit glabrous. Northern three-fourths of Ohio; wet roadside ditches, marshes and bogs, wet woods. July–early October.

Ligusticum canadense

17. Ligusticum L. LOVAGE

A genus of the North Temperate Zone, with some 30 or more species, one of them in Ohio.

1. **Ligusticum canadense** (L.) Britt. AMERICAN LOVAGE. NONDO

Perennial, to ca. 1.5 m tall; leaves 2–4 times ternate, margins of leaflets serrate; flowers white, in a compound umbel; fruit glabrous. A southeastern species reaching a northern boundary of its range in Lawrence County, in southernmost Ohio; dry woodlands. May.

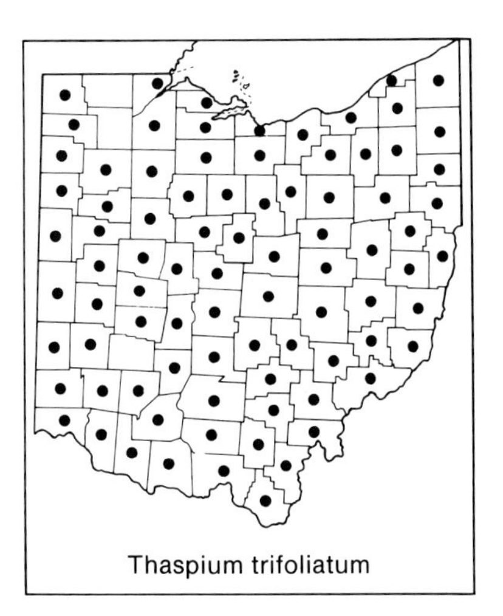

18. Thaspium Nutt.

Perennial herbs; flowers yellow or purple, in compound umbels, central flower or fruit in each umbellet on a short pedicel 1 mm or longer; fruits glabrous, ribs of fruit prominently winged. A North American genus of three species, two of which are members of the Ohio flora. The third, *Thaspium pinnatifidum* (Buckl.) A. Gray, CUT-LEAVED MEADOW-PARSNIP, was attributed to Ohio in error by Schaffner (1917, 1932) and other authors, as discussed by Cooperrider (1985). Lindsey (1984) and Lindsey and Bell (1985) did research on the reproductive biology of *Thaspium*. The text below is based on my previous study (Cooperrider, 1985).

a. Basal leaves entire or trifoliolate (rarely in the latter case with one or all of the leaflets deeply lobed or nearly divided); medial leaves usually divided into 3–6 leaflets, their margins shallowly serrate-dentate . . . . 1. *Thaspium trifoliatum*

aa. Basal leaves twice-pinnately compound or ternately decompound; medial leaves usually divided into 9 or more leaflets, their margins deeply serrate or lacerate . . . . 2. *Thaspium barbinode*

1. **Thaspium trifoliatum** (L.) A. Gray MEADOW-PARSNIP

Perennial, to 7 dm tall; basal leaves entire or trifoliolate (rarely in the latter case with one or all of the leaflets deeply lobed or nearly divided); medial leaves usually divided into 3–6 leaflets, their margins shallowly serrate-dentate. Throughout Ohio; rich woods, wooded slopes and floodplains, thickets. May–early June.

Often confused with *Zizia aptera* (p. 222; Ball (1979) discusses the characters separating the two), and sometimes difficult to distinguish from the next species, *T. barbinode*, especially those specimens without basal leaves.

I follow Mathias and Constance (1944–45) and Fernald (1950) in recognizing two varieties in this species. Variety *trifoliatum*, more eastern and Appalachian in distribution, is known from only two Ohio counties: Columbiana and Scioto, both in Ohio's southeastern Appalachian region; var. *flavum*, more common in the western part of the species' total range, occurs throughout Ohio. The two might be regarded as mere color forms, and are so treated by some authors, were it not for their differing ranges.

a. Flowers purple. . . . . var. *trifoliatum*
PURPLE MEADOW-PARSNIP

aa. Flowers yellow. . . . . var. *flavum* S. F. Blake
YELLOW MEADOW-PARSNIP

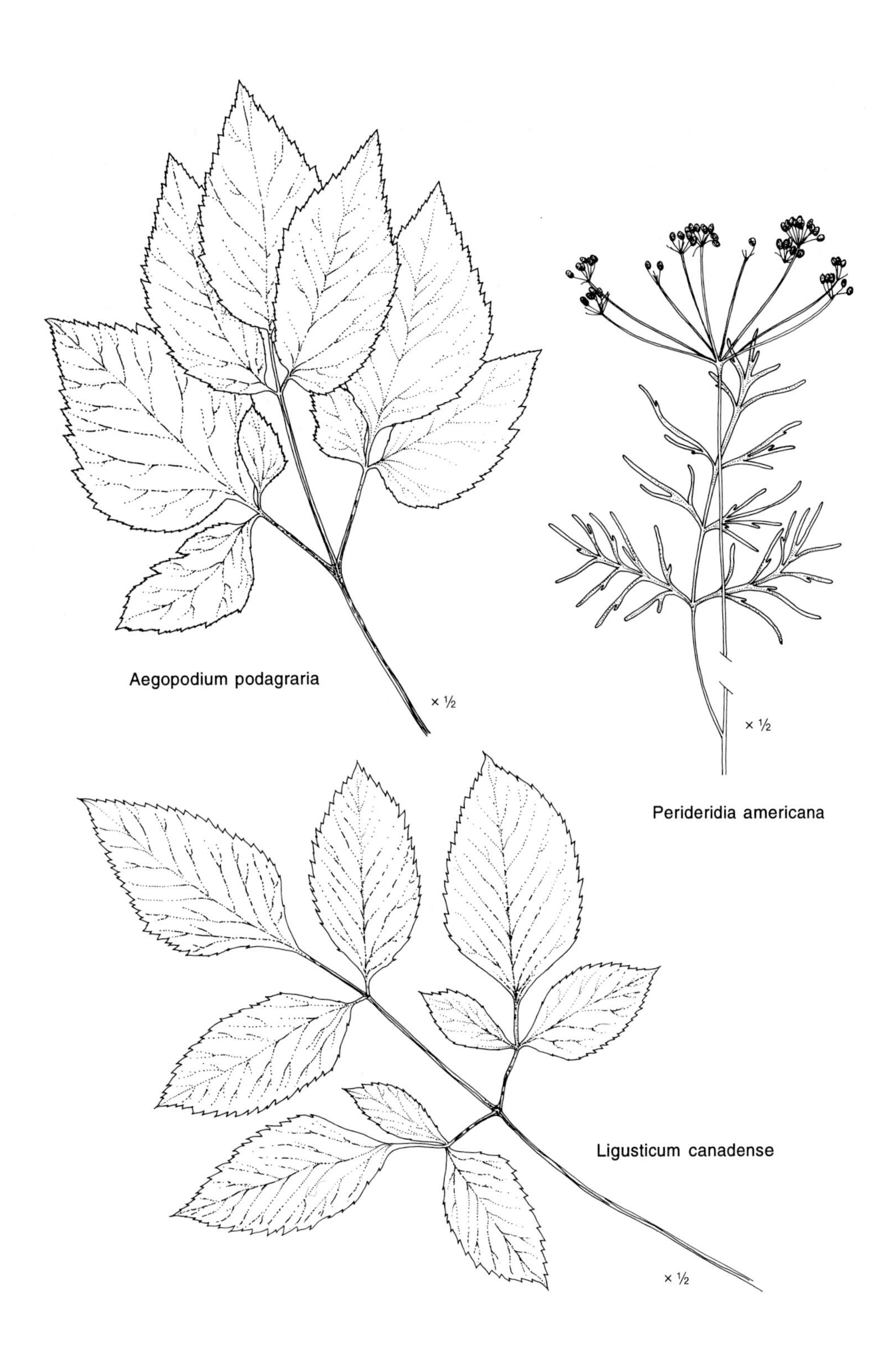
Aegopodium podagraria
× ½
Perideridia americana
× ½
Ligusticum canadense
× ½

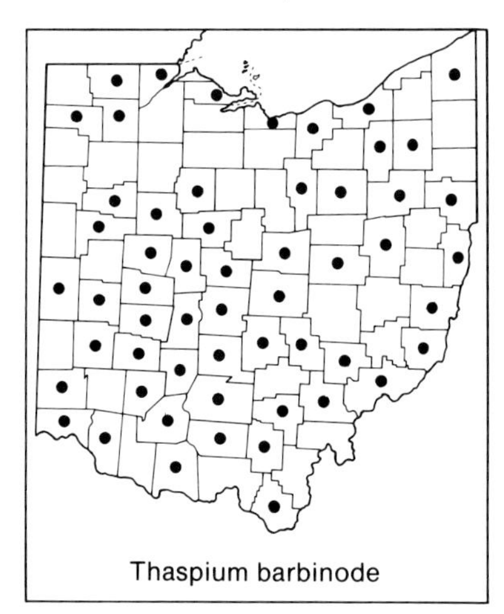
Thaspium barbinode

Conioselinum chinense

2. **Thaspium barbinode** (Michx.) Nutt. Hairy-jointed Meadow-parsnip

incl. var. *barbinode* and var. *angustifolium* C. & R.

Perennial, to 9 dm tall; basal leaves twice pinnately compound or ternately decompound; medial leaves usually divided into 9 or more leaflets, their margins deeply serrate or lacerate. Southeastern counties and scattered northward, more or less throughout the state; wooded slopes, bottoms, thickets, openings, prairie patches. May–June.

This species is sometimes confused with *Zizia aurea* (p. 224).

Following Mathias and Constance (1944–45), the species is defined here to include, without rank, plants that Fernald (1950) and other authors have segregated as var. *angustifolium*. While some Ohio plants, mostly from southwestern counties, could be assigned to var. *angustifolium*, others show a complete series of intergrades toward plants determined as var. *barbinode*; see Cooperrider (1985) for further discussion.

19. Conioselinum Hoffm. HEMLOCK-PARSLEY

A genus of the North Temperate and Arctic zones with about ten species, one of which extends south to Ohio.

1. **Conioselinum chinense** (L.) BSP. Hemlock-parsley

Perennial, to ca. 1 m tall; leaves 2–3 times pinnately compound, leaflets serrate or lobed; flowers white, in a compound umbel; fruit glabrous. A boreal species, ranging from Labrador and Newfoundland to Alaska and Siberia and southward; native to northeastern Ohio (despite the specific epithet), where it was collected from wet cliffs in Summit County, in 1891. Schaffner (1932) reports it also from Lake County, but I have found no supporting specimen to document this report. July–September.

20. Angelica L. ANGELICA

Tall perennial herbs; flowers white, in large compound umbels. A genus of some 50 species, widespread in the Northern Hemisphere and also in New Zealand.

a. Ovary and fruit pubescent; peduncles and primary and secondary rays of inflorescence pubescent; leaflets elliptic or oblong to lanceolate, their margins rather evenly low serrate-dentate, mostly unlobed . . . . 1. *Angelica venenosa*

aa. Ovary and fruit glabrous; peduncles and primary and secondary rays of inflorescence glabrous and usually spiculate; leaflets ovate to ovate-lanceolate, their margins more or less irregularly lacerate-serrate and often with one or two lobes . . . . 2. *Angelica atropurpurea*

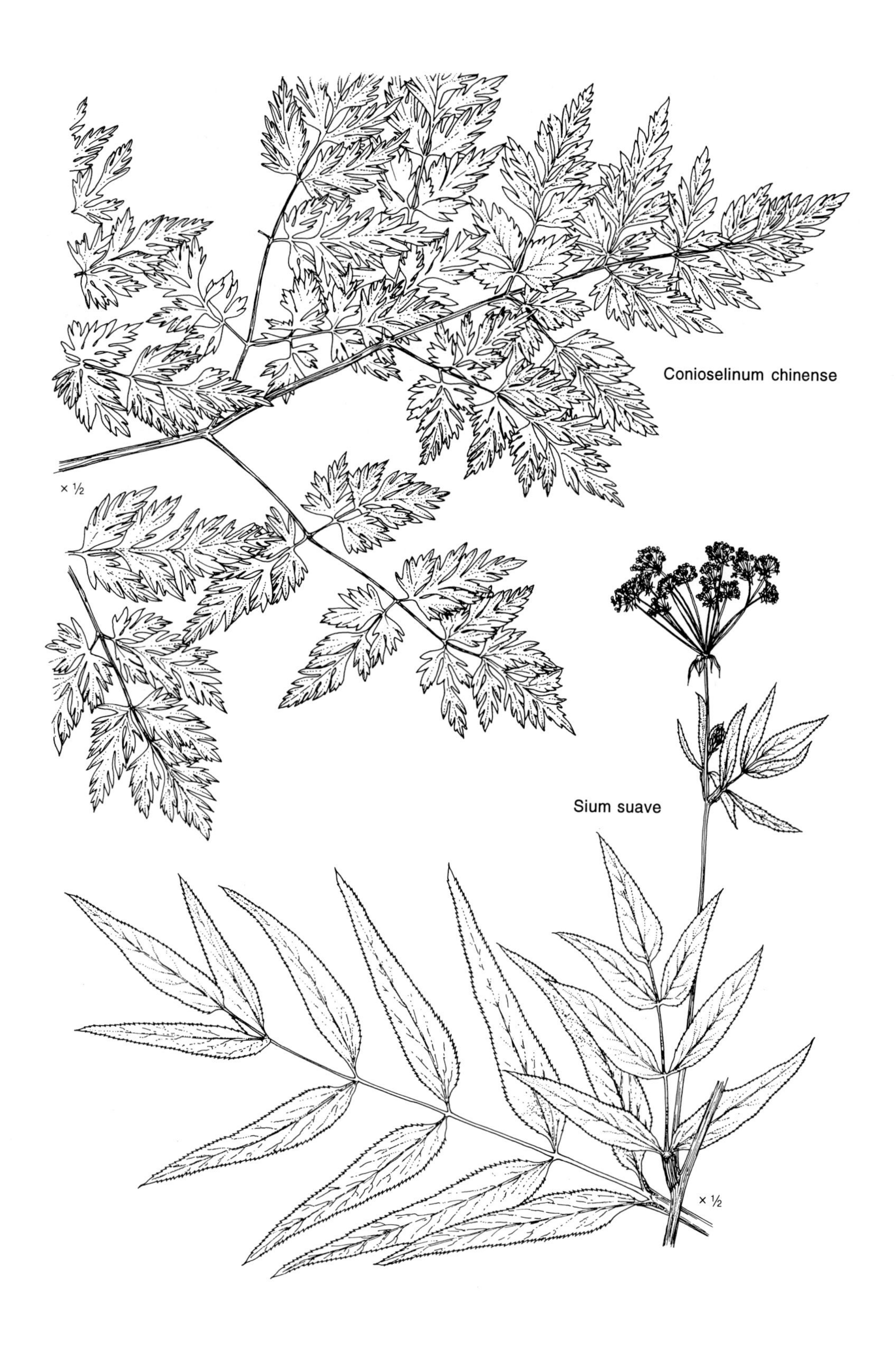
Conioselinum chinense
× ½
Sium suave
× ½

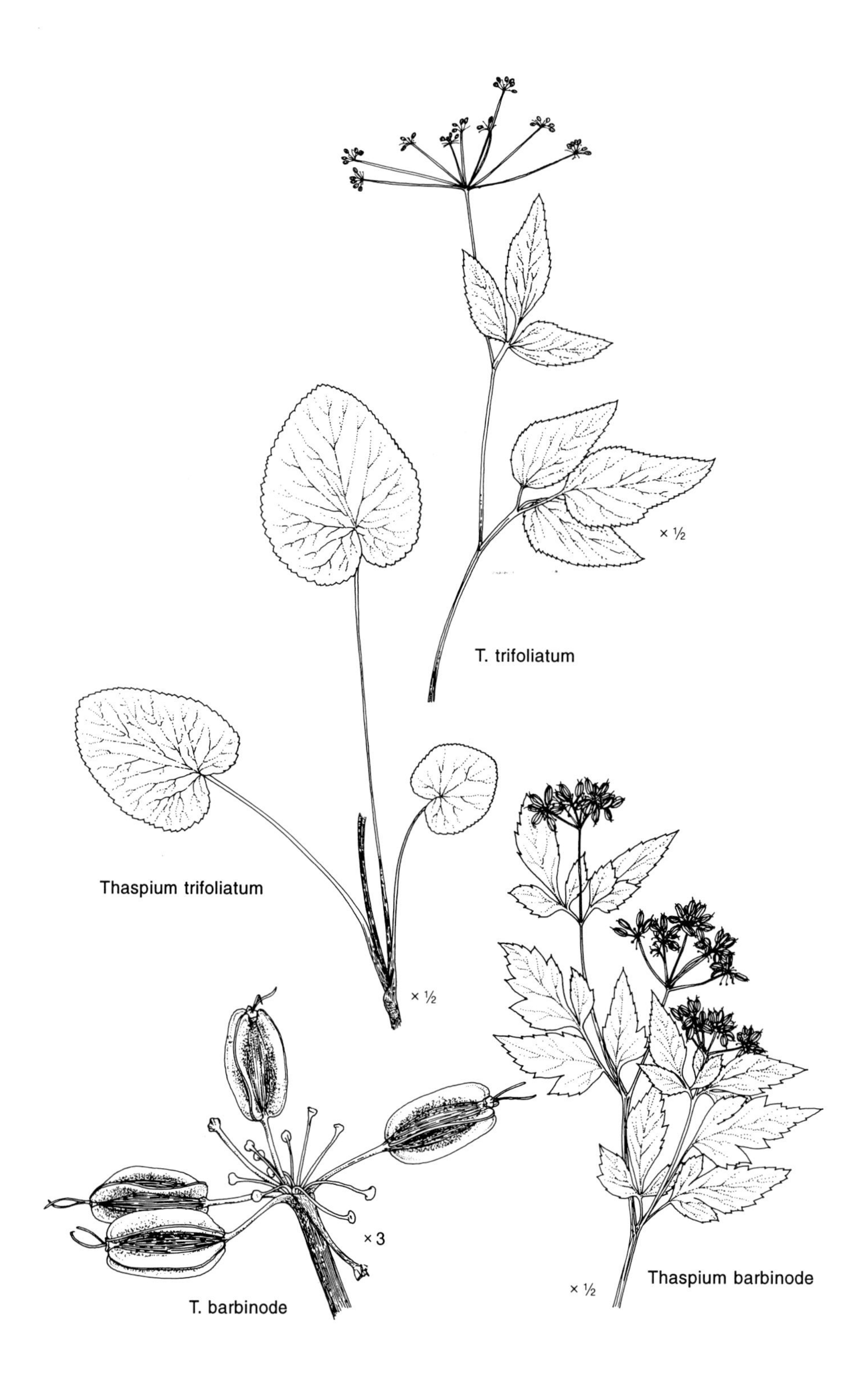
× ½
T. trifoliatum
Thaspium trifoliatum
× ½
× 3
T. barbinode
× ½
Thaspium barbinode

1. **Angelica venenosa** (Greenway) Fern. Hairy Angelica

Perennial, to 1.5 m tall; petioles very broad and sheathing; leaves ternate, the divisions once- or twice-pinnately compound, leaflets elliptic or oblong to lanceolate, their margins rather evenly low serrate-dentate, mostly unlobed; peduncles and primary and secondary rays of inflorescence pubescent; ovary and fruit pubescent. South-central and southeastern counties, and at scattered northward sites; woodland openings and borders, fields, roadside banks, usually in dry habitats. July–September.

2. **Angelica atropurpurea** L. Purple-stemmed Angelica

Perennial, to 2 m tall; stems usually purple at least in part; petioles very broad and sheathing, the sheath 1 cm or more in width when flattened; leaves twice-pinnately compound, leaflets ovate to ovate-lanceolate, their margins more or less irregularly lacerate-serrate and often with one or two lobes; peduncles and primary and secondary rays of inflorescence glabrous and usually spiculate; ovary and fruit glabrous. Chiefly in northeastern and central counties; wet fields, ditches, bogs. May–July.

21. Oxypolis Raf.

A North American genus of five to seven species, one of which occurs in Ohio.

1. **Oxypolis rigidior** (L.) Raf. Cowbane

Perennial, to 1.5 m tall; leaves once-pinnately compound, margins of leaflets entire or with 1–3 (–6) teeth on a side, leaflets varying greatly from plant to plant, ranging from linear to lanceolate to ovate; flowers white, in a compound umbel; fruit glabrous. Northern, south-central, and southwestern counties; wet ditches and thickets, marshes, swamps, bogs. August–September.

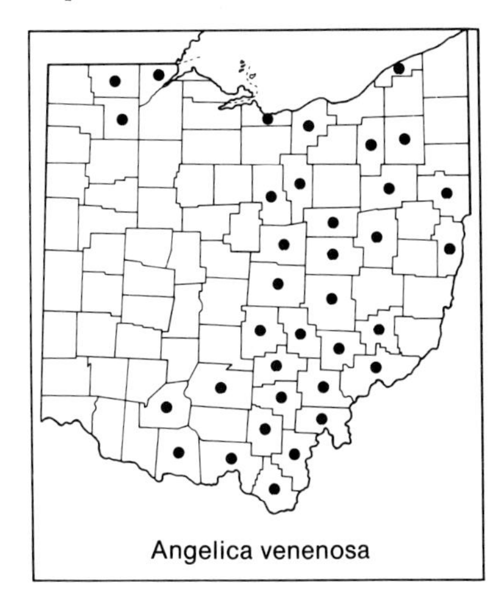
Angelica venenosa

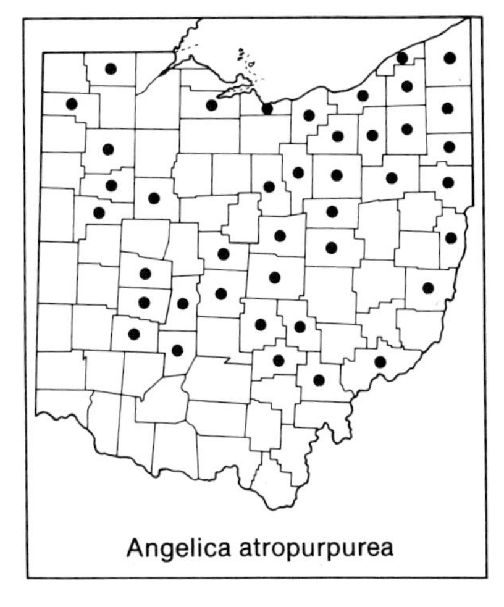
Angelica atropurpurea

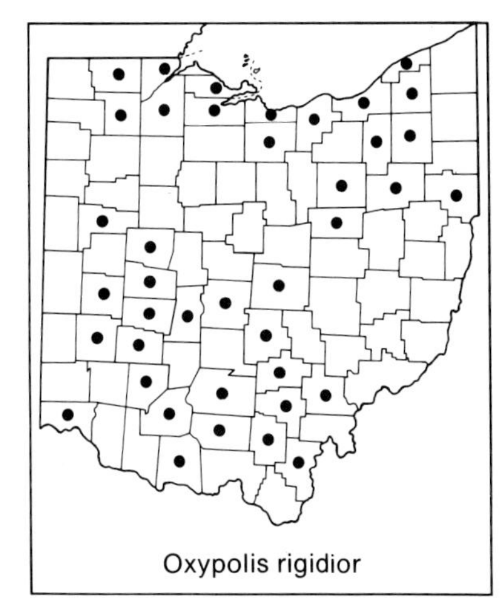
Oxypolis rigidior

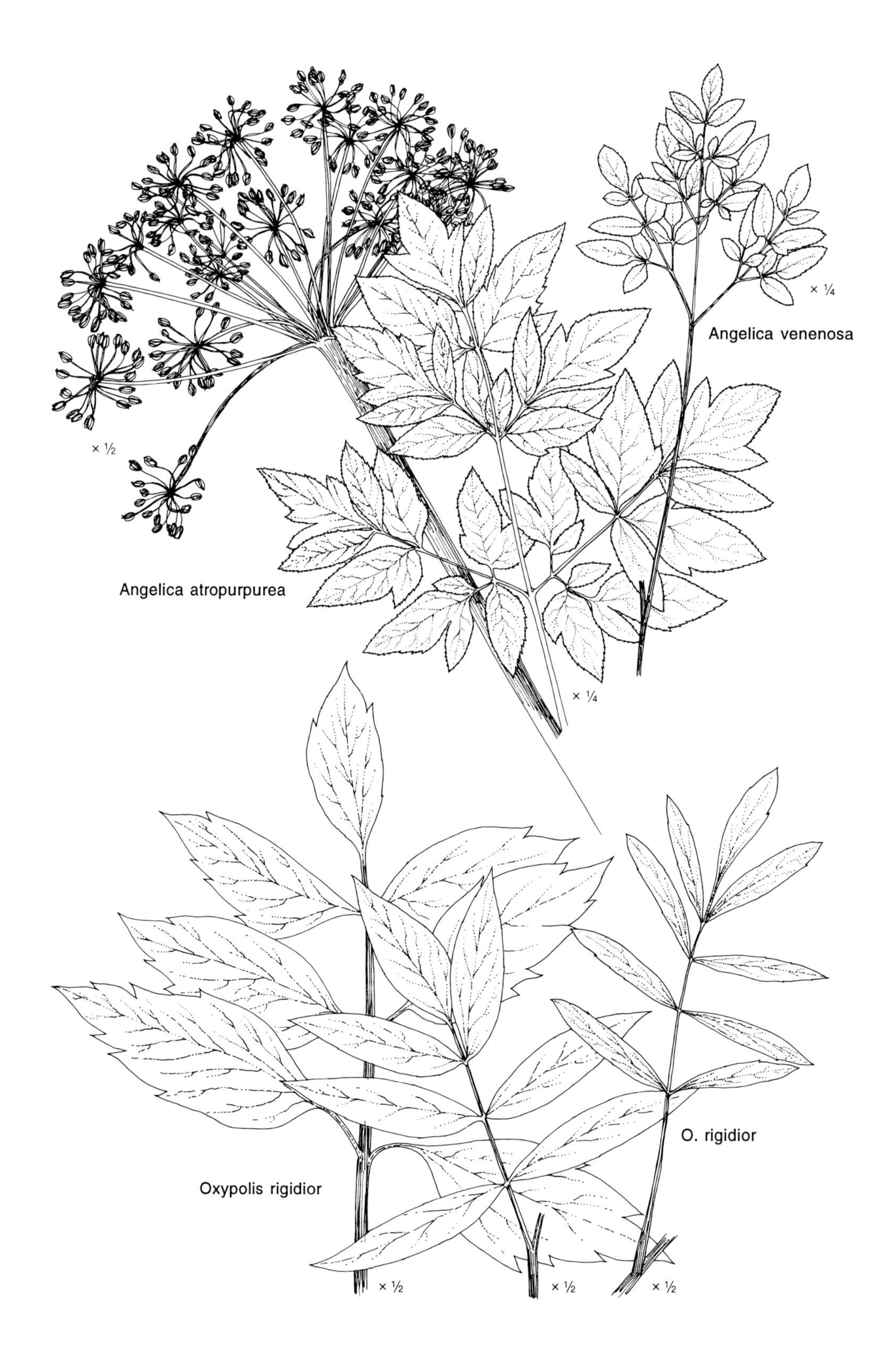
Angelica venenosa
× ¼
× ½
Angelica atropurpurea
× ¼
O. rigidior
Oxypolis rigidior
× ½
× ½
× ½

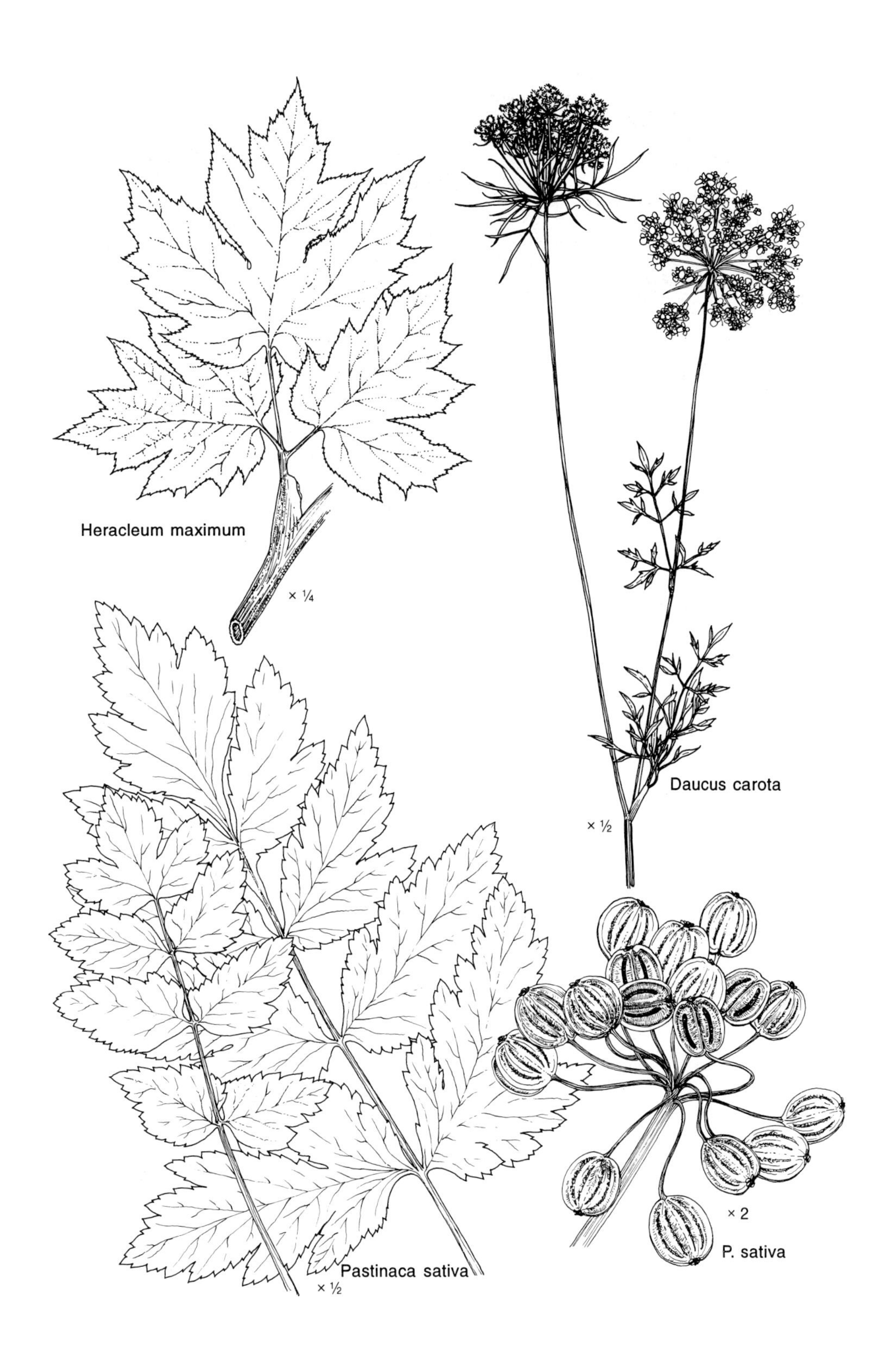
Heracleum maximum
× 1/4
Daucus carota
× 1/2
Pastinaca sativa
× 1/2
× 2
P. sativa

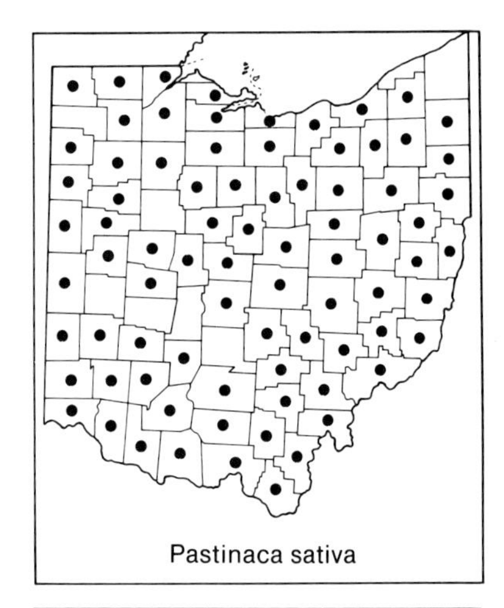
Pastinaca sativa

22. Pastinaca L. PARSNIP

A Eurasian genus of about 14 species, one of them naturalized in Ohio.

1. Pastinaca sativa L. Wild Parsnip

Stout biennial or perennial, to ca. 1.5 m tall; leaves once-pinnately compound, the leaflets mostly oblong to ovate, coarsely serrate-dentate, and sometimes lobed as well; flowers yellow, in a compound umbel; fruit glabrous, flattened, winged. A Eurasian species, naturalized throughout Ohio; along roads and railroads, and in other disturbed habitats. Baskin and Baskin (1979) studied the autecology and population biology of the species. May–July (–October).

Garden Parsnip belongs to this species.

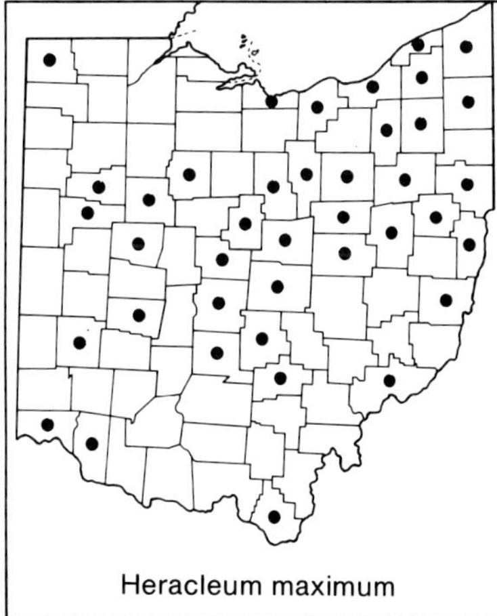
Heracleum maximum

23. Heracleum L. COW-PARSNIP

A genus of some 60 species in temperate regions of North America and Eurasia. One species is native to Ohio.

1. **Heracleum maximum** Bartr. Cow-parsnip. Masterwort
   H. lanatum Michx., *H. sphondylium* L. subsp. *montanum* (Gaudin) Briq.

Stout perennial, to 2 m tall; petioles broad and sheathing, petioles and petiolules white-woolly; leaves trifoliolate, the leaflets large, to 4 dm long, palmately or less often pinnately lobed, their margins serrate to dentate; flowers white, in a compound umbel; fruit pubescent to glabrous, flattened, winged. Northeastern and central counties, and at scattered sites elsewhere in Ohio; floodplains, stream banks, wet fields, wet ditches. May–June.

Brummitt (1971) considers *H. maximum* conspecific with the European species *H. sphondylium*.

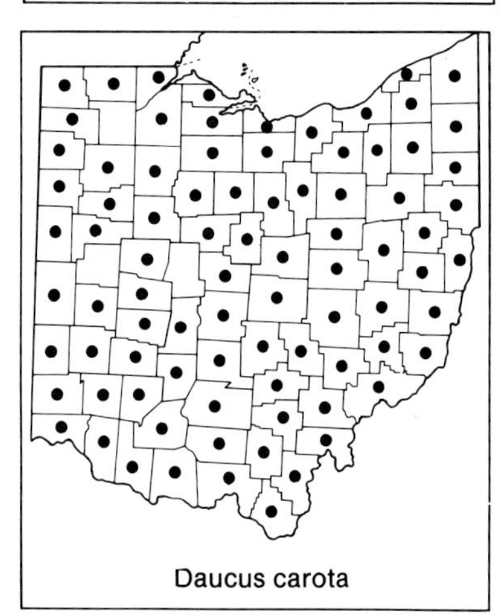
Daucus carota

24. Daucus L. CARROT

A cosmopolitan genus of 25 or more species, one of them naturalized and abundant in Ohio.

1. Daucus carota L. Wild Carrot. Queen Anne's Lace

Biennial, to ca. 1 m tall; leaves 2 or 3 times pinnately compound, finely dissected, the ultimate divisions linear to narrowly lanceolate; flowers mostly white, sometimes rose or purple, in a compound umbel subtended by conspicuous, deeply lobed bracts; fruit bristly. A Eurasian species, naturalized throughout Ohio; dry weedy habitats along roads and railroads, fields, fallow ground. Lacey (1981) studied seed dispersal mechanisms of this species. Small (1978) made a numerical taxonomic analysis of the *Daucus carota* complex. June–November.

GARDEN CARROT is derived from this species.

Three forms can be distinguished among Ohio plants. Forma *carota* and forma *epurpuratus* are common throughout Ohio; forma *roseus* is rare.

a. All or most flowers white.
  b. Inflorescence with all flowers white, except the dark red or dark purple central flower . . . . forma *carota*
  bb. Inflorescence with all flowers white . . . . forma *epurpuratus* Farw.
aa. All flowers rose-colored to purplish . . . . forma *roseus* Millsp.
  Harrison County.

Cornaceae. DOGWOOD FAMILY

A small family, chiefly of the North Temperate Zone, represented in the Ohio flora by a single genus. Rickett (1945) studied the family in North America, as did Ferguson (1966c) in southeastern United States.

1. Cornus L. DOGWOOD

Shrubs or small trees or rarely perennial herbs; leaves simple, opposite or rarely alternate; flowers small and whitish to yellowish or greenish, perfect, actinomorphic, and tetramerous, borne in cymose, paniculate, umbellate, or capitate inflorescences, in two species the inflorescence subtended by showy bracts; fruit a drupe. A genus of about 45 species, sometimes fragmented into smaller genera (discussed by Ferguson, 1966b). Eyde (1987, 1988) presents the case for recognition of *Cornus* in the broad inclusive sense, the classification followed here.

Cornus kousa Hance, KOUSA or ORIENTAL DOGWOOD, is a frequent Ohio ornamental. Reminiscent of the native *Cornus florida*, it blooms after the flowers of the native species have passed; unlike *C. florida*, its four greenish-white to creamy-white bracts taper to an acute tip.

A more southerly species, *Cornus foemina* Mill. (*C. foemina* subsp. *foemina* sensu J. S. Wilson; *C. stricta* Lam.), STIFF DOGWOOD, here excluded from the state's flora, was attributed to Ohio by Braun (1961) on the basis of a misidentified specimen of *C. racemosa* (Cooperrider, 1980). J. S. Wilson (1965) studied variation in three taxonomic complexes of *Cornus* in the eastern United States; they include species nos. 5–7, below. The treatment here is based on that of Rickett (1945) and on the study of Ohio species by Braun (1961).

a. Plants herbaceous, low, 0.7–2 dm tall; the uppermost leaves subtending the inflorescence crowded and appearing whorled; flowers in a dense cluster or head, the inflorescence closely subtended by a corolla-like involucre of 4 conspicuous white bracts mostly 1–2 cm long . . . . 1. *Cornus canadensis*

aa. Plants woody, shrubs or trees, 1–11 m tall; the leaves either opposite or alternate throughout, rarely crowded and appearing whorled; flowers in dense heads or in open cymes, the inflorescences with or without bracts.

b. Leaves alternate, mostly crowded near tips of branchlets and sometimes appearing whorled, abruptly acuminate at apex, pale on undersurface; mature petioles (1–) 2–5 cm long; pith white; fruit dark blue . . . . 8. *Cornus alternifolia*

bb. Leaves opposite, not markedly crowded near tips of branchlets, acuminate or acute at apex, either pale or green on undersurface; mature petioles 0.3–3 cm long; pith white or brown; fruit red, white, or blue.

c. Trees (sometimes shrubby); flowers and fruits in dense clusters or heads, the flowers yellow or yellowish; inflorescences closely subtended by a corolla-like involucre of 4 large white (rarely pink), showy bracts, the bracts mostly 3–7 cm long; fruit red; principal leaves ovate to broadly ovate, with 5–6 (–7) pairs of lateral veins . . . . 2. *Cornus florida*

cc. Shrubs; flowers and fruits in open cymes, the flowers white or cream-colored; bracts minute or lacking; fruit blue to white; principal leaves lanceolate to ovate or orbicular, with 3–9 pairs of lateral veins.

d. Principal leaves broadly ovate to suborbicular, 4–12 cm long and nearly as wide, with 6–9 pairs of lateral veins, abruptly short acuminate at apex; first-year branchlets greenish, often with dark reddish or purplish blotches; pith white; fruit light blue. . . . . 4. *Cornus rugosa*

dd. Principal leaves lanceolate to elliptic or ovate, 3–10 cm long and 1/4–2/3 as wide, with 3–7 pairs of lateral veins, tapering to an acute or acuminate apex; first-year branchlets reddish or reddish-brown to gray; pith white, tan, or brown; fruit blue to white.

e. Cymes paniculate, somewhat elongated, from 2/3 as long as wide to longer than wide; fruit white; pedicels red; leaves with 3–4 pairs of lateral veins, lanceolate to elliptic or narrowly ovate, acuminate at apex, acuminate to acute at base, glabrous to short-pubescent with appressed hairs on both upper surface and undersurface, the undersurface glaucous; older branches light to medium gray; first-year branchlets light reddish-brown; pith tan (or white in young branchlets); calyx lobes 0.3–0.5 mm long . . . . 7. *Cornus racemosa*

ee. Cymes broad and flattish to slightly convex, most only half as long as wide; fruit white to blue; pedicels red or some other color; leaves with 3–7 pairs of lateral veins, lanceolate to broadly elliptic, ovate, or broadly ovate, abruptly to gradually acuminate at apex, cordate to truncate or broadly rounded or broadly acute to (infrequently) acute at base, glabrous or pubescent on upper surface, pubescent with either appressed or erect hairs, and either green or whitened on undersurface; older branches gray, red, or brown; first-year branchlets red, reddish-brown, or reddish-purple; pith white to brown; calyx lobes 0.4–2.0 mm long.

f. Pith of one-year-old and two-year-old branches white; branchlets bright red, some lower branches prostrate for part of their length; leaves with 5–7 pairs of lateral veins, pubescence on undersurface appressed or mostly appressed; fruit white or dull white. . . . . 3. *Cornus stolonifera*

ff. Pith of one-year-old and two-year-old branches brown or brownish (rarely white in *C. drummondii*); branchlets gray to reddish, reddish-brown, or reddish-purple, all branches erect; leaves with 3–6 pairs of lateral veins, pubescence on undersurface spreading or appressed; fruit white or blue.

g. Leaves scabrous on upper surface, the undersurface covered with spreading and often curving hairs, narrowly to broadly ovate, abruptly long-acuminate at apex, and with 4–5 pairs of lateral veins; branches gray (but first-year branchlets often reddish); fruit white; pedicels becoming dark red; calyx lobes less than 1 mm long . . . . 5. *Cornus drummondii*

gg. Leaves smooth on upper surface, glabrous or nearly so, the undersurface with few to many spreading hairs along the veins and with numerous appressed hairs to nearly glabrous between the veins, lanceolate to ovate or broadly elliptic, variously acute or acuminate at apex, and with 3–6 pairs of lateral veins; branches and branchlets dark red to reddish-brown or reddish-purple; fruit blue; pedicels not red; calyx lobes mostly 1–2 mm long . . . . 6. *Cornus amomum*

1. **Cornus canadensis** L. BUNCHBERRY. DWARF CORNEL

Perennial herb with creeping rhizomes, erect stems to 2 dm tall; lower leaves small or scale-like, in flowering individuals the full-sized leaves lo-

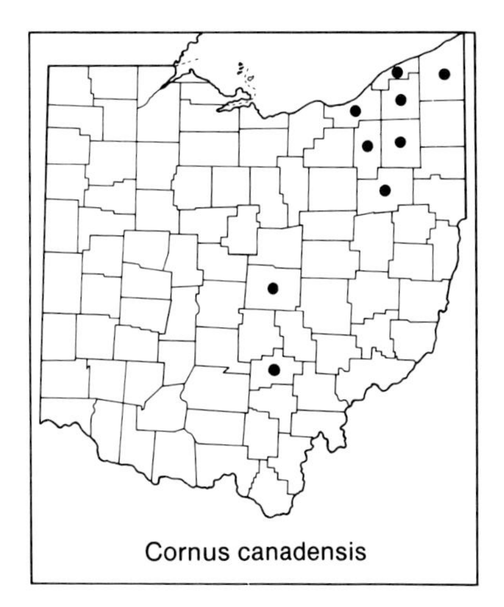

Cornus canadensis

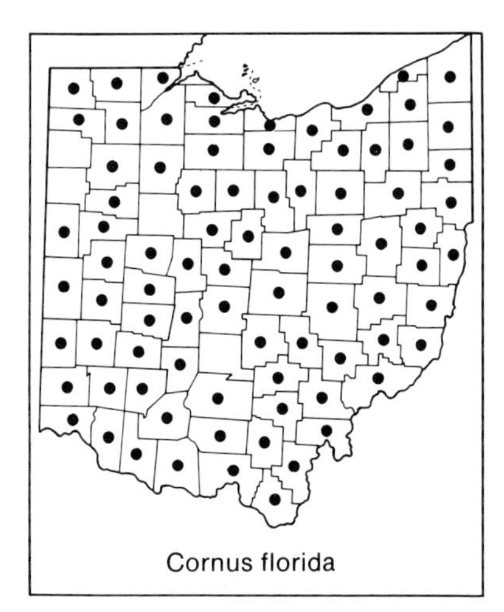

Cornus florida

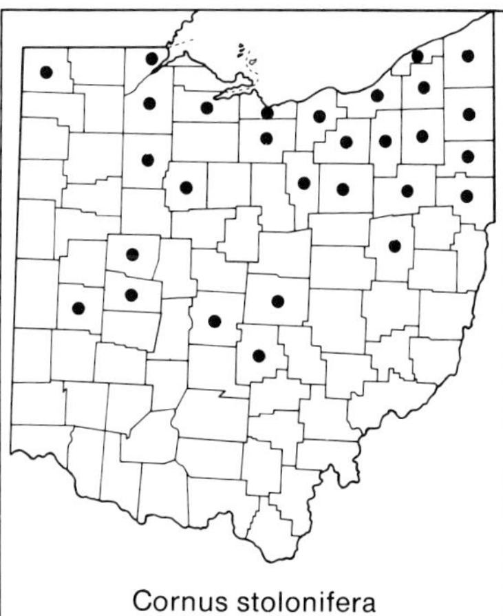

Cornus stolonifera

cated immediately beneath the inflorescence, crowded and appearing whorled; flowers cream-colored to yellowish, in a dense cluster or head closely subtended by a corolla-like involucre of 4 conspicuous white bracts mostly 1–2 cm long; fruit red. In northeastern counties and a few central Ohio stations; tamarack bogs, white pine swamps, wet woodlands. Teeri (1968) discusses variation within the species. May–July.

2. **Cornus florida** L. FLOWERING DOGWOOD

Tree, sometimes shrubby, to 9 m (29 feet) tall; leaves opposite, principal leaves ovate to broadly ovate, with 5–6 (–7) pairs of lateral veins; flowers yellowish, in dense clusters or heads, closely subtended by a corolla-like involucre of 4 white (rarely pink) bracts, the bracts large and showy, mostly 3–7 cm long, notched at apex; fruit red. Throughout Ohio, forming in spring a showy white understory in many woodlands; woods, woodland openings and borders, old fields, fencerows. April–May.

Some Ohio towns hold "dogwood festivals" in May to celebrate the flowering of this widely cultivated and much-admired tree. Ohio trees may be threatened by the spread of dogwood anthracnose disease (Herrick, 1990).

3. **Cornus stolonifera** Michx. RED OSIER. RED-OSIER DOGWOOD
incl. var. *stolonifera* and var. *baileyi* (Coult. & W. H. Evans) Drescher; ?*C. sericea* L.

Shrub, to 3 m (10 feet) tall; upper stems bright red, pith of first- and second-year branches white, some lower branches prostrate for part of their length and sometimes rooting; leaves opposite, ovate to elliptic-ovate, with 5–7 pairs of lateral veins, whitened beneath, pubescence on undersurface appressed or mostly appressed; flowers white, in a wide flattish cyme; fruits white or dull white. Northern two-thirds of Ohio, especially in northeastern counties; wet woods and thickets, shores, bogs, wet ditches. May–June.

Fosberg (1942), the Bailey Hortorium Staff (1976), and others have called this species *Cornus sericea* L. Rickett (1944), however, reiterated his earlier (Rickett, 1934) decision not to use *C. sericea* because of the ambiguity surrounding that name, a reservation recently voiced again by Voss (1985b, p. 679) and heeded here.

Following Rickett (1945), Gleason (1952a), and Braun (1961), no varieties are recognized within this species. Several forms have been named, one of which is forma *baileyi* (Coult. & Evans) Rickett, a pubescence form accepted by Braun (1961) as applicable to some Ohio plants. Forma *flaviramea* (Spaeth) Rickett, GOLDEN-TWIG DOGWOOD, a plant with bright yellow twigs and known only in cultivation, is seen frequently near Ohio homes.

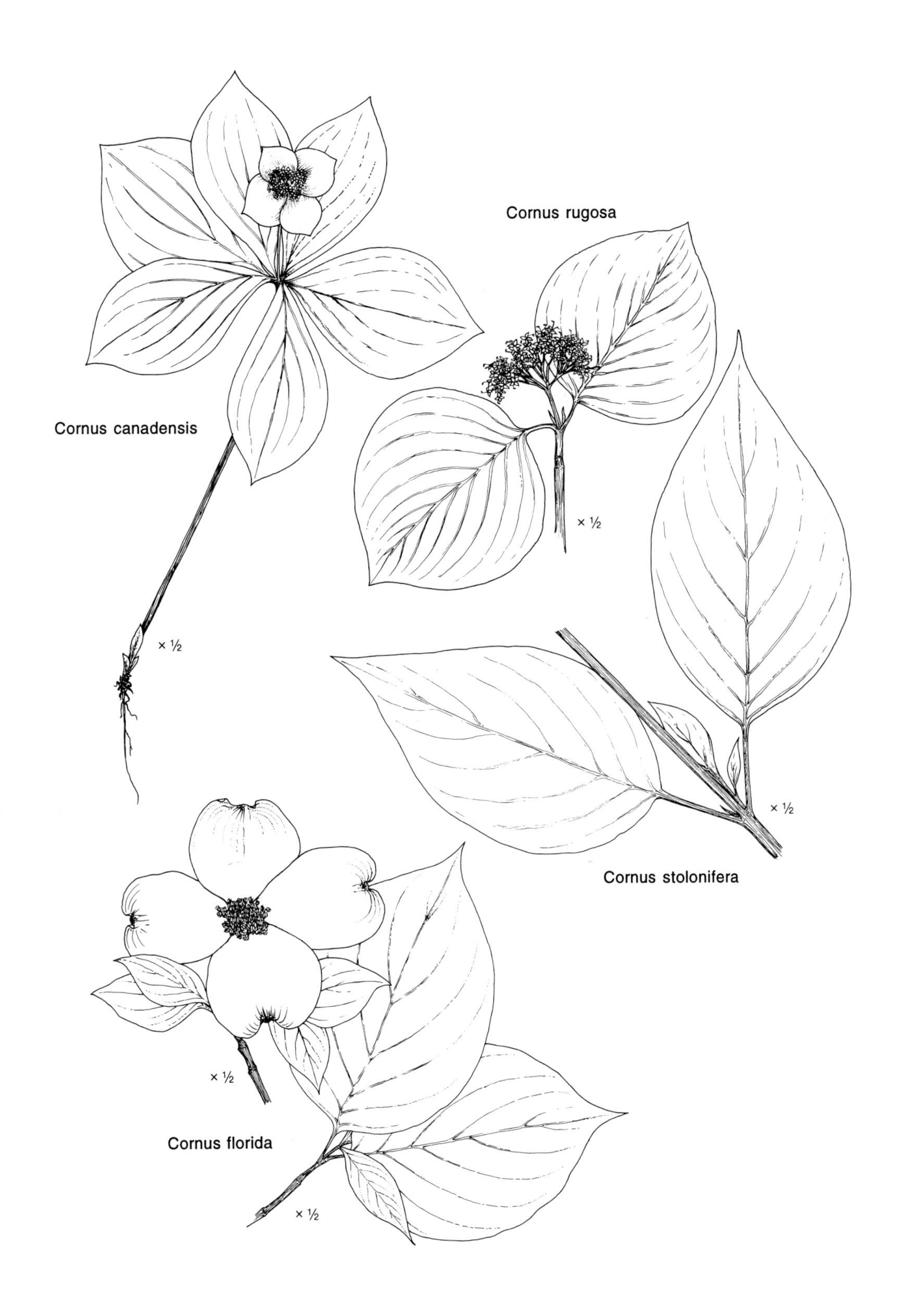
Cornus rugosa
Cornus canadensis
× ½
× ½
× ½
Cornus stolonifera
× ½
Cornus florida
× ½

4. **Cornus rugosa** Lam. ROUND-LEAVED DOGWOOD
Shrub, to 3 m (10 feet) tall; first-year branchlets greenish, often with dark reddish or purplish blotches; pith white; leaves opposite, principal leaves broadly ovate to suborbicular, 4–12 cm long and nearly as wide, with 6–9 pairs of lateral veins, pale and with white pubescence beneath; flowers white, in flat-topped cymes; fruit light blue. Northern and a few central counties; rocky slopes, woodland openings and borders, thickets. June–July.

5. **Cornus drummondii** C. Meyer ROUGH-LEAVED DOGWOOD
Shrub, to 3 m (10 feet) tall; branches gray (first-year branchlets often reddish), pith brown or rarely white; leaves opposite, narrowly to broadly ovate, abruptly long-acuminate at apex, with 4–5 pairs of lateral veins, upper surface scabrous, undersurface somewhat paler and covered with spreading and often curving hairs; flowers white, in a more or less flat-topped cyme; pedicels becoming dark red; fruit white. Western Ohio and at scattered sites eastward; woodland borders, thickets, old fields, fencerows. June–early July.

6. **Cornus amomum** Mill. SILKY DOGWOOD
incl. C. *obliqua* Raf. and C. *purpusii* Koehne
Shrub, to 3 m (10 feet) tall; branches and branchlets dark red to reddish-brown or reddish-purple, first-year branchlets usually densely pubescent, especially near stem apex; pith brown (white in youngest stems); leaves opposite, lanceolate to narrowly ovate-oblong to ovate or broadly elliptic, acute or acuminate at apex, with 3–6 pairs of lateral veins, upper surface glabrous or nearly so, undersurface with few to many spreading hairs along

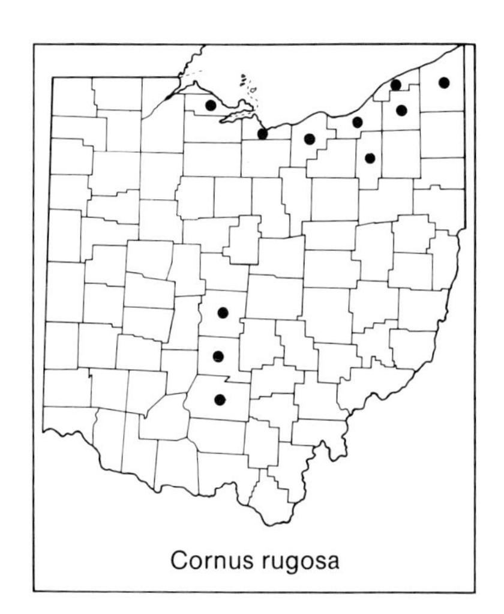
Cornus rugosa

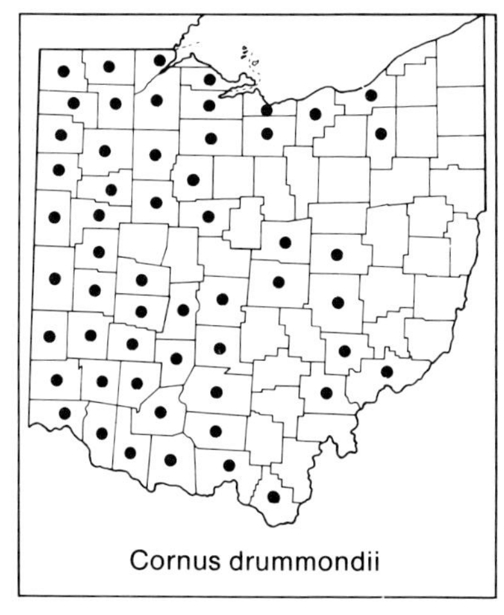
Cornus drummondii

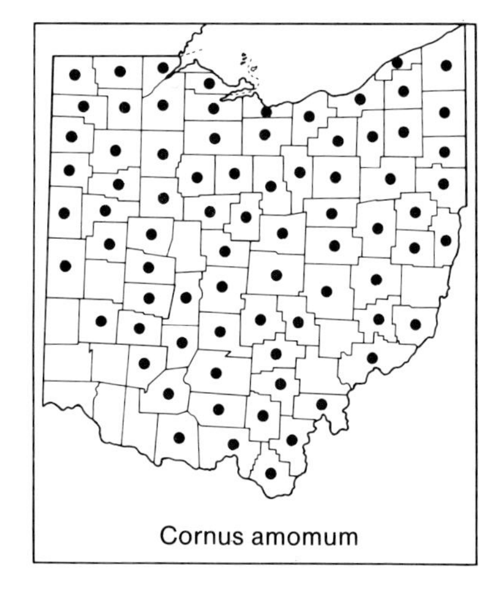
Cornus amomum

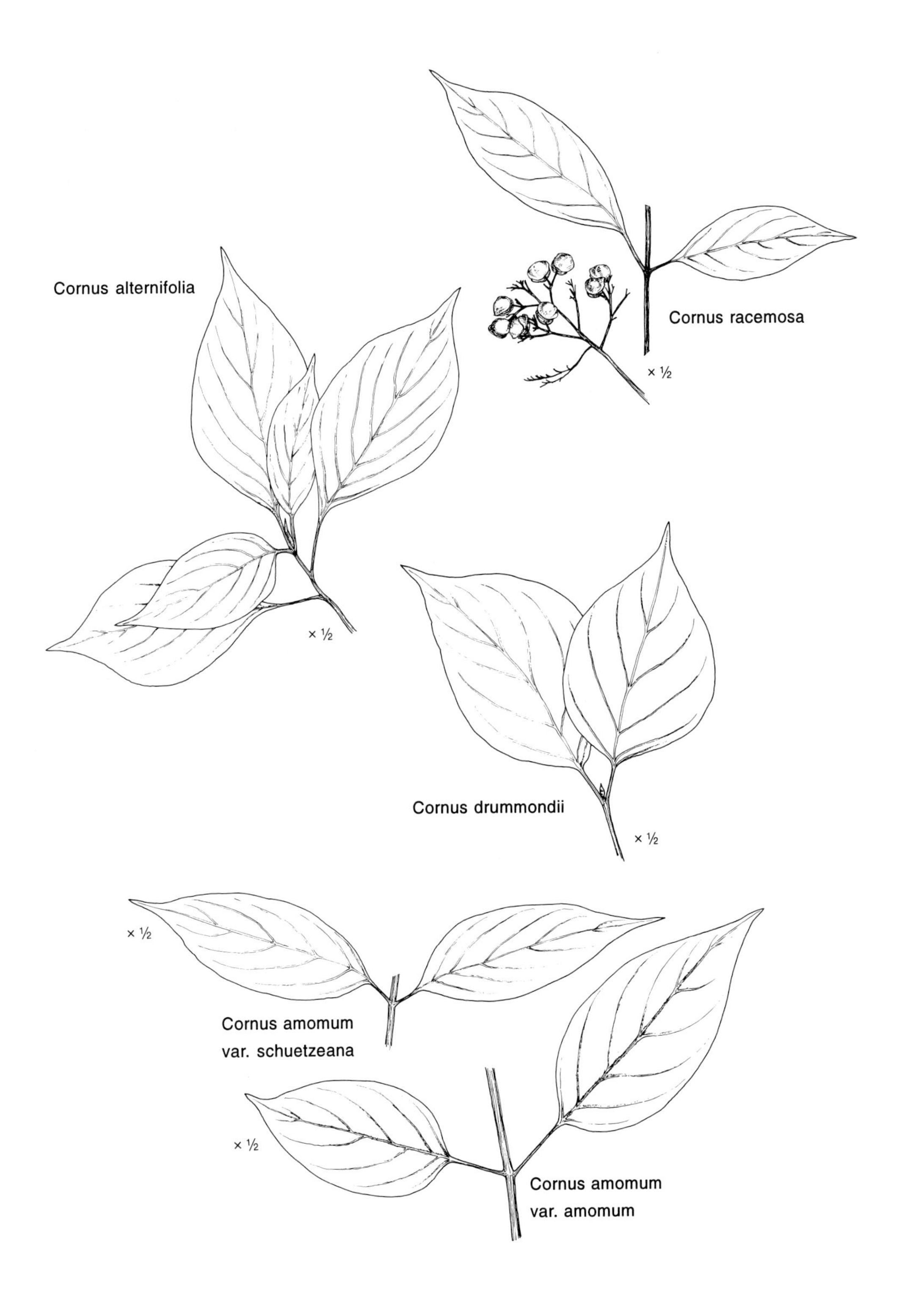
Cornus racemosa
× ½
Cornus alternifolia
× ½
Cornus drummondii
× ½
× ½
Cornus amomum
var. schuetzeana
× ½
Cornus amomum
var. amomum

the veins and either with numerous appressed hairs or nearly glabrous between the veins; flowers whitish, in more or less flat-topped cymes; fruit blue, sometimes with cream-colored areas. Throughout Ohio; moist thickets, stream banks, pond and lake shores, bogs, marshes, swamps, moist fields. June–early July.

The two varieties keyed below have often been treated at the rank of species, one called *C. amomum* and the other called either *C. obliqua* (e.g., Fernald, 1950; Braun, 1961) or *C. purpusii* (e.g., Rickett, 1945; Gleason, 1952a). Rickett (1934, 1936), the leading student of North American dogwoods, at first treated the two taxa as varieties, publishing for the latter the combination *C. amomum* var. *schuetzeana*, and defending its nomenclatural correctness, but later (Rickett, 1945) treated the two as species. They are combined here in one species following Rickett's earlier decision and the more recent work of J. S. Wilson (1965), who treated the two as subspecies of *C. amomum* sensu lato. Braun (1961) treated them as species, but recorded hybrids between the two throughout much of Ohio, reflecting the great amount of intergradation between the two taxa in this state.

a. Undersurface of leaves pubescent usually with both spreading and appressed usually brownish or reddish hairs; leaves ovate to broadly elliptic, acute to nearly rounded at base and abruptly acuminate at apex, 6–12 cm long and ca. 1/2 as wide as long, with 4–6 pairs of lateral veins; branches reddish-brown . . . . var. *amomum*

aa. Undersurface of leaves finely pubescent with appressed white or colorless hairs; leaves lanceolate to narrowly ovate-oblong, tapering to an acute base and more or less gradually to an acuminate apex, 3–9 cm long and 1/4–1/3 as wide as long, with 3–5 pairs of lateral veins; branches reddish to purple . . . . . var. *schuetzeana* (C. Meyer) Rickett
?*C. obliqua* Raf.; *C. purpusii* Koehne; *C. amomum* subsp. *obliqua* (Raf.) J. S. Wilson

7. **Cornus racemosa** Lam. Gray Dogwood. Panicled Dogwood
C. foemina Mill. subsp. *racemosa* (Lam.) J. S. Wilson; *C. paniculata* L'Hér.

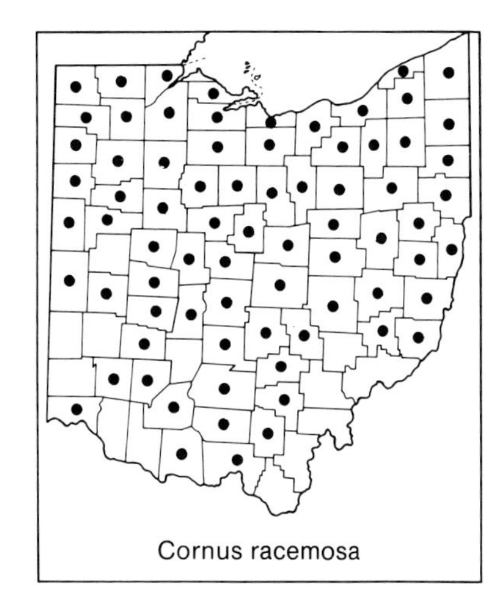
Cornus racemosa

Shrub, to 4.5 m (15 feet) tall, sometimes forming thickets; branches light to medium gray, but first-year branchlets light reddish-brown; pith tan (or white in young branchlets); leaves opposite, lanceolate to elliptic or narrowly ovate, acuminate at apex, acuminate to acute at base, with 3–4 pairs of lateral veins, glabrous to short-pubescent with appressed hairs on both upper and undersurfaces, the under also glaucous; flowers whitish, flowers and fruits in somewhat elongated paniculate cymes, the cymes from 2/3 as long as wide to longer than wide; pedicels red; fruit white. Throughout most of Ohio, but absent from some southeastern and some southwestern counties; fields, thickets, bogs and marshes, open woods, usually in moist, but sometimes in dry, situations. Late May–early July.

As indicated in the synonymy above, J. S. Wilson (1965) treats this taxon as a subspecies of the otherwise more southern *Cornus foemina*. I follow Rickett (1945) and most authors since 1945 in treating it at the rank of species. See note above, in the introduction to this genus, excluding C. *foemina* sensu stricto from the Ohio flora.

8. **Cornus alternifolia** L. f. Alternate-leaved Dogwood. Pagoda Dogwood

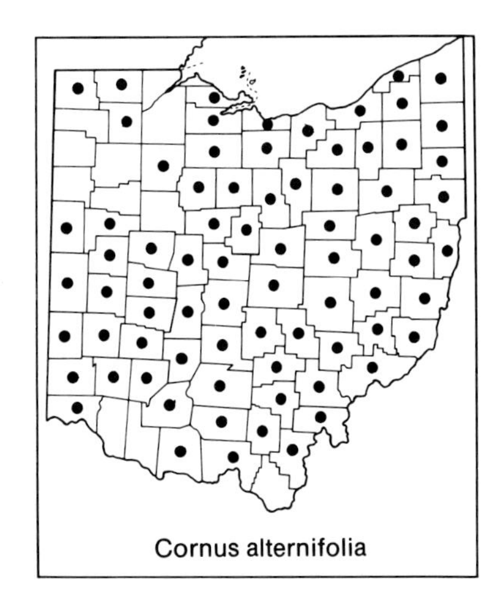
Cornus alternifolia

Usually a large shrub, sometimes a small tree, to 11 m (36 feet) tall; younger branches and branchlets greenish; pith white; leaves alternate, mostly crowded near tips of branchlets and sometimes appearing whorled, petioles (1–) 2–5 cm long, blades ovate to broadly elliptic, abruptly acuminate at apex, with 4–5 pairs of lateral veins, undersurface pale; flowers cream-colored, in a wide, somewhat rounded to flat-topped cyme; fruit dark blue. Throughout Ohio, except some northwestern and a few southwestern counties; rich woods, especially wooded slopes. May–June.

Clethraceae. CLETHRA FAMILY. PEPPERBUSH FAMILY

Clethra alnifolia L., Sweet Pepperbush or Summersweet, was attributed to Ohio by Braun (1961) on the basis of two specimens in the Oberlin College Herbarium, one dated 1850–1870, the other 1871. She noted that this summer-blooming shrub is often cultivated for its "attractive fragrant white flowers," and that the specimens "may have been taken from planted material," but thought this doubtful in view of the early dates. The closest boundary of the species' continuous range is in eastern Pennsylvania. Roberts and Cooperrider (1982), in excluding the species from the state's flora, speculated that because of its appeal and popularity as an ornamental, it may indeed have been brought westward and cultivated in Ohio as early as 1850. The species had also been excluded previously from the Ohio flora by Kellerman and Werner (1893) because "the specimens [perhaps the same ones cited in 1961 by Braun] were transplanted from another state."

Although this shrub continues to be as pleasingly fragrant as ever and is available in both white and pink corolla color forms, it is cultivated only occasionally in Ohio.

Pyrolaceae. PYROLA FAMILY. SHINLEAF FAMILY

Herbs; leaves simple; flowers perfect, solitary or in racemose or corymbose inflorescences; calyx of 5 sepals, separate or nearly so; corolla actinomorphic or nearly so, of 5 separate petals; stamens mostly 10; carpels 5, fused to form a compound pistil with a single style; ovary superior; fruit a capsule. A small family of some 50 species, chiefly in temperate regions of the Northern Hemisphere.

The circumscriptions of this family and the next two, the Monotropaceae and the Ericaceae, are problematic. Gleason (1952b), Gleason and Cronquist (1963), Thorne (1968, 1976, 1983), and Wood (1961) place all three in the Ericaceae sensu lato. Other workers, such as the Bailey Hortorium Staff (1976), Fernald (1950), Lawrence (1951), and Tutin et al. (1964–1980), recognize two families: Pyrolaceae (incl. Monotropaceae) and Ericaceae. The recognition here of three families follows Cronquist (1968, 1981).

a. Leaves cauline, their margins sharply serrate-dentate; flowers either solitary or 2–8 in an umbel or corymb; style minute and inconspicuous . . . . 1. *Chimaphila*

aa. Leaves basal, their margins subentire, crenate, or low-serrate; flowers 3–20 in a raceme; style elongate and conspicuous.

b. Flowers solitary . . . . 2. *Moneses*

bb. Flowers 3–20 in a raceme . . . . 3. *Pyrola*

1. Chimaphila Pursh

Evergreen, subshrubby, perennial herbs; leaves cauline, more or less whorled, margins sharply serrate-dentate; flowers either solitary or few in an umbel or corymb. A Northern Hemisphere genus of about seven species, mostly in temperate and boreal regions. Standley, Kim, and Hjersted (1988) studied the reproductive biology of the two species below.

a. Leaves oblanceolate, tapering at base, acute to obtuse at apex, green throughout, not variegated . . . . 1. *Chimaphila umbellata*

aa. Leaves lanceolate to ovate-lanceolate or the lowest ovate, obtuse to rounded at base, acute at apex, variegated with white along the midrib . . . . 2. *Chimpahila maculata*

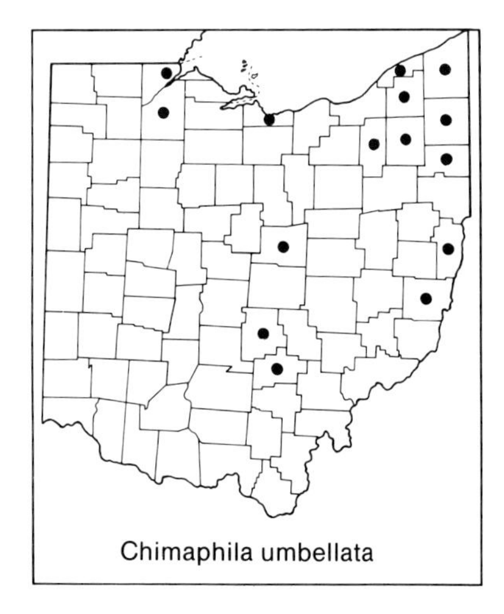
Chimaphila umbellata

1. **Chimaphila umbellata** (L.) Bart. PRINCE'S-PINE. PIPSISSEWA

Evergreen perennial; erect stems to 2.5 dm tall, arising from a creeping rhizome; leaves green throughout, mostly subwhorled, oblanceolate, tapering at base, acute to obtuse at apex; flowers 3–8, arranged in an umbel or umbellate cluster; corolla pink. Mostly in northeastern counties, but also at scattered sites in northwestern and central Ohio; dry woodlands. June–July.

Ohio plants belong to var. *cisatlantica* Blake; var. *umbellata* is a Eurasian taxon.

2. **Chimaphila maculata** (L.) Pursh   SPOTTED PIPSISSEWA. SPOTTED WINTERGREEN

Evergreen perennial; erect stems to 2.5 dm tall, arising from a creeping rhizome; leaves variegated on upper surface: whitish along the midrib and some of the lateral veins, opposite or whorled or subwhorled, lanceolate to ovate-lanceolate or the lowest ovate, obtuse to rounded at base, acute at apex; flowers (1–) 2–4 in an umbellate cluster; corolla white. Mostly in the acid soils of the eastern half of Ohio; mesic to dry woodlands. June–early August.

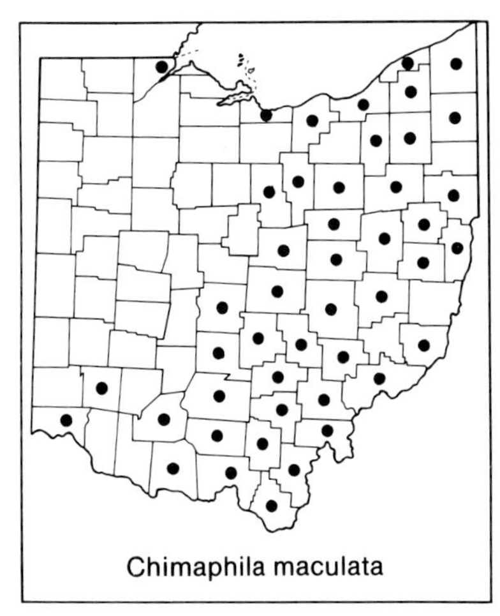
Chimaphila maculata

Moneses uniflora

2. Moneses Salisb.

A genus with a single species.

1. **Moneses uniflora** (L.) A. Gray   ONE-FLOWERED WINTERGREEN

Low perennial, to 1 dm tall; leaves basal or on lower part of stem, blades 1.0–1.5 cm long, mostly longer than petioles, broadly elliptic to orbicular, margins very low crenate to very low serrate; flowers solitary, nodding; corolla white, 1–2 cm across; style straight. A circumboreal species at a southern boundary of its range in northern Ohio. Its recent discovery in Lucas County (Cusick, 1990) gives credibility to the nineteenth-century report (Newberry, 1860) of a record from nearby Lorain County. The Lucas County population was growing under planted pines. June.

3. Pyrola L.   SHINLEAF

Perennial herbs; leaves all basal or nearly so, margins subentire to low-crenate or low-serrate; flowers in racemes. A genus of perhaps 40 species of temperate and cool regions of the Northern Hemisphere. Haber and Cruise (1974) define *Pyrola* in a more restricted sense than the treatment here, excluding *P. secunda* and assigning it to the genus *Orthilia* Raf.

a. Racemes secund; styles straight; blades of principal leaves 2–3 (–4) cm long . . . . 1. *Pyrola secunda*

aa. Racemes with spirally arranged flowers, not secund; styles declined; blades of principal leaves 1–8 cm long.

b. Blades of principal leaves 1–2 (–3) cm long, shorter than petioles; sepals about as wide as long; corolla greenish-white . . . . 2. *Pyrola chlorantha*

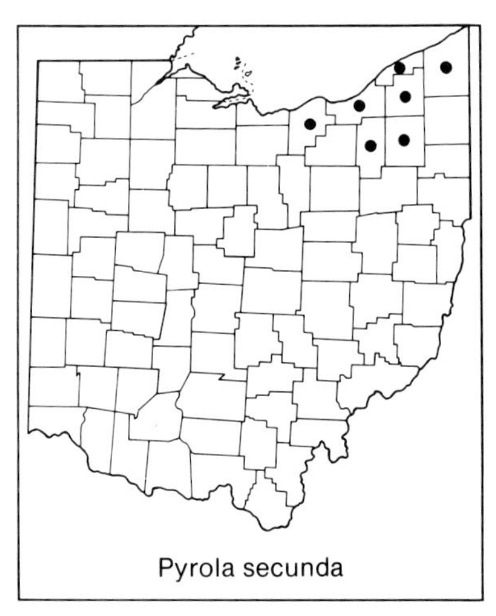
Pyrola secunda

Pyrola chlorantha

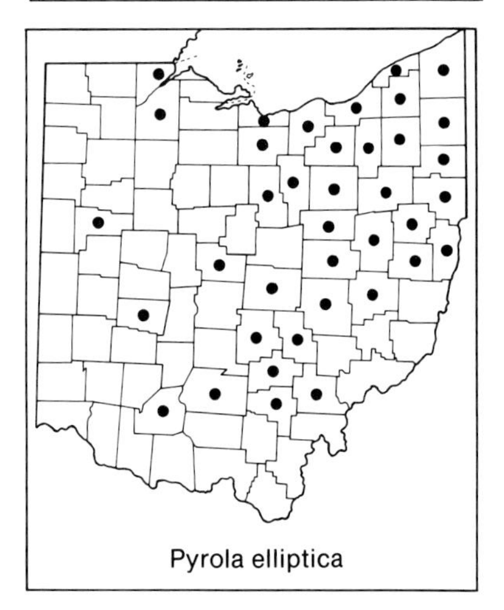
Pyrola elliptica

bb. Blades of principal leaves (3–) 4–8 cm long, about equalling to longer than petioles; sepals as wide as long or longer than wide; corolla white to cream-colored.

c. Sepals triangular or broadly triangular, as wide as or wider than long; blades of most leaves elliptic to broadly oblong or somewhat obovate, thin and deciduous . . . . 3. *Pyrola elliptica*

cc. Sepals oblong-lanceolate, approximately twice as long as wide; blades of most leaves orbicular to suborbicular or broadly elliptic, leathery and evergreen. . . . . 4. *Pyrola rotundifolia*

1. **Pyrola secunda** L. ONE-SIDED WINTERGREEN
Orthilia secunda (L.) House

Perennial; leaves basal or nearly so, blades mostly 2–3 cm long, longer than petioles, elliptic to ovate or suborbicular, margins low crenate-serrate; racemes secund; corolla greenish; styles straight. A circumboreal species at a southern boundary of its range in northeastern Ohio (a 1979 specimen reported to be from Coshocton County is probably a mislabeled collection from elsewhere); moist woods. June–July.

2. **Pyrola chlorantha** Sw. GREEN-FLOWERED SHINLEAF. GREEN-FLOWERED WINTERGREEN
P. virens Schreb.

Perennial; leaves basal or nearly so, blades mostly 1–2 cm long, shorter than petioles, broadly elliptic to suborbicular or broadly ovate, margins very low-crenate; racemes with spirally arranged flowers; sepals about as wide as long; corolla greenish-white; styles declined. A circumboreal species, the only Ohio record is of a small population discovered in 1986, in sandy soil in a pine plantation in Lucas County, near a southern boundary of the species' range. Not illustrated. June–July.

Haber (1972) concluded that the correct name for this species is *P. chlorantha*, which antedates the name *P. virens* by several years. Following Haber (1971), no infraspecific taxa are recognized in this species.

3. **Pyrola elliptica** Nutt. SHINLEAF

Perennial; leaves basal or nearly so, blades mostly 4–8 cm long, about equalling to longer than petioles, mostly elliptic to broadly oblong or somewhat obovate, margins low-crenate to low-serrate; racemes with spirally arranged flowers; sepals triangular to broadly triangular, as wide as or wider than long; corolla white to cream-colored; styles declined. Chiefly in the eastern half of Ohio; mesic to dry woodlands. June–July.

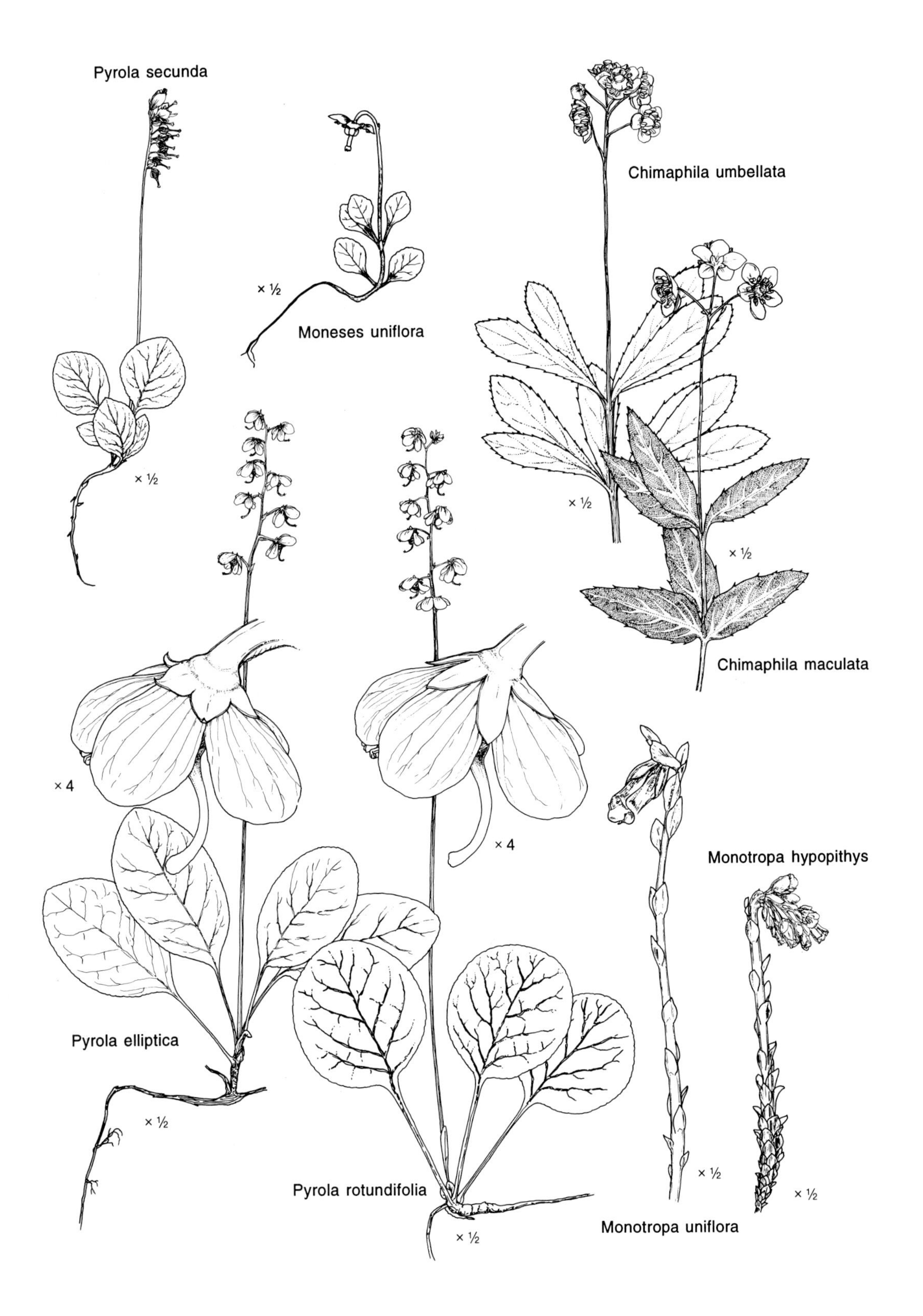
Pyrola secunda
× ½
Moneses uniflora
× ½
Chimaphila umbellata
× ½
Chimaphila maculata
× ½
× 4
× 4
Pyrola elliptica
× ½
Pyrola rotundifolia
× ½
Monotropa hypopithys
× ½
Monotropa uniflora
× ½

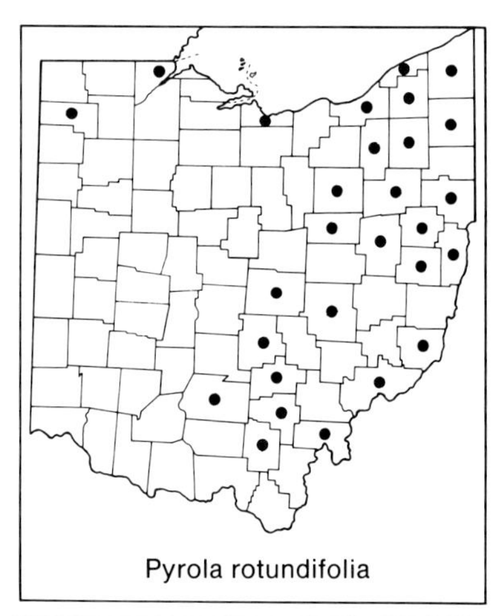
Pyrola rotundifolia

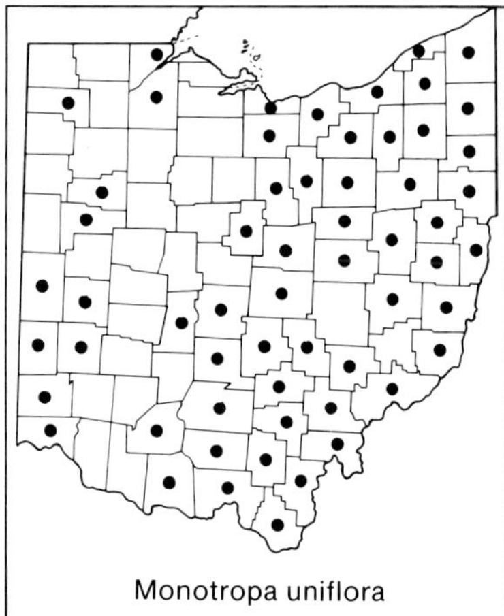
Monotropa uniflora

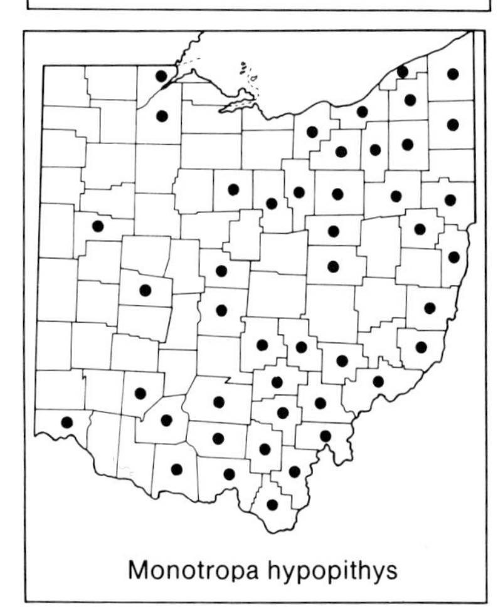
Monotropa hypopithys

4. **Pyrola rotundifolia** L. ROUND-LEAVED WINTERGREEN. WILD LILY-OF-THE-VALLEY

Perennial; leaves basal or nearly so, blades mostly 3–7 cm long, about equalling petioles, mostly orbicular to suborbicular or broadly elliptic, margins very low-crenate to subentire; racemes with spirally arranged flowers; sepals oblong-lanceolate, approximately twice as long as wide; corolla white to creamy-white; styles declined. Mostly in eastern counties; mesic to dry woodlands. June–July.

Ohio plants belong to var. *americana* (Sweet) Fern. (*P. americana* Sweet). Haber (1971) treated this taxon as *P. asarifolia* Michx. subsp. *americana* (Sweet) Krisa.

Monotropaceae. INDIAN PIPE FAMILY

A small family of about 12 species, native mostly to the Northern Hemisphere.

1. Monotropa L.

Fleshy, nongreen, perennial herbs, parasitic on fungi; stems and leaves white to brownish-yellow or reddish, becoming dark with age or when dried; leaves small, scale-like; flowers perfect, solitary or in a raceme; calyx of 2–5 separate sepals or absent; corolla actinomorphic or nearly so, of 4 or 5 separate petals; stamens 8 or 10; carpels 4 or 5, fused to form a compound pistil with a single style; ovary superior; fruit a capsule. A genus of two or more species of temperate regions of the Northern Hemisphere.

a. Each stem with a single terminal flower . . . . . . 1. *Monotropa uniflora*
aa. Each stem with several flowers arranged in a raceme . . . . 2. *Monotropa hypopithys*

1. **Monotropa uniflora** L. INDIAN PIPE

White or very pale pink perennial, to 2.5 dm tall; stems with conspicuous scale-like leaves; each stem with a single, terminal, nodding, white flower to ca. 2 cm long; the peduncle and fruit later becoming erect. Eastern half of Ohio and at scattered sites in western counties; woodlands. June–August.

Although sometimes called a "saprophyte," studies by Herrick (1957), Riley and Eichenmuller (1970), and Campbell (1971) have shown this species to be parasitic on fungal mycelia.

2. **Monotropa hypopithys** L. PINESAP

Brownish-yellow to reddish perennial, to 3 dm tall; stems with conspicuous scale-like leaves; each stem with several flowers in a terminal nodding raceme, the flowers also brownish-yellow to reddish; the raceme later straightening and the fruits becoming erect on erect or ascending pedicels. Eastern and southern counties and scattered elsewhere, not so frequent as M. *uniflora*; woodlands. June–September.

The specific epithet is sometimes spelled *hypopitys*.

Ericaceae. HEATH FAMILY

Shrubs or small trees of acid soils, the shrubs sometimes quite small and in some cases prostrate; leaves simple, alternate; flowers perfect, actinomorphic or nearly so, often showy, solitary or in various types of inflorescences; sepals usually 5; petals usually 5 and usually fused, rarely separate; stamens usually 5 or 10; carpels usually 5, fused to form a compound pistil with a single style; ovary superior or inferior; fruit usually a capsule or berry. A widely distributed family of 1,900 species, chiefly of temperate regions. Wood (1961) studied the family in southeastern United States, and Blauch (1970) prepared a taxonomic study of the family in the Southern Appalachian Highlands—an area including the unglaciated section (east-central and southeastern counties) of Ohio. See discussion under previous family, the Pyrolaceae, regarding the separation of that family from this.

The European species *Erica tetralix* L., Cross-leaved Heath, was collected once, under a Norway spruce in Mahoning County, where it was presumably an adventive escape from plantings. Braun (1961) notes a report of *Calluna vulgaris* (L.) Hull, Heather, being "spontaneous and persistent" at a locality in Trumbull County. She evidently saw no specimens, and neither have I.

a. Ovary superior; fruit a berry, drupe, or capsule.
 b. Leaves in whorls of 4, linear, 2–5 mm long; see above. . . . . . . *Erica*
 bb. Leaves alternate or opposite, linear or broader, 5 mm or longer.
  c. Petals 5, separate; leaves lanceolate to narrowly elliptic, with revolute margins, their undersurface with dense tan or rusty, woolly pubescence . . . . 1. *Ledum*
  cc. Petals 4 or 5, some or all united for at least part of their length; leaves of various shapes, not revolute-margined (except in *Andromeda*), variously pubescent to glabrous beneath.
   d. Stems prostrate.
    e. The largest leaf blades 3–6 (–10) cm long, on petioles 2–5 cm long; petals 5, corolla 15–20 mm long . . . . 8. *Epigaea*
    ee. The largest leaf blades 0.5–3 cm long, on petioles 3 mm or less in length; petals 5 or 4, corolla 2–6 mm long.

f. Leaves 0.5–1 cm long, with veins on undersurface scarcely visible; petals 4; corolla campanulate, 2–3 mm long . . . . 9. *Gaultheria*

ff. Leaves 1–3 cm long, with veins on undersurface dark and conspicuous; petals 5; corolla ovoid, 4–6 mm long . . . . 10. *Arctostaphylos*

dd. Stems erect or at least partly erect.

e. Flowers with rotate or saucer-shaped corollas; each of the 10 anthers prior to anthesis held in a separate depression on the corolla tube; leaves mostly alternate, some subopposite, all evergreen . . . . 3. *Kalmia*

ee. Flowers with corollas of various shapes, but not rotate or saucer-shaped; anthers not held in depressions; leaves all alternate, either evergreen or deciduous.

f. Corolla funnelform to campanulate, widest at mouth, the tube 15 mm or more in length . . . . 2. *Rhododendron*

ff. Corolla tubular, ovoid, or globose, more or less narrowed at mouth, the tube 3–9 mm long.

g. Leaves deciduous; shrubs or trees, 0.4–22.5 m tall.

h. Shrubs, 0.4–3 m tall; principal leaves 3–8 cm long, their margins minutely serrulate to subentire; corolla more or less globose, 3–4 (–5) mm long . . . . 5. *Lyonia*

hh. Trees, to 22.5 m tall; principal leaves 10–15 (–20) cm long, their margins usually with fine but conspicuous serrations; corolla ovoid, 6–7 mm long . . . . 6. *Oxydendrum*

gg. Leaves evergreen; shrubs, 0.1–1.5 m tall.

h. Leaves linear or very narrowly elliptic, their margins entire and revolute, white or white-pubescent on undersurface . . . . 4. *Andromeda*

hh. Leaves narrowly oblong or elliptic to broadly elliptic or suborbicular, their margins low crenate-serrate or rarely entire, not revolute, green or brown on undersurface.

i. Leafy stems 10–15 dm tall, from shrubby base; leaves narrowly oblong to elliptic, 2–5 times as long as wide. . . . . 7. *Chamaedaphne*

ii. Leafy stems 0.5–2 dm tall, from creeping rhizome; leaves broadly elliptic to suborbicular, 1–2 times as long as wide . . . . 9. *Gaultheria*

aa. Ovary inferior; fruit a berry or drupe.

b. Stems prostrate, with or without low erect branches; leaves mostly 0.5–1.5 cm long.

c. Undersurface of leaves with appressed bristles; stems trailing, without erect branches . . . . 9. *Gaultheria*

cc. Undersurface of leaves glabrous; stems trailing, with erect branches . . . . 12. *Vaccinium*

bb. Stems erect; principal leaves (1–) 2–10 cm long.

c. Undersurface of younger leaves with numerous resinous dots, these in some instances glistening, in others dull, and in older leaves these sites usually merely translucent; fruit with 10 pits. . . . . 11. *Gaylussacia*

cc. Undersurface of leaves without resinous or translucent dots; fruit with more than 10 seeds. . . . . 12. *Vaccinium*

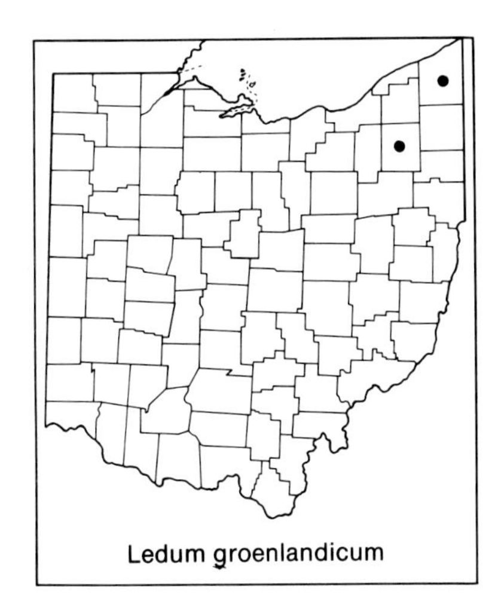

1. Ledum L.

A genus of temperate and cold regions of the Northern Hemisphere, with three or four species, one of which occurs in northern Ohio.

1. **Ledum groenlandicum** Oeder LABRADOR-TEA

L. palustre L. subsp. *groenlandicum* (Oeder) Hultén; *Rhododendron groenlandicum* (Oeder) Kron & Judd

Shrub, to 1 m (3 feet) tall; leaves lanceolate to narrowly elliptic, with revolute margins, the undersurface covered with dense woolly pubescence, the pubescence white the first year, then becoming tan or rust-colored; flowers in a terminal, somewhat umbellate raceme; petals 5, white, separate; stamens 5–7; ovary superior; fruit an ovoid capsule. A boreal species at a southern boundary of its range in northeastern Ohio; bogs. Jennings (1909) describes an early Portage County record of this species. Several collections were made in the 1930s from shrubs planted by Frederica Detmers at what is now Cranberry Bog State Nature Preserve in central Ohio, but the species apparently no longer persists there; see related comment under *Chamaedaphne calyculata* (p. 260). June.

Kron and Judd (1990) have recently merged the genus *Ledum* under *Rhododendron*.

2. Rhododendron L. RHODODENDRON. AZALEA

Shrubs; leaves evergreen or deciduous; flowers large, showy, in umbellate clusters; petals 5, corolla funnelform to campanulate; stamens 5 or 10; ovary superior; fruit a capsule. A large genus of perhaps 800 species, chiefly of temperate and cold regions of the Northern Hemisphere. Many cultivars of this genus are popular ornamentals, especially in eastern Ohio, where the acid soils provide a substrate in which these plants flourish.

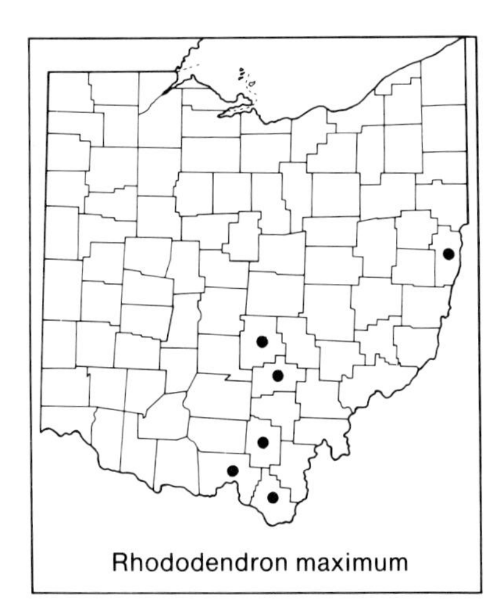

Rhododendron maximum

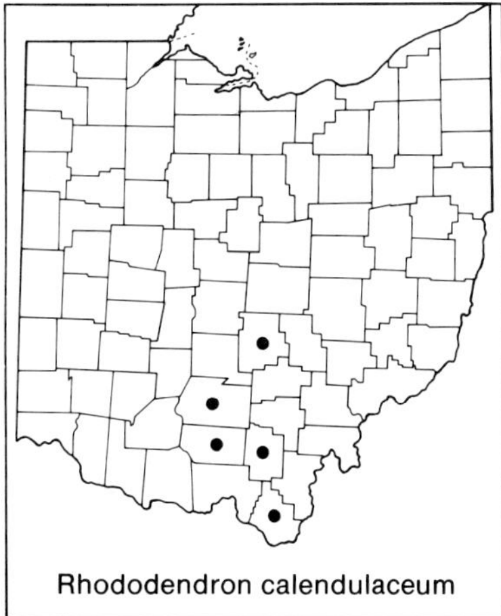

Rhododendron calendulaceum

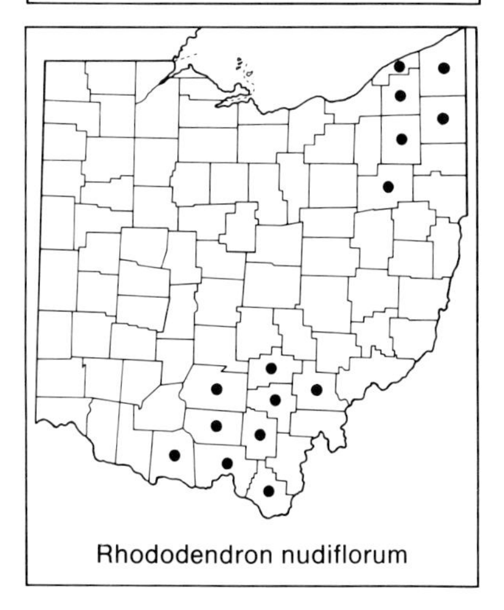

Rhododendron nudiflorum

Rhododendron canescens (Michx.) Sweet, Hoary Azalea, was credited to Ohio by both Schaffner (1933) and Fernald (1950). Braun (1961) concluded that all such attributions were based on misidentified specimens of *R. roseum* (*R. nudiflorum* var. *roseum*). *Rhododendron viscosum* (L.) Torr., Clammy Azalea, was attributed to Ohio by Fernald (1950) and Gleason (1952b), but excluded from the flora by Braun (1961) for lack of substantiating specimens.

a. Leaves evergreen, thick and leathery, 1–2 dm long, margins not ciliate; stamens 10 . . . . 1. *Rhododendron maximum*

aa. Leaves deciduous, thin and not leathery, to 1 dm long at maturity, margins ciliate; stamens 5.

b. Flowers yellow to orange; leaves acute at apex, short-pubescent on undersurface, chiefly along the veins; calyx lobes (1–) 2–4 mm long . . . . 2. *Rhododendron calendulaceum*

bb. Flowers pink to white; leaves acute or obtuse at apex, pubescent to nearly glabrous on undersurface; calyx lobes to 1 mm long . . . . 3. *Rhododendron nudiflorum*

1. **Rhododendron maximum** L. Great Rhododendron. Rosebay Rhododendron

Shrub, to 3 m (10 feet) tall; leaves evergreen, thick and leathery, 1–2 dm long; flowers pink to less often white; stamens 10. South-central counties and one east-central county (Jefferson); more or less open ravines, usually with sandstone exposures. Iltis' (1956) map shows the Ohio sites to be along the northwestern boundary of the species' range. Frequently cultivated, this attractive shrub is the only native Ohio woody dicotyledon with large evergreen leaves. Some seedlings originating from cultivated plants may become established, but these remain in close proximity to the parent plants and do not spread. June–July.

2. **Rhododendron calendulaceum** (Michx.) Torr. Flame Azalea

Shrub, to 3 m (10 feet) tall; leaves deciduous, to 1 dm long; apex acute, undersurface short-pubescent, chiefly along the veins, margins ciliate; flowers yellow to orange, calyx lobes (1–) 2–4 mm long; stamens 5. An attractive shrub of the southern Appalachians, at a northern boundary of the species' range in south-central Ohio; dry woodlands; often cultivated. Willingham (1976) studied variation in the species. May–early June.

3. **Rhododendron nudiflorum** (L.) Torr. Pink Azalea

Shrub, to 2.5 m (8 feet) tall; leaves deciduous, to 8 cm long; apex acute or obtuse, undersurface pubescent to nearly glabrous, margins ciliate; flowers pink to white, calyx lobes to 1 mm long; stamens 5. A species of northeastern North America, extending southward in the Appalachians; north-

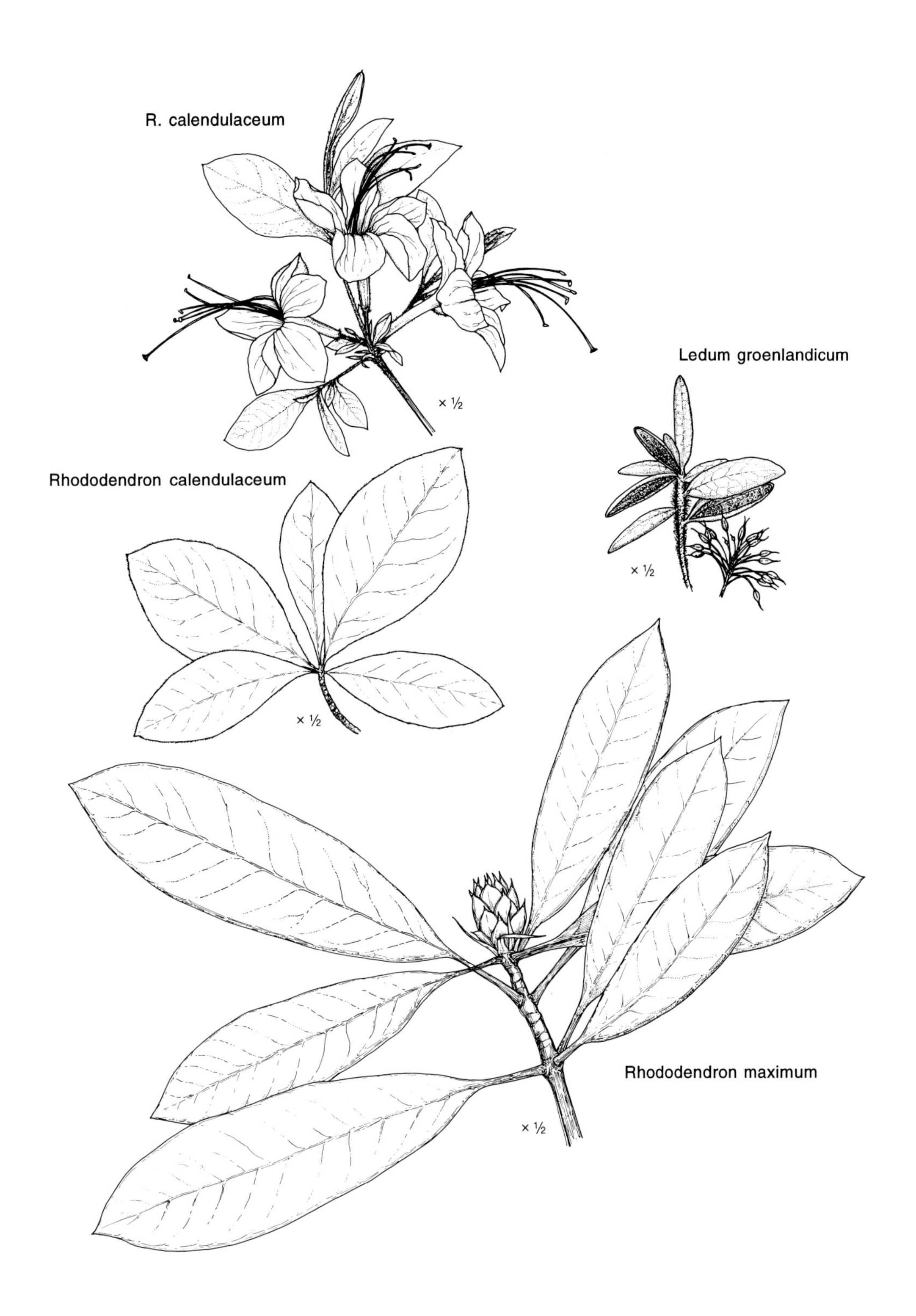
R. calendulaceum
× ½
Ledum groenlandicum
× ½
Rhododendron calendulaceum
× ½
Rhododendron maximum
× ½

eastern and south-central Ohio counties; dry woods and woodland borders, thickets, bogs; sometimes cultivated. Late April–early June.

Shinners (1962c), Roane (1975), Wilbur (1976), Uttal (1988), and Kron (1989) have commented on the nomenclature of this species. The species is represented in Ohio by two intergrading varieties. Among Ohio plants, the two taxa "hybridize freely; most Ohio populations of pink azaleas are mixed populations with many of the individuals intermediate" (Braun, 1961).

a. Leaves glabrous on undersurface except along midribs; stamens about 3 times as long as corolla tube; style 4.5–7 cm long. . . . . var. *nudiflorum*
   R. periclymenoides (Michx.) Shinners
   Pinxter-flower

aa. Leaves pubescent across all or much of the undersurface; stamens about 2 times as long as corolla tube; style 4–5 cm long . . . . var. *roseum* (Loisel.) Wieg.
   R. prinophyllum (Small) Millais, *R. roseum* (Loisel.) Rehd.
   Roseshell Azalea

3. Kalmia L. LAUREL

A North American genus of about seven species, one of which is native to Ohio. Ebinger (1974) and Southall and Hardin (1974) prepared taxonomic studies of the genus.

1. **Kalmia latifolia** L. Mountain Laurel

Shrub, to 3 m (10 feet) tall; leaves evergreen, mostly alternate but some subopposite, elliptic, to 12 cm long; flowers pinkish, showy, in few- to several-flowered terminal corymbs; corolla rotate to saucer-shaped; stamens 10, each of the anthers held when young in a separate depression (sac) on the corolla tube, from which they are released suddenly (in response to touch) soon after anthesis; ovary superior; fruit a capsule. A species of eastern United States; in Ohio, sharply limited to the acid soils of the unglaciated Allegheny Plateau (east-central and southeastern counties), where the species is at a northwestern boundary of its range (Kurmes, 1967); oak woodlands, hemlock woodlands, swampy habitats. Called by E. Lucy Braun (1961): "One of America's most beautiful shrubs," this species is frequently cultivated in Ohio. Late May–June.

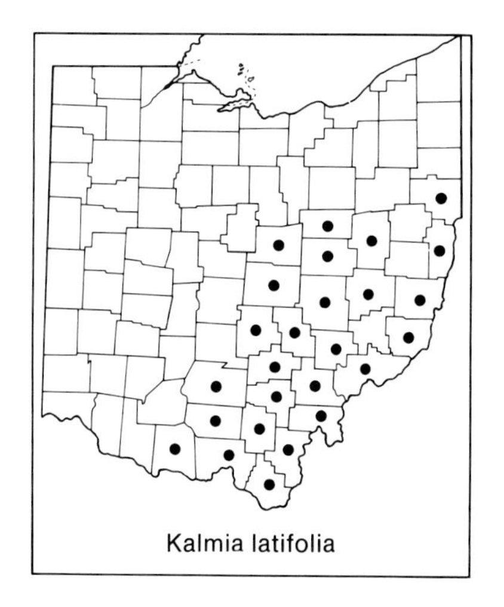
Kalmia latifolia

4. Andromeda L.

A genus of two species native to temperate and colder regions of the Northern Hemisphere, one of the two species native to northern Ohio.

1. **Andromeda glaucophylla** Link BOG-ROSEMARY

Shrub, to 6 dm (2 feet) tall; leaves evergreen, very narrowly elliptic to linear, white or white-pubescent on undersurface, the margins entire and revolute; flowers small, pinkish, in terminal racemes; corolla globose-urceolate, 5–6 mm long; stamens 10; ovary superior; fruit a capsule. A boreal species of northeastern North America; in northeastern Ohio counties, at a boundary of the species' range; bogs. May–June.

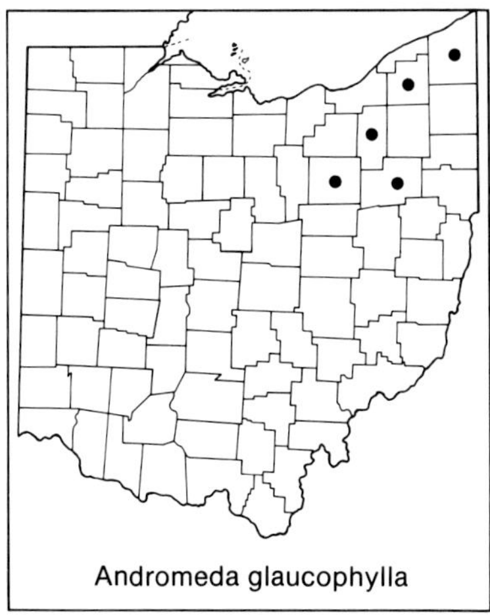
Andromeda glaucophylla

5. Lyonia Nutt.

A genus of 40 or more species native to North America and eastern Asia; a single species occurs in the Ohio flora. Judd (1981) monographed the genus.

1. **Lyonia ligustrina** (L.) DC. MALE-BERRY

Shrub, to 3 m (10 feet) tall; leaves deciduous, usually obovate or oblanceolate, 3–8 cm long, margins minutely serrulate to subentire; flowers small, white, in racemose or paniculate inflorescences; corolla more or less globose, 3–4 (–5) mm long; stamens usually 10; ovary superior; fruit a capsule. A species of eastern United States; in south-central Ohio counties, at a western boundary of the species' range; wet thickets, swampy habitats. June–July.

Following Judd (1981), Ohio plants are assigned to var. *ligustrina.*

Lyonia ligustrina

6. Oxydendrum DC.

A genus with a single species.

1. **Oxydendrum arboreum** (L.) DC. SOURWOOD. SORREL-TREE

Tree, to 22.5 m (74 feet) tall; leaves deciduous, 10–15 (–20) cm long, oblong-elliptic, acuminate at apex, margins usually with fine but conspicuous serrations; flowers small, white, in showy terminal panicles; corolla ovoid, 6–7 mm long; stamens 10; ovary superior; fruit a capsule. Chiefly a tree of southeastern United States, this species is at a northern boundary of its range in southeastern and south-central Ohio; acid woodlands. Often cultivated either for its showy inflorescences and bright red autumnal leaf coloration or for its flowers—which are the source of a distinctive honey. July.

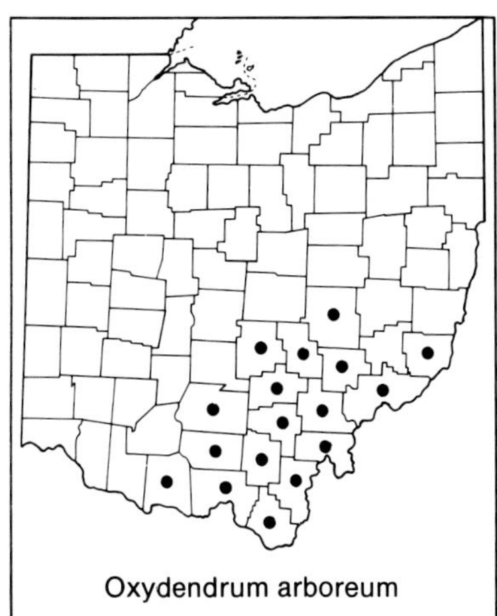
Oxydendrum arboreum

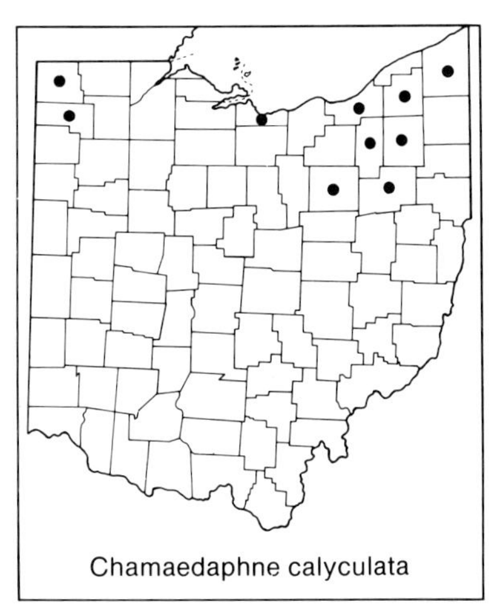

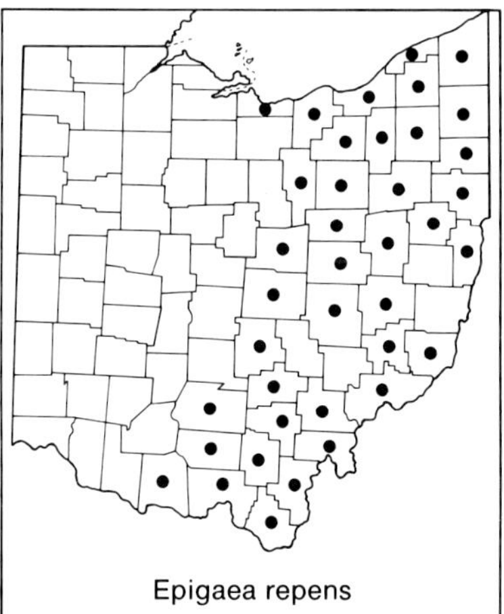

7. Chamaedaphne Moench

A genus with a single species.

1. **Chamaedaphne calyculata** (L.) Moench LEATHER-LEAF
   incl. var. *angustifolia* (Ait.) Rehd.

Shrub, to 1.5 m (5 feet) tall; leaves evergreen, narrowly oblong to elliptic, to 5 cm long, margins crenulate, undersurface rusty-scurfy; flowers small, white, solitary in axils of reduced leaves but in all appearing more or less racemose; corolla cylindric-urceolate, 6–7 mm long; stamens 10; ovary superior; fruit a capsule. A circumboreal species ranging in North America from Newfoundland to Alaska and southward, at a southern boundary of its range in northern Ohio; bogs. Braun (1961) presumed that a [1933] collection from Cranberry Bog in Licking County, in central Ohio, was taken from a plant introduced there, and that it did not represent a native population (cf. comment under *Ledum groenlandicum* above, p. 255). The species evidently did not become established at what is now Cranberry Bog State Nature Preserve. April–June.

Following Gleason and Cronquist (1963, 1991), no varieties are recognized within this species.

8. Epigaea L.

A genus of two species, the other native to Japan.

1. **Epigaea repens** L. TRAILING ARBUTUS. MAYFLOWER
   incl. var. *repens* and var. *glabrifolia* Fern.

Prostrate shrub; stems conspicuously hairy, to 4 dm long; leaves evergreen, ovate to oblong, more or less cordate at base, 1.5–6 cm long, petioles hairy, 0.5–5 cm long; flowers small, pink, fragrant, in short, crowded spikes; corolla salverform, to 1.5 cm long; stamens 10; ovary superior; fruit a capsule. In acid soils of the eastern half of Ohio; dry woods, openings, borders, banks; one of the state's earliest showy wildflowers. March–May.

Following Braun (1961), no varieties are recognized within this species.

9. Gaultheria L.

Low prostrate subshrubs; flowers solitary in leaf axils, white; fruit berry-like: a capsule closely surrounded by a fleshy persistent calyx. A widely distributed genus of 100 or more species.

a. Stems prostrate, but with erect leafy branches; leaves 2–5 cm long; corolla 7–10 mm long, petals 5; fruits red. . . . . 1. *Gaultheria procumbens*

aa. Stems and branches prostrate; leaves 0.5–1 cm long; corolla 2–3 mm long, petals 4; fruits white. . . . . 2. *Gaultheria hispidula*

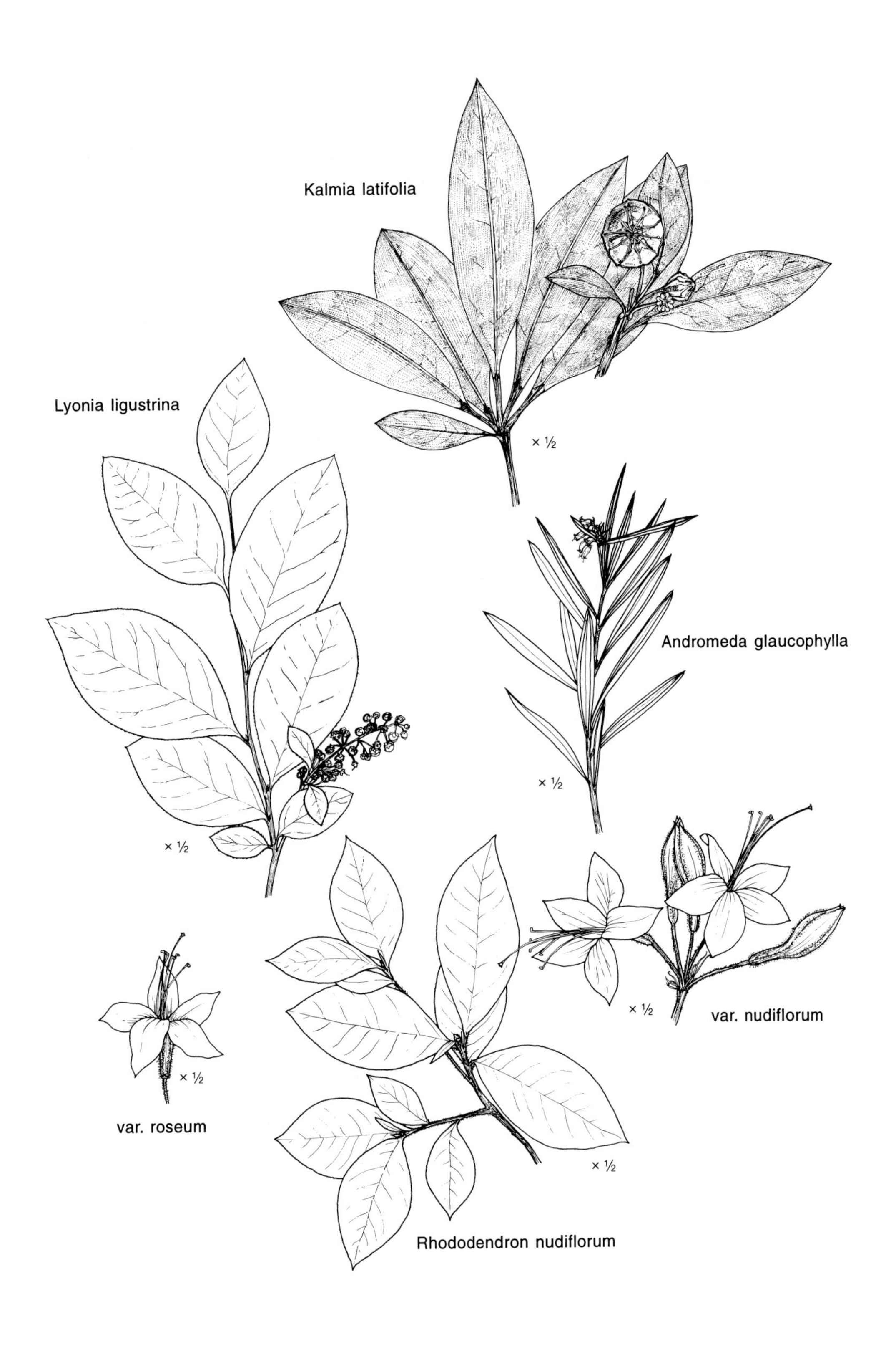
Kalmia latifolia
× ½
Lyonia ligustrina
Andromeda glaucophylla
× ½
× ½
var. nudiflorum
× ½
var. roseum
× ½
× ½
Rhododendron nudiflorum

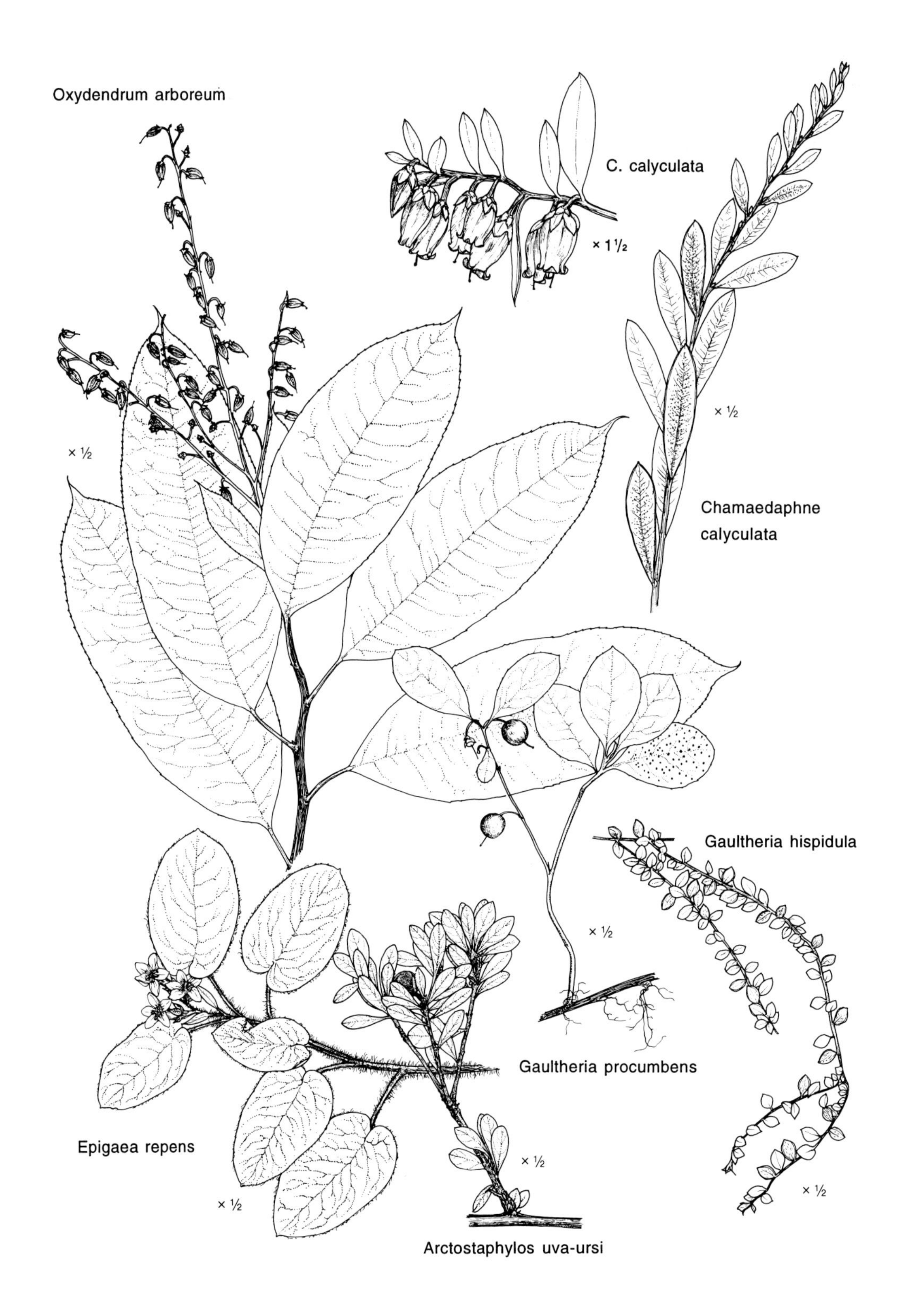
Oxydendrum arboreum
C. calyculata
× 1½
× ½
× ½
Chamaedaphne
calyculata
Gaultheria hispidula
× ½
Gaultheria procumbens
Epigaea repens
× ½
× ½
× ½
Arctostaphylos uva-ursi

1. **Gaultheria procumbens** L. WINTERGREEN. CHECKERBERRY. TEABERRY

Low herb-like shrub; principal stems (rhizomes) prostrate, their leafy branches erect, to 2 dm tall; leaves evergreen, 2–5 cm long, broadly elliptic to suborbicular, margins subentire to low crenate-serrate, undersurface often with dark dots; flowers small, white, solitary in leaf axils, nodding; corolla of 5 petals, urceolate-cylindric, 7–10 mm long; stamens 8–10; ovary superior; fruit red. Chiefly in the eastern half of Ohio; acid woodlands. The leaves have been used as a minor source of natural wintergreen. They emit that pleasant fragrance when broken. Late June–August.

2. **Gaultheria hispidula** (L.) Bigel. CREEPING SNOWBERRY

Low, trailing, herb-like shrub, both the principal stems and the branches prostrate; leaves evergreen, tiny, 0.5–1.0 cm long, broadly elliptic to suborbicular, margins slightly revolute, undersurface with appressed bristles; flowers small, white, solitary in leaf axils, nodding; corolla of 4 petals, campanulate, 2–3 mm long; stamens 8; ovary partly inferior; fruit white. A boreal species at a southern boundary of its range in northeastern Ohio; in bogs and in clumps of wet woodland moss. May.

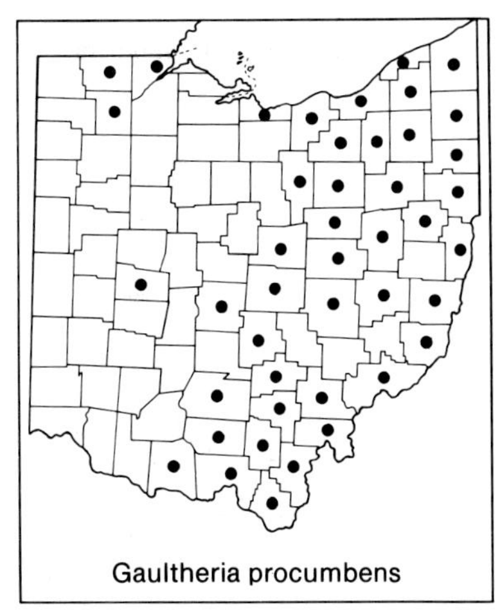

Gaultheria procumbens

Gaultheria hispidula

10. Arctostaphylos Adans.

A genus of about 50 species, chiefly North American; one species is native to Ohio.

1. **Arctostaphylos uva-ursi** (L.) Spreng. BEARBERRY. KINNIKINICK

Trailing mat-forming shrub; leaves evergreen, 1–3 cm long, obovate or oblanceolate to oblong, margins entire, undersurface with dark conspicuous veins; flowers small, white to pinkish, in terminal racemes; corolla ovoid, 4–6 mm long; stamens 10; ovary superior; fruit a red drupe. A boreal species at a southern boundary of its range in northern Ohio; sandy Lake Erie beaches; sometimes cultivated as a ground cover. May.

Ohio plants have often been assigned to var. *coactilis* Fern. & Macbr., most recently by Packer and Denford (1974). However, accepting the conclusions of Rosatti (1987), no varieties are recognized within this species.

Arctostaphylos uva-ursi

11. Gaylussacia HBK. HUCKLEBERRY

A Western Hemisphere genus of from 40 to 50 species. The one species native to Ohio is covered in the taxonomic study by W. H. Duncan and N. E. Brittain (1966) of *Gaylussacia* in Georgia.

Gaylussacia frondosa (L.) Torr., DANGLEBERRY, was attributed to Ohio by Gleason (1952b), but Braun (1961) excluded it for lack of substantiating specimens.

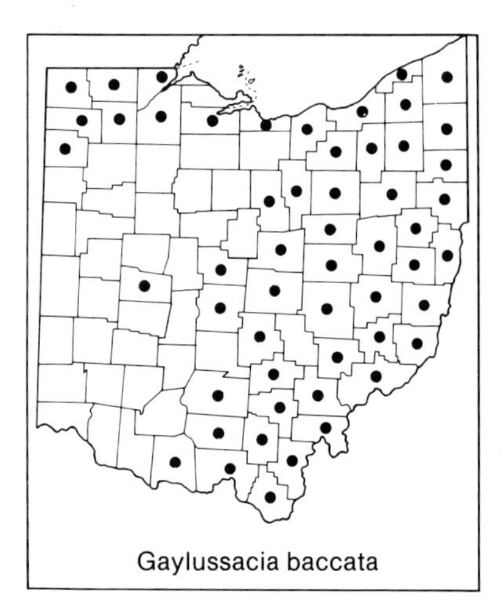
Gaylussacia baccata

1. **Gaylussacia baccata** (Wang.) K. Koch Huckleberry. Black Huckleberry

Shrub, to 1 m (3 feet) tall; leaves deciduous, mostly oblanceolate to elliptic, 3–6 cm long, entire, the undersurface (and to a lesser extent the upper) with numerous resinous dots, these in some instances glistening, in others dull, and in older leaves usually merely translucent; flowers greenish to reddish, in lateral racemes; corolla more or less cylindric, 4–6 mm long; stamens 10; ovary inferior; fruit a black berry-like drupe with 10 pits. Eastern half of Ohio and a few western, especially northwestern, counties; dry acid woods and openings, bogs. May–June.

12. Vaccinium L. BLUEBERRY

Shrubs, erect to low and trailing; leaves deciduous or less often evergreen; flowers white or greenish to pink; corolla varying from shallowly 5-lobed and usually cylindric to deeply 4-lobed; stamens 8 or 10; ovary inferior; fruit a many-seeded berry. A Northern Hemisphere genus of about 150 species. *Vaccinium corymbosum*, our principal edible blueberry, is noted for its fruit production both in wild plants and in several cultivars grown in Ohio. A few wild stands of *V. macrocarpon* produce sufficient cranberries for local harvesting.

The taxonomy adopted here is that of Vander Kloet (1988).

a. Low to tall, erect shrubs with sturdy branches; leaves deciduous; calyx and corolla 5-lobed; stamens 10.

b. Flowers and fruits borne on filiform pedicels 5–15 (–20) mm long, each solitary in the axil of a much reduced leaf 3–20 (–25) mm long, the flowers clustered with the reduced leaves on branches separate from the main vegetative branches; backs of anthers with 2 awns . . . . 1. *Vaccinium stamineum*

bb. Flowers and fruits borne on thicker pedicels 1–5 mm long, each subtended by a bract less than 3 mm long, or bracts apparently none, the flowers and bracts forming an inflorescence; backs of anthers without awns.

c. Shrubs low, to 0.6 m tall; blades of principal leaves 1–4 (–5) cm long.

d. Leaves entire, more or less densely pubescent on undersurface and pubescent along midrib on upper surface . . . . 2. *Vaccinium myrtilloides*

dd. Leaves entire or serrulate, glabrous on both surfaces or minutely pubescent on veins of undersurface.

e. Blades of leaves 2–3 times as long as wide, serrulate to entire . . . . 3. *Vaccinium pallidum*

ee. Blades of leaves 3 or more times as long as wide, minutely serrulate. . . . . 4. *Vaccinium angustifolium*

cc. Shrubs 1–3 m tall; blades of principal leaves 3–8 cm long . . . . 5. *Vaccinium corymbosum*

aa. Low trailing shrubs with delicate ascending branches; leaves evergreen; calyx and corolla 4-lobed; stamens 8.

b. Flowers 1–3 (–4) on each branch, mostly originating near tip of stem, each pedicel bearing near or below its midpoint two linear to narrowly lanceolate bractlets 1–2.5 mm long; berry 5–8 (–10) mm in diameter. . . . . 6. *Vaccinium oxycoccos*

bb. Flowers 2–4 (–6) on each branch, originating from axils of reduced leaves at sites below stem tip, each pedicel bearing well above its midpoint two elliptic to ovate bractlets 3–5 mm long; berry 10–20 mm in diameter. . . . . 7. *Vaccinium macrocarpon*

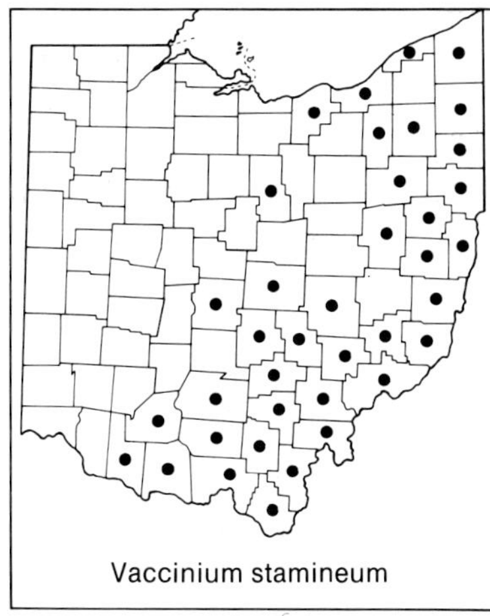
Vaccinium stamineum

1. **Vaccinium stamineum** L. Deerberry. Squaw-huckleberry
incl. var. *melanocarpum* C. Mohr and var. *neglectum* (Small) Deam; and incl. also *V. caesium* E. Greene

Shrub, to 1.5 m (5 feet) tall; leaves deciduous, 3–8 cm long, variable in shape, mostly ovate, elliptic, or oblong, margins entire or nearly so; flowers white to greenish, borne on filiform pedicels, the pedicels 5–15 (–20) mm long, each solitary in the axil of a much reduced leaf 3–20 (–25) mm long, the flowering branches separate from vegetative branches; sepals and petals 5; corolla more or less campanulate, 4–6 mm long; stamens 10, exserted, backs of anthers with 2 awns. Eastern and south-central counties, the Ohio stations along a northern boundary of the species' range; dry woodlands, openings, borders, thickets. Cane et al. (1985) studied the species' pollination ecology. May–June.

Vaccinium myrtilloides

Following P. C. Baker (1970) and Vander Kloet (1988), *Vaccinium caesium* is merged without rank under this species.

2. **Vaccinium myrtilloides** Michx. Velvet-leaved Blueberry

Low shrub, to ca. 0.5 m (1½ feet) tall; leaves 1–3 cm long, narrowly elliptic, margins entire, blades more or less densely pubescent on undersurface and pubescent along midrib on upper surface; flowers white to pinkish, in short racemes; sepals and petals 5; corolla more or less cylindric, 4–5 mm long; stamens 10. A northern species ranging across Canada and south into northern United States, at a boundary of its range in northeasternmost counties of Ohio; wet woodlands. Vander Kloet and Hall (1981) studied the biology of this species. May–June.

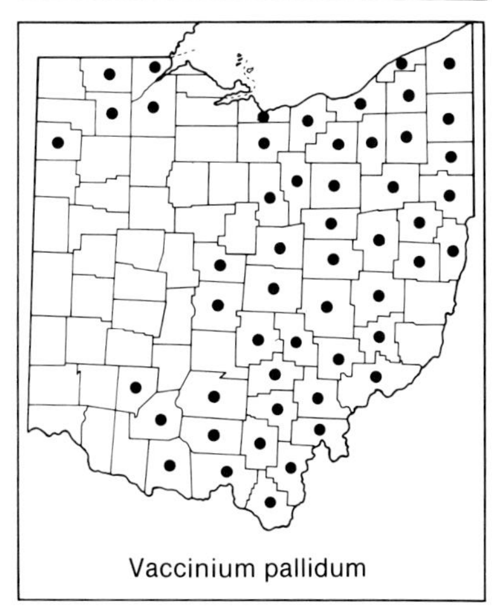
Vaccinium pallidum

3. **Vaccinium pallidum** Ait. Low Blueberry
incl. *V. alto-montanum* Ashe and *V. vacillans* Torr.

Low to medium-sized shrub, to ca. 0.6 m (2 feet) tall; leaves deciduous, 2–

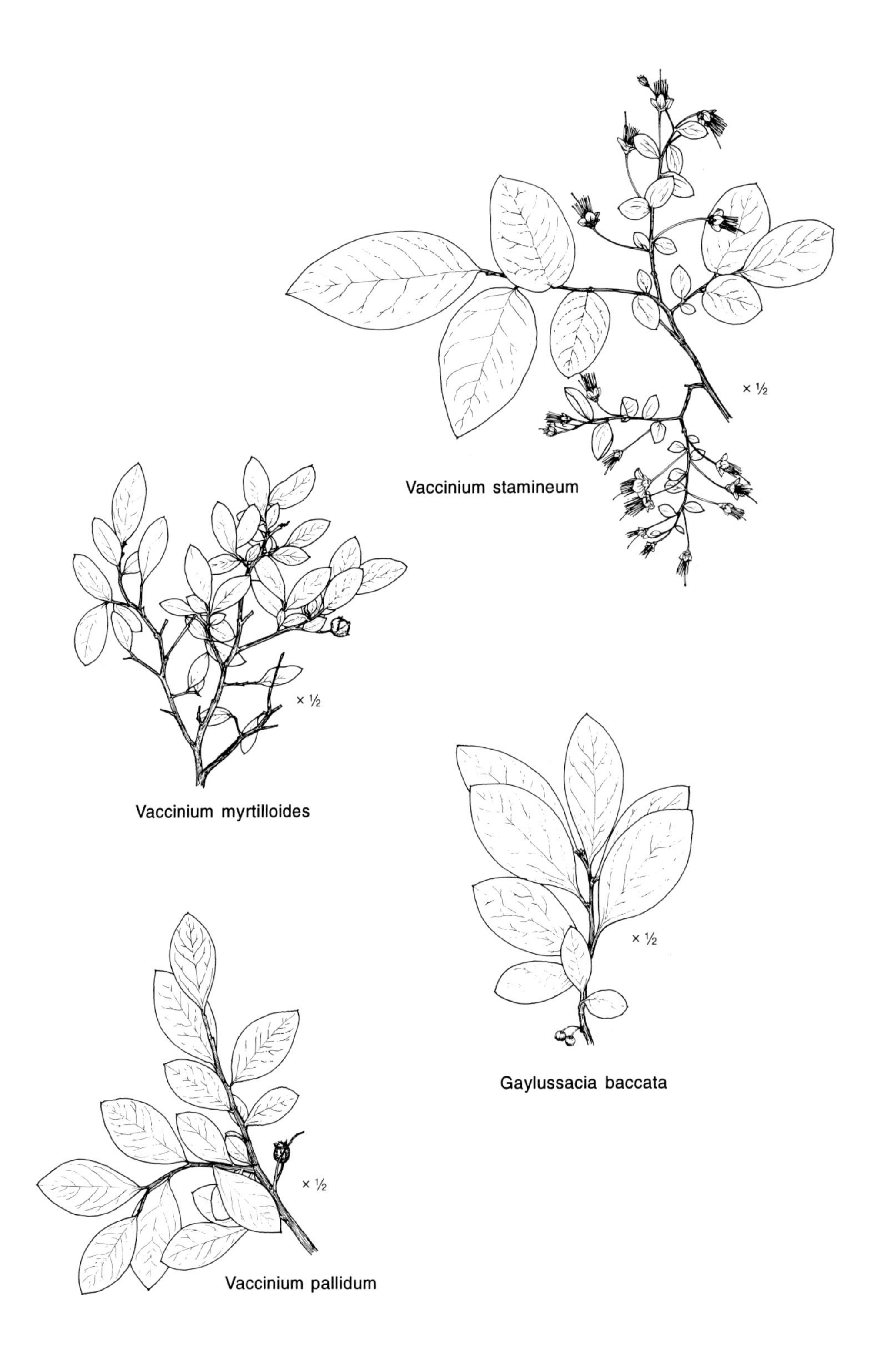
× ½
Vaccinium stamineum
× ½
Vaccinium myrtilloides
× ½
Gaylussacia baccata
× ½
Vaccinium pallidum

4 (–5) cm long, narrowly to broadly elliptic, 2–3 times as long as wide, margins serrulate to entire; flowers greenish to white, sometimes tinged with pink, in short racemes; sepals and petals 5; corolla cylindric, 4–8 mm long; stamens 10. Counties in the eastern half of Ohio, and a few western ones; dry woodlands and thickets, less often in mesic to moist woodlands, sometimes in marshes and bogs. April–June.

A few specimens can only arbitrarily be separated from the next species, *V. angustifolium*.

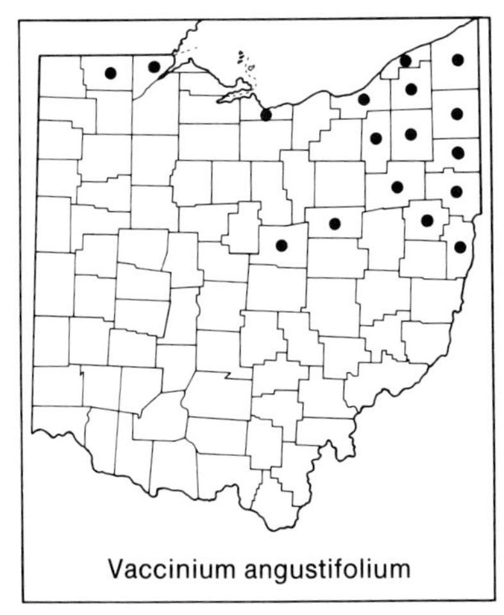

Vaccinium angustifolium

4. **Vaccinium angustifolium** Ait. Low Sugarberry. Sweet Lowbush Blueberry

incl. var. *angustifolium*, var. *hypolasium* Fern., var. *laevifolium* House, and var. *nigrum* (Wood) Dole; and incl. also *V. brittonii* Bickn. and *V. lamarckii* Camp

Low shrub, to ca. 0.3 m (1 foot) tall; leaves deciduous, 1.5–3.5 cm long, narrowly elliptic, 3 or more times as long as wide, margins minutely serrulate; flowers white, in short racemes; sepals and petals 5; corolla cylindric, 4–6 mm long; stamens 10. A plant of northeastern North America, ranging southward into northern Ohio and farther south in the mountains; dry woodlands, borders and openings, infrequently at bog margins. Hall et al. (1979) studied the biology of this species. April–May.

A few specimens can only arbitrarily be separated from *V. pallidum*, above.

Vaccinium corymbosum

5. **Vaccinium corymbosum** L. Highbush Blueberry

incl. var. *corymbosum*, var. *albiflorum* (Hook.) Fern., and var. *glabrum* A. Gray; and incl. also *V. atrococcum* (A. Gray) Heller and *V. simulatum* Small

Shrub, to ca. 3 m (10 feet) tall; leaves deciduous, 3–8 cm long, elliptic to subovate, margins entire to low-serrate; flowers white to pinkish, in racemes; sepals and petals 5; corolla cylindric, 6–11 mm long; stamens 10. A species of eastern United States, extending northward into southeastern Canada; northeastern counties and at a few sites in northwestern and southern Ohio; dry to moist thickets. April–May.

The treatment here, involving the synonymy given above, is based on the study of Vander Kloet (1980). Uttal (1987) has recently taken issue with this broad interpretation of *V. corymbosum*.

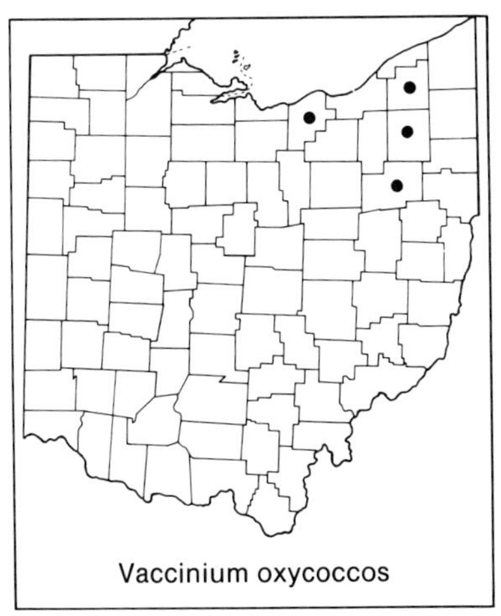

Vaccinium oxycoccos

6. **Vaccinium oxycoccos** L. Small Cranberry

incl. var. *oxycoccos* and var. *ovalifolium* Michx.

Slender trailing shrub, with delicate ascending branches to 2 dm long; leaves evergreen, tiny, 5–10 mm long, lance-elliptic to oblong, margins entire and revolute; flowers pink, solitary or few at stem tip; corolla deeply 4-lobed, ca. 6 mm long; stamens 8, exserted; berry small, 5–8 (–10) mm

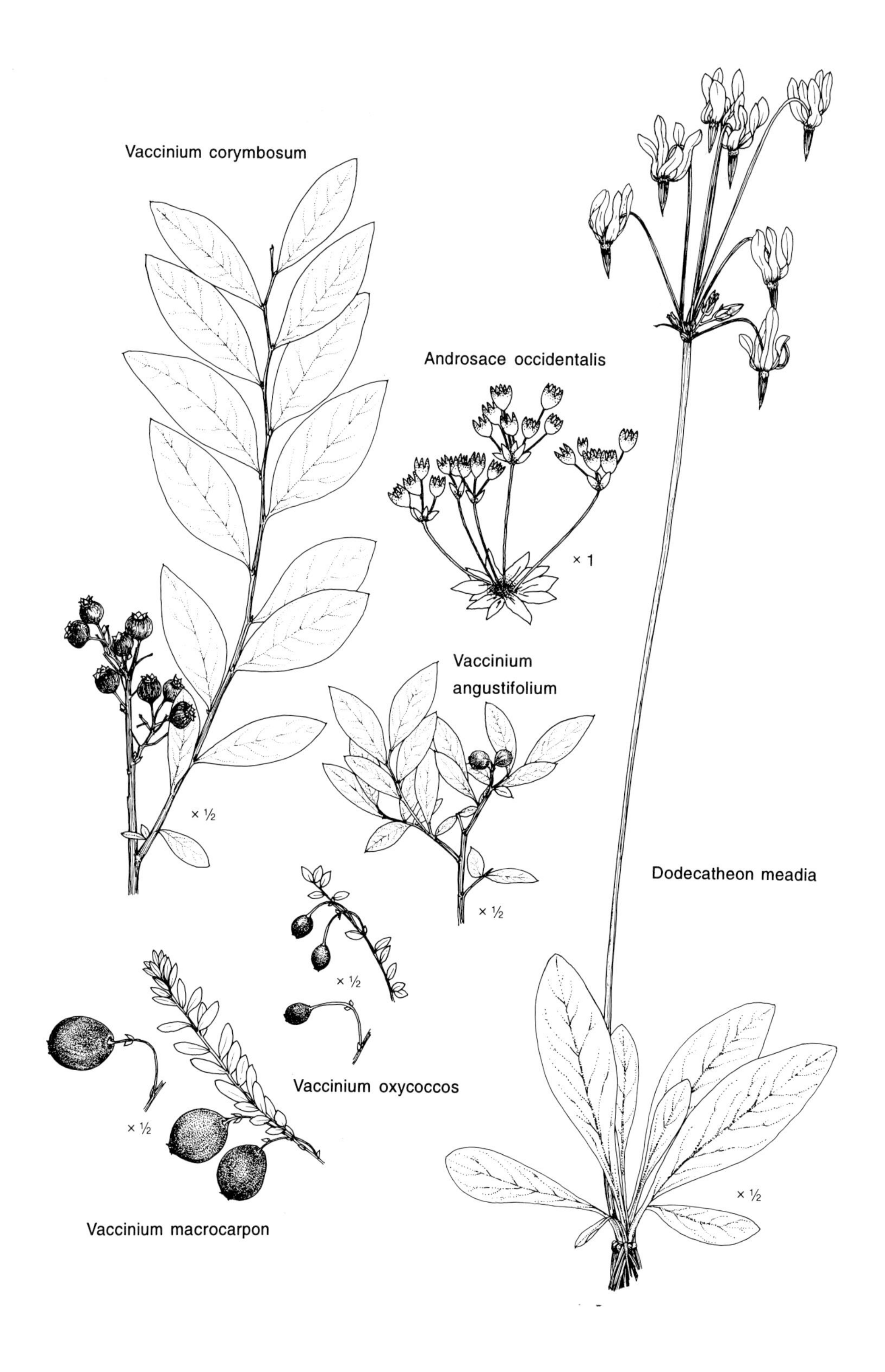
Vaccinium corymbosum
Androsace occidentalis
× 1
Vaccinium
angustifolium
× ½
× ½
Dodecatheon meadia
× ½
Vaccinium oxycoccos
× ½
Vaccinium macrocarpon
× ½

in diameter, the pedicel bearing near or below its midpoint two linear to narrowly lanceolate bractlets, each 1–2.5 mm long. A circumboreal species near a southern boundary of its range in northeastern Ohio; bogs. Additional record: Summit County (Cusick, 1991). June.

The epithet is sometimes misspelled *oxycoccus*.

7. **Vaccinium macrocarpon** Ait. CRANBERRY. LARGE CRANBERRY

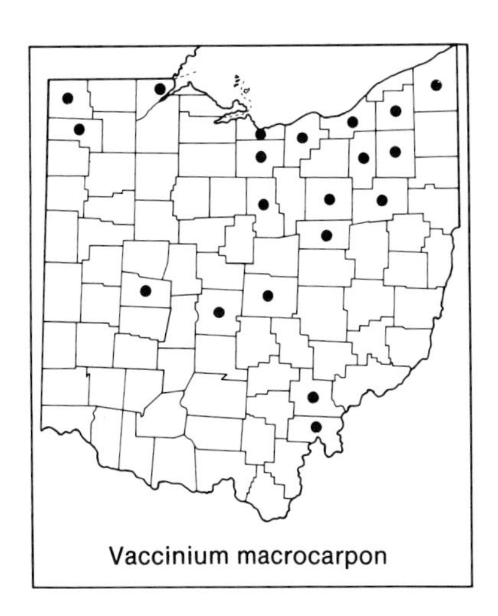
Vaccinium macrocarpon

Trailing shrub, with delicate ascending branches to 2 dm high; leaves evergreen, 6–15 mm long, narrowly to broadly elliptic, margins entire and revolute; flowers pink, originating from axils of reduced leaves at points below stem tip; corolla deeply 4-lobed, ca. 10 mm long; stamens 8, exserted; berry large, 1–2 cm in diameter, its pedicel bearing well above the midpoint two elliptic to ovate bractlets, each 3–5 mm long. A boreal species at a southern boundary of its native range in Ohio, chiefly in northeastern counties; naturalized at the southernmost stations, in Athens and Meigs counties, but native elsewhere; bogs. June–July.

The epithet is sometimes spelled *macrocarpum*.

Diapensiaceae. DIAPENSIA FAMILY

Galax urceolata (Poir.) Brummitt (*G. aphylla* of authors, not L.), WANDFLOWER, an ornamental from southeastern United States, was recently collected from Little Mountain, in Lake County (Bradt-Barnhart, 1987), where it has evidently escaped from plantings. It is an evergreen perennial with long-petiolate basal leaves and small white flowers borne in long-pedunculate, spikelike racemes. Wood and Channell (1959) studied the Diapensiaceae in southeastern United States. Brummitt (1972) discussed the nomenclature of this species.

Primulaceae. PRIMROSE FAMILY

Perennial or annual herbs; leaves simple; flowers perfect, solitary in leaf axils or in various types of inflorescences; calyx and corolla actinomorphic, (4–) 5 (–8) -lobed, often deeply so; stamens (4–) 5 (–8); carpels usually 5, fused to form a compound pistil with a single style; fruit a capsule. A family of about 1,000 species, with distribution centered in the North Temperate Zone. Channell and Wood (1959) studied the family in southeastern United States. Brockett and Cooperrider's (1983) study of the family in the Ohio flora is the basis for the treatment here.

Cyclamen coum Mill. and *C. persicum* Mill., CYCLAMEN, are popular house plants. In winter they bear numerous showy flowers, each separately long-stalked from a broad flattish tuber that sits at the surface of the soil; the flower colors range from white to pink, red, purple, and dark purple. Flowers of *Primula*, PRIMROSE, come in a host of bright colors and are popular ornamentals in flower gardens.

a. Plants aquatic, with axis of inflorescence conspicuously inflated; leaves deeply pinnately lobed, the numerous lobes linear . . . . 8. *Hottonia*

aa. Plants terrestrial although sometimes rooted in shallow water or marshy ground, axis of inflorescence not inflated; leaves entire or subentire, not deeply lobed.

b. Leaves all in a basal rosette; flowers in an umbel at the tip of a scape.

c. Plants minute, to 5 cm tall when in flower, annual; leaves mostly 0.2–1.0 cm long; scapes 2–5 cm long. . . . . 1. *Androsace*

cc. Plants larger, to 50 cm tall when in flower, perennial; leaves mostly 6–12 cm long; scapes 20–50 cm long . . . . 2. *Dodecatheon*

bb. Leaves cauline; flowers either solitary or in racemes or panicles.

c. Leaves (5–) 7 (–10) in a single terminal whorl; corolla lobes (6–) 7 (–8), white . . . . 4. *Trientalis*

cc. Leaves few to many, alternate to opposite or whorled, but not in a single terminal whorl; corolla lobes 4–6, white, yellow, pink, red, or blue.

d. Flowers and fruits sessile or subsessile, solitary in leaf axils; corolla pink. . . . . 6. *Centunculus*

dd. Flowers and fruits stalked, either solitary in leaf axils or in inflorescences; corolla white, yellow, red, or blue.

e. Leaves alternate; blades of principal leaves obovate; flowers in racemes; corolla white. . . . . 7. *Samolus*

ee. Leaves opposite, subopposite, whorled, or subwhorled; blades of principal leaves linear to ovate or orbicular; flowers in racemes or panicles or solitary; corolla yellow, red, or blue, rarely white.

f. Corolla either orangish-red to white or blue; leaves 1–2 cm long, sessile. . . . . 5. *Anagallis*

ff. Corolla yellow, rarely whitish; leaves either more than 2 cm long, or petiolate if less than 2 cm long. . . . . 3. *Lysimachia*

1. Androsace L. ROCK-JASMINE

A genus of about 100 species, mostly Eurasian. The species below is one of only a few that are native to North America.

1. **Androsace occidentalis** Pursh Western Rock-jasmine

Annual; leaves mostly 2–10 mm long, all in a basal rosette; flowers minute, in an umbel terminating a scape 2–5 cm long, the umbel subtended by tiny bracts; corolla salverform, white. A species of western United States and Canada, at an eastern boundary of its range in northwestern Ohio—where it is evidently native, although discovered only recently (Cusick, 1990); open sandy sites. May.

Androsace occidentalis

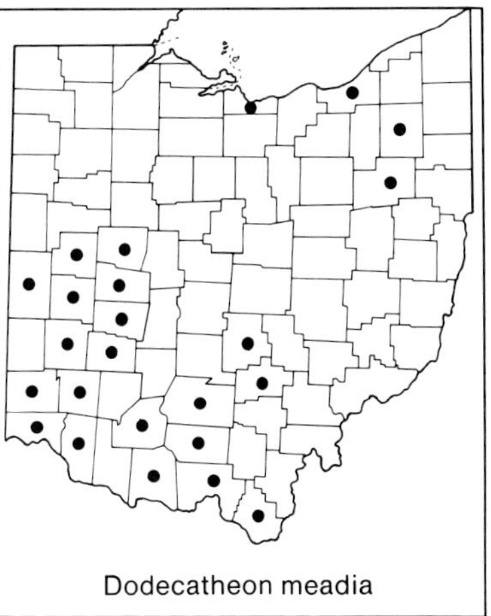

Dodecatheon meadia

2. Dodecatheon L. SHOOTING STAR

A chiefly North American genus of 14 species, one native to the Ohio flora. Fassett (1944) published a study of the taxa in eastern North America, and H. J. Thompson (1953) a biosystematic study of the entire genus.

1. **Dodecatheon meadia** L. Shooting Star. Pride of Ohio

Perennial; leaves mostly 6–12 cm long, all in a basal rosette; flowers nodding, in an umbel terminating a scape 2–5 dm long; corolla deeply 5-lobed, the lobes white to lavender, reflexed (therefore pointing skyward); anthers elongate, yellow, connivent; fruit erect. A species of the interior United States, ranging eastward through the southwestern quarter of Ohio, and with a few outlying stations in northeastern counties, the latter at a northern boundary of the species' range; moist to dry prairie remnants, woodland openings, dolomitic cliffs. April–June.

3. Lysimachia L. LOOSESTRIFE

Perennials; leaves opposite or whorled; flowers yellow or rarely whitish; calyx and corolla deeply 5- (6-) lobed, corolla more or less rotate. A cosmopolitan genus of 150 or more species. J. D. Ray's (1956) taxonomic study of the New World species includes all the Ohio taxa. Ingram (1960) has notes on the cultivated species of the United States. V. J. Coffey and S. B. Jones' (1980) biosystematic study of *Lysimachia* section *Seleucia* includes species nos. 8–10 below.

a. Stems creeping; principal leaves orbicular or suborbicular in outline; calyx valvate in bud. . . . . 7. *Lysimachia nummularia*

aa. Stems erect; principal leaves linear to lanceolate or ovate; calyx imbricate in bud.

b. Leaves at or near medial position on stem distinctly petiolate.

c. Petioles ciliate, otherwise glabrous or essentially so; leaves opposite.

d. Leaves at or near medial position on stem ovate-lanceolate to ovate . . . . 8. *Lysimachia ciliata* dd. Leaves at or near medial position on stem narrowly lanceolate or narrowly elliptic to nearly linear . . . . 9. *Lysimachia lanceolata*

cc. Petioles pubescent, but not ciliate; leaves opposite or whorled.

d. Flowers in terminal and axillary panicles; calyx lobes 3–4 mm long, green but bordered by a dark red submarginal line; corolla lobes with glabrous margins . . . . 1. *Lysimachia vulgaris*

dd. Flowers 1–3 in leaf axils; calyx lobes 7–9 mm long, green throughout; corolla lobes glandular-ciliate. . . . . . . 2. *Lysimachia punctata*

bb. Leaves at or near medial position on stem sessile or nearly so.

c. Leaves opposite or subopposite.

d. Flowers in pedunculate axillary racemes; corolla lobes narrowly oblanceolate to linear. . . . . 6. *Lysimachia thyrsiflora*

dd. Flowers solitary or in terminal racemes; corolla lobes lanceolate or wider.

e. Flowers in terminal racemes . . . . 5. *Lysimachia terrestris*

ee. Flowers solitary in leaf axils.

f. Leaves at or near medial position on stem linear, their margins often revolute . . . . 10. *Lysimachia quadriflora*

ff. Leaves at or near medial position on stem lanceolate to elliptic, their margins not revolute.

g. Leaves rarely opposite, mostly in whorls of 3–5; base of blade (and petiole if present) not ciliate, or at most with few scattered cilia . . . . 4. *Lysimachia* × *producta*

gg. Leaves opposite; base of blade (and petiole if present) conspicuously and regularly ciliate . . . . 9. *Lysimachia lanceolata*

cc. Leaves whorled or subwhorled.

d. Leaves pubescent across both surfaces; flowers in clusters of 1–3 in leaf axils; petals glandular-ciliate . . . . 2. *Lysimachia punctata*

dd. Leaves glabrous on both surfaces or pubescent toward base; flowers solitary in leaf axils or each subtended by a bract in a raceme; petals glabrous.

e. Flowers in an extended raceme . . . . . . 4. *Lysimachia* × *producta*

ee. Flowers solitary in leaf axils.

f. Blades of leaves at medial position on stem mostly more than 15 mm wide; leaves in whorls of (3–) 4–7; plant glabrous or pubescent . . . . 3. *Lysimachia quadrifolia*

ff. Blades of leaves at medial position on stem mostly less than 15 mm wide; leaves 3–4 (–5) per node; plant glabrous . . . . 4. *Lysimachia* × *producta*

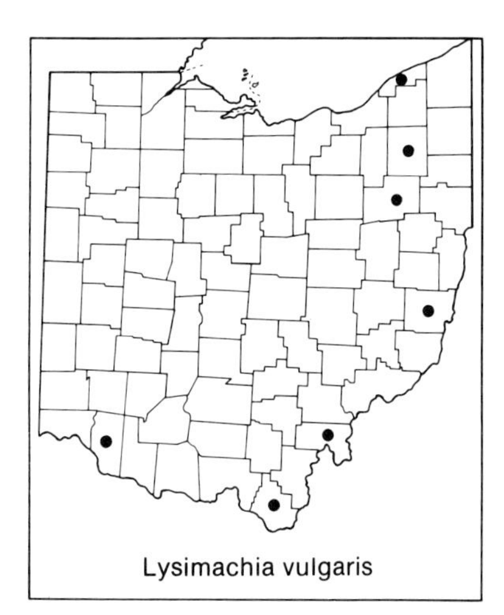
Lysimachia vulgaris

1. Lysimachia vulgaris L. Garden Loosestrife

Coarse perennial, to ca. 1 m tall; stems pubescent above; leaves in whorls of 3 (or 4) or opposite, medial leaves petiolate, the petioles pubescent; flowers ca. 2 cm across, in terminal and axillary panicles; calyx lobes 3–4 mm long, green but bordered by a dark red submarginal line; corolla lobes with glabrous margins. A Eurasian species occasionally cultivated in Ohio, and locally naturalized in a few eastern and southern counties; moist disturbed habitats. July–August.

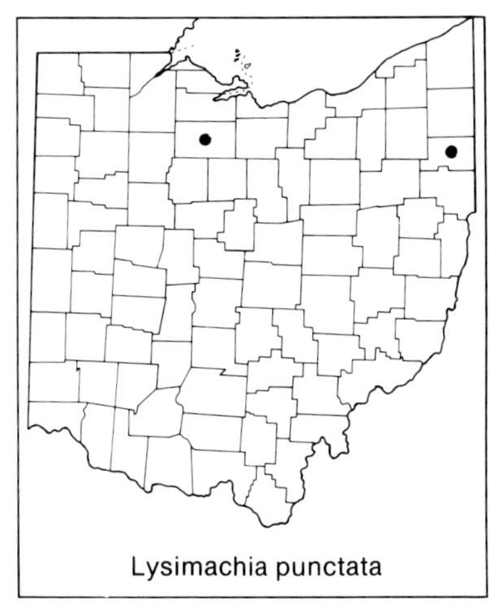
Lysimachia punctata

2. Lysimachia punctata L. Garden Loosestrife

Perennial, to 1 m tall; stems pubescent; leaves in whorls of 3 or 4, or the lower sometimes opposite; medial leaves petiolate, the petioles pubescent; flowers ca. 3 cm across, axillary, 1–3 in each axil; calyx lobes 7–9 mm long, green throughout; corolla lobes glandular-ciliate. A Eurasian species only infrequently cultivated in Ohio, and rarely escaped; locally naturalized in disturbed ground. Additional record: Hamilton County (J. D. Ray, 1956). June–August.

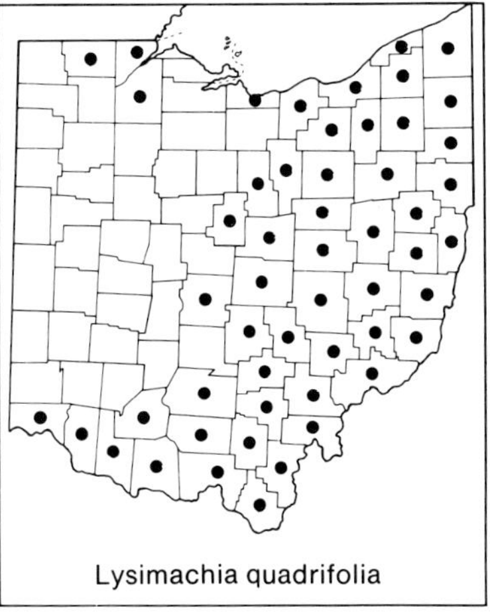
Lysimachia quadrifolia

3. **Lysimachia quadrifolia** L. Whorled Loosestrife

Perennial, to 8 dm tall; stems glabrous to pubescent, usually unbranched; leaves in whorls of (3–) 4 (–7), blades of medial leaves mostly more than 15 mm wide; flowers ca. 1.2 cm across, solitary in leaf axils. A species of east-central and northeastern United States and parts of adjacent southeastern Canada; common in counties of the eastern half of Ohio, with westward extensions across the northern and southern border counties, but absent from most of western Ohio; mesic to dry sites in woodland openings and borders, weedy thickets, roadside banks. Additional record: Greene County (J. D. Ray, 1956). June–July.

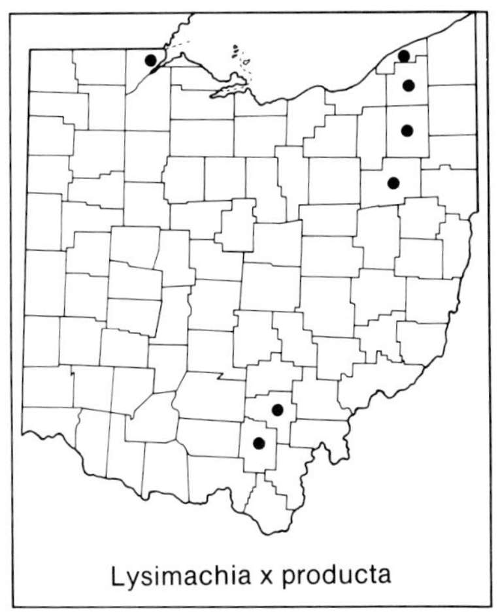
Lysimachia x producta

4. **Lysimachia** × **producta** (A. Gray) Fern.

Perennial, to 9 dm tall; stems usually glabrous, branched or unbranched; leaves mostly in whorls of 3, sometimes opposite or in whorls of 4 or 5, blades of medial leaves mostly less than 15 mm wide; flowers ca. 1.2 cm across, the lowest solitary in leaf axils, the leaves gradually reduced upward and the uppermost flowers bracteate, in all often with a more or less racemose appearance. The fertile hybrid of *Lysimachia quadrifolia* × *L. terrestris*, its morphology is generally intermediate betweeen those two species. Some specimens, presumably backcrosses or post-F_1 hybrid segregates, intergrade with one parental species or the other. Cooperrider and Brockett (1974, 1976) confirmed the hybrid status of this taxon, which was at one time treated as a taxonomic species. Known from a few scattered localities in Ohio, most in northeastern counties; disturbed, mostly mesic to moist sites near one or both parental species. June–July.

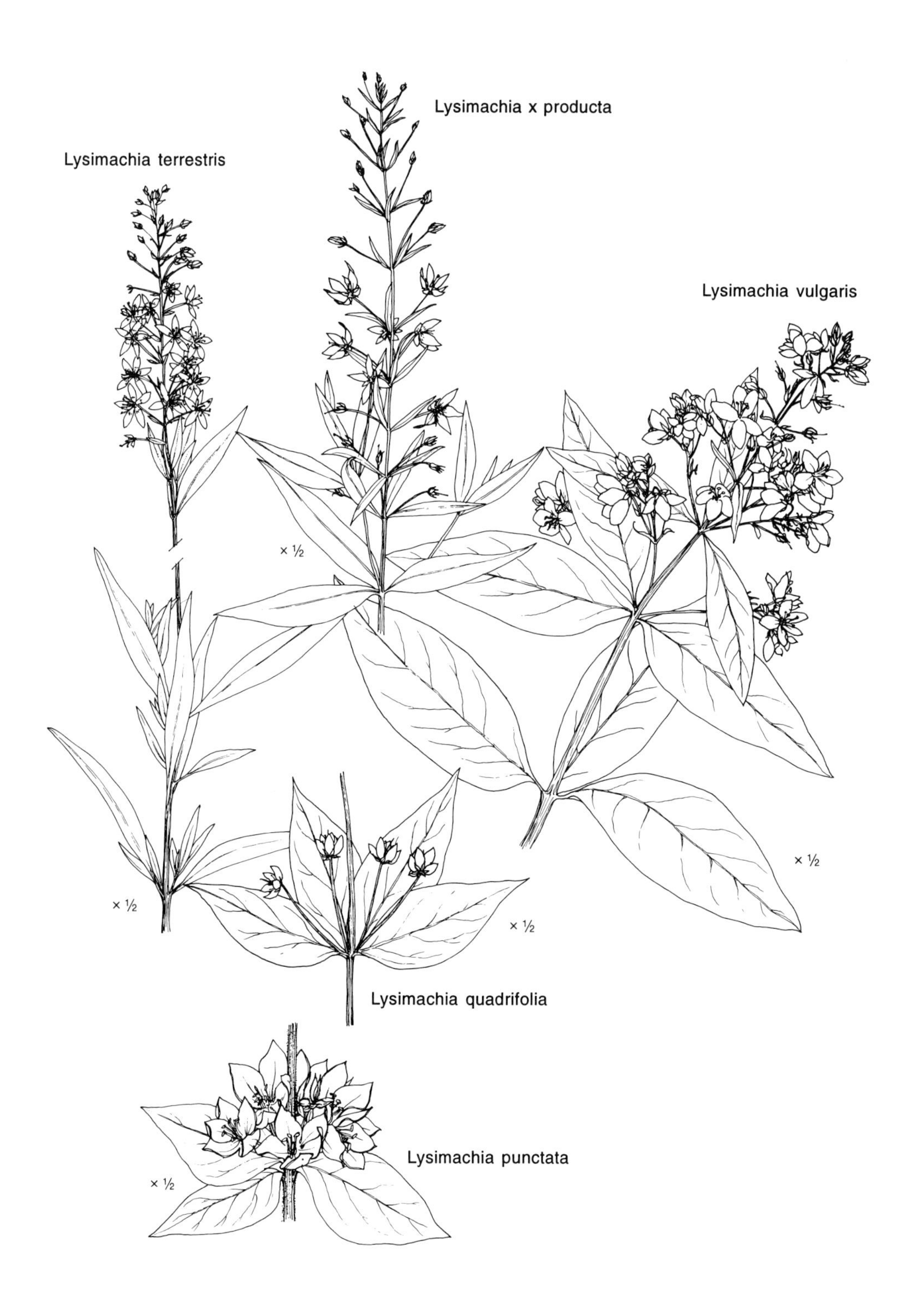
Lysimachia x producta
Lysimachia terrestris
Lysimachia vulgaris
× ½
× ½
× ½
× ½
Lysimachia quadrifolia
Lysimachia punctata
× ½

5. **Lysimachia terrestris** (L.) BSP. Swamp Candles
Perennial, to 1 m tall; stems glabrous, branched, sometimes in autumn producing narrow conelike bulblets in leaf axils, these functioning in asexual reproduction; leaves opposite; flowers ca. 1.2 cm across, in a terminal raceme. A species mainly of northeastern North America, but extending southward to east-central United States, and at a western boundary of its range in Ohio; chiefly in eastern counties; marshes, bogs, swamps, stream and lake margins, and other low moist sites. Additional records: Butler and Hamilton counties (J. D. Ray, 1956). June–July.

6. **Lysimachia thyrsiflora** L. Tufted Loosestrife
Naumburgia thyrsiflora (L.) Reichenb.
Perennial, to 7 dm tall; stems glabrous or slightly pubescent above; leaves opposite; flowers in axillary racemes; corolla ca. 8 mm across, usually 6-lobed. A circumboreal species ranging across northern United States and southern Canada, at a southern boundary of its range in Ohio; chiefly in northeastern counties; marshes, bogs, swamps, lake margins. May–July.

Although both presumed parental species are present, the putative hybrid *Lysimachia* × *commixta* Fern. (*L. terrestris* × *L. thyrsiflora*) has not been found in Ohio.

7. Lysimachia nummularia L. Moneywort
Creeping perennial; stems glabrous, prostrate, sometimes forming mats, the mats showy when many flowers are in bloom; leaves opposite, orbicular or suborbicular in outline; flowers 2–3 cm across, solitary in leaf axils. A European species sometimes cultivated and now naturalized throughout Ohio; moist woodlands, stream and lake margins, wet ditches, moist grassy sites, cemeteries, disturbed moist habitats. Late May–August.

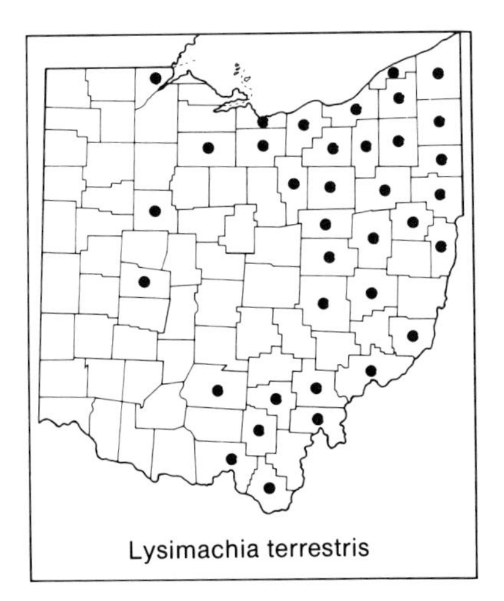
Lysimachia terrestris

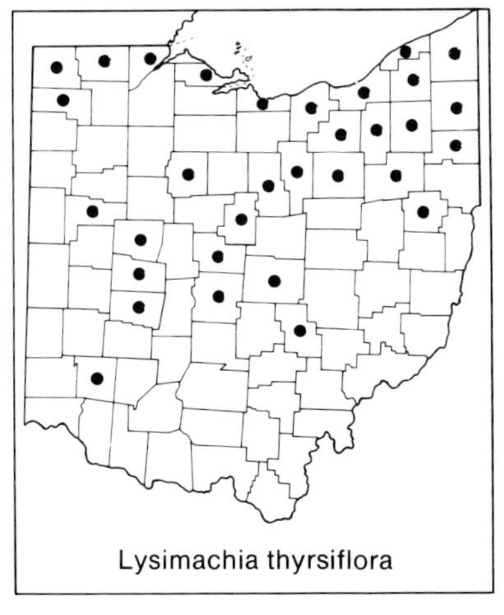
Lysimachia thyrsiflora

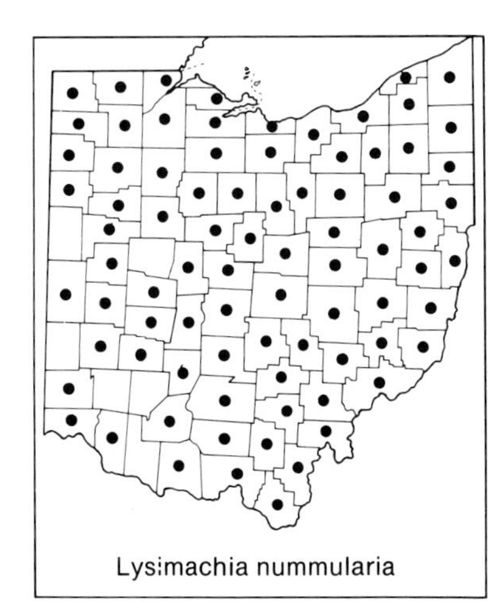
Lysimachia nummularia

8. **Lysimachia ciliata** L. Fringed Loosestrife
Steironema ciliatum (L.) Raf.

Perennial, to ca. 1 m tall; stems glabrous or nearly so; leaves opposite, those at medial position on stem petiolate, the petioles conspicuously ciliate, the blades ovate-lanceolate to ovate; flowers 2–2.5 cm across, solitary in axils of upper leaves. Throughout Ohio; damp fields, ditches, stream and lake margins, marshes, moist thickets, rich woodlands. June–August.

9. **Lysimachia lanceolata** Walt. Lance-leaved Loosestrife
Steironema lanceolatum (Walt.) A. Gray

Perennial, to 7 dm tall; stems glabrous or nearly so; leaves opposite, those at medial position on stem sessile to short-petiolate, the petioles when present and/or the base of blades ciliate, the blades narrowly lanceolate or narrowly elliptic to nearly linear; flowers 1.5–2 cm across, solitary in axils of upper leaves. A species of east-central, southeastern, and the interior United States, reaching in Ohio a northern boundary of its range; southern half of the state and northwestern counties; moist to dryish sites in woodland openings and borders, hillsides, grassy thickets, roadsides. June–August.

10. **Lysimachia quadriflora** Sims Linear-leaved Loosestrife
Steironema quadriflorum (Sims) Hitchc.

Perennial, to 7 dm tall; stems glabrous; leaves opposite, those at medial position on stem sessile, the blades linear, margins often revolute; flowers ca. 2 cm across, solitary in axils of upper leaves, but sometimes collectively appearing racemose or paniculate. Chiefly in western Ohio, with a few scattered localities in eastern counties; fens and other moist calcareous sites, limestone exposures. July–August.

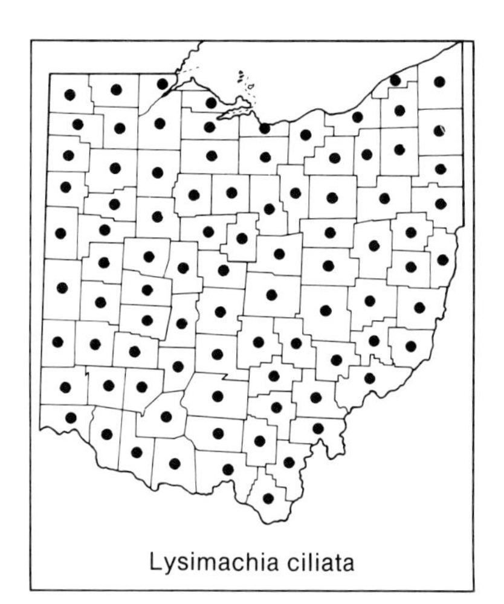
Lysimachia ciliata

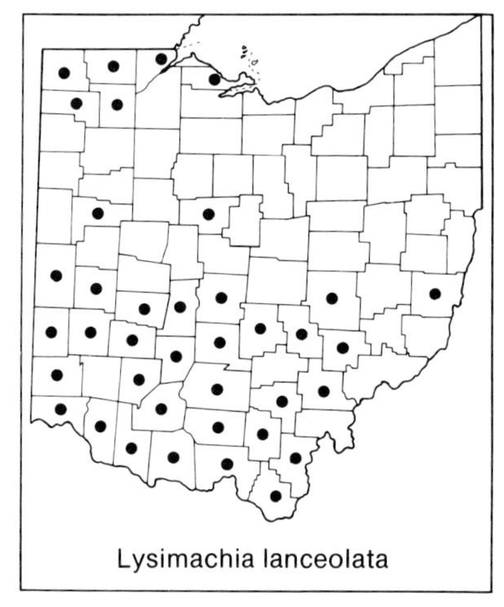
Lysimachia lanceolata

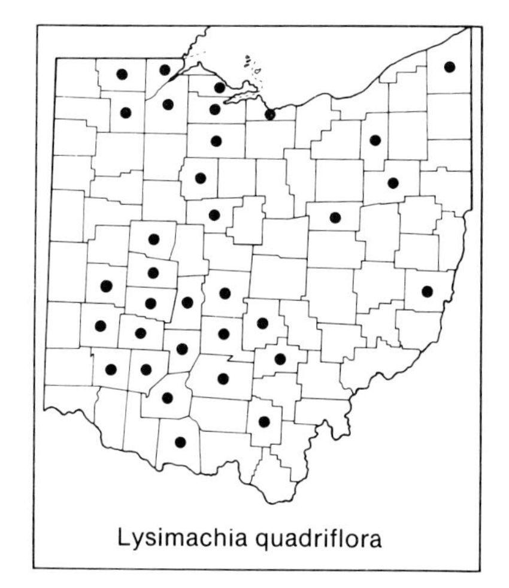
Lysimachia quadriflora

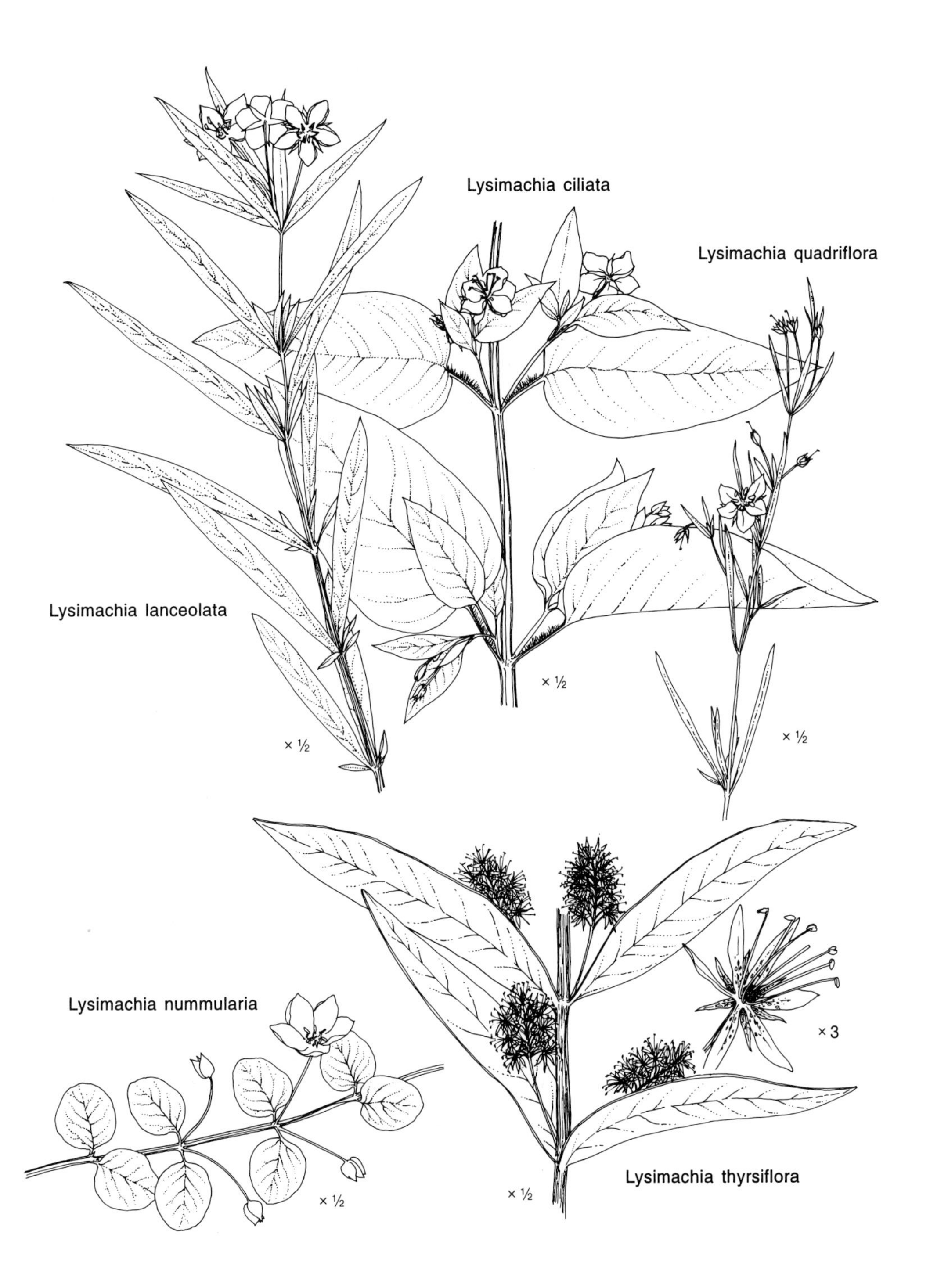
Lysimachia ciliata
Lysimachia quadriflora
Lysimachia lanceolata
× ½
× ½
× ½
Lysimachia nummularia
× 3
× ½
× ½
Lysimachia thyrsiflora

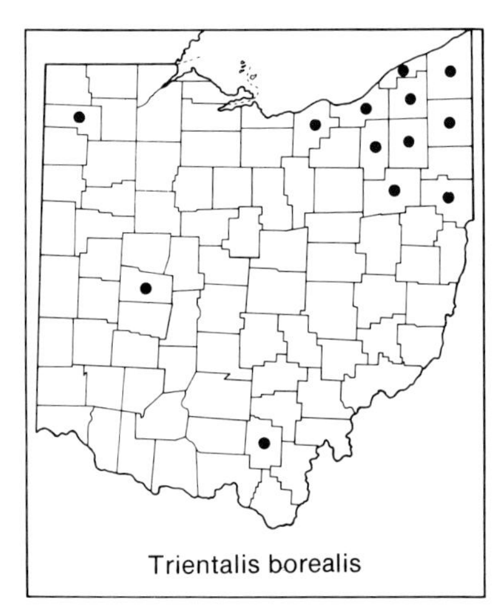
Trientalis borealis

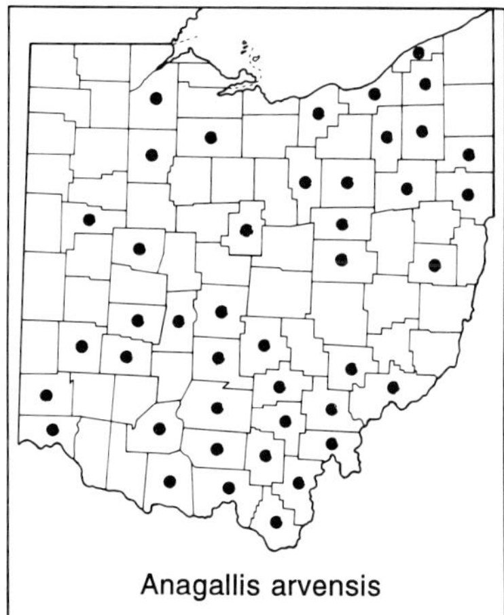
Anagallis arvensis

Centunculus minimus

4. Trientalis L. CHICKWEED-WINTERGREEN

A genus of three or four species native to temperate and colder parts of the Northern Hemisphere, with one species present in the Ohio flora.

1. **Trientalis borealis** Raf. STARFLOWER

Small perennial, to ca. 2 dm tall; leaves (5–) 7 (–10), in a single terminal whorl; flowers penduncluate, solitary in leaf axils; calyx and corolla deeply 6–8-lobed; corolla rotate, white. A northern species ranging across eastern and central Canada, southward into the northern states, and farther south in the mountains; chiefly in northeastern counties of Ohio, and with a few outliers elsewhere in the state; bogs, marshes, swamps, wet woodlands. Wilbur H. Duncan (1970) studied the species' distribution in the United States. Anderson and Loucks (1973) and Anderson and Beare (1983) studied aspects of the species' biology. May–early June.

5. Anagallis L. PIMPERNEL

A cosmopolitan genus of some 25 species, one of which is naturalized in the Ohio flora.

1. ANAGALLIS ARVENSIS L. PIMPERNEL

Simple to much-branched annual, 1–3.5 dm tall; leaves opposite, sessile; flowers solitary in leaf axils, the flowers on ascending peduncles but the fruits conspicuously nodding; calyx and corolla 5-lobed; corolla rotate. A weed of Eurasian origin, naturalized more or less throughout Ohio, especially in eastern and southern counties; grassy fields, lawns, gardens, roadside ditches. May–August (–October).

Two color forms, keyed below, are found in Ohio; the first is much the more frequent of the two.

a. Petals orangish-red to rarely white . . . . forma *arvensis*
   A. arvensis var. *arvensis*
   SCARLET PIMPERNEL

aa. Petals blue . . . . forma *caerulea* (Schreb.) Baumg.
   A. arvensis var. *caerulea* (Schreb.) Gren. & Godr.
   BLUE PIMPERNEL
   Adams, Auglaize, Butler, and Highland counties.

6. Centunculus L. CHAFFWEED

A genus with a single species. P. Taylor (1955) merges this genus under *Anagallis*.

1. **Centunculus minimus** L. CHAFFWEED

Anagallis minima (L.) E. H. Krause

Small annual, to ca. 1 dm tall; leaves alternate; flowers tiny, sessile or subsessile, borne in leaf axils; calyx and corolla 4- or 5-lobed; corolla rotate,

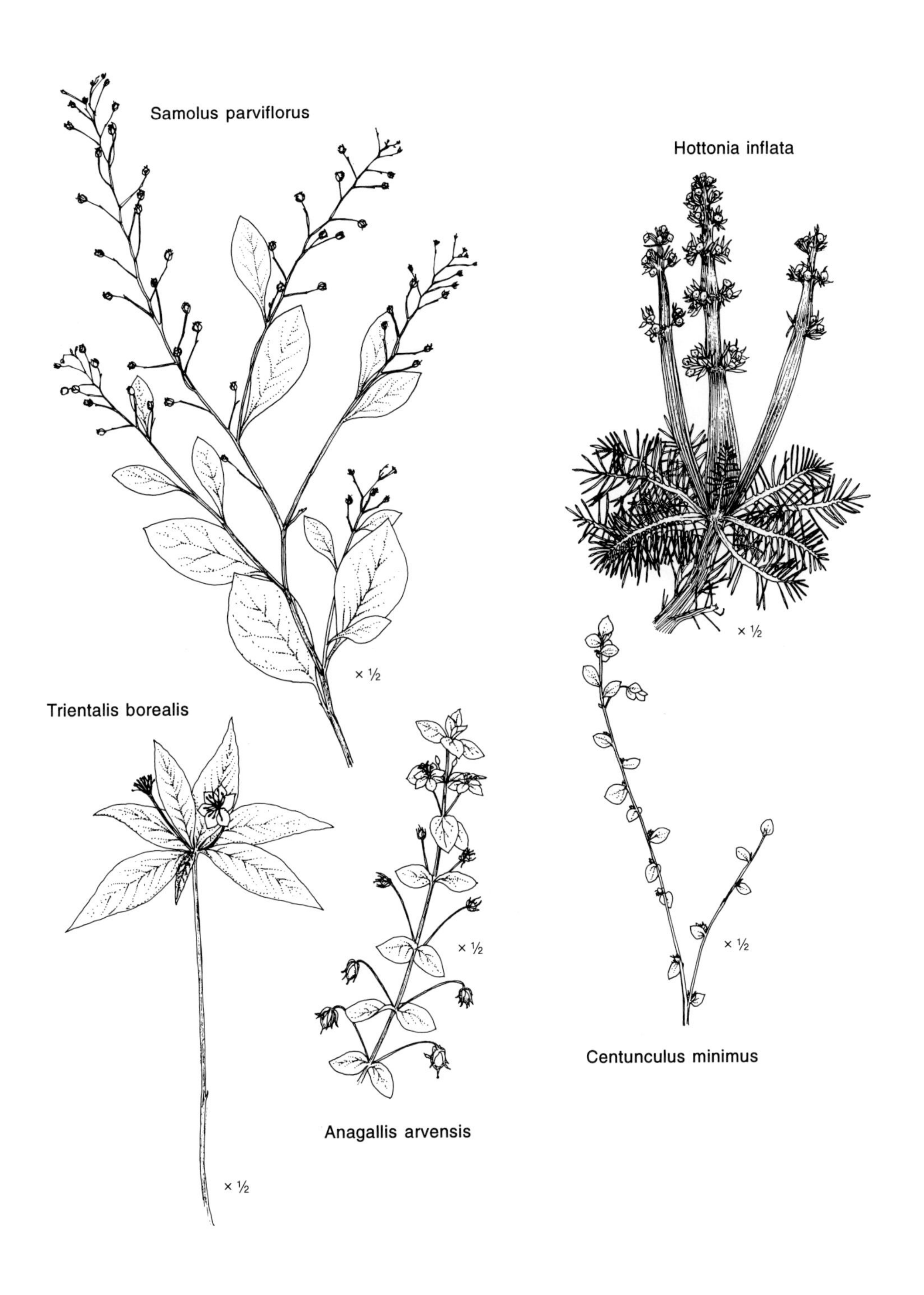
Samolus parviflorus
Hottonia inflata
× ½
× ½
Trientalis borealis
× ½
× ½
Centunculus minimus
Anagallis arvensis
× ½

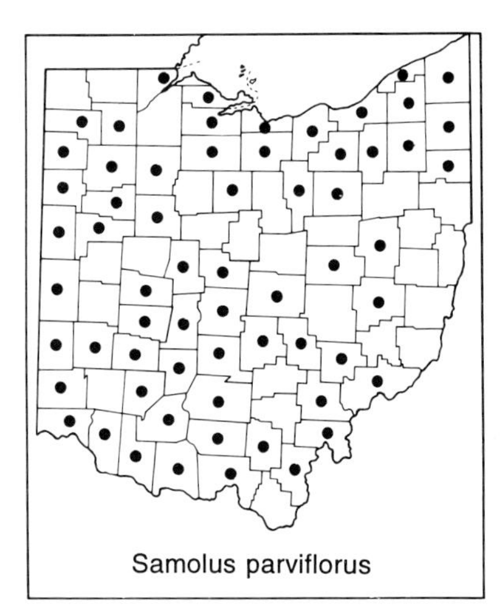
Samolus parviflorus

Hottonia inflata

pink. A rather cosmopolitan species, in Ohio known only from two collections in south-central counties, where the species is perhaps native; moist open sites. June–August.

Stuckey and Roberts (1977) treated the species as an alien member of the Ohio flora. Because of its small size and inconspicuous aspect, the species is easily overlooked.

7. Samolus L. WATER-PIMPERNEL. BROOKWEED

A widespread genus of about 10 species, one of which is native to Ohio.

1. **Samolus parviflorus** Raf. WATER-PIMPERNEL
 S. floribundus HBK.

Perennial, to 4 dm tall; leaves alternate; flowers small, in racemes (but these sometimes appearing paniculate); calyx and corolla 5-lobed; corolla more or less campanulate, white. Throughout much of Ohio; stream banks, lake and pond shores, moist woods, swamps. June–early September.

8. Hottonia L. FEATHERFOIL. WATER-VIOLET

A genus of two species, the other native to Eurasia.

1. **Hottonia inflata** Ell. FEATHERFOIL. WATER-VIOLET

An unusual-looking aquatic winter-annual, the stems hollow, swollen, jointed, inflated, and floating; leaves whorled to alternate, deeply pinnately dissected, with numerous linear lobes; flowers in racemes, the peduncle and axis of the raceme inflated; calyx and corolla 5-lobed; corolla shorter than the calyx, white. A species of eastern United States, at a northern boundary of its range in Ohio; a few localities in northern and southern counties; in shallow water of ponds and in wet ditches. May–early June.

Plumbaginaceae. LEADWORT FAMILY

Ceratostigma plumbaginoides Bunge, a blue-flowered perennial grown as a ground cover or edging plant, was reported from Ohio by Schaffner (1932) as a waif in Lake County, and the attribution repeated by Gleason (1952b). I have seen no specimens and know of no other reports of this species' appearance in the Ohio flora.

Ebenaceae. EBONY FAMILY

A family of a few hundred species, chiefly of the tropics and subtropics. Wood and Channell (1960) studied the family in southeastern United States.

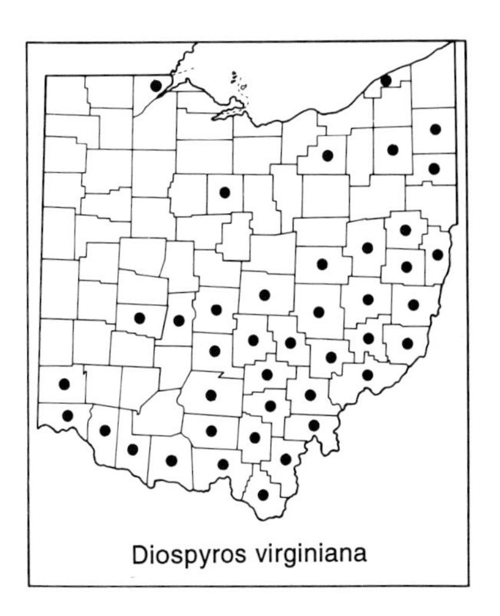
Diospyros virginiana

1. Diospyros L. PERSIMMON

A genus of some 250 or more species, one of which—known for its edible fruit—reaches Ohio. Two other important members of the genus are *Diospyros kaki* L. f., the large JAPANESE PERSIMMON sold in local markets, and the tropical tree *D. ebenum* J. Koenig, EBONY, famous for its heavy black heartwood.

1. **Diospyros virginiana** L. PERSIMMON
Tree, to 23 m (75 feet) tall; trunks of mature trees with checkered bark (the squares not so deep nor so large as those in mature specimens of *Nyssa sylvatica*), somewhat orangish in the grooves; leaves alternate, simple, entire, oblong-ovate; flowers minute, greenish to yellowish, solitary or in few-flowered clusters; flowers mostly imperfect, the trees functionally dioecious; fruit a subglobose berry, 3–4 cm in diameter, dull yellowish-orange, edible only when mature (after frost), the immature fruits noted for an astringency that causes the mouth to pucker. A species of southeastern United States, reaching a northern boundary of its native range in the southern half of Ohio. Braun (1961) concluded that specimens from northern counties were from adventive or naturalized individuals (see Skallerup, 1953, for a study of the species' distribution); dry woods and openings. June–July.

Herbarium specimens of this species are sometimes confused with those of *Nyssa sylvatica*; see notes under that species' text (p. 173) describing how the two may be distinguished from one another.

Styracaceae. STORAX FAMILY

Trees or shrubs; leaves alternate, simple; flowers perfect, white, showy, actinomorphic, solitary or in more or less racemose inflorescences; calyx and corolla 4- or 5-lobed. A small, widely distributed family of about 10 genera and 125–150 species, with distribution centered in eastern Asia. Wood and Channell (1960) studied the family in southeastern United States; Spongberg (1976) described those species hardy in temperate North America, including all three species represented in the Ohio flora.

Halesia carolina

Styrax grandifolius

a. Calyx and corolla lobes 4; fruit ellipsoid, 25–35 mm long, conspicuously 4-winged . . . . 1. *Halesia*
aa. Calyx and corolla lobes 5; fruit subglobose, 6–9 mm long, not 4-winged . . . . 2. *Styrax*

1. Halesia L. SILVER BELL

A genus of four or five species native to southeastern United States and eastern China, and one of them native to southernmost Ohio.

1. **Halesia carolina** L. CAROLINA SILVER BELL
H. tetraptera Ellis

Tree, to ca. 10 m (33 feet) tall; leaves alternate, simple, elliptic to oblong, margins low-serrate to subentire; flowers white, bell-shaped; calyx and corolla 4-lobed; fruit elongated, 2.5–3.5 cm long, conspicuously 4-winged. A southeastern species, reaching a northern boundary of its range in south-central Ohio (Chester, 1967); woodlands. Frequently cultivated in Ohio, the tallest cultivated trees attaining heights of ca. 15 m (50 feet). April–May.

Reveal and Seldin (1976) maintain that the correct name for this species is *H. tetraptera.*

2. Styrax L. STORAX

Shrubs or small trees; calyx and corolla 5-lobed; fruit subglobose, 6–9 mm long. A genus of about 100 species, chiefly in the tropical and warmer regions of the Northern Hemisphere. The most recent revision of the genus is that by Gonsoulin (1974). *Styrax japonicus* Sieb. and Zucc., JAPANESE SNOWBELL, is cultivated in Ohio.

a. Leaves broadly obovate to broadly ovate; leaf apices broadly obtuse to short-acuminate; undersurface of leaves gray with dense stellate pubescence; axes of inflorescences and pedicels also densely stellate-pubescent . . . . 1. *Styrax grandifolius*
aa. Leaves elliptic or narrowly elliptic; leaf apices acute to acuminate; undersurface of leaves glabrous or sparsely stellate-pubescent; axes of inflorescences and pedicels more or less scurfy with sparse stellate pubescence . . . . 2. *Styrax americanus*

1. **Styrax grandifolius** Ait. BIG-LEAVED SNOWBELL

Shrub or small tree, to 8 m (26 feet) tall; leaves broadly obovate to ovate; leaf apices broadly obtuse to short-acuminate, undersurface of leaves gray with dense stellate pubescence; flowers usually in racemes, rarely solitary, axes of inflorescences and also the pedicels densely stellate-pubescent. A

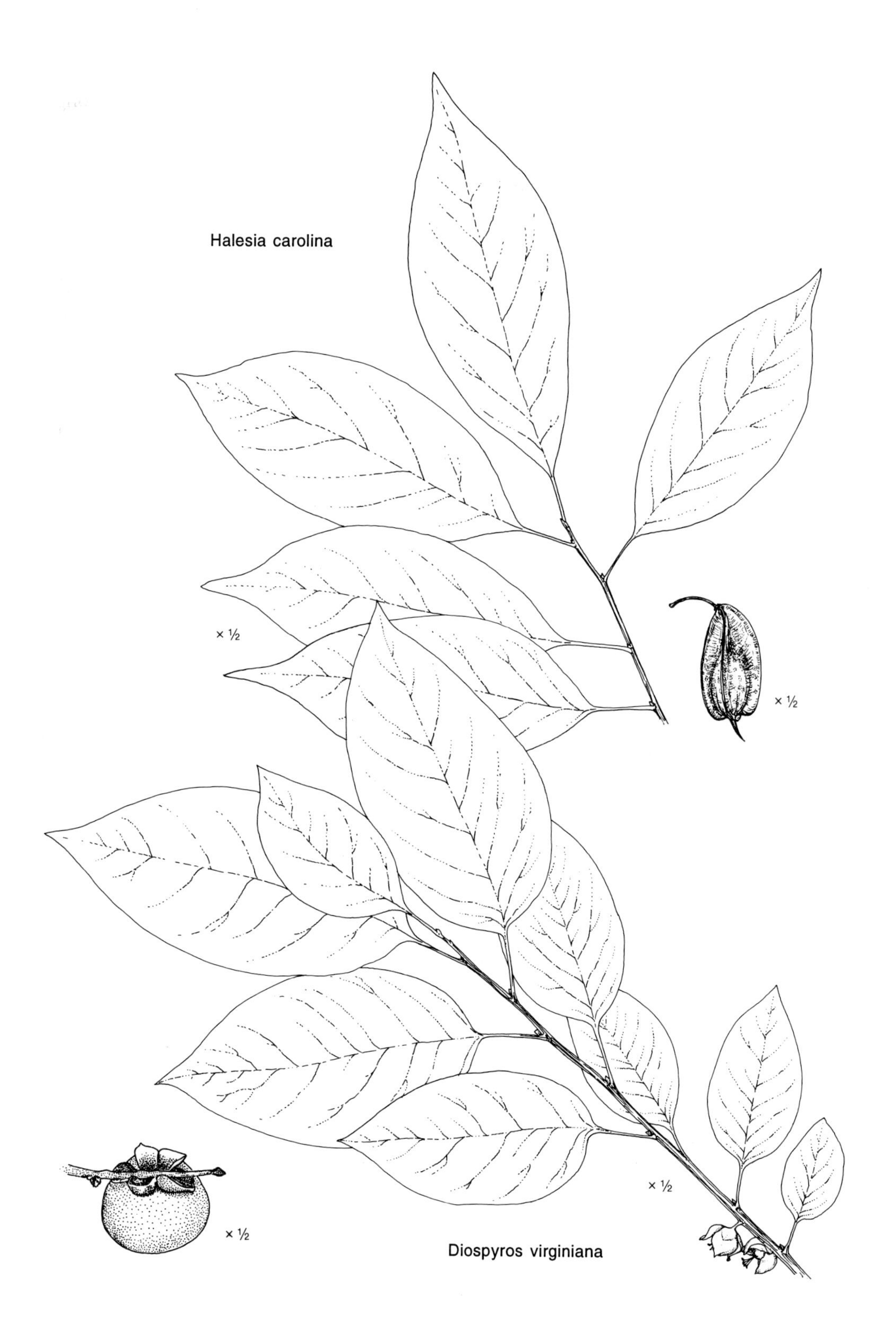
Halesia carolina
× ½
× ½
× ½
× ½
Diospyros virginiana

species of southeastern United States, at a northern boundary of its range in Athens County, in southeastern Ohio; woodland openings. May.

2. **Styrax americanus** Lam. SNOWBELL

Shrub, to 3 m (10 feet) tall; leaves elliptic or narrowly elliptic, leaf apices acute to acuminate, undersurface of leaves glabrous or sparsely stellate-pubescent; flowers solitary or in racemes, axes of inflorescences and also the pedicels more or less scurfy with sparse stellate pubescence. A species chiefly of southeastern United States, ranging to northern Indiana, but also at a boundary of its range in Pike County, in southern Ohio; alluvial woods. May.

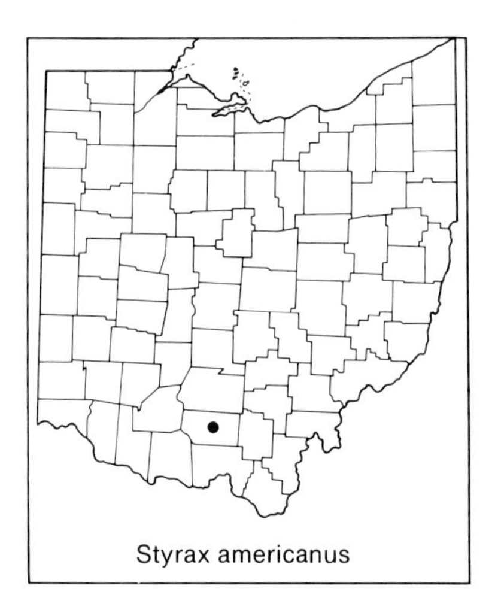
Styrax americanus

Oleaceae. OLIVE FAMILY

Trees and shrubs; leaves opposite, simple or compound; flowers perfect or less often imperfect, actinomorphic, in various types of sometimes showy inflorescences; calyx and corolla 4-lobed; fruits various. A family of about 600 species, chiefly of temperate regions. K. A. Wilson and C. E. Wood (1959) and J. W. Hardin (1974) studied the family in southeastern United States.

The fruits of the Mediterranean OLIVE, *Olea europaea* L., are the source of olive oil. Cultivars of *Forsythia* × *intermedia* Zabel, FORSYTHIA or GOLDENBELLS, and of its parental species, *F. suspensa* (Thunb.) Vahl and *F. viridissima* Lindl., all of Asian origin, are among Ohio's most admired flowering shrubs, producing numerous bright yellow flowers in early spring, often before the last snows have melted. Adventives of the hybrid have been collected from Fairfield, Meigs, and Ottawa counties.

a. Trees; leaves pinnately compound; corolla none, flowers inconspicuous; fruit an elongated samara . . . . 1. *Fraxinus*

aa. Shrubs or small trees to 10 m tall; leaves simple; corolla present, flowers conspicuous; fruit a capsule or drupe.

b. Corolla yellow; flowers in lateral clusters, appearing before the leaves; leaves serrate; fruit a capsule; see above . . . . *Forsythia*

bb. Corolla white, lilac, purple, or blue; flowers in lateral or terminal panicles, appearing with or after the leaves; leaves entire; fruit a capsule or drupe.

c. Leaves broadly elliptic to oblong or slightly oblong-ovate and 8–20 cm long; corolla white, lobed nearly to its base, the lobes elongate and narrowly linear; flowers in lateral panicles; fruit a drupe . . . . 3. *Chionanthus* cc. Leaves either elliptic to oblong and to 7 cm long, or ovate and to 12 cm long; corolla white, lilac, or purple to blue, lobed halfway to the base or less, the lobes oblong to ovate; flowers in terminal panicles; fruit a drupe or capsule.

d. Leaves ovate to broadly ovate, subcordate at base; flowers mostly lilac to purple, infrequently blue or white; fruit a capsule . . . . 2. *Syringa*

dd. Leaves elliptic to oblong, not subcordate at base; flowers white; fruit a dark drupe . . . . 4. *Ligustrum*

1. Fraxinus L. ASH

Trees; leaves pinnately compound; fruit a distinctive elongated paddle-shaped samara. Most of Ohio's species are dioecious, but *F. quadrangulata* has perfect flowers. A genus of some 65 species, chiefly of the North Temperate Zone. G. N. Miller (1955) made a taxonomic study of the genus in North America, north of Mexico. The text below is based on Braun (1961).

a. Branchlets acutely 4-angled and sometimes narrowly 4-winged; leaflets on short stalks 1–5 mm long; samaras 3–4 cm long, elliptic to narrowly oblong-ovate, 8–11 mm wide, the wing decurrent almost to base of fruit, the samaras usually notched at apex . . . . 4. *Fraxinus quadrangulata*

aa. Branchlets terete or nearly so; leaflets stalked or sessile; samaras 3–8 cm long, linear to spatulate or oblong, 3–13 mm wide, the wing decurrent to base of fruit or not, the samaras sometimes notched at apex, usually not.

b. Leaflets 7–11, the lateral ones sessile; undersurface of leaflets with swatch of red to pale pubescence at base of blade and extending onto the rachis; samaras narrowly oblong to oblong-elliptic, 3–5 cm long, 5–9 mm wide, the wing decurrent to base of fruit; calyx none . . . . 5. *Fraxinus nigra*

bb. Leaflets 5–9, the lateral ones with stalks 1 mm or longer; undersurface of leaflets pubescent or glabrous at base; samaras linear to linear-oblong or spatulate, 3–8 cm long, 4–13 mm wide, the wing at most decurrent to slightly below middle of fruit; calyx present, persistent in fruit.

c. Samaras 5–8 cm long, 7–13 mm wide; leaflets lanceolate to broadly lanceolate-elliptic or oblong-elliptic, pubescent beneath, on stalks 5–13 mm long . . . . 2. *Fraxinus profunda* cc. Samaras 3–4.5 cm long, 4–7 mm wide; leaflets lanceolate to ovate, pubescent or glabrous beneath, on stalks 1–15 mm long.

d. Stalks of lateral leaflets 3–15 mm long; the leaflets ovate to ovate-lanceolate or oblong-lanceolate, their undersurfaces pale green to whitish and more or less glaucous, and their margins somewhat serrate or dentate to often subentire; samaras linear, the wing not or only slightly decurrent along body of the fruit, samaras sometimes shallowly notched at apex . . . . 1. *Fraxinus americana*

dd. Stalks of lateral leaflets 1–5 mm long; the leaflets lanceolate to elliptic-lanceolate or subelliptic, their undersurfaces green, and their margins mostly serrate except toward base; samaras linear to lanceolate or spatulate, the wing narrowly decurrent along body of fruit for about 1/3–1/2 its length, the samaras not notched at apex . . . . 3. *Fraxinus pennsylvanica*

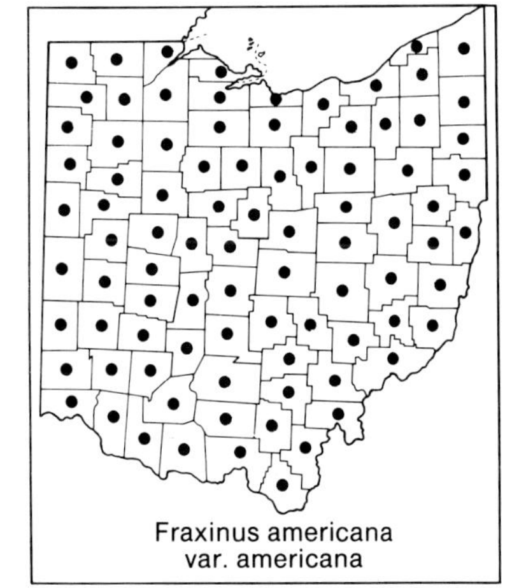
Fraxinus americana var. americana

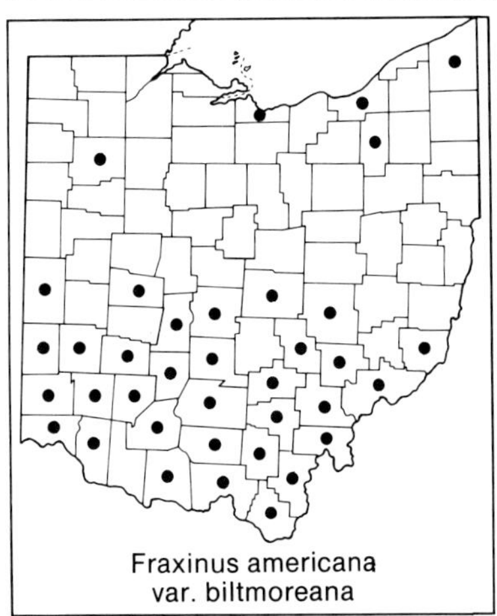
Fraxinus americana var. biltmoreana

1. **Fraxinus americana** L. White Ash

Tree, to 30 m (100 feet) tall; leaflets 5–9, ovate to ovate-lanceolate to oblong-lanceolate, undersurface pale green to whitish and more or less glaucous, margins somewhat serrate to dentate to often subentire, lateral leaflets with stalks 3–15 mm long; flowers imperfect, the plants dioecious; samaras linear, 3–4.5 cm long, 4–7 mm wide, the wing not or only slightly decurrent along the body of the fruit, the apex sometimes shallowly notched. Throughout Ohio; woodlands, and occurring occasionally as an isolated individual tree in open areas. May.

Most Ohio specimens can be assigned to one of the two varieties keyed below, but there is some intergradation. Taxonomic treatments of the two taxa differ widely. In *Hortus Third* (Bailey Hortorium Staff, 1976) the two are treated as separate species. At the other extreme, J. W. Hardin (1974) gives them no separate rank, writing that "there is neither a real morphological discontinuity . . . nor a clear ecological or geographical separation" between the two. In Ohio, the two taxa are somewhat different in geographical distribution; var. *americana* grows throughout the state, while var. *biltmoreana* occurs mostly in southern counties.

a. Young branches, petioles, leaf rachises, and undersurface of leaflets glabrous or nearly so . . . . var. *americana*
incl. var. *microcarpa* A. Gray

aa. Young branches, petioles, leaf rachises, and undersurface of leaflets pubescent . . . . var. *biltmoreana* (Beadle) J. S. Wright
F. biltmoreana Beadle
Biltmore Ash

In preparing the distribution maps, intermediate forms having pubescence on the petioles and rachises, but not on the branchlets, were as-

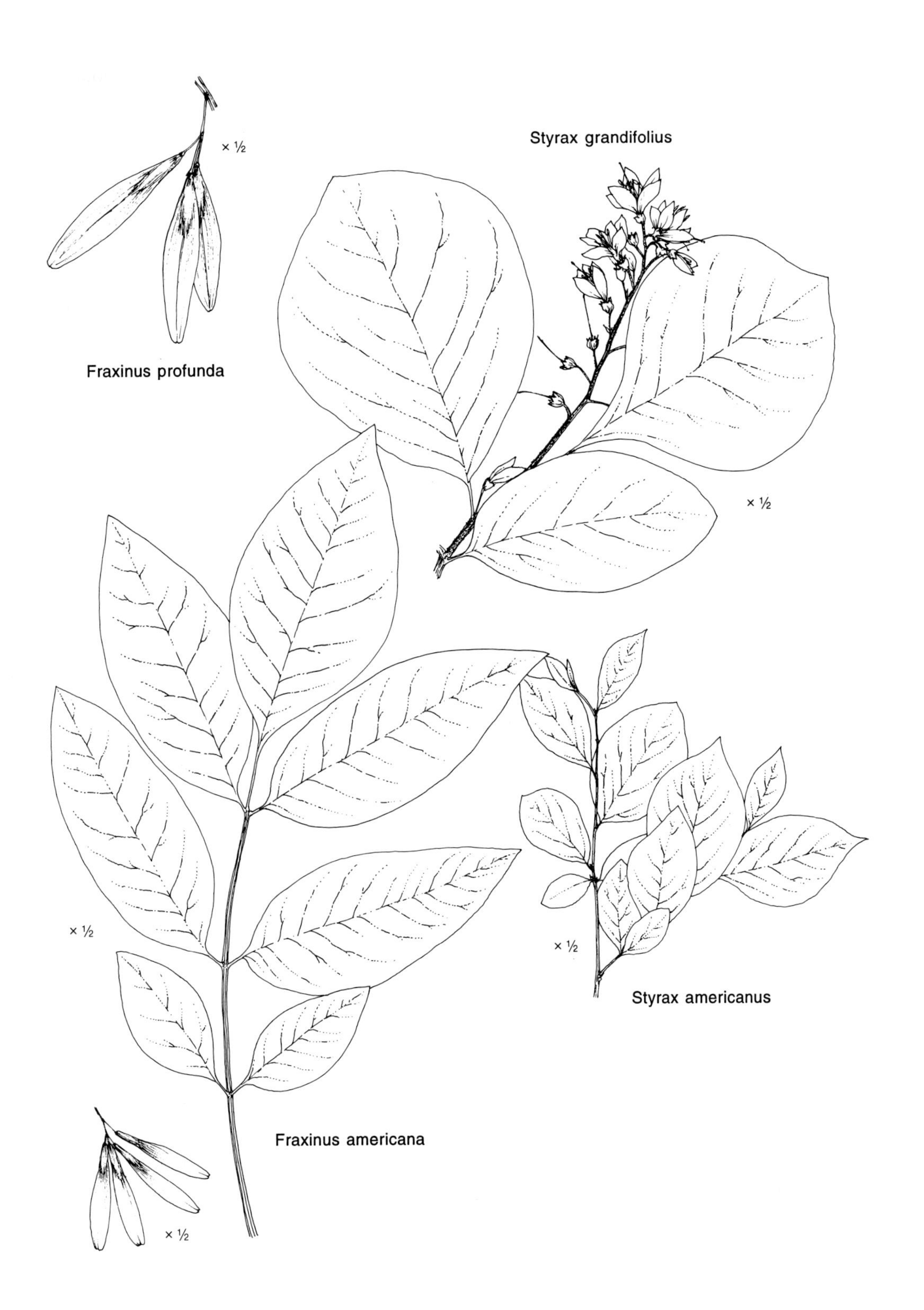
× ½
Styrax grandifolius
Fraxinus profunda
× ½
× ½
× ½
Styrax americanus
Fraxinus americana
× ½

signed to var. *biltmoreana*. Those specimens with pubescence only on the under-leaf surface were assigned to var. *americana*.

2. **Fraxinus profunda** (Bush) Bush Pumpkin Ash

   F. tomentosa Michx. f.

Tree, to ca. 15 m (50 feet) tall; leaflets 7–9, lanceolate to broadly lanceolate-elliptic or oblong-elliptic, undersurface pubescent, margins subentire, the lateral leaflets with stalks 5–13 mm long; flowers imperfect, the plants dioecious; samaras linear-oblong, 5–8 cm long, 7–13 mm wide, wing decurrent to below the middle of the fruit body, the apex blunt and sometimes notched. A species of eastern United States, at a northern boundary of its range in Ohio; swampy woods in western counties. May.

Wilson and Wood (1959) write that this "is a species of very questionable status." J. W. Hardin (1974) accepts it as a species, but observes that it is "possibly either an autopolyploid of *F. pennsylvanica* or an allopolyploid derivative of *F. americana* × *F. pennsylvanica*."

For the past several decades, this species has been called *F. tomentosa*.

3. **Fraxinus pennsylvanica** Marsh.

Tree, to 30 mm (100 feet) tall; leaflets 5–7 (–9), lanceolate to elliptic-lanceolate or subelliptic, undersurface green, margins mostly serrate (sometimes subentire) except toward base, the lateral leaflets with short stalks 1–5 mm long; flowers imperfect, the plants dioecious; samaras linear to lanceolate or spatulate, 3–4.5 cm long, 4–7 mm wide, the wing narrowly decurrent along body of fruit for about 1/3–1/2 its length, the apex not notched. Throughout Ohio; usually in low moist woods along streams, but seen occasionally in other wooded habitats or as an isolated tree in somewhat open areas. April-May.

J. W. Hardin (1974) recognizes no varieties within this species, writing "there appears to be no real discontinuity between any of the forms." I found among Ohio specimens very few intermediates between the two varieties keyed below, although both occur throughout the state.

a. Young branches, petioles, leaf rachises, and usually undersurface of leaflets pubescent .... var. *pennsylvanica*
   incl. var. *austinii* Fern.
   Red Ash.

aa. Young branches, petioles, leaf rachises, and undersurface of leaflets glabrous.................... var. *subintegerrima* (M. Vahl) Fern.
   F. pennsylvanica var. *lanceolata* (Borkh.) Sarg.; *F. lanceolata* Borkh.
   Green Ash.

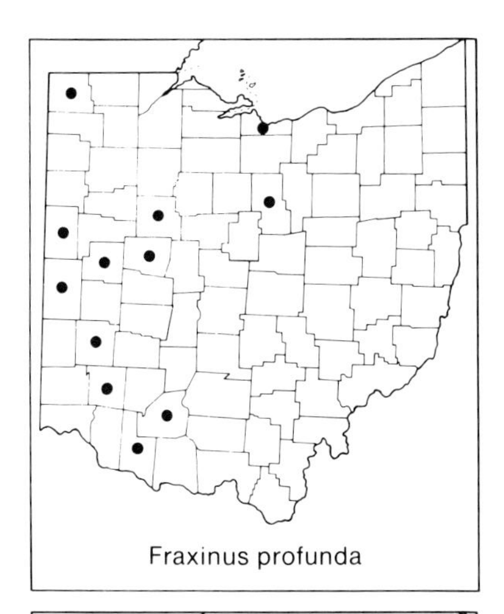
Fraxinus profunda

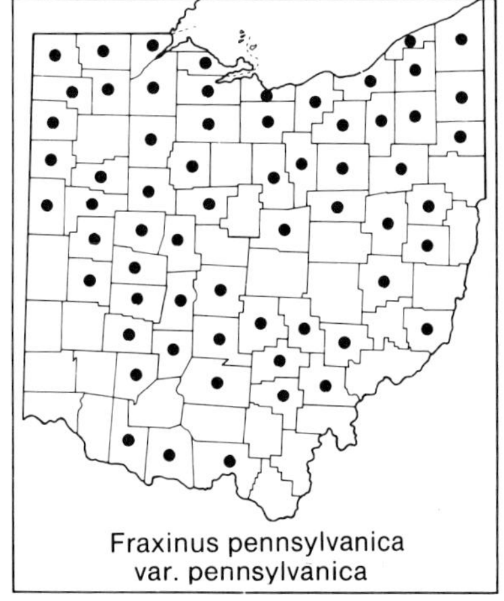
Fraxinus pennsylvanica var. pennsylvanica

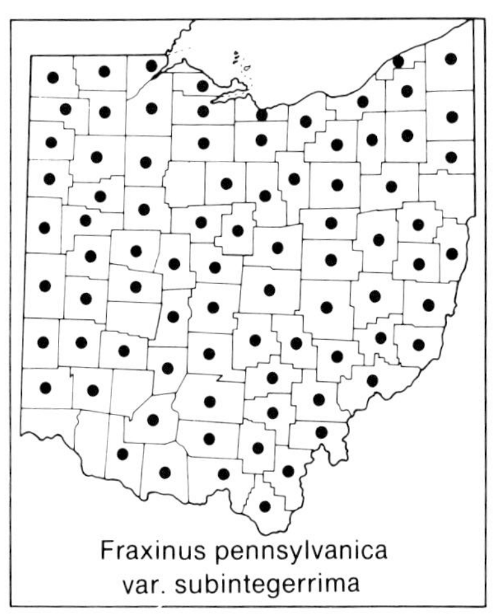
Fraxinus pennsylvanica var. subintegerrima

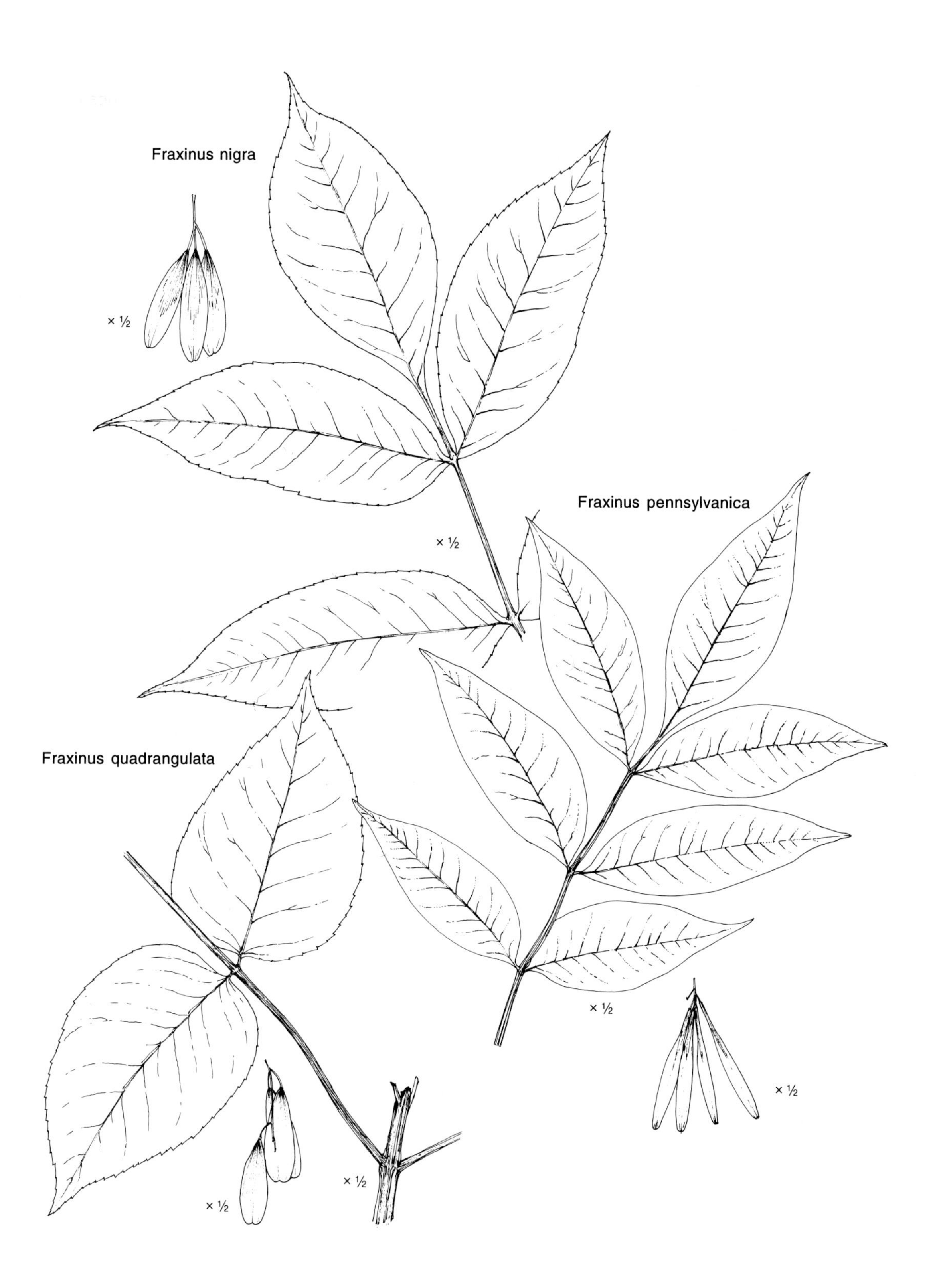
Fraxinus nigra
× ½
× ½
Fraxinus pennsylvanica
Fraxinus quadrangulata
× ½
× ½
× ½
× ½

4. **Fraxinus quadrangulata** Michx. BLUE ASH

Tree, to 26 m (85 feet) tall; branchlets acutely 4-angled and sometimes narrowly 4-winged; leaflets 7–11, ovate to ovate-lanceolate, undersurface pubescent along midrib and sometimes elsewhere as well, margins serrate, the lateral leaflets on short stalks 1–5 mm long; flowers perfect; samaras elliptic to narrowly oblong-ovate, 3–4 cm long, 8–11 mm wide, the wing decurrent almost to base of fruit, the apex usually notched. Throughout the western half of Ohio; dry to moist calcareous woodlands. April.

5. **Fraxinus nigra** Marsh. BLACK ASH

Tree, to 26 m (85 feet) tall; leaflets 7–11, oblong to oblong-lanceolate, undersurface with a swatch of red to pale pubescence at base, margins serrate, the lateral leaflets sessile; flowers imperfect, the plants dioecious; samaras narrowly oblong to oblong-elliptic, 3–5 cm long, 5–9 mm wide, the wing decurrent to base of fruit, the apex blunt or slightly notched. Throughout the northern two-thirds of Ohio but not common, absent from many southern counties; low moist woods, swamps. May.

2. Syringa L. LILAC

A Eurasian genus of about 30 species. In addition to *S. vulgaris*, described below, several other species and hybrids are cultivated. Notable among them is *Syringa reticulata* (Blume) Hara (*S. amurensis* Rupr. var. *japonica* (Maxim.) Franch. & Sav.), JAPANESE TREE-LILAC, a small tree with cherry-like bark bearing great numbers of yellowish-white flowers in panicles to 3 dm long; their fragrance is at best only marginally pleasant.

1. SYRINGA VULGARIS L. LILAC

Shrub, to 5 m (16 feet) tall; leaves simple, ovate to broadly ovate, subcordate to subtruncate at base, acute to acuminate at apex, margins entire, to

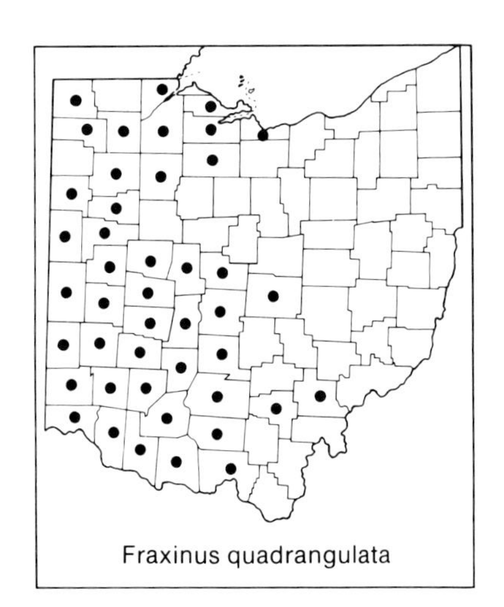
Fraxinus quadrangulata

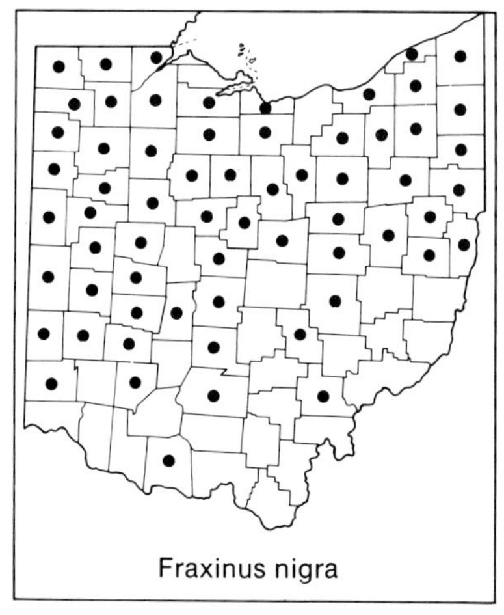
Fraxinus nigra

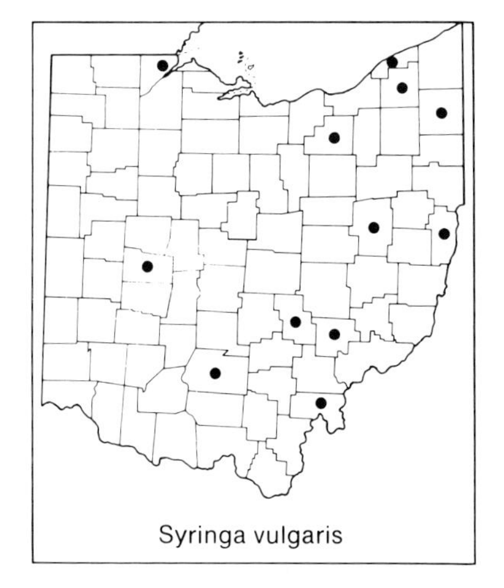
Syringa vulgaris

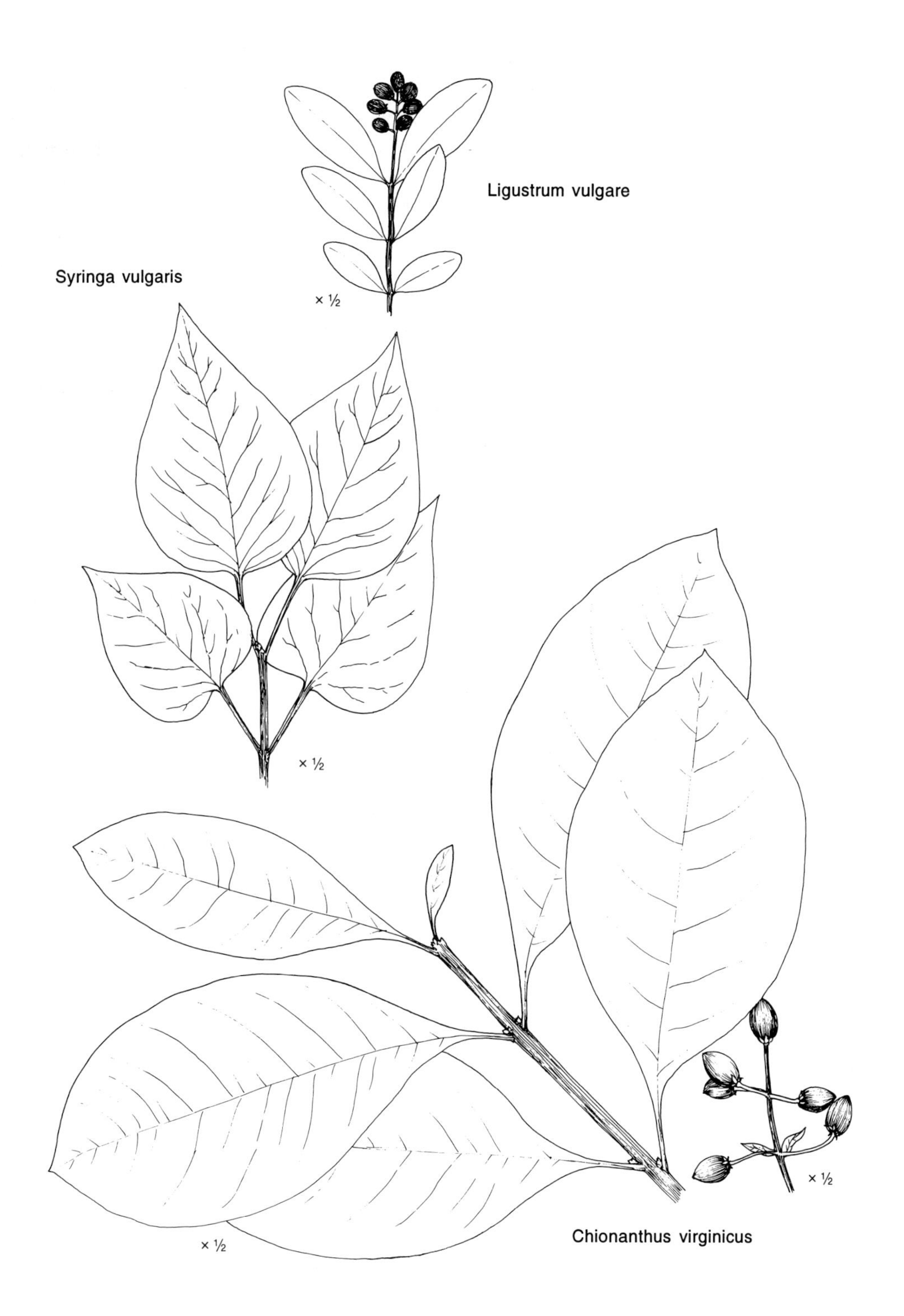
Ligustrum vulgare
Syringa vulgaris
× ½
× ½
× ½
× ½
Chionanthus virginicus

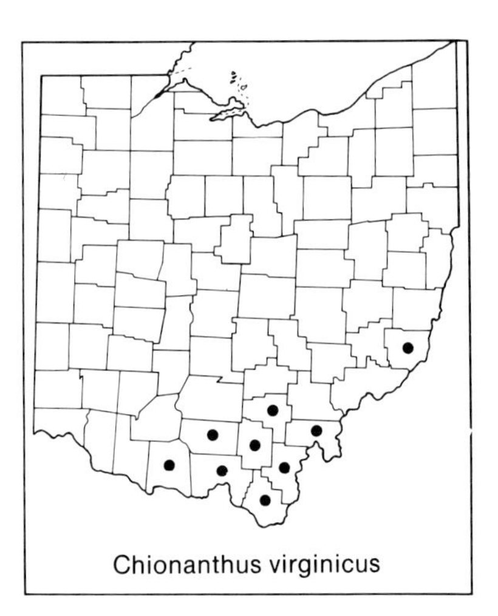
Chionanthus virginicus

12 cm long; flowers mostly white, lilac, or purple (blue or pink in some cultivars), pleasantly fragrant, in long terminal panicles; fruit a capsule. A European species with a host of named cultivars, widely planted and much admired throughout Ohio; persisting at former home sites and occasionally adventive in disturbed habitats. April–May.

3. Chionanthus L. FRINGE-TREE

A genus of three or four species native to eastern North America and eastern Asia; one of the species is native to southern Ohio.

1. **Chionanthus virginicus** L. FRINGE-TREE
Large shrub or small tree, to 6 m (20 feet) tall; leaves simple, broadly elliptic to oblong or slightly oblong-ovate, tapering at base, mostly obtuse to acute at apex, margins entire, 8–20 cm long; flowers white, numerous, in open drooping axillary panicles, with a strong sweet fragrance especially in the evening hours; corolla deeply 4-lobed, the lobes narrowly linear, elliptic, and graceful; fruit a dark blue drupe. A species of southeastern United States, at a northern boundary of its range in counties of southern Ohio; openings, thickets, mesic woodlands; often cultivated for its fragrant flowers and showy inflorescences. May.

4. Ligustrum L. PRIVET

Shrubs; leaves opposite, simple, entire-margined; flowers white, in terminal panicles; fruit a dark blue-black drupe. A genus of about 50 species native chiefly to Asia, with two species naturalized in the Ohio flora. An adventive escape of a third species, *Ligustrum ovalifolium* Hassk., CALIFORNIA PRIVET, was collected in 1960, in Geauga County.

a. Branchlets glabrous; leaves partially evergreen, sometimes persistent; length of anthers about equalling length of corolla lobes; see above . . . . *Ligustrum ovalifolium*

aa. Branchlets minutely to strongly pubescent; leaves deciduous; length of anthers less than length of corolla lobes.

b. Branchlets minutely puberulent to nearly glabrous; leaves glabrous on undersurface, or sometimes with a few scattered hairs along midrib; corolla tube about equalling length of corolla lobes . . . . 1. *Ligustrum vulgare*

bb. Branchlets strongly pubescent; leaves pubescent on undersurface, especially along midrib; corolla tube nearly twice as long as corolla lobes . . . . 2. *Ligustrum obtusifolium*

1. Ligustrum vulgare L. Common Privet

Shrub, to 3 m (10 feet) tall; branchlets minutely puberulent to nearly glabrous; leaves glabrous on undersurface, sometimes with a few scattered hairs along midrib; corolla tube about equalling length of corolla lobes. A Mediterranean species widely planted in Ohio, often used as a box-trimmed hedge bordering lawns and sidewalks, escaped and naturalized in thickets and disturbed woods, in most eastern and some western counties. Late May–early July.

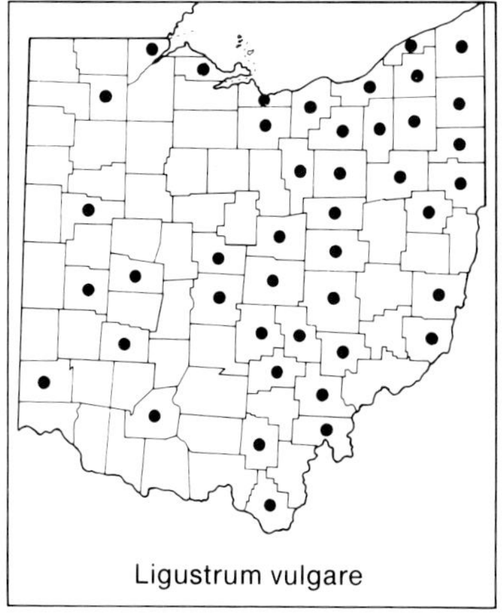
Ligustrum vulgare

2. Ligustrum obtusifolium Sieb. & Zucc.

Shrub, to 3 m (10 feet) tall; branchlets strongly pubescent; leaves pubescent on undersurface, especially along the midrib; corolla tube nearly twice as long as corolla lobes. A species of Japanese origin sometimes cultivated in Ohio, and infrequently naturalized in thickets and disturbed woodland borders; chiefly in northeastern counties. Ebinger (1983) reports the recent spread of this species in Illinois. Not illustrated. June–July.

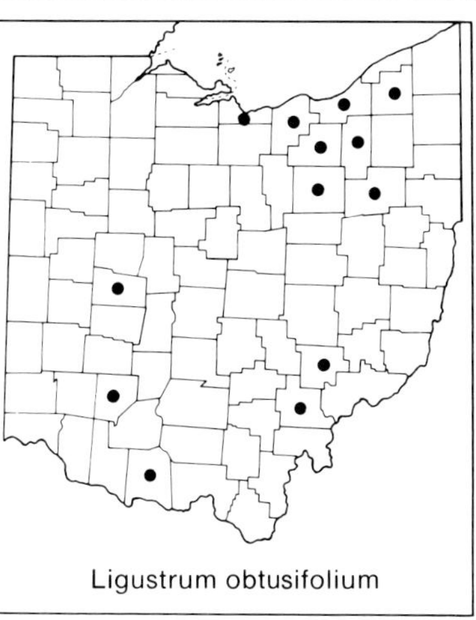
Ligustrum obtusifolium

Loganiaceae. LOGANIA FAMILY

Spigelia marilandica L., Indian Pink or Pinkroot, a red-flowered (yellow within), ornamental perennial from southeastern United States, was attributed by Schaffner (1932) to Lake County, in northeastern Ohio. The species was excluded from the Ohio flora by Roberts and Cooperrider (1982), who believed that the sole specimen was most likely taken from a cultivated plant. Fernald (1950) also attributed the species to Ohio, probably because of its inclusion by Schaffner. Deam (1940) reports the species from the southwesternmost county of Indiana.

Buddlejaceae. BUTTERFLY-BUSH FAMILY

The Chinese plant *Buddleja davidii* Franch., Butterfly-bush, an ornamental shrub with showy, pink to purple or blue, or white, flowers is frequently cultivated in Ohio. Specimens were collected from escapes of this species in Highland County in 1937, and in Lawrence County in 1971. Fernald (1950) writes that the species, sometimes escaping in southeastern United States, is becoming naturalized as far north as Maryland. I am treating the Ohio collections as adventives, although the Lawrence County population of "a few scattered plants" may be marginally naturalized.

Gentianaceae. GENTIAN FAMILY

Annual, biennial, or perennial herbs; leaves opposite or whorled, simple; flowers perfect, often showy, solitary or in cymose inflorescences; calyx with 4 or 5 lobes; corolla actinomorphic, with 4 or 5 lobes; stamens 4 or 5; carpels 2, fused to form a compound pistil with a single style; ovary superior; fruit a capsule. A family of about 1,000 species, widely distributed around the earth but most abundant in temperate regions. J. M. Gillett (1963) studied the family in Canada, as did Wood and Weaver (1982) in southeastern United States. The text here is based on Andreas and Cooperrider's (1981) treatment of Ohio taxa.

a. Leaves all scale-like, 1–3 mm long, ca. 0.5 mm wide; stems slender, 0.5–1 (–2) mm in diameter, unbranched below inflorescence. . . . . 7. *Bartonia*

aa. Leaves all broad (or the lower scale-like in *Obolaria*, but these 2 mm or more in width); stems more than 2 mm in diameter, branched or unbranched.

b. Plants with lower leaves mostly scale-like, 2–6 mm long and 2–6 mm wide, upper leaves longer and wider; the plants fleshy and small, 4–15 cm tall . . . . 8. *Obolaria*

bb. Plants with all leaves broad, none scale-like; the plants fleshy or not, 10–200 cm tall.

c. Leaves in whorls of 3–5, usually 4; plants 10–20 dm tall; corolla greenish-yellow with purple dots, each lobe bearing on its inner surface a large fringe-bordered gland . . . . 6. *Swertia*

cc. Leaves opposite; plants 1–10 dm tall; corolla of various colors, but the lobes without glands.

d. Corolla rotate, pink or rarely white, with a central greenish-white star within; stems markedly 4-angled. . . . . 1. *Sabatia*

dd. Corolla salverform, funnelform, or cylindric, blue, white, greenish-yellow, or pink, if pink—then lacking central greenish-white star within; stems more or less terete.

e. Corolla pink, salverform; calyx lobes keeled; style filiform . . . . 2. *Centaurium*

ee. Corolla blue, white, or greenish-yellow, funnelform or cylindric; calyx lobes keeled or not; style short and thick.

f. Corolla 5-lobed, with plaits between the lobes . . . 5. *Gentiana*

ff. Corolla 4- or 5-lobed, without plaits in sinuses between the lobes.

g. Plants bearing many-flowered cymes; calyx and corolla 5-lobed; corolla 1–2 cm long, its lobes not fringed . . . . 4. *Gentianella*

gg. Plants with flowers solitary; calyx and corolla 4-lobed; corolla 2–6 cm long, its lobes fringed . . . . 3. *Gentianopsis*

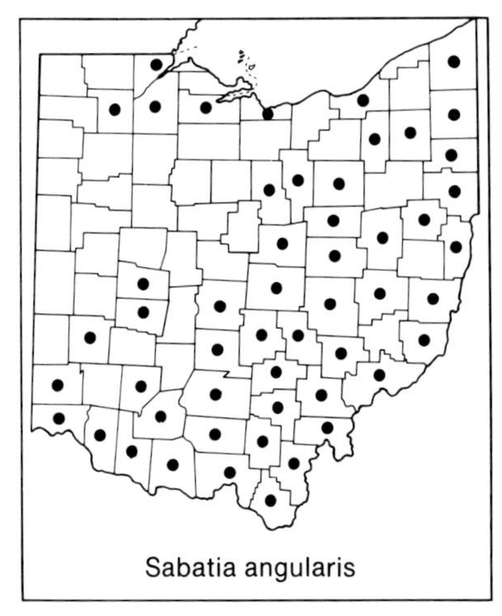
Sabatia angularis

1. Sabatia Adans.

A North American genus of about 17 species, one of which is native to Ohio. Wilbur (1955) revised *Sabatia*, and Perry (1971) made biosystematic studies in the genus.

1. **Sabatia angularis** (L.) Pursh Rose-pink. Marsh-pink

Erect biennial, to ca. 0.5 m tall; stems markedly 4-angled; leaves opposite, ovate, clasping at base; flowers pink or rarely white, with a central greenish-white star within, borne in showy cymes of several–many flowers; calyx and corolla deeply lobed, the lobes usually 5, infrequently 4 or 6; corolla rotate. Frequent in the unglaciated counties of east-central and southeastern Ohio, infrequent to rare or absent elsewhere; dry to moist fields, roadsides, pastures, woodland borders and openings. July–September (–early November).

Centaurium erythraea

2. Centaurium Hill CENTAURY

Annual or biennial herbs; leaves opposite; flowers pink, in terminal cymes; calyx and corolla 4- or 5-lobed; calyx lobes keeled; corolla salverform; style filiform. A genus of about 30 species, chiefly of the Northern Hemisphere; two of the species are alien members of the Ohio flora.

a. Plants 2–5 dm tall, stems with no or few branches below the inflorescence, not branched near base; flowers sessile or nearly so (although the terminal ones often appearing short-pedicellate), in congested inflorescences . . . . 1. *Centaurium erythraea*

aa. Plants low, 1–2 (–3) dm tall, stems usually with few to several branches below the inflorescence, sometimes much-branched from near base; flowers on pedicels 2–10 mm long, in open inflorescences . . . . 2. *Centaurium pulchellum*

1. Centaurium erythraea Rafn European Centaury
   C. umbellatum of authors, including Fernald (1950) and Gleason (1952b).

Biennial, to 5 dm tall; stems with few or no branches below the inflorescence, not branched near base; flowers sessile or nearly so, in congested cymes. A Eurasian species, adventive or perhaps locally naturalized in northeasternmost Ohio; moist to mesic disturbed sites. July–August.

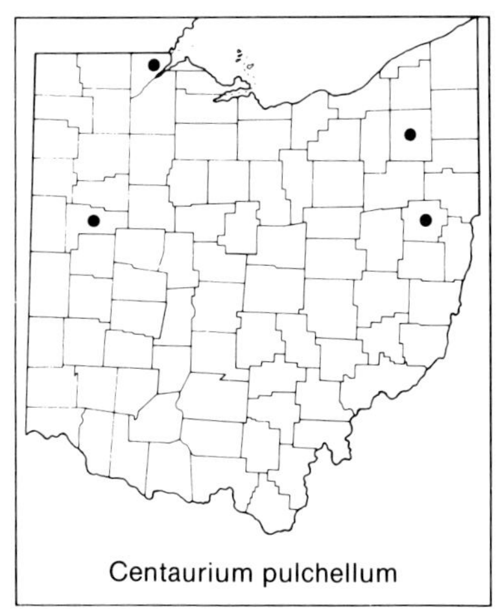
Centaurium pulchellum

2. Centaurium pulchellum (Sw.) Druce

Low annual, to 2 (–3) dm tall; stems usually with few to several branches below inflorescence, often branched near base; flowers in open inflorescences, on pedicels 2–10 mm long. A Eurasian species naturalized at a few scattered sites in Ohio; disturbed habitats. June–September.

3. Gentianopsis Ma FRINGED GENTIAN

Annuals or winter annuals; leaves opposite; flowers solitary at stem tips; calyx 4-lobed, the tube 4-angled and slightly keeled; corolla 4-lobed, funnelform, the lobes fringed. A Northern Hemisphere genus of ca. 20 species, traditionally included within *Gentiana* sensu lato, included by Gillett (1957) within the segregate genus *Gentianella*, but treated here as a segregate genus in its own right following Iltis (1965) and most other current American floristic workers. Two species, sometimes merged, occur in Ohio.

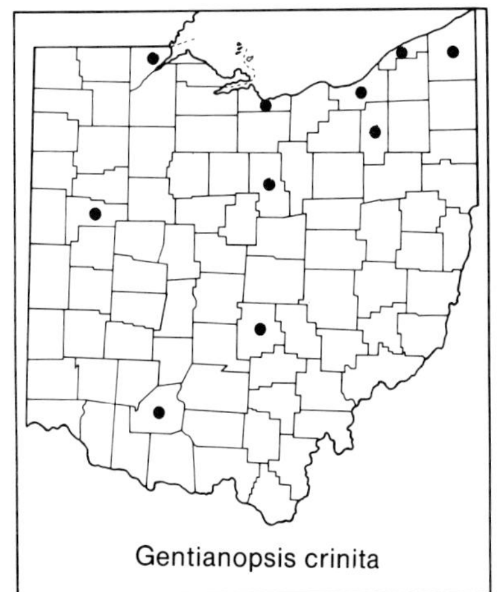
Gentianopsis crinita

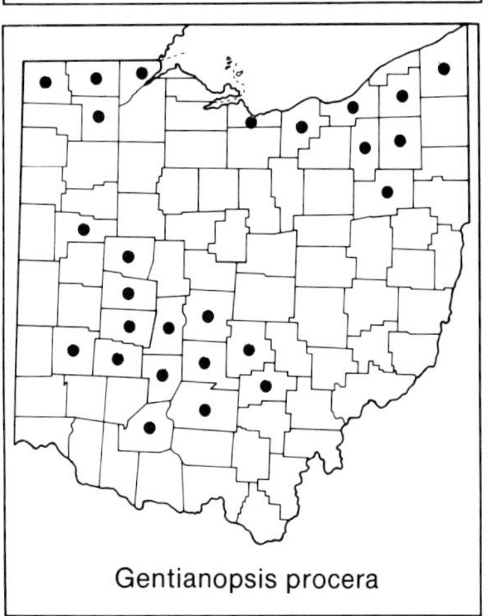
Gentianopsis procera

a. Corolla lobes deeply fringed across apex, the teeth 1.1–2.5 mm long, teeth of fringe along sides of corolla lobes 1.0–2.5 (–3.0) mm long, ½–1½ times as long as those of apical fringe; keels of calyx lobes smooth or only slightly scabrous; leaves mostly lanceolate to ovate-lanceolate, (0.9–) 1.1–2.5 cm wide, 2.1–4.5 cm long. . . . . 1. *Gentianopsis crinita*

aa. Corolla lobes coarsely dentate across apex, the teeth 0.1–1.1 (–1.4) mm long, teeth of fringe along sides of corolla lobes 1.0–2.5 (–3.0) mm long, 2–4 times as long as the apical teeth; keels of calyx lobes strongly scabrous; leaves mostly linear-lanceolate to lanceolate, 0.3–1.2 cm wide, 1.5–6.0 (–9.0) cm long. . . . . . . 2. *Gentianopsis procera*

1. **Gentianopsis crinita** (Froel.) Ma EASTERN FRINGED GENTIAN
Gentiana crinita Froel., *Gentianella crinita* (Froel.) G. Don

Annual or winter annual, to 5 (–6) dm tall; leaves mostly lanceolate to ovate-lanceolate (0.9–) 1.1–2.5 cm wide, 2.1–4.5 cm long; flowers bright blue; calyx and corolla 4-lobed; keels of calyx lobes smooth or only slightly muricate-scabrous; corolla funnelform, the lobes deeply fringed across apex, the teeth 1.1–2.5 mm long, teeth of fringe along sides of corolla lobes 1.0–2.5 (–3.0) mm long, ½–1½ times as long as those of apical fringe. Northern counties and a few scattered locations elsewhere; moist to mesic sites on calcareous substrates in open fields, meadows, marshy areas. September–October (–November).

2. **Gentianopsis procera** (Holm) Ma WESTERN FRINGED GENTIAN.
Gentiana procera Holm, *Gentianella procera* (Holm) Hiitonen, *Gentianella crinita* subsp. *procera* (Holm) J. M. Gillett, *Gentianopsis virgata* (Raf.) Holub

Annual or winter annual, to 6 (–7) dm tall; leaves mostly linear-lanceolate to lanceolate, 0.3–1.2 cm wide, 1.5–6.0 (–9.0) cm long; flow-

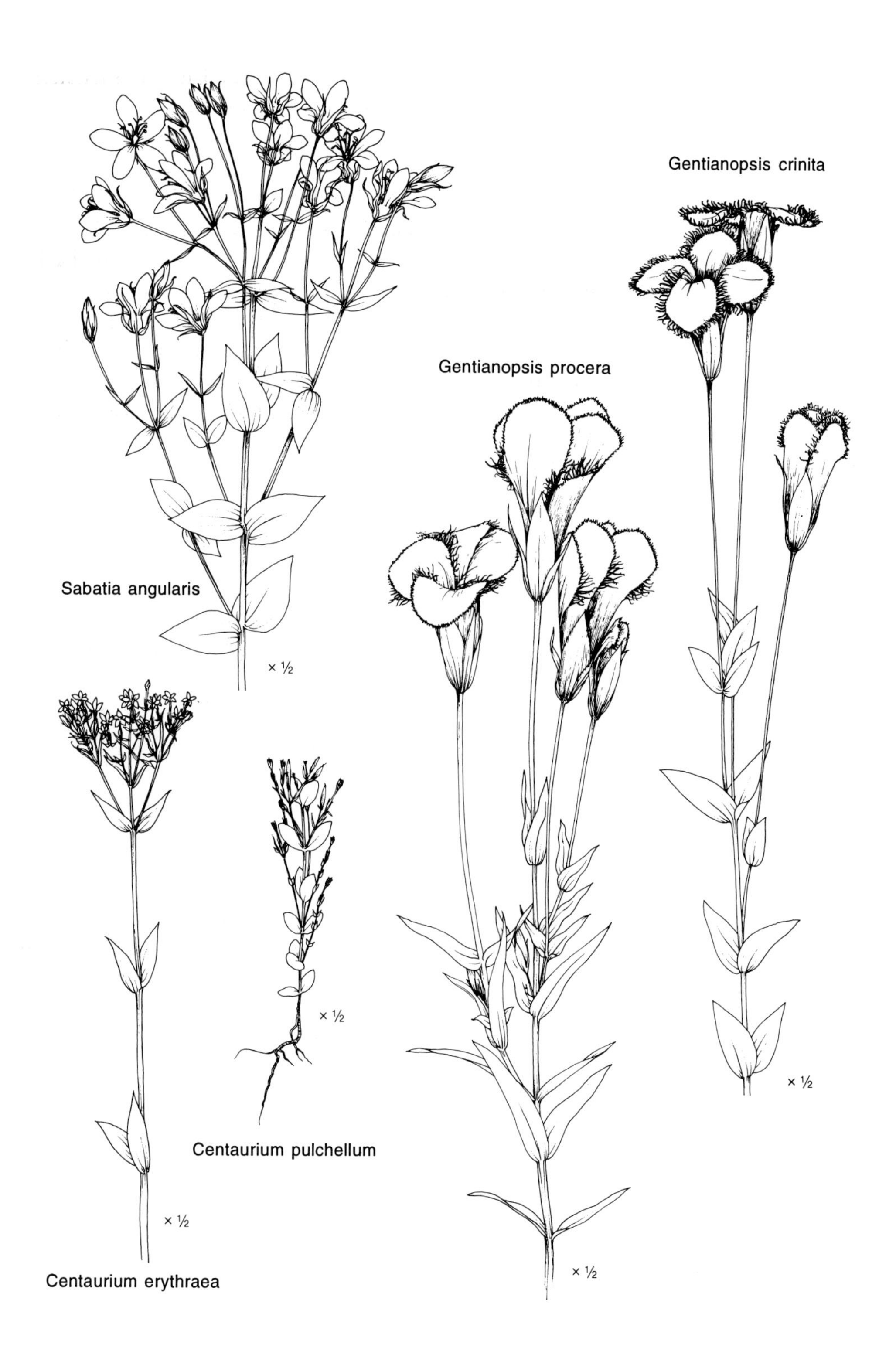
Gentianopsis crinita
Gentianopsis procera
Sabatia angularis
× ½
× ½
Centaurium pulchellum
× ½
× ½
Centaurium erythraea
× ½

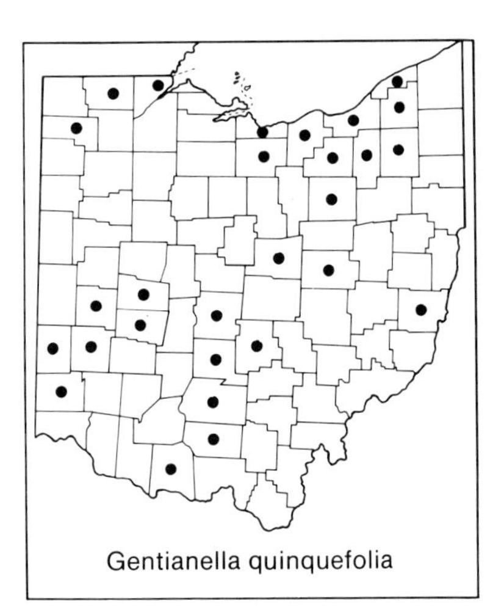
Gentianella quinquefolia

ers bright blue; calyx and corolla 4-lobed; keels of calyx lobes strongly papillose-scabrous; corolla funnelform, the lobes coarsely dentate across apex, the teeth 0.1–1.1 (–1.4) mm long, teeth of fringe along the sides of corolla lobes 1.0–2.5 (–3.0) mm long, 2–4 times as long as the apical teeth. Mostly in northern and western counties; fens, swamps, and other moist calcareous sites. September–October.

J. M. Gillett (1957) treats this taxon as a subspecies of the previous species, regarding the differences between them as clinal in nature.

4. Gentianella Moench GENTIAN

A widely distributed genus, chiefly of the Northern Hemisphere, with more than 200 species, one of them a member of the Ohio flora. This genus was traditionally included within *Gentiana* sensu lato, but is segregated here following Gillett (1957) and most other recent American floristic workers.

1. **Gentianella quinquefolia** (L.) Small STIFF GENTIAN. AGUE-WEED
Biennial, to ca. 6 dm tall, often much-branched above; leaves opposite, the medial and upper leaves ovate and more or less clasping; flowers in cymes, pale blue; calyx and corolla 5-lobed; corolla narrowly funnelform to cylindric, 1–2 cm long. Mostly in northern and southwestern counties; dry to moist woodland openings, stream banks, roadsides, prairies. September–early November.

Ohio plants are assigned to var. *occidentalis* (A. Gray) Small (*Gentiana quinquefolia* L. var. *occidentalis* (A. Gray) Hitchc.).

5. Gentiana L. GENTIAN

Perennials; leaves opposite; flowers in cymose inflorescences; calyx and corolla 5-lobed; corolla funnelform or cylindric, with plaits between the lobes. A widely distributed genus of some 300 or more species, chiefly of the Northern Hemisphere. In the past *Gentiana* has usually been defined in the broad sense so as to include the two segregate genera above, *Gentianella* and *Gentianopsis*. Pringle's (1967) taxonomic study of the species of *Gentiana* section *Pneumonanthae* native to eastern United States, includes all the Ohio species. Pringle (1965b, 1971) studied hybridization in *Gentiana*.

a. Corolla blue to bluish-purple.
  b. Flowers open wide at anthesis; the plaits 2.5–6.0 mm shorter than corolla lobes; anthers separate or separating at anthesis; stems minutely puberulent . . . . 1. *Gentiana puberulenta*

bb. Flowers closed or only slightly open at anthesis; the plaits varying from longer than to no more than 2 mm shorter than corolla lobes; anthers connate at anthesis; stems glabrous.

c. Corolla usually slightly opened at anthesis, its plaits narrower than and slightly shorter than to nearly equalling the lobes; calyx lobes linear to elliptic-linear; leaves linear-elliptic to elliptic, acute at apex, 1.5–10.0 cm long, 0.5–1.5 cm wide .... 4. *Gentiana saponaria*

cc. Corolla closed at anthesis, its plaits as wide as and as long as or longer than the lobes; calyx lobes lanceolate to elliptic, obovate, or orbicular; leaves lanceolate to ovate-lanceolate or oblong-obovate, acuminate at apex, 4–15 cm long, 1.0–5.5 cm wide.

d. Calyx lobes lanceolate to elliptic; corolla lobes truncate to broadly obtuse at apex, terminating in a single minute subapiculate tooth; corolla plaits finely dentate across apex, markedly longer than the lobes .... 2. *Gentiana andrewsii*

dd. Calyx lobes obovate to orbicular; corolla lobes rounded to obtuse at apex, entire; corolla plaits 2- or 3-cleft across apex, as long as or only slightly longer than the lobes.......... 3. *Gentiana clausa*

aa. Corolla greenish to yellow, white, or purplish-white.

b. Calyx lobes linear to narrowly oblanceolate, 3–35 mm long, 0.5–13 mm wide; corolla greenish-, yellowish-, or purplish-white .... 5. *Gentiana villosa*

bb. Calyx lobes ovate-lanceolate to broadly triangular, 3.5–15 mm long, 2–7 mm wide; corolla pale yellow or somewhat greenish-yellow .... 6. *Gentiana alba*

1. **Gentiana puberulenta** J. Pringle PRAIRIE GENTIAN. DOWNY GENTIAN

G. *puberula* of authors, incl. Fernald (1950) and Gleason (1952b), not Michx.

Perennial, to 5 dm tall; stems minutely puberulent; leaves narrowly lanceolate to oblong-lanceolate; flowers deep bluish-purple, in few-flowered terminal clusters; calyx lobes linear; corolla funnelform with spreading lobes, open at anthesis, the plaits 2.5–6.0 mm shorter than the lobes; anthers separate or separating at anthesis. A few western counties; prairies and open fields. Moseley (1931) believed that this species may have been introduced into Ohio by Native Americans. Additional record: Hamilton County (Pringle, 1968). September–October.

Pringle (1966) discusses the change in name, from *Gentiana puberula* to *G. puberulenta*.

Gentiana puberulenta

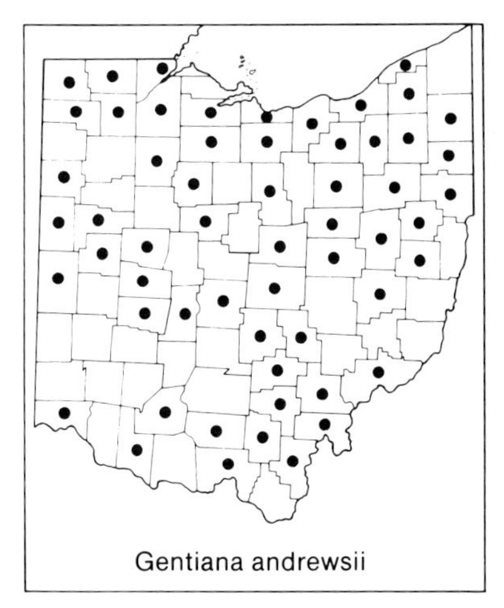
Gentiana andrewsii

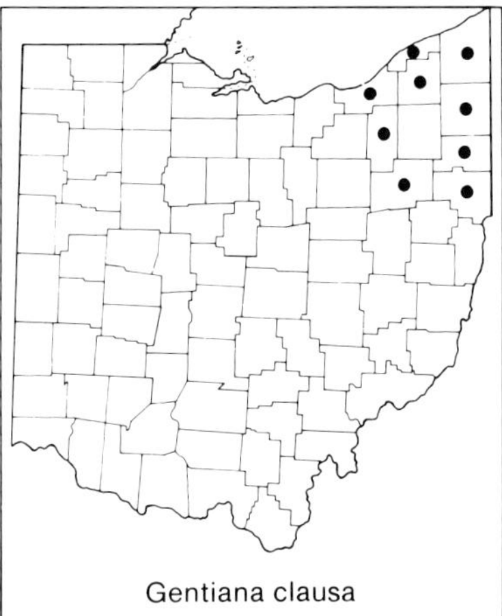
Gentiana clausa

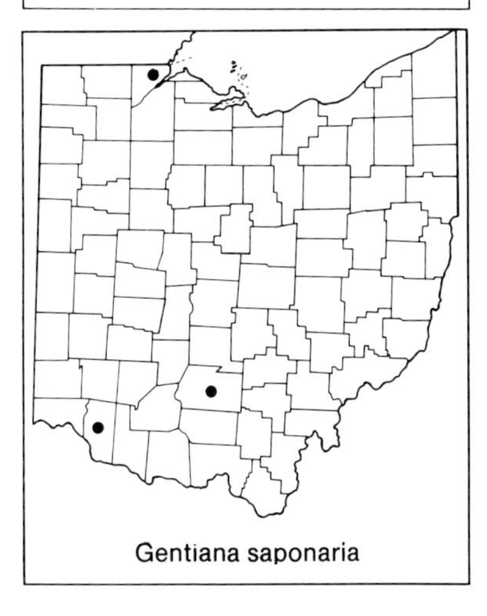
Gentiana saponaria

Gentiana × billingtonii Farw., the naturally occurring hybrid of *Gentiana andrewsii* × *G. puberulenta* has recently been collected from Wood County in northwestern Ohio.

2. **Gentiana andrewsii** Griseb. BOTTLE GENTIAN. CLOSED GENTIAN

Perennial, to 1 m tall, but often much shorter; stems glabrous; leaves narrowly lanceolate to lanceolate, elliptic, or ovate; flowers blue, in few–several-flowered terminal, cymose clusters, and often with one or more additional flowers in upper axils; calyx lobes lanceolate to elliptic, ciliate; corolla cylindric and tapering at apex, closed at anthesis, the lobes truncate to broadly obtuse at apex, terminating in a single minute subapiculate tooth, the plaits finely dentate across the summit, as wide as and markedly longer than the lobes; anthers connate at anthesis. Throughout much of Ohio, but scarce or absent in some southern counties; moist, usually open sites, floodplains, swamps, bogs, fields, roadsides. Costelloe (1988), working with Ohio populations, studied the pollination ecology of this species. Late August–October.

See note under G. *puberulenta*, above, on the hybrid between that species and this.

3. **Gentiana clausa** Raf. BLIND GENTIAN. CLOSED GENTIAN

Perennial, to 7 dm tall; stems glabrous; leaves elliptic-lanceolate to ovate; flowers blue, in few–several-flowered terminal clusters, and sometimes with one or more additional flowers in upper axils; calyx lobes obovate to orbicular; corolla cylindric and tapering at apex, closed at anthesis, the lobes rounded to obtuse at apex, entire, the plaits 2- or 3-cleft across apex, as wide as and as long as or slightly longer than the lobes; anthers connate at anthesis. A species chiefly of northeastern United States, at a western boundary of its range in northeastern Ohio; moist openings, fields, thickets. Additional record: Portage County (Pringle, 1968). August–October.

4. **Gentiana saponaria** L. SOAPWORT GENTIAN

Perennial, to 6 dm tall; stems glabrous or nearly so; leaves linear-elliptic to elliptic; flowers purplish-blue, in few-flowered terminal clusters and usually with additional flowers in the upper axils; calyx lobes linear to elliptic-linear, ciliate; corolla cylindric, usually slightly opened at anthesis, the lobes more or less erect, the plaits narrower than and slightly shorter than to nearly equalling the lobes; anthers connate at anthesis. A species chiefly of southeastern United States, known from a few sites in southwestern and northwestern Ohio, the latter at a northern boundary of the species' range; moist to mesic usually open sites. Windus (1986) reports that fruits and seeds of a population in Lucas County are severely damaged by moth larvae. Additional record: Jackson County (Pringle, 1968). September–October.

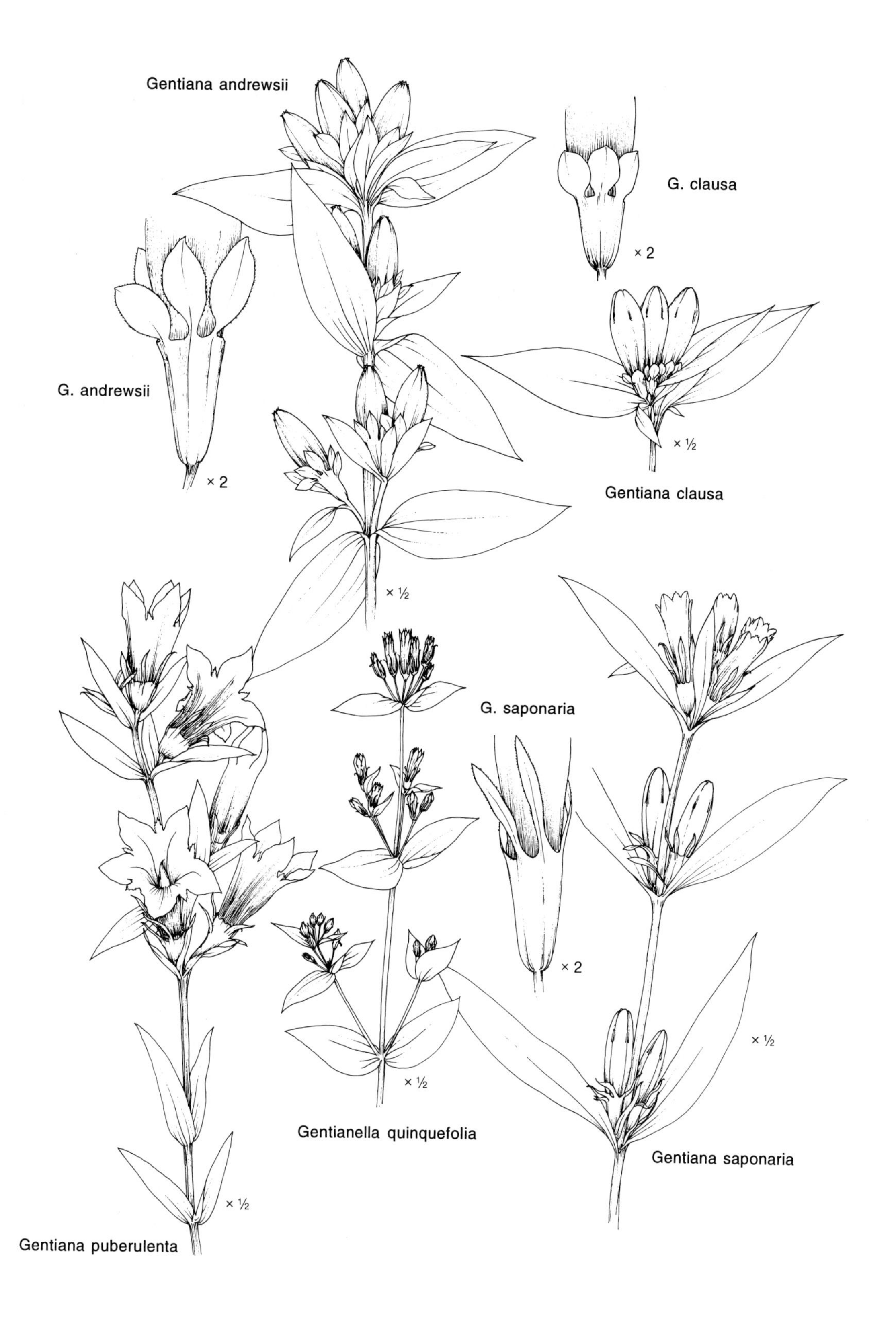
Gentiana andrewsii
G. clausa
× 2
G. andrewsii
× 2
× ½
Gentiana clausa
× ½
G. saponaria
× 2
× ½
× ½
Gentianella quinquefolia
Gentiana saponaria
× ½
Gentiana puberulenta

Gentiana villosa

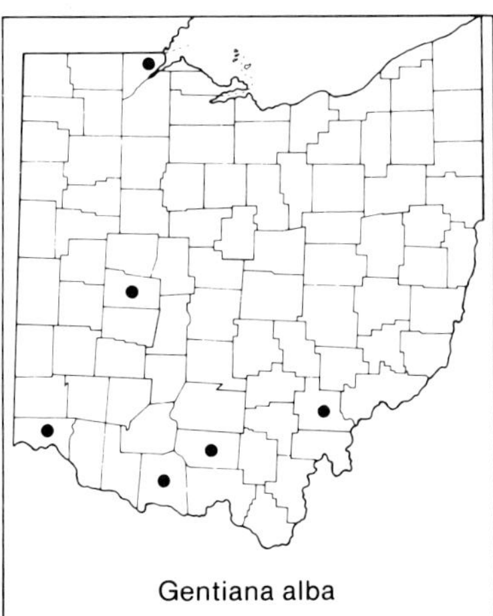
Gentiana alba

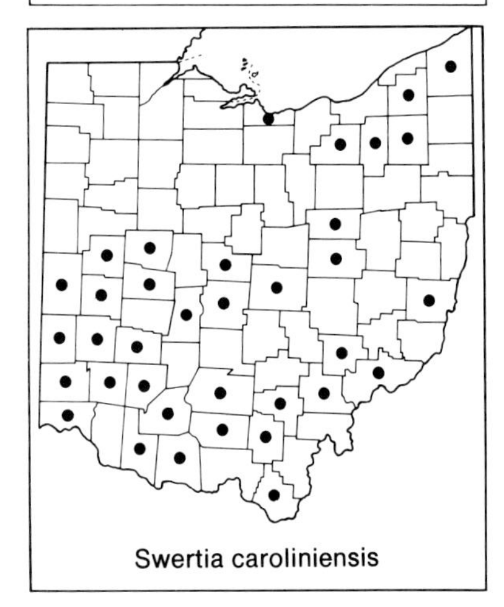
Swertia caroliniensis

5. **Gentiana villosa** L. Sampson's Snakeroot

Perennial, to 6 dm tall; stems glabrous; leaves elliptic, widest at or above the middle; flowers greenish-, yellowish-, or purplish-white, in few-flowered terminal clusters and in lateral clusters from upper axils; calyx lobes linear to narrowly oblanceolate, 3–35 mm long, 0.5–13 mm wide; corolla cylindric, tapering at apex and often slightly opened at anthesis, the plaits about as wide as, but much shorter than the lobes; anthers either connate or separate at anthesis. A species chiefly of southeastern United States, at a northern boundary of its range in the southernmost counties of Ohio; mesic woodlands. Late August–October.

6. **Gentiana alba** Muhl. Yellowish Gentian

G. flavida A. Gray

Perennial, to 1 m tall; stems glabrous; leaves lanceolate to broadly lanceolate, widest near the base; flowers pale yellow or somewhat greenish-yellow, in few–several-flowered terminal clusters, and sometimes also in axils of upper leaves; calyx lobes ovate-lanceolate to broadly triangular, 3.5–15 mm long, 2–7 mm wide; corolla cylindric, tapering at apex, but more or less open at anthesis, the plaits narrower than and shorter than the lobes; anthers connate or sometimes separate at anthesis. Chiefly in western counties; prairies, openings, grassy fields. Mid August–early October.

Accepting Pringle's (1965a) conclusions, this species is called *Gentiana alba*. However, Wilbur (1988) has recently expressed the belief that *G. flavida* is the correct name.

6. Swertia L.

A widespread genus of 50 or more species, one of which is native to Ohio. St. John (1941) revised the genus and reduced *Frasera* Walt. to synonymy under *Swertia*.

1. **Swertia caroliniensis** (Walt.) Ktze. American Columbo

Frasera caroliniensis Walt.

Rosette-producing perennial, to ca. 2 m tall; leaves in whorls of 3–5, usually 4, lanceolate, elliptic, or oblanceolate; flowers greenish-yellow with purple dots, in large paniculate or cymose clusters, terminal and from axils of upper leaves; calyx and corolla deeply 4-lobed; corolla rotate, each lobe bearing on its inner surface a large fringe-bordered gland. Counties of the southern half of Ohio and a few northeastern ones; dry, or less often mesic, woodland openings and borders, roadside banks, thickets, fields. Threadgill, Baskin, and Baskin (1979, 1981) studied the species' geographical ecology and life cycle. June–early July.

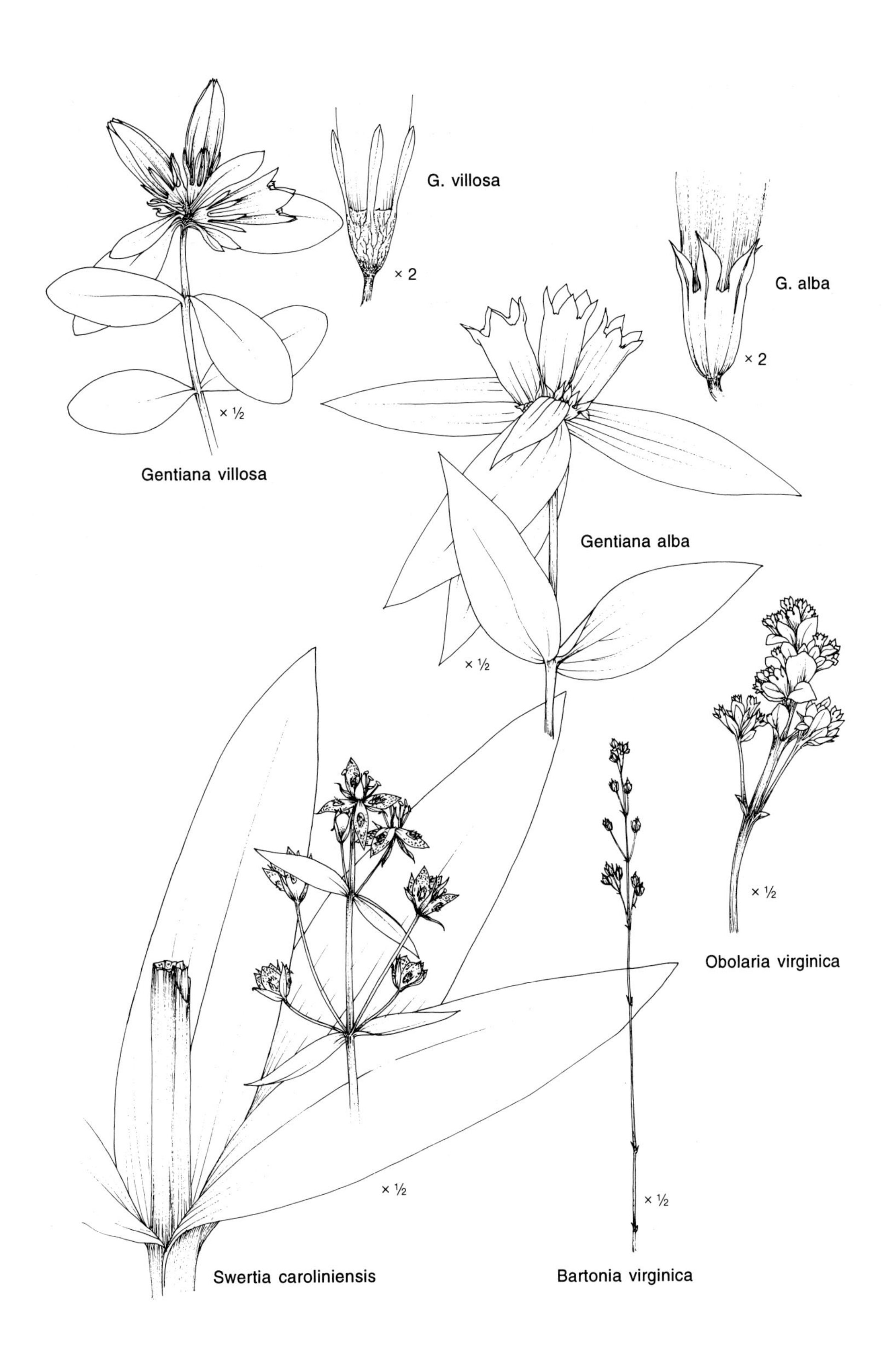
G. villosa
× 2
G. alba
× 2
× ½
Gentiana villosa
Gentiana alba
× ½
× ½
Obolaria virginica
× ½
× ½
Swertia caroliniensis
Bartonia virginica

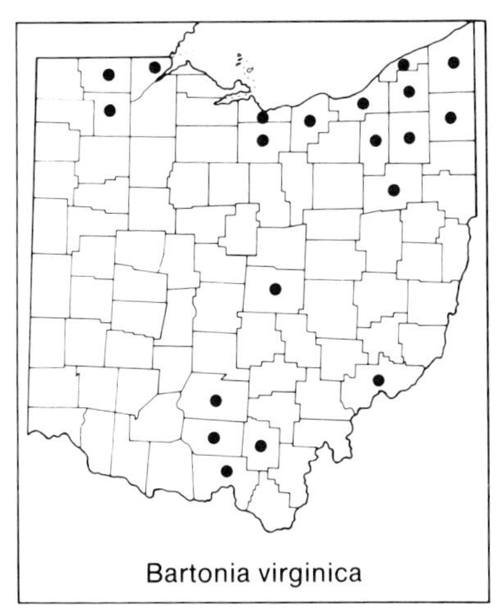

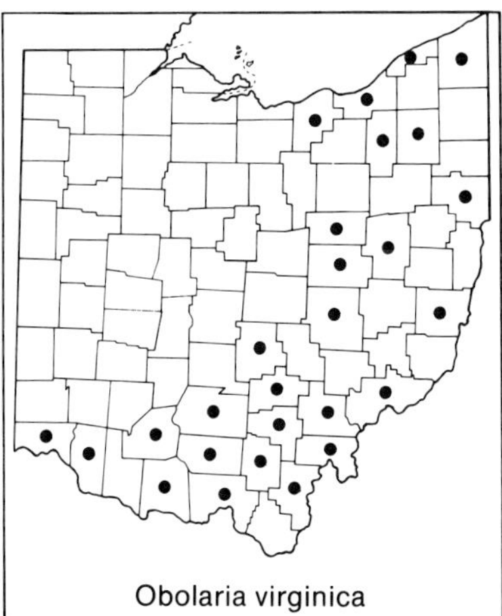

This species is placed in the genus *Swertia* following conclusions of St. John (1941) and Pringle (1990); Threadgill and Baskin (1978) and Wood and Weaver (1982) have favored its classification in *Frasera*.

7. Bartonia Willd.

A genus of eastern North America with three or four species, one of which is native to Ohio. J. M. Gillett (1959) prepared the most recent revision of the genus.

1. **Bartonia virginica** (L.) BSP. YELLOW BARTONIA

Annual, to ca. 3 dm tall; stems slender, 0.5–1.0 (–2.0) mm in diameter, unbranched below inflorescence; leaves opposite or subopposite, all scale-like, 1–3 mm long, ca. 0.5 mm wide; flowers greenish-yellow, either in few-flowered cymes or solitary, terminating the stem and from axils of upper leaves; calyx and corolla deeply 4-lobed; corolla campanulate. Northern and a few southern counties; usually in wet acid habitats, especially sphagnum bogs and meadows, occasionally in dry fields. July–September.

8. Obolaria L.

A genus with a single species. The most recent revision is by J. M. Gillett (1959).

1. **Obolaria virginica** L. PENNYWORT

Low fleshy perennial, to 1.5 dm tall; stems simple or branched; upper leaves obovate to broadly spatulate; lower leaves similar but smaller, mostly scale-like, about as wide as long, 2–6 mm long and 2–6 mm wide; flowers whitish, solitary or in few-flowered cymose clusters in axils of upper leaves and terminating the stem; calyx deeply 2-lobed; corolla 4-lobed, campanulate. A species chiefly of southeastern United States, at a northern boundary of its range in Ohio's eastern and southern counties; mesic woodlands. April–May.

Menyanthaceae. BUCKBEAN FAMILY

A small but widely distributed family of some 30 species. These species were placed within the family Gentianaceae sensu lato by Fernald (1950), Gleason (1952b), and Gleason and Cronquist (1963), but following the classifications systems of both Cronquist (1968, 1981) and Thorne (1968, 1976, 1983), are assigned here to the segregate family Menyanthaceae. Wood (1983) studied the family in southeastern United States. Andreas and Cooperrider (1981) made a study of the family in Ohio, upon which this treatment is based.

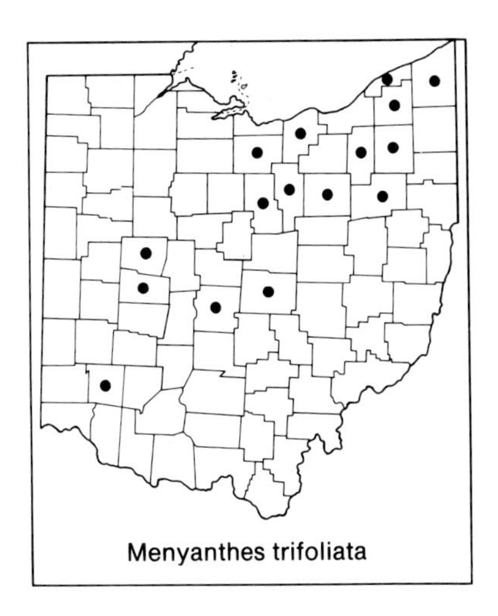
Menyanthes trifoliata

Nymphoides peltata (Gmel.) Ktze., YELLOW FLOATING-HEART, an aquatic European perennial, has been collected twice in Ohio. The first collection was from an apparently adventive stand at the mouth of the Conneaut River in Ashtabula County, in 1930. In 1971, a population was discovered in an artificial pond in Delaware County, where the species was thought to have been introduced when the pond was stocked with catfish. Stuckey (1974) described the introduction and spread of this species in North America.

a. Plants terrestrial; leaves compound, trifoliolate . . . . . . 1. *Menyanthes*
aa. Plants aquatic; leaves simple, floating; see above . . . . *Nymphoides*

1. Menyanthes L.

A genus with a single species.

1. **Menyanthes trifoliata** L. BUCKBEAN. BOGBEAN
   incl. var. *trifoliata* and var. *minor* Raf.

Perennial, 1–3 dm tall, with long creeping rhizomes; leaves alternate, trifoliolate, broadly sheathing at the base; flowers perfect, borne in racemes; calyx deeply 5-lobed; corolla actinomorphic, white to pinkish, 5-lobed, the lobes conspicuously bearded; stamens 5, attached to the corolla tube; carpels 2, fused to form a compound pistil with a single style and a 2-lobed stigma; ovary superior; fruit a capsule. Chiefly in northeastern counties, with scattered occurrences in central and southwestern Ohio; bogs and fens. Late April–May.

Apocynaceae. DOGBANE FAMILY

Perennial herbs with milky sap; leaves simple, mostly opposite; flowers perfect, solitary or in cymose inflorescences; calyx 5-lobed; corolla actinomorphic, 5-lobed; stamens 5, attached to the corolla tube; carpels 2, unusual in being separate at the base (each flower thus with two ovaries) but fused apically to form a single style, the style capped by a single and often somewhat ornate stigma (called the "clavuncle"); ovary superior; fruit a follicle. A large, chiefly tropical family of about 2,000 species. Rosatti (1989) studied the family in southeastern United States. The treatment here is based in large part on that of Andreas and Cooperrider (1979).

Catharanthus roseus (L.) G. Don, MADAGASCAR PERIWINKLE, an herbaceous ornamental with rose-colored or white flowers, is cultivated as an annual in Ohio gardens. A poisonous plant native to the Old World Tropics, it is famous as a source of several anticancer alkaloids. *Amsonia taber-*

naemontana Walt. var. *salicifolia* (Pursh) Woodson, Blue-star, an ornamental perennial with light blue flowers, native to southeastern United States, is infrequently seen in cultivation in Ohio. An adventive plant was collected in Pickaway County in 1964.

The family includes several common tropical ornamental shrubs and small trees seen as greenhouse plants in Ohio. Among them are *Nerium oleander* L., Oleander, with red to pink or white flowers; *Allamanda cathartica* L., Allamanda, with bright yellow funnel-shaped flowers to 15 cm wide; and *Plumeria rubra* L., Plumeria or Frangipani, with sweetly fragrant flowers of various colors, familiar also in tropical leis. All are highly poisonous, as are many other members of the family.

a. Plants trailing, with or without short erect branches; leaves evergreen, opposite; flowers solitary in leaf axils, blue . . . . 1. *Vinca*
aa. Plants erect or leaning; leaves not evergreen, alternate to opposite or rarely whorled; flowers in bracteate inflorescences, blue, white, greenish-white, or pink.
b. Leaves alternate to subopposite or subwhorled; corolla salverform, blue; see above . . . . *Amsonia*
bb. Leaves opposite to rarely whorled; corolla campanulate to urceolate, white, greenish-white, or pink . . . . 2. *Apocynum*

1. Vinca L. PERIWINKLE

An Old World genus of about ten species. In addition to the familiar *Vinca minor*, described below, a vigorous colony of *Vinca major* L., Greater Periwinkle or Large Periwinkle, was discovered in 1977 in Brown County, near the Ohio River (Andreas and Cooperrider, 1979). This species, native to the Mediterranean region of Europe and occasionally cultivated in Ohio, is frequently naturalized in southeastern United States. The Ohio stand seems firmly established and there are no dwellings in the immediate vicinity, but it may nevertheless represent a colony persisting from cultivation.

a. Corolla 2–3 cm wide; leaf blades 1–2.5 (–3) cm wide, not ciliate; calyx lobes not ciliate . . . . 1. *Vinca minor*
aa. Corolla 4–5 cm wide; leaf blades 3–6 cm wide, ciliate; calyx lobes ciliate; see above . . . . *Vinca major*

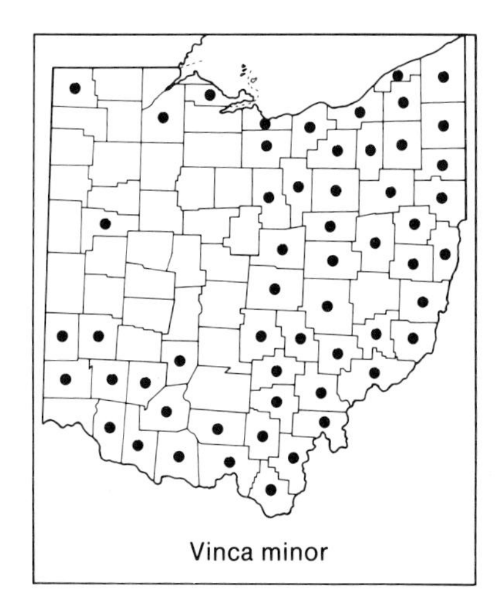

1. Vinca minor L. Common Periwinkle. Myrtle

Perennial, forming extensive mats; stems long-trailing, but with erect or ascending branches to 3 dm high; leaves opposite, evergreen, elliptic, entire-margined, 1.0–2.5 (–3.0) cm wide; flowers solitary in leaf axils; corolla blue or purplish-blue, salverform, 2–3 cm across. An ornamental of European origin, widely cultivated as a border plant or ground cover

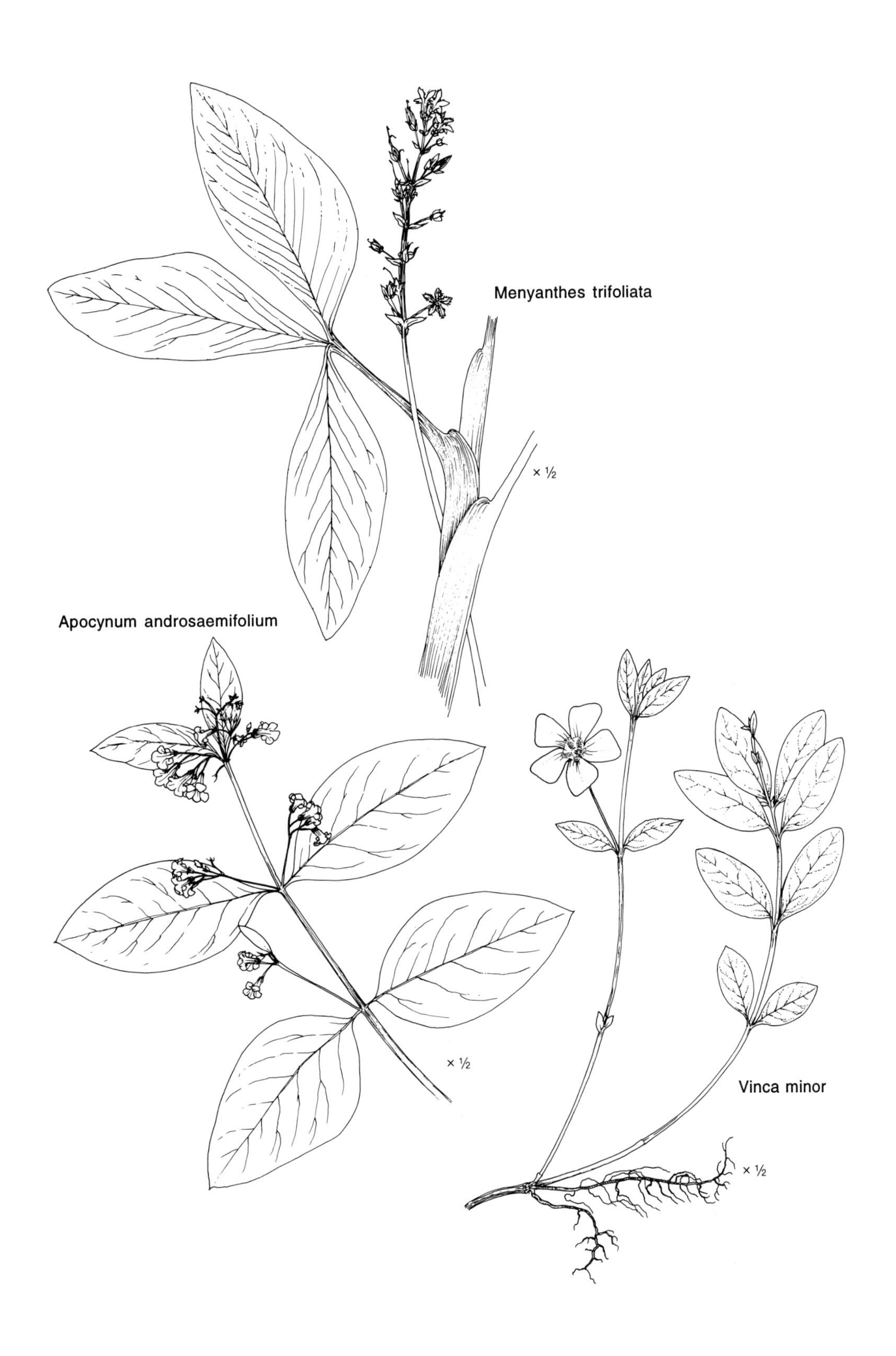
Menyanthes trifoliata
× ½
Apocynum androsaemifolium
× ½
Vinca minor
× ½

throughout much of Ohio; escaped and naturalized, mostly in eastern and southern counties, on roadsides and along railroads, near cemeteries, in disturbed woodland openings and other disturbed habitats. (February–) March–June (–January).

2. Apocynum L. DOGBANE

Erect or leaning perennials; leaves simple, opposite or rarely whorled; flowers small, in cymose inflorescences; corolla campanulate to urceolate; fruit a long follicle. A genus of seven species, all native to temperate regions of North America (Woodson, 1930).

a. Leaves of the main stem all or mostly sessile, clasping to rounded at base, leaves of the primary branches sometimes short-petiolate; corolla white, 2–4 mm long, 2–3 mm wide; bracts of the inflorescence green; follicles 4–10 cm long; seeds 3.5–4 mm long, coma of seeds 0.8–1.2 cm long . . . . 4. *Apocynum sibiricum*

aa. Leaves of main stem with distinct petioles 2–10 mm long, acute to obtuse or rounded at base, leaves of branches petiolate or rarely subsessile; corolla pink to white or greenish, 2–10 mm long, 2–10 mm wide; bracts of the inflorescence mostly scarious; follicles 5–25 cm long; seeds 2–6 mm long, coma of seeds 1.5–3 cm long.

b. Calyx lobes nearly as long as to slightly exceeding corolla tube; corolla greenish-white, mostly 3–4 mm long, 2.0–2.5 mm wide; leaves ascending or widely spreading, terminal inflorescence surpassed by lateral branches; bracts of inflorescence linear-attenuate; follicles (10–) 12–20 (–25) cm long; seeds 4–6 mm long, coma of seeds 2.5–3.0 cm long . . . . 3. *Apocynum cannabinum*

bb. Calyx lobes 1/4 to 1/2 as long as corolla tube; corolla white to pink, 4–10 mm long, 5–10 mm wide; leaves reflexed or widely spreading, terminal inflorescence rarely surpassed by lateral branches; bracts of inflorescence linear-attenuate to oblong-lanceolate; follicles 5–20 (–25) cm long; seeds 2–4 mm long, coma of seeds 1.5–2.0 cm long.

c. Calyx lobes about 1/4–1/3 the length of corolla tube; flowers soon nodding; corolla 4–10 mm long, 5–10 mm wide, pinkish with deeper pink stripes in the tube, the lobes recurved; inflorescence bracts oblong-lanceolate; seeds 2–3 mm long . . . . 1. *Apocynum androsaemifolium*

cc. Calyx lobes about 1/2 the length of corolla tube; flowers ascending or spreading; corolla 4–6 mm long, 5–8 mm wide, white tinged with pink, the lobes not or only slightly recurved; inflorescence bracts linear-attenuate; seeds 2–4 mm long . . . . 2. *Apocynum × floribundum*

1. **Apocynum androsaemifolium** L. SPREADING DOGBANE
Perennial, to 5 dm tall; terminal inflorescence not or rarely surpassed by lateral branches; leaves of main stem spreading to reflexed, elliptic to oblong-ovate, with petioles 2–6 mm long; bracts of inflorescence scarious, oblong-lanceolate; flowers soon nodding; calyx lobes about 1/4–1/3 the length of corolla tube; corolla 4–10 mm long, 5–10 mm wide, pinkish with deeper pink stripes in the tube, the lobes recurved; follicles 6–15 cm long; seeds 2–3 mm long, coma of seeds 1.5–2.0 cm long. Northeastern, northwestern, central, and some southern counties; woodland borders and openings, thickets, weedy fields, roadsides. June–August.

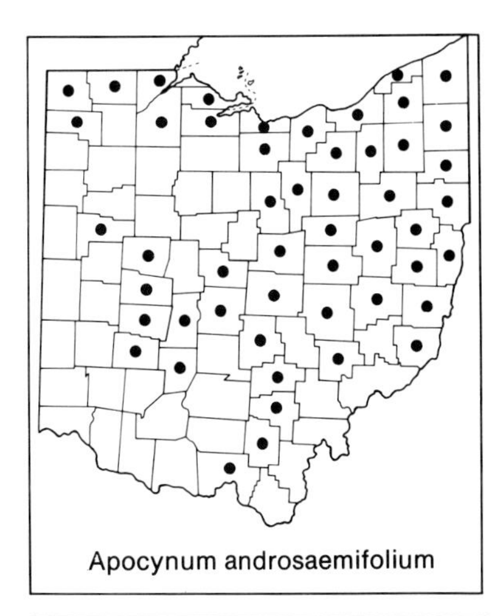
Apocynum androsaemifolium

2. **Apocynum × floribundum** E. Greene INTERMEDIATE DOGBANE
Apocynum × medium E. Greene
The spontaneous hybrid of *Apocynum androsaemifolium × A. cannabinum*, in morphological features intermediate between the two parental species; disturbed habitats at scattered sites in Ohio, mostly in the eastern half of the state. Not illustrated. June–August.

Woodson (1930) and Fernald (1950) treated these plants as a taxonomic species. The treatment here follows Anderson (1936), who concluded on the basis of experimental work that the "actual creation of a new intermediate species . . . does not seem to have been effected." Some Ohio specimens show intergradation toward the parental species and may represent backcrosses or post-F_1 hybrid segregates.

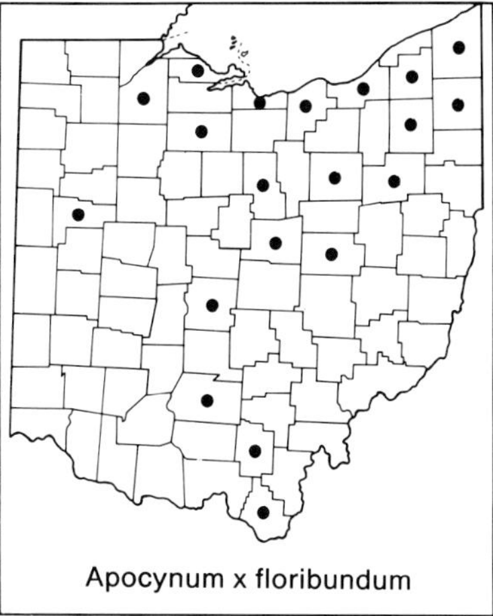
Apocynum x floribundum

3. **Apocynum cannabinum** L. INDIAN HEMP
incl. var. *cannabinum* and var. *pubescens* (R. Br.) A. DC.
Perennial, to 6 dm tall; terminal inflorescence surpassed by lateral branches; leaves of main stem ascending or widely spreading, lanceolate to elliptic or oblong, with petioles 4–10 mm long; bracts of inflorescence scarious, linear-attenuate; flowers ascending or spreading; calyx lobes nearly as long as to slightly longer than the corolla tube; corolla usually 3–4 mm long, 2.0–2.5 mm wide, greenish-white, the lobes erect; follicles (10–) 12–20 (–25) cm long; seeds 4–6 mm long, coma of seeds 2.5–3.0 cm long. Throughout Ohio; moist to mesic fields and thickets, pond and stream margins, moist to mesic woodland openings and borders, along railroads, at roadsides, and in other waste places. June–early September.

As defined here, the species includes without rank plants with white pubescence assigned by Fernald (1950) to var. *pubescens*. As described by Fernald, that variety is sympatric with the glabrous var. *cannabinum* over much of eastern North America, and occupies the same range of habitats. Most Ohio plants are glabrous or nearly so at maturity; the remainder vary more or less continuously from slightly to markedly pubescent.

Apocynum cannabinum

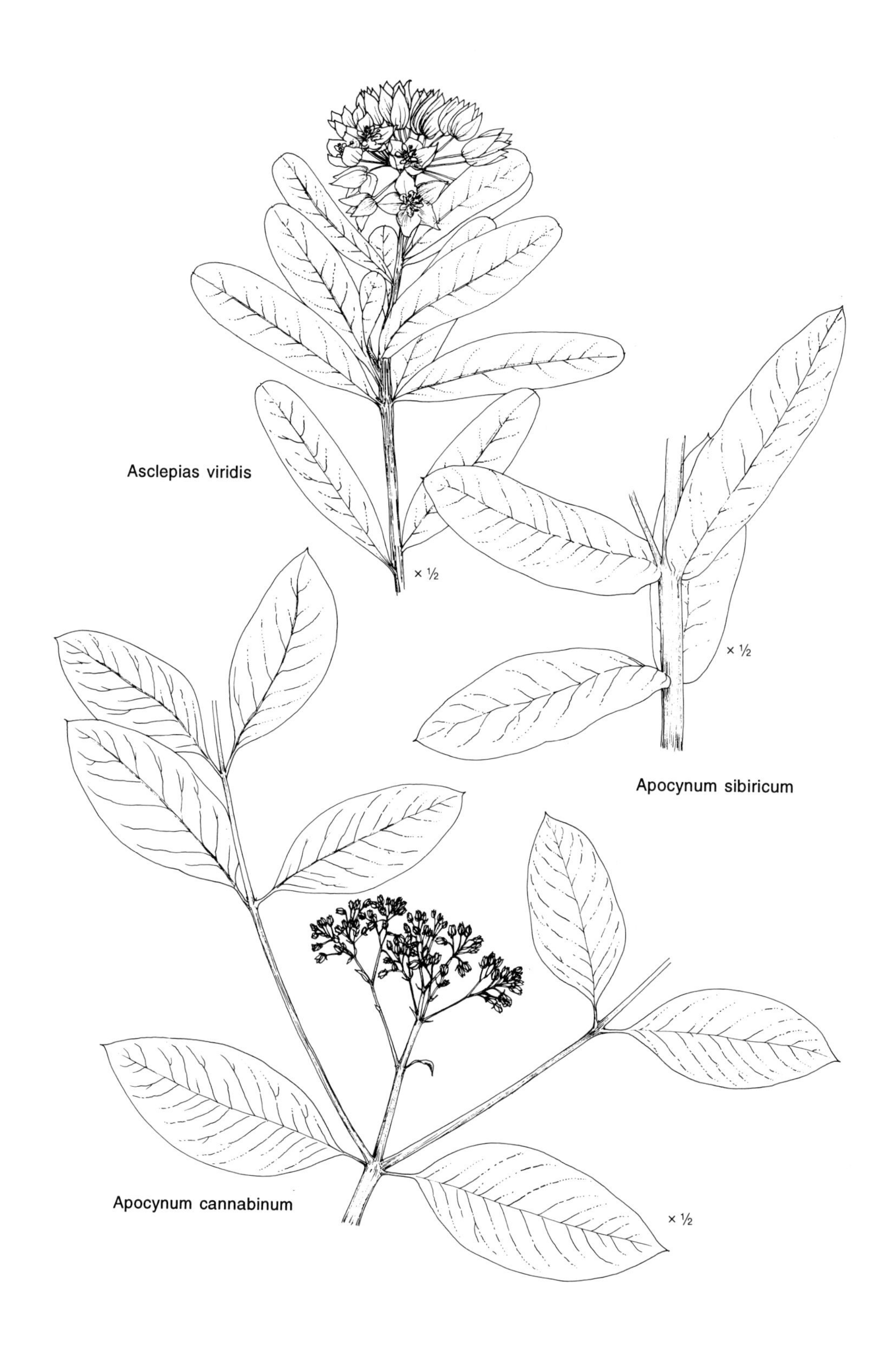
Asclepias viridis
× ½
× ½
Apocynum sibiricum
Apocynum cannabinum
× ½

4. **Apocynum sibiricum** Jacq. CLASPING-LEAVED DOGBANE

A. cannabinum var. *hypericifolium* A. Gray

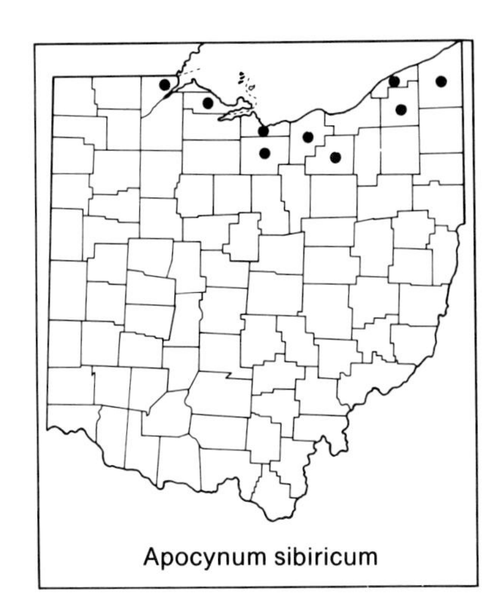

Apocynum sibiricum

Perennial, to 5 dm tall; leaves of main stem spreading or ascending, oblong-lanceolate to oblong or oblong-ovate, all or mostly sessile, rounded to cordate at base, leaves of main branches sometimes short-petiolate; bracts of inflorescence green; flowers erect; corolla 2–4 mm long, 2–3 mm wide, white, the lobes erect; follicles 4–10 cm long; seeds 3.5–4.0 mm long, coma of seeds 0.8–1.2 cm long. In northernmost Ohio counties; sandy or gravelly beaches, railroad ballast, roadsides. Additional record: Cuyahoga County (Woodson, 1930). June–July.

Ohio plants may, with some difficulty, be separated into the two varieties keyed below; but the two are essentially sympatric in Ohio, and their significance is questionable.

a. Leaves of the main stem oblong to oblong-lanceolate, rounded to slightly cordate at base . . . . var. *sibiricum*

aa. Leaves of the main stem oblong to oblong-ovate, markedly cordate at base. . . . . var. *cordigerum* (E. Greene) Fern.

Asclepiadaceae. MILKWEED FAMILY

Perennial herbs, most with milky sap; stems erect or twining; leaves simple, mostly opposite; flowers perfect, in umbellate or cymose inflorescences; calyx 5-lobed; corolla actinomorphic, 5-lobed; corona, immediately inward from the corolla and appearing to be attached to the corolla tube, of 5 or 10 appendages in 1 or 2 whorls resembling an additional set of perianth parts; stamens 5, fused to the stigma; carpels 2, separate at base (each flower with 2 ovaries and 2 styles), but fused apically to form a single stigma; the focal point of the flower a large, unique, central structure (called the "gynostegium") formed by the fusion of the 5 stamens to the single enlarged and ornate stigma; the two ovaries superior, in fruit each becoming a follicle ("pod") filled with numerous seeds bearing long silky hairs. In addition to the corona and the gynostegium, another unique aspect of the flower is that the pollen grains, instead of being separate at the time of dispersal, are held together in a waxy, sharply defined, wing-shaped structure, the pollinium. A large chiefly tropical family of about 2,000 species. Woodson (1941) published a perspective of the genera, and Rosatti (1989) a study of the family in southeastern United States. The treatment here is based mainly on Andreas and Cooperrider's (1980) study of the family in Ohio.

Several species of *Hoya*, WAXPLANT, are popular, succulent-leaved house plants.

a. Stems erect or leaning, not twining . . . . 1. *Asclepias*
aa. Stems twining, usually climbing.
b. Leaves ovate to deltoid or lance-ovate, plants either with leaves 5–7 (–10) cm long and purplish-brown to dark purple flowers *or* with leaves 5–15 (–17) cm long and white flowers.
c. Corolla white, corolla lobes ca. 6 mm long; leaves 5–15 (–17) cm long, deeply cordate at base . . . . 2. *Cynanchum*
cc. Corolla purplish-brown or purple, corolla lobes ca. 3 mm long; leaves 5–7 (–10) cm long, rounded to subcordate at base . . . . 3. *Vincetoxicum*
bb. Leaves ovate to orbicular, (7–) 10–20 (–25) cm long; flowers maroon to purplish-brown . . . . 4. *Matelea*

1. Asclepias L. MILKWEED

Erect or leaning perennial herbs, most with milky sap; leaves mostly opposite, entire-margined; corolla lobes markedly reflexed (spreading in *A. viridis*) and concealing the calyx; the corona consisting of a whorl of 5 curved, shield-shaped appendages, called "hoods," the hoods of most species bearing on their inner side a second whorl of 5 long pointed processes, called "horns." A genus of about 200 species, widely distributed and especially numerous in North America. The broad circumscription of the genus here, including the segregate genera *Acerates* Ell. and *Asclepiodora* A. Gray, is based on the revision of Woodson (1954). Kephart (1981) studied the reproductive systems in *A. incarnata*, *A. syriaca*, and *A. verticillata*.

Asclepias floridana (Lam.) Hitchc., FLORIDA MILKWEED, was attributed to Ohio by Schaffner (1932) on the basis of misidentified specimens of *A. hirtella*.

a. Corolla lobes 12–15 mm long, widely spreading, green; leaves oblong to elliptic-oblong, apex blunt . . . . 1. *Asclepias viridis*
aa. Corolla lobes 3–10 mm long, reflexed, green, white, pink, red, purple, or orange.
b. Flowers with both hoods and horns; corolla greenish-white, white, pink, red, purple, or orange.
c. Corolla orange to yellowish-orange; leaves of main stem alternate; plants with watery sap; stems rough-hairy . . . 2. *Asclepias tuberosa*
cc. Corolla greenish-white, white, pink, red, or purple; leaves of main stem opposite or whorled; plants with milky sap.

d. Leaves linear, 1–2 mm wide and 2–5 cm long, in whorls of 3–6; corolla greenish-white . . . . 11. *Asclepias verticillata*

dd. Leaves wider, 10 mm or more in width, opposite or in whorls of 4 (rarely in whorls or 3 or 5); corolla white, pink, red, purple, greenish-purple, or greenish-white.

e. Leaves either sessile and cordate-clasping at base or cordate to subcordate or rounded at base and with small flattened petioles 1–3 (–5) mm long.

f. Leaves broadly cordate-clasping and sessile, glaucous and/or minutely pubescent on under surface, margins conspicuously undulate; corolla greenish-purple or cream-colored, 7–10 mm long; hoods pale pink to tan; horns exserted and clearly visible beyond hoods . . . . 10. *Asclepias amplexicaulis*

ff. Leaves cordate to subcordate or rounded at base, not clasping, with petioles 1–3 (–5) mm long, glabrous on undersurface and green or sometimes slightly glaucous, especially toward base, margins not or only slightly undulate; corolla purple to pinkish, 9–12 mm long; hoods pink to pinkish-purple; horns largely concealed by hoods . . . . 8. *Asclepias sullivantii*

ee. Leaves petiolate, all or most with petioles 5 mm or more in length, the leaves neither clasping nor cordate at base.

f. Corolla 4–5 (–6) mm long, light pink; most plants with at least some leaves in whorls of 4 (rarely 3 or 5) and the other leaves opposite, occasional plants with all leaves opposite . . . . 5. *Asclepias quadrifolia*

ff. Corolla 6–12 mm long, white to pink, red, purple, greenish-purple, or greenish-white; leaves opposite.

g. The corolla greenish-white to white, very pale purple, or very pale pink.

h. Umbels open and lax, usually with 15–25 flowers, the flowers on weak, often nodding pedicels 2–5 cm long; corolla white to greenish-white or very pale purple; the hoods more or less oblong, truncate and toothed at the apex, the horn extending beyond the hood . . . . 6. *Asclepias exaltata*

hh. Umbels more dense, usually with 25 or more flowers, the flowers on firm stiff pedicels 1–2 cm long; corolla white or very pale pink; the hoods more or less orbicular, rounded and entire at the apex, the horn hidden behind the hood . . . . 7. *Asclepias variegata*

gg. The corolla dark pink to reddish-purple, greenish-purple, or purple, rarely white.

h. Leaves with lateral veins directed apically, diverging from midrib at an acute angle; the leaf undersurface glabrous or short-hispid along major veins; corolla 3–6 mm long; follicles on ascending pedicels; horns exserted and clearly visible beyond hoods . . . . 4. *Asclepias incarnata*

hh. Leaves with lateral veins directed horizontally, diverging from midrib at an 80°–90° angle; the leaf undersurface finely to densely pubescent; corolla usually 7–10 mm long; follicles on deflexed pedicels; horns concealed by hoods.

i. Hoods pink to tan, ovate and obtuse at apex, 3–4 (–5) mm long, only slightly surpassing the gynostegium; corolla pale purple to greenish-purple, or rarely white; umbels numerous, mostly axillary; follicles with large soft spines . . . . 9. *Asclepias syriaca*

ii. Hoods dull red to purple, oblong and abruptly narrowed at apex, 5–7 mm long, much surpassing the gynostegium; corolla reddish-purple; umbels one or few, terminal or from uppermost axils; follicles minutely pubescent, without spines . . . . 3. *Asclepias purpurascens*

bb. Flowers with hoods but without horns; corolla green or greenish.

c. Leaves all or mostly alternate, linear to narrowly linear-lanceolate, most 8–15 cm long and 0.5–1 cm wide; flowers 8–8.5 mm long; hoods 2–2.5 cm long . . . . 12. *Asclepias hirtella*

cc. Leaves all or mostly opposite, ovate to narrowly oblong, broadly oblong, or obovate, most 5–8 cm long and 1–3 cm wide; flowers 9–13 mm long; hoods 4.5–5.5 mm long . . . . . 13. *Asclepias viridiflora*

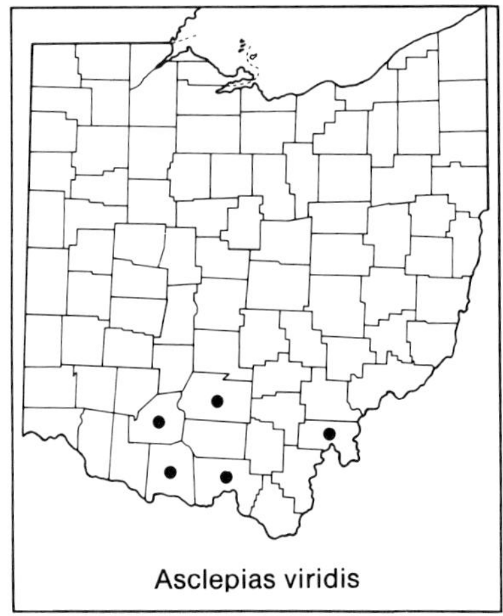
Asclepias viridis

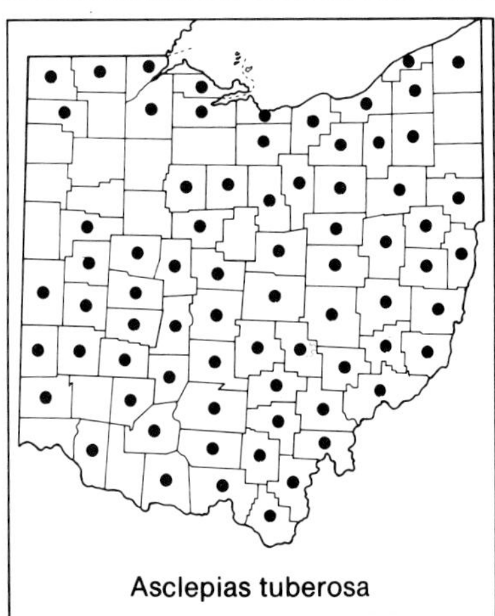
Asclepias tuberosa

1. **Asclepias viridis** Walt. Spider Milkweed. Antelope-horn
Asclepiodora viridis (Walt.) A. Gray

Perennial, more or less erect, to 6 dm tall; leaves alternate to subopposite or opposite, oblong to elliptic-oblong, apex blunt and apiculate, sometimes also slightly retuse, base acute to subtruncate, the leaves short-petiolate, glabrous to slightly puberulent on both surfaces; corolla green, 12–15 mm long, widely spreading; hoods purplish, 4–6 mm long. A species of south-central and southeastern United States, at a northern boundary of its range in counties of south-central Ohio; prairie patches, dry fields, dry roadside banks. May–June.

2. **Asclepias tuberosa** L. Butterfly-weed. Pleurisy-root
incl. var. *tuberosa* and var. *interior* (Woodson) Shinners

Perennial, to 9 dm tall, with watery rather than milky sap; stems conspicuously rough-hairy; leaves scattered, mostly alternate, linear-oblong to lanceolate, apex obtuse to more often acute or acuminate, base tapering to truncate, the leaves short-petiolate, pubescent on both surfaces, especially the under; flowers showy; corolla orange to reddish-orange or yellowish-orange; hoods orange, 4–5 mm long. Frequent in east-central and southern counties, scattered elsewhere more or less throughout Ohio; dry fields, roadside banks, and other dry to mesic open habitats. Frequently cultivated for its masses of bright orange flowers. Woodson (1962, 1964) made detailed studies of intraspecific variation. Mid-June–early September.

In Woodson's (1954) taxonomy, some Ohio plants would be assigned to the eastern var. *tuberosa*, others to the western var. *interior* (Woodson) Shinners, the two being distinguished by differences in leaf shape. Woodson's (1954) map shows that the two taxa meet in Ohio and that both occur throughout the state. Gleason and Cronquist (1991) note that intermediates are abundant, and I have found that to be particularly true of Ohio plants. The taxonomy used here is that of Fernald (1950) in which no major infraspecific taxa are recognized within this highly variable species.

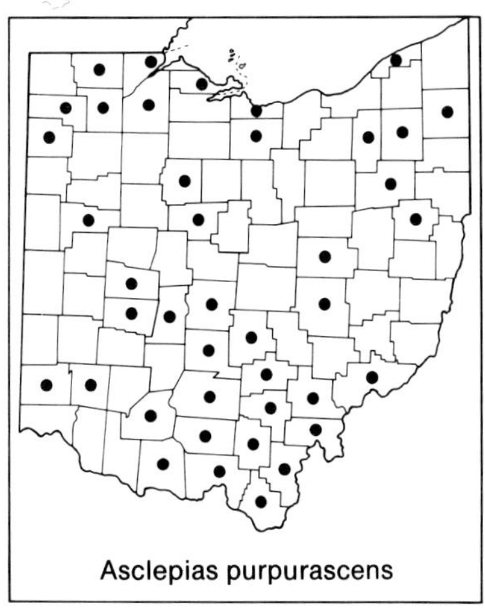
Asclepias purpurascens

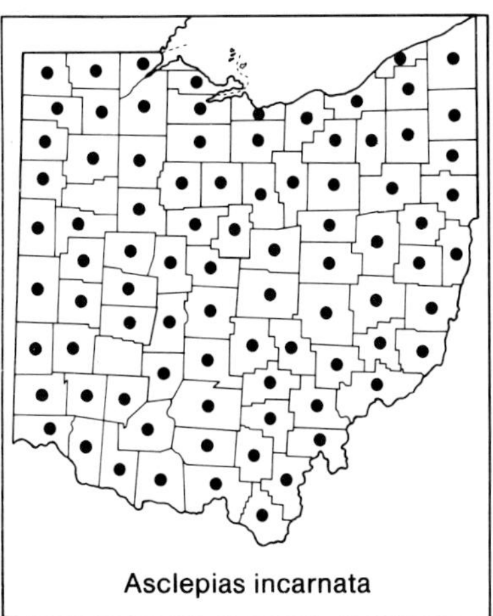
Asclepias incarnata

3. **Asclepias purpurascens** L. PURPLE MILKWEED

Perennial, to 1 m tall; leaves opposite, lanceolate to lance-oblong or ovate-oblong, apex acute to subobtuse and usually short-apiculate, base obtuse to acute, the leaves petiolate, upper surface more or less glabrous, undersurface puberulent, lateral veins diverging at right angles from the midrib; umbels one or few, terminal or from uppermost axils; corolla reddish-purple, 8–10 mm long; hoods dull red to purple, oblong and abruptly narrowed at apex, 5–7 mm long, much surpassing the gynostegium; follicles on deflexed pedicels. Chiefly in south-central counties, but also scattered elsewhere throughout Ohio; open woods, fields, thickets, roadsides. June–August.

Some specimens are difficult to separate from A. *syriaca*.

4. **Asclepias incarnata** L. SWAMP MILKWEED

Perennial, to 1.5 dm tall; leaves opposite, linear-lanceolate to linear-elliptic, oblong-elliptic, or ovate-elliptic, apex acute to acuminate, base acute to subtruncate, the leaves petiolate, both surfaces glabrous or the undersurface short-hispid along major veins, lateral veins directed apically; corolla pink to pinkish-purple, 3–6 mm long; hoods pale pink to whitish, 1.5 mm long; follicles on ascending pedicels. Throughout Ohio; marshes, swamps, wet ditches, and other moist sites. July–August.

Ohio plants belong to var. *incarnata*. *Asclepias incarnata* var. *pulchra* (Willd.) Pers. was credited to Ohio as an adventive by Woodson (1954). The sole specimen upon which his attribution was based may have been collected from a cultivated plant (Andreas and Cooperrider, 1980), and the variety is here excluded as a member of the Ohio flora.

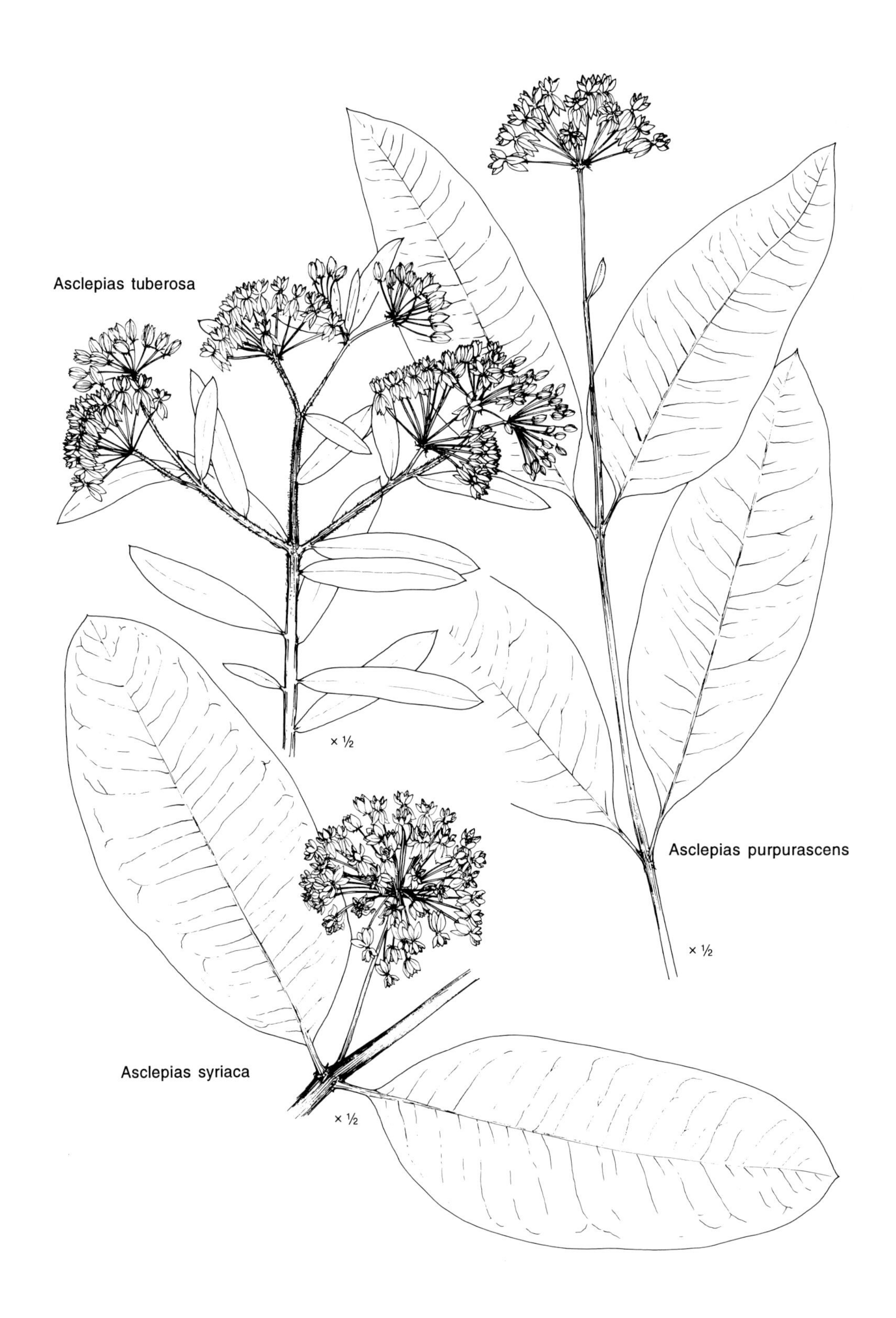
Asclepias tuberosa
× ½
Asclepias purpurascens
× ½
Asclepias syriaca
× ½

5. **Asclepias quadrifolia** Jacq. FOUR-LEAVED MILKWEED
Perennial, to 5 dm tall; leaves in whorls of 4 (rarely 3 or 5) or opposite, lanceolate to ovate, apex acute to acuminate, base broadly acute to obtuse, the leaves petiolate, both surfaces glabrous or nearly so; corolla light pink, 4–5 (–6) mm long; hoods whitish, 4–5 mm long. Throughout eastern and southern Ohio, scarce or absent in northwestern counties; dry to mesic woodlands, woodland openings and borders. Late May–June.

6. **Asclepias exaltata** L. POKE MILKWEED
Perennial, to 1 m tall; leaves opposite, elliptic to oblong-elliptic to ovate or broadly ovate, apex acuminate, sometimes abruptly so, base acute to obtuse, the leaves petiolate, both surfaces glabrous or nearly so; umbels open and lax, usually with 15–25 flowers on weak, often nodding, pedicels 2–5 cm long; corolla white to greenish-white or very pale purple, 8–12 mm long; hoods whitish, 3.5–4.0 mm long, more or less oblong, truncate and toothed at the apex, the horns extending beyond the hoods. Eastern and southern counties, and at scattered sites elsewhere; rich woodlands, or occasionally at woodland borders. Shannon and Wyatt (1986) studied this species' reproductive biology. Late May–mid-July.

Kephart et al. (1988) report natural hybridization between this species and A. *syriaca* in Michigan, Virginia, and West Virginia.

7. **Asclepias variegata** L. WHITE MILKWEED
Perennial, to 1 m tall; leaves opposite, elliptic to broadly elliptic, apex obtuse to rounded and apiculate, base obtuse to rounded, the leaves petiolate, both surfaces glabrous; umbels dense, with 25 or more flowers, the flowers on firm stiff pedicels 1–2 cm long; corolla white or very pale pink, 7–8 mm long; hoods purple at the base, white above, 2.5 mm long, more

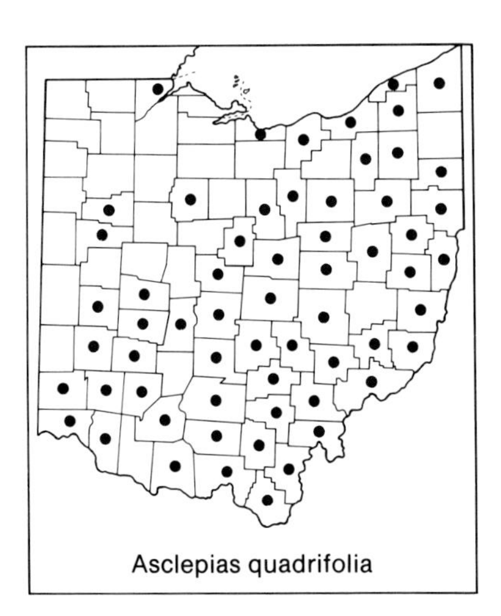
Asclepias quadrifolia

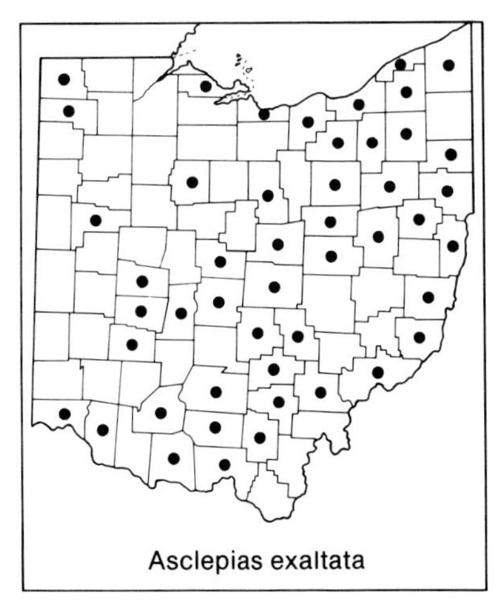
Asclepias exaltata

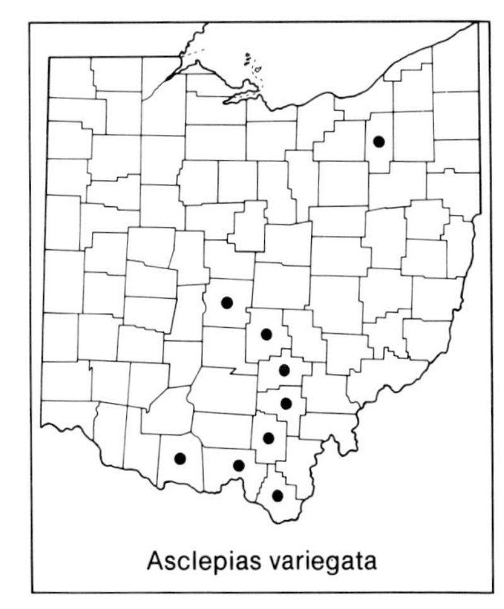
Asclepias variegata

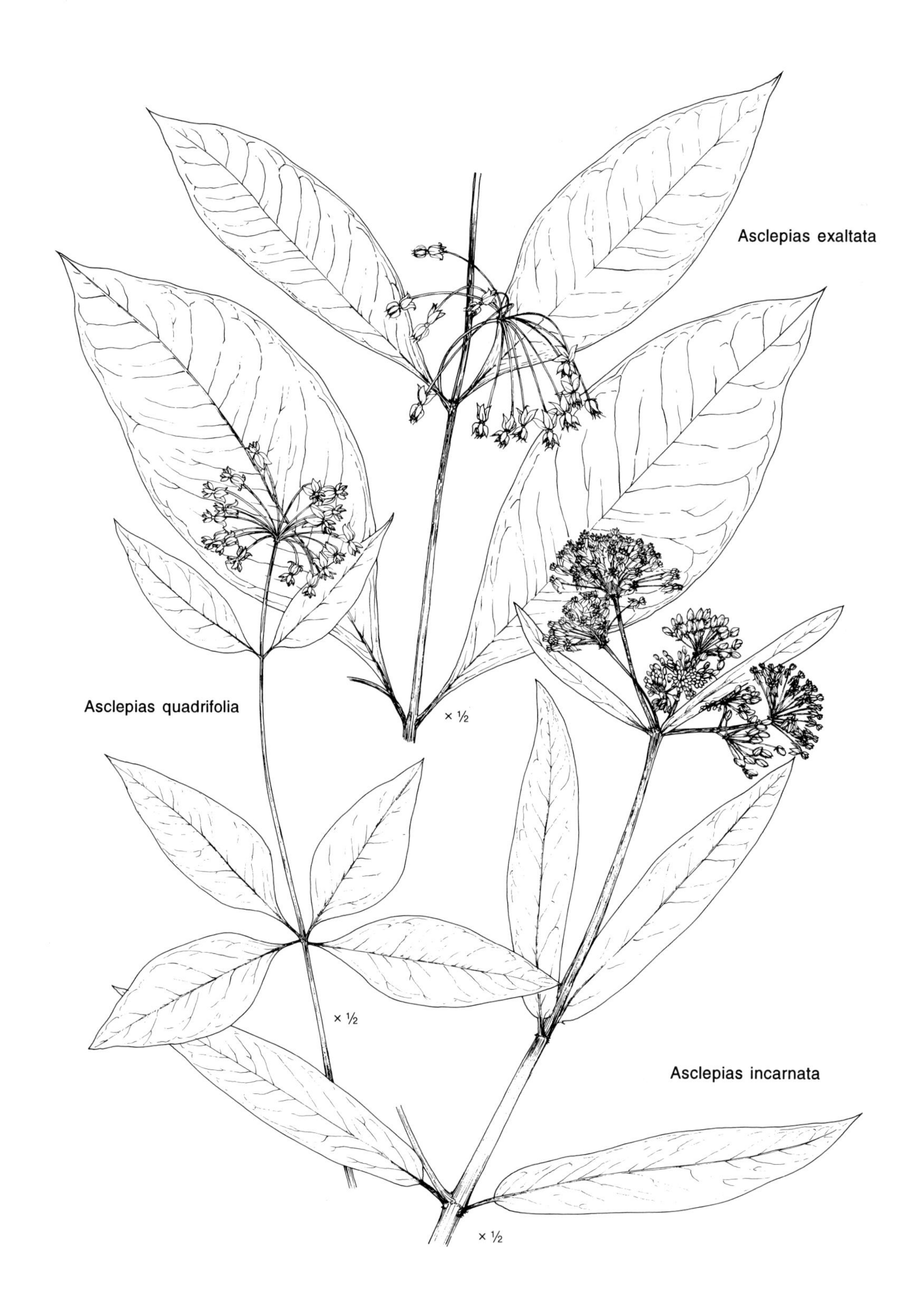
Asclepias exaltata
Asclepias quadrifolia
× ½
× ½
Asclepias incarnata
× ½

or less orbicular, rounded and entire at the apex, the horns hidden behind the hoods. South-central counties and one northeastern county (Summit); sandy soil in woodlands and thickets. Late May–early July.

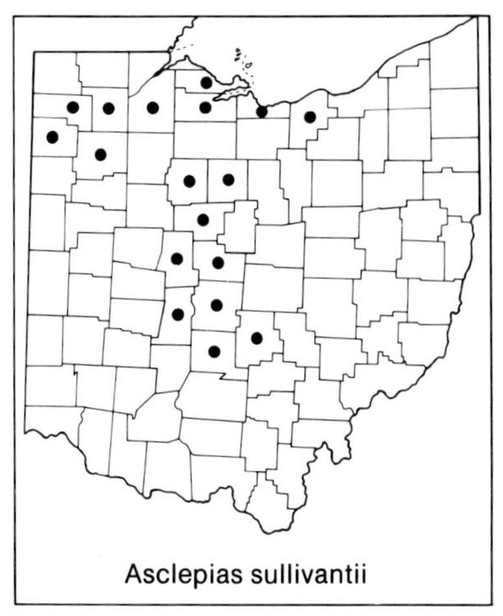

Asclepias sullivantii

8. **Asclepias sullivantii** A. Gray Prairie Milkweed. Sullivant's Milkweed

Perennial, to 1.2 m tall; leaves opposite, oblong to oblong-lanceolate or oblong-ovate, apex broadly obtuse to rounded, apiculate and sometimes retuse, base rounded to subcordate or cordate, not clasping, the leaves subsessile with petioles 1–3 (–5) mm long, margins not or only slightly undulate, both surfaces glabrous, the undersurface green and sometimes slightly glaucous, especially toward base; corolla purple to pinkish, 9–12 mm long; hoods pink to pinkish-purple, 5–6 mm long; horns largely concealed by the hoods. A species of the interior United States, at an eastern boundary of its range in Ohio; northwestern, north-central, and central counties; prairie patches, fencerows, roadside banks, dry fields. Late June–mid-August.

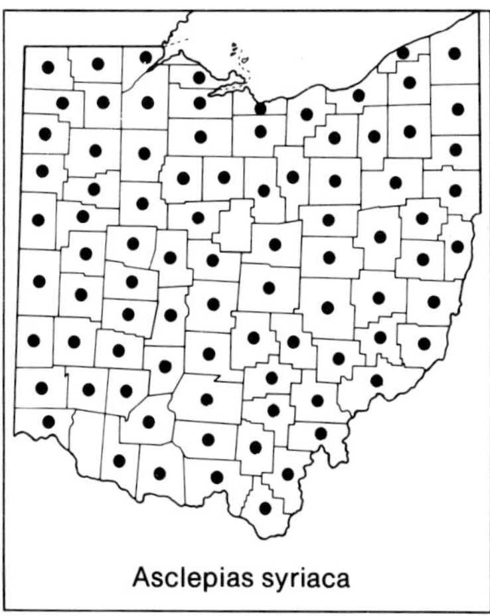

Asclepias syriaca

9. **Asclepias syriaca** L. Common Milkweed

Perennial, to 1.5 m tall; leaves opposite, elliptic-oblong to oblong, apex obtuse to rounded and often apiculate, base obtuse to rounded or subtruncate, the leaves petiolate, upper surface glabrous, the under pubescent, lateral veins diverging at right angles from the midrib; umbels numerous, mostly axillary; corolla pale purple to greenish-purple or rarely white, 7–9 mm long; hoods pink to tan, ovate and obtuse at apex, 3–4 (–5) mm long, only slightly surpassing the gynostegium; follicles softly spinulose, on deflexed pedicels. Common throughout Ohio; river banks, roadsides, railways, open fields, and many types of disturbed habitats. The empty pods are often collected for use as dry decorative ornaments. Doyon (1959) discusses the native range of this species. Baskin and Baskin (1977) report on seed germination. Mid-June–August.

Forma *leucantha* Dore, with a white or whitish corolla, has been collected from Athens, Hocking, Lawrence, Warren, Washington, and Wayne counties, most in southern Ohio.

Kephart et al. (1988) report naturally occurring hybrids between this species and A. *exaltata* from Michigan, Virginia, and West Virginia.

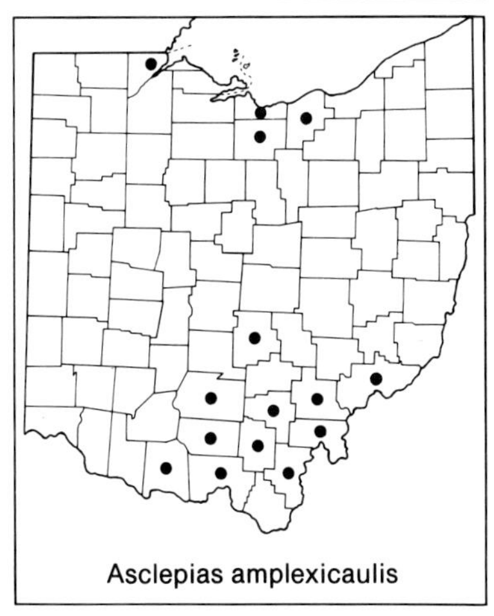

Asclepias amplexicaulis

10. **Ascelpias amplexicaulis** Sm. Clasping-leaved Milkweed. Blunt-leaved Milkweed

Perennial, to 1 m tall; leaves opposite, ovate-oblong, apex broadly obtuse to rounded or subtruncate, apiculate and sometimes retuse, base broadly cordate-clasping and sessile, margins conspicuously undulate, both surfaces glabrous, or the under minutely tomentulose, undersurface glaucous; corolla greenish-purple to cream-colored, 7–10 mm long; hoods pale pink

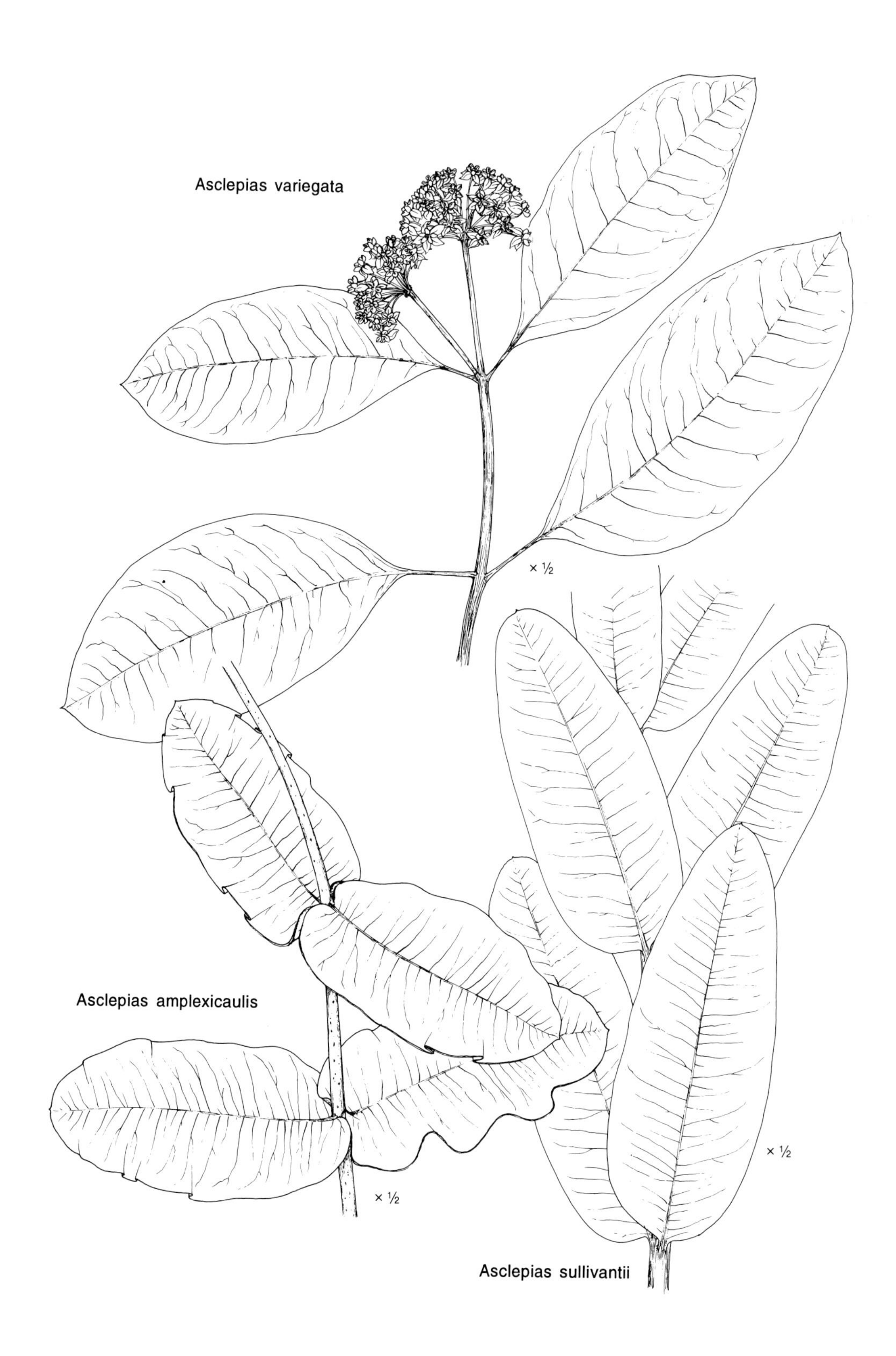
Asclepias variegata
× ½
Asclepias amplexicaulis
× ½
× ½
Asclepias sullivantii

to tan, 5 mm long; horns exserted and clearly visible beyond the hoods. South-central and north-central counties; open dry sandy sites, dry fields. Late May–July.

11. **Asclepias verticillata** L. Whorled Milkweed. Horsetail Milkweed

Perennial, to 9 dm tall; leaves in whorls of 3–6, linear, 1–2 mm wide, 2–5 cm long, sessile or subsessile, glabrous or nearly so; corolla greenish-white, 3.5 mm long; hoods greenish-white, 1.5 mm long. South-central and northwestern counties, and scattered elsewhere; dry fields, roadsides, openings. Late June–early September.

12. **Asclepias hirtella** (Pennell) Woodson Green Milkweed
    Acerates hirtella Pennell

Perennial, to 1 m tall; leaves all or mostly alternate, linear to narrowly linear-lanceolate, mostly 8–15 cm long, 0.5–1.0 cm wide, slightly scabrous beneath and along the margins; flowers 8.0–8.5 mm long; corolla green or greenish, 4–5 mm long; hoods pale green, 2.0–2.5 mm long; flowers lacking horns. A species of the interior United States, at an eastern boundary of its range in south-central and north-central counties of Ohio; mesic to damp fields, roadsides, waste places. Late June–August.

13. **Asclepias viridiflora** Raf. Green-flowered Milkweed
    incl. var. *viridiflora*, var. *lanceolata* (Ives) Torr., and var. *linearis* (A. Gray) Fern.; *Acerates viridiflora* (Raf.) Eat.

Perennial, to 9 dm tall; leaves all or mostly opposite, variable in shape: ovate to narrowly oblong, broadly oblong, or obovate, mostly 5–8 cm long and 1–3 cm wide, both surfaces, especially the under surface, short-pubescent; flowers 9–13 mm long; corolla green or greenish, 6–7 mm long;

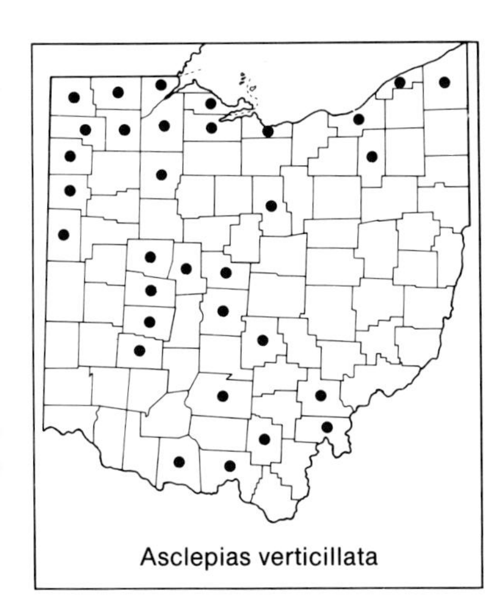
Asclepias verticillata

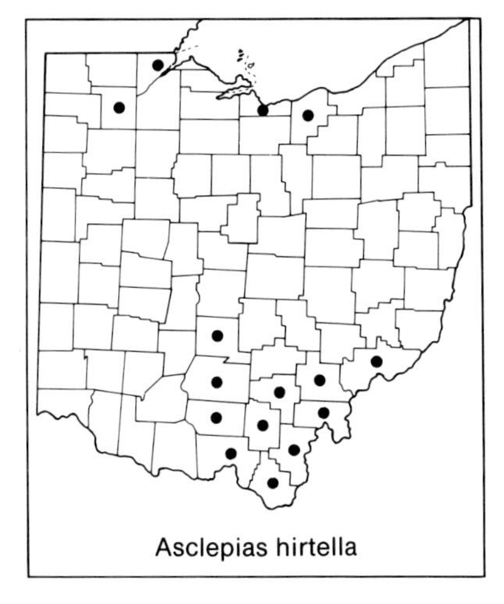
Asclepias hirtella

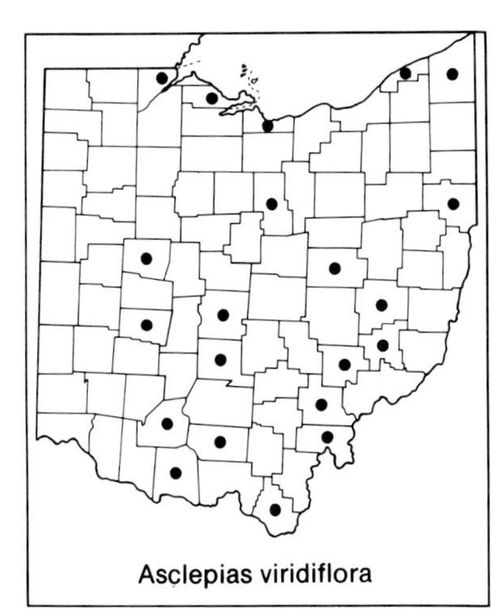
Asclepias viridiflora

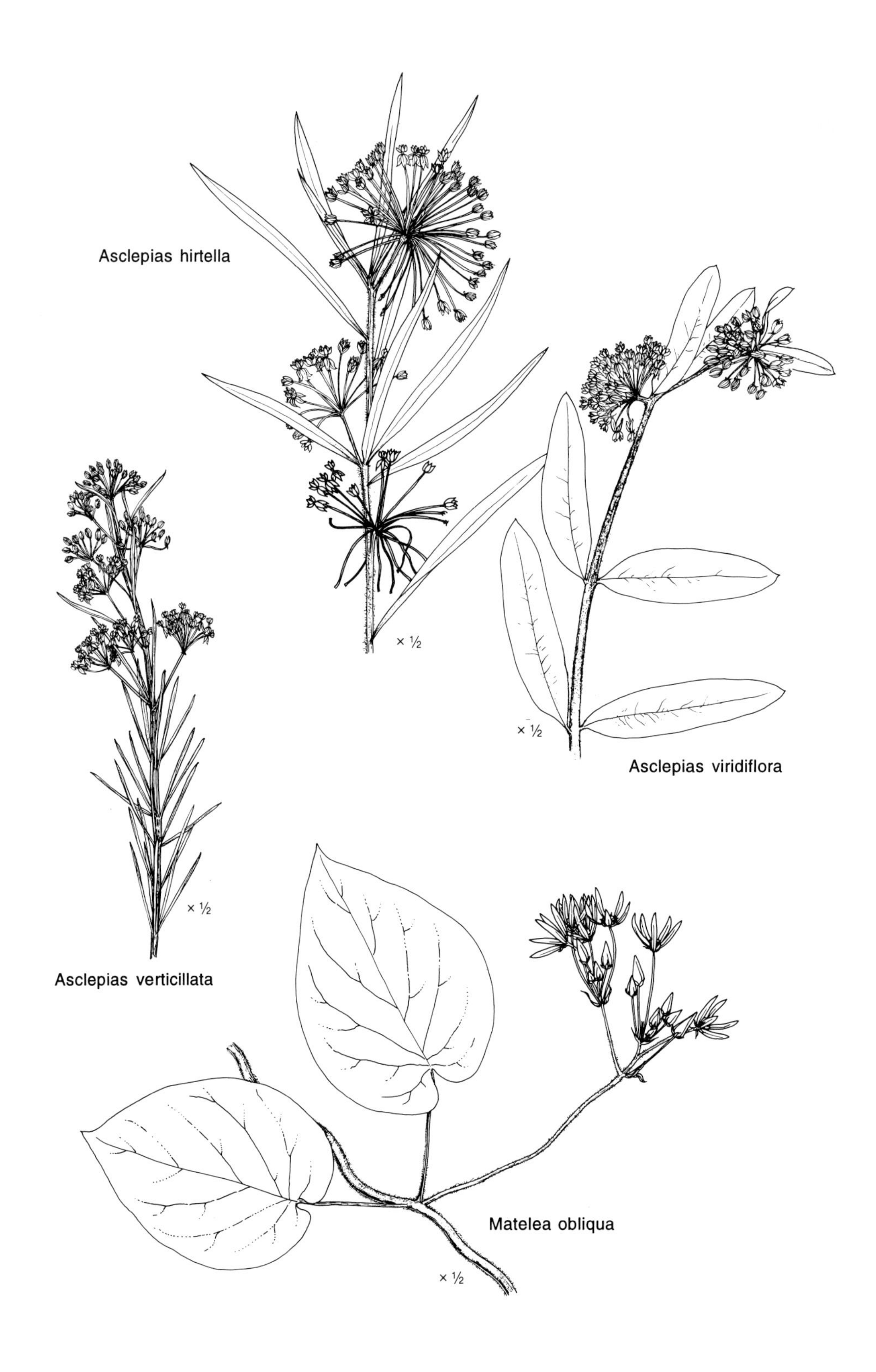
Asclepias hirtella
× ½
Asclepias viridiflora
× ½
Asclepias verticillata
× ½
Matelea obliqua
× ½

hoods pale green, 4–5 mm long; flowers lacking horns. At scattered sites throughout much of Ohio; dry grassy fields, slopes, pastures, roadside banks. June–August.

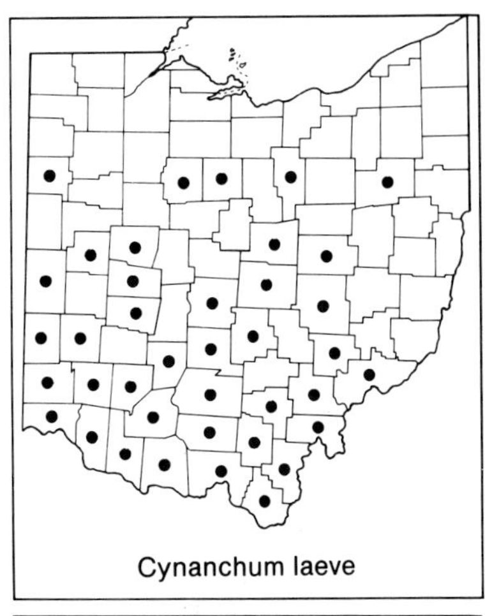
Cynanchum laeve

2. Cynanchum L.

A genus of some 100 or more species, widely distributed in tropical and temperate regions. One species is native to Ohio.

1. **Cynanchum laeve** (Michx.) Pers. SAND-VINE. HONEY-VINE
   Ampelamus albidus (Nutt.) Britt.; *Gonolobus laevis* Michx.

Twining, climbing perennial; leaves opposite, 5–15 (–17) cm long, long-triangular to ovate, apex acute to acuminate, base deeply and markedly cordate—the notch broadly to very broadly U-shaped, long-petiolate; flowers white, in axillary, umbelliform to racemose clusters; corolla lobes ca. 6 mm long. Common in southern counties, scarce or absent northward; river banks, alluvial thickets, roadside thickets, fencerows, and other disturbed habitats. July–early September.

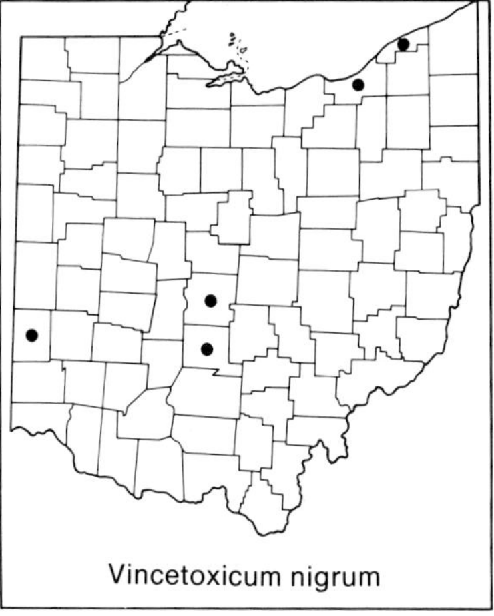
Vincetoxicum nigrum

3. Vincetoxicum Wolf

An Old World genus represented in the Ohio flora by a single, sparingly naturalized species. *Vincetoxicum rossicum* (Kleopow) Barbar. (*Cynanchum medium* of North American authors) is naturalized in Ontario (McNeill, 1981; R. J. Moore, 1959; Pringle, 1973a), but I have seen no Ohio specimens.

1. VINCETOXICUM NIGRUM (L.) Moench BLACK SWALLOW-WORT
   Cynanchum louiseae Kartesz & Gandhi, *C. nigrum* (L.) Pers.

Scrambling, twining, or climbing perennial; leaves opposite, 5–7 (–10) cm long, ovate, apex acute to acuminate, base rounded to subcordate, short-petiolate; flowers purplish-brown or dark purple, in axillary cymes; corolla lobes ca. 3 mm long. A European species, naturalized at a few scattered sites in Ohio; disturbed habitats. July–September.

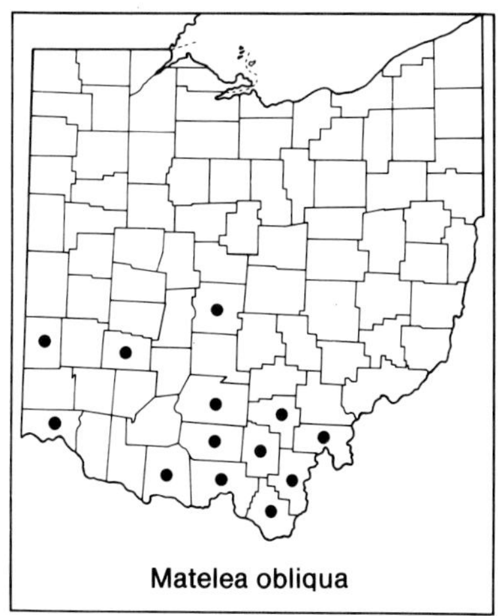
Matelea obliqua

4. Matelea Aubl.

A North American genus of perhaps 200 species, one of which is native to Ohio. Drapalik (1970) prepared a biosystematic study of the genus in southeastern United States.

1. **Matelea obliqua** (Jacq.) Woodson ANGLE-POD
   Gonolobus obliquus (Jacq.) R. & S.

Twining perennial; stems hirsute; leaves opposite, (7–) 10–20 (–25) cm long, ovate to orbicular, apex acute to short-acuminate, base cordate, pe-

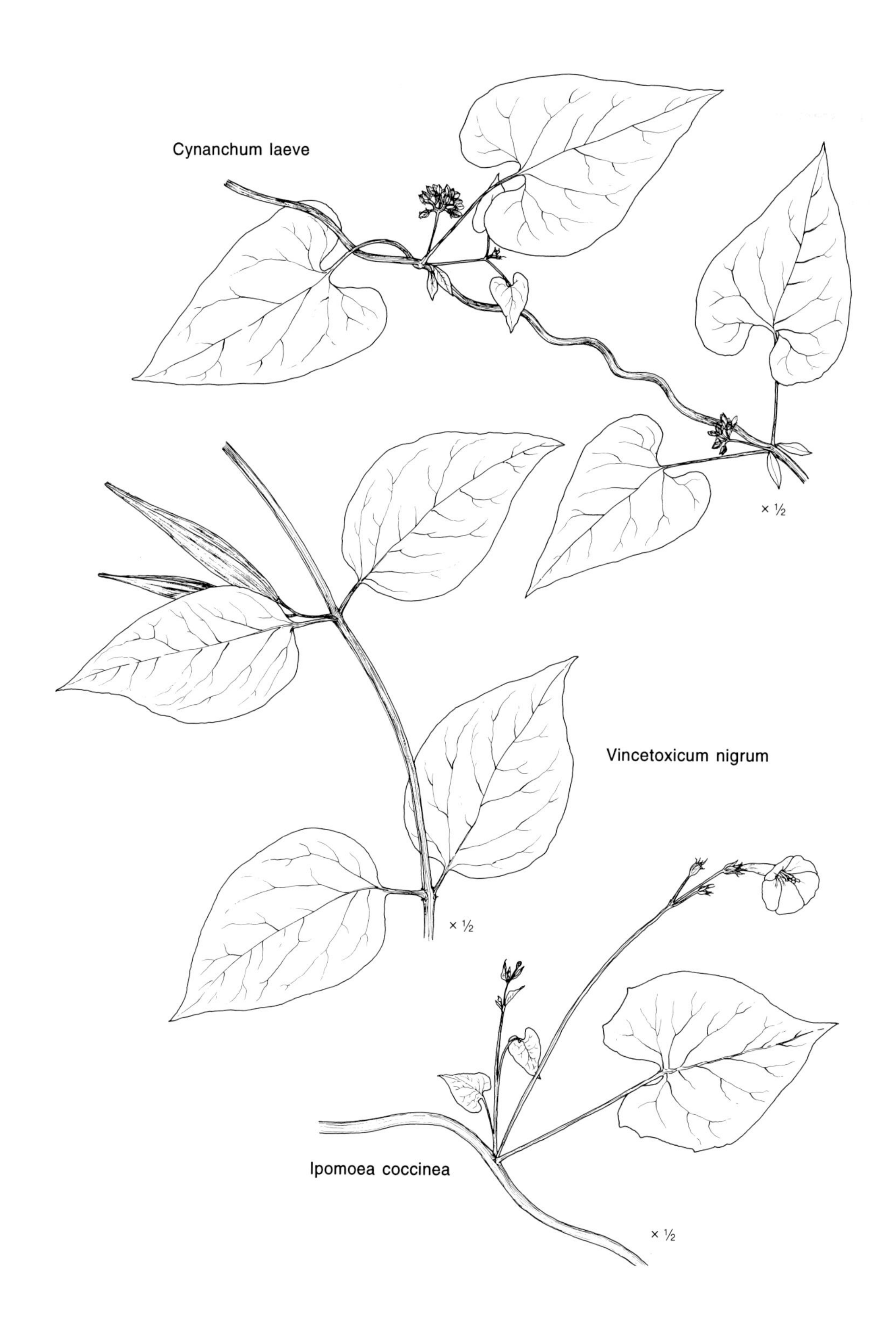
Cynanchum laeve
× ½
Vincetoxicum nigrum
× ½
Ipomoea coccinea
× ½

tiolate; flowers maroon to purplish-brown, in axillary, cymose or umbelliform clusters; corolla lobes ascending, to 1.5 cm long. A species of southeastern United States, at a northern boundary of its range in south-central and southwestern counties of Ohio; open woodlands, woodland borders, thickets. June–July.

Convolvulaceae. MORNING-GLORY FAMILY

Annual or perennial herbs, mostly climbing or twining vines; leaves alternate, simple; flowers perfect and often showy, solitary or in few-flowered inflorescences; calyx of 5 more or less separate sepals; corolla actinomorphic, of 5 partially to almost wholly fused petals; stamens 5, attached to the corolla tube; carpels 2 or 3, fused to form a compound pistil with 1 style and 1–3 stigmas; ovary superior; fruit a capsule. A family of some 1,500 species, chiefly tropical. K. A. Wilson (1960b) studied the family in southeastern United States.

Jacquemontia tamnifolia (L.) Griseb., an erect to twining weed with small blue flowers, native to tropical America, was collected once as an adventive in Hamilton County.

a. Flowers in dense clusters of 5 or more; calyx with long reddish pubescence; see above . . . . *Jacquemontia*

aa. Flowers solitary or in open inflorescences of 2–4 flowers; calyx glabrous or variously pubescent.

b. Flower stalk with 2 large sepaloid bracts closely subtending and covering and obscuring the calyx . . . . 2. *Calystegia*

bb. Flower stalk with or without bracts, if present not covering and obscuring the calyx.

c. Sepals at anthesis 6–22 mm long, usually acuminate, if blunt and rounded (as in *I. pandurata*) more than 10 mm long; leaves more or less cordate at base, acute to acuminate at apex; flower stalks with or without bracts . . . . 1. *Ipomoea*

cc. Sepals at anthesis 3–5 mm long, blunt and rounded at apex; leaves hastate to sagittate at base or rarely merely truncate, usually rounded to obtuse at apex; flower stalks bearing 2 minute bracts 1–4 mm long. . . . . 3. *Convolvulus*

1. Ipomoea L. MORNING-GLORY

Twining and scrambling vines; leaves cordate at base; flowers conspicuous and often showy. A widely distributed, chiefly tropical genus of some 500 species. Various cultivars of *Ipomoea tricolor* Cav., MORNING-GLORY, are grown as ornamentals, the striking 'Heavenly Blue' perhaps the best known. *Ipomoea batatas* (L.) Lam., SWEET POTATO, is frequently grown in Ohio gardens. Austin's (1978) taxonomic study of the *Ipomoea batatas* complex includes species no. 5 below; and Austin's (1986) discussion of the nomenclature of the *Ipomoea nil* complex covers species nos. 2 and 3. O'Donell's (1959) revision of the American species of *Ipomoea* section *Quamoclit* (Moench) Griseb. includes species no. 1. Stucky (1985) and Stucky and Beckman (1982) studied the pollination systems of the genus.

a. Corolla scarlet, salverform, the length of the tube more than twice the width of the limb; stamens and style exserted . . . . 1. *Ipomoea coccinea*

aa. Corolla pink or rose-red (not scarlet), to purple, blue, or white, funnelform to subcampanulate, the length of the tube less than twice the width of the limb; stamens and style included.

b. Peduncles and pedicels pubescent with reflexed hairs; stigma 3-lobed.

c. Sepals lanceolate, attenuate at apex, mostly 15–25 mm long; leaves 3-lobed . . . . 2. *Ipomoea hederacea*

cc. Sepals oblong to oblong-lanceolate, acute to subobtuse at apex, mostly 10–15 mm long; leaves usually unlobed, rarely 3-lobed . . . . 3. *Ipomoea purpurea*

bb. Peduncles and pedicels glabrous, smooth or spiculate; stigma unlobed or 2-lobed.

c. Sepals thick and somewhat leathery in texture, mostly obtuse or sometimes acute at apex, glabrous; corolla 5–8 cm long; peduncles smooth. . . . . 4. *Ipomoea pandurata*

cc. Sepals thin in texture, sharply acuminate at apex, usually ciliate; corolla 1.5–2.5 cm long; peduncles spiculate . . . . 5. *Ipomoea lacunosa*

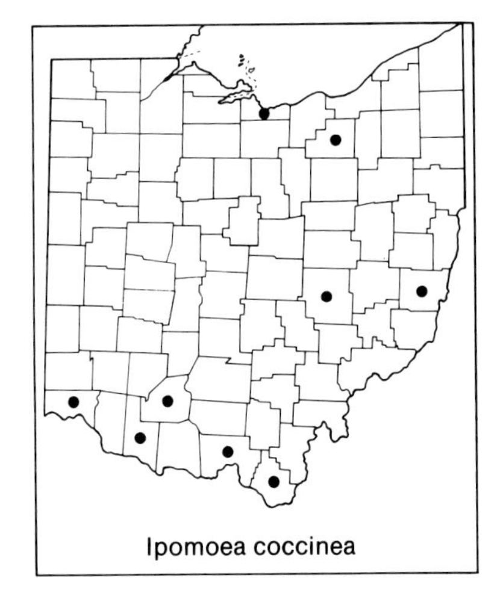
Ipomoea coccinea

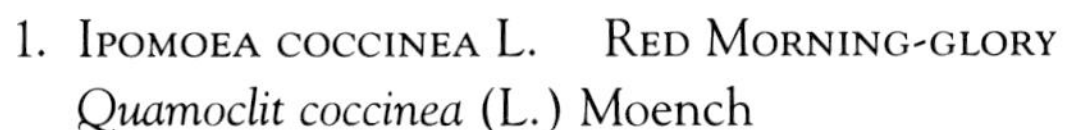

1. IPOMOEA COCCINEA L. RED MORNING-GLORY
Quamoclit coccinea (L.) Moench

Annual vine, twining and climbing; leaves ovate, entire or with a few coarse teeth or shallow angular lobes at base; flowers in axillary, few-flowered clusters or sometimes solitary, peduncles and pedicels glabrous; corolla scarlet, salverform, 2–3 cm long, the length of the tube more than twice the width of the limb; stamens and style exserted, the stigma unlobed, capitate. A species of southeastern United States; naturalized or perhaps native in a few southern counties, adventive northward; stream banks and other moist disturbed habitats. August–October.

2. Ipomoea hederacea Jacq. Ivy-leaved Morning-glory

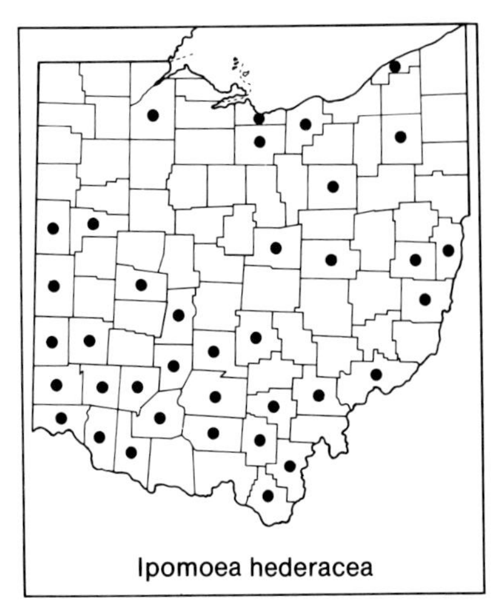
Ipomoea hederacea

Annual vine, twining and climbing; stems and petioles pubescent with reflexed hairs; leaves markedly 3-lobed; flowers solitary or in 2- or 3-flowered clusters in leaf axils; peduncles and pedicels pubescent with reflexed hairs; sepals lanceolate, attenuate at apex, mostly 15–25 mm long, conspicuously long-hirsute at base with spreading straw-colored hairs; corolla purplish-blue, white at base, funnelform, 3.5–4.5 cm long, the length of the tube less than twice the width of the limb; stamens and style included, the stigma 3-lobed. Naturalized from tropical America and sometimes casually cultivated; southern counties and at scattered sites northward; stream banks, roadsides, railways, fencerows, and other weedy disturbed habitats. August–October.

Ohio plants belong to var. *hederacea*. Variety *integriuscula* A. Gray, with entire and unlobed or nearly unlobed leaves, occurs to the south of Ohio.

3. Ipomoea purpurea (L.) Roth Common Morning-glory

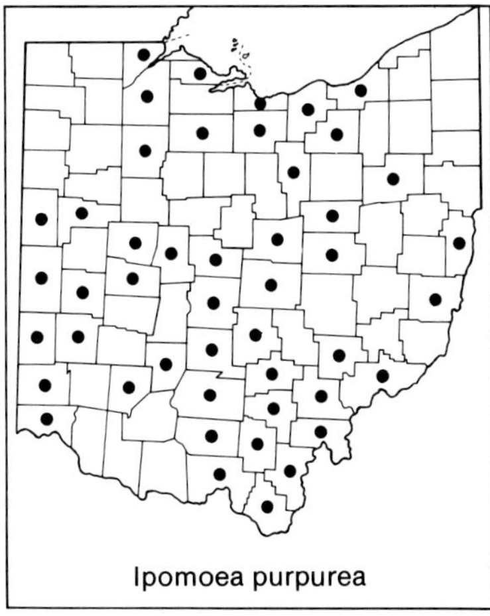
Ipomoea purpurea

Annual vine, twining and climbing; stems and petioles pubescent with reflexed hairs to nearly glabrous; leaves broadly ovate-cordate, unlobed or rarely 3-lobed; flowers solitary or in few-flowered clusters in leaf axils; peduncles and pedicels pubescent with reflexed hairs; sepals oblong to oblong-lanceolate, acute to subobtuse at apex, mostly 10–15 mm long, hirsute at base with spreading or ascending hairs; corolla rose-red to purple, blue, or white, pale below, funnelform, 4.5–6.5 cm long, ca. 4.5–5.0 cm across, the length of the tube less than twice the width of the limb; stamens and style included, the stigma 3-lobed. Naturalized from tropical America, and sometimes casually cultivated; southern counties and at scattered sites northward; cultivated fields, weed lots, stream banks, roadsides, weedy thickets, and other disturbed habitats. July–early November.

4. **Ipomoea pandurata** (L.) G. F. W. Mey. Wild Potato-vine

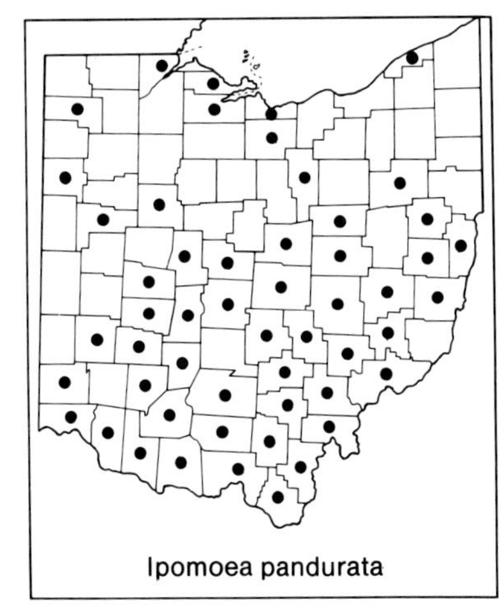
Ipomoea pandurata

Perennial vine with large tuberous root; stems trailing or twining, stems and petioles glabrous or nearly so; leaves broadly cordate-ovate or occasionally with two rounded basal lobes; flowers in few- to several-flowered clusters in leaf axils; peduncles and pedicels glabrous and smooth, their sap milky; sepals thick and somewhat leathery, mostly obtuse or sometimes acute at apex, glabrous; corolla white with reddish-purple center, funnelform, 5–8 cm long, the length of the tube less than twice the width of the limb; stamens and style included, the style 2-lobed. Southern counties and at scattered sites northward; river banks, fencerows, weedy fields, and other disturbed habitats. July–September.

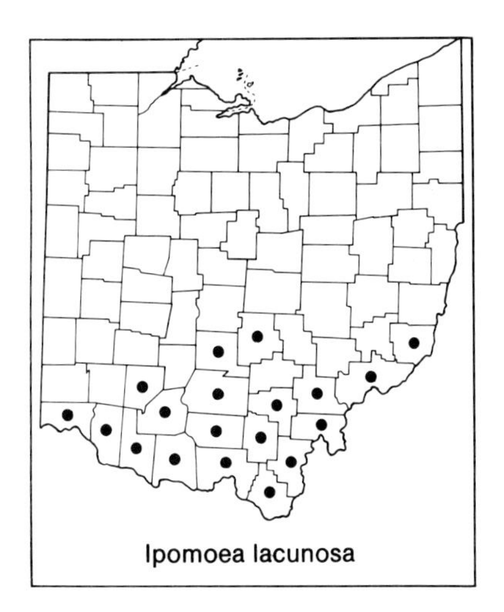

5. **Ipomoea lacunosa** L. SMALL-FLOWERED MORNING-GLORY

Annual vine, twining or occasionally somewhat erect; stems and petioles glabrous or nearly so; leaves broadly ovate-cordate, entire or sometimes 3-lobed; flowers solitary or in few-flowered clusters in leaf axils; peduncles and pedicels glabrous but often more or less spiculate; sepals thin in texture, sharply acuminate at apex, usually ciliate; corolla white or tinged with pink, funnelform, 1.5–2.5 cm long, the length of the tube less than twice the width of the limb; stamens and style included, the stigma unlobed. A species chiefly of southeastern United States, at a northern boundary of its range in southern counties of Ohio; mud flats, river banks, moist weedy habitats. August–October.

2. Calystegia R. Br.

Twining or scrambling to suberect herbs; leaves alternate, the base usually truncate to sagittate or hastate; flowers large and showy, usually solitary, each flower closely subtended by two large sepaloid bracts covering and obscuring the calyx. A widespread genus of about 25 species. *Calystegia hederacea* Wallich (*Convolvulus wallichianus* Spreng.) was collected once, in 1978, as a weed in fill dirt on the campus of Oberlin College in Lorain County. The identity of that specimen was determined by R. K. Brummitt, who also gave helpful advice on the nomenclature and identities of other *Calystegia* species occurring in Ohio.

The treatment here, recognizing this genus as one segregated from *Convolvulus*, is based on the study of W. H. Lewis and R. L. Oliver (1965); other helpful papers were those by Brummitt (1965) and Mohlenbrock (1982b).

a. Corolla 2–4 cm long, pink, and "double". . . . 3. *Calystegia pubescens*

aa. Corolla 2.5–8.0 cm long, pink to white, "single."

b. Stems erect, 1–4 (–6) dm tall and with no or few branches, not twining, or the upper stem declined and slightly twining; length of petioles half or less that of blades. . . . . 1. *Calystegia spithamaea*

bb. Stems creeping or twining, often more than 4 dm long and with many branches; length of petioles less than half to more than half that of blades.

c. Peduncles terete or nearly so; bracts at anthesis 1.7–3.0 cm long . . . . 2. *Calystegia sepium*

cc. Peduncles prominently winged; bracts at anthesis 1.2–1.5 cm long; see above. . . . . *Calystegia hederacea*

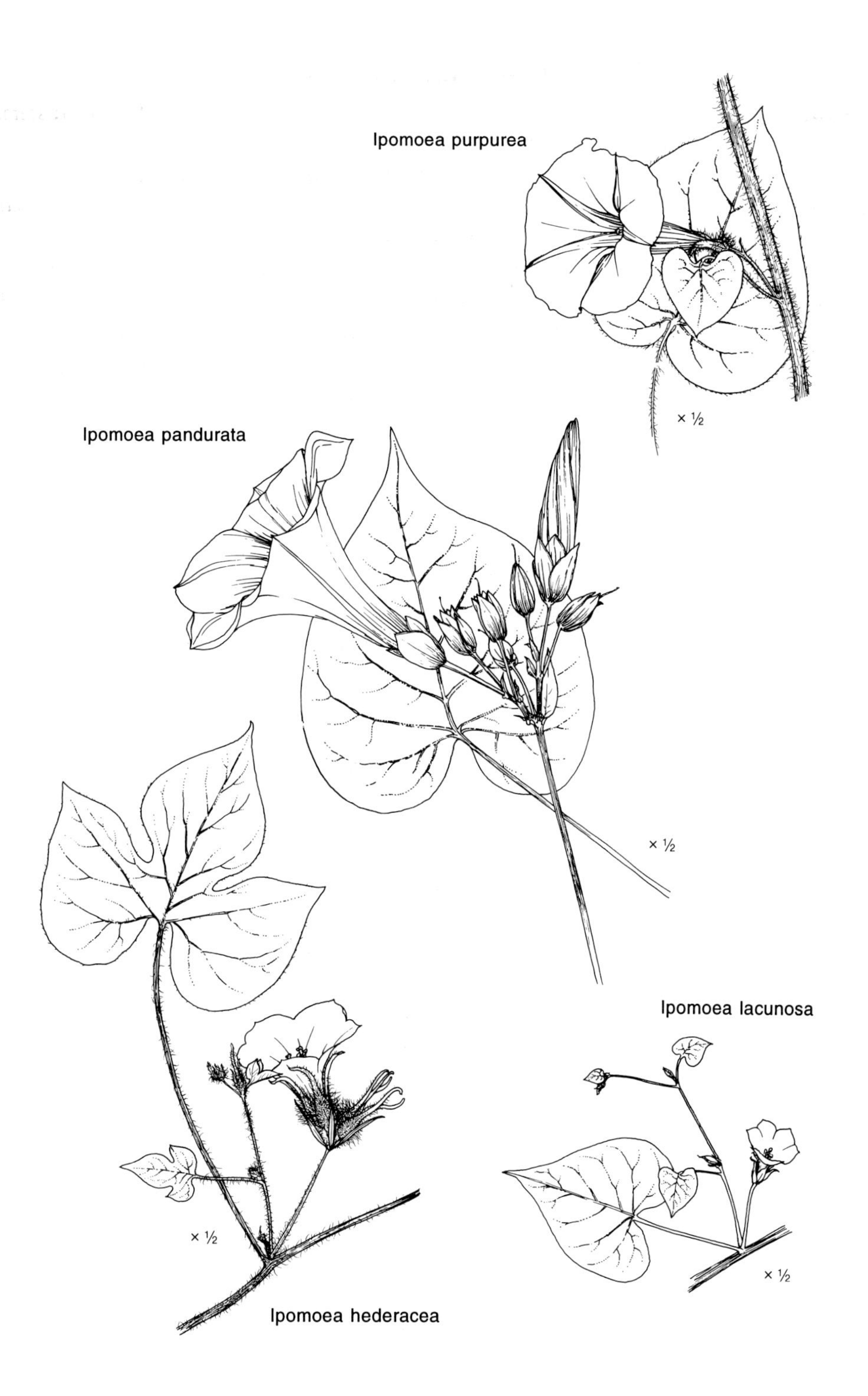
Ipomoea purpurea
× ½
Ipomoea pandurata
× ½
Ipomoea lacunosa
× ½
× ½
Ipomoea hederacea

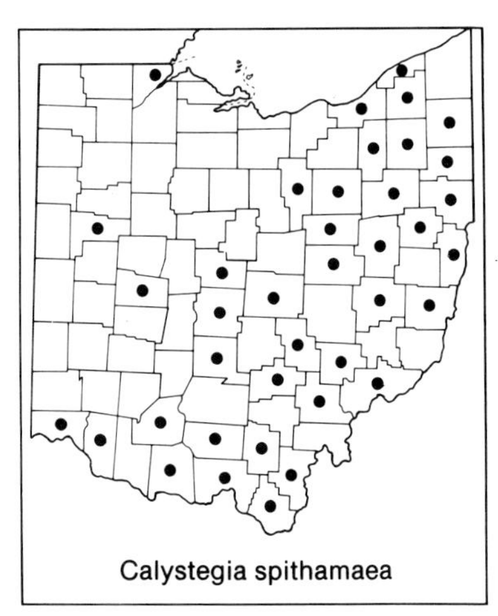
Calystegia spithamaea

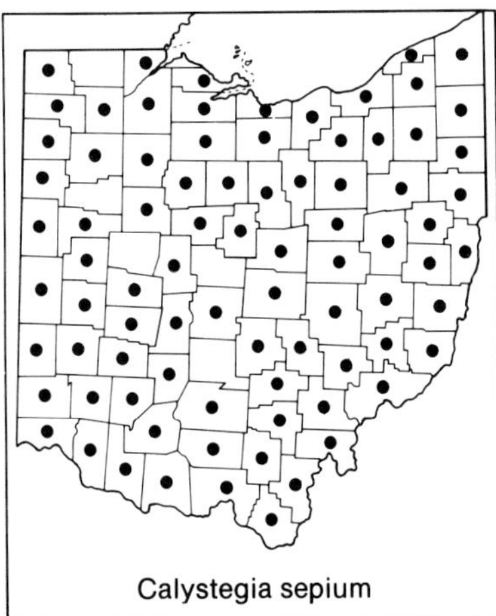
Calystegia sepium

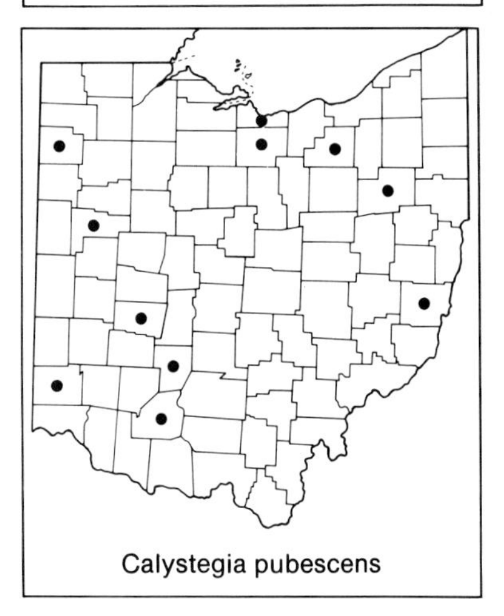
Calystegia pubescens

1. **Calystegia spithamaea** (L.) Pursh UPRIGHT BINDWEED. LOW BINDWEED
   Convolvulus spithamaeus L.

Perennial; stems erect, 1–4 (–6) dm long, with no or few branches, either not twining or the upper stem declined and twining slightly; leaves more or less oblong, the base truncate to slightly cordate, petioles less than half as long as the blades; corolla 4–6 cm long, white or tinged with pink. Chiefly in eastern and southern counties; dry fields, roadside banks, and other dry rocky or sandy habitats. May–June.

2. **Calystegia sepium** (L.) R. Br. HEDGE BINDWEED
   Convolvulus sepium L., incl. var. *sepium*, var. *fraterniflorus* Mackenz. & Bush, and var. *repens* (L.) A. Gray

Creeping or twining and climbing perennial, with many branches; leaves long-triangular to deltoid-ovate, the base mostly sagittate to hastate; peduncles terete or nearly so; bracts at anthesis 1.7–3.0 cm long; corolla 4–8 cm long, white to pink or rose-colored. Throughout Ohio; river banks, beaches, fields, fencerows, roadsides, railways, and other disturbed open habitats. June–October.

A complex species with many named infraspecific taxa, the taxa based chiefly on variation in leaf shape, sometimes in combination with bract characters and variation in flower color (Fernald, 1950; Gleason, 1952b; Brummitt, 1965, 1980; Mohlenbrock, 1982b), but intergrading more or less insensibly in Ohio. Brummitt writes from England (in personal communication), "Ohio seems to be very much the meeting point of my various subspecies, and the distinctions may be less clear there than either further east or further west." Pending clarification which may result from Brummitt's scheduled revision of the genus, I treat the species here as a polymorphic one and assign no infraspecific ranks to Ohio plants. I tentatively include within the species those plants called *Convolvulus sepium* var. *fraterniflorus* by Fernald (1950) and Gleason (1952b), but transferred to another species by Brummitt (1980) and called *Calystegia silvatica* (Kit.) Griseb. subsp. *fraterniflorus* (Mackenz. & Bush) Brummitt.

3. CALYSTEGIA PUBESCENS Lindl. CALIFORNIA-ROSE
   C. *hederacea* of authors, incl. Bailey Hortorium Staff (1976), not Wallich; *Convolvulus japonicus* of authors, incl. Gleason (1952b), not Thunb.; *Convolvulus pellitus* [forma *anestius* Fern.] of authors, incl. Fernald (1950), not Ledeb.

Twining and climbing perennial; leaves narrowly triangular, base sagittate to hastate; corolla 2–4 cm long, pink and "double." A native of eastern China, cultivated and sometimes adventive or perhaps locally naturalized at scattered sites throughout Ohio; disturbed habitats. July–September.

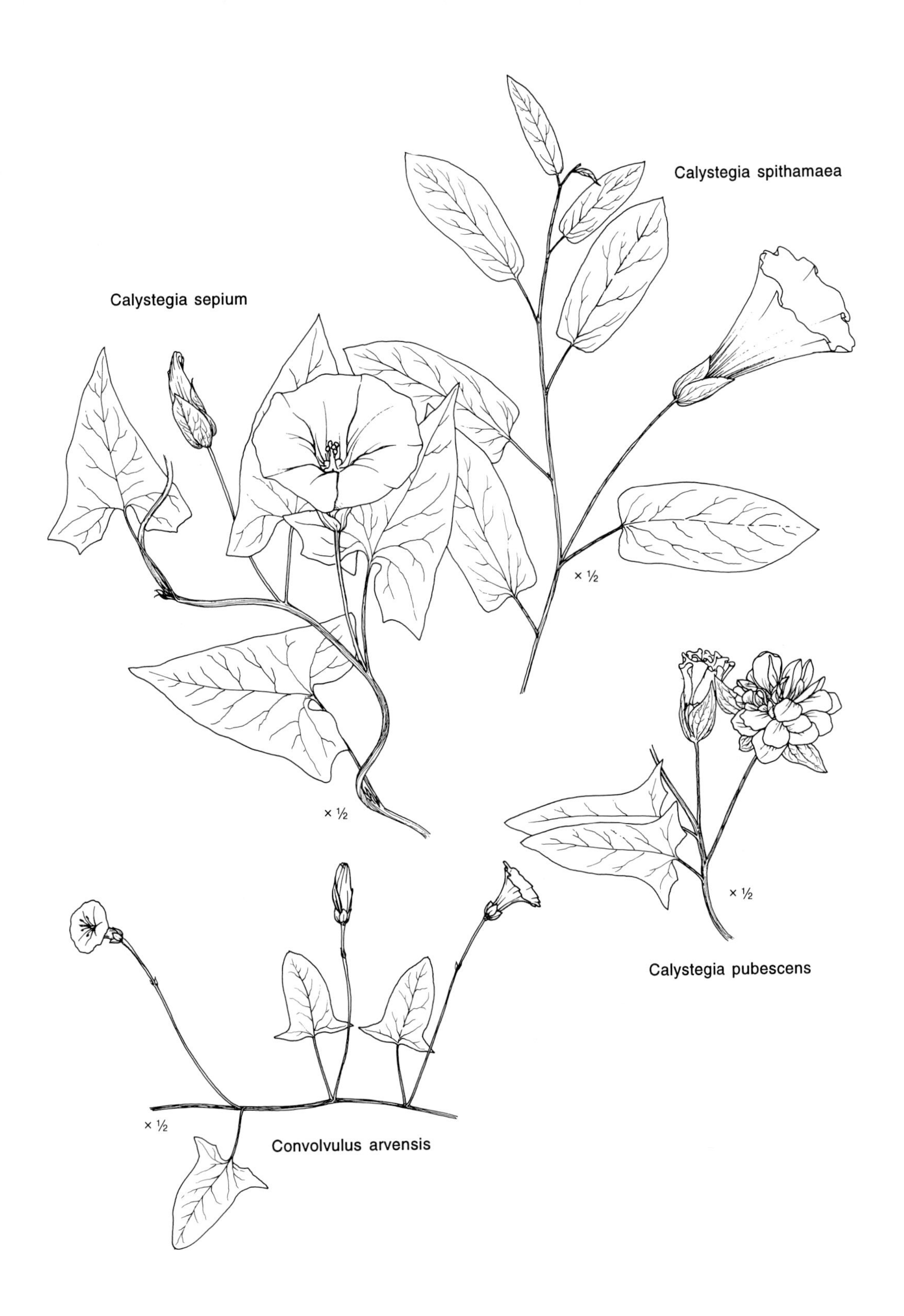
Calystegia spithamaea
Calystegia sepium
× ½
× ½
× ½
Calystegia pubescens
× ½
Convolvulus arvensis

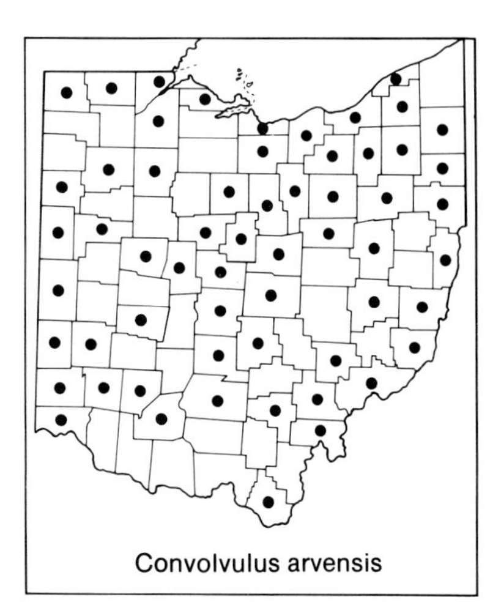

The nomenclature and synonymy listed above follow the conclusions of Brummitt (personal communication), who determined the identity of a representative Ohio specimen of this species.

3. Convolvulus L.

A widely distributed genus of more than 200 species, one of which is well established in the Ohio flora.

1. Convolvulus arvensis L. Bindweed. Field Bindweed

Trailing or twining and climbing perennial; leaves narrowly triangular to elliptic or oblong, apex rounded to obtuse, base hastate to sagittate or rarely merely truncate; flowers axillary, usually solitary, sometimes in pairs from a single peduncle; the stalk bearing 2 minute bracts 1–4 mm long; sepals at anthesis 3–5 mm long, blunt and rounded at apex; corolla 1.5–2.0 cm long, funnelform, white to pinkish; stigmas 2. A Eurasian species naturalized throughout much of Ohio, and sometimes a troublesome weed in cultivated fields and gardens; also in other disturbed habitats, especially along roads and railroads. June–September.

Cuscutaceae. DODDER FAMILY

A family of a single genus traditionally treated as a part of the Convolvulaceae, but segregated here following the classification of Cronquist (1981).

1. Cuscuta L. DODDER. LOVE-VINE. STRANGLE-WEED

Annual, yellow to orange twining vines with masses of filiform stems, parasitic by means of invasive haustoria on various genera of flowering plants; leaves lacking or represented only by minute scales; flowers perfect, whitish, few to more often numerous in open to dense cymose clusters; calyx of (3 or) 4 or 5 separate or fused sepals; corolla actinomorphic, of (3 or) 4 or 5 fused petals; stamens (3 or) 4 or 5, fused to the corolla tube; carpels 2, fused to form a compound pistil with 2 styles; ovary superior; fruit a capsule. A cosmopolitan genus of more than 150 species. Yuncker (1965) prepared a comprehensive revision of the genus as it is represented in the North American flora, Gandhi et al. (1987) a study of the genus in Louisiana, and Kuijt (1969) a study of the biology of parasitic plant genera, of which this is one.

Cuscuta epilinum Weihe, Flax Dodder, a European species, was collected once in Ohio, as an adventive, from a flax plant in 1895 in Wayne

County. Adventive specimens of *Cuscuta suaveolens* Ser., a South American species associated with alfalfa, have been collected twice: from Auglaize County in 1910, and from Lucas County in 1929. Both specimens were originally identified as *Cuscuta indecora* Choisy, PRETTY DODDER, a species of western United States, of which I have seen no bona fide Ohio records. I have been unable to locate the Montgomery County specimen of *C. indecora* cited by Schaffner (1932), but assume that it also was most likely one of *C. suaveolens*.

The key and text below are adapted from Fernald (1950), Gleason (1952b), and Yuncker (1965).

a. Stigma globose-capitate; capsules not circumscissile, closed at maturity or breaking open irregularly.
 b. Sepals fused for part of their length; inflorescences moderately dense to open.
  c. Capsules depressed-globose to globose or ovoid, rarely longer than wide, thin-walled apically at base of styles.
   d. Lobes of corolla usually 3 or 4, rarely 5, acute or obtuse at apex, erect.
    e. Corolla tube covering only basal part of capsule, the lobes 3 or 4, acute at apex; styles shorter than the ovary .... 1. *Cuscuta polygonorum*
    ee. Corolla tube covering nearly the entire capsule, the lobes 4 or rarely 5, obtuse to broadly acute at apex; styles nearly equalling to longer than the ovary .... 2. *Cuscuta cephalanthi*
   dd. Lobes of corolla usually 5, acute and inflexed at apex .... 3. *Cuscuta pentagona*
  cc. Capsules depressed-globose to globose or elongate-globose, sometimes longer than wide, thickened apically at base of styles.
   d. Tips of corolla lobes obtuse and spreading; capsules elongate-globose; calyx and corolla lobes mostly 5 .... 5. *Cuscuta gronovii*
   dd. Tips of corolla lobes acute and inflexed; capsules depressed globose, as wide as or wider than long.
    e. Flowers mostly 2 mm or more in length; calyx and corolla lobes mostly 5; see above .... *Cuscuta suaveolens*
    ee. Flowers mostly 2 mm or less in length; calyx and corolla lobes mostly 4 .... 4. *Cuscuta coryli*
 bb. Sepals separate to the base; inflorescences usually quite dense.
  c. Bracts subtending calyx appressed, their tips erect-appressed .... 6. *Cuscuta compacta*
  cc. Bracts subtending calyx ascending, their tips spreading-recurved .... 7. *Cuscuta glomerata*

aa. Stigma filiform or linear; capsules more or less globose, opening along a circumscissile suture.

b. Calyx lobes ovate; length of style plus stigma about equalling length of ovary; style and stigma included; see above . . . . *Cuscuta epilinum*

bb. Calyx lobes triangular; length of style plus stigma longer than length of ovary; style and stigma exserted . . . . 8. *Cuscuta epithymum*

1. **Cuscuta polygonorum** Engelm. SMARTWEED DODDER

Annual; inflorescences moderately dense; sepals fused basally; corolla tube covering only the basal part of the capsule, corolla lobes 3 or 4, acute at apex, erect; styles shorter than the ovary, the stigmas globose-capitate; capsule depressed-globose, thin-walled at base of styles. At scattered sites from southeastern to north-central Ohio; host genera: *Aster*, *Bidens*, *Polygonum*, *Salix*, *Trifolium*. July–September.

2. **Cuscuta cephalanthi** Engelm. BUTTONBUSH DODDER

Annual; inflorescences open to moderately dense; sepals fused basally; corolla tube covering nearly the entire capsule, corolla lobes 4 or rarely 5, obtuse to broadly acute at apex, erect; styles nearly equalling to longer than the ovary, the stigmas globose-capitate; capsule depressed-globose to globose, thin-walled at base of styles. At scattered sites, mostly in western Ohio; host genera: *Cephalanthus*, *Salix*, *Solidago*. August–September.

3. **Cuscuta pentagona** Engelm. FIVE-ANGLED DODDER

incl. C. *campestris* Yuncker

Annual; inflorescences open to moderately dense; sepals fused basally, calyx lobes as wide as to wider than long; corolla lobes usually 5, triangular to lanceolate, spreading to reflexed, acute and inflexed at apex; styles

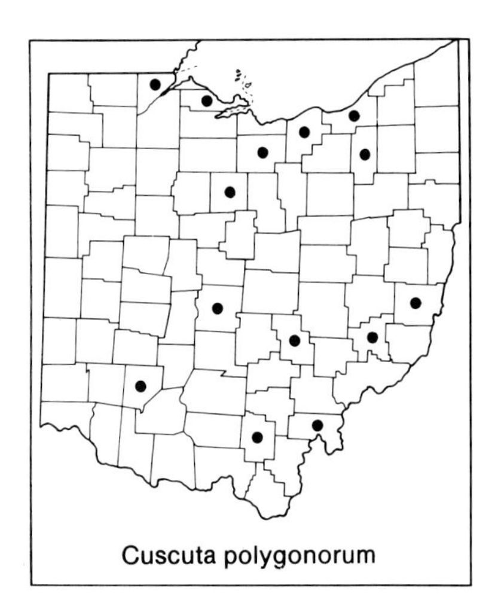

Cuscuta polygonorum

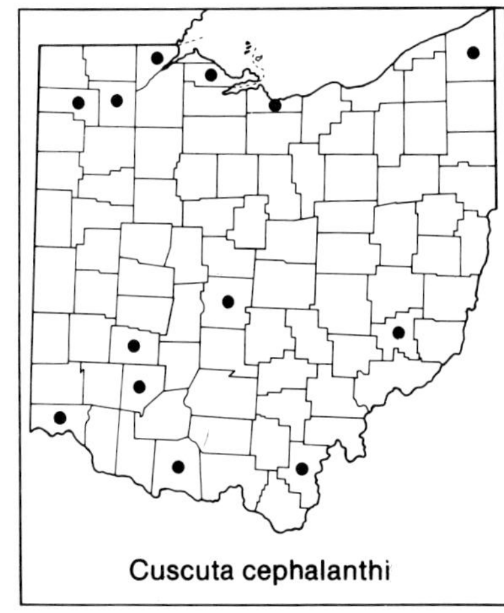

Cuscuta cephalanthi

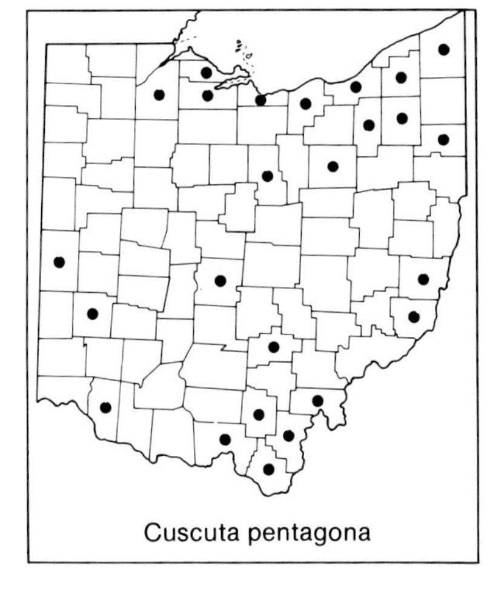

Cuscuta pentagona

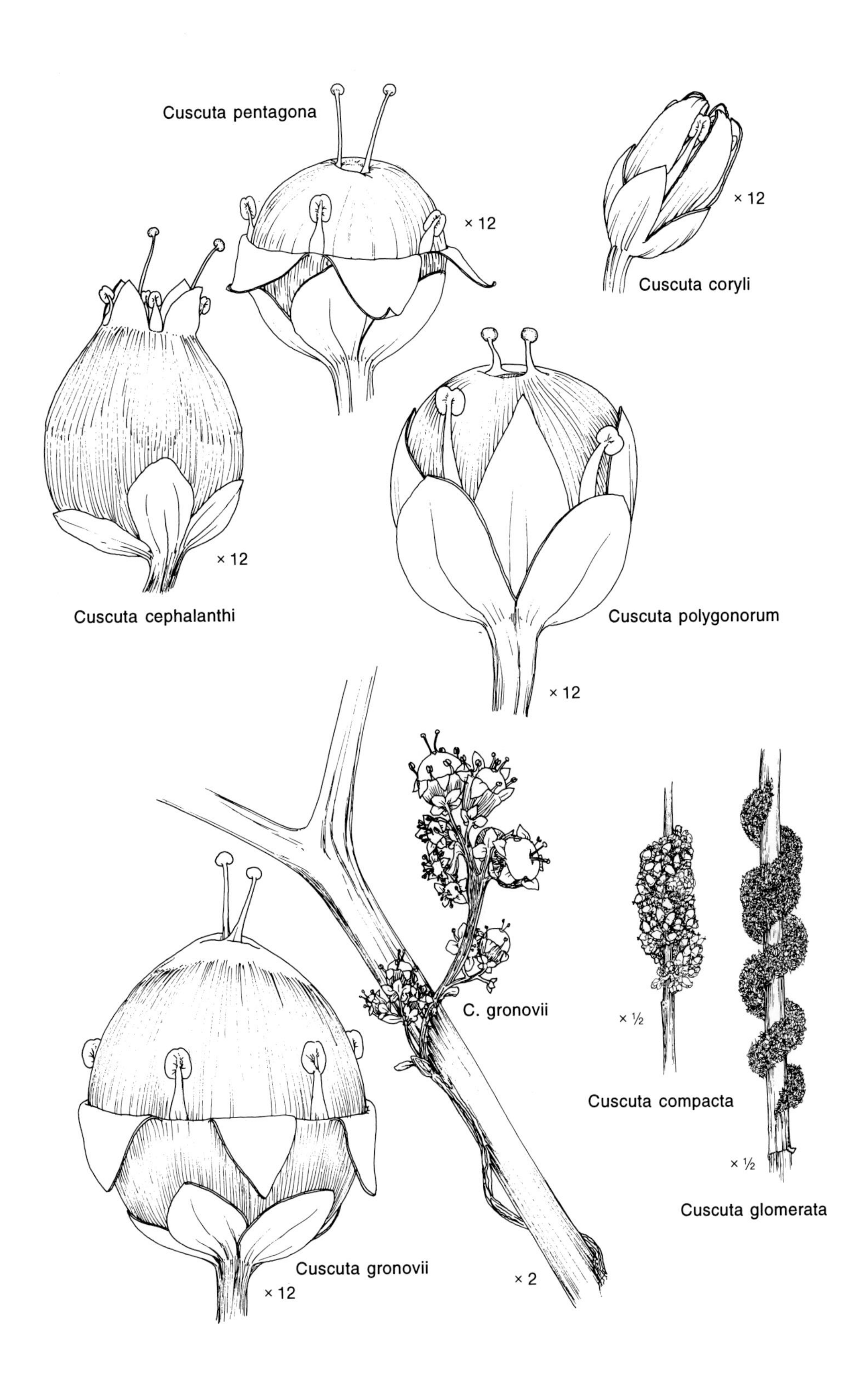
Cuscuta pentagona
× 12
× 12
Cuscuta coryli
× 12
Cuscuta cephalanthi
Cuscuta polygonorum
× 12
C. gronovii
× ½
Cuscuta compacta
× ½
Cuscuta glomerata
Cuscuta gronovii
× 12
× 2

shorter than to longer than the ovary, the stigmas globose-capitate; capsule depressed-globose to ovoid and longer than wide, thin-walled at base of styles. At scattered sites mostly in the eastern half of Ohio; host genera: *Amphicarpaea*, *Impatiens*, *Trifolium*, *Aster* and various other genera of the Asteraceae. June–October.

Cuscuta campestris is merged without rank under this species following Gleason and Cronquist (1963, 1991).

4. **Cuscuta coryli** Engelm. HAZEL DODDER

Annual; inflorescences moderately dense; sepals fused basally, calyx lobes mostly 4, triangular-ovate; corolla lobes mostly 4, longer than wide, erect, acute and inflexed at apex; styles about as long as the ovary, the stigmas globose-capitate; capsule depressed-globose, thickened apically at base of styles. At a few scattered sites; host genera: *Corylus*, *Phyla*, *Polygonum*, *Solidago*. July–September.

5. **Cuscuta gronovii** Willd. COMMON DODDER

Annual; inflorescences open to moderately dense; sepals fused basally, calyx lobes mostly 5, ovate to oblong; corolla lobes mostly 5, spreading, more or less obtuse at apex; styles shorter than the ovary, the stigmas globose-capitate; capsule elongate-globose, thickened apically at base of styles. The most frequent dodder in Ohio, throughout much of the state; host genera: *Aster*, *Bidens*, *Convolvulus*, *Eupatorium*, *Hypericum*, *Impatiens*, *Ipomoea*, *Laportea*, *Lespedeza*, *Lobelia*, *Mimulus*, *Parthenocissus*, *Rhus*, *Salix*, *Saururus*, *Solanum*, *Solidago*, *Verbena*, *Verbesina*. July–October.

6. **Cuscuta compacta** Juss. SESSILE DODDER

Annual; inflorescences thick and dense; calyx subtended by two or more appressed bracts, tips of bracts erect-appressed; sepals separate or nearly so;

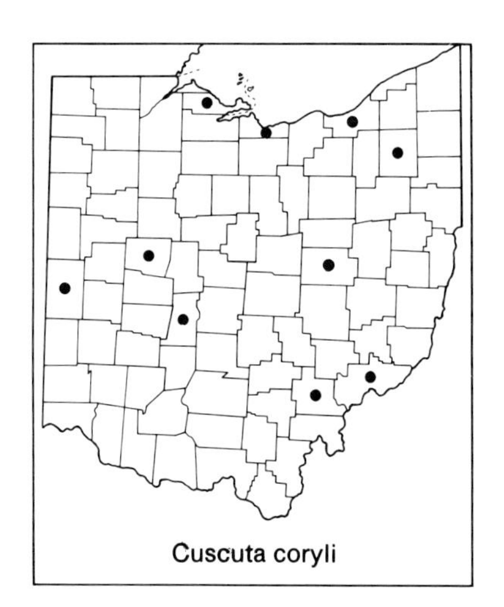

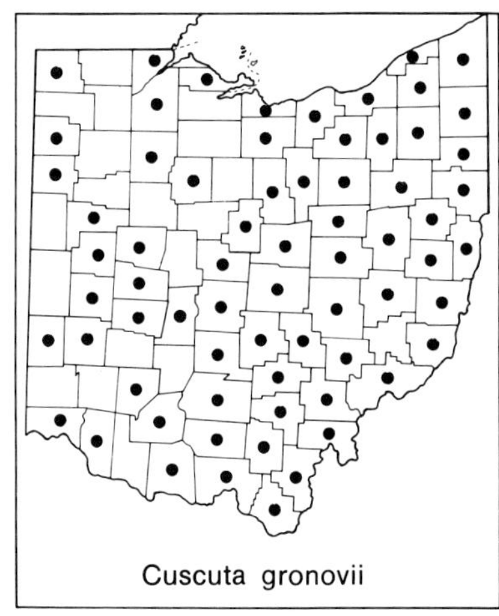

corolla lobes spreading to reflexed; styles longer than the ovary, the stigmas globose-capitate; capsule elongate-globose, thickened apically at base of styles. Jackson County in south-central Ohio; host genus: *Cephalanthus*. July–October.

7. **Cuscuta glomerata** Choisy Composite Dodder

Annual; inflorescences dense, thick and rope-like; calyx subtended by several ascending bracts, tips of bracts spreading-recurved; sepals separate; corolla lobes spreading to reflexed; styles longer than the ovary, the stigmas globose-capitate; capsule globose, thickened apically at base of styles. A species chiefly of the plains, at an eastern boundary of its range in central Ohio (Peskin, 1990); host genera: *Helianthus*, *Solidago*. July–September.

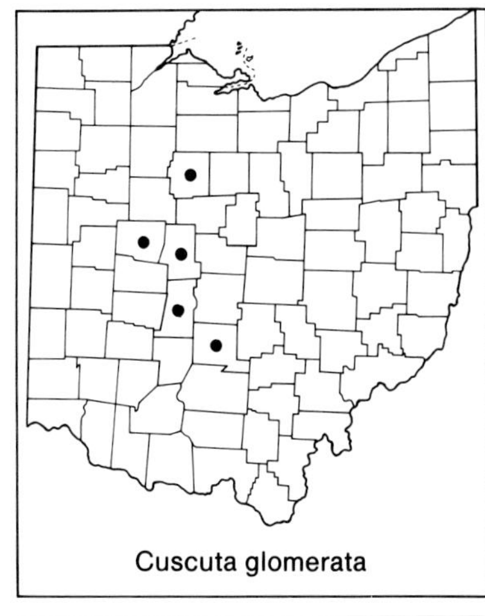

Cuscuta glomerata

8. Cuscuta epithymum (L.) L. Clover Dodder

Annual; inflorescences thick and dense; sepals fused below to form a calyx tube, the lobes triangular; corolla lobes triangular, spreading; stigmas linear, the length of styles and stigmas combined exceeding length of ovary; capsule globose, opening along a circumscissile suture. A European species, naturalized at a few Ohio sites; host genera: *Medicago*, *Trifolium*. Not illustrated. July–October.

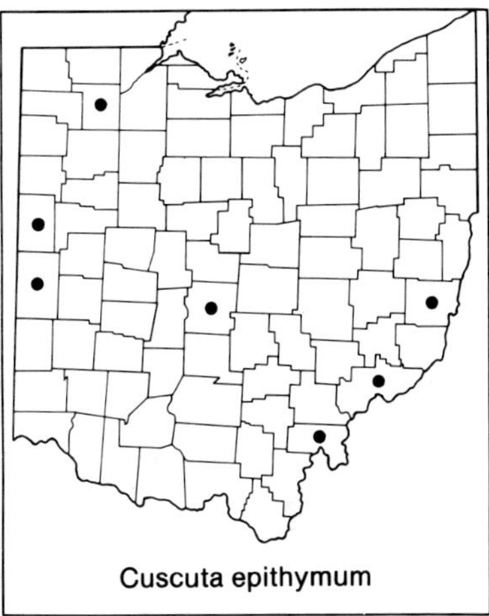

Cuscuta epithymum

Polemoniaceae. PHLOX FAMILY

Perennial or infrequently biennial or annual herbs; leaves simple or compound; flowers perfect, in cymose or paniculate or racemose, often showy, inflorescences; calyx 5-lobed, persistent; corolla actinomorphic, 5-lobed; stamens 5, attached to the corolla tube; carpels 3, fused to form a compound pistil with a single style; ovary superior; fruit a capsule. A small family with distribution centered in the temperate region of North America. V. Grant (1959) studied the family's natural history, V. Grant and K. A. Grant (1965) the pollination ecology, and K. A. Wilson (1960a) the family as it is represented in the flora of southeastern United States.

Three annuals native to western North America have appeared as adventive weeds in Ohio. *Collomia linearis* Nutt., with blue flowers and narrow, simple, alternate leaves, was collected along a railroad in Pickaway County in 1955. *Gilia capitata* Sims, with white to blue flowers crowded in a dense head and leaves dissected into narrow segments, was collected in a Highland County pasture in 1933. *Navarretia intertexta* (Benth.) Hook., with small purple flowers in dense crowded heads and leaves dissected into filiform segments, was collected in Allen County in 1920 and again in 1928; the 1920 collection was of a plant thought to have been introduced with grass seed.

a. Leaves all or mostly deeply lobed, deeply dissected, or pinnately compound, but other leaves sometimes simple and linear.

b. Leaves pinnately compound, leaflets broad: lanceolate, oblong-lanceolate, oblong, or elliptic. 1. *Polemonium*

bb. Leaves all or mostly deeply pinnately lobed, the lobes linear or filiform, other leaves sometimes simple and linear.

c. Calyx regular, the lobes approximately equal; plants either with calyx 3–4 mm long and the flowers congregated in a dense head, or with calyx 8–10 mm long and the flowers arranged in a narrow raceme or panicle.

d. Inflorescence a dense head; corolla 6–10 mm long; see above. . . . . *Gilia*

dd. Inflorescence elongated, narrowly racemose or narrowly paniculate; corolla 20–35 mm long . . . . 2. *Ipomopsis*

cc. Calyx irregular, the lobes unequal; plants with calyx 8–10 mm long and the flowers congregated into dense head-like cymes; see above . . . . *Navarretia*

aa. Leaves all simple and unlobed, their margins entire.

b. Leaves alternate; corolla lobes 1–2 mm long; see above . . . *Collomia*

bb. Leaves all or mostly opposite; corolla lobes 6–15 mm long . . . . 3. *Phlox*

1. Polemonium L. GREEK VALERIAN

A genus of some 20 or more species, rather widespread but chiefly in North America, one of the species native to the Ohio flora. The most recent revision is by Davidson (1950). *Polemonium caeruleum* L., Garden Jacob's Ladder or Greek Valerian, an ornamental from Eurasia, is sometimes cultivated in Ohio. It was collected once as an adventive in Ashtabula County, in 1931.

a. Principal lower leaves with 7–17 leaflets . . . . . 1. *Polemonium reptans*

aa. Principal lower leaves with 19–29 leaflets; see above . . . . *Polemonium caeruleum*

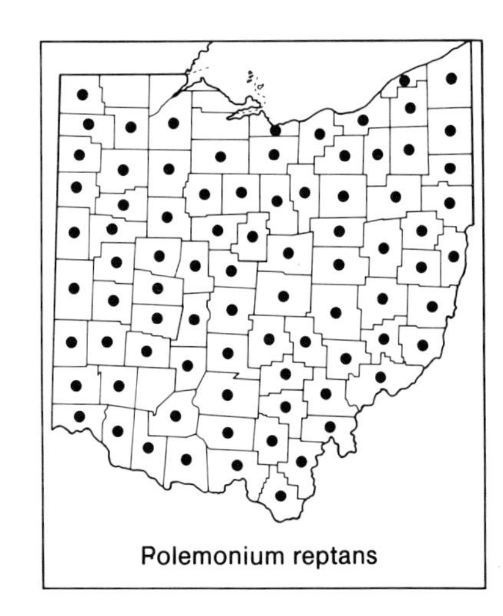

1. **Polemonium reptans** L. Jacob's Ladder

Perennial; stems erect or weakly erect, to 4 (–5) dm long; leaves alternate, pinnately compound, the leaflets lanceolate to oblong-lanceolate, oblong, or elliptic; flowers nodding, in few-flowered, more or less elongate clusters; calyx 5–10 mm long; corolla 12–16 mm long, blue, campanulate to funnelform, the lobes ascending to spreading. An attractive spring wildflower occurring throughout Ohio; rich moist wooded bottomlands, terraces, and slopes. April–mid-June.

Plants from Adams, Highland, Pike, and Scioto counties, in south-central Ohio, and from a few adjacent counties of Kentucky, have been segregated as var. *villosum* (Braun 1940, 1956). Despite its lack of acceptance by Davidson (1950), the taxon seems worthy of varietal rank. Ohio plants may be separated by the key below, but as noted by Spooner et al. (1983), "Apparent intergrades between the two varieties are frequent at the fringes of the range of var. *villosum* and determination [of specimens] sometimes is difficult."

a. Inflorescence short spiculate-pubescent, the hairs less than 0.1 mm long; stems glabrous or with few scattered hairs . . . . var. *reptans*

aa. Inflorescence densely glandular-villous, the hairs more than 0.1 mm long; stems glandular-villous, especially near nodes . . . . var. *villosum* E. Braun BRAUN'S JACOB'S LADDER

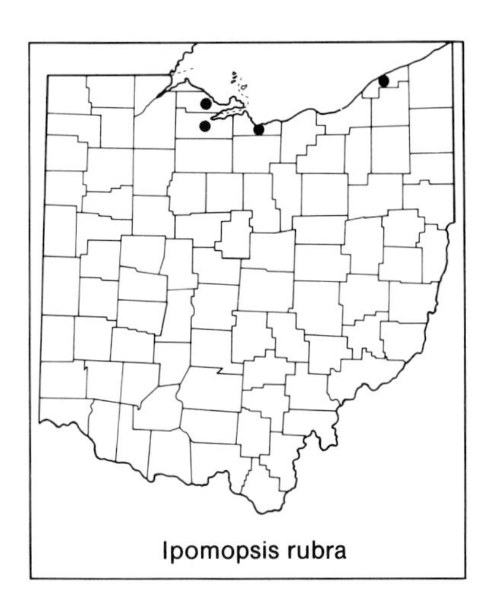
Ipomopsis rubra

2. Ipomopsis Michx.

A New World genus of 20–25 species with distribution centered in western North America; a single species is marginally established in the Ohio flora.

1. IPOMOPSIS RUBRA (L.) Wherry STANDING-CYPRESS
Gilia rubra (L.) A. A. Heller

Biennial, to 1 m tall; leaves alternate, deeply pinnately lobed, the lobes linear to filiform; flowers several to many in an elongated racemose or narrowly paniculate inflorescence; calyx 8–10 mm long; corolla orangish-red, tubular-salverform, 20–35 mm long. A native of southeastern United States, escaped from cultivation and marginally naturalized at a few sites in the northernmost tier of counties; disturbed habitats, sometimes near cemeteries. Estes and Hall (1975) describe the pollination of this species by ruby-throated hummingbirds. July–August.

3. Phlox L. PHLOX

Erect to prostrate perennials, the former often with prostrate to suberect, sterile basal shoots in addition to erect flowering stems; leaves simple, mostly opposite; inflorescences showy; corollas blue, purple, pink, or white, salverform, often fragrant. A chiefly North American genus of some 60 species (Wherry, 1955). The text below follows my recent study (Cooperrider, 1986a) of the Ohio taxa.

Several *Phlox* cultivars are seen frequently in Ohio gardens. Many are derived from *Phlox paniculata* (discussed below), a tall ornamental with large clusters of sweet- but musky-scented purple to white flowers. Cultivars derived from *Phlox bifida* Beck, *P. subulata* (discussed below), and their hybrid, are showy, low, mat-forming ornamentals flowering profusely in early spring.

a. Main stems prostrate to ascending, but producing erect flowering branches; principal leaves subulate to linear, 0.5–2.0 (–2.5) cm long . . . . 1. *Phlox subulata*

aa. Main stems either erect or erect from a decumbent base; principal leaves linear or broader, if linear then more than 2.5 cm long.

b. Tips of anthers reaching at most ⅔ the length of the corolla tube, not visible at top of tube; styles mostly united for half or less their total length; the styles short, the stigmas reaching at most to midpoint of corolla tube, often less; calyx and pedicels with glandular pubescence.

c. Principal leaves linear-oblong to lanceolate or narrowly ovate, acute to obtuse or rounded at apex; corolla bluish-purple or white, glabrous . . . . 2. *Phlox divaricata*

cc. Principal leaves linear to lanceolate, acuminate at apex; corolla pinkish-violet or white, the tube usually pilose, infrequently glabrous . . . . 3. *Phlox pilosa*

bb. Tips of anthers reaching or nearly reaching, and visible at, top of corolla tube; styles mostly united for more than half their total length; the styles longer, the stigmas reaching well above midpoint of corolla tube; calyx and pedicels either glandular-pubescent, or short-pubescent, or spiculate, or glabrous.

c. Principal leaves 8–15 cm long, their lateral veins conspicuous, their margins bearing numerous, minute, thick-based, rigid, antrorse cilia . . . . 8. *Phlox paniculata*

cc. Principal leaves 1.5–12 cm long, their lateral veins obscure, their margins eciliate or bearing few to several divergent, usually flexuous, cilia.

d. Inflorescence longer than wide; upper stems and axis of inflorescence densely spiculate to very short crispy-pubescent; stems often finely mottled or blotched with purple; leaves very long-acuminate at apex; calyx 5–7 (–8) mm long . . . . 7. *Phlox maculata*

dd. Inflorescence wider than long; upper stems and axis of inflorescence glabrous, slightly pubescent, or strongly glandular-pubescent; stems not mottled or blotched with purple; leaves rounded, obtuse, acute, or slightly acuminate at apex; calyx (5–) 7–12 mm long.

e. Plants with conspicuous basal stolons, the stolons with spatulate leaves, rounded to broadly obtuse at apex; upper flowering stems, axes of inflorescences, and pedicels strongly glandular-villous; corolla 2.5–3.0 cm wide. . . . . 4. *Phlox stolonifera*

ee. Plants usually without basal stolons, when present the stolons with lanceolate to ovate leaves, acute at apex; upper flowering stems glabrous or slightly pubescent, axes of inflorescences and pedicels glabrous or minutely glandular-pubescent; corolla 1.5–3.0 cm wide.

f. Flowering stems with (3–) 4 (–5) pairs of leaves below inflorescence . . . . 5. *Phlox ovata*

ff. Flowering stems with 6 or more pairs of leaves below inflorescence . . . . 6. *Phlox glaberrima*

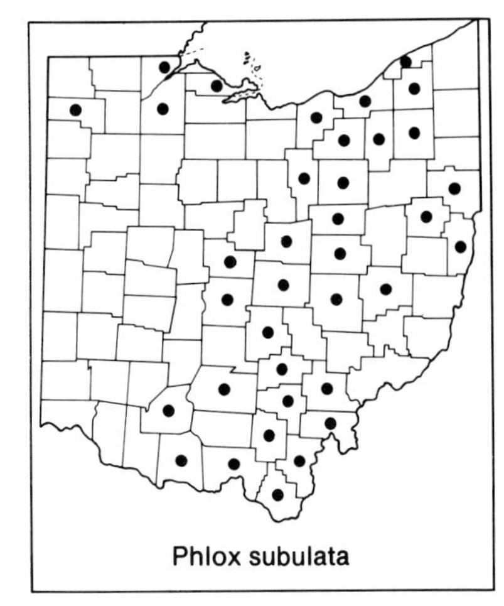
Phlox subulata

1. **Phlox subulata** L. MOSS PHLOX

Perennial; main stems prostrate to ascending, with erect flowering branches to 15 cm long; principal leaves subulate to linear, 0.5–2.0 (–2.5) cm long; calyx 4–8 mm long; corolla ca. 2 cm across, reddish-purple to lavender or white; lobes shallowly notched. Chiefly in the eastern half of Ohio; dry openings and open woodlands, roadside banks and other disturbed habitats, often near cemeteries. Frequently cultivated in Ohio, some wild populations in disturbed areas probably escapes. April–May.

Most fertile specimens can rather easily be separated into the two varieties keyed below. The two show some geographical difference, var. *subulata* being found more often in northern counties, var. *brittonii* more often in southern ones. Some intermediates occur. Older nonfertile specimens cannot be assigned with certainty to either variety.

a. Hairs of the inflorescence glandless . . . . var. *subulata*

aa. Hairs of the inflorescence glandular. . . . var. *brittonii* (Small) Wherry incl. var. *australis* Wherry; Gleason and Cronquist (1991) call this taxon var. *setacea* (L.) A. Brand.

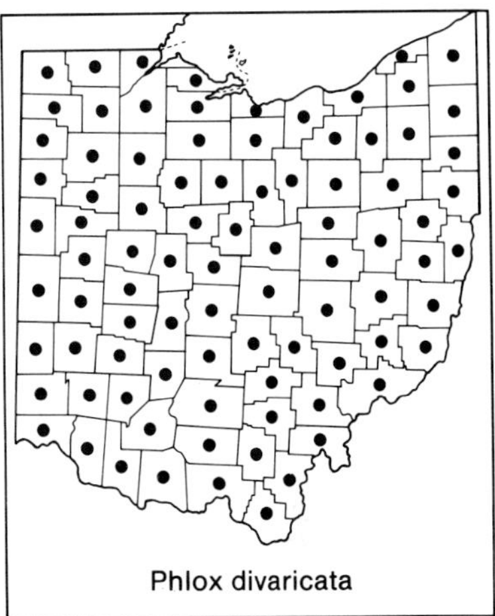
Phlox divaricata

2. **Phlox divaricata** L. BLUE PHLOX. WILD SWEET WILLIAM

Perennial, to 4 dm tall; flowering stems erect; principal leaves linear-oblong to lanceolate or narrowly ovate, acute to obtuse or rounded at apex, 2–4 cm long (leaves on prostrate basal shoots larger); calyx 5–10 mm long, glandular-pubescent; corolla 2–3 cm across, blue-purple or white, tube glabrous, lobes often shallowly notched; tips of anthers reaching at most ⅔ the length of the corolla tube, not visible at top of tube; styles mostly united for half or less their total length, short—the stigmas reaching at most to midpoint of corolla tube, often less. An attractive spring wildflower; throughout Ohio; woodlands, woodland borders and

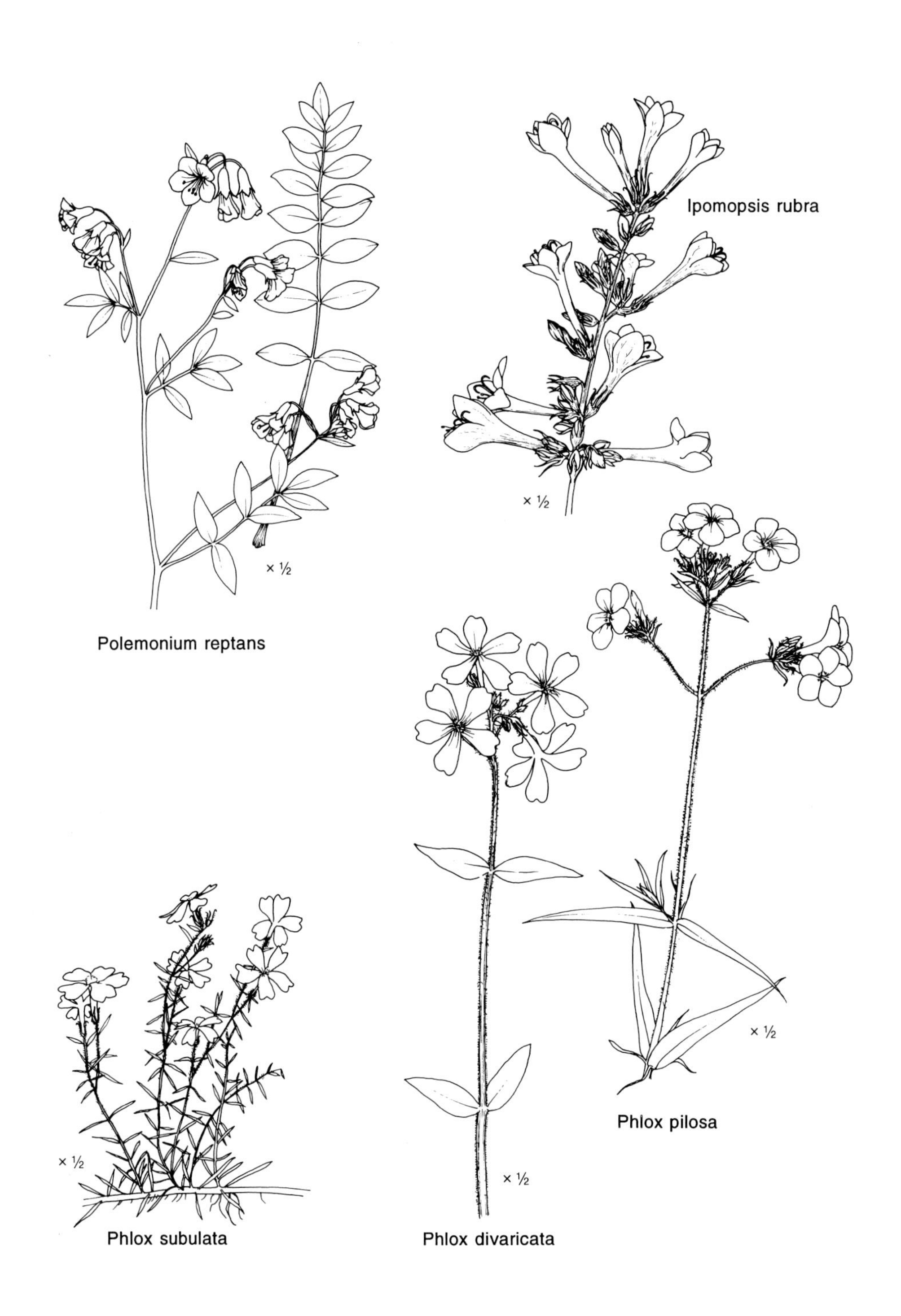
Ipomopsis rubra
× ½
Polemonium reptans
× ½
× ½
Phlox pilosa
× ½
Phlox subulata
× ½
Phlox divaricata

openings. Levin (1967) studied variation within this species. Late March–early June.

3. **Phlox pilosa** L. Downy Phlox

Perennial, to 5 dm tall; flowering stems erect; principal leaves linear to lanceolate, acuminate at apex, 3–6 cm long; calyx 5–8 mm long, glandular-pubescent; corolla 1.5–2.0 cm across, pinkish-violet or white, tube usually pilose, infrequently glabrous, lobes unnotched; tips of anthers reaching at most ²/₃ the length of the corolla tube, not visible at top of tube; styles mostly united for half or less their total length, short—the stigmas reaching at most to midpoint of corolla tube, often less. Scattered throughout much of Ohio, especially in south-central and northern counties; openings, roadside banks, fields. Levy (1983) studied variation within the species. May–early July.

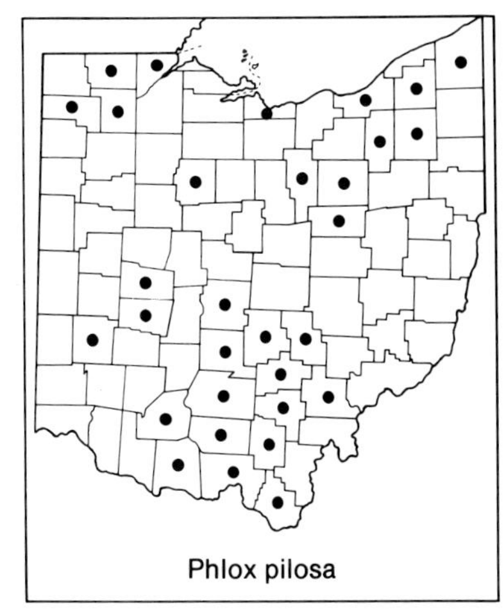
Phlox pilosa

4. **Phlox stolonifera** Sims Creeping Phlox

Perennial, to 3 dm tall, with conspicuous, sterile basal stolons to 2 dm long, leaves of the stolons to 7 cm long, spatulate and rounded to broadly obtuse at apex; flowering stems erect; principal leaves of flowering stems few, ovate to elliptic or lanceolate, acute to broadly obtuse at apex, 1.5–2.5 cm long; inflorescence wider than long, axes of inflorescence and pedicels strongly glandular-villous; calyx 7–12 mm long; corolla 2.5–3.0 cm across, reddish-purple to pink, lobes not or scarcely notched; tips of anthers reaching or nearly reaching, and visible at, top of corolla tube; styles mostly united for more than half their total length, the stigmas reaching well above midpoint of corolla tube. An Appalachian species at a western boundary of its range in a few southeastern counties; woodlands and woodland openings. Mid-April–May.

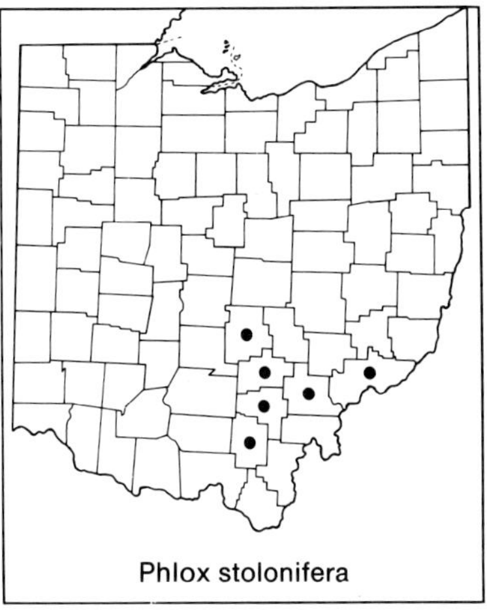
Phlox stolonifera

5. **Phlox ovata** L. Mountain Phlox. Allegheny Phlox

Phlox latifolia Michx.

Perennial, to 6 dm tall; flowering stems erect, with (3–) 4 (–5) pairs of leaves below the inflorescence; principal leaves lanceolate to elliptic- or ovate-lanceolate, acuminate at apex, 4–8 cm long; inflorescence wider than long, axes of inflorescence and pedicels minutely glandular-pubescent; calyx 8–12 mm long; corolla (1.5–) 2.5–3.0 cm across, pinkish-purple, lobes unnotched; tips of anthers reaching or nearly reaching, and visible at, top of corolla tube; styles mostly united for more than half their total length, the stigmas reaching well above midpoint of corolla tube. A species of extreme northwestern counties, the Ohio and Indiana populations disjunct from the main distribution range in the Appalachians (Wherry, 1955); woodland openings and borders. Mid-May–early June.

Phlox ovata

These plants might perhaps be better treated as a part of the next species, *P. glaberrima*; the features separating the two taxa are somewhat arbi-

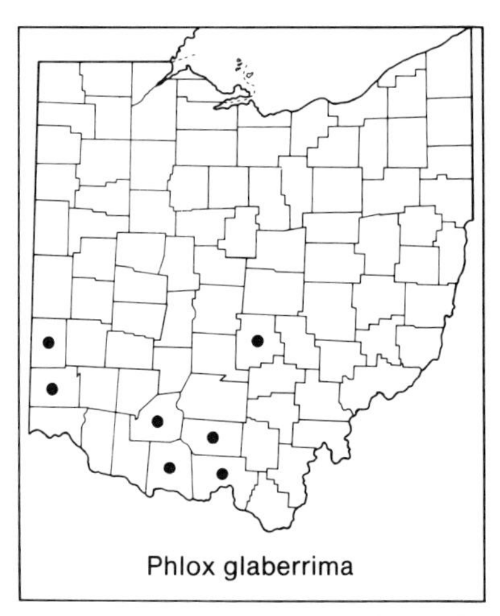

Phlox glaberrima

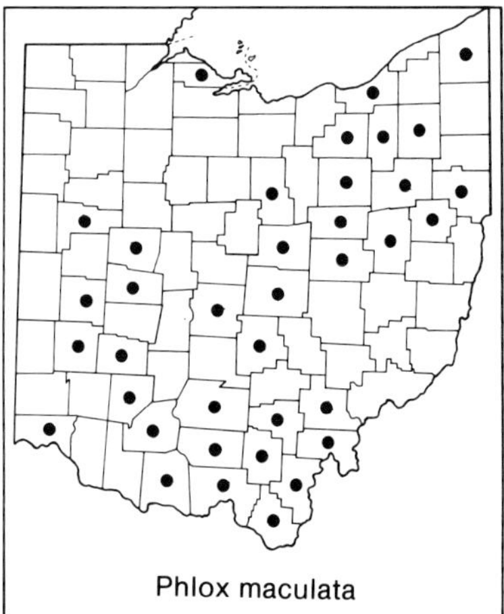

Phlox maculata

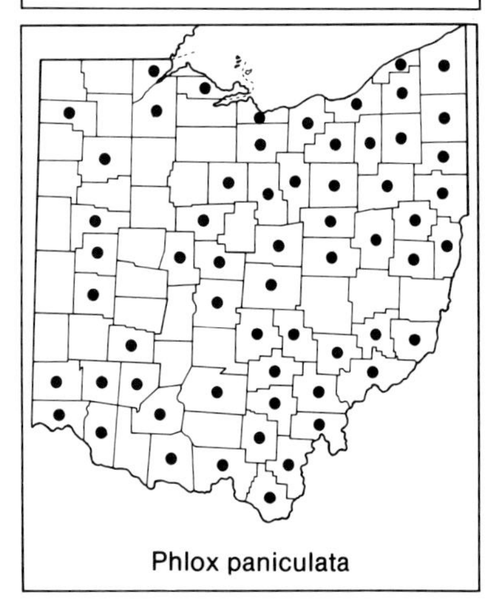

Phlox paniculata

trary (Cooperrider 1986a). Porter (1987), Reveal et al. (1982), and Wilbur (1987) discuss the nomenclature of this species.

6. **Phlox glaberrima** L. Smooth Phlox

Perennial, to 7 dm tall; flowering stems erect, with 6 or more pairs of leaves below the inflorescence; principal leaves broadly linear to elliptic or lanceolate, apex acute to acuminate, 4–7 cm long, inflorescence wider than long, axes of inflorescence and pedicels glabrous or nearly so; calyx 8–12 mm long; corolla 2–3 cm across, reddish-purple, lobes unnotched; tips of anthers reaching or nearly reaching, and visible at, top of corolla tube; styles mostly united for more than half their total length, the stigmas reaching well above midpoint of corolla tube. A taxon of a few southwestern counties; openings and woodland borders. Late May–June.

Ohio plants belong to var. *triflora* (Michx.) Reveal & Broome (*Phlox carolina* L. var. *triflora* (Michx.) Wherry).

Phlox glaberrima var. *glaberrima* (incl. var. *melampyrifolia* (Salisb.) Wherry), attributed (as var. *melampyrifolia*) to southern Ohio by several authors, was excluded from the Ohio flora by Cooperrider (1986a).

Phlox glaberrima var. *interior* Wherry (which seems to me worthy of varietal status) was attributed to northwestern Ohio by Schaffner (1934), Fernald (1950), and Wherry (1955), but no correctly identified specimens of the variety have been found in this study of Ohio *Phlox* (Cooperrider, 1986a).

7. **Phlox maculata** L. Spotted Phlox. Wild Sweet William

incl. var. *maculata* and var. *purpurea* Michx.

Perennial, to ca. 1 m tall; flowering stems erect, often finely mottled or blotched with purple; principal leaves linear to lanceolate or ovate, long-acuminate at apex, 4–12 cm long; inflorescence longer than wide, upper stem and axis of inflorescence densely spiculate to very short crispy-pubescent; calyx 5–7 (–8) mm long; corolla 1–2 cm across, purple, lobes unnotched; tips of anthers reaching or nearly reaching, and visible at, top of corolla tube; styles mostly united for more than half their total length, the stigmas reaching well above midpoint of corolla tube. Ranging from northeastern to southwestern counties; bogs, wet meadows, wet fields. June–August (–October).

Following the classifications of Gleason (1952b) and Wherry (1955), no infraspecific taxa are recognized. Levin (1963) reports naturally occurring hybrids between this species and *P. glaberrima* in Indiana.

8. **Phlox paniculata** L. Garden Phlox. Sweet William

Perennial, to 1.5 m tall; flowering stems erect; principal leaves narrowly to broadly elliptic to subrhombic, acuminate at apex, 8–15 cm long, their lateral veins conspicuous, their margins bearing numerous, minute, thick-

Phlox paniculata
Phlox stolonifera
× ½
Phlox ovata
× ½
× ½
× ½
Phlox maculata
× ½
Phlox glaberrima

based, rigid, antrorse cilia; inflorescences often dense with many flowers; calyx 5–9 mm long; corolla 2.0–2.5 cm across, purple to white (cultivars with these and other colors), lobes not or very slightly notched; tips of anthers reaching or nearly reaching, and visible at, top of corolla tube; styles mostly united for more than half their total length, the stigmas reaching well above midpoint of corolla tube. Throughout much of Ohio; woodlands, thickets, disturbed grassy habitats; probably native at some sites and naturalized at others. Garbutt, Bazzaz, and Levin (1985) studied population structure in this species. July–August.

Hydrophyllaceae. WATERLEAF FAMILY

Annual, biennial, or perennial herbs; leaves alternate, simple or compound; flowers perfect, in cymes or racemose cymes or rarely solitary; calyx 5-lobed; corolla actinomorphic, 5-lobed; stamens 5, attached to the corolla tube; carpels 2, fused to form a compound pistil with a single style; ovary superior; fruit a capsule. A small family with distribution centered in western North America. Constance (1963) studied the chromosome numbers and classification of the family, and K. A. Wilson (1960a) the genera in southeastern United States. The treatment here is based in part on Dumke's (1968) study of Ohio members of the family.

Ellisia nyctelea (L.) L., WATER-POD, a small weak-stemmed annual with pinnately lobed leaves and small white flowers, native to the west of Ohio (Constance, 1940), has been collected twice as an adventive weed in western counties bordering Indiana: in Hamilton County in 1954, and in Van Wert County in 1961.

a. Flowers white, solitary, borne opposite the upper leaves; stamens included; see above . . . . *Ellisia*

aa. Flowers white to pink or purple or blue, borne in cymes or racemose cymes; stamens included or exserted.

b. Blades (including all leaflets if compound) of principal cauline leaves (6–) 8–20 cm long, palmately or pinnately lobed or divided; flowers in cymes . . . . 1. *Hydrophyllum*

bb. Blades (including all leaflets if compound) of principal cauline leaves 2–6 (–8) cm long, pinnately lobed or divided; flowers in scorpioid cymes appearing more or less racemose at maturity. . . . . 2. *Phacelia*

1. Hydrophyllum L. WATERLEAF

Perennial or biennial herbs; leaves pinnately or palmately lobed or compound; flowers in long-pedunculate cymes. A small genus of eight species, the four below in eastern North America, the other four in western North America. Beckmann (1979) made a biosystematic study of *Hydrophyllum*. The treatment here is based on the taxonomic study of the genus by Constance (1942).

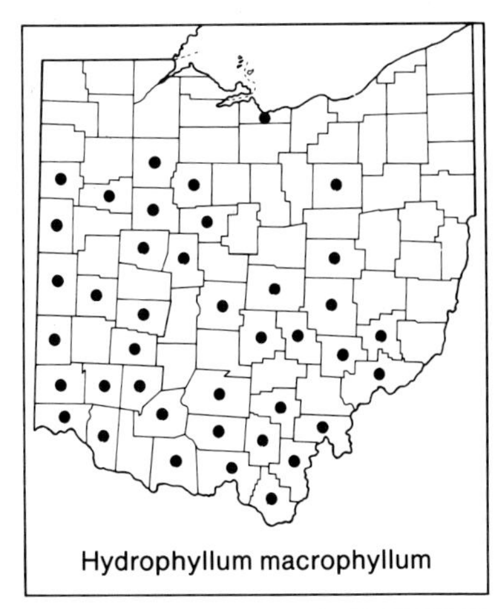
Hydrophyllum macrophyllum

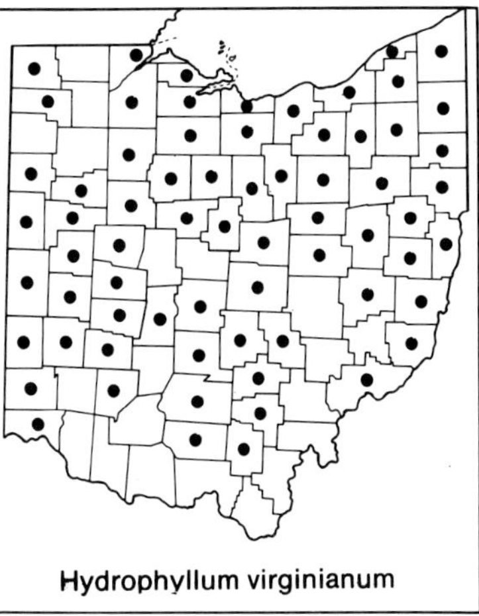
Hydrophyllum virginianum

a. Cauline leaves pinnately lobed or pinnately compound, the leaves elliptic, oblong, ovate, or deltoid in outline.
  b. Stems hispid; cauline leaves elliptic to oblong in outline, with 5–13 lobes and/or leaflets, the apices of lobes and leaflets broadly acute, obtuse, or rounded, and apiculate . . . . 1. *Hydrophyllum macrophyllum*
  bb. Stems sparsely pubescent to glabrous; cauline leaves ovate to deltoid in outline, with 3–7 lobes and/or leaflets, the apices of lobes and leaflets acute or acuminate . . . . 2. *Hydrophyllum virginianum*
aa. Cauline leaves palmately lobed and circular to reniform in outline.
  b. Perennial, with rhizome; stems sparsely hispid to glabrous; calyx glabrous or with few hairs; stamens conspicuously exserted . . . . 3. *Hydrophyllum canadense*
  bb. Biennial, tap-rooted, without rhizome; stems pubescent; calyx conspicuously hispid; stamens equalling corolla or slightly exserted . . . . 4. *Hydrophyllum appendiculatum*

1. **Hydrophyllum macrophyllum** Nutt. LARGE-LEAVED WATERLEAF
Perennial, to 9 dm tall; stems hispid; cauline leaves elliptic to oblong in outline, pinnately divided: on the lower part of the leaf cut to the midrib, on the upper part deeply pinnately lobed, in all with 5–13 lobes and/or leaflets, these deeply and coarsely toothed along the margins and broadly acute to obtuse or rounded at apex, apex apiculate; calyx conspicuously hispid and also short-pubescent; corolla whitish; stamens exserted. A species ranging from northern Alabama to Ohio, where it grows chiefly in the southern half of the state; rich moist woods and woodland borders. Baskin and Baskin (1983b) described the germination ecophysiology of this species. May–June.

2. **Hydrophyllum virginianum** L. VIRGINIA WATERLEAF
Perennial, to 7 dm tall; stems sparsely pubescent to glabrous; cauline leaves ovate to deltoid in outline, pinnately divided, the lowest sinus extending to the midrib, the upper part of the leaf deeply lobed, in all with 3–7 lobes and/or leaflets, these (especially the lowest) sometimes again lobed, the margins of all coarsely and sharply serrate or dentate, their apices acute to acuminate; calyx somewhat hispid and also with short appressed pubescence; corolla white to pinkish to pale or dark purple; stamens exserted. Throughout most of Ohio; woods, woodland borders, thickets. May–mid-June.

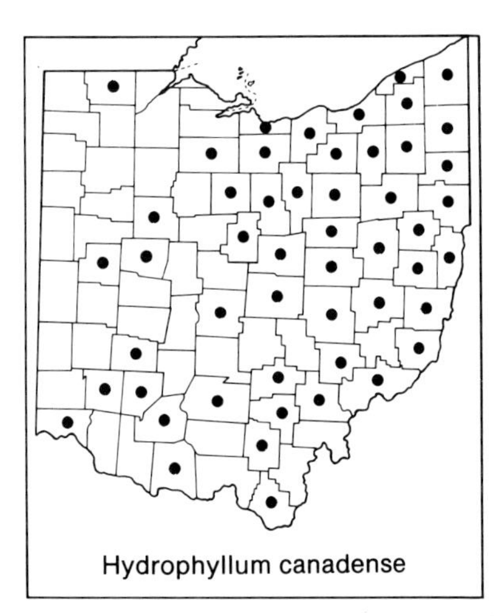
Hydrophyllum canadense

3. **Hydrophyllum canadense** L. BROAD-LEAVED WATERLEAF
Perennial, to 6 dm tall; stems sparsely hispid to glabrous; cauline leaves more or less reniform in outline, palmately lobed (the basal leaves either pinnately or palmately lobed); calyx glabrous or with few hairs; corolla whitish; stamens exserted. Throughout the eastern half of Ohio, becoming scarce or absent westward; steep rich wooded slopes, wooded floodplains. Late May–mid-July.

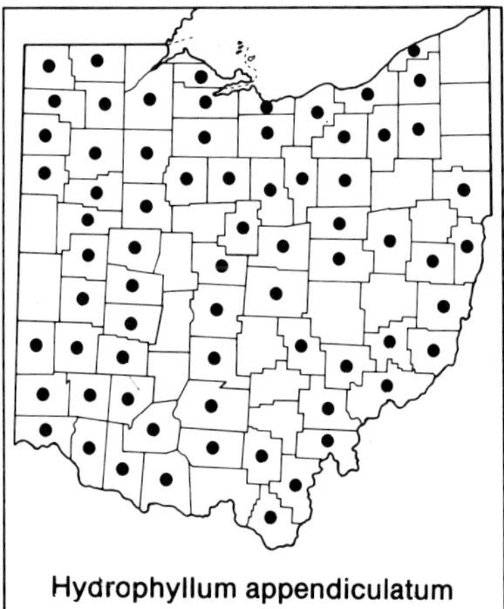
Hydrophyllum appendiculatum

4. **Hydrophyllum appendiculatum** Michx. APPENDAGED WATERLEAF
Biennial, to 6 dm tall; stems pubescent; cauline leaves more or less orbicular in outline, palmately lobed (the basal leaves often pinnately lobed or divided); calyx conspicuously hispid; corolla pale lavender to purple; stamens equalling corolla or slightly exserted. Throughout most of Ohio; moist to mesic woodlands. Baskin and Baskin (1985) studied the germination ecophysiology, and Morgan (1971) the species' life history and energy relationships. Late April–June (–September).

2. Phacelia Juss. SCORPION-WEED

Annual or biennial herbs; leaves pinnately lobed or divided; flowers in racemose scorpioid cymes. A genus of perhaps 200 species native to the New World. The text below is based on Constance's (1949) revision of *Phacelia* subgenus *Cosmanthus*, which includes all the Ohio species. G. W. Gillett (1968) studied the systematic relationships of the species in this same subgenus.

a. Uppermost cauline leaves sessile and sometimes clasping.
  b. Corolla lobes strongly fringed; calyx lobes linear to narrowly oblong; cauline leaves with (5–) 7–11 lobes, the terminal lobe acute to obtuse at apex . . . . 1. *Phacelia purshii*
  bb. Corolla lobes entire; calyx lobes oblong to lanceolate or ovate; cauline leaves with 3–5 (–9) lobes, the terminal lobe acute to obtuse or rounded at apex . . . . 2. *Phacelia dubia*
aa. Uppermost cauline leaves petiolate, never clasping.
  b. Plants 1–3 dm tall; lobes of leaves mostly obtuse to rounded at apex; corolla 2–4 mm wide; stamens 1.5–2.0 mm long, included; style 1.5–2.0 mm long . . . . 3. *Phacelia ranunculacea*
  bb. Plants 3–5 dm tall; lobes of leaves mostly acute at apex; corolla 10–15 mm wide; stamens 8–12 mm long, exserted beyond the corolla; style 8–15 mm long . . . . 4. *Phacelia bipinnatifida*

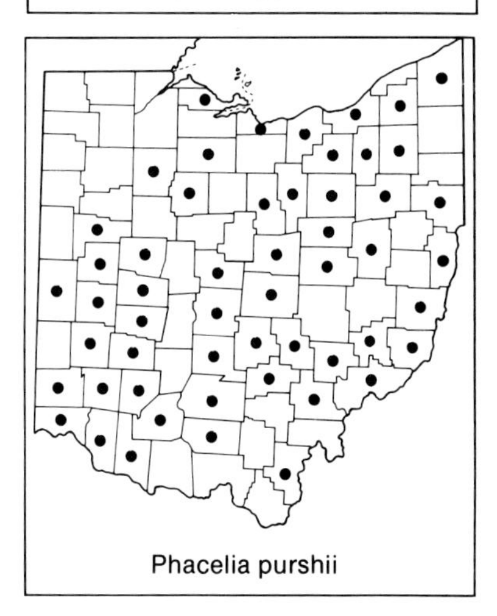
Phacelia purshii

1. **Phacelia purshii** Buckl. MIAMI-MIST
Winter annual, to 5 dm tall; cauline leaves with (5–) 7–11 lobes, the terminal lobe acute to obtuse at apex, the uppermost leaves sessile and some-

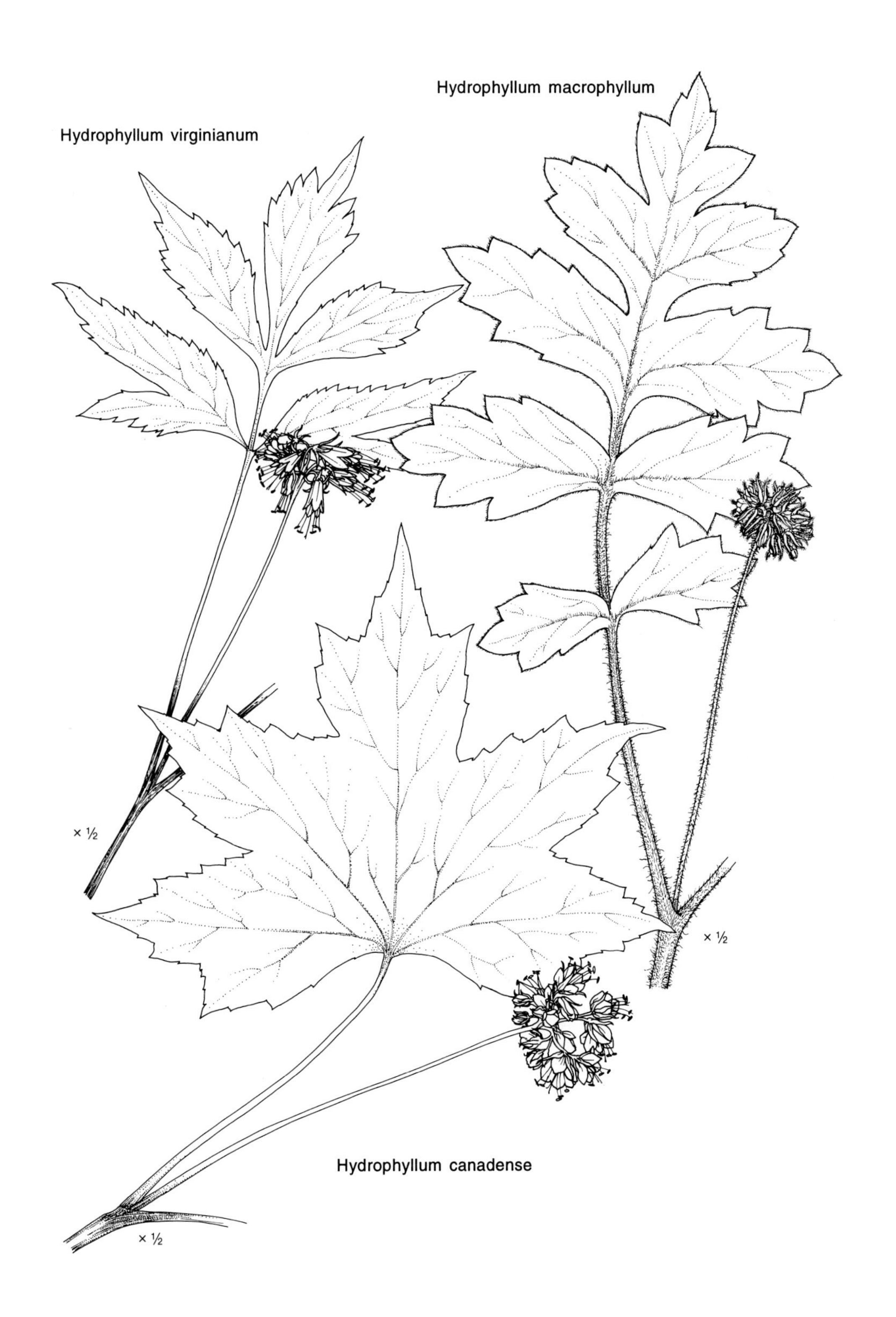
Hydrophyllum macrophyllum
Hydrophyllum virginianum
× ½
× ½
Hydrophyllum canadense
× ½

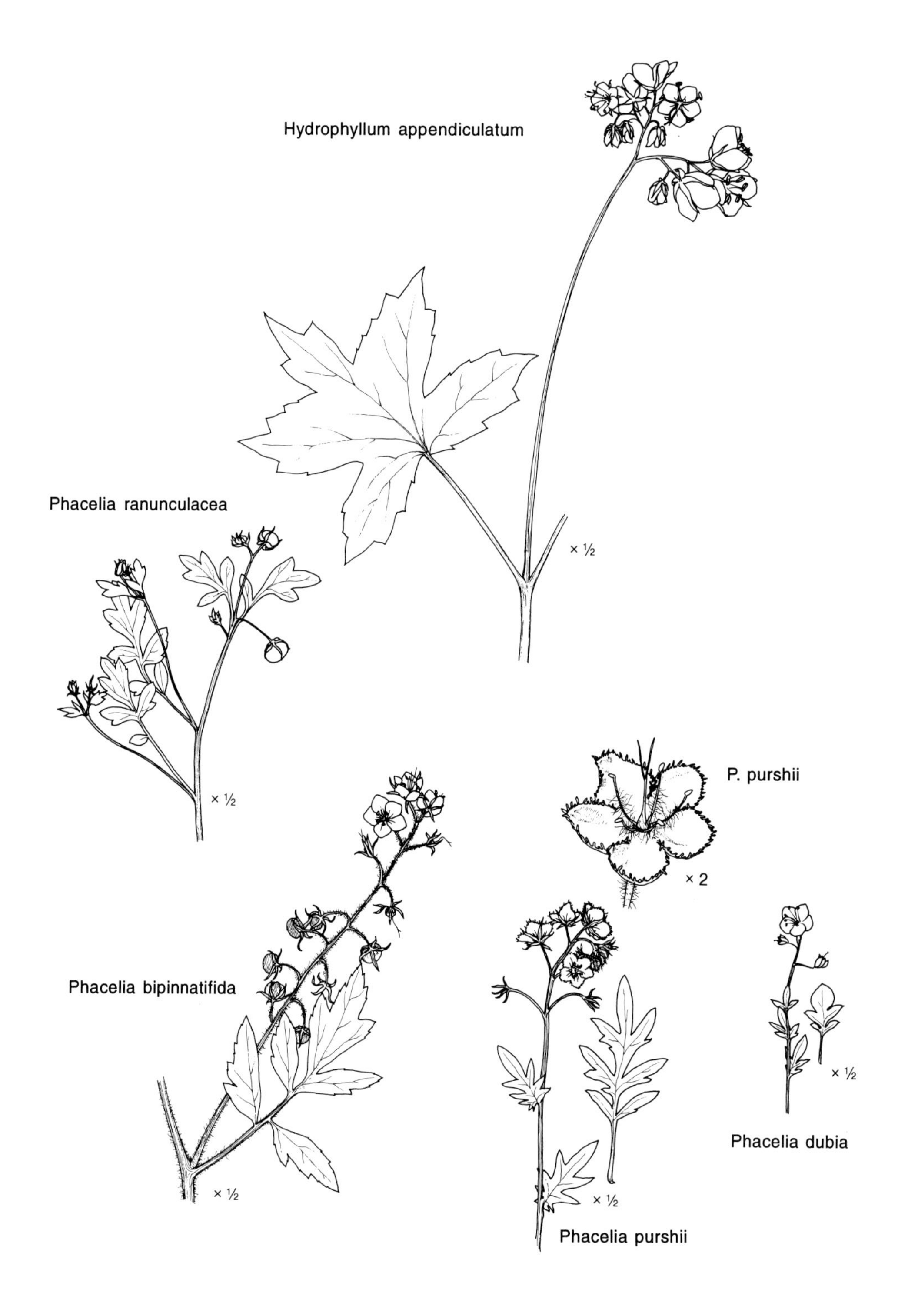
Hydrophyllum appendiculatum
× ½
Phacelia ranunculacea
× ½
P. purshii
× 2
Phacelia bipinnatifida
× ½
× ½
Phacelia dubia
× ½
Phacelia purshii

times clasping at base; calyx lobes linear to narrowly oblong; corolla 8–12 mm across, blue with white center, the lobes strongly fringed; stamens slightly exserted. Frequent throughout much of Ohio, but reaching a northern boundary of the species' range in northwestern Ohio; fields, thickets, pastures, roadsides, floodplains. Baskin and Baskin (1976a) studied the species' autecology and population biology. Late April–mid-June.

2. **Phacelia dubia** (L.) Trel. SMALL-FLOWERED SCORPION-WEED

Winter annual, to 3 dm tall; cauline leaves with 3–5 (–9) lobes, the terminal lobe acute to obtuse or rounded at apex, the uppermost cauline leaves sessile; calyx lobes oblong to lanceolate or ovate; corolla 5–10 mm across, lavender to bluish, with a white center, the lobes entire; stamens about equalling corolla. A species of the mid-Atlantic and southeastern states, with a single disjunct population known from Fairfield County, in south-central Ohio, at the northwestern boundary of the species' range; thickets and fields. Baskin and Baskin (1978) discuss aspects of the life cycle. Late April–May.

Phacelia dubia

3. **Phacelia ranunculacea** (Nutt.) Constance WEAK-STEMMED SCORPION-WEED

Winter annual, to 3 dm tall; leaves with 3–7 lobes, these mostly rounded to obtuse at apex, the uppermost leaves petiolate; corolla 2–4 mm across, lavender to pale blue, the lobes entire; stamens 1.5–2.0 mm long, included; style 1.5–2.0 mm long. A species ranging from the mid-Atlantic states westward to Missouri and Arkansas, with a disjunct population in southernmost Ohio, at a northern boundary of its range; dry woodlands. The species' cytogeography is described by Chuang and Constance (1977). Late April–May.

Phacelia ranunculacea

4. **Phacelia bipinnatifida** Michx. FERN-LEAVED SCORPION-WEED

Biennial, to 5 dm tall; leaves with 3–7 lobes, these mostly acute at apex, the uppermost leaves petiolate; corolla 10–15 mm across, lavender-blue, the lobes very slightly crenate to subentire; stamens 8–12 mm long, exserted; style 8–15 mm long. A species of the eastern interior United States, at a northeastern boundary of its range in southwestern Ohio; low woods, mesic woods, moist roadside ditches. Late April–early June.

Phacelia bipinnatifida

Boraginaceae. BORAGE FAMILY

Annual, biennial, or perennial herbs; leaves alternate, simple; flowers perfect, usually borne in a helicoid cyme; calyx of 5 sepals, separate or fused basally; corolla actinomorphic or slightly zygomorphic, 5-lobed; stamens 5, attached to the corolla tube; carpels 2, fused to form a compound pistil with a single style; the ovary superior, deeply lobed and indented into 4 (rarely 2) nearly separate segments, the style gynobasic, inserted at the base of the ovary in the midst of the segments; fruit a cluster of 4 (rarely 2) nutlets. A family of some 2,000 species, cosmopolitan in distribution. I. M. Johnston (1924) prepared a synopsis of the American genera and species, both native and alien, in the subfamily Boraginoideae. Ingram (1961a) published a key to most cultivated genera. In the text below, the treatment of alien species is based largely on the work of Valentine and Chater (1972) in *Flora Europaea*.

Several species of annuals have appeared in Ohio as adventive weeds. *Amsinckia intermedia* Fisch. & Mey., Coast Fiddle-neck, with stamens attached above the middle of the corolla tube, was attributed to Franklin County by Schaffner (1932), but I have seen no specimens. The closely related *Amsinckia lycopsoides* (Lehm.) Lehm. (*A. barbata* E. Greene), Tarweed, with stamens attached below the middle of the corolla tube, was collected from Belmont County in 1918, and from Lake County in 1930. Both species of *Amsinckia* are native to western United States; see P. M. Ray and H. F. Chisaki (1957) for a discussion of the genus. *Asperugo procumbens* L., Madwort, was collected from Lake County in 1901. *Borago officinalis* L., Talewort or Cooltankard, sometimes cultivated in Ohio, has been collected from Erie, Highland, and Jefferson counties. Several species of *Heliotropium* collected in, or reported for, Ohio are discussed in the text below. *Lycopsis arvensis* L. (*Anchusa arvensis* (L.) Bieb.), Small Bugloss, has been collected from Ashtabula, Lake, and Stark counties. Two adventive perennials of the genus *Anchusa* are discussed in the text below.

The family's most frequently cultivated species in Ohio is *Pulmonaria officinalis* L., Blue Lungwort, a perennial with pinkish-lavender to deep blue flowers, blooming in early spring.

a. Flowers solitary or in groups of 2–4 in small, tight, axillary clusters.
- b. Corolla white to yellow; stems erect, pubescent but without prickles . . . . 6. *Lithospermum*
- bb. Corolla blue; stems low and procumbent, harsh with small prickles; see above . . . . *Asperugo*

aa. Flowers in more well defined inflorescences.
- b. Corolla blue, rotate; each filament bearing a conspicuous spur-like appendage; see above . . . . *Borago*
- bb. Corolla blue or some other color, not rotate; filaments without spur-like appendages.
  - c. Stems rough and bristly-hispid; corolla blue or sometimes pale blue, somewhat irregular, the tube not straight, the lobes slightly unequal.
    - d. Stamens included, equal in length, attached near base of corolla tube; see above . . . . *Lycopsis*

dd. Stamens exserted, unequal in length, attached near middle of corolla tube. . . . . 4. *Echium*

cc. Stems rough to smooth; corolla blue or some other color, regular, the tube straight, the lobes equal.

d. Corolla lobes erect or only slightly spreading.

e. The corolla white, its lobes acute to acuminate at the apex; stems hispid . . . . 5. *Onosmodium*

ee. The corolla blue, pale purple, pinkish, yellow or rarely white, its lobes obtuse to rounded at the apex; stems hirsute or scabrous to glabrous.

f. Stems more or less hirsute to scabrous; corolla 10–17 mm long . . . . 2. *Symphytum*

ff. Stems glabrous; corolla 18–25 mm long. . . . . 9. *Mertensia*

dd. Corolla lobes spreading.

e. Fruits bearing apically barbed bristles.

f. Leaves all linear, 2–6 mm wide; corolla pale blue . . . . 10. *Lappula*

ff. Leaves on lower half of stem oblong to broadly elliptic, 10 mm or more in width; corolla white, red, purple, or blue.

g. Corolla red, purple, blue, or rarely white, 6–12 mm wide; middle and upper leaves sessile or subsessile, rarely narrowed at base, often clasping . . . . 7. *Cynoglossum*

gg. Corolla white or pale blue, ca. 2 mm wide; middle and upper leaves narrowed to a petiole 1–5 mm long, not clasping . . . . 11. *Hackelia*

ee. Fruits lacking apically barbed bristles.

f. Inflorescences without bracts.

g. Corolla yellow; stems hispid; see above . . . . *Amsinckia*

gg. Corolla blue or white; stems hirsute to glabrous.

h. Flowers sessile or subsessile . . . . 1. *Heliotropium*

hh. Flowers pedicellate. . . . . 8. *Myosotis*

ff. Inflorescences with bracts throughout or at least below.

g. Corolla pale yellow to orange-yellow, 10–25 mm wide . . . . 6. *Lithospermum*

gg. Corolla white or blue, 1–20 mm wide.

h. Flowers blue, 6–20 mm wide; inflorescences bracteate throughout . . . . 3. *Anchusa*

hh. Flowers blue or white, 1–9 mm wide; inflorescences not bracteate throughout, the upper flowers bractless . . . . 8. *Myosotis*

1. Heliotropium L. HELIOTROPE

A widespread genus of some 200–250 species. Three weedy species have been collected or reported as rare adventives in Ohio. All are small, more or less hirsute herbs, with small, white to purple or blue, sessile or subsessile flowers, and with spreading corolla lobes, the flowers borne in a bractless, well-defined, spike-like scorpioid cyme. The South American perennial, *Heliotropium amplexicaule* M. Vahl, CLASPING HELIOTROPE, was attributed to Ohio by Fernald (1950), but I have seen no specimens. The annual *Heliotropium europaeum* L., EUROPEAN HELIOTROPE, has been collected recently (since 1979) from Columbiana, Lucas, and Mahoning counties, all in northern Ohio. *Heliotropium indicum* L., INDIAN HELIOTROPE, an annual of South American origin, has been collected from Hamilton and Scioto counties in southern Ohio, most recently in 1944. *Heliotropium tenellum* (Nutt.) Torr., SLENDER HELIOTROPE, was attributed to Ohio by Cusick and Silberhorn (1977), based on misidentified specimens.

a. Ovary 4-lobed, fruit at maturity separating into 4 nutlets, each with 1 seed; see above . . . . *Heliotropium europaeum*

aa. Ovary 2-lobed, fruit at maturity separating into 2 nutlets, each with either 1 seed or 2.

b. Leaves petiolate, blades ovate, 4–10 cm long; see above . . . . *Heliotropium indicum*

bb. Leaves sessile or subsessile, blades lanceolate to oblong-lanceolate, 2–8 cm long; see above . . . . *Heliotropium amplexicaule*

2. Symphytum L.

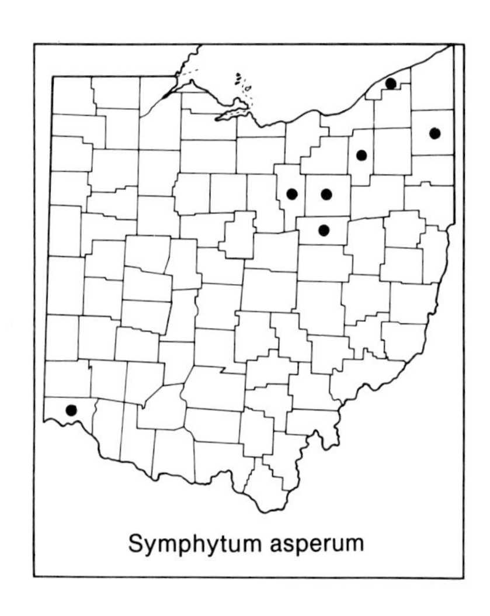

Symphytum asperum

Erect, coarse, more or less hirsute perennials; flowers borne in nodding, bractless, racemose, scorpioid cymes; corolla medium-sized, 10–17 mm long, the lobes erect; nutlets wrinkled or smooth, glabrous. A Eurasian genus of 20–25 species, two of them naturalized in the Ohio flora. Gadella (1984) has notes on the morphology and chromosome numbers of the two species below and on their hybrid, *S.* × *uplandicum* Nyman; Ingram (1961b) has additional notes on these two species.

a. Bases of leaves not or only slightly decurrent, stem slightly or not at all winged; nutlets wrinkled-reticulate . . . . . . 1. *Symphytum asperum*

aa. Bases of leaves markedly decurrent along stem, stems conspicuously winged; nutlets smooth . . . . 2. *Symphytum officinale*

1. SYMPHYTUM ASPERUM Lepechin PRICKLY COMFREY

Perennial, to ca. 1 m tall; stems with recurved, almost prickle-like hairs; bases of leaves not or only slightly decurrent, the stems slightly or not at all winged; corolla 10–14 mm long, pinkish-red in bud, blue at maturity; nutlets wrinkled-reticulate. A Eurasian species sometimes cultivated in Ohio, naturalized at a few sites, mostly in northeastern counties; roadsides and other disturbed habitats. Late May–July.

2. Symphytum officinale L. Common Comfrey

Perennial, to 8 dm tall; stems hirsute to scabrous; bases of leaves markedly decurrent, the stems conspicuously winged; corolla 14–17 mm long, whitish to red or purple; nutlets smooth and shiny. A Eurasian species, sometimes cultivated in Ohio, naturalized at scattered sites, mostly in northern counties; roadsides and other disturbed habitats. Late May–July.

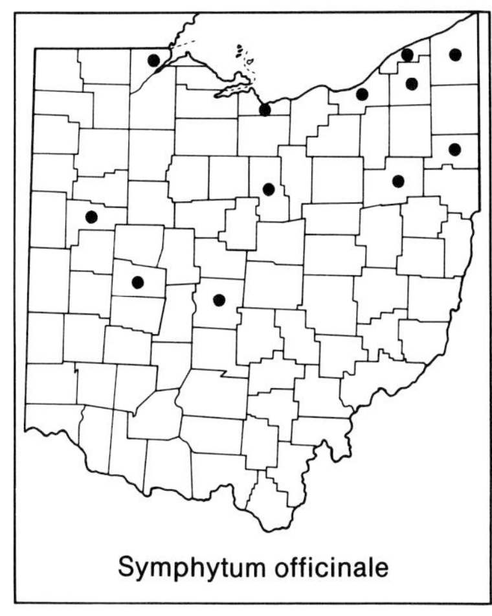
Symphytum officinale

3. Anchusa L. ALKANET

A genus of 30–40 species of the Mediterranean region. Two species, both sometimes cultivated, have been collected in Ohio as adventive escapes. They are erect hirsute perennials with bright blue flowers and spreading corolla lobes, the flowers borne in bracteate, racemose, scorpioid cymes. *Anchusa azurea* Mill., Garden Bugloss, has been collected from Ashtabula and Auglaize counties, both in northern Ohio, most recently in 1930. *Anchusa officinalis* L., Common Bugloss or Common Alkanet, was collected in 1931 near a Lake County nursery, also in northern Ohio.

a. Calyx usually divided for only 1/2–2/3 its length; corolla 5–10 mm wide, violet to red, rarely yellow or white; nutlets wider than long; see above . . . . *Anchusa officinalis*

aa. Calyx divided to base; corolla 12–20 mm wide, violet to deep blue; nutlets longer than wide; see above . . . . *Anchusa azurea*

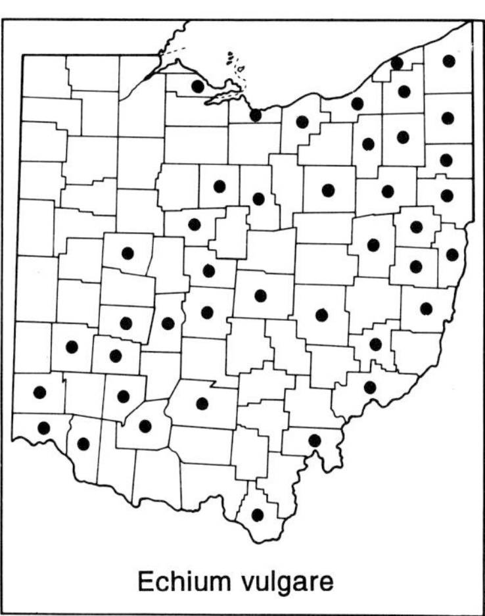
Echium vulgare

4. Echium L. VIPER'S BUGLOSS

A genus of 30 or more species, most native to Eurasia, with a single species naturalized in Ohio.

1. Echium vulgare L. Blue-weed. Blue-devil. Viper's Bugloss
incl. var. *vulgare* and var. *pustulatum* (Sibth. & Sm.) Coincy

Biennial, to 1 m tall; stems rough-hispid; corolla blue, slightly zygomorphic, the tube somewhat curved and the lobes slightly unequal; stamens exserted, unequal in length, attached near middle of corolla tube; nutlets rough-wrinkled, glabrous. A Eurasian species naturalized from northeastern to southwestern Ohio, but absent from most northwestern counties; railroad ballast, calcareous fields. June–October.

Following the conclusions reached by Pusateri and Blackwell (1979), no major infraspecific taxa are recognized within this variable species.

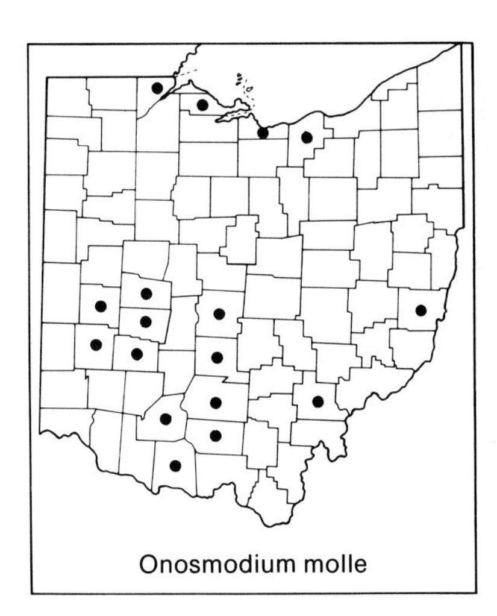

5. Onosmodium Michx. FALSE GROMWELL

A North American genus of about 12 species, one native to Ohio.

1. **Onosmodium molle** Michx. FALSE GROMWELL. MARBLE-SEED

Perennial, to 1 m tall; stems hispid; corolla white, actinomorphic, the tube straight, the lobes equal, acute to acuminate at apex; nutlets pitted, glabrous. Chiefly in southwestern and a few north-central counties; calcareous fields, roadsides. June–August.

Ohio plants belong to var. *hispidissimum* (Mackenz.) Cronq. (*O. hispidissimum* Mackenz.). This taxonomy follows that of Gleason and Cronquist (1963, 1991) and is supported by the research of Cochrane (1976).

6. Lithospermum L. PUCCOON. GROMWELL

Erect pubescent annuals or perennials; flowers solitary in leaf axils or in bracteate scorpioid cymes; corolla whitish to yellow or orangish, the lobes spreading; nutlets wrinkled or smooth, glabrous. A widely distributed genus of 40–50 species. A taxonomic study of the genus was prepared by I. M. Johnston (1952), and a study of the genus as it is represented in the Ohio flora by Cusick (1985).

Lithospermum officinale L., EUROPEAN GROMWELL, was collected as an adventive weed in Hamilton County in 1924, and in Ross County in 1976; both counties are in southwestern Ohio.

a. Corolla white, yellowish-white, greenish-white, or bluish-white, 4–8 mm long, only slightly exceeding the calyx.
  b. Plants annual with slender taproot; principal leaves linear to narrowly lanceolate, 3–7 (–10) mm wide; nutlets wrinkled .... 1. *Lithospermum arvense*
  bb. Plants perennial with thick root; principal leaves lanceolate to narrowly ovate, (6–) 8–45 mm wide; nutlets smooth.
    c. Leaves narrowly ovate to ovate, 20–45 mm wide .... 2. *Lithospermum latifolium*
    cc. Leaves lanceolate, 6–15 (–20) mm wide; see above .... *Lithospermum officinale*
aa. Corolla yellow to yellowish-orange, 8–18 mm long, exceeding the calyx.
  b. Leaves hispid; mature calyx lobes 8–10 (–13) mm long, hirsute to hispid; corolla 15–25 mm wide....... 3. *Lithospermum caroliniense*
  bb. Leaves densely soft-pilose; mature calyx lobes 3–7 mm long, appressed-pubescent; corolla 11–15 mm wide .... 4. *Lithospermum canescens*

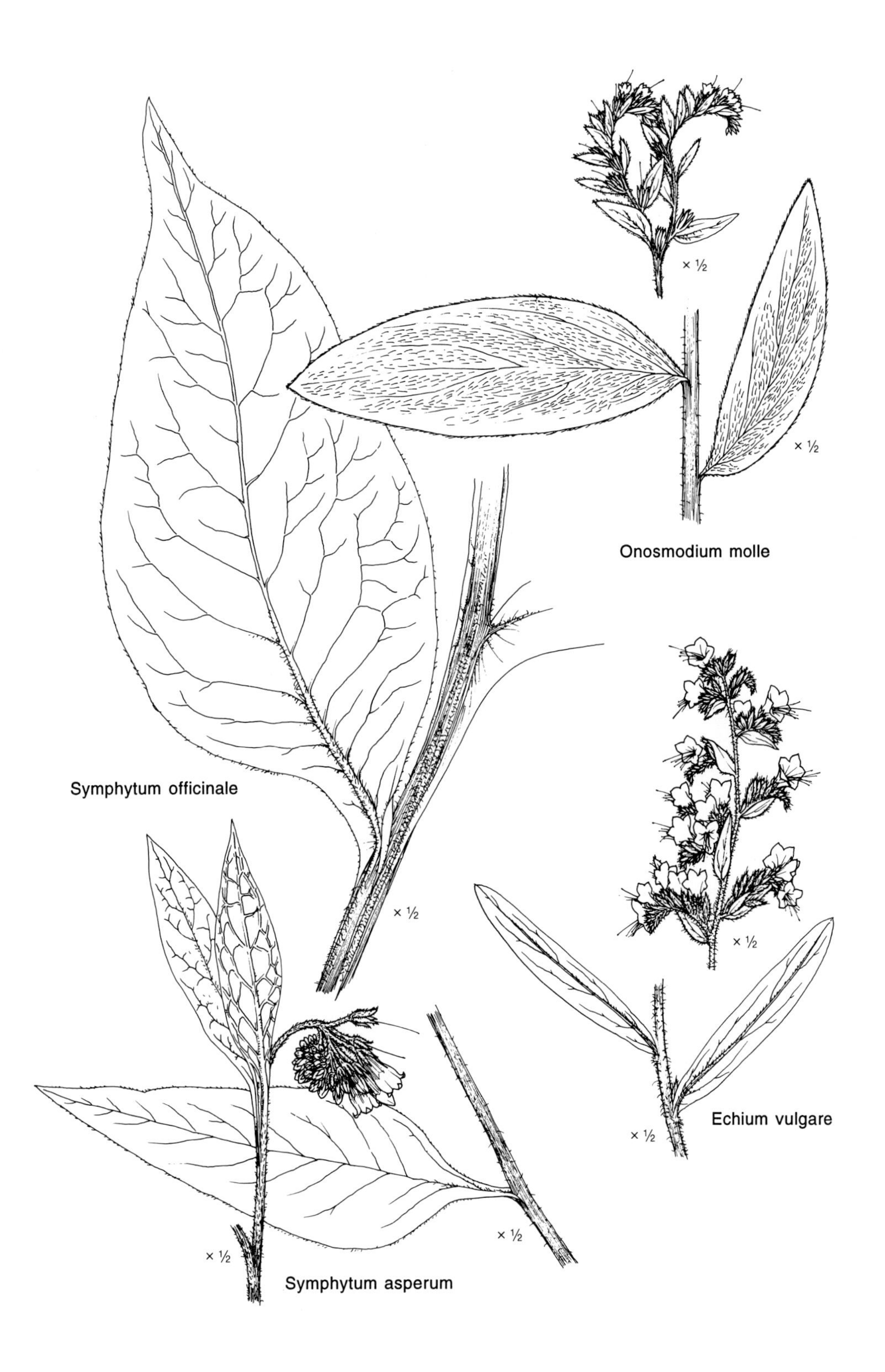
× ½
× ½
Onosmodium molle
Symphytum officinale
× ½
× ½
× ½
Echium vulgare
× ½
× ½
Symphytum asperum

1. Lithospermum arvense L. Corn Gromwell

Buglossoides arvensis (L.) I. M. Johnston

Annual, to 6 dm tall; principal leaves linear to narrowly lanceolate, 3–7 (–10) mm wide; corolla whitish, very small, ca. 4 mm across, slightly exceeding calyx; nutlets grayish-tan, wrinkled. A Eurasian species naturalized throughout Ohio; roadsides, fields, disturbed openings, and other weedy sites. May–July.

Fernandes (in Valentine and Chater, 1972) and I. M. Johnston (1952, 1954) favor placement of this species in a segregate genus, *Buglossoides* Moench. Gleason and Cronquist (1991) continue to treat it as a part of *Lithospermum*, as it is here.

2. **Lithospermum latifolium** Michx. American Gromwell

Perennial, to 8 dm tall; leaves narrowly ovate to ovate, 20–45 mm wide; corolla whitish to pale yellow, small—ca. 6 mm across, slightly exceeding calyx; nutlets shining, whitish, smooth. Chiefly in the southern half of Ohio; woods, woodland openings. May–June.

3. **Lithospermum caroliniense** (Walt.) MacM. Hairy Puccoon

Perennial, to 6 dm tall; leaves linear-oblong to lanceolate or lance-ovate, hispid; calyx lobes 5–7 mm long at anthesis, mostly 8–10 (–13) mm long when mature, hirsute to hispid; corolla yellowish-orange, 8–15 mm long, exceeding calyx, 15–25 mm across; nutlets shining, whitish, and smooth. Northern counties; open, usually sandy, habitats. Levin (1972) studied the species' reproductive system. Late May–July.

Ohio plants belong to var. *croceum* (Fern.) Cronq. (*L. croceum* Fern.); var. *caroliniense* occurs to the south of Ohio.

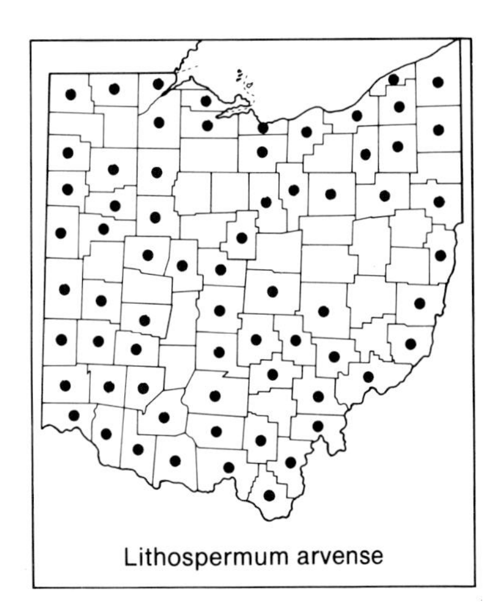
Lithospermum arvense

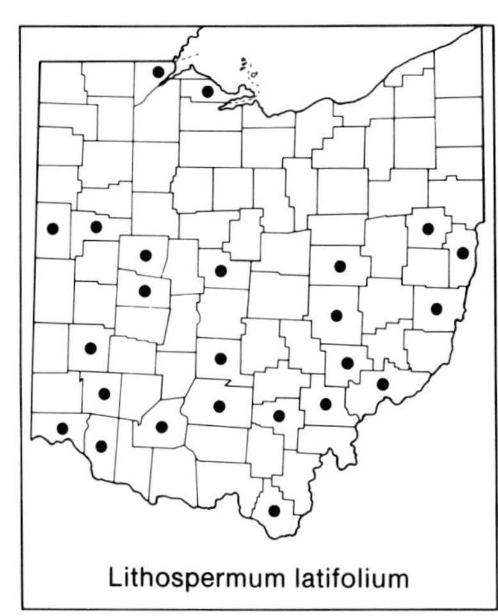
Lithospermum latifolium

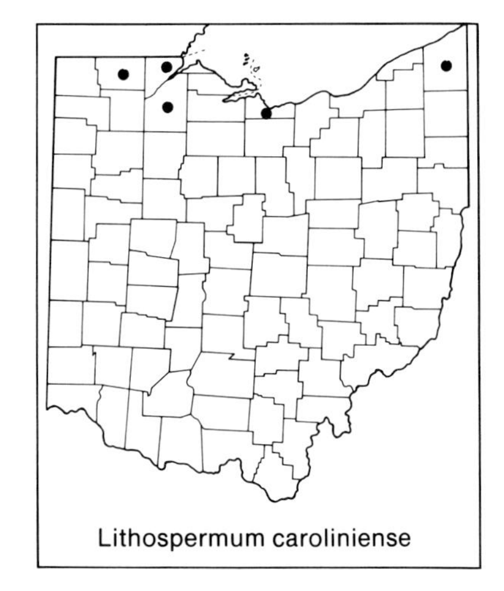
Lithospermum caroliniense

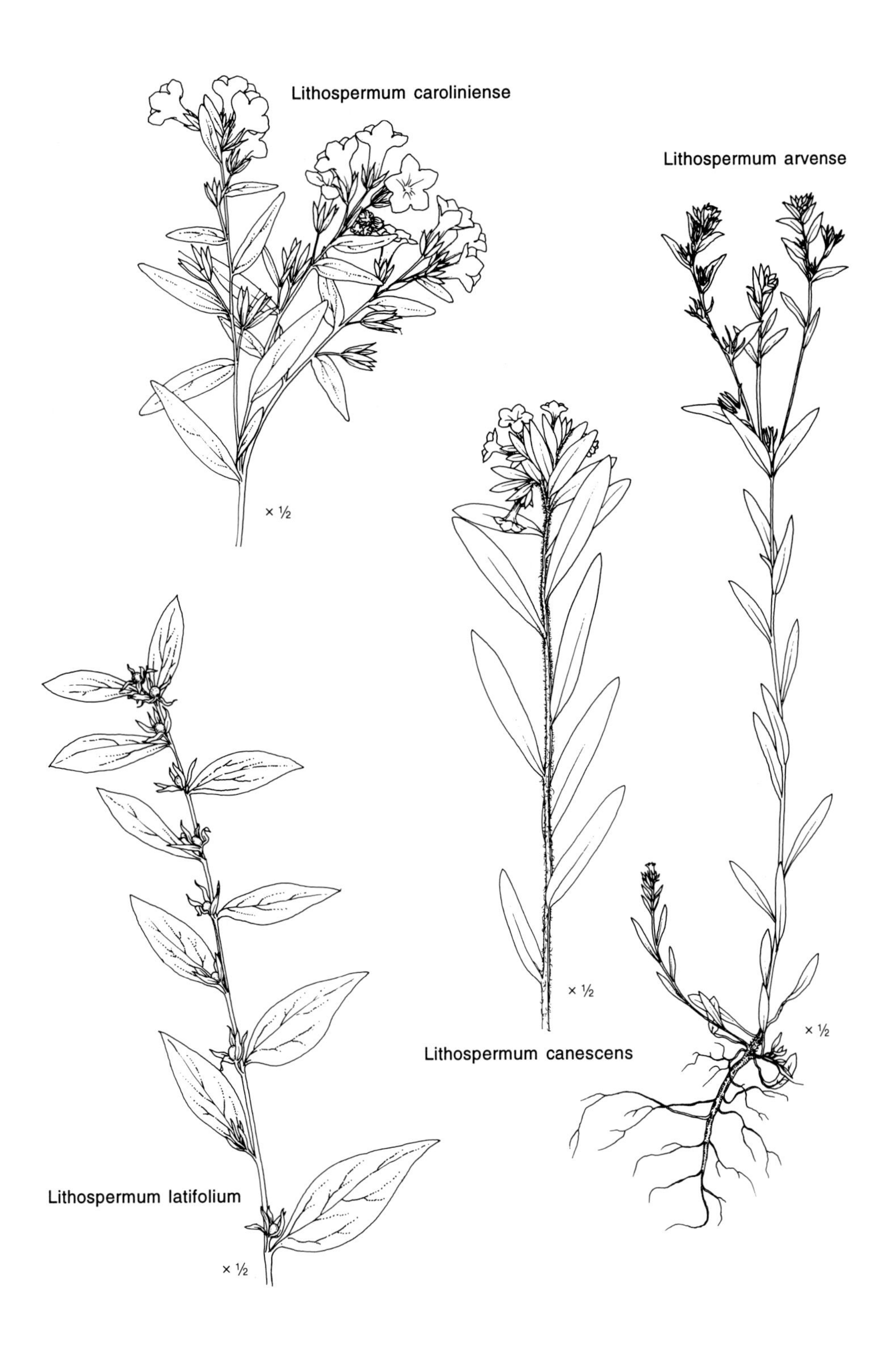
Lithospermum caroliniense
Lithospermum arvense
× ½
Lithospermum latifolium
× ½
× ½
Lithospermum canescens
× ½

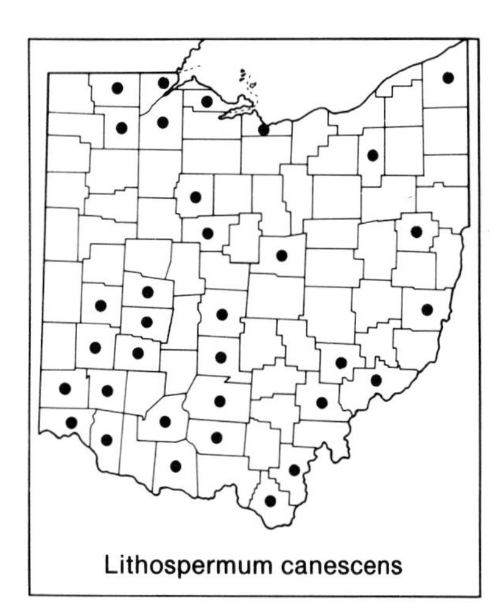
Lithospermum canescens

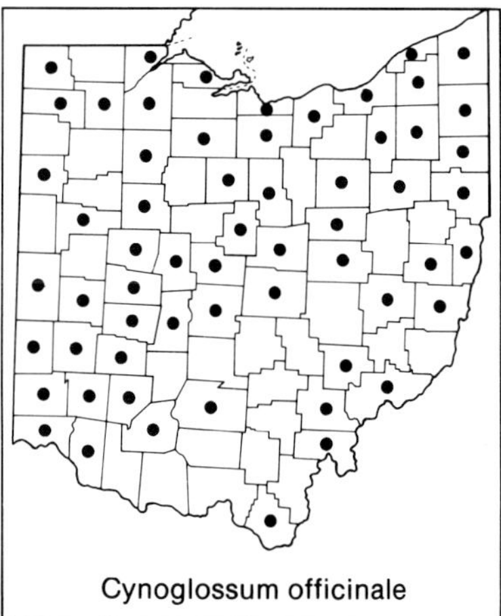
Cynoglossum officinale

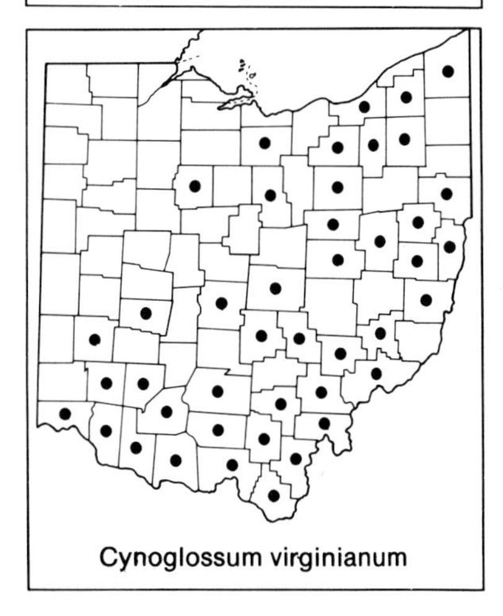
Cynoglossum virginianum

4. **Lithospermum canescens** (Michx.) Lehm. HOARY PUCCOON

Perennial, to 4 dm tall; leaves narrowly oblong to elliptic-oblong or lance-oblong, densely soft-pilose; calyx lobes 2–3 mm long at anthesis, 3–7 mm long when mature; corolla yellow, 10–18 mm long, exceeding calyx, 11–15 mm across; nutlets shining, whitish, smooth. Southwestern counties and at scattered sites elsewhere in Ohio; prairies, grassy openings, fields, dry roadside banks. Heterostyly and homostyly in this species were studied by H. G. Baker (1961). Late April–May.

7. Cynoglossum L. HOUND'S-TONGUE. BEGGAR'S-LICE

Erect pubescent biennials or perennials; flowers borne in bractless, racemose or paniculate, scorpioid cymes; corollas red, purple, or blue, the lobes spreading; nutlets bearing barbed bristles. A widely distributed genus of temperate regions, with some 80 or more species.

a. Stems with numerous full-sized leaves on upper half as well as lower half of stem; inflorescences axillary; nutlets somewhat concave on the sides but with an elevated border . . . . 1. *Cynoglossum officinale*

aa. Stems with large full-sized leaves only on lower half of stem; inflorescences terminal; nutlets rounded on sides and without an elevated border . . . . 2. *Cynoglossum virginianum*

1. CYNOGLOSSUM OFFICINALE L. HOUND'S-TONGUE.

Biennial, to ca. 6 dm tall; stems with numerous full-sized leaves on the upper half as well as the lower; inflorescences axillary; corolla dull red-purple, 6–9 mm across; nutlets obovoid, somewhat concave on the sides, but with an elevated border. A Eurasian species, naturalized throughout much of Ohio, but scarce or absent from many south-central counties; fields, disturbed sites. May–early July.

2. **Cynoglossum virginianum** L. WILD COMFREY

Perennial, to 8 dm tall; stems with large full-sized leaves only on the lower half, those of the upper half few and reduced in size; inflorescences terminal; corollas blue to pale blue, 6–12 mm across; nutlets subglobose to obovoid, rounded on the sides and without an elevated border. Mostly in eastern and southern counties; woodlands. May–early June.

Two varieties of this species occur in the Ohio flora (Cooperrider, 1984b; Gleason and Cronquist, 1991). Using the key below, based chiefly on that of Fernald (1950), most Ohio material will key to var. *virginianum*; a few specimens from northern counties will key to var. *boreale*, and a significant number will be intermediate between the two.

a. Corolla (8–) 10–12 mm wide, the lobes somewhat overlapping; calyx at anthesis (3.0–) 3.5–4.5 mm long; nutlets 5.5–7 (–9) mm long . . . . var. *virginianum*
SOUTHERN WILD COMFREY

aa. Corolla 6–8 mm wide, the lobes not overlapping; calyx at anthesis 2–2.5 mm long; nutlets 3.5–5 mm long . . . . var. *boreale* (Fern.) Cooperr.
C. *boreale* Fern.
NORTHERN WILD COMFREY

8. Myosotis L. FORGET-ME-NOT. SCORPION-GRASS

Erect, ascending, or decumbent annuals, biennials, or perennials; flowers borne in racemose or paniculate scorpioid cymes; corollas usually blue, sometimes white, yellow, or pink, the lobes spreading; nutlets smooth. A widely distributed genus of temperate regions, with 40 or more species.

a. Pubescence on calyx consisting of short straight hairs, appressed and pointing apically.
b. Corolla 5–9 mm wide; calyx lobes mostly shorter than the calyx tube; style exceeding nutlets. . . . . 1. *Myosotis scorpioides*
bb. Corolla 3–5 (–6) mm wide; calyx lobes mostly equalling or longer than the calyx tube; style shorter than nutlets . . . . 2. *Myosotis laxa*

aa. Pubescence on calyx consisting, at least in part, of spreading and/or hooked hairs.
b. Calyx slightly two-lipped, 2 of the lobes longer than the other 3; corolla white, 1–2 mm wide.
c. Principal leaves linear to narrowly oblong or narrowly spatulate, 2–6 (–10) mm wide; fruiting pedicels erect to appressed, but slightly divergent at tip, 1–5 mm long; the pedicels not widely separated: at maturity, the lowest less than 2 cm apart, and the others less than 1 cm apart. . . . . 3. *Myosotis verna*
cc. Principal leaves oblong to oblanceolate, (5–) 8–20 mm wide; fruiting pedicels ascending to divergent, 3–10 mm long; the pedicels widely separated: at maturity, the lowest more than 2 cm apart, the others more than 1 cm apart . . . . 4. *Myosotis macrosperma*
bb. Calyx lobes more or less equal in length; corolla blue, pink, yellow, or rarely white, 1.5–8 mm wide.
c. Corolla 5–6 (–8) mm wide, lobes spreading; fruiting pedicels 5–7 (–10) mm long, exceeding length of mature calyx. . . . . 5. *Myosotis sylvatica*

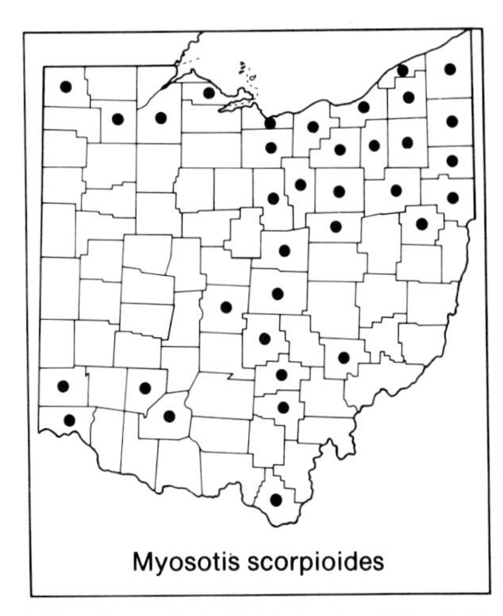
Myosotis scorpioides

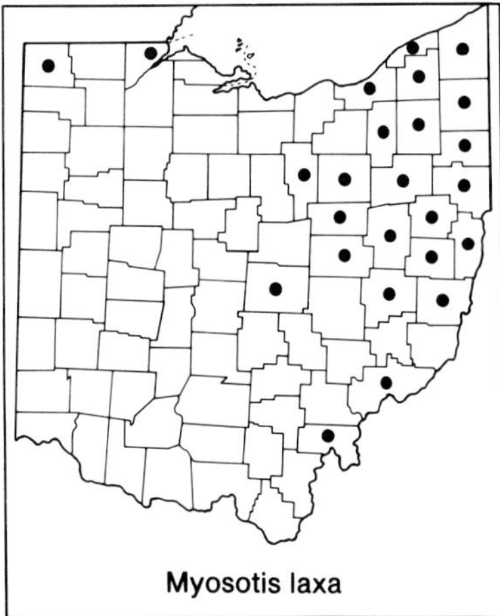
Myosotis laxa

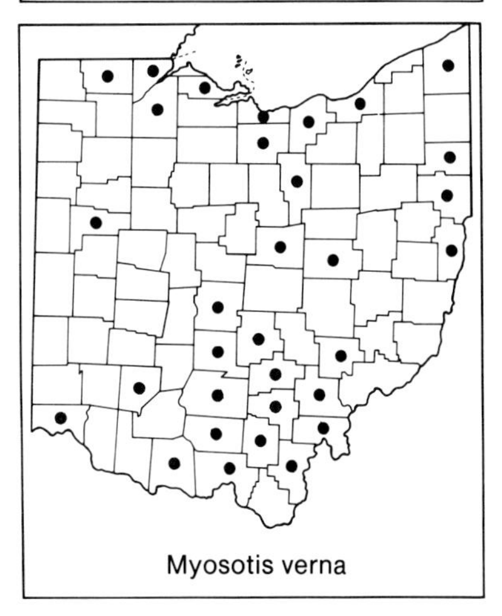
Myosotis verna

cc. Corolla 1.5–3 mm wide, lobes erect; fruiting pedicels 1–9 mm long, exceeding or shorter than mature calyx.

d. Fruiting pedicels 5–9 mm long, exceeding length of mature calyx; corolla 2–3 mm wide . . . . 6. *Myosotis arvensis*

dd. Fruiting pedicels 1–2.5 mm long, less than length of mature calyx; corolla 1–2 (–2.5) mm wide.

e. Veins on undersurface of leaves with some hooked hairs; axis of inflorescence with some hairs ascending to divergent; corolla blue, scarcely exceeding calyx; style shorter than nutlets . . . . 7. *Myosotis stricta*

ee. Veins on undersurface of leaves without hooked hairs; axis of inflorescence with all hairs appressed; corolla at first pale yellow, then becoming pink, violet, or blue, exceeding the calyx; style exceeding nutlets . . . . 8. *Myosotis discolor*

1. Myosotis scorpioides L. True Forget-me-not

Erect, ascending, or decumbent perennial; stems to 6 dm long; calyx pubescent with short, straight, appressed and apically pointing hairs, the calyx lobes approximately equal in length, mostly shorter than the tube; corolla blue with a golden yellow center, 5–9 mm across; style exceeding nutlets. A Eurasian species naturalized in the northeastern quarter of Ohio, and at scattered sites in other parts of the state; river and lake shores, wet ditches and banks, and other wet muddy habitats. May–October.

2. **Myosotis laxa** Lehm. Smaller Forget-me-not

Erect, ascending, or decumbent perennial; stems to 5 dm long; calyx pubescent with short, straight, appressed and apically pointing hairs, the calyx lobes approximately equal in length, mostly equalling or longer than the tube; corolla pale blue with a golden yellow center, 3–5 (–6) mm across; style shorter than nutlets. Chiefly in northeastern counties; lake and stream shores, marshes, swamps, and other wet habitats. May–August.

3. **Myosotis verna** Nutt. Spring Forget-me-not

Annual, to 3 dm tall; stems erect; principal leaves linear to narrowly oblong or narrowly spatulate, 2–6 (–10) mm wide; fruiting pedicels erect to appressed, but slightly divergent at tip, 1–5 mm long; the pedicels not widely separated at maturity, the lowest less than 2 cm apart and the others less than 1 cm apart; calyx unequally cleft and slightly two-lipped, 2 of the lobes longer than the other 3, the tube with some hairs spreading, of these some straight but with a small hook at the tip; corolla white, 1–2 mm across. Southern and northern counties and at scattered sites between; dry fields. May–June.

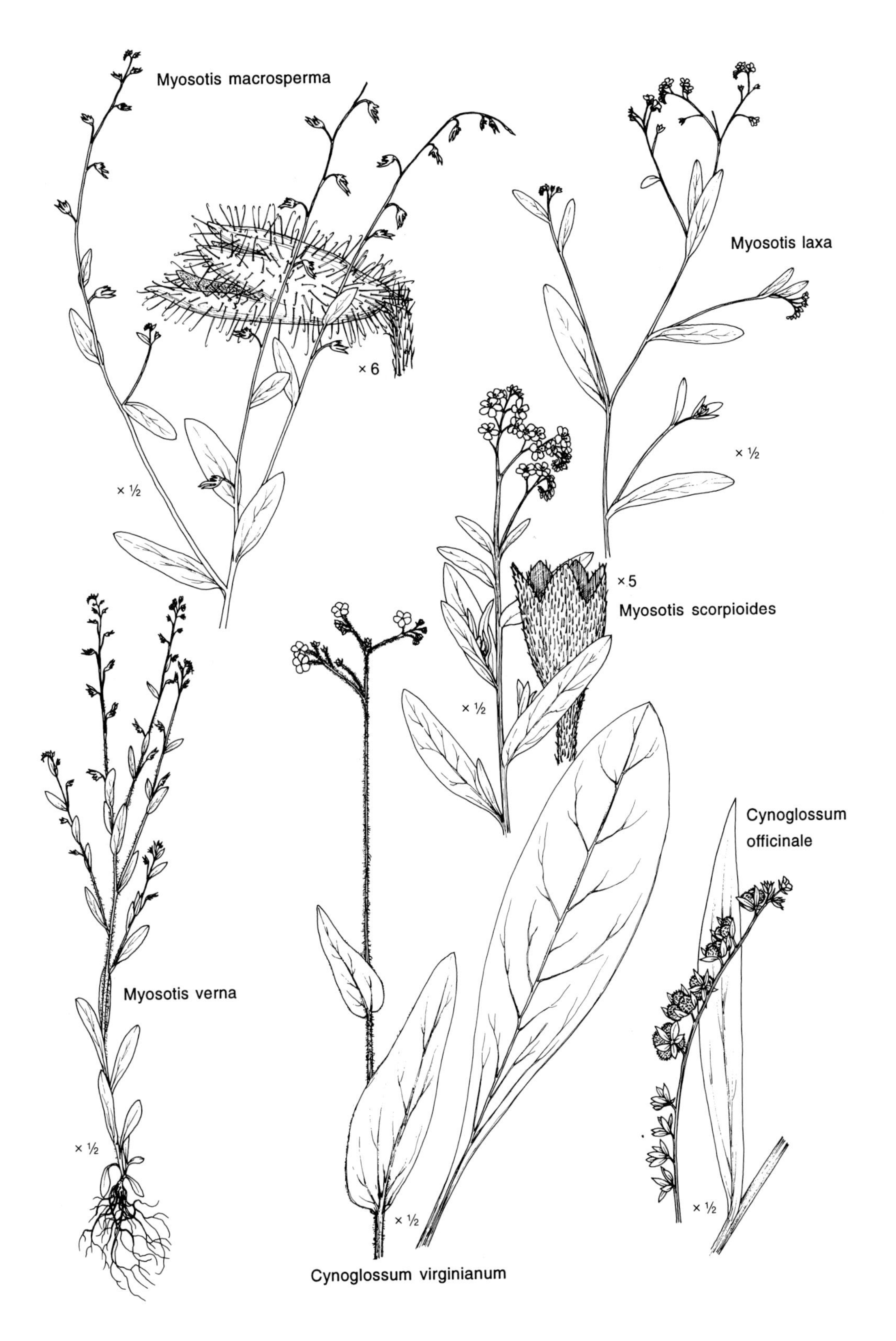
Myosotis macrosperma
×6
× ½
Myosotis laxa
× ½
× ½
×5
Myosotis scorpioides
Myosotis verna
× ½
Cynoglossum officinale
× ½
× ½
Cynoglossum virginianum

4. **Myosotis macrosperma** Engelm. Bristly Scorpion-grass

Annual, to 6 dm tall; stems erect or mostly so; principal leaves oblong to oblanceolate, (5–) 8–20 mm wide; fruiting pedicels ascending to divergent, 3–10 mm long; the pedicels widely separated at maturity, the lowest more than 2 cm apart, the others more than 1 cm apart; calyx unequally cleft and slightly two-lipped, 2 of the lobes longer than the other 3, the tube with all or nearly all hairs spreading and strongly hooked at the tip; corolla white, 1–2 mm across. Chiefly in southern counties, also at scattered northern sites where the species is perhaps adventive; disturbed woods on floodplains and terraces, woodland borders, thickets. Late April–May.

5. Myosotis sylvatica Hoffm. Garden Forget-me-not

Biennial or perennial, to 5 dm tall; stems erect; calyx lobes more or less equal in length; corolla blue (or rarely white) with a yellow center, 5–6 (–8) mm across, the lobes spreading; fruiting pedicels 5–7 (–10) mm long, exceeding length of the mature calyx. A Eurasian species, sometimes cultivated, escaped and naturalized or perhaps merely adventive, chiefly in northeastern counties; disturbed habitats. June.

6. Myosotis arvensis (L.) Hill Field Forget-me-not. Field Scorpion-grass

Annual or biennial, to 5 dm tall; stems more or less erect; calyx lobes more or less equal in length, the tube with spreading hairs; corolla whitish to blue, 2–3 mm across; fruiting pedicels 5–9 mm long, exceeding length of the mature calyx. A Eurasian species, naturalized at scattered sites in Ohio; disturbed habitats. June–July.

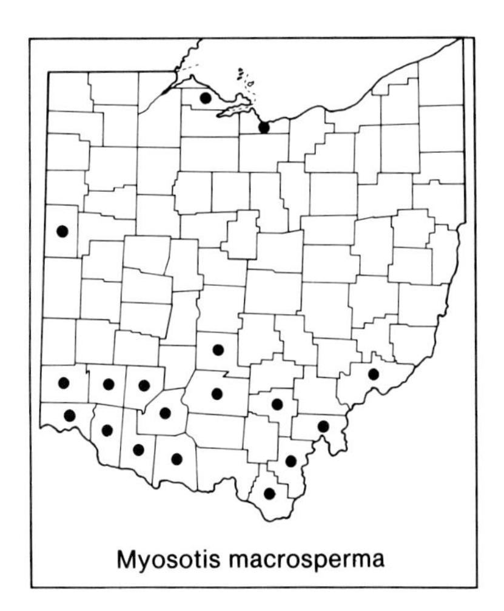
Myosotis macrosperma

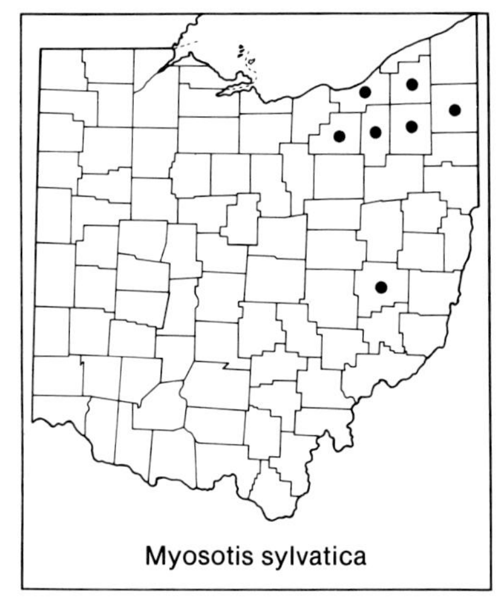
Myosotis sylvatica

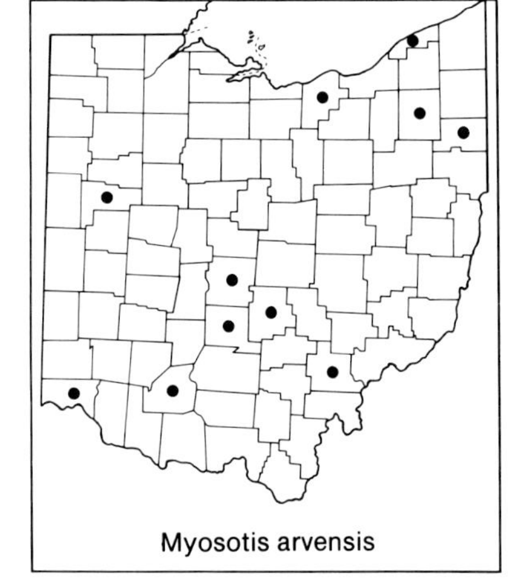
Myosotis arvensis

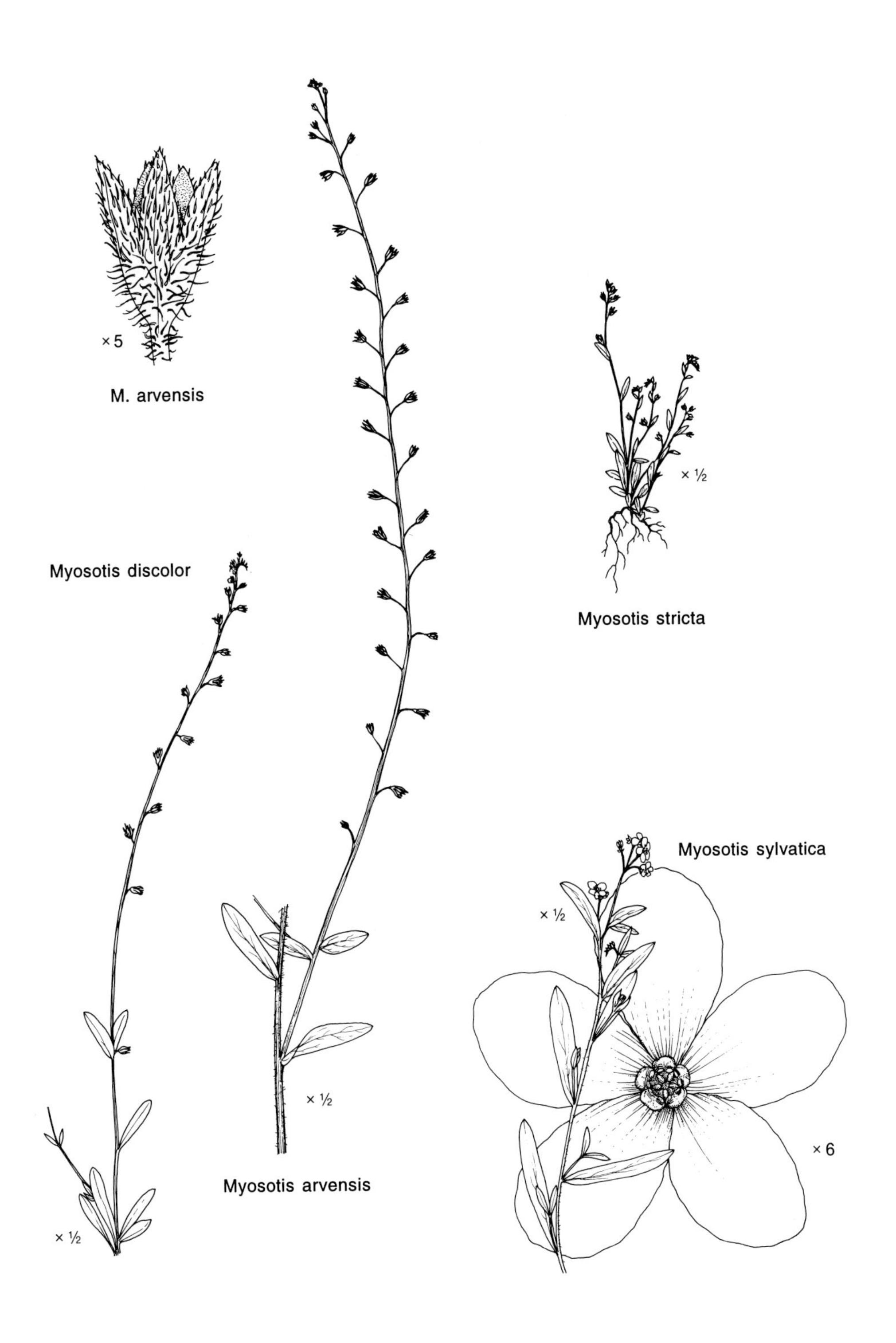
×5
M. arvensis
Myosotis discolor
× ½
× ½
Myosotis arvensis
× ½
Myosotis stricta
Myosotis sylvatica
× ½
× 6

7. Myosotis stricta R. & S.   Small-flowered Forget-me-not
   M. micrantha of authors, not Pall.

Annual, to 2 dm tall, but often shorter; stems erect; veins on undersurface of leaves with some hooked hairs; axis of inflorescence with some hairs ascending to divergent; calyx lobes more or less equal in length; corolla blue, 1–2 mm across, scarcely exceeding calyx; fruiting pedicels 1.0–1.5 mm long, less than length of mature calyx; style shorter than nutlets. A Eurasian species, naturalized or perhaps merely adventive at scattered Ohio sites; fields, dry disturbed habitats. May–July.

8. Myosotis discolor Pers.   Two-colored Forget-me-not
   M. versicolor (Pers.) Sm.

Annual, to 3 dm tall; stems more or less erect; veins on undersurface of leaves without hooked hairs; axis of inflorescence with all hairs appressed; calyx lobes more or less equal in length; corolla at first pale yellow, then becoming pink, violet, or blue, 1.5–2.0 (–2.5) mm across, exceeding the calyx; fruiting pedicels 1.5–2.5 mm long, less than length of mature calyx; style exceeding nutlets. A Eurasian species, naturalized or perhaps merely adventive at a few widely scattered sites in Ohio; lawns, cemeteries, weedy grassy areas. April–May.

9. Mertensia Roth   bluebells. lungwort

A genus of the Northern Hemisphere with some 40 species, one of which is native to Ohio.

1. **Mertensia virginica** (L.) Link   Bluebells. Virginia Cowslip

Glabrous subfleshy perennial; stems erect or ascending, to 7 dm long; flow-

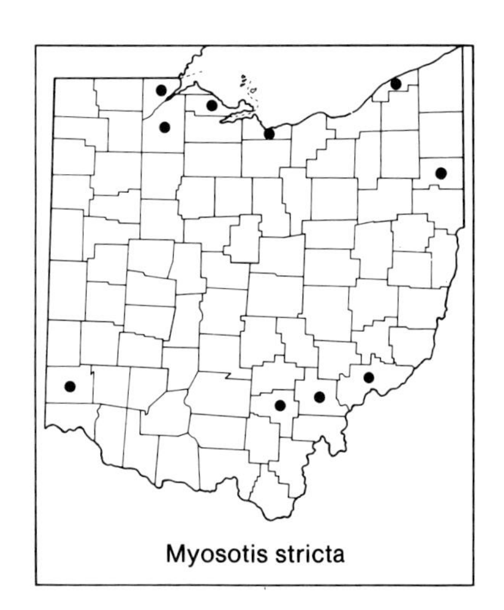
Myosotis stricta

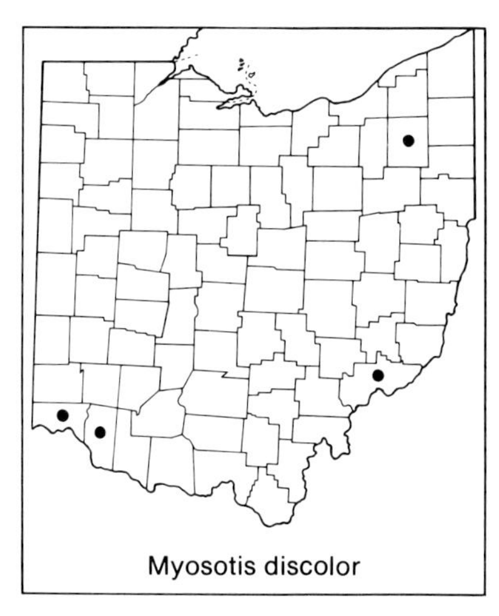
Myosotis discolor

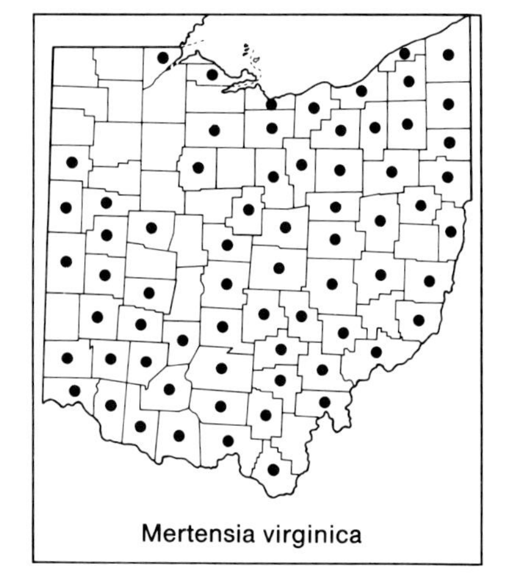
Mertensia virginica

ers in racemose cymes; corolla pink in bud, sky-blue (rarely white) in anthesis, (1.8–) 2.0–2.5 cm long, 1.0–1.5 cm across; nutlets wrinkled, glabrous. A showy spring wildflower; throughout much of Ohio, but absent from most northwestern counties; wooded floodplains and other rich moist woodlands. April–mid-May.

The rare white-flowered plant, forma *berdii* Moldenke (forma *alba* Allard, a later synonym), has been collected from Athens and Miami counties.

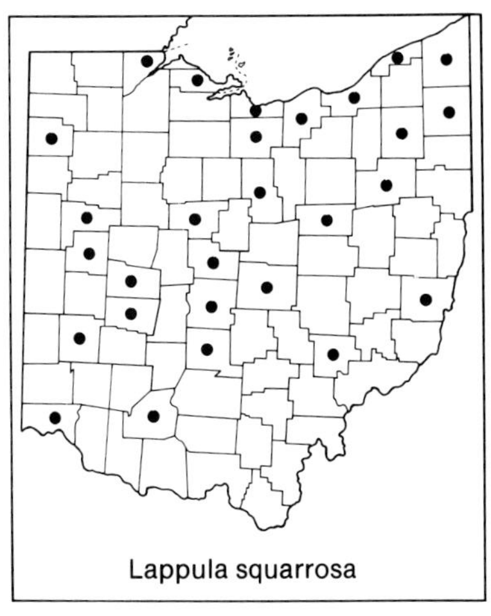
Lappula squarrosa

10. Lappula Moench STICKSEED

A widespread genus of about 14 species, one of which is a naturalized member of the Ohio flora.

1. LAPPULA SQUARROSA (Retz.) Dumort. EUROPEAN STICKSEED
   L. echinata Gilib.

Pubescent annual, to 7 dm tall; stems erect; leaves linear, 2–6 mm wide; corolla pale blue, minute—3–4 mm long and 2–3 mm across; nutlets bearing apically barbed bristles. A Eurasian species, naturalized at scattered sites across Ohio; railroad ballast and other disturbed dry habitats. Frick (1984) studied the biology of this weed. June–July.

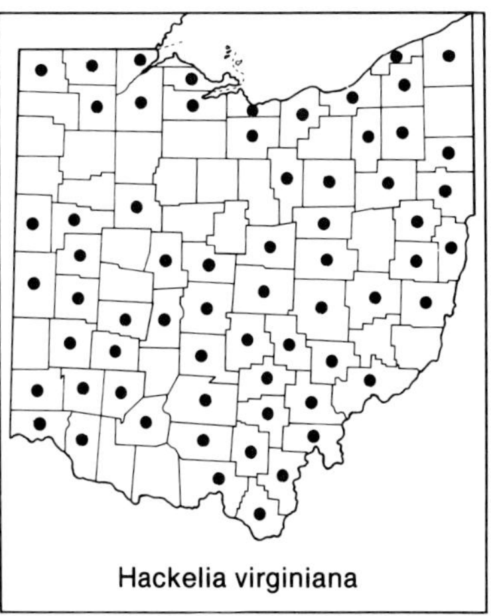
Hackelia virginiana

11. Hackelia Opiz STICKSEED. BEGGAR'S-LICE

Erect pubescent biennials or annuals; flowers borne in bractless, racemose, scorpioid cymes; corolla minute, white or pale blue, the lobes spreading; nutlets with barbed bristles. A genus of 30 or more species, widespread but chiefly in temperate regions of the Northern Hemisphere, merged under *Lappula* by Chater (in Valentine and Chater, 1972). Gentry and Carr (1976) revised *Hackelia* in North America, north of Mexico.

a. Prickles of approximately equal length on both the margin and dorsal face of nutlets . . . . 1. *Hackelia virginiana*
aa. Prickles only on margins of nutlets, or those on the margins much longer than the few on the dorsal face . . . . 2. *Hackelia deflexa*

1. **Hackelia virginiana** (L.) I. M. Johnston COMMON STICKSEED

Biennial, to ca. 1 m tall; stems erect; corolla white, ca. 2 mm long and 2 mm across; nutlets with prickles of approximately equal length on both margin and dorsal face. Throughout much of Ohio, but absent from some western counties; open woodlands, woodland borders, fields. Late May–August.

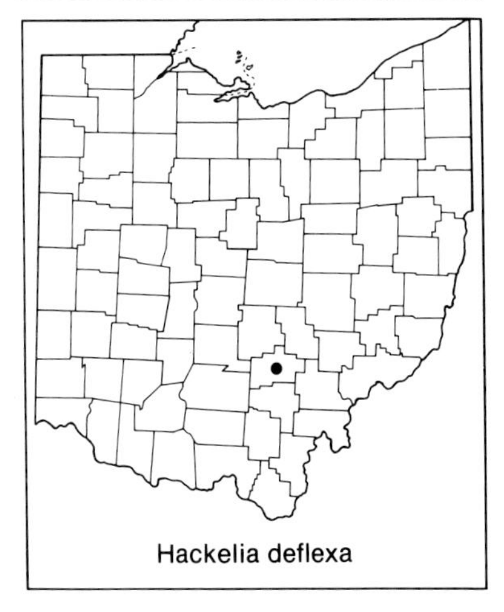
Hackelia deflexa

2. **Hackelia deflexa** (Wahlenb.) Opiz NORTHERN STICKSEED

Annual or biennial, to 5 dm tall; stems erect; corolla pale blue, ca. 2 mm long and 2 mm across; nutlets either with prickles only on margins or those

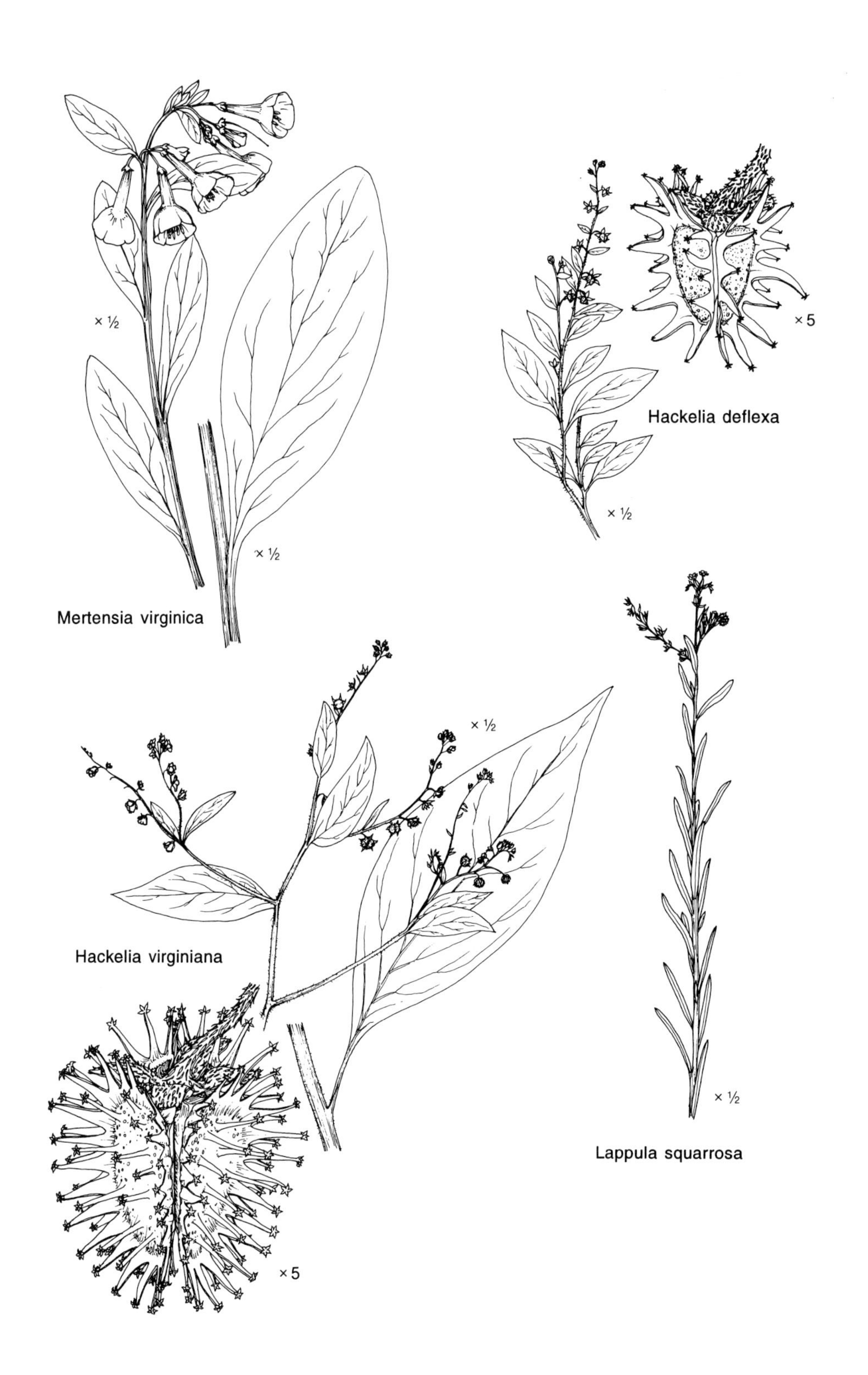
× ½
× ½
Mertensia virginica
× 5
Hackelia deflexa
× ½
× ½
Hackelia virginiana
× 5
× ½
Lappula squarrosa

on the margins much longer than the few on the dorsal face. A northern species ranging across Canada and northern United States, with a single Ohio collection at a southern boundary of the species' range, made in 1899 from Hocking County, in the southeastern quarter of the state (Cooperrider, 1982b); moist rocky woodlands. Late June–July.

Ohio plants belong to var. *americana* (A. Gray) Fern. & I. M. Johnston (*H. americana* (A. Gray) Fern.).

Verbenaceae. VERVAIN FAMILY. VERBENA FAMILY

Perennial herbs; leaves opposite, simple; flowers perfect, in spicate, racemose, or paniculate inflorescences; calyx 2- or 5-lobed; corolla zygomorphic —although often not markedly so, 4- or 5-lobed; stamens 4, sometimes didynamous, attached to the corolla tube; carpels 2, fused to form a compound pistil with a single terminal style; ovary superior; fruit 1–4 nutlets. A large, chiefly tropical family of some 2,600 species. Following Cronquist (1981), Thorne (1968, 1976, 1983), and Whipple (1972), the family is defined here to include the genus *Phryma*. The treatment below is based in part on studies of the Ohio taxa made by Dumke (1967, 1968).

Lantana camara L., LANTANA or YELLOW-SAGE, a native of the American tropics producing clusters of small bright orange or yellow-orange flowers, is seen occasionally as a house or patio plant. *Vitex negundo* L. var. *heterophylla* (Franch.) Rehd., CUT-LEAVED CHASTE-TREE, is a small tree rarely seen in cultivation in Ohio. There is a single 1929 collection from an adventive escape in Preble County.

a. Leaves palmately compound with 5 (rarely 3) leaflets; see above .... *Vitex*
aa. Leaves simple or rarely pinnately lobed.
  b. Flowering calyx widely divergent from inflorescence axis, fruiting calyx strongly deflexed .... 3. *Phryma*
  bb. Flowering and fruiting calyx ascending or only slightly divergent from inflorescence axis.
    c. Calyx 5-lobed; corolla 5-lobed, not 2-lipped; inflorescenses axillary or terminal, when axillary in the form of elongate spikes somewhat open at base .... 1. *Verbena*

cc. Calyx 2-lobed; corolla 4-lobed and 2-lipped; inflorescences axillary, long-pedunculate, globose to short-cylindric, head-like spikes . . . . 2. *Phyla*

1. Verbena L. VERVAIN

Erect or infrequently prostrate to ascending perennials; calyx 5-lobed; corolla 5-lobed, white to pink, purple, or blue. A genus of some 200 or more species, chiefly of the New World. Barber (1982) made taxonomic studies of the *Verbena stricta* complex, which includes species nos. 1–5 below. W. H. Lewis and R. L. Oliver (1961) studied the cytogeography and phylogeny of the North American species of *Verbena*. Moldenke (1958) and Poindexter (1962) studied hybridization within the genus.

The cultivated *Verbena* × *hybrida* Voss, Garden Vervain or Garden Verbena, was reported by Moldenke (1980) as an escape from Ashtabula County, but I have seen no specimens.

a. Corolla 2–9 mm wide; stems prostrate to erect.
 b. Plants prostrate to weakly ascending; bracts of inflorescence conspicuously exserted and spreading, twice or more as long as the calyx . . . . 5. *Verbena bracteata*
 bb. Plants ascending to, more often, erect; bracts of inflorescence mostly ascending, much less than to only slightly exceeding length of calyx.
  c. Inflorescences in groups of 5–many, forming paniculate clusters at tips of main stem and branches.
   d. Corolla white, ca. 2 mm wide; flowers and especially the fruits remote, all or least those on lower half of inflorescence axis separated and not overlapping; leaves unlobed . . . . 1. *Verbena urticifolia*
   dd. Corolla blue to purple or rarely pink, 3–4 mm wide; flowers and fruits congested, all or all except the lowest few on the inflorescence axis crowded and overlapping; leaves unlobed or hastately 3-lobed . . . . 2. *Verbena hastata*
  cc. Inflorescences solitary or in groups of 3 at tips of main stem and branches.
   d. Stems and leaves slightly scabrous to glabrous; leaves linear to narrowly lanceolate or narrowly oblanceolate; corolla 5–6 mm wide . . . . 3. *Verbena simplex*
   dd. Stems and leaves densely soft-pubescent; leaves elliptic to oblong or ovate; corolla 8–10 mm wide . . . . 4. *Verbena stricta*
aa. Corolla 10–25 mm wide; stems decumbent to ascending, not totally erect.

b. Calyx lobes 2–4 mm long, lanceolate with attenuate apex; corolla 10–15 mm wide; leaves moderately pubescent to glabrous. . . . . 6. *Verbena canadensis*

bb. Calyx lobes ca. 1 mm long, deltoid to triangular with short-acuminate apex; corolla 10–25 mm wide; leaves densely pubescent; see above . . . . *Verbena* × *hybrida*

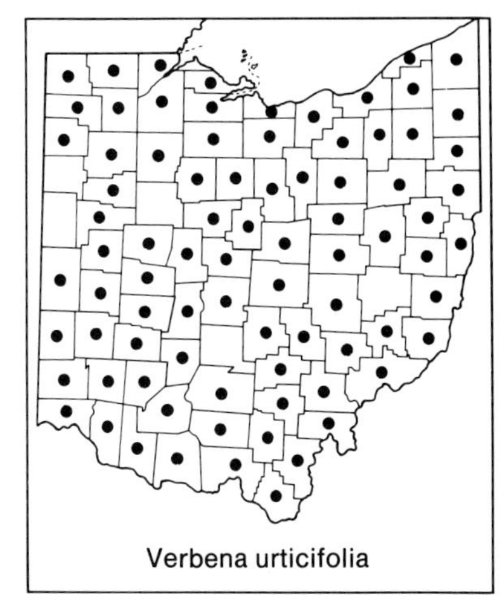
Verbena urticifolia

1. **Verbena urticifolia** L. White Vervain

Perennial, to 1.5 m tall; stems erect, slightly pubescent to nearly glabrous; leaves serrate to dentate, sometimes doubly so, but unlobed; flowers in several to many, long, spicate inflorescences, these forming paniculate clusters at tips of main stem and axillary branches; flowers and especially the fruits remote, all or at least those on the lower half of the inflorescence axis separated and not overlapping; corolla white, ca. 2 mm across. Throughout Ohio; roadsides, pastures, weedy fields and thickets, and a variety of other disturbed habitats. Additional record: Pickaway County (Moldenke, 1980). July–September.

Ohio plants belong to var. *urticifolia*. *Verbena urticifolia* var. *leiocarpa* Perry & Fern. was attributed to Coshocton and Hamilton counties by Moldenke (1980). I have found no specimens in this study, although there is a considerable amount of variation within Ohio specimens and some may approach var. *leiocarpa*. The range given by Fernald (1950) for this variety lies to the east and south of Ohio.

Verbena hastata

Verbena × **engelmannii** Moldenke, the naturally occurring hybrid of *Verbena hastata* × *V. urticifolia*, has been collected in Auglaize, Columbiana, Lorain, Lucas, Portage, and Wood counties. Moldenke (1980) reports it also from Champaign County.

2. **Verbena hastata** L. Blue Vervain

Perennial, to 1.5 m tall; stems erect, short-pubescent; leaves coarsely serrate, sometimes doubly so, the lower leaves sometimes hastately lobed; flowers in 5–many (fewer in depauperate individuals) spicate inflorescences, these forming paniculate clusters at tips of main stem and axillary branches; flowers and fruits congested, all—or all except the lowest few on the inflorescence axis—crowded and overlapping; corolla blue to purple or rarely pink, 3–4 mm across. Throughout Ohio; open wet fields, ditches, shores, and other wet disturbed habitats. Additional records: Brown and Miami counties (Moldenke, 1980). Late June–September.

See comments under *V. urticifolia* above, regarding naturally occurring hybrids between that species and this.

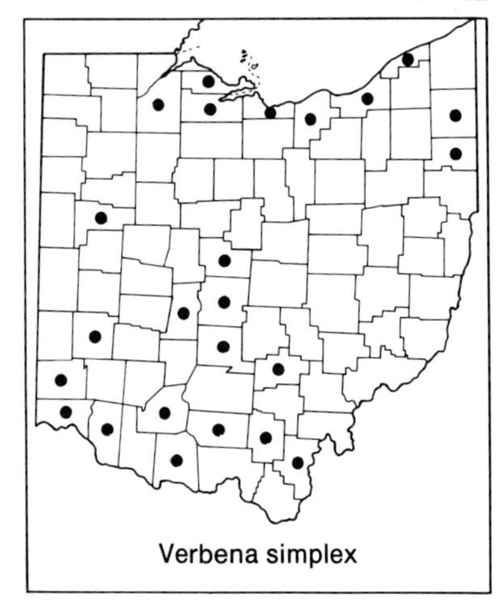
Verbena simplex

3. **Verbena simplex** Lehm. Narrow-leaved Vervain

Perennial, to 6 dm tall; stems erect, slightly scabrous to glabrous; leaves linear to narrowly lanceolate or narrowly oblanceolate; flowers in one to

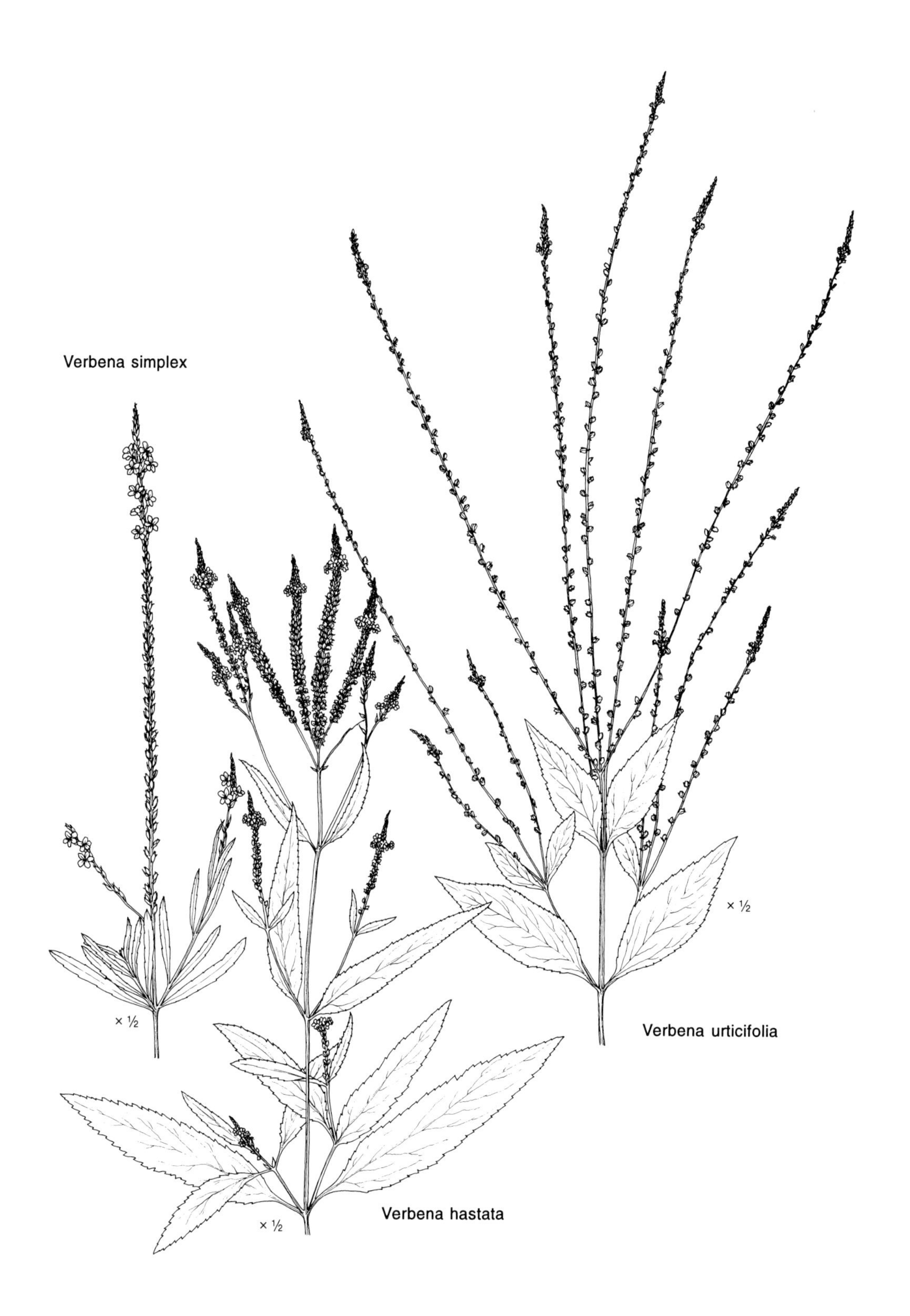
Verbena simplex
× ½
Verbena hastata
× ½
Verbena urticifolia
× ½

several spicate inflorescences, these solitary or in groups of 3 at tips of main stem and branches; corolla purple, 5–6 mm across. Northern and southwestern counties and at scattered sites elsewhere; calcareous fields, railroad ballast, dry disturbed habitats. Additional record: Champaign County (Moldenke, 1980). Mid-May–mid-September.

Verbena × moechina Moldenke, the naturally occurring interspecific hybrid of *Verbena simplex* × *V. stricta*, was reported from Erie County, by Moldenke (1980).

4. **Verbena stricta** Vent. HOARY VERVAIN

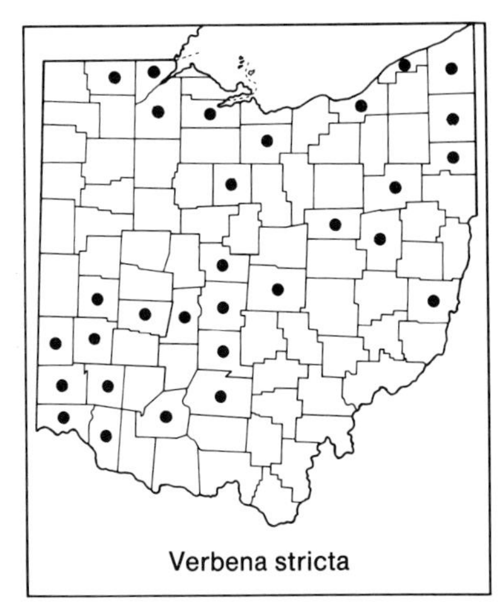

Verbena stricta

Perennial, to 1 m tall; stems erect, densely soft-pubescent; leaves elliptic to oblong or ovate; flowers in one to several spicate inflorescences, these solitary or in groups of 3 at tips of main stem and branches; corolla purple or bluish-purple, 8–10 mm across. Southwestern counties and scattered elsewhere, collections from eastern counties are probably of adventives; dry rocky open sites, railroad ballast. Additional record: Champaign County (Moldenke, 1980). July–September.

See note under *V. simplex* above, regarding the report of a hybrid between that species and this.

5. **Verbena bracteata** Lag. & Rodr. PROSTRATE VERVAIN. BRACTED VERVAIN

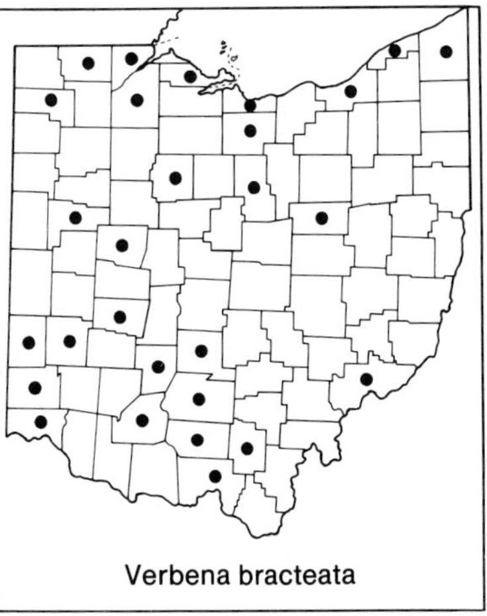

Verbena bracteata

Perennial; stems hirsute, prostrate to weakly ascending, to 4 dm long; leaves coarsely incised, sometimes 3-lobed; flowers in 1–several dense spicate inflorescences, the bracts of the inflorescence spreading and conspicuously exserted, twice or more as long as the calyx; corolla purplish-blue, 2–3 mm across. Chiefly in western counties, with a few scattered occurrences eastward; along roads, railroads, and sidewalks, and other similar open weedy habitats. Additional record: Gallia County (Moldenke, 1980). June–September.

6. VERBENA CANADENSIS (L.) Britt. ROSE VERVAIN
Glandularia canadensis (L.) Nutt.

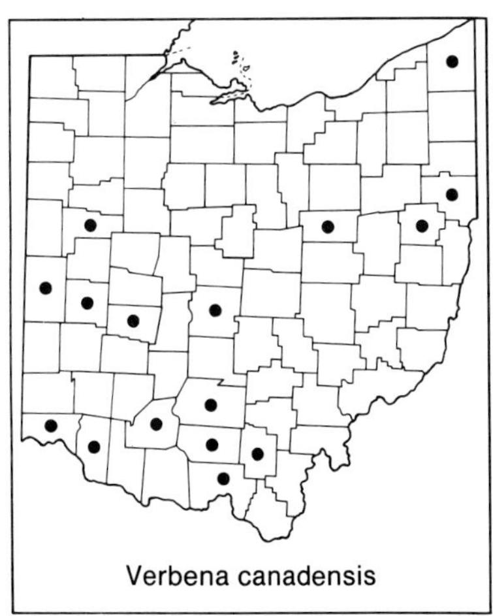

Verbena canadensis

Perennial; stems more or less hirsute, ascending, to 4 dm long; leaves deeply incised to shallowly lobed, moderately pubescent to glabrous; calyx lobes 2–4 mm long; corolla reddish-purple to rose or white, 10–15 mm across. A species of southern and western United States, cultivated for its summer flowers, naturalized or at some Ohio sites merely adventive, chiefly in southwestern counties; fields, cemeteries, and other open disturbed habitats. May–early September.

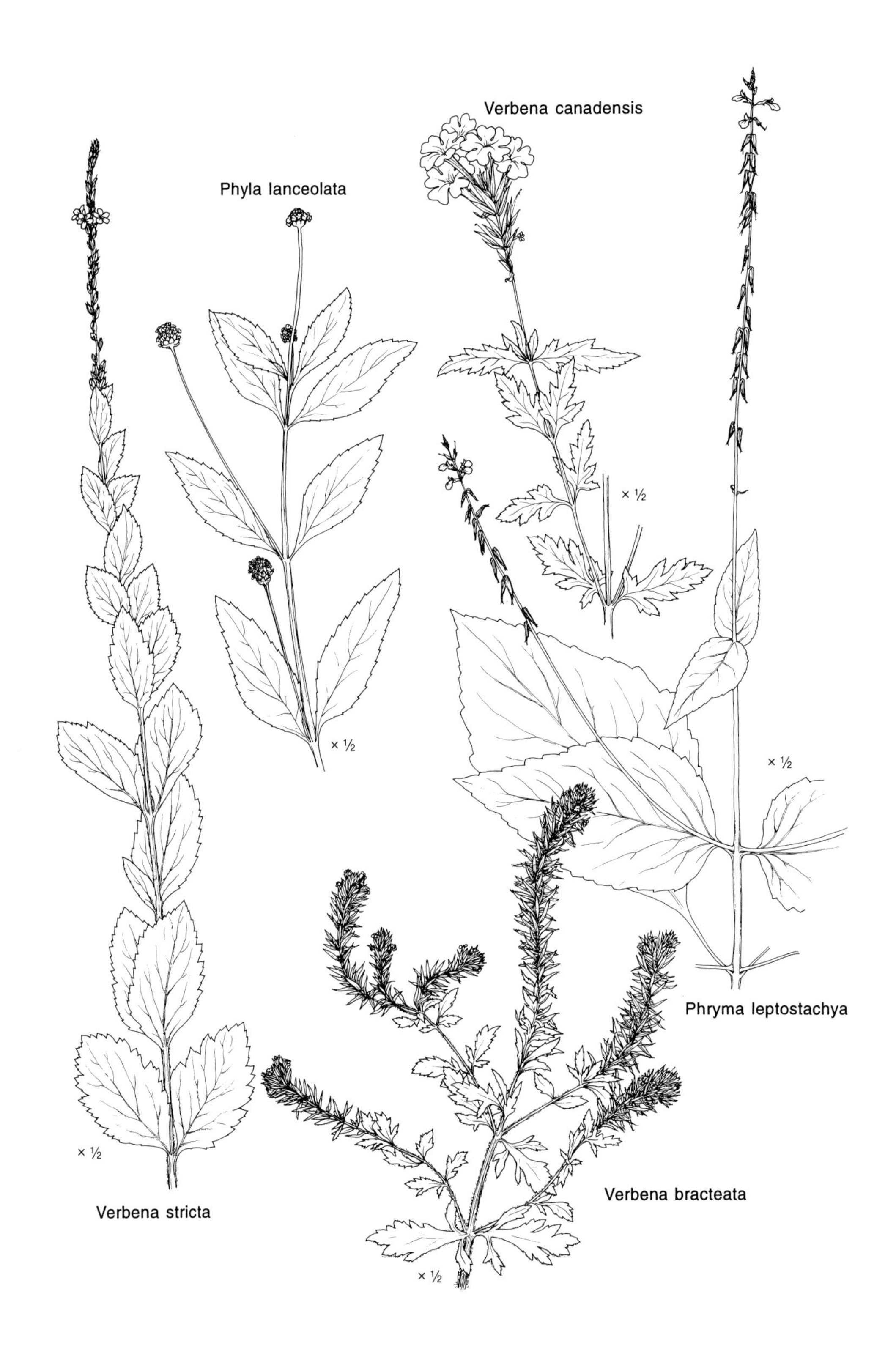
Verbena canadensis
Phyla lanceolata
× ½
× ½
× ½
Phryma leptostachya
× ½
Verbena stricta
Verbena bracteata
× ½

2. Phyla Lour.

A genus of some 15 species, chiefly of tropical and warm regions, with a single species native to the Ohio flora. *Phyla* is merged under the genus *Lippia* L. by Fernald (1950).

1. **Phyla lanceolata** (Michx.) E. Greene FROG-FRUIT. FOG-FRUIT.
   incl. var. *lanceolata* and var. *recognita* (Fern. & Grisc.) Soper; *Lippia lanceolata* Michx.

Procumbent to ascending or suberect perennial; stems short-pubescent to nearly glabrous, squarish, sometimes purplish-red in places, to ca. 6 dm long; leaves lanceolate to lance-elliptic, serrate to shallowly dentate; flowers in axillary, long-pedunculate, globose to short-cylindric, head-like spikes; calyx 2-lobed; corolla 4-lobed and 2-lipped, whitish to pinkish or pale blue, ca. 2 mm across. Throughout much of Ohio, but rare or absent from some eastern counties; open moist fields, banks, and shores, marshy areas. June–September.

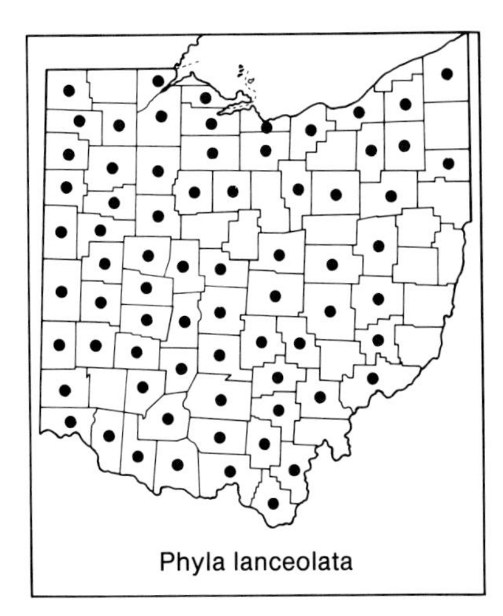
Phyla lanceolata

3. Phryma L.

A genus of one or two species of eastern Asia and eastern North America, placed in the segregate family, Phrymaceae, by Fernald (1950) and Gleason (1952b). Thieret (1972) studied the genus in southeastern United States.

1. **Phryma leptostachya** L. LOPSEED

Erect perennial, to 8 dm tall; stems more or less pubescent; leaves ovate to narrowly ovate, margins shallowly dentate; flowers in 1–several elongated spicate inflorescences, the flowers widely divergent from the inflorescence axis at anthesis, becoming strongly and conspicuously deflexed in fruit; calyx 5-lobed, 2-lipped; corolla 5-lobed, 2-lipped, white to pinkish, ca. 8 mm long, ca. 3 mm across. Throughout Ohio; woodlands. Whipple (1972) discusses the plant structure and systematics. July–August.

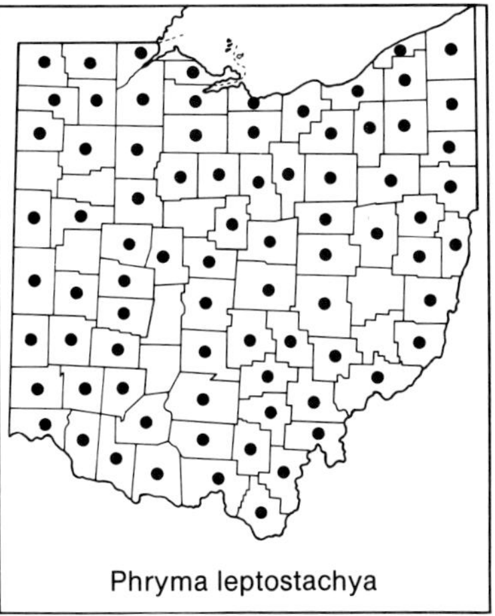
Phryma leptostachya

Lamiaceae or Labiatae. MINT FAMILY

Aromatic annual or perennial herbs; stems square in cross-section; leaves opposite, simple; flowers perfect, most often in clusters formed from 1–several consecutive whorls (verticils) of several flowers each; calyx 5-lobed; corolla 5-lobed, usually zygomorphic and bilabiate; stamens attached to the corolla tube, either 4 and didynamous or only 2; carpels 2, united to form a compound pistil with a single style; the ovary superior, deeply lobed and indented into 4 nearly separate segments, the style gynobasic, inserted at the base of the ovary in the midst of the 4 segments; fruit a group of 4 nutlets. A major plant family of some 3,200 species, widely distributed around the earth. The treatment below is based in part on a study of Ohio mints by Sabo (1965); the treatment of alien species is based largely on that of Heywood and Richardson (1972) in *Flora Europaea*. The Canadian mints were studied by Gill (1981) and the Michigan mints by Waterman (1960). The family has two alternate scientific names; the use of either is equally correct.

Acinos arvensis (Lam.) Dandy (*A. thymoides* (Lam.) Moench, *Satureja acinos* (L.) Scheele), MOTHER-OF-THYME or BASIL-THYME, a weed of European origin, was collected once, in 1892, as an adventive in Lake County. *Ballota nigra* L. var. *alba* (L.) Sm., BLACK HOREHOUND, an adventive weed from Eurasia, was collected in 1963 from a parking lot in Brown County. *Dracocephalum parviflorum* Nutt. (*Moldavica parviflora* (Nutt.) Britt.), AMERICAN DRAGONHEAD, native to Canada and western United States, has been collected as an adventive weed from disturbed habitats in Belmont, Henry, Highland, Perry, Pickaway, and Trumbull counties. *Satureja hortensis* L., SAVORY or SUMMER SAVORY, used for seasoning foods, has been collected as an adventive escape in Athens, Lake, and Ottawa counties. *Coleus blumei* Benth., PAINTED NETTLE, was attributed to Ohio by Cusick and Silberhorn (1977) on the basis of a specimen later determined to be *Perilla frutescens*.

Several members of this family are cultivated for use as flavorings, seasonings, and fragrances. In addition to *Satureja hortensis* noted above and other species discussed in the text below, the following are often seen in Ohio herb gardens: *Lavandula angustifolia* Mill., ENGLISH LAVENDER; *Ocimum basilicum* L., COMMON BASIL or SWEET BASIL, and *Rosmarinus officinalis* L., ROSEMARY.

a. Corolla actinomorphic or nearly so, the lobes appearing approximately equal.
  b. Flowers solitary or few in loose clusters, on pedicels 3–10 mm long; leaves entire.
    c. Calyx actinomorphic or subactinomorphic, the lobes equal or nearly so; stamens about equalling length of corolla lobes, included or slightly exserted . . . . 1. *Isanthus*
    cc. Calyx zygomorphic, the lobes strongly unequal; stamens much longer than corolla lobes, long-exserted. . . . . 2. *Trichostema*
  bb. Flowers several to many in dense clusters or spicate inflorescences, sessile or on pedicels to 2 mm long; leaves finely serrate to serrate, dentate, or deeply incised.
    c. Fertile stamens 2; flowers white, sessile in axillary clusters . . . . 29. *Lycopus*

cc. Fertile stamens 4; flowers white to pale purple or pale blue, short-pedicelled in either axillary clusters or spicate inflorescences . . . . 30. *Mentha*

aa. Corolla zygomorphic, the lobes unequal.

b. Anther-bearing stamens 2.

c. Calyx actinomorphic or nearly so, not bilabiate.

d. Corolla yellow, the lower median corolla lobe fringed across its summit; the flowers in open terminal panicles . . . 31. *Collinsonia*

dd. Corolla white, pink, red, lavender, or purple (yellow in *Monarda punctata* only), the lower median corolla lobe not fringed; the flowers in more or less congested clusters.

e. Corolla 15–45 mm long, markedly bilabiate; the stamens arching closely under the upper lip. 19. *Monarda*

ee. Corolla 5–8 mm long, only moderately bilabiate; the stamens free from the upper lip. . . . . 28. *Cunila*

cc. Calyx bilabiate.

d. Principal cauline leaves 1–3 cm long; corolla 3–5 mm long, scarcely or not exceeding the calyx . . . . 21. *Hedeoma*

dd. Principal cauline leaves 3–12 cm long; corolla 6–25 mm long, exceeding the calyx.

e. Flowers in few-flowered verticils, these together forming a terminal spicate raceme . . . . 18. *Salvia*

ee. Flowers in crowded, compact, many-flowered clusters . . . . 20. *Blephilia*

bb. Anther-bearing stamens 4.

c. Calyx tube with a low ridge or upright protuberance on upper side . . . . 5. *Scutellaria*

cc. Calyx-tube without ridge or protuberance.

d. Corolla having or appearing to have only a lower lip, the upper lip very short or modified.

e. Leaves deeply lobed, the lobes linear or narrowly oblong . . . . 4. *Teucrium*

ee. Leaves unlobed.

f. Stems 1–2 (–3) dm tall; flowers in a congested leafy spike. . . . . 3. *Ajuga*

ff. Stems 3–10 dm tall; flowers in racemes . . . . 4. *Teucrium*

dd. Corolla appearing bilabiate.

e. Most flowers solitary, or in clusters in leaf axils, or in leafy terminal inflorescences.

f. Calyx markedly zygomorphic, the lobes unequal.

g. Corolla 30–35 mm long. 13. *Synandra* gg. Corolla 5–15 mm long.

h. Leaves with petioles 1–4 cm long . . . . 22. *Melissa*

hh. Leaves sessile or with petioles to 1 cm long.

i. Leaves 0.6–1.5 cm long; flowers few in open clusters; calyx tube gibbous at base; see above. . . . . *Acinos*

ii. Leaves 2–4 (–6) cm long; flowers numerous in dense clusters; calyx tube not gibbous . . . . 24. *Clinopodium*

ff. Calyx actinomorphic or nearly so, the lobes approximately equal.

g. Flowers on evident pedicels.

h. Leaves linear or sublinear . . . . 23. *Calamintha*

hh. Leaves ovate to orbicular or reniform.

i. Leaf blades 6–20 cm long; corolla yellow. . . 31. *Collinsonia*

ii. Leaf blades 1.5–6 cm long; corolla pink, purple, or blue.

j. Stems prostrate, extensively creeping; leaves orbicular to reniform, 1.5–4.5 cm long. . . . . 10. *Glechoma*

jj. Stems erect; leaves ovate to broadly ovate, 3–6 cm long; see above. . . . . *Ballota*

gg. Flowers sessile or nearly so.

h. Calyx lobes acuminate, but not terminating in spines.

i. Corolla more than twice as long as calyx; upper lip of corolla helmet-shaped . . . . 16. *Lamium*

ii. Corolla about 1.5 times as long as calyx; upper lip of corolla not helmet-shaped; see above. . . . . *Satureja*

hh. Calyx lobes terminating in spines.

i. Calyx terminating in 10 spines, each spine more or less hooked at the tip. . . . . 6. *Marrubium*

ii. Calyx terminating in 5 spines.

j. Spines on calyx lobes about 1 mm or less in length; plants perennial or biennial, mostly 10–15 dm tall. . . . . 14. *Leonurus*

jj. Spines on calyx lobes about 2 mm or more in length; plants annual, 1–6 dm tall . . . . 15. *Galeopsis*

ee. Most flowers in terminal, bracteate but not leafy inflorescences, or in nonleafy axillary spikes or racemes.

f. Calyx markedly zygomorphic, the lobes unequal.

g. Principal leaves 0.5–1 cm long . . . . 27. *Thymus*

gg. Principal leaves 2–20 cm long.

h. Inflorescences somewhat open at maturity, 5–15 cm long; corolla 4–10 mm long.

i. Leaves linear to elliptic or elliptic-lanceolate, entire or crenate; calyx pubescent with nonseptate hairs; corolla 5–10 mm long . . . . 2. *Trichostema*

ii. Leaves ovate to oblong-ovate, coarsely serrate to dentate; calyx bearing septate hairs; corolla 4–6 mm long . . . . 32. *Perilla*

hh. Inflorescences usually dense, 2–10 cm long; corolla 7–16 mm long.

i. Calyx, bracts, and undersurface of leaves gray- to white-canescent; corolla white to pink or purple.

j. Leaves truncate to cordate at base, the leaf margins coarsely serrate-dentate; corolla 10–12 mm long. . . . . 9. *Nepeta*

jj. Leaves tapering to acute or obtuse base, the leaf margins remotely low-serrate; corolla 7–9 mm long . . . . 25. *Pycnanthemum*

ii. Calyx, bracts, and undersurface of leaves green, either pubescent or glabrous; corolla blue, purple, pink, or rarely white.

j. Margins of leaves coarsely and sharply serrate; margins of inflorescence bracts spinulose-serrate; see above . . . . *Dracocephalum*

jj. Margins of leaves low serrulate, crenate, or entire; margins of inflorescence bracts ciliate but not spinulose-serrate.

k. Inflorescence longer than wide, the bracts broad and reniform . . . . 11. *Prunella*

kk. Inflorescence wider than long, the bracts linear and bristle-like. . . . . 24. *Clinopodium*

ff. Calyx actinomorphic, the lobes approximately equal.

g. Corolla 25–30 mm long.

h. Leaves long-petiolate, the blades broadly ovate, with crenate margins, and rounded to a usually cordate base . . . . 8. *Meehania*

hh. Leaves sessile or subsessile, the blades narrowly lanceolate to oblong, with sharply serrate margins, and tapered to an acute base . . . . 12. *Physostegia*

gg. Corolla 3–15 mm long.

h. Stamens not exserted beyond corolla . . . . 17. *Stachys*

hh. Stamens, or at least some of them, longer than corolla and exserted.

i. Inflorescences longer than wide, spicate; leaves ovate, 5–15 cm long . . . . 7. *Agastache*

ii. Inflorescences wider than long, corymbose or cymose; leaves linear to ovate, 1.5–12 cm long, less than 5 cm long if ovate.

j. Flowers in dense head-like cymes; calyx lobes acuminate, 1–2.5 mm long. . . . . 25. *Pycnanthemum*

jj. Flowers in somewhat open corymbs or broad panicles; calyx lobes acute, 0.5–1 mm long. . . . . 26. *Origanum*

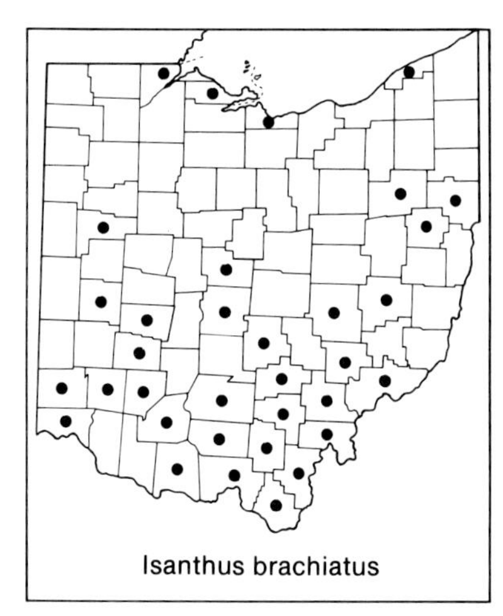
Isanthus brachiatus

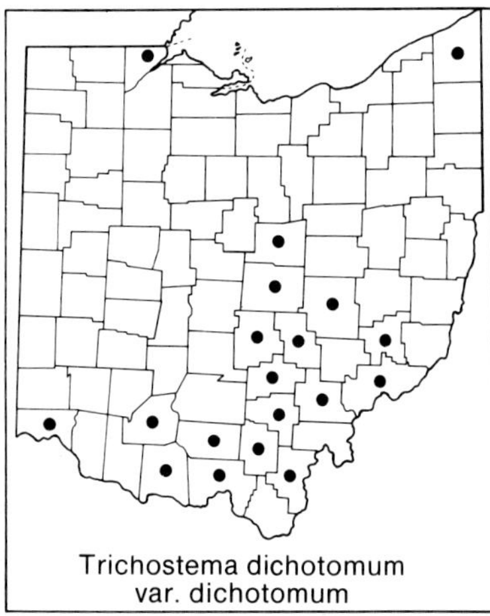
Trichostema dichotomum var. dichotomum

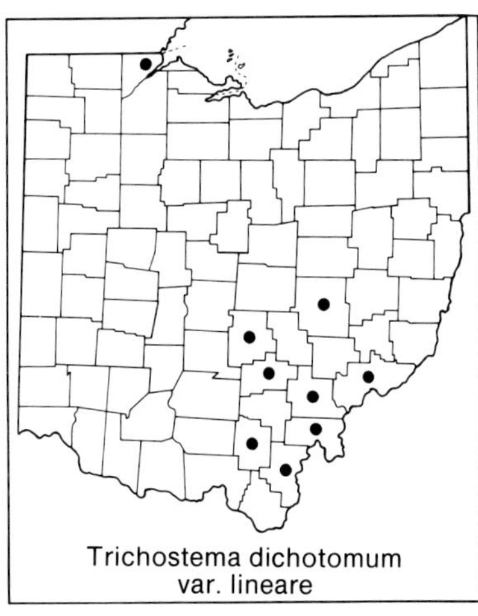
Trichostema dichotomum var. lineare

1. Isanthus Michx. FALSE PENNYROYAL

A genus with but a single species. H. Lewis (1945, 1960) merged *Isanthus* under the related genus *Trichostema*, a merger supported by Abu-Asab and Cantino (1989). Concurring with Gleason and Cronquist (1991), I continue to recognize the segregation of *Isanthus*.

1. **Isanthus brachiatus** (L.) BSP. FALSE PENNYROYAL
Trichostema brachiatum L.

Annual, to 4 dm tall; stems erect, short-pubescent; leaves entire; flowers either solitary or few in loose clusters, on pedicels 3–10 mm long; calyx actinomorphic or subactinomorphic, the lobes equal or nearly so; corolla blue, 4–5 mm long, actinomorphic or nearly so, the lobes appearing approximately equal; stamens 4, equalling length of corolla lobes, included or slightly exserted. Southern counties and at scattered sites northward; open, rocky, calcareous areas, and railroad ballast. August–mid-September.

2. Trichostema L. BLUECURLS

A genus of two species, the other in southeastern United States. H. Lewis (1945, 1960) studied the taxonomy of the genus.

1. **Trichostema dichotomum** L. BLUECURLS. BASTARD PENNYROYAL

Annual, to 6 dm tall; stems erect, short-pubescent; leaves entire or slightly crenate; flowers either solitary or few in loose clusters, on pedicels 3–10 mm long; calyx markedly zygomorphic, the lobes strongly unequal; corolla blue, 5–10 mm long, actinomorphic to slightly zygomorphic; stamens 4, much longer than corolla lobes, long-exserted. Mostly in southern counties, with a few outlying stations northward; open sandy ground, dry fields and woodland borders. Late August–September.

Most Ohio specimens can be separated into the varieties keyed below, and the maps show some difference in their Ohio distribution. However, at a few sites the two occur together and intergrade.

a. Leaves narrowly elliptic or rhombic-elliptic to suboblong, mostly 4 mm or more in width, with both a midrib and evident lateral veins . . . . var. *dichotomum* aa. Leaves linear to linear-lanceolate, mostly less than 4 mm in width, with a midrib but either no or scarcely evident lateral veins . . . . var. *lineare* (Walt.) Pursh

T. setaceum Houtt.

NARROW-LEAVED BLUECURLS

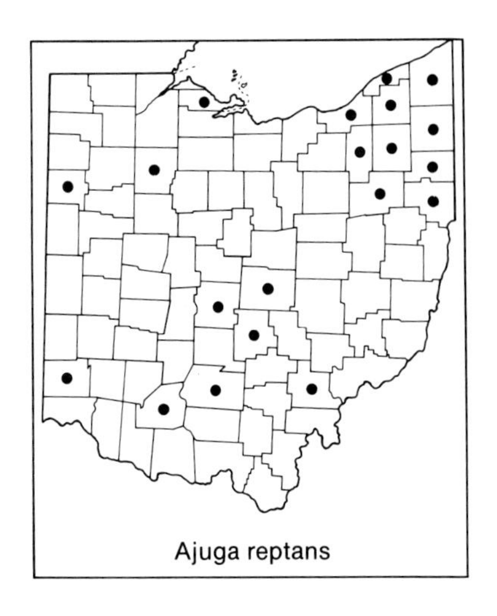

Ajuga reptans

3. Ajuga L. BUGLE-WEED

An Old World genus of 40 or more species. *Ajuga genevensis* L., ERECT BUGLE-WEED, is cultivated in Ohio. It was collected once in Cincinnati (Hamilton County) as a lawn weed, in 1917, where its occurrence was most likely adventive.

a. Erect stems glabrous or slightly pubescent; plants with numerous stolons, the stolons bearing leaves whose bases are narrowed to a short petiole, most leaves of flowering stems sessile . . . . . . 1. *Ajuga reptans*

aa. Erect stems densely long-pubescent; plants without stolons, all stems erect, all leaves broadly sessile; see above . . . . *Ajuga genevensis*

1. AJUGA REPTANS L. CARPET BUGLE-WEED

Glabrous to slightly pubescent perennial; plants with numerous long stolons to ca. 4 dm long, and erect flowering stems to 2 (–3) dm tall; leaves shallowly crenate, those of the stolons with bases narrowed to a short petiole, those of the flowering stems mostly sessile; flowers in a congested leafy spike; calyx actinomorphic or nearly so; corolla purplish-blue, ca. 1.5 cm long, bilabiate but the upper lip quite short; stamens 4, slightly exserted. A Eurasian species, naturalized at several sites in northeastern counties and at scattered sites elsewhere in Ohio; roadsides, fields, lawns, and other open disturbed habitats. April–May.

4. Teucrium L. GERMANDER

A genus of 200 or more species, widely distributed in temperate regions and frequent in the Mediterranean area. McClintock and Epling (1946) revised the genus. *Teucrium botrys* L., CUT-LEAVED GERMANDER, a weed of European origin, was reported from Ohio by Fernald (1950), but I have seen no specimens. *Teucrium scorodonia* L., WOOD GERMANDER, a cultivated ornamental shrub, also of European origin, was attributed to Ohio by Schaffner (1932), Fernald (1950), and Gleason (1952b), all probably on the basis of two specimens in The Ohio State University Herbarium. Both specimens were collected from a Lake County nursery on 19 July 1900; one is clearly marked "cult."

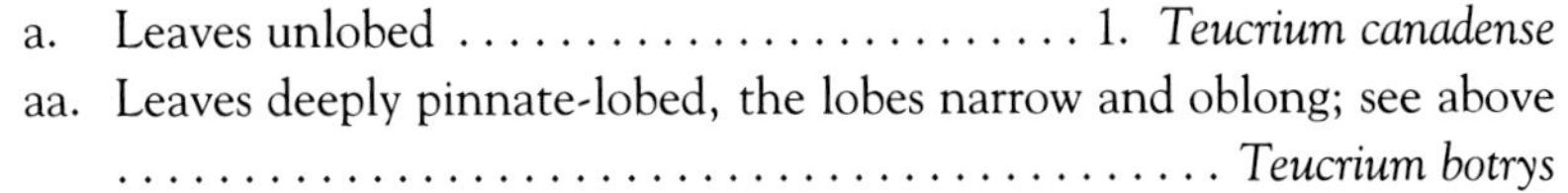

a. Leaves unlobed . . . . 1. *Teucrium canadense*

aa. Leaves deeply pinnate-lobed, the lobes narrow and oblong; see above . . . . *Teucrium botrys*

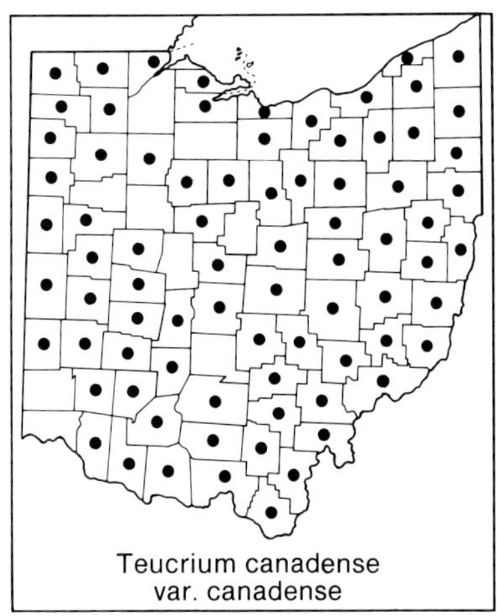
Teucrium canadense
var. canadense

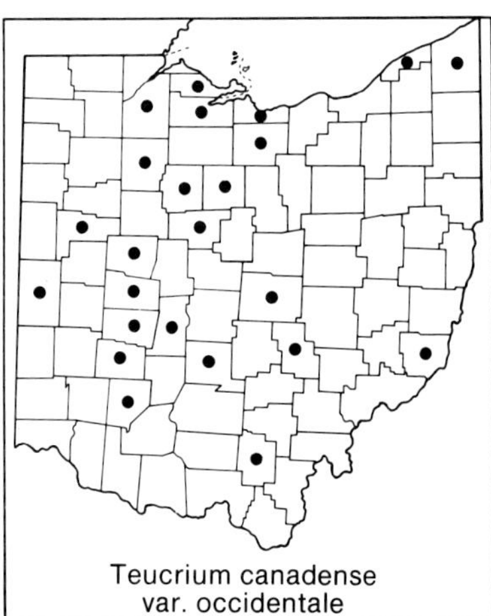
Teucrium canadense
var. occidentale

1. **Teucrium canadense** L. AMERICAN GERMANDER

Perennial, to 1 m tall; stems erect, pubescent; leaves toothed; flowers numerous, in rather dense spicate racemes; calyx regular or nearly so; corolla purplish-pink to yellowish-pink, 12–18 mm long, appearing to have only a lower lip, the upper lip quite short; stamens 4, appearing exserted because of the tiny upper corolla lip, but shorter than the lower lip. Marshes, wet ditches, low wet ground along streams and lakes, and to a lesser extent in dry fields and other dry habitats. Late June–August.

Most Ohio plants can be assigned to one of the two varieties keyed below, but some are intergrades between the two. Variety *canadense* occurs throughout Ohio, var. *occidentale* chiefly in western counties. The treatment of the two taxa at the rank of variety rather than species follows the classification of McClintock and Epling (1946).

a. Axis of inflorescence and calyx with only short, incurved, eglandular hairs . . . . var. *canadense*
incl. var. *boreale* (Bickn.) Shinners, var. *virginicum* (L.) Eat., and incl. *T. boreale* Bickn., *T. occidentale* A. Gray var. *boreale* (Bickn.) Fern. Shinners (1963) erred in believing *T. boreale* to be synonymous with *T. canadense* var. *occidentale*.

aa. Axis of inflorescence and calyx with long spreading hairs as well as short incurved hairs, many of both types glandular . . . . var. *occidentale* (A. Gray) McClintock & Epl.
T. occidentale A. Gray var. *occidentale*

5. Scutellaria L. SKULLCAP

Erect perennial herbs; the calyx bilabiate and bearing on the upper lip a distinctive low transverse ridge or upright protuberance (called a "crest"); corolla blue or white, bilabiate, the upper lip usually arched and concave; stamens 4, included, arching under the upper corolla lip. A widely distributed genus of 250–300 species. Epling's (1942) revision covered the American species of *Scutellaria*. More recently, J. L. Collins (1976) revised section *Annulatae*, which includes species nos. 2–5 below.

a. Flowers in axillary and/or terminal racemes, each flower subtended by a bract or much-reduced leaf; principal leaves on petioles 5 mm or more in length (except *S. integrifolia* with leaves sometimes sessile).

b. Corolla 5–9 mm long; flowers mostly in axillary racemes originating at medial as well as upper nodes, a few in terminal racemes . . . . 7. *Scutellaria lateriflora*

bb. Corolla 10–30 mm long; flowers all or mostly in terminal racemes or in axillary racemes originating only at uppermost nodes.

c. Medial and upper leaves entire (the lowest leaves sometimes crenate-dentate), linear-lanceolate to oblanceolate, long-tapering nearly or quite to base of petiole . . . . 5. *Scutellaria integrifolia*

cc. Medial and upper leaves crenate, serrate, or dentate, mostly ovate, not tapering or only short-tapering at base.

d. Upper leaves cordate to truncate at base; inflorescence glandular-pubescent.

e. Principal leaves with blades 6–13 cm long, petioles 2.5–8 cm long; upper internodes densely pubescent; inflorescence densely glandular-pubescent. . . . . 1. *Scutellaria ovata*

ee. Principal leaves with blades 2–5 cm long, petioles 1–2.5 cm long; upper internodes glabrous or slightly pubescent especially along the angles; inflorescence sparingly glandular-pubescent . . . . 6. *Scutellaria saxatilis*

dd. Upper leaves obtuse to rounded or truncate at base; inflorescence glandular-pubescent or not.

e. Stems with spreading hairs either on lower half of stem or throughout; medial and upper leaves rounded or obtuse at apex; calyx with divergent glandular hairs. . . . . . 2. *Scutellaria elliptica*

ee. Stems with either short curved hairs or glabrous; medial and upper leaves acute at apex; calyx with eglandular, mostly ascending or appressed hairs.

f. Stems below inflorescence glabrous or nearly so; racemes usually 1, sometimes 3; calyx green and glabrous or with few hairs . . . . 3. *Scutellaria serrata*

ff. Stems pubescent; racemes usually 5 or more; calyx usually gray with dense pubescence . . . . 4. *Scutellaria incana*

aa. Flowers mostly solitary, in axils of reduced or nonreduced leaves; leaves sessile or on short petioles to 3 mm long.

b. Corolla 15–25 mm long; principal leaves oblong-lanceolate to oblong-ovate, 3–8 cm long . . . . 8. *Scutellaria epilobiifolia*

bb. Corolla 6–9 mm long; principal leaves oblong-ovate to broadly ovate, 1–5 cm long, or if oblong-ovate then less than 2 cm long.

c. Principal leaves broadly ovate to ovate, 2–5 cm long, 1.5–3 cm wide, the margins of the lower principal leaves crenate-dentate or crenate-serrate; rhizomes without tubers . . . . 9. *Scutellaria nervosa*

cc. Principal leaves ovate to oblong-ovate, 1–2 cm long, 0.3–1 cm wide, the margins entire or nearly so; rhizomes producing small tubers . . . . 10. *Scutellaria parvula*

1. **Scutellaria ovata** Hill Heart-leaved Skullcap

Perennial, to 7 dm tall; stems with upper internodes densely pubescent; principal leaves with petioles 2.5–8.0 cm long, blades 6–13 cm long and

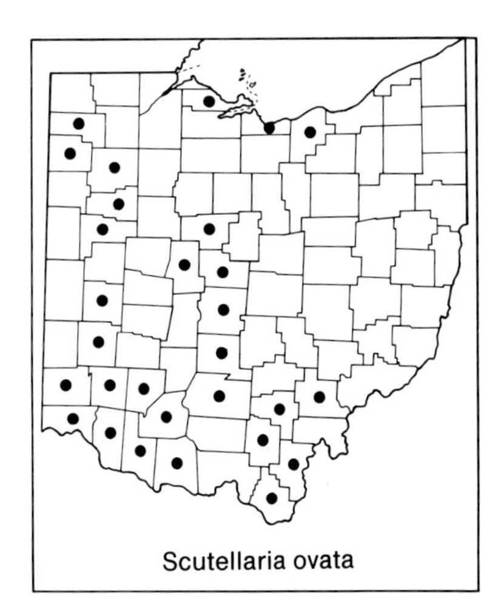
Scutellaria ovata

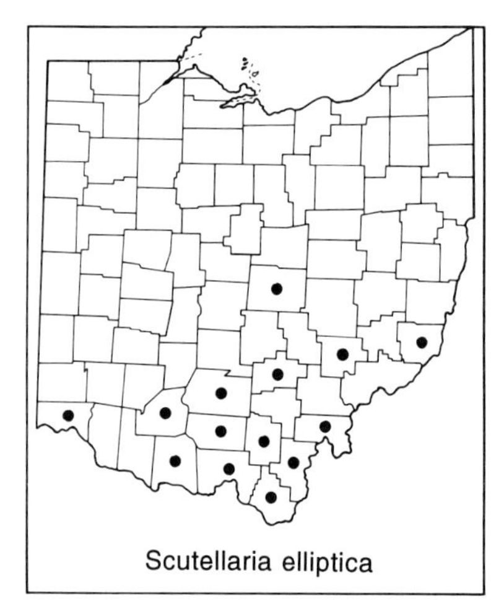
Scutellaria elliptica

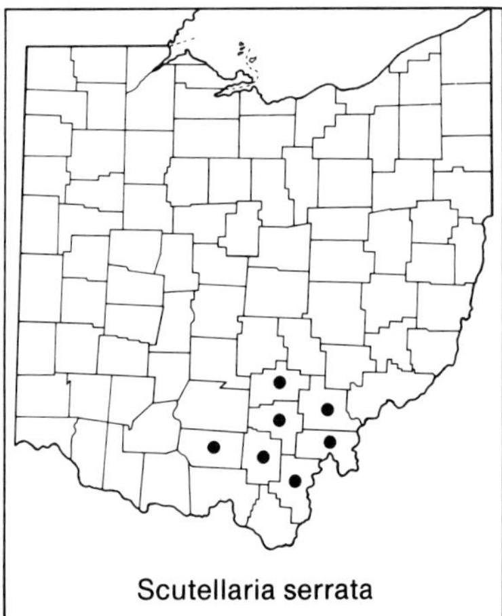
Scutellaria serrata

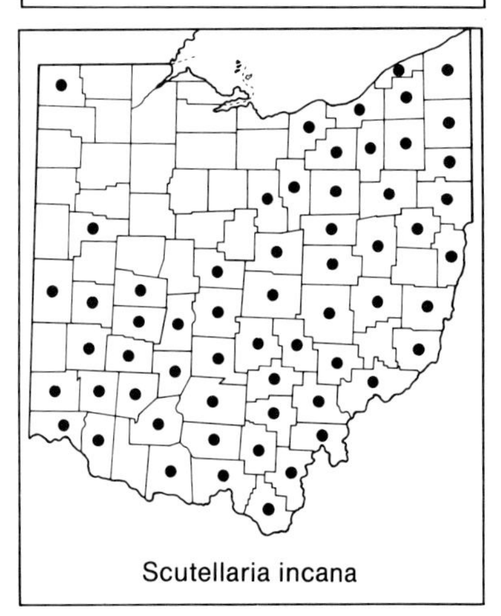
Scutellaria incana

markedly to slightly cordate at base, margins crenate-dentate; flowers in a terminal raceme, plants sometimes also with a few additional racemes from upper axils, axis of inflorescence densely glandular-pubescent; corolla blue, 15–20 mm long. Western, especially southwestern, counties; open woodlands. June–July.

Ohio plants belong to var. *versicolor* (Nutt.) Fern. (*S. versicolor* Nutt.).

2. **Scutellaria elliptica** Muhl. HAIRY SKULLCAP

Perennial, to 5 dm tall; stems pubescent with spreading hairs either on lower half or throughout; medial and upper leaves rounded to obtuse at apex, obtuse to broadly obtuse at base, margins crenate; inflorescence a terminal raceme, sometimes with additional smaller racemes from upper axils, axis of inflorescence densely pubescent; calyx with divergent glandular hairs; corolla purplish-blue, 14–19 mm long. A species of southeastern United States, ranging northward into southern counties of Ohio; woodlands. Ohio specimens can, if desired, be separated into the two varieties keyed below. June.

a. Stems with upwardly incurved hairs, most eglandular, the longest to 1 mm in length . . . . var. *elliptica*

aa. Stems with spreading hairs, most minutely glandular at tip, the longest 1–2 mm in length . . . . var. *hirsuta* (Short) Fern.

3. **Scutellaria serrata** Andr. SHOWY SKULLCAP

Perennial, to 6 dm tall; stems glabrous or nearly so below the inflorescence; medial and upper leaves acute at apex, obtuse at base, margins serrate; inflorescence a terminal raceme, plants sometimes with two smaller racemes from upper axils, axis of inflorescence slightly pubescent; calyx green and glabrous or with few hairs; corolla blue, 22–26 mm long. A species of the mid-Atlantic states, at a northwestern boundary of its range in south-central counties of Ohio; woodlands. Late May–June.

4. **Scutellaria incana** Biehler DOWNY SKULLCAP

Perennial, to ca. 1 m tall; stems pubescent; medial and upper leaves acute at apex, obtuse to rounded or subtruncate at base, margins crenate; inflorescence a terminal raceme, plants usually with several additional racemes from upper axils—in all sometimes appearing paniculate, axis of inflorescence densely pubescent; calyx usually gray with dense pubescence; corolla blue (rarely white), 15–23 mm long. A species of the southeastern states and the Ohio Valley, throughout southern and eastern Ohio, and in scattered western counties; woodlands, woodland openings and borders, roadside banks. July–August.

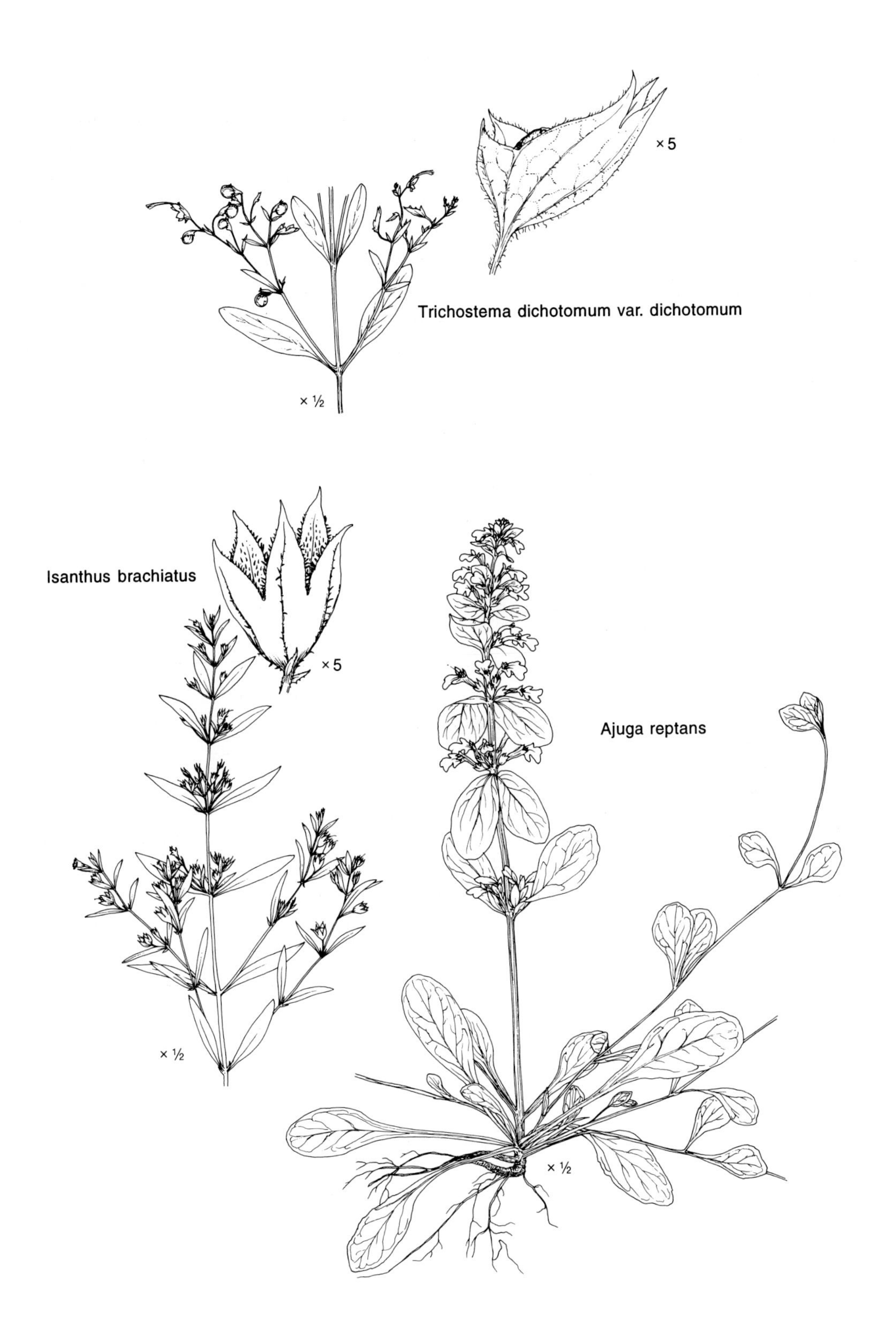
×5
Trichostema dichotomum var. dichotomum
× ½
Isanthus brachiatus
×5
Ajuga reptans
× ½
× ½

5. **Scutellaria integrifolia** L. Hyssop Skullcap
Perennial, to 7 dm tall; stems pubescent; medial and upper leaves linear-lanceolate to oblanceolate, acute to blunt at apex, acute and long-tapering at base, margins entire (margins of the lowest leaves sometimes crenate-dentate); inflorescence a terminal raceme with few to several additional racemes from upper axils, axis of inflorescence pubescent; corolla blue to purplish-blue, 18–27 mm long. A species chiefly of the southeastern United States, but extending northward along the Atlantic seaboard to New England, at a northwestern boundary of its range in southern counties of Ohio; fields, shores, open sandy habitats. June–July.

6. **Scutellaria saxatilis** Riddell Rock Skullcap
Perennial, to 5 dm tall; stems with upper internodes glabrous or slightly pubescent, especially along the angles; principal leaves with petioles 1.0–2.5 cm long, blades 2–5 cm long and truncate to slightly cordate at base, margins crenate-serrate to deeply crenate; flowers in a terminal raceme and in other racemes from upper axils, axis of inflorescence sparingly glandular-pubescent; corolla violet, 12–18 mm long. A species chiefly of the Appalachians, at a northern boundary of its range in east-central and south-central Ohio; dry woodlands. Additional record: Ross County (Epling, 1942). Late June–August.

7. **Scutellaria lateriflora** L. Mad-dog Skullcap
Perennial, to 8 dm tall; stems glabrous or very slightly pubescent on the angles; principal leaves with petioles 1–3 cm long, blades 3–9 cm long, broadly obtuse to rounded at base, margins serrate to coarsely serrate-dentate; flowers mostly in axillary, often spreading racemes originating from mid and upper nodes, a few in terminal racemes, axis of inflorescence

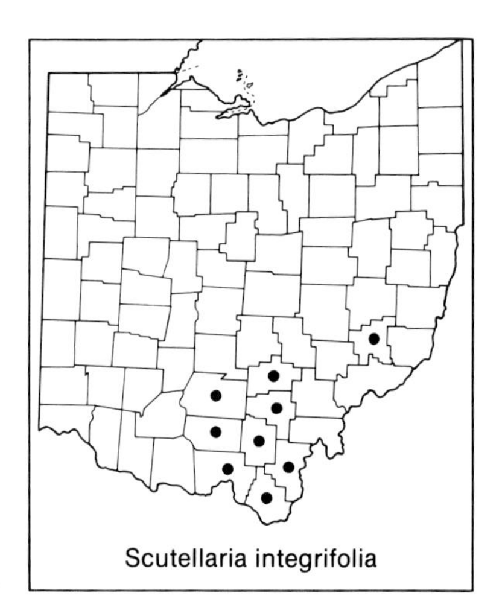
Scutellaria integrifolia

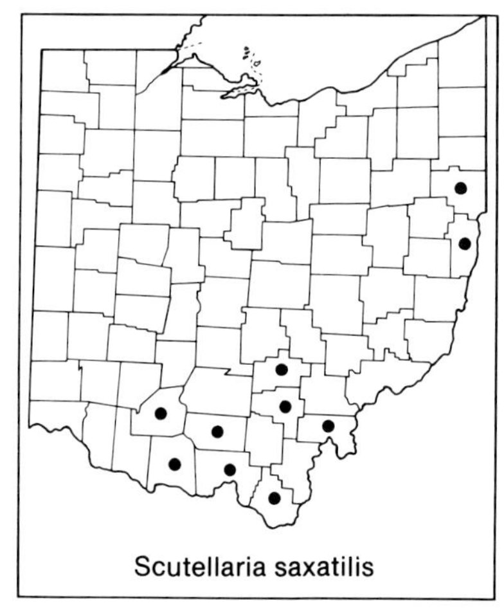
Scutellaria saxatilis

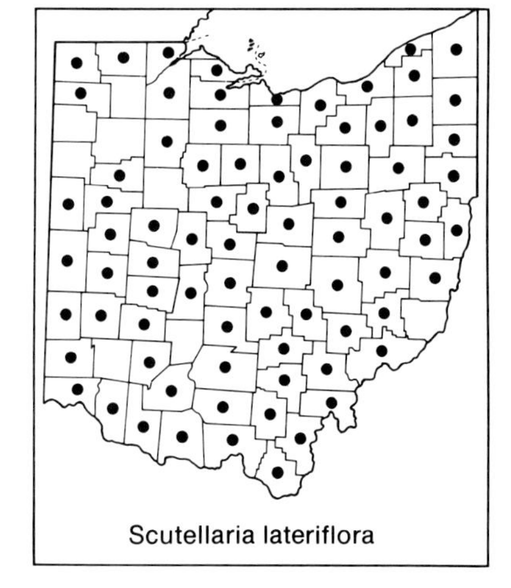
Scutellaria lateriflora

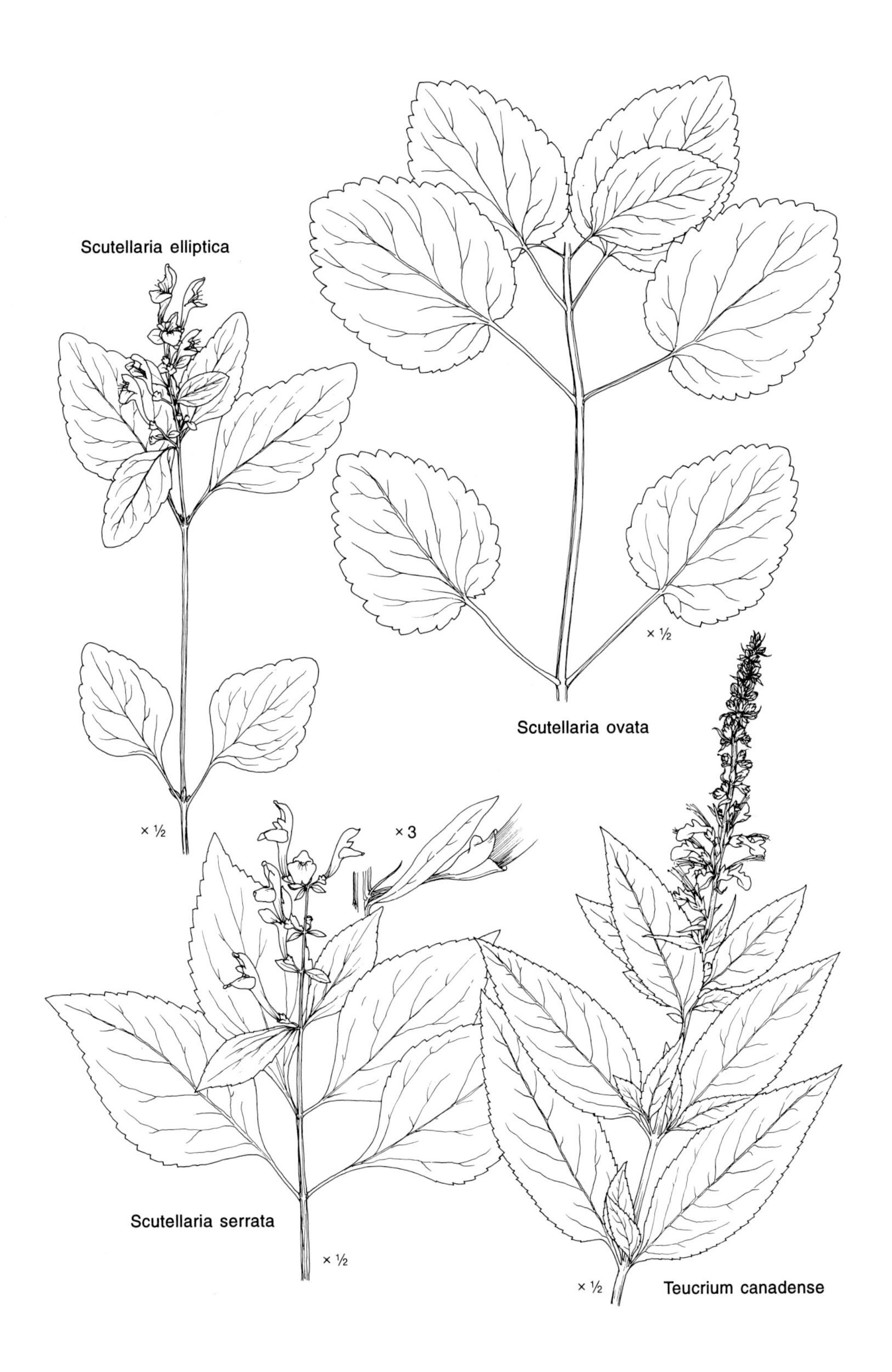
Scutellaria elliptica
× ½
Scutellaria ovata
× ½
× 3
Scutellaria serrata
× ½
× ½
Teucrium canadense

glabrous or low-puberulent on the angles; corolla blue, 5–8 mm long. Throughout Ohio; moist ditches and shores, marshes, wooded floodplains, moist wooded slopes. July–early October.

8. **Scutellaria epilobiifolia** A. Hamilton MARSH SKULLCAP

Perennial, to 8 dm tall; stems glabrous or low puberulent on the angles; principal leaves oblong-lanceolate to oblong-ovate, sessile or subsessile with petioles to 3 mm long, blades 3–8 cm long, rounded to truncate or slightly cordate-clasping at base, margins mostly low serrate to low crenate-serrate; flowers solitary in the axils of full-sized or reduced leaves; corolla violet-blue, 15–25 mm long. A northern species, transcontinental across North America, at a southern boundary of its range in Ohio, where it grows chiefly in northeastern and central counties; marshes, bogs, pond and lake margins, wet ditches. June–early September.

This species is often merged under the otherwise Eurasian *S. galericulata* L., with which it is perhaps conspecific.

9. **Scutellaria nervosa** Pursh VEINED SKULLCAP

Perennial, to 5 dm tall; stems erect or weakly erect, glabrous or nearly so; principal leaves broadly ovate to ovate, sessile or subsessile with petioles to 2 mm long, blades 2–5 cm long, 1.5–3.0 cm wide, rounded to cordate and sometimes slightly clasping at base, margins of lower leaves crenate-dentate to crenate-serrate, the upper nearly entire; flowers solitary in the axils of full-sized or reduced leaves; corolla pale blue, 7–9 mm long. Throughout much of Ohio, but chiefly in east-central and southeastern counties; moist woodlands and thickets. Late May–June.

Although there are some intermediates and the distribution pattern is somewhat clinal, most Ohio plants of this species can be assigned to one of

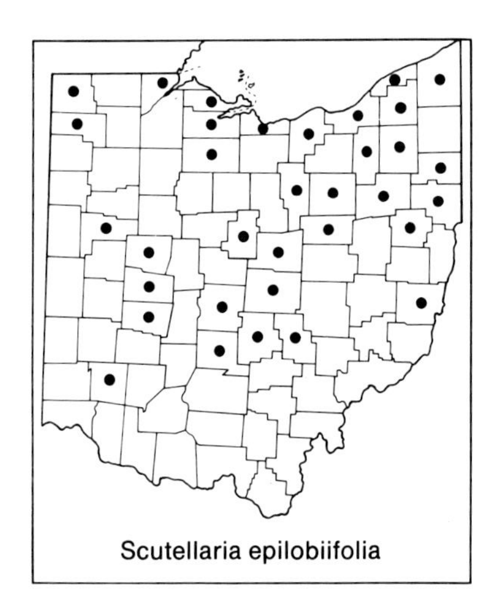
Scutellaria epilobiifolia

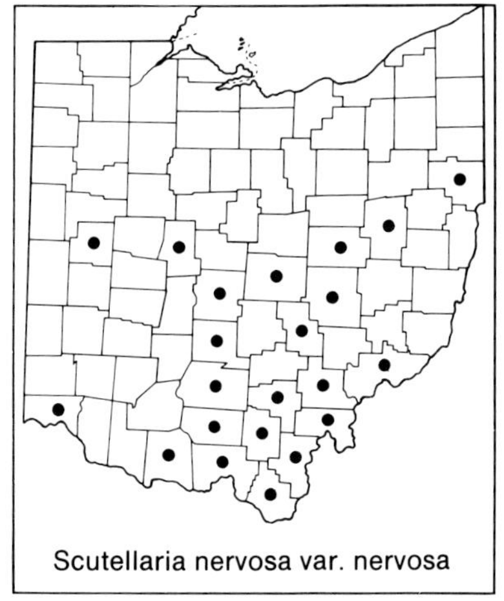
Scutellaria nervosa var. nervosa

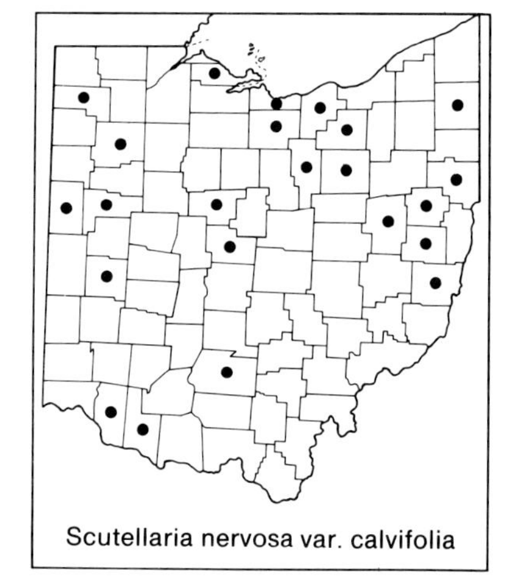
Scutellaria nervosa var. calvifolia

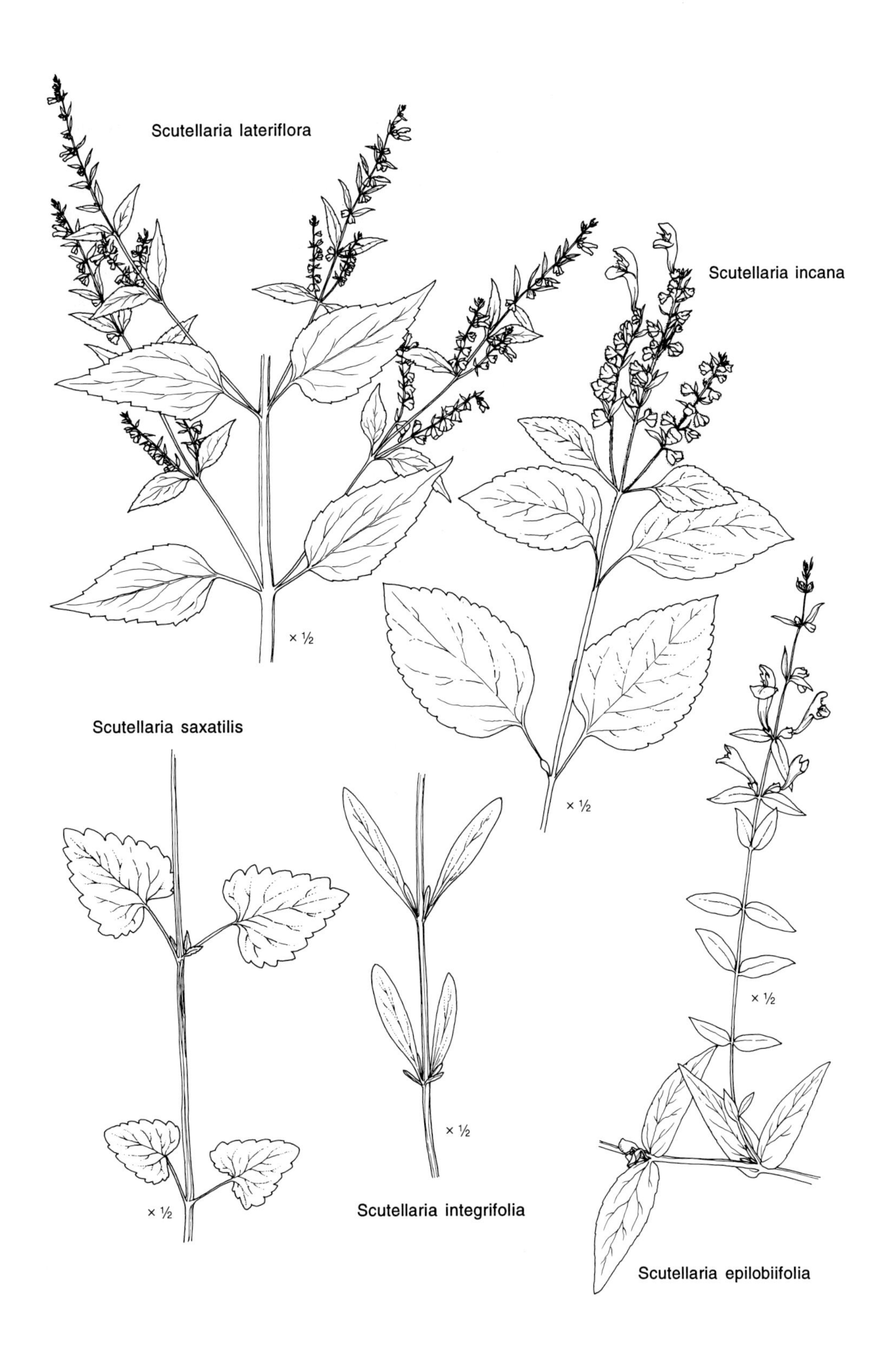
Scutellaria lateriflora
× ½
Scutellaria incana
× ½
Scutellaria saxatilis
× ½
× ½
× ½
Scutellaria integrifolia
× ½
Scutellaria epilobiifolia

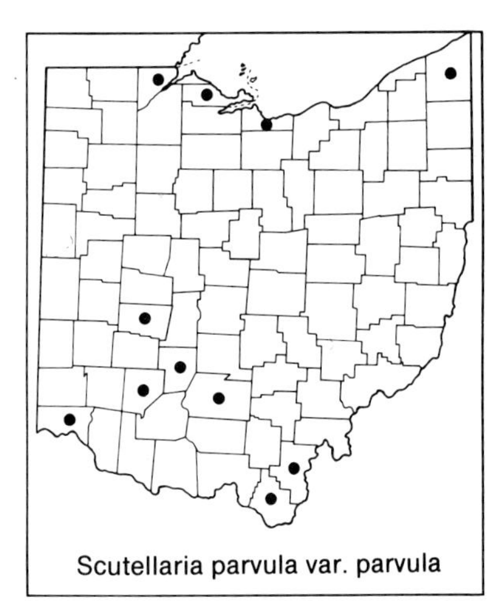
Scutellaria parvula var. parvula

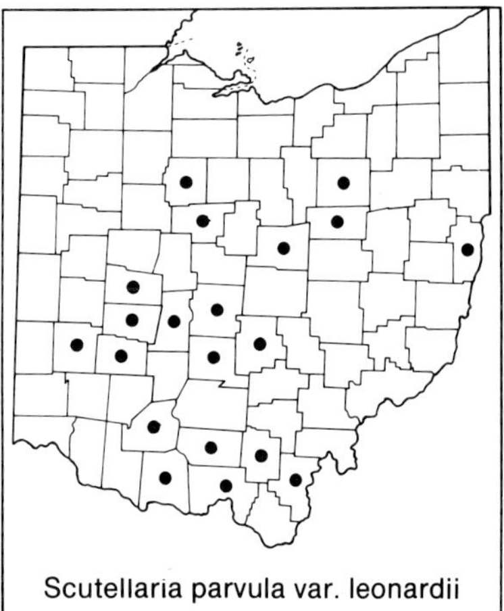
Scutellaria parvula var. leonardii

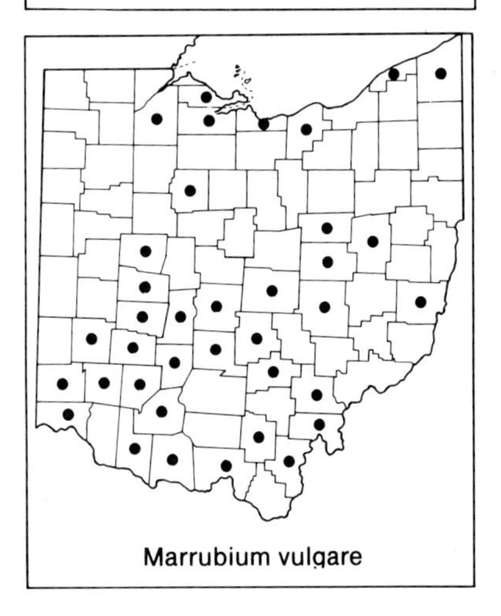
Marrubium vulgare

the two varieties keyed below. The two show some geographical separation in Ohio, var. *nervosa* occurring chiefly in central and south-central counties, var. *calvifolia* in other parts of the state.

a. Upper leaf surface with stout, more or less evenly distributed hairs, the longest hairs to 1 mm long. . . . . var. *nervosa* incl. var. *ambigua* (Nutt.) Fern.

aa. Upper leaf surface glabrous . . . . var. *calvifolia* Fern.

10. **Scutellaria parvula** Michx. SMALL SKULLCAP

Perennial, to 2 dm tall, with rhizomes producing small tubers; stems pubescent; principal leaves ovate to ovate-oblong, sessile, 1–2 cm long, 0.3–1.0 cm wide, rounded at base, margins entire or nearly so; flowers solitary in leaf axils; corolla purplish-blue, 6–9 mm long. Southwestern counties and scattered elsewhere; calcareous woodlands and open areas. Baskin and Baskin (1982a) studied the life cycle. June–July.

The two varieties keyed below are usually well marked and with few intergrades. The key is adapted from Gleason and Cronquist (1963).

a. Stems and calyx with glandular pubescence; principal leaves mostly with 3 or more pairs of lateral veins. . . . . var. *parvula*

aa. Stems and calyx with eglandular pubescence; principal leaves mostly with 2 pairs of lateral veins . . . . var. *leonardii* (Epl.) Fern. *S. leonardii* Epl.

6. Marrubium L. HOREHOUND

A Eurasian genus of some 30 species, one of which was once widely cultivated in Ohio.

1. MARRUBIUM VULGARE L. COMMON HOREHOUND. WHITE HOREHOUND

Perennial, to 5 dm tall; stems erect, densely white woolly-pubescent; leaves petiolate, ovate to orbicular to subreniform, margins markedly crenate; flowers in dense clusters in the upper axils, sessile or nearly so; calyx actinomorphic or nearly so, terminating in ten spines, each more or less hooked at the tip; corolla white, ca. 5 mm long, bilabiate; stamens 4, included. A Eurasian species, cultivated and escaped to disturbed areas near gardens throughout southern Ohio and at scattered sites northward. Most Ohio collections of this plant date prior to 1910, and very few since 1940—reflecting its decreasing use as a medicinal grown in dooryards and home gardens. Not illustrated. June–August.

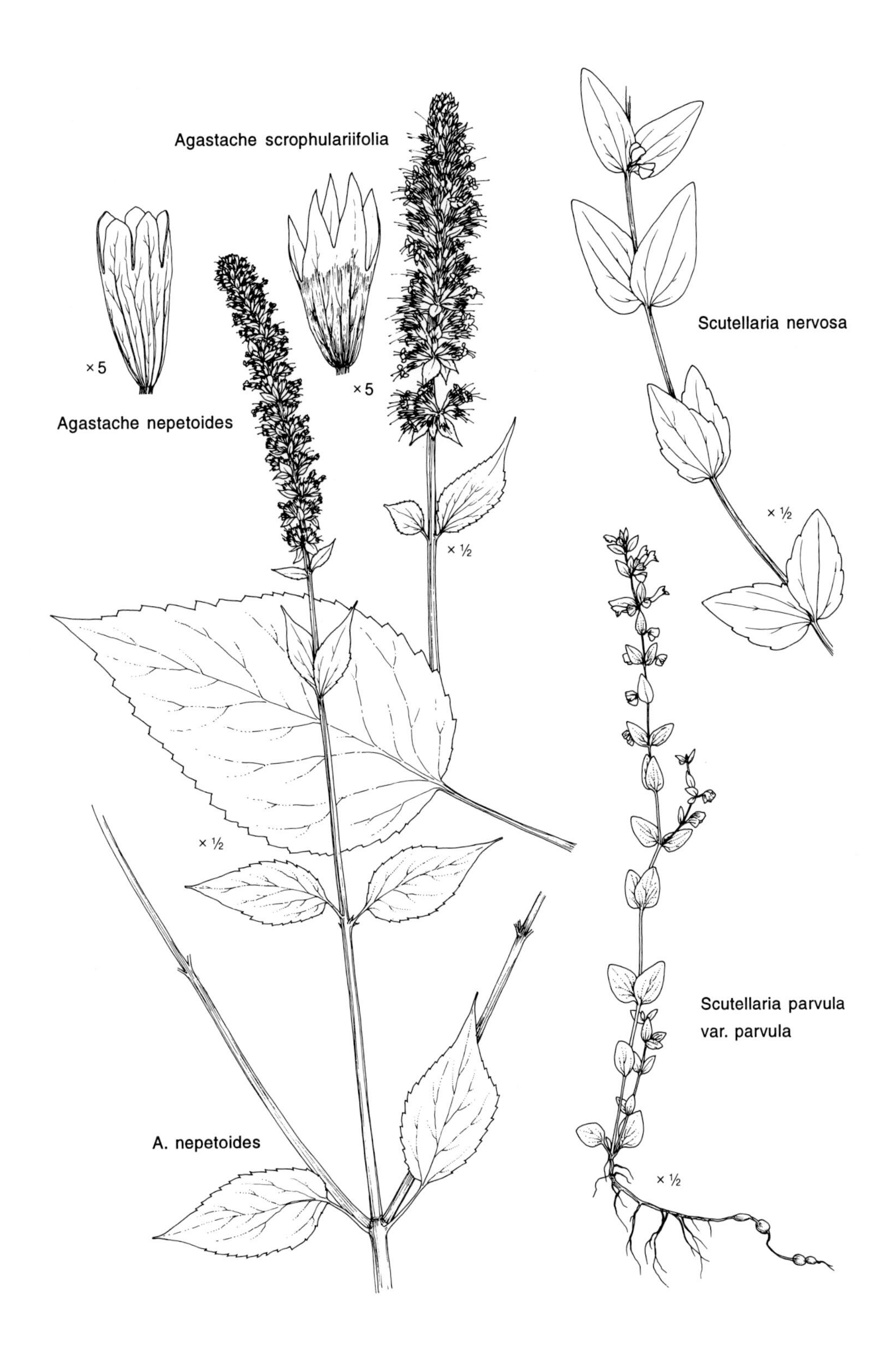
Agastache scrophulariifolia
×5
×5
Agastache nepetoides
× ½
× ½
A. nepetoides
Scutellaria nervosa
× ½
Scutellaria parvula
var. parvula
× ½

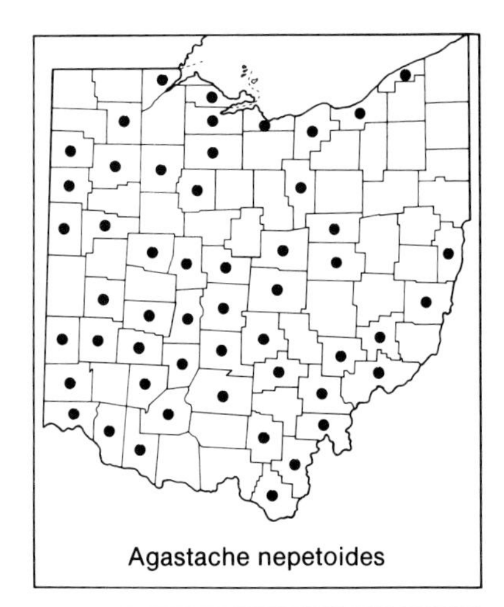
Agastache nepetoides

7. Agastache Gronov. GIANT-HYSSOP

Erect perennial herbs; flowers in dense spikes; calyx actinomorphic or nearly so; corolla bilabiate, yellowish to greenish-yellow or pale pink to pale purple; stamens 4, exserted. A genus of about 30 species, most native to North America, but some to eastern Asia. The most recent revision of the genus was by Lint and Epling (1945); a study of crossing relationships between the American and Asian taxa was made by Vogelmann (1985).

a. Calyx-lobes ovate to elliptic, acute to obtuse at apex, green throughout; corolla pale yellow to greenish-yellow; stems glabrous or slightly pubescent . . . . 1. *Agastache nepetoides*

aa. Calyx-lobes lanceolate to broadly lanceolate, acute to subacuminate at apex, green below, whitish apically; corolla pale pink to pale purple; stems glabrous, or more often with a few long hairs on the angles . . . . 2. *Agastache scrophulariifolia*

1. **Agastache nepetoides** (L.) Ktze. YELLOW GIANT-HYSSOP

Perennial, to 1.5 m tall; stems glabrous to slightly pubescent; leaves petiolate, ovate; calyx lobes ovate to elliptic, acute to obtuse at apex, green throughout; corolla pale yellow to greenish, 6–7 mm long. Southern counties and some northern ones, scarce or absent from most of northeastern Ohio; stream banks, wet thickets, woodlands, roadside ditches. July–August.

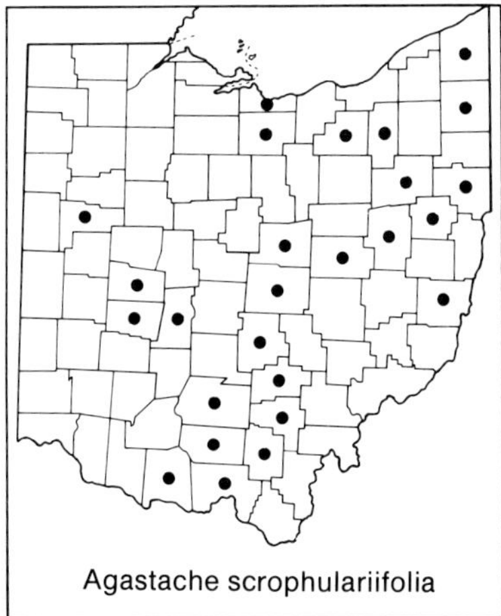
Agastache scrophulariifolia

2. **Agastache scrophulariifolia** (Willd.) Ktze. PURPLE GIANT-HYSSOP

incl. var. *scrophulariifolia* and var. *mollis* (Fern.) A. A. Heller

Perennial, to 1.5 m tall; stems glabrous or more often with a few long hairs on the angles; leaves petiolate, ovate; calyx lobes lanceolate to broadly lanceolate, acute to subacuminate at apex, green below, whitish apically; corolla pale pink to purple, 7–9 mm long. Mostly in northeastern and south-central counties; woods, thickets, banks and ditches along roads and railroads. Late July–September.

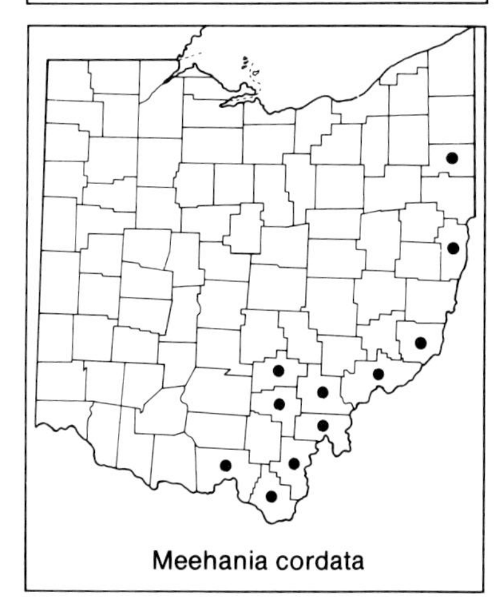
Meehania cordata

8. Meehania Britt.

A genus with a single species.

1. **Meehania cordata** (Nutt.) Britt. MEEHANIA

Perennial, with trailing or creeping stems and erect pubescent flowering branches to 2 dm high; leaves long-petiolate, the blades broadly ovate, with crenate margins and a rounded to usually cordate base; flowers in a terminal raceme; calyx actinomorphic or nearly so; corolla bilabiate, purplish-blue, 2.5–3.2 cm long; stamens 4, included. A mid-Appalachian species at a northern boundary of its range in southeastern and east-central Ohio; wooded slopes, ravines, and floodplains. May–early July.

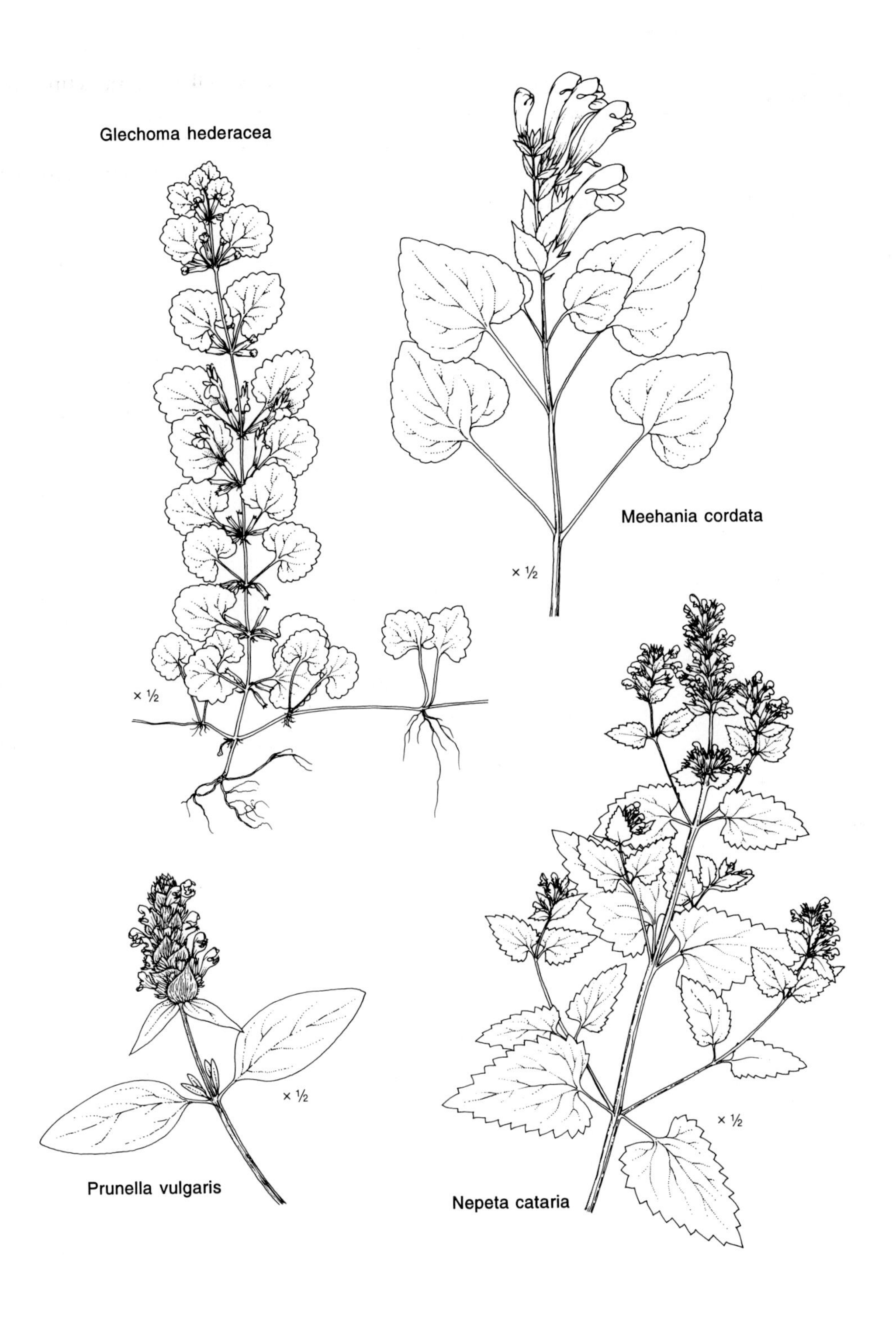
Glechoma hederacea
× ½
Meehania cordata
× ½
Prunella vulgaris
× ½
Nepeta cataria
× ½

Nepeta cataria

9. Nepeta L. CATMINT

A Eurasian genus of 200 or more species, one of them naturalized in Ohio.

1. NEPETA CATARIA L. CATNIP

Perennial, to 1 m tall; stems erect, densely gray-pubescent; leaves ovate, truncate to cordate at base, margins coarsely serrate-dentate; flowers in dense clusters terminating main stem and branches; calyx zygomorphic; corolla bilabiate, whitish and with pink or purplish dots, 10–12 mm long; stamens 4, included or very slightly exserted. A Eurasian species, once frequently cultivated and now naturalized throughout Ohio; along roads and railroads, and in thickets and disturbed weedy habitats. Late June–early October.

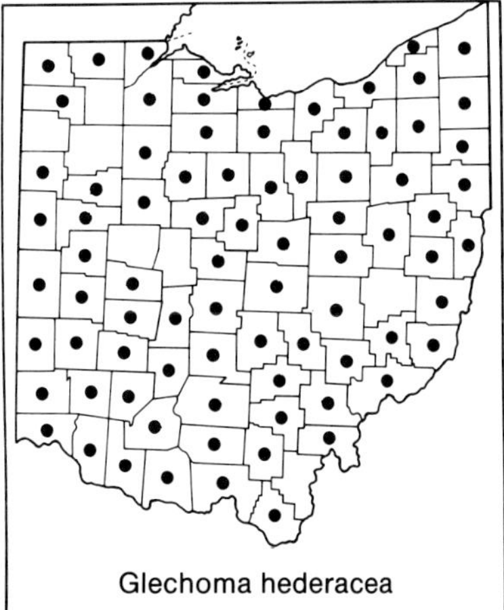
Glechoma hederacea

10. Glechoma L.

A Eurasian genus of five or more species, one of them naturalized in Ohio. The generic name has sometimes been spelled *Glecoma*.

1. GLECHOMA HEDERACEA L. GROUND IVY. GILL-OVER-THE-GROUND
 incl. var. *hederacea* and var. *micrantha* Moricand

Perennial; stems extensively creeping and forming mats, with erect or ascending flowering stems to 2 dm high; stems hispid to nearly glabrous; leaves petiolate, the blades orbicular to reniform, cordate at base, margins markedly crenate; flowers few to solitary in leaf axils; calyx actinomorphic or nearly so; corolla bilabiate, purple to purplish-blue, 1.2–2.2 cm long; stamens 4, included. A Eurasian species naturalized throughout Ohio; in a variety of moist habitats in disturbed woods, fields, cemeteries, dooryards, waste places. March–July.

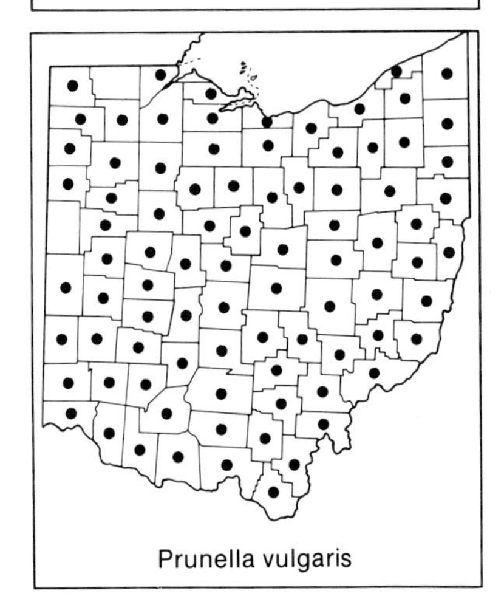
Prunella vulgaris

11. Prunella L. SELF-HEAL

A genus of the Northern Hemisphere with six or seven species, one of them a member of the Ohio flora.

1. **Prunella vulgaris** L. SELF-HEAL. HEAL-ALL
 incl. var. *lanceolata* (Bart.) Fern.

Perennial, to 6 dm tall; stems erect or ascending, glabrous to slightly pubescent; leaves petiolate, blades lanceolate to elliptic or lance-ovate; flowers in a dense, terminal, head-like spike with broad reniform bracts; calyx zygomorphic; corolla bilabiate, blue or purple to white, 10–15 mm long; stamens 4, included. Throughout Ohio; woods, fields, lawns and other disturbed habitats. A. P. Nelson (1964, 1965) made population studies of this species. May–October.

Fernald (1950), Gleason (1952b), and others recognize var. *lanceolata*, the native element, as well as var. *vulgaris*, naturalized from Eurasia. The high frequency of intergradation found among Ohio plants makes the distinction generally meaningless; the treatment here recognizing no major infraspecific taxa follows that of Ahles (1968). However, the two color forms keyed below can be distinguished. Both occur throughout Ohio; the blue- to purple-flowered plants are common, the white-flowered plants are rare.

a. Corolla blue to purple or lavender . . . . forma *vulgaris*
aa. Corolla white . . . . forma *albiflora* Britt.
incl. *P. vulgaris* var. *lanceolata* forma *candida* Fern.

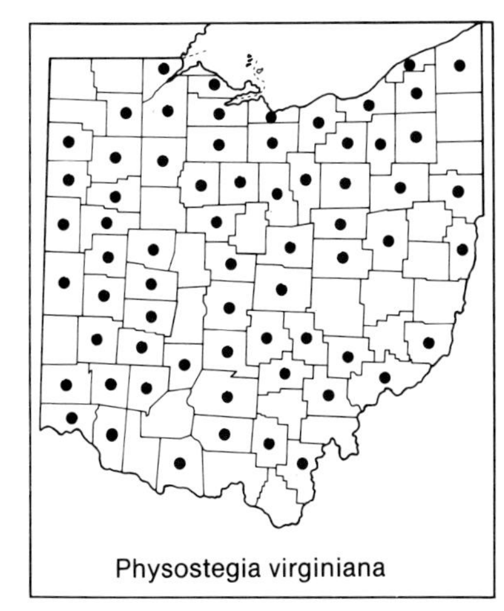
Physostegia virginiana

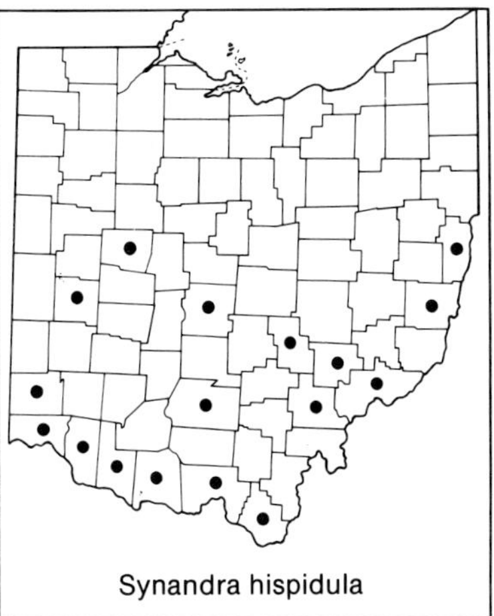
Synandra hispidula

12. Physostegia Benth. FALSE DRAGONHEAD

A North American genus of about 15 species, one of them a member of the Ohio flora.

1. **Physostegia virginiana** (L.) Benth. OBEDIENCE. FALSE DRAGONHEAD
incl. var. *virginiana*, var. *arenaria* Shimek, and var. *speciosa* (Sweet) A. Gray; *Dracocephalum virginianum* L.

Perennial, to ca. 1 m tall; stems erect, glabrous; leaves sessile or subsessile, narrowly lanceolate to oblong, tapering to an acute base, margins sharply serrate; flowers borne in a dense terminal raceme and usually also in a few to several additional racemes from upper axils, axis of inflorescence more or less puberulent; calyx actinomorphic or nearly so; corolla bilabiate, rose-purple to white, 1.5–3.0 cm long; stamens 4, slightly exserted. Throughout Ohio and also often cultivated, native at some sites, escaped from cultivation and naturalized at others; prairies, moist fields and ditches, shores. Additional records: Scioto and Williams counties (Cantino, personal communication). Late July–early October.

Cantino (1982) has divided the native plants into subsp. *virginiana* and subsp. *praemorsa* (Shinners) Cantino (var. *arenaria* Shimek). His map shows Ohio specimens to have a considerable amount of diversity, with four elements represented in the state: (1) subsp. *virginiana*, (2) subsp. *praemorsa*, (3) garden escapes, and (4) plants of "uncertain subspecific affinities." Other papers dealing with this species are those by Cantino and Lammers (1983) and Mohlenbrock (1963).

13. Synandra Nutt.

A genus with a single species.

1. **Synandra hispidula** (Michx.) Britt. SYNANDRA

Biennial, to 6 dm tall; stems erect, pubescent; leaves petiolate, ovate to broadly ovate, cordate at base, margins crenate to crenate-dentate; flowers solitary in the axils of large to progressively smaller bracts, in all forming

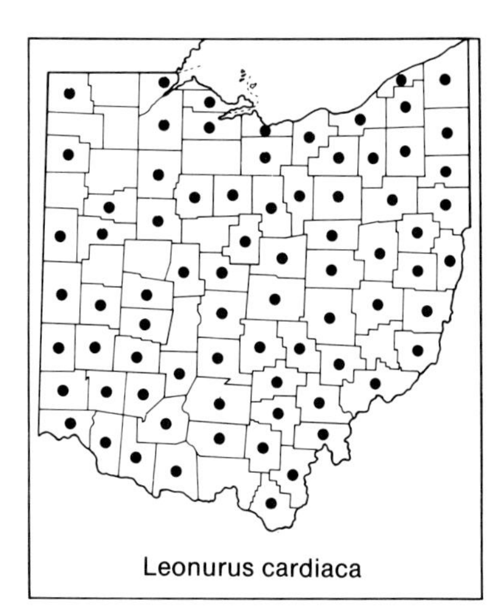

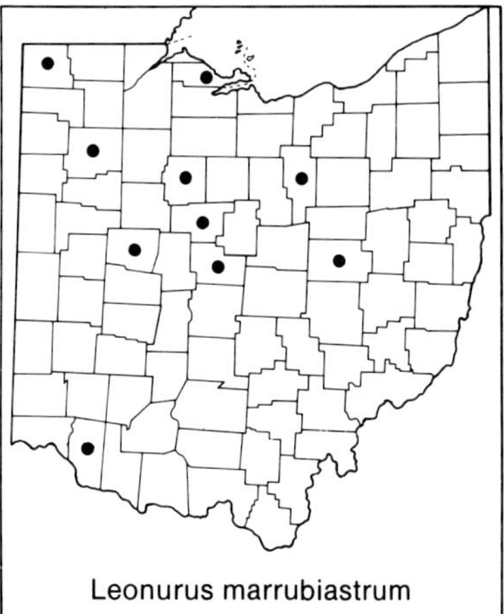

an interrupted spike; calyx actinomorphic or nearly so; corolla bilabiate, yellowish-white to pale greenish-yellow, 3.0–3.5 cm long; stamens 4, included. A species of the mid-Appalachians and the Ohio River Valley, at a northern boundary of its range in the southern half of Ohio; wooded floodplains and terraces, moist wooded slopes. Baskin, Bryan, and Baskin (1986), Cantino (1985), and Moran (1986) have studied the life history and ecology of this species. May–mid-June.

14. Leonurus L. MOTHERWORT

Biennial or perennial herbs; flowers sessile, clustered in axils of upper leaves and bracts; calyx actinomorphic or nearly so, the lobes terminating in spines; corolla bilabiate; stamens 4, included or slightly exserted. A Eurasian genus or four or more species, two of them naturalized in Ohio.

a. Leaves all, or at least the lower, 3-lobed; stems pubescent on the angles; bracts one-half the length of the calyx or less; corolla pink to pale purple, exceeding the calyx, the upper lip long villous with dense tangled hairs; plants perennial . . . . 1. *Leonurus cardiaca*

aa. Leaves all coarsely toothed but unlobed; stems pubescent on surface as well as angles; bracts one-half to fully as long as or longer than the calyx; corolla whitish, not or only slightly exceeding the calyx, the upper lip short-pubescent; plants biennial. . . . . 2. *Leonurus marrubiastrum*

1. LEONURUS CARDIACA L. MOTHERWORT. COMMON MOTHERWORT

Perennial, to ca. 1.5 m tall; stems erect, pubescent on the angles; leaves petiolate, the principal ones 3-lobed; corolla pink to pale purple, 8–11 mm long, exceeding the calyx, the upper lip long villous with dense tangled hairs. A Eurasian species naturalized throughout Ohio; roadsides, weedy fields, disturbed woods, and other disturbed habitats. June–September.

2. LEONURUS MARRUBIASTRUM L. HOREHOUND MOTHERWORT

Biennial, to ca. 1 m tall; stems erect, pubescent on angles and surface; leaves petiolate, coarsely toothed; corolla whitish, 5–7 mm long, not or only slightly exceeding the calyx, the upper lip short-pubescent. A Eurasian species, naturalized chiefly in the northwestern quarter of Ohio; disturbed weedy sites. Not illustrated. July–September.

15. Galeopsis L. HEMP-NETTLE

A Eurasian genus of seven or more species. *Galeopsis ladanum* L., RED HEMP-NETTLE, was collected once, in 1928, as an adventive railroad track weed in Highland County.

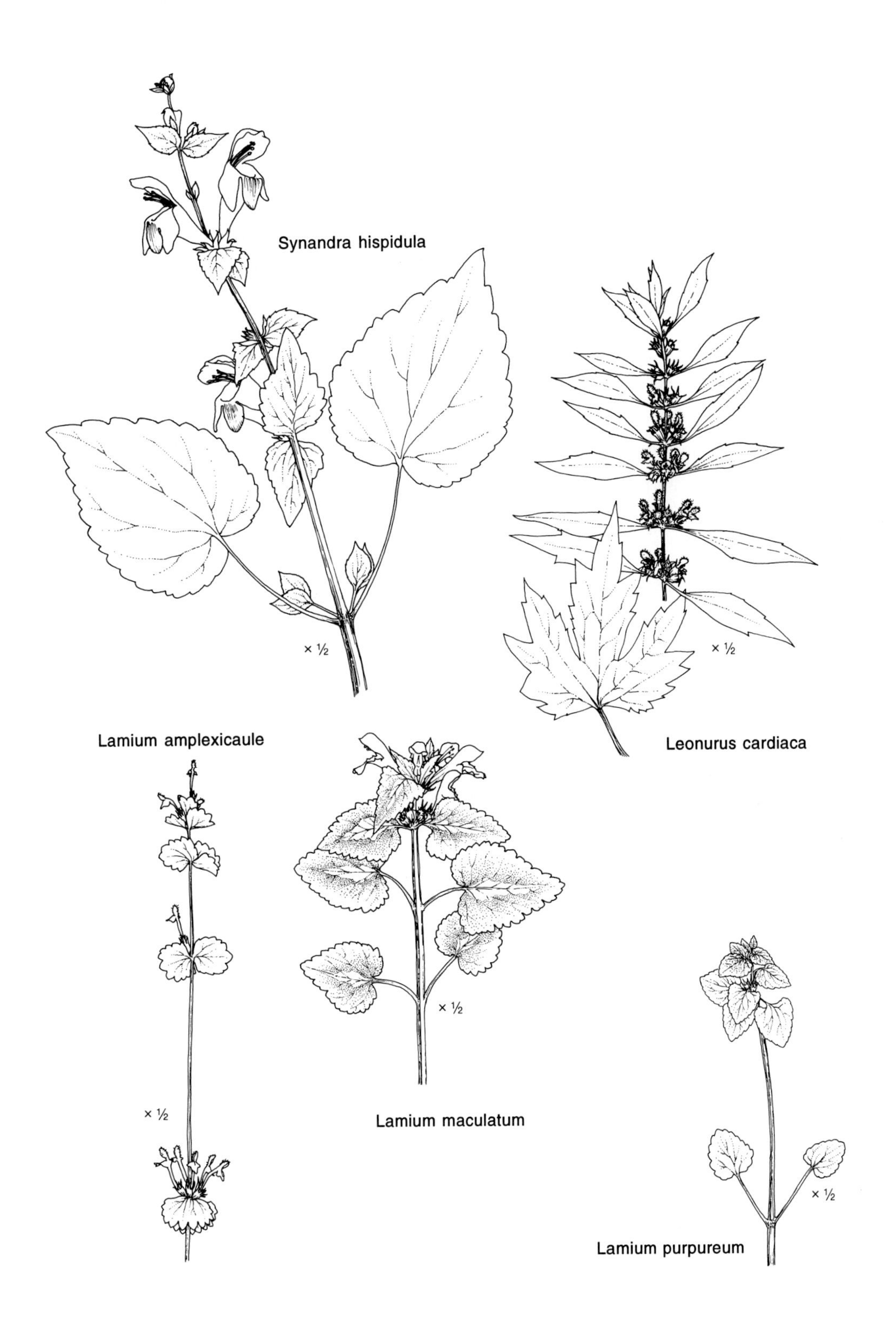
Synandra hispidula
× ½
× ½
Leonurus cardiaca
Lamium amplexicaule
× ½
× ½
Lamium maculatum
× ½
Lamium purpureum

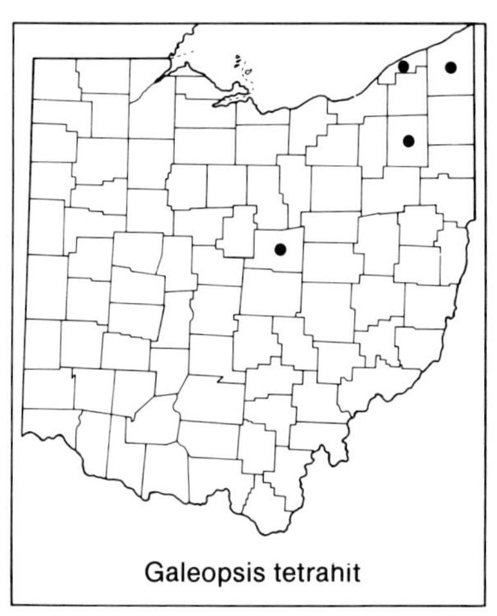
Galeopsis tetrahit

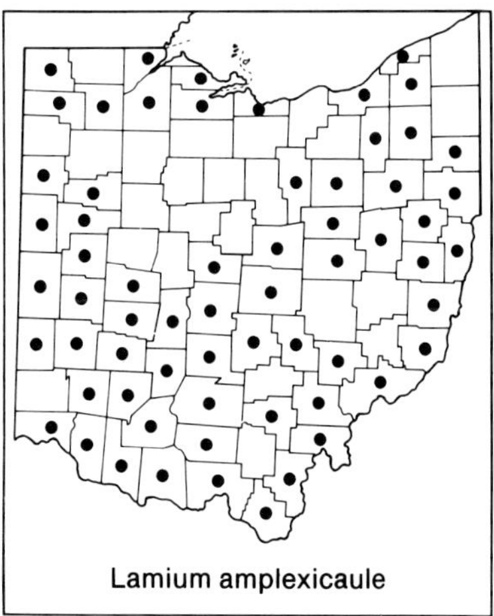
Lamium amplexicaule

a. Stem hispid with bristly, spreading, and often slightly reflexed hairs; leaves coarsely serrate. . . . . 1. *Galeopsis tetrahit*

aa. Stem pubescent with short, soft, recurved or appressed hairs; leaves serrate to usually entire; see above. . . . . *Galeopsis ladanum*

1. GALEOPSIS TETRAHIT L. COMMON HEMP-NETTLE

Annual, to 6 dm tall; stems erect, hispid with bristly, spreading, and often slightly reflexed hairs; leaves petiolate, broadly lanceolate to ovate, coarsely serrate; flowers sessile, clustered in axils of upper leaves; calyx actinomorphic or nearly so, the lobes terminating in sharp spines 2 mm or more in length; corolla bilabiate, pinkish, 12–17 mm long; stamens 4, included or slightly exserted. A Eurasian species, naturalized at a few sites in northeastern counties; roadsides and other disturbed habitats. O'Donovan and Sharma (1987) studied the biology of this species. June–August.

16. Lamium L. DEAD-NETTLE

Annual or perennial herbs; flowers clustered in axils of upper leaves or bracts; calyx actinomorphic or nearly so; corolla bilabiate; stamens 4, included or slightly exserted. A Eurasian genus of about 40 species, three of which are naturalized members of the Ohio flora. In addition, *Lamium album* L., SNOWFLAKE, was attributed to Ohio by Schaffner (1932), perhaps on the basis of misidentified specimens of *L. maculatum*; I have seen no Ohio collections of *L. album*.

a. Corolla 10–18 mm long, the tube straight at base; leaves without a whitish blotch along midrib.

b. Upper leaves and bracts sessile, broadly orbicular to reniform, wider than long, green; calyx lobes not or only slightly spreading . . . . 1. *Lamium amplexicaule*

bb. Upper leaves and bracts petiolate, suborbicular to ovate, longer than wide, usually red or purple; calyx lobes somewhat spreading. . . . . 2. *Lamium purpureum*

aa. Corolla 20–30 mm long, the tube curved at the base; leaves usually with a whitish blotch along midrib . . . . 3. *Lamium maculatum*

1. LAMIUM AMPLEXICAULE L. HENBIT

Annual or winter annual, to 3 dm tall; stems ascending to suberect, glabrous or slightly pubescent; middle and upper leaves sessile, broadly orbicular to reniform, wider than long, margins crenate; calyx lobes not or only slightly spreading; corolla reddish-purple, 14–18 mm long, the tube straight at the base. A Eurasian species naturalized throughout much of Ohio; roadsides, grassy fields, lawns, cemeteries, open woodlands. Baskin and Baskin (1981, 1984c) made seed germination studies of this species. (February–) March–May (–July).

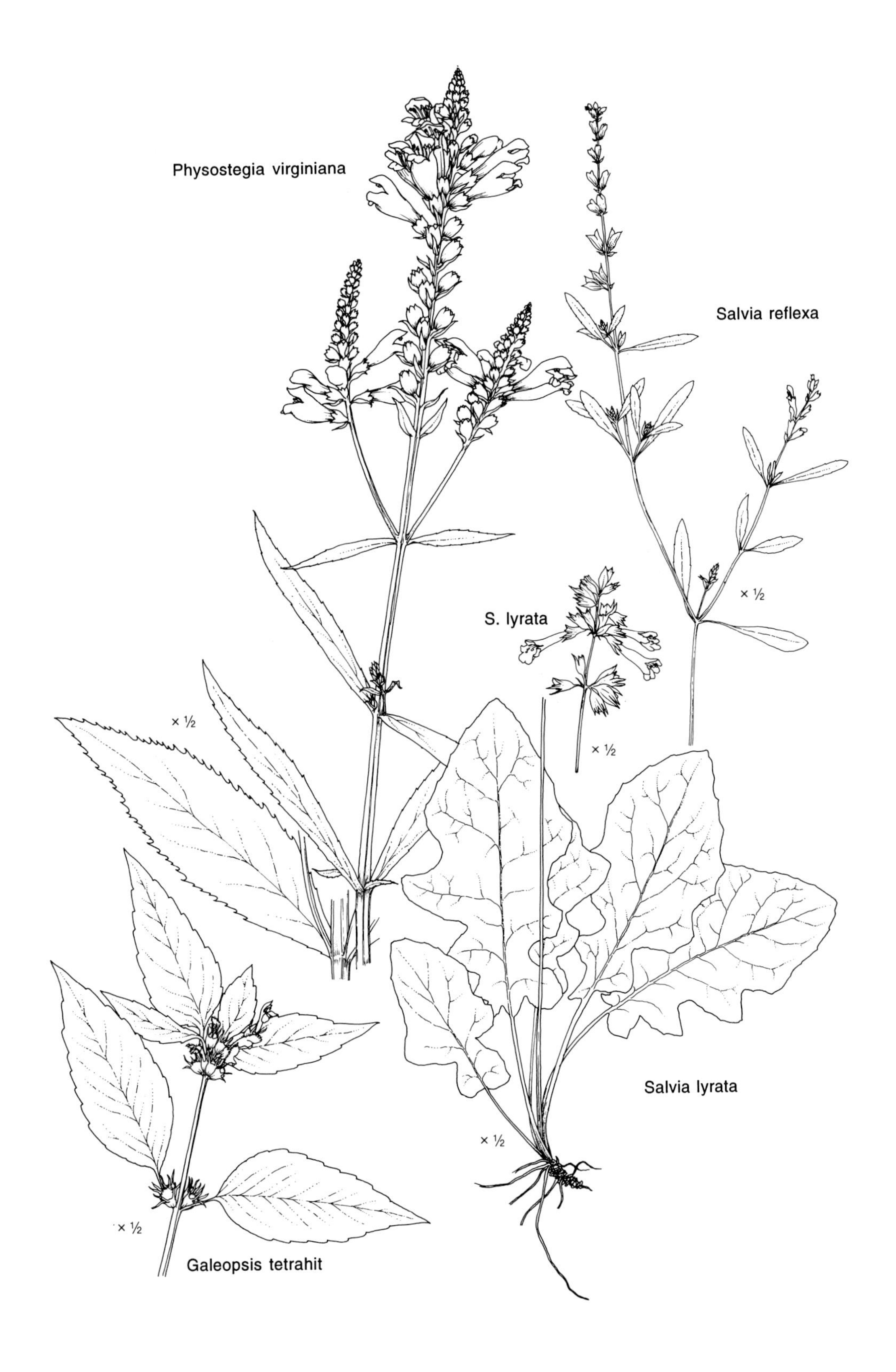
Physostegia virginiana
Salvia reflexa
× ½
S. lyrata
× ½
× ½
× ½
Salvia lyrata
× ½
Galeopsis tetrahit

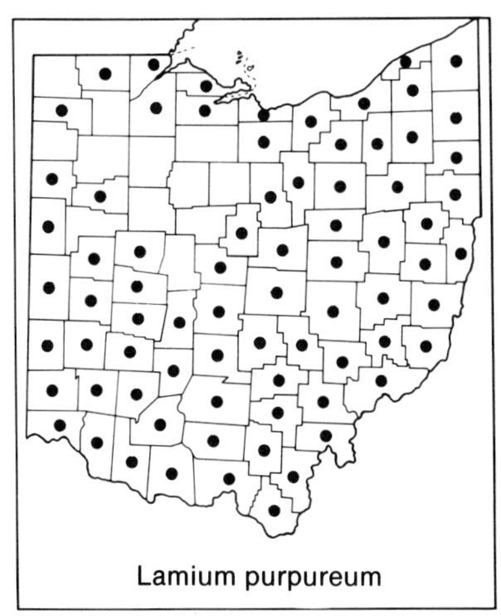
Lamium purpureum

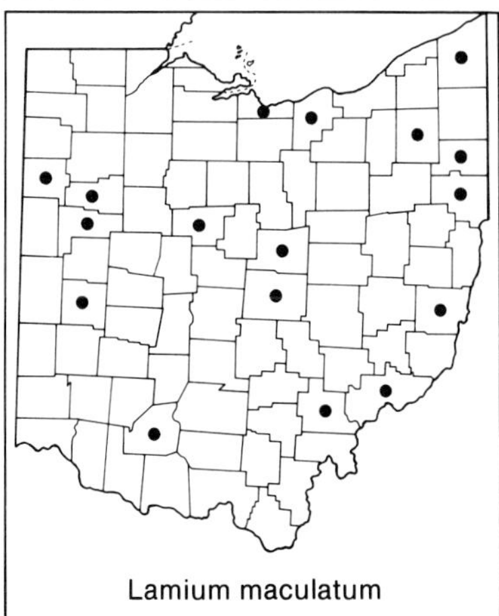
Lamium maculatum

2. Lamium purpureum L. Purple Dead-nettle
Annual or winter annual, to 3 dm tall; stems ascending to suberect, low-pubescent; mid and upper leaves petiolate, suborbicular to ovate, longer than wide, margins crenate, the upper leaves and the bracts usually reddish or purplish; calyx lobes somewhat spreading; corolla pinkish-purple (rarely white), 10–14 mm long, the tube straight at the base. A Eurasian species naturalized throughout much of Ohio; roadsides, lawns, gardens, cultivated fields, cemeteries, thickets, open woods. Baskin and Baskin (1984d) studied seed germination in this species. (February–) March–May.

A white-flowered form has been collected in Brown and Fairfield counties.

3. Lamium maculatum L. Spotted Dead-nettle. Spotted Henbit
Perennial, to 4 dm tall; stems ascending to suberect, pubescent; middle and upper leaves petiolate, triangular-ovate, usually with a whitish blotch along the midrib, margins crenate to crenate-dentate; corolla reddish-purple or rarely white, 20–30 mm long, the tube curved at the base. A Eurasian species, cultivated and occasionally escaped, adventive or perhaps locally naturalized at scattered sites throughout Ohio; roadsides, fields, open woods, disturbed habitats near gardens. May–October.

17. Stachys L. hedge-nettle. woundwort

Erect perennial or biennial herbs; flowers in several clusters forming a rather open spike; calyx actinomorphic or nearly so; corolla bilabiate; stamens 4, included or slightly exserted. A widely distributed genus of 200–300 species. Epling (1934) made a preliminary revision of the American species, Gill (1980b) a cytotaxonomic study of Canadian species, J. B. Nelson (1981) a study of the species in southeastern United States, and Mulligan and Munro (1989) a taxonomic study of the North American species north of Mexico. *Stachys germanica* L., Downy Woundwort, was collected as an adventive from Adams County in 1935, Ross County in 1955, and Stark County in 1898. The key below is adapted from Sabo (1965).

a. Stems and leaves with dense, soft, white pubescence; plants biennial; see above . . . . *Stachys germanica*

aa. Stems and leaves pubescent or glabrous, but not with dense, soft, white pubescence; plants perennial.

b. Stems with at least some pubescence on both the sides and the angles.

c. Lower leaves sessile or with petioles to 1 (–1.5) cm in length; calyx lobes two-thirds as long as to equalling the length of calyx tube . . . . 1. *Stachys palustris*

cc. Lower leaves with petioles 2 cm or more in length; calyx lobes less than half as long as calyx tube .... 2. *Stachys nuttallii*

bb. Stems with pubescence only on the angles, the sides glabrous.

c. Petioles of medial leaves (3–) 6–30 mm long.. 3. *Stachys tenuifolia*

cc. Petioles of medial leaves less than 3 (–5) mm long or the leaves sessile .... 4. *Stachys aspera*

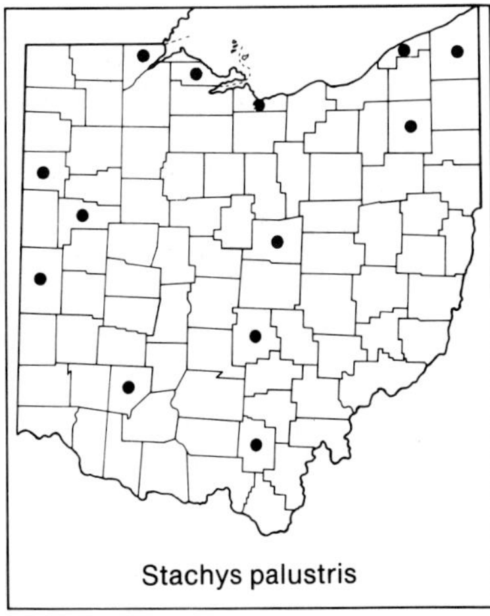
Stachys palustris

1. **Stachys palustris** L. MARSH HEDGE-NETTLE. MARSH WOUNDWORT
incl. var. *homotricha* Fern., var. *nipigonensis* Jennings, and var. *phaneropoda* Weath.; and including *S. pilosa* Nutt. var. *pilosa* and var. *arenicola* (Britt.) G. Mulligan and D. Munro

Perennial, to 1 m tall; stems pubescent on both the sides and the angles; lower leaves sessile or with petioles to 1 (–1.5) cm in length; calyx lobes two-thirds as long as to equalling the length of calyx tube; corolla pink to purple. A northern and western species, at a southern boundary of its range at scattered sites in Ohio; low wet fields, shores, and sometimes in drier habitats; probably native at some sites and naturalized at others. July–August.

A highly variable species with poorly defined infraspecific taxa. Gill (1980a) and Mulligan, Munro, and McNeill (1983) studied variation within North American populations of this species. Mulligan and Munro (1989) treat the tetraploid native plants as *S. pilosa* and the hexaploid aliens as *S. palustris* sensu stricto; I found these entities morphologically indistinguishable among Ohio plants. In Ohio, a few rare intergrades between this species and *S. nuttallii* have been collected.

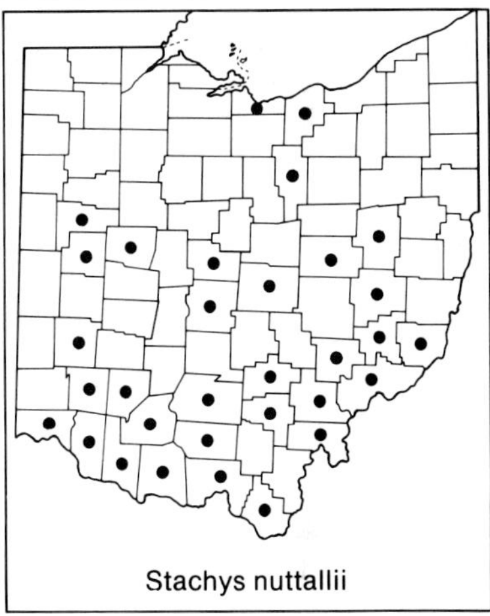
Stachys nuttallii

2. **Stachys nuttallii** Benth. HEART-LEAVED HEDGE-NETTLE
S. cordata Riddell, *S. riddellii* House

Perennial, to 7 dm tall; stems pubescent on both the sides and the angles; lower leaves with petioles 2 cm or more in length, mostly cordate at base, sometimes rounded; calyx lobes less than half as long as calyx tube; corolla pale pinkish to purplish, 11–14 mm long. A species of the mideastern states, at a northern boundary of its range in Ohio; southern counties and at scattered sites northward; woodlands. June–early August.

J. B. Nelson and J. E. Fairey (1979) discuss the nomenclature of this species.

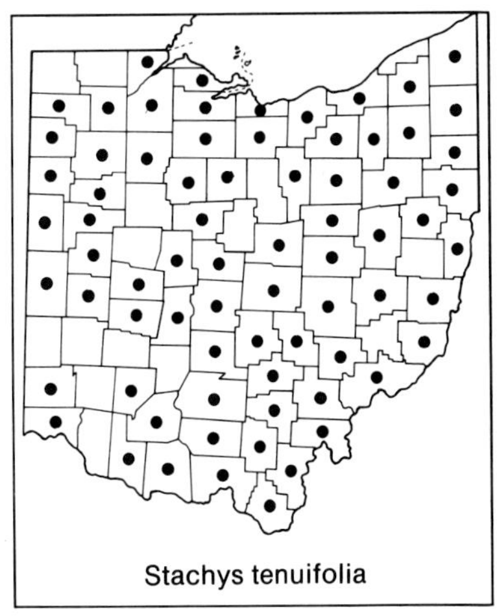
Stachys tenuifolia

3. **Stachys tenuifolia** Willd. COMMON HEDGE-NETTLE
incl. var. *hispida* (Pursh) Fern. (*S. hispida* Pursh)

Perennial, to ca. 9 dm tall; stems with pubescence only on the angles, the sides glabrous; medial leaves with petioles (3–) 6–30 mm long; corolla pinkish, 10–12 mm long. Throughout Ohio; wet fields, floodplains, wet ditches, wet woodlands, stream banks. June–October.

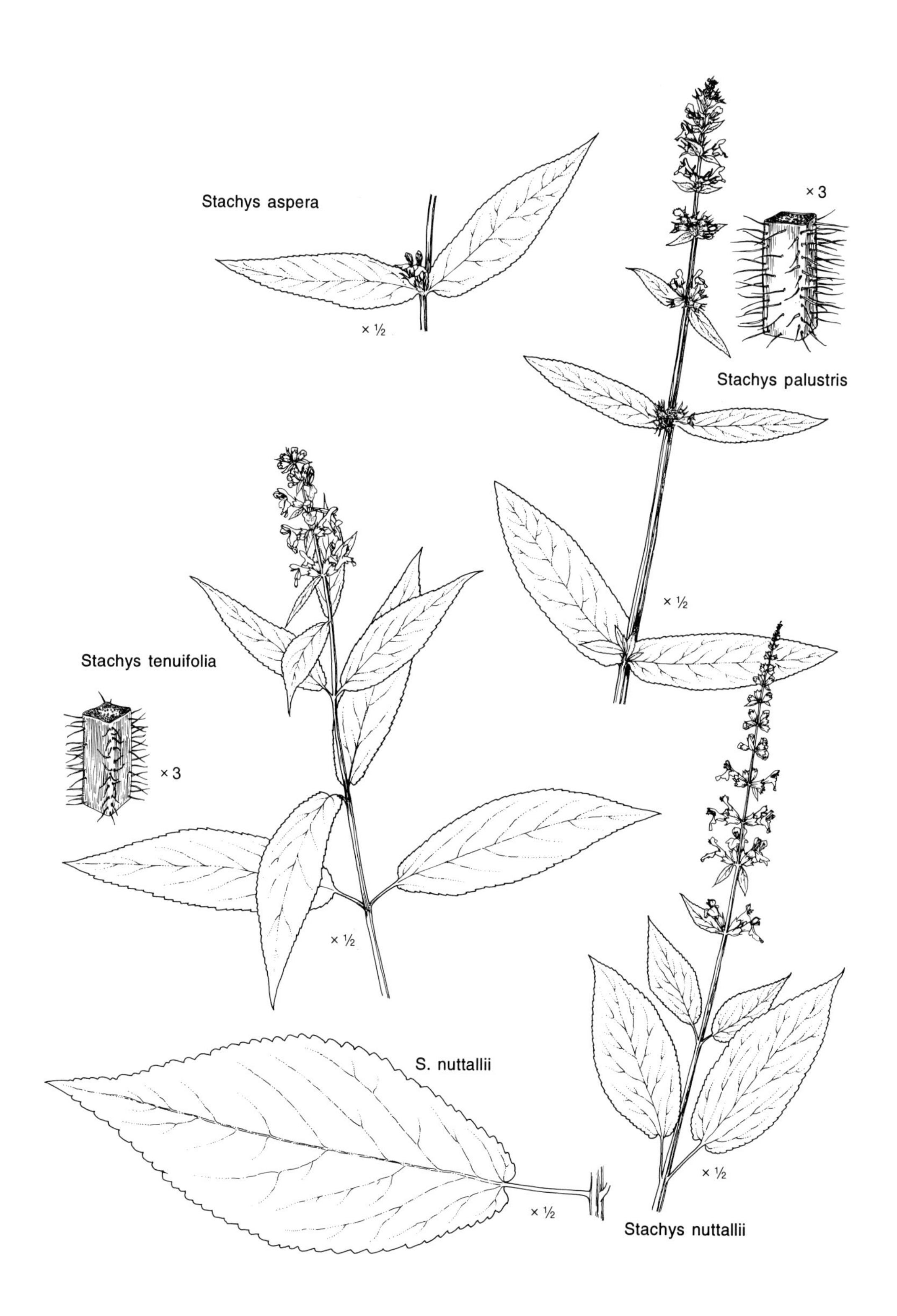
Stachys aspera
× ½
× 3
Stachys palustris
× ½
Stachys tenuifolia
× 3
× ½
S. nuttallii
× ½
× ½
Stachys nuttallii

This complex has often been divided into two species: *S. hispida* and *S. tenuifolia* sensu stricto on the basis of differences in petiole length and leaf and calyx pubescence. The most recent proponents of this separation were Mulligan and Munro (1989), who found only tetraploids among plants they called *S. hispida*, but both diploids and tetraploids among those they called *S. tenuifolia*—all elements having the same base number. The large percentage of intermediates found among Ohio plants (Cooperrider and Sabo, 1969) leads me to favor the treatment of J. B. Nelson (1981) in which *S. hispida* is merged without rank under *S. tenuifolia* var. *tenuifolia*. A few Ohio specimens were encountered which show a tendency toward intergradation with *S. aspera*.

4. **Stachys aspera** Michx. Rough Hedge-nettle
S. hyssopifolia Michx. var. *ambigua* A. Gray

Perennial, to 7 dm tall; stems with pubescence only on the angles, the sides glabrous; medial leaves sessile or with petioles less than 3 (–5) mm long; corolla pinkish to purplish, 10–12 mm long. North-central counties, and at scattered sites southward; moist banks, ditches, borders; probably native, but perhaps naturalized. June–August.

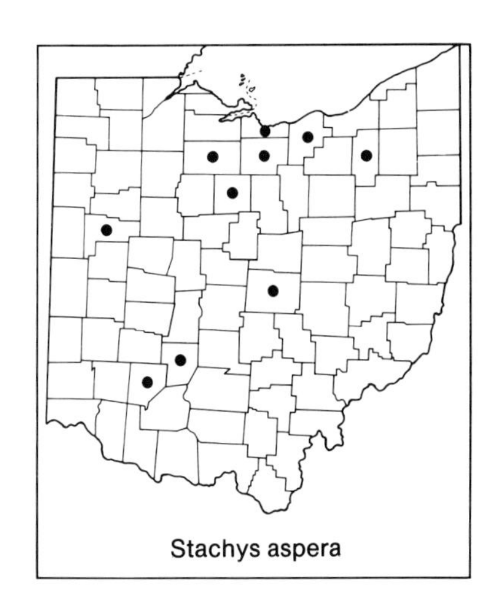

Stachys aspera

18. Salvia L. SAGE

Erect annual or perennial herbs; flowers in clusters forming spikes or racemes; calyx bilabiate; corolla bilabiate, usually blue, violet, or purple, but in some species white, red, or pink; stamens 2, included. A cosmopolitan genus of some 750 species. Several species are cultivated in Ohio and have been collected occasionally as adventive escapes. *Salvia azurea* Lam. var. *grandiflora* Benth. (*S. pitcheri* Torr.), Blue Sage, a species from the plains to the west of Ohio, has been collected from disturbed sites in Auglaize, Franklin, Highland, and Lucas counties, all in western Ohio. *Salvia coccinea* Murr., Crimson Sage or Texas Sage, from southeastern United States and Mexico, was collected once in Highland County, in 1934. *Salvia officinalis* L., Common Sage or Garden Sage, a plant of the Mediterranean region grown for use as a food seasoning, has been collected from Ashtabula, Highland, Hocking, and Ottawa counties. *Salvia pratensis* L., Meadow Sage, a weed of European origin, has been collected from Pickaway and Portage counties. The much-cultivated *Salvia splendens* Roem. & Schult., Scarlet Sage, an ornamental from Brazil, with bright red calyx and corolla, was collected as an adventive in Highland County in 1917. *Salvia × superba* Stapf, also an ornamental, was collected in Fulton County in 1930.

Salvia urticifolia L. was attributed to Ohio by Fernald (1950) and Gleason (1952b), although Ohio lies beyond the range of the species given by Epling (1940). No specimens have been found in this study.

a. Corolla bright red.
  b. Calyx purple; corolla 2–2.5 cm long; see above. . . . . *Salvia coccinea*
  bb. Calyx red; corolla 3–4 cm long; see above . . . . *Salvia splendens*
aa. Corolla blue, violet, purple, white, or rarely pink.
  b. Most leaves in a basal rosette, cauline leaves none or at most 1–3 pairs and smaller than the basal leaves.
    c. Leaves mostly pinnately lobed, the unlobed leaves tapering to an acute base. . . . . 1. *Salvia lyrata*
    cc. Leaves unlobed but with crenate margins, all leaves cordate at base; see above. . . . . *Salvia pratensis*
  bb. Most leaves cauline, plants with or without basal leaves.
    c. Leaves sessile, truncate at base; see above . . . . 7. *Salvia* × *superba*
    cc. Leaves petiolate and tapering, rounded, or subtruncate at base.
      d. Calyx 10–14 mm long; leaves lanceolate to elliptic; stems pubescent; see above . . . . *Salvia officinalis*
      dd. Calyx 6–8 mm long; leaves linear to lanceolate; stems minutely pubescent to glabrous.
        e. The calyx densely pubescent throughout; corolla 13–15 mm long, its tube much exceeding length of calyx; principal leaves with blades 3–12 cm long; see above . . . . *Salvia azurea*
        ee. The calyx pubescent only on the nerves; corolla 6–8 mm long, its tube scarcely or not exceeding length of calyx; principal leaves with blades 3–5 cm long . . . . 2. *Salvia reflexa*

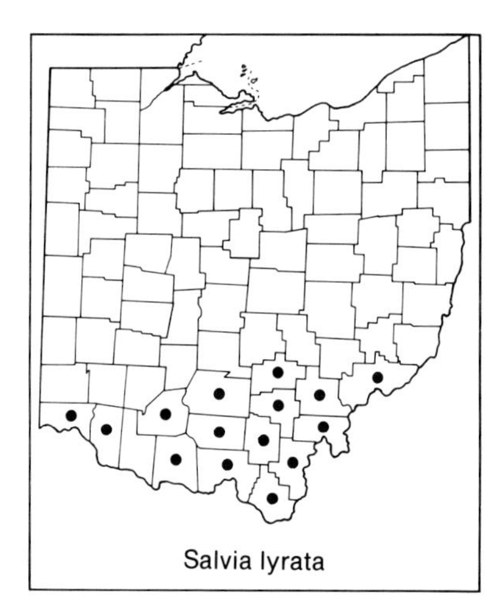
Salvia lyrata

1. **Salvia lyrata** L. LYRE-LEAVED SAGE

Perennial, to 5 dm tall; stems pubescent; plants with most leaves in a basal rosette, these petiolate, their blades pinnately lobed to merely crenate, 4–12 cm long, much larger than the single pair of usually sessile cauline leaves; corolla violet to blue-violet, 1.5–3.0 cm long. A species chiefly of southeastern United States, at a northern boundary of its range in southern counties of Ohio; open sandy soil, sandy woodland openings and borders. There is also a 1933 collection of an adventive from Portage County in northeastern Ohio. May–June.

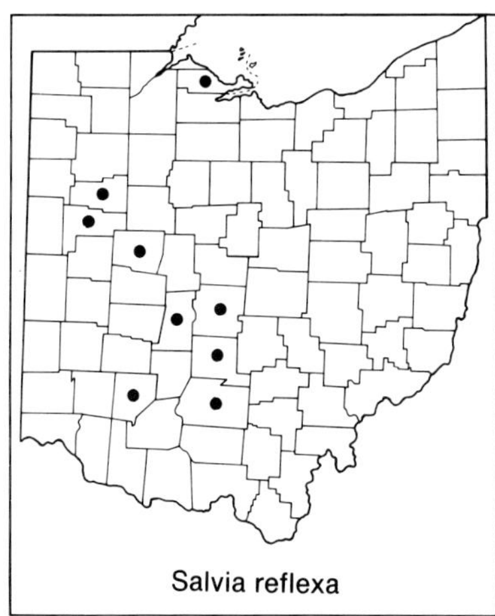
Salvia reflexa

2. **Salivia reflexa** Hornem. ROCKY MOUNTAIN SAGE

Annual, to 6 dm tall; stems pubescent; leaves cauline, petiolate, the principal ones with blades 3–5 cm long; corolla blue, 6–8 mm long, the tube scarcely or not exceeding length of calyx. Probably native in west-central counties; prairies, fields, open sites. There is also a 1918 collection of an adventive from a pigeon-yard in Belmont County in eastern Ohio. August–October.

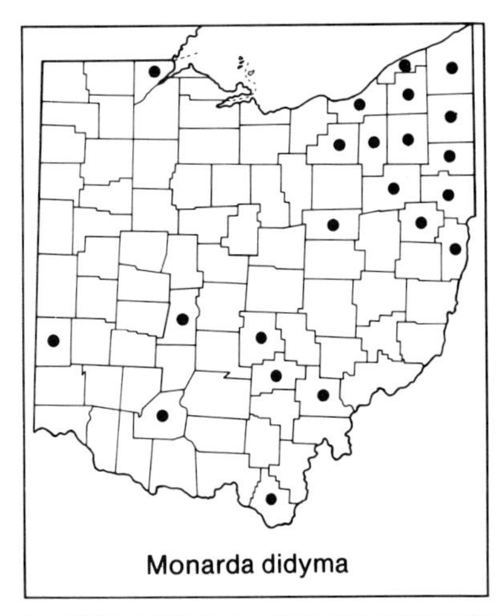
Monarda didyma

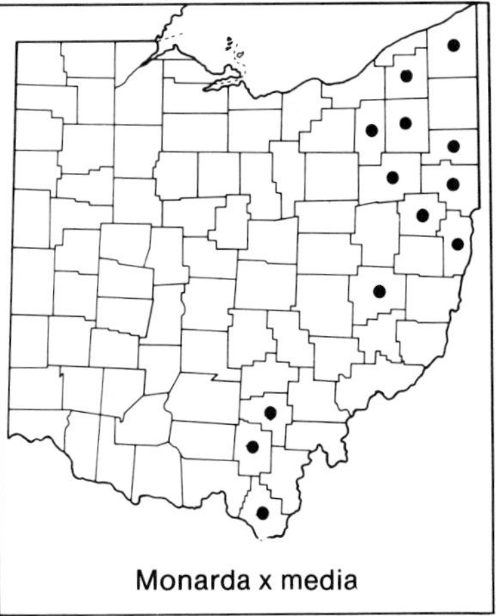
Monarda x media

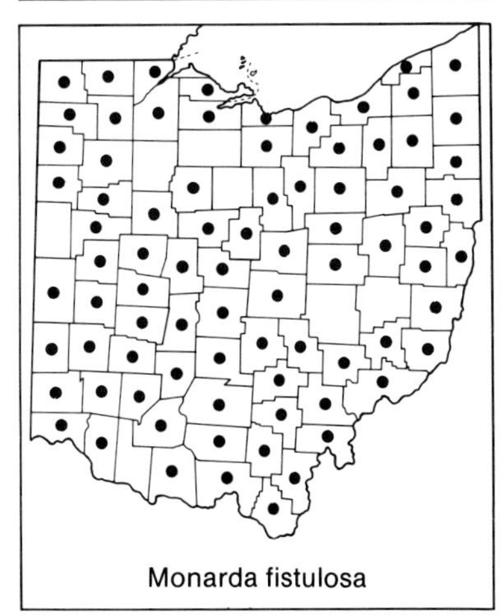
Monarda fistulosa

19. Monarda L. HORSEMINT

Erect perennial herbs; flowers in dense terminal or axillary clusters; calyx actinomorphic or nearly so; corolla bilabiate, purple, red, pink, white, or yellow; stamens 2, exserted or not. A North American genus of 12 or more species. Taxonomic studies of the genus have been made by McClintock and Epling (1942) and Scora (1967).

a. Corolla yellow with purple spots; stamens not exserted beyond corolla; flowers in both terminal and axillary clusters . . . . 4. *Monarda punctata*

aa. Corolla purple, red, lavender, pink, or white; stamens exserted beyond corolla; flowers in terminal clusters only.

b. Corolla bright red or reddish-purple.

c. Corolla bright red, its upper lip glabrous at apex . . . . 1. *Monarda didyma*

cc. Corolla reddish-purple, its upper lip with few hairs at apex . . . . 2. *Monarda* × *media*

bb. Corolla pale lavender to pink or white . . . . 3. *Monarda fistulosa*

1. **Monarda didyma** L. OSWEGO-TEA. BEE-BALM
Perennial, to 1.5 m tall; stems slightly pubescent; leaves petiolate, broadly lanceolate to ovate; flowers in terminal clusters; corolla bright red, 3–4 cm long, the upper lip glabrous at apex; stamens exserted. Northeastern counties and at scattered sites westward and southward; wet open woodlands, thickets, stream margins, ditches. Late June–early August.

2. **Monarda × media** Willd. PURPLE BERGAMOT
Perennial, to 1.2 m tall; stems glabrous or with very few hairs; leaves petiolate, broadly lanceolate to ovate; flowers in terminal clusters; corolla reddish-purple, 2–3 cm long, the upper lip with a few hairs at apex; stamens exserted. Northeastern to south-central Ohio; roadsides, fields, thickets. The spontaneous hybrid of *Monarda didyma* × M. *fistulosa*; papers of W. H. Duncan (1959), Egler (1973), and Whitten (1981) describe aspects of its hybrid nature; some workers, however, accord it species status. Not illustrated. July–early August.

3. **Monarda fistulosa** L. WILD BERGAMOT
Perennial, to 1 m tall; stems glabrous to pubescent; leaves petiolate, broadly lanceolate to ovate; flowers in terminal clusters; corolla pale lavender to pink or white, 1.5–3 cm long; stamens exserted. Throughout Ohio; moist

Monarda punctata

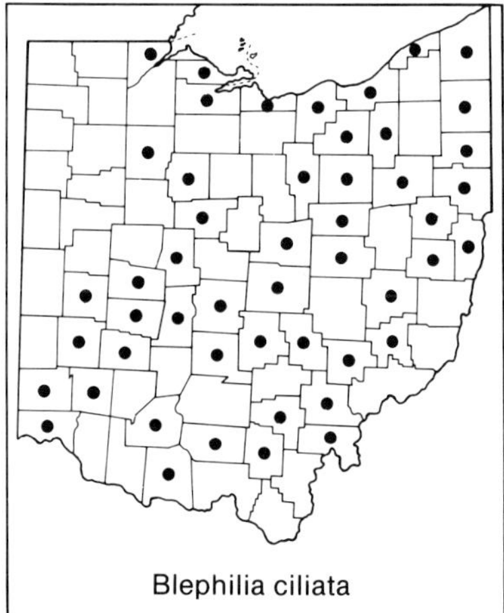
Blephilia ciliata

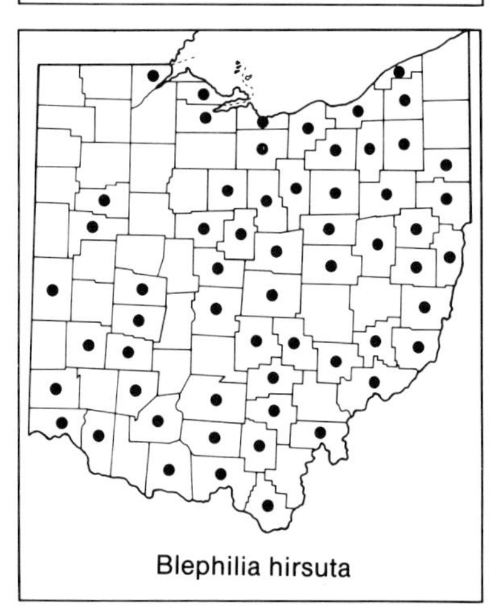
Blephilia hirsuta

woodland openings and borders, thickets, ditches, fields. (Mid-June–) July–August (–mid-September).

The two varieties keyed below, often treated as distinct species, show a continuum of intergradation in Ohio (Cooperrider, 1984b).

a. Orifice of calyx tube densely hirsute; upper corolla lip villous at apex . . . . var. *fistulosa* incl. var. *mollis* (L.) Benth.

aa. Orifice of calyx tube moderately to slightly hirsute; upper corolla lip not villous at apex . . . . var. *clinopodia* (L.) Cooperr. *M. clinopodia* L.

4. **Monarda punctata** L. Dotted Horsemint

Perennial, to 5 dm tall; stems densely pubescent; leaves tapering to a short petiole, lanceolate to linear-lanceolate; flowers in both terminal and axillary clusters; corolla yellow with purple spots, 1.2–1.6 mm long; stamens not exserted beyond corolla. Known from a few scattered collections; open sandy habitats; probably native at some Ohio sites and naturalized or adventive at others. July–September.

Ohio plants are assigned to var. *villicaulis* (Pennell) Shinners.

20. Blephilia Raf. woodmint

Erect perennial herbs; flowers in dense axillary clusters; calyx bilabiate; corolla bilabiate, white to purple; stamens 2, slightly exserted. An eastern North American genus of two species, both present in the Ohio flora.

a. Stems with most hairs recurved, none or only a few hairs spreading; upper leaves narrowly lanceolate to lanceolate or oblong, sessile or with petioles to 1 cm long; corolla usually purplish with darker purple spots, rarely white with dark purple spots . . . . 1. *Blephilia ciliata*

aa. Stems with all hairs spreading, sometimes matted; upper leaves mostly ovate or narrowly ovate, with petioles 1–3 cm long; corolla whitish with purple spots . . . . 2. *Blephilia hirsuta*

1. **Blephilia ciliata** (L.) Benth. Downy Woodmint

Perennial, to 7 dm tall; stems pubescent with most hairs recurved, none or only a few hairs spreading; upper leaves narrowly lanceolate to lanceolate or oblong, sessile or with petioles to 1 cm long; corolla usually purplish with darker purple spots, rarely white with dark purple spots, 8–12 mm long. Mostly in eastern and southern counties, absent from northwestern and west-central Ohio; woodland openings and borders, thickets, fields. Mid-June–July.

2. **Blephilia hirsuta** (Pursh) Benth. Hairy Woodmint

Perennial, to ca. 1 m tall; stems pubescent with all hairs spreading and sometimes matted; upper leaves mostly ovate or narrowly ovate, with

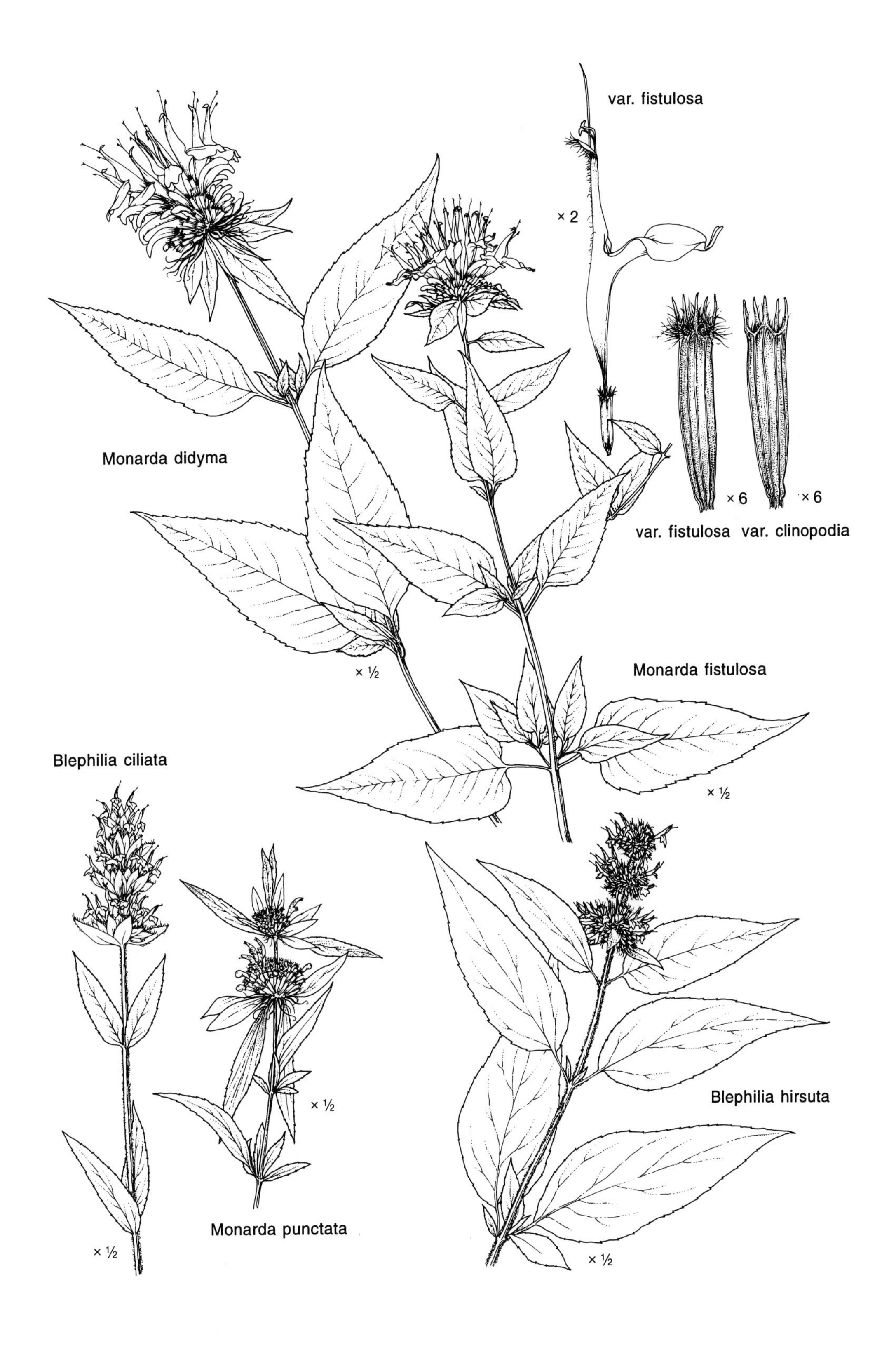
var. fistulosa
× 2
Monarda didyma
× 6
× 6
var. fistulosa
var. clinopodia
× ½
Monarda fistulosa
Blephilia ciliata
× ½
× ½
Blephilia hirsuta
Monarda punctata
× ½
× ½

petioles 1–3 cm long; corolla whitish with purple spots, 8–10 mm long. Throughout Ohio, except some northwestern and west-central counties; woodlands. Late June–early September.

Herbarium specimens of this species are sometimes confused with those of *Clinopodium vulgare* (p. 410), but can be separated by these characters: *B. hirsuta* has white flowers and long-petiolate upper leaves; *C. vulgare* has pink to reddish-purple flowers and subsessile or short-petiolate upper leaves.

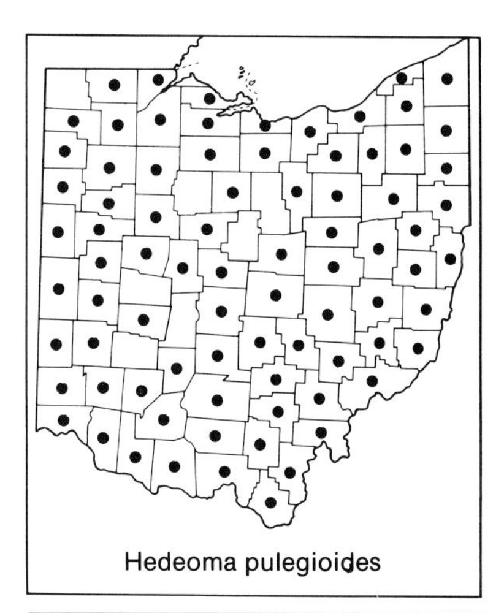
Hedeoma pulegioides

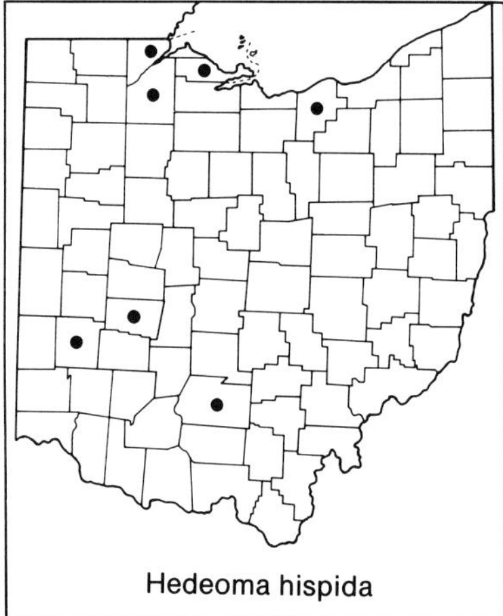
Hedeoma hispida

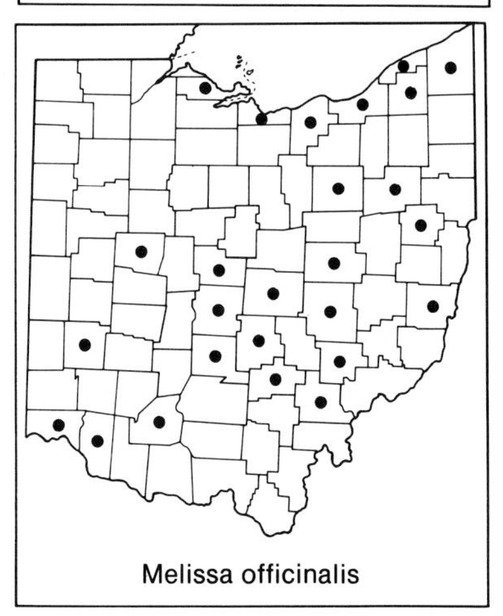
Melissa officinalis

21. Hedeoma Persoon

Erect annual herbs; flowers in few-flowered axillary clusters; calyx bilabiate; corolla bilabiate, blue; stamens 2, not or scarcely exserted. A New World genus of 25–30 species, two of them native to the Ohio flora. Irving (1980) prepared a systematic study of the genus.

a. Principal leaves lanceolate to rhombic-lanceolate, short-petiolate, margins of most leaves with a few low teeth; the two lower calyx lobes subulate and ciliate, the three upper triangular and eciliate . . . . 1. *Hedeoma pulegioides*

aa. Principal leaves linear, sessile, margins entire; the calyx lobes all subulate and ciliate . . . . 2. *Hedeoma hispida*

1. **Hedeoma pulegioides** (L.) Pers. American Pennyroyal. Mock Pennyroyal

Annual, to 4 dm tall, with a strong but pleasant fragrance; stems simple to much-branched, pubescent; principal leaves lanceolate to rhomboid-lanceolate, short-petiolate, margins of most leaves with a few low teeth; calyx with the lower two lobes subulate and ciliate, the upper three triangular and eciliate; corolla blue, 3–5 mm long, scarcely longer than calyx. Throughout Ohio; open woodlands, thickets, fields, weed lots. July–September.

2. **Hedeoma hispida** Pursh Rough Mock Pennyroyal

Annual or winter annual, to 2.5 dm tall; stems simple or with few to several branches, pubescent; principal leaves linear, sessile, margins entire; calyx lobes all subulate and ciliate; corolla blue, 5–7 mm long, slightly exserted from calyx. Scattered sites in the western half of Ohio; open sandy habitats. June–July.

22. Melissa L. Balm

A Eurasian genus of three species, one of them once widely cultivated in Ohio.

1. Melissa officinalis L. Common Balm

Lemon-scented perennial, to ca. 6 dm tall; stems erect, glandular-puberulent; leaves petiolate, crenate-serrate; flowers in clusters of few to

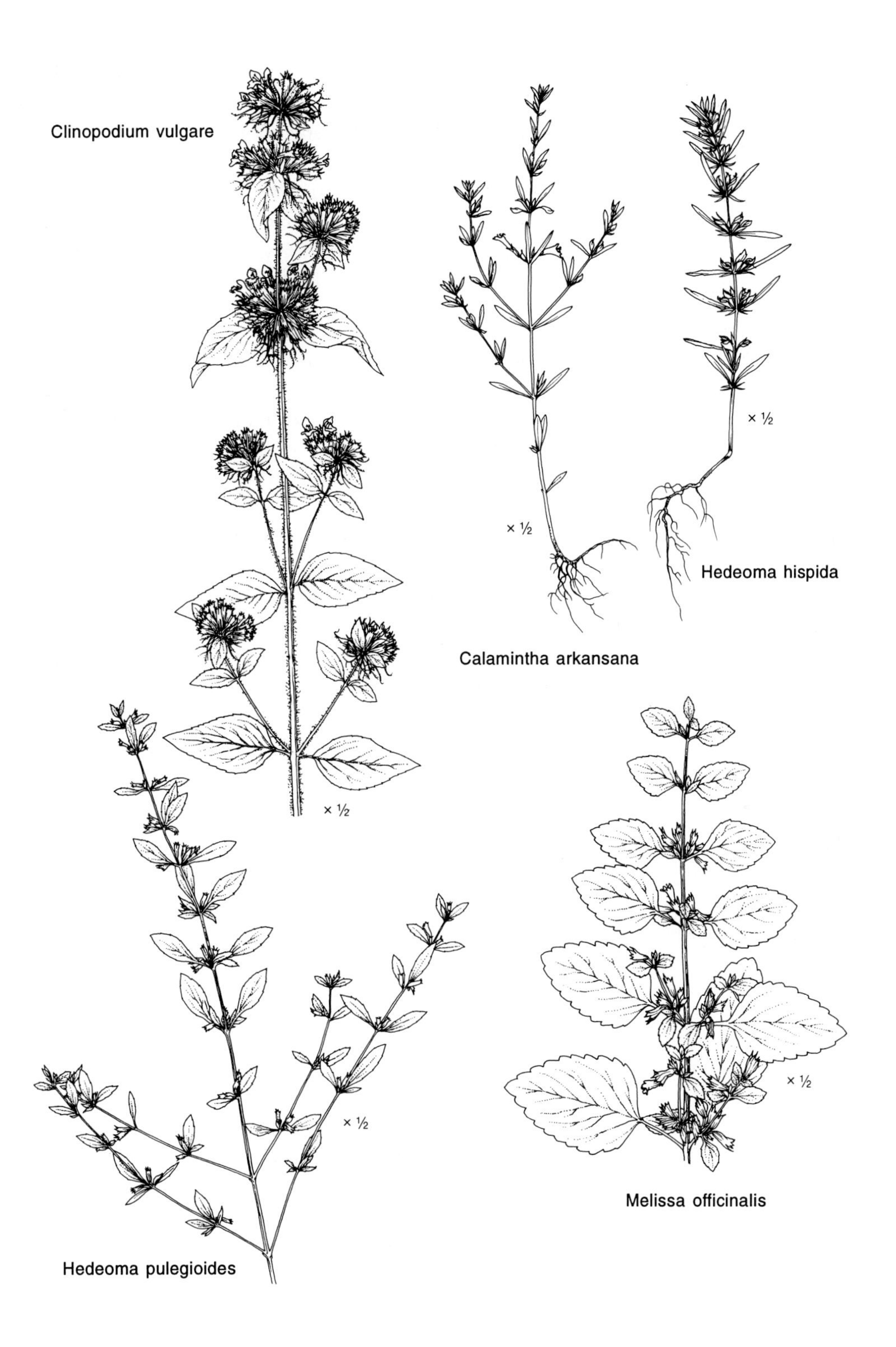
Clinopodium vulgare
× ½
Calamintha arkansana
× ½
Hedeoma hispida
× ½
Hedeoma pulegioides
× ½
Melissa officinalis
× ½

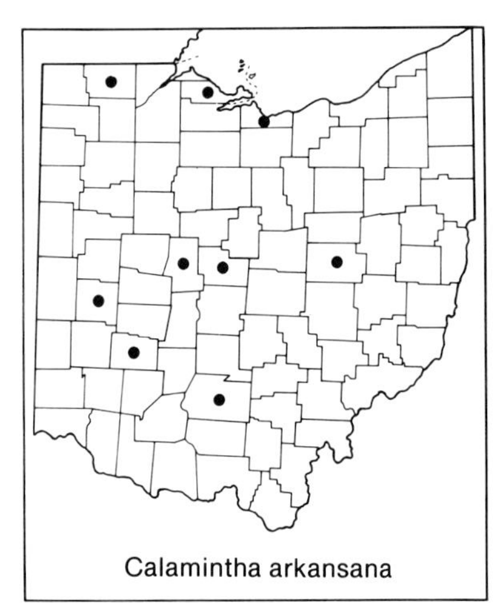
Calamintha arkansana

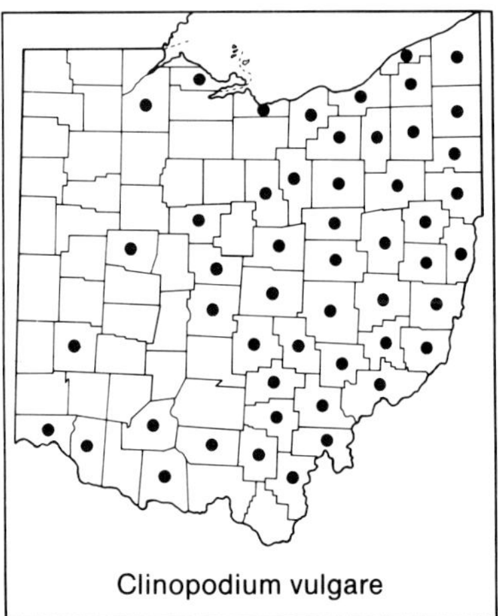
Clinopodium vulgare

several; calyx bilabiate; corolla whitish, 10–15 mm long, bilabiate; stamens 4, not or scarcely exserted. A European species, frequently cultivated (more so in the past than at present) and escaped into disturbed habitats, adventive—or perhaps in some cases naturalized—at scattered sites throughout much of Ohio. Most Ohio specimens were collected between 1890–1910, the lack of recent collections due probably to the species' declining popularity as a kitchen garden plant, once used for culinary and medicinal purposes, now usually grown merely as a fragrant ornamental. July–September.

23. Calamintha Mill. CALAMINT

A Northern Hemisphere genus of about twelve species, one of them native to Ohio.

1. **Calamintha arkansana** (Nutt.) Shinners LIMESTONE SAVORY. LOW CALAMINT
   Satureja arkansana (Nutt.) Briq.; *S. glabella* (Michx.) Briq. var. *angustifolia* (Torr.) Svenson

Perennial with leafy stolons, the tiny stolon leaves roundish; erect flowering stems glabrous, to 3 dm tall; leaves linear or sublinear; flowers pedicellate, solitary or few in leaf axils; calyx actinomorphic or nearly so; corolla bilabiate, lavender to purple, 8–10 mm long; stamens 4, slightly exserted. At scattered sites, mostly in western Ohio; dry, rocky, open, calcareous habitats. Shinners (1962a) discusses the members of this genus known from southern United States. Mid-June–August.

24. Clinopodium L.

A small genus of four species, widely distributed in temperate regions, with one species present in the Ohio flora. DeWolf (1955) comments on the segregation of this genus from *Satureja*.

1. **Clinopodium vulgare** L. WILD BASIL
   Satureja vulgaris (L.) Fritsch

Perennial, to 9 dm tall; stems erect, white-pubescent; leaves short-petiolate, blades ovate to lance-ovate; flowers in dense terminal and axillary clusters; calyx zygomorphic; corolla bilabiate, pink to reddish-purple, 12–14 mm long; stamens 4, included. Throughout the eastern half of Ohio and at scattered sites westward; open woods, fields, roadsides, open wet habitats; native but perhaps naturalized at some sites. June–September.

See notes under *Blephilia hirsuta* (p. 408) on separating that species from this.

25. Pycnanthemum Michx. MOUNTAIN-MINT

Erect perennial herbs; flowers in dense heads or cymes; calyx actinomorphic to zygomorphic; corolla bilabiate, white to pinkish-purple; stamens 4, usually exserted. A North American genus of about 20 species. E. Grant and C. Epling (1943) published a systematic treatment of the genus, and Chambers (1961) and Chambers and Chambers (1971) reports of chromosome numbers and hybridization studies. As Ahles (1968) observed, *Pycnanthemum* is a genus "in much need of critical study." Yetter (1989) recently made a systematic study of the "Virginianum species complex," which includes species nos. 1–4 below. Fernald (1950) reports *Pycnanthemum torrei* Benth. from Ohio, but I have seen no specimens in this study.

a. Cymes dense and head-like, with no branches or only the primary branches of the inflorescence visible; leaves ovate-lanceolate to lanceolate or linear, blades of principal leaves 1–18 (–20) mm wide; calyx actinomorphic with all lobes equal.

b. Leaves linear to linear-lanceolate, 1–5 (–7) mm wide, the lateral veins, if any, originating from low on midrib.

c. Internodes glabrous; calyx lobes 1–1.5 mm long, aristate-acuminate at apex. . . . . 1. *Pycnanthemum tenuifolium*

cc. Internodes pubescent on the angles; calyx lobes 0.7–1 mm long, acute at apex . . . . 2. *Pycnanthemum virginianum*

bb. Leaves lanceolate to ovate-lanceolate, principal ones 8–18 (–20) mm wide, the lateral veins originating from upper as well as lower half of midrib.

c. Leaves lanceolate, four or more times as long as wide, the margins finely serrate to entire. . . . . 3. *Pycnanthemum verticillatum*

cc. Leaves ovate to ovate-lanceolate, less than three times as long as wide, the margins serrate. . . . . 4. *Pycnanthemum muticum*

aa. Cymes somewhat open at base, with the primary, secondary, and sometimes tertiary branches of the inflorescence visible; leaves ovate to ovate-oblong or ovate-lanceolate, blades of principal leaves 15–55 mm wide; calyx more or less zygomorphic.

b. Tips of bracts and calyx-lobes, and usually the sides of the upper internodes, with numerous, long, spreading, flexuous hairs in addition to abundant, short, incurved hairs; calyx lobes only slightly unequal; corolla white with purple dots. . . . . 5. *Pycnanthemum pycnanthemoides*

bb. Tips of bracts and calyx-lobes, and usually the sides of the upper internodes, with abundant, short, incurved hairs only—or also with a very few long flexuous hairs; calyx lobes of two different lengths; corolla white to more often dark pink, with or without purple dots . . . . 6. *Pycnanthemum incanum*

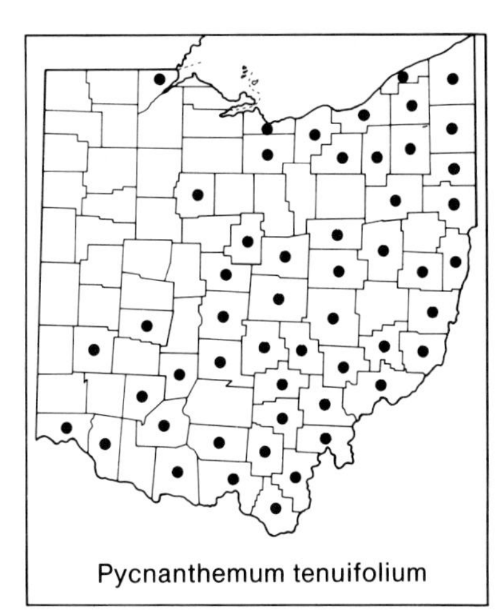

Pycnanthemum tenuifolium

1. **Pycnanthemum tenuifolium** Schrad. Narrow-leaved Mountain-mint

Perennial, to ca. 8 dm tall; stems glabrous; leaves linear, the lateral veins, if any, originating from low on the midrib; cymes dense and head-like, with no branches or only the primary branches of the inflorescence visible; calyx actinomorphic, the lobes 1.0–1.5 mm long, aristate-acuminate at apex; corolla whitish, ca. 5 mm long. Throughout the eastern half of Ohio and extending into some western parts of the state, but absent from west-central and northwestern counties; dry fields, roadside banks, thickets, woodland openings and borders. Late June–mid-September.

This species is called *P. flexuosum* (Walt.) BSP. by E. Grant and C. Epling (1943) and others.

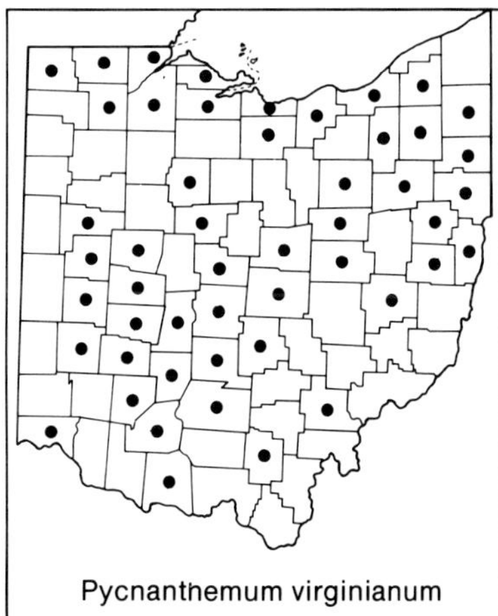

Pycnanthemum virginianum

2. **Pycnanthemum virginianum** (L.) Robins. & Fern. Virginia Mountain-mint

Perennial, to ca. 1 m tall; stems pubescent on the angles; leaves linear-lanceolate, the lateral veins originating from low on the midrib; cymes dense and head-like, with either no branches or only the primary branches of inflorescence visible; calyx actinomorphic, the lobes 0.7–1.0 mm long, acute at apex; corolla whitish, ca. 5 mm long. Throughout much of Ohio, but scarce or absent in southeastern and westernmost counties; open moist fields and thickets, bogs, marshes. July–mid-September.

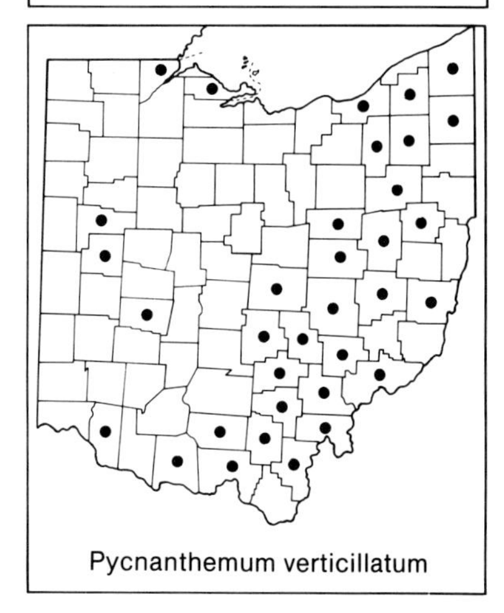

Pycnanthemum verticillatum

3. **Pycnanthemum verticillatum** (Michx.) Pers.

Perennial, to ca. 1 m tall; stems more or less pubescent on sides and angles; leaves lanceolate, four or more times as long as wide, margins finely serrate to entire; cymes dense and head-like, with no or only the primary branches of inflorescence visible; calyx actinomorphic; corolla whitish, 5–7 mm long. Eastern counties and at scattered sites westward; moist fields, open woods, woodland borders. July–early September.

Two intergrading varieties of this species occur in Ohio (Cooperrider, 1984b). Yetter (1989) proposed merging both taxa (each to be treated at the rank of subspecies) under *P. virginianum* sensu lato.

a. Leaves with a few hairs along the midrib on undersurface, otherwise usually glabrous, leaf margins sometimes with a few low teeth . . . . var. *verticillatum*
Verticillate Mountain-mint

aa. Leaves pubescent across undersurface, leaf margins entire . . . . var. *pilosum* (Nutt.) Cooperr.
P. pilosum Nutt.
HAIRY MOUNTAIN-MINT

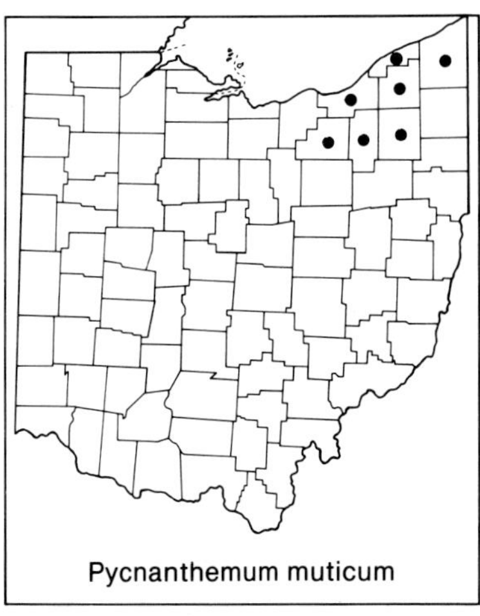
Pycnanthemum muticum

4. **Pycnanthemum muticum** (Michx.) Pers. BLUNT MOUNTAIN-MINT
Perennial, to ca. 8 dm tall; stems pubescent on sides and angles; leaves ovate to ovate-lanceolate, less than three times as long as wide, the margins serrate; cymes dense and head-like, with no or only the primary branches visible; calyx actinomorphic; corolla whitish, ca. 5 mm long. A few northeasternmost counties; open rocky areas, fens. Late July–mid-September.

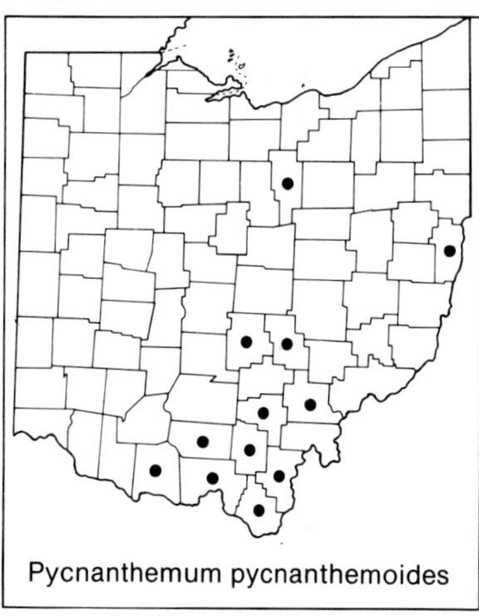
Pycnanthemum pycnanthemoides

5. **Pycnanthemum pycnanthemoides** (Leavenw.) Fern. SOUTHERN MOUNTAIN-MINT
Perennial, to 1 m tall; stems pubescent on sides and angles; leaves ovate-lanceolate to ovate; cymes somewhat open at base, with the primary, secondary, and sometimes tertiary branches of inflorescence visible; calyx slightly zygomorphic, tips of lobes (and tips of bracts and usually the sides of the upper internodes) with numerous, long, spreading flexuous hairs; corolla white with purple dots, 5–7 mm long. A species of the southeastern states, at a northern boundary of its range in Ohio; chiefly in south-central counties; dry roadside banks, woodland openings and borders. July–August.

Among Ohio specimens, there are a few intergrades between this species and the next, *P. incanum*. Ahles (1968) merged this species without rank under *P. incanum*.

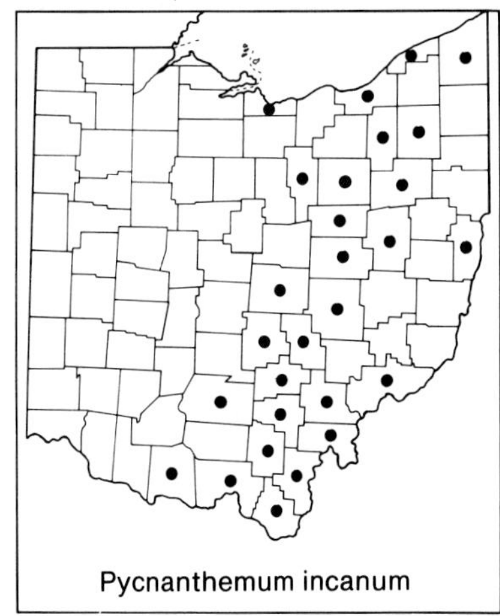
Pycnanthemum incanum

6. **Pycnanthemum incanum** (L.) Michx. HOARY MOUNTAIN-MINT
incl. var. *incanum* and var. *puberulum* (Grant & Epl.) Fern.
Perennial, to ca. 1 m tall; stems pubescent on sides and angles; leaves ovate to elliptic or ovate-oblong; cymes somewhat open at base, the primary, secondary, and sometimes tertiary branches of inflorescence visible; calyx zygomorphic, tips of lobes (and tips of bracts and usually the sides of the upper internodes) with abundant short incurved hairs only—or also with a very few long flexuous hairs; corolla white to more often pink, with or without purple dots, 6–9 mm long. Eastern half of Ohio; dry banks, slopes, thickets, and fields. July–early September.

26. Origanum L. WILD MARJORAM

A Eurasian genus of 15 or more species, one of which is widely cultivated in Ohio. A. O. Tucker and E. D. Rollins (1989) list and discuss the taxa of this genus that are cultivated in the United States.

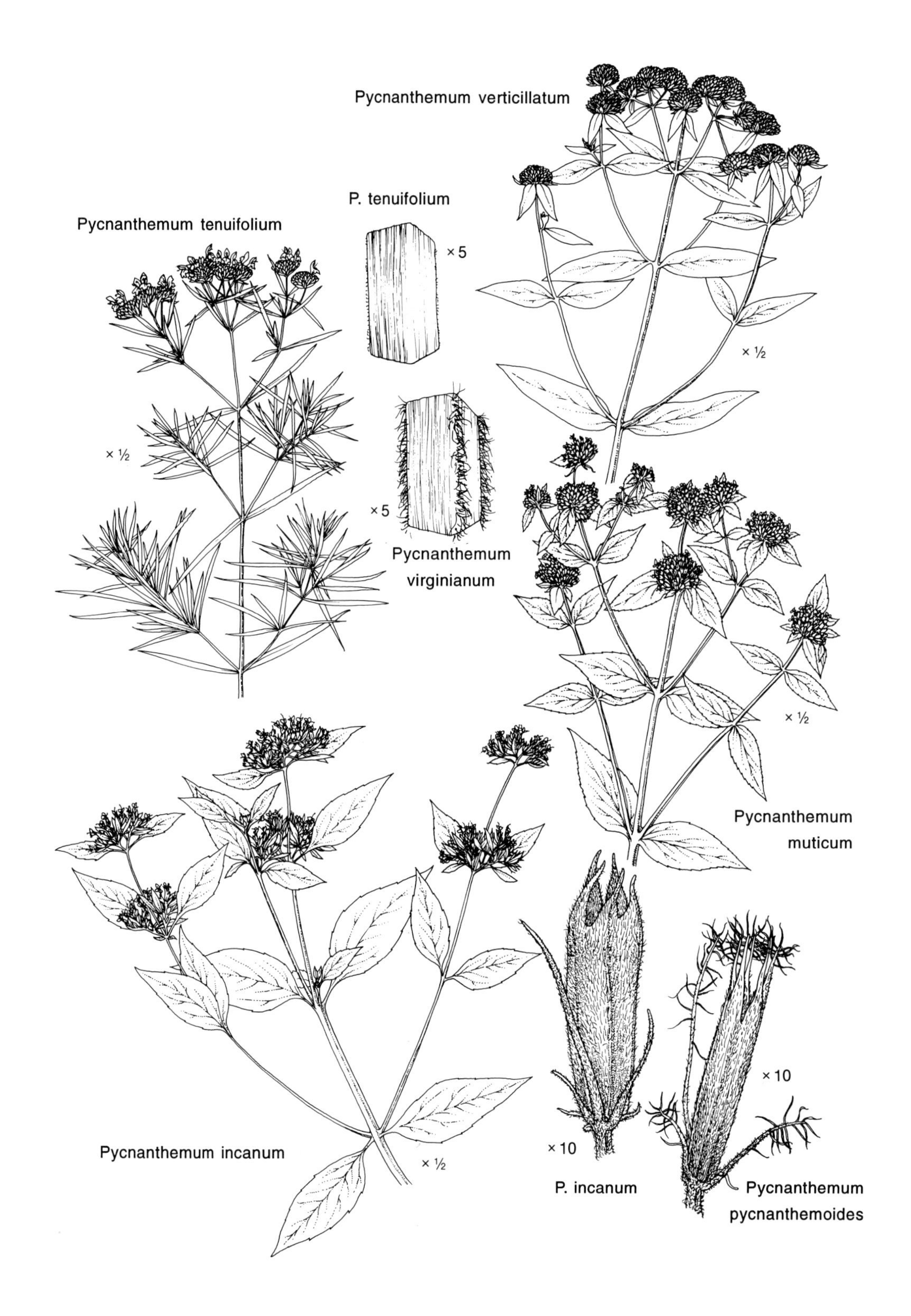
Pycnanthemum verticillatum
P. tenuifolium
Pycnanthemum tenuifolium
× 5
× ½
× ½
× 5
Pycnanthemum
virginianum
× ½
Pycnanthemum
muticum
× 10
Pycnanthemum incanum
× ½
× 10
P. incanum
Pycnanthemum
pycnanthemoides

1. Origanum vulgare L. Marjoram. Pot Marjoram. Oregano
Perennial, to 1 m tall; stems erect, pubescent; leaves petiolate, ovate; flowers in somewhat open corymbs or broad panicles, bracts of the inflorescence purple; calyx actinomorphic or nearly so; corolla bilabiate, rose-purple, 6–8 mm long; stamens 4, slightly exserted. A Eurasian species, cultivated for a food seasoning, escaped and adventive or perhaps locally naturalized at a few scattered sites in eastern Ohio. Not illustrated. July–September.

Origanum vulgare

27. Thymus L. thyme

A Eurasian genus of 300 or more species, one of them cultivated and naturalized in Ohio.

1. Thymus pulegioides L. Creeping Thyme. Wild Thyme
Perennial, to 2.5 dm tall; stems ascending, pubescent; leaves linear to elliptic, sessile or short-petiolate; flowers in terminal clusters and clusters from upper axils, in all forming a spike-like inflorescence; calyx zygomorphic; corolla bilabiate, purplish, ca. 5 mm long; stamens 4, slightly exserted. A cultivated Eurasian species, escaped and naturalized in disturbed habitats from northeastern to south-central Ohio. July–September.

Following the lead of past regional floras (e.g., Fernald, 1950; Gleason, 1952b), most Ohio collections of this species were originally misidentified as *T. serpyllum* L.

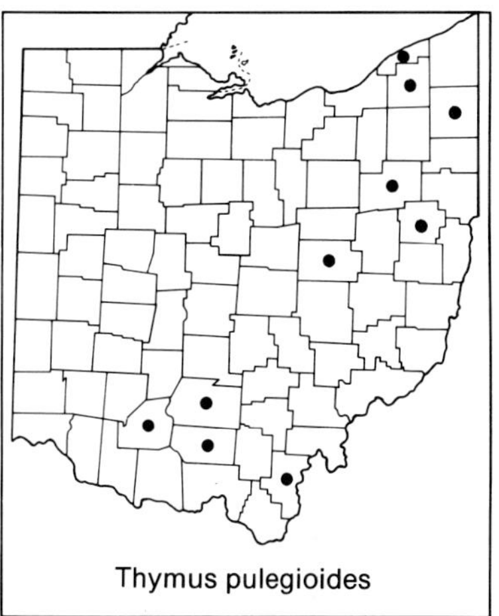
Thymus pulegioides

28. Cunila L. dittany

A New World genus of 15 species, one of them native to Ohio.

1. **Cunila origanoides** (L.) Britt. Common Dittany
Branched perennial herb, somewhat woody near base, to 5 dm tall; stems erect, slightly pubescent; leaves ovate, sessile or nearly so; flowers in small clusters from upper axils and at stem apex; calyx actinomorphic or nearly so; corolla pinkish-purple, slightly zygomorphic, only moderately bilabiate, 5–8 mm long; stamens 2, exserted. A species chiefly of the Appalachians, at a northern boundary of its range in east-central and southeastern counties of Ohio; open woods, woodland openings and borders, roadside banks. Mid-August–early October.

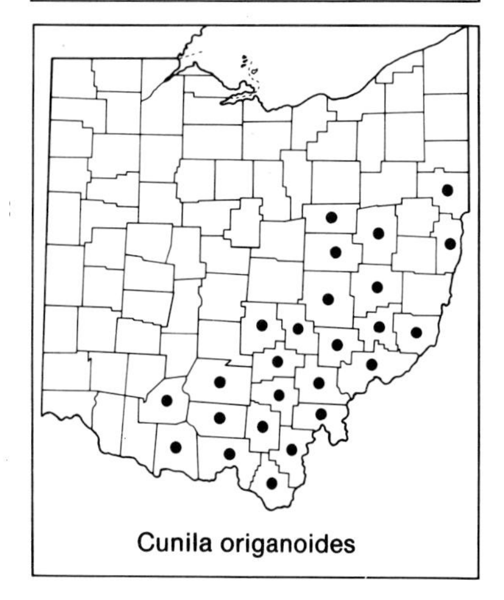
Cunila origanoides

29. Lycopus L. bugle-weed. water-horehound

Erect perennial herbs; flowers in dense axillary clusters; calyx actinomorphic or nearly so; corolla white or whitish, tiny, actinomorphic or nearly so; stamens 2, not or only slightly exserted. A genus of ten or more species of temperate regions of the Northern Hemisphere, revised by Henderson (1962). Several characters in the key below are adapted from Gleason (1952b).

a. Calyx lobes triangular or oblong-triangular, and rounded, obtuse, or acute at apex, at maturity shorter than to equalling height of nutlets . . . . 1. *Lycopus virginicus*

aa. Calyx lobes lanceolate, and acuminate to attenuate-cuspidate at apex, at maturity exceeding height of nutlets.

b. Lower and medial leaves sessile, margins of basal part of blade (from lowest tooth to blunt base) straight to slightly convex; leaf margins uniformly low serrate . . . . 5. *Lycopus asper*

bb. Lower and medial leaves more or less distinctly petiolate, margins of basal part of blade (from lowest tooth to acute or acuminate base) slightly to markedly concave; leaf margins uniformly low serrate to more often serrate, dentate, or lobed, sometimes irregularly so.

c. Margins of leaves serrate; corolla tube 1½–2 times as long as calyx . . . . 4. *Lycopus rubellus*

cc. Margins of leaves, especially the lower, deeply serrate, dentate, or lobed; corolla tube only slightly longer than calyx.

d. Lower and upper leaves evenly broad-dentate; upper surface of leaves with appressed hairs near base of blade, the undersurface with numerous hairs on veins; calyx lobes 1.5–2 mm long . . . . 2. *Lycopus europaeus*

dd. Lower leaves deeply lobed to irregularly serrate; upper surface of leaves glabrous, the undersurface glabrous except for few hairs on veins; calyx lobes 1–1.5 mm long. . . . . 3. *Lycopus americanus*

1. **Lycopus virginicus** L. VIRGINIA WATER-HOREHOUND. VIRGINIA BUGLE-WEED

incl. *L. uniflorus* Michx.

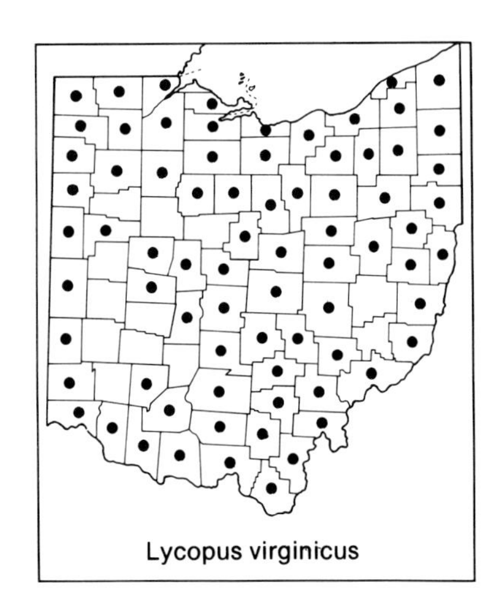

Perennial, to 7 dm tall; stems glabrous to pubescent; leaves short-petiolate to subsessile; calyx lobes triangular or oblong-triangular, and rounded, obtuse, or acute at apex, at maturity shorter than to equalling length of nutlets; corolla 2–3 mm long, conspicuously exceeding calyx. Throughout Ohio; shores, wet meadows, bogs, wet disturbed fields, sometimes in dry sites. July–early October.

Some Ohio plants can be assigned to one of the two varieties keyed below; others are intermediate between the two. Plants of this complex are treated by Henderson (1962) and others as two species, *L. virginicus* sensu stricto and *L. uniflorus*, and the intermediates as interspecific hybrids called *L.* × *sherardii* E. Steele. The complex is treated here as a single species composed of two intergrading varieties, with var. *virginicus* occupying

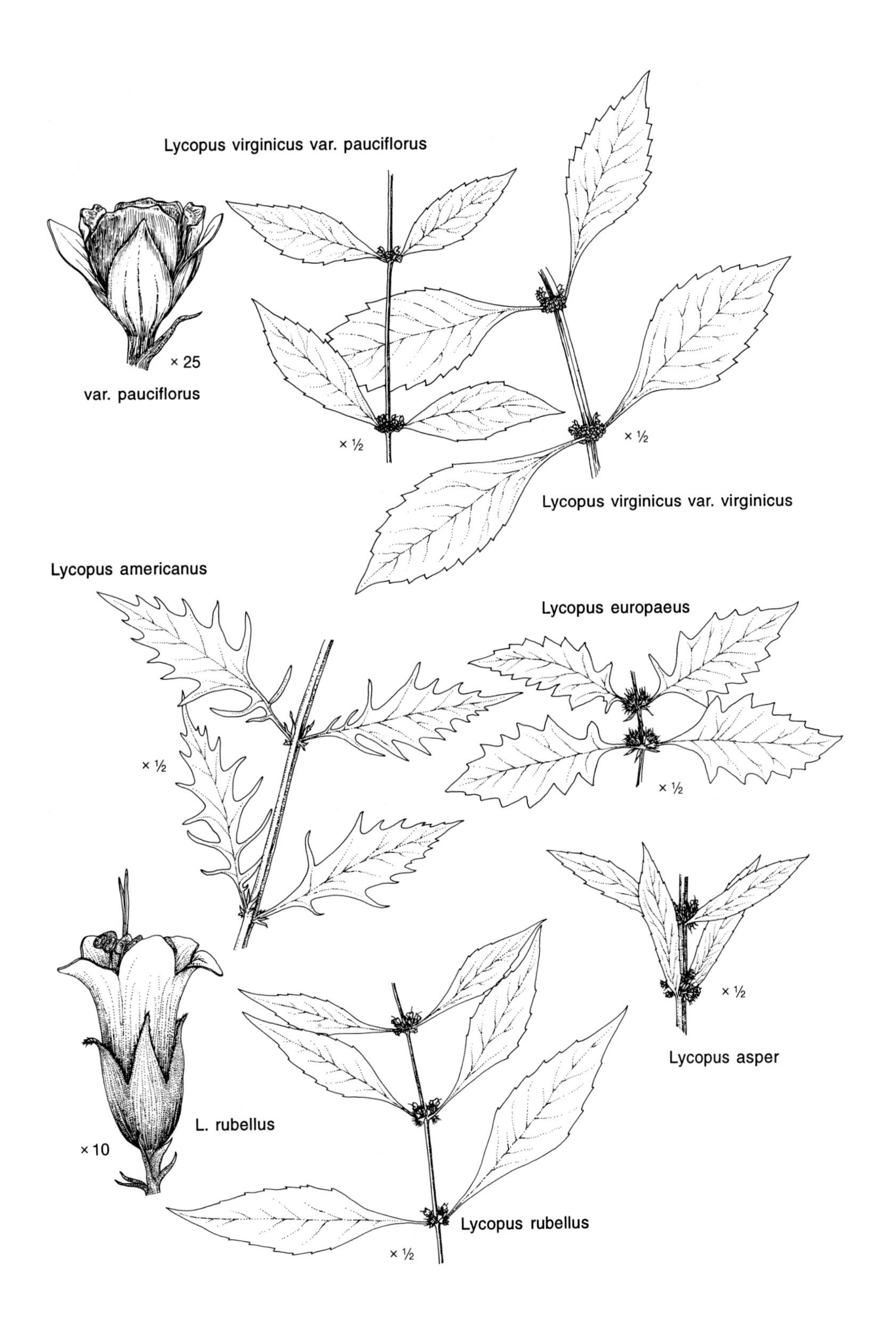
Lycopus virginicus var. pauciflorus
× 25
var. pauciflorus
× ½
× ½
Lycopus virginicus var. virginicus
Lycopus americanus
Lycopus europaeus
× ½
× ½
× ½
Lycopus asper
L. rubellus
× 10
Lycopus rubellus
× ½

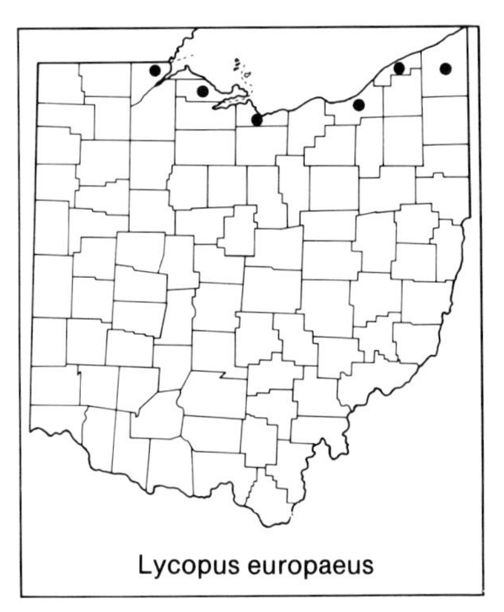

Lycopus europaeus

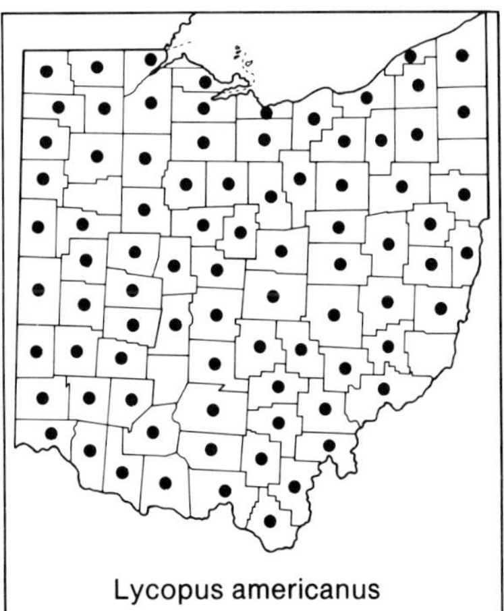

Lycopus americanus

the southern part of the species' range, var. *pauciflorus* the northern. Henderson's (1962) map shows the intermediates ("*L.* × *sherardii*") occurring throughout a vast area running diagonally across the eastern half of the continent from Maine south to Maryland, from Maine southwestward to Nebraska, and from Maryland southwestward to Louisiana. In Ohio, which lies near the center of this area, var. *virginicus* occurs in southern counties, var. *pauciflorus* in northern and western counties, and the intermediates throughout the state.

a. Base of leaves concave, making leaves more or less petiolate; stolons usually without tubers . . . . var. *virginicus*

aa. Base of leaves straight or convex, making leaves scarcely or not petiolate; stolons bearing small tubers . . . . var. *pauciflorus* Benth. *L. uniflorus* Michx.

2. Lycopus europaeus L. European Bugle-weed
incl. var. *europaeus* and var. *mollis* (Kern.) Briq.

Perennial, to 6 dm tall; stems puberulent to nearly glabrous; leaves short-petiolate, margins evenly broad-dentate, upper surface with appressed hairs near base of blade, undersurface with numerous hairs on veins; calyx lobes 1.5–2.0 mm long, attenuate-cuspidate at apex, at maturity exceeding nutlets; corolla 3–4 mm long, equalling calyx. At several Ohio sites, chiefly in the northernmost tier of counties bordering Lake Erie; beaches, shores, marshes. Stuckey and Phillips (1970) recorded the spread of this alien species in eastern North America; the earliest Ohio records they found are from the 1960s. August–September.

Andrus and Stuckey (1981) and Webber and Ball (1980) report introgressive hybridization between this species and the next, *L. americanus*.

3. **Lycopus americanus** Muhl. Cut-leaved Water-horehound
incl. var. *americanus* and var. *scabrifolius* Fern.

Perennial, to 1 m tall; stems slightly pubescent to nearly glabrous; leaves short-petiolate, margins of lower leaves deeply lobed to irregularly serrate, upper leaves with a few low teeth, upper surface of leaves glabrous, the undersurface glabrous except for a few hairs on veins; calyx lobes 1.0–1.5 mm long, attenuate-cuspidate at apex, at maturity exceeding nutlets; corolla 3–4 mm long, equalling or slightly exceeding calyx. Throughout Ohio; shores, marshy ground, wet fields and ditches. July–September.

A few specimens are intermediate between this species and the next, *L. rubellus*. See notes above under *L. europaeus* reporting hybridization between that species and this.

4. **Lycopus rubellus** Moench STALKED WATER-HOREHOUND
incl. var. *rubellus* and var. *arkansanus* (Fresn.) Benner
Perennial, to 1 m tall; stems glabrous or nearly so; leaves short-petiolate, margins serrate; calyx lobes acute to acuminate; corolla 3.5–4 mm long, conspicuously exceeding calyx. Northern counties and at scattered sites southward; shores, marshes, wet fields and thickets. Late July–September.

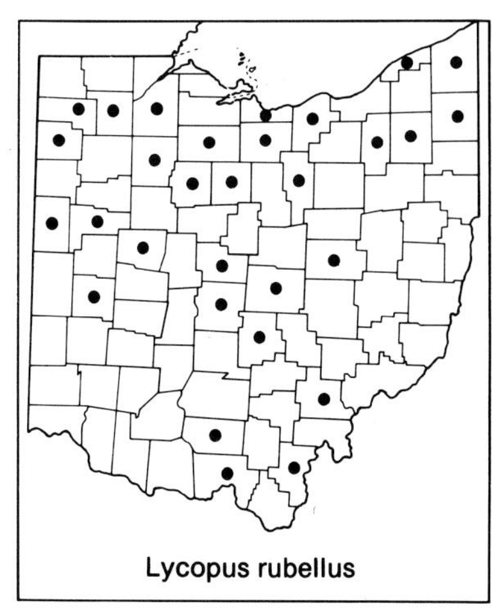
Lycopus rubellus

5. LYCOPUS ASPER E. Greene ROUGH WATER-HOREHOUND
Perennial, to 7 dm tall; stems pubescent, especially on the angles; leaves sessile, margins of basal part of blade (from lowest tooth to blunt base) straight or slightly convex; calyx lobes sharply acuminate; corolla 4–5 mm long, exceeding calyx. A few sites in north-central counties near Lake Erie; beaches, shores, mud flats, wet ditches. Stuckey (1969) concluded that this species, native to western North America, is a naturalized member of the Ohio flora. Late July–September.

Lycopus asper

30. Mentha L. MINT

Erect, pleasantly fragrant, perennial herbs; calyx actinomorphic or nearly so; corolla actinomorphic or nearly so; stamens 4, exserted. Chiefly an Old World genus, of some 25 species and several interspecific hybrids. The treatment of the naturalized taxa below is based primarily on Harley (1972), in *Flora Europaea*, and on personal communication from Harley, elaborating on parts of that treatment.

a. Leaves mostly on petioles 5 mm or more in length.
  b. Flowers all or mostly in a terminal head or spike . . . . 5. *Mentha* × *piperita*
  bb. Flower clusters axillary.
    c. Leaves subtending flower clusters somewhat reduced in size, most only 2–3 times as long as flower clusters; calyx pubescent only toward apex . . . . 6. *Mentha* × *gracilis*
    cc. Leaves subtending flower clusters full-sized, 4 or more times as long as flower clusters, or the uppermost leaves sometimes slightly reduced; calyx usually pubescent from apex to base. . . . . 7. *Mentha arvensis*
aa. Leaves sessile or on petioles to 2 mm long.
  b. Leaves glabrous on undersurface, or with a few hairs along the veins . . . . 4. *Mentha spicata*
  bb. Leaves pubescent across the undersurface.
    c. Leaves oblong-ovate to oblong-elliptic or oblong-lanceolate, the apex acute. . . . . 1. *Mentha longifolia*
    cc. Leaves orbicular-ovate or orbicular-oblong to suborbicular, the apex obtuse or rounded.

d. Spikes 8–15 mm in diameter . . . . 3. *Mentha* × *villosa*
dd. Spikes 5–8 mm in diameter . . . . 2. *Mentha* × *rotundifolia*

1. Mentha longifolia (L.) L. European Horsemint
incl. var. *longifolia*, var. *mollissima* (Borkh.) Rouy, and var. *undulata* Fiori & Paoletti

Perennial, to ca. 1 m tall; stems pubescent; leaves sessile or subsessile with petioles to 2 mm long, pubescent across the undersurface, oblong-ovate to oblong-elliptic or oblong-lanceolate, apex acute; flowers in elongated, crowded, terminal spikes; corolla white to lavender, 3–4 mm long. A Eurasian species, sometimes cultivated, locally naturalized at several sites in Ohio; disturbed weedy habitats. Ohio specimens are highly variable in leaf size and shape and in the amount of pubescence on leaves and stems. Not illustrated. July–September.

2. Mentha × rotundifolia (L.) Huds. Apple Mint

Perennial, to 1 m tall; stems pubescent; leaves sessile or on petioles to 2 mm long, pubescent on both surfaces, especially the undersurface, suborbicular to orbicular-ovate or orbicular-oblong, apex obtuse or rounded; flowers in a terminal spike 5–8 mm in diameter; corolla white to pinkish. A European taxon, the hybrid of *Mentha longifolia* × *M. suaveolens* Ehrh., adventive or perhaps naturalized at several scattered Ohio sites; disturbed weedy ground. Not illustrated. July–October.

3. Mentha × villosa Huds. Foxtail Mint

Perennial, to ca. 1 m tall; stems pubescent; leaves sessile, pubescent on both upper and under surfaces, suborbicular to orbicular-ovate or orbicular-oblong, apex obtuse or rounded; flowers in a terminal spike 8–15 mm in diameter; corolla white to pink or lavender. A European taxon, the hybrid

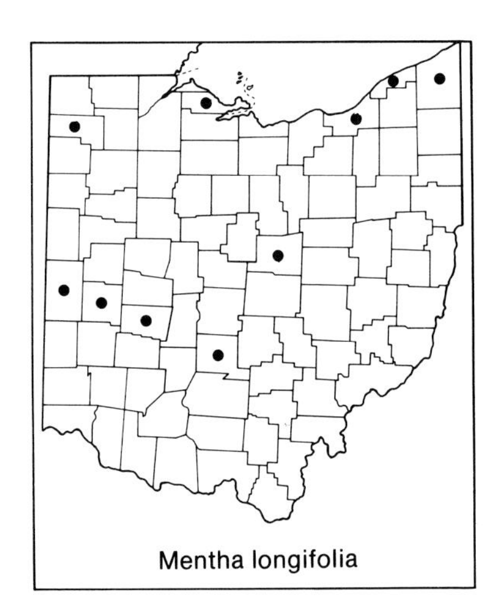
Mentha longifolia

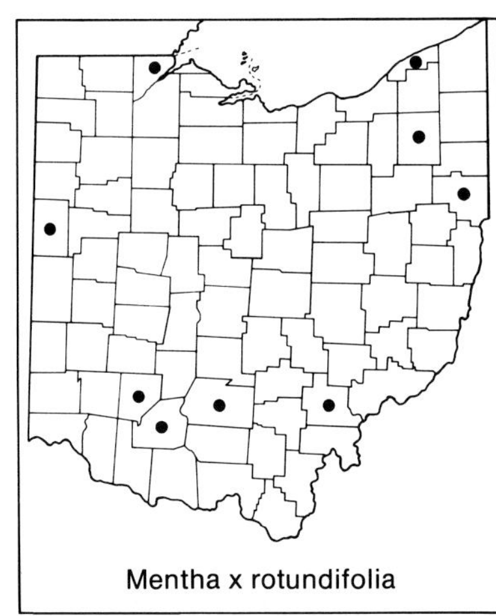
Mentha x rotundifolia

Mentha x villosa

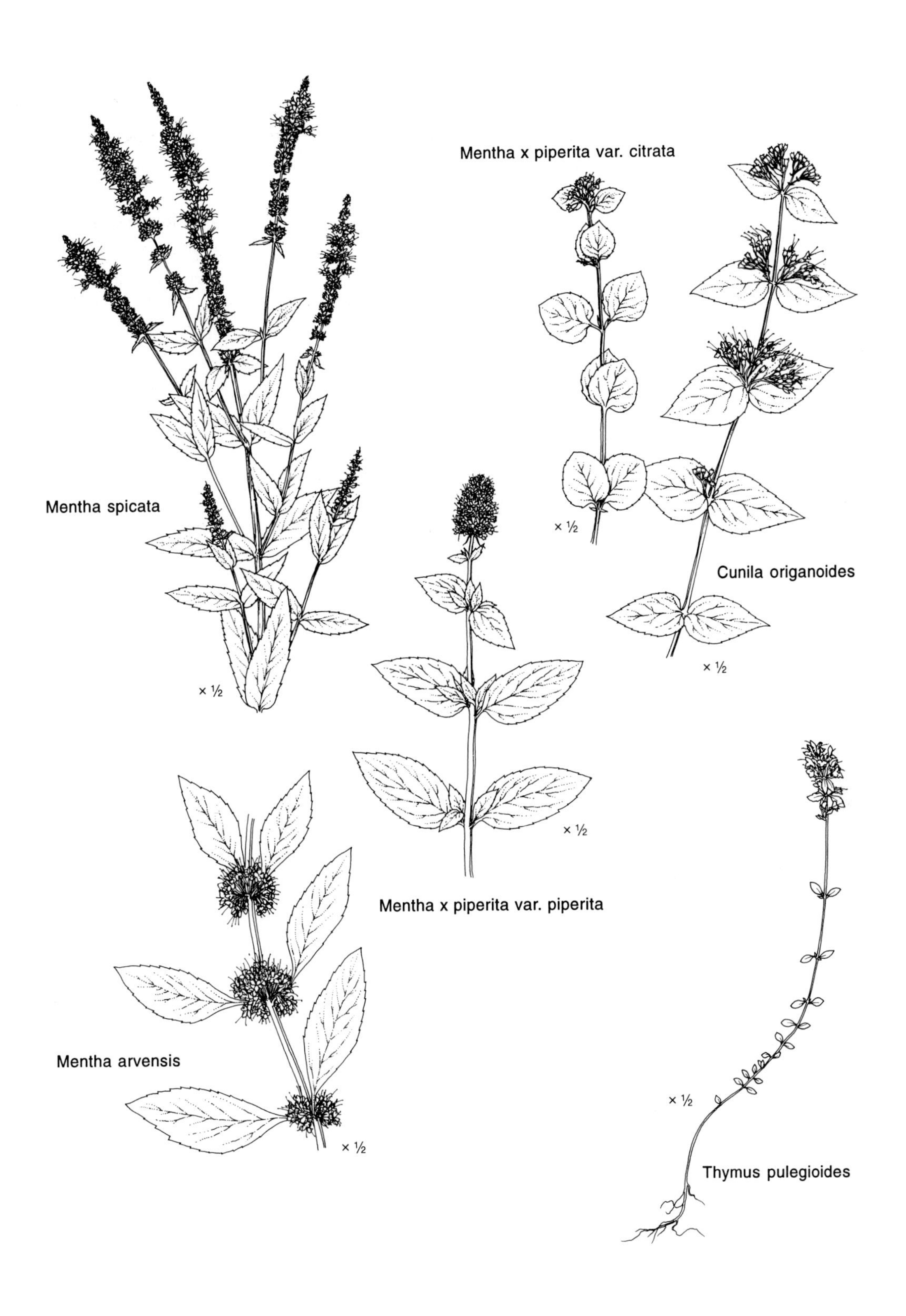
Mentha x piperita var. citrata
Mentha spicata
× ½
× ½
Cunila origanoides
× ½
× ½
Mentha x piperita var. piperita
Mentha arvensis
× ½
× ½
Thymus pulegioides

of *Mentha spicata* × M. *suaveolens* Ehrh., adventive or perhaps naturalized at a few Ohio sites; disturbed weedy ground. Not illustrated. July–October.

Ohio plants are assigned to var. *alopecuroides* (Hull) R. Harley (M. *alopecuroides* Hull).

4. Mentha spicata L. Spearmint

Perennial, to 1 m tall; stems glabrous or nearly so; leaves sessile, glabrous on undersurface or with few hairs along the veins; flowers in elongated interrupted spikes, terminating main stem and sometimes originating from upper axils; corolla pale purple to pinkish or whitish, 4–5 mm long. A frequently cultivated Eurasian species, naturalized throughout all of Ohio except some northwestern counties; weedy roadside banks, ditches, thickets, fields. July–September.

5. Mentha × piperita L.

Perennial, to 1 m tall; stems glabrous or nearly so; leaves with petioles 5 mm or longer; flowers all or mostly in a terminal head or spike; corolla pinkish to rose-purple, 4–5 mm long. Two varieties of this Eurasian hybrid, *Mentha aquatica* L. × M. *spicata*, have been introduced and become naturalized in the Ohio flora; var. *piperita* is found more or less throughout the state, var. *citrata* in only a few scattered sites; both are established in moist disturbed habitats. July–September.

a. Flowers in an elongated spike, the spike usually composed of 4 or more verticils; leaves oblong-lanceolate .... var. *piperita* Peppermint

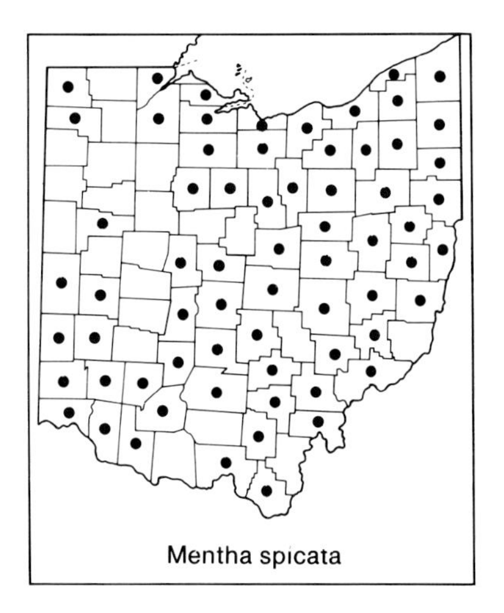
Mentha spicata

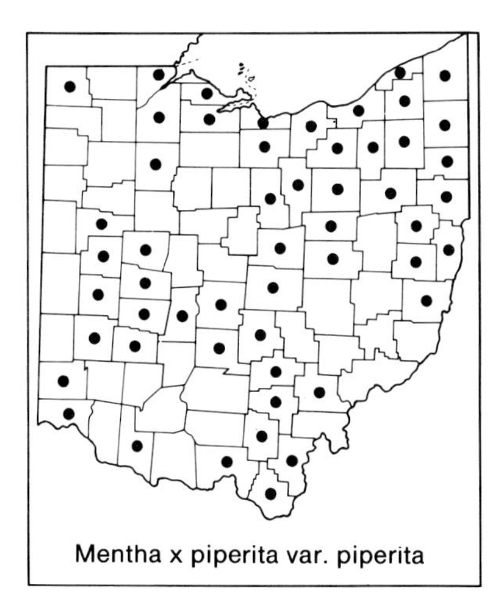
Mentha x piperita var. piperita

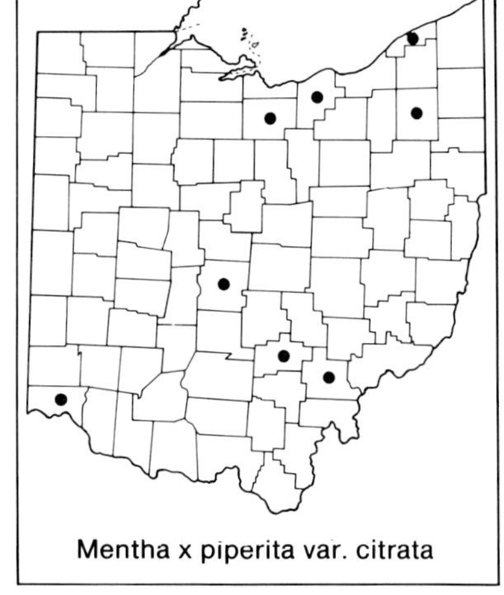
Mentha x piperita var. citrata

aa. Flowers in a globose to ovoid head composed of 1–3 compact verticils, or the lowest verticil sometimes slightly separated; leaves ovate to broadly ovate . . . . var. *citrata* (Ehrh.) Briq.
M. *citrata* Ehrh.
BERGAMOT MINT

6. MENTHA × GRACILIS Sole RED MINT
M. *cardiaca* (S. F. Gray) Baker, M. *gentilis* of authors, not L.

Perennial, to 1 m tall; stems glabrous or nearly so; leaves with petioles 5 mm or more in length; flowers in axillary clusters subtended by reduced leaves; calyx glabrous below, pubescent toward apex; corolla pink to purplish. A European taxon, the hybrid of M. *arvensis* × M. *spicata*, adventive or perhaps naturalized at several Ohio sites, mostly in western counties. Not illustrated. July–September.

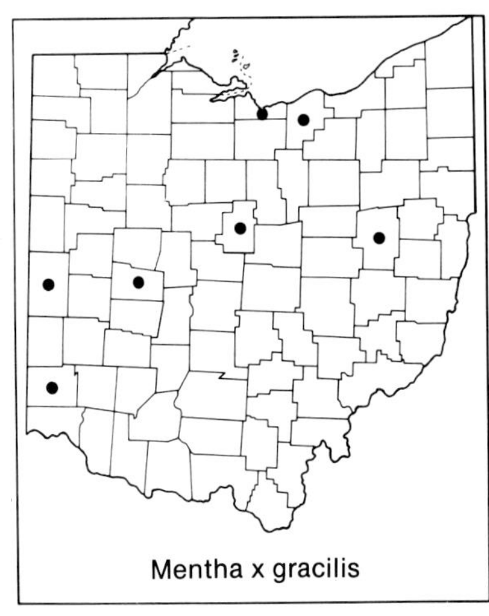
Mentha x gracilis

7. **Mentha arvensis** L. FIELD MINT
incl. M. *gentilis* L. (see A. O. Tucker, Harley, and Fairbrothers, 1980)

Perennial, to 1 m tall; stems pubescent; leaves with petioles 5 mm or more in length; flowers in axillary clusters subtended by full-sized leaves, or the uppermost leaves sometimes slightly reduced; corolla white to lavender or pinkish-purple. Throughout most of Ohio; shores, marshes, bogs, moist woods, disturbed moist habitats. Gill, Lawrence, and Morton (1973) studied variation in North American populations of this species. July–October.

Ohio plants are assignable to M. *arvensis* var. *glabrata* (Benth.) Fern. (incl. var. *villosa* (Benth.) S. F. Stewart). *Mentha arvensis* var. *arvensis* was attributed to Ohio by Schaffner (1932) on the basis of a specimen probably collected from cultivation.

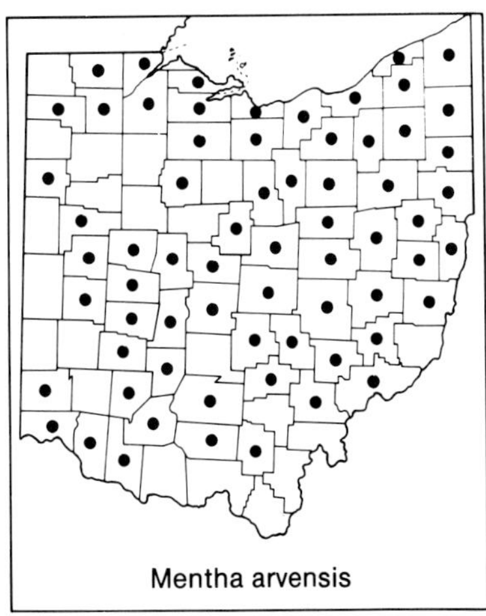
Mentha arvensis

31. Collinsonia L. HORSE-BALM

Erect perennial herbs; flowers in terminal panicles or racemes; calyx actinomorphic or slightly zygomorphic; corolla bilabiate; stamens 2 or 4, exserted. An eastern North American genus of five species. Shinners (1962b) published a synopsis of the genus.

a. Leaves opposite, several pairs scattered along the stem, not closely approximate; fertile stamens 2; corolla yellow . . . . 1. *Collinsonia canadensis*

aa. Leaves appearing whorled, 2 or 3 pairs closely approximate on the stem; fertile stamens 4; corolla dull yellowish-pink . . . . 2. *Collinsonia verticillata*

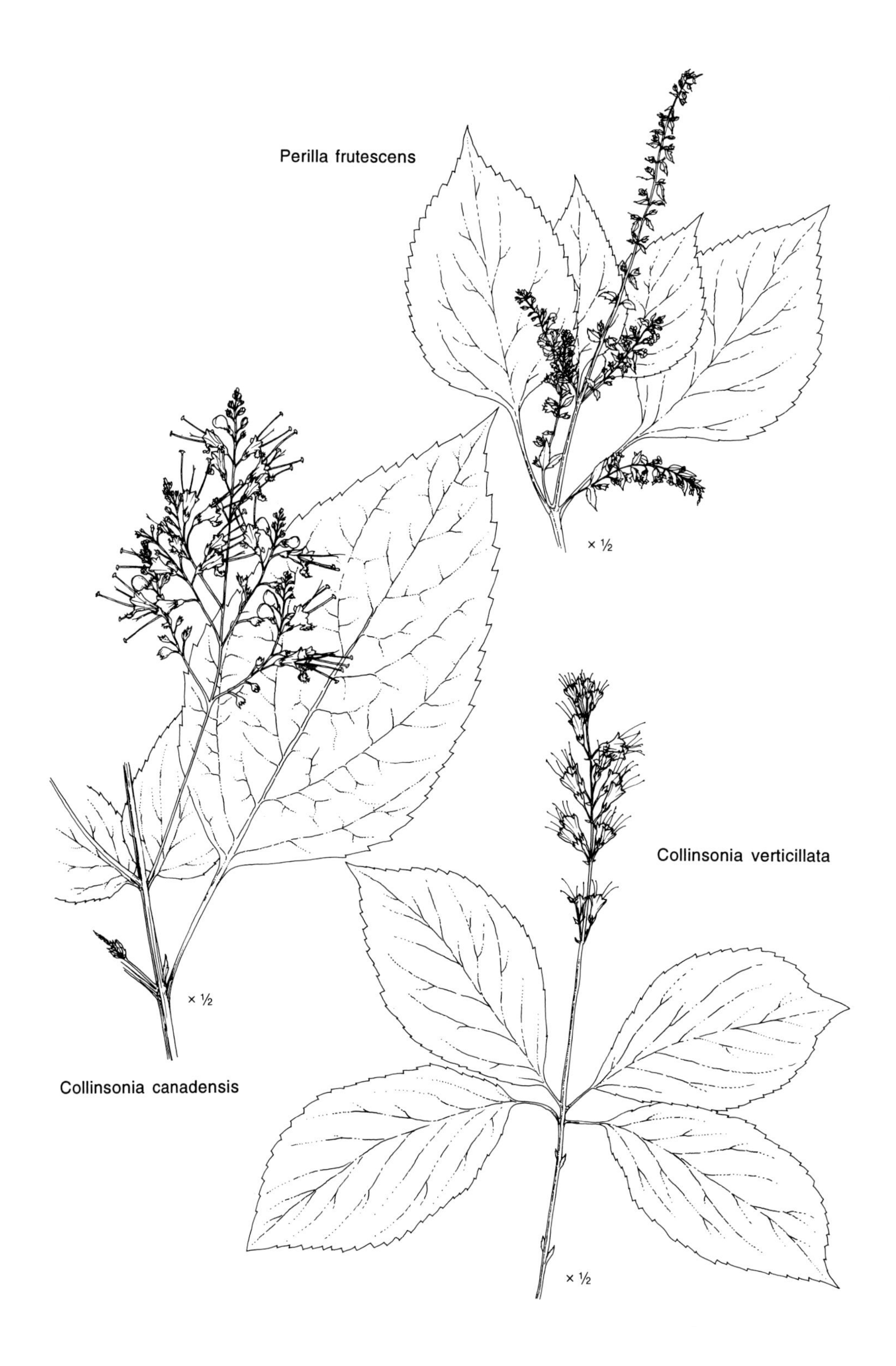

Perilla frutescens
× ½
× ½
Collinsonia canadensis
Collinsonia verticillata
× ½

1. **Collinsonia canadensis** L. RICHWEED. STONEROOT

Perennial, to 1 m tall; stems glabrous or nearly so; leaves petiolate, opposite, in several pairs scattered along the stem; corolla yellow, 10–14 mm long; fertile stamens 2. Throughout Ohio; woodlands. Mid-July–September.

2. **Collinsonia verticillata** Ell. EARLY STONEROOT

   Micheliella verticillata (Ell.) Briq.

Perennial, to 8 dm tall; stems slightly pubescent; leaves petiolate, appearing whorled, the two or three pairs closely approximate; corolla dull yellowish-pink, 12–18 mm long. Known only from Scioto County, in southernmost Ohio, where it is disjunct and at a northern limit of the species' range (Braun, 1964); woodlands. May–June.

32. Perilla L.

A small genus of southeastern Asia with six species, one of them occurring in Ohio.

1. PERILLA FRUTESCENS (L.) Britt. BEEFSTEAK-PLANT

   incl. var. *frutescens* and var. *crispa* (Benth.) W. Deane

Annual, to 1 m tall; stems erect, pubescent to nearly glabrous; stems and leaves sometimes purple; leaves petiolate; flowers in terminal and axillary panicles and racemes; calyx zygomorphic; corolla bilabiate, white, 5–6 mm long. An Asian plant, sometimes cultivated and escaped, adventive or perhaps locally naturalized, mostly in eastern and southern counties; disturbed habitats. September–October.

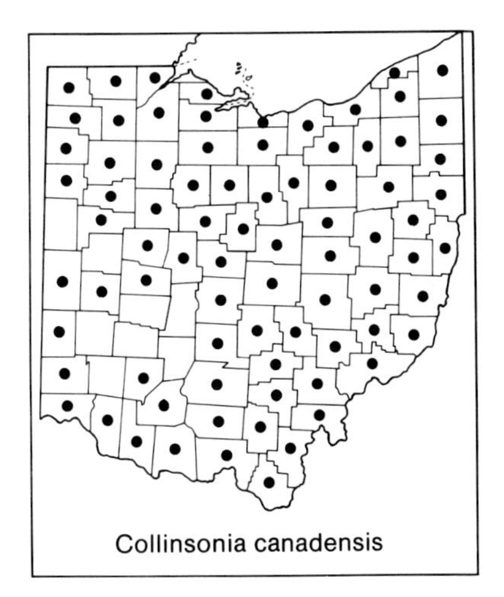

Collinsonia canadensis

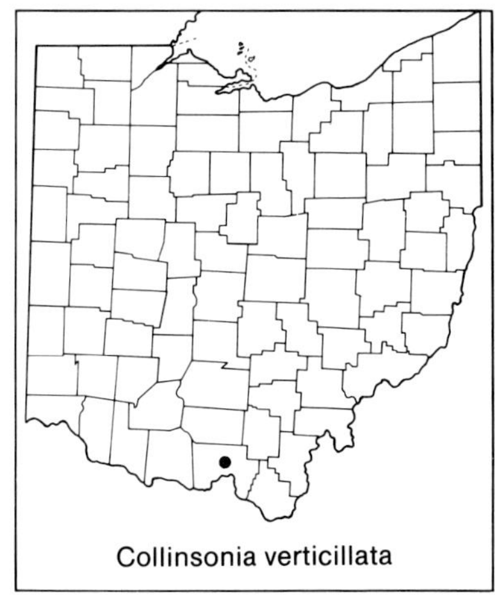

Collinsonia verticillata

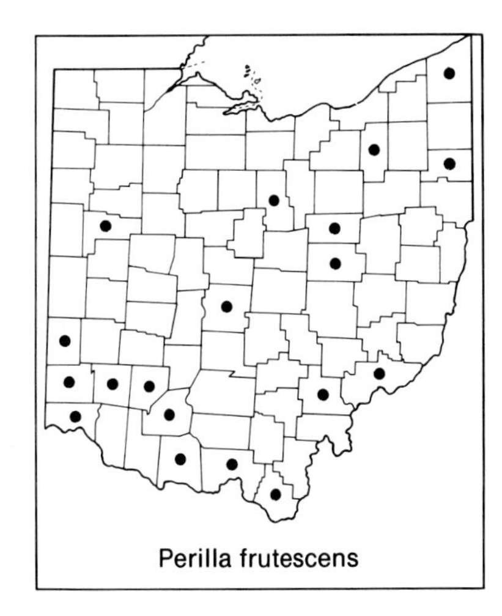

Perilla frutescens

Solanaceae. NIGHTSHADE FAMILY

Annual or perennial herbs, or in a few genera shrubs or woody vines; leaves petiolate, alternate or occasionally opposite, simple or compound; flowers perfect, solitary or in various types of inflorescences, usually cymose; calyx 5-lobed; corolla 5-lobed, actinomorphic or nearly so; stamens 5, attached to the corolla tube; carpels 2, united to form a compound pistil with a single style; ovary superior; fruit a berry. A chiefly tropical family with more than 2,500 species. The treatment below is based in part on Jacobs and Eshbaugh's (1983) study of the family as it is represented in the Ohio flora.

Several members of the family are cultivated in Ohio, the best known being petunia, potato, tobacco, and tomato, all discussed in the text below. Other cultivated species include *Capsicum annuum* L., Bell Pepper or Sweet Pepper, *Capsicum frutescens* L., Tabasco Pepper, and the flowering ornamental, *Schizanthus pinnatus* Ruiz & Pavón, Poor-man's Orchid. Specimens of *Browallia americana* L., Bush-violet, an ornamental from tropical America, with blue to purple flowers, have been collected in Belmont and Highland counties from adventive escapes or perhaps from individuals persisting from cultivation.

a. Shrubs or partially woody vines, the stems often arching, scrambling, or climbing.
  b. Leaves all or mostly with 1–4 basal lobes or small leaflets, a few leaves sometimes unlobed; corolla rotate, purple or purple-blue .... 1. *Solanum*
  bb. Leaves all unlobed and simple; corolla more or less funnelform, dull purple, dull violet, or dull greenish-pink .... 5. *Lycium*
aa. Herbaceous plants, the stems erect or ascending, not climbing.
  b. Corolla 0.3–3 cm long, rotate to campanulate; fruit a berry.
    c. Leaves once- or-twice pinnately compound, the stems clammy-pubescent; stems without spines .... 2. *Lycopersicon*
    cc. Leaves simple and unlobed, or if deeply pinnately lobed or pinnately compound the stems not clammy-pubescent; stems with or without spines.
      d. Corolla rotate; calyx not inflated and bladder-like at maturity .... 1. *Solanum*
      dd. Corolla rotate to campanulate; calyx inflated and bladder-like at maturity.
        e. Corolla white to yellow or yellowish-green; calyx lobed half its length or less .... 3. *Physalis*
        ee. Corolla blue; calyx lobed nearly to the base....... 4. *Nicandra*

bb. Corolla (1–) 4–20 cm long, funnelform or salverform; fruit a capsule.

c. Calyx less than 1 cm long; corolla 1–2 cm long; see above .... *Browallia*

cc. Calyx 1–10 cm long; corolla 4–20 cm long.

d. Calyx 3–10 cm long; fruit bearing many spines ...... 6. *Datura*

dd. Calyx 1–2.5 cm long; fruit without spines.

e. Leaves more than 10 cm long; calyx lobed less than half its length .... 7. *Nicotiana*

ee. Leaves less than 10 cm long; calyx lobed nearly to the base .... 8. *Petunia*

1. Solanum L. NIGHTSHADE

Annuals or perennials; leaves alternate, simple or compound; flowers in cymose or racemose clusters; corollas white or purple. A large genus of 1,700 or more species, cosmopolitan in distribution, but especially numerous in the New World tropics. Several species have been collected or reported as adventives in Ohio. The most frequent, *Solanum tuberosum*, POTATO, is discussed below. *Solanum elaeagnifolium* Cav., SILVER-LEAVED NIGHTSHADE, native to southwestern United States and found occasionally as a weed in states to the east of Ohio, was reported for Ohio by Schaffner (1932) and Fernald (1950), but I have seen no specimens. Similarly, *Solanum pseudocapsicum* L., JERUSALEM-CHERRY, cultivated as an ornamental for its brightly colored (but poisonous) fruits, was listed for Ohio by C. H. Jones (1942); I have seen no specimens other than ones collected from cultivation. The South American plant *Solanum physalifolium* Rusby (*S. sarrachoides* of authors, not Sendtner), HAIRY NIGHTSHADE, was collected as a garden weed in Trumbull County, in 1951. *Solanum sisymbriifolium* Lam., STICKY NIGHTSHADE, a tall prickly weed from the American tropics, was attributed to Ohio by Fernald (1950), but I have seen no specimens; Gleason (1952b) reports it only from southern and Atlantic coastal states. *Solanum triflorum* Nutt., CUT-LEAVED NIGHTSHADE, a weed from western United States, was collected once in a Belmont County pigeon yard in 1918.

a. Undersurface of leaves with simple hairs or glabrous; stems and leaves without prickles or spines.

b. Stems somewhat woody at base, plants often scrambling or climbing; corolla purple to purple-blue or rarely white; fruit red at maturity .... 1. *Solanum dulcamara*

bb. Stems herbaceous, plants erect or ascending; corolla white or whitish; fruit black, yellow, or greenish at maturity.
c. Leaves deeply pinnately lobed or pinnately compound.
d. Leaves pinnately compound; corolla 20–25 mm wide . . . . 2. *Solanum tuberosum*
dd. Leaves simple, pinnately lobed; corolla 8–9 mm wide; see above . . . . *Solanum triflorum*
cc. Leaves simple, unlobed but sometimes with a few large teeth.
d. Stems with a few appressed or incurved hairs; calyx subtending fruit, 1–3 mm long; fruit purplish to blackish. . . . . 3. *Solanum nigrum*
dd. Stems, especially above, with numerous spreading hairs; calyx enlarging and nearly or quite surrounding fruit, becoming 3–6 mm long; fruit brownish-green to yellowish; see above . . . . *Solanum physalifolum*
aa. Undersurface of leaves with stellate hairs and sometimes with spines; stems often with prickles or spines.
b. Leaves entire to toothed or shallowly pinnately lobed, the sinuses less than half the distance to midrib; calyx-tube stellate-pubescent, bearing no or only a few stubby prickles.
c. Principal leaves green, ovate to elliptic-ovate, 3–6 cm wide, margins mostly with a few large broad teeth or shallow lobes, undersurface usually with a few large spines along midrib . . . . 4. *Solanum carolinense*
cc. Principal leaves silvery-pubescent, linear-lanceolate to oblong-lanceolate, 0.5–2 (–3) cm wide, margins entire or nearly so, undersurface usually without spines; see above . . . . *Solanum elaeagnifolium*
bb. Leaves deeply pinnately lobed, the sinuses more than half the distance to midrib; calyx-tube stellate-pubescent and bearing also several to many sharp spines.
c. Corolla blue to white; inflorescence with both glandular and stellate pubescence; see above. . . . . *Solanum sisymbriifolium*
cc. Corolla yellow; inflorescence with stellate pubescence only . . . . 5. *Solanum rostratum*

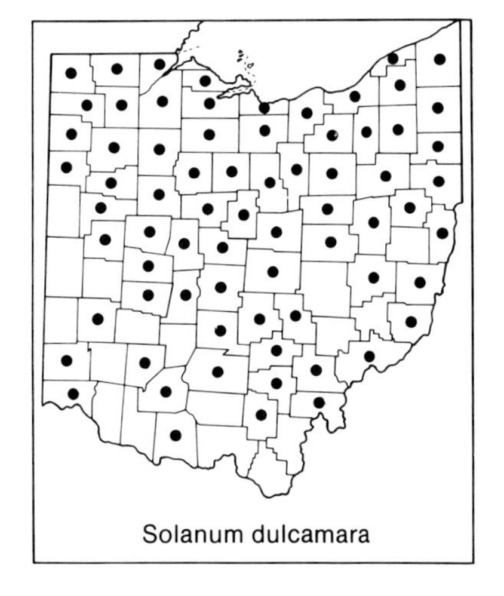
Solanum dulcamara

1. Solanum dulcamara L. Bittersweet Nightshade. Deadly Nightshade. Bittersweet
incl. var. *dulcamara* and var. *villosissimum* Desv.

Scrambling or climbing perennial; stems somewhat woody at base, to ca. 2 m in length; leaves usually with two (or sometimes four or more) deep opposite basal lobes or sometimes leaflets, or the leaves occasionally unlobed, pubescent to glabrous on undersurface; inflorescence a several-flowered cyme; corolla purple to purple-blue (rarely white), 10–14 mm across; the connivent anthers forming a bright yellow cone-shaped beak in the center of the flower; berries bright red, juicy. A poisonous Eurasian species naturalized throughout Ohio; weed lots, thickets, fencerows, wet ground along marshes and streams, and other disturbed habitats. All parts of the plant are poisonous, and the death of at least one Ohio child has been attributed to ingestion of the attractive red berries. Mid-May–early September.

Plants of the purple-flowered forma *dulcamara* are common; plants of the rare white-flowered forma *albiflorum* House have been collected from Ashland, Erie, Geauga, Ottawa, and Stark counties.

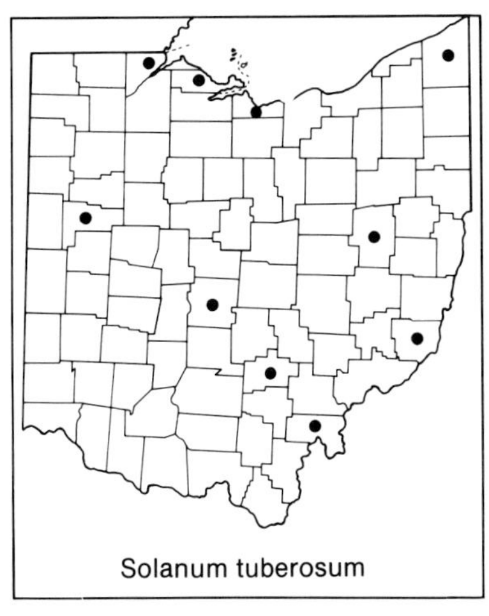
Solanum tuberosum

2. Solanum tuberosum L. Potato. Irish Potato

Annual; stems ascending to weakly erect, to 1 m long; leaves pinnately compound, more or less pubescent on undersurface; inflorescence a few-flowered cyme; corolla white to whitish and tinged with purple or blue, 20–25 mm across; berries yellowish to greenish, produced infrequently. A South American species cultivated throughout Ohio; adventive individuals probably originating from discarded plant material have been collected from scattered sites around the state. Green parts of the plant and the whitish sprouts are poisonous. Not illustrated. June–July.

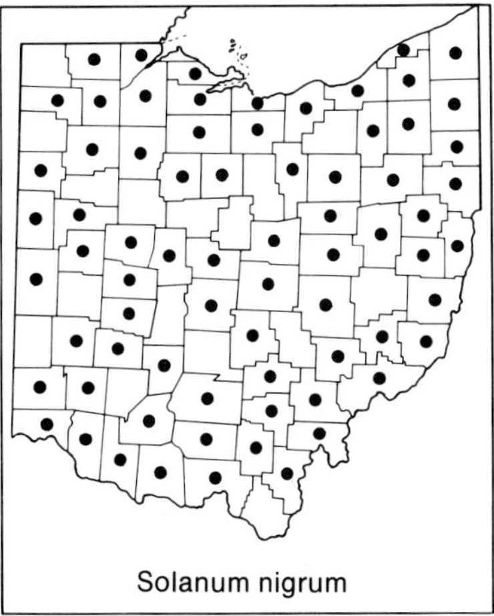
Solanum nigrum

3. **Solanum nigrum** L. Black Nightshade. Common Nightshade
incl. *S. americanum* Mill. and *S. ptychanthum* DC.

Annual, to ca. 7 dm tall; stems erect; leaves entire to coarsely and shallowly toothed, slightly pubescent to glabrous on undersurface; inflorescence a few–several-flowered cyme; corolla white to whitish with a lavender tinge, 6–10 mm across; berries purplish to blackish. Throughout Ohio; stream banks, thickets, roadsides, and other open disturbed habitats. The two taxa listed in synonymy above are often treated as segregate species of this complex (Bassett and Munro, 1985; Schilling, 1981). All plants of the complex are poisonous. July–early October.

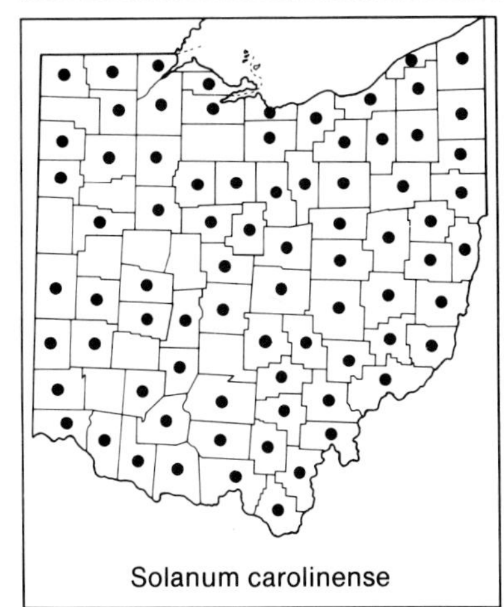
Solanum carolinense

4. **Solanum carolinense** L. Horse-nettle

Perennial, to 1 m tall; stems erect, pubescent and with scattered gold-colored prickles or spines; leaves mostly with a few large broad teeth or shallow lobes, the undersurface stellate-pubescent and usually with a few large spines along the midrib; inflorescence few-flowered and racemose; corolla violet to whitish, 2.5–3.0 cm across; berries yellow. Probably native in part and in part naturalized, the species grows throughout Ohio; stream banks, along roads and railroads, cultivated fields, barnyards, and other disturbed habitats. The berries are poisonous. Bassett and Munro (1986) studied the biology of this weed. Late May–mid-September.

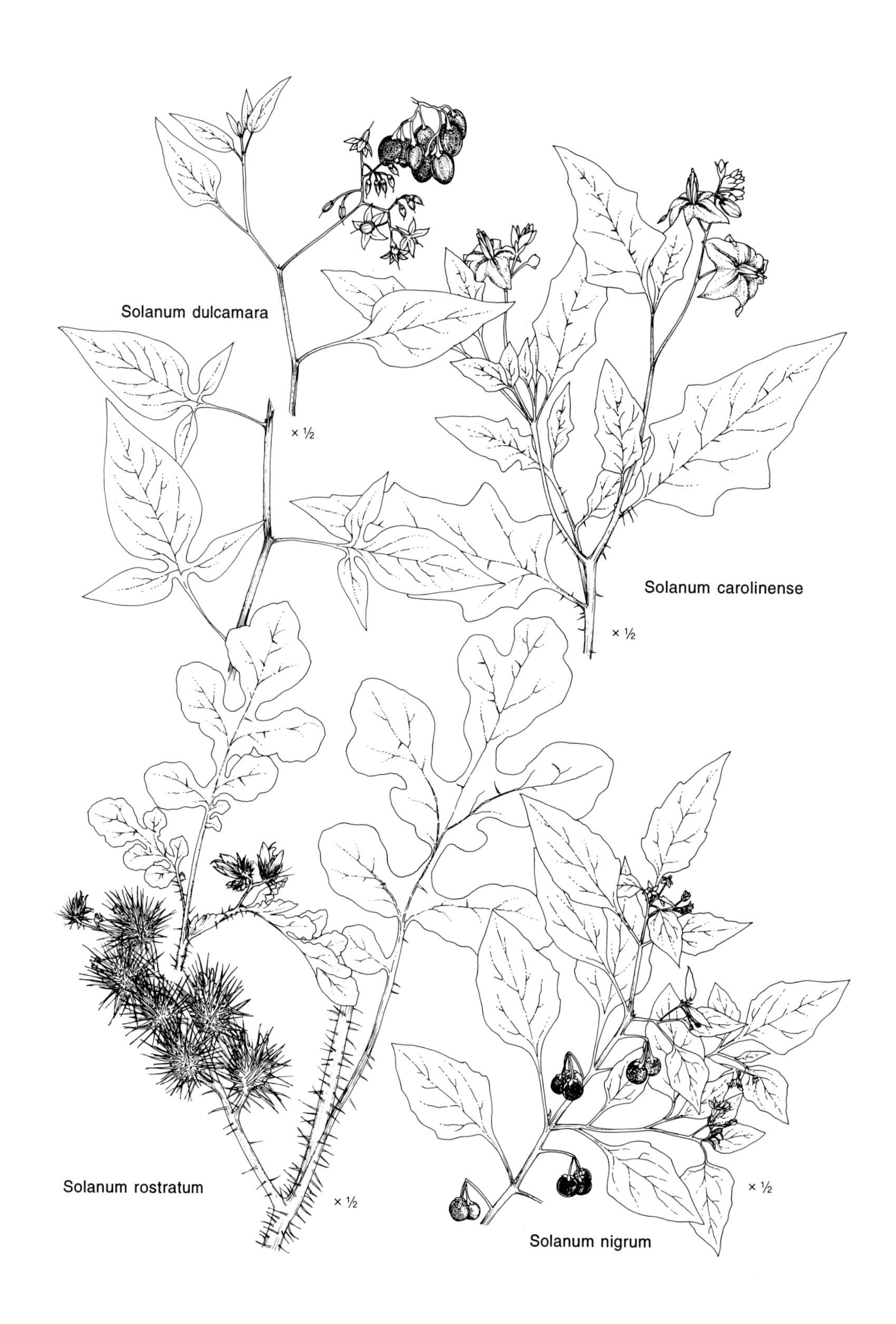
Solanum dulcamara
× ½
Solanum carolinense
× ½
Solanum rostratum
× ½
Solanum nigrum
× ½

5. SOLANUM ROSTRATUM Dunal   BUFFALO-BUR

?*S. cornutum* Lam.

Annual, to 6 dm tall; stems erect, stellate-pubescent and with numerous straw-colored prickles or spines; leaves deeply pinnately lobed, the undersurface stellate-pubescent and with a few large spines along the main veins; inflorescence few-flowered and racemose; corolla yellow, 1.5–2.5 cm across; berries enclosed by the spine-covered calyx. An American species native to the west of Ohio; naturalized at scattered sites in the state; fields, cultivated ground, and other disturbed habitats. The plant is poisonous but seldom ingested because of its numerous spines. Bowers (1975) studied the pollination ecology of this species; Bassett and Munro (1986) made a general study of its biology. August–September.

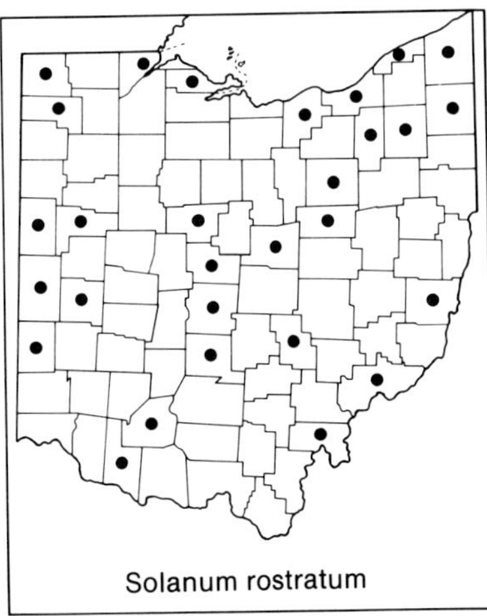
Solanum rostratum

2. Lycopersicon Mill.   TOMATO

A South American genus of about seven species, the fruit of one of them a familiar garden vegetable in Ohio. The generic name is sometimes misspelled *Lycopersicum*.

1. LYCOPERSICON ESCULENTUM Mill.   TOMATO

L. lycopersicum (L.) Karst.

Annual, to ca. 2 m tall; stems ascending, clammy-pubescent; leaves once- or twice-pinnately compound; inflorescence a few-flowered cyme; corolla yellow, 1–2 cm across; fruit red or sometimes yellow, juicy. A South American species widely cultivated throughout Ohio, and adventive at scattered sites around the state; dumps, weed lots, and other disturbed habitats. Not illustrated. July–October.

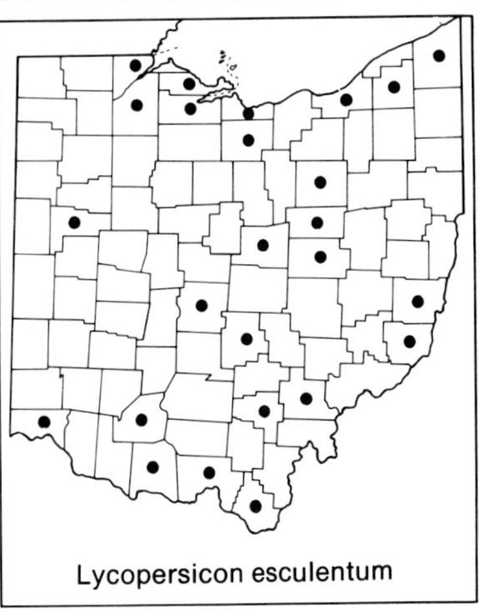
Lycopersicon esculentum

3. Physalis L.   GROUND-CHERRY

Annuals or perennials; leaves alternate to subopposite; flowers axillary, solitary or in few-flowered clusters; calyx lobed half its length or less; corolla white to yellow or greenish-yellow; berries enclosed in a persistent, inflated, bladderlike calyx. A genus of 80 or more species of the New World tropics. Menzel (1951) made a study of the cytotaxonomy and genetics of *Physalis*, and Waterfall (1958) a taxonomic study of the genus in North America, north of Mexico.

In addition to those covered in the text below, two other species have been collected as adventives in Ohio. *Physalis philadelphica* Lam. (*P. ixocarpa* of authors, not Hornem.), TOMATILLO, cultivated for its edible fruits in Mexico and southwestern United States, and occasionally in Ohio, has been collected from Auglaize, Belmont, and Franklin counties. *Physalis hispida* (Waterfall) Cronq. (*P. virginiana* var. *hispida* Waterfall; and incl. *P.*

lanceolata of authors, not Michx.), PRAIRIE GROUND-CHERRY, was collected as a railroad track weed in Stark County, in 1960.

The keys below are based in part on those of Gleason and Cronquist (1963).

a. Calyx reddish-orange at maturity; corolla whitish, 5-lobed .... 5. *Physalis alkekengi*

aa. Calyx green or greenish at maturity; corolla yellow or yellowish-green, not to very shallowly lobed.

b. Pedicels 3–6 mm long at anthesis, not or only slightly longer in fruit; fruit purplish, filling or nearly filling calyx; see above .... *Physalis philadelphica*

bb. Pedicels 5–20 mm long at anthesis, longer in fruit; fruit purple, red, orange, or yellow, filling or not filling calyx.

c. Fruiting pedicels mostly 1 cm or less in length; corolla 5–10 mm wide; filaments slender .... 1. *Physalis pubescens*

cc. Fruiting pedicels 1–3 cm long; corolla 12–20 mm wide; filaments broad and flat.

d. Upper internodes with numerous soft spreading, often glandular, hairs .... 4. *Physalis heterophylla*

dd. Upper internodes glabrous or with stiff nonglandular hairs.

e. Calyx tube pubescent on surface as well as on veins, the hairs ascending or spreading; leaves with hairs on surface as well as on veins; upper internodes and pedicels with recurved hairs .... 3. *Physalis virginiana*

ee. Calyx tube pubescent only along the veins, the hairs appressed, ascending, or spreading; leaves glabrous or with pubescence only along the veins.

f. Upper internodes, pedicels, and calyx glabrous or with hairs appressed or ascending; leaves ovate to rarely lanceolate .... 2. *Physalis longifolia*

ff. Upper internodes, pedicels, and calyx with spreading hairs; leaves elliptic to oblanceolate; see above .... *Physalis hispida*

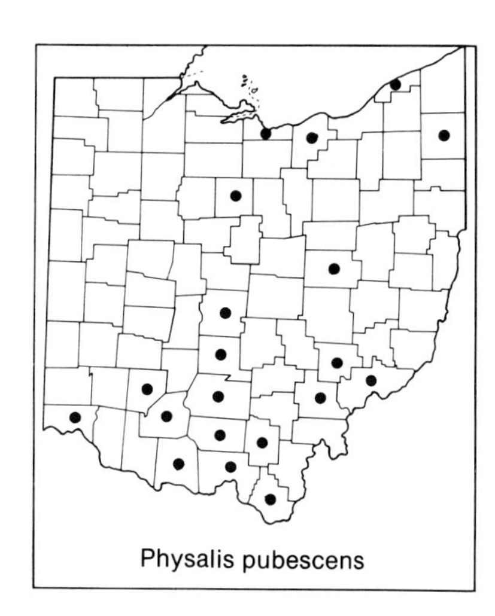
Physalis pubescens

1. **Physalis pubescens** L. DOWNY GROUND-CHERRY

Annual, to 5 dm tall; stems erect, pubescent; undersurface of leaves pubescent, at least on the veins; pedicels pubescent, in fruit mostly 1 cm or less in length; calyx tube pubescent on surface as well as on veins; corolla yellow with dark blotches at center, 5–10 mm across; filaments slender. Probably native in southern counties—where the species is at a northern boundary of its range, and presumably naturalized or adventive at scattered sites into northeastern Ohio; margins of streams and lakes, woodland openings, disturbed habitats. August–October.

Ohio plants can, if desired, be separated into the three varieties keyed below, but the varieties are of questionable significance.

a. Leaves entire or nearly so, with 4 or fewer teeth on each side of blade . . . . var. *integrifolia* (Dunal) Waterfall

aa. Leaves more evidently dentate, with 5 or more teeth on each side of blade.

b. Upper internodes, petioles, and undersurface of leaves pubescent but green . . . . var. *pubescens*

bb. Upper internodes, petioles, and undersurface of leaves gray-pubescent. . . . . var. *grisea* Waterfall
P. pruinosa of authors, not L.

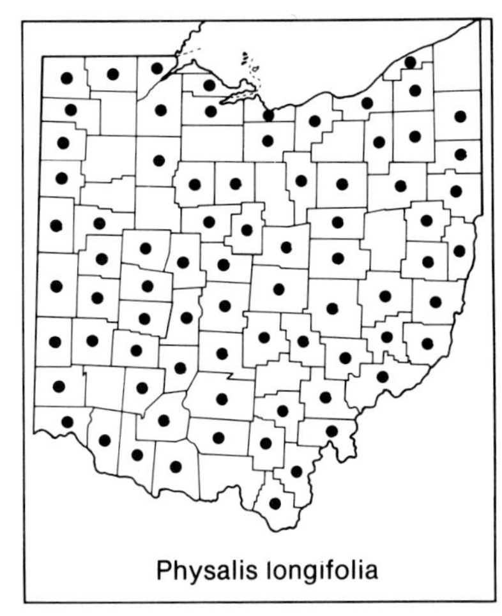
Physalis longifolia

Physalis virginiana

2. **Physalis longifolia** Nutt.
Perennial, to ca. 1 m tall; stems erect, upper internodes glabrous or with hairs appressed and ascending; undersurface of leaves glabrous or with hairs along the veins; pedicels with hairs appressed and ascending, 1.5–2.0 cm long in fruit; calyx tube pubescent only along the veins, the hairs appressed, ascending, or spreading; corolla yellowish with purple-brown blotches at center, 12–17 mm across; filaments broad and flat. Throughout Ohio, but more common westward; stream banks, shores of lakes and ponds, fields, along roads and railroads, and in other disturbed habitats. Mid-June–September.

Most Ohio plants can be separated into the two varieties keyed below, but the leaf shape is quite variable and there are intergrades between the two. Variety *longifolia* is an alien plant, naturalized from areas to the west of Ohio; var. *subglabrata* is native.

a. Leaves lanceolate to linear, more or less entire . . . . var. *longifolia*
Long-leaved Ground-cherry

aa. Leaves ovate to broadly lanceolate, shallowly toothed (illustrated) . . . . var. *subglabrata* (Mackenz. & Bush) Cronq.
P. subglabrata Mackenz. & Bush
Smooth Ground-cherry

3. **Physalis virginiana** Mill. Virginia Ground-cherry
Perennial, to 5 dm tall; stems erect, upper internodes with stiff recurved hairs; undersurface of leaves with hairs on surface as well as on veins; pedicels pubescent, 2–3 cm long in fruit; calyx tube pubescent on surface as well as on veins, the hairs ascending or spreading; corolla yellow with purple-brown blotches in center, ca. 1.5 cm across; filaments broad and flat. A few counties in southern Ohio; fields, openings. Hinton (1975) reported hybridization between this species and the next, *P. heterophylla*. June–August.

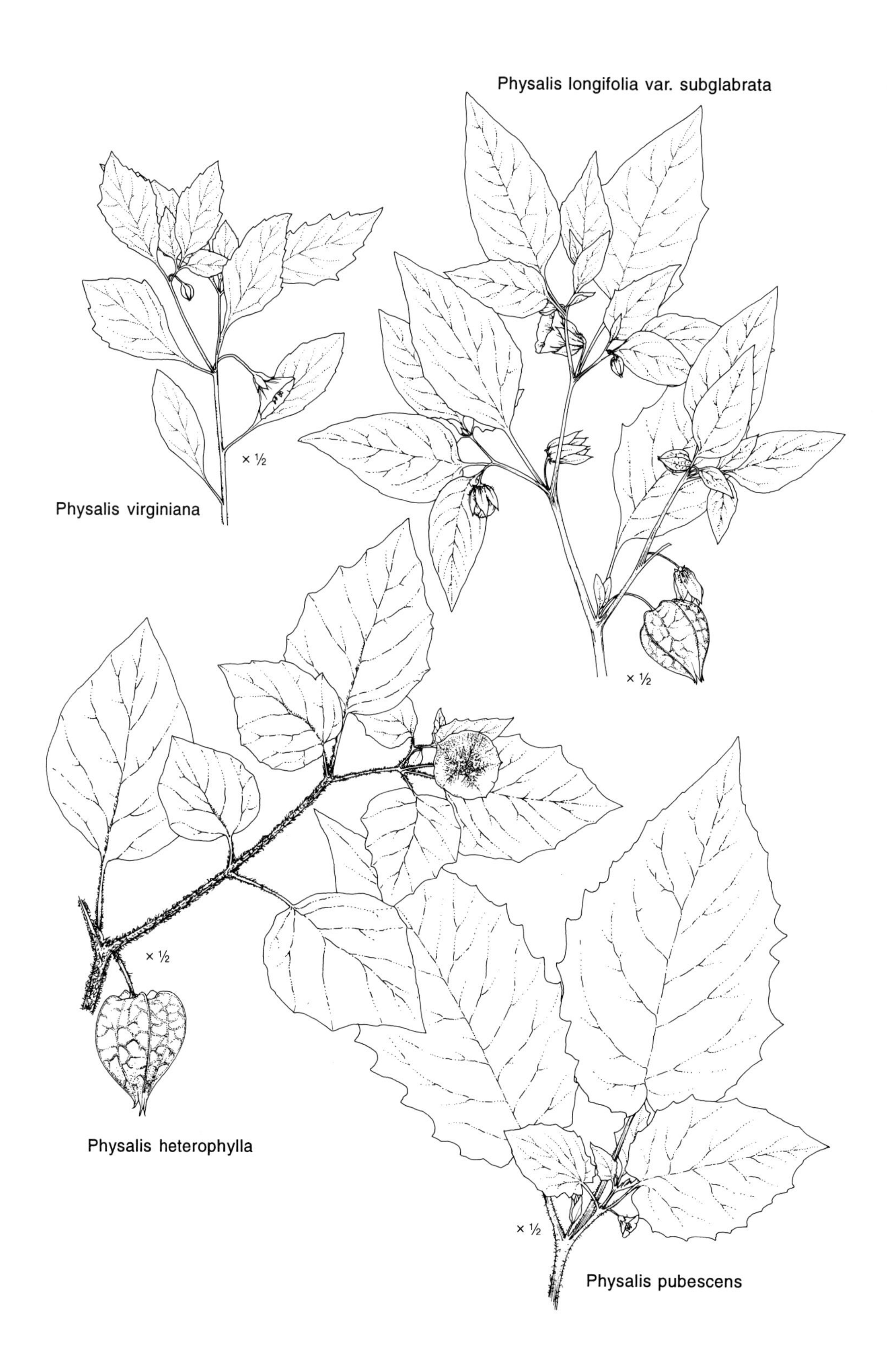
Physalis longifolia var. subglabrata
× ½
Physalis virginiana
× ½
× ½
Physalis heterophylla
× ½
Physalis pubescens

4. **Physalis heterophylla** Nees CLAMMY GROUND-CHERRY
incl. var. *heterophylla*, var. *ambigua* (A. Gray) Rydb., and var. *nyctaginea* (Dunal) Rydb.

Perennial, to 8 dm tall; stems erect, upper internodes with numerous soft, spreading, often glandular hairs; undersurface of leaves pubescent on surface as well as on veins; pedicels pubescent, 1–3 cm long in fruit; calyx tube pubescent on surface as well as on veins, the hairs spreading; corolla yellow with purple-brown blotches in center, 1–2 cm across; filaments broad and flat. Throughout Ohio; stream banks, woodland openings and borders, and especially along roads and railroads. Mid-June–September.

5. PHYSALIS ALKEKENGI L. CHINESE LANTERN

Perennial, to 6 dm tall; stems glabrous, more or less erect; mature calyx showy, inflated, papery, and bright orange; corolla whitish, 5-lobed, ca. 2 cm across. A frequently cultivated Eurasian species, escaped and perhaps locally established, mostly in northern counties; disturbed habitats. Not illustrated. June–August.

4. Nicandra Adans.

A genus with a single species.

1. NICANDRA PHYSALODES (L.) Gaertn. APPLE OF PERU

Annual, to ca. 1.5 m tall; stems erect, glabrous; leaves more or less ovate, margins irregularly dentate to shallowly lobed; flowers solitary in upper axils; calyx somewhat inflated at maturity, lobed nearly to the base, the lobes large, overlapping, and broadly ovate-cordate; corolla blue, campanulate, 2–3 cm across. A South American species sometimes cultivated in Ohio;

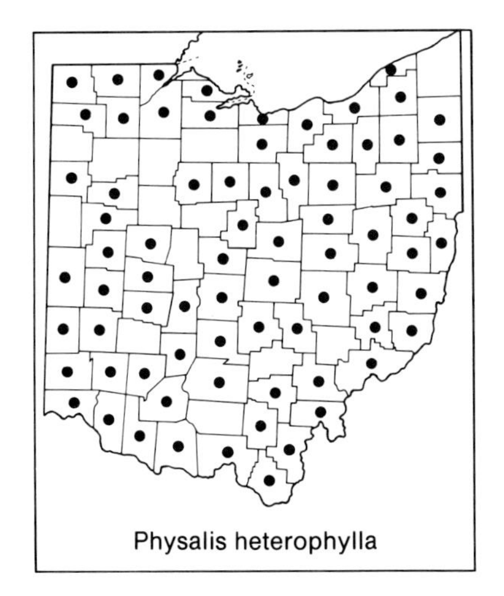

Physalis heterophylla

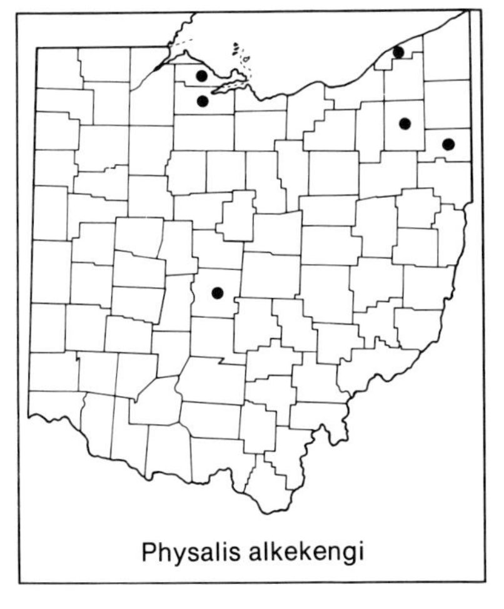

Physalis alkekengi

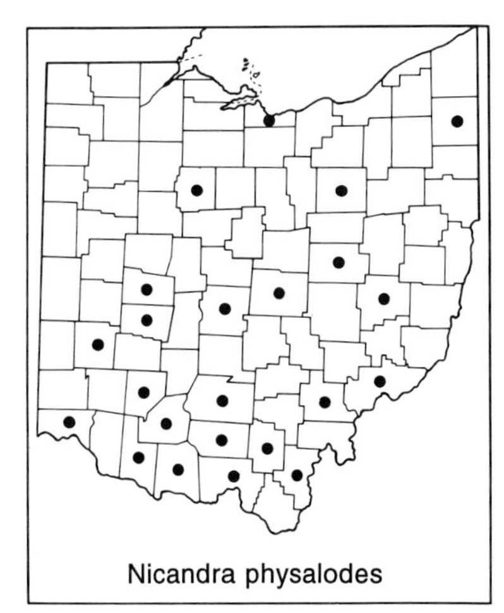

Nicandra physalodes

adventive or perhaps locally naturalized in southern counties and at scattered sites northward; cultivated fields, roadsides, waste places. July–September.

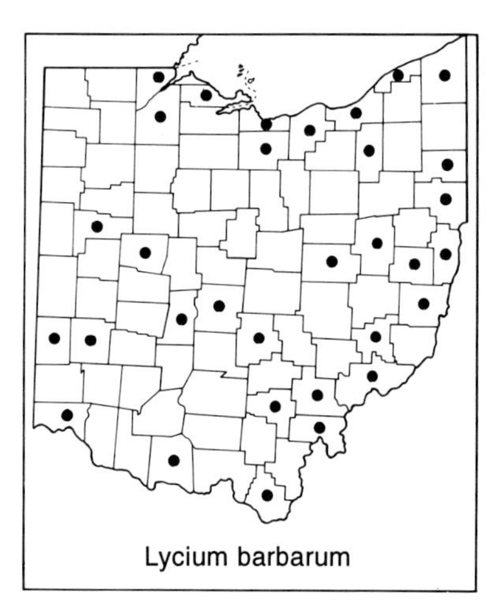

5. Lycium L. MATRIMONY-VINE

A genus widely distributed in tropical and temperate regions, with about 100 species, one of which is a naturalized member of the Ohio flora.

1. LYCIUM BARBARUM L. COMMON MATRIMONY-VINE
 incl. *L. chinense* Mill. and *L. halimifolium* Mill.

Scrambling shrub with arching branches; leaves lanceolate to ovate, margins entire or nearly so, petiolate to very short-petiolate; flowers solitary or few in leaf axils; corolla dull purple, dull violet, or dull greenish-pink, more or less funnelform, 8–12 mm across. A Eurasian species sometimes cultivated in Ohio; escaped and naturalized at scattered sites, mostly in eastern counties; river banks, weed lots, and other disturbed habitats. July–mid-October.

The merger of *L. chinese* and *L. halimifolium* under this species follows Gleason and Cronquist (1991).

6. Datura L. THORN-APPLE

A genus of about 10 species, chiefly of the New World tropics and subtropics, one of them a common Ohio weed. The plants are poisonous, but also a source of medicinal alkaloids. In addition to the species described below, *Datura inoxia* Mill. (incl. *D. meteloides* Dunal and *D. wrightii* Regel), DOWNY THORN-APPLE or ANGEL'S TRUMPET, a pubescent ornamental cultivated for its huge flowers—to 20 cm long and to 14 cm across, has been collected as an adventive from Cuyahoga, Franklin, Hamilton, Henry, Lake, and Ottawa counties. *Datura metel* L., HORN-OF-PLENTY, similar to *D. inoxia* but glabrous or becoming so, is also cultivated in Ohio, but I have seen no specimens of escapes. Schaffner's (1932) records of *D. metel* from Franklin and Lake counties were based on misidentified specimens of *D. inoxia*. *Datura inoxia* is native to southwestern United States and Mexico, *D. metel* to southern Asia.

a. Upper internodes, petioles, and undersurface of leaves glabrous or nearly so; corolla 6–10 cm long; calyx strongly 5-angled; capsules erect; leaves coarsely toothed to shallowly lobed .... 1. *Datura stramonium*

aa. Upper internodes, petioles, and undersurface of leaves short-pubescent; corolla 11–19 cm long; calyx terete or weakly 5-angled; capsules nodding; leaves entire or with few irregular teeth; see above .... *Datura inoxia*

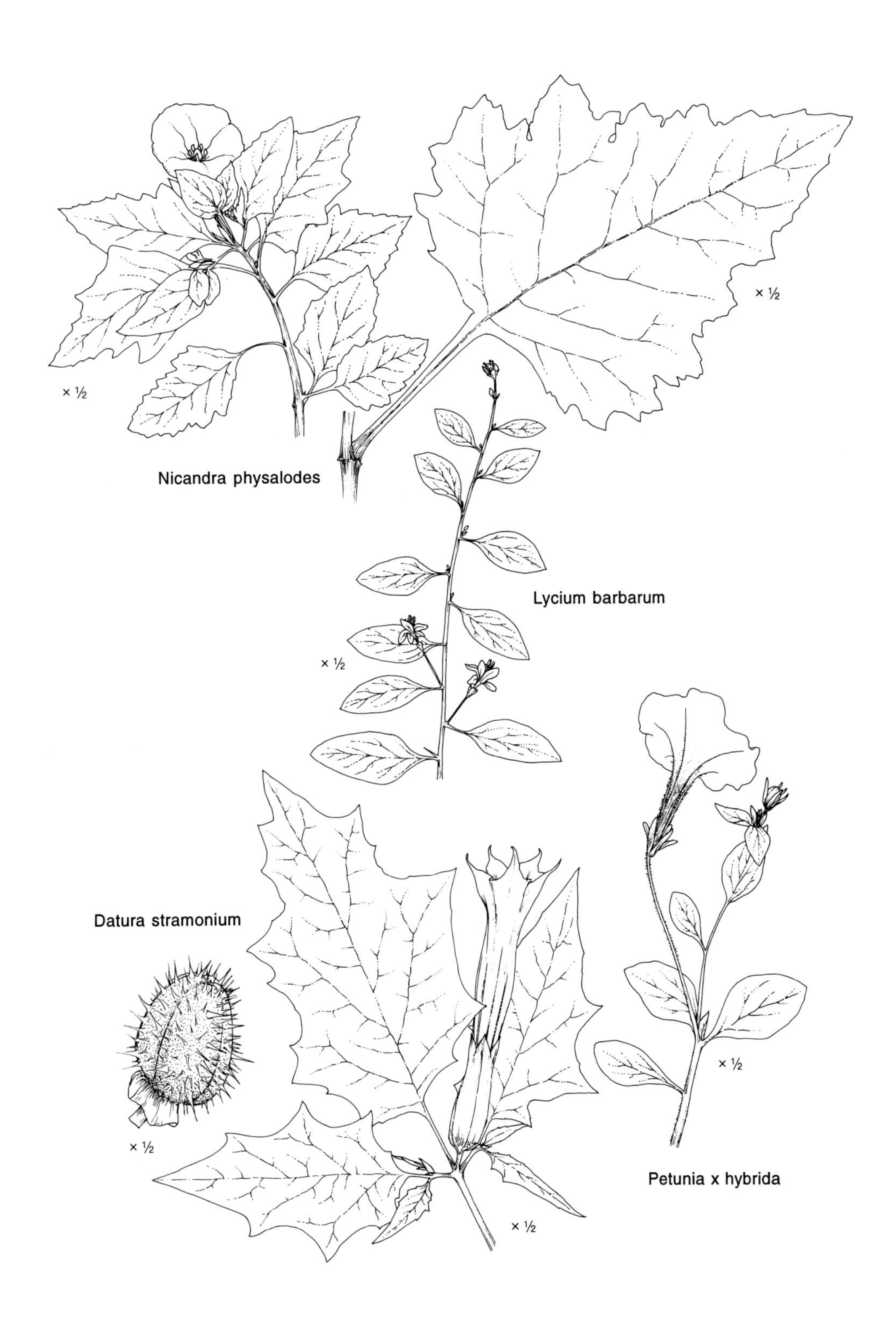
× ½
× ½
Nicandra physalodes
Lycium barbarum
× ½
Datura stramonium
× ½
× ½
Petunia x hybrida
× ½

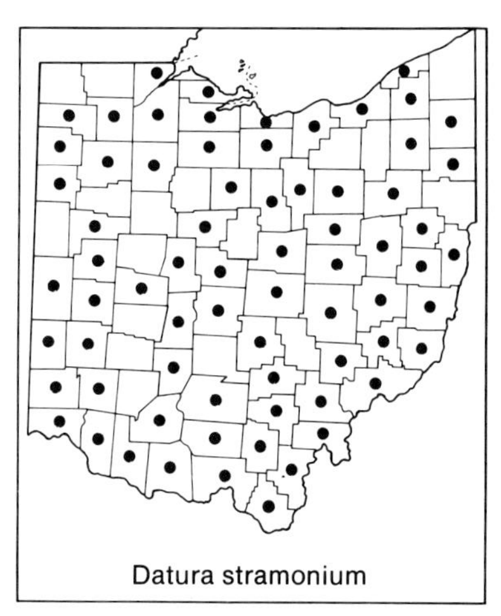

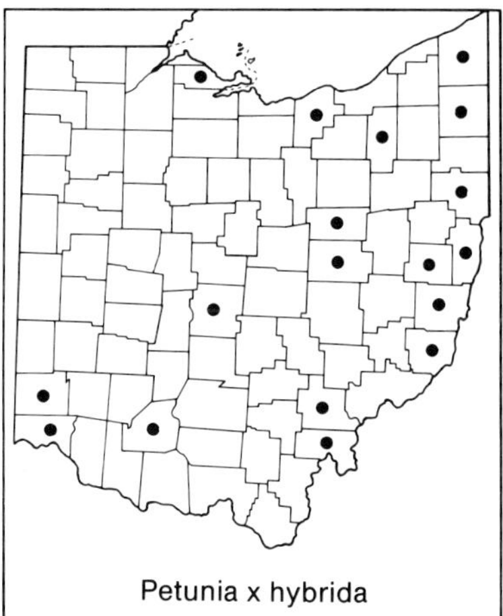

1. Datura stramonium L. Jimson-weed
incl. var. *stramonium* and var. *tatula* (L.) Torr.

Ill-smelling coarse annual, to 1.5 m tall; stems erect, glabrous to slightly pubescent; leaves irregularly and coarsely toothed to very shallowly lobed; flowers solitary; calyx tubular, 3–5 cm long, strongly 5-angled; corolla white to purple, cylindrical-funnelform, 6–10 cm long, 2–3 cm across; fruit an erect spine-covered capsule. Native to warmer parts of the New World, but now naturalized throughout Ohio; barnyards and feedlots, cultivated fields, along roads and railroads, and in other disturbed habitats. Late June–early October.

7. Nicotiana L. TOBACCO

A genus of some 70 species, native to the Americas and Australia, two of the species are cultivated and rarely adventive in Ohio. Both are perennials, cultivated as annuals, with clammy viscid-pubescent stems and white to dark red, salverform flowers. *Nicotiana alata* Link & Ott, Flowering Tobacco, is grown for its showy flowers that emit a sweet fragrance, especially in the evening hours. Escapes have been collected from Fairfield and Highland counties. *Nicotiana tabacum* L., Tobacco, is cultivated on a commercial scale in parts of southern Ohio (Jencks, 1972), and grown locally for personal use elsewhere in the state. Escapes have been collected from Adams and Huron counties. In addition, *Nicotiana rustica* L., Tobacco or Wild Tobacco, a species cultivated by Native Americans, was once presumably cultivated and escaped in Ohio (Bailey Hortorium Staff, 1976), but I have encountered no specimens. Gleason (1952b) notes that in northeastern United States, *N. rustica* was last reported as an escape "from roadsides in Indian settlements in western New York." All species of the genus are poisonous.

a. Flowering plants 1–3 m tall; flowers in a panicle; corolla limb 2–3 cm wide; see above . . . . *Nicotiana tabacum*

aa. Flowering plants mostly less than 1 m tall; flowers in a raceme; corolla limb ca. 5 cm wide; see above . . . . *Nicotiana alata*

8. Petunia Juss. PETUNIA

A New World genus of about 30 species.

1. Petunia × hybrida Vilmorin Garden Petunia

Perennial, cultivated in Ohio as an annual; stems erect to ascending, clammy viscid-pubescent, to 6 dm long; leaves less than 10 cm long; flowers solitary in leaf axils; calyx lobed nearly to the base; corolla of various colors, funnelform, 4–10 cm across. Probably the most commonly cultivated flowering plant in Ohio, popular for its fragrant flowers and for the ease with which it may be grown in a variety of garden situations. Adventives have been collected from scattered sites, chiefly in eastern Ohio; dumps and waste places, and other disturbed habitats. June–October.

Scrophulariaceae. FIGWORT FAMILY

Annual, biennial, or perennial herbs; leaves simple, but sometimes deeply lobed; flowers perfect, solitary in leaf axils or in a variety of bracteate inflorescences; calyx 5- or 4-lobed; corolla 5- or 4-lobed, usually zygomorphic and often bilabiate, infrequently subactinomorphic; stamens 5 or more often 4 and didynamous, and with or without a staminode attached to the corolla tube; carpels 2, united to form a compound pistil with a single style, ovary superior; fruit a capsule. A major, widely distributed family of more than 3,000 species, many of them in temperate regions of the earth. Pennell's (1935) taxonomic study of the family as it is represented in the flora of eastern temperate North America is the basis for much of the treatment here. The treatment of alien members of the flora is based in part on that of Tutin (1972) in *Flora Europaea*. Sutton's (1988) revision of the tribe Antirrhineae covers genera nos. 2–6 below. The text is also based in part on two previous studies of Ohio members of the family: McCready and Cooperrider (1978) on the subfamily Scrophularioideae, which includes genera nos. 1–15, and Bentz and Cooperrider (1978) on the subfamily Rhinanthoideae, which includes genera nos. 16–26. Special studies on the semiparasitic members of subfamily Rhinanthoideae include those of Kondo, Musselman, and Mann (1978) on karyomorphology, Kuijt (1969) on biological aspects of the life cycle, and Musselman and Mann (1977) on host plants.

The family's two most widely planted ornamentals, foxglove and snapdragon, are discussed in the text below. Also popular in Ohio gardens are cultivars of *Calceolaria crenatifolia* Cav., SLIPPERWORT, a South American plant in which the corolla's lower lip forms an inflated, brightly colored pouch. *Mazus pumilus* (Burm. f.) van Stennis (*M. japonicus* (Thunb.) Ktze.), a small blue-flowered ornamental of Asian origin, has been collected twice as an adventive escape: from Pickaway County in 1953, and from Scioto County in 1951.

a. Leaves either alternate or mostly alternate, or in a basal rosette.
  b. Stems glabrous.
    c. Leaves pinnately lobed; calyx lobes 2 . . . . 26. *Pedicularis*

cc. Leaves not pinnately lobed; calyx lobes 4 or 5.
d. Calyx lobes 4; stamens 2; corolla rotate . . . . 18. *Veronica*
dd. Calyx lobes 5; stamens 4 or 5; corolla rotate or not.
e. Leaves linear or narrowly linear-lanceolate; corolla spurred; anthers 4 . . . . 4. *Linaria*
ee. Leaves broader, not linear or narrowly linear-lanceolate; corolla spurred or not; anthers 4 or 5.
f. Plant a trailing vine; leaves palmately veined and shallowly palmately lobed; corolla spurred; stamens 4. . . . . . . 2. *Cymbalaria*
ff. Plant erect; leaves pinnately veined, not lobed; corolla spurred or not; stamens 4 or 5.
g. Corolla spurred; flowers yellow; leaves entire, ovate-lanceolate or ovate . . . . 4. *Linaria*
gg. Corolla not spurred; flowers yellow or some other color; leaves not entire or if so then not ovate-lanceolate or ovate.
h. Bracts subtending flowers shorter than pedicel length; corolla rotate; stamens 5 . . . . 1. *Verbascum*
hh. Bracts subtending flowers equalling or exceeding pedicel length; corolla tubular to campanulate; stamens 4 . . . . 15. *Digitalis*
bb. Stems more or less pubescent.
c. Leaves deeply cleft or lobed for 1/3 or more of way to midrib; stamens 4.
d. Leaves uniformly pinnately lobed, the lobes oblong or ovate, lobe margins crenate or serrate or again lobed; bracts subtending flowers not showy . . . . 26. *Pedicularis*
dd. Some or all leaves deeply cleft into 3–5 linear segments; bracts subtending flowers, especially the upper ones, showy and usually red, sometimes orange or yellow . . . . 24. *Castilleja*
cc. Leaves entire or variously toothed but not deeply cleft or lobed; stamens 4 or 2.
d. Calyx lobes 4; stamens 2.
e. Basal leaves long-petiolate, ovate or cordate, the cauline leaves sessile and much smaller than the basal . . . . 16. *Besseya*
ee. Leaves various but more or less uniform throughout plant. . . . . 18. *Veronica*
dd. Calyx lobes 5; stamens 4 or 5.
e. Leaves linear or very narrowly lanceolate or narrowly elliptic; stamens 4.
f. Corolla with slender spur . . . . 6. *Chaenorrhinum*
ff. Corolla saccate at base but without spur . . . . . . 5. *Antirrhinum*

ee. Leaves broader than narrowly lanceolate or narrowly elliptic; stamens 4 or 5.

f. Stems prostrate; leaves ovate or oblong, usually with a cordate or hastate base . . . . 3. *Kickxia*

ff. Stems erect; leaves various.

g. Leaves velvety pubescent. . . . . 1. *Verbascum*

gg. Leaves not velvety pubescent.

h. Bracts subtending flowers shorter than pedicel length; corolla rotate; stamens 5 . . . . 1. *Verbascum*

hh. Bracts subtending flowers equalling or exceeding pedicel length; corolla tubular; stamens 4 . . . . 15. *Digitalis*

aa. Leaves all or mostly opposite or whorled, the upper leaves sometimes alternate, or some leaves in a basal rosette and the cauline leaves opposite.

b. Leaves whorled, or some whorled and some opposite.

c. Leaves throughout in whorls of 4–6; corolla white throughout or partly pinkish . . . . 17. *Veronicastrum*

cc. Leaves mostly opposite, but some in whorls of 3 or rarely 4; corolla usually not white throughout.

d. Leaves deeply pinnately lobed . . . . 13. *Leucospora*

dd. Leaves entire, dentate, serrate or doubly serrate, but not lobed.

e. Principal cauline leaves sessile, their margins entire or remotely serrate; corolla variegated blue and white, 10–15 mm long . . . . 7. *Collinsia*

ee. Leaves petiolate, their margins dentate, serrate, or doubly serrate; corolla blue or violet, 6 mm long or less . . . . . 18. *Veronica*

bb. Leaves not whorled, all or mostly opposite, or the upper sometimes alternate.

c. Flowers borne in an inflorescence.

d. Leaves, at least the basal ones, deeply crenate to lobed, sinuses extending 1/3 or more of way to midrib; stamens 4; corolla yellow.

e. Calyx deeply 2-lobed, each lobe with a leafy auriculate appendage at apex; corolla bilabiate . . . . 26. *Pedicularis*

ee. Calyx 5-lobed, the lobes not appendaged; corolla more or less campanulate, not bilabiate.

f. Flowers sessile or nearly so; corolla tube 1 cm long or less; stamens nearly equal in length . . . . 19. *Dasistoma*

ff. Flowers on pedicels 2 mm long or more; corolla tube 1.5 cm long or more; stamens didynamous . . . . 22. *Aureolaria*

dd. Leaves entire or variously toothed, but not lobed; stamens 4 or 2; corolla yellow or not.

e. Leaves linear-lanceolate, linear, or filiform.

f. Corolla with slender spur; stamens 4.

g. Stems glandular-pubescent throughout; flowers solitary and axillary . . . . 6. *Chaenorrhinum*

gg. Stems glabrous throughout or slightly puberulent; flowers borne in a loosely defined raceme . . . . 4. *Linaria*

ff. Corolla without spur; stamens 4 or 2.

g. Calyx lobes 4; stamens 4 or 2.

h. Stamens 2; corolla rotate . . . . 18. *Veronica*

hh. Stamens 4; corolla bilabiate . . . . 25. *Melampyrum*

gg. Calyx lobes 5; stamens 4; corolla subcampanulate or markedly bilabiate.

h. Leaves serrate; flowers in dense spike-like inflorescences; corolla bilabiate . . . . 9. *Chelone*

hh. Leaves entire; flowers solitary in axils of upper leaves or in open, loosely defined inflorescences; corolla bilabiate or subcampanulate.

i. Corolla subcampanulate, open at apex; calyx lobes shorter than calyx tube. . . . . 20. *Agalinis*

ii. Corolla bilabiate, closed at apex; calyx lobes longer than calyx tube . . . . 5. *Antirrhinum*

ee. Leaves broader than linear-lanceolate.

f. Calyx lobes 4.

g. Corolla bilabiate, the tube longer than lobes; stamens 4; calyx lobed about half way to base, the lobes about equalling length of calyx tube. . . . . 25. *Melampyrum*

gg. Corolla rotate, the lobes much longer than tube; stamens 2; calyx lobed nearly to base. . . . . 18. *Veronica*

ff. Calyx lobes 5.

g. Calyx lobes very short, less than 1/2 length of calyx tube.

h. Stem scabrous; calyx 6–7 mm long; corolla salverform, nearly actinomorphic. . . . . 23. *Buchnera*

hh. Stem glabrous; calyx 10–16 mm long; corolla bilabiate. . . . . 11. *Mimulus*

gg. Calyx lobes longer, usually equalling or exceeding length of calyx tube.

h. Plants small, 0.5–2.5 dm tall.

i. Stems creeping to erect, leaves mostly basal; see above . . . . *Mazus*

ii. Stems erect, leafy throughout.

j. Calyx lobed nearly to base; corolla spurred, purple or bluish-purple . . . . 6. *Chaenorrhinum*

jj. Calyx lobed a little more than half-way to the base, the lobes only slightly longer than calyx tube; corolla not spurred.

k. Upper leaves petiolate; corolla white to purplish, and yellow-tipped, the lips shorter than corolla tube . . . . 25. *Melampyrum*

kk. Upper leaves sessile or clasping; lower lip of corolla bright blue, upper lip white or pale blue, the lips longer than corolla tube . . . . 7. *Collinsia*

hh. Plants 2.5 dm tall or more.

i. Flowers in spikes or spike-like inflorescences, the flowers sessile or with pedicels less than 2 mm long.

j. Stems glabrous or puberulent; bracts much shorter than flowers; corolla tubular, nearly closed at apex, white, greenish, or tinged with pink or purple . . . . 9. *Chelone*

jj. Stems harshly scabrous; bracts longer than flowers; corolla campanulate, open at apex, purple throughout . . . . 21. *Tomanthera*

ii. Flowers in racemes and with pedicels 2 mm long or more, or in panicles or umbel-like clusters.

j. Inflorescence a raceme.

k. Corolla bilabiate, closed at apex; stamens 4. . . . . 5. *Antirrhinum*

kk. Corolla rotate or campanulate, open at apex; stamens 4 or 2.

l. Leaves 2–18 cm long; corolla yellow; stamens 4. . . . . 22. *Aureolaria*

ll. Leaves 1.5–4 cm long; corolla blue; stamens 2. . . . . 18. *Veronica*

jj. Inflorescence a panicle, or a panicle-like or umbel-like cluster.

k. Flowers few or solitary in axils of upper leaves, the uppermost in a terminal umbel-like cluster; corolla variegated blue and white . . . . 7. *Collinsia*

kk. Flowers in a panicle or panicle-like inflorescence; corolla not variegated blue and white.

l. Corolla 5–11 mm long, greenish- or reddish-brown; staminode small, much shorter than fertile stamens, gland-like, glabrous . . . . 8. *Scrophularia* ll. Corolla 12–35 mm long, white to purple; staminode long and bearded, nearly equalling to exceeding length of fertile stamens . . . . 10. *Penstemon*

cc. Flowers usually solitary in leaf axils, not borne in a distinct inflorescence.

d. Leaves, or at least basal ones, lobed, the sinuses extending 1/3 or more of way to midrib; stamens 4.

e. Plants small, 1–2 dm tall; leaves deeply pinnately lobed, the lobes linear or oblong; corolla light purple to greenish-white, bilabiate . . . . 13. *Leucospora*

ee. Plants 3 dm tall or more; leaves variously lobed but the lobes not linear or oblong; corolla yellow, campanulate.

f. Flowers sessile or nearly so; corolla tube 1 cm long or less; stamens nearly equal in length . . . . 19. *Dasistoma*

ff. Flowers on pedicels 2 mm long or more; corolla tube 1.5 cm long or more; stamens didynamous . . . . 22. *Aureolaria*

dd. Leaves entire or variously dissected but not lobed; stamens 4 or 2.

e. Leaves linear-lanceolate, linear, or filiform; stamens 4.

f. Calyx lobes shorter than calyx tube; corolla campanulate, open at apex, without a spur . . . . 20. *Agalinis*

ff. Calyx lobes as long as to much longer than calyx tube; corolla bilabiate, closed at apex, with or without a spur.

g. Calyx 2/3 as long as corolla to longer than corolla; corolla without a spur. . . . . 5. *Antirrhinum*

gg. Calyx half as long as corolla or less; corolla with or without a spur.

h. Calyx 4-lobed; corolla without a spur . . . . 25. *Melampyrum*

hh. Calyx 5-lobed; corolla spurred.

i. Stems glandular-pubescent throughout; flowers distinctly axillary . . . . 6. *Chaenorrhinum*

ii. Stems glabrous throughout, or slightly puberulent; flowers in a loosely defined raceme . . . . 4. *Linaria*

ee. Leaves various, broader than linear-lanceolate; stamens 4 or 2.

f. Calyx lobes 4.

g. Corolla bilabiate, tube much longer than lobes; stamens 4; calyx lobes 1–2 times as long as calyx tube . . . 25. *Melampyrum*

gg. Corolla rotate, lobes much longer than tube; stamens 2; calyx lobed nearly to base. . . . . 18. *Veronica*

ff. Calyx lobes 5.

g. Calyx lobes very short, appearing as small teeth on apex of tube; flowers bilabiate; corollas purple-blue to pinkish or rarely white; stamens 4 .... 11. *Mimulus*

gg. Calyx lobes nearly as long as tube or longer; flowers bilabiate or not; corollas of various colors; stamens 4 or 2.

h. Plants large; stems 3 dm tall or more, stiff, erect or leaning; stamens 4.

i. Flowers few or solitary in axils of upper leaves, the uppermost in a terminal umbel-like cluster; corolla bilabiate, variegated blue and white .... 7. *Collinsia*

ii. Flowers solitary in axils of upper leaves; corolla campanulate, yellow or purple.

j. Flowers and fruits on pedicels 2 mm long or more; corolla yellow............................... 22. *Aureolaria*

jj. Flowers and fruits sessile or on pedicels less than 2 mm long; corolla purple........................ 21. *Tomanthera*

hh. Plants small; stems less than 3 dm tall, or if more than 3 dm the stems prostrate or creeping to ascending, but not stiffly erect; stamens 4 or 2.

i. Stems prostrate or creeping to erect; stamens 4.

j. Stems creeping to erect, leaves mostly basal; corolla not spurred; see above .... *Mazus*

jj. Stems prostrate, leafy throughout; corolla spurred........ .... 3. *Kickxia*

ii. Stems mostly upright; stamens 4 or 2.

j. Calyx deeply lobed, lobes scarcely united at base; stamens 4 or 2.

k. Corolla spurred; fertile stamens 4 .... 6. *Chaenorrhinum*

kk. Corolla not spurred; fertile stamens 2.

l. Calyx closely subtended by two sepal-like bracts; capsule symmetric........................... 12. *Gratiola*

ll. Calyx not subtended by bracts; capsule somewhat asymmetric............................. 14. *Lindernia*

jj. Calyx tube half as long as to equalling calyx lobes; stamens 4.

k. Corolla white to purplish, yellow-tipped; corolla lips shorter than corolla tube; upper leaves short-petiolate .... 25. *Melampyrum*

kk. Lower lip of corolla blue, upper lip usually white; corolla lips longer than corolla tube; upper leaves sessile or clasping .... 7. *Collinsia*

1. Verbascum L. MULLEIN

Biennials, forming a basal rosette the first year and erect flowering stems the second; basal leaves more or less petiolate, cauline leaves alternate, sessile; flowers in racemose or spicate inflorescences; calyx deeply 5-lobed; corolla yellow, white, or purple, 5-lobed and nearly actinomorphic; stamens 5. An Old World genus of some 250 species, two of them naturalized throughout Ohio. In addition, three other European species have been collected, once each, as adventives in Ohio: *Verbascum phlomoides* L., CLASPING-LEAVED MULLEIN, from Adams County in 1937; *V. phoeniceum* L., PURPLE MULLEIN, from Lawrence County in 1984; and *V. virgatum* Stokes, TWIGGY MULLEIN, from Franklin County in 1891—the last species recently reported also from Indiana (Vernon and Schoknecht, 1981).

a. Plants densely pubescent with stellate, nonglandular hairs; leaf blades entire to crenate; corolla yellow; the two lower filaments mostly glabrous, the three upper ones with white or pale yellow hairs; capsules stellate-pubescent.

b. Leaves decurrent down stem to next leaf below; inflorescence dense or sometimes interrupted below, usually with 1 flower in each axil; corolla 12–22 mm wide. . . . . 1. *Verbascum thapsus*

bb. Leaves not or only shortly decurrent; inflorescence dense above, somewhat open and interrupted below, usually with 3–7 flowers in each axil; corolla 25–40 mm wide; see above. . . . . *Verbascum phlomoides*

aa. Plants either glabrous or pubescent with glandular, simple, and/or a few stellate hairs, the last not dense; leaf blades toothed, crenate, or entire; corolla yellow, white, or purple; the filaments all with long purple hairs; capsules glabrous or pubescent with simple hairs, or rarely with a few stellate hairs.

b. Pedicels 5–20 mm long; corolla yellow, white, or purple.

c. Corolla yellow or white (often becoming red-violet upon drying); pedicels 5–15 mm long; undersurface of leaves glabrous . . . . 2. *Verbascum blattaria*

cc. Corolla purple; pedicels 10–20 mm long; undersurface of leaves slightly pubescent; see above . . . . *Verbascum phoeniceum*

bb. Pedicels 2–5 mm long; corolla yellow; see above . . . . *Verbascum virgatum*

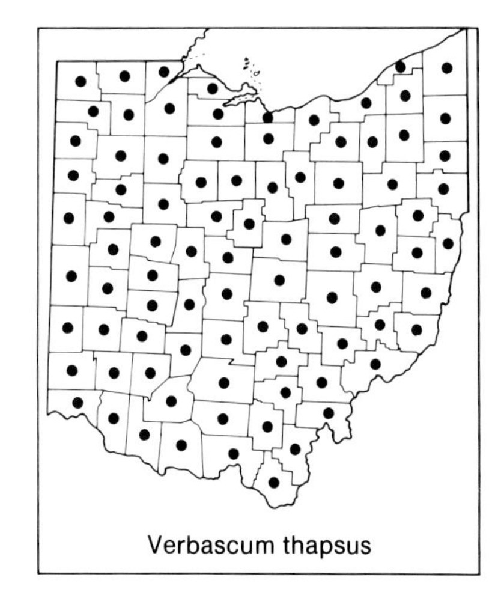

Verbascum thapsus

1. VERBASCUM THAPSUS L. COMMON MULLEIN. VELVET-PLANT

Biennial, the plants densely stellate-pubescent throughout; erect flowering stems to 2.5 m tall; leaves entire to crenate, those of the flowering stem decurrent down the stem to the next leaf below; flowers in a dense spicate inflorescence; corolla yellow, 12–22 mm across; filaments of the three upper stamens shorter and pubescent with white or pale yellow hairs, those of the two lower stamens glabrous or nearly so; capsules stellate-pubescent. A European species naturalized throughout Ohio; along roads and railroads (where the velvety-leaved, first-year rosettes are conspicuous), old fields, and other disturbed habitats. Late June–August (–early October).

Wagner, Daniel, and Hansen (1980) report a Michigan population of hybrids between this species and *V. phlomoides*.

2. VERBASCUM BLATTARIA L. MOTH MULLEIN

Biennial, the plants glabrous or nearly so below the inflorescence; erect flowering stems to 1.5 m tall; leaves irregularly toothed to crenate or subentire; flowers in an open terminal raceme, and sometimes in one or more smaller additional axillary racemes, the inflorescence axis glandular-pubescent; corollas either all white (the bases of the petals somewhat purplish) or all yellow, 20–30 mm across; all filaments densely glandular-villous with purple hairs; capsules glandular-pubescent. A European species naturalized throughout Ohio; along roads and railroads, in old fields and other disturbed habitats. Late May–early October.

The white-flowered plants, forma *erubescens* Brügger (forma *albiflorum* (Don) House), are somewhat more frequent in Ohio than the yellow-flowered plants, forma *blattaria*, but both are common. Flowers of the former often become red-violet upon drying, sometimes causing them to be confused with *V. phoeniceum*.

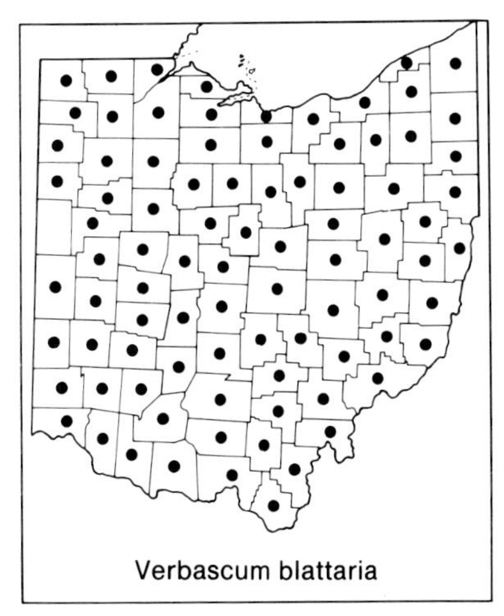

Verbascum blattaria

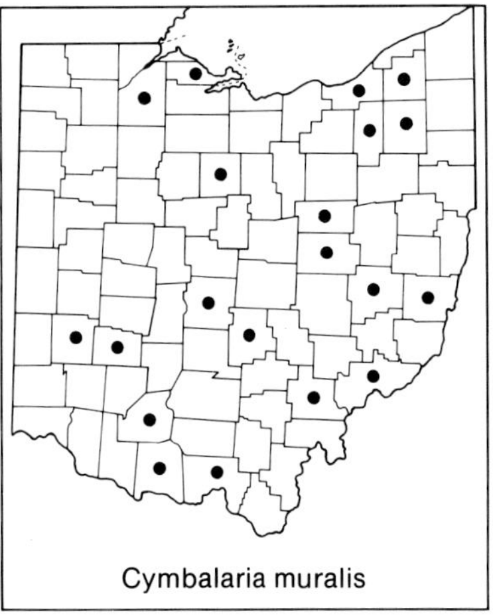

Cymbalaria muralis

2. Cymbalaria Hill

A European genus of about 10 species.

1. CYMBALARIA MURALIS Gaertn., Mey. & Scherb. KENILWORTH IVY

Annual, trailing vine; stems glabrous, to 7 dm long; leaves alternate, palmately veined and shallowly palmately lobed; flowers solitary in leaf axils, their peduncles becoming recurved in fruit; calyx deeply 5-lobed; corolla pale purple to bluish, spurred, 6–14 mm long, bilabiate, the throat closed by a yellow palate; stamens 4, staminode minute. An attractive, frequently cultivated, trailing Eurasian plant with ivy-like leaves, escaped and locally naturalized at scattered sites in Ohio; usually in towns and often near gardens, especially in crevices of walls and sidewalks, and in similar habitats. April–early November (–January).

3. Kickxia Dumort. CANCERWORT

Villous annuals; leaves alternate; flowers solitary, axillary; calyx deeply 5-lobed; corolla yellow to purplish, bilabiate and with the throat closed by

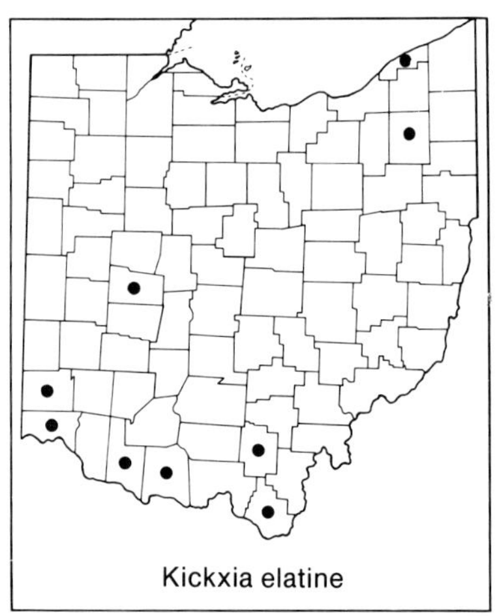
Kickxia elatine

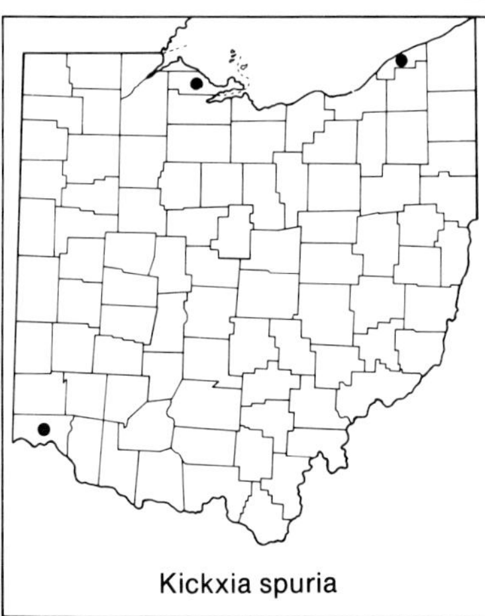
Kickxia spuria

a palate, spurred; stamens 4, staminode minute. An Old World genus of ca. 50 species, two of them naturalized in Ohio.

a. Leaf blades ovate to broadly triangular or obovate, their bases hastate, truncate, or sagittate; margins of lower leaves with 1–several teeth near base; peduncles glabrous or with pubescence thicker at apex and base than in middle; calyx lobes scarious-margined toward base . . . . 1. *Kickxia elatine*

aa. Leaf blades ovate to suborbicular, their bases rounded or cordate; margins of all leaves entire; peduncles uniformly pubescent throughout their length; calyx lobes scarcely or not scarious-margined . . . . 2. *Kickxia spuria*

1. Kickxia elatine (L.) Dumort. Sharp-pointed Cancerwort
Annual; stems pubescent, prostrate to ascending, to 6 dm long; leaves ovate to broadly triangular or obovate, hastate, truncate, or sagittate at base, the margins of lower leaves with 1–several teeth near the base, the upper without teeth; peduncles glabrous or with pubescence heavier at apex and base of peduncles than in the middle; calyx lobes scarious-margined toward base; corolla cream-colored, upper lip purple within, lower lip and especially the palate yellow within, 5–13 mm long. A European species naturalized at a few Ohio sites, chiefly in southwestern counties; mud flats, railways, stony parking lots, and similar waste places. June–September.

2. Kickxia spuria (L.) Dumort. Round-leaved Cancerwort
Annual; stems pubescent, prostrate to ascending, to 6 dm long; leaves ovate to suborbicular, rounded or cordate at base, the margins entire; peduncles uniformly pubescent throughout their length; calyx lobes scarcely or not at all scarious-margined; corolla yellow, upper lip purple within, 10–15 mm long. A European species naturalized or perhaps merely adventive at a few sites in northern and southern Ohio counties; open disturbed habitats. June–September.

4. Linaria Mill. toadflax

Annual or perennial herbs; leaves opposite or alternate; flowers in racemes or solitary in the axils of reduced upper leaves; calyx deeply 5-lobed; corolla yellow or blue, bilabiate and with the throat closed by a palate, spurred; stamens 4, staminode none. A genus of perhaps 100 species, most native to Eurasia. The treatment here is based in part on Sutton (1988). Two European species have been collected as adventives in Ohio: *Linaria dalmatica* (L.) Mill. (*L. genistifolia* subsp. *dalmatica* (L.) Marie & Petitmengin), Dalmatian Toadflax, from Franklin County in 1935 and from Ottawa County in 1967; and *Linaria genistifolia* (L.) Mill. sensu stricto from Van Wert County in 1925.

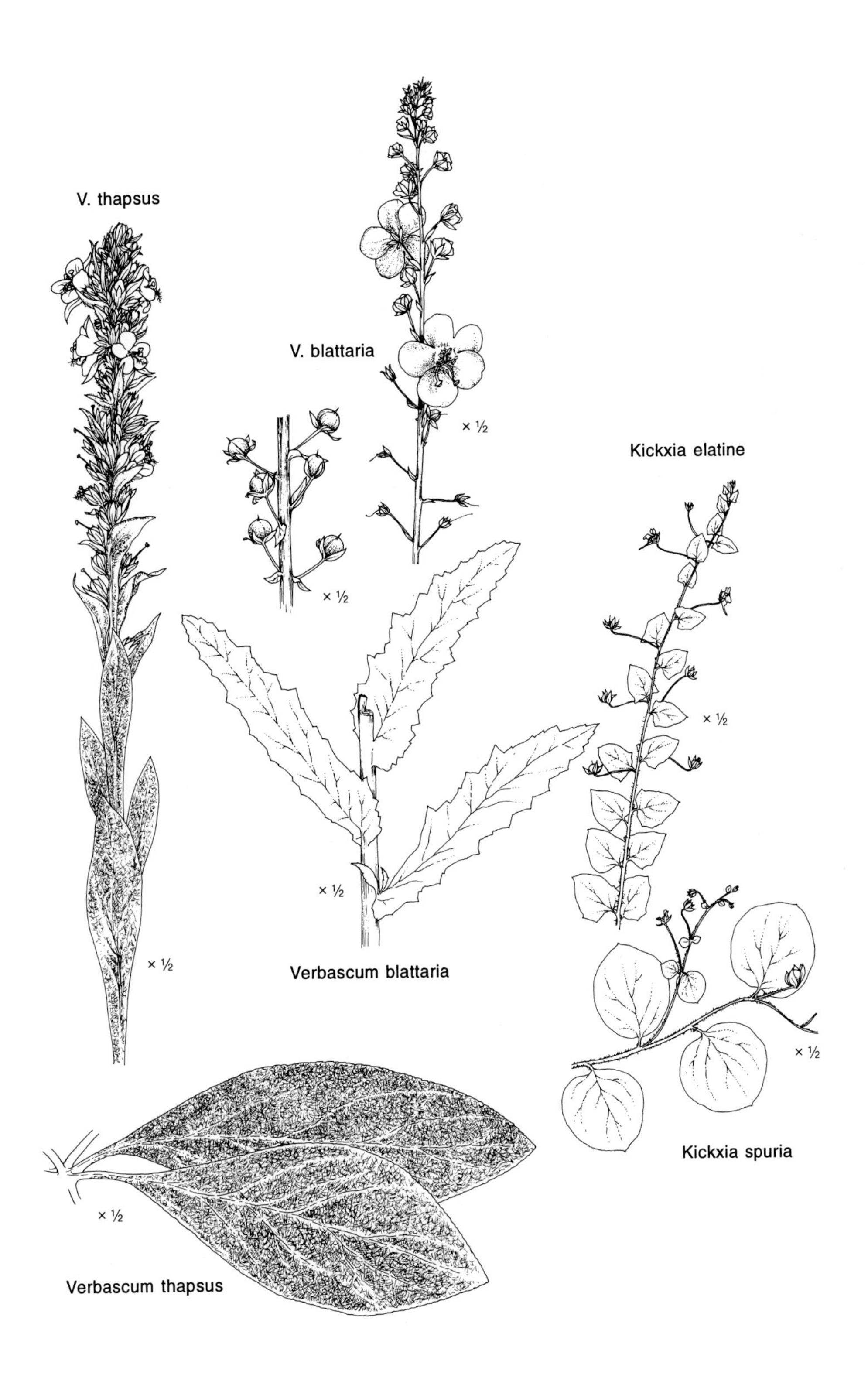
V. thapsus
V. blattaria
× ½
× ½
Kickxia elatine
× ½
× ½
× ½
Verbascum blattaria
× ½
× ½
Kickxia spuria
Verbascum thapsus

a. Corolla (including spur) 2–5 cm long, yellow with orange palate; calyx 3–10 mm long; capsules 4–12 mm long; leaves linear to ovate.

b. Principal leaves linear to oblong-lanceolate, 1–5 mm wide; calyx 3–5 mm long; corolla (including spur) 3 cm long or less; capsules longer than wide.

c. Corolla (including spur) 2–3 cm long; leaves linear to narrowly linear-lanceolate, tapering at base . . . . 1. *Linaria vulgaris*

cc. Corolla (including spur) 2 cm long or less; leaves lanceolate or oblong-lanceolate, narrowly rounded at base; see above . . . . *Linaria genistifolia*

bb. Principal leaves ovate to oblong-lanceolate, 6–25 (–40) mm wide; calyx 5–10 mm long; corolla (including spur) 3–5 cm long; capsules wider than long; see above. *Linaria dalmatica*

aa. Corolla (including spur) 0.5–1.5 cm long, blue or violet, with white palate; calyx 1–4 mm long; capsules 2–3 mm long; leaves narrowly linear. . . . . 2. *Linaria canadensis*

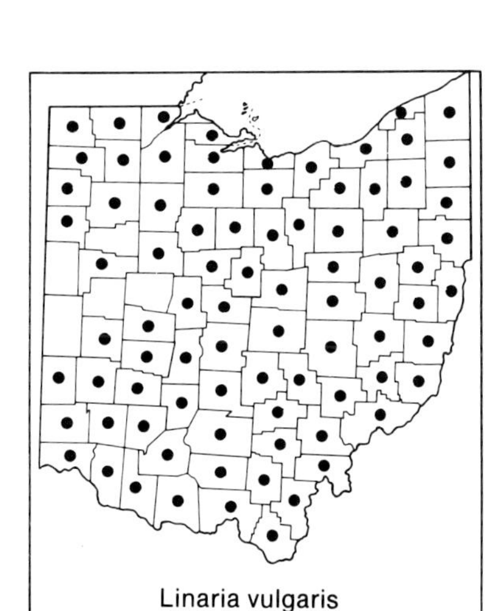
Linaria vulgaris

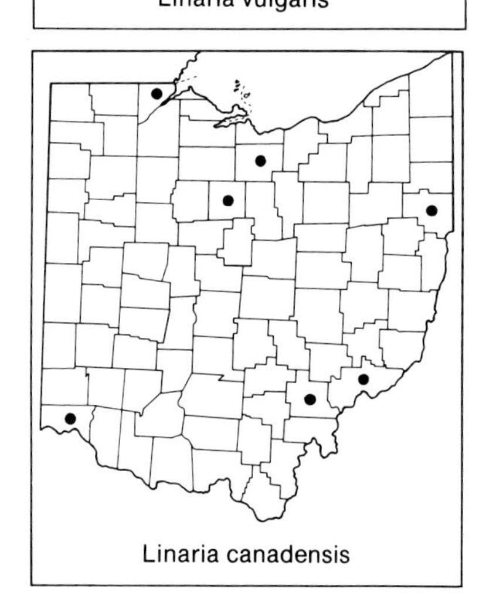
Linaria canadensis

1. LINARIA VULGARIS Mill. BUTTER-AND-EGGS. YELLOW TOADFLAX

Perennial, to 10 dm tall; stems erect, glabrous or slightly puberulent, sometimes glaucous; leaves linear to narrowly linear-lanceolate; corolla yellow with an orange palate, 20–30 mm long; capsules 7–12 mm long. A European species naturalized throughout Ohio, but scarce in west-central counties; along roads and railroads, in cultivated fields—where it is sometimes a troublesome weed, old fields, waste lots, and other disturbed habitats. The species' reproductive biology has been studied by Arnold (1982a) and Bakshi and Coupland (1960). June–October.

2. **Linaria canadensis** (L.) Dumont OLD-FIELD TOADFLAX. BLUE TOADFLAX

Nuttallanthus canadensis (L.) D. Sutton

Annual or biennial, with basal rosette of short trailing stems; flowering stems erect, glabrous, to 7 dm tall; leaves narrowly linear; corolla blue or violet, the palate white, 5–15 mm long; capsules 2–3 mm long. Known from a few widely scattered sites in Ohio, native at some and probably merely adventive at others; sandy fields, sterile soils. Late April–July (–October).

Sutton (1988) places this species in the segregate genus *Nuttallanthus* D. Sutton.

5. Antirrhinum L. SNAPDRAGON

A genus of the temperate regions of the Northern Hemisphere with about 40 species, one of them a widely planted ornamental. DeWolf (1956) has notes on cultivated members of this genus. *Antirrhinum orontium* L. (*Misopates orontium* (L.) Raf.), LESSER SNAPDRAGON, was collected as an adventive weed in Tuscarawas County in 1970.

a. Corolla 25–45 mm long; calyx lobes broadly ovate, shorter than capsules and shorter than corolla-tube; bracts subtending flowers much shorter than foliage leaves; leaf blades lanceolate to elliptic, 5–30 mm wide . . . . 1. *Antirrhinum majus*

aa. Corolla 10–12 mm long; calyx lobes linear, longer than capsules and longer than corolla-tube; bracteal leaves scarcely or not shorter than foliage leaves; leaf blades linear to narrowly elliptic-lanceolate, 2–6 mm wide; see above. . . . . *Antirrhinum orontium*

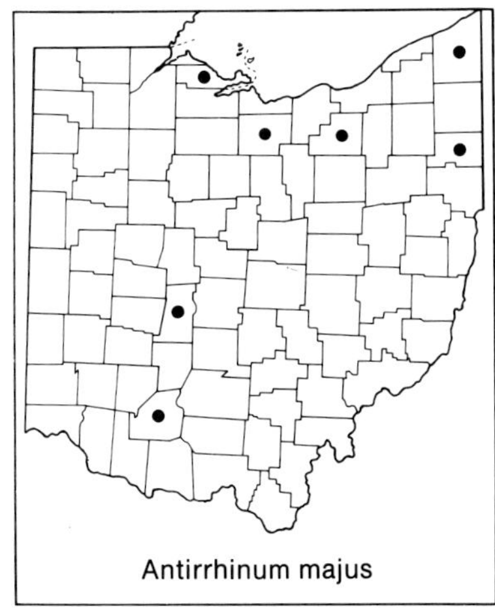

Antirrhinum majus

1. ANTIRRHINUM MAJUS L. SNAPDRAGON. GARDEN SNAPDRAGON

Perennial, cultivated as an annual; stems erect, usually glabrous below, glandular-pubescent above, to 9 dm tall; leaves opposite to subopposite, lanceolate to elliptic; flowers in an elongate, terminal, bracteate raceme; calyx deeply 5-lobed; corolla bilabiate with the throat closed by a palate (or subactinomorphic with an open throat in one group of cultivars), 25–45 mm long; stamens 4; staminode minute. A European species, from which many cultivars with a wide assortment of corolla colors have been derived, grown in flower gardens throughout Ohio. Adventive escapes have been collected from a few scattered Ohio sites, the species not persisting out of cultivation. Not illustrated. April–November.

Chaenorrhinum minus

6. Chaenorrhinum (DC.) Reichenb.

A genus of some 20 species, most of the Mediterranean region, and one naturalized in the Ohio flora.

1. CHAENORRHINUM MINUS (L.) Lange DWARF SNAPDRAGON. LESSER TOADFLAX

Annual, to 3 dm tall, glandular-pubescent throughout; stems erect; leaves linear to narrowly elliptic or spatulate, the lowermost opposite, the middle and upper alternate; flowers solitary in leaf axils; calyx deeply 5-lobed; corolla pale purple or bluish-purple, bilabiate and the throat closed or nearly closed by a palate, spurred, 5–8 mm long; stamens 4; staminode none. A European species naturalized throughout most of Ohio, it was formerly common on railroad ballast but is now scarce along main railways because of herbicide use; railroad sidetracks, abandoned railroads, open calcareous rocky habitats, rocky (limestone) parking lots. For studies on floral biology and seed dispersal of this species, see Arnold (1981, 1982b). Widrlechner (1983) details the history of its spread across North America. Late May–October.

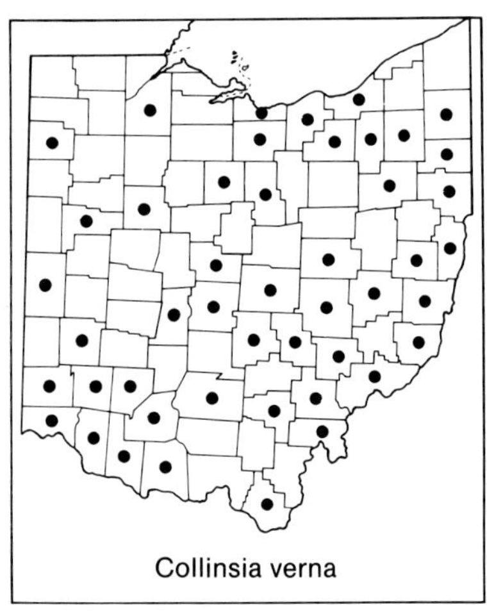
Collinsia verna

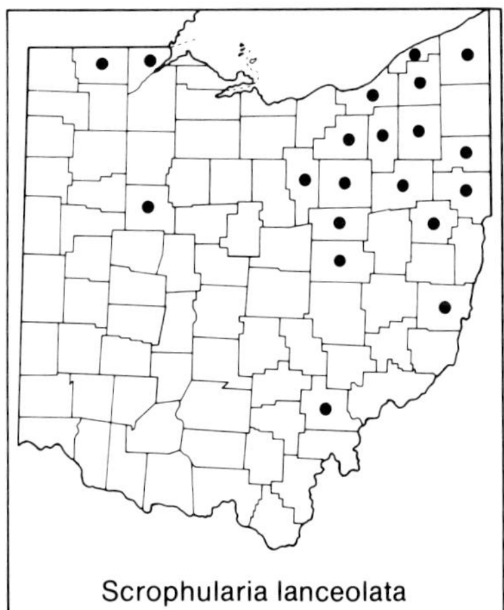
Scrophularia lanceolata

7. Collinsia Nutt.

An American genus of about 20 species, all but the one below native to western North America.

1. **Collinsia verna** Nutt. BLUE-EYED MARY

Winter annual; stems weakly erect or decumbent, puberulent, to 5 dm long; leaves opposite or the uppermost in whorls of 3, the principal leaves sessile, their margins entire or remotely serrate, truncate to cordate-clasping at base; flowers solitary or few in axils of upper leaves, the uppermost group forming an umbel-like cluster; calyx 5-lobed; corolla 10–15 mm long, bilabiate, variegated: the upper lip white or pale blue, the lower lip with its two lateral lobes bright blue and the shorter and keeled middle lobe purplish-blue; stamens 4, staminode tiny, green-tipped. One of Ohio's most attractive spring wildflowers; mostly in southern counties, and at scattered sites northward; low rich woodlands, open areas along streams. Baskin and Baskin (1983a) studied the germination ecology of this species. Additional records: Gallia, Lake, and Seneca counties (Pennell, 1935). Mid-April–early June.

8. Scrophularia L. FIGWORT

Tall perennials; leaves opposite; flowers in large terminal paniculate inflorescences; calyx 5-lobed; corolla reddish-brown to greenish-brown, bilabiate; stamens 4, a staminode borne on the upper corolla lip. A genus of 150–200 species of the Northern Hemisphere, two of them native to Ohio. Creasy (1953) discusses the classification history of these two species.

a. Staminode yellowish-green; principal leaves broadly cuneate or broadly rounded to subtruncate at base, their margins usually coarsely serrate to dentate or sharply incised; petioles 1–3 (–4) cm long; corolla (5–) 7–11 mm long, shiny reddish- or greenish-brown; plants flowering from April to early July . . . . 1. *Scrophularia lanceolata*

aa. Staminode purplish or brown; principal leaves usually rounded to subcordate at base, their margins serrate or dentate; petioles 3–7 cm long; corolla 5–9 mm long, dull reddish-brown; plants flowering from July to October . . . . 2. *Scrophularia marilandica*

1. **Scrophularia lanceolata** Pursh LANCE-LEAVED FIGWORT. EARLY FIGWORT

Perennial, to 2.5 m tall; stems erect; glandular-puberulent throughout or glabrous below; principal leaves broadly cuneate or broadly rounded to subtruncate at base, their margins usually coarsely serrate, dentate, or sharply incised, petioles 1–3 (–4) cm long; corolla reddish- or greenish-

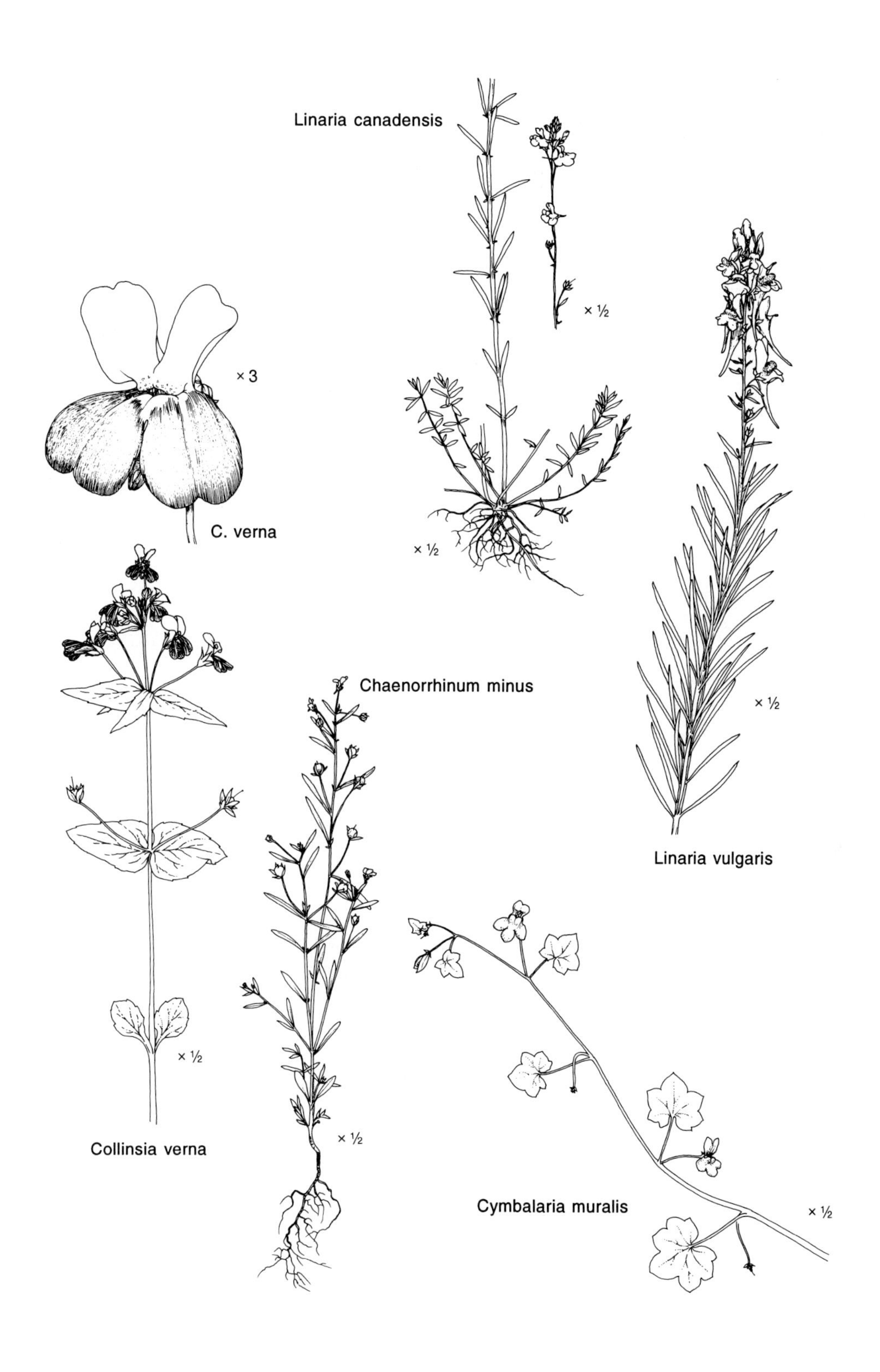
Linaria canadensis
× ½
× 3
C. verna
× ½
Linaria vulgaris
× ½
Chaenorrhinum minus
× ½
Collinsia verna
× ½
Cymbalaria muralis
× ½

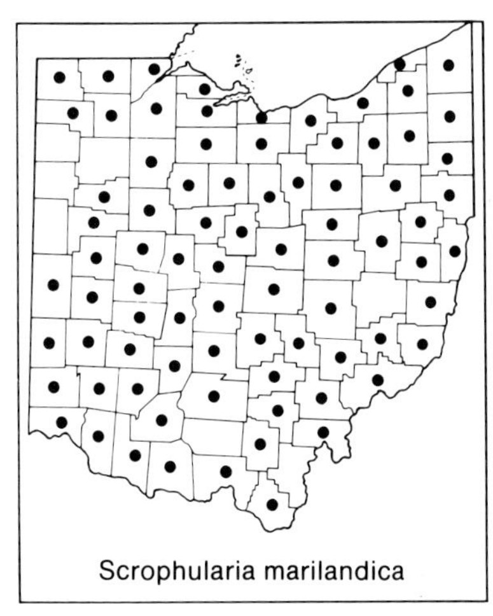
Scrophularia marilandica

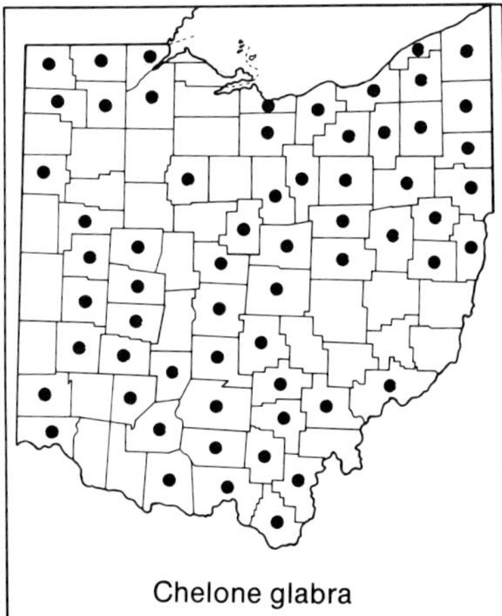
Chelone glabra

brown, shiny, (5–) 7–11 mm long; staminode yellowish-green. Chiefly in northeastern counties; woodland openings and borders, along roads and railroads. Additional records: Licking and Morgan counties (Pennell, 1935). Late April–early July.

2. **Scrophularia marilandica** L. MARYLAND FIGWORT. CARPENTER'S-SQUARE

Perennial, to 2 m tall; stems erect, glabrous below, glabrous or often minutely glandular-puberulent above; principal leaves usually rounded to subcordate at base, their margins serrate or dentate, petioles 3–7 mm long; corolla reddish-brown, dull, 5–9 mm long; staminode purplish or brown. Throughout Ohio, but scarce in northwesternmost counties; woodland openings and borders, pastures, and along roads and railroads. July–October.

9. Chelone L. TURTLEHEAD

An eastern North American genus of four species, one of them a member of the Ohio flora. Cooperrider and McCready (1970) reported the chromosome numbers of taxa in this genus.

1. **Chelone glabra** L. TURTLEHEAD

Perennial, to ca. 1.5 m tall; stems erect, glabrous or rarely puberulent; leaves opposite, subsessile or short-petiolate, margins serrate; flowers subsessile in dense spicate inflorescences, terminal on main stem and branches; calyx deeply 5-lobed; corolla 15–35 mm long, bilabiate and with villous palate nearly closing the throat; stamens 4, filaments flat and densely woolly, staminode conspicuous but smaller than filaments. Throughout much of Ohio; moist fields, bogs, swamps, borders of lakes and streams. Cooperrider (1967) studied the reproductive systems in this species. Additional records: Brown and Muskingum counties (Pennell, 1935). Late July–early October.

Many Ohio specimens can be assigned to one of the three varieties keyed below; others are intergrades. The three are somewhat distinct, both ecologically and geographically: var. *elatior* occurs mostly at stream margins and in wet woods in southeastern Ohio; var. *glabra* in many types of moist habitats, mostly in eastern and southern counties; var. *linifolia* in bogs, fens, and wet meadows mostly in northern, especially northwestern, counties.

a. Corolla basally white but pink or purplish at apex for 1/4–1/2 its total length, the lips purplish within; leaves broadly lanceolate to ovate . . . . var. *elatior* Frick

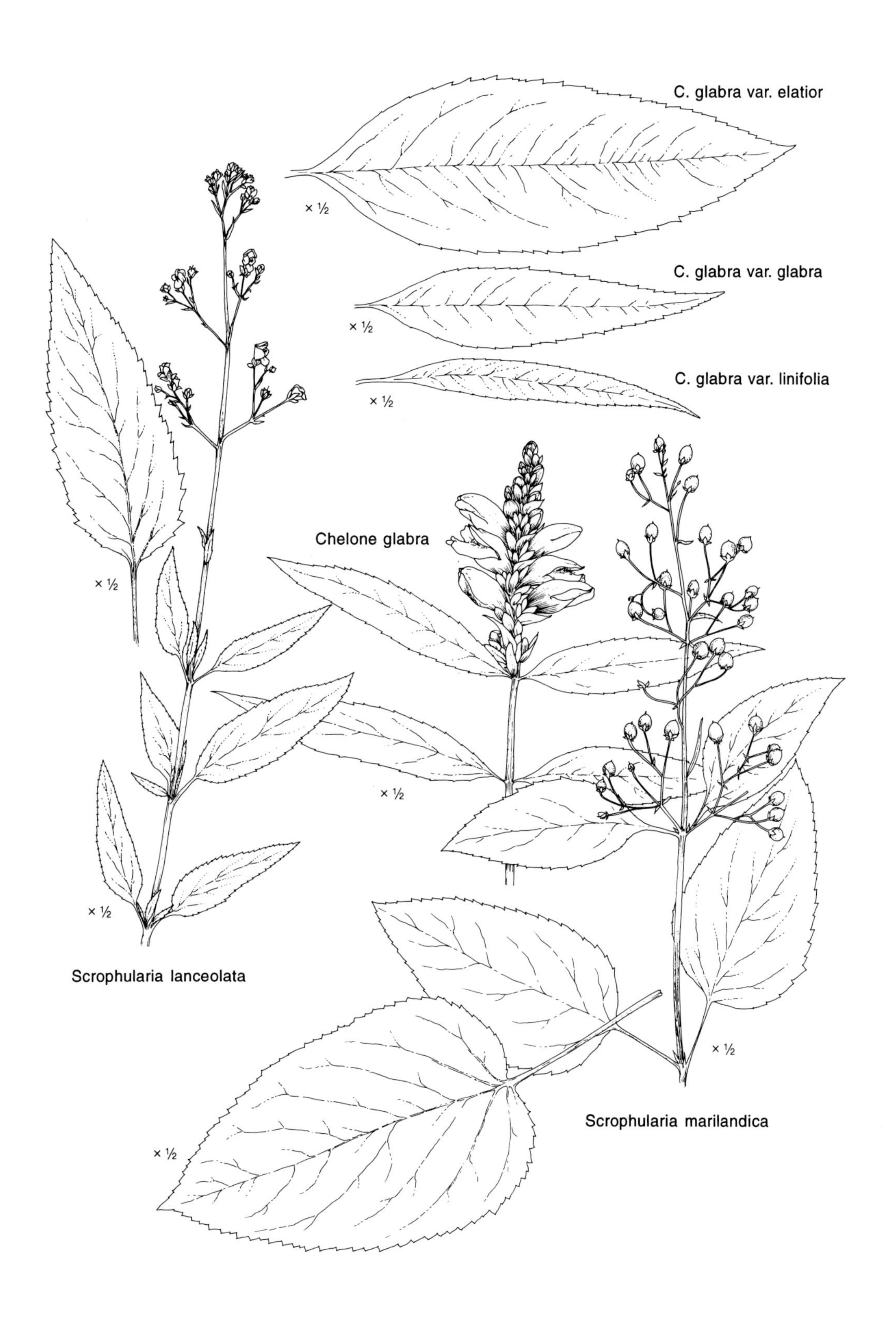
C. glabra var. elatior
× ½
C. glabra var. glabra
× ½
C. glabra var. linifolia
× ½
× ½
Chelone glabra
× ½
× ½
Scrophularia lanceolata
× ½
× ½
Scrophularia marilandica

aa. Corolla white throughout or slightly tinged with pink, green, or yellowish-green at apex, the lips purplish or white within; leaves linear to lanceolate or ovate.

b. Corolla white throughout or slightly tinged with pink or pinkish-green at apex, the lips purplish within; leaves lanceolate to narrowly ovate . . . . var. *glabra*
incl. var. *elongata* Pennell & Wherry

bb. Corolla white throughout but tinged with green or yellowish-green at apex, the lips whitish within; leaves linear to narrowly lanceolate . . . . var. *linifolia* N. Coleman

10. Penstemon Mitch. BEARD-TONGUE

Perennials; leaves opposite; flowers in a paniculate, thyrsoid, or rarely racemose inflorescence; calyx deeply 5-lobed; corolla white to purple or blue, varying from bilabiate and with a palate closing the throat to subactinomorphic and open; stamens 4 and didynamous, staminode 1 and bearded, equalling or exceeding length of stamens. A genus of 250–300 species, all but one native to North and Central America. The treatment of species nos. 2–4 below is based on the taxonomic studies of Koelling (1964) and on my discussion (Cooperrider, 1976) of the problems surrounding this same complex.

Two species native to central United States, both sometimes cultivated as ornamentals, have been collected as adventive escapes in Ohio: *Penstemon cobaea* Nutt. from Lake County in 1902, and *Penstemon grandiflorus* Nutt., LARGE-FLOWERED BEARD-TONGUE, from Hamilton County in 1917. *Penstemon brevisepalus* Pennell, SHORT-SEPALED BEARD-TONGUE, was attributed to Ohio by Schaffner (1938) and Gleason (1952b), but I have seen no specimens (Cooperrider, 1976). The point may be moot, since Mohlenbrock and Wallace (1963) merged *P. brevisepalus* without rank under *P. pallidus*.

a. Corolla 1–3 cm long; calyx 2–8 mm long.

b. Corolla with glandless hairs on inner surface of lower lobes; corolla tubular to funnelform or campanulate, not trumpet-shaped; leaves entire to distinctly toothed.

c. Stems glabrous to finely puberulent; corolla with upper and lower lips subequal or lower lip only slightly the longer; corolla tube strongly inflated below and moderately ridged within; staminode moderately to sparsely bearded.

d. Corolla white to pale pink or pale purple, often with pink or purple lines internally, 15–30 mm long; anther surface with few to several, stiff, white or tan hairs; calyx lobes 3–8 mm long at anthesis, ovate to lanceolate; capsules 6–12 mm long . . . . 2. *Penstemon digitalis*

dd. Corolla pale lavender to purple, 10–25 mm long; anther surface glabrous or slightly spiculate; calyx lobes 2–8 mm long at anthesis, ovate-lanceolate to linear; capsules 3–8 mm long.

e. Calyx lobes 4–8 mm long, narrowly lanceolate to linear-attenuate; corolla 13–25 mm long; cauline leaves broadly lanceolate to oblong- or ovate-lanceolate, usually sharply low-serrate . . . . 3. *Penstemon calycosus*

ee. Calyx lobes 2–5 mm long, oblong-lanceolate to ovate-lanceolate; corolla 10–17 mm long; cauline leaves lanceolate to narrowly oblong, entire or remotely low-serrate . . . . 4. *Penstemon laevigatus*

cc. Stems puberulent to densely pubescent; lower lip of corolla longer and projecting well beyond the upper; corolla tube flattened below, strongly ridged within; staminode densely bearded.

d. Lower lip of corolla not arching upward, orifice of tube open; anther sacs longer than wide, cup- or trough-shaped when open at maturity; corolla 1.7–3 cm long, lined with deeper color; calyx lobes obtusish to attenuate, not recurved; calyx less than 1/2 as long as to equalling length of capsule.

e. Corolla 18–30 mm long, pale purple to pinkish, the tube with darker purple lines within; cauline leaves ovate to broadly oblong, broadly lanceolate, or broadly oblanceolate, sparsely hirsute, but the midrib usually densely hirsute on undersurface . . . . 5. *Penstemon canescens*

ee. Corolla 17–22 mm long, white, the tube lined with purple within; cauline leaves narrowly lanceolate to oblong-lanceolate, pubescent to densely pubescent on both surfaces . . . . 6. *Penstemon pallidus*

dd. Lower lip of corolla arching upward against the upper lip, closing or nearly closing orifice to tube; anther sacs as wide as long, saucer-shaped when open at maturity; corolla 2–3 cm long, with no or only indistinct lines of deeper color within; calyx lobes usually acuminate to attenuate at apex and becoming recurved; calyx often greater than 1/2 the length of capsule . . . . 7. *Penstemon hirsutus*

Penstemon tubaeflorus

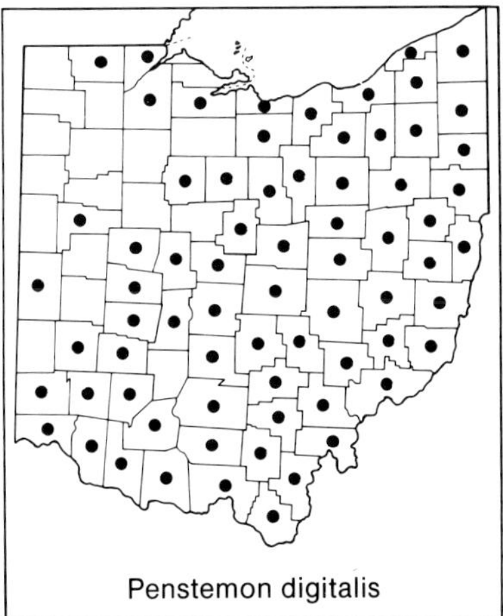
Penstemon digitalis

bb. Corolla with minute glandular pubescence on all of inner surface; corolla trumpet-shaped: the tube gradually widened from base to apex, nearly actinomorphic, the limb abruptly flared; leaves entire to crenate or obscurely toothed. . . . . 1. *Penstemon tubaeflorus*

aa. Corolla 3.5–5 cm long; calyx 7–15 mm long.

b. Upper and bracteal leaves lanceolate and often dentate, not glaucous; calyx and pedicels pubescent, the corolla glandular-pubescent; see above . . . . *Penstemon cobaea*

bb. Upper and bracteal leaves orbicular to orbicular-ovate, entire, glaucous; calyx, pedicels, and corolla glabrous; see above . . . . *Penstemon grandiflorus*

1. **Penstemon tubaeflorus** Nutt. White-wand Beard-tongue

Perennial, to ca. 1 m tall; stems erect, glabrous; cauline leaves elliptic to oblong or oblong-lanceolate, margins entire to crenate or obscurely toothed; calyx 2–5 mm long; corolla white or whitish, 1.5–2.3 cm long, trumpet-shaped: the tube gradually widening from base to apex, subactinomorphic, the limb abruptly flared, with minute glandular pubescence on all the inner surface of the corolla; staminode slightly bearded; capsule 5–7 mm long. A species of south-central United States and the lower Midwest, escaped from cultivation and naturalized in the northeastern states. The only known Ohio record is a 1916 collection from "fields, on gravel terrace" from Clermont County in southwesternmost Ohio, where the species—as shown in Pennell's (1935, p. 244) distribution map—was presumably at a northeastern boundary of its native range. Perhaps merely adventive in Ohio. Not illustrated. June.

Crosswhite (1969) discussed the spelling of this specific epithet.

2. **Penstemon digitalis** Sims Foxglove Beard-tongue
incl. *P. alluviorum* Pennell

Perennial, to 1.5 m tall; stems erect, glabrous and shining to less often finely puberulent and dull, sometimes purple-glaucous; cauline leaves lanceolate or narrowly oblong to triangular, narrowly truncate to rounded or cordate-clasping at base, margins entire to denticulate; calyx 3–8 mm long, the lobes lanceolate to ovate, usually attenuate-acuminate at apex; corolla white to pale pink or pale purple, and often with pink or purple lines internally, 15–30 mm long, the tube funnelform, narrowed at base, the upper 2/3 abruptly dilated; anther surface with few to several stiff, white or tan hairs, staminode moderately bearded at apex with white or tan hairs; capsule 6–12 mm long. Nearly throughout the state, but absent from west-central and northwestern counties; sometimes locally abundant; along roads and railroads, in old fields, and in woodland openings and borders. Late May–mid-July.

As treated here, *P. alluviorum* Pennell is included in *P. digitalis* without rank, following the conclusions of Koelling (1964). *Penstemon digitalis* and the next species, *P. calycosus*, are merged under *Penstemon laevigatus* sensu lato by Farwell (1941) and Bennett (1963). Some intermediate Ohio specimens give support to that classification (Cooperrider, 1976).

3. **Penstemon calycosus** Small

Perennial, to 1 m tall; stems erect, glabrous to puberulent; cauline leaves broadly lanceolate to oblong-lanceolate or ovate-lanceolate, broadly rounded at base, margins usually sharply low-serrate—rarely entire; calyx 4–8 mm long, the lobes narrowly lanceolate to linear-attenuate; corolla more or less purple externally but nearly white within, 13–25 mm long, the throat open and inflated; anthers glabrous or slightly spiculate, staminode moderately bearded above; capsule 5–8 mm long. Chiefly in counties of the southwestern quarter of Ohio; open fields, along roads and railroads, woodland openings and borders. Additional records: Clinton and Greene counties (Pennell, 1935). Late May–mid-July.

This species is merged under *Penstemon laevigatus* sensu lato by Farwell (1941), Bennett (1963), and Gleason and Cronquist (1991). In the eastern part of Ohio, there are some specimens that seem to represent intergrades between this species and *P. digitalis* and others seemingly intergrades between this species and *P. laevigatus* (Cooperrider, 1976).

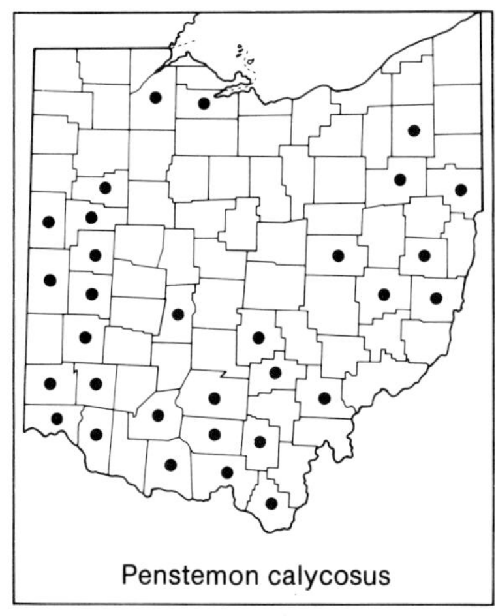
Penstemon calycosus

4. **Penstemon laevigatus** Ait. Smooth Beard-tongue

Perennial, to 1 m tall; stems erect, glabrous to minutely puberulent; cauline leaves lanceolate to narrowly oblong, rounded and clasping at base, margins entire or remotely low-serrate; calyx 2–5 mm long, the lobes oblong-lanceolate to ovate-lanceolate; corolla lavender, 10–17 mm long, the throat open and inflated; anthers glabrous to slightly spiculate, staminode with long, pale yellow hairs on the upper half; capsule 3–7 mm long. A species of the mid-Atlantic and southeastern states, at a northwestern boundary of its range at a few scattered sites in southern counties; fields, woodland openings and borders. May–June.

The two previous species, *P. digitalis* and *P. calycosus*, are merged under this one by Farwell (1941) and Bennett (1963). As I have noted earlier (Cooperrider, 1976), a few Ohio specimens are intergrades between each of those species and this one. Gleason and Cronquist (1991) merge *P. calycosus* under this species but treat *P. digitalis* as distinct.

Penstemon laevigatus

5. **Penstemon canescens** (Britt.) Britt. Gray Beard-tongue

Perennial, to 8 dm tall; stems erect, grayish-puberulent; cauline leaves ovate to broadly oblong or broadly lanceolate or broadly oblanceolate, rounded to cordate-clasping at base, margins unevenly serrate-dentate or rarely subentire, sparsely hirsute on both surfaces, but the midrib usually

Penstemon canescens

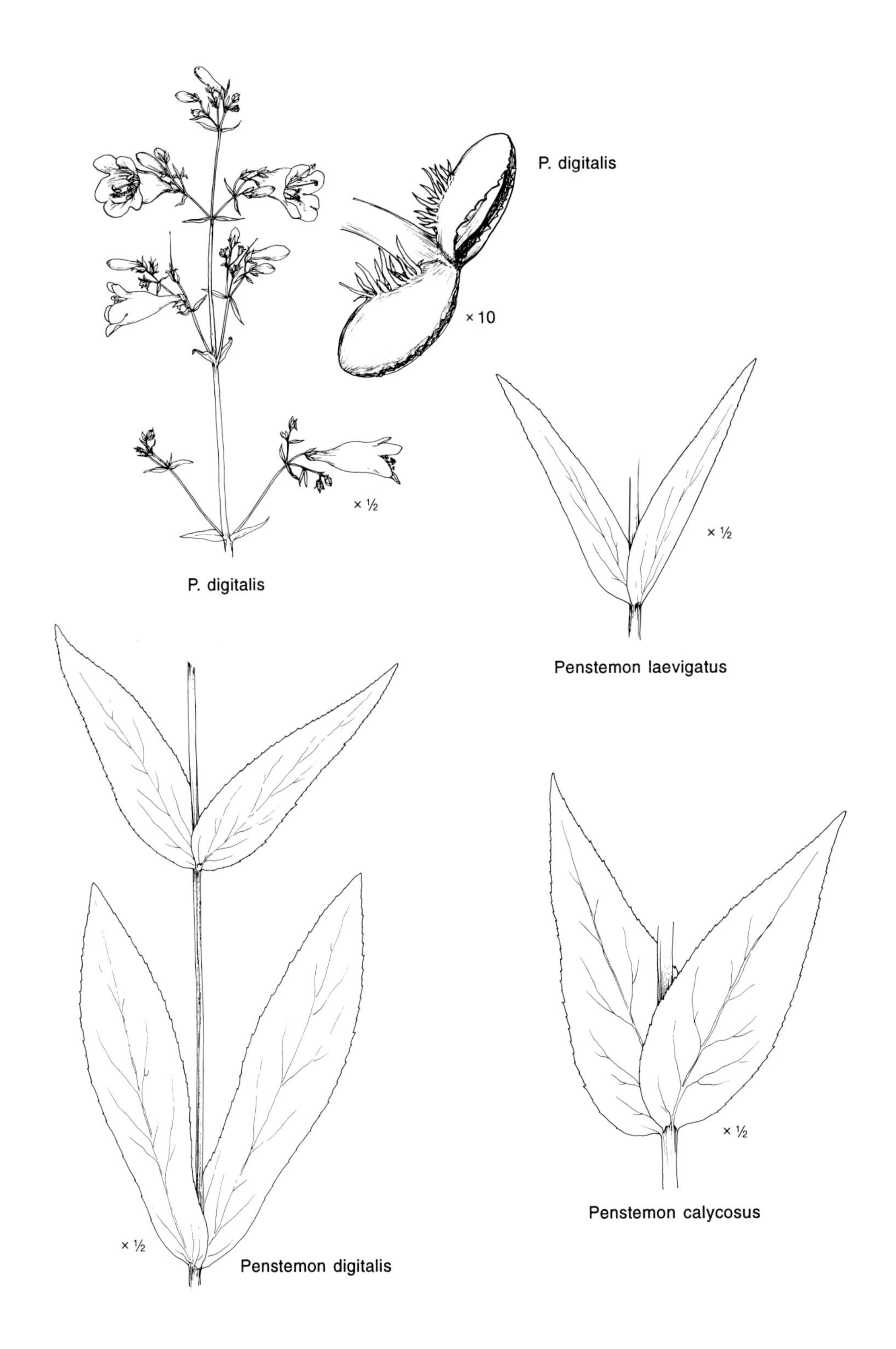
P. digitalis
× 10
× ½
P. digitalis
× ½
Penstemon laevigatus
× ½
Penstemon digitalis
× ½
Penstemon calycosus

densely hirsute on undersurface; calyx 2.5–5 mm long, the lobes ovate-lanceolate; corolla pale purple to pinkish, with darker purple lines in throat, 18–30 mm long, the lower lip longer than the upper and projecting forward, the throat flattened but open; anthers glabrous, spiculate along the valve edges, anther sacs longer than wide, staminode densely bearded; capsule 5–8 mm long. A species of southeastern United States, in Ohio known only from Lawrence County, the southernmost in the state, where the species is at a northern boundary of its range; rocky woodland openings and borders. May–early July.

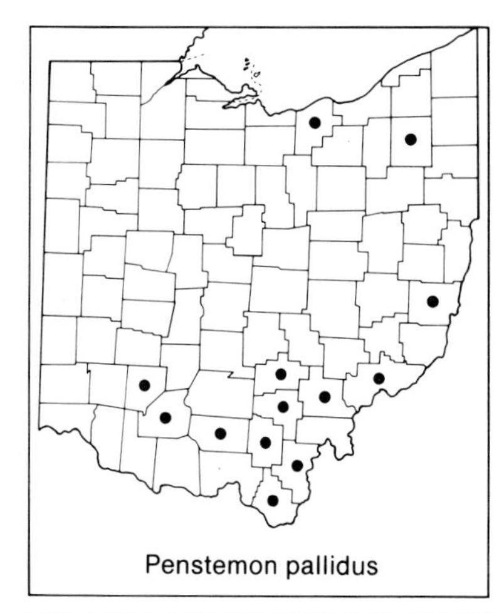
Penstemon pallidus

6. **Penstemon pallidus** Small DOWNY WHITE BEARD-TONGUE
Perennial, to 8 dm tall; stems erect, puberulent to densely pubescent; cauline leaves narrowly lanceolate to lanceolate or oblong-lanceolate, mostly rounded-clasping at base, margins remotely serrate-dentate to subentire, pubescent to densely pubescent on both surfaces; calyx 2.5–4.5 mm long, the lobes ovate to elliptic; corolla white, the tube lined with purple within, 17–22 mm long, the lower lip longer than the upper, the throat open but flattened; anthers glabrous, spiculate along the valve edges, anther sacs longer than wide, staminode densely bearded with yellow hairs for most of its length; capsule 4–7 mm long. Chiefly in south-central counties; woodland openings and borders, roadsides. Additional records: Franklin, Greene, and Hamilton counties (Pennell, 1935). May–June.

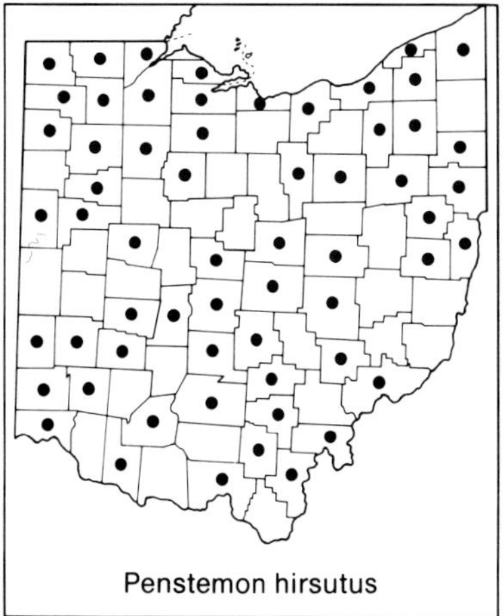
Penstemon hirsutus

7. **Penstemon hirsutus** (L.) Willd. HAIRY BEARD-TONGUE
Perennial, to 8 dm tall; stems erect, short-puberulent and usually also with longer glandular or nonglandular hairs; cauline leaves narrowly lanceolate to oblong-lanceolate or ovate-lanceolate, narrowly truncate to rounded-clasping at base, margins serrate to subentire; calyx 4–6 mm long, the lobes usually acuminate to attenuate at apex and becoming recurved; corolla tube purple to whitish, lips white, with no or only indistinct lines of deeper color within, 20–30 mm long, the tube wide and flat, the lower lip longer than the upper, projecting forward and arching upward and nearly closing the throat; anthers glabrous, anther sacs as broad as long, saucer-shaped when open at maturity, upper half of staminode densely bearded with yellow or pale hairs; capsule 6–9 mm long. A distinctive species growing throughout most of Ohio; dry woodlands and openings, slopes, fields, roadside banks. Additional record: Trumbull County (Pennell, 1935). May–mid-July.

11. Mimulus L. MONKEY-FLOWER

Perennials; stems 4-angled; leaves opposite; the flowers solitary in the axils of upper leaves; calyx shallowly 5-lobed; corolla usually purplish-blue, bi-

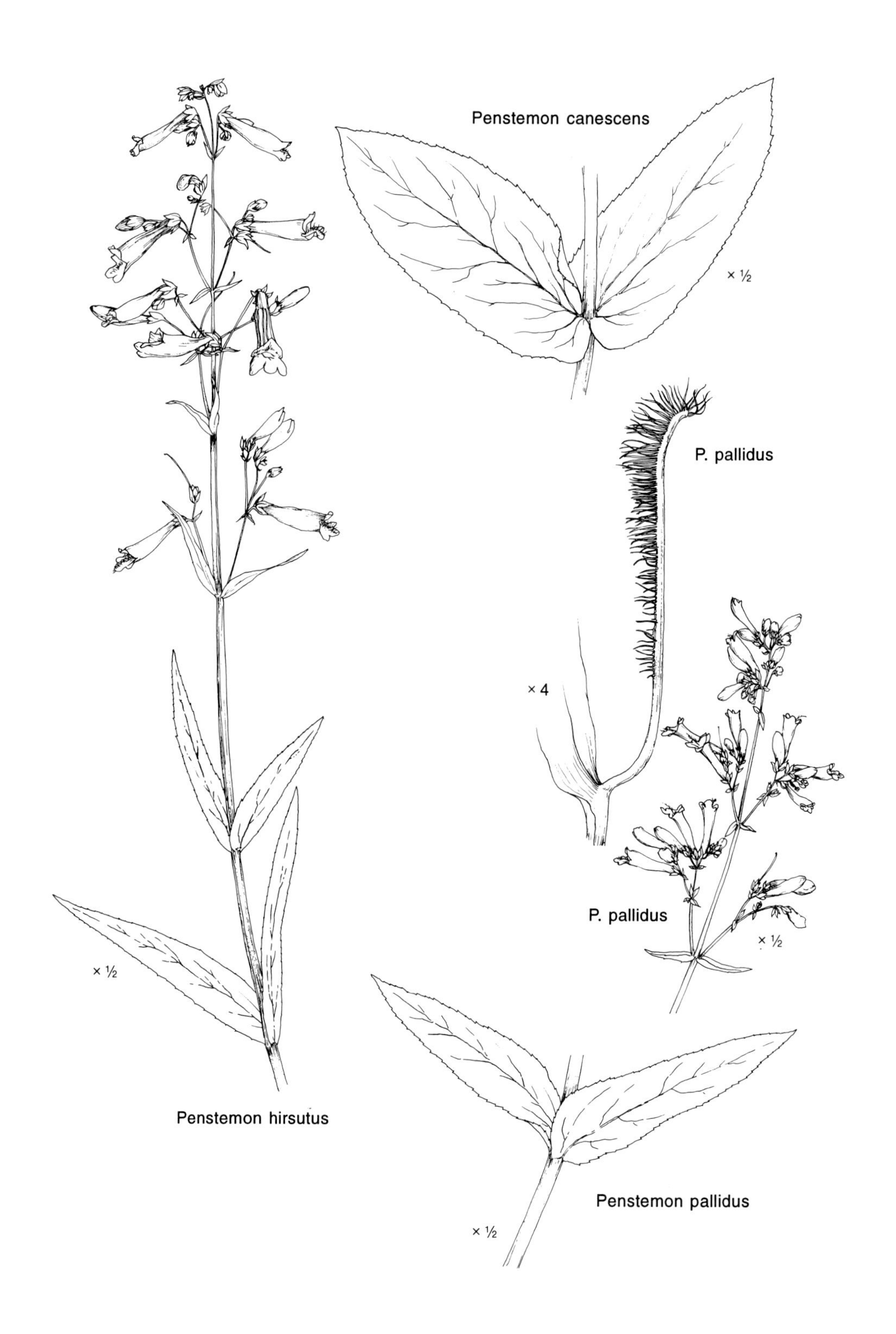
Penstemon canescens
× ½
P. pallidus
× 4
P. pallidus
× ½
× ½
Penstemon hirsutus
Penstemon pallidus
× ½

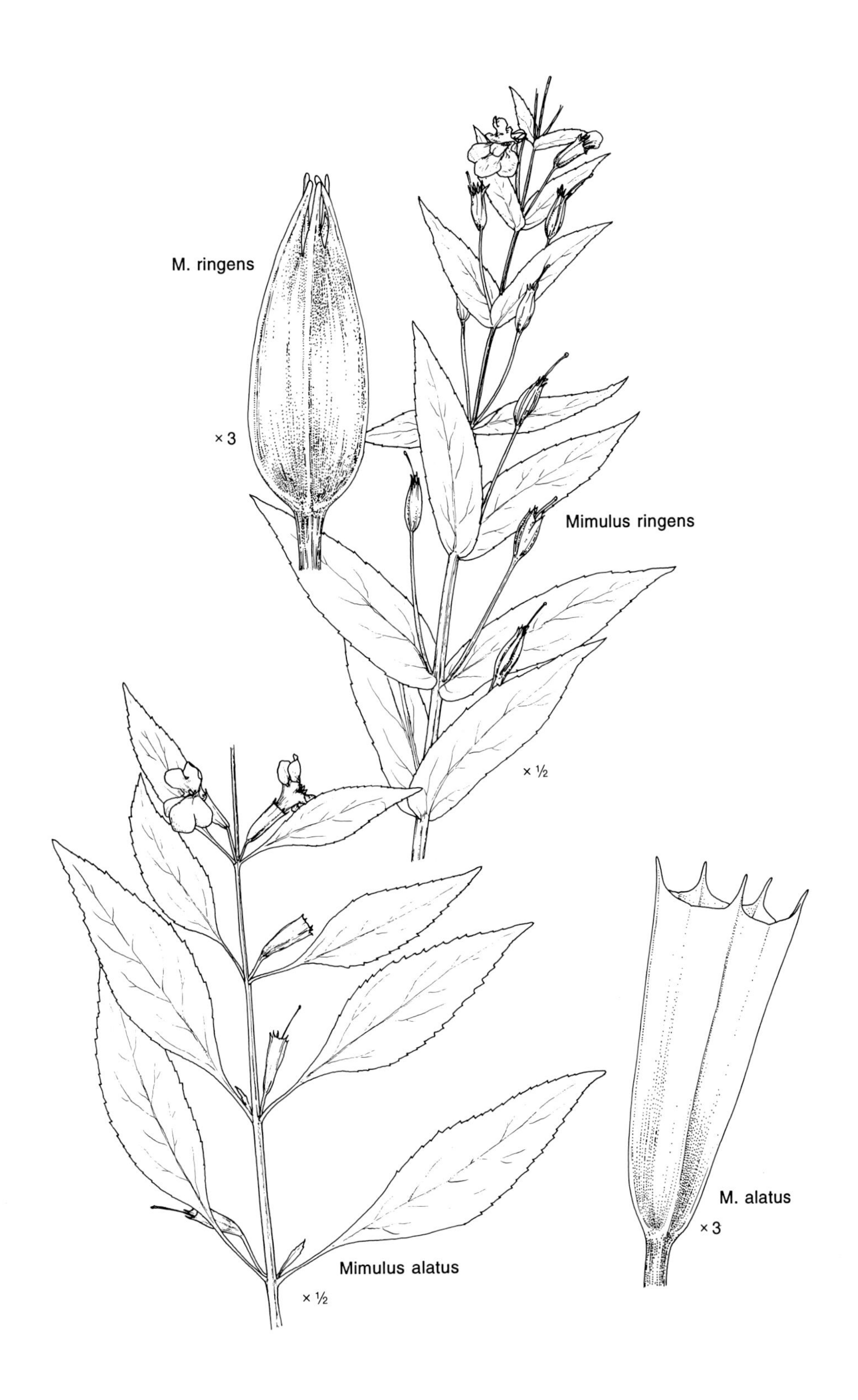
M. ringens
× 3
Mimulus ringens
× ½
M. alatus
× 3
Mimulus alatus
× ½

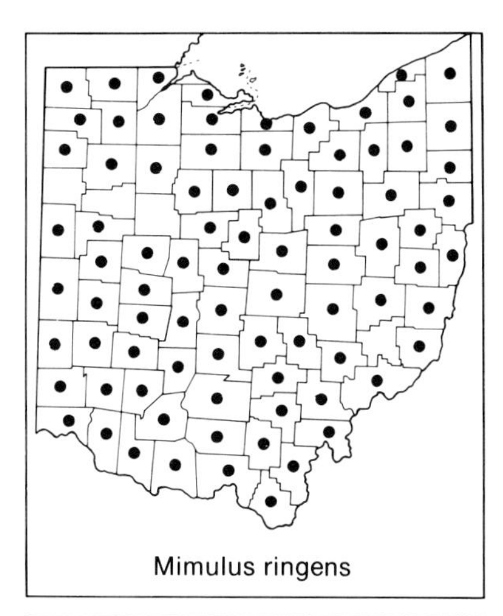
Mimulus ringens

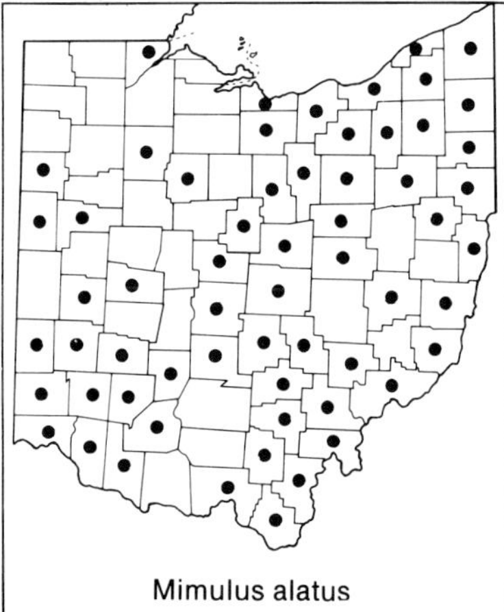
Mimulus alatus

labiate and with a palate closing the throat; stamens 4 and didynamous, staminode none. A cosmopolitan genus of 100 or more species. Among Maryland plants, Windler, Wofford, and Bierner (1976) reported naturally occurring hybrids between the two species listed below, but no hybrids have been found in Ohio.

a. Leaves sessile, broad at base and usually auriculate-clasping; peduncles 1–5 cm long, longer than the calyx; calyx lobes acutely triangular-attenuate, 3–8 mm long . . . . 1. *Mimulus ringens*

aa. Leaves petiolate, tapering at base, not clasping; peduncles 0.2–1.0 (–1.8) cm long, mostly shorter than the calyx; calyx lobes low, broad, obtuse, terminated by a narrow tooth, lobes (incl. teeth) 2–3.5 mm long . . . . 2. *Mimulus alatus*

1. **Mimulus ringens** L. COMMON MONKEY-FLOWER

Perennial, to ca. 1 m tall; stems erect, angles of the stem not or only narrowly winged; leaves sessile, broad at base and usually auriculate-clasping; peduncles 1–5 cm long, longer than the calyx; calyx lobes acutely triangular-attenuate, 3–8 mm long; corolla purplish-blue or rarely white, 2.0–3.5 cm long. Throughout Ohio; stream banks, margins of lakes and ponds, swamps, wet meadows. June–September.

Ohio plants belong to var. *ringens* (incl. var. *minthodes* (E. Greene) A. L. Grant).

2. **Mimulus alatus** Ait. SHARP-WINGED MONKEY-FLOWER

Perennial, to 1 m tall; stems erect, angles of the stem more or less winged; leaves petiolate, narrowed at the base, not clasping; peduncles 0.2–1.0 (–1.8) cm long, mostly shorter than the calyx; calyx lobes low, broad, and obtuse, terminated by a narrow tooth, lobes (incl. teeth) 2.0–3.5 mm long; corolla purplish-blue or rarely white, 2.0–3.5 cm long. Eastern, central, and southern Ohio, absent from most of the northwestern quarter of the state; wet fields, stream banks, swamps. Additional record: Hardin County (Pennell, 1935). Late June–September.

12. Gratiola L. HEDGE-HYSSOP

Annuals or perennials; stems erect to sprawling; leaves opposite; flowers solitary in leaf axils; calyx deeply 5-lobed, closely subtended by two sepaloid bracts; corolla white, sometimes tinged with green or yellow or purple, bilabiate; stamens 2, staminodes 2, 1, or none; capsules symmetric. A genus of about 20 species, widely distributed but chiefly in temperate regions.

a. Leaves widest at the base and somewhat clasping; capsules 2–3 mm long, much shorter than calyx lobes; plants perennial, with creeping rhizomes; stems puberulent or glandular . . . . 3. *Gratiola viscidula* aa. Leaves widest well above, and tapering at, the base, scarcely or not clasping; capsules (2–) 3–7 mm long, about equalling or slightly longer than calyx lobes; plants annual, without creeping rhizomes; stems glabrous to short glandular-puberulent.

b. Upper stem short glandular-puberulent; peduncles slender, (5–) 10–30 mm long; corolla 8–11 mm long. . . . . 1. *Gratiola neglecta*

bb. Stem usually glabrous throughout; peduncles stout, 1–5 (–10) mm long; corolla 10–14 mm long . . . . 2. *Gratiola virginiana*

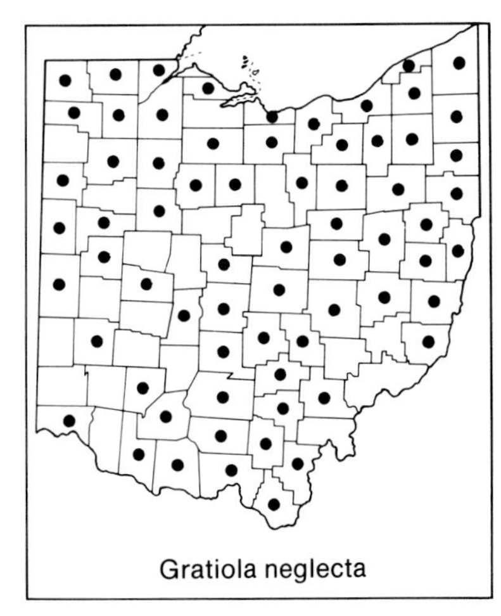
Gratiola neglecta

1. **Gratiola neglecta** Torr. COMMON HEDGE-HYSSOP

Annual; stems reclining to ascending or suberect, short glandular-pubescent above, to 4 dm long; leaves widest at midpoint or above, tapering and not or scarcely clasping at base; peduncles slender (5–) 10–30 mm long; corolla white or white tinged with green or yellow, 8–11 mm long; capsules (2–) 3–5 mm long, about equalling calyx lobes. Throughout much of Ohio, inconspicuous and perhaps under-collected; wet habitats, often in muddy ground at margins of streams, lakes, and ponds, and especially at such sites when they are becoming desiccated. Late May–early October.

Herbarium specimens are often confused with those of *Lindernia dubia* (p. 467). They can be distinguished by these two characters: flowers and fruits of *Gratiola* are closely subtended by two sepaloid bracts, and the capsules are symmetric; flowers and fruits of *Lindernia* are without closely subtending bracts, and the capsules are somewhat asymmetric.

Gratiola virginiana

2. **Gratiola virginiana** L. ROUND-FRUITED HEDGE-HYSSOP

Annual, to 4 dm tall; stems more or less erect, usually glabrous throughout; leaves widest at midpoint or above and tapering to base; peduncles stout, 1–5 (–10) mm long; corolla white above, the tube tinged with yellowish-green, 10–14 mm long; capsules 4–7 mm long, about equalling calyx lobes. A species of the southeastern states at a northern boundary of its range in Ohio; known from a few sites in southern counties and from Erie County in the northernmost tier; shady stream margins and wet ditches. Late May–October.

Gratiola viscidula

3. **Gratiola viscidula** Pennell

incl. var. *viscidula* and var. *shortii* (Pennell) Gleason; Spooner (1984) erred in concluding that the latter combination was not validly published.

Perennial, with creeping rhizomes; stems creeping, 1–5 dm long, with a few erect to sprawling branches, these puberulent or glandular; leaves widest at the base and somewhat clasping; peduncles slender, 7–30 mm long; corolla white, tinged with pale purple above, the tube yellowish and with many reddish-brown lines, 11–14 mm long; capsules 2–3 mm long,

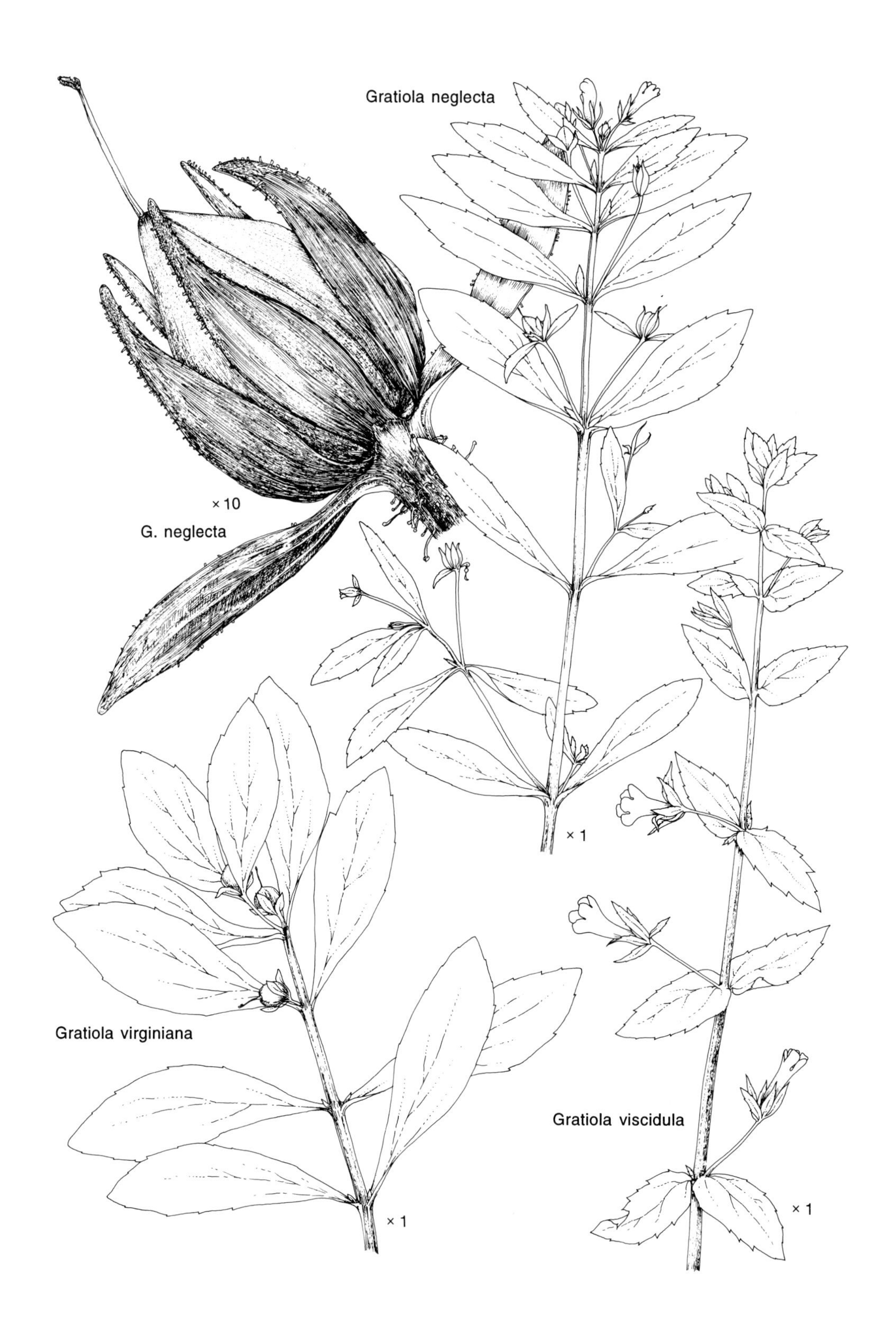
Gratiola neglecta
×10
G. neglecta
×1
Gratiola virginiana
×1
Gratiola viscidula
×1

much shorter than the calyx lobes. A species chiefly of the mid-Atlantic states and areas inland from them, at a northern boundary of its range in a few southernmost counties of Ohio; margins of ponds and lakes, moist roadside ditches, and similar habitats. Late June–October.

No infraspecific taxa are recognized within this species, Spooner (1984) having shown the intraspecific variation pattern to be continuous and not organized geographically.

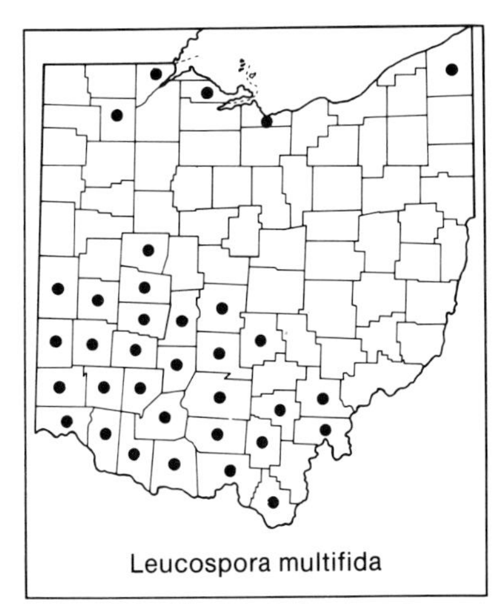
Leucospora multifida

13. Leucospora Nutt.

A genus with a single species.

1. **Leucospora multifida** (Michx.) Nutt.
   Conobea multifida (Michx.) Benth.

Annual; stems erect or decumbent, to 2 dm long, puberulent or glandular-puberulent; leaves opposite or in whorls of 3, deeply pinnately lobed with 3–7 linear to oblong segments; flowers solitary in leaf axils; calyx deeply 5-lobed; corolla pale purple to greenish-white, 3–5 mm long, bilabiate; stamens 4. A species of the central and lower Mississippi Valley, in Ohio at a northeastern boundary of the species' range; south-central and southwestern parts of the state, and at a few sites in northern counties; open calcareous habitats, often at margins of ponds and streams, and adventive on railroad ballast. Additional record: Shelby County (Pennell, 1935). Mid-July–early October.

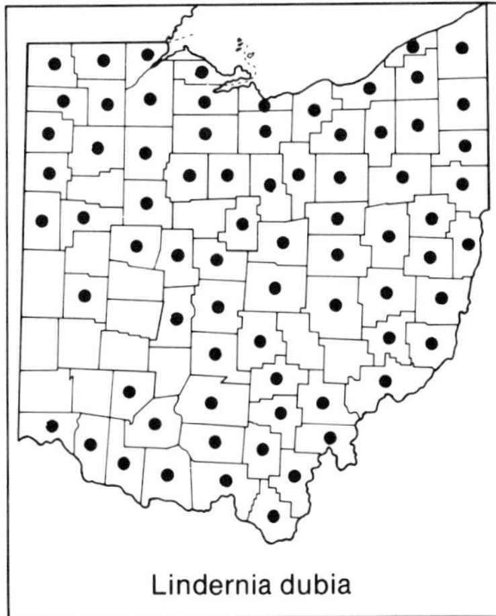
Lindernia dubia

14. Lindernia All. FALSE PIMPERNEL

A widespread genus of about 70 species, one of them native to Ohio.

1. **Lindernia dubia** (L.) Pennell FALSE PIMPERNEL

Annual, to 3 dm tall; stems erect to ascending, glabrous; leaves opposite; flowers solitary in leaf axils; calyx deeply 5-lobed; corolla pale purple to white, 5–10 mm long, bilabiate; fertile stamens 2, staminodes 2; capsules somewhat asymmetric. Throughout most of Ohio, except some western counties; mud flats, margins of lakes, ponds, and streams. See comments under *Gratiola neglecta* (p. 465) on features distinguishing that species from this. Late June–early October.

Some members of this species can be identified as belonging to one of the varieties keyed below, but many Ohio plants are intermediate between the two (Cooperrider and McCready, 1975; Cooperrider, 1976). Plants assignable to var. *dubia* occur throughout the species' Ohio range; the few assignable to var. *anagallidea* are chiefly from east-central and southeastern counties. The intermediates are widespread.

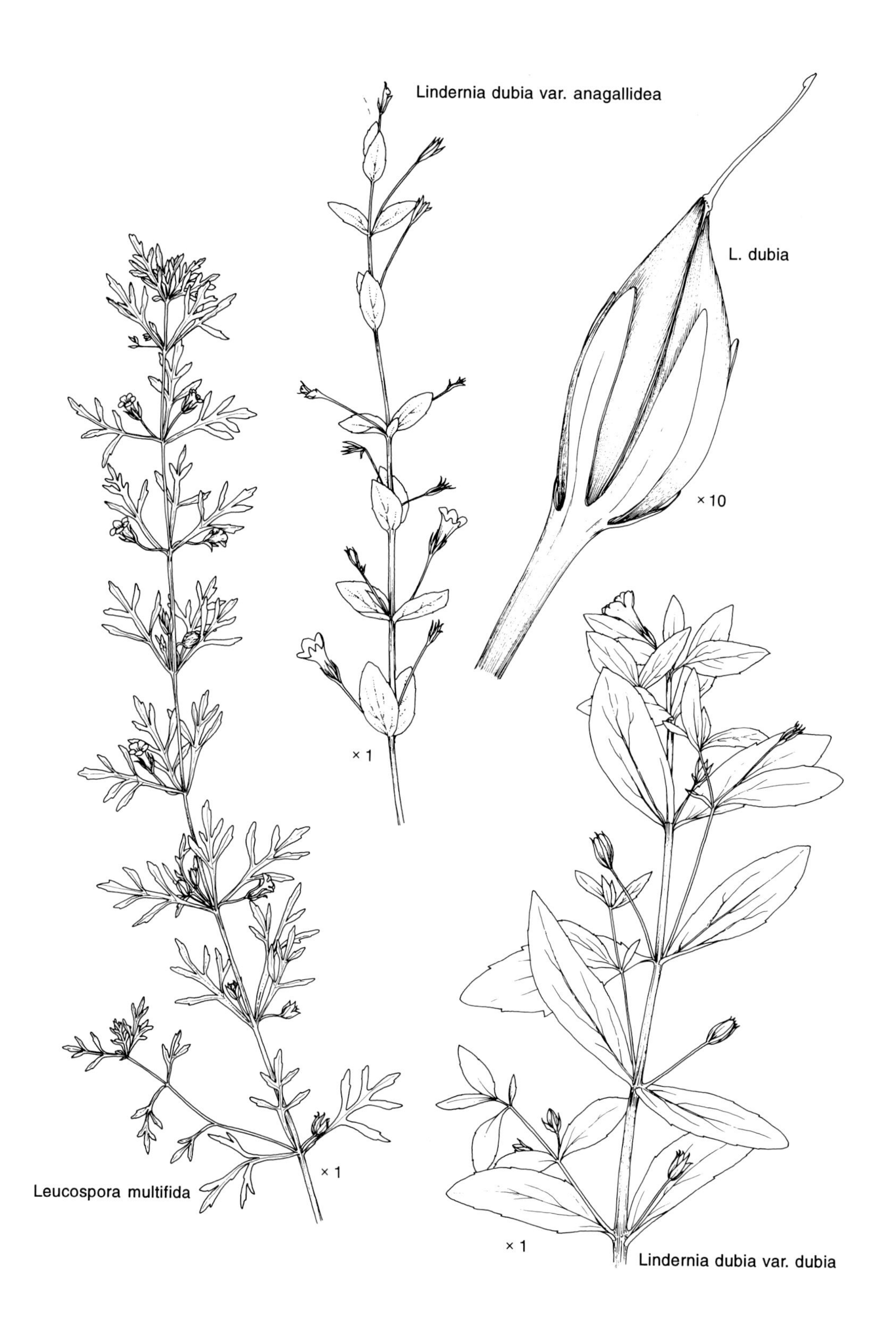
Lindernia dubia var. anagallidea
L. dubia
×10
×1
×1
Leucospora multifida
×1
Lindernia dubia var. dubia

a. Leaves 10–32 mm long, tapering at base; peduncles shorter than to slightly exceeding subtending leaves. . . . . var. *dubia*
incl. var. *riparia* (Raf.) Fern.

aa. Leaves 5–20 mm long, somewhat rounded and clasping at base; peduncles slightly longer than to much exceeding length of subtending leaves . . . . var. *anagallidea* (Michx.) Cooperr.
L. anagallidea (Michx.) Pennell

15. Digitalis L. FOXGLOVE

A Eurasian genus of 20 or more species, several of them cultivated as ornamentals. Adventive escapes of four of the species have been collected in Ohio. *Digitalis grandiflora* Mill., YELLOW FOXGLOVE, was collected from Portage County in 1974 and 1978, and from Scioto County in 1938. *Digitalis lanata* Ehrh., GRECIAN FOXGLOVE, was collected from Pickaway County in 1936 and 1952, and from Trumbull County in 1952. *Digitalis lutea* L., STRAW FOXGLOVE, was collected from Cuyahoga County in 1890. *Digitalis purpurea* L., COMMON FOXGLOVE, the showiest and most widely cultivated of the species (and the source of the medicinal drug "digitalis"), was collected from Cuyahoga County in 1896, and from Lake County in 1901 and 1985. The key below is based in part on the study of cultivated *Digitalis* species by Werner (1962).

a. Calyx lobes ovate to broadly elliptic; lower leaves long-petiolate; leaf margins crenate to crenate-dentate; corolla 4–6 cm long, purple, red, or white; see above . . . . *Digitalis purpurea*

aa. Calyx lobes linear to narrowly ovate-lanceolate; lower leaves short-petiolate or sessile; leaf margins entire, undulate, or finely serrate; corolla 4 cm long or less, yellow to white.

b. Inflorescence villous or glandular-puberulent; corolla 20–40 mm long, brown-veined.

c. Inflorescence glandular-pubescent; margins of leaves finely serrate; lower median lobe of corolla not prolonged; see above . . . . *Digitalis grandiflora*

cc. Inflorescence downy-villous; margins of leaves entire or undulate; lower median lobe of corolla much prolonged, equalling length of corolla tube; see above . . . . *Digitalis lanata*

bb. Inflorescence glabrous; corolla 10–20 mm long, without colored veins; see above . . . . *Digitalis lutea*

16. Besseya Rydb. KITTEN-TAILS

A genus of seven species, all but the one below limited to western North America. Kruckeberg and Hedglin (1963) found evidence supporting the

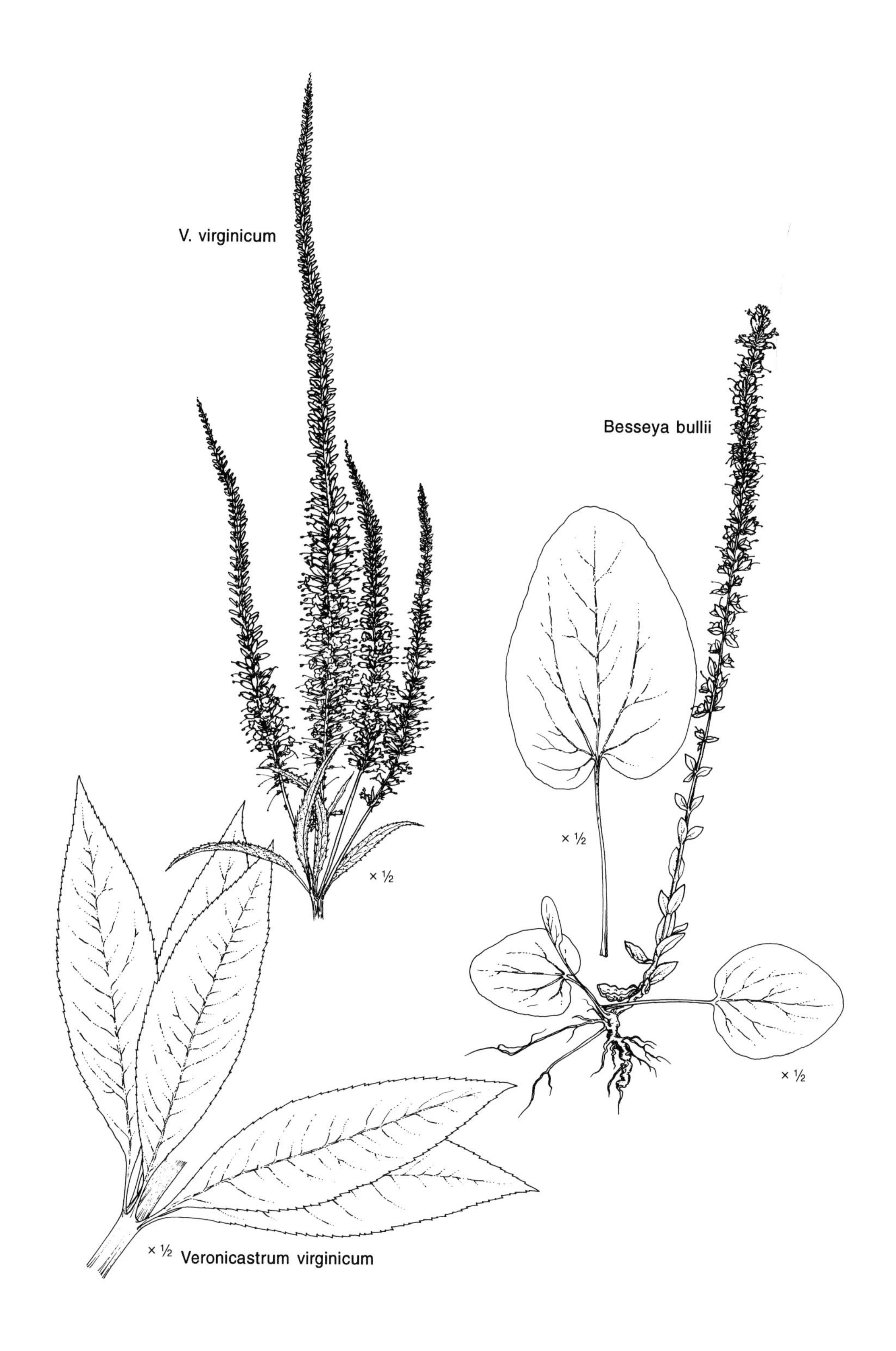
V. virginicum
Besseya bullii
× ½
× ½
× ½
× ½ Veronicastrum virginicum

merger of *Besseya* under *Synthyris* Benth. However, following the conclusions of Schaack (1983), who also kindly discussed the matter in personal correspondence, it is treated here as a separate genus.

1. **Besseya bullii** (Eat.) Rydb. KITTEN-TAILS

Synthyris bullii (Eat.) Heller; *Wulfenia bullii* (Eat.) Barnh.

Perennial, to 4 dm tall, with a rosette of large, petiolate, basal leaves; stems erect, pubescent; cauline leaves alternate, sessile or nearly so, much smaller than the basal leaves; flowers very short-pedicellate, in a terminal spicate raceme; calyx deeply 4-lobed; corolla yellow to pale greenish, 6–8 mm long, bilabiate; stamens 2. A species of the upper Mississippi valley, at an eastern border of its range in southwestern Ohio, where it has been collected only from Montgomery County; prairies, barren fields. May–early June.

Besseya bullii

17. Veronicastrum Fabricius CULVER'S-PHYSIC

A genus of two species, the other in eastern Asia.

1. **Veronicastrum virginicum** (L.) Farw. CULVER'S-ROOT

Perennial, to 2 m tall; stems erect, glabrous or somewhat villous—especially at base; leaves in whorls of 4–6; flowers very short-pedicellate, in a terminal spicate raceme and usually 1–several additional racemes originating from upper axils; calyx deeply 5-lobed; corolla white to partly pinkish, 4–7 mm long, 4-lobed and nearly actinomorphic; stamens 2, conspicuously exserted. Throughout eastern Ohio, becoming less frequent westward; woodland openings, thickets, along roads and railroads. Additional record: Pickaway County (Pennell, 1935). July–early October.

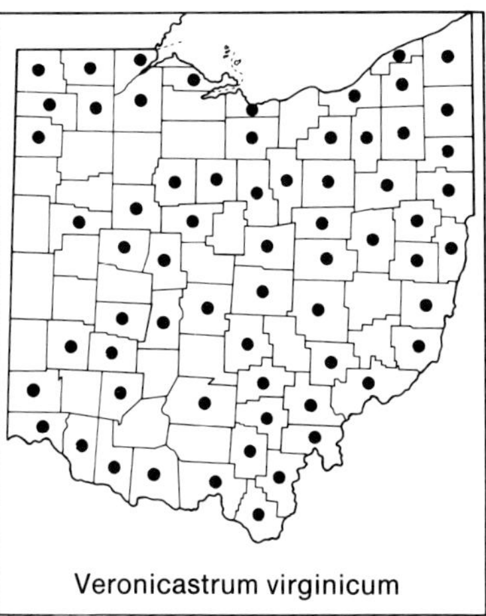
Veronicastrum virginicum

18. Veronica L. SPEEDWELL

Annual or perennial herbs; leaves opposite or partly alternate; flowers in racemes or solitary in leaf axils; calyx usually deeply 4-lobed; corolla blue or purple to white, often with darker colored lines, usually 4-lobed and only slightly zygomorphic; stamens 2, staminodes none; capsules flattened, often notched at the apex. A genus of the North Temperate Zone, with about 250 species. Three Eurasian species have been collected as adventives in Ohio: *Veronica agrestis* L., FIELD SPEEDWELL, from Summit County in 1955; *V. latifolia* L. (*V. austriaca* L., ? *V. teucrium* L.), BROAD-LEAVED SPEEDWELL, from Medina County in the 1890s; and the frequently cultivated *V. longifolia* L., GARDEN SPEEDWELL, from Cuyahoga, Hamilton, Lake, and Ottawa counties, several times during the present century.

a. Flowers in axillary or terminal racemes, each flower subtended by a small narrow bract.

b. Main stem terminated by an inflorescence, additional axillary inflorescences present or not.

c. Stem erect, 0.3–1 m tall; leaves opposite or in whorls of 3, petioles 5–15 mm long, blades 3–10 cm long; see above. . . . . *Veronica longifolia*

cc. Stems decumbent to more or less erect, 5–25 cm long; leaves opposite, petioles absent or to 3 mm long, blades 0.5–1.7 cm long . . . . 1. *Veronica serpyllifolia*

bb. Main stem not terminated by an inflorescence, racemes 1–several, all axillary.

c. Stems pubescent; pedicels less than 1½ times as long as calyx; capsules pubescent.

d. Stems creeping, ascending at apex; leaves elliptic to obovate; corolla pale purple with darker purple lines, ca. 5–7 mm wide; capsules much exceeding calyx . . . . 2. *Veronica officinalis*

dd. Stems erect or ascending; leaves lanceolate to ovate or subcordate; corolla blue, 6–12 mm wide; capsules shorter than to slightly exceeding calyx.

e. Leaves lanceolate to ovate, 1.5–4 cm long; stems 3–7 dm tall, more or less uniformly pubescent; calyx lobes 5, unequal; see above . . . . *Veronica latifolia*

ee. Leaves ovate to subcordate, 1–3 cm long; stems 1–3 dm tall, with pubescence in two lines; calyx lobes 4, nearly equal . . . . 3. *Veronica chamaedrys*

cc. Stems usually glabrous, rarely with scattered pubescence; pedicels equalling to much longer than calyx; capsules glabrous.

d. Leaves linear to lanceolate, thin, entire to remotely denticulate; racemes with flexuous rachis; pedicels filiform, divergent to more or less deflexed; capsules exceeding calyx . . . . 4. *Veronica scutellata*

dd. Leaves lanceolate to oblong, elliptic, or ovate, somewhat fleshy, subentire to crenate or serrate; racemes with straight or arching rachis; pedicels divergent to ascending; capsules shorter than to only slightly exceeding calyx.

e. Leaves of flowering stems petiolate.

f. Principal leaves lanceolate to oblong-lanceolate or narrowly ovate, acute to subobtuse at apex, their margins serrate to sharply serrate; pedicels (6–) 7–12 mm long. . 5. *Veronica americana*

ff. Principal leaves ovate to oblong-orbicular, obtuse to rounded at apex, their margins subentire to shallowly crenate-serrate; pedicels 2–4 (–5) mm long . . . . 6. *Veronica beccabunga*

ee. Leaves of flowering stems sessile and sometimes cordate-clasping (those of sterile autumnal branches sometimes petiolate).

f. Corolla light blue-purple; capsules broadly elliptic to suborbicular, slightly longer than wide, scarcely or not notched. . . . . 7. *Veronica anagallis-aquatica*

ff. Corolla white or pale pinkish-purple; capsules suborbicular to orbicular-reniform or broadly obcordate, wider than long, shallowly to deeply notched . . . . 8. *Veronica catenata*

aa. Flowers solitary in leaf axils.

b. Flowers sessile or on peduncles to 1 mm (rarely 2 mm) long, borne in axils of ordinary or much reduced leaves; stems erect to ascending.

c. Lower leaves oblong to linear-oblong or spatulate, 1–3 cm long, glabrous and somewhat succulent; calyx lobes glabrous, nearly equal, 3–7 mm long; corolla white; capsules 4–5 mm wide, wider than long . . . . 9. *Veronica peregrina*

cc. Lower leaves ovate, 0.5–1.5 cm long, pubescent, not succulent; calyx lobes pubescent, unequal, the lower pair 3–5 mm long, the upper pair 2–3.5 mm long; corolla bright blue; capsules 3–4 mm wide and equally as long . . . . 10. *Veronica arvensis*

bb. Flowers on peduncles 2 mm long or more, borne in axils of ordinary leaves; stems prostrate to erect.

c. Leaves of two types: blades of middle and upper leaves 3–5-palmately lobed, wider than long, orbicular to reniform, blades of lowest leaves entire and elliptic; calyx lobes broadly ovate to deltoid . . . . 14. *Veronica hederifolia*

cc. Leaves on any one plant more or less alike: entire to serrate or coarsely serrate, not palmately lobed, blades longer than wide or not; calyx lobes ovate, ovate-lanceolate, or narrowly elliptic.

d. Peduncles 2–4 mm long; leaf margins entire or slightly crenate . . . . 1. *Veronica serpyllifolia*

dd. Peduncles 6–25 mm long; leaf margins mostly crenate, serrate, coarsely serrate, or coarsely dentate.

e. Leaf blades wider than long, reniform to cordate-orbicular, 3–7 mm long . . . . 12. *Veronica filiformis*

ee. Leaf blades mostly longer than wide, broadly ovate to oblong or suborbicular, 4–30 mm long.

f. Fruiting peduncles 13–25 mm long; capsules 5–8 mm wide, the capsule lobes rather widely spreading and more or less acutish . . . . 13. *Veronica persica*

ff. Fruiting peduncles 6–18 mm long; capsules 3–5.5 mm wide, the capsule lobes rounded, not widely spreading.

g. Leaf blades ovate to narrowly elliptic, 10–15 mm long, narrow to broadly tapering at base; style shorter than to equalling summit of capsule lobes; capsules slightly pubescent to glabrous; see above. . . . . *Veronica agrestis*

gg. Leaf blades ovate to suborbicular, 4–11 (–17) mm long, rounded to subcordate or truncate at base; style surpassing capsule lobes; capsules rather densely and uniformly pubescent . . . . 11. *Veronica polita*

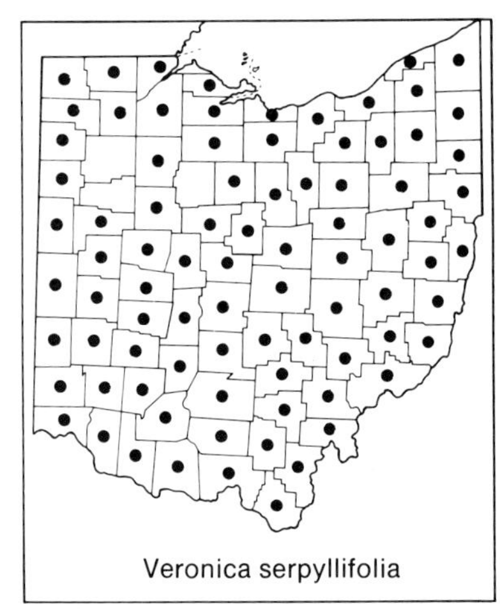
Veronica serpyllifolia

1. Veronica serpyllifolia L. Thyme-leaved Speedwell

Perennial, often forming small mats; stems creeping at base, the upper stem more or less erect, to 25 cm long, puberulent to glabrous; leaves opposite, round-ovate to ovate, elliptic, or oblong, sessile or with petioles to 3 mm long; flowers in an elongated terminal raceme or some appearing solitary in the axils of upper leaves, pedicels 2–4 mm long; corolla whitish to pale blue and with darker purple lines, ca. 3 mm long, 4–6 mm across; capsules sparingly glandular-pubescent. A Eurasian species naturalized throughout Ohio, but more abundant in the eastern half of the state; lawns, pastures, moist open fields. April–June (–October).

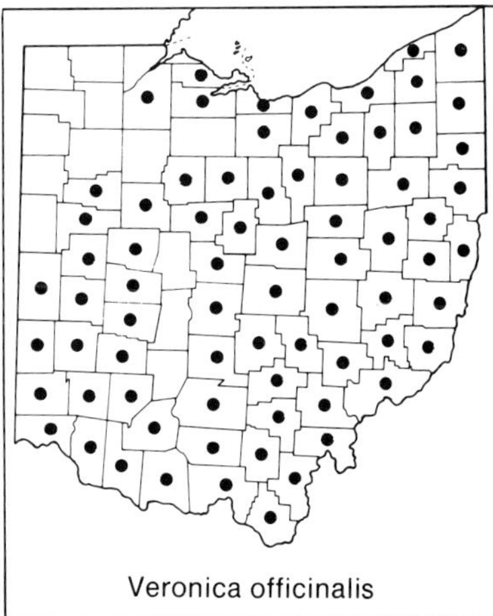
Veronica officinalis

2. Veronica officinalis L. Common Speedwell

Perennial; stems creeping, rooting at nodes, to 5 dm long, the apex and branches ascending, pubescent; leaves opposite, elliptic to obovate, with petioles 1–4 mm long; racemes 1-several, axillary, pedicels 1–2 mm long; corolla pale purple with darker lines or rarely white, 5–7 mm across; capsules glandular-pubescent. A Eurasian species naturalized throughout much of Ohio, but absent from northwestern counties; lawns, fields, pastures, roadsides, disturbed woodlands, waste places. Late April–June (–early August).

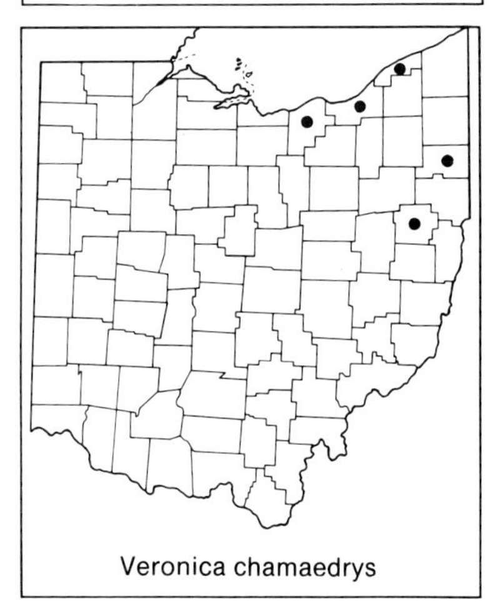
Veronica chamaedrys

3. Veronica chamaedrys L. Germander Speedwell. Bird's-eye Speedwell

Perennial; stems ascending from a creeping base, to 3 dm long, with pubescence in two lines; leaves opposite, subsessile or with petioles to 3 mm long, blades ovate to triangular-ovate or subcordate; flowers in 1–4 racemes originating from upper axils, pedicels 3–8 mm long; corolla blue with darker lines, 6–10 mm across; capsules pubescent. A Eurasian species naturalized, or perhaps merely adventive, at a few sites in northeastern Ohio; lawns, fields, waste places. Late April–June.

4. **Veronica scutellata** L. Marsh Speedwell

Perennial; stems ascending or weakly erect, to 8 (or 10) dm long, glabrous or nearly so; leaves opposite, sessile, linear to lanceolate; racemes several, originating from alternate axils, pedicels at maturity ca. 1 cm long, filiform and more or less deflexed; corolla pale red-purple to blue, 5–6 mm across; capsules glabrous. A northern species at a southern boundary of its range in Ohio; counties of the northeastern quarter and at scattered sites elsewhere in the state; marshes, swamps, shores, and other wet places. Mid-June–early September.

5. **Veronica americana** Benth. American Brooklime

Perennial; stems creeping or sprawling at base and rooting at the nodes, upper stem ascending to erect, to 7 dm long, glabrous and somewhat fleshy; leaves opposite, lanceolate to oblong-lanceolate or narrowly ovate, acute to subobtuse at apex, margins serrate to sharply serrate, petioles 4–10 mm long; racemes several from upper or most axils, pedicels (6–) 7–12 mm long; corolla blue to white, 4–8 mm across; capsules glabrous. Counties of the northeastern quarter of Ohio and at scattered sites elsewhere; shallow streams, margins of springs, marshes. Additional records: Franklin and Hamilton counties (Pennell, 1935). Late May–early September.

6. Veronica beccabunga L. European Brooklime

Perennial; stems ascending, to 5 dm long, glabrous and somewhat fleshy; leaves opposite, the principal ones ovate to oblong-orbicular, obtuse to rounded at apex, margins subentire to shallowly crenate-serrate, petioles 5–10 mm long; racemes several from upper axils, pedicels 2–4 (–5) mm long; corolla blue, 5–7 mm across; capsules glabrous. A Eurasian species, naturalized at a few sites mostly in central Ohio; stream margins, shallow

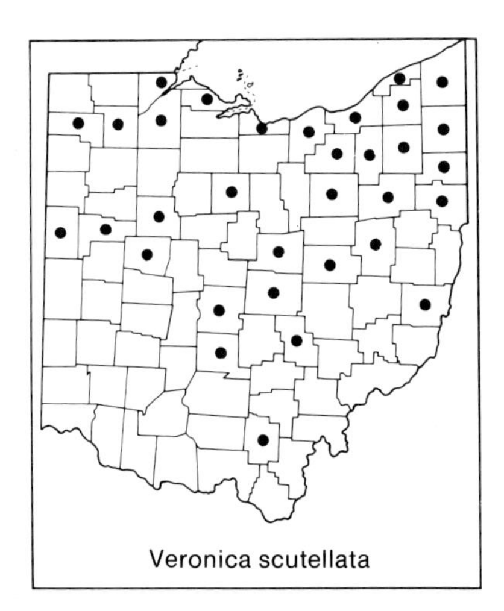
Veronica scutellata

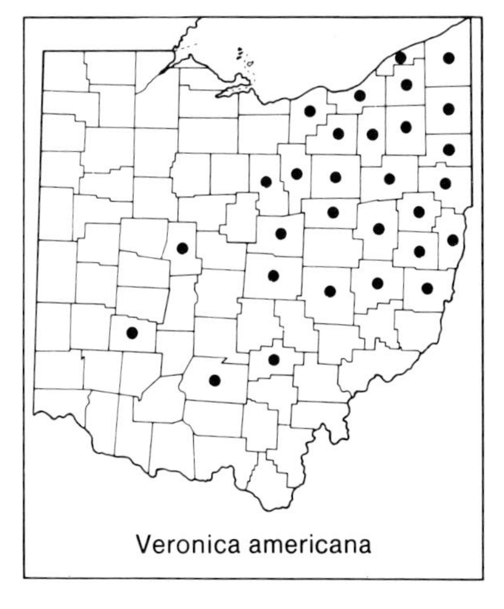
Veronica americana

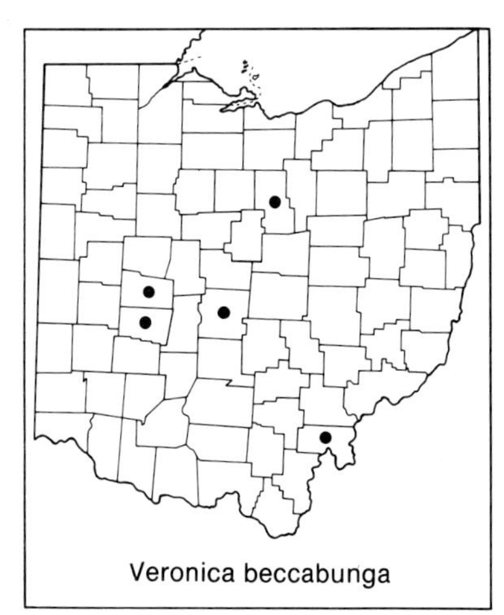
Veronica beccabunga

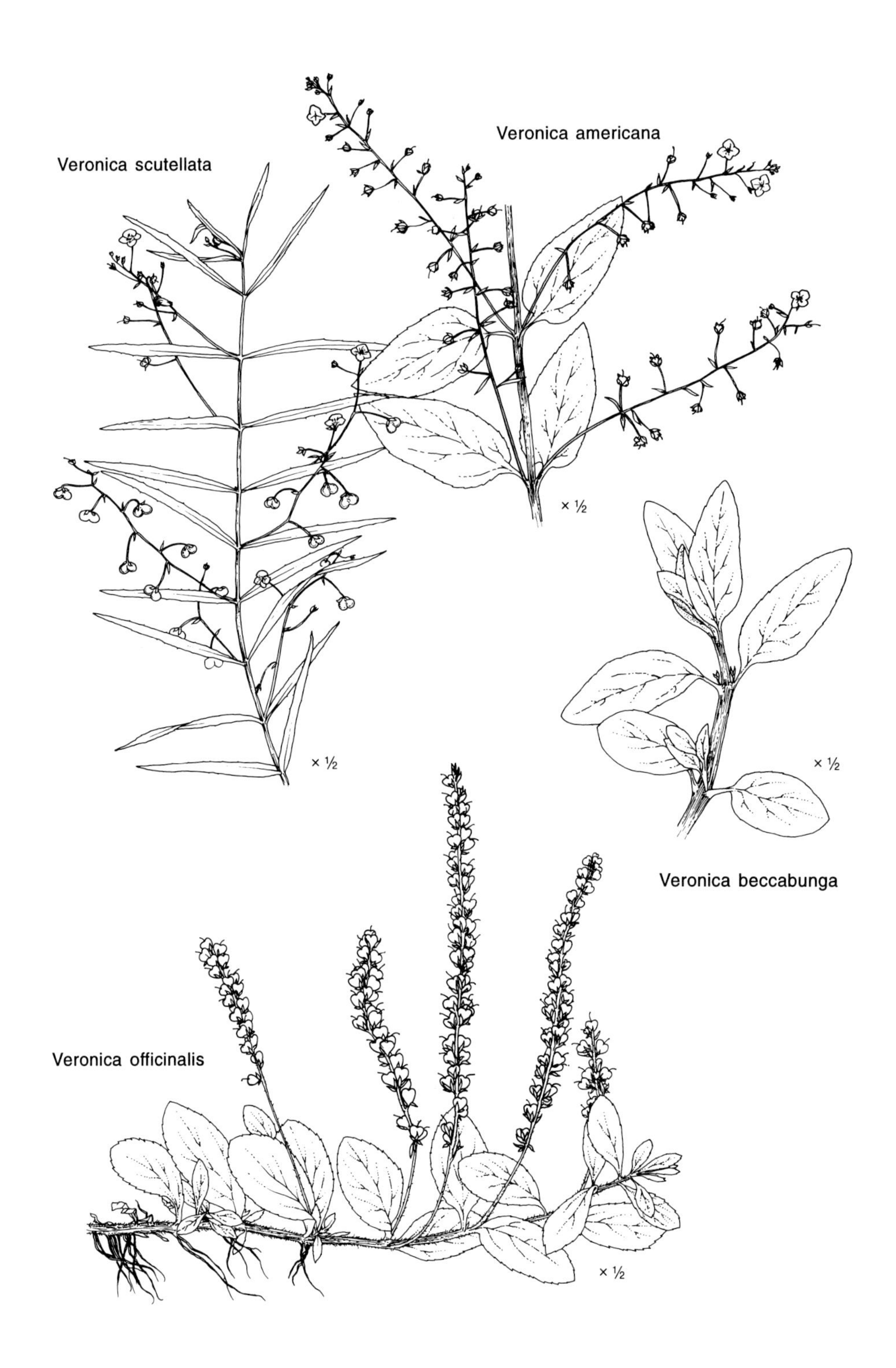
Veronica americana
Veronica scutellata
× ½
× ½
× ½
Veronica beccabunga
Veronica officinalis
× ½

water around springs. Les and Stuckey (1985) described the introduction and spread of this species in eastern North America. The earliest Ohio specimen I have seen was collected from Richland County, in 1967. Additional record: Delaware County (Les and Stuckey, 1985). June–August.

7. **Veronica anagallis-aquatica** L. WATER SPEEDWELL
   incl. *V. glandifera* Pennell

Perennial; stems ascending to erect, to 6 dm long, glabrous or somewhat glandular-pubescent above; leaves opposite, narrowly oblong-lanceolate to elliptic or ovate, sessile (the leaves on sterile autumnal branches petiolate and ovate to orbicular); flowers in axillary racemes, pedicels 3–6 mm long; corolla light bluish-purple with darker purple lines, 4–6 mm across; capsules glabrous, slightly longer than wide, broadly elliptic to suborbicular, scarcely or not notched at apex. Mostly in the northeastern and southwestern quarters of the state, presumably in part native to Ohio and in part naturalized from Eurasia; streams, wet ditches. Late May–early October.

Brooks (1976) and Heckard and Rubtzoff (1977) discuss the naturally occurring hybrid, *V. anagallis-aquatica* × *V. catenata*.

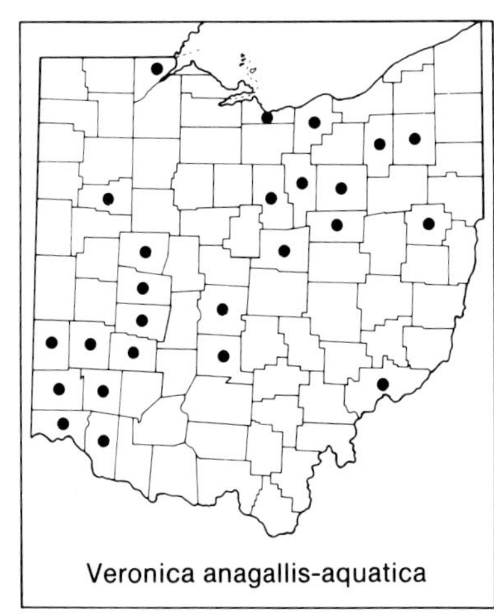
Veronica anagallis-aquatica

8. **Veronica catenata** Pennell WATER SPEEDWELL
   V. comosa of authors, incl. Fernald (1950), not K. Richter; *V. connata* of authors, not Raf.; *V. salina* of authors, incl. Gleason (1952b), not Schur.

Perennial; stems creeping to ascending and nearly erect, to 6 dm long, glabrous or nearly so; leaves opposite, lanceolate to oblong-lanceolate, sessile; flowers in axillary racemes, pedicels 2.5–5.0 mm long; corolla white or pale pinkish-purple with a few darker lines, 4–6 mm across; capsules glabrous, wider than long, suborbicular to orbicular-reniform or broadly obcordate, shallowly to deeply notched at apex. Western counties; calcareous waters in slow streams and wet ditches. Burnett (1950) discusses the nomenclature of this species. June–September.

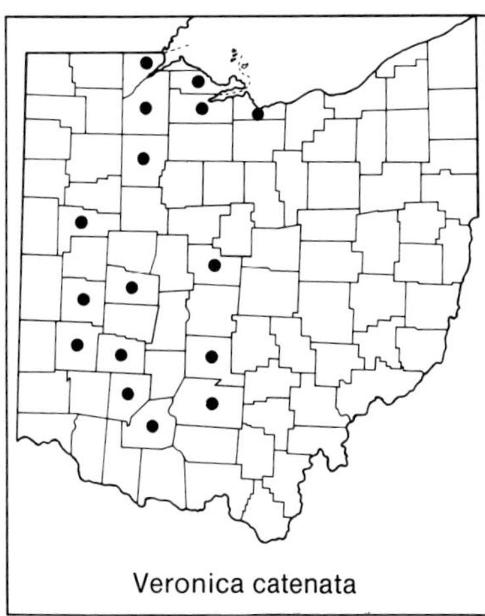
Veronica catenata

9. **Veronica peregrina** L. PURSLANE SPEEDWELL

Annual; stems erect or ascending, to 3 dm long; lower leaves opposite and petiolate, the upper alternate and sessile, oblong to linear-oblong or the lowest somewhat spatulate; flowers solitary in axils of middle and upper leaves, peduncles ca. 1 mm (rarely 2 mm) long; corolla white, 1–2 mm across, exceeded by calyx; capsules glabrous. Throughout Ohio; moist shaded areas at borders of lawns and gardens, and other moist disturbed habitats. Baskin and Baskin (1983c) studied seed germination in this species. Mid-April–early June (–September).

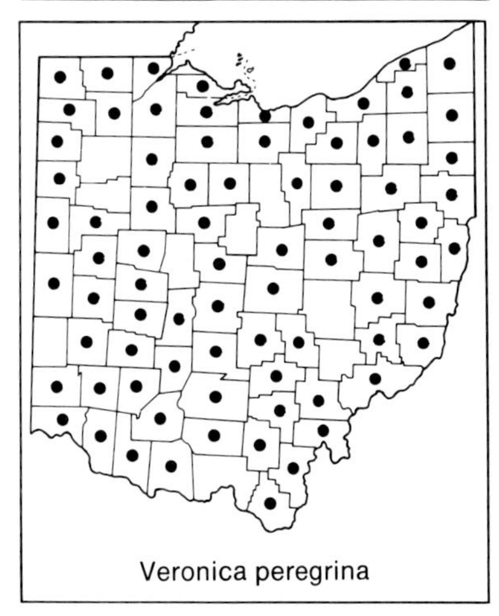
Veronica peregrina

Veronica anagallis-aquatica
× 12
× ½
× 12
Veronica catenata
× ½

10. Veronica arvensis L. Corn Speedwell

Annual; stems erect or ascending, to 4 dm long, villous-puberulent with hairs of two lengths; lower leaves opposite and short-petiolate, upper leaves and bracts sessile or nearly so, principal leaves ovate; flowers solitary in axils of bracts and all but the lowest leaves, peduncles to 1 mm long; corolla blue with a few darker purple lines, 2–3 mm across, exceeded by calyx; capsules pubescent. A Eurasian species naturalized throughout Ohio; open fields, cultivated ground, lawns, roadsides, and similar disturbed habitats. Baskin and Baskin (1983d) studied this species' germination ecology. Mid-April–early June (–early August).

Crins, Sutherland, and Oldham (1987) report the Ontario distribution of the similarly naturalized and closely related *Veronica verna* L.

11. Veronica polita Fries Wayside Speedwell

Annual; stems erect to ascending, to 3 dm long; puberulent to pilose; lowest leaves opposite, the others alternate, ovate to suborbicular, petiolate; flowers solitary in leaf axils, mature peduncles 8–18 mm long, often arched; corolla with upper lip blue and lower lip paler blue to white, both with darker blue or purple veins, ca. 8 mm across; capsules uniformly and densely pubescent, lobed at apex, the lobes rounded. A Eurasian species naturalized in south-central and southwestern counties and at scattered sites northward; lawns, waste places. Mid-March–May (–October).

12. Veronica filiformis Sm. Blue-eyed Speedwell. Slender Speedwell

Perennial, sometimes forming mats; stems slender, prostrate and sometimes ascending at tips, to 3 dm long, puberulent; lower leaves opposite, the others alternate, cordate-orbicular to reniform, petiolate, wider than

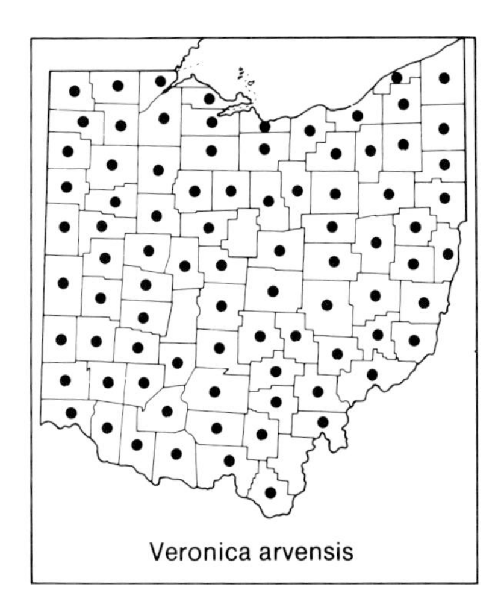
Veronica arvensis

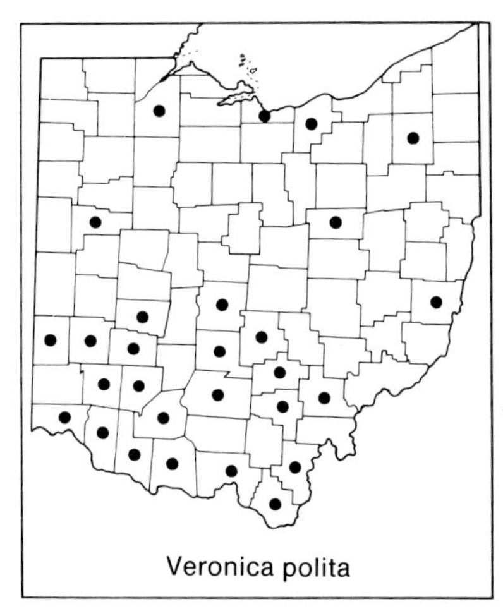
Veronica polita

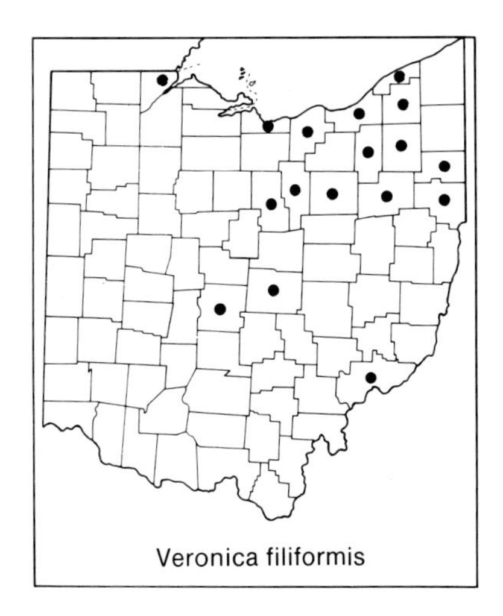
Veronica filiformis

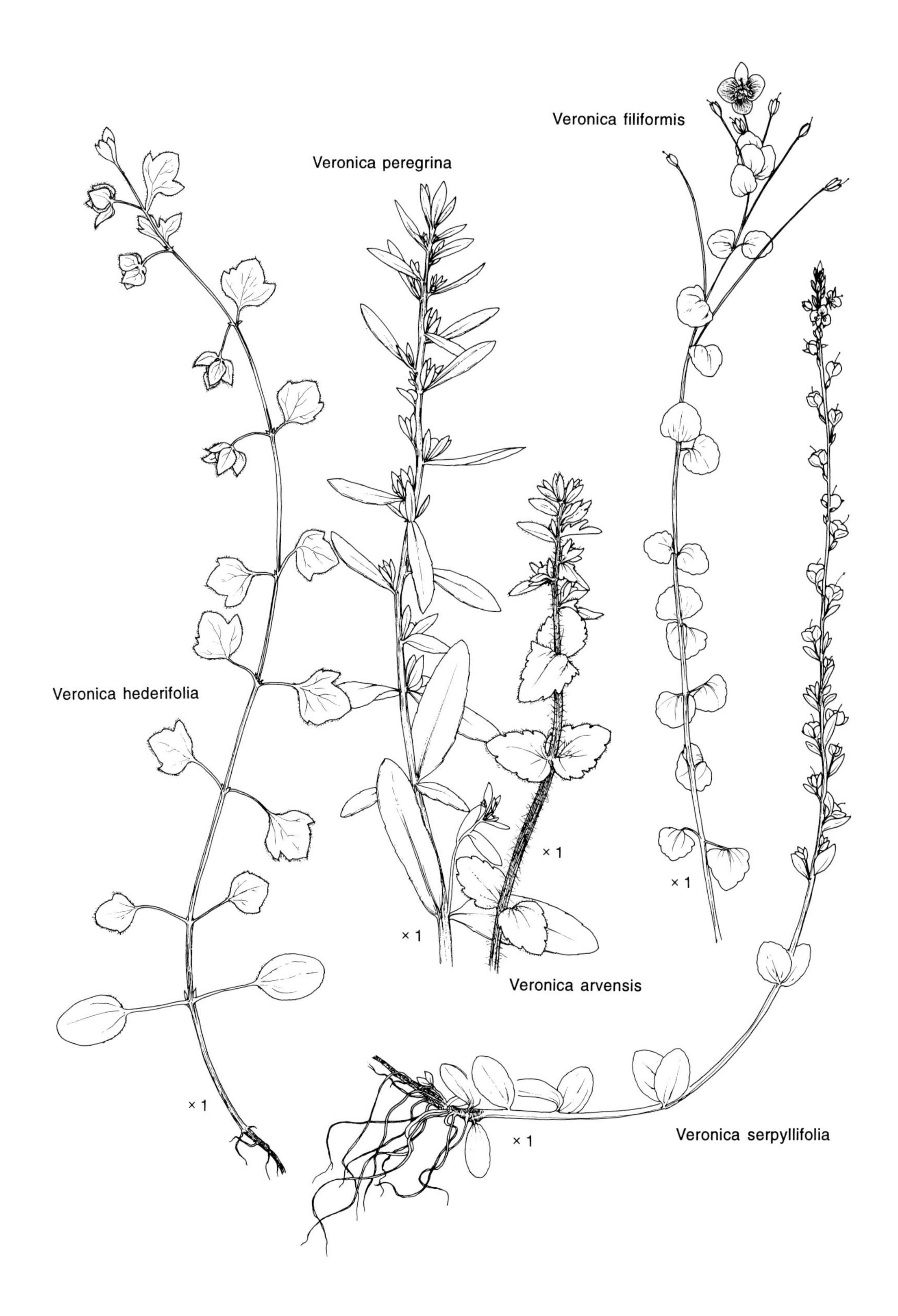
Veronica filiformis
Veronica peregrina
Veronica hederifolia
× 1
× 1
Veronica arvensis
× 1
× 1
× 1
× 1
Veronica serpyllifolia

long; flowers solitary in axils of middle and upper leaves, mature peduncles 15–25 mm long, filiform, becoming arched; corolla blue with darker stripes, lower lip paler than the upper, 8–9 mm across; capsules rarely, if ever, produced. A Eurasian species sometimes cultivated in Ohio, escaped and naturalized chiefly in northeastern counties; lawns, waste places. J. W. Adams (1976), Bentz and Cooperrider (1978), and Muenscher (1949) comment on the aggressiveness of this species, now spreading in parts of northeastern Ohio. April–mid-June.

13. Veronica persica Poir. Cat's-eye Speedwell. Bird's-eye Speedwell

Annual; stems prostrate to ascending, to 6 dm long, more or less hirsute; leaves ovate to broadly elliptic or suborbicular, the lowest opposite, the others usually alternate, petioles longest on lowest leaves and becoming progressively shorter above, the uppermost leaves subsessile; flowers solitary in axils of all but lowest leaves, mature peduncles 13–25 mm long, more or less recurved; corolla blue with darker lines, 6–9 mm across; capsules reticulately veined, ciliate along the upper margin, broadly notched at apex, the lobes mostly acute and widely spreading. An Asian species, naturalized mostly in northeastern and southern counties; gardens and other cultivated ground, along roads and railroads, and other waste places. (Late January–) March–June (–November).

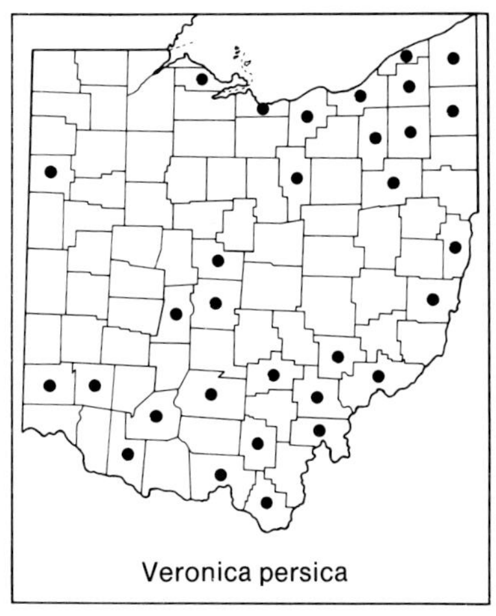
Veronica persica

14. Veronica hederifolia L. Ivy-leaved Speedwell

Annual; stems reclining or weakly ascending, to 4 dm long, more or less hirsute; leaves petiolate, of two forms, the lowest: opposite, entire, and elliptic, the others: mostly alternate, orbicular-reniform to reniform, wider than long, and mostly 3–5-palmately lobed with the central lobe the largest; flowers solitary in axils of middle and upper leaves, peduncles 6–15 mm long; corolla pale purple-blue, 2.5–4.0 mm across; capsules glabrous. A Eurasian species, locally naturalized at a few sites in northernmost and southernmost counties; lawns, fields, and similar places. Late March–early June.

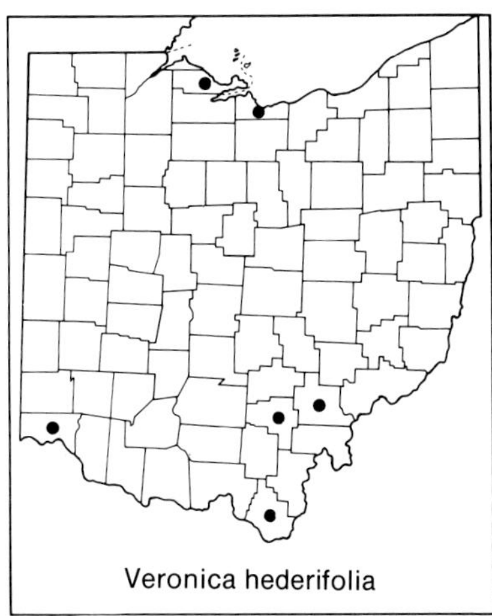
Veronica hederifolia

19. Dasistoma Raf.

A genus with a single species.

1. **Dasistoma macrophylla** (Nutt.) Raf. Mullein Foxglove
   Seymeria macrophylla Nutt.

Annual or perhaps biennial or perennial, to 2 m tall, semiparasitic on other flowering plants, sometimes blackening on drying; stems erect, retrorsely pubescent; leaves opposite, the lowest petiolate, broadly ovate in outline, deeply pinnatifid or twice-pinnatifid, the middle and upper leaves progressively shorter-petioled, reduced in size, and less dissected, the upper-

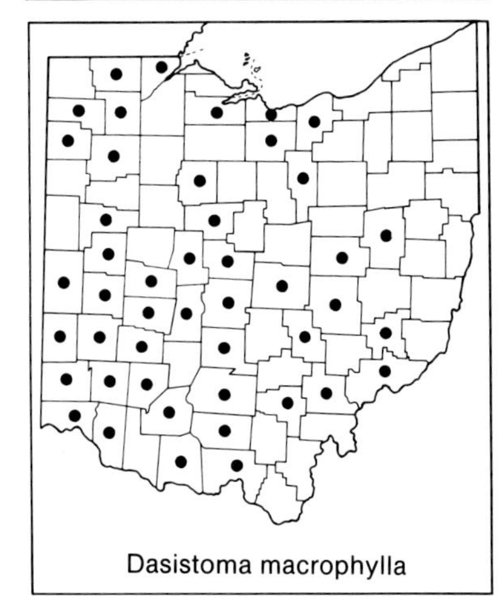
Dasistoma macrophylla

most subsessile, narrowly lanceolate, and with margins crenate to subentire; flowers solitary in axils of reduced upper leaves, sessile or on peduncles to 2 mm long; calyx deeply 5-lobed; corolla yellow, 12–15 mm long, 5-lobed and nearly actinomorphic; stamens 4; capsules glabrous. A species of the Ohio and mid–Mississippi River valleys, in Ohio at an eastern boundary of the species' range; chiefly in western counties; stream banks, rich woodlands. Piehl (1962b) reports research on the parasitism of this species. Additional record: Meigs County (Pennell, 1935). Mid-July–early September.

20. Agalinis Raf. GERARDIA. AGALINIS

Annuals, semiparasitic on other flowering plants; leaves opposite or subopposite, linear to filiform; flowers solitary in the axils of upper leaves; calyx shallowly 5-lobed; corolla purple to pink, rarely white, often spotted within, shallowly 5-lobed and only slightly zygomorphic; stamens 4. A New World genus of about 60 species. Canne (1984) reports chromosome numbers in the genus. The treatment below is based in part on the studies of Pennell (1928, 1929).

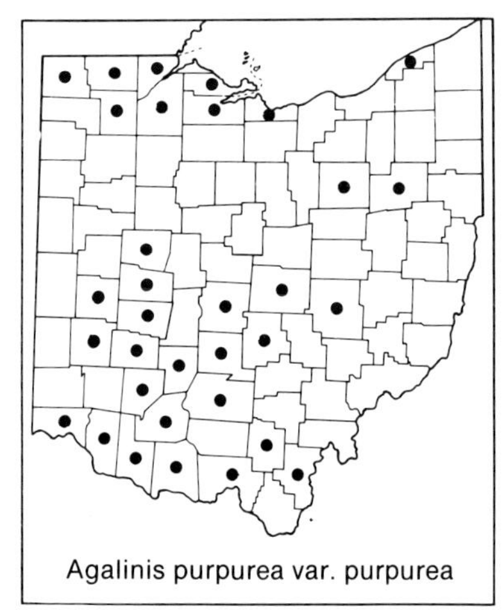

Agalinis purpurea var. purpurea

a. Peduncles shorter than to equalling length of calyx . . . . 1. *Agalinis purpurea*

aa. Peduncles twice or more than twice as long as calyx.

b. Plants deep-green to purple-tinged when fresh, sometimes blackening in drying; calyx tube at anthesis moderately thick, not membranous, with inconspicuous reticulate venation; leaves 1–3 mm wide, 1–4 cm long; peduncles spreading; stems usually glabrous . . . . 2. *Agalinis tenuifolia*

bb. Plants yellow-green or pale when fresh, mostly not blackening in drying; calyx tube at anthesis membranous and with conspicuous reticulate venation; leaves 0.3–1 mm wide, 1–3 cm long; peduncles ascending; stems glabrous to evidently scabrous . . . . 3. *Agalinis skinneriana*

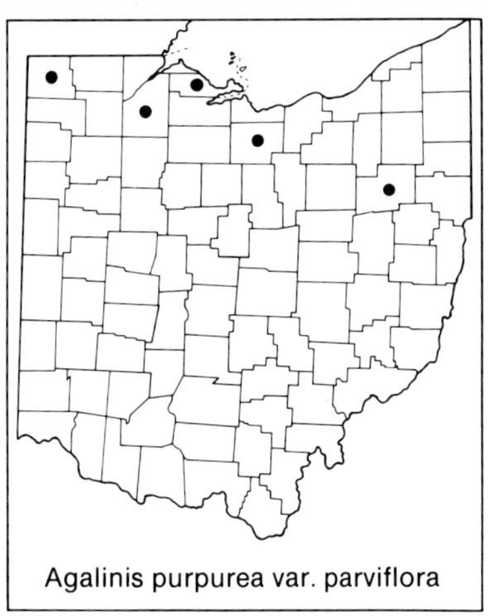

Agalinis purpurea var. parviflora

1. **Agalinis purpurea** (L.) Pennell PURPLE GERARDIA. PURPLE AGALINIS.

Annual, to 8 dm tall, plants usually darkening upon drying; stems erect, glabrous to puberulent or minutely scabrous; peduncles 2–5 mm long, shorter than to equalling length of calyx; corolla pinkish-purple. Chiefly in western counties, and at scattered sites eastward. August–early October.

Using the key below, most Ohio plants will be identified as members of the rather widespread var. *purpurea*, a plant of bogs, wet meadows, moist to dry sandy fields, and sometimes dry barren soils. Others will be identi-

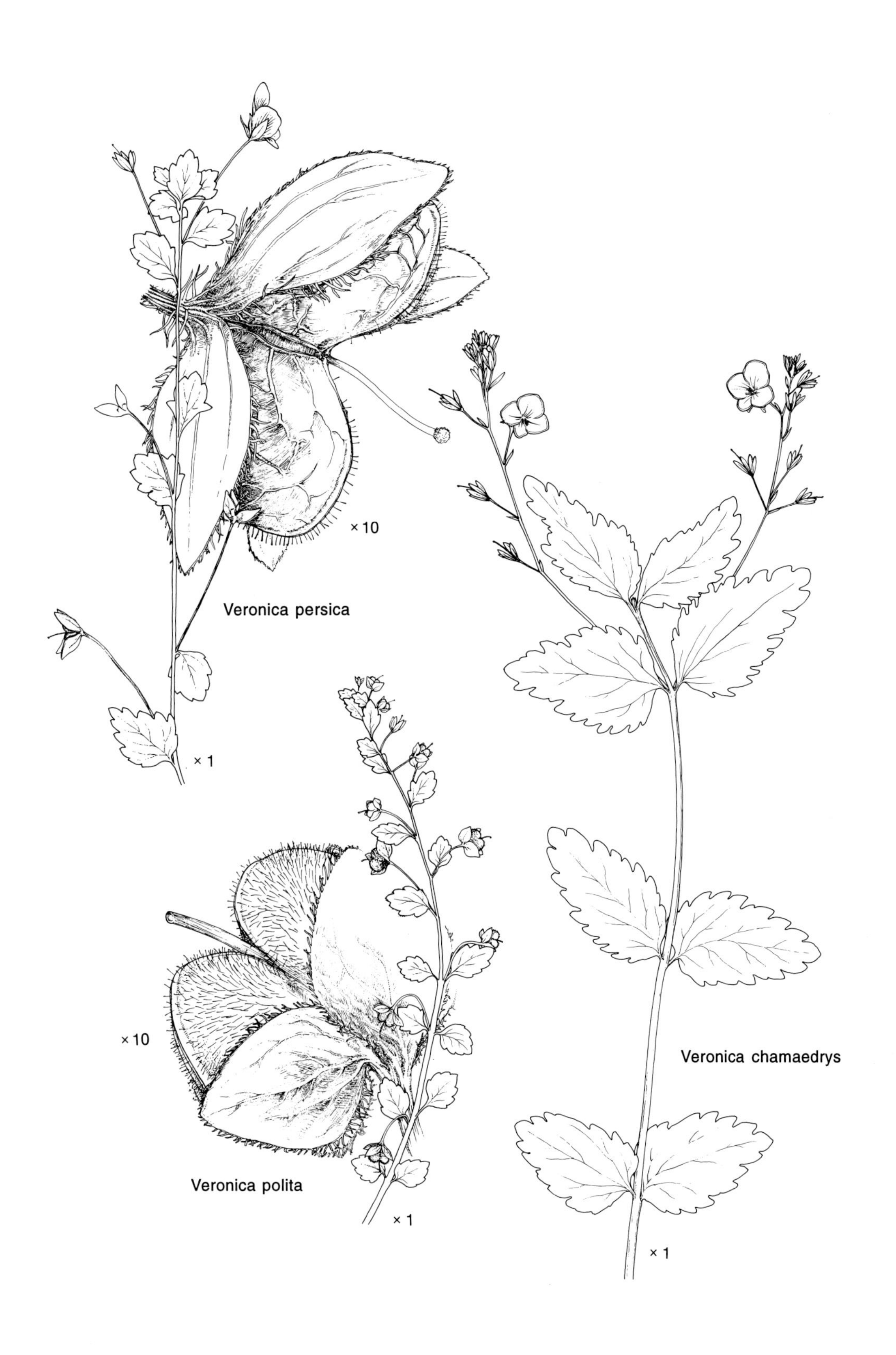
×10
Veronica persica
×1
×10
Veronica polita
×1
Veronica chamaedrys
×1

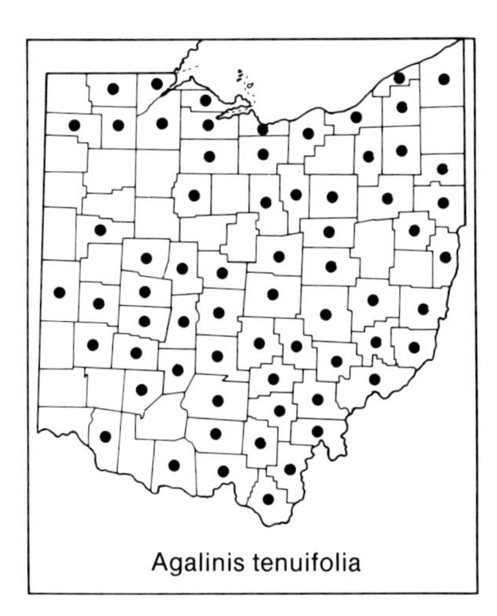
Agalinis tenuifolia

fied as var. *parviflora*, known only from a few northern counties, where it grows in damp open ground at stream and lake margins, and in bogs. A few other Ohio plants represent intergrades between the two varieties.

a. Corolla 1.8–2.7 cm long; style 13–20 mm long; calyx lobes 0.5–1.5 mm long, sinuses between the lobes broadly U-shaped; the largest leaves to 3 mm wide . . . . var. *purpurea*
Gerardia purpurea L. var. *purpurea*

aa. Corolla 1–2 cm long; style 8–10 mm long; calyx lobes 1–2.5 mm long, sinuses between the lobes V-shaped or narrowly U-shaped; the largest leaves 1.8 mm wide. . . . . var. *parviflora* (Benth.) J. Boivin
incl. *A. paupercula* (A. Gray) Britt. var. *paupercula* and var. *borealis* Pennell; *Gerardia paupercula* (A. Gray) Britt. var. *paupercula* and var. *borealis* (Pennell) Deam; G. *purpurea* var. *parviflora* Benth. Additional records: Erie and Lucas counties (Pennell, 1929).

2. **Agalinis tenuifolia** (M. Vahl) Raf. Slender Gerardia. Slender Agalinis
Gerardia tenuifolia M. Vahl

Annual, to 6 dm tall, plants often darkening upon drying; stems erect, usually glabrous; peduncles 7–18 mm long, spreading; calyx tube at anthesis moderately thick, not membranous, and with inconspicuous reticulate venation; corolla pinkish to purplish, 10–17 mm long. Throughout eastern Ohio, but infrequent or absent in some western counties; dry to moist open ground, woodland openings and borders. Additional records: Brown and Hamilton counties (Pennell, 1935). Mid-August–early October.

Three intergrading varieties occur in Ohio, the assignment of individual specimens sometimes difficult. Plants of var. *tenuifolia* are more frequent in the eastern half of the state, those of var. *macrophylla* more frequent in the western half. The relatively few plants assignable to var. *parviflora* are from the lake plains in northern Ohio.

a. Leaves and branches somewhat spreading to more often stiffly ascending; calyx lobes 1–2 mm long. . . . . var. *parviflora* (Nutt.) Pennell
Gerardia tenuifolia var. *parviflora* Nutt.

aa. Leaves and branches spreading.

b. Calyx lobes 0.2–1 mm long; leaves 1–3.5 mm wide. . . . . var. *tenuifolia*
Gerardia tenuifolia var. *tenuifolia*

bb. Calyx lobes 1–2 mm long; leaves (1–) 3–6 mm wide. . . . . var. *macrophylla* (Benth.) S. F. Blake
Gerardia tenuifolia var. *macrophylla* Benth.

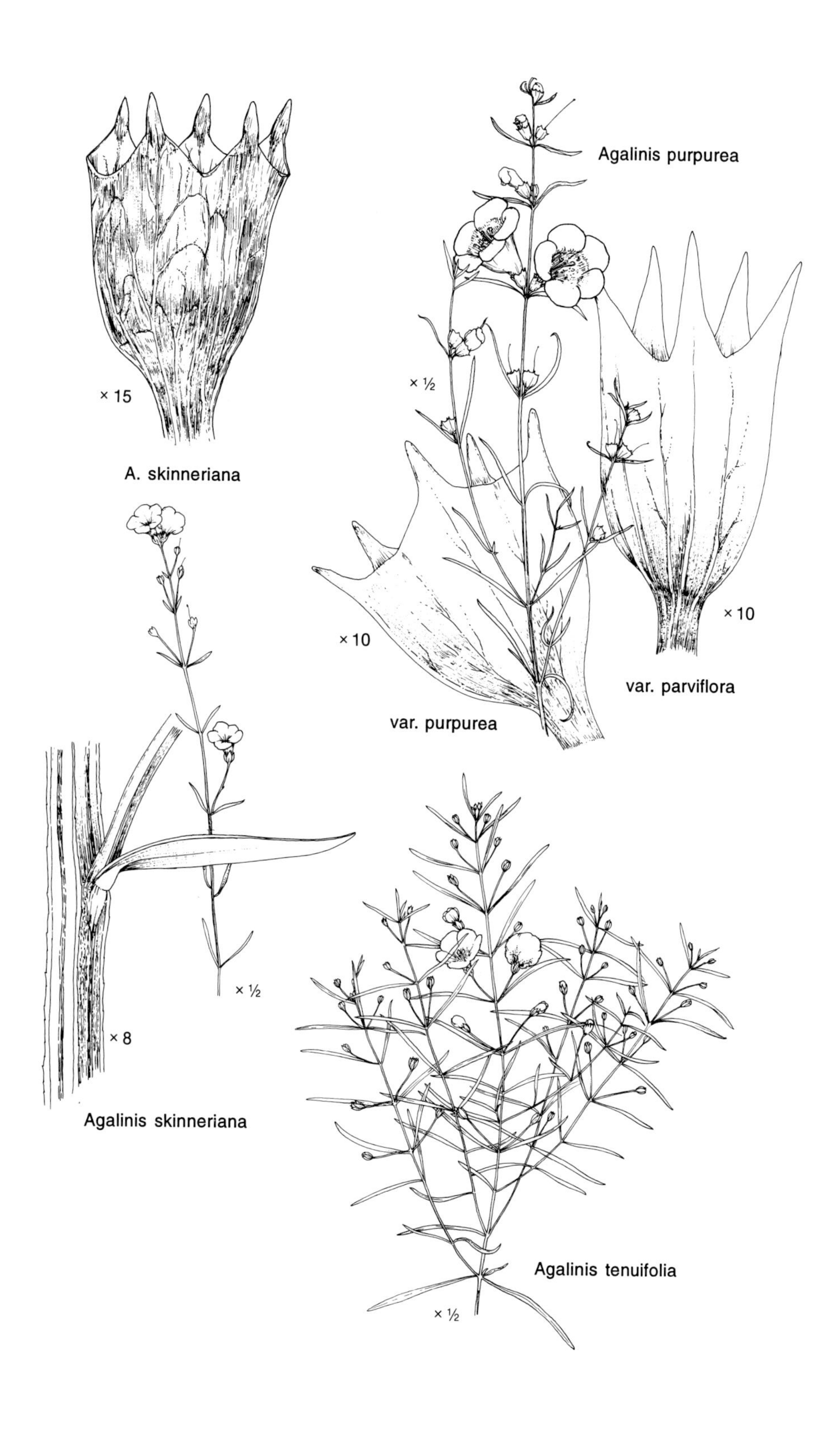
× 15
A. skinneriana
Agalinis purpurea
× ½
× 10
× 10
var. parviflora
var. purpurea
× ½
× 8
Agalinis skinneriana
Agalinis tenuifolia
× ½

Agalinis skinneriana

Tomanthera auriculata

3. **Agalinis skinneriana** (Alph. Wood) Britt. Skinner's Gerardia. Skinner's Agalinis

Gerardia skinneriana Alph. Wood; and incl. *A. gattingeri* (Small) Small, *Gerardia gattingeri* Small

Annual, to 5 dm tall, plants darkening little or not at all upon drying; stems erect, glabrous to scabrous; peduncles 5–25 mm long, ascending; calyx tube at anthesis membranous and with conspicuous reticulate venation; corolla purplish-pink or white, 8–15 mm long. A rare species of the interior United States, at an eastern boundary of its range in Ohio; known from only a few sites in western counties; sandy woods, woodland openings and borders, thickets. August–September.

Agalinis gattingeri is merged without rank under this species, following Holmgren (1986).

21. Tomanthera Raf.

A genus of two species, one of which is native to Ohio. D'Arcy (1979) and Gleason and Cronquist (1991) merge this genus under *Agalinis*.

1. **Tomanthera auriculata** (Michx.) Raf.

Agalinis auriculata (Michx.) S. F. Blake; *Gerardia auriculata* Michx.

Annual, to 8 dm tall, semiparasitic on other flowering plants; stems erect, roughly scabrous with hairs both spreading and retrorse; leaves opposite, sessile or nearly so, narrowly lanceolate to ovate-lanceolate, upper leaves often with two small basal lobes; flowers solitary in axils of upper leaves, sessile or on peduncles to 2 mm long; calyx 5-lobed; corolla purple, 18–28 mm long, 5-lobed and subactinomorphic; stamens 4, didynamous. Known from a few scattered sites in Ohio; prairies, open woodlands. Knoop (1988) describes a recently discovered colony in Adams County. Additional records: Butler County (Pennell, 1935); Erie and Muskingum counties (Pennell, 1928). Cunningham and Parr (1990) describe culture techniques, Baskin et al. (1991) its seed germination ecology. July–September.

22. Aureolaria Raf. FALSE FOXGLOVE

Perennials or biennials, semiparasitic on oaks; leaves opposite or subopposite, the upper sometimes alternate; flowers solitary in the axils of upper leaves; calyx 5-lobed; corolla yellow, shallowly 5-lobed and only slightly zygomorphic; stamens 4 and didynamous, staminode none. A North American genus of about ten species. The treatment here is based in part on that of Pennell (1928).

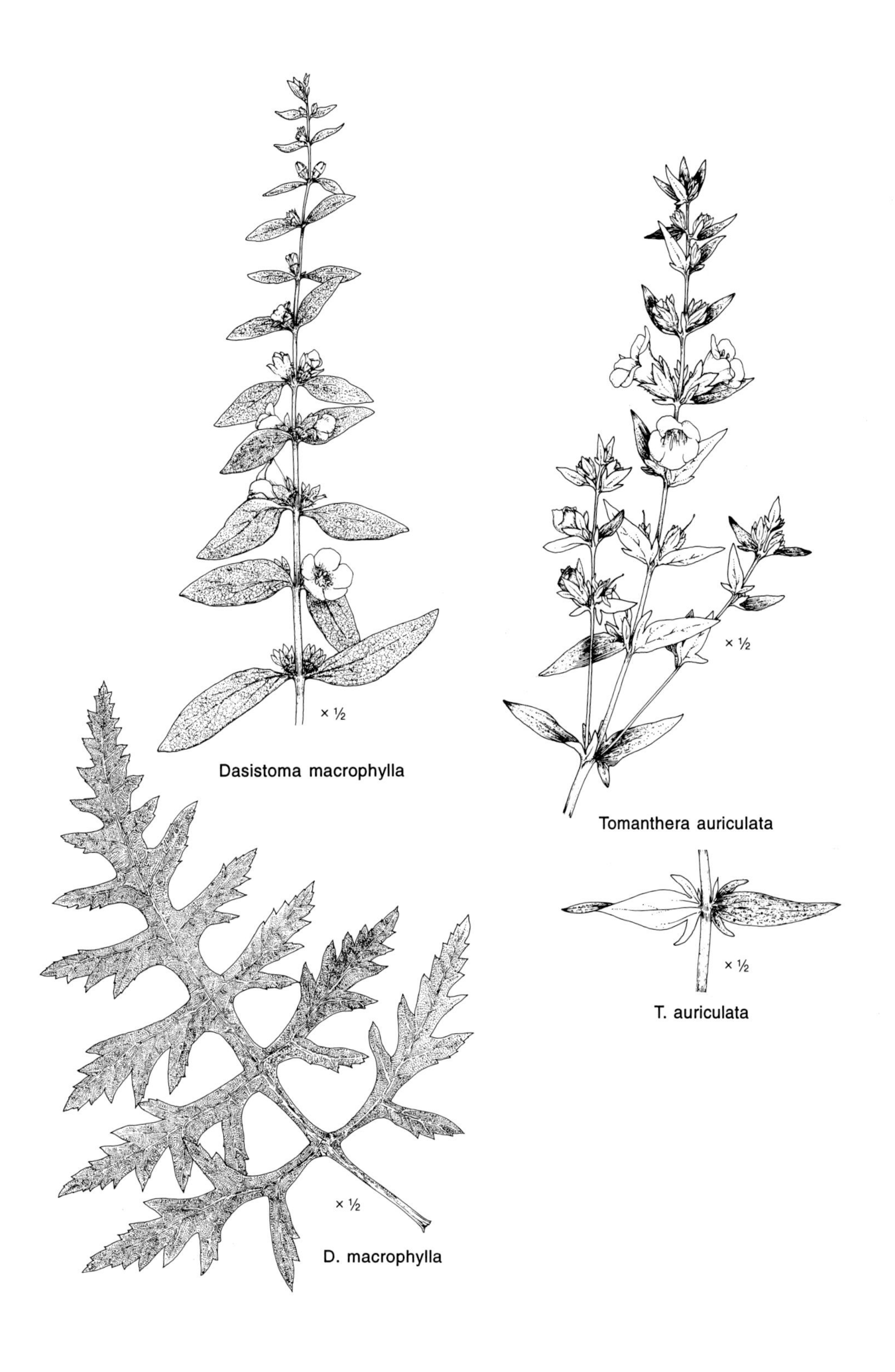
× ½
Dasistoma macrophylla
× ½
Tomanthera auriculata
× ½
T. auriculata
× ½
D. macrophylla

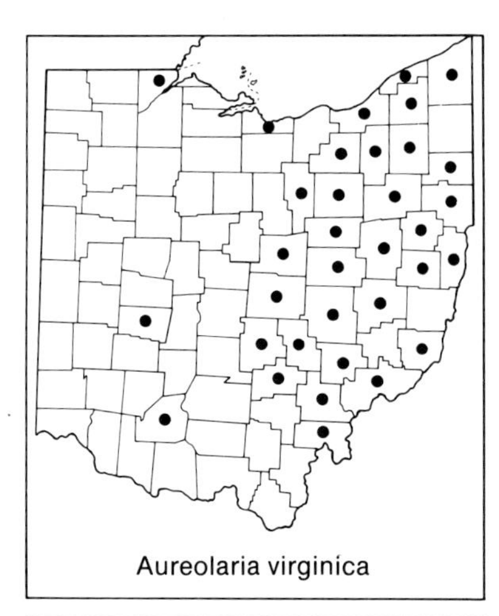
Aureolaria virginica

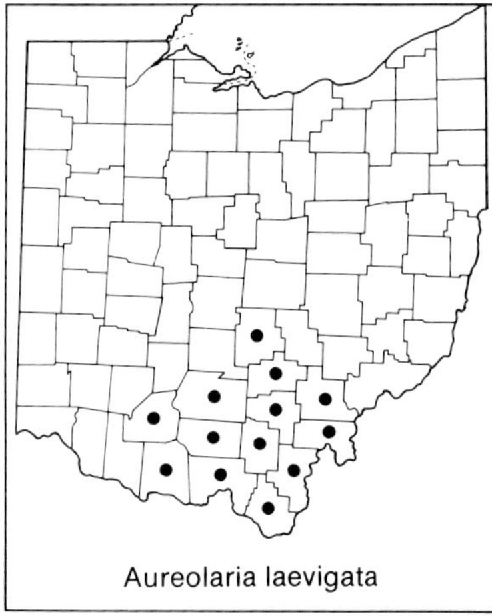
Aureolaria laevigata

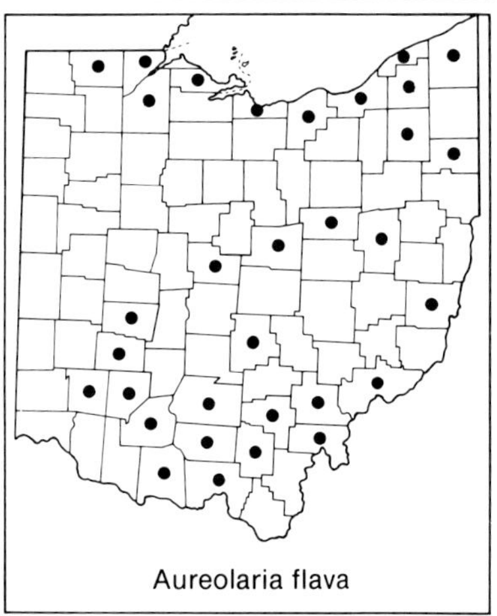
Aureolaria flava

a. Stem, calyx, and capsules pubescent or glandular-pubescent.
  b. Lower leaves usually with 1–3 pairs of large obtuse teeth or rounded lobes on lower half of blade; peduncles straight at anthesis, 2–6 mm long; calyx densely short-pubescent, the calyx lobes not spreading and usually entire; corolla yellow throughout. . . . . 1. *Aureolaria virginica*
  bb. Lower leaves once- or twice-pinnately lobed, often deeply so; peduncles curved upward at anthesis, 7–25 mm long; calyx glandular-pubescent, the calyx lobes spreading and usually crenate to pinnately lobed; corolla yellow and also usually tinged with purple . . . . 4. *Aureolaria pedicularia*
aa. Stem, calyx, and capsules glabrous.
  b. Blades of lower leaves mostly entire, rarely lobed; stems not glaucous; peduncles mostly 1–5 mm long; corolla 2.5–3.5 cm long . . . . 2. *Aureolaria laevigata*
  bb. Blades of lower leaves deeply lobed or incised; stems glaucous; peduncles mostly 4–10 mm long, sometimes longer; corolla 3.5–6 cm long. . . . . 3. *Aureolaria flava*

1. **Aureolaria virginica** (L.) Pennell   DOWNY FALSE FOXGLOVE
   Gerardia virginica (L.) BSP.

Perennial, to ca. 1 m tall; stems erect, puberulent; lower leaves usually with 1–3 pairs of large obtuse teeth or rounded lobes on lower half of blade; peduncles straight at anthesis, 2–6 mm long; calyx densely short-pubescent, the lobes not spreading and usually entire; corolla yellow throughout, 3.0–4.5 cm long; capsules pubescent. Eastern Ohio and a few scattered western sites; sandy acid soils, dry open woodlands. Late June–early September.

2. **Aureolaria laevigata** (Raf.) Raf.   ENTIRE-LEAVED FALSE FOXGLOVE
   Gerardia laevigata Raf.

Perennial, to ca. 1 m tall; stems erect, glabrous, green or rarely purplish but not glaucous; blades of lower leaves mostly entire, rarely lobed; peduncles mostly 1–5 mm long; calyx glabrous; corolla yellow, 2.5–3.5 cm long; capsules glabrous. An Appalachian species at a northern boundary of its range in south-central Ohio; rocky woodlands, woodland openings and borders. June–early October.

3. **Aureolaria flava** (L.) Farw.   SMOOTH FALSE FOXGLOVE
   Gerardia flava L.

Perennial, to 2 m tall; stems erect, glabrous, glaucous and often purplish or bluish; blades of lower leaves deeply lobed or incised; peduncles 4–10 (–15) mm long; calyx glabrous; corolla yellow, 3.5–6.0 cm long; capsules glabrous. Scattered across southern and northern counties, but generally

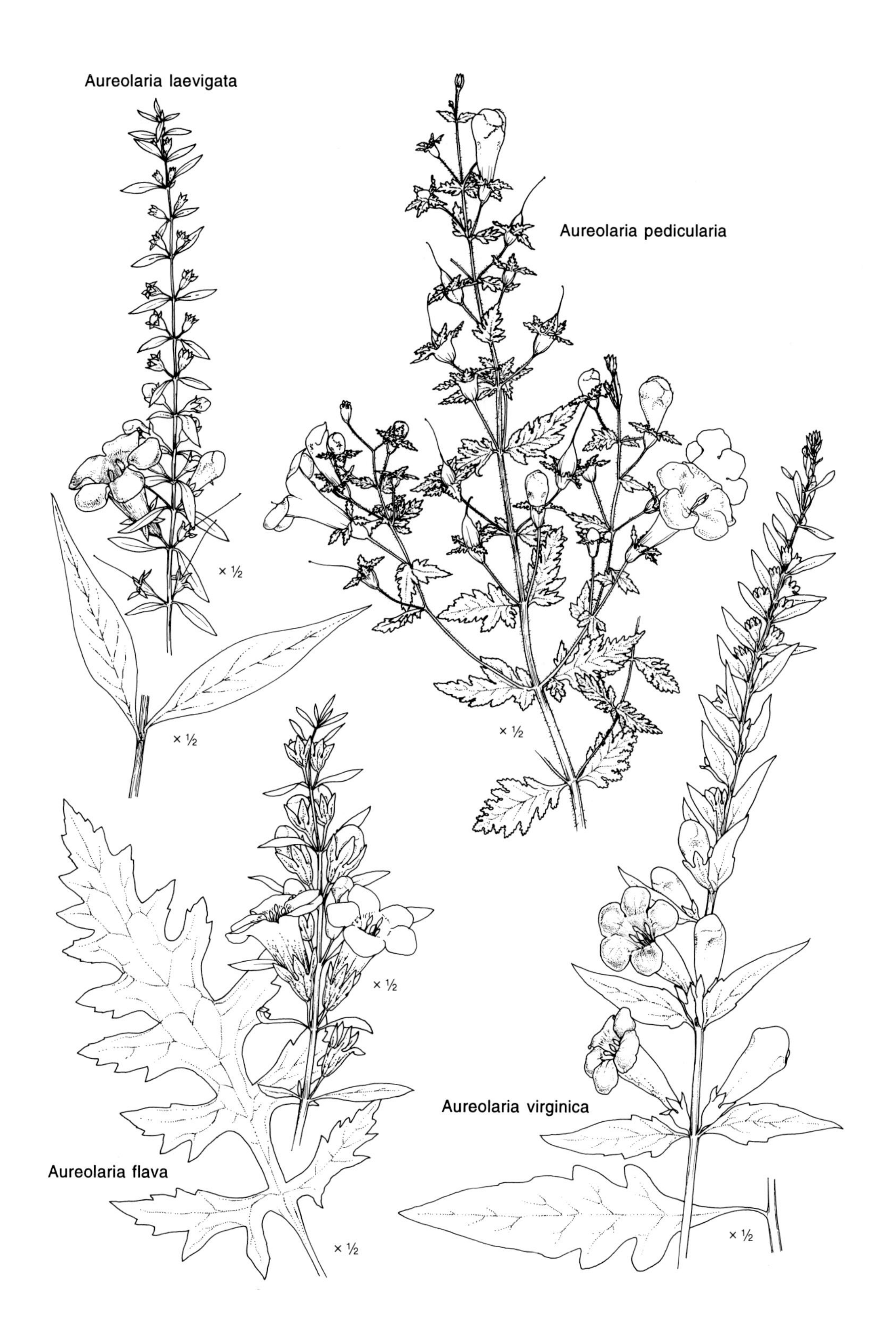

Aureolaria laevigata
Aureolaria pedicularia
× ½
× ½
× ½
× ½
Aureolaria virginica
Aureolaria flava
× ½
× ½

Aureolaria pedicularia var. pedicularia

Aureolaria pedicularia var. ambigens

absent from westernmost Ohio; dry often sandy soils in woodland openings and at woodland borders. Additional records: Brown County (Pennell, 1935), Hamilton, Medina, Montgomery, and Stark counties (Pennell, 1928). August–September.

Two intergrading varieties, keyed below, occur in Ohio, var. *flava* mostly in northern counties, var. *macrantha* mostly in southern counties.

a. Calyx lobes 2–5 (–7) mm long; corolla 35–40 mm long; capsules 12–16 mm long . . . . var. *flava*
Gerardia flava var. *flava*

aa. Calyx lobes 5–24 mm long; corolla (35-) 40–60 mm long; capsules 15–20 mm long . . . . var. *macrantha* Pennell
Gerardia flava var. *macrantha* (Pennell) Fern.

Ballard and Pippen (1991) report the discovery, in Michigan, of a hybrid between this species and the next, *A. pedicularia*.

4. **Aureolaria pedicularia** (L.) Raf. FERN-LEAVED FALSE FOXGLOVE
Gerardia pedicularia L.

Biennial, 4–12 dm tall; stems erect, pubescent and sometimes densely glandular-pubescent; lower leaves once- or twice-pinnately lobed, often deeply so; peduncles curved upward at anthesis, 7–25 mm long; calyx glandular-pubescent, the lobes spreading and usually crenate-margined to pinnately lobed; corolla yellow, usually tinged with purple, 2.5–4.0 cm long; capsules glandular-pubescent. The life history of this species has been studied by Musselman (1969) and Werth and Riopel (1979). August–September.

The species is represented in Ohio by the two rather distinct varieties keyed below; var. *pedicularia* grows in dry woods and woodland openings and borders in eastern counties, var. *ambigens* in moist prairies, sandy woods, and dunes in northwestern counties.

a. Upper stems and leaves with all or nearly all pubescence nonglandular (but that of the peduncles and calyces glandular), the hairs ca. 0.2–0.4 mm long, rarely septate; capsules narrowly ellipsoid, 9–11 mm long. . . . . var. *pedicularia*
Gerardia pedicularia var. *pedicularia*
WOODLAND FERN-LEAVED FALSE FOXGLOVE

aa. Upper stems and leaves with some short nonglandular hairs and many longer glandular hairs, the latter mostly 0.6–0.8 mm long, flat, and with 2–3 septations, the glandular hairs twice as long as the nonglandular; capsules ellipsoid or broadly ellipsoid, 11–15 mm long. . . . . var. *ambigens* (Fern.) Farw.
Gerardia pedicularia var. *ambigens* Fern.
PRAIRIE FERN-LEAVED FALSE FOXGLOVE

23. Buchnera L. BLUEHEARTS

A widespread genus of about 100 species, one of them native to Ohio. The most recent revision of the genus is that of Philcox (1965).

1. **Buchnera americana** L. BLUEHEARTS

Perennial, to 8 dm tall, probably semiparasitic on other flowering plants, plants darkening upon drying; stems erect, rough-pubescent; leaves opposite, sessile; flowers sessile, in a terminal spike and sometimes in additional spikes originating from upper axils; calyx shallowly 5-lobed; corolla pinkish-purple to purple (becoming dark purple when dried), nearly actinomorphic, 10–15 mm long; stamens 4. Known in Ohio from a few northwestern and south-central counties; prairies, woodland openings and borders. Late June–mid-September.

Buchnera americana

24. Castilleja L.f. PAINTED CUP

A genus chiefly of the Western Hemisphere with about 200 species, one of them native to Ohio. Heckard (1962) described root parasitism in this genus.

1. **Castilleja coccinea** (L.) Spreng. INDIAN PAINTBRUSH. PAINTED CUP

Annual or biennial, to 8 dm tall, semiparasitic on other flowering plants; stems erect, pubescent with hairs of two lengths; some leaves in a basal rosette, cauline leaves alternate, sessile, usually cleft into 3–5 linear segments; flowers subsessile in dense, terminal, spicate inflorescences, flowers borne in axils of showy bracts, the bracts 3–5-lobed, green at the base, and scarlet, orange, yellow, or white above; calyx 2-lobed, often orange or scarlet at apex; corolla whitish to pale yellow or greenish-yellow, bilabiate; stamens 4. South-central counties and at scattered sites northward; prairies, old fields, grassy roadside banks, sphagnum bogs. Additional records: Erie and Stark counties (Pennell, 1935). Malcolm (1962) gives techniques for pot-culture of this species, including a list of the best "hosts." May–June (–August).

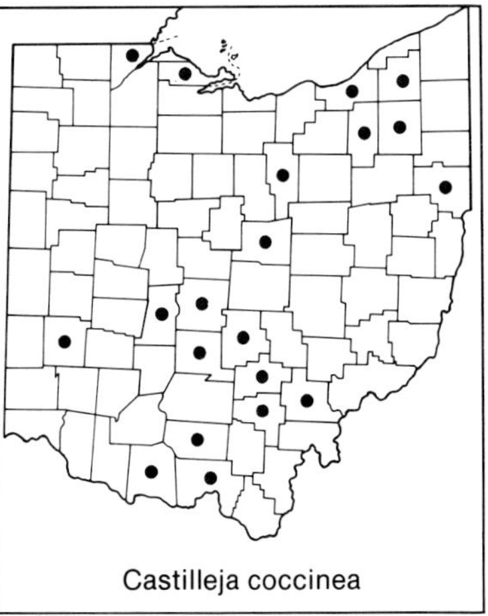

Castilleja coccinea

25. Melampyrum L. COW-WHEAT

A genus of about 25 species, all but the one below native to Eurasia.

1. **Melampyrum lineare** Desr. COW-WHEAT

Annual, to 4 dm tall, semiparasitic on other flowering plants; stems erect, somewhat short-puberulent; leaves opposite, subsessile to short-petiolate; flowers solitary in axils of upper leaves, peduncles 1–5 mm long; calyx 4-lobed; corolla 10–13 mm long, bilabiate, the tube white or sometimes purplish, the lips pale yellow; stamens 4. Northern counties and a few sites

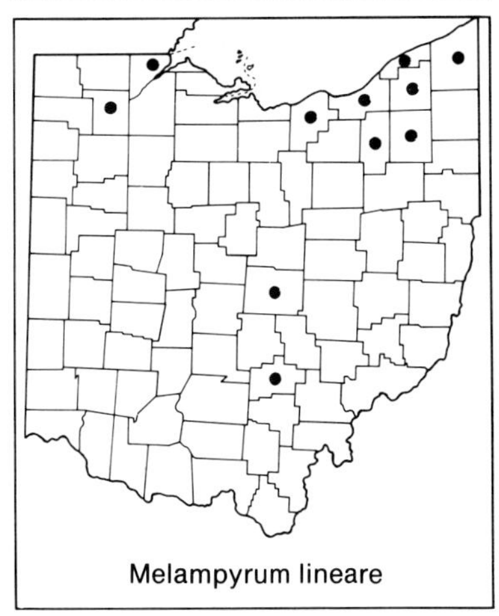

Melampyrum lineare

southward; rocky woodland openings and borders, stream banks. Piehl (1962a) and Cantlon, Curtis, and Malcolm (1963) studied parasitism and other aspects of this species' life history. Late May–early August.

Ohio plants may, when desired, be separated into the two varieties keyed below. There are, however, intergrades and the varieties are perhaps of only marginal significance. The range of var. *lineare* lies to the north of Ohio.

a. Principal leaves lanceolate to ovate, more or less rounded at base, 10–30 mm wide; the bracteal leaves all or mostly without teeth . . . . var. *latifolium* Bart.

aa. Principal leaves lanceolate to broadly linear, narrowed at base, 3–10 mm wide; the bracteal leaves mostly with several sharp narrow teeth . . . . var. *americanum* (Michx.) Beauv.

26. Pedicularis L. LOUSEWORT. WOOD-BETONY

Perennials, semiparasitic on other flowering plants; leaves alternate or opposite; flowers in terminal spike-like racemes; calyx obliquely 2-lobed or 2-cleft; corolla yellow, red, or purple, bilabiate, the upper lip more or less arched and hooded; stamens 4. A genus of some 350 species, all but one native to the Northern Hemisphere, and two native to the flora of Ohio.

a. Stems erect, 3–8 dm tall; leaves opposite or alternate, blades shallowly pinnately lobed, depth of sinuses half the way to midrib or less, each lobe sharply crenately toothed; calyx markedly 2-lobed, each half with a terminal, leafy appendage; corolla pale yellow to greenish-yellow; capsules oblong to subovoid . . . . 1. *Pedicularis lanceolata*

aa. Stems erect or ascending, 1–5 dm tall; leaves chiefly basal, cauline leaves alternate, blades deeply pinnately lobed, depth of sinuses more than halfway to the midrib, the lobes sometimes again lobed, otherwise their margins crenate to serrate; calyx shallowly 2-cleft, without appendages; corolla yellow or partly or wholly red or purple; capsules oblong-lanceolate . . . . 2. *Pedicularis canadensis*

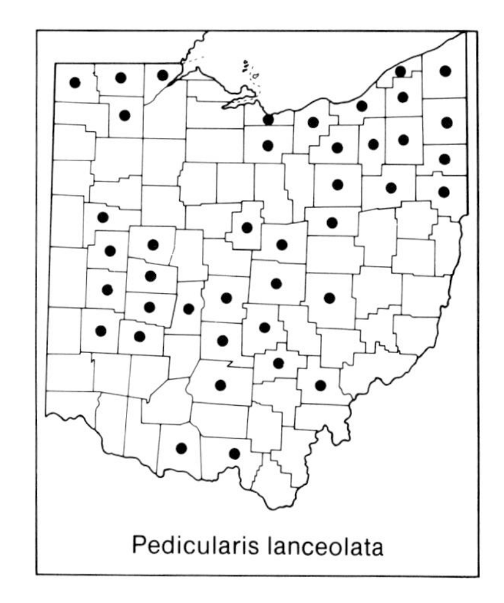
Pedicularis lanceolata

1. **Pedicularis lanceolata** Michx. SWAMP LOUSEWORT

Perennial, to 8 dm tall; stems erect, glabrous or pubescent; leaves opposite or alternate, the blades shallowly pinnately lobed—the depth of the sinuses halfway to the midrib or less, each lobe sharply crenately toothed; pedicels 1–2 mm long; calyx markedly 2-lobed, each half with a terminal leafy appendage; corolla pale yellow to greenish-yellow, 17–22 mm long; capsules oblong to subobovoid. Chiefly in northeastern and central counties; low wet fields, bogs, swamps. Lackney (1981) and Piehl (1965) studied root parasitism in this species. Additional record: Hamilton County (Pennell, 1935). Mid-August–early October.

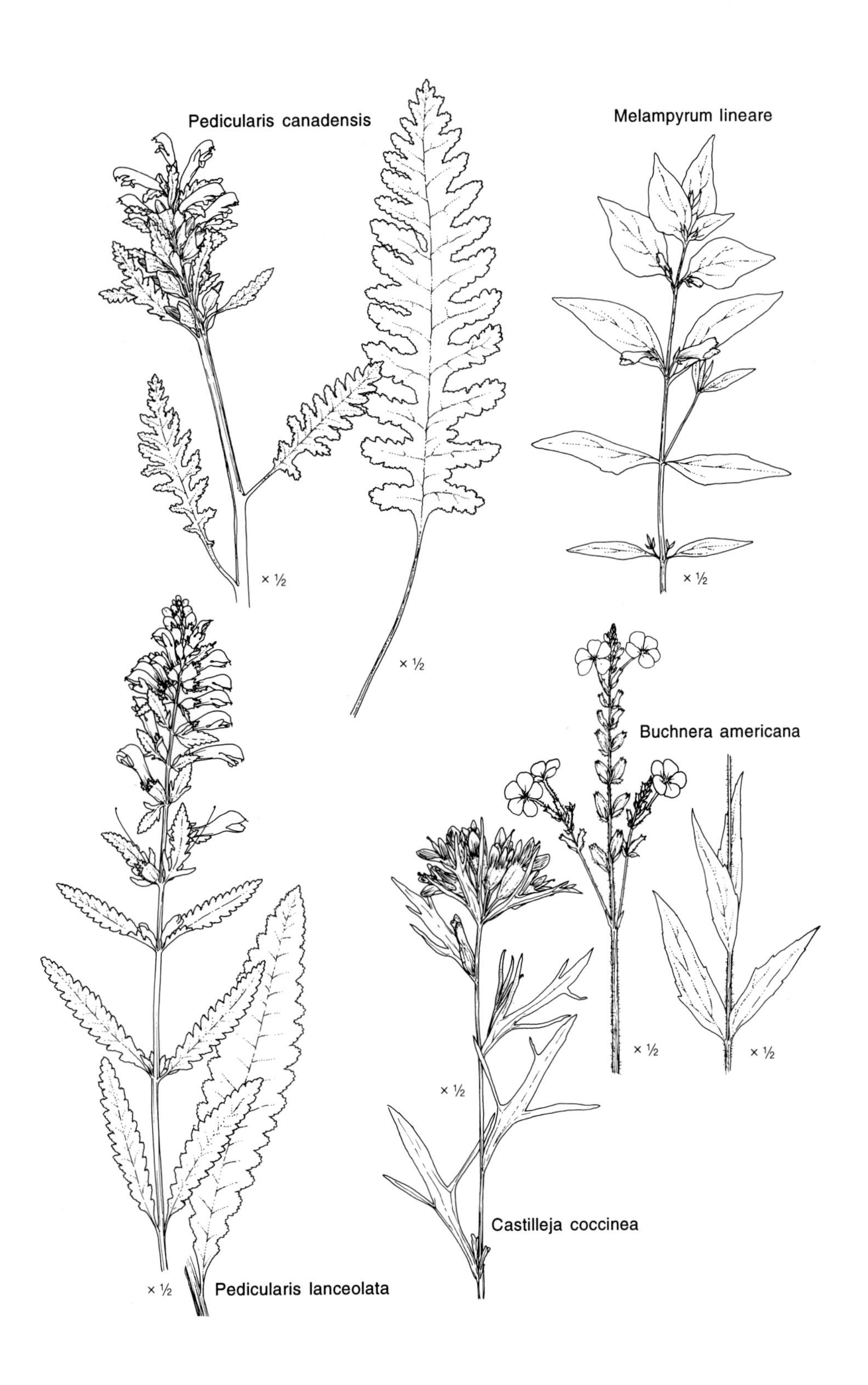
Pedicularis canadensis
Melampyrum lineare
× ½
× ½
× ½
Buchnera americana
× ½
× ½
× ½
Castilleja coccinea
× ½
Pedicularis lanceolata

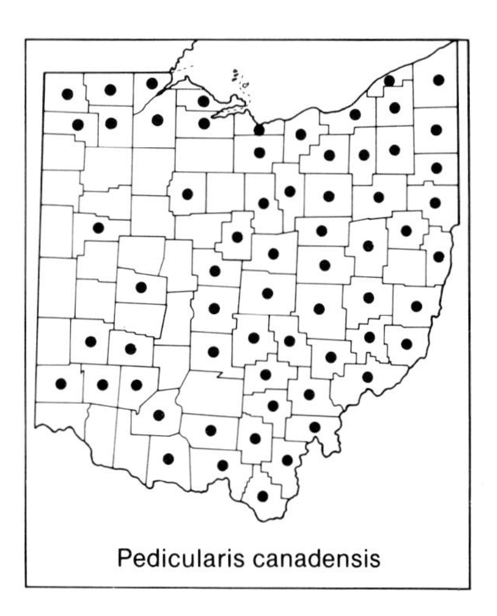
Pedicularis canadensis

2. **Pedicularis canadensis** L. COMMON LOUSEWORT. WOOD-BETONY
incl. var. *canadensis* and var. *dobbsii* Fern.

Perennial, to 5 dm tall; stems erect or ascending; leaves chiefly basal, cauline leaves alternate, the blades deeply pinnately lobed—the depth of the sinuses more than halfway to the midrib, the lobes sometimes again lobed, otherwise their margins crenate to serrate; pedicels 1–3 (–5) mm long; calyx shallowly 2-cleft, without appendages; corolla yellow to partly or wholly red or purple, 15–25 mm long; capsules oblong-lanceolate. An attractive spring wildflower growing throughout much of Ohio, but absent from most west-central counties; woodland openings and borders. Piehl (1963) studied root parasitism in this species. Additional record: Clark County (Pennell, 1935). Mid-April–early June.

Bignoniaceae. TRUMPET-CREEPER FAMILY. BIGNONIA FAMILY

Trees or woody vines; leaves opposite or whorled; flowers perfect, in cymes or panicles; calyx 2- or 5-lobed; corolla 5-lobed, showy, zygomorphic to subactinomorphic; stamens either 2 and equal or 4 and didynamous, attached to the corolla tube; carpels 2, fused to form a compound pistil with a single style; ovary superior; fruit a capsule, sometimes elongated, bearing winged seeds. A family of some 750 species, chiefly tropical, with distribution centered in northern South America. The treatment below is based in part on Dumke's (1967) study of Ohio members of this family.

a. Trees; leaves simple.
  b. Leaves opposite, their undersurface velvety with dense stellate pubescence; corolla 5–7 cm long, violet to bluish, yellow-striped within; fruits ovoid, 3–4 cm long . . . . 1. *Paulownia*
  bb. Leaves mostly in whorls of 3, sometimes opposite, their undersurface with simple hairs, not velvety; corolla 2.5–5 cm long, white to yellow, with yellow, orange, or purple spots; fruits linear-cylindric, 15–50 cm long . . . . 4. *Catalpa*
aa. Vines; leaves compound.
  b. Leaves with (5–) 7–11 (–13) leaflets, lacking a terminal tendril; corolla 6–8 cm long . . . . 2. *Campsis*
  bb. Leaves with (1 or) 2 leaflets and a terminal tendril; corolla 4–5 cm long . . . . 3. *Bignonia*

1. Paulownia Sieb. & Zucc.

A genus of six species of eastern Asia, one species naturalized in Ohio. This genus is placed in the family Scrophulariaceae by Fernald (1950),

Gleason (1952b), and Gleason and Cronquist (1963). Armstrong (1985) presents evidence for its retention in the Scrophulariaceae on the basis of floral anatomy. Cronquist (1981), Gleason and Cronquist (1991), and others place it in the Bignoniaceae.

1. Paulownia tomentosa (Thunb.) Steud. Princess-tree. Royal Paulownia

Tree, to 15 m (50 feet) tall; leaves simple, usually opposite, sometimes in whorls of 3 on young branches, long-petiolate, the blades broadly ovate, large—to ca. 3 dm long, undersurface velvety with dense stellate pubescence, margins entire to shallowly few-lobed; flowers showy, fragrant, in large panicles, appearing before the leaves; calyx 5-lobed; corolla violet to bluish, yellow-striped within, 5–7 cm long; stamens 4, included; capsules ovoid, 3–4 cm long. An Asian tree, cultivated in southernmost counties, escaped and locally naturalized along Ohio River slopes. May.

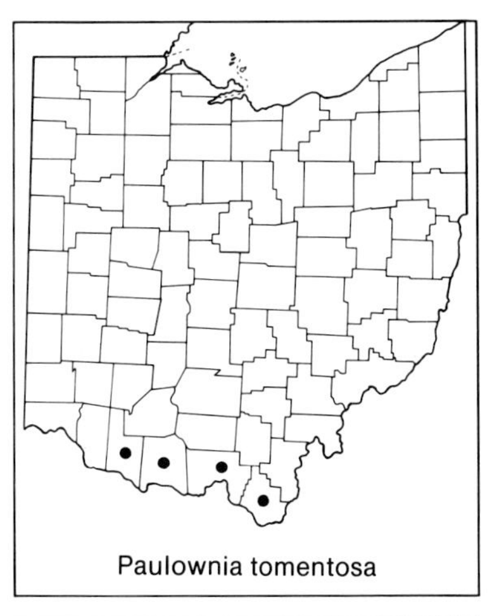
Paulownia tomentosa

2. Campsis Lour. TRUMPET-FLOWER

A genus of two species, the other native to eastern Asia.

1. **Campsis radicans** (L.) Bureau Trumpet-vine. Trumpet-creeper

Woody vine, climbing or sometimes trailing; leaves opposite, pinnately compound, with (5–) 7–11 (–13) leaflets; flowers showy, in terminal paniculate clusters; calyx shallowly 5-lobed; corolla with an orange tube and red limb, 6–8 cm long; stamens 4, included; capsule a thick legume-shaped pod, to 15 cm long. Native and common throughout southern Ohio, and naturalized at scattered sites northward; low moist woodlands, thickets, old fields, and along roadsides—where it often climbs fence-posts and trees. The species is frequently cultivated, sometimes only casually, especially in southern counties. A yellow-flowered form is occasionally seen in cultivation. Mid-June–early September.

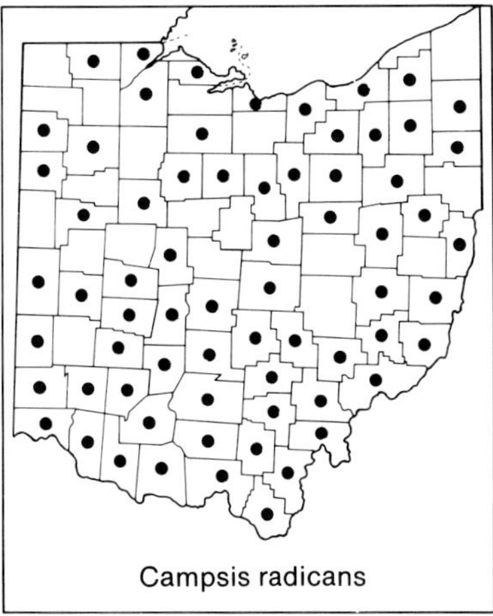
Campsis radicans

3. Bignonia L.

A genus with a single species.

1. **Bignonia capreolata** L. Cross-vine
   Anisostichus capreolata (L.) Bureau

Woody vine, climbing or trailing; leaves opposite, compound: the leaf composed of 2 leaflets (or sometimes only 1) and a terminal tendril; flowers showy, in few-flowered axillary cymes; calyx very shallowly 5-lobed to subtruncate; corolla 4–5 cm long, the tube orange-yellow to reddish, the lips yellow and reflexed; stamens 4, included; capsules linear-cylindric, to 2 dm long. A species of southeastern United States, at a northern boundary of its range in several southernmost Ohio counties at sites near the Ohio River; cliffs, bluffs, open-wooded slopes. May.

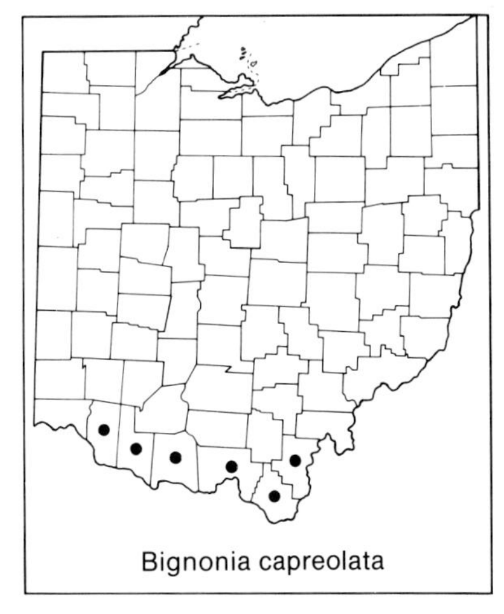
Bignonia capreolata

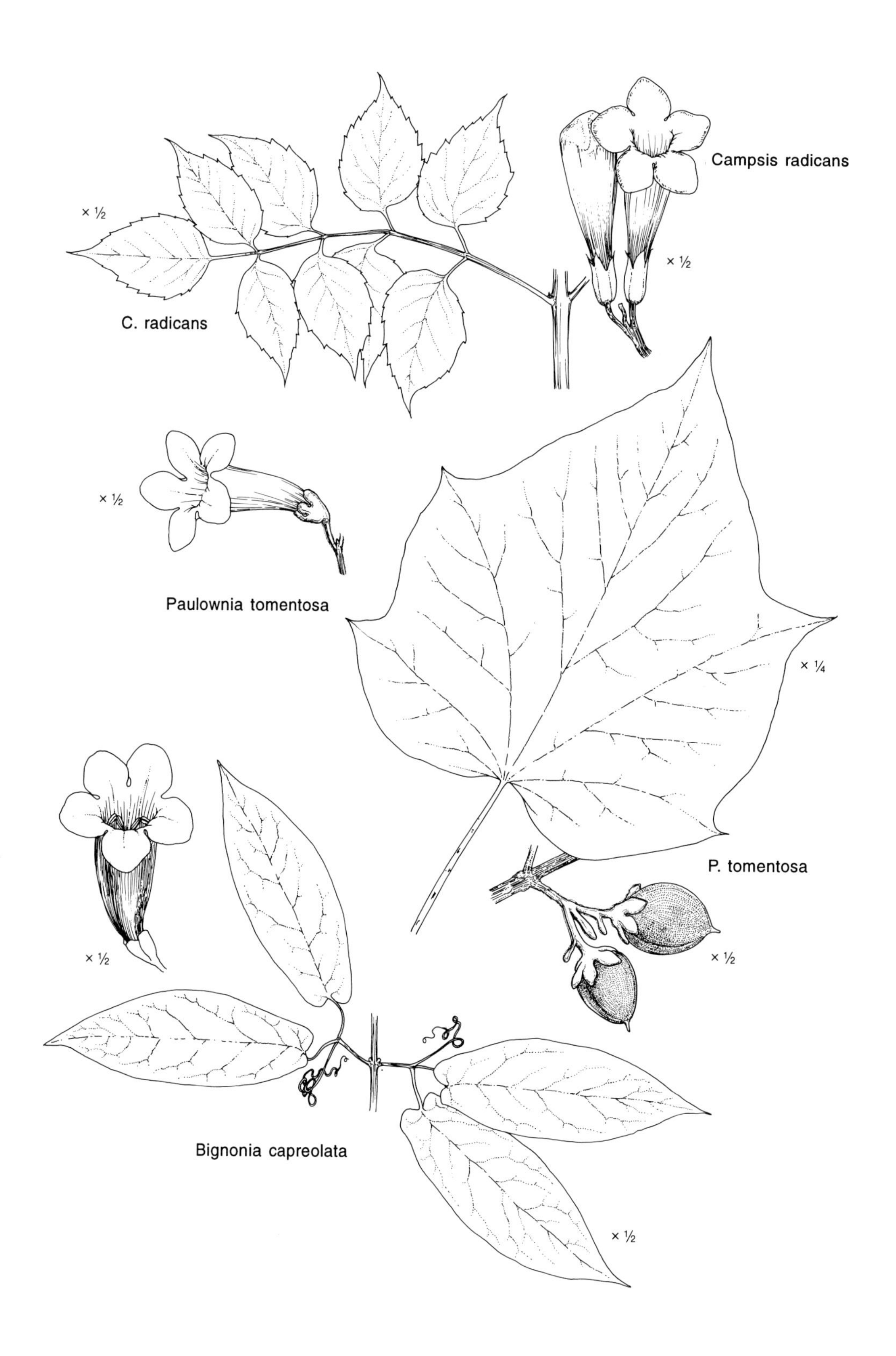
Campsis radicans
× ½
× ½
C. radicans
× ½
Paulownia tomentosa
× ¼
P. tomentosa
× ½
× ½
Bignonia capreolata
× ½

4. Catalpa Scop. CATAWBA. INDIAN BEAN

Trees; leaves simple, in whorls of three or opposite, long-petiolate, blades large, ovate; flowers in large terminal panicles; calyx more or less 2-lobed; corolla subcampanulate and zygomorphic, 5-lobed and somewhat bilabiate, large but delicate in texture; stamens 2; capsules long, linear-cylindric. A genus of some 12 species native to eastern North American and eastern Asia, three of the species naturalized in Ohio. The characters in the key below are mostly from Fernald (1950).

a. Undersurface of leaves glabrous at maturity; leaves frequently lobed; corolla 1–2 cm wide, yellow, with orange stripes and purple spots within; capsules 5–8 mm in diameter . . . . 3. *Catalpa ovata*

aa. Undersurface of leaves pubescent at maturity; leaves rarely lobed; corolla 2–6 cm wide, white or whitish with yellow stripes and purple-brown spots within; capsules 8–15 mm in diameter.

b. Leaves long attenuate-acuminate at apex; corolla limb 4–6 cm wide; capsules 2–5 dm long, 12–15 mm in diameter . . . . 1. *Catalpa speciosa*

bb. Leaves abruptly short-acuminate at apex; corolla limb 2–3 cm wide; capsules 1.5–4 dm long, 8–10 (-12) mm in diameter. . . . . 2. *Catalpa bignonioides*

1. CATALPA SPECIOSA Engelm. NORTHERN CATALPA. CIGAR-TREE

Tree, to 24 m (80 feet) tall; leaves not or rarely lobed, long-attenuate at apex, pubescent on undersurface; corolla white with darker spots and yellow stripes within, 4–6 cm across; capsules 2–5 dm long, 12–15 mm in diameter. A species native to regions south of Ohio, often cultivated and in some places escaped and naturalized; roadsides, disturbed habitats along lakes and streams, and other disturbed sites. Braun (1961) suggests that some Ohio herbarium specimens may have been collected from cultivated plants without that having been known or noted on the label. Late May–early August.

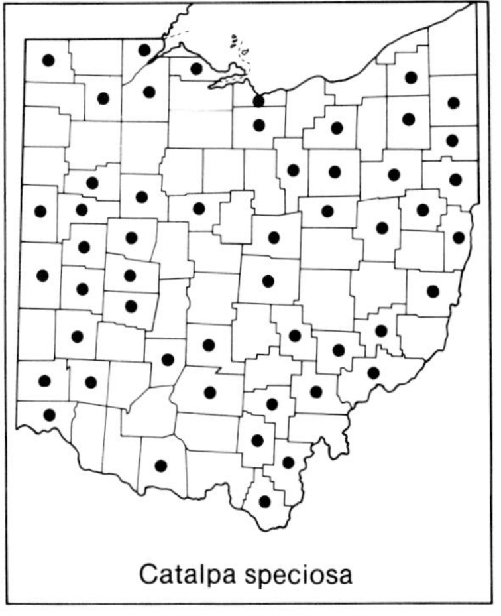
Catalpa speciosa

2. CATALPA BIGNONIOIDES Walt. SOUTHERN CATALPA. INDIAN BEAN

Tree, to 21 m (68 feet) tall; leaves not or rarely lobed, abruptly short-acuminate at apex, pubescent on undersurface; corolla white with darker spots and yellow stripes within, 2–3 cm across; capsules 1.5–4.0 dm long, 8–10 (–12) mm in diameter. A species of southeastern United States, cultivated, and escaped and naturalized at a few sites in Ohio; disturbed woodland borders, roadside thickets. June–early July.

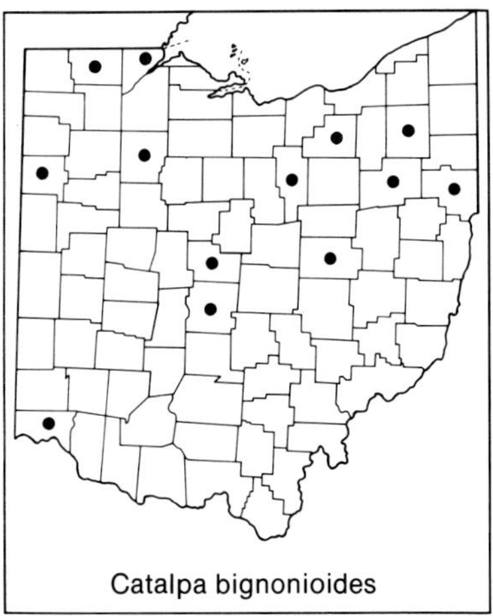
Catalpa bignonioides

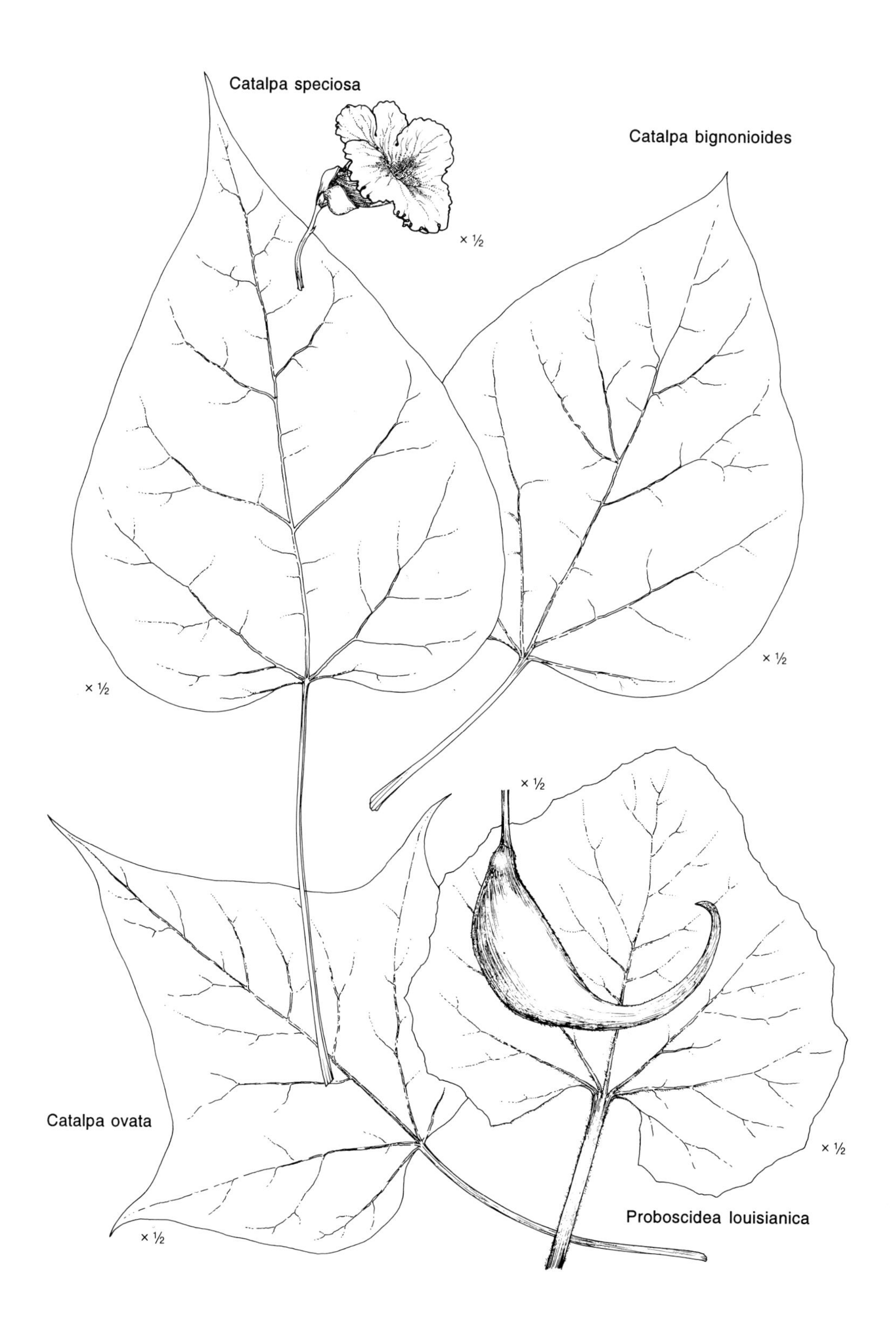
Catalpa speciosa
× ½
Catalpa bignonioides
× ½
× ½
× ½
Catalpa ovata
× ½
Proboscidea louisianica
× ½

3. Catalpa ovata G. Don Chinese Catalpa

Tree, to 9 m (30 feet) tall; leaves often lobed, acuminate at apex, at maturity glabrous on undersurface; corolla yellow with orange stripes and purple spots within, 1–2 cm across; capsules 5–8 mm in diameter. An Asian tree infrequently cultivated in Ohio, escaped and locally naturalized at a few sites, mostly in northeastern counties. Burk and McMaster (1988) describe the spread of this species in western Massachusetts. June–early July.

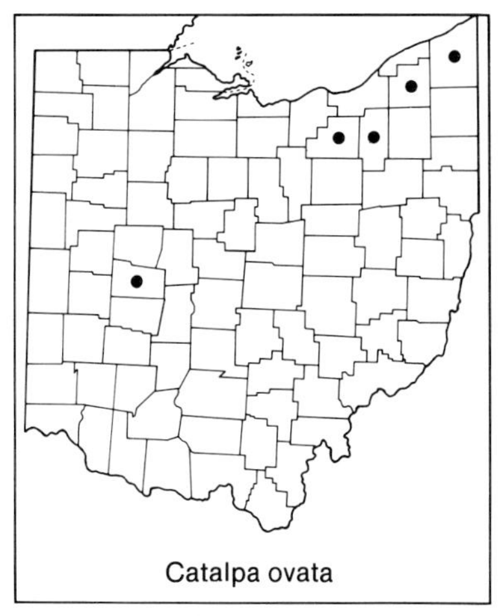
Catalpa ovata

Pedaliaceae. SESAME FAMILY

A chiefly tropical family of some 80 species. Following Cronquist (1981), the family is defined here to include the segregate family Martyniaceae. Thieret (1977) studied the Martyniaceae in southeastern United States.

1. Proboscidea Schmidel

A genus of about 14 species, one of which has been found in Ohio. The treatment here is based in part on the study by Dumke (1967).

1. Proboscidea louisianica (Mill.) Thell. Unicorn-plant. Proboscis-flower

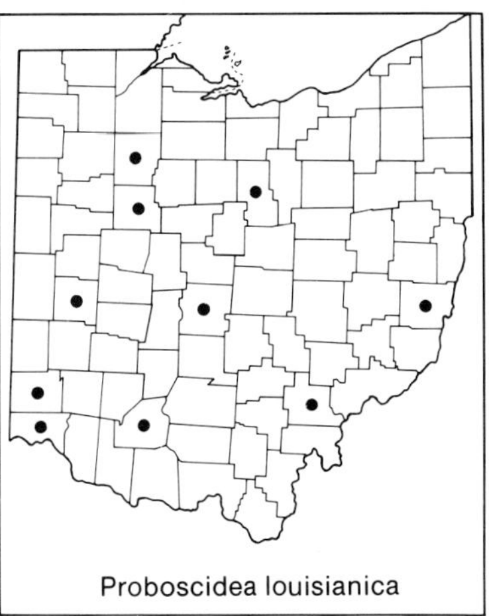
Proboscidea louisianica

Low, spreading, glandular-hairy, aromatic annual; leaves simple, mostly opposite; flowers perfect, borne in racemes; calyx 5-lobed, the tube split to the base on one side; corolla white—tinged with pink, zygomorphic, 5-lobed, to 5 cm long; stamens 4, didynamous, attached to the corolla tube; carpels 2, fused to form a compound pistil with a single style; ovary superior; fruit a distinctively beaked, firm drupe to 10 cm or more in length. Cultivated as an ornamental—now much less frequently than in the past, this plant is also grown for its young fruits, which can be pickled and eaten. At scattered Ohio sites, mostly in southern counties; low disturbed ground. Thieret (1976b) studied the species' floral biology. June–September.

Proboscidea louisianica may be native to southern Ohio, but that seems doubtful. Thieret (1977) questions whether it is in fact native anywhere in southeastern United States. If alien in Ohio, as seems likely from its habitat, it is perhaps merely an adventive.

Orobanchaceae. BROOM-RAPE FAMILY

Nongreen fleshy annual or perennial herbs, mostly tan to brown, parasitic on the roots of other flowering plants; leaves reduced and scale-like; flowers perfect, solitary or in spicate, racemose, or paniculate inflorescences; calyx 4- or 5-lobed, sometimes bilabiate; corolla 5-lobed, zygomorphic;

stamens 4 and didynamous, attached to the corolla tube; carpels 2, fused to form a compound pistil with a single style; ovary superior; fruit a capsule. A small family of about 150 species, mostly native to the Old World. Kuijt's (1969) book on the biology of parasitic plants includes members of this family. Thieret (1971) studied the family in southeastern United States. The text here is based in part on Valley and Cooperrider's (1966) study of the family as it is represented in the flora of Ohio.

a. Plants with underground stems, bearing 1–4 flowers and fruits on aerial scape-like stalks many times longer than calyx . . . . . 3. *Orobanche*

aa. Plants with aerial stems bearing 10–many flowers and fruits, the pedicels shorter than or scarcely exceeding length of calyx.

b. Flowers and fruits openly spaced along all branches, not or scarcely overlapping . . . . 1. *Epifagus*

bb. Flowers and fruits crowded and overlapping in dense spicate inflorescences.

c. Plants glabrous; calyx shallowly lobed except for one deep sinus, most lobes much shorter than tube; stamens more or less exserted . . . . 2. *Conopholis*

cc. Plants glandular-pubescent; calyx deeply lobed, all lobes much longer than tube; stamens included . . . . 3. *Orobanche*

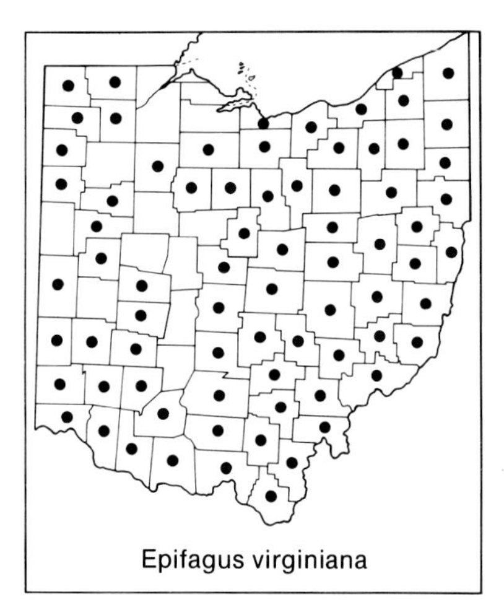
Epifagus virginiana

1. Epifagus Nutt.

A genus with a single species.

1. **Epifagus virginiana** (L.) Bart. BEECH-DROPS

Annual, to ca. 4 dm tall; stems erect, tan to brown or sometimes purplish, puberulent to nearly glabrous, with small, widely spaced bracts; flowers solitary in the axils of bracts, each flower-bearing branch appearing racemose, the branches collectively appearing paniculate; calyx shallowly 5-lobed; corolla pale tan with purplish stripes, 8–12 mm long. Parasitic on the roots of beech (*Fagus*) trees; beech woods throughout Ohio, scarce in some western counties. Mid-August–early October.

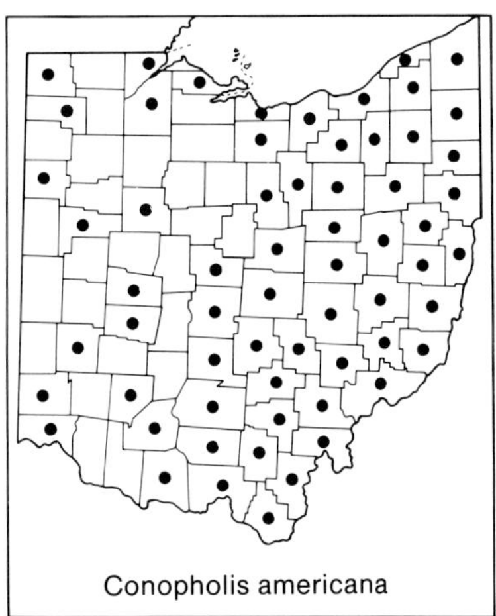
Conopholis americana

2. Conopholis Wallr.

A North American genus of two species, one of them native to Ohio. The treatment here is based in part on the monograph of the genus by Haynes (1971).

1. **Conopholis americana** (L.) Wallr. SQUAW-ROOT

Perennial, to ca. 2 dm tall; stems erect, brown, thick, usually unbranched, glabrous, covered with numerous overlapping brown scales; flowers crowded in a dense, thick, spicate raceme; calyx 4- or 5-lobed, the lobes shallow

Orobanche uniflora
×1
×1
Orobanche ludoviciana
×1
Conopholis americana
Epifagus virginiana
×1

except for one deep sinus, most lobes shorter than the tube; corolla pale yellow or cream-colored, 8–14 mm long. Parasitic on roots of oak (*Quercus*) trees; throughout the eastern half of Ohio, but scarce or absent from much of the western half; oak woodlands. W. V. Baird and J. L. Riopel (1986) studied this species' life history. Mid-April–June.

3. Orobanche L. BROOM-RAPE

Perennials; stems underground or aerial. A widely distributed genus of about 100 species.

a. Plants with an aerial stem, bearing numerous flowers and fruits in a dense spicate inflorescence; flowers subsessile, the pedicels, if present, shorter than to scarcely exceeding length of calyx . . . . 1. *Orobanche ludoviciana*

aa. Plants with an underground stem, bearing 1–4 flowers, each on a separate aerial scape-like stalk many times length of calyx . . . . 2. *Orobanche uniflora*

Orobanche ludoviciana

1. **Orobanche ludoviciana** Nutt. LOUISIANA BROOM-RAPE

Perennial, to ca. 3 dm tall; stems erect, whitish—tinged with purple, glandular-pubescent, with small scattered scales; flowers subsessile to sessile, crowded in dense spikes; calyx deeply 5-lobed; corolla tube whitish—tinged with purple, the lips purplish, 2–2.5 cm long. Parasitic on a number of flowering plant hosts; known from only two sites in southern Ohio, one a tobacco field in Clermont County, the other a weed lot in Gallia County. A species of western United States, at an eastern boundary of its range in Ohio. L. T. Collins (1973) made a systematic study of *Orobanche* section *Myzorrhiza*, to which this species belongs. July–August.

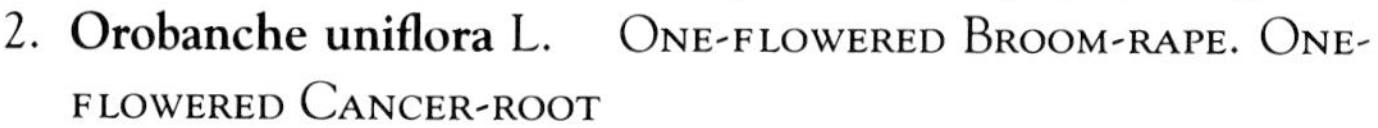

2. **Orobanche uniflora** L. ONE-FLOWERED BROOM-RAPE. ONE-FLOWERED CANCER-ROOT

Perennial; stems completely or largely underground, producing 1–4 aerial, scape-like peduncles, each bearing a single flower, the peduncles 5–20 cm tall, glandular-puberulent; calyx 5-lobed; corolla whitish to lavender, 2.0–2.5 cm long. Parasitic on the roots of various plants; mostly in the eastern half of Ohio; woodlands. May–early June.

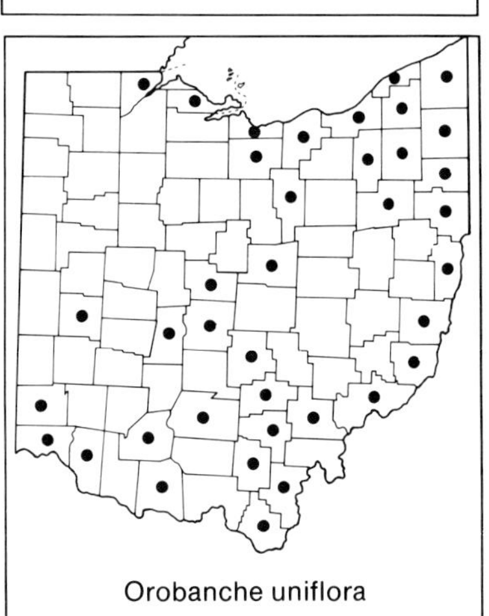

Orobanche uniflora

Lentibulariaceae. BLADDERWORT FAMILY

A small cosmopolitan family of aquatic and wetland plants, represented in the Ohio flora by a single genus. Crow and Hellquist (1985) studied the Lentibulariaceae in the New England states, and Tans (1987) studied the family in Wisconsin.

1. Utricularia L. BLADDERWORT

Rootless perennial herbs; leaves alternate or basal, either simple and linear or more often compound and highly dissected into linear-filiform segments, the leaves with numerous tiny bladder-traps, by means of which the plant is carnivorous, capturing and digesting minute aquatic animals; flowers perfect, in mostly bracteate, few-flowered racemes or infrequently the flowers solitary; calyx deeply 2-lobed; corolla zygomorphic and spurred, bilabiate with the throat closed by a palate, the upper lip more or less 2-lobed, the lower 3-lobed; stamens 2, attached to the corolla tube; carpels 2, fused to form a compound pistil with a single, short to minute style; ovary superior; fruit a capsule. A widely distributed genus of more than 200 species. P. Taylor (1989) prepared a monograph of the genus; Ceska and Bell (1973) studied the species of the Pacific Northwest. The text below is based in part on Dumke's (1967) treatment of the genus as it is represented in the flora of Ohio.

a. Plants terrestrial; leaves undivided; flowering branches arising from subterranean stems growing in wet soil, peat, or sand, and bearing 1–3 flowers . . . . 5. *Utricularia cornuta*

aa. Plants aquatic or semiaquatic; leaves divided into 3 or more linear-filiform segments; flowering branches arising from floating or submersed stems, and bearing 1 to many flowers.

b. Leaves mostly divided 4 or more times; flowering branches 10–50 cm long, bearing 6–20 flowers; calyx 3–5 mm long; corolla 15–25 mm long . . . . 1. *Utricularia vulgaris*

bb. Leaves mostly divided 1–3 times; flowering branches 2–20 cm long, bearing 1–6 flowers; calyx 1.5–3 mm long; corolla 4–15 mm long.

c. Lower lip of corolla not or only slightly longer than upper lip; final divisions of leaves terete, much less than 1 mm in width. . . . . 2. *Utricularia gibba*

cc. Lower lip of corolla approximately twice the length of upper lip; final divisions of leaves flat (except at tip), 1–2 mm wide.

d. Margins of leaf segments without bristle-like hairs; bladders 1.0–1.5 mm long, borne on ordinary leaves . . . . . 3. *Utricularia minor*

dd. Margins of leaf segments with bristle-like hairs; bladders 2–3 mm long, borne in racemose or paniculate fashion on separate leafless branches . . . . 4. *Utricularia intermedia*

1. **Utricularia vulgaris** L. GREATER BLADDERWORT
   incl. *U. macrorhiza* LeConte

Perennial; stems floating, to ca. 1 m long; leaves mostly divided 4 or more times, bearing numerous bladders; erect flowering branches to 5 dm tall, with 6–20 flowers in a terminal raceme; corolla yellow, 15–25 mm long.

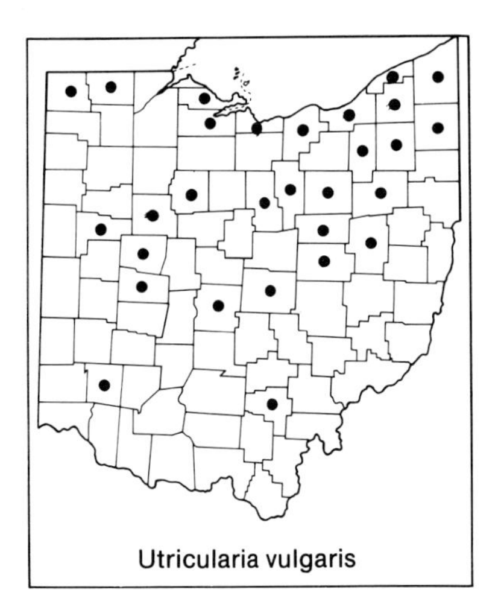
Utricularia vulgaris

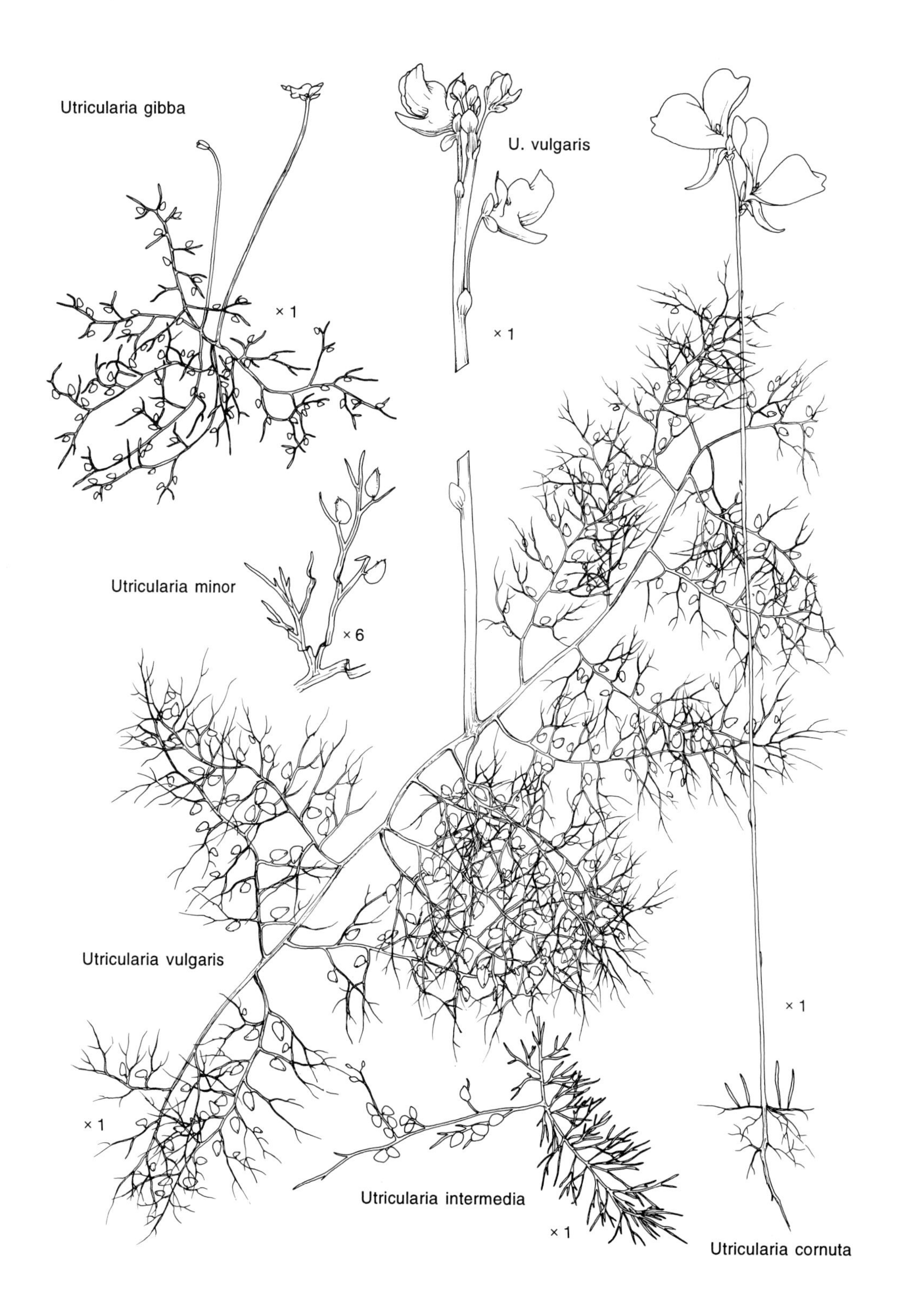
Utricularia gibba
U. vulgaris
× 1
× 1
Utricularia minor
× 6
Utricularia vulgaris
× 1
× 1
Utricularia intermedia
× 1
Utricularia cornuta

Northeastern counties and at scattered sites elsewhere in Ohio; lakes and ponds, standing water in marshes and wet ditches. Late April–September.

P. Taylor (1989) places the American and eastern Asian elements of this species in the segregate species *U. macrorhiza*.

2. **Utricularia gibba** L. HUMPED BLADDERWORT
Perennial; stems floating or somewhat creeping; leaves bearing numerous bladders, and 1 (or 2) times divided, the final divisions terete and much less than 1 mm in width; erect flowering branches to 1 dm tall, with 1 or 2 flowers; corolla yellow, 4–6 mm long, the lower lip not or only slightly longer than the upper. Chiefly in northeastern and south-central counties; shallow water in marshes, swamps, bogs, and ponds. July–September.

3. **Utricularia minor** L. LESSER BLADDERWORT
Perennial; stems floating or somewhat creeping; leaves bearing numerous bladders 1.0–1.5 mm long, mostly 1–3 times divided, the final divisions flat (except at the tip) and 1–2 mm wide, margins of leaf segments lacking bristle-like hairs; erect flowering branches to 1.5 dm tall, with 1–6 flowers; corolla dull pale yellow, 5–8 mm long, the lower lip approximately twice as long as the upper. A circumboreal species transcontinental across North America, at a southern boundary of its range at a few scattered sites in central and northeastern Ohio; shallow water of marshes and bogs. June–August.

4. **Utricularia intermedia** Hayne FLAT-LEAVED BLADDERWORT
Perennial; stems floating or somewhat creeping; bladders 2–3 mm long, borne in racemose or paniculate fashion on separate leafless branches; leaves mostly 1–3 times divided, the final divisions flat (except at the tip) and 1–2 mm wide, margins of leaf segments with bristle-like hairs; erect

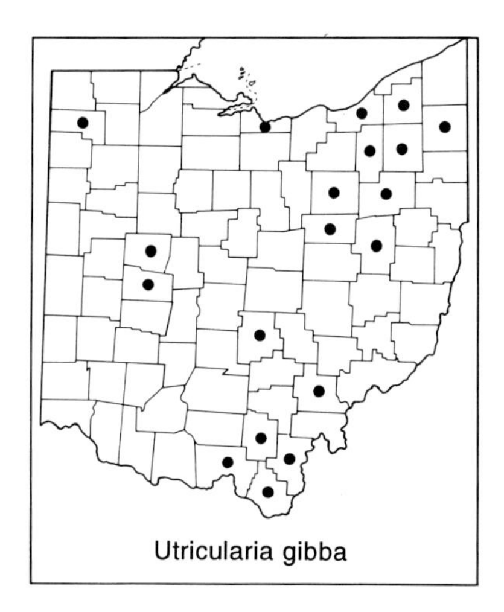
Utricularia gibba

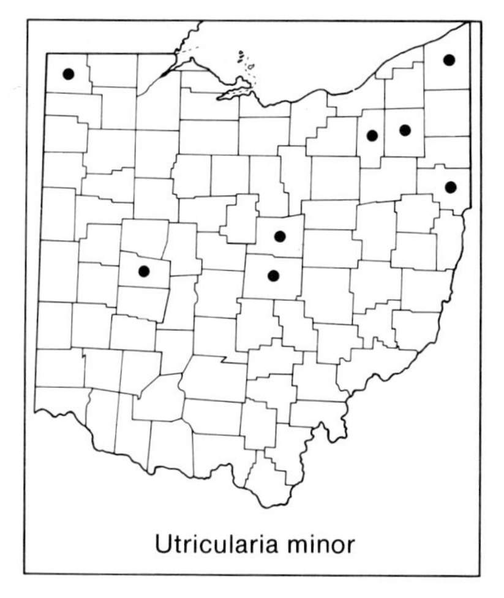
Utricularia minor

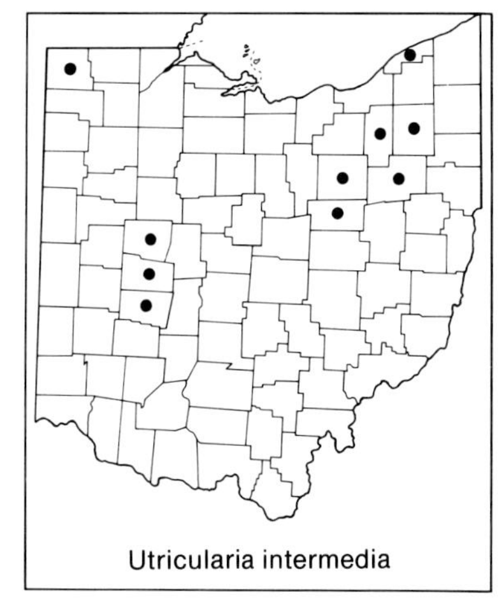
Utricularia intermedia

flowering branches to 2 dm tall, with 1–5 flowers; corolla yellow, 10–15 mm long, the lower lip approximately twice as long as the upper. A circumboreal species, transcontinental across North America, at a southern boundary of its range in northern and central counties of Ohio; shallow water of bogs and fens; marshy pond margins. June–September.

5. **Utricularia cornuta** Michx. HORNED BLADDERWORT

Perennial; stems subterranean, creeping; leaves subterranean, linear, undivided, with few bladders; flowering branches erect, to 3 dm tall, with 1–3 or more flowers; corolla yellow, 20–25 mm long. A few sites in northeastern and central counties; quiet shallow waters of bogs and fens. Kondo (1972) describes the variability in this species. June–September.

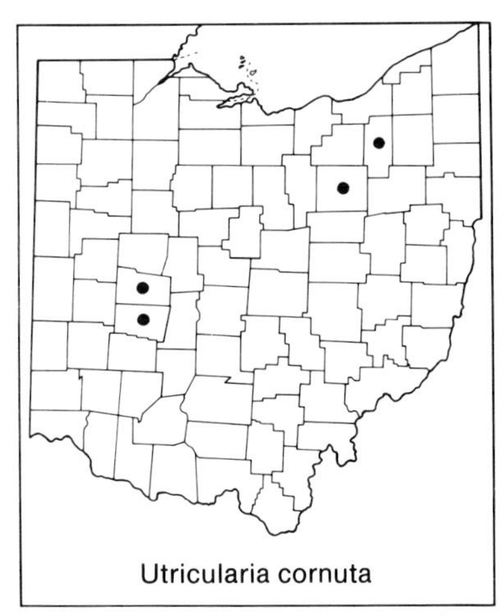
Utricularia cornuta

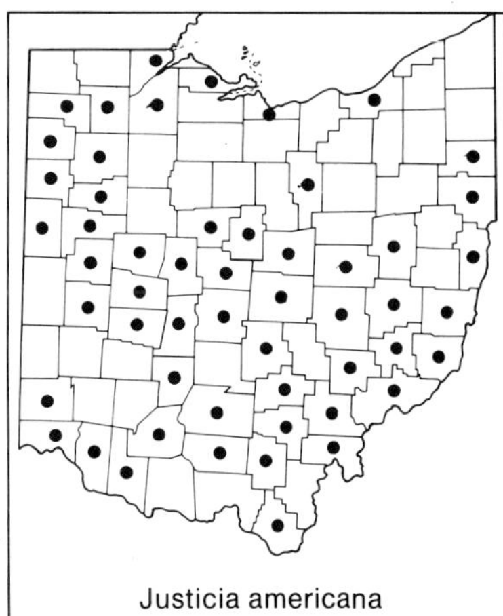
Justicia americana

Acanthaceae. ACANTHUS FAMILY

Perennial herbs; leaves opposite, simple; flowers perfect, solitary or in cymes or dense spikes; calyx 5-lobed; corolla 5-lobed, slightly to markedly zygomorphic; stamens 4 and didynamous or 2, attached to the corolla tube; carpels 2, fused to form a compound pistil with a single style; ovary superior; fruit a capsule. A chiefly tropical family of some 2,500 species. Long (1970) studied the family in southeastern United States. The treatment here is based in part on Dumke's (1967) study of Ohio members of the family.

a. Leaves and upper internodes glabrous; leaves linear to lanceolate or narrowly elliptic; flowers in dense head-like spikes borne on axillary peduncles, the peduncles 5–12 cm long; corolla 1–1.5 cm long; stamens 2 . . . . 1. *Justicia*

aa. Leaves and upper internodes slightly to markedly pubescent; leaves lanceolate to elliptic or ovate; flowers either in axillary cymes on peduncles less than 1 cm long or solitary in leaf axils; corolla 2–7 cm long; stamens 4 . . . . 2. *Ruellia*

1. Justicia L. WATER-WILLOW

A widely distributed genus chiefly of the tropics and subtropics with 300 or more species, one of them native to Ohio.

1. **Justicia americana** (L.) M. Vahl WATER-WILLOW

Perennial, to 1 m tall, with conspicuous stolons and/or rhizomes; stems erect, all or at least the middle and upper internodes glabrous; leaves glabrous, linear to lanceolate or narrowly elliptic; flowers in dense head-like spikes borne on axillary peduncles, the peduncles 5–12 cm long; corolla bilabiate, whitish to rose-purple, 1.0–1.5 cm long; stamens 2. Chiefly in southern and western counties; shallow water of streams and lakes. June–September.

2. Ruellia L. RUELLIA

Perennials; stems with upper internodes slightly to markedly pubescent; leaves more or less pubescent, opposite, lanceolate to elliptic or ovate; flowers mostly in axillary cymes, sometimes solitary in leaf axils; corolla showy, pinkish-lavender to purple, funnelform with 5 spreading lobes, only slightly zygomorphic; stamens 4. A widely distributed genus, chiefly of the tropics and subtropics, with ca. 250 species. Long (1966) did research on artificial interspecific hybridization in *Ruellia.*

a. Calyx lobes lanceolate, 2–3 mm wide at widest point; blades of principal leaves 8–13 cm long, 3–5 cm wide . . . . 1. *Ruellia strepens*

aa. Calyx lobes linear, 1–1.5 mm wide at widest point; blades of principal leaves 2–8 (–10) cm long, 1–3 (–3.5) cm wide.

b. Leaves sessile or subsessile with petioles to 3 mm long, blades mostly ovate to elliptic; stem usually branching only from lower nodes, often giving the plant a low bushy appearance . . . . . 2. *Ruellia humilis*

bb. Leaves with petioles 5–12 mm long, blades ovate-lanceolate to lanceolate; stem with few scattered branches or none, giving the plant a strict erect appearance . . . . 3. *Ruellia caroliniensis*

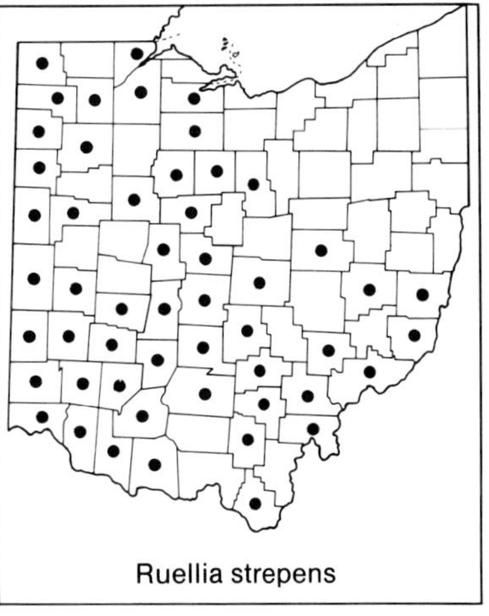
Ruellia strepens

1. **Ruellia strepens** L. SMOOTH RUELLIA

Perennial, to 1 m tall; stems erect; leaves petiolate, blades of principal leaves 8–13 cm long, 3–5 cm wide; calyx lobes lanceolate, 2–3 mm wide at widest point; corolla violet, 4–5 cm long and ca. 4–5 cm across. A species chiefly of southeastern and south-central United States, at a northern boundary of its range in Ohio, where it occurs throughout southern and western counties; woodland openings and borders, cliffs, roadside banks, thickets. Late May–early August.

Intergrades between this species and *R. humilis* have been collected from Champaign, Clermont, Logan, and Miami counties.

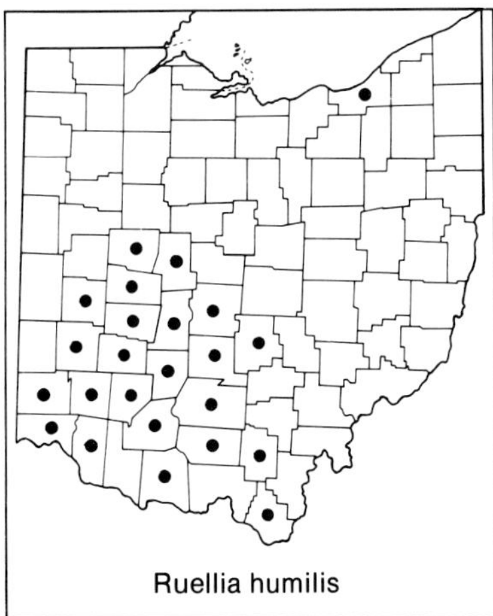
Ruellia humilis

2. **Ruellia humilis** Nutt. WILD PETUNIA

Perennial, to 6 dm tall; stems usually branching only from lower nodes, often giving the plant a low bushy appearance; leaves sessile or subsessile with petioles to 3 mm long, blades mostly ovate to elliptic, blades of principal leaves 2–6 cm long, 1–3 cm wide; calyx lobes linear, 1 mm wide at widest point; corolla pinkish-lavender, 2–3 cm long, 3–4 cm across. A species chiefly of southeastern and south-central United States, at a northern boundary of its range in Ohio, mostly in counties of the southwestern quarter; prairies, fields, roadsides. The disjunct 1896 collection (in The

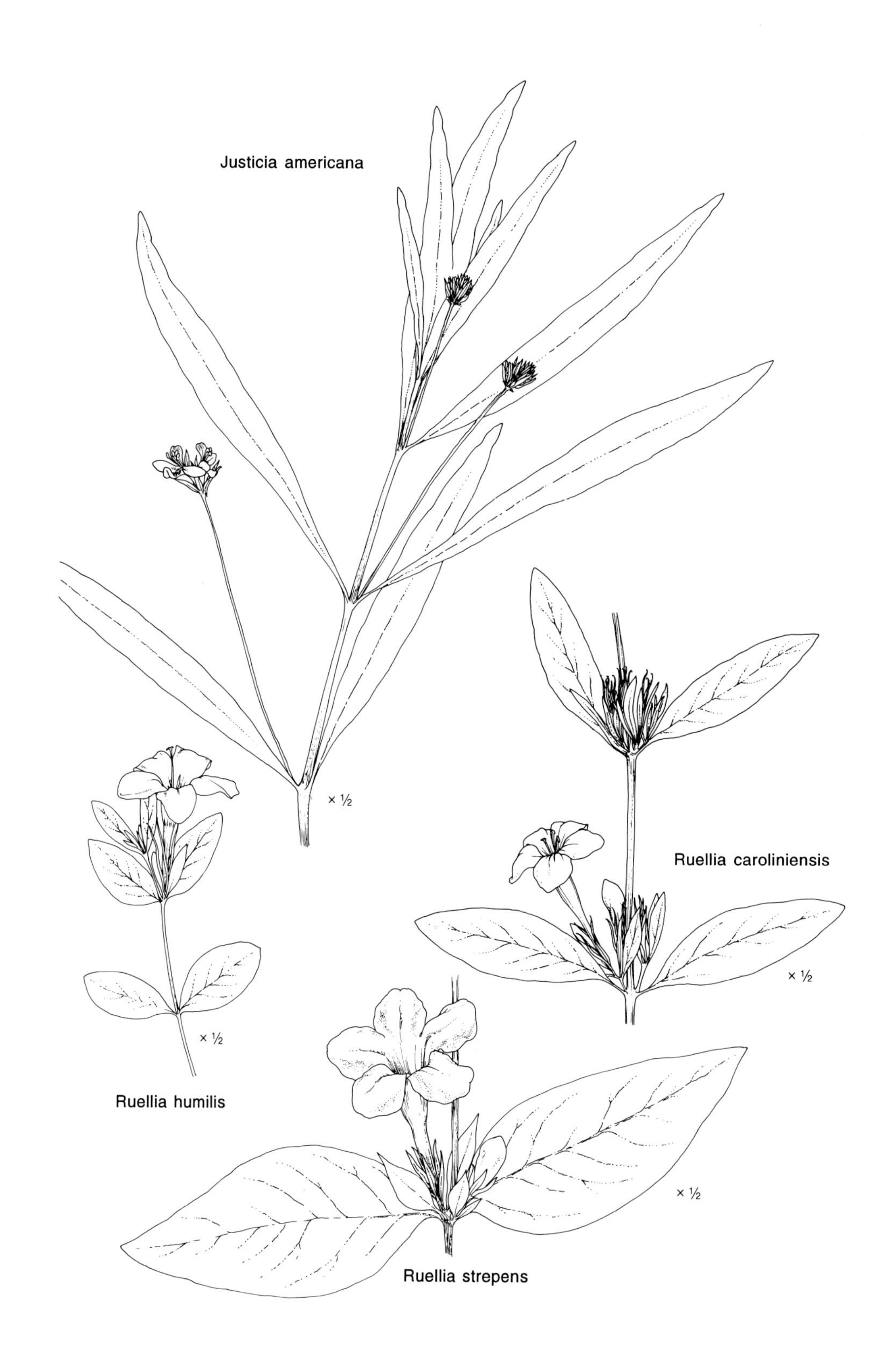
Justicia americana
× ½
Ruellia caroliniensis
× ½
× ½
Ruellia humilis
× ½
Ruellia strepens

Ohio State University Herbarium) attributed to "Cleveland" in northern Ohio is questionable, but does lie within the species' range drawn by Long (1961). Baskin and Baskin (1982b) discuss seed germination in this species. Mid-June–September.

Ohio plants belong to var. *calvescens* Fern.

See notes above under *R. strepens* on intergrades between that species and *R. humilis*. Long (1961) describes convergent patterns of variation between *R. humilis* and the next species, *Ruellia caroliniensis*; intergrades, treated as hybrids between the two species, have been collected from Lawrence County—one of only a few counties, all in southernmost Ohio—where the two species are sympatric.

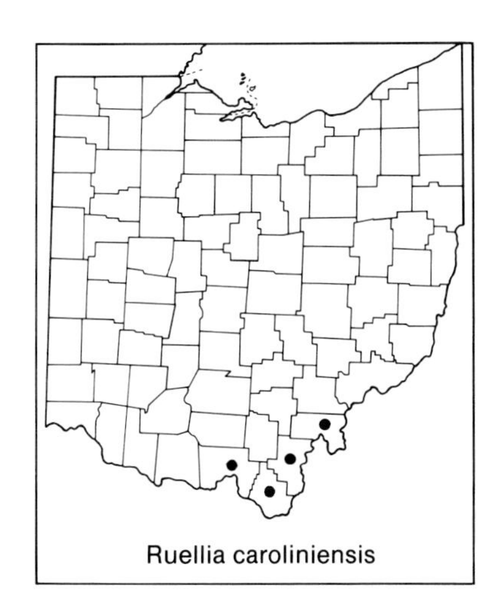
Ruellia caroliniensis

3. **Ruellia caroliniensis** (J. F. Gmel.) Steud. CAROLINA RUELLIA
   incl. var. *caroliniensis*, var. *dentata* (Nees) Fern., and var. *membranacea* Fern.

Perennial, to 8 dm tall; stems with few scattered branches or none, giving the plant a strict erect appearance; leaves with petioles 5–12 mm long, blades ovate-lanceolate to lanceolate, those of principal leaves 4–8 (–10) cm long, 1.5–3.0 (–3.5) cm wide; calyx lobes linear, 1.5 mm wide at widest point; corolla lavender to purplish, 2.5–4.5 cm long, 3–4 cm across. Known from a few southernmost Ohio counties, where this species of southeastern United States is at a northern boundary of its range; woodland openings, grassy slopes, dry fields. Long (1974) studied variation within this species. July–August.

See notes above under *R. humilis* on intergrades between that species and this.

Plantaginaceae. PLANTAIN FAMILY

A small cosmopolitan family of three genera. Nearly all the species belong to *Plantago*, the only genus represented in the Ohio flora. Rosatti (1984) studied the family in southeastern United States.

1. Plantago L. PLANTAIN

Perennial or annual herbs; leaves basal in most species, cauline and opposite in one species; flowers small, perfect, or rarely imperfect, borne in pedunculate spikes or heads; calyx of 4 separate or nearly separate sepals, often appearing as 3 due to fusion of one pair; corolla actinomorphic, 4-lobed; stamens 4, attached to the corolla tube, often exserted on long weak filaments; carpels 2, fused to form a compound pistil with a single style; ovary superior; fruit a capsule. A genus of about 250 species, cosmopolitan in distribution but most abundant in temperate regions. Bassett (1973) studied the genus in Canada.

a. Leaves cauline, borne on aerial stems, opposite, linear . . . . 8. *Plantago psyllium*

aa. Leaves in basal rosettes, alternate, linear to lanceolate to broadly ovate or suborbicular.

b. Blades of principal leaves narrowly lanceolate, narrowly oblanceolate, lanceolate, elliptic, or ovate, (0.5–) 1–19 cm wide.

c. Peduncles downy with soft spreading hairs . . . 7. *Plantago virginica*

cc. Peduncles glabrous or with short appressed-ascending or sometimes with diverging stiffish hairs.

d. Leaves narrowly lanceolate or oblanceolate, blades of principal leaves 4–10 or more times as long as wide . . . . 4. *Plantago lanceolata*

dd. Leaves broadly elliptic to ovate, broadly ovate, or suborbicular, blades of principal leaves 1–2.5 times as long as wide.

e. Uppermost pair of lateral veins diverging from midrib just below midpoint of leaf blade; peduncles 3–4 mm in diameter, hollow . . . . 1. *Plantago cordata*

ee. Uppermost pair of lateral veins diverging from midrib near base of leaf blade; peduncles 1–2 mm in diameter, solid.

f. Capsules 2–3 (–4) mm long, circumscissile near midpoint; sepals rounded to obtuse at apex, without an aristate tip . . . . 2. *Plantago major*

ff. Capsules 3–5 (–6) mm long, circumscissile well below midpoint; sepals acute at apex, often with an aristate tip. . . . . 3. *Plantago rugelii*

bb. Blades of principal leaves linear, 0.1–0.6 (–0.8) cm wide.

c. Bracts conspicuously exserted from inflorescence, some or all bracts 3–10 times longer than flowers . . . . 5. *Plantago aristata*

cc. Bracts not exserted from inflorescence, mostly hidden by dense pubescence, scarcely or not exceeding length of flowers . . . . 6. *Plantago patagonica*

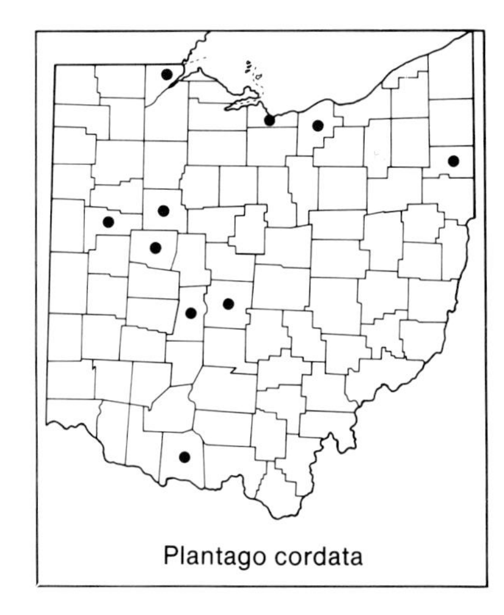
Plantago cordata

1. **Plantago cordata** Lam. HEART-LEAVED PLANTAIN

Perennial; leaves in a basal rosette, long-petiolate, blades of principal leaves ovate to broadly ovate or suborbicular, 1–2 times as long as wide, rounded to cordate at base, the uppermost pair of lateral veins diverging from midrib just below midpoint of blade; peduncles 3–4 mm in diameter, hollow, glabrous. A rare species at scattered Ohio sites; shallow water of slow-moving streams, wet woodlands. Genetic variation was studied by Meagher, Antonovics, and Primack (1978), and the systematics and ecology by Tessene (1969). Spooner et al. (1983) have notes on extant Ohio populations. April–May.

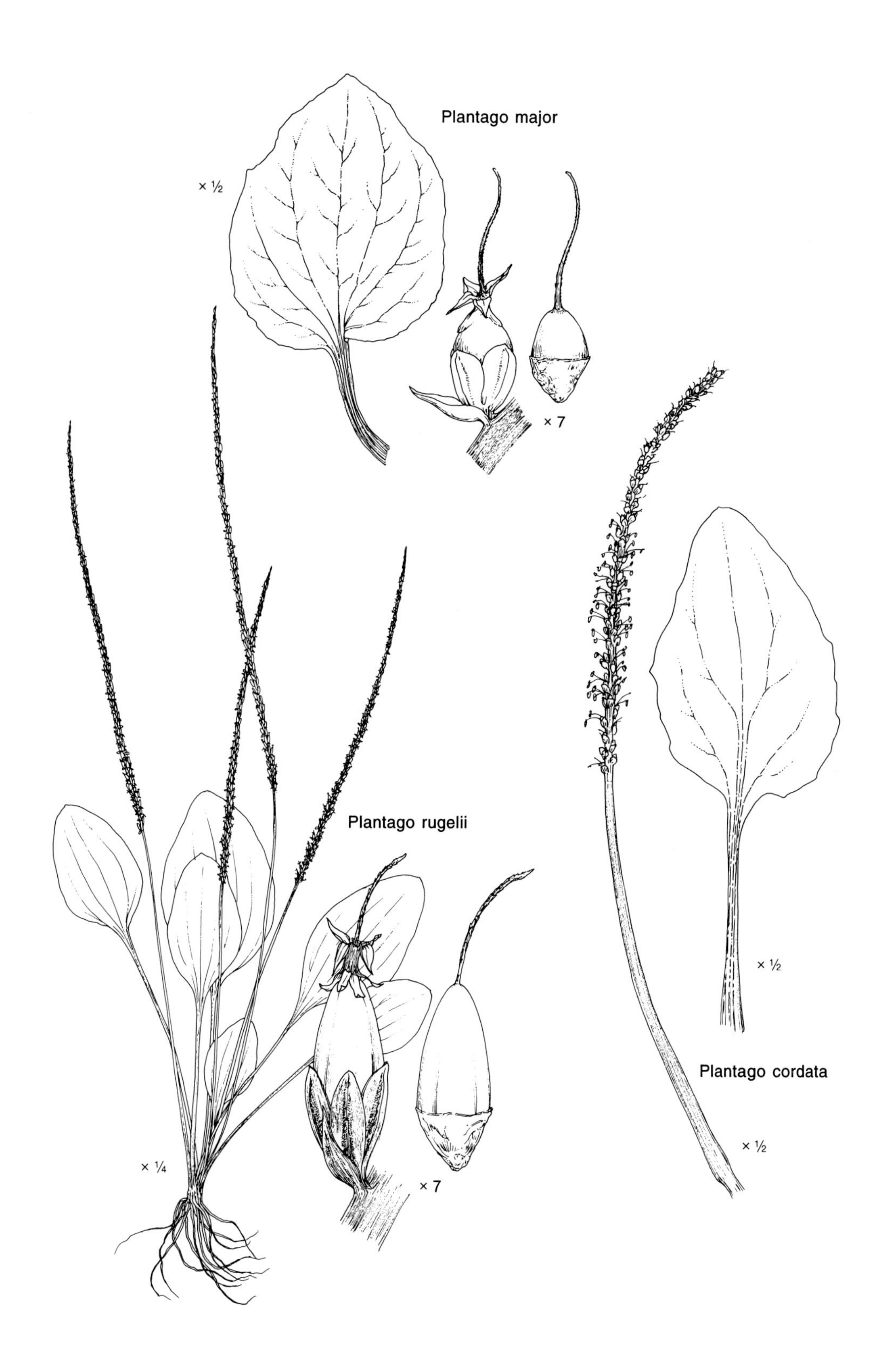
Plantago major
× ½
× 7
Plantago rugelii
× ¼
× 7
× ½
Plantago cordata
× ½

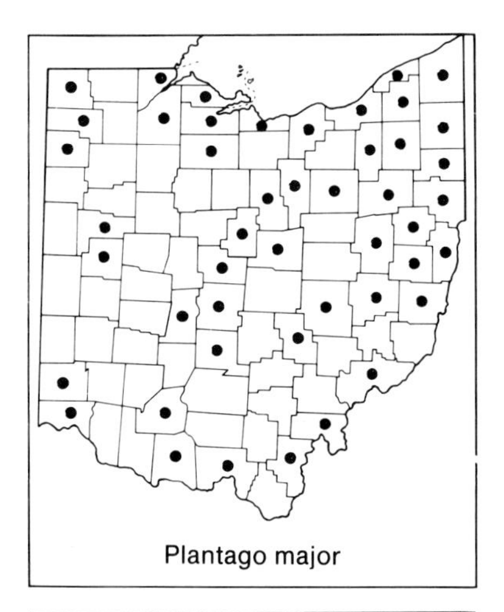

Plantago major

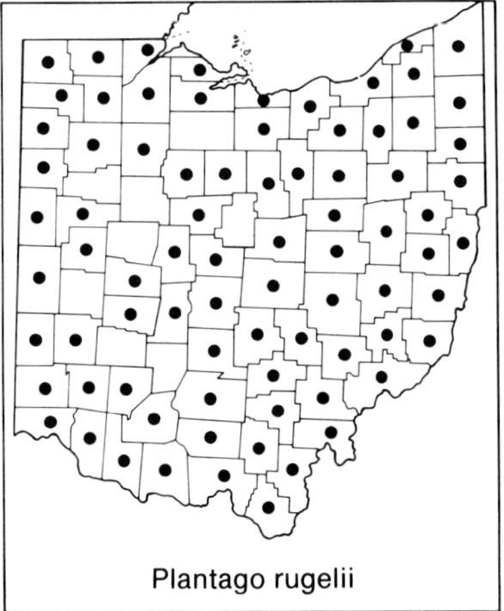

Plantago rugelii

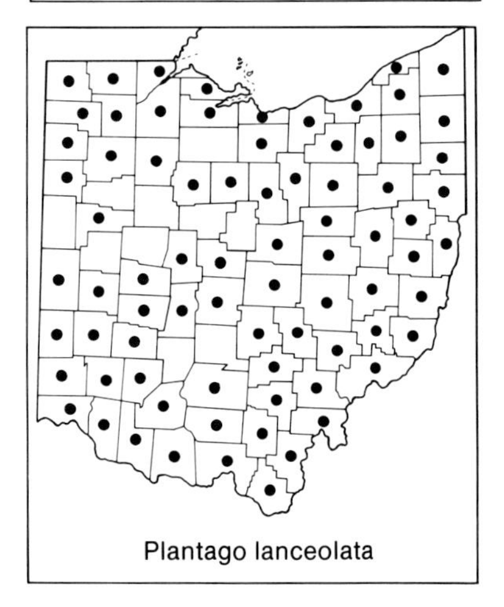

Plantago lanceolata

2. Plantago major L. Common Plantain

Perennial, sometimes functioning as an annual; leaves in a basal rosette, long-petiolate, pubescent, blades of principal leaves elliptic to ovate, broadly ovate, or suborbicular, 1–2.5 times as long as wide, obtuse to rounded or slightly cordate at base, the uppermost pair of lateral veins diverging from midrib near base of blade; peduncle 1–2 mm in diameter, solid, more or less pubescent with stiffish ascending to somewhat spreading hairs; sepals rounded to obtuse at apex; capsules 2–3 (–4) mm long, circumscissile near midpoint. A Eurasian species, naturalized throughout much of Ohio, but infrequent in western counties; roadsides, disturbed ground along railroads and sidewalks, weed lots and weedy fields. Hawthorn and Cavers (1976) describe the species' population dynamics. July–September.

3. **Plantago rugelii** Dcne. Rugel's Plantain

Perennial, sometimes functioning as an annual; leaves in a basal rosette, long-petiolate, slightly pubescent to glabrous, blades of principal leaves elliptic to ovate or suborbicular, 1–2.5 times as long as wide, broadly acute to rounded at base, uppermost pair of lateral veins diverging from midrib near base of blade; peduncles 1–2 mm in diameter, solid, more or less pubescent with stiffish, ascending to somewhat spreading hairs or nearly glabrous; sepals acute at apex and often with an aristate tip; capsules 3–5 (–6) mm long, circumscissile well below midpoint. Throughout Ohio; open ground along streams, roadsides, along railroads and sidewalks, parking lots, lawns, fields, cultivated ground, and similar disturbed habitats. Hawthorn and Cavers (1976) describe the species' population dynamics. Mid-June–September.

4. Plantago lanceolata L. English Plantain. Buckhorn Plantain. Narrow-leaved Plantain

Perennial, sometimes functioning as an annual; leaves in a basal rosette, tapering at base to a short petiole, glabrous or with a few long hairs, blades of principal leaves narrowly lanceolate or narrowly oblanceolate, 4–10 or more times as long as wide; peduncles very long, with appressed hairs to nearly glabrous; flowers in a dense spike. A European species naturalized throughout most of Ohio; along roads, railroads, and sidewalks, in gardens and other cultivated ground, lawns, fields, weed lots, and similar disturbed habitats. Cavers, Bassett, and Crompton (1980) discuss the weedy aspect of this species in Canada. May–August.

Plantago psyllium
× ½
Plantago virginica
× ½
× ½
× ½
× ½
Plantago aristata
× ¼
Plantago lanceolata
Plantago patagonica

5. Plantago aristata Michx. Bracted Plantain

Annual; leaves in a basal rosette, linear and scarcely petiolate, glabrous to villous; peduncles with appressed and/or spreading hairs; flowers in a dense spike with bracts conspicuously exserted, some or all bracts 3–10 times longer than flowers. Native to western United States, naturalized in eastern, southern, and parts of northern Ohio; dry woodland openings and borders, dry fields and roadside banks, rock and gravel quarries. Mid-June–August.

6. **Plantago patagonica** Jacq. Salt-and-pepper-plant

P. purshii of authors, not R. & S.

Annual; leaves in a basal rosette, linear and scarcely petiolate, white-villous; peduncles densely gray-pubescent; flowers in a dense spike, bracts not exserted from inflorescence, mostly hidden by dense pubescence, scarcely or not exceeding length of flowers. Known from a few sites in northwestern Ohio, where the species is at an eastern boundary of its range; dry sandy fields. Mid-June–August.

7. **Plantago virginica** L. Dwarf Plantain. Hoary Plantain

Annual; leaves in a basal rosette, short-petiolate, pubescent, blades of principal leaves elliptic to broadly oblanceolate, 2–4 times as long as wide, narrowly tapering at base; peduncles downy with soft spreading hairs. Chiefly in southern and eastern counties; dry woodland openings and borders, sandy fields, dry weedy fields and roadsides, and similar dry disturbed habitats. Rahn (1974) revised *Plantago* section *Virginica*, which includes this species. Mid-April–mid-June.

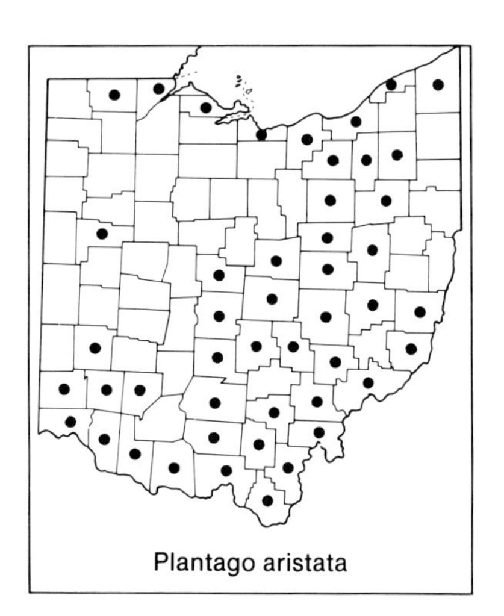

Plantago aristata

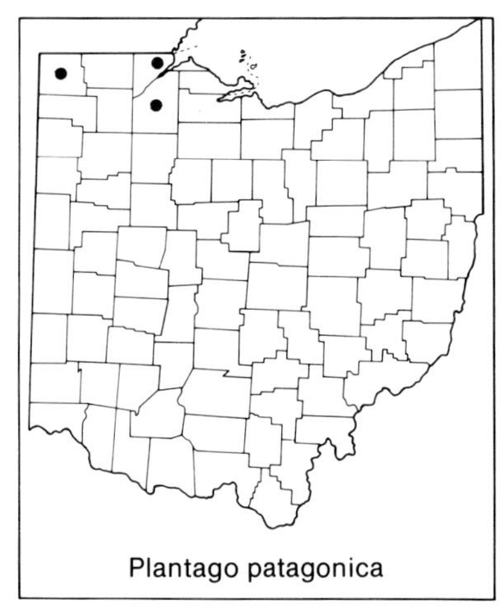

Plantago patagonica

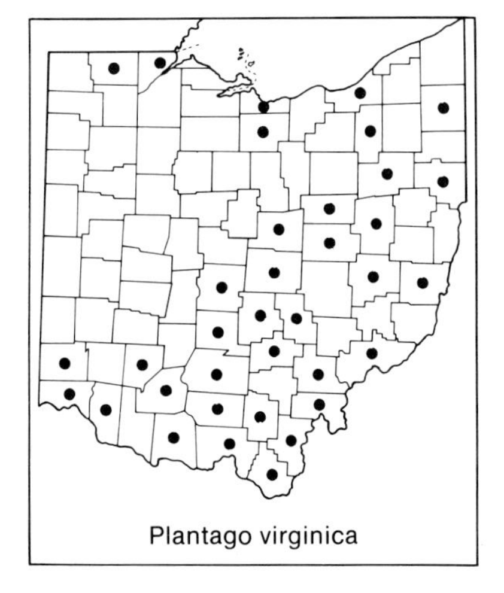

Plantago virginica

8. Plantago psyllium L. Fleawort. Sand Plantain
incl. *P. indica* L.; *P. arenaria* Waldst. & Kit.

Annual, to 3 dm tall; leaves cauline, borne on aerial stems, opposite, linear; flowers few to numerous in dense heads on pubescent peduncles arising from axils of upper leaves. A Eurasian species locally naturalized or perhaps merely adventive at scattered Ohio sites, mostly in northern counties; dry cinders and ballast along railroads, and similar disturbed habitats. The plant is the source of psyllium, used in laxatives. July–September.

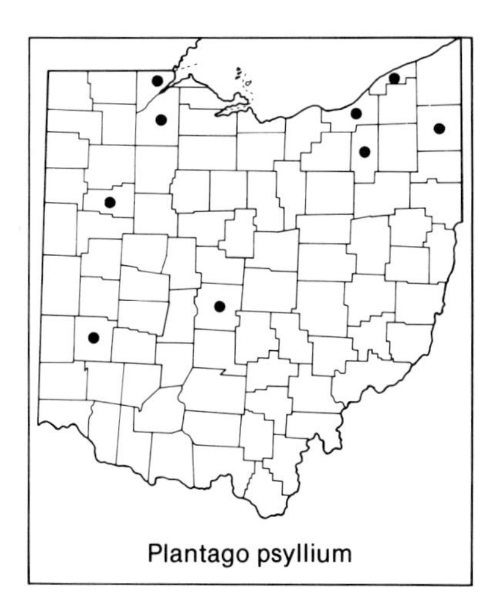
Plantago psyllium

Rubiaceae. MADDER FAMILY

Annual or perennial herbs or shrubs; leaves simple, opposite or whorled; flowers perfect, solitary or more often in various types of few- to many-flowered inflorescences; calyx usually 4-lobed, sometimes absent; corolla usually 4-lobed, actinomorphic; stamens usually 4, attached to the corolla tube; carpels 2, fused to form a compound pistil with a single style; ovary inferior; fruit usually a capsule or berry. A major, chiefly tropical family of some 6,000 species. G. K. Rogers (1987) studied the subfamily Cinchonoideae in southeastern United States, which includes genera nos. 6 and 7 below. The text here is based in part on Hauser's (1964) study of the family in Ohio.

Coffee is made from the dried seeds (called "beans") of *Coffea* species from Africa, principally *C. arabica* L. Quinine, used in the treatment of malaria, is extracted from the bark of South American trees of the genus *Cinchona*. The sweetly fragrant Gardenia of florists, sometimes raised as a houseplant, is the Chinese species, *Gardenia jasminoides* Ellis.

a. Plants woody, shrubs; flowers and fruits in dense, globose, pedunculate heads . . . . 6. *Cephalanthus*

aa. Plants herbaceous or essentially so; flowers solitary or in various types of inflorescences, if in heads, the heads not globose.

b. Stems trailing; leaves evergreen, orbicular to orbicular-ovate, less than twice as long as wide; fruit a fleshy red berry. . . . . 5. *Mitchella*

bb. Stems erect to procumbent or reclining, but not trailing; leaves not evergreen, more than twice as long as wide; fruit dry or leathery, not red.

c. Leaves all or mostly whorled, rarely opposite.

d. Inflorescence a head, subtended by an involucre of leaves equalling or nearly equalling length of flowers; calyx lobes triangular-lanceolate . . . . 1. *Sherardia*

dd. Inflorescence cymose, paniculate, or umbellate, subtended by small bracts or none; calyx lobes none . . . . 2. *Galium*

cc. Leaves all or mostly opposite.

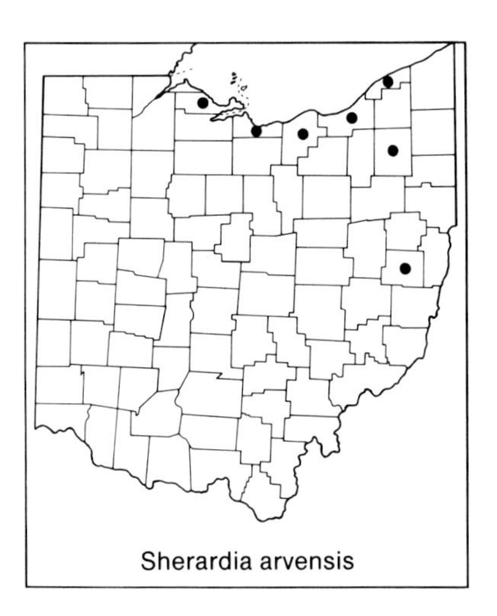

d. Flowers and fruits stalked (except sometimes in *Houstonia nigricans*), either in terminal or axillary pedunculate cymes or solitary on long peduncles . . . . 7. *Houstonia*

dd. Flowers and fruits sessile or nearly so, all or mostly axillary.

e. Upper internodes glabrous; leaf margins glabrous; flowers and fruits numerous in dense axillary and terminal clusters . . . . 3. *Spermacoce*

ee. Upper internodes more or less pubescent; leaf margins scabrous-ciliate; flowers and fruits solitary in leaf axils or in axillary groups of two or three . . . . 4. *Diodia*

1. Sherardia L.

A genus with a single species.

1. Sherardia arvensis L. Field Madder

Annual; stems to 4 dm long, procumbent or prostrate with erect branches; leaves sessile, mostly in whorls of 6, sometimes fewer; flowers in terminal heads closely subtended by an involucre of leaves equalling or nearly equalling length of flowers; calyx lobes 4–6, triangular-lanceolate; corolla lilac, 3–5 mm long, funnelform, 4-lobed; stamens 4; fruit a dryish berry. A European species, adventive or perhaps locally naturalized at a few sites in northern and eastern counties; lawns, disturbed ground near homes and gardens. May–June.

2. Galium L. BEDSTRAW. CLEAVERS

Annuals, winter annuals, or perennials; stems squarish, slender, erect or often weak and reclining; leaves whorled; flowers small, in axillary and/or terminal, pedunculate cymes or panicles; calyx lobes absent; corolla small to minute, white or whitish or sometimes yellow or greenish, mostly rotate and 3- or 4-lobed; stamens usually 4, rarely 3; fruit a dryish, smooth or bristly berry, the berries usually borne in pairs. A widely distributed genus of some 300 species.

Gleason (1952b) noted a report of *Galium vernum* Scop. (*Cruciata glabra* (L.) Ehrh.) from Ohio, presumably based on Schaffner's (1925) list in which G. *vernum* was included—evidently by clerical error. The specimen cited by Schaffner was originally, and correctly, determined to be G. *verum* (discussed in the text below), and had not been annotated as G. *vernum*.

a. Fruits bearing hairs or bristles.

b. Leaves all or mostly in whorls of 6–8 (–10), with only one main vein visible at base of blade.

c. Plants annual, their stems with evident retrorse bristles; leaves linear to narrowly oblanceolate, bristles at leaf margin mostly divergent or retrorse. . . . . 1. *Galium aparine*

cc. Plants perennial, their stems with few or no retrorse bristles; leaves oblanceolate to elliptic-lanceolate or elliptic, bristles at leaf margin mostly ascending.

d. Flowers in axillary or terminal inflorescences; corolla rotate or nearly so, the tube 1/5–1/4 as long as the lobes; fruits 1–2 mm long . . . . 2. *Galium triflorum*

dd. Flowers in terminal inflorescences; corolla funnel-form, the tube about half as long as the lobes; fruits 3–4 mm long . . . . 17. *Galium odoratum*

bb. Leaves all or mostly in whorls of 4, all or most with 3 main veins visible at base of blade.

c. Leaves linear to narrowly linear-lanceolate, the principal leaves 2–5 cm long and 2–6 mm wide . . . . 6. *Galium boreale*

cc. Leaves lanceolate to elliptic or ovate, the principal leaves 1–8 cm long and 5–25 mm wide, if less than 10 mm wide then also less than 2 cm long.

d. Leaves ovate to elliptic, most less than 2 (–2.5) cm long; flowers and fruits all separately stalked, on pedicels more than 1 mm long . . . . 3. *Galium pilosum*

dd. Leaves mostly elliptic to lanceolate, rarely ovate, most more than 2.5 cm long; some or all flowers and fruits sessile, or on pedicels to only 1 mm long, arranged along axes of inflorescence.

e. Leaves ovate-lanceolate to ovate-elliptic or elliptic, rarely ovate, widest near midpoint, acute to obtuse at apex . . . . 4. *Galium circaezans*

ee. Leaves lanceolate to broadly lanceolate, widest near base, acute to acuminate at apex. . . . . 5. *Galium lanceolatum*

aa. Fruits smooth or roughened, but without hairs or bristles.

b. Leaves in whorls of 2–4.

c. Plants erect, 3–9 dm tall; leaves linear to narrowly linear-lanceolate, the principal leaves 2–5 cm long; inflorescences paniculate with numerous flowers . . . . 6. *Galium boreale*

cc. Plants matted, reclining, or ascending, 0.5–6 (–8) dm long; leaves linear to oblanceolate or oblong, the principal leaves 0.3–2.5 cm long; inflorescences cymose with few to many flowers, or flowers solitary.

d. Stems with long divergent hairs and also with retrorse bristles on the angles; leaves elliptic to oblong, 3–6 (–8) mm long, about twice as long as wide; corolla yellow . . . . 16. *Galium pedemontanum*

dd. Stems glabrous or with only retrorse bristles on the angles; leaves linear to oblanceolate or oblong, the principal leaves 5–20 mm long, 2–5 or more times as long as wide; corolla white or whitish.

e. Inflorescences with numerous flowers; corolla 4-lobed . . . . 9. *Galium palustre*

ee. Inflorescences with 2–6 flowers or flowers solitary; corolla 4-lobed or 3-lobed.

f. Corolla 0.5–1.0 mm long, 3-lobed or rarely 4-lobed; upper internodes scabrous.

g. All flowers and fruits axillary, borne singly at the tips of flexuous, arching, scabrous peduncles 8–15 mm long, the peduncles solitary or in groups of 2 or 3 from a node; leaves in whorls of 4 . . . . 10. *Galium trifidum*

gg. All or nearly all flowers and fruits borne in groups of 3 (sometimes 2 or 4, rarely 5) at the tip of a usually straight peduncle 5–15 mm long, the flowers and fruits on short smooth pedicels 1–6 mm long, occasionally with 1 or more solitary flowers on same plant; leaves in whorls of 4 or 5 (or 6) . . . . 11. *Galium tinctorium*

ff. Corolla 1.2–1.5 mm long, 4-lobed; upper internodes smooth or scabrous.

g. Leaves broadly linear to lanceolate, oblanceolate, or elliptic, spreading or ascending, the principal leaves 15–30 mm long, 2–7 mm wide; upper internodes smooth . . . . 12. *Galium obtusum*

gg. Leaves linear-spatulate or linear-oblanceolate, deflexed, the principal leaves 5–10 mm long, 1–2 mm wide; upper internodes smooth or scabrous . . . . 13. *Galium labradoricum*

bb. Leaves on main stem, and often those of the branches, in whorls of 5–8 (–9).

c. Corolla yellow; upper internodes pubescent with short divergent hairs on surfaces as well as on angles . . . . 7. *Galium verum*

cc. Corolla white or pinkish; upper internodes glabrous or with retrorse bristles only on the angles.

d. Stems firm and erect or nearly so; leaves of main stem all or mostly in whorls of 8. . . . . 8. *Galium mollugo* dd. Stems weak and leaning, reclining, or ascending; leaves of main stem in whorls of 4–6.

e. Leaves of main stem all or mostly in whorls of 4 or 5 (sometimes 6), rounded or obtuse at apex, without cuspidate or mucronate tip.

f. Inflorescences of numerous (usually more than 6) flowers; corolla 4-lobed . . . . 9. *Galium palustre*

ff. Inflorescences of 2–5 (–6) flowers; corolla 3-lobed or 4-lobed.

g. Corolla 3-lobed; upper internodes scabrous . . . . 11. *Galium tinctorium*

gg. Corolla 4-lobed; upper internodes smooth . . . . 12. *Galium obtusum*

ee. Leaves of main stem all or mostly in whorls of 6, acute at apex and with cuspidate or mucronate tip.

f. Leaves linear to linear-elliptic, their margins either entire and smooth or scabrous with fine antrorse bristles . . . . 14. *Galium concinnum*

ff. Leaves narrowly elliptic to ovate-lanceolate or oblanceolate, their margins scabrous with usually retrorse spinules . . . . 15. *Galium asprellum*

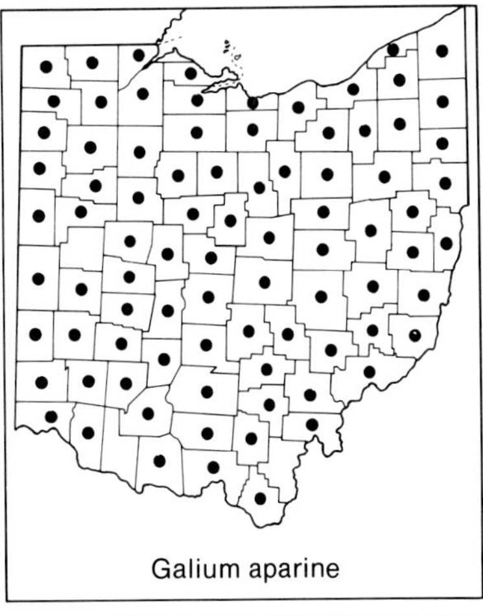

Galium aparine

1. **Galium aparine** L. CLEAVERS

incl. G. *spurium* L.

Annual or winter annual; stems suberect to reclining, to ca. 1 m long, with retrorse bristles; leaves mostly in whorls of 6–8, linear to narrowly oblanceolate, with bristles at leaf margin mostly divergent or retrorse; flowers on axillary peduncles; corolla whitish, 4-lobed; fruits covered with bristles. Throughout Ohio, in part native and in part naturalized from Eurasia; woodlands, thickets, grassy roadsides, weed lots. Mid-April–mid-June.

The treatment here, in which G. *spurium* L. is included without rank as a part of G. *aparine* sensu lato, follows that of Fernald (1950) and of Gleason and Cronquist (1991). Some recent workers, including Ehrendorfer (1976), Malik and Vanden Born (1988), and R. J. Moore (1975), treat G. *spurium* as a distinct species.

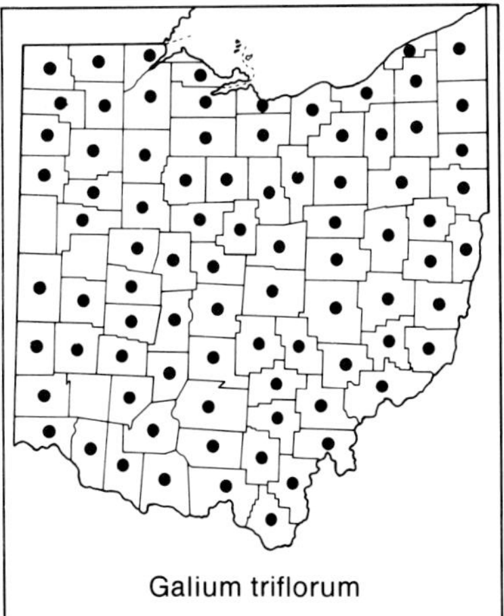

Galium triflorum

2. **Galium triflorum** Michx. SWEET-SCENTED BEDSTRAW. FRAGRANT BEDSTRAW

Perennial; stems suberect to reclining, to ca. 1 m long, either with few retrorse bristles or glabrous; leaves in whorls of 6, elliptic to narrowly elliptic or oblanceolate, cuspidate at apex, bristles on leaf margins mostly ascending; corolla greenish-white, 4-lobed, rotate or nearly so, the tube 1/5–1/4 as long as the lobes; fruits 1–2 mm long, covered with bristles.

Throughout Ohio; woodlands, thickets. Leyendecker (1941) discussed variation in this species. Late May–September.

3. **Galium pilosum** Ait. HAIRY BEDSTRAW

Perennial; stems more or less erect, to 8 dm long, with spreading pubescence; leaves in whorls of 4, elliptic to narrowly ovate or ovate, most less than 2 (or 2.5) cm long; flowers on pedicels longer than 1 mm; corolla greenish-white to purple, 4-lobed; fruits covered with bristles. Eastern half of Ohio, and at a few western sites; dry woodland openings and borders, grassy fields and roadsides. Mid-June–early September.

4. **Galium circaezans** Michx. WILD LICORICE
incl. var. *circaezans* and var. *hypomalacum* Fern.

Perennial; stems more or less erect, to 5 dm long, glabrous throughout or pubescent at the nodes; leaves in whorls of 4, ovate-lanceolate to ovate-elliptic or elliptic, rarely ovate, widest near midpoint, acute to obtuse at apex, more or less pubescent; some or all flowers either sessile or subsessile on pedicels 1 mm long or less; corolla greenish-white, 4-lobed; fruits covered with bristles. Throughout Ohio; woods, woodland borders. Late May–July.

Occasional specimens are difficult to distinguish from G. *lanceolatum*.

5. **Galium lanceolatum** Torr. LANCE-LEAVED BEDSTRAW. LANCE-LEAVED WILD LICORICE

Perennial; stems more or less erect, to 6 dm long, glabrous or nearly so; leaves in whorls of 4, lanceolate to broadly lanceolate, widest near base, acute to acuminate at apex; some or all flowers either sessile or subsessile on pedicels 1 mm long or less; corolla purple, 4-lobed; fruits covered with

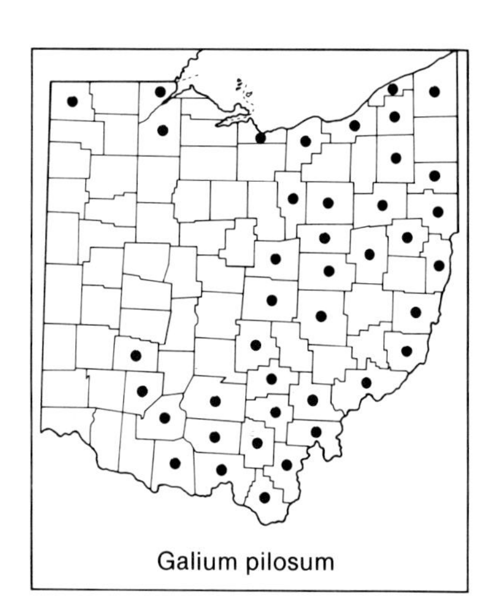
Galium pilosum

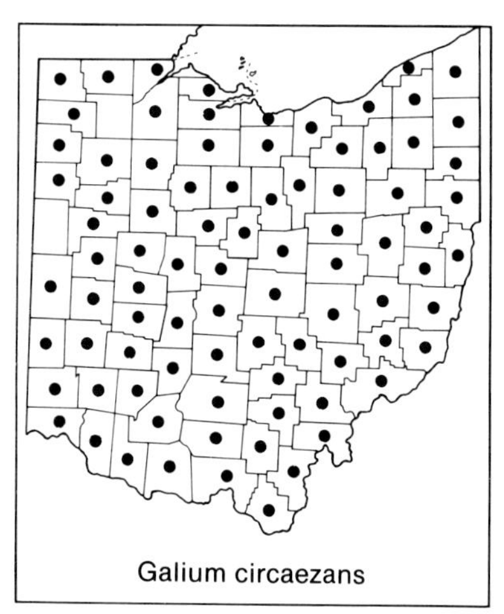
Galium circaezans

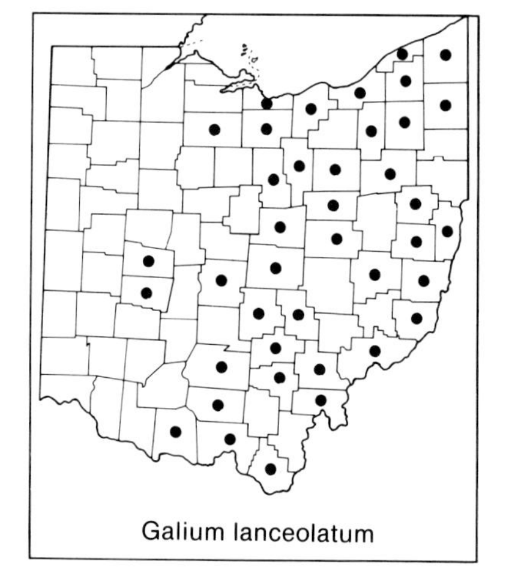
Galium lanceolatum

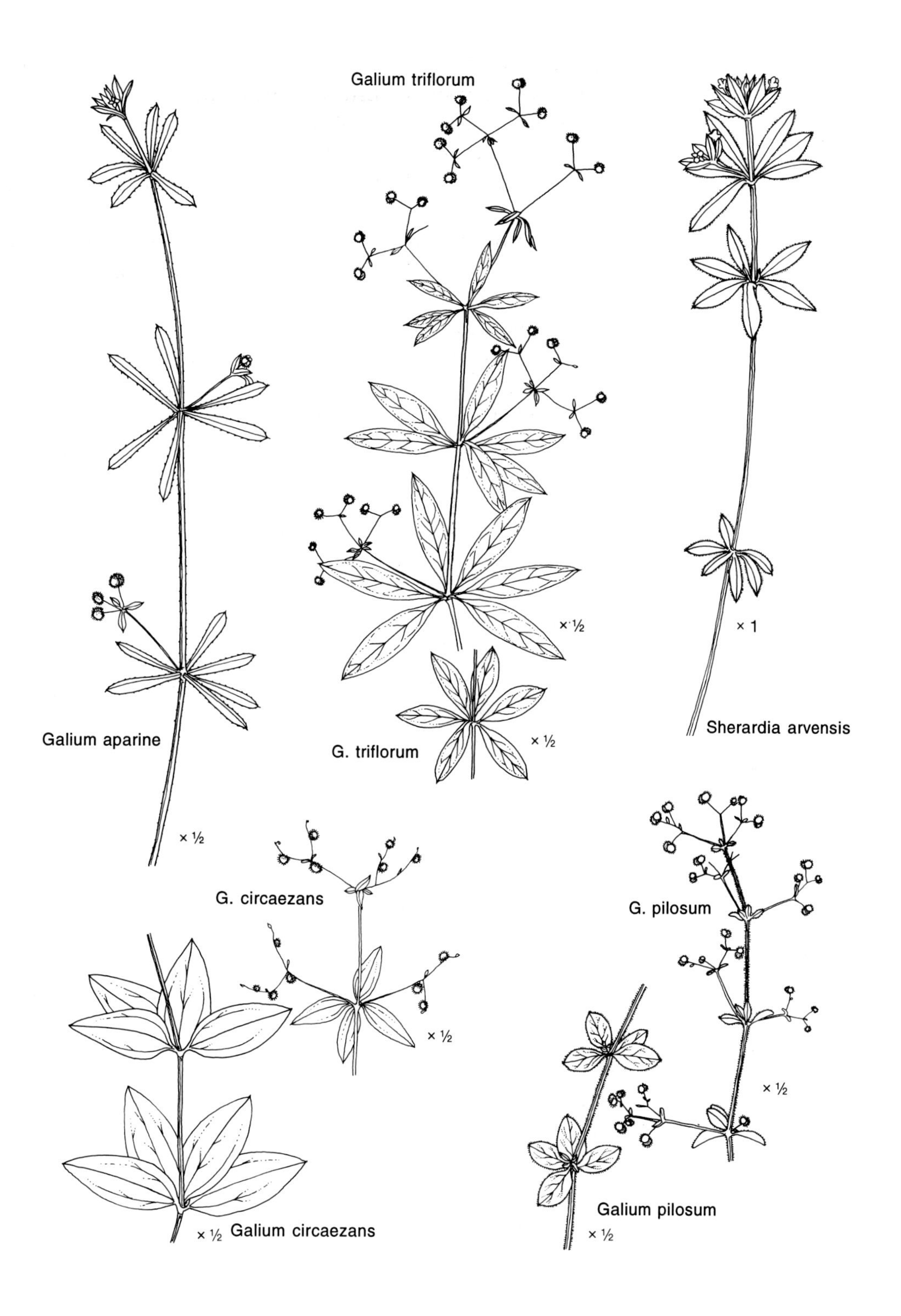
Galium triflorum
Galium aparine
× ½
× ½
G. triflorum
× ½
× 1
Sherardia arvensis
G. circaezans
× ½
G. pilosum
× ½
× ½ Galium circaezans
Galium pilosum
× ½

bristles. Eastern half of Ohio, and a few western sites; woodlands. June–July.

Occasional specimens are difficult to distinguish from G. *circaezans*.

6. **Galium boreale** L. NORTHERN BEDSTRAW

incl. var. *boreale*, var. *hyssopifolium* (Hoffm.) DC., and var. *intermedium* DC.

Perennial, to 1 m tall; stems erect, glabrous or nearly so; leaves in whorls of 4, linear to narrowly linear-lanceolate, the principal leaves 2–5 cm long and 2–6 mm wide; flowers numerous; corolla white, 4-lobed; fruits glabrous or covered with very short bristles. Northern and a few western counties; fens, moist thickets. June–July.

The treatment here in which the three varieties listed above are included in the species without rank, follows the conclusions of Leyendecker (1941), who thought them at most mere forms of the species.

7. GALIUM VERUM L. YELLOW BEDSTRAW

Perennial; stems erect or suberect, to 1 m long, upper internodes pubescent with short divergent hairs on surface as well as on angles; leaves mostly in whorls of 8, linear; flowers numerous; corolla bright yellow, 4-lobed, fragrant; fruits glabrous. A Eurasian species naturalized at scattered sites, mostly in southern counties; roadsides, fields, and other weedy habitats. Mid-May–July.

8. GALIUM MOLLUGO L. WHITE BEDSTRAW

Perennial; stems firm and erect or nearly so, to ca. 1 m long, glabrous; leaves of main stem all or mostly in whorls of 8, narrowly elliptic to oblanceolate; flowers numerous; corolla white, 4-lobed; fruits glabrous. A Eurasian species naturalized in counties of the eastern half of Ohio, and at

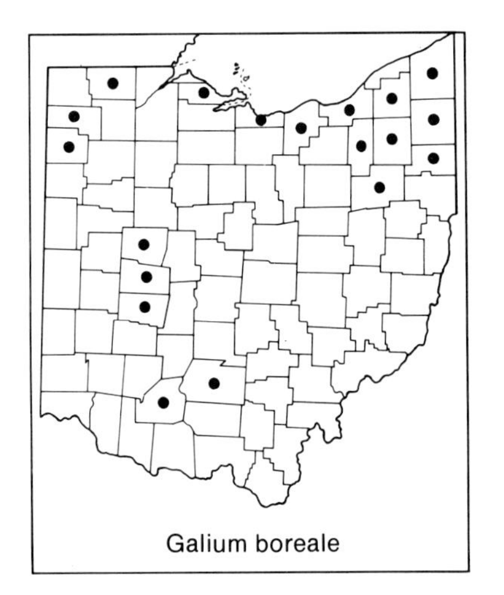
Galium boreale

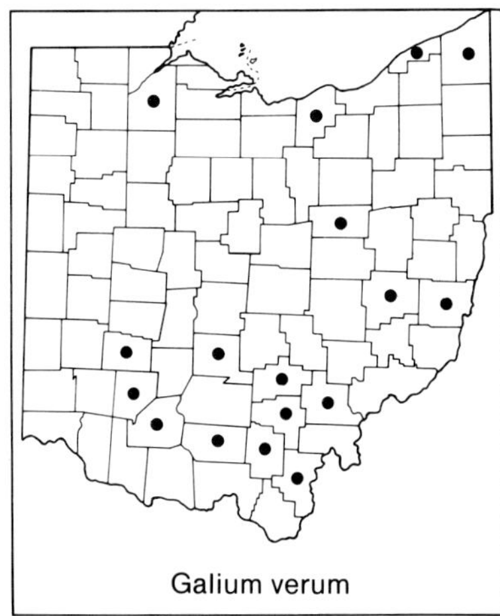
Galium verum

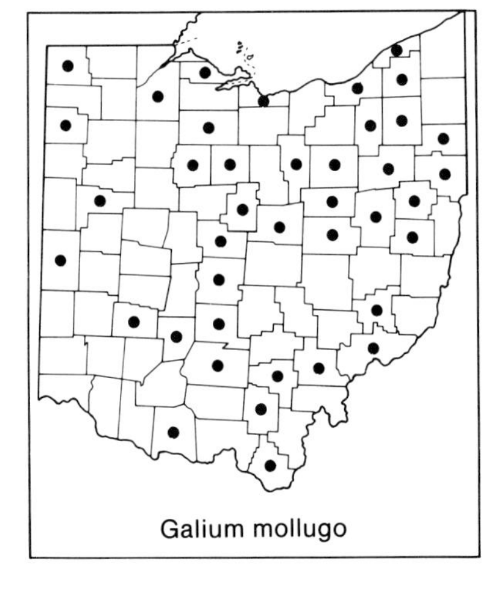
Galium mollugo

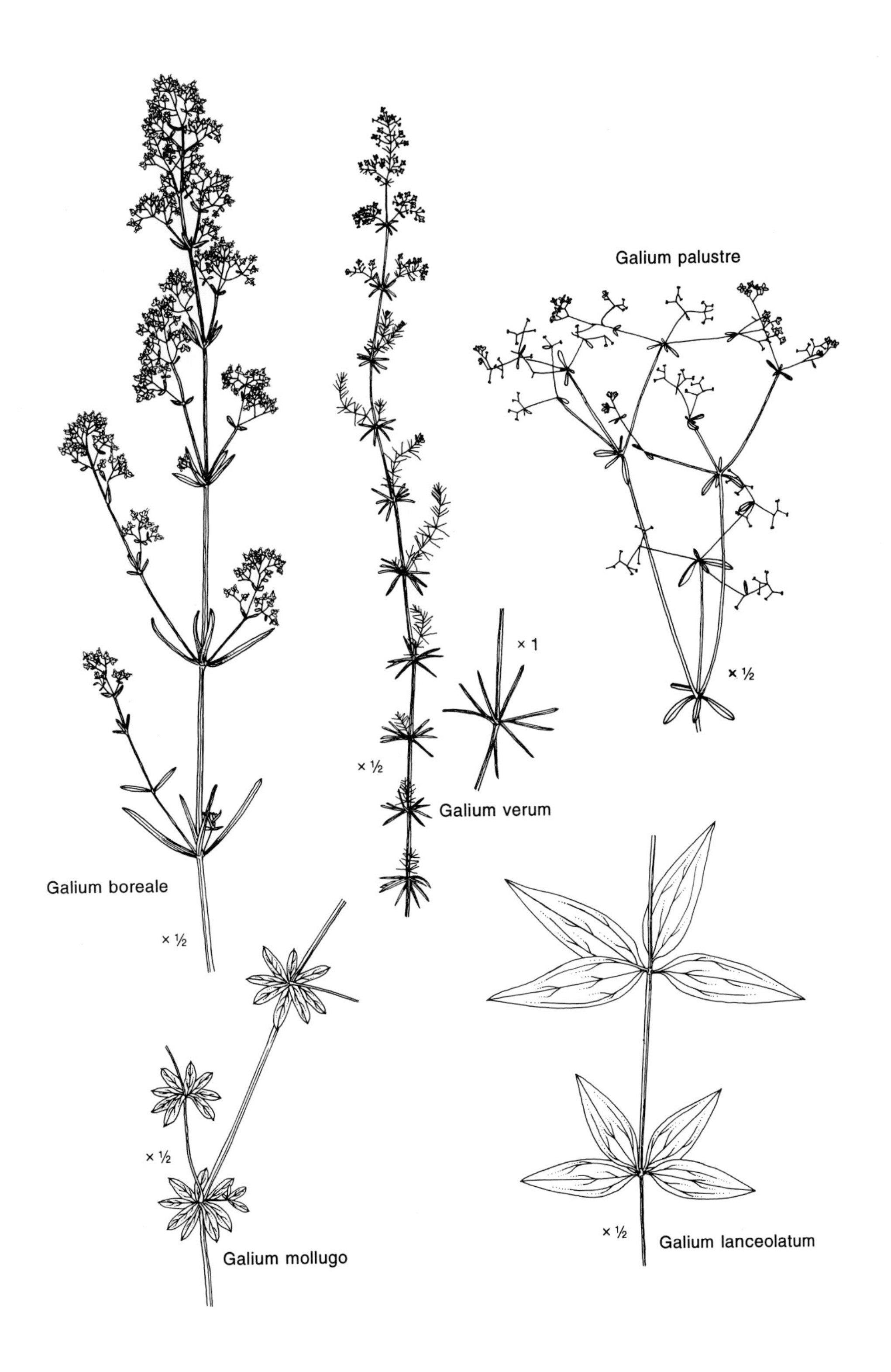
Galium palustre
× ½
× 1
× ½
Galium verum
Galium boreale
× ½
× ½
Galium mollugo
× ½
Galium lanceolatum

scattered sites elsewhere in the state; roadsides, fields, and other open disturbed habitats. June–July (–October).

9. **Galium palustre** L. NORTHERN MARSH BEDSTRAW

Perennial; stems weak, suberect to reclining, to 8 dm long, glabrous or nearly so; leaves mostly in whorls of 4–6, linear-elliptic to narrowly oblanceolate; flowers numerous; corolla white to pinkish, 4-lobed; ovaries sometimes with a few hairs at anthesis, but the mature fruits glabrous. A species of northeastern North America, at a southwestern boundary of its range in Ashtabula County, in northeasternmost Ohio; wet marshy ground, wet fields and thickets. June–July.

10. **Galium trifidum** L. SMALL BEDSTRAW

Perennial; stems weak, to 4 dm long, reclining and sometimes forming mats, minutely retrorsely scabrous on the angles, otherwise glabrous; leaves in whorls of 4, elliptic-linear to linear-oblanceolate; flowers few; corolla whitish, 3-lobed; flowers and fruits axillary, borne singly at the tips of flexuous, arching, scabrous peduncles 8–15 mm long, the peduncles solitary at the nodes or in groups of 2 or 3; fruits glabrous. A circumboreal species at a southern boundary of its range in Ohio; swamps, marshes, and other wetlands. Puff's (1976) systematic study of the *Galium trifidum* group, includes this species and the next, *G. tinctorium*. July–September.

Herbarium specimens of this species are often confused with those of *G. tinctorium*.

11. **Galium tinctorium** L.

G. trifidum L. var. *tinctorium* (L.) T. & G.

Perennial; stems suberect to reclining, to 6 dm long, sometimes forming mats, very minutely scabrous on the angles, otherwise glabrous; leaves in

Galium palustre

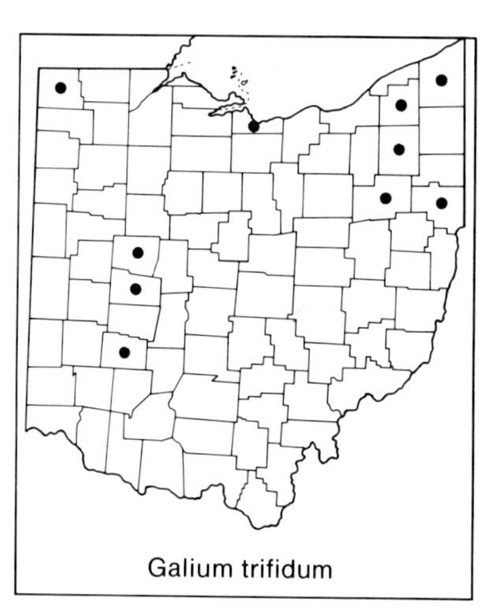
Galium trifidum

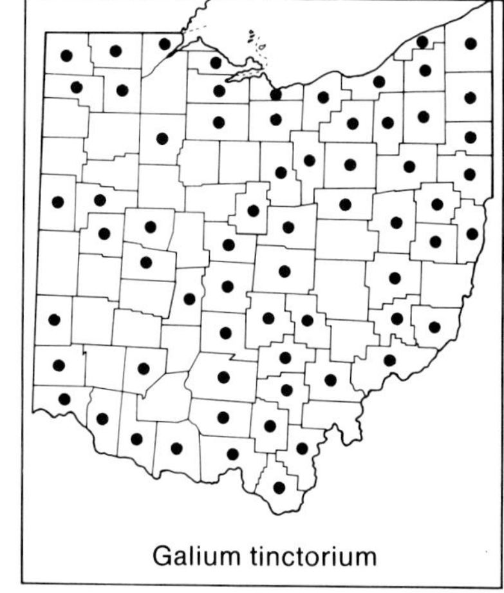
Galium tinctorium

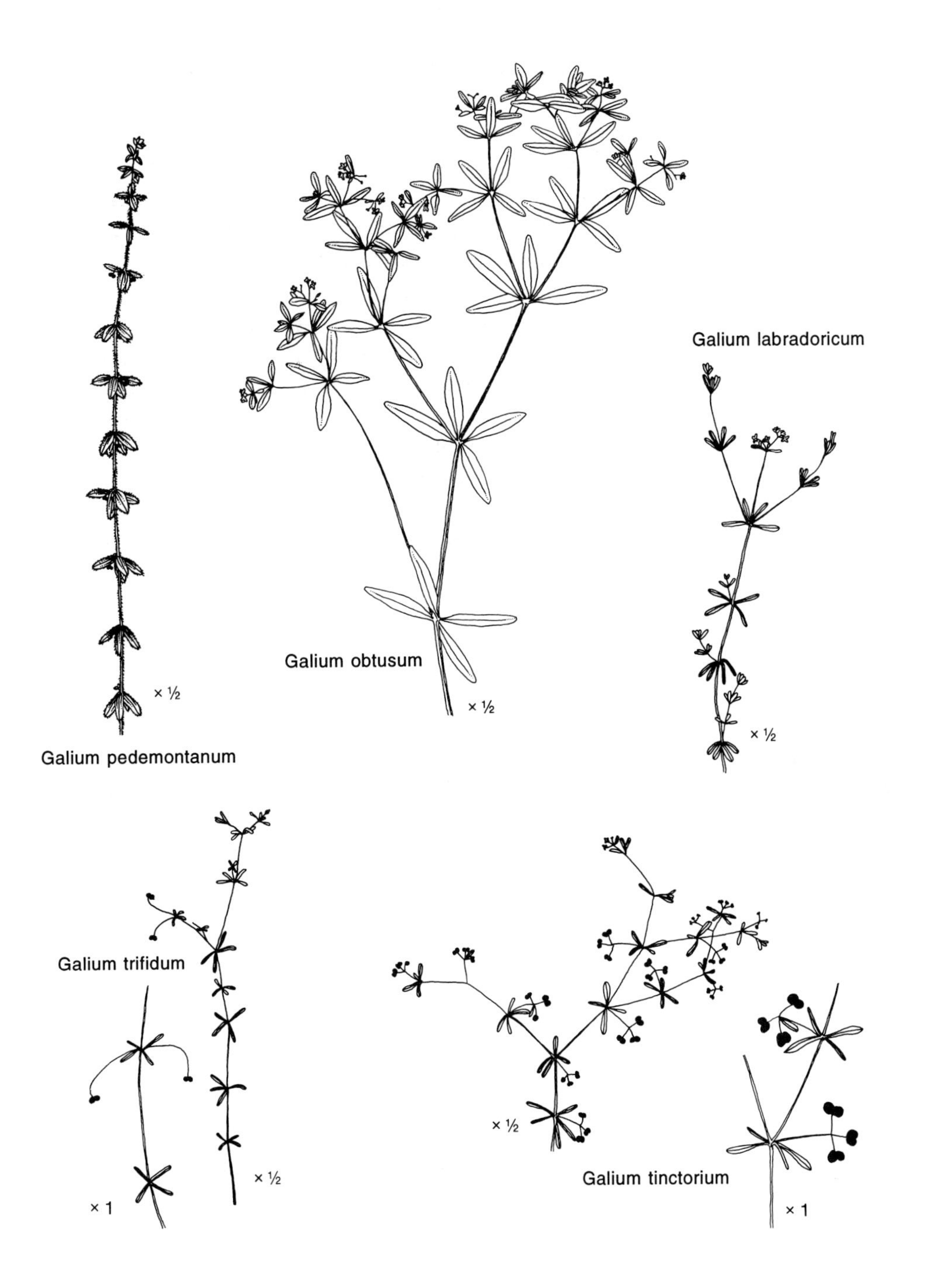
Galium labradoricum
Galium obtusum
× ½
× ½
× ½
Galium pedemontanum
Galium trifidum
× ½
× 1
× ½
Galium tinctorium
× 1

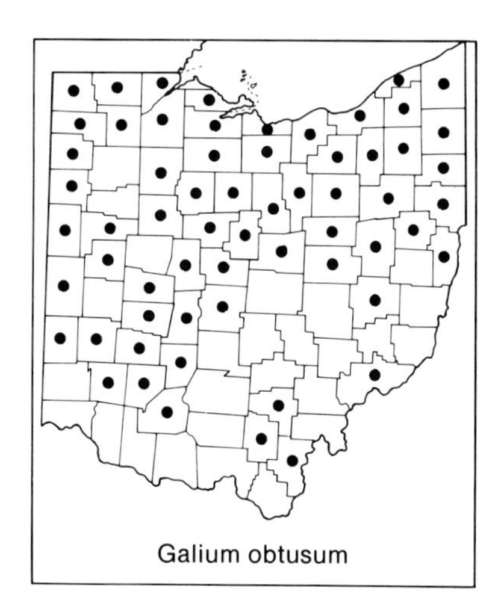
Galium obtusum

Galium labradoricum

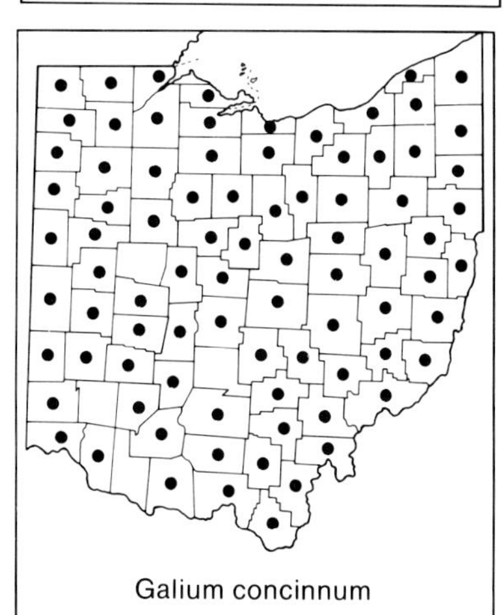
Galium concinnum

whorls of 4 or 5 (or 6), linear to narrowly oblanceolate or narrowly elliptic; flowers numerous; corolla whitish, 3- (or rarely 4-) lobed; flowers and fruits borne in groups of 3 (sometimes 2, 4, or 5) at the tip of a usually straight peduncle 5–15 mm long, the flowers and fruits each on separate, short, smooth pedicels 1–6 mm long originating at the peduncle tip, occasionally with 1 or more solitary flowers on the same plant; fruits glabrous. Throughout eastern Ohio, and at scattered sites in western counties; marshes, bogs, wet fields, borders of lakes and ponds, and other wetlands. Puff's (1976) systematic study of the G. *trifidum* complex, includes that species and this. Mid-May–September.

Herbarium specimens of this species are often confused with those of G. *obtusum*.

12. **Galium obtusum** Bigel. Stiff Marsh Bedstraw
Perennial; stems suberect to reclining, to 7 dm long, glabrous; leaves in whorls of 4 or 5, broadly linear to lanceolate, oblanceolate, or elliptic, spreading or ascending, the principal leaves 15–30 mm long, 2–7 mm wide; flowers few to several; corolla whitish, 4-lobed; fruits glabrous. Chiefly in northern and western counties; low wet woodlands, wet fields and ditches. Puff's (1977) systematic study of the *Galium obtusum* complex includes this species and the next, G. *labradoricum*. Late May–July.

Ohio plants belong to var. *obtusum* (incl. var. *ramosum* Gleason).

13. **Galium labradoricum** (Wieg.) Wieg. Bog Bedstraw
Perennial; stems weak, reclining, to 3 dm long, sometimes forming mats, glabrous or sometimes minutely scabrous on the angles; leaves in whorls of 4, linear-spatulate to linear-oblanceolate, deflexed, the principal leaves 5–10 mm long, 1–2 mm wide; flowers few; corolla white, 4-lobed; fruits glabrous. A species of northeastern North America, at a southern boundary of its range in northeastern Ohio; bogs, wet meadows. Puff's (1977) systematic study of the *Galium obtusum* complex includes that species and this. May–June.

14. **Galium concinnum** T. & G. Shining Bedstraw
Perennial; stems reclining to suberect, to 5 dm long, sometimes minutely scabrous on the angles, otherwise glabrous; leaves of main stem all or mostly in whorls of 6, linear to linear-elliptic, becoming deflexed at maturity, the margins either entire and smooth or scabrous with fine antrorse bristles; flowers many; corolla white, 4-lobed; fruits glabrous. Throughout Ohio; woods, woodland borders, thickets. May–early September.

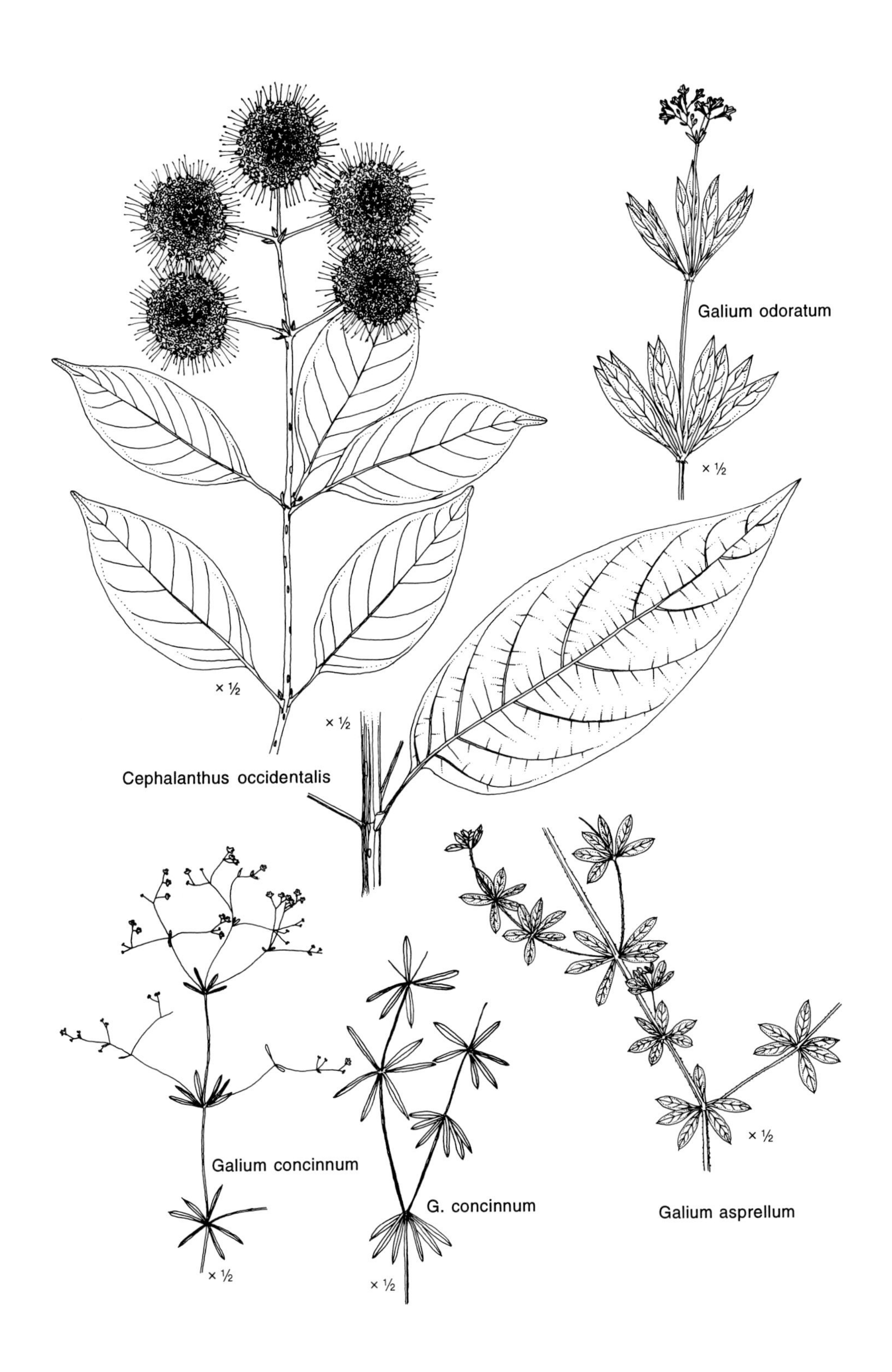
Galium odoratum
× ½
× ½
× ½
Cephalanthus occidentalis
Galium concinnum
G. concinnum
Galium asprellum
× ½
× ½
× ½

15. **Galium asprellum** Michx. ROUGH BEDSTRAW
Perennial; stems reclining, to ca. 1 m long, scabrous on the angles; leaves of main stem all or mostly in whorls of 6, narrowly elliptic to ovate-lanceolate or oblanceolate, often darkening upon drying, the margins scabrous with usually retrorse spinules; flowers numerous; corolla white, 4-lobed; fruits glabrous. Counties of the northeastern quarter of Ohio, and at scattered sites elsewhere; wooded floodplains, marshes, swamps, wet fields and ditches. Mid-June–September.

16. GALIUM PEDEMONTANUM (Bellardi) All. PIEDMONT BEDSTRAW
Cruciata pedemontana (Bellardi) Ehrh.
Annual; stems suberect or reclining, to 2 dm long, with long divergent hairs and also retrorse bristles on the angles; leaves in whorls of 4, elliptic; flowers few; corolla yellow, 4-lobed; fruits glabrous. A European species, naturalized in southern counties; lawns, cemeteries, grassy fields and banks. Sanders (1976) described the first North American collection, made in 1933 near Lexington, Kentucky, and the species' subsequent spread to surrounding areas. The earliest known Ohio collection was made in 1960. Braun (1976) has notes on this species. May–June.

17. GALIUM ODORATUM (L.) Scop. SWEET WOODRUFF
Asperula odorata L.
Perennial, to 2 dm tall, the herbage fragrant upon drying; stems erect, with a few hairs at the nodes, otherwise glabrous; leaves in whorls of 6–8, oblanceolate to narrowly oblong-elliptic, with antrorse marginal bristles; flowers few to several, in terminal inflorescences; corolla white, 4-lobed, funnelform, the tube about half as long as the lobes; fruits hairy. A species of Eurasia and northern Africa, cultivated as a ground cover and some-

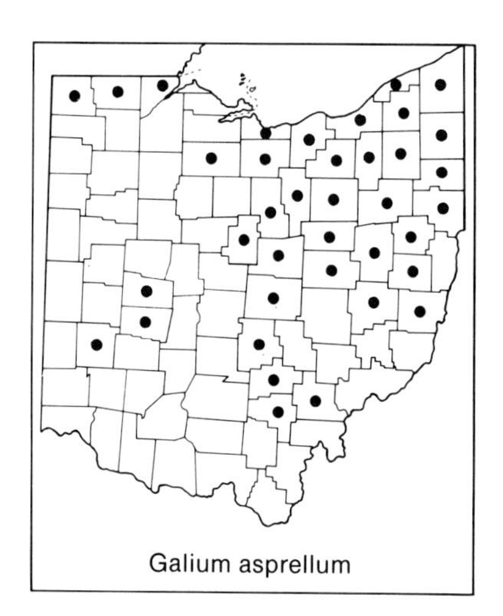
Galium asprellum

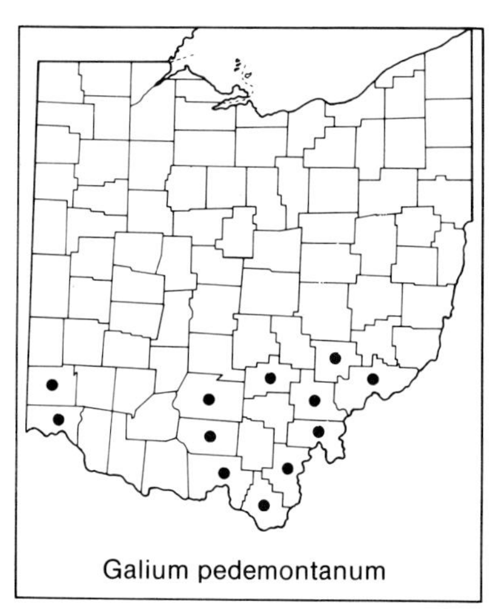
Galium pedemontanum

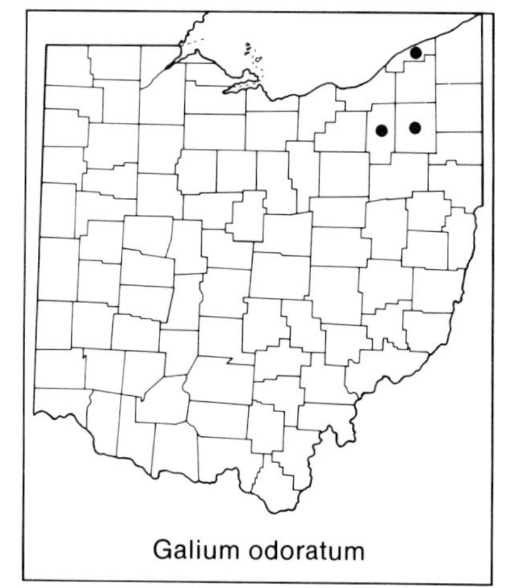
Galium odoratum

times escaped; adventive or perhaps locally naturalized at a few sites in northeastern Ohio. May–June.

3. Spermacoce L.

An American genus of five species, one of them native to Ohio.

1. **Spermacoce glabra** Michx. SMOOTH BUTTON-WEED

Perennial; stems decumbent to suberect, to 6 dm long, glabrous; leaves opposite, tapering and joined at the base by a membranous, fringed, stipular sheath, lanceolate to elliptic, the margins glabrous; flowers sessile in dense axillary and terminal clusters; calyx 4-lobed; corolla whitish, funnelform, 4-lobed, 3–4 mm long; fruits small, dry, glabrous. A species of southeastern United States, at a northern boundary of its range in southern Ohio; mud flats and shores along the Ohio River. July–September.

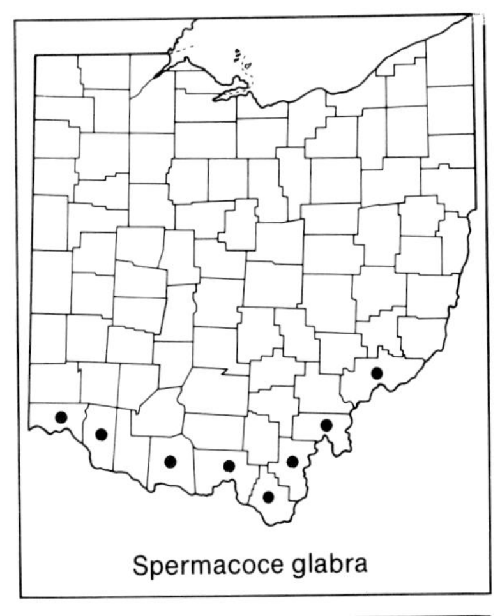
Spermacoce glabra

4. Diodia L.

Annuals or perennials; stems spreading, decumbent to suberect; leaves opposite, tapering and joined at base by a membranous stipular sheath; flowers solitary or in groups of 2 or 3 in leaf axils; calyx 2- or 4-lobed; corolla salverform or funnelform, with 4 lobes; stamens 4; fruit a schizocarp splitting into 2 nutlets. A genus of some 35 species, native to the Western Hemisphere and Africa.

a. Fruits 6–8 mm long; stipular bristles few and inconspicuous, shorter than fruits; calyx lobes 2 or rarely 4; corolla salverform, 7–10 mm long; plants perennial. . . . . 1. *Diodia virginiana*

aa. Fruits 3–4 mm long; stipular bristles several and conspicuous, exceeding length of fruits; calyx lobes 4; corolla funnelform, 3–5 mm long; plants annual . . . . 2. *Diodia teres*

1. **Diodia virginiana** L. SOUTHERN BUTTON-WEED

Perennial; stems with hairs only on the angles, branches spreading, to 6 dm long; flowers usually solitary in leaf axils; calyx usually 2-lobed, rarely 4-lobed; corolla whitish to pinkish, salverform, 7–10 mm long; fruits 6–8 mm long, the stipular bristles few and inconspicuous, shorter than the fruits. Mud flats and shores of the Ohio River in the southernmost tier of counties, and along a pond shore in one county (Jackson) of the second tier. July–September.

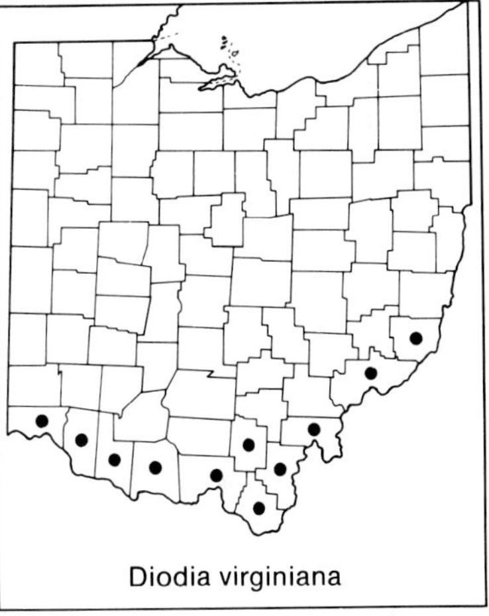
Diodia virginiana

2. **Diodia teres** Walt. ROUGH BUTTON-WEED
   incl. var. *setifera* Fern. & Grisc.

Annual; stems pubescent, to ca. 4 dm long; flowers 1–3 in leaf axils; calyx 4-lobed; corolla whitish to pinkish, funnelform, 3–5 mm long; fruits 3–4 mm long, the stipular bristles several and conspicuously exceeding length

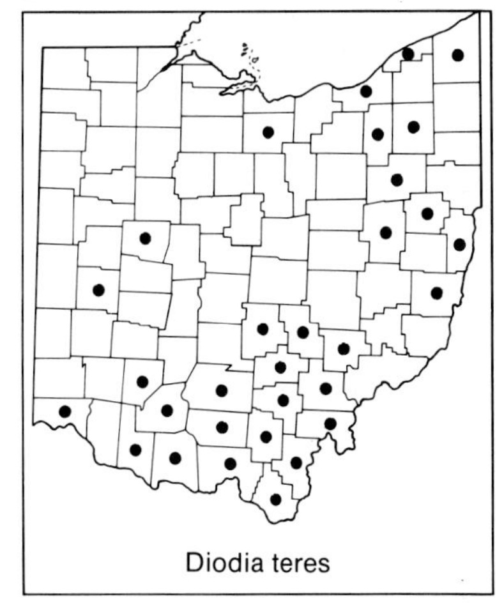
Diodia teres

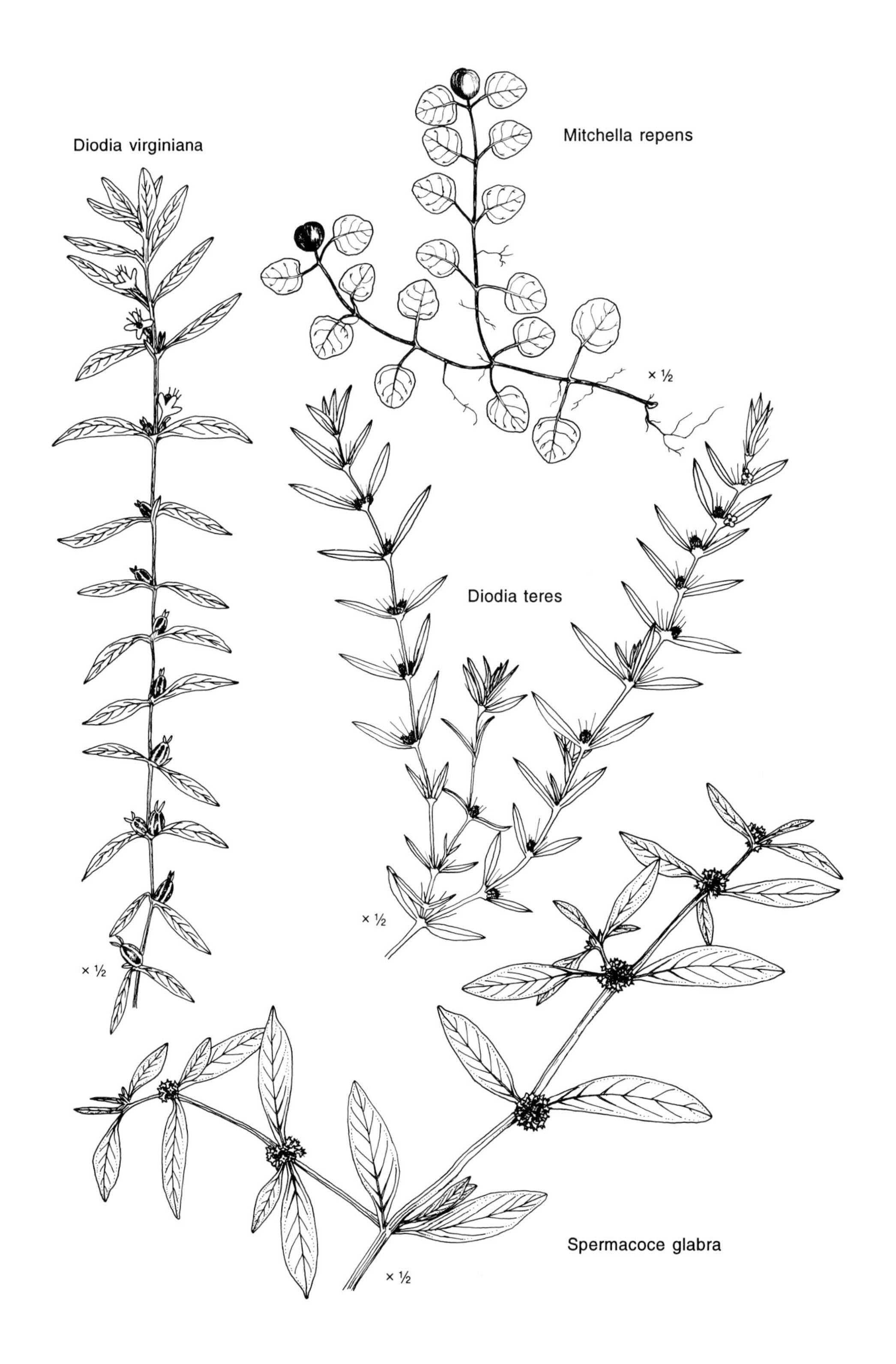
Diodia virginiana
Mitchella repens
× ½
Diodia teres
× ½
× ½
Spermacoce glabra
× ½

of fruits. In eastern and southern counties and at scattered sites elsewhere in the state; dry fields, roadsides, railways, pastures, thickets. July–September.

Following the revision of Bruza (1982), no infraspecific taxa are recognized within this species.

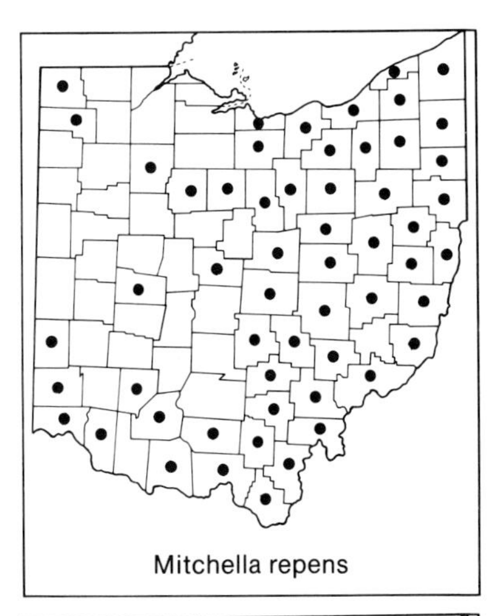
Mitchella repens

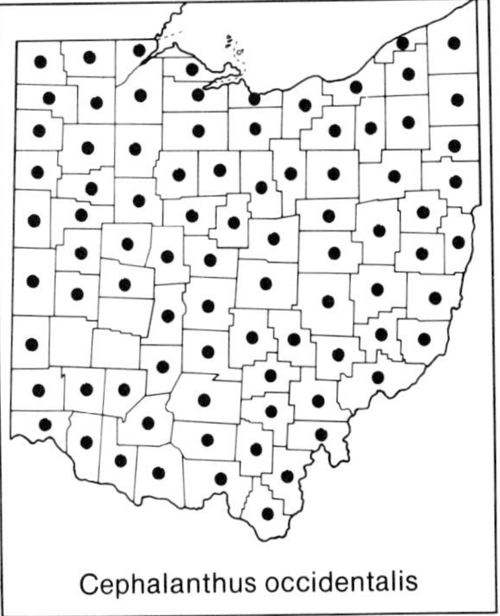
Cephalanthus occidentalis

5. Mitchella L.

A genus of two species, the other in eastern Asia.

1. **Mitchella repens** L. PARTRIDGE-BERRY

Perennial; stems trailing, rooting at the nodes, to ca. 3 dm in length, with ascending branches to 1.5 dm tall; leaves opposite, evergreen, orbicular to orbicular-ovate; flowers sessile, in pairs at the tip of a single peduncle, the peduncles terminal on stems or sometimes axillary; calyx 4-lobed; corolla white—tinged with red, fragrant, funnelform, 4-lobed, 12–16 mm long; the ovaries of the paired flowers partially fused to form a single, fleshy, red berry. Chiefly in eastern and southern counties; woodlands, woodland openings, open ridges. Hicks, Wyatt, and Meagher (1985) studied the reproductive biology of this species. June–early July.

6. Cephalanthus L.

A widely distributed genus of six or seven species, one of which is native to Ohio.

1. **Cephalanthus occidentalis** L. BUTTONBUSH
   incl. var. *occidentalis* and var. *pubescens* Raf.

Shrub, to 6 m (20 feet) tall; leaves either in whorls of 3 or opposite, sometimes both conditions on the same branch, ovate to elliptic or oblong-elliptic; flowers in dense globose heads; calyx 4-lobed; corolla creamy white, 7–10 mm long, 4-lobed; styles conspicuously long-exserted; fruits dry. Throughout Ohio; marshes, margins of lakes and ponds, stream banks, low moist ground, "buttonbush bogs." Late June–mid-August.

Variety *pubescens*, included in synonymy above, is recognized by Fernald (1950) and by Braun (1961), the latter reporting it from Hamilton County, in southwestern Ohio. Gleason (1952b) describes it as part of a vestiture cline, in which the more northerly plants are glabrous, the more southerly ones pubescent. The treatment here follows that of Bell (1968), in which no infraspecific ranks are recognized.

7. Houstonia L. BLUETS

Perennial herbs; leaves opposite; flowers solitary or more often in cymes; calyx 4-lobed and persistent; corolla blue or purple to white, salverform or funnelform, with 4 usually spreading lobes; stamens 4; fruit a capsule. A North American genus of about 35 species.

Houstonia has been merged under *Hedyotis* L. by W. H. Lewis (1961, 1962) and others, including G. K. Rogers (1987) and Gleason and Cronquist (1991). The treatment here follows the judgment of Fernald (1950) and Gleason (1952b) as well as that of Terrell (1975), the leading American student of the genus, who noted the heterogeneity of *Hedyotis* sensu lato, and wrote, "I have continued to recognize *Houstonia* as distinct on the grounds that no firm evidence was ever presented to back up the merger."

For a study of the phylogenetic implications of data from seeds, pollen, and chromosome numbers of several *Houstonia* species, including all the Ohio species described below, see Terrell et al. (1986). Terrell's (1959) revision of the *Houstonia purpurea* group, includes species nos. 2–4 below. As noted by Hauser (1964), during Terrell's preliminary work for this revision, he annotated some specimens at The Ohio State University Herbarium as *Houstonia tenuifolia* Nutt. and others as intergrades between *H. tenuifolia* and *H. purpurea*. In the published revision, however, the range of both these elements was presented as lying outside and to the south of Ohio.

a. Flowers few, solitary on long peduncles; corolla salverform, blue, pale blue, or purplish-blue, yellow-centered within .... 1. *Houstonia caerulea*

aa. Flowers numerous, in cymes; corolla funnelform, purple to white.

b. Most flowers and fruits sessile or on pedicels less than 2 mm long .... 5. *Houstonia nigricans*

bb. Most or all flowers and fruits on pedicels longer than 2 mm.

c. Cauline leaves lanceolate to oblong-lanceolate or ovate, blades of the principal leaves (0.7–) 1–2.5 (–3) cm wide, with 3–5 main veins, rounded or slightly cordate at base; stems with few to many divergent hairs .... 2. *Houstonia purpurea*

cc. Cauline leaves linear, lanceolate, oblanceolate, or narrowly oblong, blades of the principal leaves less than 1 cm wide, usually with only one main vein, narrowed and tapering at base; stems usually spiculate or smooth, rarely with divergent hairs.

d. Basal leaves few or absent at anthesis, if present not ciliate; stems slightly to densely short-spiculate........ 3. *Houstonia longifolia*

dd. Basal leaves few to many at anthesis, ciliate; stems smooth or rarely with diverging hairs .... 4. *Houstonia canadensis*

1. **Houstonia caerulea** L. BLUETS. QUAKER LADIES
Hedyotis caerulea (L.) Hook.

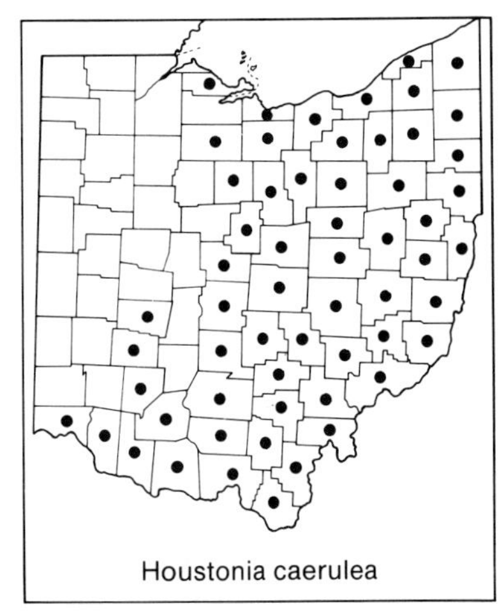
Houstonia caerulea

Perennial, to 1.5 dm tall; stems erect, usually in clusters; leaves tiny, mostly in a basal tuft, spatulate to obovate or oblanceolate or the uppermost linear; flowers solitary on long peduncles; corolla blue, pale blue, or purplish-blue, yellow-centered within, salverform, 8–10 mm long, 10–15 mm across. An attractive spring wildflower; mostly in eastern and southern Ohio; grassy fields, open woods, woodland openings and borders, and other open sites. April–June (–October).

2. **Houstonia purpurea** L. LARGE SUMMER BLUETS
incl. var. *purpurea* and var. *calycosa* A. Gray (*H. lanceolata* (Poir.) Britt.); *Hedyotis purpurea* (L.) T. & G.

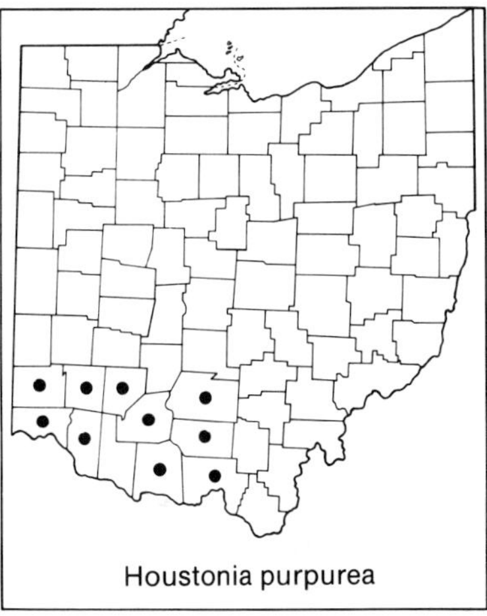
Houstonia purpurea

Perennial, to 4 dm tall; stems erect, with few to many divergent hairs; cauline leaves lanceolate to oblong-lanceolate or ovate, blades of principal leaves (0.7–) 1–2.5 (–3.0) cm wide, with 3–5 main veins, rounded or slightly cordate at base; flowers in cymes terminal on main stem or on axillary branches, pedicels 3–6 mm long; corolla purplish-white to purplish, funnelform, 5–9 mm long. A species of east-central and southeastern United States, at a northern boundary of its range in southwestern counties of Ohio; woodland openings and borders, open slopes, roadside banks. Late May–June.

Terrell (1991) recognizes both var. *purpurea* and var. *calycosa*, noting that they intergrade in Ohio. The two are merged here because of difficulty in assigning the Ohio plants to either variety.

Terrell (1959) reports intergrades between this species and the next, *H. longifolia*, from several southwestern Ohio counties: Adams, Clermont, Hamilton, Preble, Ross, and Warren, as well as from Auglaize County in west-central Ohio, and from Coshocton County in east-central Ohio. This large number of intergrades lends support to the classification adopted for *Hortus Third* (Bailey Hortorium Staff, 1976), in which *H. longifolia* is treated as a variety of *H. purpurea*.

3. **Houstonia longifolia** Gaertn. LONG-LEAVED SUMMER BLUETS
Hedyotis longifolia (Gaertn.) Hook.; *Houstonia purpurea* var. *longifolia* (Gaertn.) A. Gray

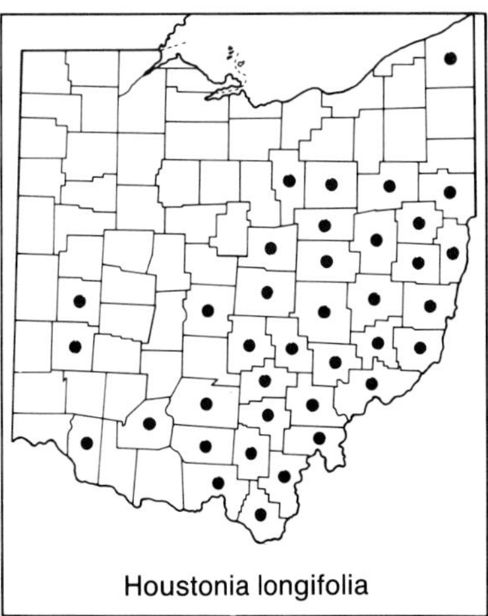
Houstonia longifolia

Perennial, to 3.5 dm tall; stems erect, slightly to densely short-spiculate; basal leaves few if any at anthesis; cauline leaves narrowly elliptic to narrowly lanceolate or linear, 2–6 mm wide, 10–30 mm long; flowers in terminal or axillary cymes, pedicels 3–6 mm long; corolla whitish to purplish, funnelform, 5–9 mm long. Chiefly in the unglaciated section of Ohio, the east-central and southeastern counties; dry fields and roadside banks, open slopes, woodland openings. Mid-June–August.

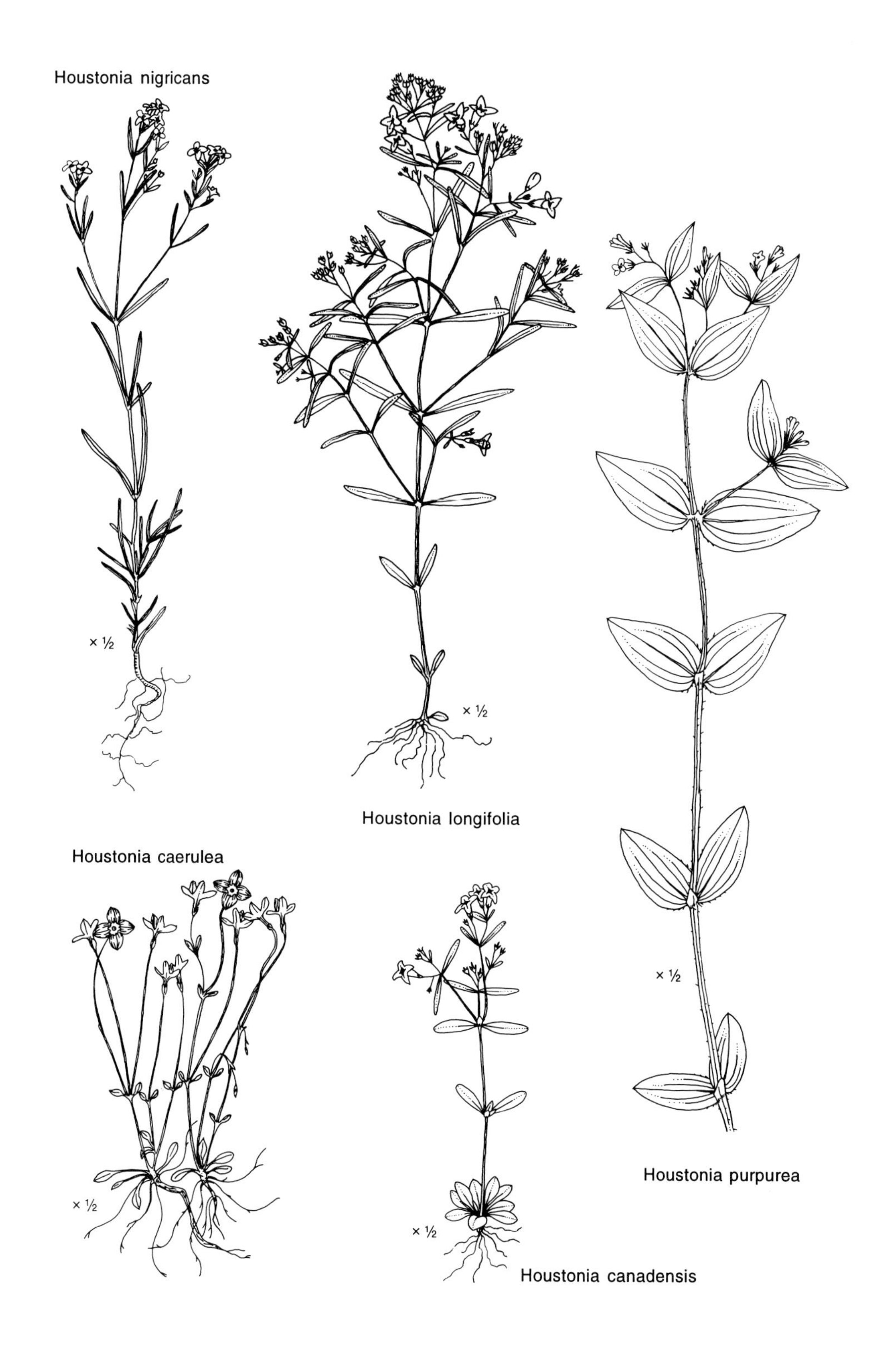
Houstonia nigricans
× ½
× ½
Houstonia longifolia
Houstonia caerulea
× ½
× ½
Houstonia purpurea
× ½
Houstonia canadensis

Ohio plants belong to var. *compacta* Terrell. See notes above under *H. purpurea* regarding intergrades between that species and this.

4. **Houstonia canadensis** Willd. CANADIAN BLUETS

incl. *H. setiscaphia* L. Carr; *Hedyotis canadensis* (Willd.) Fosberg

Perennial, to 2.5 dm tall; stems erect, glabrous or rarely with diverging hairs; basal leaves few to many at anthesis; cauline leaves elliptic to elliptic-oblong or linear-elliptic, 2–6 mm wide, 10–30 mm long; flowers in terminal or axillary cymes, pedicels 2–5 mm long; corolla whitish to purplish, funnelform, 5–9 mm long. At scattered Ohio sites; dry woodland openings, open sandy or rocky ground, roadsides. May–June.

Braun (1976) reported *Houstonia setiscaphia* from Adams County in southwestern Ohio, but Terrell (1959) thought these plants to be merely a "local extreme or local race" not worthy of taxonomic rank.

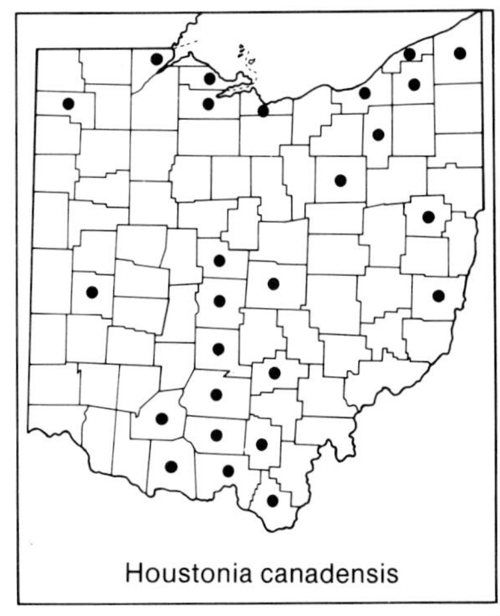
Houstonia canadensis

5. **Houstonia nigricans** (Lam.) Fern. NARROW-LEAVED SUMMER BLUETS

Hedyotis nigricans (Lam.) Fosberg

Perennial, to 4 dm tall; stems erect, very low spiculate-pubescent to glabrous; cauline leaves linear to narrowly linear-elliptic, 1.5–3 mm wide, 20–35 mm long; flowers in terminal and axillary cymes, either sessile or subsessile on pedicels less than 2 mm long; corolla whitish to purplish, salverform, 5–8 mm long. At a few sites each in the northernmost and southernmost tiers of counties; open dry habitats. Terrell (1986) has taxonomic and nomenclatural notes on this species; recently (Terrell, 1991) he assigned it to the genus *Hedyotis* sensu stricto. June–September.

Houstonia nigricans

Caprifoliaceae. HONEYSUCKLE FAMILY

Shrubs, woody vines, or perennial herbs; leaves opposite, simple or rarely compound; flowers perfect, solitary or in pairs or in cymose inflorescences; calyx (4- or) 5-lobed and often very small; corolla (4- or) 5-lobed, zygomorphic to actinomorphic; stamens usually 5, attached to the corolla tube; carpels 3–5, fused to form a compound pistil with a single style; ovary inferior; fruit a berry, drupe, capsule, or achene. A family of some 400 species, with distribution centered in the North Temperate Zone. Ferguson (1966a) studied the family in southeastern United States. The treatment here is based in part on Hauser (1965).

Abelia × *grandiflora* (André) Rehd., GLOSSY ABELIA, an Asian shrub with profuse numbers of glossy white to pinkish flowers, is a popular ornamental, hardy in the southern half of Ohio and in northern areas near Lake Erie. Another Asian shrub, *Weigela florida* (Bunge) A. DC.,

Weigela (also spelled Weigelia), with rose to white flowers, is seen in cultivation throughout Ohio.

a. Plants erect herbs . . . . 5. *Triosteum*

aa. Plants woody or mostly woody shrubs, small trees, or vines, the last sometimes partly herbaceous.

b. Leaves pinnately compound. . . . . 7. *Sambucus*

bb. Leaves simple.

c. Stems erect or ascending and arching, shrubs or small trees.

d. All leaves serrate, toothed, or lobed.

e. Calyx lobes linear, 2–7 mm long; corolla yellow, turning reddish in age, 12–20 mm long; style exserted; fruit a capsule . . . . 1. *Diervilla*

ee. Calyx lobes none or to 1 mm long; corolla white, pink, purplish, or greenish, in fertile flowers 3–10 mm long, larger in sterile flowers when present; style included or at most equalling length of corolla; fruit a berry or drupe.

f. Flowers in axillary and/or terminal clusters or spikes; shrubs to 1 (–1.5) m tall . . . . 3. *Symphoricarpos*

ff. Flowers in terminal cymes; shrubs or small trees, 1–8 m tall . . . . 6. *Viburnum*

dd. All or most leaves entire or slightly undulate.

e. Flowers in pairs, borne at the tip of an axillary peduncle; corolla (7–) 10–25 mm long, varying from nearly actinomorphic to markedly zygomorphic; fruits borne in pairs . . . . 2. *Lonicera*

ee. Flowers in axillary and/or terminal clusters or spikes (of usually more than two flowers); corolla 3–9 mm long, actinomorphic or nearly so; fruits borne in clusters of few to several . . . . 3. *Symphoricarpos*

cc. Stems trailing, creeping, twining, climbing, or scrambling.

d. Leaves 3 cm or more in length, deciduous or partially evergreen; peduncles to 2 cm long; flowers in clusters of 3-several, erect or divergent, not nodding; corolla 15–50 mm long . . . . 2. *Lonicera*

dd. Leaves 1–2 cm long, evergreen; peduncles 2–10 cm long; flowers in pairs, nodding; corolla 10–15 mm long . . . . 4. *Linnaea*

1. Diervilla Mill. BUSH-HONEYSUCKLE

A genus of eastern North America with one, two, or perhaps three species. J. W. Hardin (1968) studied the genus in southeastern United States.

Diervilla hybrida Dippel (= *D.* × *splendens* (Carr.) Kirch.?), Garden Bush-honeysuckle, was listed as an escape by Schaffner (1935), but no specimens have been found in this study.

1. **Diervilla lonicera** Mill. BUSH-HONEYSUCKLE

Low shrub, to ca. 1 m (3 feet) tall; leaves simple, lanceolate to oblong-lanceolate or ovate, long-acuminate and often slightly curved at apex, margins serrate; flowers in few-flowered cymes; calyx lobes 5, linear, 2–7 mm long, conspicuously persistent in fruit; corolla yellow, turning orange or reddish in age, 10–15 mm long, 5-lobed; stamens 5; style exserted; fruit a capsule. Counties of the northeastern quarter and at a few scattered sites elsewhere in Ohio; wooded slopes, rocks and cliffs. Schoen (1977) studied the floral biology of this species, Thomson (1985) its pollination and seed set. June–early July.

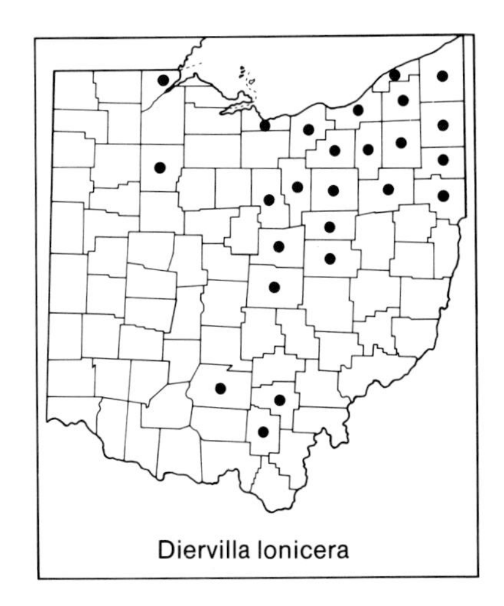
Diervilla lonicera

2. Lonicera L. HONEYSUCKLE

Shrubs or climbing or reclining woody vines; leaves opposite, simple, entire (except for the infrequently encountered form of *L. japonica* with pinnately lobed leaves); flowers either in pairs borne at tip of a single axillary peduncle, or in terminal several-flowered heads or spikes; calyx tiny, 5-toothed; corolla 5-lobed; stamens 5; fruit a berry. A Northern Hemisphere genus of 150 or more species. Green (1966) reports research on the species and hybrids of the *Lonicera tatarica* complex, which includes taxa nos. 2–5 below. W. H. Duncan (1975) has notes on the honeysuckle vines of southeastern United States, including species nos. 8–13 below.

Several members of the genus are cultivated in Ohio. An escape of *Lonicera fragrantissima* Lindl. & Paxt., SWEET-BREATH-OF-SPRING, an Asian shrub with sweetly fragrant whitish flowers appearing in early spring before the leaves, was recently collected from Franklin County. *Lonicera caprifolium* L., ITALIAN WOODBINE, a twining vine with fragrant white or purplish flowers, was listed for Cuyahoga County by Schaffner (1932)—presumably as an escape from cultivation. No specimens, other than ones clearly marked "cultivated," have been found in this study. *Lonicera hirsuta* Eat., HAIRY HONEYSUCKLE, a species native to the north of Ohio, was attributed to the state by Schaffner (1932) and Fernald (1950), but excluded from the flora by Braun (1961) for lack of substantiating specimens.

a. All leaves distinct; flowers in pairs on a single axillary peduncle.
  b. Stems twining, the plants trailing or climbing vines; corolla 3–4.5 cm long, very fragrant; fruits black (produced infrequently) .... 9. *Lonicera japonica*
  bb. Stems not twining, the plants erect or ascending shrubs; corolla 0.7–2.2 cm long, not or only slightly fragrant; fruits yellow, red, purple, or blue.
    c. Young branches with pith solid and white between the nodes.

d. Leaves widest above the middle; peduncles short, 3–7 (–10) mm long; the two ovaries fused into a single blue fruit . . . . 1. *Lonicera villosa*

dd. Leaves all or mostly widest at or below the middle; peduncles longer, (10–) 20–40 mm long; the two ovaries separate or fused, red or purple to yellowish in fruit.

e. Leaves with petioles 5–8 mm long, the blades ovate to ovate-lanceolate, acute to rounded or cordate at base; corollas nearly actinomorphic; the two berries of each pair separate and strongly divergent . . . . 7. *Lonicera canadensis*

ee. Leaves sessile or with petioles to 3 mm long, the blades oblong to ovate-lanceolate, acute at base; corollas strongly bilabiate; the two berries of each pair separate to more often partially fused, not strongly divergent . . . . 8. *Lonicera oblongifolia*

cc. Young branches with pith hollow between the nodes, the lining of the cavity usually brown to tan, rarely white.

d. Peduncles 2–4 (–5) mm long, shorter than to equalling petioles of subtending leaves . . . . 6. *Lonicera maackii*

dd. Peduncles 10–15 mm long, longer than petioles of subtending leaves.

e. Leaves pubescent beneath; flowers white, yellowish sometimes tinged with red, or pink.

f. Corolla bilabiate, yellowish sometimes tinged with red; young branches short-puberulent with few or no long spreading hairs . . . . 2. *Lonicera xylosteum*

ff. Corolla actinomorphic or nearly so, white to yellowish, or pink to pinkish-yellow; young branches varying from pubescent with both short hairs and longer spreading hairs to nearly glabrous.

g. Corolla white to yellowish; peduncles pubescent . . . . 3. *Lonicera morrowii*

gg. Corolla pink to pinkish-yellow; peduncles sparsely pubescent . . . . 4. *Lonicera* × *bella*

ee. Leaves glabrous beneath or soon becoming glabrous or essentially so; flowers pink to reddish.

f. Petioles, peduncles, and bracts sparsely pubescent; corolla pink to pinkish-yellow. . . . . 4. *Lonicera* × *bella*

ff. Petioles, peduncles, and bracts glabrous or nearly so; corolla red . . . . 5. *Lonicera tatarica* aa. Uppermost pair of leaves (and sometimes one or more of the next lower pairs) fused at base; flowers several in each cluster, the clusters either terminal or axillary above the uppermost leaves and thereby appearing terminal.

b. Corolla 3.5–5.5 cm long, red to orange-red, not or only slightly bilabiate, the lobes approximately equal; stamens and style included or only slightly exserted . . . . 10. *Lonicera sempervirens*

bb. Corolla 1.5–3 cm long, greenish, pale yellow, or yellow, tinged with or varying to red or purple, strongly bilabiate; stamens and style exserted.

c. Uppermost pair of fused leaves together forming an orbicular or nearly orbicular disk, its ends entire or retuse; the nonfused leaves broadly ovate to suborbicular; corolla pale yellow, often marked with purple. . . . . 13. *Lonicera prolifera*

cc. Uppermost pair of fused leaves together forming a rhombic, elliptic, or sometimes suborbicular structure or one that is elliptic and narrowed at the middle, the ends retuse, entire, obtuse, or more often acute or mucronate; the nonfused leaves oblong or elliptic to obovate or very broadly ovate; corolla greenish, yellow, red, or purple.

d. The nonfused leaves oblong to elliptic, 1½–2 times as long as wide, narrowed to a sessile base or to a short petiole; corolla greenish or yellowish to red or purple, the tube slightly gibbous at base. . . . . 11. *Lonicera dioica*

dd. The nonfused leaves ovate to very broadly ovate or obovate, only slightly longer than wide, abruptly narrowed at base to a short petiole; corolla yellow, the tube not gibbous at base. . . . . 12. *Lonicera flava*

1. **Lonicera villosa** (Michx.) R. & S. MOUNTAIN FLY-HONEYSUCKLE
incl. var. *villosa* and var. *tonsa* Fern.; *L. caerulea* L. var. *villosa* (Michx.) T. & G.

Lonicera villosa

Low shrub, to 1 m (3 feet) tall; young branches with pith solid between the nodes; leaves sessile or subsessile, oblong to oblong-oblanceolate, usually widest above the middle, more or less rounded at the apex; flowers in axillary pairs on a single peduncle, the peduncles short, 3–7 (–10) mm long; the two ovaries fused to form a single blue fruit. A boreal species of northeastern North America, at a southern boundary of its range in Ashtabula County, in northeasternmost Ohio; bogs, wet thickets. June.

The sole Ohio specimen of this species was reported by Braun (1961, p. 324) and cited more fully by Hauser (1965). Braun assigned the Ohio plant to var. *tonsa*, but that variety is placed in synonymy above because

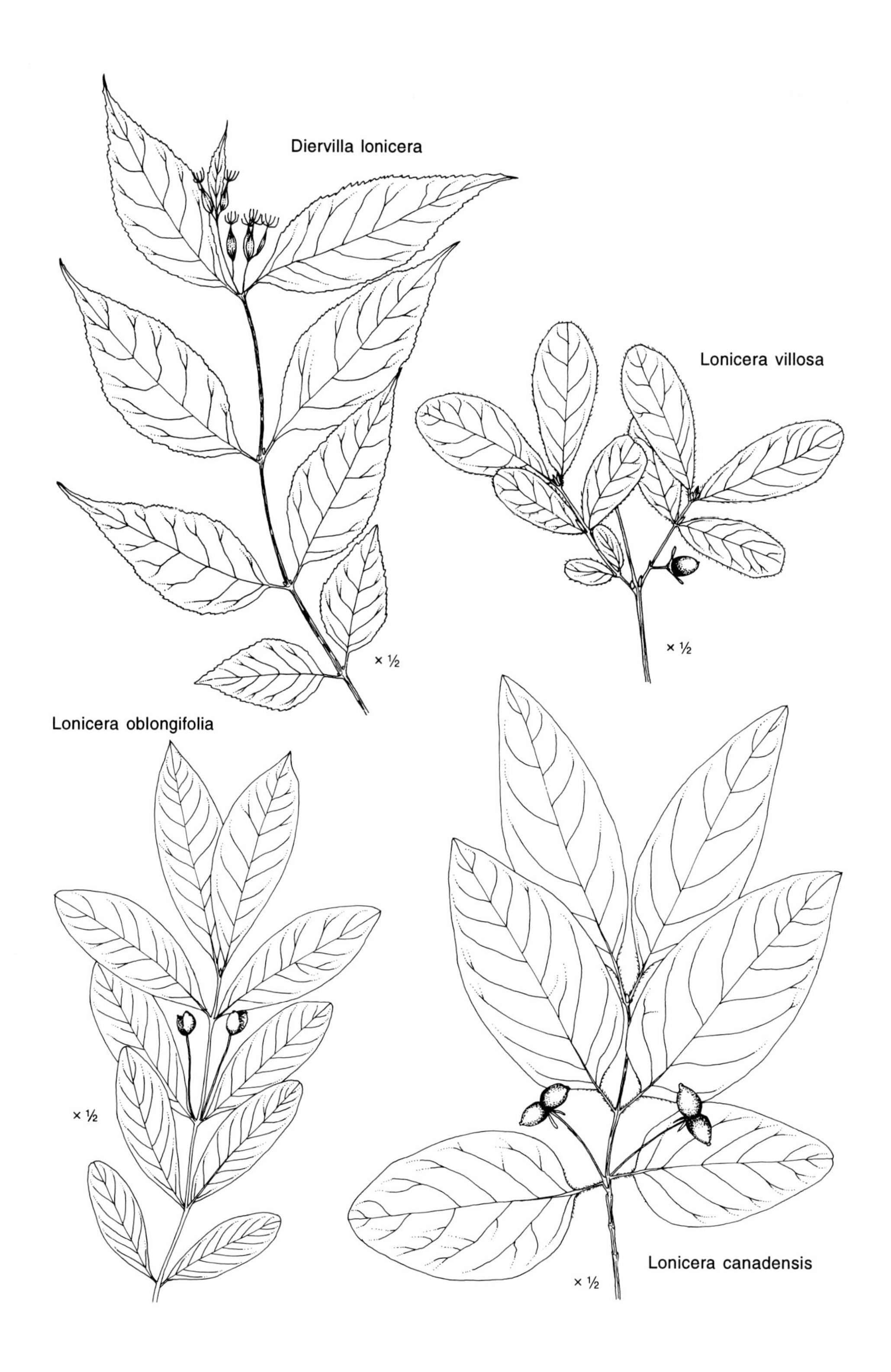
Diervilla lonicera
Lonicera villosa
× ½
× ½
Lonicera oblongifolia
× ½
Lonicera canadensis
× ½

of Gleason's (1952b) reservations about the recognition of any infraspecific taxa in the species.

2. LONICERA XYLOSTEUM L. EUROPEAN FLY-HONEYSUCKLE

Shrub, to 3 m (10 feet) tall; young branches with pith hollow between the nodes, and externally short-puberulent with few or no long spreading hairs; leaves ovate or obovate, rounded at base, pubescent on under surface; flowers in axillary pairs on a single peduncle, the peduncles 10–15 mm long; corolla bilabiate, yellowish and sometimes tinged with red; fruits red. A cultivated species from Eurasia, adventive at a few sites in Ohio, mostly in northeastern counties. Not illustrated. May–June.

3. LONICERA MORROWII A. Gray MORROW'S HONEYSUCKLE

Shrub, to 2.5 m (8 feet) tall; young branches with pith hollow between the nodes, and externally pubescent with many short hairs and some long spreading hairs; leaves oblong to lance-elliptic, pubescent on under surface, obtuse to rounded or subtruncate at base; flowers in axillary pairs on a single peduncle, the peduncles 10–12 mm long and pubescent; corolla white or yellowish, actinomorphic or nearly so; fruits red. A cultivated species from Asia, naturalized in northern counties and at scattered sites southward; openings and borders of disturbed woodlands, thickets. April–early June.

4. LONICERA × BELLA Zabel

Shrub, to 3 m (10 feet) tall; young branches with pith hollow between the nodes, and externally slightly pubescent to nearly glabrous; leaves narrowly ovate to elliptic-oblong, pubescent to glabrate on under surface, obtuse to rounded or slightly subcordate at base; flowers in axillary pairs on a single peduncle, the peduncles 10–15 mm long and slightly pubescent; co-

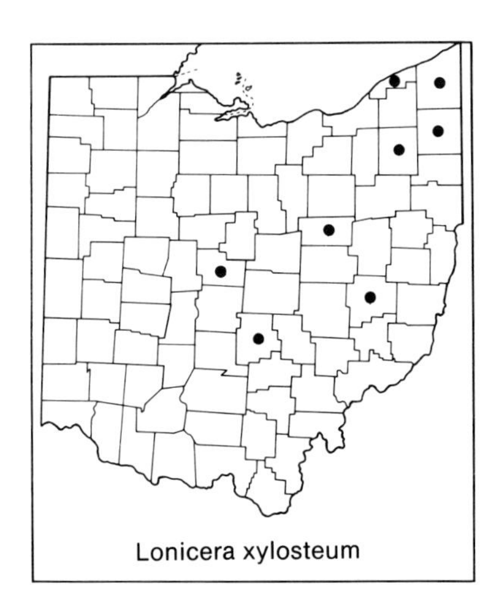

Lonicera xylosteum

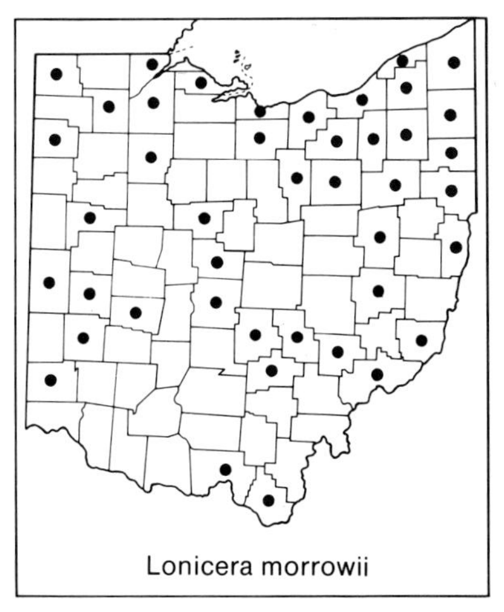

Lonicera morrowii

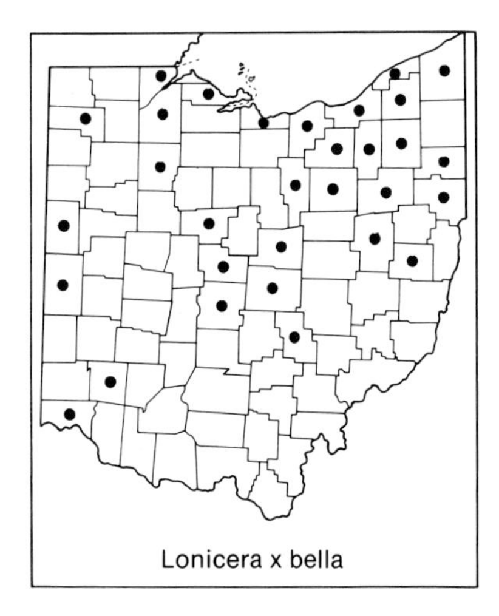

Lonicera x bella

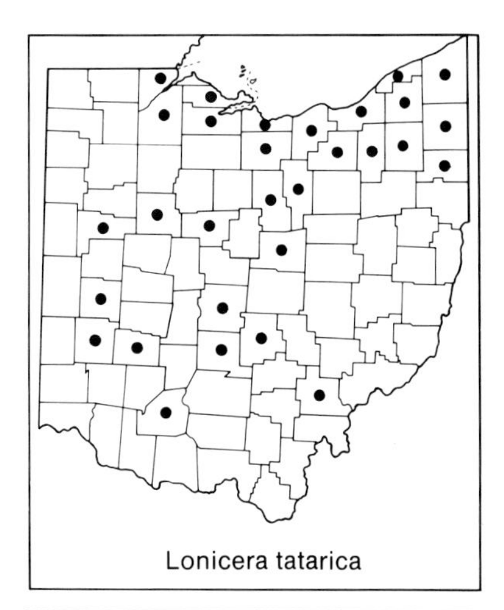
Lonicera tatarica

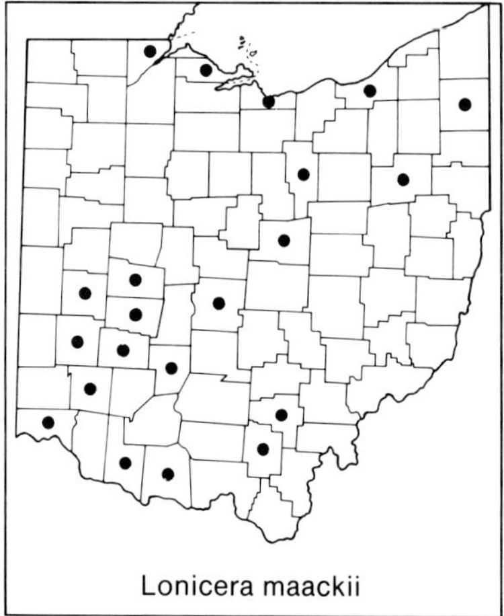
Lonicera maackii

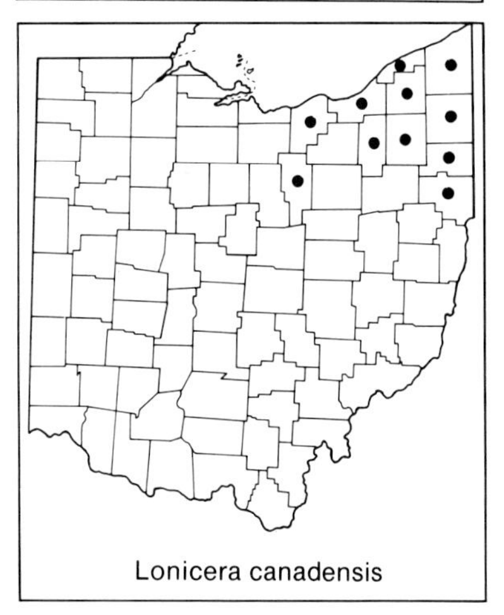
Lonicera canadensis

rolla pink to pinkish-yellow, actinomorphic or nearly so; fruits red. This is the hybrid of *Lonicera morrowii* × *L. tatarica*, treated as a taxonomic species by Gleason (1952b) and Gleason and Cronquist (1963). Hauser (1966) studied the occurrence of this hybrid in Ohio and Michigan; both he and Green (1966) speculate that it may backcross to one or both parents. Naturalized in northern counties and at scattered sites southward; fields, thickets, fencerows, disturbed woodland openings and borders. Not illustrated. May–early June.

5. Lonicera tatarica L. Tatarian Honeysuckle

Shrub, to 3 m (10 feet) tall; young branches with pith hollow between the nodes, and externally glabrous or nearly so; leaves elliptic to oblong or ovate, glabrous on undersurface, rounded to truncate or subcordate at base; flowers in axillary pairs on a single peduncle, the peduncles 10–15 mm long and glabrous; corolla red, actinomorphic or nearly so; fruits red. A cultivated species from Asia, naturalized in northern counties and at scattered sites southward; disturbed fields, pastures, thickets. May–mid-June (–September).

6. Lonicera maackii (Rupr.) Maxim. Amur Honeysuckle

Shrub (or small tree), to 4.5 m (15 feet) tall; young branches with pith hollow between the nodes, and externally pubescent; leaves ovate to elliptic or rhombic-elliptic, pubescent on veins of undersurface, broadly acute to obtuse at base, long-acuminate at apex; flowers in axillary pairs on a single peduncle, the peduncles pubescent and very short—2–4 (–5) mm long, shorter than to equalling petioles of subtending leaves; corolla white to yellowish-white, bilabiate; fruits dark red. An Asian species, cultivated and sometimes escaped and naturalized in southern counties and at scattered sites northward. Pringle (1973b) notes its occurrence as an adventive in southern Ontario. Luken (1988) studied its population structure. May–early June.

7. **Lonicera canadensis** Marsh. Canadian Fly-honeysuckle

Shrub, to 1.5 m (5 feet) tall; young branches with pith solid between the nodes; leaves ovate to ovate-lanceolate, acute to rounded or cordate at base, petioles ciliate, 5–8 mm long; flowers in axillary pairs on a single peduncle; corolla pale yellowish, nearly actinomorphic; fruits red, the two berries of each pair separate and strongly divergent. A northern species at a southern boundary of its range in northeastern counties of Ohio; steep wooded slopes, swamp forests. April–May.

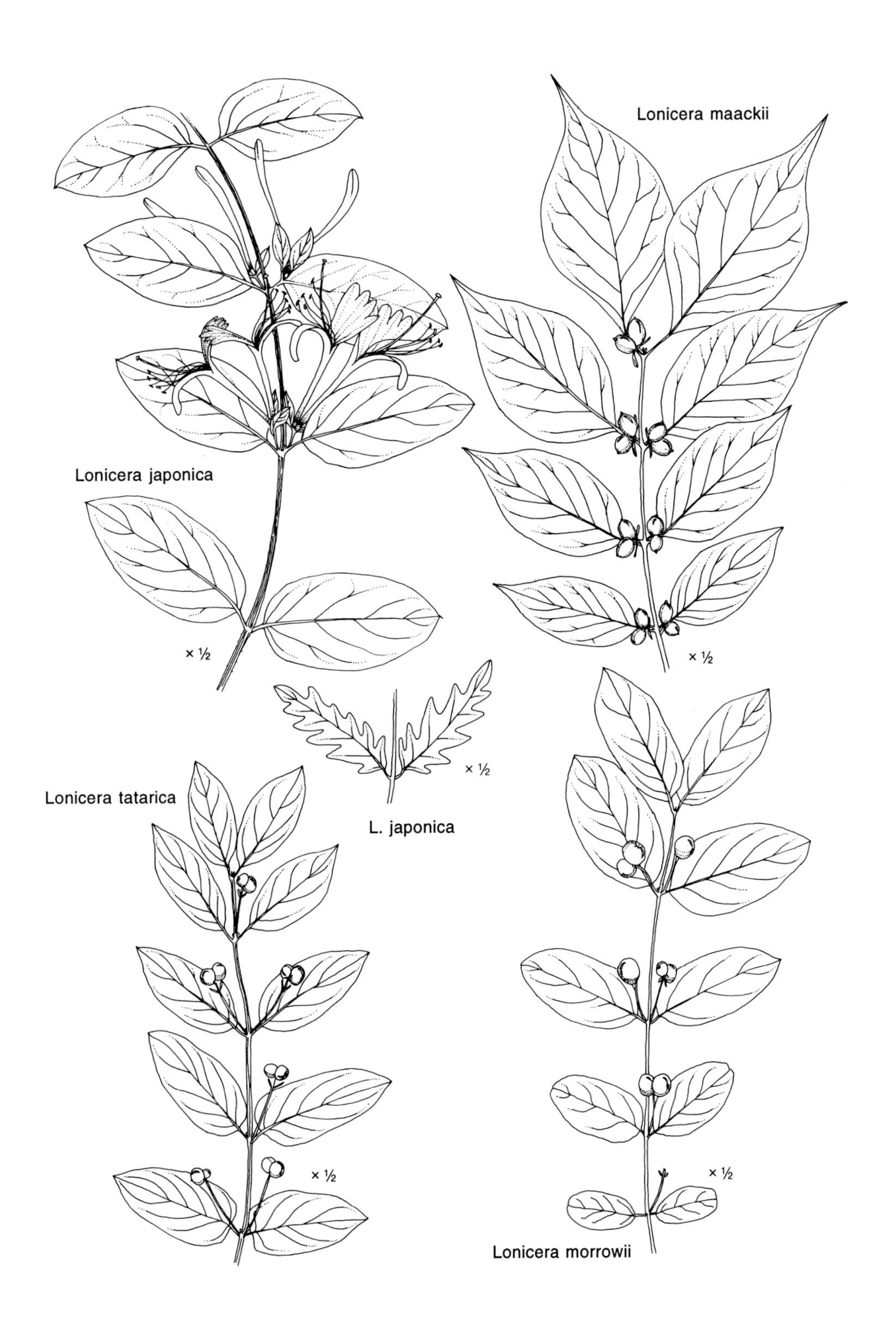
Lonicera maackii
Lonicera japonica
× ½
× ½
× ½
L. japonica
Lonicera tatarica
× ½
× ½
Lonicera morrowii

Lonicera oblongifolia

8. **Lonicera oblongifolia** (Goldie) Hook. Swamp Fly-honeysuckle
Shrub, to 1.5 m (5 feet) tall; young branches with pith solid between the nodes; leaves oblong to ovate-lanceolate, acute at base, sessile or with petioles to 3 mm long; flowers in axillary pairs on a single peduncle; corolla yellowish-white, bilabiate; fruits reddish, the two berries of each pair separate or more often partially fused. A northern species at a southern boundary of its range in extreme northeastern Ohio, where it is known only from Ashtabula County; bogs. Late May–June.

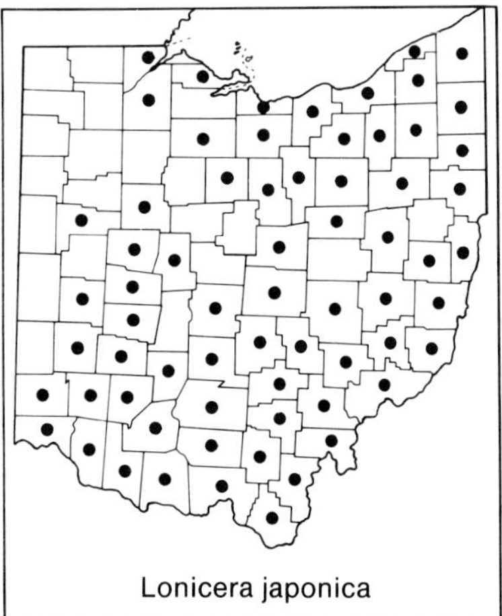

Lonicera japonica

9. Lonicera japonica Thunb. Japanese Honeysuckle
incl. var. *japonica* and var. *chinensis* (P. W. Wats.) Baker
Trailing, twining, and climbing vine, sometimes forming dense tangled thickets; stems pubescent, pith hollow between the nodes; leaves partially evergreen, ovate to oblong or elliptic, usually entire but sometimes pinnately lobed, more or less pubescent on the undersurface; flowers in axillary pairs on a single peduncle; corolla white when young and turning dull golden yellow with age, bilabiate, sweetly fragrant, 3.0–4.5 cm long; fruits black but produced only infrequently. An Asian species widely cultivated, escaped and naturalized in all except some northwestern and west-central counties, abundant in parts of southern Ohio, forming dense tangles and becoming a troublesome woodland weed; roadsides, thickets, waste lots, disturbed woodlands. Mid-May–July (–November).

Plants approaching var. *chinensis*, having the corolla infused with red and the leaves and young branches infused with purple, are known from extreme southern Ohio. W. H. Duncan (1975), writing of this species in southeastern United States, argues that the two varieties (var. *japonica* and var. *chinensis*) "probably cannot be maintained" because of the abundance of intermediates; I concur.

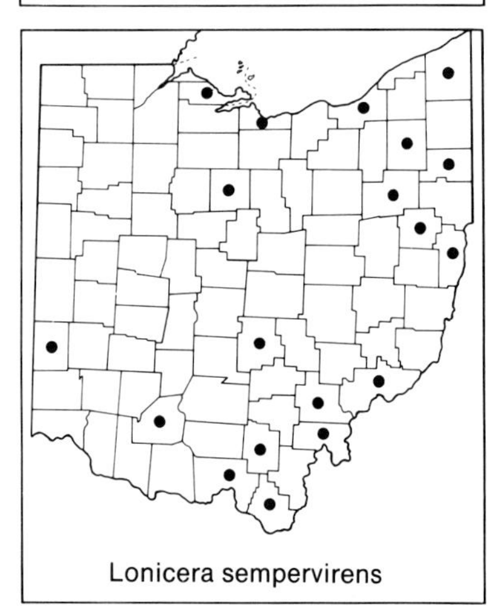

Lonicera sempervirens

10. Lonicera sempervirens L. Trumpet Honeysuckle
Climbing, twining, trailing, or scrambling vine; uppermost pair of leaves fused at base to form an elliptic-oblong double-leaf, sometimes indented at the point of fusion, lower nonfused leaves elliptic to elliptic-ovate; inflorescence an open terminal spike with flowers in 2–4 whorls; corolla bright red or orange-red, 3.5–5.5 cm long and narrowly trumpet-shaped, the lobes approximately equal, not or only slightly bilabiate; stamens and style included or only slightly exserted. A species chiefly of the southeastern states, often cultivated there and farther north; although regarded by Braun (1961) as native to southernmost counties, I am inclined to believe that all Ohio records are of naturalized plants escaped from cultivation. At scattered sites mostly in the eastern half of the state; thickets, roadsides, woodland borders. May–July.

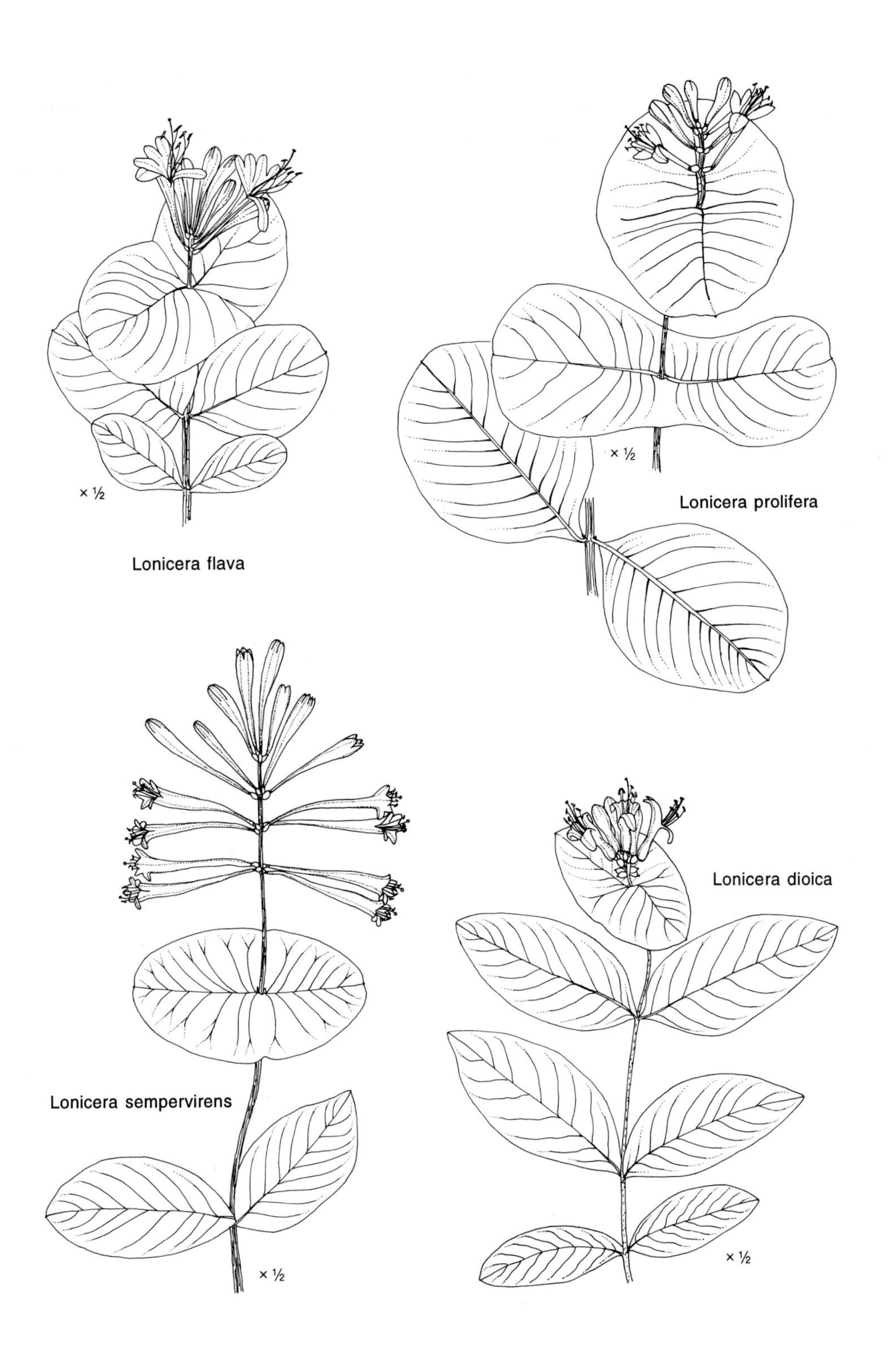
× ½
Lonicera flava
× ½
Lonicera prolifera
Lonicera dioica
Lonicera sempervirens
× ½
× ½

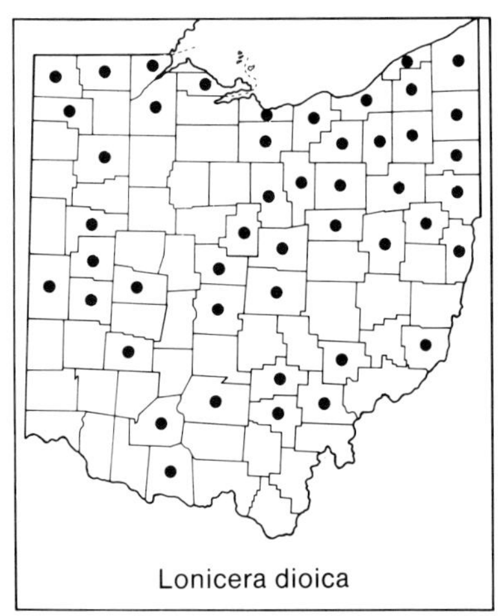
Lonicera dioica

Lonicera flava

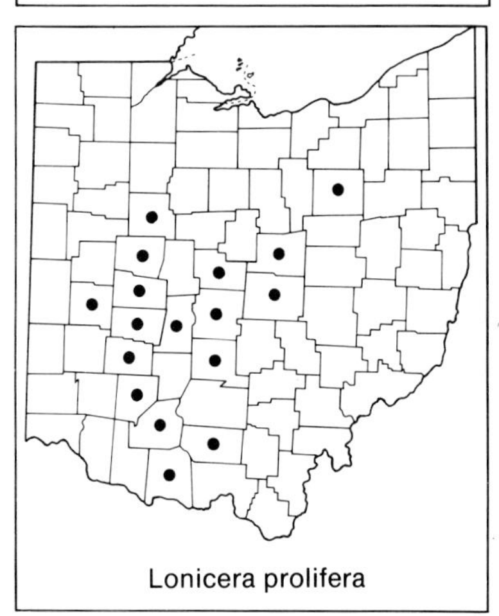
Lonicera prolifera

11. **Lonicera dioica** L. WILD HONEYSUCKLE
incl. var. *dioica*, var. *dasygyna* (Rehd.) Gleason, and var. *glaucescens* (Rydb.) Butters

Climbing, twining, or scrambling shrubby vine; uppermost pair of leaves fused at base to form a rhombic, elliptic, or suborbicular double-leaf, sometimes indented at the point of fusion, at the ends retuse, entire, obtuse, or more often acute and mucronate, the second pair of leaves sometimes also fused basally, the lower nonfused leaves oblong to elliptic, 1½–2 times as long as wide, narrowed to a sessile base or short petiole; corolla greenish or yellowish to red or purple, 1.5–2.2 cm long, bilabiate, the tube slightly gibbous at base; stamens and style exserted. Northeastern counties and at scattered sites elsewhere in Ohio; woodlands, thickets. May–June.

Fernald (1950), Gleason (1952b), and Braun (1961) recognized several varieties within this species, two or more of which would occur in Ohio. However, following the treatment of W. H. Duncan (1975) who wrote, "only one taxon seems justified since characters intergrade over a wide area as well as form different combinations," no varieties are recognized here among Ohio members of this species.

12. **Lonicera flava** Sims YELLOW HONEYSUCKLE

Twining and climbing vine; uppermost pair of leaves fused at base to form a rhombic or elliptic double-leaf, sometimes indented at the point of fusion, the second pair of leaves sometimes also fused basally, the lower nonfused leaves ovate to very broadly ovate or obovate, only slightly longer than wide, abruptly narrowed at base to a short petiole; corolla yellow, 2.5–3.0 cm long, bilabiate; stamens and style exserted. A species of southeastern and south-central United States, known from a 1911 collection made in Clermont County in southwestern Ohio, where the species is at a northern boundary of its range; rocky thickets, calcareous gravelly bluffs. May.

Ohio plants belong to var. *flavescens* (Small) Gleason (*L. flavida* Cockerell).

13. **Lonicera prolifera** (Kirchn.) Rehd. GRAPE HONEYSUCKLE

Twining and reclining or sometimes climbing vine; the uppermost pair of leaves fused at base to form an orbicular or nearly orbicular disk-shaped double-leaf, its ends entire or retuse, the nonfused leaves broadly ovate to suborbicular; corolla pale yellow and often marked with purple, 2.5–3.0 cm long, bilabiate; stamens and style exserted. A species of the interior United States, at an eastern boundary of its range in counties of western and central Ohio; rocky calcareous woodlands and openings. May–July.

W. H. Duncan (1975) observed: "This taxon is closely related to and intergrades with *L. dioica* and may not be separable from it."

3. Symphoricarpos Duham.

Shrubs; leaves opposite, simple; flowers in short, axillary or terminal spikes or clusters; calyx very short, 4- or 5-lobed; corolla white or greenish-white to pinkish or purplish, campanulate with 4 or 5 erect lobes, actinomorphic or nearly so; stamens 4 or 5; fruit a berry. A genus of 17 species, one native to China, the others to North America. All the Ohio species below are sometimes cultivated.

a. Fruits reddish, 4–7 mm long; upper internodes more or less densely pubescent with several to many divergent hairs as well as shorter curved hairs; corolla 2.5–4 mm long, greenish or purplish . . . . 3. *Symphoricarpos orbiculatus*

aa. Fruits whitish, 8–14 mm long; upper internodes mostly either with short curved hairs or glabrous, rarely with a few divergent hairs; corolla 6–8 mm long, pinkish.

b. Style and anthers included; upper internodes glabrous or minutely puberulent; leaves mostly entire, infrequently crenate to rounded-dentate . . . . 1. *Symphoricarpos albus*

bb. Style and anthers slightly exserted; upper internodes puberulent; leaves entire or more often crenate to markedly rounded-dentate . . . . 2. *Symphoricarpos occidentalis*

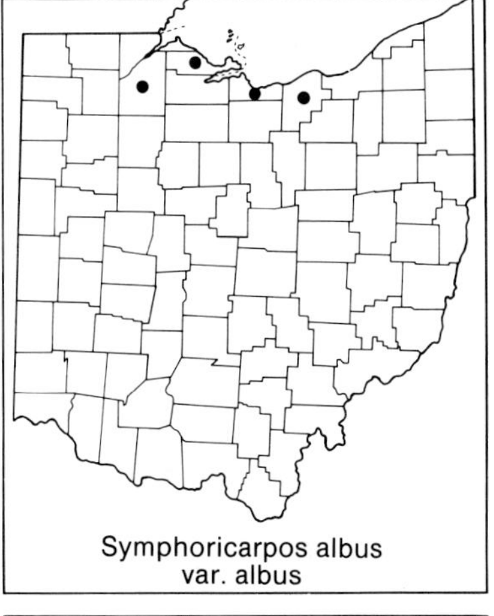

Symphoricarpos albus var. albus

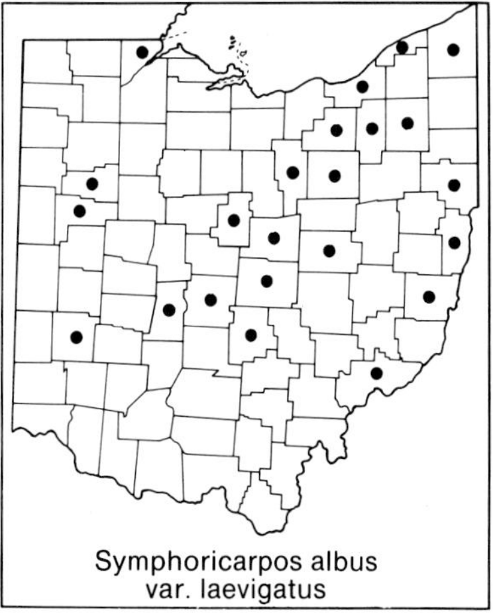

Symphoricarpos albus var. laevigatus

1. **Symphoricarpos albus** (L.) S. F. Blake SNOWBERRY

Shrub, to 1 m (3 feet) tall; upper internodes either minutely puberulent or glabrous; leaves ovate to broadly elliptic to nearly orbicular, margins mostly entire, infrequently irregularly crenate to rounded-dentate; corolla pinkish, 6–8 mm long; anthers and style included; fruit whitish, more or less globose, 6–15 mm long. This species is represented in the Ohio flora by the two varieties keyed below. Variety *albus* is known from a few counties of the northern tier, where it grows in open rocky habitats. A boreal plant, transcontinental across North America, it is presumably at or near a southern boundary of its range at these Ohio sites. Although the taxon is generally thought to be indigenous in northern Ohio, and is so regarded here, Moseley (1931) suggested that it may have been introduced there by Native Americans. Variety *laevigatus*, from the Pacific Northwest, is cultivated as an ornamental and has escaped and become naturalized in northeastern and central counties and at scattered sites in other parts of Ohio; roadsides, disturbed ground along railroad tracks, and other waste places. May–July.

a. Undersurface of leaves usually pubescent; upper internodes usually minutely puberulent; fruits 6–10 mm in diameter . . . . var. *albus*

aa. Undersurface of leaves glabrous; upper internodes glabrous; fruits 10–15 mm in diameter .... var. *laevigatus* (Fern.) S. F. Blake
S. rivularis Suksd.

2. Symphoricarpos occidentalis Hook. Wolfberry

Shrub, to 1 m tall (3 feet) tall; upper internodes puberulent; leaves mostly elliptic to broadly elliptic, margins entire to more often crenate or markedly rounded-dentate; corolla pinkish, 6–8 mm long; anthers and style slightly exserted; fruit whitish, more or less globose, 8–14 mm long. A species of western North America, naturalized or perhaps merely adventive at a few sites in northeastern Ohio; along railroad tracks and in other disturbed habitats. June–August.

3. **Symphoricarpos orbiculatus** Moench Coralberry. Indian Currant

Shrub, to 1.5 m (5 feet) tall; upper internodes more or less densely pubescent with several to many divergent hairs as well as shorter curved hairs; leaves elliptic to ovate or suborbicular, margins entire; corolla greenish to purplish, 2.5–4.0 mm long; stamens and style included; fruit dull red, more or less ellipsoid, 4–7 mm long. Throughout most of Ohio; woodland openings and borders, prairies, fields, thickets, pastures, roadsides. Mid-July–mid-September.

4. Linnaea L.

A genus of a single species.

1. **Linnaea borealis** L. Twinflower

Low shrub; stems trailing and creeping, with erect or ascending branches to 2 dm tall; leaves opposite, simple, evergreen, 1–2 cm long, elliptic to

Symphoricarpos occidentalis

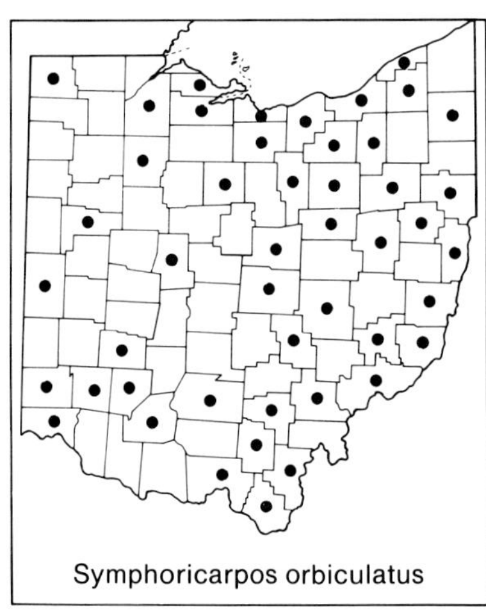
Symphoricarpos orbiculatus

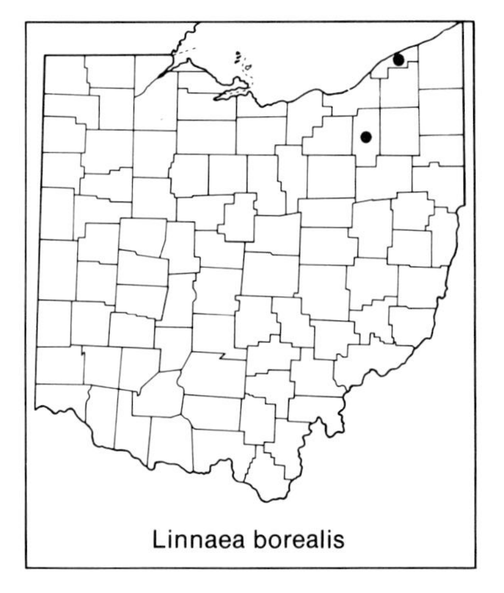
Linnaea borealis

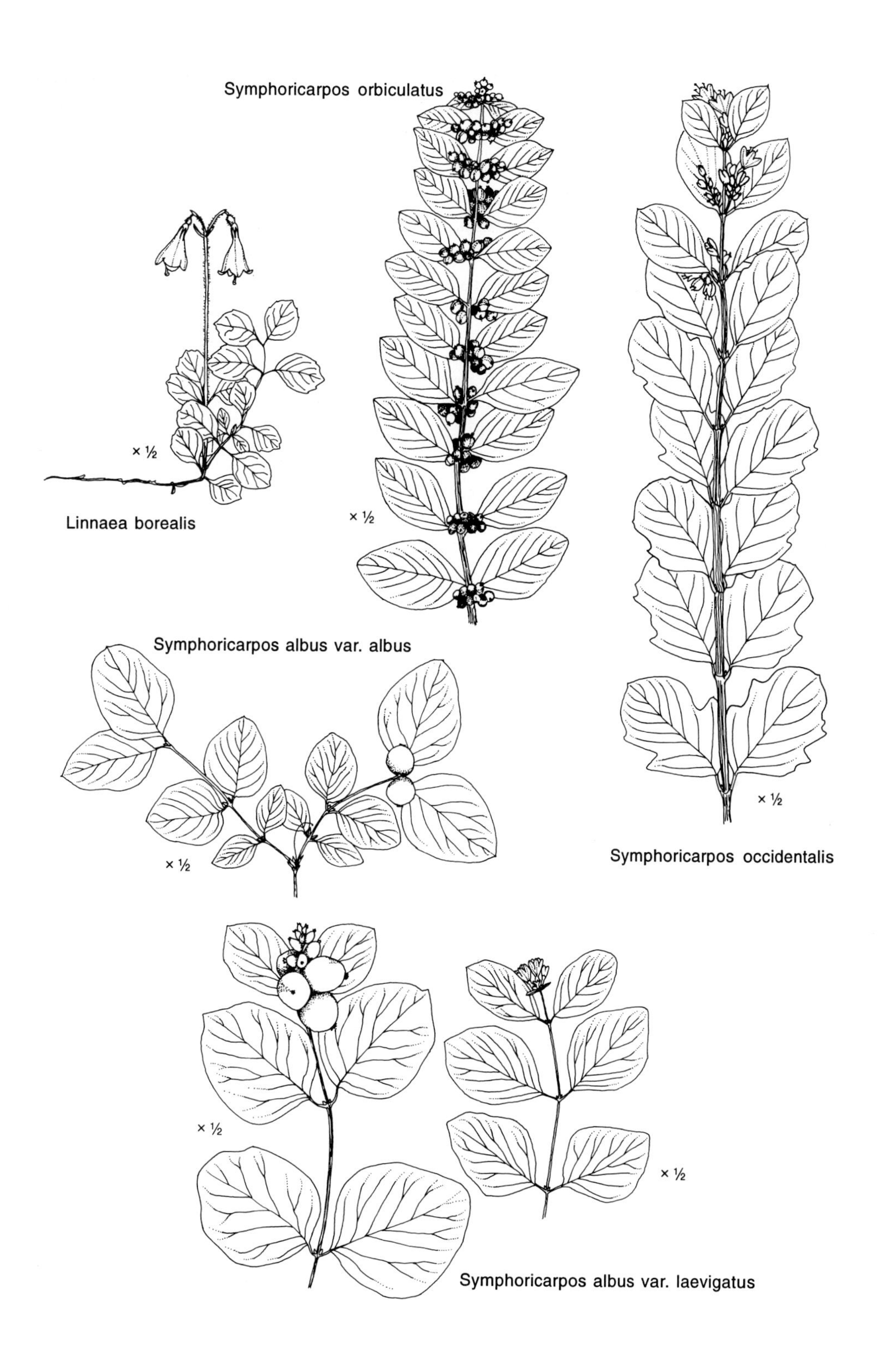
Symphoricarpos orbiculatus
× ½
Linnaea borealis
× ½
× ½
Symphoricarpos albus var. albus
× ½
× ½
Symphoricarpos occidentalis
× ½
× ½
Symphoricarpos albus var. laevigatus

obovate or orbicular, margins low-serrate; flowers nodding, fragrant, short-pedicellate, in pairs on a single peduncle 2–10 cm long; corolla pinkish, 10–15 mm long, campanulate-funnelform, 5-lobed, actinomorphic; fruit an achene. A boreal species at a southern boundary of its range in northeasternmost Ohio; wet woodlands, bogs. July–August.

Ohio plants are assigned to var. *americana* (J. Forbes) Rehd. Gleason and Cronquist (1991) merge var. *americana* under var. *longiflora* Torr.

5. Triosteum L. FEVERWORT. HORSE-GENTIAN

Coarse, perennial herbs; leaves simple, opposite and sometimes connate-perfoliate, stipulate; flowers axillary, solitary or in few-flowered clusters; calyx deeply 5-lobed, cut nearly or quite to the base, the lobes linear or lanceolate, persistent; corolla yellow to greenish or purplish, tubular with 5 erect lobes, subactinomorphic; stamens 5; fruit a dryish drupe. A genus of about six species of eastern North America and eastern Asia, revised by Lane (1954).

a. Calyx lobes with cilia 0.7–1.5 (–2.0) mm long, the outer surface of the sepals otherwise glabrous or with a few shorter hairs; stipules usually extending beyond tips of sepals; stems with longest hairs of upper internodes 1.5–2.5 mm long; none of the leaves connate-perfoliate . . . . 3. *Triosteum angustifolium*

aa. Calyx lobes more or less uniformly pubescent across outer surface, cilia none or only ca. 0.5 mm long; stipules not extending to tips of sepals; stems with longest hairs of upper internodes rarely exceeding 1.5 mm in length; leaves separate or connate-perfoliate.

b. Some or all leaves, especially the principal ones located near medial position on stem, strongly connate-perfoliate; stems densely glandular-pubescent, with few or no longer glandless hairs . . . . 1. *Triosteum perfoliatum*

bb. None of the leaves connate-perfoliate or only a few of the upper ones very slightly so; stems with all or most hairs long and glandless, these sometimes intermixed with shorter, often glandular, hairs . . . . 2. *Triosteum aurantiacum*

1. **Triosteum perfoliatum** L. TINKER'S-WEED

Perennial, to ca. 1 m tall; stems erect, densely glandular-pubescent, with few or no longer glandless hairs; some or all leaves, especially the principal ones located near the medial section of the stem, strongly connate-perfoliate, stipules not extending to tips of sepals; calyx lobes more or less uniformly pubescent across outer surface; corolla yellow, greenish, or purplish, 10–15 mm long; fruit dull orange. Throughout much of Ohio,

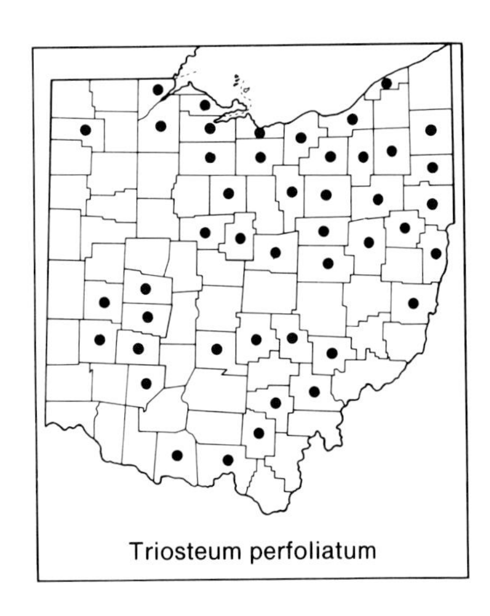
Triosteum perfoliatum

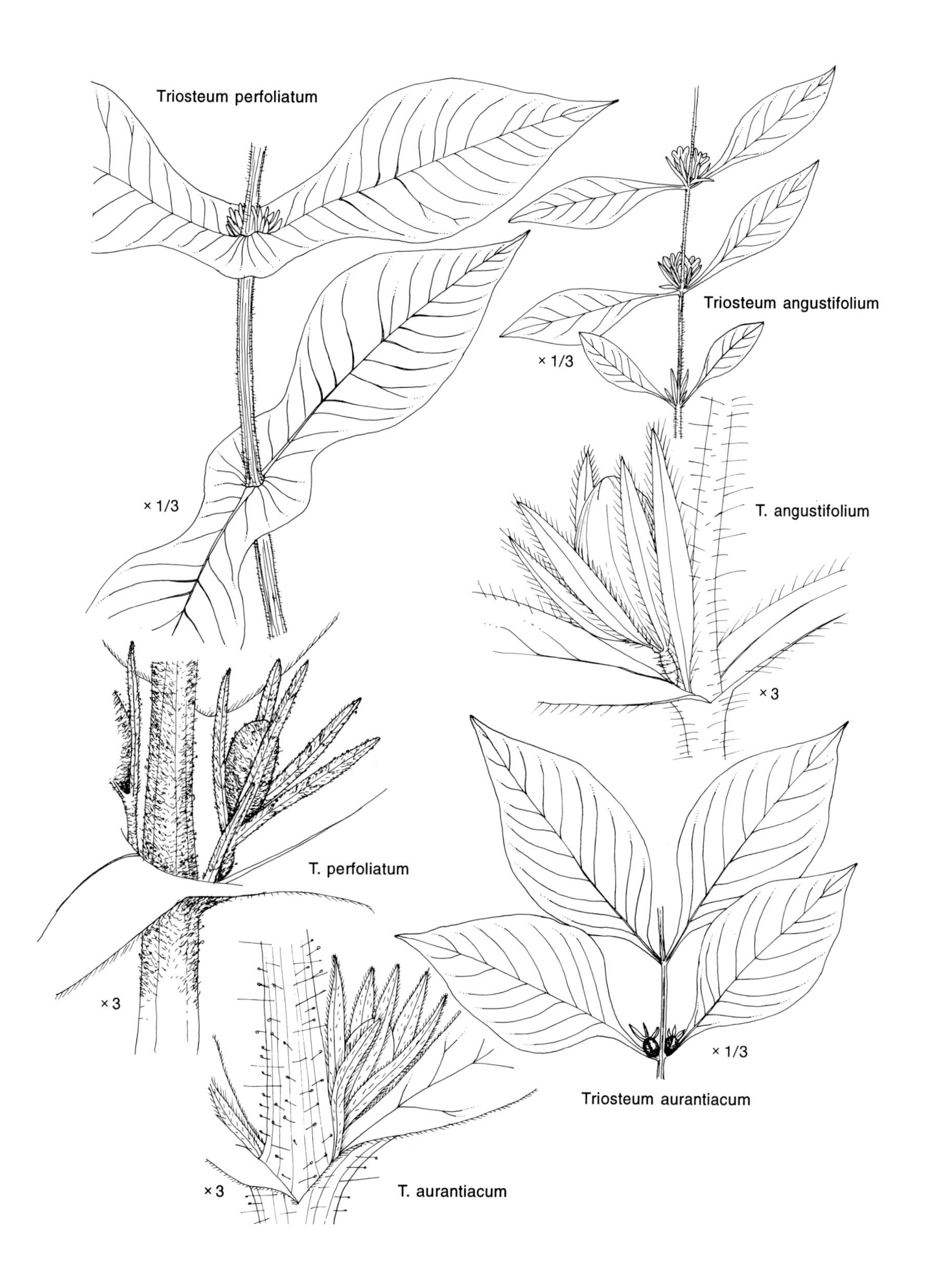
Triosteum perfoliatum
× 1/3
Triosteum angustifolium
× 1/3
T. angustifolium
× 3
T. perfoliatum
× 3
× 1/3
Triosteum aurantiacum
× 3
T. aurantiacum

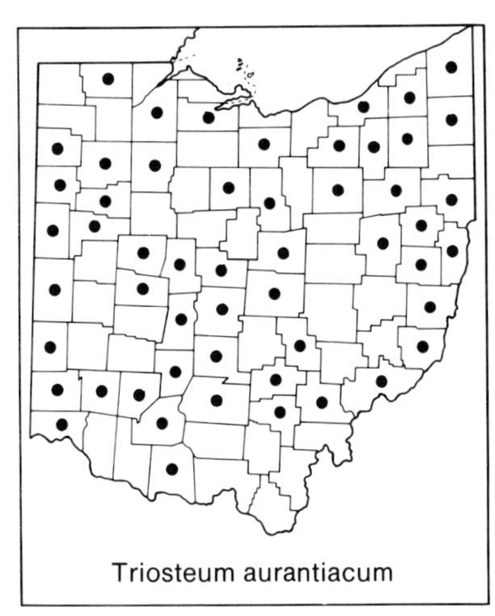
Triosteum aurantiacum

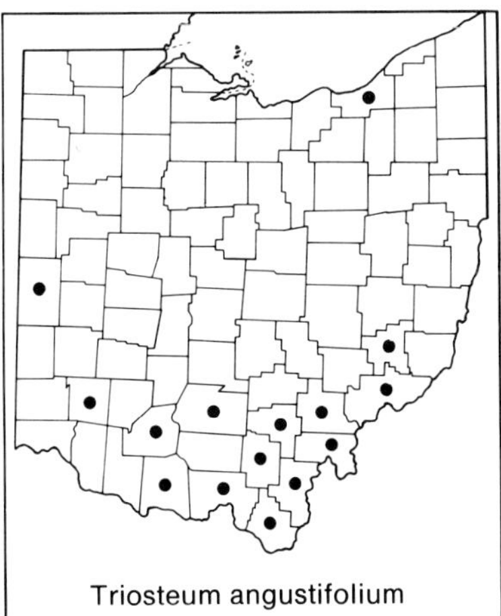
Triosteum angustifolium

except the westernmost counties; woodland borders, fields, thickets, roadsides. June.

2. **Triosteum aurantiacum** Bickn. WILD COFFEE
incl. var. *aurantiacum*, var. *glaucescens* Wieg., and var. *illinoense* (Wieg.) E. J. Palmer & Steyerm. (*T. illinoense* (Weig.) Rydb.); *T. perfoliatum* L. var. *aurantiacum* (Bickn.) Wieg.

Perennial, to ca. 1 m tall; stems erect, with all or most hairs long and glandless, but the hairs rarely exceeding 1.5 m in length, and sometimes intermixed with shorter often glandular hairs; plants either with none of the leaves connate-perfoliate or with only a few of the upper ones very slightly so, stipules not extending to tips of sepals; calyx lobes more or less uniformly pubescent across outer surface, usually ciliate—the cilia ca. 0.5 mm long; corolla reddish-purple, 10–15 mm long; fruit orangish-red. Throughout Ohio; woodland openings and borders, thickets, roadsides. Mid-May–mid-June.

Gleason (1952b) treated this species as a variety of the previous species, *T. perfoliatum*. Fernald (1950), Gleason and Cronquist (1991), and others accord it specific rank, as is done here. Those authors recognize within the species three varieties, listed above in synonymy, all based on differences in vestiture, and all presumably occurring in Ohio. Because of intergradation in the diagnostic character states, I find little merit in attempting to separate Ohio plants into these varieties.

3. **Triosteum angustifolium** L. YELLOW HORSE-GENTIAN

Perennial, to 8 dm tall; stems erect, with most hairs of the upper internodes 1.5–2.5 mm long, and with few if any short glandular hairs; none of the leaves connate-perfoliate, stipules usually extending beyond tips of sepals; calyx lobes with cilia 0.7–1.5 (–2.0) mm long, the outer surface of the sepals otherwise glabrous or with a few shorter hairs; corolla yellowish, 12–17 mm long; fruit orangish-red. A species of the southeastern and east-central United States, at a northern boundary of its range in Ohio; woods, woodland openings and borders. The disjunct 1896 record from Cuyahoga County, in northern Ohio, is supported by the more recent discovery of the species on Pelee Island, Ontario, in western Lake Erie (T. Duncan, 1973). Late April–May.

6. Viburnum L. ARROW-WOOD

Shrubs or rarely small trees; leaves opposite, simple, petiolate; flowers usually small and numerous, in terminal cymes; calyx of 5 tooth-like lobes; corolla 5-lobed, actinomorphic; stamens 5; fruit a drupe. A Northern Hemisphere genus of perhaps 200 species. Donoghue (1983) made a preliminary numerical analysis of phylogenetic relationships in the genus, and Egolf (1962) a study of chromosome numbers. T. H. Jones (1983) revised section *Lentago*, which includes species nos. 3–6 below.

Several of the species described in the text below are cultivated in Ohio as ornamentals. A few other species and hybrids also are cultivated, among them the Asian shrubs *Viburnum dilatatum* Thunb., notable for large clusters of scarlet fruits produced in late summer and autumn, and *V. plicatum* Thunb., JAPANESE SNOWBALL. A specimen of the latter species, perhaps escaped or perhaps merely persisting from cultivation, was collected from Geauga County in 1985.

a. Plants with all leaves unlobed, the blades pinnately or palmately veined.

b. Upper internodes densely stellate-pubescent; leaves cordate or rounded at base.

c. Leaves mostly gray stellate-pubescent on undersurface, the leaf blades oblong to ovate, 5–10 (–14) cm long, their margins singly serrate with teeth terminating in stout hyaline tips; inflorescences without marginal sterile flowers. . . . . 1. *Viburnum lantana*

cc. Leaves mostly rusty brown stellate-pubescent on undersurface, the leaf blades broadly ovate to suborbicular, 10–20 cm long, their margins slightly doubly serrate with teeth mostly terminating in green tissue, not in hyaline tips; inflorescences with large marginal sterile flowers. . . . . 2. *Viburnum alnifolium*

bb. Upper internodes sparingly stellate-pubescent, or with simple hairs, or glabrous; leaves cordate, rounded, or most often tapering at base.

c. Leaves entire, crenate, finely dentate, or finely serrate, the lateral veins curving and anastomosing, not reaching margin of blade.

d. Leaf margins entire to crenate or bluntly low-dentate; buds reddish to cinnamon-brown; cymes on peduncles 1–3 cm long . . . . 3. *Viburnum cassinoides*

dd. Leaf margins finely and sharply serrate; buds not reddish to cinnamon-brown; cymes sessile.

e. Petioles of young leaves and the midribs on their undersurfaces with dense, spreading, rusty-brown, stellate tomentum; leaf apices retuse, rounded, obtuse, or rarely shortly acute . . . . 6. *Viburnum rufidulum*

ee. Petioles of young leaves and the midribs on their undersurfaces with at most sparse, appressed, rusty-brown tomentum to more often glabrous, or with simple whitish pubescence; leaf apices acuminate, acute, or obtuse.

f. Margins of all or some petioles usually broad and slightly winged and wavy-edged; leaf apices usually acuminate, often abruptly so . . . . 4. *Viburnum lentago*

ff. Margins of petioles none or narrow and not wavy-edged; leaf apices acute to obtuse . . . . 5. *Viburnum prunifolium*

cc. Leaves coarsely dentate or serrate-dentate, each lateral vein and vein-branch reaching margin of blade, ending in a tooth.

d. Petioles 0.1–1.0 (–1.2) cm long; leaves with linear stipules . . . . 7. *Viburnum rafinesquianum*

dd. Petioles (0.5–) 1–8 cm long; leaves with or without linear stipules.

e. Leaves markedly cordate at the base, the 2 or 3 lowest pairs of lateral veins converging at or near midrib; petioles 1.5–6 (–8) cm long, with linear stipules. . . . . 8. *Viburnum molle*

ee. Leaves rounded, truncate, or rarely subcordate at the base, the lowest lateral veins not converging; petioles mostly 1.0–2.5 cm long, with or more often without linear stipules.

f. Petioles and undersurface of leaves pubescent, especially along the major veins; principal leaves mostly suborbicular to broadly ovate, as wide as to 3/4 as wide as long. . . . . 9. *Viburnum dentatum*

ff. Petioles glabrate or glabrous, undersurface of leaves glabrous or with pubescence in vein axils only; principal leaves mostly broadly ovate to ovate, 3/4 to 1/4 as wide as long . . . . 10. *Viburnum recognitum*

aa. Plants with all or most leaves palmately lobed, the blades palmately veined.

b. Petioles without glands near base of blades; undersurface of blades stellate-pubescent and with numerous yellow to orange or brown dots; inflorescences 4–7 cm wide, without marginal sterile flowers; fruit at first red, then becoming dark purple to black at maturity . . . . 11. *Viburnum acerifolium*

bb. Petioles with 2 or more large glands near base of blades; undersurface of blades with simple hairs and lacking dots; inflorescences 7–10 cm wide, with large marginal sterile flowers; fruit red at maturity. . . . . 12. *Viburnum opulus*

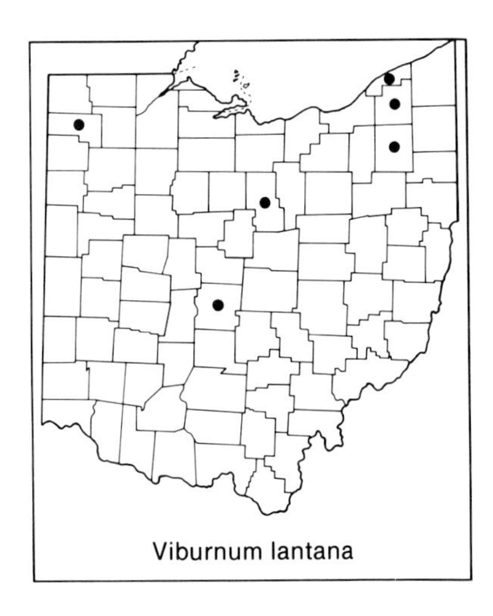
Viburnum lantana

1. Viburnum lantana L. Wayfaring-tree

Shrub, to 4.5 m (15 feet) tall; upper internodes densely stellate-pubescent; leaves oblong to ovate, 5–10 (–14) cm long, rounded to cordate at base, margins singly serrate with teeth terminating in stout hyaline tips, undersurface usually gray stellate-pubescent; inflorescences without marginal sterile flowers; corolla white; fruits red, changing to black. A Eurasian species frequently cultivated in Ohio; escaped and adventive or perhaps locally naturalized at a few sites in central and northern counties; roadsides, disturbed ground along railroads; disturbed woodlands. Not illustrated. May–June.

2. **Viburnum alnifolium** Marsh. Hobblebush
V. lantanoides Michx.

Shrub, to ca. 2 m (6.5 feet) tall; upper internodes densely stellate-pubescent; leaves broadly ovate to suborbicular, 10–20 cm long, cordate at base, margins slightly double-serrate with teeth mostly terminating in green tissue, not in hyaline tips, undersurface mostly rusty brown stellate-pubescent; inflorescences with large white marginal sterile flowers; corolla of fertile flowers white; fruits red, darkening to nearly black. A northern species ranging southward in the Appalachians, at a southwestern border of its range in counties of northeasternmost Ohio; wooded slopes and ravines, wet woodlands. May.

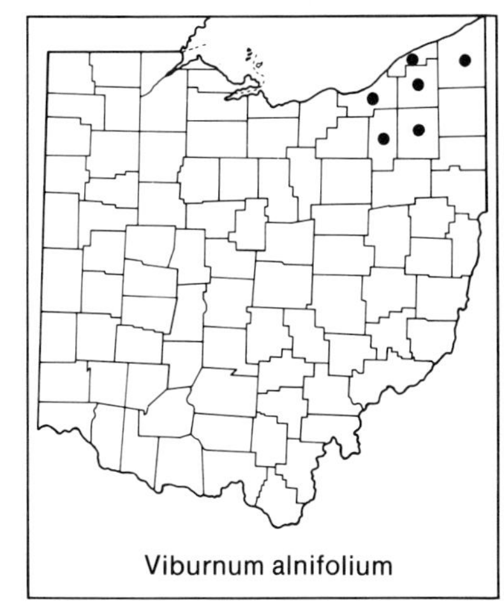
Viburnum alnifolium

3. **Viburnum cassinoides** L. Withe-rod. Wild Raisin
V. nudum L. var. *cassinoides* (L.) T. & G.

Shrub, to ca. 2 m (6.5 feet) tall; upper internodes with patches of scurfy, dark brown stellate pubescence, buds oblong-linear, reddish to cinnamon-brown; leaves quite variable: lanceolate to elliptic, ovate, or obovate, 4–15 cm long, the margins entire to crenate or bluntly low-dentate, undersurface with small to large amounts of brown stellate pubescence especially conspicuous along the midrib, lateral veins curving and anastomosing and failing to reach leaf margin; cymes on peduncles 1–3 cm long; corolla white; fruits bluish-black at maturity. Northeastern and south-central counties, and in Williams County in northwesternmost Ohio; wet woodlands, swamp margins. June.

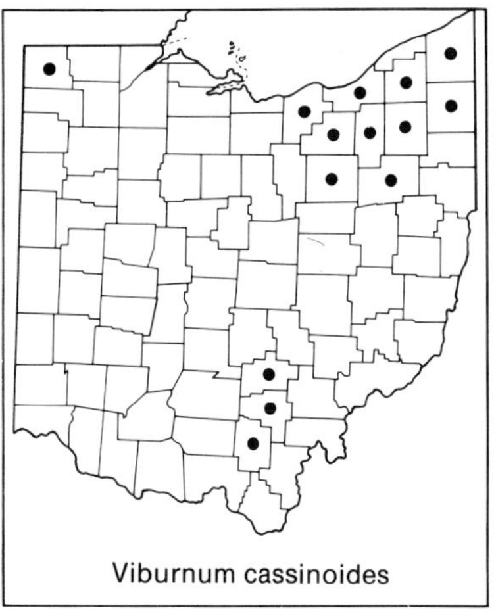
Viburnum cassinoides

T. H. Jones (1983) and Gleason and Cronquist (1991) treat this species as *Viburnum nudum* var. *cassinoides*. Jones also reports a specimen of *V. nudum* var. *nudum* from Cuyahoga County, in northernmost Ohio, greatly disjunct from its more southerly range, but I have not seen the specimen upon which this attribution is based.

4. **Viburnum lentago** L. Nannyberry. Sheepberry

Shrub, to ca. 4.5 m (15 feet) tall; terminal flower bud distinctively shaped: swollen and globose-ellipsoid at base and conic above; margins of some or all petioles usually broad, slightly winged and wavy-edged, leaf blades quite variable in size and shape: lanceolate to elliptic, oblong, ovate, suborbicular, or obovate, 5–15 cm long, apices usually acuminate, often abruptly so, lateral veins curving and anastomosing and failing to reach leaf margin; cymes sessile; corolla white; fruits bluish-black, glaucous. A

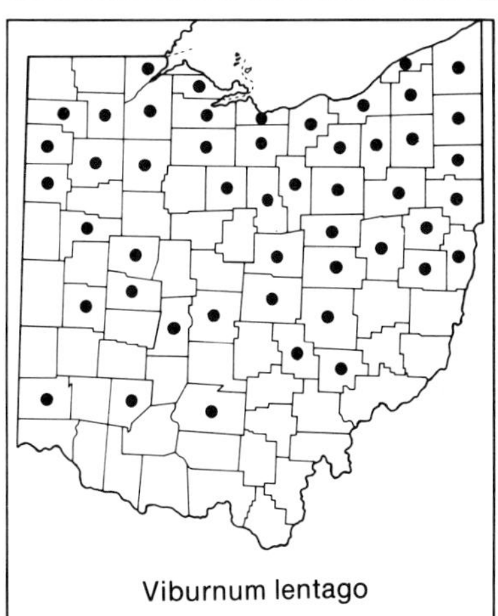
Viburnum lentago

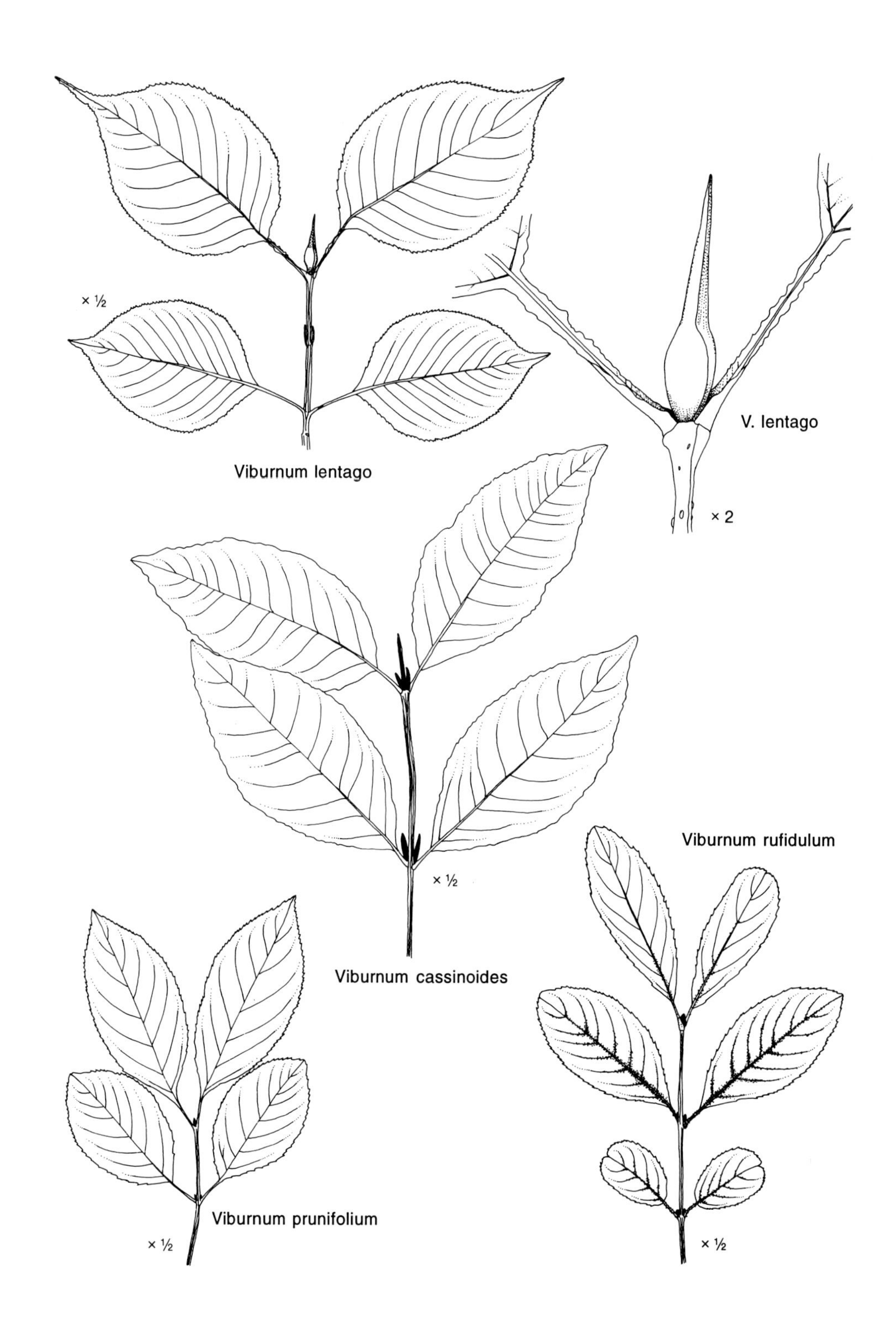
× ½
Viburnum lentago
V. lentago
× 2
× ½
Viburnum cassinoides
Viburnum rufidulum
Viburnum prunifolium
× ½
× ½

species of northeastern North America, at a southern boundary of its range in Ohio; northern 3/4 of the state; wet thickets, wet woodland borders. Mid-April–early June.

Some Ohio specimens are difficult to separate from *V. prunifolium.*

5. **Viburnum prunifolium** L. Black-haw

Shrub or small tree, to 8 m (26 feet) tall; petiole margins either none or narrow and not wavy-edged; leaf blades mostly narrowly to broadly elliptic or elliptic-obovate, 4–8 cm long, apices acute or obtuse, lateral veins curving and anastomosing and failing to reach leaf margin; cymes sessile; corolla white; fruits bluish-black, glaucous. Throughout Ohio; open woods, woodland openings and borders, thickets. Mid-April–early June.

Some Ohio specimens are difficult to separate from *V. lentago.*

6. **Viburnum rufidulum** Raf. Southern Black-haw

Shrub or small tree, to 8 m (26 feet) tall; petioles very narrowly winged, the petioles of young leaves and the midrib on undersurface with dense, spreading, rusty-brown, stellate tomentum; leaf blades elliptic to broadly elliptic or obovate, the apices retuse, rounded, obtuse, or rarely shortly acute, the lateral veins curving and anastomosing and failing to reach leaf margin; cymes sessile; corolla white; fruits bluish-black, glaucous. A species of southeastern United States, at a northern boundary of its range in a few southwestern counties of Ohio; open rocky slopes, grassy fields, thickets. May.

7. **Viburnum rafinesquianum** Schultes Downy Arrow-wood

Shrub, to 1.5 m (5 feet) tall; petioles short, 0.1–1.0 (–1.2) cm long, stipules linear; leaf blades ovate, rounded to cordate at base, margins serrate-

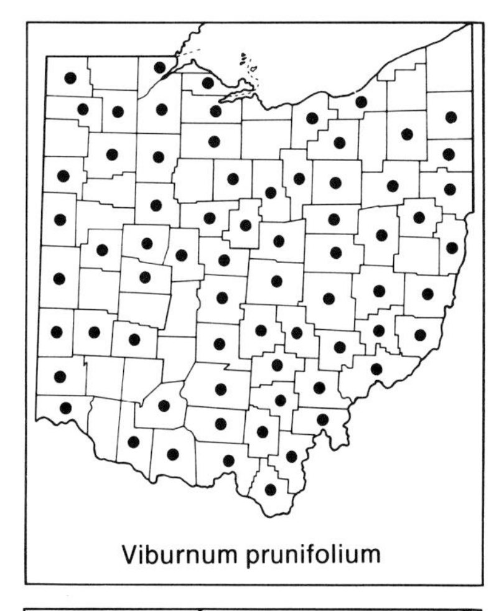

Viburnum prunifolium

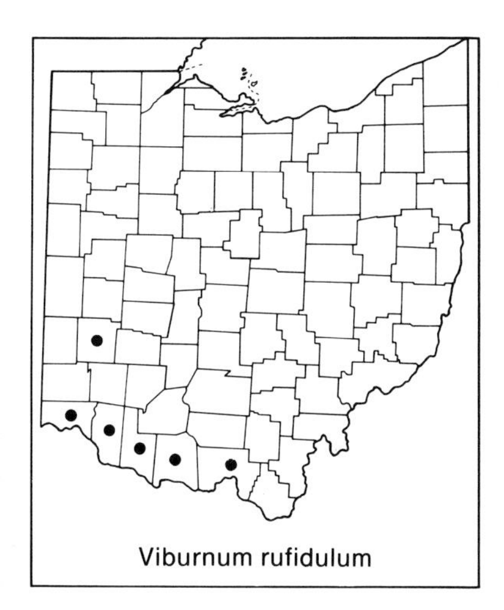

Viburnum rufidulum

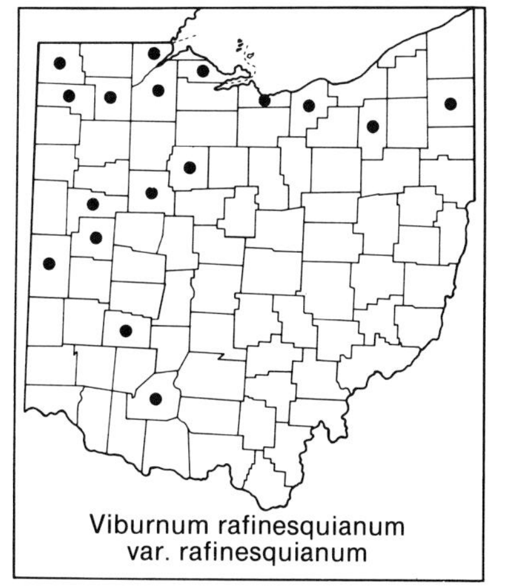

Viburnum rafinesquianum var. rafinesquianum

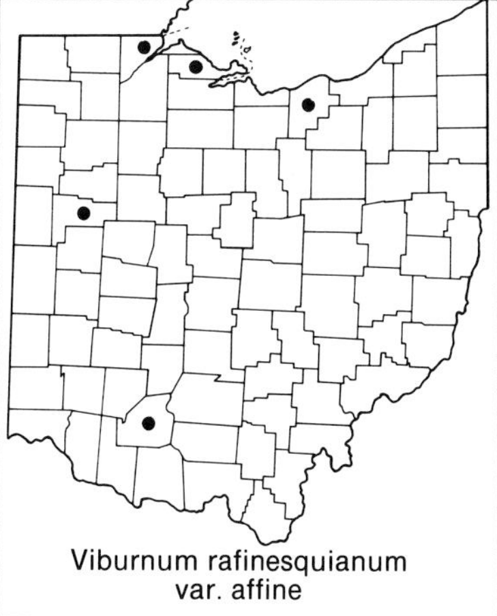

Viburnum rafinesquianum var. affine

Viburnum molle

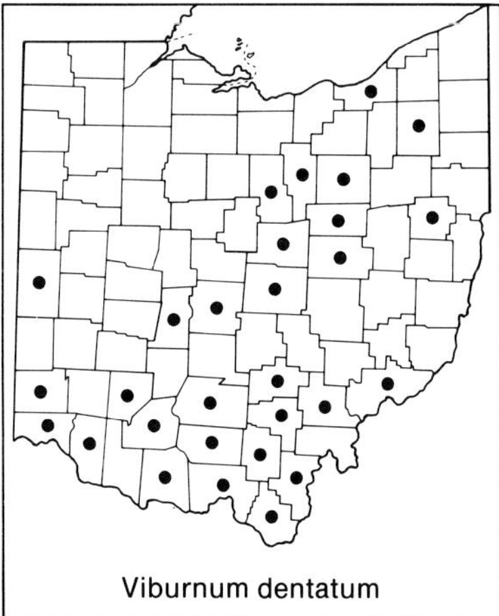
Viburnum dentatum

dentate, each lateral vein and vein-branch reaching margin, ending in a tooth; corolla whitish; fruits bluish-black. Chiefly in western counties; open woodlands. Mid-May–June.

Ohio plants may be separated into the two varieties keyed below, but their recognition is of questionable value because of the great amount of intergradation.

a. Leaves pubescent across all the undersurface; petioles ca. 5 mm long or less. . . . . var. *rafinesquianum*

aa. Leaves pubescent only on the veins on undersurface; petioles ca. 10 mm long . . . . var. *affine* (Bush) House

8. **Viburnum molle** Michx. SOFT-LEAVED ARROW-WOOD

Shrub, to 3 m (10 feet) tall; petioles 1.5–6 (–8) cm long, stipules linear; leaf blades ovate to suborbicular, markedly cordate at base, the 2 or 3 lowest pairs of lateral veins converging at or near midrib, margins coarsely dentate, each lateral vein and vein-branch reaching margin of blade and ending in a tooth; corolla white; fruits bluish-black. A species of the interior United States, at a northeastern boundary of its range at a few sites in counties of southwestern Ohio; wooded floodplains, rocky woodlands. Blackwell et al. (1981) note a vigorous stand of this species in Butler County. May–June.

9. **Viburnum dentatum** L. SOUTHERN ARROW-WOOD

incl. var. *dentatum*, var. *deamii* (Rehd.) Fern., var. *indianense* (Rehd.) Gleason, var. *scabrellum* T. & G., and var. *venosum* (Britt.) Gleason

Shrub, to 3 m (10 feet) tall; petioles pubescent, 1.0–2.5 cm long, stipules linear or none; leaf blades ovate to suborbicular, mostly as wide as to 3/4 as wide as long, rounded to truncate or rarely subcordate at base, undersurface pubescent—especially along the major veins, margins dentate, each lateral vein and vein-branch reaching margin of blade and ending in a tooth; corolla white; fruits bluish-black. In southern and central counties and at scattered sites northward; woodlands, woodland openings and borders, thickets. June–July.

Braun (1961), following Gleason (1952b), recognizes five varieties (listed in synonymy above) among Ohio plants of this species, but presents only one map for the lot. All are based on varying amounts and types of vestiture on differing parts of the plant body. The degree of intergradation I have observed in Ohio plants has led to their treatment here as part of a polymorphic species not divided into varieties.

There are among Ohio specimens some intergrades between this species and the next, *V. recognitum*.

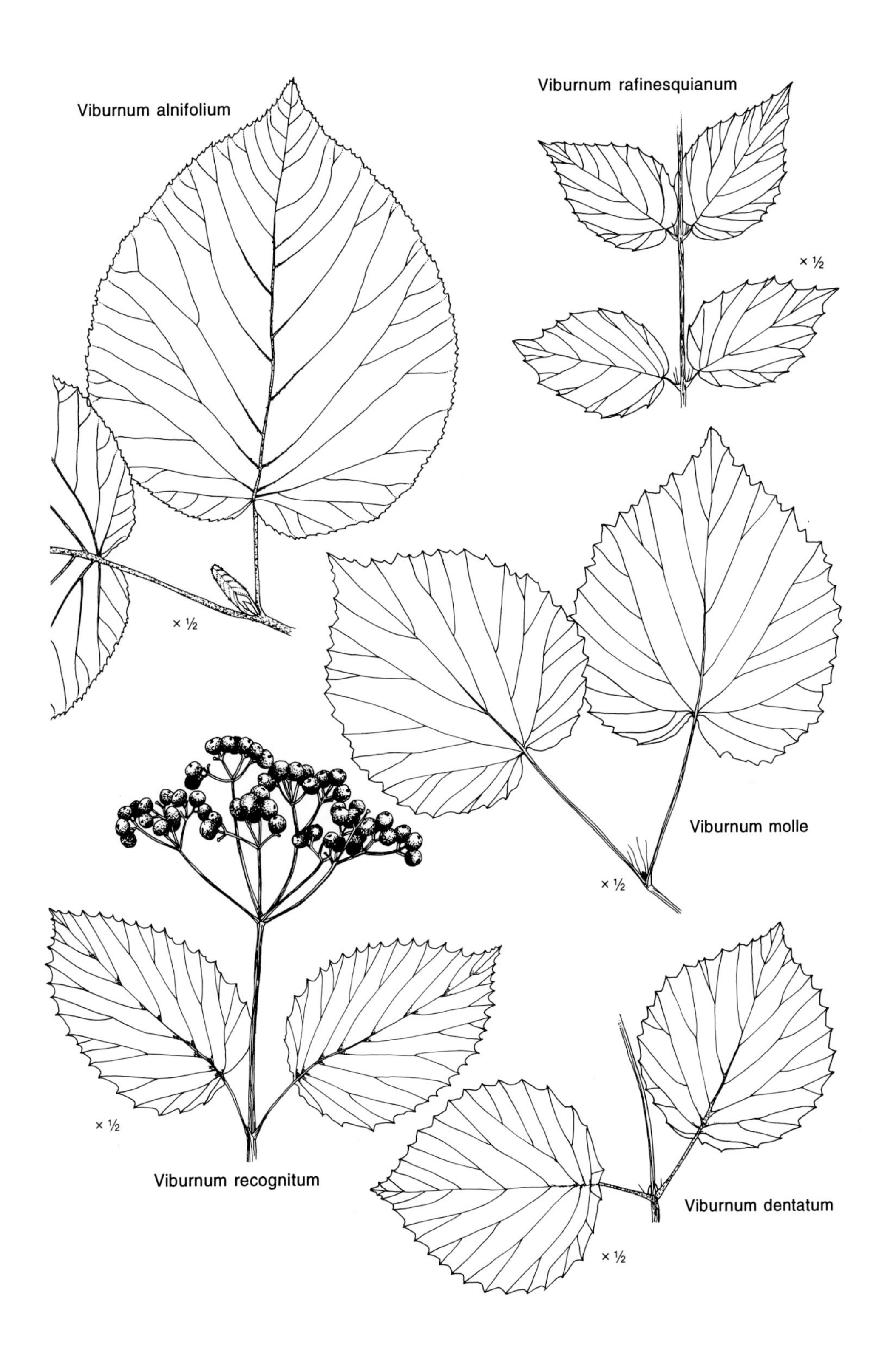
Viburnum alnifolium
× ½
Viburnum rafinesquianum
× ½
Viburnum molle
× ½
× ½
Viburnum recognitum
Viburnum dentatum
× ½

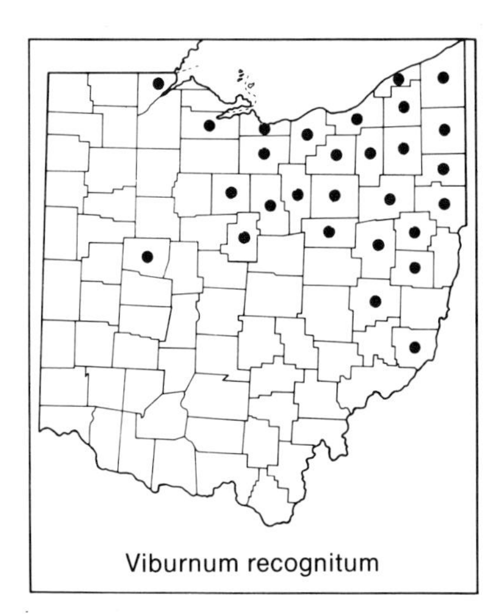
Viburnum recognitum

10. **Viburnum recognitum** Fern. Northern Arrow-wood

V. dentatum L. var. *lucidum* Ait.

Shrub, to 3 m (10 feet) tall; petioles glabrate or glabrous, 1.0–2.5 cm long, stipules usually none; leaf blades ovate to broadly ovate, mostly 1/4 to 3/4 as wide as long, rounded at base, margins dentate, undersurface glabrous or with pubescence only in vein axils, each lateral vein or vein-branch reaching margin of blade and ending in a tooth; corolla white; fruits bluish-black. Counties of the northeastern quarter of Ohio and at a few sites elsewhere in the state; woodland borders, borders of bogs and ponds, thickets. Mid-May–June.

There are among Ohio specimens some intergrades between this species and the last, *V. dentatum*. Gleason and Cronquist (1991) treat this species as a part of *V. dentatum*, assigning it the varietal name listed above in synonymy.

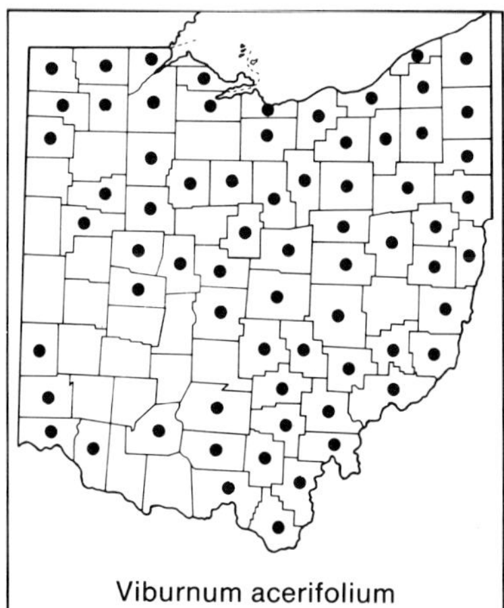
Viburnum acerifolium

11. **Viburnum acerifolium** L. Maple-leaved Viburnum. Maple-leaved Arrow-wood

incl. var. *acerifolium* and var. *glabrescens* Rehd.

Shrub, to 2 m (6.5 feet) tall; all or most leaves palmately lobed and palmately veined, occasional specimens with some leaves unlobed, undersurface stellate-pubescent and with numerous yellow to orange or brown dots; cymes 4–7 cm across, without marginal sterile flowers; corolla whitish; fruits at first red, then becoming dark purple to black at maturity. Throughout most of Ohio, but absent from some west-central and southwestern counties; woodlands. Mid-May–June.

12. **Viburnum opulus** L.

Shrub, to 3 m (10 feet) tall; petioles with 2 or more large glands near base of blade; leaf blades palmately lobed and palmately veined, undersurface of

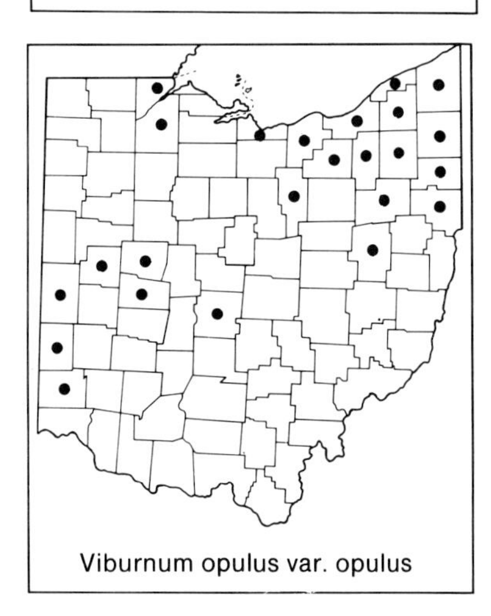
Viburnum opulus var. opulus

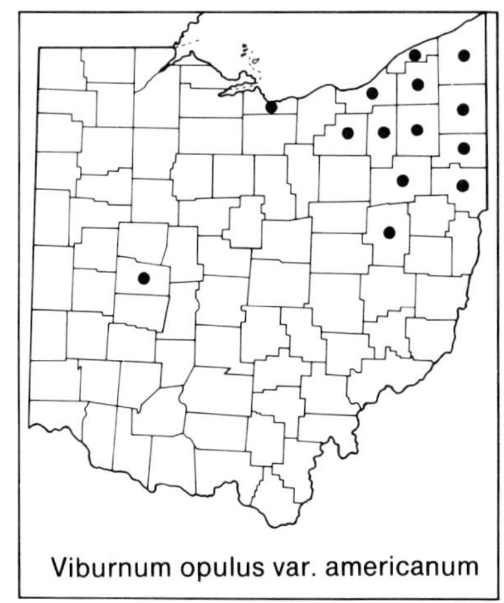
Viburnum opulus var. americanum

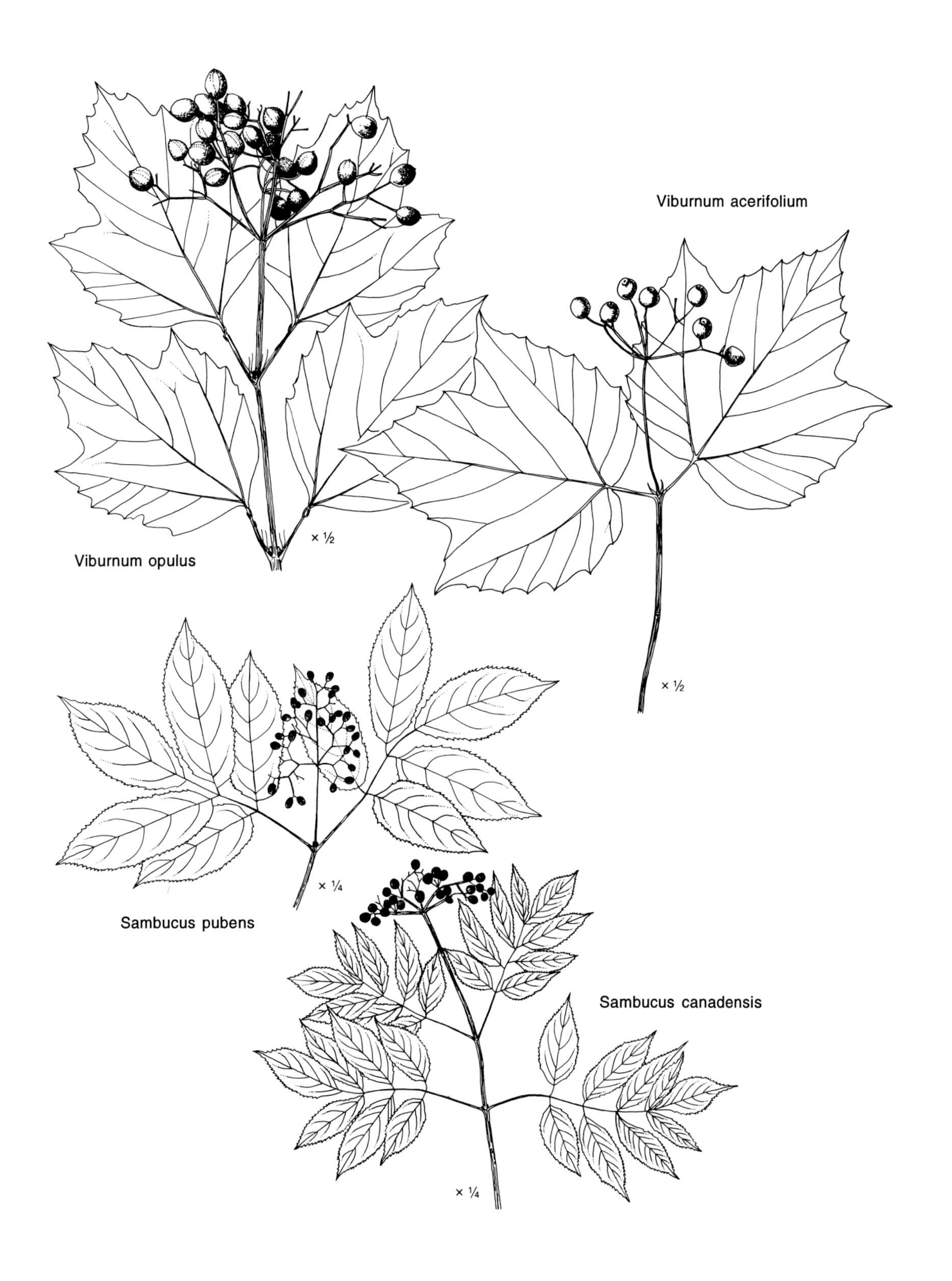
Viburnum acerifolium
× ½
Viburnum opulus
× ½
× ¼
Sambucus pubens
Sambucus canadensis
× ¼

blades with simple hairs and lacking dots; cymes 7–10 cm across, with large marginal white sterile flowers as well as white fertile flowers; fruits red at maturity. Counties of the northeastern quarter of Ohio and at scattered sites elsewhere; bogs, wet thickets, wet woodlands. May–mid-June.

This species contains two elements in Ohio; var. *americanum* is native, var. *opulus* is naturalized from Eurasia. The two are sometimes regarded as separate species. Because of the great amount of intergradation, presumably the result of hybridization, they are treated here as varieties of the same species. Some Ohio specimens cannot be satisfactorily assigned to either.

a. Glands on petiole (just below junction with blade) low and more or less saucer-shaped; stipules attenuate at apex . . . . var. *opulus*
European Cranberry-bush. Guelder-rose

aa. Glands on petiole (just below junction with blade) longer than wide and more or less cylindrical; some or all stipules thickened and blunt at apex . . . . var. *americanum* Ait.
V. trilobum Marsh.
Highbush-cranberry

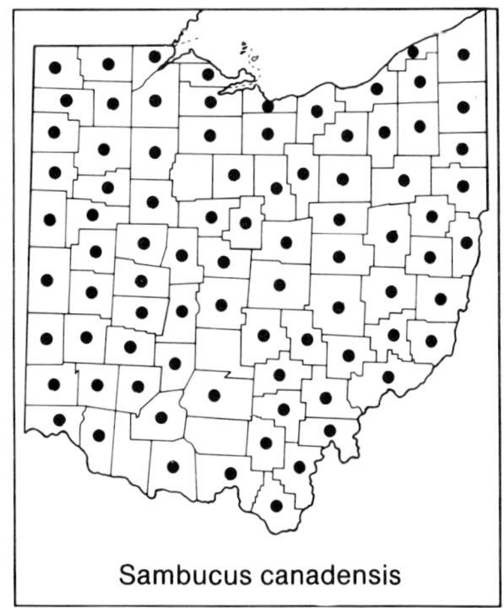
Sambucus canadensis

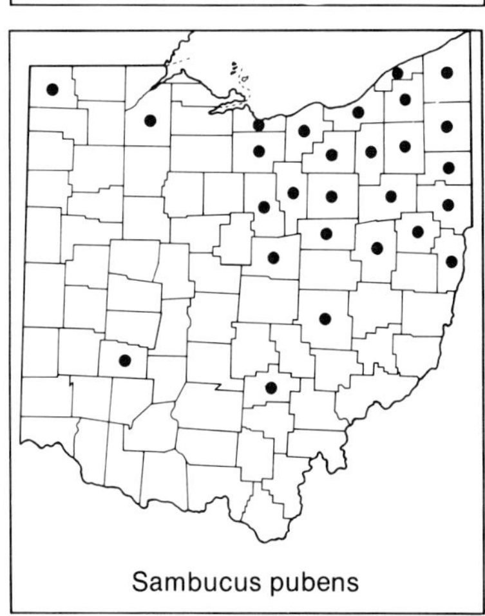
Sambucus pubens

7. Sambucus L. ELDERBERRY. ELDER

Shrubs; leaves opposite, pinnately compound, the leaflets serrate-margined; flowers numerous, tiny, in terminal cymes; calyx of 5 tooth-like lobes; corolla white, 5-lobed, actinomorphic, and more or less rotate; stamens 5; fruit a berry-like drupe. A genus of about 20 species widely distributed in temperate regions, two of the species native to Ohio.

a. Inflorescence a flat-topped and more or less umbellate cyme; fruits purple-black at maturity; pith white . . . . 1. *Sambucus canadensis*

aa. Inflorescence an ovoid to conical and more or less paniculate cyme; fruits red at maturity; pith brown . . . . 2. *Sambucus pubens*

1. **Sambucus canadensis** L. Common Elder. Elderberry

Shrub, to 3 m (10 feet) tall; stems with large white or whitish pith; leaflets 5–11; inflorescence a flat-topped and more or less umbellate cyme; fruits purple-black at maturity, juicy. Throughout Ohio; wet ground at margins of swamps, marshes, bogs, ponds, and streams, fencerows, thickets, woodland borders. The tangy fruits are often collected from the wild in late summer and autumn for pies, jelly, and wine. June–July.

2. **Sambucus pubens** Michx. Red-berried Elder

S. racemosa L. var. *pubens* (Michx.) Koehne

Shrub, to 3 m (10 feet) tall; stems with large brown or reddish-brown pith; leaflets 5–7; inflorescence an ovoid to conical, more or less paniculate cyme; fruits red at maturity. Northeastern counties and at a few scattered sites elsewhere in Ohio; rich woodlands, often on moist wooded north-facing slopes. Mid-April–May.

This species is regarded by some authors, including Gleason and Cronquist (1991), as a variety of the European species, *S. racemosa.*

Valerianaceae. VALERIAN FAMILY

Annuals, winter annuals, or perennials; leaves opposite, simple or compound; flowers perfect or imperfect, borne in regular or paniculate cymes; calyx either small—sometimes nearly absent—or modified to form bristles; corolla 5-lobed, actinomorphic to somewhat zygomorphic; stamens 3 (or 4), attached to the corolla tube; carpels 3, but only one functional, fused to form a compound pistil with a single style; ovary inferior; fruit an achene. A small, widely distributed family of about 300 species. Ferguson (1965) studied the family in southeastern United States. The text here is based in part on Hauser's (1963) study of the family in Ohio.

a. Cauline leaves pinnately compound or pinnately lobed, the basal leaves either pinnately lobed or simple and unlobed; calyx composed of bristles, these inrolled at anthesis but later expanded and plumose . . . . 1. *Valeriana*

aa. All leaves simple and unlobed, usually entire, occasionally few-toothed; calyx minute or absent . . . . 2. *Valerianella*

1. Valeriana L. VALERIAN

Perennial herbs; leaves opposite or basal, pinnately lobed or pinnately compound; flowers in cymes or paniculate cymes; calyx inrolled at anthesis, later expanding into 5–15 plumose bristles; corolla white or pinkish, funnelform to tubular or conic, slightly zygomorphic, with 5 erect or spreading lobes. A widely distributed genus of 200 or more species, chiefly of temperate regions. Meyer (1951) revised the genus in North America.

a. Aerial stems arising from a taproot; the principal leaves basal, usually unlobed and broadly to elliptically spatulate with entire margins, sometimes lobed; the basal leaves with petioles broadened and clasping at the base. . . . . 1. *Valeriana edulis*

aa. Aerial stems arising from rhizomes or stolons; the principal leaves cauline, pinnately lobed or pinnately compound, their margins usually crenate, serrate, or dentate; the basal leaves, if any, with petioles narrowed at the base, not clasping.

Valeriana edulis

Valeriana uliginosa

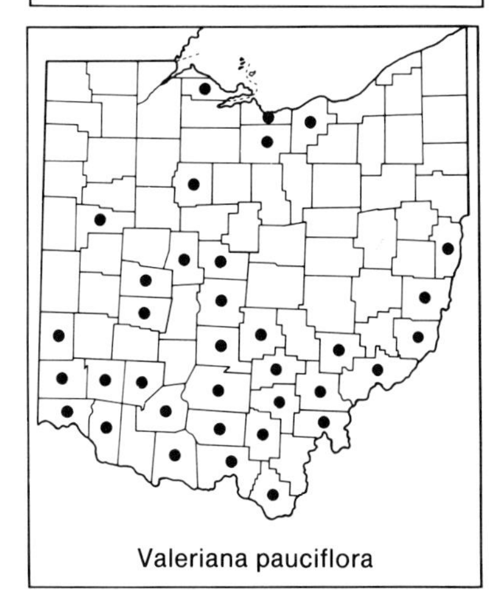
Valeriana pauciflora

b. Corolla 13–19 mm long; cauline leaves with 3–5 (–7) lobes or leaflets, the terminal leaflet ovate to suborbicular and 2–4 cm wide; plants with a short rhizome and also stoloniferous . . . . 3. *Valeriana pauciflora*

bb. Corolla 3–6 mm long; cauline leaves with (5–) 7–19 lobes or leaflets, the terminal leaflet broadly linear, lanceolate, elliptic, or oblanceolate—0.5–2.0 (–2.5) cm wide; plants with a rhizome but not stoloniferous.

c. Cauline leaves with (5–) 7–9 (–13) lobes or leaflets, their margins glabrous or with spreading cilia . . . . 2. *Valeriana uliginosa*

cc. Cauline leaves with 11–19 lobes or leaflets, their margins with stiff antrorse cilia . . . . 4. *Valeriana officinalis*

1. **Valeriana edulis** T. & G. PRAIRIE VALERIAN

Perennial, to 1 m tall; stems erect, arising from a taproot; basal leaves large—to 4 dm long, with petioles broadened and clasping at base, blades broadly to elliptically spatulate, entire-margined, sometimes lobed but usually unlobed; cauline leaves smaller, pinnately compound to deeply pinnately lobed; flowers numerous, tiny, borne in a paniculate cyme; corolla white, ca. 3 mm long. A species of western and southwestern United States and Mexico, at an eastern boundary of its range in west-central Ohio; fens, wet prairies. Meyer (1951) reported a record from Franklin County, but Stuckey (1966) concluded that this collection was in fact from Champaign County. June–October.

Ohio plants are assigned to var. *ciliata* (T. & G.) Cronq. (*V. ciliata* T. & G.).

2. **Valeriana uliginosa** (T. & G.) Rydb. SWAMP VALERIAN

V. sitchensis Bong. subsp. *uliginosa* (T. & G.) F. Meyer

Rhizomatous perennial; aerial stems erect, to 1 m tall; principal leaves cauline, opposite, pinnately compound to pinnately lobed, with (5–) 7–9 (–13) leaflets or lobes, their margins low-serrate to subentire and glabrous or with spreading cilia; flowers in 1–3 flat-topped cymes; corolla white to pinkish, 5–6 mm long. A species of northeastern North America, at a southern boundary of its range in northeastern Ohio, where it is known only from a "swamp" in Stark County. June–July.

3. **Valeriana pauciflora** Michx. LARGE-FLOWERED VALERIAN

Stoloniferous and rhizomatous perennial; aerial stems erect, to 1 m tall; principal leaves cauline, opposite, pinnately lobed or pinnately compound, with 3–5 (–7) lobes or leaflets, the terminal one ovate to suborbicular and 2–4 cm wide; flowers in a rounded terminal cyme; corolla purplish-pink, 13–19 mm long. A species chiefly of the Ohio Valley, at a northern boundary of its range in Ohio; southern counties and at scattered

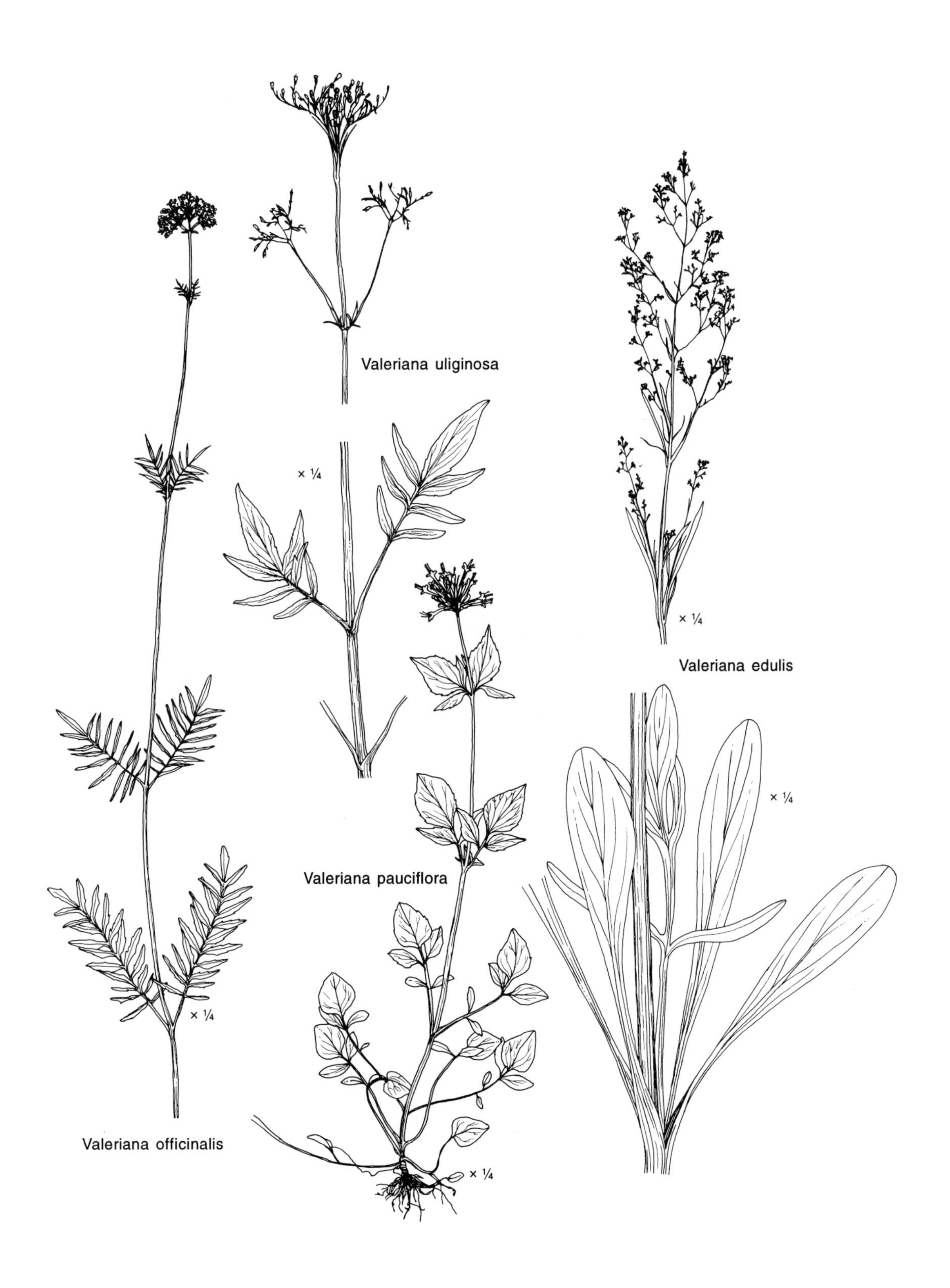
Valeriana uliginosa
× ¼
× ¼
Valeriana edulis
× ¼
Valeriana pauciflora
× ¼
Valeriana officinalis
× ¼

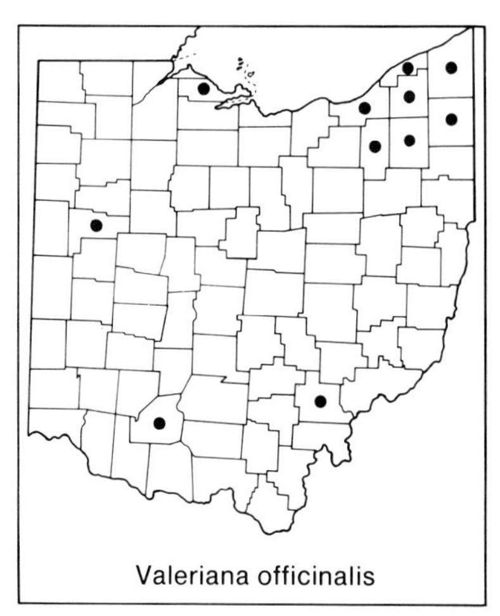

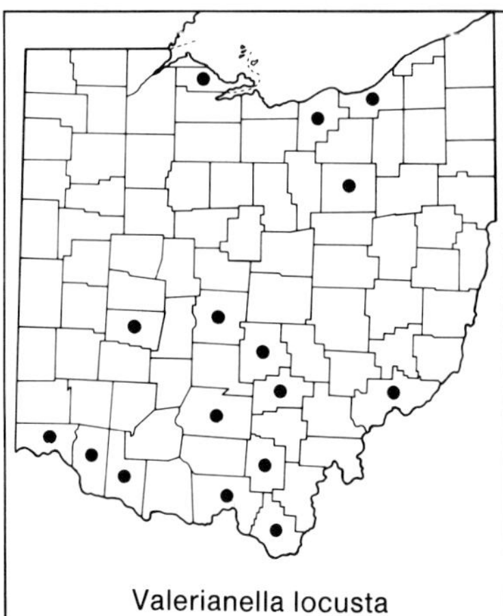

sites northward; moist wooded floodplains and terraces. Mid-May–mid-June.

4. VALERIANA OFFICINALIS L. GARDEN VALERIAN

Rhizomatous perennial; aerial stems erect, to 1 m tall; principal leaves cauline, opposite, pinnately lobed or pinnately compound, with 11–19 lobes or leaflets, their margins with a few scattered teeth and with stiff antrorse cilia; flowers in a rounded terminal cyme; corolla white to pinkish, 3–5 mm long. A Eurasian species, sometimes cultivated in Ohio, escaped and naturalized at a few scattered sites, mostly in northeastern counties. Mid-May–June.

2. Valerianella Mill. CORN-SALAD. LAMB'S-LETTUCE

Annuals or winter annuals; leaves opposite, simple; flowers in bracteate cymes; calyx minute or absent; corolla white or pale blue, more or less funnelform, subactinomorphic, with 5 spreading lobes. A Northern Hemisphere genus of perhaps 60 species. Ware (1983) studied fruit polymorphism and its taxonomic implications. In addition to the species treated below, an adventive specimen of *V. radiata* (L.) Dufr., a plant of more southerly states, was collected from railroad ballast in Holmes County in 1971, and Ware (1969) also reports records of this species from Franklin and Lorain counties.

a. Bracts and bractlets ciliate at apex; the fruit asymmetric, enlarged on one side by a thick spongy wall; corolla 1–2 mm long, the lobes pale blue . . . . 1. *Valerianella locusta*

aa. Bracts and bractlets glabrous or with few cilia at apex; the fruit more or less symmetric, not enlarged on one side by thick spongy wall; corolla 1.5–5 mm long, the lobes white.

b. Corolla 1.5–2 mm long; the outer bracts slightly ciliate, the inner eciliate; see above . . . . *Valerianella radiata*

bb. Corolla 3–5 mm long; the outer and inner bracts eciliate.

c. Fruits 2–3 (–3.25) mm long, plants without an early-maturing fruit at base of cyme. . . . . 2. *Valerianella umbilicata*

cc. Fruits 3–4 mm long, the single fruit at base of cyme maturing early, reaching its maximum length while flowers on upper branches of the same cyme are in anthesis . . . . . . 3. *Valerianella chenopodiifolia*

1. VALERIANELLA LOCUSTA (L.) Betcke EUROPEAN CORN-SALAD. LAMB'S-LETTUCE

incl. *V. olitoria* (L.) Poll.

Winter annual, to 4 dm tall; stems erect; bracts and bractlets of inflorescence ciliate at apex; corolla 1–2 mm long, the lobes pale blue; achenes

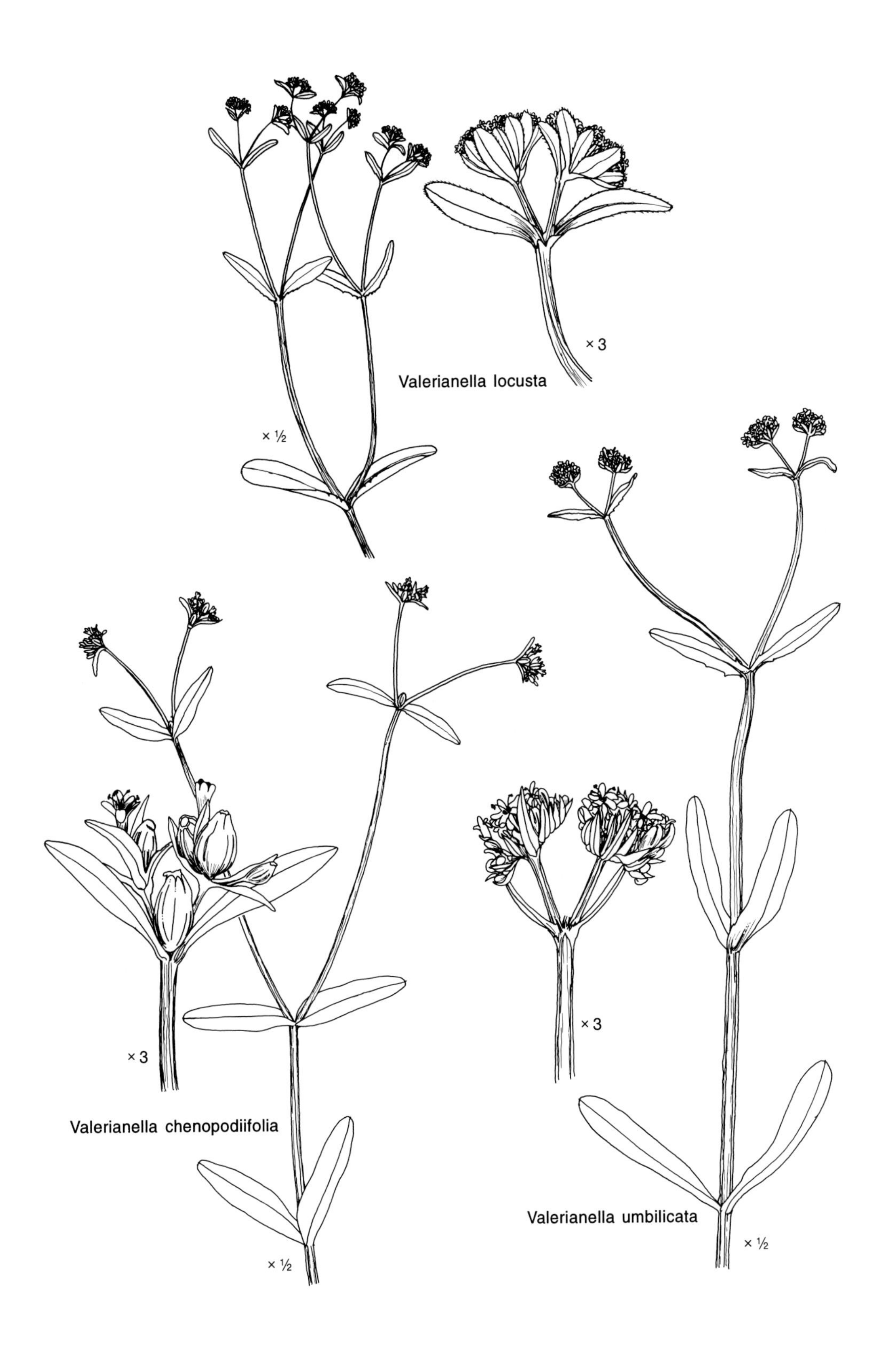
×3
Valerianella locusta
× ½
×3
×3
Valerianella chenopodiifolia
Valerianella umbilicata
× ½
× ½

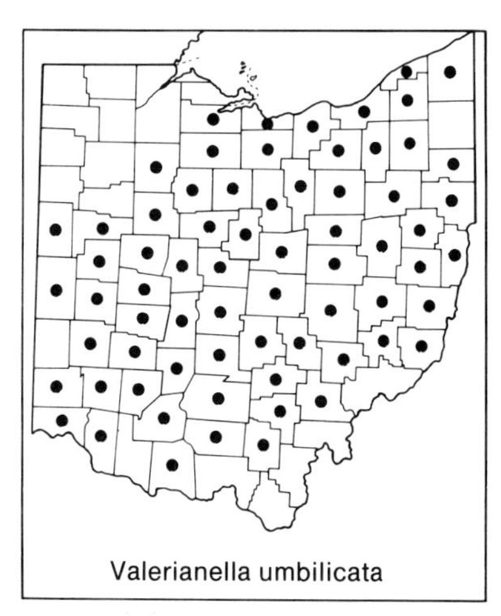
Valerianella umbilicata

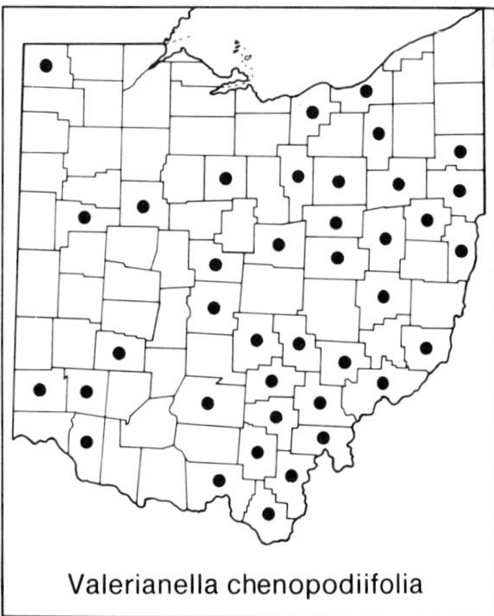
Valerianella chenopodiifolia

asymmetric, enlarged on one side by a thick spongy wall. A Eurasian species, naturalized locally at scattered Ohio sites; moist ground along streams, moist fields and roadsides. Late April–May.

2. **Valerianella umbilicata** (Sulliv.) Alph. Wood BEAKED CORN-SALAD
   incl. *V. intermedia* Dyal (*V. radiata* var. *intermedia* (Dyal) Gleason)
   and *V. patellaria* (Sulliv.) Alph. Wood

Winter annual, to 7 dm tall; stems erect; bracts and bractlets of inflorescence glabrous and without cilia; corolla 3–5 mm long, the lobes white; achenes more or less symmetric, 2–3 (–3.25) mm long, plants lacking an early-maturing achene at base of cyme. Throughout Ohio except extreme southeastern and northwestern counties; floodplains, stream banks, marshes, wet fields and roadsides. Baskin and Baskin (1976b) studied this species' germination ecology. Late April–mid-June.

The taxonomy adopted here is that of Ware (1983). She recognizes three forms within this species: *Valerianella umbilicata* forma *intermedia* (Dyal) Eggers, forma *patellaria* (Sulliv.) Eggers, and forma *umbilicata*—all of which occur in Ohio. Because the characters separating them are minor and often obscure, many Ohio specimens cannot be assigned with certainty to any of the three.

3. **Valerianella chenopodiifolia** (Pursh) DC. GOOSEFOOT CORN-SALAD

Annual, to 6 dm tall; stems erect; bracts and bractlets of inflorescence glabrous and without cilia; corolla 3–5 mm long, the lobes white; achenes more or less symmetric, 3–4 mm long, the one at the base of the cyme maturing early, reaching its maximum length while flowers on upper branches of the same cyme are in anthesis. Chiefly in the unglaciated counties of east-central and southeastern Ohio, and at scattered sites elsewhere in the state; moist wooded floodplains, terraces, and slopes, and other moist woodlands. Additional record: Miami County (Ware, 1969). May–early June.

Dipsacaceae. TEASEL FAMILY

A small family of fewer than 200 species, native to the Old World and represented in the Ohio flora by a single genus. Ferguson (1965) studied the family in southeastern United States. The treatment below is based in part on Hauser's (1963) study of the family in Ohio.

An adventive plant of the Old World ornamental, *Scabiosa columbaria* L., SMALL SCABIOUS, was collected from Highland County in 1926. At first misidentified, this specimen was the basis for Schaffner's (1932) record of *Scabiosa atropurpurea* L. (see Hauser, 1963).

1. Dipsacus L. TEASEL

Biennial herbs; stems prickly; leaves opposite, sessile and often connate-perfoliate; flowers perfect, borne in dense many-flowered heads, the heads subtended by prickly involucral bracts; calyx more or less 4-lobed; corolla 4-lobed, zygomorphic; stamens 4, attached to the corolla tube; carpels 2, but only one functional, fused to form a compound pistil with a single style; ovary inferior; fruit an achene. A genus of about 12–15 species, native to Eurasia and northern Africa. The nomenclature adopted here is that of Hansen (1976) in *Flora Europaea.*

Dipsacus sativus (L.) Honck. (*D. fullonum* of authors, not L.), FULLER'S TEASEL, has been collected as a rare adventive from Clinton and Ross counties, both in southwestern Ohio.

a. Cauline leaves irregularly pinnately lobed or divided; involucral bracts curved upward, but the longest shorter than to equalling the top of the inflorescence; corolla white to very pale pink. . . . . 2. *Dipsacus laciniatus*

aa. Cauline leaves unlobed, entire to crenate-serrate or crenate-dentate; involucral bracts spreading or curved upward, if all bracts curved upward, the longest surpassing the inflorescence; corolla lavender to pinkish-purple.

b. Receptacular scales slightly exceeding length of flowers at anthesis, the scales terminated by a straight spine; the involucral bracts curved upward, the longest surpassing the inflorescence . . . . 1. *Dipsacus fullonum*

bb. Receptacular scales about equalling length of flowers at anthesis, the scales terminated by a recurved spine; the involucral bracts mostly spreading; see above . . . . *Dipsacus sativus*

1. DIPSACUS FULLONUM L. COMMON TEASEL. WILD TEASEL
incl. *D. sylvestris* Huds.

Biennial, to 2 m tall; stems erect; leaf-margins entire to crenate-serrate or crenate-dentate, but not lobed; the involucral bracts curved upward, the longest surpassing the inflorescence; corolla lavender to pinkish-purple. A Eurasian species naturalized throughout Ohio, the tough dead stems with their conspicuous fruiting heads standing through the winter (and often cut for ornamental arrangements of dried plant materials); roadsides, fields, waste lots, disturbed ground along railroad tracks, and other disturbed open habitats. Late June–August (–October).

Ferguson and Brizicky (1965) and Brummitt (1986) have notes on the scientific name of this species.

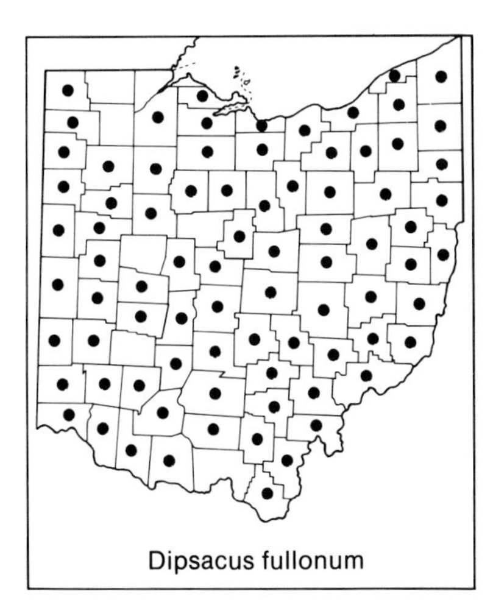
Dipsacus fullonum

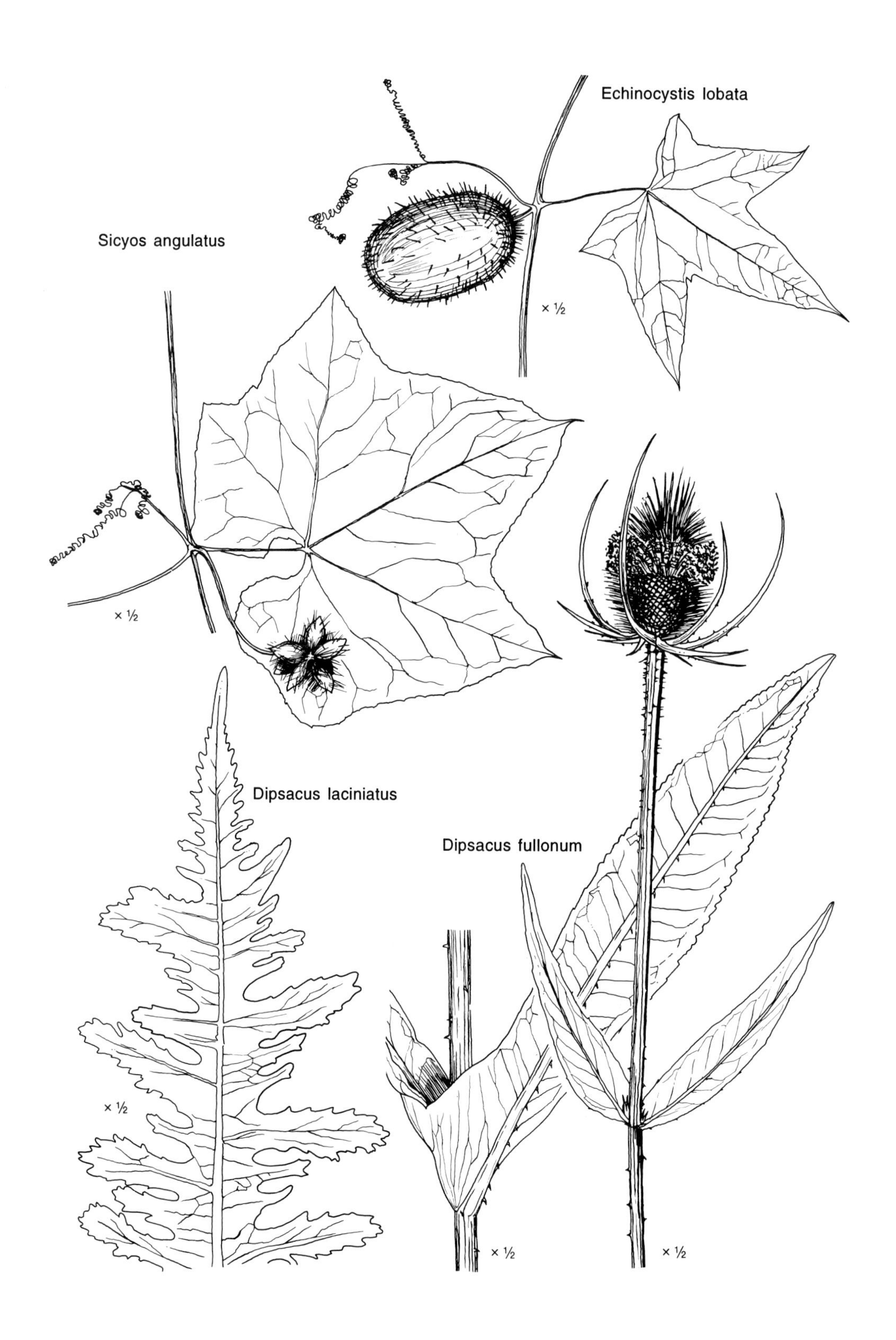
Echinocystis lobata
Sicyos angulatus
× ½
× ½
Dipsacus laciniatus
Dipsacus fullonum
× ½
× ½
× ½

2. Dipsacus laciniatus L.   Cut-leaved Teasel

Biennial, to 2 m tall; stems erect; leaf margins irregularly pinnately lobed or divided; the involucral bracts curved upward, but the longest shorter than or at most equalling the top of the inflorescence; corolla white to very pale pink. A Eurasian species, naturalized in western and a few eastern counties; roadsides, weedy fields, disturbed ground along railroads, and similar open disturbed habitats. Mid-July–October.

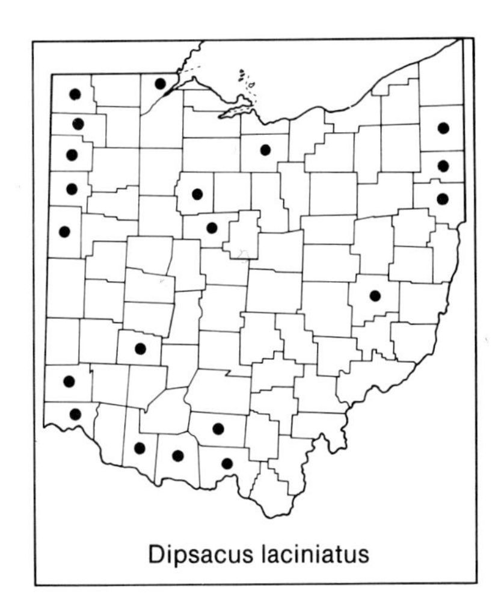
Dipsacus laciniatus

Cucurbitaceae. CUCUMBER FAMILY. GOURD FAMILY

Annual herbaceous vines, climbing by tendrils; leaves alternate, simple; flowers imperfect, the plants monoecious; flowers solitary or borne in various types of inflorescences; calyx 5-lobed; corolla 5- or 6-lobed, actinomorphic; staminate flowers with 5 stamens cohering by their anthers, and sometimes with a vestigial pistil; pistillate flowers with 2–4 carpels, fused to form a compound pistil with a single style and 2–4 stigmas; ovary inferior; fruit a firm or inflated berry (called a "pepo"), often bristly, sometimes becoming dry and dehiscent. A small chiefly tropical family with two species native to the Ohio flora.

Several species of the family are commonly cultivated in Ohio, almost all of which have been collected as, and are to be expected as, adventives in disturbed areas, especially near cultivated plots. They include the following, listed with the counties from which they have been collected: *Citrullus lanatus* (Thunb.) Matsumura & Nakai (*C. vulgaris* Schrad.), Watermelon, from Erie, Jackson, Jefferson, Lorain, Lucas, Meigs, Mercer, Ottawa, Portage, and Ross counties; *Cucumis melo* L., Muskmelon, from Ashtabula, Auglaize, Clinton, Erie, Franklin, Holmes, Jefferson, Knox, and Lucas counties; *Cucurbita foetidissima* HBK., Missouri Gourd or Wild Pumpkin, from Fairfield and Franklin counties; *Cucurbita maxima* Duchesne, Squash, from Erie, Franklin, Holmes, and Wood counties; and *Cucurbita pepo* L., Pumpkin, from Brown, Butler, Holmes, Jackson, Knox, and Licking counties. I have seen no collections of *Cucumis sativus* L., Cucumber, but have observed adventive plants in Ohio. In addition, *Cucurbita lagenaria* L. (now generally called *Lagenaria siceraria* (Molina) Standl.), Gourd, was reported for Ohio by Schaffner (1932), but I have seen no specimens.

a. Flowers yellow; fruits smooth at maturity, but sometimes slightly prickly when immature.
  b. Leaves pinnately lobed; flowers 1.5–4 cm wide; see above. . . . . *Citrullus*

bb. Leaves unlobed or palmately lobed.
c. Flowers 5–12 cm wide; tendrils branched; see above .... *Cucurbita*
cc. Flowers 1–2.5 cm wide; tendrils unbranched; see above .... *Cucumis*
aa. Flowers white or greenish-white; fruits prickly at maturity.
b. Corolla 5-lobed; staminate flowers in corymbose racemes; pistillate flowers in long-peduncled clusters; fruits ovoid, 1.0–1.5 cm long, not inflated, the wall thick, firm, and indehiscent .... 1. *Sicyos*
bb. Corolla 6-lobed; staminate flowers in narrow elongated panicles or racemes; pistillate flowers solitary or in short-peduncled clusters, axillary; fruits ellipsoid, 4–5 cm long, inflated, the wall thin and papery, dehiscing at apex .... 2. *Echinocystis*

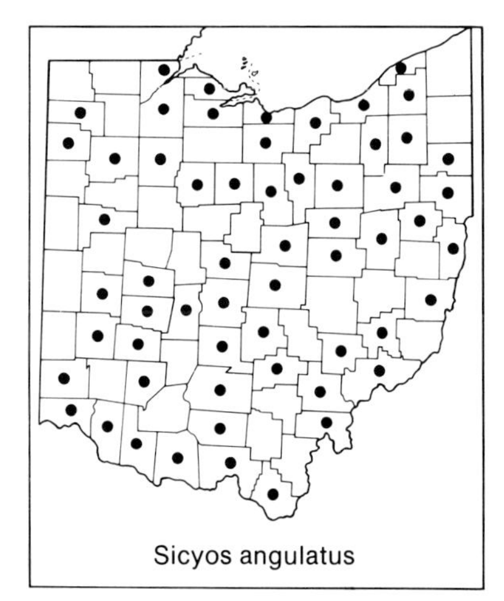
Sicyos angulatus

1. Sicyos L.

A genus of some 30 species, of the New World, Australia, and southeastern Asia, with one species native to Ohio.

1. **Sicyos angulatus** L. BUR-CUCUMBER. STAR-CUCUMBER

Annual herbaceous vine, climbing by tendrils; leaves palmately veined and shallowly palmately lobed; staminate flowers in corymbose racemes; pistillate flowers in long-pedunculed clusters; corolla white to greenish-white, 5-lobed; fruits indehiscent, ovoid, 1.0–1.5 cm long, covered with prickly spines. Throughout most of Ohio; river banks, floodplains, wet thickets, fields, roadsides. Mid-August–mid-September.

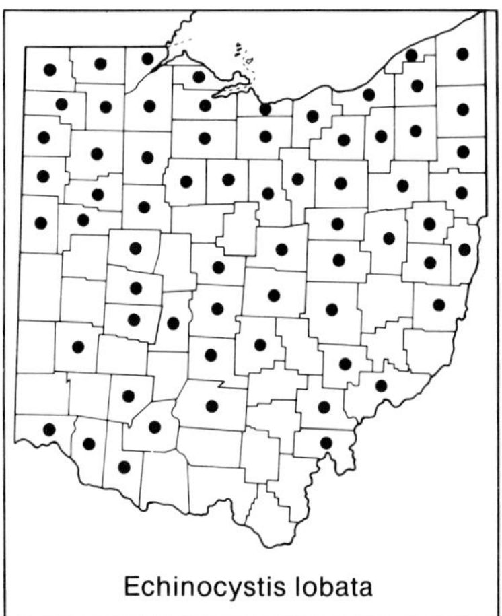
Echinocystis lobata

2. Echinocystis T. & G.

A genus of a single species.

1. **Echinocystis lobata** (Michx.) T. & G. WILD CUCUMBER. BALSAM-APPLE

Annual herbaceous vine, climbing by tendrils; leaves palmately veined and moderately to deeply palmately lobed; corolla white to greenish-white, 6-lobed; staminate flowers in narrow elongated panicles or racemes; pistillate flowers solitary or in short-peduncled clusters, corolla white to greenish-white, 6-lobed; fruits ellipsoid, inflated, the wall thin and papery, dehiscing at apex, 4–5 cm long, covered with weak prickles. Throughout northern Ohio and also in many southern counties; river banks, wet thickets and woodlands, moist roadsides. August–mid-September.

Campanulaceae. BELLFLOWER FAMILY

Annuals, winter annuals, or perennials; leaves alternate, simple; flowers perfect, solitary or in various types of inflorescences; calyx usually 5-lobed; corolla 5-lobed, actinomorphic or zygomorphic; stamens 5, separate or fused apically, attached or not to the corolla tube; carpels 2 or 3, fused to form a compound pistil with a single style; ovary inferior or half-inferior; fruit a capsule. A widely distributed family of about 2,000 species. Shetler and Morin (1986) described the seed morphology of North American members of this family. Rosatti (1986) studied the family in southeastern United States, Cruden (1962) the Ohio members.

Platycodon grandiflorus (Jacq.) A. DC., BALLOON-FLOWER, is a showy, blue-flowered, summer-blooming perennial from eastern Asia, often grown in Ohio flower gardens. Some of its cultivars have white or pinkish flowers.

The genus *Lobelia* was placed in a segregate family, the Lobeliaceae, by Gleason (1952b) and Gleason and Cronquist (1963), but is retained here within the Campanulaceae sensu lato, following the classifications of both Cronquist (1968, 1981) and Thorne (1968, 1976, 1983).

a. Corolla zygomorphic; anthers and terminal half of filaments fused . . . . 3. *Lobelia*

aa. Corolla actinomorphic or nearly so; filaments and anthers free.

  b. Cauline leaves about as wide as long, sessile and cordate-clasping at base; corolla more or less rotate . . . . 1. *Triodanis*

  bb. Cauline leaves all or mostly two or more times as long as wide, petiolate or tapering but not clasping at base; corolla rotate, campanulate, or funnelform. . . . . 2. *Campanula*

1. Triodanis Raf.

A genus of nine species, one of the Mediterranean region, the others of the Western Hemisphere, and one a member of the flora of Ohio. Fernald (1946) favored merging this genus under *Specularia* Fabr., but McVaugh (1945, 1948) argued convincingly that *Triodanis* is worthy of generic rank.

1. **Triodanis perfoliata** (L.) Nieuwl. VENUS' LOOKING-GLASS
   Specularia perfoliata (L.) A. DC.

Winter annual, to 8 dm tall; stems erect to ascending; cauline leaves 1–3 cm long, ovate to orbicular or reniform, about as wide as or wider than long, sessile and cordate-clasping at base; corolla purplish-blue, actinomorphic and more or less rotate, 1–2 cm across. Mostly in eastern and southern Ohio; woodland openings and borders, fields, roadsides. Mid-May–early July.

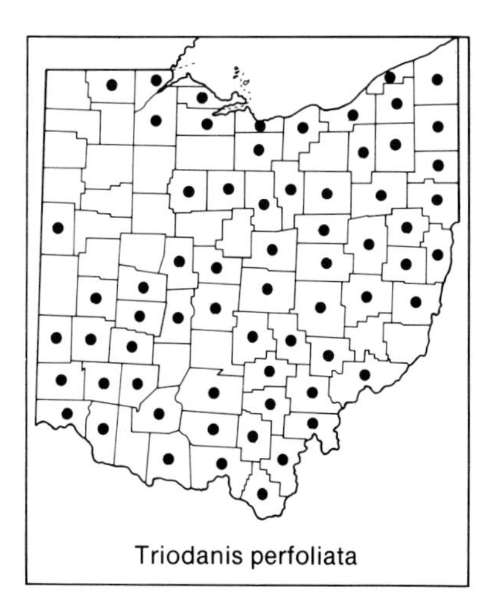
Triodanis perfoliata

2. Campanula L. BELLFLOWER. HAREBELL

Annual, biennial, or perennial herbs; leaves alternate, simple; corolla blue to less often white, actinomorphic, campanulate to funnelform or rotate; stamens separate; carpels 3, stigmas 3. A genus of the Northern Hemisphere with about 300 species. Shetler (1963) published a checklist and key for the North American species. In addition to the species covered below, adventive specimens of *Campanula trachelium* L., NETTLE-LEAVED BELLFLOWER, a weed of Eurasian origin, have been collected from Lorain and Pickaway counties.

a. Lower cauline leaves lanceolate to ovate or oblong-ovate, 1.5–7 cm wide, serrate or crenate-serrate.

b. Most flowers sessile or nearly so, in a terminal spike or spicate raceme; corolla rotate. . . . . 2. *Campanula americana*

bb. Most flowers on peduncles or pedicels (0.3–) 5–25 (–40) mm long, either in a terminal raceme or in few-flowered clusters or some solitary; corolla campanulate or campanulate-funnelform.

c. Flowers in loose clusters of (1–) 2–4 flowers each, these terminal on both main stem and axillary branches; calyx lobes ascending at anthesis; see above . . . . *Campanula trachelium*

cc. Most or all flowers in a terminal raceme, the others solitary; calyx lobes spreading or reflexed at anthesis. . . . . 1. *Campanula rapunculoides*

aa. Lower cauline leaves linear to narrowly lanceolate or narrowly elliptic, 0.1–0.7 (–1.2) cm wide, entire or low-serrate.

b. Corolla bright purplish-blue, 1.5–2.5 cm long; calyx lobes linear, 4–7 (–10) mm long; stem glabrous. . . . . 3. *Campanula rotundifolia*

bb. Corolla pale blue to whitish, 0.4–1.2 cm long; calyx lobes lanceolate-triangular, 0.7–3 mm long; stem minutely and retrorsely scabrous . . . . 4. *Campanula aparinoides*

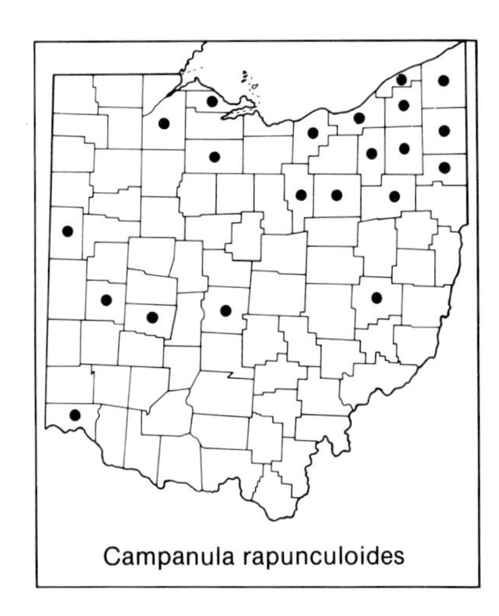
Campanula rapunculoides

1. CAMPANULA RAPUNCULOIDES L. EUROPEAN BELLFLOWER. CREEPING BELLFLOWER

incl. var. *rapunculoides* and var. *ucranica* (Bess.) K. Koch

Perennial, to 1 m tall; stems erect; lower cauline leaves lanceolate to ovate, margins crenate-serrate; most or all flowers in a terminal raceme, the others solitary in axils of uppermost leaves, pedicels or peduncles 2–25 mm long; corolla bluish-purple, campanulate-funnelform, 2–3 cm long, 2–3 cm across, nodding at anthesis. A cultivated Eurasian species, escaped and naturalized in northeastern counties and at scattered sites elsewhere in Ohio; along roadsides and railroad tracks, and in other disturbed habitats. Mid-June–October.

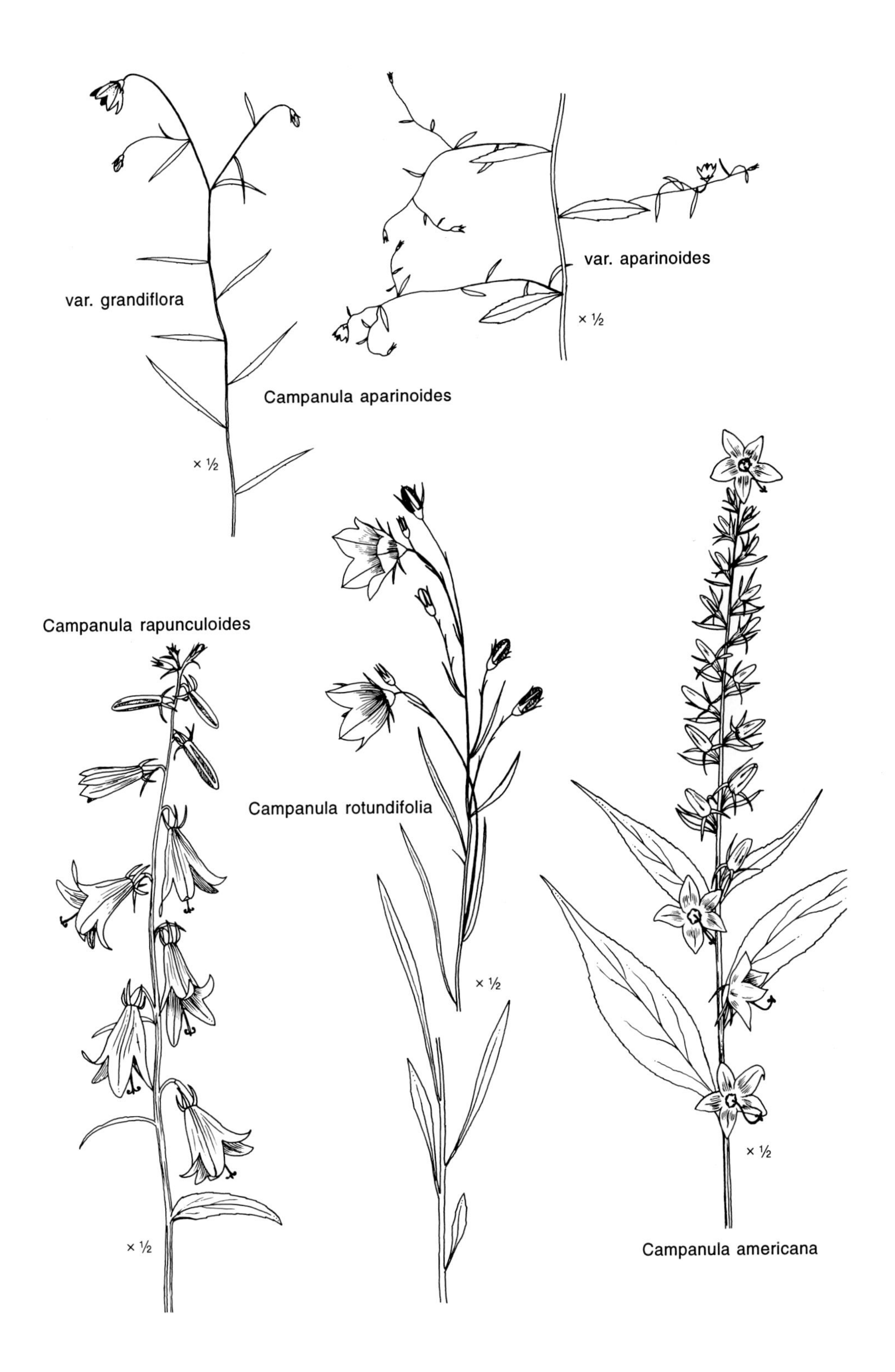
var. grandiflora
var. aparinoides
× ½
Campanula aparinoides
× ½
Campanula rapunculoides
Campanula rotundifolia
× ½
× ½
× ½
Campanula americana

2. **Campanula americana** L. TALL BELLFLOWER

incl. var. *americana* and var. *illinoensis* (Fresn.) Farw.

Annual or biennial, to 2 m tall; stems erect; lower cauline leaves lanceolate to ovate or oblong-ovate, margins serrate to crenate-serrate; most flowers sessile or nearly so in a terminal spike or spicate-raceme, the others short-pedunculate and solitary or few in the axils of uppermost leaves, frequently the plant's uppermost flower opening at the same time as the lowest; corolla blue to pale blue, rotate, 2–3 cm across. Throughout Ohio; woods, woodland openings and borders, thickets, roadsides. Baskin and Baskin (1984a) studied the ecological life cycle of this species. June–mid-September.

3. **Campanula rotundifolia** L. HAREBELL

Perennial; stems ascending to erect, to 5 dm long, glabrous; lower cauline leaves linear to narrowly elliptic, margins entire to low-serrate; flowers in open terminal racemes and/or solitary in axils of upper leaves; calyx lobes linear, 4–7 (–10) mm long; corolla purplish-blue, campanulate, 15–25 mm long, 15–20 mm across. A boreal species ranging across North America, along a southern boundary of its range at a few scattered sites in western Ohio; calcareous cliffs and rock exposures. Late June–early September.

4. **Campanula aparinoides** Pursh MARSH BELLFLOWER

Perennial; stems weak, suberect to ascending or reclining, to 1 m long, minutely and retrorsely scabrous; lower cauline leaves linear to lanceolate or narrowly elliptic, entire to low-serrate; flowers solitary in the axils of upper leaves; calyx lobes lanceolate-triangular, 0.7–3.0 mm long; corolla pale blue to whitish, campanulate, 4–12 mm long, 4–12 mm across. Bogs,

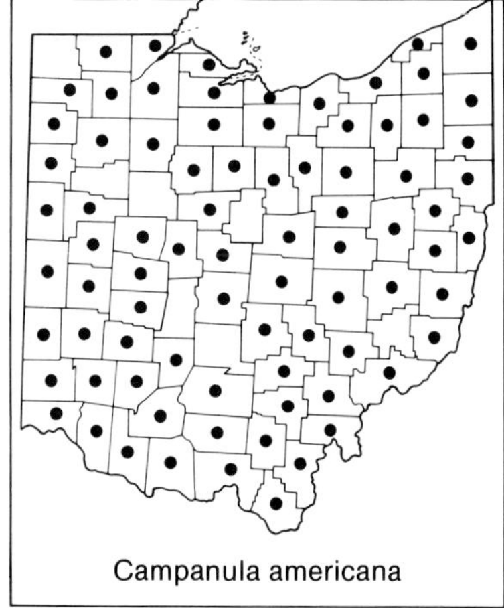

Campanula americana

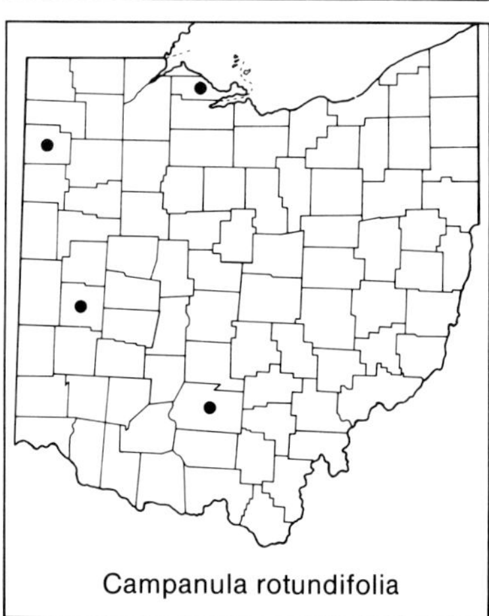

Campanula rotundifolia

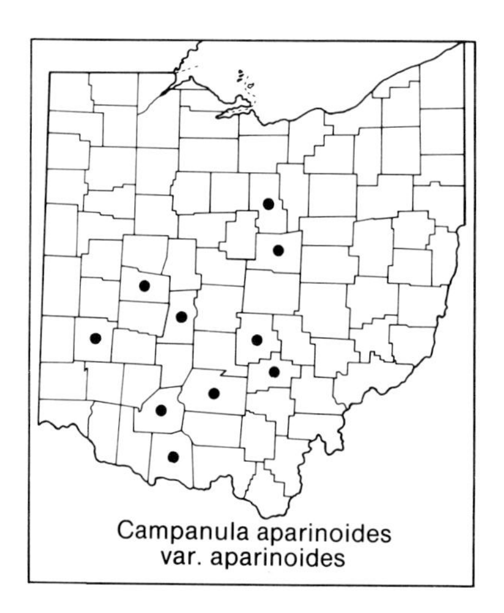

Campanula aparinoides var. aparinoides

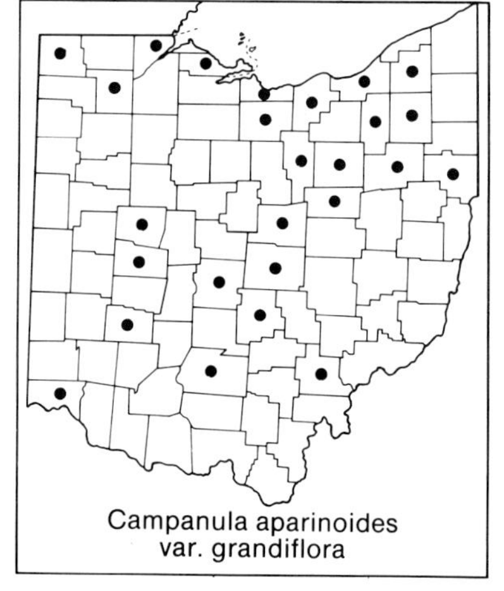

Campanula aparinoides var. grandiflora

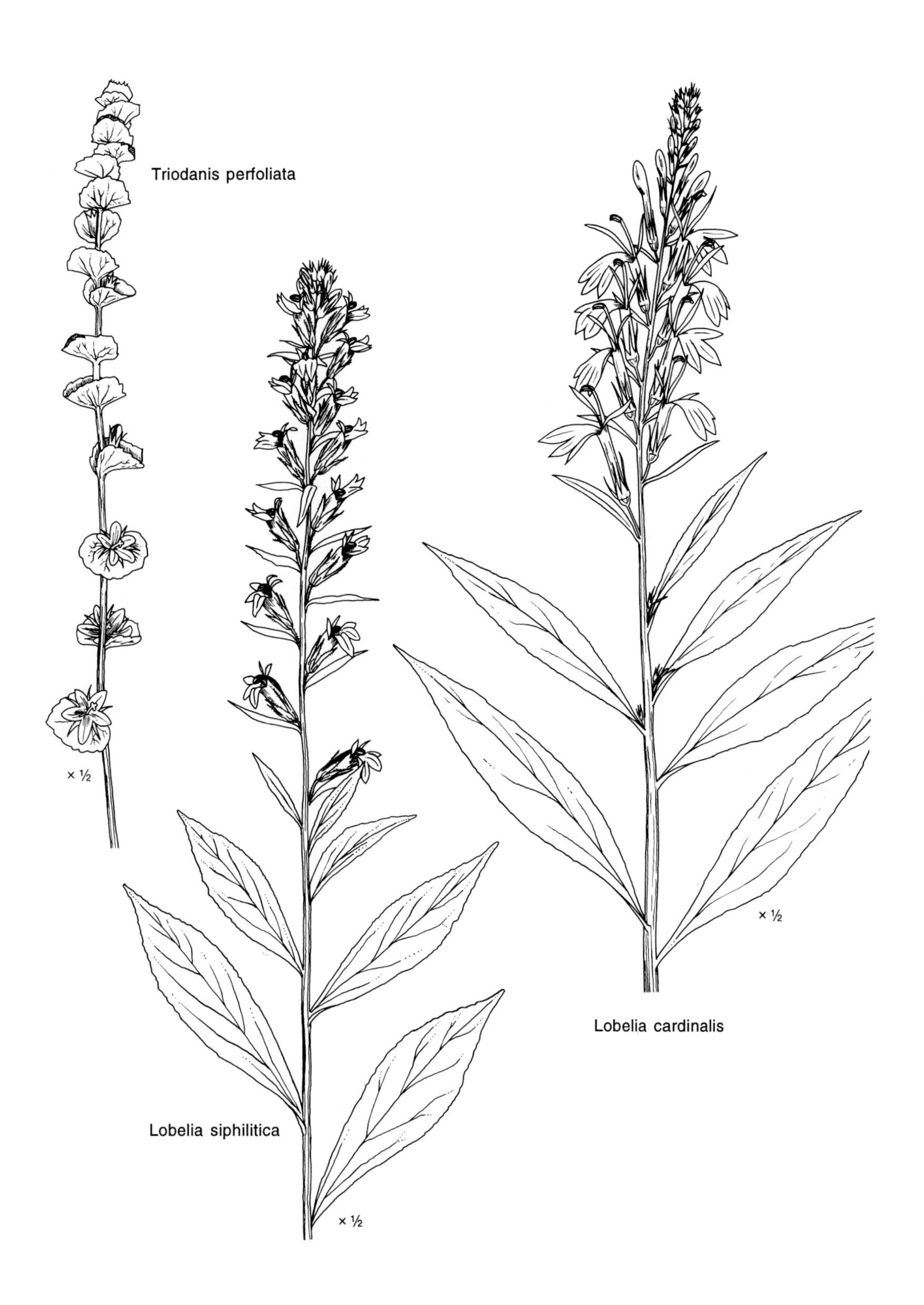
Triodanis perfoliata
× ½
Lobelia siphilitica
× ½
× ½
Lobelia cardinalis

marshes, swamps, shores, and other wet habitats. Late June–early September.

Two intergrading varieties, treated by Fernald (1950) as separate species, occur in Ohio. Variety *aparinoides* is known from central and southwestern counties, var. *grandiflora* from northeastern and some southern, and some western counties. The length of the calyx lobes, often used to separate the two taxa, seems to have little or no value in separating Ohio plants.

a. Peduncles (resembling thin lateral branches) strongly and irregularly divergent and bearing leafy bracts; corolla 4–9 mm long, whitish . . . . var. *aparinoides*

aa. Peduncles (not resembling lateral branches) more or less ascending and bractless; corolla 8–12 mm long, pale blue . . . . var. *grandiflora* Holz. *C. uliginosa* Rydb.

3. Lobelia L. LOBELIA

Annual or perennial herbs; leaves alternate, simple; flowers in bracteate racemes or rarely solitary in axils of upper leaves; corolla red or blue to white, rarely pinkish, zygomorphic and bilabiate, the tube split to the base on one side; stamens fused above; carpels 2. A widely distributed genus of 350 or more species. *Lobelia erinus* L., Edging Lobelia, a small annual less than 20 cm tall, is often used as a border plant in flower gardens; it blooms profusely with blue, violet, or white flowers throughout the growing season.

a. Flowers 3–4 cm long; corolla bright red (rarely white); calyx lobes 8–12 mm long, glabrous. . . . . 1. *Lobelia cardinalis*

aa. Flowers 0.7–3 cm long; corolla blue to white or rarely pinkish; calyx lobes either 5–12 mm long and more or less hirsute or 1–5 mm long and glabrous.

b. Flowers 1.5–3 cm long; calyx lobes more or less hirsute, 5–12 mm long.

c. Flowers 2.3–3 cm long; calyx with broad conspicuous basal auricles; stems glabrous to sparsely hirsute . . . . 2. *Lobelia siphilitica*

cc. Flowers 1.5–2 cm long; calyx with inconspicuous basal auricles or none; stems densely short-hirsute . . . . 3. *Lobelia puberula*

bb. Flowers 0.7–1.2 (–1.5) cm long; calyx lobes glabrous, 1–5 mm long.

c. Medial (and upper) cauline leaves linear to broadly linear, 0.5–3 mm wide; lower lip of corolla glabrous at base . . . 6. *Lobelia kalmii* cc. Medial cauline leaves lanceolate to narrowly elliptic, elliptic, oblanceolate, oblong, obovate, or ovate, 5–35 mm wide; lower lip of corolla pubescent at base.

d. Lowest internodes of main stem stiffly short-hirsute with hairs divergent to deflexed; leaves mostly entire to subentire; calyx not inflated . . . . 4. *Lobelia spicata*

dd. Lowest internodes of main stem pubescent with hairs divergent and variously bent; leaves crenate-serrate; calyx becoming conspicuously inflated at maturity . . . . 5. *Lobelia inflata*

1. **Lobelia cardinalis** L. CARDINAL-FLOWER

Perennial, to ca. 1 m tall; stems erect, glabrous to pubescent; leaves mostly lanceolate to elliptic-lanceolate; flowers showy, in a congested terminal raceme; calyx lobes 8–12 mm long, glabrous; corolla bright red (rarely white), 3–4 cm long. One of Ohio's most resplendent wildflowers; throughout much of the state; stream and river banks, marshes, wet fields. Bowden (1982) discusses the infraspecific taxonomy of this species, and Devlin and Stephenson (1985) its pollination. Mid-July–mid-September.

The rare white-flowered forma *alba* (McNab) St. John has been collected from Franklin County, in central Ohio; Pringle (1988) and E. C. Nelson (1988) comment on the authorship of this name. A more or less purple-flowered interspecific hybrid between the red-flowered forms of this species and blue-flowered forms of the next species, *L. siphilitica*, has been reported from Illinois (Bowden, 1982; Ebinger, 1985).

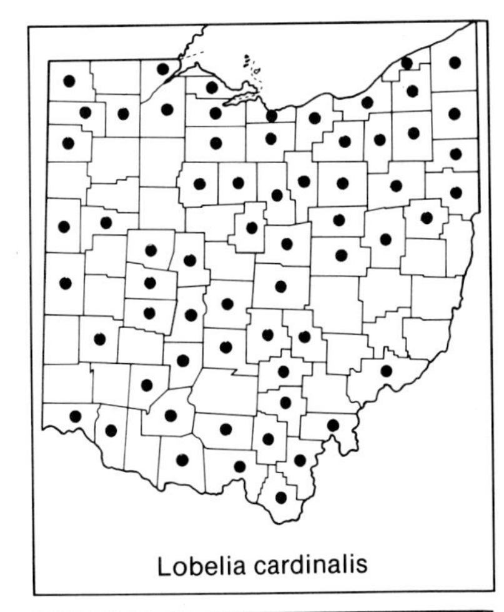
Lobelia cardinalis

2. **Lobelia siphilitica** L. GREAT BLUE LOBELIA. GREAT LOBELIA

Perennial, to ca. 1 m tall; stems erect, glabrous to sparsely hirsute; leaves mostly oblanceolate to elliptic; flowers showy, in a congested terminal raceme; calyx with broad conspicuous basal auricles, the calyx lobes 8–12 mm long, more or less hirsute; corolla bright blue (rarely white), 2.3–3.0 cm long. A showy wildflower of late summer and early autumn growing throughout Ohio; low damp ground along streams, rivers, ponds, and lakes; and in marshes, swamps, wet fields, and wet roadside ditches; sometimes cultivated. Bowden (1982) studied the infraspecific taxonomy of this species. Mid-July–October.

The white-flowered forma *albiflora* Britt. has been collected from Clinton, Columbiana, Coshocton, Highland, Hocking, Lawrence, and Wayne counties. See notes above, under *L. cardinalis*, on the hybrid between that species and this.

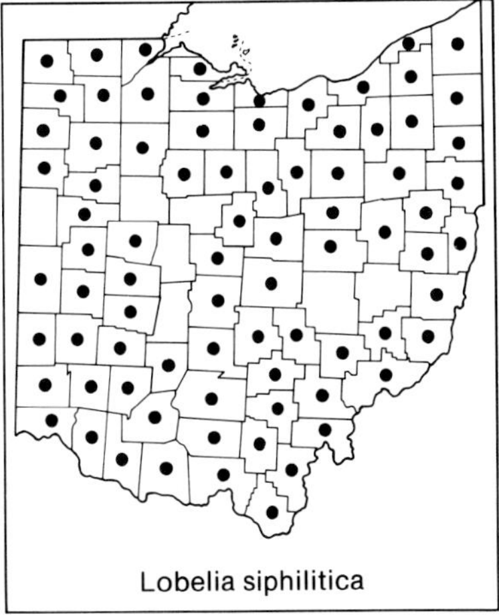
Lobelia siphilitica

3. **Lobelia puberula** Michx. DOWNY LOBELIA

incl. var. *puberula* and var. *simulans* Fern.

Perennial, to 1 m tall; stems erect, densely short-hirsute; leaves oblanceolate to elliptic; flowers somewhat showy, in a congested terminal raceme;

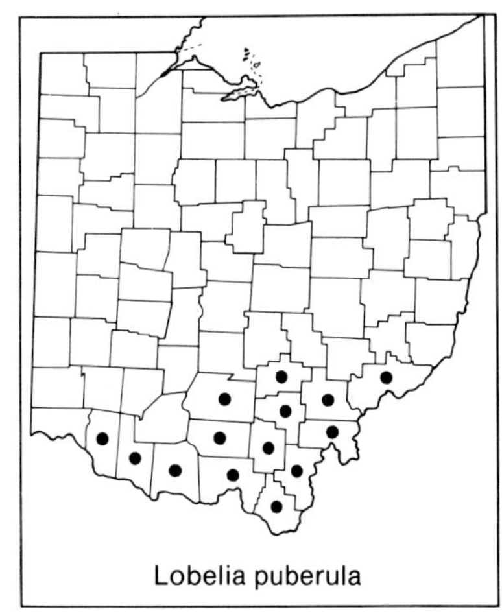
Lobelia puberula

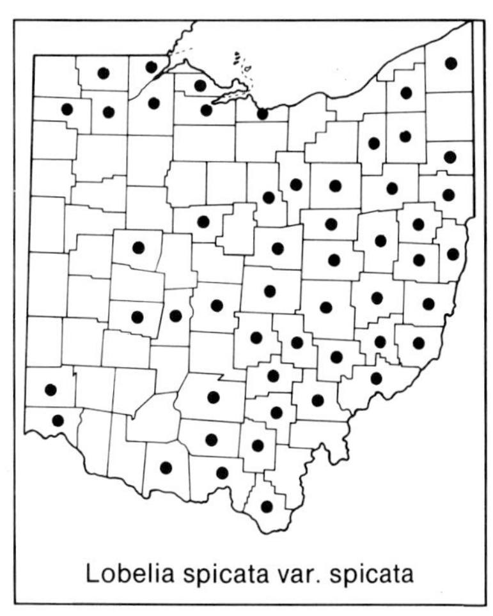
Lobelia spicata var. spicata

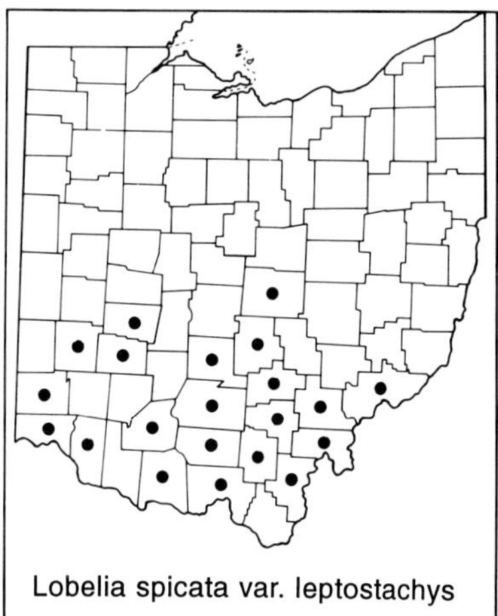
Lobelia spicata var. leptostachys

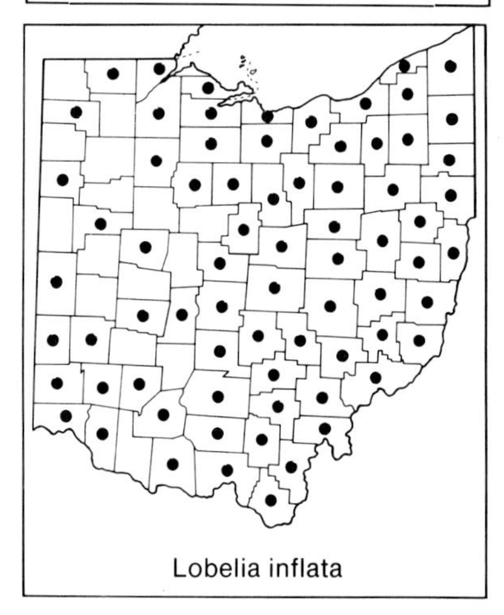
Lobelia inflata

calyx with inconspicuous basal auricles or none, the calyx lobes 5–10 mm long, slightly hirsute; corolla blue, 1.5–2.0 cm long. Southern counties; dry woodland openings and borders, dry roadsides. Mid-July–mid-October.

Following the taxonomy of Gleason (1952b), no varieties are recognized within this species.

4. **Lobelia spicata** Lam. Pale-spike Lobelia

Perennial, to 1 m tall; stems erect, the lowest internodes stiffly short-hirsute with hairs divergent to deflexed, the upper internodes glabrous or nearly so; lower leaves oblanceolate, middle and upper leaves lanceolate to elliptic, all with margins mostly entire to subentire; flowers in a crowded terminal raceme; calyx with or without auricles, the calyx lobes 3–4 mm long, glabrous; corolla blue to white, 7–12 mm long, lower lip pubescent at base. Throughout most of Ohio, but absent from west-central counties; dry woods, woodland openings and borders, rock exposures, dry roadside banks, grassy fields. June–August.

Most Ohio specimens can rather readily be assigned to one of the two varieties keyed below. The two have somewhat differing geographical ranges in Ohio: var. *spicata* occurs throughout eastern and most of northern Ohio, var. *leptostachys* is found only in the southern half of the state. However, there is some overlap in range, and I found specimens from Butler and Hamilton counties, in southwestern Ohio, with individuals of both varieties on the same herbarium sheet, suggesting that the two were growing together in a mixed population. Furthermore, a few specimens are intermediate in regard to auricle formation and length, the main diagnostic character separating the two. Following McGregor's (1985) conclusions, var. *campanulata* is merged here without rank under var. *spicata*.

a. Calyx lacking auricles or the auricles minute, less than 1 mm long . . . . var. *spicata*
incl. var. *campanulata* McVaugh

aa. Calyx with evident filiform auricles, mostly 2–3 mm long. . . . . var. *leptostachys* (A. DC.) Mackenz. & Bush.

5. **Lobelia inflata** L. Indian Tobacco

Annual, to 1 m tall; stems erect, pubescent with hairs divergent to variously bent; leaves oblanceolate to elliptic, oblong, ovate, or obovate, with margins crenate-serrate; flowers in terminal and axillary racemes, and occasional flowers solitary in axils of upper leaves; calyx lobes 3–5 mm long, glabrous; corolla pale blue to whitish or sometimes pinkish, 7–10 mm long, lower lip pubescent at base; calyx becoming conspicuously inflated at maturity. Throughout most of Ohio, absent from some northwestern counties; woods, woodland openings and borders, marshy areas, grassy fields, cultivated ground, weed lots. July–October.

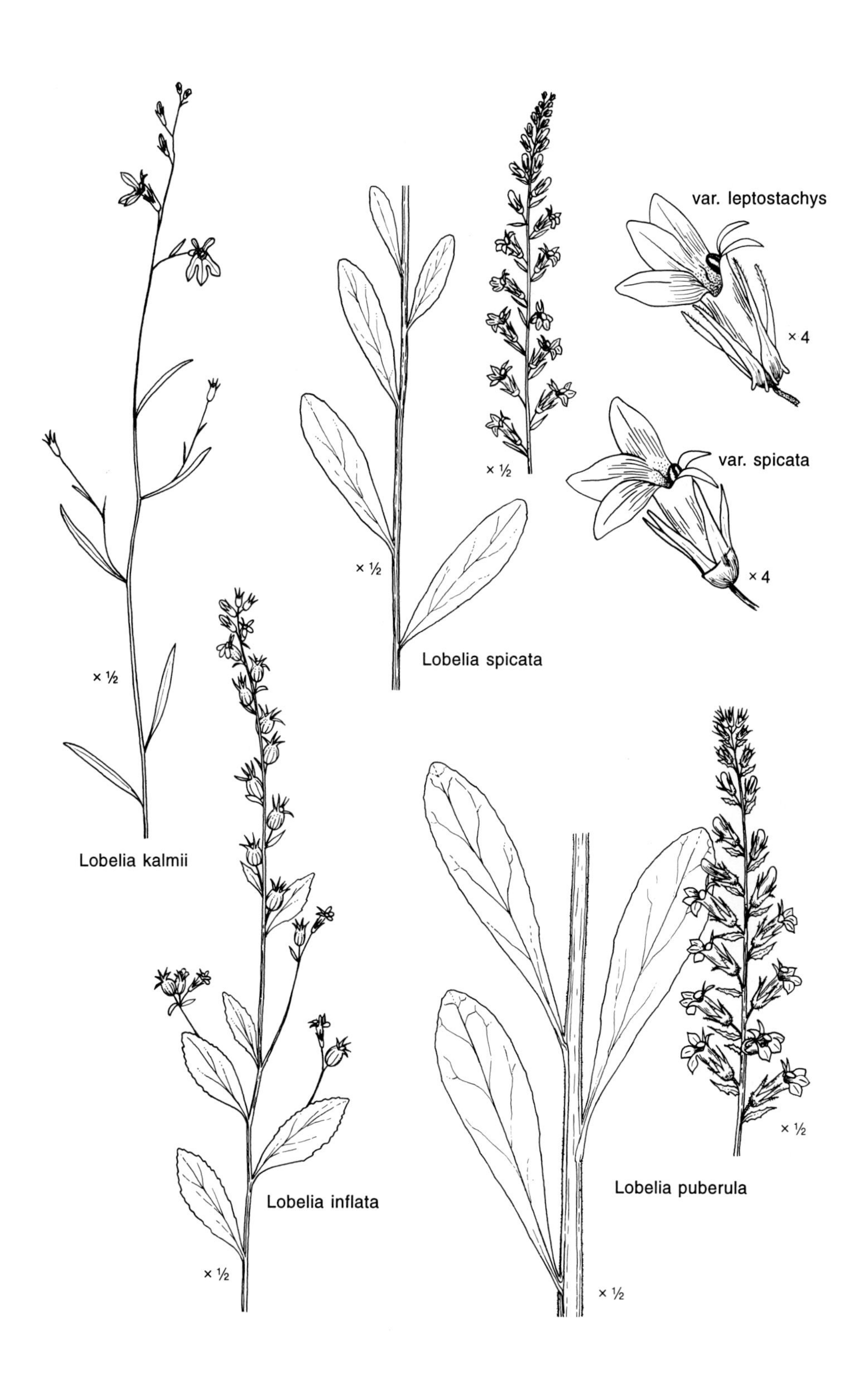
var. leptostachys
× 4
var. spicata
× 4
× ½
× ½
Lobelia spicata
× ½
Lobelia kalmii
Lobelia inflata
× ½
Lobelia puberula
× ½
× ½

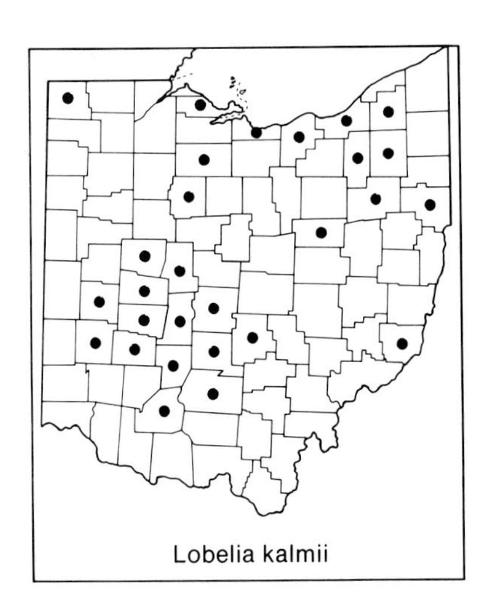
Lobelia kalmii

6. **Lobelia kalmii** L.   Kalm's Lobelia

Perennial, to 6 dm tall; stems erect or sometimes weak and suberect, glabrous or minutely puberulent below; basal leaves when present tiny and elliptic-spatulate, lower cauline leaves linear-elliptic and with a few shallow teeth or crenations, middle and upper cauline leaves linear to broadly linear, 0.5–3 mm wide; flowers in open terminal racemes or solitary in leaf axils; calyx lobes 1–3 mm long, glabrous; corolla pale blue and white-centered, 10–15 mm long, lower lip of corolla glabrous at base. A boreal species, transcontinental across North America, at a southern boundary of its range in Ohio; mostly in northern and central counties; fens, marl bogs, wet shale cliffs. Mid-July–mid-October.

LITERATURE CITED

ABU-ASAB, M. S., and P. D. CANTINO. 1989. Pollen morphology of *Trichostema* (Labiatae) and its systematic implications. Syst. Bot. **14:** 359–369.

ADAMS, J. W. 1976. *Veronica filiformis*: a beautiful weed or a blue-eyed nuisance? The Green Scene **4:** 15–17.

ADAMS, W. P. 1959. The status of *Hypericum prolificum*. Rhodora **61:** 249–251.

———. 1962. Studies in the Guttiferae. I. A synopsis of *Hypericum* section *Myriandra*. Contr. Gray Herb. **189:** 1–51.

———. 1973. Clusiaceae of the southeastern United States. J. Elisha Mitchell Sci. Soc. **89:** 62–71.

ADAMS, W. P., and N. K. B. ROBSON. 1961. A re-evaluation of the generic status of *Ascyrum* and *Crookea* (Guttiferae). Rhodora **63:** 10–16.

AGREN, J., and M. F. WILLSON. 1991. Gender variation and sexual differences in reproductive characters and seed production in gynodioecious *Geranium maculatum*. Amer. J. Bot. **78:** 470–480.

AHLES, H. E. 1968. Linaceae, pp. 645–646; Aquifoliaceae, pp. 697–684; Violaceae, pp. 723–733; Labiatae, pp. 895–925. *In* A. E. Radford, H. E. Ahles, and C. R. Bell. Manual of the vascular flora of the Carolinas. Univ. of North Carolina Press, Chapel Hill. lxi + 1183 pp.

AIKEN, S. G. 1981. A conspectus of *Myriophyllum* (Haloragaceae) in North America. Brittonia **33:** 57–69.

AIKEN, S. G., P. R. NEWROTH, and I. WILE. 1979. The biology of Canadian weeds. 34. *Myriophyllum spicatum* L. Canad. J. Pl. Sci. **59:** 210–215.

AIKEN, S. G., and K. F. WALZ. 1979. Turions of *Myriophyllum exalbescens*. Aquatic Bot. **6:** 357–363.

ANDERSON, E. 1936. An experimental study of hybridization in the genus *Apocynum*. Ann. Missouri Bot. Gard. **23:** 159–167.

ANDERSON, R. C., and M. H. BEARE. 1983. Breeding system and pollination ecology of *Trientalis borealis* (Primulaceae). Amer. J. Bot. **70:** 408–415.

ANDERSON, R. C., and O. L. LOUCKS. 1973. Aspects of the biology of *Trientalis borealis* Raf. Ecology **54:** 798–808.

ANDREAS, B. K. 1989. The flora of the glaciated Allegheny Plateau region of Ohio. Ohio Biol. Surv. Bull. N.S. **8:** i-viii, 1–191.

ANDREAS, B. K., and T. S. COOPERRIDER. 1979. The Apocynaceae of Ohio. Castanea **44:** 238–241.

———. 1980. The Asclepiadaceae of Ohio. Castanea **45:** 51–55.

———. 1981. The Gentianaceae and Menyanthaceae of Ohio. Castanea **46:** 102–108.

ANDRUS, M. E., and R. L. STUCKEY. 1981. Introgressive hybridization and habitat separation in *Lycopus americanus* and *L. europaeus* at the southwestern shores of Lake Erie. Michigan Bot. **20:** 127–135.

ANONYMOUS. 1989. Ohio's big trees. 19th ed. Ohio Forestry Association, Columbus. 27 pp.

ARMSTRONG, J. E. 1985. The delimitation of Bignoniaceae and Scrophulariaceae based on floral anatomy, and the placement of problem genera. Amer. J. Bot. **72:** 755–766.

Arnold, R. M. 1981. Population dynamics and seed dispersal of *Chaenorrhinum minus* on railroad cinder ballast. Amer. Midl. Naturalist **106:** 80–91.

———. 1982a. Pollination, predation and seed set in *Linaria vulgaris* (Scrophulariaceae). Amer. Midl. Naturalist **107:** 360–369.

———. 1982b. Floral biology of *Chaenorrhinum minus* (Scrophulariaceae) a self-compatible annual. Amer. Midl. Naturalist **108:** 317–324.

Ashby, W. C. 1964. A note on basswood nomenclature. Castanea **29:** 109–116.

Austin, D. F. 1978. The *Ipomoea batatas* complex—I. Taxonomy. Bull. Torrey Bot. Club **105:** 114–129.

———. 1986. Nomenclature of the *Ipomoea nil* complex (Convolvulaceae). Taxon **35:** 355–358.

Baas, P. 1984. Vegetative anatomy and the taxonomic status of *Ilex collina* and *Nemopanthus* (Aquifoliaceae). J. Arnold Arbor. **65:** 243–250.

Bailey Hortorium Staff. 1976. Hortus Third. A concise dictionary of plants cultivated in the United States and Canada. Macmillan Publishing Co., New York. xiv + 1290 pp.

Bailey, V. L. 1962. Revision of the genus *Ptelea* (Rutaceae). Brittonia **14:** 1–45.

Bailey, V. L., S. B. Herlin, and H. E. Bailey. 1970. *Ptelea trifoliata* subsp. *trifoliata* (Rutaceae) in deciduous forest regions of eastern North America. Brittonia **22:** 346–358.

Baird, V. B. 1942. Wild violets of North America. Univ. of California Press, Berkeley. xv + 225 pp.

Baird, W. V., and J. L. Riopel. 1986. Life history studies of *Conopholis americana* (Orobanchaceae). Amer. Midl. Naturalist **116:** 140–151.

Baker, H. G. 1961. Heterostyly and homostyly in *Lithospermum canescens* (Boraginaceae). Rhodora **63:** 229–235.

Baker, M. S. 1953a. Studies in western violets. VII. Madroño **12:** 8–18.

———. 1953b. A correction in the status of *Viola Macloskeyi*. Madroño **12:** 60.

Baker, P. C. 1970. A systematic study of the genus *Vaccinium* L. subgenus *Polycodium* (Raf.) Sleumer, in the southeastern United States. Ph.D. dissertation. Univ. of North Carolina, Chapel Hill. xvi + 248 pp.

Bakshi, T. S., and R. T. Coupland. 1960. Vegetative propagation in *Linaria vulgaris*. Canad. J. Bot. **38:** 243–249.

Ball, P. W. 1979. *Thaspium trifoliatum* (meadow-parsnip) in Canada. Canad. Field-Naturalist **93:** 306–307.

Ballard, H. E., Jr. 1989. *Viola* × *eclipes*, a new hybrid violet. Michigan Bot. **28:** 216–219.

———. 1990. Hybrids among three caulescent violets, with special reference to Michigan. Michigan Bot. **29:** 43–54.

———. 1993. Three new rostrate violet hybrids from Appalachia. Castanea **58:** 1–9.

Ballard, H. E., Jr., and R. W. Pippen. 1991. An intersubgeneric hybrid of *Aureolaria flava* and *A. pedicularia*. Michigan Bot. **30:** 59–63.

Balogh, G. R., and T. A. Bookhout. 1989. Purple loosestrife (*Lythrum salicaria*) in Ohio's Lake Erie marshes. Ohio J. Sci. **89:** 62–64.

Barber, S. C. 1982. Taxonomic studies in the *Verbena stricta* complex (Verbenaceae). Syst. Bot. **7:** 433–456.

Barkley, F. A. 1937. A monographic study of *Rhus* and its immediate allies in

North and Central America, including the West Indies. Ann. Missouri Bot. Gard. **24:** 265–466.
Baskin, J. M., and C. C. Baskin. 1972. Physiological ecology of germination of *Viola rafinesquii.* Amer. J. Bot. **59:** 981–988.
———. 1976a. Some aspects of the autecology and population biology of *Phacelia purshii.* Amer. Midl. Naturalist **96:** 431–442.
———. 1976b. Germination ecology of winter annuals: *Valerianella umbilicata,* f. *patellaria* and f. *intermedia.* J. Tennessee Acad. Sci. **51:** 138–141.
———. 1977. Germination of common milkweed (*Asclepias syriaca* L.) seeds. Bull. Torrey Bot. Club **104:** 167–170.
———. 1978. Role of temperature in the regulation of the life cycle of the winter annual *Phacelia dubia* var. *dubia* in Tennessee cedar glades. J. Tennessee Acad. Sci. **53:** 118–120.
———. 1979. Studies on the autecology and population biology of the weedy monocarpic perennial, *Pastinaca sativa.* J. Ecol. **67:** 601–610.
———. 1981. Seasonal changes in the germination responses of buried *Lamium amplexicaule* seeds. Weed Res. **21:** 299–306.
———. 1982a. Ecological life cycle and temperature relations of seed germination and bud growth of *Scutellaria parvula.* Bull. Torrey Bot. Club **109:** 1–6.
———. 1982b. Temperature relations of seed germination in *Ruellia humilis,* and ecological implications. Castanea **47:** 119–131.
———. 1983a. Germination ecology of *Collinsia verna,* a winter annual of rich deciduous woodlands. Bull. Torrey Bot. Club **110:** 311–315.
———. 1983b. Germination ecophysiology of eastern deciduous forest herbs: *Hydrophyllum macrophyllum.* Amer. Midl. Naturalist **109:** 63–71.
———. 1983c. Seasonal changes in the germination responses of seeds of *Veronica peregrina* during burial, and ecological implications. Canad. J. Bot. **61:** 3332–3336.
———. 1983d. Germination ecology of *Veronica arvensis.* J. Ecol. **71:** 57–68.
———. 1984a. The ecological life cycle of *Campanula americana* in northcentral Kentucky. Bull. Torrey Bot. Club **111:** 329–337.
———. 1984b. Germination ecophysiology of the woodland herb *Osmorhiza longistylis* (Umbelliferae). Amer. J. Bot. **71:** 687–692.
———. 1984c. Effect of temperature during burial on dormant and non-dormant seeds of *Lamium amplexicaule* L. and ecological implications. Weed Res. **24:** 333–339.
———. 1984d. Role of temperature in regulating timing of germination in soil seed reserves of *Lamium purpureum* L. Weed Res. **24:** 341–349.
———. 1984e. Environmental conditions required for germination of prickly sida (*Sida spinosa*). Weed Sci. **32:** 786–791.
———. 1985. Germination ecophysiology of *Hydrophyllum appendiculatum,* a mesic forest biennial. Amer. J. Bot. **72:** 185–190.
———. 1988. The ecological life cycle of *Cryptotaenia canadensis* (L.) DC. (Umbelliferae), a woodland herb with monocarpic ramets. Amer. Midl. Naturalist **119:** 165–173.
———. 1991. Nondeep complex morphophysiological dormancy in seeds of *Osmorhiza claytonii* (Apiaceae). Amer. J. Bot. **78:** 588–593.
Baskin, J. M., C. C. Baskin, and M. T. McCann. 1988. A contribution to the

germination ecology of *Floerkea proserpinacoides* (Limnanthaceae). Bot. Gaz. **149:** 427–431.

Baskin, J. M., C. C. Baskin, P. D. Parr, and M. Cunningham. 1991. Seed germination ecology of the rare hemiparasite *Tomanthera auriculata* (Scrophulariaceae). Castanea **56:** 51–58.

Baskin, J. M., H. D. Bryan, and C. C. Baskin. 1986. Ecological life cycle of *Synandra hispidula* in northcentral Kentucky. Bull. Torrey Bot. Club **113:** 168–172.

Bassett, I. J. 1973. The plantains of Canada. Canada Dept. Agr. Monogr. 7, Ottawa. 47 pp.

Bassett, I. J., and D. B. Munro. 1985. The biology of Canadian weeds. 67. *Solanum ptycanthum* Dun., *S. nigrum* L., and *S. sarrachoides* Sendt. Canad. J. Pl. Sci. **65:** 401–414.

———. 1986. The biology of Canadian weeds. 78. *Solanum carolinense* L. and *Solanum rostratum* Dunal. Canad. J. Pl. Sci. **66:** 977–991.

Bates, D. M. 1965. Notes on cultivated Malvaceae. 1. *Hibiscus*. Baileya **13:** 56–130.

Baum, B. R. 1967. Introduced and naturalized tamarisks in the United States and Canada (Tamaricaceae). Baileya **15:** 19–25.

Beal, E. O., and J. W. Thieret. 1986. Aquatic and wetland plants of Kentucky. Kentucky Nature Preserves Commission, Frankfort. vii + 313 pp.

Beatley, J. C. 1979. Distribution of buckeyes (*Aesculus*) in Ohio. Castanea **44:** 150–163.

Beckmann, R. L., Jr. 1979. Biosystematics of the genus *Hydrophyllum* L. (Hydrophyllaceae). Amer. J. Bot. **66:** 1053–1061.

Bell, C. R. 1963. The genus *Eryngium* in the southeastern United States. Castanea **28:** 73–79.

———. 1968. Rubiaceae, pp. 978–988. *In* A. E. Radford, H. E. Ahles, and C. R. Bell. Manual of the vascular flora of the Carolinas. Univ. of North Carolina Press, Chapel Hill. lxi + 1183 pp.

Bennett, R. W. 1963. Notes on eastern American species of *Penstemon*. Phytologia **9:** 57–58.

Benson, L. 1982. The cacti of the United States and Canada. Stanford Univ. Press, Stanford. ix + 1044 pp.

Bentz, G. D., and T. S. Cooperrider. 1978. The Scrophulariaceae subfamily Rhinanthoideae of Ohio. Castanea **43:** 145–154.

Best, K. F., G. G. Bowes, A. G. Thomas, and M. G. Maw. 1980. The biology of Canadian weeds. 39. *Euphorbia esula* L. Canad. J. Pl. Sci. **60:** 651–663.

Blackwell, W. H., Jr. 1970. The Lythraceae of Ohio. Ohio J. Sci. **70:** 346–352.

Blackwell, W. H., Jr., D. M. Brandenburg, M. D. Baechle, and P. D. Doran. 1981. Checklist of vascular plants of the "Highbanks," an oak-hickory stand in southwestern Ohio. Castanea **46:** 300–310.

Blanchard, O. J., Jr. 1976. A revision of species segregated from *Hibiscus* sect. *Trionum* (Medicus) De Candolle sensu lato (Malvaceae). Ph.D. dissertation. Cornell Univ., Ithaca. viii + 356 pp.

Blauch, D. S. 1970. A taxonomic study of the Ericaceae of the southern Appalachian highlands. Ph.D. dissertation. West Virginia Univ., Morgantown. vii + 230 pp.

Bouchard, A., and P. F. Maycock. 1970. A phytogeographical and phytosociological study of *Viola rotundifolia* in eastern Canada. Canad. J. Bot. **48:** 2285–2303.

Boufford, D. E. 1982. The systematics and evolution of *Circaea* (Onagraceae). Ann. Missouri Bot. Gard. **69:** 804–994.

Boufford, D. E., J. V. Crisci, H. Tobe, and P. C. Hoch. 1990. A cladistic analysis of *Circaea* (Onagraceae). Cladistics **6:** 171–182.

Bowden, W. M. 1982. The taxonomy of *Lobelia* × *speciosa* s.l. and its parental species, *L. siphilitica* and *L. cardinalis* s.l. (Lobeliaceae). Canad. J. Bot. **60:** 2054–2070.

Bowers, K. A. W. 1975. The pollination ecology of *Solanum rostratum* (Solanaceae). Amer. J. Bot. **62:** 633–638.

Bradt-Barnhart, J. S. 1987. A floristic survey of the vascular plants of Little Mountain and other peaks in northeastern Ohio. M.S. thesis. Kent State Univ., Kent. vii + 152 pp.

Brainerd, E. 1907. The older types of North American violets.I. Rhodora **9:** 93–98.

———. 1924. Some natural violet hybrids of North America. Vermont Agric. Exp. Sta. Bull. **239:** 1–205.

Brant, A. E. 1987. Additions to the flora of West Virginia. Sida **12:** 425–428.

Braun, E. L. 1940. New plants from Kentucky. Rhodora **52:** 47–51.

———. 1956. Variation in *Polemonium reptans*. Rhodora **58:** 103–116.

———. 1960. The genus *Tilia* in Ohio. Ohio. J. Sci. **60:** 257–261.

———. 1961. The woody plants of Ohio. Ohio State Univ. Press, Columbus. 362 pp.; reprinted 1989.

———. 1964. *Micheliella verticillata* in Ohio. Rhodora **66:** 275–277.

———. 1967. The Monocotyledoneae [of Ohio]. Cat-tails to orchids. With the Gramineae by Clara G. Weishaupt. Ohio State Univ. Press, Columbus. ix + 464 pp.

———. 1976. Two members of the Rubiaceae new to Ohio. Rhodora **78:** 549–551.

Brizicky, G. K. 1961a. The genera of Turneraceae and Passifloraceae in the southeastern United States. J. Arnold Arbor. **42:** 204–218.

———. 1961b. The genera of Violaceae in the southeastern United States. J. Arnold Arbor. **42:** 321–333.

———. 1962a. The genera of Rutaceae in the southeastern United States. J. Arnold Arbor. **43:** 1–22.

———. 1962b. The genera of Simaroubaceae and Burseraceae in the southeastern United States. J. Arnold Arbor. **43:** 173–186.

———. 1962c. The genera of Anacardiaceae in the southeastern United States. J. Arnold Arbor. **43:** 359–375.

———. 1963a. Taxonomic and nomenclatural notes on the genus *Rhus* (Anacardiaceae). J. Arnold Arbor. **44:** 60–80.

———. 1963b. The genera of Sapindales in the southeastern United States. J. Arnold Arbor. **44:** 462–501.

———. 1964a. The genera of Celastrales in the southeastern United States. J. Arnold Arbor. **45:** 206–234.

———. 1964b. The genera of Cistaceae in the southeastern United States. J. Arnold Arbor. **45:** 346–357.

———. 1964c. The genera of Rhamnaceae in the southeastern United States. J. Arnold Arbor. **45:** 439–463.

———. 1964d. A further note on *Ceanothus herbaceus* versus *C. ovatus*. J. Arnold Arbor. **45:** 471–473.

———. 1965a. The genera of Vitaceae in the southeastern United States. J. Arnold Arbor. **46:** 48–67.

———. 1965b. The genera of Tiliaceae and Elaeocarpaceae in the southeastern United States. J. Arnold Arbor. **46:** 286–307.

Brockett, B. L., and T. S. Cooperrider. 1983. The Primulaceae of Ohio. Castanea **48:** 37–40.

Brooks, R. E. 1976. A new *Veronica* (Scrophulariaceae) hybrid [*V. anagallis-aquatica* × *catenata*] from Nebraska. Rhodora **78:** 773–775.

Brummitt, R. K. 1965. New combinations in North American *Calystegia*. Ann. Missouri Bot. Gard. **52:** 214–216.

———. 1971. Relationship of *Heracleum lanatum* Michx. of North America to *H. sphondylium* [L.] of Europe. Rhodora **73:** 578–584.

———. 1972. Nomenclatural and historical considerations concerning the genus *Galax*. Taxon **21:** 303–317.

———. 1980. Further new names in the genus *Calystegia* (Convolvulaceae). Kew Bull. **35:** 327–334.

———. 1986. Report of the Committee for Spermatophyta:30. Taxon **35:** 556–563.

Bruza, J. D. 1982. A revision of the *Diodia teres* complex (Rubiaceae). Ph.D. dissertation. Mississippi State Univ., Mississippi State. 165 pp.

Burch, D. 1966. The application of the Linnaean names of some New World species of *Euphorbia* subgenus *Chamaesyce*. Rhodora **68:** 155–166.

Burk, C. J., and N. D. McMaster. 1988. The spread of *Catalpa ovata* G. Don in western Massachusetts and its biogeographic implications. Rhodora **90:** 461–464.

Burnett, J. H. 1950. The correct name for *Veronica aquatica* Bernhardi. Watsonia **1:** 349–353.

Burns, J. F. 1986. The Polygalaceae of Ohio. Castanea **51:** 137–144.

Cain, S. A. 1967. Studies on the stemmed yellow violets of eastern North America. II. Mass collections of *Viola pubescens* and *V. eriocarpa* in the Michigan area. Naturaliste Canad. **94:** 79–129.

Campbell, E. O. 1971. Notes on the fungal association of two *Monotropa* species in Michigan. Michigan Bot. **10:** 63–67.

Cane, J. H., G. C. Eickwort, F. R. Wesley, and J. Spielholz. 1985. Pollination ecology of *Vaccinium stamineum* (Ericaceae: Vaccinioideae). Amer. J. Bot. **72:** 135–142.

Canne, J. M. 1984. Chromosome numbers and the taxonomy of North American *Agalinis* (Scrophulariaceae). Canad. J. Bot. **62:** 454–456.

Cantino, P. D. 1982. A monograph of the genus *Physostegia* (Labiatae). Contr. Gray Herb. **211:** 1–105.

———. 1985. Facultative autogamy in *Synandra hispidula* (Labiatae). Castanea **50:** 105–111.

Cantino, P. D., and T. G. Lammers. 1983. *Physostegia virginiana* var. *arenaria* Shimek, an overlooked name. Rhodora **85:** 263–264.

Cantlon, J. E., E. J. C. Curtis, and W. M. Malcolm. 1963. Studies of *Melampyrum lineare*. Ecology **44:** 466–474.

Carr, B. L., J. V. Crisci, and P. C. Hoch. 1990. A cladistic analysis of the genus *Gaura* (Onagraceae). Syst. Bot. **15:** 454–461.

Cavers, P. B., I. J. Bassett, and C. W. Crompton. 1980. The biology of Canadian weeds. 47. *Plantago lanceolata* L. Canad. J. Pl. Sci. **60:** 1269–1282.

Ceska, A., and M. A. M. Bell. 1973. *Utricularia* (Lentibulariaceae) in the Pacific Northwest. Madroño **22:** 74–84.

Ceska, A., and O. Ceska. 1986. Notes on *Myriophyllum* (Haloragaceae) in the Far East: the identity of *Myriophyllum sibiricum* Komarov. Taxon **35:** 95–100.

Chambers, H. L. 1961. Chromosome numbers and breeding systems in *Pycnanthemum* (Labiatae). Brittonia **13:** 116–128.

Chambers, H. L., and K. L. Chambers. 1971. Artificial and natural hybrids in *Pycnanthemum* (Labiatae). Brittonia **23:** 71–88.

Channell, R. B., and C. E. Wood, Jr. 1959. The genera of the Primulales of the southeastern United States. J. Arnold Arbor. **40:** 268–288.

Chester, E. W. 1967. *Halesia carolina* L. in Kentucky, Indiana, and Ohio. Rhodora **69:** 380–382.

Chuang, T. I., and L. Constance. 1969. A systematic study of *Perideridia* (Umbelliferae-Apioideae). Univ. Calif. Publ. Bot. **55:** 1–74.

———. 1977. Cytogeography of *Phacelia ranunculacea* (Hydrophyllaceae). Rhodora **79:** 115–122.

Clausen, J., R. B. Channell, and U. Nur. 1964. *Viola rafinesquii*, the only *Melanium* violet native to North America. Rhodora **66:** 32–46.

Clement, I. D. 1957. Studies in *Sida* (Malvaceae). I. Contr. Gray Herb. **180:** 1–91.

Cochrane, T. S. 1976. Taxonomic status of the *Onosmodium molle* complex (Boraginaceae) in Wisconsin. Michigan Bot. **15:** 103–110.

Coffey, B. T., and C. D. McNabb. 1974. Eurasian water-milfoil in Michigan. Michigan Bot. **13:** 159–165.

Coffey, V. J., and S. B. Jones, Jr. 1980. Biosystematics of *Lysimachia* section *Seleucia* (Primulaceae). Brittonia **32:** 309–322.

Collins, J. L. 1976. A revision of the annulate *Scutellaria* (Labiatae). Ph.D. dissertation. Vanderbilt Univ., Nashville. x + 294 pp.

Collins, L. T. 1973. Systematics of *Orobanche* section *Myzorrhiza* (Orobanchaceae). With emphasis on *Orobanche ludoviciana*. Ph.D. dissertation. Univ. of Wisconsin, Milwaukee. viii + 219 pp.

Constance, L. 1940. The genus *Ellisia*. Rhodora **42:** 33–39.

———. 1942. The genus *Hydrophyllum* L. Amer. Midl. Naturalist **27:** 710–731.

———. 1949. A revision of *Phacelia* subgenus *Cosmanthus* (Hydrophyllaceae). Contr. Gray Herb. **168:** 1–48.

———. 1963. Chromosome number and classification in Hydrophyllaceae. Brittonia **15:** 273–285.

Cooperrider, M. K., and T. S. Cooperrider. 1994. History and computerization of the Kent State University Herbarium. Ohio J. Sci. **94:** 24–28.

Cooperrider, T. S. 1961. Ohio floristics at the county level. Ohio J. Sci. **61:** 318–320.

———. 1962. The occurrence and hybrid nature of an enchanter's nightshade in Ohio. Rhodora **64:** 63–67.

———. 1967. Reproductive systems in *Chelone glabra* var. *glabra*. Proc. Iowa Acad. Sci. **74:** 32–35.

———. 1976. Notes on Ohio Scrophulariaceae. Castanea **41:** 223–226.

———. 1980. *Cornus foemina* excluded from the known Ohio flora. Castanea **42:** 216–217.

———, ed. 1982a. Endangered and threatened plants of Ohio. Ohio Biol. Surv. Biol. Notes, No. 16. Ohio State Univ., Columbus. iv + 92 pp.

———. 1982b. *Geranium bicknellii* and *Hackelia deflexa* added to the known Ohio flora. Castanea **47:** 408.

———. 1984a. Ohio's herbaria and the Ohio Flora Project. Ohio J. Sci. **84:** 189–196.

———. 1984b. Some species mergers and new combinations in the Ohio flora. Michigan Bot. **23:** 165–168.

———. 1985. *Thaspium* and *Zizia* (Umbelliferae) in Ohio. Castanea **50:** 116–119.

———. 1986a. The genus *Phlox* (Polemoniaceae) in Ohio. Castanea **51:** 145–148.

———. 1986b. *Viola* × *brauniae* (*Viola rostrata* × *V. striata*). Michigan Bot. **25:** 107–109.

———. 1989. The Clusiaceae (or Guttiferae) of Ohio. Castanea **54:** 1–11.

———. 1992. Changes in knowledge of the vascular plant flora of Ohio, 1860–1991. Ohio J. Sci. **92:** 73–76.

Cooperrider, T. S., and B. K. Andreas. 1991. Noteworthy collections. Ohio. *Epilobium parviflorum* Schreber (Onagraceae). Small-flowered hairy willow-herb. Michigan Bot. **30:** 69–70.

Cooperrider, T. S., and B. L. Brockett. 1974. The nature and status of *Lysimachia* × *producta* (Primulaceae). Brittonia **26:** 119–128.

———. 1976. The nature and status of *Lysimachia* × *producta* (Primulaceae)— II. Brittonia **28:** 76–80.

Cooperrider, T. S., and G. A. McCready. 1970. Chromosome numbers in *Chelone* (Scrophulariaceae). Brittonia **22:** 175–183.

———. 1975. On separating Ohio specimens of *Lindernia dubia* and *L. anagallidea* (Scrophulariaceae). Castanea **40:** 191–197.

Cooperrider, T. S., and R. F. Sabo. 1969. *Stachys hispida* and *S. tenuifolia* in Ohio. Castanea **34:** 432–435.

Corbett, G. A. 1973. Euphorbiaceae of West Virginia. Proc. West Virginia Acad. Sci. **45:** 44–52.

Costelloe, B. H. 1988. Pollination ecology of *Gentiana andrewsii*. Ohio J. Sci. **88:** 132–138.

Creasy, W. D. 1953. Taxonomy of *Scrophularia marilandica* L. and *S. lanceolata* Pursh. Castanea **18:** 65–68.

Crins, W. J., D. A. Sutherland, and M. J. Oldham. 1987. *Veronica verna* (Scrophulariaceae), an overlooked element of the naturalized flora of Ontario. Michigan Bot. **26:** 161–165.

Cronquist, A. 1968. The evolution and classification of flowering plants. Houghton Miflin Co., Boston. x + 396 pp.

———. 1981. An integrated system of classification of flowering plants. Columbia Univ. Press, New York. xviii + 1262 pp.

———. 1982. Reduction of *Pseudotaenidia* to *Taenidia* (Apiaceae). Brittonia **34:** 365–367.

Crosswhite, F. S. 1969. An incorrect correction. Rhodora **71:** 480.

Crow, G. E., and C. B. Hellquist. 1983. Aquatic vascular plants of New England: Part 6. Trapaceae, Haloragaceae, Hippuridaceae. New Hampshire Agric. Exp. Sta. Bull. **524:** i-ii, 1–26.

———. 1985. Aquatic vascular plants of New England: Part 8. Lentibulariaceae. New Hampshire Agric. Exp. Sta. Bull. **528:** i-ii, 1–22.

Cruden, R. W. 1962. The Campanulaceae of Ohio. Ohio J. Sci. **62:** 142–149.

Cunningham, M., and P. D. Parr. 1990. Successful culture of the rare annual hemiparasite *Tomanthera auriculata* (Michx.) Raf. (Scrophulariaceae). Castanea **55:** 266–271.

Cusick, A. W. 1985. *Lithospermum* (Boraginaceae) in Ohio, with a new taxonomic rank for *Lithospermum croceum* Fernald. Michigan Bot. **24:** 63–69.

———. 1990. The best plant discoveries of 1989. Ohio Department of Natural Resources, Division of Natural Areas & Preserves Newsletter **12** (1): 1, 5.

———. 1991. The best plant discoveries of 1990. Ohio Department of Natural Resources, Division of Natural Areas & Preserves Newsletter **13** (2): 1–2.

Cusick, A. W., and G. M. Silberhorn. 1977. The vascular plants of unglaciated Ohio. Ohio Biol. Surv. Bull. N.S. **5:** i–x, 1–153.

Daoud, H. S., and R. L. Wilbur. 1965. A revision of the North American species of *Helianthemum* (Cistaceae). Rhodora **67:** 63–82, 201–216, 255–312.

D'Arcy, W. G. 1979. Proposal to conserve the name *Agalinis* Raf. (1837) against *Virgularia* Ruiz & Pavon (1794) (Scrophulariaceae). Taxon **28:** 419–422.

Davidson, J. F. 1950. The genus *Polemonium* [Tournefort] L. Univ. Calif. Publ. Bot. **23:** 209–282.

Deam, C. C. 1940. Flora of Indiana. Indiana Department of Conservation, Division of Forestry, Indianapolis. 1236 pp.

Denton, M. F. 1973. A monograph of *Oxalis* section *Ionoxalis* (Oxalidaceae) in North America. Publ. Mus. Michigan State Univ., Biol. Ser. **4:** 455–615.

Deschamp, P. A., and T. J. Cooke. 1985. Leaf dimorphism in the aquatic angiosperm *Callitriche heterophylla*. Amer. J. Bot. **72:** 1377–1387.

Desmarais, Y. 1952. Dynamics of leaf variation in the sugar maples. Brittonia **7:** 347–387.

Devlin, B., and A. G. Stephenson. 1985. Sex differential floral longevity, nectar secretion, and pollinator foraging in a protandrous species. Amer. J. Bot. **72:** 303–310.

DeWolf, G. P., Jr. 1955. Notes on cultivated labiates. 4. *Satureja* and some related genera. Baileya **2:** 142–150.

———. 1956. Notes on cultivated Scrophulariaceae. 2. *Antirrhinum* and *Asarina*. Baileya **4:** 55–68.

Dietrich, W., and W. L. Wagner. 1988. Systematics of *Oenothera* section *Oenothera* subsection *Raimannia* and subsection *Nutantigemma* (Onagraceae). Syst. Bot. Monographs **24:** 1–91.

Donoghue, M. J. 1983. A preliminary analysis of phylogenetic relationships in *Viburnum* (Caprifoliaceae s.l.). Syst. Bot. **8:** 45–58.

DOYON, D. 1959. Les aires géographiques actuelle et primitive d'*Asclepias syriaca* L. en Amérique. Rapport Société Québec Protection des Plantes **41:** 43–68.

DRAPALIK, D. J. 1970. A biosystematic study of the genus *Matelea* in southeastern United States. Ph.D. dissertation. Univ. of North Carolina, Chapel Hill. iii + 220 pp.

DRESSLER, R. L. 1961. A synopsis of *Poinsettia* (Euphorbiaceae). Ann. Missouri Bot. Gard. **48:** 329–341.

DUMKE, J. D. 1967. The Polemoniales of Ohio. I. Acanthaceae, Bignoniaceae, Lentibulariaceae, Martyniaceae, and Phrymaceae. Typewritten manuscript. 34 pp. [In the possession of Tom S. Cooperrider.]

———. 1968. The Polemoniales of Ohio. II. Hydrophyllaceae and Verbenaceae. Typewritten manuscript. 25 pp. [In the possession of Tom S. Cooperrider.]

DUNCAN, T. 1973. Three plant species new to Canada on Pelee Island: *Triosteum angustifolium* L., *Valerianella umbilicata* (Sull.) Wood, and *Valerianella intermedia* Dyal. Canad. Field-Naturalist **87:** 261–265.

DUNCAN, W. H. 1959. A naturally occurring F_1 hybrid of *Monarda media* and M. *fistulosa*. Rhodora **61:** 302–305.

———. 1964. New *Elatine* (Elatinaceae) populations in the southeastern United States. Rhodora **66:** 47–53.

———. 1970. The southern limits of *Trientalis borealis*. Rhodora **72:** 489–492.

———. 1975. Woody vines of the southeastern United States. Univ. of Georgia Press, Athens. 76 pp.

DUNCAN, W. H., and N. E. BRITTAIN. 1966. The genus *Gaylussacia* (Ericaceae) in Georgia. Bull. Georgia Acad. Sci. **24**. 14 pp.

EBINGER, J. E. 1974. A systematic study of the genus *Kalmia* (Ericaceae). Rhodora **76:** 315–398.

———. 1983. Exotic shrubs. A potential problem in natural area management in Illinois. Natural Areas J. **3:** 3–6.

———. 1985. *Lobelia cardinalis* × *L. siphilitica* in Illinois. Trans. Illinois State Acad. Sci. **78:** 29–31.

EBINGER, J. E., and L. LEHNEN. 1981. Naturalized autumn olive in Illinois. Trans. Illinois State Acad. Sci. **74:** 83–85.

EGLER, F. E. 1973. The hybrid nature of "*Monarda media* Willd." Castanea **38:** 209–214.

EGOLF, D. R. 1962. A cytological study of the genus *Viburnum*. J. Arnold Arbor. **43:** 132–172.

EHRENDORFER, F. 1976. *Galium*, pp. 14–36. *In* T. G. Tutin et al., eds. Flora Europaea. Vol. 4. Cambridge Univ. Press, London. xxix + 505 pp.

EITEN, G. 1955. The typification of the names "*Oxalis corniculata* L." and "*Oxalis stricta* L." Taxon **4:** 99–105.

———. 1963. Taxonomy and regional variation of *Oxalis* section *Corniculatae*. I. Introduction, keys and synopsis of the species. Amer. Midl. Naturalist **69:** 257–309.

ELLIS, W. H. 1963. Revision of section *Rubra* of *Acer* in eastern North America, excluding *Acer saccharinum* L. Ph.D. dissertation. Univ. of Tennessee, Knoxville. vii + 195 pp.

EPLING, C. 1934. Preliminary revision of American *Stachys*. Feddes Repert. Spec. Nov. Beih. **80:** 1–75.

———. 1940. A revision of *Salvia* subgenus *Calosphace*. Publ. Biol. Sci. Univ. Calif. Los Angeles **2:** 1–383.

———. 1942. The American species of *Scutellaria*. Univ. Calif. Publ. Bot. **20:** 1–146.

Estes, J. R., and P. M. Hall. 1975. Pollination of *Ipomopsis rubra* (Polemoniaceae) by ruby-throated hummingbirds. Bull. Torrey Bot. Club **102:** 413–415.

Eyde, R. H. 1966. The Nyssaceae in the southeastern United States. J. Arnold Arbor. **47:** 117–125.

———. 1987. The case for keeping *Cornus* in the broad Linnaean sense. Syst. Bot. **12:** 505–518.

———. 1988. Comprehending *Cornus*: puzzles and progress in the systematics of dogwoods. Bot. Rev. (Lancaster) **54:** 233–251.

Farwell, O. A. 1941. Notes on the Michigan flora. VIII. Pap. Michigan Acad. Sci. **26:** 3–20.

Fassett, N. C. 1939. Notes from the herbarium of the University of Wisconsin—XVII. *Elatine* and other aquatics. Rhodora **41:** 367–377.

———. 1944. *Dodecatheon* in eastern North America. Amer. Midl. Naturalist **31:** 455–486.

———. 1951. *Callitriche* in the New World. Rhodora **53:** 137–155, 161–194, 209–222.

Ferguson, I. K. 1965. The genera of Valerianaceae and Dipsacaceae in the southeastern United States. J. Arnold Arbor. **46:** 218–231.

———. 1966a. The genera of Caprifoliaceae in the southeastern United States. J. Arnold Arbor. **47:** 33–59.

———. 1966b. Notes on the nomenclature of *Cornus*. J. Arnold Arbor. **47:** 100–105.

———. 1966c. The Cornaceae in the southeastern United States. J. Arnold Arbor. **47:** 106–116.

Ferguson, I. K., and G. K. Brizicky. 1965. Nomenclatural notes on *Dipsacus fullonum* and *Dipsacus sativus*. J. Arnold Arbor. **46:** 362–365.

Fernald, M. L. 1946. Identifications and reidentifications of North American plants. Rhodora **48:** 207–216.

———. 1950. Gray's manual of botany. 8th ed. American Book Co., New York. lxiv + 1632 pp.

Fisher, T. R. 1988. The Dicotyledoneae of Ohio. Part 3. Asteraceae. Ohio State Univ. Press, Columbus. x + 280 pp.

Flanagan, L. B., and J. F. Bain. 1988. The biological flora of Canada 8. *Aralia nudicaulis* L., Wild Sarsaparilla. Canad. Field-Naturalist **102:** 45–59.

Frick, B. 1984. The biology of Canadian weeds. 62. *Lappula squarrosa* (Retz.) Dumort. Canad. J. Pl. Sci. **64:** 375–386.

Fosberg, F. R. 1942. *Cornus sericea* L. (*Cornus stolonifera* Michx.). Bull. Torrey Bot. Club **69:** 583–589.

Fryxell, P. A. 1985. Sidus sidarum—V. The North and Central American species of *Sida*. Sida **11:** 62–91.

Gadella, T. W. J. 1984. Notes on *Symphytum* (Boraginaceae) in North America. Ann. Missouri Bot. Gard. **71:** 1061–1067.

Gandhi, K. N., R. D. Thomas, and S. L. Hatch. 1987. Cuscutaceae of Louisiana. Sida **12:** 361–379.

Garbutt, K., F. A. Bazzaz, and D. A. Levin. 1985. Population and genotype niche width in clonal *Phlox paniculata*. Amer. J. Bot. **72:** 640–648.

Gentry, J. L., and R. L. Carr. 1976. A revision of the genus *Hackelia* (Boraginaceae) in North America, north of Mexico. Mem. New York Bot. Gard. **26:** 121–227.

Gill, L. S. 1980a. A study of the *Stachys palustris* L. complex (Labiatae) in northern North America. Phytologia **46:** 231–245.

———. 1980b. Cytotaxonomy of the genus *Stachys* L. in Canada. Caryologia **33:** 473–481.

———. 1981. Taxonomy, distribution and ecology of the Canadian Labiatae. Feddes Repert. **92:** 33–93.

Gill, L. S., B. M. Lawrence, and J. K. Morton. 1973. Variation in *Mentha arvensis* L. (Labiatae). I. The North American populations. J. Linn. Soc., Bot. **67:** 213–232.

Gillett, G. W. 1968. Systematic relationships in the *Cosmanthus* phacelias (Hydrophyllaceae). Brittonia **20:** 368–374.

Gillett, J. M. 1957. A revision of the North American species of *Gentianella* Moench. Ann. Missouri Bot. Gard. **44:** 195–269.

———. 1959. A revision of *Bartonia* and *Obolaria* (Gentianaceae). Rhodora **61:** 43–62.

———. 1963. The gentians of Canada, Alaska, and Greenland. Canada Dept. of Agriculture, Research Branch Publ. 1180. Ottawa. 99 pp.

———. 1968. The milkworts of Canada. Canada Dept. of Agriculture, Monograph No. 5. Ottawa. 24 pp.

———. 1979. New combinations in *Hypericum*, *Triadenum*, and *Gentianopsis*. Canad. J. Bot. **57:** 185–186.

Gillett, J. M., and N. K. B. Robson. 1981. The St. John's-worts of Canada (Guttiferae). Natl. Mus. Canada Publ. Bot. **11:** 1–40.

Gillis, W. T. 1971. The systematics and ecology of poison-ivy and the poison-oaks (*Toxicodendron*, Anacardiaceae). Rhodora **73:** 72–159, 161–237, 370–443, 465–540.

Gleason, H. A. 1947. Notes on some American plants. Phytologia **2:** 281–291.

———. 1952a. The new Britton and Brown illustrated flora of the northeastern United States and adjacent Canada. Vol. 2. New York Botanical Garden, Bronx. iv + 655 pp.

———. 1952b. The new Britton and Brown illustrated flora of the northeastern United States and adjacent Canada. Vol. 3. New York Botanical Garden, Bronx. iii + 595 pp.

Gleason, H. A., and A. Cronquist. 1963. Manual of vascular plants of northeastern United States and adjacent Canada. D. Van Nostrand Co., Princeton. li + 810 pp.

———. 1991. Manual of vascular plants of northeastern United States and adjacent Canada. 2d ed. New York Botanical Garden, Bronx. lxxv + 910 pp.

Gonsoulin, G. J. 1974. A revision of *Styrax* (Styracaceae) in North America, Central America, and the Caribbean. Sida **5:** 191–258.

Graham, S. A. 1964a. The genera of Lythraceae in the southeastern United States. J. Arnold Arbor. **45:** 235–250.

———. 1964b. The Elaeagnaceae in the southeastern United States. J. Arnold Arbor. **45:** 274–278.
———. 1966. The genera of Araliaceae in the southeastern United States. J. Arnold Arbor. **47:** 126–136.
———. 1975. Taxonomy of the Lythraceae in the southeastern United States. Sida **6:** 80–103.
———. 1979. The origin of *Ammannia* × *coccinea* Rottboell. Taxon **28:** 169–178.
———. 1985. A revision of *Ammannia* (Lythraceae) in the Western Hemisphere. J. Arnold Arbor. **66:** 395–420.
———. 1988. Revision of *Cuphea* section *Heterodon* (Lythraceae). Syst. Bot. Monographs **20:** 1–168.
Grant, E., and C. Epling. 1943. A study of *Pycnanthemum* (Labiatae). Univ. Calif. Publ. Bot. **20:** 195–240.
Grant, V. 1959. Natural history of the phlox family. I. Systematic botany. Nijhoff, The Hague. 280 pp.
Grant, V., and K. A. Grant. 1965. Flower pollination in the phlox family. Columbia Univ. Press, New York. xi + 180 pp.
Green, P. S. 1966. Identification of the species and hybrids in the *Lonicera tatarica* complex. J. Arnold Arbor. **47:** 75–88.
Grover, F. O. 1939. Reports on the flora of Ohio. I. Notes on the Ohio violets with additions to the state flora. Ohio J. Sci. **39:** 144–154.
Guire, K. E., and E. G. Voss. 1963. Distributions of distinctive shoreline plants in the Great Lakes region. Michigan Bot. **2:** 99–114.
Haber, E. 1971. A biosystematic study of the eastern North American species of the genus *Pyrola*. Ph.D. dissertation. Univ. Toronto, Toronto. v + 172 pp.
———. 1972. Priority of the binomial *Pyrola chlorantha*. Rhodora **74:** 396–397.
———. 1977. *Circaea* × *intermedia* in eastern North America with particular reference to Ontario. Canad. J. Bot. **55:** 2919–2935.
Haber, E., and J. E. Cruise. 1974. Generic limits in the Pyroloideae (Ericaceae). Canad. J. Bot. **52:** 877–883.
Hall, I. V., L. E. Aalders, N. L. Nickerson, and S. P. Vander Kloet. 1979. The biological flora of Canada 1. *Vaccinium angustifolium* Ait., Sweet Lowbush Blueberry. Canad. Field-Naturalist **93:** 415–430.
Hanks, L. T., and J. K. Small. 1907. Geraniaceae. North American Flora **25:** 3–24.
Hansen, A. 1976. *Dipsacus*, pp. 58–59. *In* T. G. Tutin et al., eds. Flora Europaea. Vol. 4. Cambridge Univ. Press, London. xxix + 505 pp.
Hardin, E. D. 1985. *Geranium bicknellii* extant in Ohio. Castanea **50:** 123.
Hardin, J. W. 1957a. A revision of the American Hippocastanaceae. Brittonia **9:** 145–171,—II. 173–195.
———. 1957b. Studies in the Hippocastanaceae. IV. Hybridization in *Aesculus*. Rhodora **59:** 185–203.
———. 1960. Studies in the Hippocastanaceae. V. Species of the Old World. Brittonia **12:** 26–38.
———. 1968. *Diervilla* (Caprifoliaceae) of the southeastern U.S. Castanea **33:** 31–36.
———. 1974. Studies of the southeastern United States flora. IV. Oleaceae. Sida **5:** 274–285.

———. 1990. Variation patterns and recognition of varieties of *Tilia americana* s.l. Syst. Bot. **15:** 33–48.

Hardin, J. W., and L. L. Phillips. 1985. Hybridization in eastern North American *Rhus* (Anacardiaceae). Assoc. Southeast. Biol. Bull. **32:** 99–106.

Harley, R. M. 1972. *Mentha*, p. 183–186. *In* T. G. Tutin et al., eds. Flora Europaea. Vol. 3. Cambridge Univ. Press, London. xxix + 370 pp.

Harris, B. D. 1968. Chromosome numbers and evolution in North American species of *Linum*. Amer. J. Bot. **55:** 1197–1204.

Hauser, E. J. P. 1963. The Dipsacaceae and Valerianaceae of Ohio. Ohio J. Sci. **63:** 26–30.

———. 1964. The Rubiaceae of Ohio. Ohio J. Sci. **64:** 27–35.

———. 1965. The Caprifoliaceae of Ohio. Ohio J. Sci. **65:** 118–129.

———. 1966. The natural occurrence of a hybrid honeysuckle (*Lonicera* × *bella*) in Ohio and Michigan. Michigan Bot. **5:** 211–217.

Hawthorn, W. R., and P. B. Cavers. 1976. Population dynamics of the perennial herbs *Plantago major* L. and *P. rugelii* Decne. J. Ecol. **64:** 511–527.

Haynes, R. R. 1971. A monograph of the genus *Conopholis* (Orobanchaceae). Sida **4:** 246–264.

Heckard, L. R. 1962. Root parasitism in *Castilleja*. Bot. Gaz. **124:** 21–29.

Heckard, L. R., and P. Rubtzoff. 1977. Additional notes on *Veronica anagallis-aquatica* × *catenata* (Scrophulariaceae). Rhodora **79:** 579–582.

Henderson, N. C. 1962. A taxonomic revision of the genus *Lycopus* (Labiatae). Amer. Midl. Naturalist **68:** 95–138.

Henry, L. K. 1953. The Violaceae in western Pennsylvania. Castanea **18:** 37–59.

Herrick, J. A. 1957. The nutrition of Indian pipe. Turtox News **35:** 188–190.

———. 1990. Is the dogwood next? On the Fringe **8:** 15.

Heywood, V. H., and I. B. K. Richardson, eds. 1972. Labiatae, pp. 126–192. *In* T. G. Tutin et al., eds. Flora Europaea. Vol. 3. Cambridge Univ. Press, London. xxix + 370 pp.

Hickok, L. G., and J. C. Anway. 1972. A morphological and chemical analysis of geographical variation in *Tilia* L. of eastern North America. Brittonia **24:** 2–8.

Hicks, D. J., R. Wyatt, and T. R. Meagher. 1985. Reproductive biology of distylous partridgeberry, *Mitchella repens*. Amer. J. Bot. **72:** 1503–1514.

Hinton, W. F. 1975. Natural hybridization and extinction of a population of *Physalis virginiana* (Solanaceae). Amer. J. Bot. **62:** 198–202.

Hoch, P. C. 1978. Systematics and evolution of the *Epilobium ciliatum* complex in North America (Onagraceae). Ph.D. dissertation. Washington Univ., St. Louis. vii + 176 pp.

Hodgdon, A. R. 1938. A taxonomic study of *Lechea*. Rhodora **40:** 29–69, 87–131.

Holmgren, N. H. 1986. Scrophulariaceae, pp. 751–797. *In* Great Plains Flora Association, Flora of the Great Plains. Univ. Press of Kansas, Lawrence. vii + 1392 pp.

Hou, D. 1955. A revision of the genus *Celastrus*. Ann. Missouri Bot. Gard. **42:** 215–302.

House, H. D. 1924. Annotated list of the ferns and flowering plants of New York State. Bull. 254. New York State Museum, Albany. 759 pp.

Howell, J. A., and W. H. Blackwell, Jr. 1977. The history of *Rhamnus frangula* (glossy buckthorn) in the Ohio flora. Castanea **42:** 111–115.

Iltis, H. H. 1956. Studies in Virginia plants. II. *Rhododendron maximum* in the Virginia coastal plain and its distribution in North America. Castanea **21:** 114–124.

———. 1963. *Napaea dioica* (Malvaceae): whence came the type? Amer. Midl. Naturalist **70:** 90–109.

———. 1965. The genus *Gentianopsis* (Gentianaceae): transfers and phytogeographic comments. Sida **2:** 129–154.

Ingram, J. 1960. Notes on the cultivated Primulaceae. 1. *Lysimachia*. Baileya **8:** 84–97.

———. 1961a. Studies in the cultivated Boraginaceae. 4. A key to the genera. Baileya **9:** 1–12.

———. 1961b. Studies in the cultivated Boraginaceae. 5. *Symphytum*. Baileya **9:** 92–99.

Irving, R. S. 1980. The systematics of *Hedeoma* (Labiatae). Sida **8:** 218–295.

Jacobs, B. F., and W. H. Eshbaugh. 1983. The Solanaceae of Ohio: a taxonomic and distributional study. Castanea **48:** 239–249.

James, C. W. 1956. A revision of *Rhexia* (Melastomataceae). Brittonia **8:** 201–230.

———. 1957. Notes on the cleistogamous species of *Polygala* in southeastern United States. Rhodora **59:** 51–56.

Jasieniuk, M., and M. J. Lechowicz. 1987. Spatial and temporal variation in chasmogamy and cleistogamy in *Oxalis montana* (Oxalidaceae). Amer. J. Bot. **74:** 1672–1680.

Jencks, W. R. 1972. Ohio's burley tobacco agriculture: a primary regional cash crop. Ohio J. Sci. **72:** 285–291.

Jennings, O. E. 1909. The Labrador tea in Ohio. Ohio Naturalist **10:** 13.

Jepson, W. L. 1925. Manual of the flowering plants of California. Univ. of California, Berkeley. 1238 pp.

Johnston, I. M. 1924. Studies in the Boraginaceae. II. 1. A synopsis of the American native and immigrant borages of the subfamily Boraginoideae. Contr. Gray Herb. **70:** 1–61.

———. 1952. Studies in the Boraginaceae. XXIII. A survey of the genus *Lithospermum*. J. Arnold Arbor. **33:** 299–366.

———. 1954. Studies in the Boraginaceae. XXVI. Further revaluations of the genera of the Lithospermeae. J. Arnold Arbor. **35:** 1–81.

Johnston, L. A. 1975. Revision of the *Rhamnus serrata* complex. Sida **6:** 67–79.

Jones, C. H. 1941. Additions to the revised catalogue of Ohio vascular plants. IX. Ohio J. Sci. **41:** 328–345.

———. 1942. Additions to the revised catalogue of Ohio vascular plants. X. Ohio J. Sci. **42:** 201–210.

———. 1945. Additions to the revised catalogue of Ohio vascular plants. XIII. Ohio J. Sci. **45:** 162–166.

Jones, G. N. 1959. *Viola eriocarpa* vs. *V. pensylvanica*. Rhodora **61:** 219–220.

———. 1968. Taxonomy of American species of linden (*Tilia*). Illinois Biol. Monogr. 39. Univ. of Illinois Press, Urbana. 156 pp.

Jones, G. N., and G. D. Fuller. 1955. Vascular plants of Illinois. Univ. of Illinois Press, Urbana, and Illinois State Museum, Springfield. xii + 593 pp.

Jones, T. H. 1983. A revision of the genus *Viburnum* section *Lentago* (Caprifoliaceae). Ph.D. dissertation. North Carolina State Univ. at Raleigh, Raleigh. vii + 157 pp.

Judd, W. S. 1981. A monograph of *Lyonia* (Ericaceae). J. Arnold Arbor. **62:** 63–209, 315–436.

Kartesz, J. T., and R. Kartesz. 1980. A synonymized checklist of the vascular flora of the United States, Canada, and Greenland. Univ. of North Carolina Press, Chapel Hill. xlviii + 498 pp.

Kartesz, J. T., and K. N. Gandhi. 1990. Nomenclatural notes for the North American flora. III. Phytologia **69:** 129–137.

Kearney, T. H. 1951. The American genera of Malvaceae. Amer. Midl. Naturalist **46:** 93–131.

Kellerman, W. A., and W. C. Werner. 1893. Catalogue of Ohio plants. Rep. Geol. Surv. Ohio **7:** 56–406.

Kephart, S. R. 1981. Breeding systems in *Asclepias incarnata* L., *A. syriaca* L., and *A. verticillata* L. Amer. J. Bot. **68:** 226–232.

Kephart, S. R., R. Wyatt, and D. Parrella. 1988. Hybridization in North American species of *Asclepias*. I. Morphological evidence. Syst. Bot. **13:** 456–473.

Kerrigan, W. J., Jr., and W. H. Blackwell, Jr. 1973. The distribution of Ohio Araliaceae. Castanea **38:** 168–170.

Klaber, D. 1976. Violets of the United States. A. S. Barnes & Co., Cranbury, NJ. 208 pp.

Knoop, J. D. 1988. *Tomanthera auriculata* (Michx.) Raf. extant in Ohio. Ohio J. Sci. **88:** 120–121.

———. 1990. *Euphorbia purpurea* (Raf.) Fern. extant in Ohio. Castanea **55:** 286–288.

Knuth, R. 1930. Oxalidaceae. Das Pflanzenreich IV. 130 (Heft 95). Wilhelm Engelmann, Leipzig. 481 pp.

Koelling, A. C. 1964. Taxonomic studies in *Penstemon deamii* and its allies. Ph.D. dissertation. Univ. of Illinois, Urbana. iv + 98 pp.

Kondo, K. 1972. A comparison of variability in *Utricularia cornuta* and *Utricularia juncea*. Amer. J. Bot. **59:** 23–37.

Kondo, K., L. J. Musselman, and W. F. Mann, Jr. 1978. Karyomorphological studies in some parasitic species of the Scrophulariaceae, I. Brittonia **30:** 345–354.

Kral, R., and P. E. Bostick. 1969. The genus *Rhexia* (Malastomataceae). Sida **3:** 387–440.

Kron, K. A. 1989. *Azalea rosea* Loiseleur is a superfluous name. Sida **13:** 331–333.

Kron, K. A., and W. S. Judd. 1990. Phylogenetic relationships within the Rhodoreae (Ericaceae) with specific comments on the placement of *Ledum*. Syst. Bot. **15:** 57–68.

Kruckeberg, A. R., and F. L. Hedglin. 1963. Natural and artificial hybrids of *Besseya* and *Synthyris* (Scrophulariaceae). Madroño **17:** 109–115.

Kuijt, J. 1969. The biology of parasitic flowering plants. Univ. of California Press, Berkeley and Los Angeles. 246 pp.

Kurmes, E. A. 1967. The distribution of *Kalmia latifolia* L. Amer. Midl. Naturalist **77:** 525–526.

Lacey, E. P. 1981. Seed dispersal in wild carrot (*Daucus carota*). Michigan Bot. **20:** 15–20.

Lackney, V. K. 1981. The parasitism of *Pedicularis lanceolata* Michx., a root hemiparasite. Bull. Torrey Bot. Club **108:** 422–429.

LaFrankie, J. V., Jr. 1983. A note on the growth habit of fringed polygala. Rhodora **85:** 457–460.

Lane, F. C. 1954. The genus *Triosteum* (Caprifoliaceae). Ph.D. dissertation. Univ. of Illinois, Urbana. vi + 158 pp.

Lawrence, G. H. M. 1951. Taxonomy of vascular plants. Macmillan Co., New York. xiii + 823 pp.

Les, D. H., and R. L. Stuckey. 1985. The introduction and spread of *Veronica beccabunga* (Scrophulariaceae) in eastern North America. Rhodora **87:** 503–515.

Lévesque, L., and P. Dansereau. 1966. Études sur les violettes jaunes caulescentes de l'Est de l'Amérique du Nord. I. Taxonomie, nomenclature, synonymie et bibliographie. Naturaliste Canad. **93:** 489–569.

Levin, D. A. 1963. Natural hybridization between *Phlox maculata* and *Phlox glaberrima* and its evolutionary significance. Amer. J. Bot. **50:** 714–720.

———. 1967. Variation in *Phlox divaricata*. Evolution **21:** 92–108.

———. 1972. Plant density, cleistogamy and self-fertilization in natural populations of *Lithospermum caroliniense*. Amer. J. Bot. **59:** 71–77.

Levy, M. 1983. Flavone variation and subspecific divergence in *Phlox pilosa* (Polemoniaceae). Syst. Bot. **8:** 118–126.

Lewis, H. 1945. A revision of the genus *Trichostema*. Brittonia **5:** 276–303.

———. 1960. Chromosome numbers and phylogeny of *Trichostema*. Brittonia **12:** 93–97.

Lewis, W. H. 1961. Merger of the North American *Houstonia* and *Oldenlandia* under *Hedyotis*. Rhodora **63:** 216–223.

———. 1962. Phylogenetic study of *Hedyotis* (Rubiaceae) in North America. Amer. J. Bot. **49:** 855–865.

Lewis, W. H., and R. L. Oliver. 1961. Cytogeography and phylogeny of the North American species of *Verbena*. Amer. J. Bot. **48:** 638–643.

———. 1965. Realignment of *Calystegia* and *Convolvulus* (Convolvulaceae). Ann. Missouri Bot. Gard. **52:** 217–222.

Leyendecker, P. J., Jr. 1941. The variations in *Galium triflorum* and *Galium boreale*. Iowa State Coll. J. Sci. **15:** 179–181.

Lindsey, A. H. 1984. Reproductive biology of Apiaceae. I. Floral visitors to *Thaspium* and *Zizia* and their importance in pollination. Amer. J. Bot. **71:** 375–387.

Lindsey, A. H., and C. R. Bell. 1985. Reproductive biology of Apiaceae. II. Cryptic specialization and floral evolution in *Thaspium* and *Zizia*. Amer. J. Bot. **72:** 231–247.

Lint, H., and C. Epling. 1945. A revision of *Agastache*. Amer. Midl. Naturalist **33:** 207–230.

Long, R. W. 1961. Convergent patterns of variation in *Ruellia caroliniensis* and *R. humilis* (Acanthaceae). Bull. Torrey Bot. Club **88:** 387–396.

———. 1966. Artificial interspecific hybridization in *Ruellia* (Acanthaceae). Amer. J. Bot. **53:** 917–927.

———. 1970. The genera of Acanthaceae in the southeastern United States. J. Arnold Arbor. **51:** 257–309.

———. 1974. Variation in natural populations of *Ruellia caroliniensis* (Acanthaceae). Bull. Torrey Bot. Club **101:** 1–6.

Lourteig, A. 1979. Oxalidaceae extra-Austroamericanae II. *Oxalis* L. Section *Corniculatae* DC. Phytologia **29:** 57–198.

Lovett Doust, L., J. Lovett Doust, and P. B. Cavers. 1981. Fertility relationships in closely related taxa of *Oxalis* section *Corniculatae*. Canad. J. Bot. **59:** 2603–2609.

Lowry, P. P., II, and A. G. Jones. 1984. Systematics of *Osmorhiza* Raf. (Apiaceae: Apioideae). Ann. Missouri Bot. Gard. **71:** 1128–1171.

Luken, J. O. 1988. Population structure and biomass allocation of the naturalized shrub *Lonicera maackii* (Rupr.) Maxim. in forest and open habitats. Amer. Midl. Naturalist **119:** 258–267.

Makowski, R. M. D., and I. A. Morrison. 1989. The biology of Canadian weeds. 91. *Malva pusilla* Sm. (=*M. rotundifolia* L.). Canad. J. Pl. Sci. **69:** 861–879.

Malcolm, W. M. 1962. Culture of *Castilleja coccinea* (Indian paint-brush), a root parasitic flowering plant. Michigan Bot. **1:** 77–79.

Malik, N., and W. H. Vanden Born. 1988. The biology of Canadian weeds. 86. *Galium aparine* L. and *Galium spurium* L. Canad. J. Pl. Sci. **68:** 481–499.

Mathias, M. E., and L. Constance. 1944–45. Umbelliferae. North American Flora **28:** 43–295.

McCance, R. M., Jr., and J. F. Burns, eds. 1984. Ohio endangered and threatened vascular plants. Abstracts of state-listed taxa. Ohio Dept. Natural Resources, Division of Natural Areas and Preserves, Columbus. xi + 635 pp.

McClintock, E., and C. Epling. 1942. A review of the genus *Monarda* (Labiatae). Univ. Calif. Publ. Bot. **20:** 147–194.

———. 1946. A revision of *Teucrium* in the New World, with observations on its variation, geographical distribution and history. Brittonia **5:** 491–510.

McCready, G. A., and T. S. Cooperrider. 1978. The Scrophulariaceae subfamily Scrophularioideae of Ohio. Castanea **43:** 76–86.

———. 1984. The Geraniaceae of Ohio. Castanea **49:** 138–141.

McGregor, R. L. 1985. Studies on the validity of *Lobelia spicata* infraspecific taxa in the prairies and plains of central North America with notes on *Lobelia appendiculata*. Contr. Univ. Kansas Herb. **16:** 1–10.

McKinney, L. E. 1992. A taxonomic revision of the acaulescent blue violets (*Viola*) of North America. Sida, Botanical Miscellany, No. 7. Botanical Research Institute of Texas. 60 pp.

McNeill, J. 1981. Taxonomic, nomenclatural and distributional notes on Canadian weeds and aliens. Naturaliste Canad. **108:** 237–244.

McVaugh, R. 1945. The genus *Triodanis* Rafinesque, and its relationships to *Specularia* and *Campanula*. Wrightia **1:** 13–52.

———. 1948. Generic status of *Triodanis* and *Specularia*. Rhodora **50:** 38–49.

Meagher, T. R., J. Antonovics, and R. Primack. 1978. Experimental ecological

genetics in *Plantago*. III. Genetic variation and demography in relation to survival of *Plantago cordata*, a rare species. Biol. Conserv. **14:** 243–257.
MEIKLE, R. D., ed. 1980. Draft index of author abbreviations compiled at the herbarium, Royal Botanic Gardens, Kew. Royal Botanic Gardens, Kew. 249 pp.
MENZEL, M. Y. 1951. The cytotaxonomy and genetics of *Physalis*. Proc. Amer. Philos. Soc. **95:** 132–183.
MEYER, F. G. 1951. *Valeriana* in North America and the West Indies. Ann. Missouri Bot. Gard. **38:** 377–503.
MEYER, F. G., and J. W. HARDIN. 1987. Status of the name *Aesculus flava* Solander (Hippocastanaceae). J. Arnold Arbor. **68:** 335–341.
MEYER, M. W., and R. H. MOHLENBROCK. 1966. The Illinois species of Haloragaceae and Hippuridaceae. Trans. Illinois State Acad. Sci. **59:** 149–162.
MILLER, G. N. 1955. The genus *Fraxinus*, the ashes, in North America, north of Mexico. Cornell Univ. Agric. Exp. Sta. Mem. **335:** 1–64.
MILLER, L. A. W. 1964. A taxonomic study of the species of *Acalypha* in the United States. Ph.D. dissertation. Purdue Univ., Lafayette. x + 211 pp.
MILLER, L. D. 1976. The Violaceae of Ohio. M.S. thesis. Kent State Univ., Kent. vii + 203 pp.
MILLER, N. G. 1971. The Polygalaceae in the southeastern United States. J. Arnold Arbor. **52:** 267–284.
MOHLENBROCK, R. H. 1963. Contributions to an Illinois flora. I. The genus *Physostegia*. Rhodora **65:** 58–64.
———. 1982a. The illustrated flora of Illinois. Flowering plants, basswoods to spurges. Southern Illinois Univ. Press, Carbondale and Edwardsville. xi + 234 pp.
———. 1982b. Illinois Convolvulaceae in the Missouri Botanical Garden herbarium. Ann. Missouri Bot. Gard. **69:** 393–401.
———. 1986. Guide to the vascular flora of Illinois. Revised and enlarged ed. Southern Illinois Univ. Press, Carbondale and Edwardsville. viii + 507 pp.
MOHLENBROCK, R. H., and W. WALLACE. 1963. The taxonomic status of *Penstemon brevisepalus* Pennell. Castanea **28:** 42–44.
MOLDENKE, H. N. 1958. Hybridity in the Verbenaceae. Amer. Midl. Naturalist **59:** 333–370.
———. 1980. A sixth summary of the Verbenaceae, Avicenniaceae, Stilbaceae, Chloanthaceae, Symphoremaceae, Nyctanthaceae, and Eriocaulaceae of the world as to valid taxa, geographic distribution and synonymy. Phytologia Memoirs 2. Plainfield, NJ. 629 pp.
MOORE, M. O. 1991. Classification and systematics of eastern North American *Vitis* L. (Vitaceae) north of Mexico. Sida **14:** 339–367.
MOORE, R. J. 1959. The dog-strangling vine *Cynanchum medium*, its chromosome number and its occurrence in Canada. Canad. Field-Naturalist **73:** 144–147.
———. 1975. The *Galium aparine* complex in Canada. Canad. J. Bot. **53:** 877–893.
MOORE, R. J., and C. FRANKTON. 1969. *Euphorbia* × *pseudo-esula* (*E. cyparissias* × *E. esula*) in Canada. Canad. Field-Naturalist **83:** 243–246.
MORAN, V. S. 1986. A demographic and phenological study of *Synandra hispidula* (Labiatae): a threatened Ohio species. Ohio J. Sci. **86** (April Program Abstracts): 5.

Morgan, M. D. 1971. Life history and energy relationships of *Hydrophyllum appendiculatum*. Ecol. Monogr. **41:** 329–349.

Moseley, E. L. 1931. Some plants that were probably brought to northern Ohio from the West by Indians. Pap. Michigan Acad. Sci. **13:** 169–172.

Mosquin, T. 1966. A new taxonomy for *Epilobium angustifolium* L. (Onagraceae). Brittonia **18:** 167–188.

Muenscher, W. C. 1949. *Veronica filiformis*, a weed of lawns and gardens. Rhodora **51:** 365.

Mulligan, G. A. 1980. The genus *Cicuta* in North America. Canad. J. Bot. **58:** 1755–1767.

Mulligan, G. A., and D. B. Munro. 1989. Taxonomy of species of North American *Stachys* (Labiatae) found north of Mexico. Naturaliste Canad. **116:** 35–51.

Mulligan, G. A., D. B. Munro, and J. McNeill. 1983. The status of *Stachys palustris* (Labiatae) in North America. Canad. J. Bot. **61:** 679–682.

Munz, P. A. 1965. Onagraceae. North American Flora, Series II. **5:** 1–278.

Musselman, L. J. 1969. Observations on the life history of *Aureolaria grandiflora* and *Aureolaria pedicularia* (Scrophulariaceae). Amer. Midl. Naturalist **82:** 307–311.

Musselman, L. J., and W. F. Mann, Jr. 1977. Host plants of some Rhinanthoideae (Scrophulariaceae) of eastern North America. Plant Systematics and Evolution **127:** 45–53.

Navaro, A. M., and W. H. Blackwell. 1990. A revision of *Paxistima* (Celastraceae). Sida **14:** 231–249.

Nelson, A. P. 1964. Relationships between two subspecies in a population of *Prunella vulgaris* L. Evolution **18:** 43–51.

———. 1965. Taxonomic and evolutionary implications of lawn races in *Prunella vulgaris* (Labiatae). Brittonia **17:** 160–174.

Nelson, E. C. 1988. *Lobelia cardinalis* f. *alba* (McNab) St. John—a correction. Notes Roy. Bot. Gard. Edinburgh **45:** 375–376.

Nelson, J. B. 1981. *Stachys* (Labiatae) in southeastern United States. Sida **9:** 104–123.

Nelson, J. B., and J. E. Fairey III. 1979. Misapplication of the name *Stachys nuttallii* (Lamiaceae) to a new southeastern species. Brittonia **31:** 491–494.

Nevling, L. I., Jr. 1962. The Thymelaeaceae in the southeastern United States. J. Arnold Arbor. **43:** 428–434.

Newberry, J. S. 1860. Catalogue of the flowering plants and ferns of Ohio. Ann. Rep. Ohio State Board Agric. (1859) **14:** 235–273.

Nichols, S. A. 1975. Identification and management of Eurasian water milfoil in Wisconsin. Trans. Wisconsin Acad. Sci. **63:** 116–128.

Noelle, H. J., and W. H. Blackwell, Jr. 1972. The Cactaceae in Ohio. Castanea **37:** 119–124.

Nuttall, T. 1818. The genera of North American plants. Vol. 1. Philadelphia. viii + 312 pp.

O'Connor, M. A., and W. H. Blackwell, Jr. 1974. Taxonomy and distribution of Ohio Cistaceae. Castanea **39:** 228–239.

O'Donell, C. A. 1959. Las especies americanas de "*Ipomoea*" L. sect. "*Quamoclit*" (Moench) Griseb. Lilloa **29:** 19–86.

O'Donovan, J. T., and M. P. Sharma. 1987. The biology of Canadian weeds. 78. *Galeopsis tetrahit* L. Canad. J. Pl. Sci. **67:** 787–796.

Orchard, A. E. 1981. A revision of South American *Myriophyllum* (Haloragaceae), and its repercussions on some Australian and North American species. Brunonia **4:** 27–65.

Ostertag, C. P., and R. J. Jensen. 1980. Species and population variability of *Osmorhiza longistylis* and *Osmorhiza claytonii*. Ohio J. Sci. **80:** 91–95.

Packer, J. G., and K. E. Denford. 1974. A contribution to the taxonomy of *Arctostaphylos uva-ursi*. Canad. J. Bot. **52:** 743–753.

Patten, B. C., Jr. 1954. The status of some American species of *Myriophyllum* as revealed by the discovery of intergrade material between M. *exalbescens* Fern. and M. *spicatum* L. in New Jersey. Rhodora **56:** 213–225.

Peng, C. I. 1989. The systematics and evolution of *Ludwigia* sect. *Microcarpium* (Onagraceae). Ann. Missouri Bot. Gard. **76:** 221–302.

Pennell, F. W. 1928. *Agalinis* and allies in North America, I. Proc. Acad. Nat. Sci. Philadelphia **80:** 339–449.

———. 1929. *Agalinis* and allies in North America, II. Proc. Acad. Nat. Sci. Philadelphia **81:** 111–249.

———. 1935. The Scrophulariaceae of eastern temperate North America. Academy of Natural Sciences of Philadelphia, Monograph 1. xiv + 650 pp.

Perry, J. D. 1971. Biosystematic studies in the North American genus *Sabatia* (Gentianaceae). Rhodora **73:** 309–369.

Peskin, P. K. 1990. Knotted dodder. Ohio. *Cuscuta glomerata*: (Convolvulaceae) in Ohio. Michigan Bot. **29:** 125–127.

Petrides, G. A. 1972. A field guide to trees and shrubs. 2d ed. Houghton Miflin Co., Boston. xxxii + 428 pp.

Philbrick, C. T. 1983. Contributions to the reproductive biology of *Panax trifolium* L. (Araliaceae). Rhodora **85:** 97–113.

Philcox, D. 1965. Contributions to the flora of tropical America. LXXIV. Revision of the New World species of *Buchnera* L. (Scrophulariaceae). Kew Bull. **18:** 275–315.

Phillippe, L. R. 1978. A biosystematic study of *Sanicula* section *Sanicula*. Ph.D. dissertation. Univ. of Tennessee, Knoxville. viii + 262 pp.

Piehl, M. A. 1962a. The parasitic behavior of *Melampyrum lineare* and a note on its seed color. Rhodora **64:** 15–23.

———. 1962b. The parasitic behavior of *Dasistoma macrophylla*. Rhodora **64:** 331–336.

———. 1963. Mode of attachment, haustorium structure, and hosts of *Pedicularis canadensis*. Amer. J. Bot. **50:** 978–985.

———. 1965. Studies of root parasitism in *Pedicularis lanceolata*. Michigan Bot. **4:** 75–81.

Poindexter, J. D. 1962. Natural hybridization among *Verbena stricta*, *V. hastata*, and *V. urticifolia* in Kansas. Trans. Kansas Acad. Sci. **65:** 409–419.

Polunin, O. 1969. Flowers of Europe. Oxford Univ. Press, London. 662 pp. + 192 colour plates.

Porter, D. M. 1972. The genera of Zygophyllaceae in the southeastern United States. J. Arnold Arbor. **53:** 531–552.

———. 1976. *Zanthoxylum* (Rutaceae) in North America north of Mexico. Brittonia **28:** 443–447.

———. 1987. The need for care in attributing lectotypification. Taxon **36:** 436–437.

Pringle, J. S. 1965a. The white gentian of the prairies. Michigan Bot. **4:** 43–47.

———. 1965b. Hybridization in *Gentiana* (Gentianaceae): a resume of J. T. Curtis' studies. Trans. Wisconsin Acad. Sci. **54:** 283–293.

———. 1966. *Gentiana puberulenta* sp. nov., a known but unnamed species of the North American prairies. Rhodora **68:** 209–214.

———. 1967. Taxonomy of *Gentiana*, section *Pneumonanthae*, in eastern North America. Brittonia **19:** 1–32.

———. [1968]. List of specimens examined in the preparation of the paper: Taxonomy of *Gentiana*, Section *Pneumonanthae*, in eastern North America. Mimeographed. 44 pp.

———. 1971. Hybridization in *Gentiana* (Gentianaceae): further data from J. T. Curtis's studies. Baileya **18:** 41–51.

———. 1973a. The spread of *Vincetoxicum* species (Asclepiadaceae) in Ontario. Canad. Field-Naturalist **87:** 27–33.

———. 1973b. *Lonicera maackii* (Caprifoliaceae) adventive in Ontario. Canad. Field-Naturalist **87:** 54–55.

———. 1988. Nomenclature of the white cardinal flower, *Lobelia cardinalis* forma *alba* (Campanulaceae: Lobelioideae). P1. Press **5:** 4–5.

———. 1990. Taxonomic notes on western American Gentianaceae. Sida **14:** 179–187.

Pryer, K. M., and L. R. Phillippe. 1989. A synopsis of the genus *Sanicula* (Apiaceae) in eastern Canada. Canad. J. Bot. **67:** 694–707.

Puff, C. 1976. The *Galium trifidum* group (*Galium* sect. *Aparinoides*, Rubiaceae). Canad. J. Bot. **54:** 1911–1925.

———. 1977. The *Galium obtusum* group (*Galium* sect. *Aparinoides*, Rubiaceae). Bull. Torrey Bot. Club **104:** 202–208.

Purcell, N. J. 1976. *Epilobium parviflorum* Schreb. (Onagraceae) established in North America. Rhodora **78:** 785–787.

Pusateri, W. P., and W. H. Blackwell, Jr. 1979. The *Echium vulgare* complex in eastern North America. Castanea **44:** 223–229.

Radford, A. E., H. E. Ahles, and C. R. Bell. 1968. Manual of the vascular flora of the Carolinas. Univ. of North Carolina Press, Chapel Hill. lxi + 1183 pp.

Rahn, K. 1974. *Plantago* section *Virginica*. A taxonomic revision of a group of American plantains, using experimental, taximetric and classical methods. Dansk Bot. Ark. **30:** 1–180.

Ramey, R. E. 1992. The best plant discoveries of 1991. Ohio Dept. Natural Resources, Division of Natural Areas and Preserves Newsletter **14** (1): 1–2.

Raven, P. H., W. Dietrich, and W. Stubbe. 1979. An outline of the systematics of *Oenothera* subsect. *Euoenothera* (Onagraceae). Syst. Bot. **4:** 242–252.

Raven, P. H., and D. P. Gregory. 1972. A revision of the genus *Gaura* (Onagraceae). Mem. Torrey Bot. Club **23:** 1–96.

Ray, J. D., Jr. 1956. The genus *Lysimachia* in the New World. Univ. of Illinois Press, Urbana. v + 160 pp.

Ray, P. M., and H. F. Chisaki. 1957. Studies on *Amsinckia*. I. A synopsis of the genus, with a study of heterostyly in it. Amer. J. Bot. **44:** 529–536.

Reinartz, J. A., and J. W. Popp. 1987. Structure of clones of northern prickly ash (*Xanthoxylum americanum*). Amer. J. Bot. **74:** 415–428.

Reveal, J. L. 1986. Additional comments on Linnaean types of eastern North American plants. J. Linn. Soc. Bot. **92:** 161–176.

———. 1991. *Rhus hirta* (L.) Sudworth, a newly revived correct name for *Rhus typhina* L. (Anacardiaceae). Taxon **40:** 489–492.

Reveal, J. L., C. R. Broome, M. L. Brown, and G. F. Frick. 1982. Comments on the typification of two Linnaean species of *Phlox* (Polemoniaceae). Taxon **31:** 733–736.

Reveal, J. L., and M. J. Seldin. 1976. On the identity of *Halesia carolina* L. (Styracaceae). Taxon **25:** 123–140.

Richardson, J. W. 1968. The genus *Euphorbia* of the high plains and prairie plains of Kansas, Nebraska, South and North Dakota. Univ. Kansas Sci. Bull **48:** 45–112.

Rickett, H. W. 1934. *Cornus amomum* and *Cornus candidissima*. Rhodora **36:** 269–274.

———. 1936. On nomenclature in *Cornus*. Rhodora **38:** 51.

———. 1944. *Cornus stolonifera* and *Cornus occidentalis*. Brittonia **5:** 149–159.

———. 1945. Cornaceae. North American Flora **28:** 299–311.

———, ed. 1966. Wild flowers of the United States. Vol. 1, 2 parts. The northeastern states. McGraw Hill Book Co., New York. xvi + 559 pp.

———, ed. 1967. Wild flowers of the United States. Vol. 2, 2 parts. The southeastern states. McGraw Hill Book Co., New York. xvi + 688 pp.

Riley, R. K., and J. J. Eichenmuller. 1970. Tree species associated with *Monotropa uniflora* in West Virginia. Proc. West Virginia Acad. Sci. **42:** 16–19.

Roane, M. K. 1975. The valid names of *Rhododendron nudiflorum* (L.) Torr. and *R. roseum* (Loisel.) Rehd. Taxon **24:** 540.

Roberts, M. L., and T. S. Cooperrider. 1982. Dicotyledons, pp. 48–84. *In* T. S. Cooperrider, ed. Endangered and threatened plants of Ohio. Ohio Biological Survey Bull. Biological Notes, No. 16. Ohio State Univ., Columbus. iv + 92 pp.

Robertson, K. R. 1971. The Linaceae in the southeastern United States. J. Arnold Arbor. **52:** 649–665.

———. 1972. The genera of Geraniaceae in the southeastern United States. J. Arnold Arbor. **53:** 182–201.

———. 1975. The Oxalidaceae in the southeastern United States. J. Arnold Arbor. **56:** 223–239.

Rodríguez Jiménez, C. 1973. Recherches sur *Hypericum* L. section *Brathys* (Mutis ex L. f.) Choisy, sous-section *Spachium* Keller (Guttiferae). Mem. Soc. Ci. Nat. La Salle **33:** 1–151.

Rogers, C. M. 1957. Linaceae in Michigan. Asa Gray Bull. N. S. **3:** 199–204.

———. 1963. Yellow flowered species of *Linum* in eastern North America. Brittonia **15:** 97–122.

———. 1984. Linaceae. North American Flora, Series II. **12:** 1–58.

Rogers, G. K. 1987. The genera of Cinchonoideae (Rubiaceae) in the southeastern United States. J. Arnold Arbor. **68:** 137–183.

Rosatti, T. J. 1984. The Plantaginaceae in the southeastern United States. J. Arnold Arbor. **65:** 533–562.

———. 1986. The genera of Sphenocleaceae and Campanulaceae in the southeastern United States. J. Arnold Arbor. **67:** 1–64.

———. 1987. Field and garden studies of *Arctostaphylos uva-ursi* (Ericaceae) in North America. Syst. Bot. **12:** 61–77.

———. 1989. The genera of suborder Apocynineae (Apocynaceae and Asclepiadaceae) in the southeastern United States. J. Arnold Arbor. **70:** 307–401, 443–514.

Russell, N. H. 1954. Three field studies of hybridization in the stemless white violets. Amer. J. Bot. **41:** 679–686.

———. 1955. The taxonomy of the North American acaulescent white violets. Amer. Midl. Naturalist **54:** 481–494.

———. 1965. Violets (*Viola*) of central and eastern United States: an introductory survey. Sida **2:** 1–113.

Russell, N. H., and A. C. Risser, Jr. 1960. The hybrid nature of *Viola emarginata* (Nuttall) LeConte. Brittonia **12:** 298–305.

Rust, R. W. 1977. Pollination in *Impatiens capensis* and *Impatiens pallida* (Balsaminaceae). Bull. Torrey Bot. Club **104:** 361–367.

Sabo, R. F. 1965. The Labiatae of Ohio. M.A. thesis. Kent State Univ., Kent. 168 pp.

St. John, H. 1941. Revision of the genus *Swertia* (Gentianaceae) of the Americas and the reduction of *Frasera*. Amer. Midl. Naturalist **26:** 1–29.

Sanders, R. W. 1976. Distributional history and probable ultimate range of *Galium pedemontanum* (Rubiaceae) in North America. Castanea **41:** 73–80.

Saulmon, J. G. 1971. The genus *Polygala* in the southeastern United States. Proc. West Virginia Acad. Sci. **43:** 9–13.

Schaack, C. G. 1983. A monographic revision of the genera *Synthyris* and *Besseya* (Scrophulariaceae). Ph.D. dissertation. Univ. of Montana, Missoula. 376 pp.

Schaffner, J. H. 1917. Additions to the catalog of Ohio vascular plants for 1916. Ohio J. Sci. **17:** 132–136.

———. 1925. Additions to the catalog of Ohio vascular plants for 1924. Ohio J. Sci. **25:** 130–138.

———. 1932. Revised catalog of Ohio vascular plants. Ohio Biol. Surv. Bull. **25:** 87–215.

———. 1933. Additions to the revised catalog of Ohio vascular plants. I. Ohio J. Sci. **33:** 288–294.

———. 1934. Additions to the revised catalog of Ohio vascular plants. II. Ohio J. Sci. **34:** 165–174.

———. 1935. Additions to the revised catalog of Ohio vascular plants. III. Ohio J. Sci. **35:** 297–303.

———. 1938. Additions to the revised catalog of Ohio vascular plants. VI. Ohio J. Sci. **38:** 211–216.

Schilling, E. E. 1981. Systematics of *Solanum* sect. *Solanum* (Solanaceae). Syst. Bot. **6:** 172–185.

Schlessman, M. A. 1985. Floral biology of American ginseng (*Panax quinquefolium*). Bull. Torrey Bot. Club. **112:** 129–133.

SCHMIDT, B. L., and W. F. MILLINGTON. 1968. Regulation of leaf shape in *Proserpinaca palustris*. Bull. Torrey Bot. Club **95:** 264–286.

SCHOEN, D. J. 1977. Floral biology of *Diervilla lonicera* (Caprifoliaceae). Bull. Torrey Bot. Club **104:** 234–240.

SCORA, R. W. 1967. Interspecific relationships in the genus *Monarda* (Labiatae). Univ. Calif. Publ. Bot. **41:** 1–71.

SHAN, R. H., and L. CONSTANCE. 1951. The genus *Sanicula* (Umbelliferae) in the Old World and the New. Univ. Calif. Publ. Bot. **25:** 1–76.

SHANNON, T. R., and R. WYATT. 1986. Reproductive biology of *Asclepias exaltata*. Amer. J. Bot. **73:** 11–20.

SHETLER, S. G. 1963. A checklist and key to the species of *Campanula* native or commonly naturalized in North America. Rhodora **65:** 319–337.

SHETLER, S. G., and N. R. MORIN. 1986. Seed morphology in North American Campanulaceae. Ann. Missouri Bot. Gard. **73:** 653–688.

SHINNERS, L. H. 1962a. *Calamintha* (Labiatae) in the southern United States. Sida **1:** 69–75.

———. 1962b. Synopsis of *Collinsonia* (Labiatae). Sida **1:** 76–83.

———. 1962c. *Rhododendron nudiflorum* and *R. roseum* (Ericaceae), illegitimate names. Castanea **27:** 94–95.

———. 1963. The varieties of *Teucrium canadense* (Labiatae). Sida **1:** 182–183.

SILBERHORN, G. M. 1970. A distinct phytogeographic area in Ohio: the southeastern Allegheny Plateau. Castanea **35:** 277–292.

SKALLERUP, H. R. 1953. The distribution of *Diospyros virginiana* L. Ann. Missouri Bot. Gard. **40:** 211–225.

SKOG, J. T., and N. H. NICKERSON. 1972. Variation and speciation in the genus *Hudsonia*. Ann. Missouri Bot. Gard. **59:** 454–464.

SMALL, E. 1978. A numerical taxonomic analysis of the *Daucus carota* complex. Canad. J. Bot. **56:** 248–276.

SMITH, E. B. 1982. Juvenile and adult leaflet phases in *Aralia spinosa* (Araliaceae). Sida **9:** 330–332.

SOPER, J. H., and M. L. HEIMBURGER. 1982. Shrubs of Ontario. Royal Ontario Museum, Toronto. xxxi + 495 pp.

SOUTHALL, R. M., and J. W. HARDIN. 1974. A taxonomic revision of *Kalmia* (Ericaceae). J. Elisha Mitchell Sci. Soc. **90:** 1–23.

SPENCER, N. R. 1984. Velvetleaf, *Abutilon theophrasti* (Malvaceae), history and economic impact in the United States. Econ. Bot. **38:** 407–416.

SPONGBERG, S. A. 1971. The Staphyleaceae in the southeastern United States. J. Arnold Arbor. **52:** 196–203.

———. 1976. Styracaceae hardy in temperate North America. J. Arnold Arbor. **57:** 54–73.

SPOONER, D. M. 1984. Infraspecific variation in *Gratiola viscidula* Pennell (Scrophulariaceae). Rhodora **86:** 79–87.

SPOONER, D. M., A. W. CUSICK, B. ANDREAS, and D. ANDERSON. 1983. Notes on Ohio vascular plants previously considered for listing as federally endangered or threatened species. Castanea **48:** 250–258.

SPOONER, D. M., A. W. CUSICK, G. F. HALL, and J. M. BASKIN. 1985. Observations on the distribution and ecology of *Sida hermaphrodita* (L.) Rusby (Malvaceae). Sida **11:** 215–225.

Standley, L. A., S. S. Kim, and I. M. Hjersted. 1988. Reproductive biology of two sympatric species of *Chimaphila*. Rhodora **90:** 233–244.

Steyermark. J. A. 1963. Flora of Missouri. Iowa State Univ. Press, Ames. lxxxiii + 1725 pp.

Straley, G. B. 1977. Systematics of *Oenothera* sect. *Kneiffia* (Onagraceae). Ann. Missouri Bot. Gard. **64:** 381–424.

Strausbaugh, P. D., and E. L. Core. 1978. Flora of West Virginia. 2d ed. Seneca Books, Inc., Grantsville. xl + 1079 pp.

Stuckey, R. L. 1966. The botanical pursuits of John Samples, pioneer Ohio plant collector (1836–1840). Ohio J. Sci. **66:** 1–41.

———. 1969. The introduction and spread of *Lycopus asper* (western water hore-hound) in the western Lake Erie and Lake St. Clair region. Michigan Bot. **8:** 111–120.

———. 1970. Distributional history of *Epilobium hirsutum* (great hairy willow-herb) in North America. Rhodora **72:** 164–181.

———. 1974. The introduction and distribution of *Nymphoides peltatum* (Meny-anthaceae) in North America. Bartonia **42:** 14–23.

———. 1980. Distributional history of *Lythrum salicaria* (purple loosestrife) in North America. Bartonia **47:** 3–20.

Stuckey, R. L., and W. L. Phillips. 1970. Distributional history of *Lycopus europaeus* (European water-horehound) in North America. Rhodora **72:** 351–369.

Stuckey, R. L., and M. L. Roberts. 1977. Rare and endangered aquatic vascular plants of Ohio: an annotated list of the imperiled species. Sida **7:** 24–41.

Stucky, J. M. 1985. Pollination systems of sympatric *Ipomoea hederacea* and *I. purpurea* and the significance of interspecific pollination flow. Amer. J. Bot. **72:** 32–43.

Stucky, J. M., and R. L. Beckmann. 1982. Pollination biology, self-incompatibility, and sterility in *Ipomoea pandurata* (L.) G. F. W. Meyer (Convolvulaceae). Amer. J. Bot. **69:** 1022–1031.

Sutton, D. A. 1988. A revision of the tribe Antirrhineae. Oxford Univ. Press. 575 pp. + 344 pp. on microfiche.

Tans, W. 1987. Lentibulariaceae: the bladderwort family in Wisconsin. Michigan Bot. **26:** 52–62.

Taylor, P. 1955. The genus *Anagallis* in tropical and south Africa. Kew Bull. **[1955]**: 321–350.

———. 1989. The genus *Utricularia*—A taxonomic monograph. Kew Bull. Additional Series **14:** i-xi, 1–724 pp.

Taylor, R. L., and B. MacBryde. 1977. Vascular plants of British Columbia. Univ. of British Columbia Press, Vancouver. xxiv + 754 pp.

Teeri, J. A. 1968. Developmental variability of *Cornus canadensis* in northern New England. Rhodora **70:** 278–282.

Terrell, E. E. 1959. A revision of the *Houstonia purpurea* group (Rubiaceae). Rhodora **61:** 157–180, 188–207.

———. 1975. Relationships of *Hedyotis fruticosa* to *Houstonia* L. and *Oldenlandia* L. Phytologia **31:** 418–424.

———. 1986. Taxonomic and nomenclatural notes on *Houstonia nigricans* (Rubiaceae). Sida **11:** 471–481.

———. 1991. Overview and annotated list of North American species of *Hedyotis*,

Houstonia, *Oldenlandia* (Rubiaceae), and related genera. Phytologia **71:** 212–243.

Terrell, E. E., W. H. Lewis, H. Robinson, and J. W. Nowicke. 1986. Phylogenetic implications of diverse seed types, chromosome numbers, and pollen morphology in *Houstonia* (Rubiaceae). Amer. J. Bot. **73:** 103–115.

Tessene, M. F. 1969. Systematic and ecological studies on *Plantago cordata*. Michigan Bot. **8:** 72–104.

Thieret, J. W. 1971. The genera of Orobanchaceae in the southeastern United States. J. Arnold Arbor. **52:** 404–434.

———. 1972. The Phrymaceae in the southeastern United States. J. Arnold Arbor. **53:** 226–233.

———. 1976a. Vascular plants new to Ohio. Castanea **41:** 181–183.

———. 1976b. Floral biology of *Proboscidea louisianica* (Martyniaceae). Rhodora **78:** 169–179.

———. 1977. The Martyniaceae in the southeastern United States. J. Arnold Arbor. **58:** 25–39.

Thompson, D. Q., R. L. Stuckey, and E. B. Thompson. 1987. Spread, impact, and control of purple loosestrife (*Lythrum salicaria*) in North American wetlands. U.S. Dept. Interior. Fish and Wildlife Serv., Washington, DC. v + 55 pp.

Thompson, H. J. 1953. The biosystematics of *Dodecatheon*. Contr. Dudley Herb. **4:** 73–154.

Thomson, J. D. 1985. Pollination and seed set in *Diervilla lonicera* (Caprifoliaceae): temporal patterns of flower and ovule development. Amer. J. Bot. **72:** 737–740.

Thorne, R. F. 1968. Synopsis of a putatively phylogenetic classification of the flowering plants. Aliso **6:** 57–66.

———. 1976. A phylogenetic classification of the Angiospermae. Evol. Biol. **9:** 35–106.

———. 1983. Proposed new realignments in the angiosperms. Nord. J. Bot. **3:** 85–117.

Threadgill, P. F., and J. M. Baskin. 1978. *Swertia caroliniensis* or *Frasera caroliniensis*? Castanea **43:** 20–22.

Threadgill, P. F., J. M. Baskin, and C. C. Baskin. 1979. Geographical ecology of *Frasera caroliniensis*. Bull. Torrey Bot. Club **106:** 185–188.

———. 1981. The ecological life cycle of *Frasera caroliniensis*, a long-lived monocarpic perennial. Amer. Midl. Naturalist **105:** 277–288.

Tucker, A. O., R. M. Harley, and D. E. Fairbrothers. 1980. The Linnaean types of *Mentha* (Lamiaceae). Taxon **29:** 233–255.

Tucker, A. O., and E. D. Rollins. 1989. The species, hybrids, and cultivars of *Origanum* (Lamiaceae) cultivated in the United States. Baileya **23:** 14–27.

Tucker, G. C. 1986. The genera of Elatinaceae in the southeastern United States. J. Arnold Arbor. **67:** 471–483.

Tutin, T. G., ed. 1972. Scrophulariaceae, pp. 202–281. *In* T. G. Tutin et al., eds. Flora Europaea. Vol. 3. Cambridge Univ. Press, London. xxix + 370 pp.

Tutin, T. G., et al., eds. 1964–1980. Flora Europaea. 5 vols. Cambridge Univ. Press, London.

Ugent, D. 1963. *Epilobium* × *wisconsinense*, hybr. nov. Rhodora **65:** 274–279.

Uttal, L. J. 1987. The genus *Vaccinium* L. (Ericaceae) in Virginia. Castanea **52:** 231–255.

———. 1988. Lectotypification of *Azalea rosea* Loisel. (Ericaceae) and a new combination in *Rhododendron periclymenoides* (Michx.) Shinners. Sida **13:** 167–169.

Valentine, D. H., and A. O. Chater, eds. 1972. Boraginaceae, pp. 83–122. *In* T. G. Tutin et al., eds. Flora Europaea. Vol. 3. Cambridge Univ. Press, London. xxix + 370 pp.

Valley, K. R., and T. S. Cooperrider. 1966. The Orobanchaceae of Ohio. Ohio J. Sci. **66:** 264–265.

Vander Kloet, S. P. 1980. The taxonomy of the highbush blueberry, *Vaccinium corymbosum*. Canad. J. Bot. **58:** 1187–1201.

———. 1988. The genus *Vaccinium* in North America. Research Branch Agriculture Canada, Ottawa. xi + 201 pp.

Vander Kloet, S. P., and I. V. Hall. 1981. The biological flora of Canada. 2. *Vaccinium myrtilloides* Michx., velvet-leaf blueberry. Canad. Field-Naturalist **95:** 329–345.

Vernon, D. P., and J. D. Schoknecht. 1981. *Verbascum virgatum* Stokes: an addition to the flora of Indiana, with comparison to *Verbascum blattaria* L. Proc. Indiana Acad. Sci. **91:** 515–524.

Vincent, M. A., and J. W. Thieret. 1987. *Thymelaea passerina* (Thymelaeaceae) in Ohio. Sida **12:** 75–78.

Vogelmann, J. E. 1985. Crossing relationships among North American and eastern Asian populations of *Agastache* sect. *Agastache* (Labiatae). Syst. Bot. **10:** 445–452.

Voss, E. G. 1985a. Nomenclatural notes on some Michigan dicots. Michigan Bot. **24:** 117–124.

———. 1985b. Michigan Flora. Part II. Dicots (Saururaceae–Cornaceae). Cranbrook Institute of Science Bull. 59 and University of Michigan Herbarium, Bloomfield Hills. xix + 724 pp.

Wagner, W. H., Jr. 1975. Notes on the floral biology of box-elder (*Acer negundo*). Michigan Bot. **14:** 73–82.

Wagner, W. H., Jr., T. F. Daniel, and M. K. Hansen. 1980. A hybridizing *Verbascum* population in Michigan. Michigan Bot. **19:** 37–45.

Ware, D. M. E. 1969. A revision of *Valerianella* in North America. Ph.D. dissertation. Vanderbilt Univ., Nashville. x + 249 pp.

———. 1983. Genetic fruit polymorphism in North American *Valerianella* (Valerianaceae) and its taxonomic implications. Syst. Bot. **8:** 33–44.

Waterfall, U. T. 1958. A taxonomic study of the genus *Physalis* in North America north of Mexico. Rhodora **60:** 107–114, 128–142, 152–173.

———. 1971. New taxa, combinations and distribution records for the Oklahoma flora. Rhodora **73:** 552–555.

Waterman, A. H. 1960. The mints (family Labiatae) of Michigan. Publ. Mus. Michigan State Univ., Biol. Ser. **1:** 269–302.

Watson, M. F. 1989. Nomenclatural aspects of *Oxalis* section *Corniculatae* in Europe. J. Linn. Soc., Bot. **101:** 347–362.

Webb, D. A. 1967. What is *Parthenocissus quinquefolia* (L.) Planchon? Feddes Repert. **74:** 6–10.

Webb, D. H. 1980. A biosystematic study of *Hypericum* section *Spachium* in eastern North America. Ph.D. dissertation. Univ. of Tennessee, Knoxville. x + 331 pp.

Webber, J. M., and P. W. Ball. 1980. Introgression in Canadian populations of *Lycopus americanus* Muhl. and *L. europaeus* L. (Labiatae). Rhodora **82:** 281–304.

Weber, J. A. 1972. The importance of turions in the propagation of *Myriophyllum exalbescens* (Haloragidaceae) in Douglas Lake, Michigan. Michigan Bot. **11:** 115–121.

Weber, J. A., and L. D. Noodén. 1974. Turion formation and germination in *Myriophyllum verticillatum*: phenology and its interpretation. Michigan Bot. **13:** 151–158.

Webster, G. L. 1967. The genera of Euphorbiaceae in the southeastern United States. J. Arnold Arbor. **48:** 303–430.

———. 1970. A revision of *Phyllanthus* (Euphorbiaceae) in the continental United States. Brittonia **22:** 44–76.

Weishaupt, C. G. 1971. Vascular plants of Ohio. A manual for use in field and laboratory. 3d. ed. Kendall/Hunt Publishing Co., Dubuque. iii + 292 pp.

Werner, K. 1962. Die kultivierten *Digitalis*-Arten. Kulturpflanze Beih. **3:** 167–182.

Werth, C. R., and J. L. Riopel. 1979. A study of the host range of *Aureolaria pedicularia* (L.) Raf. (Scrophulariaceae). Amer. Midl. Naturalist **102:** 300–306.

Wheeler, L. C. 1941. *Euphorbia* subgenus *Chamaesyce* in Canada and the United States exclusive of southern Florida. Rhodora **43:** 97–154, 168–205, 223–286.

———. 1943. History and orthography of the celastraceous genus "*Pachystima*" Rafinesque. Amer. Midl. Naturalist **29:** 792–795.

Wherry, E. T. 1955. The genus *Phlox*. Morris Arboretum Monograph. III. Philadelphia. 174 pp.

Whipple, H. L. 1972. Structure and systematics of *Phryma leptostachya* L. J. Elisha Mitchell Sci. Soc. **88:** 1–17.

Whitten, W. M. 1981. Pollination ecology of *Monarda didyma*, *M. clinopodia*, and hybrids (Lamiaceae) in the southern Appalachian mountains. Amer. J. Bot. **68:** 435–442.

Widrlechner, M. P. 1983. Historical and phenological observations on the spread of *Chaenorrhinum minus* across North America. Canad. J. Bot. **61:** 179–187.

Wilbur, R. L. 1955. A revision of the North American genus *Sabatia* (Gentianaceae). Rhodora **57:** 1–33, 43–71, 78–104.

———. 1966. Notes on Rafinesque's species of *Lechea* (Cistaceae). Rhodora **68:** 192–208.

———. 1976. Illegitimate names: *Rhododendron nudiflorum* (L.) Torr. and *R. roseum* (Loisel.) Rehd. Taxon **25:** 178–179.

———. 1987. Lectotypification of *Phlox ovata* L. (Polemoniaceae). Taxon **36:** 130–132.

———. 1988. The correct name of the pale, yellow or white gentian of the eastern United States. Sida **13:** 161–165.

Wilbur, R. L., and H. S. Daoud. 1961. The genus *Lechea* (Cistaceae) in the southeastern United States. Rhodora **63:** 103–118.

Willingham, F. F., Jr. 1976. Variation and phenological forms in *Rhododendron calendulaceum* (Michx.) Torrey (Ericaceae). Castanea **41:** 215–223.

Wilson, J. S. 1965. Variation in three taxonomic complexes of the genus *Cornus* in eastern United States. Trans. Kansas Acad. Sci. **67:** 747–817.

Wilson, K. A. 1960a. The genera of Hydrophyllaceae and Polemoniaceae in the southeastern United States. J. Arnold Arbor. **41:** 197–212.

———. 1960b. The genera of Convolvulaceae in the southeastern United States. J. Arnold Arbor. **41:** 298–317.

Wilson, K. A., and C. E. Wood, Jr. 1959. The genera of Oleaceae in the southeastern United States. J. Arnold Arbor. **40:** 369–384.

Windler, D. R., B. E. Wofford, and M. W. Bierner. 1976. Evidence of natural hybridization between *Mimulus ringens* and *Mimulus alatus* (Scrophulariaceae). Rhodora **78:** 641–649.

Windus, J. L. 1986. Monitoring *Gentiana saponaria* L. (Gentianaceae), an endangered species in Ohio. Ohio J. Sci. **86** (April Program Abstracts): 3.

Wingo, C. O., Jr. 1962. The herbaceous members of the families Crassulaceae, Saxifragaceae, and Oxalidaceae in Ohio. M.S. thesis. Ohio Univ., Athens. 114 pp.

Wisniewski, R. W. 1967. The Euphorbiaceae of Ohio. M.S. thesis. Ohio Univ., Athens. 70 pp.

Wood, C. E., Jr. 1961. The genera of Ericaceae in the southeastern United States. J. Arnold Arbor. **42:** 10–80.

———. 1975. The Balsaminaceae in the southeastern United States. J. Arnold Arbor. **56:** 413–426.

———. 1983. The genera of Menyanthaceae in the southeastern United States. J. Arnold Arbor. **64:** 431–445.

Wood, C. E., Jr., and P. Adams. 1976. The genera of Guttiferae (Clusiaceae) in the southeastern United States. J. Arnold Arbor. **57:** 74–90.

Wood, C. E., Jr., and R. B. Channell. 1959. The Empetraceae and Diapensiaceae of the southeastern United States. J. Arnold Arbor. **40:** 161–171.

———. 1960. The genera of the Ebenales in the southeastern United States. J. Arnold Arbor. **41:** 1–35.

Wood, C. E., Jr., and R. E. Weaver, Jr. 1982. The genera of Gentianaceae in the southeastern United States. J. Arnold Arbor. **63:** 441–487.

Woodson, R. E., Jr. 1930. Studies in the Apocynaceae. I. A critical study of the Apocynoideae (with special reference to the genus *Apocynum*). Ann. Missouri Bot. Gard. **17:** 1–212.

———. 1941. The North American Asclepiadaceae. I. Perspective of the genera. Ann. Missouri Bot. Gard. **28:** 193–244.

———. 1954. The North American species of *Asclepias* L. Ann. Missouri Bot. Gard. **41:** 1–211.

———. 1962. Butterflyweed revisited. Evolution **16:** 168–185.

———. 1964. The geography of flower color in the butterflyweed. Evolution **18:** 143–163.

Wurdack, J. J., and R. Kral. 1982. The genera of Melastomataceae in the southeastern United States. J. Arnold Arbor. **63:** 429–439.

WYMAN, D. 1965. Trees for American gardens. Revised and enlarged edition. Macmillan Co., New York. viii + 502 pp.

YETTER, T. C. 1989. Systematic and phylogenetic relationships in the Virginianum species complex of the genus *Pycnanthemum* (Labiatae). Ph.D. dissertation. Miami University, Oxford, OH. vii + 121 pp.

YOUNG, D. P. 1958. *Oxalis* in the British Isles. Watsonia **4:** 51–69.

YUNCKER, T. G. 1965. *Cuscuta*. North American Flora, Series II. **4:** 1–51.

INDEX TO PLANT NAMES

The page number of the main text entry is printed in roman. Species distribution maps appear on the same page as the main text entry. The page on which a taxon is illustrated is in **boldface**. Synonyms are in *italics*. Forms are not indexed. Names occurring in the introductory sections are not indexed.